Lecture Notes in Computer Science 711
Edited by G. Goos and J. Hartmanis

Andrzej M. Borzyszkowski
Stefan Sokołowski (Eds.)

Mathematical Foundations of Computer Science 1993

18th International Symposium, MFCS'93
Gdańsk, Poland, August 30-September 3, 1993
Proceedings

Springer-Verlag
Berlin Heidelberg GmbH

Series Editors

Gerhard Goos
Universität Karlsruhe
Postfach 69 80
Vincenz-Priessnitz-Straße 1
D-76131 Karlsruhe, Germany

Juris Hartmanis
Cornell University
Department of Computer Science
4130 Upson Hall
Ithaca, NY 14853, USA

Volume Editors

Andrzej M. Borzyszkowski
Stefan Sokołowski
Institute of Computer Science, Polish Academy of Sciences
ul. Jaskowa Dolina 31, 80-252 Gdańsk, Poland

CR Subject Classification (1991): D.3.3, F.1.1, F.2.2, F.3.1, F.4.1, F.4.3, D.2.1-2, G.2.2

ISBN 978-3-540-57182-7 ISBN 978-3-540-47927-7 (eBook)
DOI 10.1007/978-3-540-47927-7

Originally published by Springer-Verlag Berlin Heidelberg New York in 1993
Softcover reprint of the hardcover 1st edition 1993

Typesetting: Camera-ready by authors

45/3140-543210 - Printed on acid-free paper

Preface

This volume contains the proceedings of the 18th International Symposium on Mathematical Foundations of Computer Science, MFCS'93, held in Gdańsk, Poland, August 30–September 3, 1993.

The MFCS symposia, organized annually in Poland and the former Czechoslovakia since 1972, have a long and well-established tradition. Over the years they have served as a meeting ground for specialists from all branches of theoretical computer science, in particular

> algorithms and complexity, automata theory and theory of languages,
> concurrent, distributed and real-time systems,
> the theory of functional, logic and object-oriented programming,
> lambda calculus and type theory,
> semantics and logics of programs,

and others.

The interest in MFCS symposia seems to be growing. MFCS'93 sets a new record on the number of submissions: 133. Out of them, the Programme Committee has selected 56 for presentation, strictly on the basis of their scientific merit and appropriateness for the symposium. The committee was assisted by over 210 referees. The scientific programme includes also lectures by 12 distinguished scholars.

The proceedings contain both the invited lectures and the contributions. Unfortunately, most of the latter had to be abridged due to space limitations.

We would like to thank, first of all, the authors of the papers submitted—their interest justifies the organization of the symposia. Thanks are also due to the members of the Programme Committee and to the referees for their work put into the evaluation of the papers. Finally, we thank Springer-Verlag for a long-term cooperation with MFCS symposia.

A number of institutions have financially contributed to the success of the symposium. We gratefully acknowledge the support from Stefan Batory Foundation, *Gdańsk Development Method* research project, Warsaw University, Wrocław University, Gdańsk University.

MFCS'93 is organized by the Institute of Computer Science of the Polish Academy of Sciences. The Organizing Committee consists of Marek A. Bednarczyk (chairman), Andrzej M. Borzyszkowski, Ryszard Kubiak, Wiesław Pawłowski and Stefan Sokołowski.

Gdańsk, June 1993

Andrzej M. Borzyszkowski
Stefan Sokołowski

Programme Committee

J. Adámek (Praha)
M. A. Bednarczyk (Gdańsk)
J. A. Bergstra (Amsterdam)
A. M. Borzyszkowski (Gdańsk)
B. Courcelle (Bordeaux)
H.-D. Ehrich (Braunschweig)
I. Havel (Praha)
G. Huet (Paris)
N. Jones (København)
E. Moggi (Genova)
V. Nepomniaschy (Novosibirsk)
E.-R. Olderog (Oldenburg)
F. Orejas (Barcelona)
A. Poigné (St. Augustin)
I. Sain (Budapest)
D. A. Schmidt (Manhattan, Kansas)
S. Sokołowski (Gdańsk), chair
M. Wirsing (München)

Non-PC Referees

K. Ambos-Spies
N. Andersen
H. R. Andersen
H. Andréka
M. Anselmo
P. Audebaud
J. L. Balcazar
A. Banerjee
P. Barth
M. Bauderon
T. Bétréma
U. Berger
J. Berstel
J. Biskup
A. J. Blikle
J. Bradfield
F. J. Brandenburg
R. Breu
C. Brown
F. Broy
A. Bystrov
R. Casas
I. Castellani
M. V. Cengarle
M. M. Chiara
C. Choffrut
J. Chomicki
J. Clausen
L. Colson
T. Coquand
G. Costa
L. Csirmaz
P.-L. Curien
L. Czaja
R. De Nicola
P. Dembiński
M. Dezani
R. Diaconescu
J. Diaz
V. Diekert
G. Dowek
D. Dranidis
P. Dybjer
H. Ehrig
V. Evstigneev
G.-L. Ferrari
J. L. Fiadeiro
H. Fleischhack
W. J. Fokkink
G. S. Frandsen
D. de Frutos-Escrig
J. Gabarro
A. Gammelgaard
G. Ghelli
P. Y. Gloess
L. Godo
M. Gogolla
J. Goguen
M. Grosse-Rhode
R. Grosu
C. A. Gunter
P. Hajnal
M. R. Hansen
T. Hardin
J. Hatcliff
M. Hennessy
M. Hermann
R. Herzig
L. Holenderski
R. Howell
J. Hromkovič
M. Huhn
H. Hungar
G. Jarzembski
J. Jędrzejowicz
B. Jenner
C. T. Jensen
J. Jørgensen
B. Josko
J.-P. Jouannaud
A. Jung
S. Kaes
G. Kahn
R. Kaivola
J. Kari
S. Kars
J. Katajainen
C. Kenyon
D. Kesner
N. Klarlund
J. W. Klop
S. Klusener
B. Konikowska
S. Kotov
V. Koubek
H.-J. Kreowski
B. Krieg-Brückner
M. Křivánek
F. Kröger
R. Kubiak
A. Kučera
G. Kucherov
Á. Kurucz
H. Le Verge
U. Lechner
C. Lengauer
P. Lincoln
U. W. Lipeck
I. Litovsky
G. Longo
A. Lozano
Z. Luo
T. Maibaum
K. Malmkjær
V. Manca
S. Martini

E. W. Mayr	C. Palamidessi	M. Rusinowitch	A. Sulimov
N. D. N. Measor	C. Paulin	G. Saake	A. Szałas
S. Meiser	W. Pawłowski	N. Saheb	A. Szepietowski
V. Ménissier-Morain	W. Penczek	A. Salomaa	A. Tarlecki
A. Middeldorp	H. Peterreins	D. Sands	E. Teniente
S. Mikulás	W. Phoa	M. Schenke	J. Toran
P. B. Miltersen	B. Pierce	G. Schreiber	J. L. Träff
T. Mogensen	G. M. Pinna	M. Schwartzbach	E. Upfal
B. Monien	D. Pisinger	A. Schwill	P. Urzyczyn
L. Moss	A. Pitts	G. Scollo	Y. Venema
P. D. Mosses	V. Pollara	G. Serény	R. Verbeek
A. W. Mostowski	J. Power	M. Serna	A. Vinacua
F. Mráz	T. Przytycka	H. Shì	I. B. Virbitskaite
A. Mück	Z. Qian	N. V. Shilov	S. Vorobyov
D. Murphy	A. Raspand	A. Simon	D. Wätjen
I. Németi	G. Rauzy	G. Snelting	K. Weihrauch
M. Nielsen	G. Reggio	S. Soloviev	B. Werner
R. Nieuwenhuis	J. Rehof	E. Sopena	J. Winkowski
U. Nilsson	W. Reisig	Z. Spławski	P. Winter
E. Ohlebusch	B. Reus	M. Srebrny	M. Zaionc
L. Ong	J. Roman	R. Statman	E. Zucca
E. Orłowska	M. Rosendahl	O. Stepankova	A. Zvonkin
Y. Ortega-Mallén	G. Rosolini	T. Streicher	J. Zwiers
L. Pacholski	B. Rovan	W. Struckmann	
P. Pączkowski	M. Rodriguez-Artalejo		

A couple of referees have remained anonymous to the editors of the volume. We are as thankful to them as to the ones listed above for their work.

Table of Contents

Invited lectures

Contributions

On the Unification Free Prolog Programs

Krzysztof R. Apt[1] and Sandro Etalle[2]

[1] CWI
P.O. Box 4079, 1009 AB Amsterdam, The Netherlands
and
Faculty of Mathematics and Computer Science
University of Amsterdam, Plantage Muidergracht 24
1018 TV Amsterdam, The Netherlands
[2] Dipartimento di Matematica Pura ed Applicata
Università di Padova
Via Belzoni 7, 35131 Padova, Italy

Abstract. We provide simple conditions which allow us to conclude that in case of several well-known Prolog programs the unification algorithm can be replaced by iterated matching. The main tools used here are types and generic expressions for types. As already noticed by other researchers, such a replacement offers a possibility of improving the efficiency of program's execution.

Notes. The work of the first author was partly supported by ESPRIT Basic Research Action 6810 (Compulog 2). This research was done partly during the second author's stay at Centre for Mathematics and Computer Science, Amsterdam.

1 Introduction

Unification is heralded as one of the crucial features offered by Prolog, so it is natural to ask whether it is actually used in specific programs. The aim of this paper to identify natural conditions under which unification can be replaced by iterated matching and to show that they are applicable to several well-known Prolog programs. These conditions can be statically checked without analyzing the search trees for the queries. For programs which use ground inputs they can be efficiently tested.

The problem of replacing unification by iterated matching was already studied in the literature by a number of researchers – see e.g. Deransart and Maluszynski [DM85b], Maluszynski and Komorowski [MK85] and Attali and Franchi-Zannettacci [AFZ88]. As in the previous works on this subject, we use modes, which indicate how the arguments of a relation should be used. Our results improve upon the previous ones due to the additional use of types. This allows us to deal with non-ground inputs.

We use here a simple notion of a type, which is a set of terms closed under substitution. The main tool in our approach is the concept of a *generic expression*. Intuitively, a term s is a generic expression for a type T if it is more general than all elements of T which unify with s. This simple notion turns out to be crucial here, because surprisingly often the input positions of the heads of program clauses are filled in by generic expressions for appropriate types.

We combine in our analysis the use of generic expressions with the notion of a *well-typed program*, recently introduced by Bronsard, Lakshman and Reddy [BLR92], which allows us to ensure that the input positions of the selected atoms remain correctly typed. As the table included at the end of this paper shows, our results can be applied to astonishingly many Prolog programs.

2 Preliminaries

In what follows we study logic programs executed by means of the *LD-resolution*, which consists of the SLD-resolution combined with the leftmost selection rule. An SLD-derivation in which the leftmost selection rule is used is called an *LD-derivation*. We allow in programs various first-order built-in's, like $=$, $\neq$, $>$, etc, and assume that they are resolved in the way conforming to their interpretation.

We work here with *queries*, that is sequences of atoms, instead of *goals*, that is constructs of the form $\leftarrow Q$, where Q is a query. Apart from this we use the standard notation of Lloyd [Llo87] and Apt [Apt90]. In particular, given a syntactic construct E (so for example, a term, an atom or a set of equations) we denote by $Var(E)$ the set of the variables appearing in E. Given a substitution $\theta = \{x_1/t_1, ..., x_n/t_n\}$ we denote by $Dom(\theta)$ the set of variables $\{x_1, \ldots, x_n\}$, by $Range(\theta)$ the set of terms $\{t_1, \ldots, t_n\}$, and by $Ran(\theta)$ the set of variables appearing in $\{t_1, \ldots, t_n\}$. Finally, we define $Var(\theta) = Dom(\theta) \cup Ran(\theta)$.

Recall that a substitution θ is called *grounding* if $Ran(\theta)$ is empty, and is called a *renaming* if it is a permutation of the variables in $Dom(\theta)$. Given a substitution θ and a set of variables V, we denote by $\theta|V$ the substitution obtained from θ by restricting its domain to V.

2.1 Unifiers

Given two sequences of terms $\mathbf{s} = s_1, ..., s_n$ and $\mathbf{t} = t_1, ..., t_n$ of the same length we abbreviate the set of equations $\{s_1 = t_1, ..., s_n = t_n\}$ to $\{\mathbf{s} = \mathbf{t}\}$ and the sequence $s_1\theta, ..., s_n\theta$ to $\mathbf{s}\theta$. Two atoms can unify only if they have the same relation symbol. With two atoms $p(\mathbf{s})$ and $p(\mathbf{t})$ to be unified we associate the set of equations $\{\mathbf{s} = \mathbf{t}\}$. In the applications we often refer to this set as $p(\mathbf{s}) = p(\mathbf{t})$. A substitution θ such that $\mathbf{s}\theta = \mathbf{t}\theta$ is called a *unifier* of the set of equations $\{\mathbf{s} = \mathbf{t}\}$. Thus the set of equations $\{\mathbf{s} = \mathbf{t}\}$ has the same unifiers as the atoms $p(\mathbf{s})$ and $p(\mathbf{t})$.

A unifier θ of a set of equations E is called a *most general unifier* (in short *mgu*) of E if it is more general than all unifiers of E. An mgu θ of a set of equations E is called *relevant* if $Var(\theta) \subseteq Var(E)$.

The following lemma was proved in Lassez, Marriot and Maher [LMM88].

Lemma 1. *Let θ_1 and θ_2 be mgu's of a set of equations. Then for some renaming η we have $\theta_2 = \theta_1\eta$.* □

Finally, the following well-known lemma allows us to search for mgu's in an iterative fashion.

Lemma 2. *Let E_1, E_2 be two sets of equations. Suppose that θ_1 is a relevant mgu of E_1 and θ_2 is a relevant mgu of $E_2\theta_1$. Then $\theta_1\theta_2$ is a relevant mgu of $E_1 \cup E_2$. Moreover, if $E_1 \cup E_2$ is unifiable then θ_1 exists and for any such θ_1 an appropriate θ_2 exists, as well.* □

2.2 Modes and Types

Below we extensively use modes.

Definition 3. Consider an n-ary relation symbol p. By a *mode* for p we mean a function m_p from $\{1, \ldots, n\}$ to the set $\{+, -\}$. If $m_p(i) =$ '+', we call i an *input position* of p and if $m_p(i) =$ '−', we call i an *output position* of p (both w.r.t. m_p). □

Modes indicate how the arguments of a relation should be used. The definition of moding assumes one mode per relation in a program. Multiple modes may be obtained by simply renaming the relations. When every considered relation has a mode associated with it, we can talk about input positions and output positions of an atom. In that case for an atom A we denote by $In(A)$ and $Out(A)$ the family of terms filling in, respectively, the input and the output positions of A. Given an atom A, we denote by $VarIn(A)$ (resp. $VarOut(A)$) the set of variables occurring in the input (resp. output) positions of A. Similar notation is used for sequences of atoms.

In the sequel, we also use types. The following very general definition is sufficient for our purposes.

Definition 4. A *type* is a decidable set of terms closed under substitution. □

We call a type T *ground* if all its elements are ground, and *non-ground* if some of its elements is non-ground. By a *typed term* we mean a construct of the form $s : S$ where s is a term and S is a type. Given a sequence $\mathbf{s} : \mathbf{S} = s_1 : S_1, \ldots, s_n : S_n$ of typed terms we write $\mathbf{s} \in \mathbf{S}$ if for $i \in [1, n]$ we have $s_i \in S_i$.

Certain types will be of special interest:

U — the set of all terms,

List — the set of lists,

BinTree — the set of binary trees,

Nat — the set of natural numbers,

Ground — the set of ground terms.

Of course, the use of the type *List* assumes the existence of the empty list `[]` and the list constructor `[.|.]` in the language, and the use of the type *Nat* assumes the existence of the numeral `0` and the successor function `s(.)`, etc. Throughout the paper we fix a specific set of types, denoted by *Types*, which includes the above ones.

We also associate types with relation symbols.

Definition 5. Consider an n-ary relation symbol p. By a *type* for p we mean a function t_p from $[1, n]$ to the set *Types*. If $t_p(i) = T$, we call T *the type associated with the position i of p*. Assuming a type t_p for the relation p, we say that an atom $p(s_1, \ldots, s_n)$ is *correctly typed in position i* if $s_i \in t_p(i)$. □

When every considered relation has a mode and a type associated with it, we can talk about types of input positions and of output positions of an atom. An n-ary relation p with a mode m_p and type t_p will be denoted by

$$p(m_p(1) : t_p(1), \ldots, m_p(n) : t_p(n)).$$

For example, app($+ : List,\ + : List,\ - : U$) denotes a ternary relation app with the first two positions moded as input and typed as *List*, and the third position moded as output and typed as U.

¿From the context it will be always clear whether modes and/or types are assumed for the considered relations. In this paper we shall always use types in presence of modes.

3 Solvability by (Iterated) Matching

3.1 Solvability by Matching

We begin by recalling the following concepts.

Definition 6. Consider a set of equations $E = \{\mathbf{s} = \mathbf{t}\}$.

- A substitution θ such that either $Dom(\theta) \subseteq Var(\mathbf{s})$ and $\mathbf{s}\theta = \mathbf{t}$ or $Dom(\theta) \subseteq Var(\mathbf{t})$ and $\mathbf{s} = \mathbf{t}\theta$, is called a *match* for E.
- E is called *left-right disjoint* if $Var(\mathbf{s}) \cap Var(\mathbf{t}) = \emptyset$. □

Clearly, if E is left-right disjoint, then a match for E is also a relevant mgu of E. The sets of equations we consider in this paper will always satisfy this disjointness proviso due to the standardization apart.

Definition 7. Let E be a left-right disjoint set of equations. We say that E is *solvable by matching* if E is unifiable implies that a match for E exists. □

A simple test allowing us to determine whether a given set of equations is solvable by matching is summarized in the following lemma.

Definition 8.

- We call an atom (resp. a term) a *pure atom* (resp. *pure variable term*) if it is of the form $p(\mathbf{x})$ with $\mathbf{x}$ a sequence of different variables.
- Two atoms (resp. terms) are called *disjoint* if they have no variables in common. □

Lemma 9 (Matching 1). *Consider two disjoint atoms A and H with the same relation symbol. Suppose that*

- *one of them is ground or pure.*

Then $A = H$ is solvable by matching.

Proof. Clear. □

3.2 Generic Expressions

A more interesting condition for solvability by matching can be obtained using types. For example, assume the standard list notation and consider a term $t = [x|y]$ with x and y variables. Note that whenever a list l unifies with t, then l is an instance of t, i.e $l = t$ is solvable by matching.

Thus solvability by matching can be sometimes deduced from the shape of the considered terms. This motivates the following definition.

Definition 10. Let T be a type. A term t is a *generic expression* for T if for every $s \in T$ disjoint with t, if s unifies with t then s is an instance of t. □

In other words, t is a generic expression for type T iff all left-right disjoint equations $s = t$, where $s \in T$, are solvable by matching. Note that a generic expression for type T needs not to be a member of T.

Example 1.

- 0, $s(x)$, $s(s(x))$, ... are generic expressions for the type *Nat*,
- $[\,]$, $[x]$, $[x|y]$, $[x|x]$, $[x, y|z]$, ... are generic expressions for the type *List*. □

Next, we provide some important examples of generic expressions which will be used in the sequel.

Lemma 11. *Let T be a type. Then*

- *variables are generic expressions for T,*
- *the only generic expressions for type U are variables,*
- *if T does not contain variables, then every pure variable term is a generic expression for T,*
- *if T is ground, then every term is a generic expression for T.*

Proof. Clear. □

When the types are defined by structural induction (as for example in Bronsard, Lakshman and Reddy [BLR92] or in Yardeni, T. Frühwirth and E. Shapiro [YFS92]), then it is easy to characterize the generic expressions for each type by structural induction.

We can now provide another simple test for establishing solvability by matching.

Lemma 12 (Matching 2). *Consider two disjoint typed atoms A and H with the same relation symbol. Suppose that*

- *A is correctly typed,*
- *the positions of H are filled in by mutually disjoint terms and each of them is a generic expression for its position's type.*

Then $A = H$ is solvable by matching. Moreover, if A and H are unifiable, then a substitution θ with $Dom(\theta) \subseteq Var(H)$ exists such that $A = H\theta$.

Proof. Clear. □

3.3 Solvability by Iterated Matching

Consider a selected atom A and the head H of an input clause used to resolve A. In presence of modes the input and output positions can be used to model a parameter passing mechanism as follows. First the input values are passed from the selected atom A to the head H. Then the output values are passed from H to A.

To formalize and extend this idea we introduce the following notion where passing a value is modeled by matching.

Definition 13. Let E be a left-right disjoint set of equations.

- We say that E is *solvable by iterated matching* if E is unifiable implies that for some $E_1, \ldots, E_n$ and substitutions $\theta_1, \ldots, \theta_n$
  - $E = \dot{\bigcup}_{j=1}^{n} E_j$,
  and for $i \in [1, n]$
  - $E_i\theta_1 \ldots \theta_{i-1}$ is left-right disjoint,
  - θ_i is a match for $E_i\theta_1 \ldots \theta_{i-1}$. □

We shall also call it *double matching* when $n = 2$. In fact, in this paper we shall only study this form of iterated matching.

Note that when $\theta_1, \ldots, \theta_n$ satisfy the above three conditions, then by Lemma 2 $\theta_1\theta_2 \ldots \theta_n$ is a relevant mgu of E.

A slightly less general definition of solvability by (iterated) matching was considered by Maluszynski and Komorowski [MK85], where for $E = \{s_1 = t_1, \ldots, s_n = t_n\}$ the fixed partition $E = \dot{\bigcup}_{j=1}^{n} E_j$ with $E_j = \{s_j = t_j\}$ is used.

According to this terminology the above modeling of a parameter passing mechanism amounts to solvability by double matching.

To study solvability by double matching, modes and types are useful. Again, let us consider the case of passing the input values from a selected atom A to the head H of the clause used to resolve A. In presence of types, we can expect those input values to be correctly typed. Then the Matching 2 Lemma 12 can be applicable to deal with the input positions. If we are able to combine it with the Matching 1 Lemma 9 applied to the output positions, we can then conclude that $A = H$ is solvable by double matching. This observation is at the base of the following definitions.

Definition 14. An atom is called *input safe* if

- each of its input positions is filled in with a generic expression for this position's type,
- either the types of all input positions are ground or the terms filling in the input positions are mutually disjoint. □

In particular, an atom is *input safe* if the types of all input positions are ground.

Definition 15. An atom is called *input-output disjoint* if the family of terms occurring in its input positions has no variable in common with the family of terms occurring in its output positions. □

Definition 16. An atom A is called *i/o regular* if

(i) it is correctly typed in its input positions,
(ii) it is input-output disjoint,
(iii) each of its output positions is filled in by a distinct variable. □

We now prove a result allowing us to conclude that $A = H$ is solvable by double matching.

Lemma 17 (Double Matching). *Consider two disjoint atoms A and H with the same relation symbol. Suppose that*

- *A is i/o regular,*
- *H is input safe,*

Then $A = H$ is solvable by double matching.

Proof. Assume that $A = H$ is unifiable. Take as E_1 the subset of $A = H$ corresponding to the input positions, as E_2 the subset of $A = H$ corresponding to the output positions.

By the Matching 1 Lemma 9 or the Matching 2 Lemma 12 E_1 is solvable by matching, and it determines a match θ_1 such that $Dom(\theta_1) \subseteq Var(H)$ and $Ran(\theta_1) \subseteq VarIn(A)$. But A is input-output disjoint, so $Ran(\theta_1) \cap VarOut(A) = \emptyset$. Thus $E_2\theta_1$ is left-right disjoint. Applying to $E_2\theta_1$ the Matching 1 Lemma 9 we get a match θ_2 for $E_2\theta_1$ such that $Dom(\theta_2) \subseteq VarOut(A)$. □

3.4 Unification Free Programs

Recall that the aim of this paper is to clarify for what Prolog programs unification can be replaced by iterated matching. The following definition is the key one.

Definition 18.

- Let ξ be an LD-derivation. Let A be an atom selected in ξ and H the head of the input clause selected to resolve A in ξ. Suppose that A and H have the same relation symbol. Then we say that the system $A = H$ *is considered in* ξ.
- Suppose that all systems of equations considered in the LD-derivations of $P \cup \{Q\}$ are solvable by iterated matching. Then we say that $P \cup \{Q\}$ is *unification free.*

□

The Double Matching Lemma 17 allows us to conclude when $P \cup \{Q\}$ is unification free. We need this notion.

Definition 19. We call an LD-derivation *i/o driven* if all atoms selected in it are i/o regular. □

Theorem 20. *Suppose that*

- *the head of every clause of P is input safe,*
- *all LD-derivations of $P \cup \{Q\}$ are i/o driven.*

Then $P \cup \{Q\}$ is unification free. □

In order to apply this theorem we need to find conditions which imply that all considered *LD*-derivations are i/o driven. To deal with the first condition for an atom to be i/o regular we use the concept of well-typed queries and programs.

4 Well-Typed Programs

The notion of well-typed queries and programs relies on the concept of a type judgement.

Definition 21.

- By a *type judgement* we mean a statement of the form

$$\mathbf{s} : \mathbf{S} \Rightarrow \mathbf{t} : \mathbf{T}. \tag{1}$$

- We say that a type judgement (1) *is true*, and write

$$\models \mathbf{s} : \mathbf{S} \Rightarrow \mathbf{t} : \mathbf{T},$$

if for all substitutions θ, $\mathbf{s}\theta \in \mathbf{S}$ implies $\mathbf{t}\theta \in \mathbf{T}$. □

For example, the type judgement $s(s(x)) : Nat,\ l : ListNat \Rightarrow [x \mid l] : ListNat$ is true.

To simplify the notation, when writing an atom as $p(\mathbf{u} : \mathbf{S}, \mathbf{v} : \mathbf{T})$ we now assume that $\mathbf{u} : \mathbf{S}$ is a sequence of typed terms filling in the input positions of p and $\mathbf{v} : \mathbf{T}$ is a sequence of typed terms filling in the output positions of p. We call a construct of the form $p(\mathbf{u} : \mathbf{S}, \mathbf{v} : \mathbf{T})$ a *typed atom*.

The following notion is due to Bronsard, Lakshman and Reddy [BLR92].

Definition 22.

- A query $p_1(\mathbf{i_1} : \mathbf{I_1}, \mathbf{o_1} : \mathbf{O_1}), \ldots, p_n(\mathbf{i_n} : \mathbf{I_n}, \mathbf{o_n} : \mathbf{O_n})$ is called *well-typed* if for $j \in [1, n]$

$$\models \mathbf{o_1} : \mathbf{O_1}, \ldots, \mathbf{o_{j-1}} : \mathbf{O_{j-1}} \Rightarrow \mathbf{i_j} : \mathbf{I_j}.$$

- A clause $p_0(\mathbf{o_0} : \mathbf{O_0}, \mathbf{i_{n+1}} : \mathbf{I_{n+1}}) \leftarrow p_1(\mathbf{i_1} : \mathbf{I_1}, \mathbf{o_1} : \mathbf{O_1}), \ldots, p_n(\mathbf{i_n} : \mathbf{I_n}, \mathbf{o_n} : \mathbf{O_n})$ is called *well-typed* if for $j \in [1, n+1]$

$$\models \mathbf{o_0} : \mathbf{O_0}, \ldots, \mathbf{o_{j-1}} : \mathbf{O_{j-1}} \Rightarrow \mathbf{i_j} : \mathbf{I_j}.$$

- A program is called *well-typed* if every clause of it is. □

Thus, a query is well-typed if

- the types of the terms filling in the *input* positions of an atom can be deduced from the types of the terms filling in the *output* positions of the previous atoms.

And a clause is well-typed if

- ($j \in [1, n]$) the types of the terms filling the *input* positions of a body atom can be deduced from the types of the terms filling in the *input* positions of the head and the *output* positions of the previous body atoms,
- ($j = n + 1$) the types of the terms filling in the *output* positions of the head can be deduced from the types of the terms filling in the *input* positions of the head and the types of the terms filling in the *output* positions of the body atoms.

Note that a query with only one atom is well-typed iff this atom is correctly typed in its input positions. The following lemma due to Bronsard, Lakshman and Reddy [BLR92] shows persistence of the notion of being well-typed.

Lemma 23. *An LD-resolvent of a well-typed query and a disjoint with it well-typed clause is well-typed.* □

Corollary 24. *Let P and Q be well-typed, and let ξ be an LD-derivation of $P \cup \{Q\}$. All atoms selected in ξ are correctly typed in their input positions.*

Proof. A variant of a well-typed clause is well-typed and the first atom of a well-typed query is correctly typed in its input positions. □

This shows that by restricting our attention to well-typed programs and queries we ensure that all atoms selected in the LD-derivations satisfy the first condition of i/o regularity.

5 Simply Moded Programs

To ensure that the other two conditions of i/o regularity are satisfied we introduce further syntactic restrictions. Later we shall discuss how confining these restrictions are. We need a definition first.

Definition 25. A family of terms is called *linear* if every variable occurs at most once in it. □

Thus a family of terms is linear iff no variable has two distinct occurrences in any of the terms and no two terms have a variable in common.

Definition 26.

- A query $p_1(\mathbf{s_1}, \mathbf{t_1}), \ldots, p_n(\mathbf{s_n}, \mathbf{t_n})$ is called *simply moded* if $\mathbf{t_1}, \ldots, \mathbf{t_n}$ is a linear family of variables and for $i \in [1, n]$

$$Var(\mathbf{s_i}) \cap (\bigcup_{j=i}^{n} Var(\mathbf{t_j})) = \emptyset.$$

- A clause

$$p_0(\mathbf{s_0}, \mathbf{t_0}) \leftarrow p_1(\mathbf{s_1}, \mathbf{t_1}), \ldots, p_n(\mathbf{s_n}, \mathbf{t_n})$$

  is called *simply moded* if $p_1(\mathbf{s_1}, \mathbf{t_1}), \ldots, p_n(\mathbf{s_n}, \mathbf{t_n})$ is simply moded and

$$Var(\mathbf{s_0}) \cap (\bigcup_{j=1}^{n} Var(\mathbf{t_j})) = \emptyset.$$

  In particular, every unit clause is simply moded.
- A program is called *simply moded* if every clause of it is. □

Thus, assuming that in every atom the input positions occur first, a query is simply moded if

- all output positions are filled in by variables,
- every variable occurring in an output position of an atom does not occur earlier in the query.

And a clause is simply moded if

- all output positions of body atoms are filled in by variables,
- every variable occurring in an output position of a body atom occurs neither earlier in the body nor in an input position of the head.

So, intuitively, the concept of being simply moded prevents a "speculative binding" of the variables which fill in the output positions — these variables are required to be "fresh". A similar notion of nicely moded programs and queries was introduced in Chadha and Plaisted [CP91] and further studied in Apt and Pellegrini [AP92]. The difference is that the output positions do not need there to be filled in by variables.

Note that a query with only one atom is simply moded iff it is input-output disjoint and each of its output positions is filled in by a distinct variable, i.e. so iff the conditions (ii) and (iii) of i/o regularity are satisfied. The following lemma shows the persistence of the notion of being simply moded.

Lemma 27. *An LD-resolvent of a simply moded query and a disjoint with it simply moded clause is simply moded.*

Proof. First, we establish two claims.

Claim 1 *Let θ be a substitution and* $\mathbf{A}$ *a simply moded query such that* $Var(\theta) \cap VarOut(\mathbf{A}) = \emptyset$. *Then* $\mathbf{A}\theta$ *is simply moded, as well.*

Proof. θ does not affect the variables appearing in the output positions of $\mathbf{A}$ and does not introduce these variables when applied to the terms appearing in the input positions of $\mathbf{A}$. □

Claim 2 *Suppose* $\mathbf{A}$ *and* $\mathbf{B}$ *are simply moded queries such that* $VarOut(\mathbf{A}) \cap Var(\mathbf{B}) = \emptyset$. *Then* $\mathbf{B}, \mathbf{A}$ *is a simply moded query, as well.*

Proof. Immediate by the definition of a simply moded query. □

Consider now a simply moded query $A, \mathbf{A}$ and a disjoint with it simply moded clause $H \leftarrow \mathbf{B}$, such that A and H unify. Take as E_1 the subset of $A = H$ corresponding to the input positions and as E_2 the subset of $A = H$ corresponding to the output positions. Let θ_1 be a relevant mgu of E_1. Then $Var(\theta_1) \subseteq VarIn(H) \cup VarIn(A)$, so $Var(\theta_1) \cap VarOut(A) = \emptyset$, since A is input-output disjoint. Thus $E_2\theta_1$ is left-right disjoint and by virtue of the Matching 1 Lemma 9 is solvable by matching. Let θ_2 be a match for $E_2\theta_1$. Then $Dom(\theta_2) \subseteq VarOut(A)$ and $Ran(\theta_2) \subseteq VarOut(H)$.

Let $\theta = \theta_1\theta_2$. Then $Var(\theta) \subseteq Var(A) \cup Var(H)$, so by the disjointness assumption and the definition of simply modedness $Var(\theta) \cap VarOut(\mathbf{A}) = \emptyset$. Thus by Claim 1 $\mathbf{A}\theta$ is simply moded.

Next, $\theta = \theta_1 \dot\cup \theta_2$, since A is input-output disjoint. So by the disjointness assumption $\mathbf{B}\theta = \mathbf{B}\theta_1$. But $Var(\theta_1) \cap VarOut(\mathbf{B}) \subseteq (VarIn(A) \cup VarIn(H)) \cap VarOut(\mathbf{B}) = \emptyset$, so by Claim 1 $\mathbf{B}\theta$ is simply moded.

Finally, $VarOut(\mathbf{A}\theta) = VarOut(\mathbf{A})$ and $Var(\mathbf{B}\theta) \subseteq Var(\mathbf{B}) \cup Var(A) \cup Var(H)$, so by the disjointness assumption and the definition of simply modedness we have that $VarOut(\mathbf{A}\theta) \cap Var(\mathbf{B}\theta) = \emptyset$. By Claim 2 $(\mathbf{B}, \mathbf{A})\theta$ is simply moded. Now by Lemma 2 θ is an mgu of A and H, so $(\mathbf{B}, \mathbf{A})\theta$ is a resolvent of $A, \mathbf{A}$ and $H \leftarrow \mathbf{B}$.

$\theta = \theta_1\theta_2$ is just one specific mgu of $A = H$. By Lemma 1 every other mgu of $A = H$ is of the form $\theta\eta$ for a renaming η. But a renaming of a simply moded query is simply moded, so we conclude that every LD-resolvent of $A, \mathbf{A}$ and $H \leftarrow \mathbf{B}$ is simply moded. □

It is useful to note that the above Lemma can be easily established as a consequence of Lemma 5.9 of Apt and Pellegrini [AP92] stating persistence of the notion of being nicely moded. To keep the paper self-contained we preferred to give here a direct proof.

The following immediate consequence show that the notion of being simply moded is the one we need.

Corollary 28. *Let P and Q be simply moded, and let ξ be an LD-derivation of $P \cup \{Q\}$. All atoms selected in ξ are input-output disjoint and such that each of their output positions are filled in by a distinct variable.*

Proof. A variant of a simply moded clause is simply moded and the first atom of a simply moded query is input-output disjoint and each of its output positions is filled in by a distinct variable. □

Theorem 29. *Suppose that*

- *P and Q are well-typed and simply moded,*

Then all LD-derivations of $P \cup \{Q\}$ are i/o driven.

Proof. By Corollaries 24 and 28. □

This brings us to the desired conclusion.

Theorem 30 (Main). *Suppose that*

- *P and Q are well-typed and simply moded,*
- *the head of every clause of P is input safe.*

Then $P \cup \{Q\}$ is unification free.

Proof. By Theorems 20 and 29. □

6 Examples

Let us see now how the established result can be applied to specific programs. When presenting the programs we adhere here to the usual syntactic conventions of Prolog with the exception that Prolog's ":-" is replaced by the logic programming " ← ".

(i) Consider the proverbial program append:

```
app([X | Xs], Ys, [X | Zs]) ← app(Xs, Ys, Zs).
app([], Ys, Ys).
```

with the typing app(+: *List*,+: *List*,-: *List*). First note that append is well-typed in the assumed typing. Indeed, the following type judgements are true:

$$[X|Xs] : List \Rightarrow Xs : List,$$
$$Ys : List \Rightarrow Ys : List,$$
$$Zs : List \Rightarrow [X|Zs] : List.$$

append is also obviously simply moded and the heads of all clauses are input safe. By the Main Theorem 30 we conclude that for lists s and t, and a variable u , append ∪ { app(s, t, u)} is unification free.

(ii) Examine now the program append with the typing app(-: *List*,-: *List*,+: *List*). First note that by virtue of the same type judgements as above append is well-typed. Moreover, append is also simply moded and the heads of all clauses are input safe. The Main Theorem 30 yields that for a list u and variables s,t, append ∪ { app(s, t, u)} is unification free.

(iii) Consider now the program permutation sort which is often used as a benchmark program.

```
ps(Xs, Ys) ←permutation(Xs, Ys), ordered(Ys).

permutation(Xs, [Y | Ys]) ←
    select(Y, Xs, Zs),
    permutation(Zs, Ys).
permutation([], []).

select(X, [X | Xs], Xs).
select(X, [Z | Xs], [Z | Zs]) ← select(X, Xs, Zs).

ordered([]).
ordered([X]).
ordered([X, Y | Xs]) ←X ≤ Y, ordered([Y| Xs]).
```

With the following typing: ps(+: *List*,-: *List*), permutation(+: *List*, -: *List*), select(-: *U*,+: *List*,-: *List*), ≤(+: *U*,+: *U*), ordered(+: *List*), the program is well-typed. Indeed, in addition to the above type judgements the following type judgement is true:

$$[X, Y|Xs] : List \Rightarrow [Y|Xs] : List.$$

`permutation sort` is also simply moded and the heads of all clauses are input safe. By the Main Theorem 30 we get that for a list `s` and a variable `t`, `permutation sort` ∪ { `ps(s, t)`} is unification free.

In all the examples seen before, the generic expressions which were filling in the input positions of the clauses were always either variables or pure variable terms. This is not the case with `permutation sort`. Indeed, the terms `[X]` and `[X, Y | Xs]`, filling in the input positions of, respectively, the first and the third clause defining the relation `ordered`, are generic expressions for *List*, but are not pure variable terms. In a sense we could say that `[X]` and `[X, Y | Xs]` are nontrivial generic expressions.

(iv) Finally, consider the program `in-order` which converts a (n ordered) binary tree into a (n ordered) list and consists in the following clauses:

```
in-order(tree(X, L, R), Xs) ←
   in-order(L, Ls),
   in-order(R, Rs),
   app(Ls, [X | Rs], Xs).
in-order(void, []).
```

augmented by the `append` program.

The type *BinTree* can be defined recursively as follows:
- `void` is a binary tree,
- if `l` and `r` are binary trees and `label` is a term, then `tree(label, l, r)` is a binary tree.

With the typing `in-order(+:`*BinTree*`,-:`*List*`)`, `app(+:`*List*`,+:`*List*`,-:`*List*`)`, the program `in-order` is well-typed. Indeed, in addition to the above type judgements the following type judgements are clearly true:

$$tree(X, L, R) : BinTree \Rightarrow L : BinTree,$$
$$tree(X, L, R) : BinTree \Rightarrow R : BinTree.$$

`in-order` is also simply moded and the heads of all clauses are input safe. By the Main Theorem 30 we conclude that for a binary tree `t` and a variable `s`, `in-order` ∪ { `in-order(t, s)`} is unification free.

7 Avoiding Unification Using Modes

When trying to apply the Main Theorem 30 one has to verify whether for a given typing a program and a query are well-typed. This can be inefficient and for some artificially constructed types even undecidable. However, when dealing with programs which use ground inputs the conditions of the Main Theorem 30 can be efficiently tested for a given moding, program and query.

This is due to the fact that it is possible then to formulate this theorem without explicit reference to types. The notion which is sufficient then is that of a well-moded program. The concept is essentially due to Dembinski and Maluszynski [DM85a]; we use here an elegant formulation due to Rosenblueth [Ros91]. To simplify the notation, when writing an atom as $p(\mathbf{u}, \mathbf{v})$, we now assume that $\mathbf{u}$ is a sequence of terms filling in the input positions of p and that $\mathbf{v}$ is a sequence of terms filling in the output positions of p.

Definition 31.

- A query $p_1(\mathbf{s_1}, \mathbf{t_1}), \ldots, p_n(\mathbf{s_n}, \mathbf{t_n})$ is called *well-moded* if for $i \in [1, n]$

$$Var(\mathbf{s_i}) \subseteq \bigcup_{j=1}^{i-1} Var(\mathbf{t_j}).$$

- A clause

$$p_0(\mathbf{t_0}, \mathbf{s_{n+1}}) \leftarrow p_1(\mathbf{s_1}, \mathbf{t_1}), \ldots, p_n(\mathbf{s_n}, \mathbf{t_n})$$

is called *well-moded* if for $i \in [1, n+1]$

$$Var(\mathbf{s_i}) \subseteq \bigcup_{j=0}^{i-1} Var(\mathbf{t_j}).$$

- A program is called *well-moded* if every clause of it is. □

Thus, a query is well-moded if

- every variable occurring in an input position of an atom ($i \in [1, n]$) occurs in an output position of an earlier ($j \in [1, i-1]$) atom.

And a clause is well-moded if

- ($i \in [1, n]$) every variable occurring in an input position of a body atom occurs either in an input position of the head ($j = 0$), or in an output position of an earlier ($j \in [1, i-1]$) body atom,
- ($i = n+1$) every variable occurring in an output position of the head occurs in an input position of the head ($j = 0$), or in an output position of a body atom ($j \in [1, n]$).

It is useful to note that the concept of a well-moded program (resp. query) is a particular case of that of a well-typed program. Indeed, if the only type used is *Ground*, then the notions of a well-typed program (resp. query) and a well-moded program (resp. query) coincide. All programs considered in Section 6 become well-moded when the type information is dropped.

This brings us to the following special case of the Main Theorem 30.

Corollary 32. *Suppose that*

- *P and Q are simply moded and well-moded.*

Then $P \cup \{Q\}$ is unification free.

Proof. When the only types used are ground, all atoms are input safe. □

Note that for a given moding it is easy to test whether conditions of this corollary are applicable. Indeed, assume that in every atom the input positions occur first. Then a query Q is simply moded and well-moded iff

- every first from the left occurrence of a variable in Q is within an output position,
- the output positions of Q are filled in by distinct variables.

And a clause $p(\mathbf{s}, \mathbf{t}) \leftarrow \mathbf{B}$ is simply moded and well-moded iff

- every first from the left occurrence of a variable in the sequence $\mathbf{s}, \mathbf{B}, \mathbf{t}$ is in $\mathbf{s}$ or within an output position in $\mathbf{B}$,
- the output positions of $\mathbf{B}$ are filled in by distinct variables which do not appear in $\mathbf{s}$.

This corollary allows us to deal with more restricted queries than the Main Theorem 30, but in a number of cases this is sufficient.

Example 2.

(i) Examine the following program `palindrome`:

```
palindrome(Xs) ← reverse(Xs, Xs).

reverse(X1s, X2s) ← reverse(X1s, [], X2s).
reverse([X | X1s], X2s, Ys) ← reverse(X1s, [X | X2s], Ys).
reverse([], Xs, Xs).
```

With the typing `palindrome(+:`*List*`)`, `reverse(+:`*List*`, -:`*List*`)`, `reverse(+:`*List*`, +:`*List*`, -:`*List*`)`, `reverse` is simply moded, but `palindrome` is not, as the body of the first clause does not satisfy Definition 26 (the variable in the second position of `reverse` appears "earlier" twice). Switching to the typing `palindrome(+:`*List*`)`, `reverse(+:`*List*`,+:`*List*`)`, `reverse(+:`*List*`,+:`*List*`,+:`*List*`)`, does not help as now the head of the last clause is not input safe.

On the other hand, when adopting the mode `palindrome(+)`, `reverse(+,+)`, `reverse(+,+,+)`, `palindrome` is simply moded and well-moded. Hence, by Corollary 32, when `t` is ground, `palindrome` ∪ { `palindrome(t)`} is unification free.

It is worth noticing that for non-ground inputs `palindrome(+:`*List*`)` may actually require unification in order to run properly. Indeed, consider the following query: `palindrome([f(X,a),f(b,X)])`. When evaluated it eventually leads to the equation $\mathtt{f(X, a)} = \mathtt{f(b, X)}$, which to be solved requires unification.

(ii) Applying Corollary 32 to the programs handled in Section 6 we can only draw conclusions for the case when all terms filling in the input positions are ground.

8 Discussion

To apply the established results to a program and a query, one needs to find appropriate modings and typings for the considered relations such that the conditions of the Main Theorem 30 or of Corollary 32 are satisfied. In the table below several programs taken from the book of Sterling and Shapiro [SS86] are listed. For each program it is indicated for which *nonground* typing (i.e. one with some non-ground type) the Main Theorem 30 is applicable, and for which modings Corollary 32 is. All built-in's are moded completely input with all positions typed U.

In programs which use difference-lists we replaced "\" by ",", thus splitting a position filled in by a difference-list into two positions. Because of this change in some relations additional arguments are introduced, and so certain clauses have to be modified in an obvious way. For example, in the parsing program on page 258 each clause of the form `p(X) ← r(X)` has to be replaced by `p(X,Y) ← r(X,Y)`. Such changes are purely syntactic and they allow us to draw conclusions about the original program.

In the program `dutch` we refer to a new type *CList* which consists of lists of *colored objects*, where a colored object is a term of the form *red*(*s*), *blue*(*s*) or *white*(*s*).

program	page	(nonground) typings	modings
member	45	(–:*U*,+:*List*)	(–,+)
member	45	none	(+,+)
prefix	45	(–:*List*,+:*List*)	(–,+)
prefix	45	(+:*List*,–:*List*)	(+,–)
prefix	45	none	(+,+)
suffix	45	(–:*List*,+:*List*)	(–,+)
suffix	45	(+:*List*,–:*List*)	(+,–)
suffix	45	none	(+,+)
naive reverse	48	(+:*List*,–:*List*)	(+,–)
reverse-accum.	48	r(+:*List*,–:*List*)	r(+,–)
		r(+:*List*,+:*List*,–:*List*)	r(+,+,–)
reverse-accum.	48	none	r(+,+)
			r(+,+,+)
delete	53	none	(+,+,–)
delete	53	none	(+,+,+)
select	53	none	(+,+,–)
select	53	(–:*U*,+:*List*,–:*List*)	(–,+,–)
select	53	(+:*U*,–:*List*,+:*List*)	(+,–,+)
select	53	none	(+,+,+)
insertion sort	55	s(+:*List*,–:*List*)	s(+,–)
		i(+:*List*,+:*List*,–:*List*)	i(+,+,–)
quicksort	56	q(+:*List*,–:*List*)	q(+,–)
		p(+:*List*,+:*U*,–:*List*,–:*List*)	p(+,+,–,–)
		app(+:*List*,+:*List*,–:*List*)	app(+,+,–)
tree-member	58	(–:*U*,+:*BinTree*)	(–,+)
tree-member	58	none	(+,+)

isotree	58	(–:*BinTree*,+:*BinTree*)	(–,+)
isotree	58	(+:*BinTree*,–:*BinTree*)	(+,–)
isotree	58	none	(+,+)
substitute	60	none	(+,+,+,–)
pre-order	60	(+:*BinTree*,–:*List*)	(+,–)
post-order	60	(+:*BinTree*,–:*List*)	(+,–)
polynomial	62	none	(+,+)
derivative	63	none	(+,+,–)
derivative	63	none	(+,+,+)
hanoi	64	(+:*Nat*,+:*U*,+:*U*,+:*U*,–:*List*)	(+,+,+,+,–)
flatten	243	none	f(+,–) f(+,+,–)
reverse_dl	244	r(+:*List*,–:*List*) r_dl(+:*List*,–:*List*,+:*List*)	r(+,–) r_dl(+,–,+)
reverse_dl	244	none	r(+,+) r_dl(+,+,+)
dutch	246	dutch(+:*CList*,–:*CList*) di(+:*CList*,–:*CList*,–:*CList*,–:*CList*)	dutch(+,–) di(+,–,–,–)
parsing	258	none	all (+,–)

9 Conclusions

In view of the above results it is natural to ask when unification is intrinsically needed in Prolog programs. A canonic example is the Prolog program `curry` which computes a type assignment to a lambda term, if such an assignment exists (see e.g. Reddy [Red86]). We are not aware of other natural examples, though it should be added that for complicated queries which anticipate in their ouput positions the form of computed answers, almost any program will necessitate the use of unification.

In our analysis we restricted our attention to the case of programs in which the output positions of the clause bodies are filled in by variables. This obviously limits applicability of our results, since in a number of natural programs these output positions are compound terms. An example is the following program `permutation`:

`perm(Xs, Ys)` ← `Ys` is a permutation of the list `Xs`.

```
perm(Xs, [X | Ys]) ←
    app(X1s, [X | X2s], Xs),
    app(X1s, X2s, Zs),
    perm(Zs, Ys).
perm([], []).
```

augmented by the APPEND program.

in which the first call of app uses a compound term `[X | X2s]` in an output position. We checked by hand that when s is a list and x a variable, **permutation** ∪ { perm(s, x) } is unification free. Currently we are working on extension of the obtained results to the case of non-variable outputs.

This work is naturally related to the study of conditions which guarantee that Prolog programs can be executed using unification without the occur-check. It should be noted however, that unification freedom property rests exclusively upon those considered systems of equations which *are* unifiable, whereas the property of being occur-check free rests exclusively upon those considered systems which are *not* unifiable. Indeed, the occur-check is only needed to correctly identify the non-unifiable systems of equations. Still, when comparing the outcome of this paper with our previous work on the occur-check problem (Apt and Pellegrini [AP92]) we note an astonishing similarity between both classes of identified Prolog programs.

Acknowledgements

We thank Jan Heering and Alessandro Pellegrini for interesting discussions on the subject of this paper.

References

[AFZ88] I. Attali and P. Franchi-Zannettacci. Unification-free execution of TYPOL programs by semantic attribute evaluation. In R.A. Kowalski and K.A. Bowen, editors, *Proceedings of the Fifth International Conference on Logic Programming*, pages 160–177. The MIT Press, 1988.

[AP92] K. R. Apt and A. Pellegrini. Why the occur-check is not a problem. In M. Bruynooghe and M. Wirsing, editors, *Proceeding of the Fourth International Symposium on Programming Language Implementation and Logic Programming (PLILP 92)*, Lecture Notes in Computer Science 631, pages 69–86, Berlin, 1992. Springer-Verlag.

[Apt90] K. R. Apt. Logic programming. In J. van Leeuwen, editor, *Handbook of Theoretical Computer Science*, pages 493–574. Elsevier, 1990. Vol. B.

[BLR92] F. Bronsard, T.K. Lakshman, and U.S. Reddy. A framework of directionality for proving termination of logic programs. In K.R. Apt, editor, *Proc. of the Joint International Conference and Symposium on Logic Programming*, pages 321–335. MIT Press, 1992.

[CP91] R. Chadha and D.A. Plaisted. Correctness of unification without occur check in Prolog. Technical report, Department of Computer Science, University of North Carolina, Chapel Hill, N.C., 1991.

[DM85a] P. Dembinski and J. Maluszynski. AND-parallelism with intelligent backtracking for annotated logic programs. In *Proceedings of the International Symposium on Logic Programming*, pages 29–38, Boston, 1985.

[DM85b] P. Deransart and J. Maluszynski. Relating Logic Programs and Attribute Grammars. *Journal of Logic Programming*, 2:119–156, 1985.

[Llo87] J. W. Lloyd. *Foundations of Logic Programming*. Springer-Verlag, Berlin, second edition, 1987.

[LMM88] J.-L. Lassez, M. J. Maher, and K. Marriott. Unification Revisited. In J. Minker, editor, *Foundations of Deductive Databases and Logic Programming*, pages 587–625. Morgan Kaufmann, Los Altos, Ca., 1988.

[MK85] J. Maluszynski and H. J. Komorowski. Unification-free execution of logic programs. In *Proceedings of the 1985 IEEE Symposium on Logic Programming*, pages 78–86, Boston, 1985. IEEE Computer Society Press.

[Red86] U.S. Reddy. On the relationship between logic and functional languages. In D. DeGroot and G. Lindstrom, editors, *Functional and Logic Programming*, pages 3–36. Prentice-Hall, 1986.

[Ros91] D.A. Rosenblueth. Using program transformation to obtain methods for eliminating backtracking in fixed-mode logic programs. Technical Report 7, Universidad Nacional Autonoma de Mexico, Instituto de Investigaciones en Matematicas Aplicadas y en Sistemas, 1991.

[SS86] L. Sterling and E. Shapiro. *The Art of Prolog*. MIT Press, 1986.

[YFS92] E. Yardeni, T. Frühwirth, and E. Shapiro. Polymorphically typed logic programs. In F. Pfenning, editor, *Types in Logic Programming*, pages 63–90. MIT Press, Cambridge, Massachussets, 1992.

Equivalences and preorders of transition systems

A. Arnold*, A. Dicky**

LaBRI
Université Bordeaux I

Abstract. Two transition systems are logically equivalent if they satisfy the same formulas of a given logic. For some of these logics, such as Hennessy-Milner's logic, there is an algebraic characterization of this equivalence involving particular homomorphisms of transition systems. This logical equivalence is associated with a preorder: a transition system S is less than S' if all formulas satisfied by S are satisfied by S'. For particular logics, this preorder can also be algebraically characterized, using homomorphisms and a specific notion of inclusion of transition systems.

Keywords : bisimulation, temporal logic, transition system.

1 Introduction

This paper deals with labelled transition systems, as a model of the behaviour of nondeterministic processes; a general problem is to define a criterion to determine whether a transition system represents a given process.

Such a criterion is commonly provided for by some propositional logic formalizing the properties of processes, and interpreted in transition systems: a formula is interpreted as a property of the states, or paths, of a transition system, and a transition system is a model of a formula if at least one of its states (or paths) satisfies this formula. A process is thus characterized by a set of formulas, and transition systems are equivalent when they satisfy the same formulas.

Conversely, any definition of an equivalence of transition systems leads to a semantical interpretation: the characterization of a process is a class of equivalent transition systems. In particular, the different kinds of bisimulation formalize the notion of *observational equivalence* of processes, as a structural relationship between transition systems representing the same process.

It is well-known that (strong) bisimulation equivalence may be characterized by particular homomorphisms of transition systems, *transition-preserving homomorphisms* [DDN88], so that transition systems are bisimulation-equivalent if, and only if, they have a common homomorphic image.

* Université Bordeaux I

** IUT La Rochelle

More generally, any class of transition system homomorphisms satisfying the confluence property induces a similar equivalence (transition systems are equivalent when they have a common image); this algebraic characterization is particularly interesting when dealing with finite-state transition systems, since a canonical minimal transition system can be associated with every equivalence class.

The Hennessy-Milner theorem [HM85] relating bisimulation to logical equivalence with respect to Hennessy-Milner logic can be extended to some other cases, where such a characterization may be obtained for the logical equivalence, via ***saturating homomorphisms***, as defined in [AD89, A92]: roughly, the idea that a transition system homomorphism of S onto S' is ***saturating*** with respect to a logic $\mathcal{L}$ captures the property that every state (or path) of S and its image in S' satisfy the same formulas of $\mathcal{L}$ (the precise definition, which involves an algebraic interpretation of logical operators, is slightly more restrictive).

Given a transition system S, say that two states of S are ***indistinguishable*** with respect to $\mathcal{L}$ if they satisfy the same formulas of $\mathcal{L}$. The notion extends to transitions ($q_1 \xrightarrow{a} q_1'$ and $q_2 \xrightarrow{a} q_2'$ are indistinguishable whenever q_1 and q_2 on one hand, q_1' and q_2' on the other, are indistinguishable). By merging together the indistinguishable states and transitions of S, we obtain a transition system $S\,|_{\sim_{\mathcal{L}}}$, so that the canonical mapping applying the states of S onto the states of $S\,|_{\sim_{\mathcal{L}}}$ is a transition system homomorphism.

Now, provided that for every transition system S, the canonical mapping of S onto $S\,|_{\sim_{\mathcal{L}}}$ is saturating with respect to $\mathcal{L}$, the algebraic characterization is ensured: i.e., transition systems are logically equivalent if, and only if, they have a common image under saturating homomorphisms.

In this paper, we extend the algebraic approach to preorders of transition systems. The intuitive idea, suggested by K. Larsen and B. Thomsen's *modal transition systems* [LT88], is that transition systems implement **incomplete** specifications of processes: a transition system S is finer than S' if S implements a finer specification than S'.

Consider, for instance, the preorder induced by one-way simulations: say that a transition system S is finer than S' if there is a correspondence between the states of S and the states of S' such that

- every state of S' has at least a corresponding state in S;
- if a state q of S corresponds to a state q' of S', then every path of S', starting from q', can be simulated by a corresponding path of S, starting from q.

An obvious way of refining a transition system consists in adding states or transitions to that system. Indeed, it can be shown that it is the only possible refinement, up to the bisimulation equivalence; more precisely, S is finer than S' if, and only if, there are transition systems T and T', respectively bisimulation-equivalent to S and S', such that T "contains" T'.

From the logical point of view, an incomplete specification of a process may be given by a set of logical formulas, considering that a transition system implements the specification when it satisfies each of its formulas. This induces a preorder on transition systems: given a set $\mathcal{F}$ of formulas of a logic $\mathcal{L}$, we shall say that a transition system S if *finer* than S' with respect to $\mathcal{F}$ if all formulas of $\mathcal{F}$ satisfied in S' are also satisfied in S.

For instance, a logical characterization of the preorder induced by one-way simulations is given by the *positive* formulas of the Hennessy-Milner logic, and the result we mentioned may be stated as follows: a transition system S is finer than S' iff S (resp. S') is the image of some transition system T (resp. T') under a saturating homomorphism, T' "included" in T in the sense that

- every state of T is a state of T';
- if a state of T' satifies a positive formula of the Hennessy-Milner logic, the same state also satisfies this formula in T.

The algebraic characterization

$$
\begin{array}{ccc}
 & & T \quad \text{included in} \quad T' \\
 & & \text{wrt } \mathcal{F} \\
S \text{ is finer than } S' & \Leftrightarrow & h \Big\downarrow \quad \text{saturating} \quad \Big\downarrow h' \\
\text{wrt } \mathcal{F} & & \text{wrt } \mathcal{L} \\
 & & S \qquad\qquad\qquad S'
\end{array}
$$

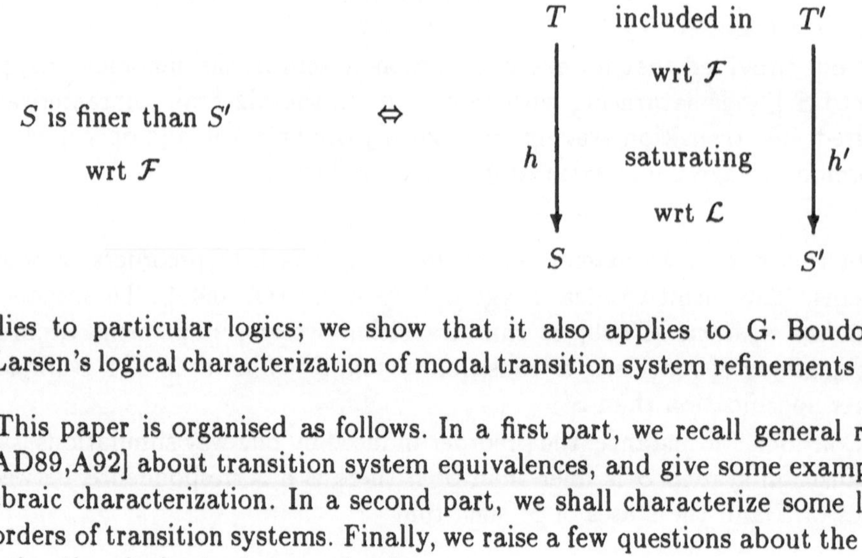

applies to particular logics; we show that it also applies to G. Boudol and K. Larsen's logical characterization of modal transition system refinements [BL92].

This paper is organised as follows. In a first part, we recall general results of [AD89,A92] about transition system equivalences, and give some examples of algebraic characterization. In a second part, we shall characterize some logical preorders of transition systems. Finally, we raise a few questions about the logics allowing the algebraic characterization.

2 Logical equivalences of transition systems

Given an set A (which elements represent the actions a process may perform), a transition system S is defined by a set Q_S of states, and a set $T_S \subseteq Q_S \times A \times Q_S$ of transitions. As usual, a transition (q_1, a, q_2) will be denoted as $q_1 \xrightarrow{a} q_2$. A path of S is a (finite or infinite) sequence of successive transitions, and will be denoted as

$$
q_0 \xrightarrow{a_1} \ldots \xrightarrow{a_n} q_n \ldots
$$

Note that our results always refer to finite transition systems[3].

A transition system *homomorphism* $h : S \longrightarrow S'$ is a mapping of Q_S into $Q_{S'}$ such that $h(T_S) \subseteq T_{S'}$, where $h(T_S) = \{h(q_1) \xrightarrow{a} h(q_2) \mid q_1 \xrightarrow{a} q_2 \in T_S\}$. The homomorphism h is said to be *surjective* if $h(Q_S) = Q_{S'}$. S' is said to be the *image* of S under h if $h(Q_S) = Q_{S'}$ and $h(T_S) = T_{S'}$.

Branching-time temporal logics, such as CTL* [ES84, EH86], applying to transition systems, may contain "state formulas" and "path formulas"; usually, the set of formulas is recursively defined using boolean connectors (disjunction, conjunction, and negation) and particular operators, each of which is given a functional interpretation in transition systems. For legibility we shall assume that logics contain only "state formulas" such as CTL [CES86] or Hennessy-Milner logic [HM85]; the results may be extended to hybrid logics [AD89,A92].

We consider that a logic $\mathcal{L}$ is defined by a set Φ_L of nonlogical operators, so that each operator $\varphi \in \Phi_L$ has a fixed arity $n \geq 0$, and is given an interpretation in every transition system S as a mapping

$$\varphi_S : (\mathcal{P}(Q_S))^n \longrightarrow \mathcal{P}(Q_S)$$

Obviously the interpretations of the logical operators $\vee, \wedge, \neg$ are defined by

$$X \vee_S Y = X \cup Y, X \wedge_S Y = X \cap Y, \neg_S X = Q_S \setminus X.$$

It follows that every closed formula f is interpreted in S as a subset f_S of Q_S.

We shall say that a transition system homomorphism $h : S \longrightarrow S'$ is *saturating* with respect to $\mathcal{L}$ if every operator $\varphi \in \Phi_L$, of arity n, satisfies the property

$$\forall (X_1, \ldots, X_n) \in \mathcal{P}(Q_{S'})^n : \varphi_S(h^{-1}(X_1), \ldots, h^{-1}(X_n)) = h^{-1}(\varphi_{S'}(X_1, \ldots, X_n))$$

Obviously, $h^{-1}(X \cup X') = h^{-1}(X) \cup h^{-1}(X')$ and $h^{-1}(X \cap X') = h^{-1}(X) \cap h^{-1}(X')$. Moreover, if h is surjective, $h^{-1}(Q_{S'} \setminus X) = Q_S \setminus h^{-1}(X)$. It follows that if h is a surjective homomorphism saturating with respect to $\mathcal{L}$, then for every closed formula f, $h^{-1}(f_{S'}) = f_S$.

Given a transition system S, say that two states q_1 and q_2 of S are *indistinguishable* with respect to $\mathcal{L}$ (notation: $q_1 \sim_{\mathcal{L}} q_2$) when they satisfy the same formulas of $\mathcal{L}$. Let h be the canonical mapping of Q_S onto the quotient set $Q_S \mid_{\sim_{\mathcal{L}}}$. Let $S \mid_{\sim_{\mathcal{L}}}$ denote the transition system whose set of states is $Q_S \mid_{\sim_{\mathcal{L}}}$, and whose transitions are the $h(q_1) \xrightarrow{a} h(q_2)$, $q_1 \xrightarrow{a} q_2$ a transition of S. By construction h is a transition system homomorphism of S onto $S \mid_{\sim_{\mathcal{L}}}$.

Proposition 1 *Let S be a transition system and $S' = S \mid_{\sim_{\mathcal{L}}}$ its quotient by indistinguishability with respect to $\mathcal{L}$. If $h : S \rightarrow S \mid_{\sim_{\mathcal{L}}}$ is saturating with respect to $\mathcal{L}$, then for any state q' of $Q_{S'}$ there is a characteristic formula f such that $f_{S'} = \{q'\}$ and $f_S = h^{-1}(q')$.*

[3] This assumption is not necessary if the logics we consider are closed under infinite conjunction and disjunction of formulas.

Proof Firstly, two distinct states of $Q_{S'}$ are always distinguishable. Let q'_1 and q'_2 be two distinct states of $Q_{S'}$ and q_1, q_2 two states of Q_S such that $q'_i = h(q_i)$. Since q_1 and q_2 have distinct images under h they are distinguishable and there is a formula f such that $q_1 \in f_S$ and $q_2 \notin f_S$. Since h is saturating, $f_S = h^{-1}(f_{S'})$, thus, $q'_1 = h(q_1) \in h(f_S) = h(h^{-1}(f_{S'})) \subseteq f_{S'}$ and $q'_2 = h(q_2) \in h((\neg f)_S) = h(h^{-1}((\neg f)_{S'}) \subseteq (\neg f)_{S'}$.

Let q' be a state of $Q_{S'}$. For any state $q'' \neq q'$ there is a formula $f_{q''}$ such that $q' \in (f_{q''})_{S'}$ and $q'' \notin (f_{q''})_{S'}$. A characteristic formula of q'' is $\bigwedge_{q''\neq q'} f_{q''}$ since $(\bigwedge_{q''\neq q'} f_{q''})_{S'} = \{q'\}$.

□

We shall say that the logic $\mathcal{L}$ is *fully adequate* if for every transition system S, the canonical homomorphism of S onto $S\,|_{\sim_\mathcal{L}}$ is saturating with respect to $\mathcal{L}$.

Example 2.1 *A non fully adequate logic*

Let us consider the logic which contains as unique nonlogical operator the nullary operator *Inf* whose interpretation Inf_S in a transition system S is the set of origins of infinite paths. In the transition system S having the unique transition $q_1 \xrightarrow{a} q_2$, $Inf_S = \emptyset$. Therefore q_1 and q_2 are indistiguishable. The quotient $S' = S\,|_{\sim_\mathcal{L}}$ contains the unique transition $q \xrightarrow{a} q$, hence, $Inf_{S'} = \{q\}$. If h is the canonical mapping from S onto S', $h^{-1}(Inf_{S'}) = \{q_1, q_2\} \neq Inf_S$, thus h is not saturating and this logic is not fully adequate.

□

Say that transition systems S_1 and S_2 are *equivalent* with respect to $\mathcal{L}$ (we shall use the same notation: $S_1 \sim_\mathcal{L} S_2$) when for every formula f of $\mathcal{L}$, some state of S_1 satisfies f iff some state of S_2 satisfies f, i.e., $f_{S_1} \neq \emptyset$ iff $f_{S_2} \neq \emptyset$.

Proposition 2 *If $\mathcal{L}$ is fully adequate, then two finite transition systems S_1 and S_2 are equivalent with respect to $\mathcal{L}$ iff there exist a transition system S and two surjective homomorphisms $h_1 : S_1 \to S$ and $h_2 : S_2 \to S$, saturating with respect to $\mathcal{L}$.*

Proof
if: Let $h_1 : S_1 \to S$ and $h_2 : S_2 \to S$ be two surjective saturating homomorphisms. For every formula f, $f_{S_1} = h_1^{-1}(f_S) \neq \emptyset$ iff $f_S \neq \emptyset$ iff $f_{S_2} = h_2^{-1}(f_S) \neq \emptyset$.
only if: Let S be the disjoint union of S_1 and S_2 and define, for $X_1, \ldots, X_n \in \mathcal{P}(Q_S)$ and $\varphi \in \Phi_L$,

$$\varphi_S(X_1, \ldots, X_n) = \varphi_{S_1}(X_1 \cap Q_{S_1}, \ldots, X_n \cap Q_{S_1}) \cup \varphi_{S_2}(X_1 \cap Q_{S_2}, \ldots, X_n \cap Q_{S_2})$$

so that

$$\varphi_S(X_1, \ldots, X_n) \cap Q_{S_i} = \varphi_{S_i}(X_1 \cap Q_{S_i}, \ldots, X_n \cap Q_{S_i}).$$

Let h_i be the restriction of the surjective saturating homorphism $h : S \rightarrow S' = S\,|_{\sim_{\mathcal{L}}}$ to S_i, i.e., $h_i^{-1}(X) = h^{-1}(X) \cap Q_{S_i}$. We have

$$\begin{aligned} h_i^{-1}(\varphi_{S'}(X_1,\ldots,X_n)) &= h^{-1}(\varphi_{S'}(X_1,\ldots,X_n)) \cap Q_{S_i} \\ &= \varphi_S(h^{-1}(X_1),\ldots,h^{-1}(X_n)) \cap Q_{S_i} \\ &= \varphi_{S_i}(h^{-1}(X_1) \cap Q_{S_i},\ldots,h^{-1}(X_n) \cap Q_{S_i}) \\ &= \varphi_{S_i}(h_i^{-1}(X_1),\ldots,h_i^{-1}(X_n)) \end{aligned}$$

hence, h_i is saturating. To prove that h_i is surjective, i.e., $\forall q \in Q_{S'}, h_i^{-1}(q) \neq \emptyset$, let us consider, for each state q the characteristic formula f such that $f_{S'} = \{q\}$ defined in Proposition 1; it follows that $h_i^{-1}(q) = f_{S_i}$ and $h^{-1}(q) = f_{S_1} \cup f_{S_2}$. Since $f_{S_1} = \emptyset$ iff $f_{S_2} = \emptyset$, we get from the fact that $h^{-1}(q) \neq \emptyset$, $f_{S_1} \neq \emptyset$ and $f_{S_2} \neq \emptyset$.

□

Example 2.2 *Characterization of the strong bisimulation equivalence by the Hennessy-Milner logic*

The Hennessy-Milner logic is algebraically defined by an operator $\langle a \rangle$ for each transition label a: $\langle a \rangle_S X$ is the set of states q such that there is a transition $q \xrightarrow{a} r$, r a state of X.

The proof of Hennessy-Milner theorem amounts to proving that this logic is fully adequate, and it is easy to see that homomorphisms saturating the operators $\langle a \rangle$ are precisely the transition-preserving homomorphisms characterizing the strong bisimulation equivalence. Indeed, if h is a homomorphism from S onto S', for any $r' \in Q_{S'}$, the equality $h^{-1}(\langle a \rangle_{S'}(\{r'\}) = \langle a \rangle_S(h^{-1}(\{r'\}))$ is equivalent to the property:

$$\forall q \in Q_S, h(q) \xrightarrow{a} r' \in T_{S'} \Rightarrow \exists r \in Q_S : q \xrightarrow{a} r \text{ and } h(r) = r'.$$

Moreover, we shall show that in this case the two homomorphisms h_1 and h_2 constructed in the second part of the proof of Proposition 1 satisfy also $h_i(T_{S_i}) = T_{S'}$, so that Proposition 1 can be stated in the following form: *For Hennessy-Milner logic, two finite transition systems are logically equivalent iff they have a common image under saturating homomorphisms.*

Let $q_1' \xrightarrow{a} q_2'$ be a transition of $T_{S'}$. We have to show that there is a transition $q_1 \xrightarrow{a} q_2$ in T_{S_i} such that $q_j' = h_i(q_j)$. Let us consider the characteristic formulas f and f' of q_1' and q_2' and the formula $g = f \wedge \langle a \rangle f'$. Obviously, $q_1' \in g_{S'}$, thus $g_{S_i} = h_i^{-1}(g_{S'}) \neq \emptyset$. Let q_1 be in g_{S_i}. By definition of g and of $\langle a \rangle_{S_i}$, there is a transition $q_1 \xrightarrow{a} q_2$ of T_{S_i} such that $q_1 \in f_{S_i}$ and $q_2 \in f'_{S_i}$, i.e., $h_i(q_1) \in f_{S'} = \{q_1'\}$ and $h_i(q_2) \in f'_{S'} = \{q_2'\}$.

□

Example 2.3 *The "Future Perfect" Logic*

A *generalized* transition system [HS84] is a transition system in which infinite paths are restricted by arbitrary fairness constraints; thus it is defined by a transition system S, and a subset I_S of the infinite paths of S.

M. Hennessy and C. Stirling have defined a logic J_T that characterizes the bisimulation equivalence of generalized transition systems; the formulas of J_T apply to finite paths of transition systems, but its restriction to states (as paths of length 0) provide a natural definition of $S \mid_{\sim_{J_T}}$. It can be shown [AD89] that J_T is fully adequate; indeed, a saturating homomorphism h of S onto S' is a transition-preserving homomorphism such that $h(I_S) = I'_S$.

□

Example 2.4 *The "until" Hennessy-Milner logic and the branching bisimulation*

Assume that the alphabet A contains a particular action ε (label of unobservable transitions). The dyadic "until" operator $\langle a \rangle$ [NV90] interpretes as follows: $\langle a \rangle_S(X, Y)$ is the set of states q such that

either $a = \varepsilon$ and $q \in Y$,
or there is a path $q = q_0 \xrightarrow{\varepsilon} \ldots \xrightarrow{\varepsilon} \ldots q_n \xrightarrow{a} r$ with $n \geq 0$, $q_0, \ldots, q_n \in X$ and $r \in Y$.

It can easily be shown, by considering $\langle a \rangle_{S'}(\{q'\}, \{r'\})$, that a transition system homomorphism $h : S \longrightarrow S'$ saturates the dyadic operators $\langle a \rangle$ iff for each transition $q' = h(q) \xrightarrow{a} r'$ of S',

either $a = \varepsilon$ and $r' = q'$,
or there is a path $q = q_0 \xrightarrow{\varepsilon} \ldots \xrightarrow{\varepsilon} \ldots q_n \xrightarrow{a} r$ with $h(q_0) = \ldots = h(q_n) = q'$ and $h(r) = r'$.

In particular, if a homomorphism saturates the unary operator $\langle a \rangle$ of the Hennessy-Milner logic, it also saturates the dyadic $\langle a \rangle$ of the "until" Hennessy-Milner logic.

The logic is fully adequate, and saturating homomorphisms provide an algebraic characterization of the branching bisimulation; indeed, the minimal equivalent of a transition system is unique up to ε-loops (i.e, transitions $q \xrightarrow{\varepsilon} q$): if S and S' are equivalent their quotients $S \mid_{\sim_\mathcal{L}}$ and $S' \mid_{\sim_\mathcal{L}}$ have the same set of states, and the only transitions which belong to one quotient and not to the other are ε-loops.

□

3 Logical preorders of transition systems

Let $\mathcal{L}$ be a logic, $\mathcal{F}$ a set of formulas of this logic. We shall say that a transition system S is *finer* than S' with respect to $\mathcal{F}$ (notation: $S' \sqsubseteq_\mathcal{F} S$) when for each formula f of $\mathcal{F}$, if some state of S' satisfies f, then some state of S satisfies f, i.e., $f_{S'} \neq \emptyset \Longrightarrow f_S \neq \emptyset$.

Obviously, this relation is a preorder and the associated equivalence $\sim_{\mathcal{F}}$ is defined by $S \sim_{\mathcal{F}} S'$ iff $\forall f \in \mathcal{F}, f_S \neq \emptyset \Leftrightarrow f_{S'} \neq \emptyset$. In case $\mathcal{F}$ is the set of all formulas, we get the logical equivalence of the previous section.

The following result is an immediate consequence of the definition of a saturating homomorphism (cf. *if* part of the proof of Proposition 2).

Proposition 3 *Let S and S' be two transition systems. If there is a surjective homomorphism from S onto S', saturating with respect to $\mathcal{L}$, then $S \sim_{\mathcal{F}} S'$.*

Now, we establish a relation between the states of S' and the states of S when $S' \sqsubseteq_{\mathcal{F}} S$.

Proposition 4 *If $\mathcal{F}$ is closed under boolean conjunction, then transition systems S and S' satisfy $S' \sqsubseteq_{\mathcal{F}} S$ iff the relation $R \subseteq Q_S \times Q_{S'}$, defined by*

> $(q, q') \in R$ *iff for each formula f of $\mathcal{F}$, if q' satisfies f in S', then q satisfies f in S*

is total on $Q_{S'}$.

Proof

if: Obvious.

only if: For any state $q' \in Q_{S'}$, let $F_{q'}$ be the set of formulas $f \in \mathcal{F}$ such that $q' \in f_{S'}$. Define R by $(q, q') \in R$ iff $q \in \bigcap_{f \in F_{q'}} f_S$. It follows that $(q, q') \in R$ iff for all f if q' satisfies f then q satisfies f. It remains to prove that R is total on $Q_{S'}$, i.e., $\bigcap_{f \in F_{q'}} f_S \neq \emptyset$. If this set is empty, for every $q \in Q_S$ there is $f_q \in F_{q'}$ such that $q \notin (f_q)_S$. Since $\mathcal{F}$ is closed under conjunction, $g = \bigwedge_{q \in Q_S} f_q$ is in $\mathcal{F}$ and $q' \in g_{S'}$. On the other hand, $g_S = \bigcap_{q \in Q_S} (f_q)_S = \emptyset$, a contradiction.

□

This applies, in particular, when $\mathcal{F}$ is the set of *positive* formulas of the logic $\mathcal{L}$ (i.e. formulas in which no boolean connector $\neg$ occurs).

We shall say that S' is *included* in S with respect to $\mathcal{F}$ (notation: $S' \subseteq_{\mathcal{F}} S$) when $Q_{S'} \subseteq Q_S$ and for each formula f of $\mathcal{F}$, if a state of S' satisfies f, it also satisfies f in S, i.e., $f_{S'} \subseteq f_S$ (indeed S is obtained by adding states to S', and modifying the set of transitions so that the properties of states of S' remain unchanged).

Although this relation $\subseteq_{\mathcal{F}}$ is defined logically, it may have a more algebraic characterization. A logic $\mathcal{L}$ is *monotonic* if for all its operators φ and all transition systems S, the interpretation φ_S of φ in S is monotonic, i.e.,

$$\forall X_1 \subseteq X_1' \subseteq Q_S, \ldots, X_n \subseteq X_n' \subseteq Q_S, \ \varphi_S(X_1, \ldots, X_n) \subseteq \varphi_S(X_1', \ldots, X_n').$$

It should be noted that all particular logics considered in this paper are monotonic.

Proposition 5 *Let $\mathcal{L}$ be a monotonic logic. Let S and S' be two transition systems such that $Q_{S'} \subseteq Q_S$. If for all operators $\varphi \in \Phi$, and for all $X_1, \ldots, X_n$ included in $Q_{S'}$, $\varphi_{S'}(X_1, \ldots, X_n) \subseteq \varphi_S(X_1, \ldots, X_n)$, then for any set $\mathcal{F}$ of positive formulas, $S' \subseteq_{\mathcal{F}} S$.*

Proof It is easy to show, by induction on any positive formula f, that $f_{S'} \subseteq f_S$.

□

Clearly, $S' \subseteq_{\mathcal{F}} S \Longrightarrow S' \sqsubseteq_{\mathcal{F}} S$. Because of Proposition 3 we get

Proposition 6 *If $S'_1 \subseteq_{\mathcal{F}} S_1$, and if there are transition systems S' and S, and surjective homomorphisms $h : S_1 \longrightarrow S$ and $h' : S'_1 \longrightarrow S'$, saturating with respect to $\mathcal{L}$, then $S' \sqsubseteq_{\mathcal{F}} S$.*

For at least two logics, we can prove the converse of this characterization:

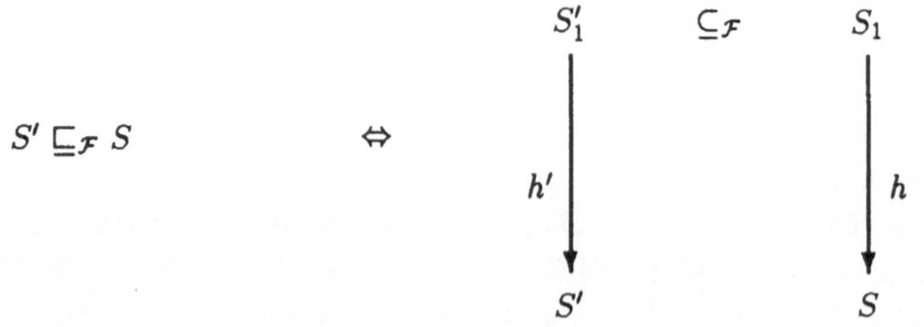

Let us show the construction of S'_1 and S_1 in the case of

Example 3.1 *The positive "until" Hennessy-Milner logic*

Let $\mathcal{F}$ denote the positive formulas of the "until" Hennessy-Milner logic given in Example 2.4. Since this set of formulas is closed under conjunction, let R be a relation as stated in Proposition 4. We define

- $Q_{S'_1} = R$, and $(q_1, q'_1) \xrightarrow{a} (q_2, q'_2)$ is a transition of S'_1 whenever $q'_1 \xrightarrow{a} q'_2$ is a transition of S'.
- $Q_{S_1} = R \cup Q_S$, and the transitions of S_1 are the transitions $(q_1, q'_1) \xrightarrow{a} q_2$ and $q_1 \xrightarrow{a} q_2$ such that $q_1 \xrightarrow{a} q_2$ is a transition of S.

By construction, the canonical projections π and π' of S_1 and S'_1 onto S and S' are transition system homomorphisms that saturate the monadic $\langle a \rangle$ (for π it is obvious, for π' we use the fact that R is total on $Q_{S'}$), thus they are saturating with respect to the "until" Hennessy-Milner logic, as explained in the Example 2.4.

Now, let us check that $S'_1 \subseteq_{\mathcal{F}} S_1$. Let f be a formula and let us prove that $f_{S'_1} \subseteq f_{S_1}$. We know that $f_{S'_1} = \pi'^{-1}(f_{S'})$, thus, $(q, q') \in f_{S'_1}$ implies $q' \in f_{S'}$. Since $(q, q') \in R$, we have $q \in f_S$. Thus, $(q, q') \in \pi^{-1}(f_S) = f_{S_1}$. □

Example 3.2 *Modal transition system refinements*

Modal transition systems were introduced by K. Larsen and B. Thomsen [LT88] as incomplete specifications of processes: a modal transition system S is defined by a set Q_S of states, a set T_S^N of *necessary* transitions, and a set T_S^P of *possible* transitions; a consistency rule specifies that any necessary transition is also a possible transition.

If S and S' are two modal transition systems, a binary relation $R \subseteq Q_S \times Q_{S'}$ is a *refinement* of S' by S if for every $(q, q') \in R$

1. if $q' \xrightarrow{a} q'_1$ is a necessary transition of S', there is a necessary transition $q \xrightarrow{a} q_1$ in S such that $(q_1, q'_1) \in R$;
2. if $q \xrightarrow{a} q_1$ is a possible transition of S, there is a possible transition $q' \xrightarrow{a} q'_1$ in S' such that $(q_1, q'_1) \in R$.

We say that a refinement R is *complete* if $\mathrm{codom}(R) = Q_{S'}$.

Following [BL92] we write $q \sqsubseteq q'$ if there is a refinement R such that $R(q, q')$. This relation is itself a refinement.

An obvious way of refining a modal transition system S' consists in specifying some of its possible (but not necessary) transitions, by either adding them to $T_{S'}^N$, or removing them from $T_{S'}^P$. Formally, $S' \trianglelefteq S$ if

$$Q_{S'} \subseteq Q_S, T_{S'}^N \subseteq T_S^N, T_S^P \subseteq T_{S'}^P.$$

It is clear that if $S' \trianglelefteq S$, the relation $\{(q', q') \mid q' \in Q_{S'}\}$ is a complete refinement of S' by S.

G. Boudol and K. Larsen [BL92] have defined a logic characterizing modal transition system refinements: the basic operators of this logic are the usual operator $\langle a \rangle$ of the Hennessy-Milner logic, interpreted over **necessary** transitions, and the operator $[a]$, interpreted as the dual of $\langle a \rangle$ over **possible** transitions. Formally, the interpretations of $\langle a \rangle$ and $[a]$ in a modal transition systems $S = (Q_S, T_S^P, T_S^N)$ are defined by

$$\langle a \rangle_S(X) = \{q \in Q_S \mid \exists q' \in Q_S : q \xrightarrow{a} q' \in T_S^N, q' \in X\},$$
$$[a]_S(X) = \{q \in Q_S \mid \forall q' \in Q_S, q \xrightarrow{a} q' \in T_S^P \Rightarrow q' \in X\},$$

for any subset X of Q_S. The formulas of the logic are the closure of $\{true, false\}$ under these operators and conjunction and disjunction.

From now on let $\mathcal{L}$ denote the logic defined by the operators $\langle a \rangle$ and $[a]$ and $\mathcal{F}$ denote the set of positive formulas of $\mathcal{L}$.

Proposition 7 *Let S and S' be two modal transition systems. If $S' \trianglelefteq S$ then $S' \subseteq_{\mathcal{F}} S$.*

Proof If $S' \trianglelefteq S$ then $Q_{S'} \subseteq Q_S$. Since this logic is monotonic, by Proposition 5 we have to prove $\langle a\rangle_{S'}(X) \subseteq \langle a\rangle_S(X)$ and $[a]_{S'}(X) \subseteq [a]_S(X)$. The first inclusion is an immediate consequence of $T^N_{S'} \subseteq T^N_S$ and the second one of $T^P_S \subseteq T^P_{S'}$.

□

Boudol and Larsen proved that $q' \sqsubseteq q$ iff for any formula f of $\mathcal{F}$, $q' \in f_{S'} \Rightarrow q \in f_S$. From our point of view, this characterization may be stated as $S' \sqsubseteq_{\mathcal{F}} S$ iff $\forall q' \in Q_{S'}, \exists q \in Q_S : q' \sqsubseteq q$ iff there is a complete refinement of S' by S. We are going to prove a more precise statement of this result.

Proposition 8 *A modal transition system S is finer than S' with respect to $\mathcal{F}$ iff there are modal transition systems S_1 and S'_1 such that $S'_1 \trianglelefteq S_1$, and that S (resp. S') is the image of S_1 (resp. S'_1) under a surjective homomorphism, saturating with respect to $\mathcal{L}$.*

Proof
if: Obvious, by Proposition 6 and 7
only if: Let R be the relation defined in Proposition 4. We only give a construction of S'_1 and S_1:

- the states of S'_1 are the elements of R, and
  - the necessary transitions of S'_1 are the $(q_1, q'_1) \xrightarrow{a} (q_2, q'_2)$ where $q_1 \xrightarrow{a} q_2$ is a necessary transition of S, and $q'_1 \xrightarrow{a} q'_2$ a necessary transition of S';
  - the possible transitions of S_1 are the $(q_1, q'_1) \xrightarrow{a} (q_2, q'_2)$ where $q'_1 \xrightarrow{a} q'_2$ is a possible transition of S';
- the states of S_1 are either elements of R or states of S, and
  - the necessary transitions of S_1 are either necessary transitions of S, or the $(q_1, q'_1) \xrightarrow{a} (q_2, q'_2)$ where $q_1 \xrightarrow{a} q_2$ is a necessary transition of S;
  - the possible transitions of S_1 are either possible transitions of S, or the $(q_1, q'_1) \xrightarrow{a} (q_2, q'_2)$ where $q_1 \xrightarrow{a} q_2$ is a possible transition of S, and $q'_1 \xrightarrow{a} q'_2$ a possible transition of S'.

By definition $S'_1 \trianglelefteq S_1$. The proof that the projections π' of S'_1 onto S' and π of S_1 onto S are saturating is straightforward.

□

4 Open questions

As a conclusion, we wish to raise a few questions.

The condition that a logic is fully adequate is sufficient to ensure the algebraic characterization of the logical equivalence; does there exist a necessary condition?

Next, it would be interesting to characterize the full adequacy by particular properties of the specific operators of the logic, so that the characterization of a fully adequate logic can be expressed in a more algebraic way.

Also, we would like to have a criterion ensuring the algebraic characterization of logical preorders. In particular, in which cases the relation $\subseteq_{\mathcal{F}}$ appearing in the converse of Proposition 6 can be replaced by the stronger relation used as hypothesis in Proposition 5. It is the case for modal transition systems, but it is open for the "until" Hennessy-Milner logic.

And finally, we may define a logical equivalence with respect to a subset of formulas of a given logic (for instance, its positive formulas), and one could wonder whether it is possible to give an algebraic characterization of such an equivalence.

References

[A92] A. Arnold. *Systèmes de transitions finis et sémantique des processus communicants.* Masson, 1992.

[AD89] A. Arnold, A. Dicky. An Algebraic Characterization of Transition System Equivalences. *Information and Computation*, 82:198–229, 1989.

[BL92] G. Boudol, K. Larsen. Graphical versus Logical Specifications. *Theor. Comput. Sci.*,106:3–20, 1992

[DDN88] P. Degano, R. De Nicola, U. Montanari. A Distributed Operational Semantics for CCS Based on Condition/Event Systems, *Acta Informatica*,26:59–91, 1988.

[DV90] R. De Nicola, F. Vaandrager. Three Logics for Branching Bisimulation. in: *Proc. of LICS '90*(1990)118-129.

[CES86] E. M. Clarke, E. A. Emerson, and A. P. Sistla. Automatic verification of finite state concurrent systems using temporal logics specifications. *ACM Trans. Prog. Lang. Syst.*, 8:244–263, 1986.

[EH86] E. A. Emerson and J. Y. Halpern. "Sometimes" and "Not Never" revisited : on branching versus linear time temporal logic. *J. Assoc. Comput. Mach.*, 33:151-178, 1986.

[ES84] E. A. Emerson and A. P. Sistla. Deciding full branching time logic. *Information and Control*, 61:175–201, 1984.

[HM85] M. Hennessy, R.Milner. Algebraic laws for nondeterminism and concurrency. *J. Assoc. Comput. Mach.*, 32:137-161, 1985.

[HS84] M. Hennessy, C. Stirling. The power of the future perfect in program logics. in: *11th Math. Foundations of Comput. Sci, 1984.* Lect. Notes Comput. Sci. 176(1984)301-311.

[LT88] K. Larsen, B. Thomsen. A Modal Process Logic. in: *Proc. of LICS '88*(1988)203–210.

Deliverables: a categorical approach to program development in type theory

James McKinna and Rod Burstall *

Laboratory for the Foundations of Computer Science
University of Edinburgh**

Abstract. We describe a method for formally developing functional programs using the "propositions as types" paradigm. The idea is that a function together with its proof of correctness forms a morphism in a category whose objects are input/output specifications. The function-proof pairs, called "deliverables", can be combined by the operations of a cartesian closed category, indeed by the same operations which are usually used to combine functions. The method has been implemented using the Lego proof assistant and tried on some examples.

1 Introduction

This paper outlines a method for constructing a program together with a proof of its correctness with respect to a given specification. The technology used is type theory, in fact an extended version of Calculus of Constructions Luo's ECC [20] as implemented in Pollack's 'Lego' system [22]; here *program* means a primitive recursive description of a function using simply typed lambda calculus with higher order functions and data types such as numbers and lists.

Given a precondition and a postcondition we consider pairs consisting of

- a program
- a proof that the program satisfies the postcondition given the precondition.

We call such a pair together with the pre- and postconditions a *deliverable*, since it is what a software house should ideally deliver to its client instead of just a program.

Now we observe that various operations, composition, pairing, iteration and so on, can be used to combine deliverables, and that these operations can be implemented in ECC. Consider for example composition. If (f, p) is a deliverable from *pre* to *post* and (f', p') is a deliverable from *post* to *post'* then their composition is a deliverable from *pre* to *post'* thus

- $f \circ f'$, the composition of the functions

* The authors gratefully acknowledge the support of the EC Logical Frameworks BRA and the SERC

** J.C.M.B., King's Buildings, Mayfield Rd., Edinburgh EH9 3JZ, UK;
e-mail `jhm@dcs.ed.ac.uk`, `rb@dcs.ed.ac.uk`

– a proof using p and $p\prime$ that $f \circ f'$ satisfies $post'$ given pre.

In fact the combining operations (excluding iteration) are exactly those of a cartesian closed category whose objects are the pre- and postconditions and whose arrows are deliverables. This is comforting since it assures us that we have a complete set of combinators. They enable us to build a complicated program together with its correctness proof. Since the resulting expression of the calculus denotes a pair we can extract the program trivially by evaluating the first element of the pair.

We have used the Lego system to define these combinators and use them to create the deliverables for several programs as examples.

However in many cases a precondition and a postcondition do not satisfactorily specify a desired program. For example when developing a sorting program the precondition might ask for an arbitrary list and the postcondition demand that the result be a sorted list. But we would get little thanks for a program which always returned the nil list, even though this is sorted. We need to specify that the output list should also be a permutation of the input one.

If we try to replace the pre- and postconditions by a single statement specifying a relation between input and output there do not seem to be convenient combination operators; so we discard that approach. Thinking syntactically, we could allow the pre- and postconditions to share a free variable. For example using a free variable l for lists

$$pre(x) = x = l$$

$$post(y) = isPermutation(y, l)$$

This ties the pre- and postconditions together, but such free variables should have a definite scope.

We can think, more semantically, that l satisfies some base condition (it must be a list) and the pre- and post conditions are relations whose first argument is l. Thus we can talk of pre- and post conditions *relative to l*. Now a deliverable relative to l is a program which takes the precondition to the postcondition for all l. We will refer to these as *second order* specifications and deliverables in contrast to the original *first order* ones.

How should we represent specifications and deliverables in type theory? We choose the following:

A first order specification is a predicate over a type, thus

– a type, s;
– a function, S, from s to $Prop$ (the type of propositions);

and a first order deliverable from specification (s, S) to specification (t, T) is a program paired with a proof thus

– a function, $f : s \rightarrow t$;
– a proof, $\phi : \forall x : s.Sx \rightarrow T(fx)$ where $\forall$ is the $\prod$ quantifier on propositions.

Now we can generalise these to work relative to a first order specification (u, U), which corresponds to the free variable l in our example. Call u the base type; now the second order specification becomes a relation over the base type and another type, and the second order deliverable must respect the pre- and postrelations. We leave the details to the body of the paper.

We might try to characterise the first order deliverables as a cartesian closed category whose objects are specifications and whose arrows are deliverables; this is almost correct, but we have to modify the CCC to a *semi*-cartesian closed category, a mild generalisation due to Hayashi [10]. The second order deliverables are then characterised using a fibration whose fibres are categories of first order deliverables. In fact this structure constitutes a model of type theory.

The paper includes some illustrative examples of program development using deliverables and brief mention of others, of which the most substantial is an algorithm for the Chinese Remainder theorem.

We also discuss the relation to earlier work on extracting programs from proofs, particularly the work of Martin-Löf as developed by Nördström, Petersson and Smith [27]. They define a translation from a type theory, say S, to an underlying type theory U such that the types in S correspond to our specifications and the unary terms in S correspond to our deliverables. A somewhat similar approach is taken by Paulin-Mohring in her work on program extraction using the realisability interpretation [28].

The virtue of our method is that it can quite easily be coded up in an existing system, and there is no difficulty about extracting the program from the proof, since they are built together but as separate components of a pair; normalising the first component gives us the program. We have mentioned operations for combining deliverables, and this might suggest a bottom-up approach; but we can just as well start with a specification of the desired program and derive the corresponding deliverable top-down by refinement in Lego.

The basic approach was proposed in an unpublished talk by Burstall at the Båstad Workshop on type theory in 1988. A brief account by the present authors appeared as [4], and a full treatment of work so far is in McKinna's thesis [23].

2 Review of ECC and a simple example

Luo's Extended Calculus of Constructions, ECC [19, 20], is a rich type theory containing Coquand and Huet's Calculus of Constructions [6, 7] as a subsystem, together with strong Σ-types and a cumulative hierarchy of predicative universes, much as in the systems considered by Martin-Löf and his collaborators [24, 25, 27]. In Martin-Löf systems, *all* types may be read as propositions, and in Coquand and Huet's original system, all propositions may be read as types. ECC avoids this blurring of distinctions, giving us access to full intuitionistic higher-order logic at a propositional level, together with a predicative environment for computation and abstract mathematics. It is precisely this ability to distinguish propositional from computational information within a single framework which underlies our approach to program development.

ECC is built out of a calculus of terms with the following official syntax

$$\begin{aligned} T ::= {} & \kappa \mid V \mid \\ & \Pi V{:}T.T \mid \lambda V{:}T.T \mid TT \mid \\ & \Sigma V{:}T.T \mid \mathbf{pair}_T(T,T) \mid \pi_1(T) \mid \pi_2(T) \end{aligned}$$

where V ranges over some infinite collection of variables, and κ ranges over *Prop* and $Type_i$ $(i \in \omega)$, the so-called *kinds* of ECC. *Prop* is an impredicative universe, as in Coquand and Huet's original systems [6], intended to contain propositions, while the $Type_i$ are predicative universes much like a set-theoretic hierarchy (and very similar to the $\mathbf{U}_i$ of some versions of Martin-Löf's theories [27, 25]). Substitution for free occurrences of variables is defined in the usual way. Terms are identified up to renaming of bound variables. The basic conversion relation $\simeq_\beta$ is defined on all terms, and as usual is the congruence closure of the familiar reductions:

(β) $(\lambda x{:}A.M)N \rhd M[N/x]$, and
(σ) $\pi_i(\mathbf{pair}_T(M_1, M_2)) \rhd M_i (i = 1, 2)$.

LEGO is Pollack's implementation of a typechecker and refinement proof system for ECC and a number of related systems, based on earlier ideas of Huet, de Bruijn and others [3, 6].

The LEGO syntax for terms in the language of ECC is given by

T ::= `Prop` | `Type(`i`)` $(i \in \omega)$ | `V`
`{V:T}T` | `[V:T]T` | `TT` |
`<V:T>T` | `(T,T)` | `T.1` | `T.2`

where the correspondence with the official syntax is given in Table 1. As usual, the non-dependent Π-types are denoted by an infix $\rightarrow$.

`{V:T}T`	corresponds to	$\Pi x{:}\ T.\ T$,
`[V:T]T`	to	$\lambda x{:}T.\ T$,
`<V:T>T`	to	$\Sigma V{:}T.T$,
`(T,T)`	to	$\mathbf{pair}_{\Sigma x:A.B}(T, T)$,
`T.1`	to	$\pi_1(T)$,
and `T.2`	to	$\pi_2(T)$.

Table 1. Comparison between syntax of LEGO and ECC

The syntax has recently been extended to provide arbitrary extensions to the conversion relation. As an example, we may add a type of natural numbers, by declaring appropriate constants in an initial context

$$\begin{aligned} \Gamma_{nat} = {} & nat{:}Type_0, 0{:}nat, S{:}nat \rightarrow nat, \\ & natrecd{:}\Pi C{:}nat \rightarrow Type.\ (C0) \rightarrow (\Pi k{:}nat.\ (Ck) \rightarrow (C(Sk))) \rightarrow \Pi n{:}nat.\ Cn \end{aligned}$$

together with reductions

$$natrecd\ C\ z\ s\ 0 \rhd z$$
$$natrecd\ C\ z\ s\ (Sk) \rhd s\ k\ (natrecd\ C\ z\ s\ k)$$

for C, z, s, k of the appropriate types.

In this context, we may derive the following term

$$\Gamma_{nat}, n{:}nat \vdash \forall\Phi{:}nat \rightarrow Prop.\Phi(0) \Rightarrow (\forall k{:}nat.\Phi(k) \Rightarrow \Phi(S(Sk))) \Rightarrow \Phi(n) : Prop$$

which represents impredicatively the (informal) proposition that n is an *even* natural number. That is, we define even numbers to be those n which satisfy *all* predicates satisfied by 0 and closed under successor of successor (the impredicativity, of course, lies in the fact that "evenness" is just such a predicate). Moreover, in this representation, the judgment

$$\begin{aligned}\Gamma_{nat} \vdash\ & \lambda\Phi{:}nat \rightarrow Prop. \\ & \quad \lambda z{:}\Phi(0)\,. \\ & \quad\quad \lambda s{:}\forall k{:}nat\,.\,\Phi(k) \Rightarrow \Phi(S(Sk))\,. \\ & \quad\quad\quad z \\ & : \ \forall\Phi{:}nat \rightarrow Prop.\Phi(0) \Rightarrow (\forall k{:}nat.\Phi(k) \Rightarrow \Phi(S(Sk))) \Rightarrow \Phi(0)\end{aligned}$$

represents a proof that 0 is even.

Example In the same way, one could define a predicate representing "oddness",

$$Odd =_{\text{def}} \lambda n{:}nat.\ \forall\Phi{:}nat \rightarrow Prop.\Phi(S0) \Rightarrow (\forall k{:}nat.\Phi(k) \Rightarrow \Phi(S(Sk))) \Rightarrow \Phi(n)$$

being one such representation. One might then show that the successor function S transforms even numbers into odds, and odds to evens, yielding terms sEO of type $\forall n{:}nat.\ Even(n) \rightarrow Odd(Sn)$ and sOE of type $\forall n{:}nat.\ Odd(n) \rightarrow Even(Sn)$ (the patient reader may care to experiment with the system by doing just such a thing).

What is the behaviour of the composite function $x \mapsto x + 2$? We may use the proofs sEO and sOE, together with *modus ponens*, to derive a term $ssEE$ of type $\forall n{:}nat.\ Even(n) \rightarrow Even(n+2)$.

Now the Σ-types of the calculus allow us to pair together these fragments of code with their associated proof terms, for example

$$(S, sEO) : \Sigma f{:}nat \rightarrow nat.\ \forall n{:}nat.\ Even(n) \rightarrow Odd(fn)$$

and we may see the *pair* $(\lambda n{:}nat.\ n{+}2,\ ssEE)$ arising as a composite construction from the *pairs* (S, sEO) and (S, sOE).

The pair $(\lambda n{:}nat.\ n + 2,\ ssEE)$ of type

$$\Sigma f{:}nat \rightarrow nat.\ \forall n{:}nat.\ Even(n) \rightarrow Even(fn)$$

is a very simple instance of the general idea below of a *first-order deliverable*, in this case from $(nat, Even)$ to $(nat, Even)$.

3 Definition of first-order deliverables

We work relative to some well-formed context Γ in ECC.

Definition 3.1 *specification*
A *specification* is a pair of terms s, S such that $\Gamma \vdash s : Type$, and $\Gamma, x{:}s \vdash Sx : Prop$.

We typically write $\mathcal{S}, \mathcal{T}, \mathcal{U}$ for specifications, and understand s, t, u (respectively S, T, U) as referring to their underlying type (resp. predicate). Formally, there is a type of specifications, namely $SPEC_1 =_{\text{def}} \Sigma s{:}Type.s \rightarrow Prop$, so that we may consider operations which construct specifications entirely within the framework of ECC . Luo, in [21], describes an approach to specification and data refinement in ECC. This idea has been further pursued by Streicher and Reus, who have shown how to capture the traditional language of algebraic specification in this setting [36].

We will consider a category whose objects are the specifications. Specifications defined by *logically* equivalent predicates in general define distinct objects. The appropriate notion of morphism between specifications consists of a pair (f, ϕ), where f is a function between the underlying types, and ϕ is a proof that f respects the predicates. We call these morphisms *first-order deliverables.*

Definition 3.2 *first-order deliverable*
Given specifications $\mathcal{S} = (s, S), \mathcal{T} = (t, T)$, a *first-order deliverable* is a term $\mathcal{F}$ such that

$$\Gamma \vdash \mathcal{F} : \Sigma f : s \longrightarrow t.\forall x : s.Sx \Longrightarrow T(fx).$$

The motivation for such a definition goes right back to Hoare's original paper on axiomatic semantics [12], and, like the logic of triples which bears his name, expresses in a *formal* system the informal notion of a program, together with a certificate of some specified input/output behaviour. Of course, we are concerned with a functional language, rather than an imperative one, so there is no confusion over program variables and logical variables. In this framework, moreover, the proof and the program are linked as a pair. This definition uses the Σ-types in an essential way to capture this idea. We may use all the features of ECC to construct such pairs, but based on our intuitions about computational *vs.* propositional information, we insist upon a trivial extraction process: first projection π_1 from the Σ-type yields the underlying algorithm f. Indeed this accounts for the name "deliverables": they are what a software house should deliver to its customers, a program plus a proof in a box with the specification printed on the lid (the Σ-type). The customer can independently check the proof and then run the program, without the need for a complicated extraction process which may yield an unusual algorithm. Indeed, we propose this method as a style for developing programs in the first place. We usually have a reasonable idea of the algorithm in advance, as opposed to its proof of termination, or even correctness, and we would like an understanding, which reflects our intuitions, of how to build up these deliverables from smaller ones, for example by refinement and composition, possibly with machine assistance.

In terms of LEGO, we define a predicate Del_1 and a type constructor $\mathbf{del}_1$ which describe first-order deliverables within ECC (compare Definition 3.2 above):

```
Lego> Del1;
value = [s,t|Type][S:Pred s][T:Pred t][f:s->t]{x|s}(S x)->T (f x)
type  = {s,t|Type}(Pred s)->(Pred t)->(s->t)->Prop
Lego> del1;
value = [s,t|Type][S:Pred s][T:Pred t]<f:s->t>Del1 S T f
type  = {s,t|Type}(Pred s)->(Pred t)->Type
```

(This is the output from the typechecker after giving the definitions). We have exploited LEGO's type inference mechanism [31] to enable us to suppress the argument types[3] in the predicate `Del1` and the type `del1`.

With an eye to the categorical aspects of this definition, we typically write

$$S \xrightarrow{\mathcal{F}} T \in \mathbf{del}_1 \quad \text{or} \quad (s,S) \xrightarrow{(f,\phi)} (t,T) \in \mathbf{del}_1$$

when $\Gamma \vdash f : s \longrightarrow t$, and $\Gamma \vdash \phi : \mathrm{Del}_1\ S\ T\ f$. Since s, t may be inferred by the typechecker, and we are in general not interested in ϕ, save to know that it exists, we may even abuse our notation and write

$$S \xrightarrow{f} T \in \mathbf{del}_1 .$$

At this stage, we need not have used the Σ-types to present these definitions. However, internalising the mathematical pair in a Σ-type allows us to represent operations which produce such function-proof pairs within the calculus. This gives us the possibility of developing a structure on these morphisms.

Definition 3.3 *equality of specifications*
Given specifications $(s,S),(t,T)$, we say $(s,S) = (t,T)$ if

$$s \simeq_{\beta\delta} t \text{ and } \lambda x{:}s.\, S\,x \simeq_{\beta\delta} \lambda x{:}t.\, T\,x.$$

Definition 3.4 *equality of deliverables*
Given

$$(s,S) \underset{(g,\psi)}{\overset{(f,\phi)}{\rightrightarrows}} (t,T) \in \mathbf{del}_1,$$

we say $(f,\phi) = (g,\psi)$ if

$$\lambda x{:}s.\, fx \simeq_{\beta\delta} \lambda x{:}s.\, gx$$

and

$$\lambda x{:}s.\, \lambda h{:}S\,x.\, \phi x h \simeq_{\beta\delta} \lambda x{:}s.\, \lambda h{:}S\,x.\, \psi x h.$$

This definition of equality of deliverables seems to be the minimal extension of the basic conversion relation which ensures good categorical properties. This should be contrasted with the notion of equality based on logical equivalence in [36].

[3] This accounts for our choice of notation: since the types `s,t` may be inferred from `S,T`, we relegate them notationally to the lower case.

3.1 Semi-structure in categories

The use of cartesian closed categories to give models of the simply-typed λ-calculus is by now familiar in computer science [8, 18, for example], as are various equational presentations of the structure of a cartesian closed category. The basic type and term constructors are defined by adjunctions. In this analysis, the unit and counit of the adjunction defining the arrow type correspond, loosely, to η and β conversion, respectively (and similarly for the product type constructor). The absence of η-conversion and surjective pairing in ECC forces some extra technical difficulty upon us. However, models of various typed λ-calculi without η-conversion and surjective pairing can be given a rigorous semantic account in terms of *semi-adjunctions*, introduced by Hayashi in [10]. Essentially, the equations defining an adjunction are relaxed sufficiently that, under suitable conditions, there is a notion of counit which corresponds appropriately to β-conversion, without a unit corresponding to η-conversion. A detailed treatment is given in McKinna's thesis [23].

Definition 3.5 *Semi-functor*
Given two categories $\mathbf{C}, \mathbf{D}$, a *semi-functor* between them is "a functor which need not preserve identities": that is to say, we are given an assignment F of objects of $\mathbf{D}$ to objects of $\mathbf{C}$, and an assignment, also called F, of arrows of $\mathbf{D}$ to arrows of $\mathbf{C}$, such that

$$A \xrightarrow{f} B \in \mathbf{C} \text{ implies } FA \xrightarrow{Ff} FB \quad \in \mathbf{D}$$

and

$$F(A \xrightarrow{f} B \xrightarrow{g} C) = FA \xrightarrow{Ff} FB \xrightarrow{Fg} FC \in \mathbf{D}.$$

In a *semi*-adjunction, we replace the usual natural bijection on homsets with a pair of maps, which need not be mutual inverses. However, we still require them to behave "naturally". The beauty of this idea lies in the following lemma.

Lemma 1 Hayashi. *Suppose F, G above are in fact functors. Then a semi-adjunction between F and G yields an adjunction between F and G.*

So this concept subsumes the whole of our ordinary understanding of adjunctions, and hence all the constructions of universal gadgets defined by adjunctions[4]. In particular, in a category $\mathbf{C}$, we may define the notions of *semi-terminal object*,*semi-product* and *semi-exponential* exactly by analogy with the usual cartesian closed structure. If $\mathbf{C}$ has all the above structure, we say $\mathbf{C}$ is a *semi-cartesian closed* category, or *semi-ccc*. Hayashi's development leads to the following main results, generalising previous accounts of models of the λ-calculus with β-conversion only.

[4] However, by contrast with adjunctions, structure defined by a semi-adjunction is *not* in general unique.

Proposition 2. *Semi-cccs are sound and complete for interpretations of the $\lambda\beta$-calculus.*

Proposition 3. *Semi-cccs can be presented algebraically.*

In [23], the second author used LEGO to prove the following theorem:

Theorem 4. *del_1 is semi-cartesian closed.*

We sketch some of the ideas in the discussion below.

3.2 Identities and composition

By considering our slight modification of the underlying conversion on the terms of ECC, to remedy the failure of surjective pairing and η-conversion in the calculus, we fixed a notion of equality on arrows. With this definition, it is now trivial to establish that specifications, together with first-order deliverables as morphisms, form a category, denoted by $\mathbf{del}_1$.

The identities are given simply by

$$(s,S) \xrightarrow{(\lambda x{:}s.\ x,\ \lambda x{:}s.\ \lambda h{:}Sx.\ h)} (s,S) \in \mathbf{del}_1;$$

in LEGO, we have

```
Lego> id_del1;
value = [s|Type][S:Pred s]([x:s]x,[x|s][p:S x]p)
type  = {s|Type}{S:Pred s}del1 S S .
```

Composition is given by

$$S \xrightarrow{(f,\phi)} T \xrightarrow{(g,\psi)} \mathcal{U} = S \xrightarrow{(\lambda x{:}s.\ g(fx),\ \lambda x{:}s.\ \lambda h{:}Sx.\ \psi(fx)(\phi xh))} \mathcal{U};$$

We may now verify that these definitions do indeed yield the structure of a category on $\mathbf{del}_1$[5].

3.3 Semi-Terminal object

$$Unit =_{\text{def}} (unit, \lambda u : unit.u = ()^6) : SPEC_1$$

defines a trivial specification, where we postulate the existence of a unit type in the same way as other inductive types. We then obtain, for any specification (s,S) the deliverable

$$!_{(s,S)} =_{\text{def}} (\lambda x{:}s.\ (),\ \lambda x{:}s.\ \lambda p{:}S\,x.\ \mathit{reflEQ}()) : \mathbf{del}_1\ S\ Unit.$$

[5] Indeed, we only need the modified equality to prove the identity laws. The conversion relation is sufficient to establish associativity of composition.

[6] Here, () is the unique term of type *unit*. We write void in LEGO.

3.4 Binary semi-products

To obtain semi-products, we use the non-dependent Σ-type as the underlying type, with conjunction at the predicate level:

$$\mathcal{S} \times \mathcal{T} =_{\text{def}} (s \times t, \lambda p{:}s \times t.\ (S(\pi_1 p)) \wedge (T(\pi_2 p))).$$

That we indeed have a semi-product structure now follows straightforwardly from this definition.

3.5 Semi-exponentials

The notion of first-order deliverable is based on the predicate Del_1 on terms of arrow type. Precisely this predicate defines the specification which yields a semi-exponential object in $\mathbf{del}_1$. λ-abstraction and evaluation then follow from those operations in the underlying type theory on values, and implication introduction and elimination at the level of proofs.

$$\mathcal{S} \Longrightarrow \mathcal{T} =_{\text{def}} (s \to t, \lambda f{:}s \to t.\ \mathrm{Del}_1\ S\ T\ f)$$

3.6 Trivial deliverables

Every function — that is to say a term of arrow type — gives rise to a deliverable, very much in the manner of the assignment rule of Hoare logic [12] or Dijkstra's predicate transformer for assignment. Namely, for

$$f : s \longrightarrow t,\ P : t \longrightarrow \mathit{Prop}$$

we obtain

$$(f, \lambda x{:}s.\lambda h{:}f^{\star}P.h){:}\mathbf{del}_1\ f^{\star}P\ P,\ \text{ where } f^{\star}P =_{\text{def}} \lambda x{:}s.P(fx).$$

We call these deliverables *trivial*, since they come with vacuous proofs of correctness.

3.7 A consequence rule

Logical implication induces a pointwise ordering $\subset$ on predicates, for which we have the following consequence rule, in the manner of Hoare logic:

$$\frac{\mathcal{S} \subset \mathcal{S}' \quad \mathcal{S}' \xrightarrow{(f,\phi')} \mathcal{T}' \quad \mathcal{T}' \subset \mathcal{T}}{\mathcal{S} \xrightarrow[(f,\phi)]{} \mathcal{T}}$$

In fact, this rule is subsumed by composition, since any logical entailment gives rise to a deliverable whose function component is the identity.

3.8 Pointwise construction

A basic combinator in the theory of deliverables constructs a function-proof pair from a function which returns value-proof pairs[7]. Mendler, in his thesis [26], calls such gadgets "pointwise designs": for each argument value x, the pointwise existence of a value y (of type t) satisfying some property Tv, yields a deliverable with codomain (t, T). In detail,

$$\frac{\mathcal{F}\colon \Pi x{:}s.\ \Sigma y{:}t.\ (Sx) \longrightarrow (Ty)}{\mathcal{S} \xrightarrow{(f,\phi)} \mathcal{T}}$$

where $f =_{\text{def}} \lambda x{:}s.\ \pi_1(\mathcal{F}x)$, and $\phi =_{\text{def}} \lambda x{:}s.\ \lambda h{:}Sx.\ \pi_2(\mathcal{F}x)h$.

3.9 Inductively defined types

Provided we accept a weak definition of inductive type, it is relatively straightforward to add inductive types to the categorical structure developed so far. Categorical accounts of inductive types, via initial algebras, impose extra equalities on the iterator, due to the uniqueness clause in the definition of initial algebra. As with the semi-structures above, we only have existence, and not uniqueness, of the relevant universal arrows.

The basic idea is very simple: we add inductive types at the *Type* level, with a strong[8] elimination rule, yielding a simply typed recursor at the *Type* level, and the usual induction principle at the *Prop* level. This type is then paired with the identically true predicate. The elimination rule for first-order deliverables is easily derived, by packaging primitive recursion at the type level with induction at the predicate level. We illustrate this general idea by considering the case of natural numbers and lists.

Natural numbers Recall from Section 2 that we may postulate the existence of a type of natural numbers, with two constructors, zero and successor. This yields the well-formed context

$$nat{:}Type,\ 0{:}nat,\ S{:}nat \longrightarrow nat$$

We typically abbreviate S to "+1" in informal mathematical language. We extend this context with a dependent elimination constant *natrecd*,

$$natrecd{:}\Pi C{:}nat \longrightarrow Type.\ \Pi z{:}C0.\ \Pi s{:}(\Pi k{:}nat.\ \Pi ih{:}Ck.\ C(k+1)).\ \Pi n{:}nat.\ Cn$$

together with reduction rules defining the δ-redexes[9] (in some context where C, z, s, n have the appropriate types):

---

[7] This just corresponds to Howard's observation that, given a strong interpretation of the existential quantifier as Σ-type, the axiom of choice becomes constructively valid [13].

[8] *Strong*, that is, because we can eliminate over *types*, and not merely propositions.

[9] *cf.* Martin-Löf type theory, or Gödel's earlier system **T** of functionals. We essentially use this language of primitive recursion in all finite types as our programming language.

– $natrecd\ C\ z\ s\ 0 \rhd z$;
– $natrecd\ C\ z\ s\ (n+1) \rhd s\ n\ (natrecd\ C\ z\ s\ n)$.

This is precisely expressed in LEGO as follows:

```
[nat:Type(0)];
[zero:nat];
[succ:nat -> nat];
[natrecd:{C:nat->Type}
        {z:C zero}{s:{k:nat}{ih:C k}C (succ k)}{n:nat}C n];
[[n:nat][C:nat->Type][z:C zero][s:{k:nat}{ih:C k}C (succ k)]
   natrecd C z s zero ==> d
|| natrecd  C z s (succ n) ==> s n (natrecd C z s n)].
```

This yields a derived iterator and primitive recursor

$$natiter{:}\Pi\alpha{:}Type.\ \alpha \longrightarrow (\alpha \longrightarrow \alpha) \longrightarrow \alpha$$

and

$$natrec{:}\Pi\alpha{:}Type.\ \alpha \longrightarrow (nat \longrightarrow \alpha \longrightarrow \alpha) \longrightarrow \alpha$$

where[10]

$$\begin{aligned} natiter\ z\ s\ 0 &=_{\text{def}} z \\ natiter\ z\ s\ (n+1) &=_{\text{def}} s\ (natiter\ z\ s\ n) \\ natrec\ z\ s\ 0 &=_{\text{def}} z \\ natrec\ z\ s\ (n+1) &=_{\text{def}} s\ n\ (natrec\ z\ s\ n), \end{aligned}$$

and an induction principle

$$natind : \Pi\Phi{:}nat \longrightarrow Type.\ \Pi z{:}\Phi(0).\ \Pi s{:}(\Pi k{:}nat.\ \Pi ih{:}\Phi(k).\ \Phi(k{+}1)).\ \Pi n{:}nat.\ \Phi n.$$

Our methodology suggests we examine derived induction principles for the iterator and recursor, since we are interested in programs, in this case of the form *natiter z s* or *natrec z s*, together with proven propositions about them. For the iterator, we obtain

$$\frac{\Phi(z)\ \forall y{:}\alpha.\ \Phi(y) \Longrightarrow \Phi(s\ y)}{\forall n{:}nat.\ \Phi(natiter\ z\ s\ n)}[z{:}\alpha, s{:}\alpha \longrightarrow \alpha]$$

and for the recursor

$$\frac{\Phi(z)\ \forall k{:}nat.\ \forall y{:}\alpha.\ \Phi(k) \Longrightarrow \Phi(s\ k\ y)}{\forall n{:}nat.\ \Phi(natrec\ z\ s\ n)}[z{:}\alpha, s{:}nat \longrightarrow \alpha \longrightarrow \alpha].$$

Since we are interested in building new deliverables from less complex ones, we could of course use the dependent eliminator *natrecd* to construct terms of type $\mathbf{del}_1$, but in doing so we violate the separation of proofs from programs

[10] The type α is, of course, inferred by the typechecker.

which distinguishes our approach. So we package recursion at the *Type* level with induction at the *Prop* level in a pair.

We introduce the predicate $Nat =_{\text{def}} \lambda n{:}nat.\ true$ on the natural numbers. For each constructor of the type, we obtain a corresponding deliverable:

$$\frac{}{Zero =_{\text{def}} (\lambda u{:}unit.\ 0,\ \lambda u{:}unit.\ \lambda h{:}\,Unit\ u.\ \top^{11}){:}\,Unit \longrightarrow Nat}$$

$$\frac{}{Succ =_{\text{def}} (\lambda n{:}nat.\ S\,n,\ \lambda n{:}nat.\ \lambda h{:}Nat\,n.\ \top){:}Nat \longrightarrow Nat}$$

For the iterator, we obtain

$$\frac{Unit \xrightarrow{(z,Z)} (t,T) \quad (t,T) \xrightarrow{(s,S)} (t,T)}{Natiter\,(z,Z)\ (s,S){:}(nat,\ Nat) \longrightarrow (t,T)}$$

and the recursor *Natrec* analogously, where the function component of *Natiter* $(z, Z)\ (s, S)$ (respectively *Natrec*) is *natiter* $(z())\ s$ (respectively *natrec*), and the proof component is obtained from the appropriate derived induction principle above.

These terms are easily obtained by refinement in LEGO. For example, here is the significant structure of *Natiter*, with the uninformative details of the proof component suppressed:

```
Lego> natiter_del1;
value = [t|Type][T|Pred t][ZZ:del1 Unit T][SS:del1 T T]
        [z=ZZ.1 void][Z=ZZ.2][s=SS.1][S=SS.2]
        (natiter z s,
         natind ([m:nat](Nat m)->T (natiter z s m)) ... )
type  = {t|Type}{T|Pred t}(del1 Unit T)->(del1 T T)->del1 Nat T
```

Example As an example of the use of these combinators, we present a correctness proof of a doubling function, given by

$$double =_{\text{def}} \lambda n{:}nat.\ natiter\,0\ (\lambda k{:}nat.\ k+2)\ n.$$

Suppose we wish to show that *double* n is even for all natural numbers n. Posed in terms of deliverables, we seek a term of type $\mathbf{del}_1$ *Nat Even*, whose function component is *double*.

Using the rule for *Natiter* above, the problem reduces to finding:

base case $Unit \xrightarrow{(z,Z)} (nat, Even)$. We take $z =_{\text{def}} \lambda u{:}unit.\ 0$, and use the proof that 0 is even from Section 2;

step case $(nat, Even) \xrightarrow{(s,S)} (nat, Even)$. We simply use the first-order deliverable which we constructed by composition at the end of Section 2, namely $(\lambda n{:}nat.\ n+2,\ ssEE)$.

We thus obtain a non-trivial recursive first-order deliverable.

Lists In much the same way as above, we may define combinators for deliverables over a type of lists. As before, we extend the context with a type constructor, in this case $list{:}\,Type_0 \longrightarrow Type_0$, together with constructors the usual *nil* and *cons*, and a dependent eliminator *listrecd*. Again we derive an iterator *listiter*, a primitive recursor *listrec* and an induction combinator *listind*.

When it comes to considering the derived induction principles for the iterator and recursor, however, we now have the freedom to specify recursions over lists of elements satisfying some predicate, rather than over all lists of the parameter type a. That is, given some specification (a, A), we obtain a derived specification $List\,(a, A) =_{\text{def}} (list\,a,\ Listof\ A)$, where

$$\begin{aligned} Listof\ A\ (nil\ a) &=_{\text{def}} true \\ Listof\ A\ (cons\ x\ l) &=_{\text{def}} (A\,x) \wedge (Listof\ A\ l) \end{aligned}$$

defines *Listof A* by primitive recursion.

We may then proceed in the same way as above, obtaining constructors

$$\frac{}{\overline{Nil =_{\text{def}} (\lambda u{:}unit.\ nil\ a,\ \lambda u{:}unit.\ \lambda h{:}\,Unit\ u.\ \top){:}\,Unit \longrightarrow Listof\ A}}$$

and

$$\frac{}{\overline{Cons{:}A \longrightarrow (Listof\ A)^{Listof\ A}}}$$

with function component,

$$\lambda x{:}a.\ \lambda l{:}list\,a.\ cons\ x\ l$$

and proof component

$$\lambda x{:}a.\ \lambda p{:}A\ x.\ \lambda l{:}list\ a.\ \lambda q{:}Listof\ A\ l.\ pair\,p\ q.$$

Likewise, we package together recursion and an appropriate derived induction principle, to obtain the following rules for constructing deliverables: an iterator,

$$\frac{Unit \xrightarrow{(n,N)} (t,T) \quad (a,A) \xrightarrow{(c,C)} (t,T)^{(t,T)}}{Listiter\,(n,N)\ (c,C){:}\ (list\ a,\ Listof\ A) \longrightarrow (t,T)};$$

and analogously a recursor.

4 Second-order deliverables

The system which we have described above amounts to a functional version of the well known invariants used in proofs of imperative programs. Unfortunately the specification makes no connection between the input and the output of the function. All we say is that if the input satisfies property S then the output satisfies property T, but there is no *relation* between them. Recalling the example in the introduction, we might specify that a sorting function takes lists to ordered lists, but we cannot specify that the output is a permutation of the input. The function might always produce the empty list, which is indeed sorted, but not very interesting. As a matter of fact the classical invariant proofs have the same weakness, masked by the tacit assumption that some variable which is carried through the computation does not change its value. To enforce the constraint that the output bear some relation to the input, we need to develop a compositional theory in which relations are the basic objects of study, with a notion of arrow which respects relations, rather than predicates. This gives us a categorical explanation of the idea that the pre- and postconditions are linked by having a free variahle in common, that is they are over the same context.

4.1 A thought experiment

Suppose we are given some specification $\forall x{:}s.\ \exists z{:}u.\ R(x,z)$, and we wish to find some function $f{:}s \longrightarrow u$ which satisfies it. In what sense may we refine such specifications by composition? Suppose we wish to instantiate f via the composition $f = g; h$ of two functions

$$s \xrightarrow{\ g\ } t \xrightarrow{\ h\ } u$$

where t is some intermediate type. Then, following our intuition in the case of predicates, we anticipate some intermediate specification $Q(x,y) \quad [x{:}s\ , y{:}t]$, such that g solves

$$\forall x{:}s.\ \exists y{:}t.\ Q(x,y),$$

and h solves

$$\forall x{:}s.\ (\exists y{:}t.\ Q(x,y)) \Longrightarrow \exists z{:}u.\ R(x,z).$$

This last is logically equivalent to

$$\forall x{:}s.\ \forall y{:}t.\ Q(x,y) \Longrightarrow \exists z{:}u.\ R(x,z).$$

But now we are left in something of a quandary: h, our intended solution, makes no reference to the intermediate value of y. Also, we have introduced an asymmetry between the rôles of g and h. A remedy, which underlies the definitions 4.2, and 4.3 below, is to separate the rôles of the independent parameter x and the dependent variables y, z.

We consider relations such as Q, R as the objects of study, for a fixed type s, but allowing the types t, u to vary. The following provides an appropriate notion of morphism which re-establishes a symmetry between Q and R.

Definition 4.1 An arrow from $Q(x,y)$ $[x{:}s\ ,y{:}t]$ to $R(x,z)$ $[x{:}s\ ,z{:}u]$ consists of the following data:

- a function $f{:}s \longrightarrow t \longrightarrow u$, that is to say a function of *two* arguments — this recovers the missing dependence we observed above;
- a proof $\phi{:}\forall x{:}s.\ \forall y{:}t.\ Q(x,y) \Longrightarrow R(x,(f\,x\,y))$.

The composition of two such arrows

$$P(x,w) \xrightarrow{(g,\phi)} Q(x,y) \xrightarrow{(h,\psi)} R(x,z)$$

is definable as

$$(\lambda x, w{:}s, r.\ h\,x\,(g\,x\,w),\ \ \lambda x, w{:}s, r.\ \lambda p{:}P(x,w).\ \psi\,x\,(g\,x\,w)\,(\phi\,x\,w\,p)).$$

We have now established a definition which respects the symmetry of source and target in our previous analysis of the decomposition of the specification $\forall x{:}s.\ \exists z{:}u.\ R(x,z)$. In so doing, we have generalised the notion of specification, and our old specification corresponds in this new setting to choosing $r =_{\text{def}}$ *unit*, $P(x,w) =_{\text{def}}$ *true*, and $f_{old} =_{\text{def}} \lambda x{:}s.\ f_{new}\,x\,()$.

4.2 Basic definitions

In view of the above discussion, we relativise specifications and first-order deliverables to depend on some input type s. Indeed, by observing that we may uniformly impose some condition S on the input parameter $x{:}s$, without affecting the notion of composition, we arrive at the definition of a "second-order" deliverable.

Definition 4.2 *relativised specification*
Suppose $\Gamma \vdash s : \mathit{Type}$, $\Gamma \vdash S : s \longrightarrow \mathit{Prop}$. Then a *relativised specification with respect to* (s,S) is given by a pair of terms t, R, such that

- $\Gamma \vdash t : \mathit{Type}$, and
- $\Gamma \vdash R : s \longrightarrow t \longrightarrow \mathit{Prop}$.

Definition 4.3 *second-order deliverable*
Suppose $\Gamma \vdash s : \mathit{Type}$, $\Gamma \vdash S : s \longrightarrow \mathit{Prop}$. Given two relativised specifications (t,Q) and (u,R), a *second-order deliverable over* (s,S) is a term $\mathcal{F}$ such that

$$\Gamma \vdash \mathcal{F} : \Sigma f : s \longrightarrow t \longrightarrow u.\forall x : s.S(x) \Longrightarrow \forall y : t.Q(x,y) \Longrightarrow R(x,(fxy)).$$

We define $\mathrm{Del}_2\, S\, Q\, R$ to be the predicate

$$\lambda f{:}s \longrightarrow t \longrightarrow u.\ \forall x : s.Sx \Longrightarrow \forall y : t.Q(x,y) \Longrightarrow R(x,(fxy)).$$

Definition 4.3 embodies the idea that, for each $x{:}s$ such that Sx holds, $(fx, \phi x)$ is a first-order deliverable from Qx to Rx, where $\mathcal{F} =_{\text{def}} (f,\phi)$. This suggests that the study of second-order deliverables amounts to the study of first-order deliverables in an extended context. In particular we expect to obtain, for a given specification $\mathcal{S} =_{\text{def}} (s, S)$, a category structure on the second-order deliverables over $\mathcal{S}$. We make a similar definition of equality on second-order deliverables to that given in Section 3 above, the details of which are left to the reader. Then we can indeed define identities and composition, given by

Identities The identity morphism from (t, Q) to itself over (s, S) is given by

$$(\lambda x{:}s.\ \lambda y{:}t.\ y,\ \lambda x{:}s.\ \lambda h{:}S\,x.\ \lambda y{:}t.\ \lambda q{:}Q\,x\,y.\ q)$$

Composition This is defined much as in the thought experiment above (4.1): the composition of

$$P(x, w) \xrightarrow{(f,\phi)} Q(x, y) \xrightarrow{(g,\psi)} R(x, z)$$

over (s, S) is definable as

$$(\lambda x{:}s.\ \lambda w{:}r.\ g\,x\,(f\,x\,w),\ \ \lambda x{:}s.\ \lambda h{:}S\,x.\ \lambda w{:}r.\ \lambda p{:}P\,x\,w).\ \psi\,x\,h\,(f\,x\,w)\,(\phi\,x\,h\,w\,p)).$$

This gives us a category $\mathbf{del_2}\mathcal{S}$.

If (f, ϕ) is a second-order deliverable over (s, S), we typically write

$$(t, Q) \xrightarrow{(f,\phi)} (u, R)\ \ [(s, S)]\ ,\ \text{or even } Q \xrightarrow{(f,\phi)} R\ \ [S],$$

since, as usual, the types s, t, u may be inferred by the typechecker. Our notation is intended to indicate that we are considering deliverables relative to some assumption defined by the specification (s, S). This notation is deliberately intended to echo the style of contexts in Martin-Löf type theory.

4.3 Each $\mathbf{del_2}\mathcal{S}$ is a semi-ccc

As in Section 3, we work relative to some context Γ. We have seen how a second-order deliverable (f, ϕ) over (s, S) in context Γ may be viewed as arising from a first-order deliverable in the extended context $\Gamma, x{:}s, h{:}Sx$. The conditions of definitions 4.2, and 4.3 are intended to enforce a hierarchy of dependencies in this extended context. The type t must not depend on x or h. The relation R, considered as a predicate on t in context $\Gamma, x{:}s$ does not depend on h. The function component f may not depend on h, but the proof component ϕ may do so.

Given these conditions, we may lift the structure in $\mathbf{del_1}$, by observing that the various constructions of Section 3 respect this hierarchy of dependencies. The predicates concerned need to be modified to include an explicit hypothesis Sx. We arrive at the following result.

Theorem 5. *For each specification $\mathcal{S}$, $del_2\mathcal{S}$ has the structure of a semi-ccc.*

Proof We merely sketch some of the constructions, on the basis of the above informal intuition. Full details are in McKinna's thesis [23].

Semi-terminal object This is given by the relativised specification

$$1_S =_{\text{def}} (\mathit{unit},\ \lambda x{:}s.\ \lambda u{:}\mathit{unit}.\ \mathit{true}).$$

The map $!_{(t,Q)}$, from some relativised specification (t, Q) to 1_S, has function component $\lambda x{:}s.\ \lambda y{:}t.\ ()$, and proof component

$$\lambda x{:}s.\lambda h{:}Sx.\lambda y{:}t.\lambda p{:}P\,x\,w.\ \top.$$

Semi-products Suppose we are given two relativised specifications $(t, Q), (u, R)$. Then we may form the relativised specification

$$(t,Q) \times (u,R) =_{\text{def}} (t \times u,\ \lambda x{:}s.\ \lambda p{:}t \times u.\ (Qx(\pi_1 p)) \wedge (Rx(\pi_2 p)))$$

This defines a semi-product object in $\mathbf{del}_2(s, S)$. The pairing map is given by

$$\frac{P \xrightarrow{(f,\phi)} Q\ [S] \quad P \xrightarrow{(g,\psi)} R\ [S]}{P \xrightarrow[(<f,g>,\ <\phi,\psi>)]{} Q \times R\ [S]}$$

where $< f, g >$ and $< \phi, \psi >$ are appropriate pairing operations on values and proofs respectively. Projections are similarly straightforward to define.

Semi-exponentials Just as the predicate Del_1 defined a semi-exponential object in the category $\mathbf{del}_1$, we may define a semi-exponential object in $\mathbf{del}_2\ (s, S)$, using the relativised specification to which Del_2 gives rise. More precisely, suppose (r, P), (t, Q), (u, R) are relativised specifications. We obtain a relativised specification

$$R^Q =_{\text{def}} (t \longrightarrow u,\ \lambda x{:}s.\ \lambda f{:}t \longrightarrow u.\ \forall y{:}t.Q\,x\,y \Longrightarrow R\,x\,(fy)).$$

If $\Gamma \vdash f : s \longrightarrow t \longrightarrow u$, and $\Gamma \vdash \phi : \mathrm{Del}_2\,S\,Q\,R\,f$, then $\Gamma, x{:}s \vdash \phi x : R^Q\,fx$. Moreover, given

$$P \times Q \xrightarrow{\mathcal{F}} R\ [S]$$

we obtain

$$P \xrightarrow{\Lambda(\mathcal{F})} R^Q\ [S]$$

by currying in the obvious way: if $\mathcal{F} =_{\text{def}} (f, \phi)$, then $\Lambda(\mathcal{F}) =_{\text{def}} (\hat{f}, \hat{\phi})$, where $\hat{f}$ and $\hat{\phi}$ are appropriate currying operations on values and proofs respectively. We may similarly define an evaluation map, details of which are left to the imaginative reader: it is rather less taxing to develop this construction by refinement in LEGO. Likewise, the proofs that these data meet Hayashi's conditions for a semi-exponential are best dealt with by refinement.

□

4.4 $\mathbf{del}_2$: an indexed category over $\mathbf{del}_1$

The categorically minded reader now asks herself what relationships exist between the various categories $\{\mathbf{del}_2\mathcal{S} | \mathcal{S} \in SPEC_1\}$, and to what extent we may elaborate upon the structure of this collection. In particular, she may ask what is the relationship between $\mathbf{del}_2\mathcal{S}$ and $\mathbf{del}_2\mathcal{T}$, given a first-order deliverable from $\mathcal{S}$ to $\mathcal{T}$. A moment's pause should convince her that composition in $\mathbf{del}_1$ should induce an operation on second-order deliverables, since they somehow are no more than first-order deliverables, except that they are defined in an extended context. In other words, we are groping towards the following theorem:

Theorem 6. *del_2 is an indexed category [17, 1] over del_1, whose fibres are semi-cccs, with semi-cc structure strictly preserved by reindexing along arrows in del_1.*

4.5 Pullback functors

The above theorem depends on the existence of pullback functors, which translate, or reindex, data between the categories $\mathbf{del}_2\mathcal{S}$. The obvious definition, which we give below, works, and moreover trivially respects the equality of objects and arrows, so we do indeed have pullback functors — and they are functors, not merely semi-functors, since identities and composition are preserved. It is then a straightforward, and tedious, task to verify that these operations compose, and strictly preserve the structure in each fibre.

Definition 4.4 *pullback along a first-order deliverable*
Suppose we are given specifications $\mathcal{S}, \mathcal{T}$, and a first-order deliverable $\mathcal{S} \xrightarrow{(k,K)} \mathcal{T}$. We define an operation of *pullback along* (k, K), where we abuse notation in the standard way by employing the same symbol for the operation on objects and arrows, as follows: given a relativised specification $\mathcal{Q} =_{\text{def}} (u, Q)$ with respect to $\mathcal{T}$, let

$$(k, K)^{\star}\mathcal{Q} =_{\text{def}} \lambda x{:}s.\ \lambda z{:}u.\ Q\,(kx)\,z;$$

moreover, given a relativised specification $\mathcal{R} =_{\text{def}} (v, R)$, and a second-order deliverable

$$\mathcal{Q} \xrightarrow{(f,\phi)} \mathcal{R}\ \ [\mathcal{T}]$$

we define $(k, K)^{\star}(f, \phi)$ to be the pair

$$(\lambda x{:}s.\ \lambda z{:}u.\ f\,(k\,x)\,z,\ \lambda x{:}s.\ \lambda h{:}Sx.\ \lambda z{:}u.\ \lambda p{:}Q\,(k\,x)\,z.\ \phi\,(k\,x)\,(K\,x\,h)\,z\,p).$$

Lemma 7. *$(k, K)^{\star}\mathcal{Q}$ is a relativised specification with respect to $\mathcal{S}$. Moreover, $(k, K)^{\star}(f, \phi)$ is a second-order deliverable from $(k, K)^{\star}\mathcal{Q}$ to $(k, K)^{\star}\mathcal{R}$.*

Lemma 8. *$(k, K)^{\star}$ preserves identities and composition.*

So, indeed, we do have the the existence of functors between the fibres $\mathbf{del}_2$. That they are a satisfactory notion of reindexing requires us to show that they obey the condition $\mathcal{H}^{\star}; \mathcal{K}^{\star} \cong (\mathcal{K}; \mathcal{H})^{\star}$. In fact, more is true. We have the following:

Lemma 9. *The reindexing is* strict, *in the sense that*

$$\mathcal{H}^\star;\mathcal{K}^\star = (\mathcal{K};\mathcal{H})^\star.$$

We now turn to the remainder of Theorem 6, namely that the pullback functors preserve the structure of a semi-ccc in each fibre. As above, we find that the structure is preserved strictly. We state only the case of exponentials, the cases of products and terminal object being exactly similar, and rather easier.

Lemma 10. *In the notation of Theorem 5 above, with* $\mathcal{V} \xrightarrow{\mathcal{K}} \mathcal{S}$, *we have*

$$\mathcal{K}^\star(R^Q) = (\mathcal{K}^\star R)^{\mathcal{K}^\star Q}.$$

Proposition 11. *Suppose* $(r, P), (t, Q), (u, R)$ *are relativised specifications with respect to* $\mathcal{S}$. *Given*

$$P \times Q \xrightarrow{\mathcal{F}} R \;\; [\mathcal{S}] \quad \textit{and} \quad \mathcal{V} \xrightarrow{\mathcal{K}} \mathcal{S}$$

we have $\Lambda(\mathcal{K}^\star\mathcal{F}) \equiv \mathcal{K}^\star\Lambda(\mathcal{F})$.

4.6 Second-order deliverables for natural numbers and lists

In the context of second-order deliverables, the situation regarding inductive types is less well understood. We do not regard this section as giving a definitive account, but the examples we have considered suggest that we have a usable set of combinators for reasoning about recursive programs.

We take as our guiding motivation the derived induction principles of Section 3.9. Since we now work in the relativised case, these will be subtly altered by the presence of the induction variable.

This means, for the case of natural numbers, that we now examine proofs of statements of the form[12]:

$$\forall n{:}nat.\ R\ n\ (natrec\ z\ s\ n)$$

where, for some type t, $z{:}t$ and $s{:}nat \longrightarrow t \longrightarrow t$. A proof of this, by induction, yields

$$R\ 0\ z \quad \text{and} \quad \forall k{:}nat.\ \forall y{:}t.\ R\ k\ y \Longrightarrow R\ (k+1)\ (s\ k\ y)$$

as the requisite hypotheses in the base and step cases. We may now recognise the second hypothesis as the logical component of some second-order deliverable, whose function component is s.

The question arises as to how to view the first hypothesis $R\ 0\ z$. Do we regard it as part of some first or second-order deliverable? In a sense, neither, in the choice we have made in the current version of deliverables. If we examine the derived rule of induction again, but this time rephrased as

$$\frac{\forall k{:}nat.\ \forall y{:}t.\ R\ k\ y \Longrightarrow R\ (k+1)\ (s\ k\ y)}{\forall n{:}nat.\ \forall z{:}t.\ R\ 0\ z \Longrightarrow R\ n\ (natrec\ z\ s\ n)}$$

[12] We only consider *natrec*, since *natiter* is a degenerate instance of it.

this captures how we currently view recursions at the second-order level. Namely, we see the function which recursively applies s to an *arbitrary* initial value z as the function component of some second-order deliverable, whose proof component is the proof by induction of the conclusion

$$\forall n{:}nat.\ \forall z{:}t.\ R\ 0\ z \Longrightarrow R\ n\ (natrec\ z\ s\ n).$$

As observed above, the hypothesis in the step case of induction arises as the proof component of a second-order deliverable

$$R \xrightarrow{(s,S)} (+1)^{\star}R \quad [1_{nat}]$$

where $(+1)^{\star}R$, otherwise written $R[(n+1)/n]$, is the relation

$$\lambda n{:}nat.\ \lambda y{:}t.\ R\ (n+1)\ y.$$

In like manner, we write $0^{\star}R$, or $R[0/n]$, for the relation

$$\lambda n{:}nat.\ \lambda y{:}t.\ R\ 0\ y.$$

We thus obtain the second-order deliverable constructor for *nat* recursions as the following derived rule

$$\frac{R \xrightarrow{(s,S)} (+1)^{\star}R \quad [1_{nat}]}{Natrec_2\ (s,S){:}0^{\star}R \rightarrow R \quad [1_{nat}]}$$

where $Natrec_2$ has function component

$$\lambda n{:}nat.\ \lambda z{:}t.\ natrec\ z\ s\ n.$$

The principle reason for making this choice of representation is a pragmatic one, based partly on experience, and on the behaviour of unification in the typechecker. If we were to mimic the construction of first-order deliverables by induction, we would expect some rule with one hypothesis for each constructor of the datatype, such as for example

$$\frac{1 \xrightarrow{(z,Z)} 0^{\star}R \quad [1_{nat}] \quad R \xrightarrow{(s,S)} (+1)^{\star}R \quad [1_{nat}]}{Natrec_2'\ (z,Z)\ (s,S){:}1 \rightarrow R \quad [1]}.$$

We would typically apply such a rule in a top-down proof, to a subgoal of the form
`del2 ?n ?m R.`
In a top-down development, where we may construct deliverables using all the constructions described above, we would like the instantiation of `?m` to be both as general as possible, to allow for subsequent development, and yet to allow unification to constrain `?m` to make the application valid. Our choice of the above rule for $Natrec_2$ seems to achieve this. We do not regard this choice as necessarily definitive, however: it merely represents our present view.

4.7 Lists

We may extend this analysis to the case of lists, where, as in the case of first-order deliverables, we find the richer structure of lists reflected in a richer collection of predicates and relations.

Firstly, if we work in the fibre $\mathbf{del}_2\ 1_{list\ a}$, then we obtain in exactly the same way as above, the following derived rule:

$$\frac{\Pi x{:}a.\ R \xrightarrow{\ \mathcal{F}\ } (cons\ x)^{\star}R \quad [1_{list\ a}]}{Listrec_2\ \mathcal{F}{:}(nil)^{\star}R \ \rightarrow\ R \quad [1_{list\ a}]}$$

where

$$(cons\ x)^{\star}R =_{\text{def}} \lambda l{:}list\ a.\ \lambda y{:}t.\ R\,(cons\ x\ l)\ y$$

and

$$(nil)^{\star}R =_{\text{def}} \lambda l{:}list\ a.\ \lambda y{:}t.\ R\,(nil\ a)\ y.$$

But already in this rule we find something new: the outermost Π binding. That is to say, the rule has as its premise a *dependent* family of second-order deliverables. This phenomenon arises from the parameter type a of the lists in question. The rule is susceptible to the same criticisms as the rule for $Natrec_2$ above, but also the criticism that we have accorded a different status to the parameter type. In particular, it does not seem to be constrained by any predicate A we might impose on a. Our justification, as above, is essentially pragmatic. We have found this rule to be a useful construction, as in the example of minimum finding in a list.

We can obtain forms of this rule in which the input list is further constrained. We may, for example, consider the predicate *Listof* A, for some predicate A on a. In fact, since we are now considering second-order deliverables, where we can take into account relations which depend on both the input variable and the result of some computation step, we may extend this predicate to a dependent version, which we call *depListof* A, defined as follows:

$$depListof\ \Phi\ (nil\ a) =_{\text{def}} true$$
$$depListof\ \Phi\ (cons\ x\ l) =_{\text{def}} (\Phi\ x\ l) \wedge (depListof\ \Phi\ l)$$

Here Φ is some *relation* between values of the variable x varying over the parameter type a, and lists over a. An example is the predicate *Sorted*, where we take $\Phi\ x\ l$ to be the relation that x is less than each element of the list l. This crops up in the example of an insert sort in the next section.

This introduces an extra hypothesis into the induction scheme we must consider. Suppose we wish to prove

$$\forall l{:}list\ a.\ (depListof\ \Phi\ l) \Longrightarrow R\ l\ (listrec\ n\ c\ l)$$

where $n{:}t$, $c{:}a \longrightarrow t \longrightarrow t$. A proof by induction generates the following hypotheses for each constructor.

base case *true* $\Longrightarrow R\ nil\ n$, which reduces logically to $R\ nil\ n$. As with the rules for $Natrec_2$, we shall fold this assumption into the rule as the initial relation in the second-order deliverable we eventually derive.

step case Formally, we obtain

$$\forall x{:}a.\ \forall l{:}list\ a.\ ((depListof\ \Phi\ l) \Rightarrow R\ l\ (listrec\ n\ c\ l)) \Longrightarrow$$
$$((depListof\ \Phi\ (x :: l)) \Rightarrow R\ (x :: l)\ (c\ x\ l\ (listrec\ n\ c\ l))).$$

Two simplifications present themselves. The first is to replace the explicit mention of $(listrec\ n\ c\ l)$ by an additional universally quantified parameter y. The second is to observe that $depListof\ \Phi\ (cons\ x\ l) \Longrightarrow depListof\ \Phi\ l$. Combining these, we obtain as an induction hypothesis in the step case

$$\forall x{:}a.\ \forall l{:}list\ a.\ \forall y{:}t.\ (depListof\ \Phi\ (x :: l)) \Longrightarrow R\ l\ y \Longrightarrow R\ (x :: l)\ (c\ x\ l\ y).$$

In this form, we see the logical part of a second-order deliverable emerge.

We thus obtain the following derived rule, which yields a second-order deliverable with function component *listrec* from a dependent family of second-order deliverables:

$$\frac{\mathcal{F}{:}\Pi x{:}a.\ R \rightarrow (cons\ x)^{\star}R\quad [(cons\ x)^{\star}depListof\ \Phi]}{depListrec_2\ \mathcal{F}{:}nil^{\star}R \rightarrow R\quad [depListof\ \Phi]}$$

5 Examples

In his thesis [23], the second author considered a number of example developments using the methodology outlined above, culminating in a machine-checked proof of the Chinese Remainder Theorem. This last example is too long to include here, although it perhaps best exhibits some of the strengths of our approach. We hope to discuss it in a forthcoming paper.

To illustrate the discussion of the previous sections, we give two examples of the use of deliverables in small-scale program development. The use of deliverables may seem heavy-handed and even counter-productive for these smaller examples. The discussion is taken from [23], where detailed accounts in LEGO of the derivations may be found.

5.1 Finding the minimum of a list

An example which is treated several times in the literature [28, 34]. In our case, it involves a somewhat unnatural reperesentation, which makes non-trivial use of the semi-cartesian closed structure in the fibres $\mathbf{del}_2$.

The mathematical specification Suppose R is a decidable total order on some type α, which for the purposes of this example contains a maximal element a_0 (this avoids having to consider exceptions for the case of the *nil* list). We distinguish between R, a *boolean* valued function, and the *relation* $\lambda a, b{:}\alpha.\ R\,a\,b = tt$, denoted $\leq_R$. Then we may specify the minimum of a list as follows:

$$\forall l{:}list\alpha . l \neq nil \Rightarrow \exists m{:}\alpha . m \in l \wedge (\forall a{:}\alpha . a \in l \Rightarrow m \leq_R a).$$

We abbreviate the second conjunct to $m \preceq_R l$. We may easily express a solution to this specification in SML as follows (:: adds an element to a list):

```
fun min a b = if (R a b) then a else b;

fun minelemaux nil = (fn a => a) |
    minelemaux (b::l) = (fn a => (min a (minelemaux l b));

fun minelem nil = a_0 |
    minelem (a::l) = minelemaux l a;
```

We have explicitly curried the function definition of [34][13]

```
fun minelem (a::nil) = a
  | minelem(a::b::l) = min a (minelem (b::l));
```

since the type system of ECC is too strict to allow such a definition based on pattern matching. Our definition is by recursion on the first argument l of **minelemaux**; the corresponding proof is by induction on l. We seek to verify that **minelemaux** meets the following specification:

$$\forall l{:}list\ \alpha.\ \exists f{:}\alpha \rightarrow \alpha.\ \forall a{:}\alpha.\ fa \in a :: l \wedge fa \preceq_R a :: l;$$

The proof for **minelem** follows by composition with a (suitably relativised) deliverable for application.

The correctness proof We just consider the verification of **minelemaux**. As indicated, the proof is by induction.

Base case

$$\exists f{:}\alpha \rightarrow \alpha.\ \forall a{:}\alpha.\ fa \in a :: nil \wedge fa \preceq_R a :: nil.$$

The condition $fa \in a :: nil$ forces us to choose $f =_{\text{def}} \lambda a{:}\alpha.\ a$. For a reflexive R, the second conjunct is then satisfied.

Step case Suppose

$$\exists f{:}\alpha \rightarrow \alpha.\ \forall a{:}\alpha.\ fa \in a :: k \wedge fa \preceq_R a :: k.$$

Then for $b \in \alpha$, taking $g =_{\text{def}} \lambda a{:}\alpha.\ min\ a\ (f\,b)$, we obtain for all $a \in \alpha$, by cases on $R\ a\ (f\,b)$:

[13] Sannella in fact treats the *maximum* of the list.

- $R\ a\ (f\ b) = true$, and hence $min\ a\ (f\ b) = a$. Now $a \in a :: b :: k$, and $a \preceq_R a :: b :: k$, since $a \leq_R a$, and $a \leq_R (f\ b) \preceq_R b :: k$, by hypothesis, and the transitivity of $\leq_R$.
- $R\ a\ (f\ b) = false$, and hence $min\ a\ (f\ b) = f\ b$. We have $fb \in b :: k$, by hypothesis, and hence $fb \in a :: b :: k$. Again, by hypothesis, $f\ b \preceq_R b :: k$, and $fb \leq_R a$. Hence $fb \preceq_R a :: b :: k$, and we are done.

The development in terms of deliverables We give an outline of the proof in the style of a derivation tree, indicating the combinators used to resolve significant subgoals.

Let

$$MinAuxSpec =_{\mathrm{def}} \lambda l{:}list\alpha.\ \lambda f{:}\alpha \longrightarrow \alpha.\ \forall a{:}\alpha.\ fa \in a :: l \wedge fa \preceq_R a :: l.$$

Abbreviating the trivial specifications $\lambda l{:}list\alpha.\ true$ and $\lambda l{:}list\alpha.\ \lambda u{:}unit.\ true$ to 1 (since they are terminal objects, respectively in $\mathbf{del}_1$, and $\mathbf{del}_2\ (list\alpha,\ 1)$), we seek

$$1 \xrightarrow{minelemaux} MinAuxSpec\ \ [1]$$

We use composition to enable us to exploit our recursion combinator for second-order deliverables over lists of Section 4.6, to package up the proof by induction:

$$\frac{1 \xrightarrow{base} ?1\ \ [1]\ ?1 \xrightarrow{step} MinAuxSpec\ \ [1]}{1 \xrightarrow{minelemaux} MinAuxSpec\ \ [1]}(compose)$$

and

$$\frac{\Pi a{:}\alpha.\ MinAuxSpec \xrightarrow{\mathcal{F}_{step}} (cons\ a)^{\star}MinAuxSpec\ \ [1_{list\,\alpha}]}{?1 \xrightarrow{step} MinAuxSpec\ \ [1]}(Listrec_2)$$

which instantiates subgoal ?1 with the specification $nil^{\star}MinAuxSpec$.

We resolve the base case of our induction using a pointwise construction, and thereafter exactly the informal reasoning above. This is sufficient to allow us to completely close this branch of the derivation.

$$\frac{\begin{array}{l} value = \lambda a{:}\alpha.\ a \\ proof : a \in [a] \wedge a \preceq_R [a] \end{array}}{1 \xrightarrow{base} nil^{\star}MinAuxSpec\ \ [1]}(pointwise)$$

This leaves the step case. We extend the context with a free variable b of type α, and then conclude with another pointwise construction. The proof follows the informal argument above.

$$\frac{\begin{array}{l} value = \lambda a{:}\alpha.\ min\ a\ (f\ b) \\ proof : \ldots \end{array}}{MinAuxSpec \xrightarrow{\mathcal{F}_{step}\ b} (cons\ b)^{\star}MinAuxSpec\ \ [1]}\,(pointwise)$$

5.2 Insert sort

We consider insert sort, since it is expressible naturally within primitive recursion, as opposed to more efficient algorithms such as Hoare's quicksort, which has a natural expression only in terms of general recursion.

The mathematical specification Informally, this is straightforward enough. For every list l, over some type α which carries a decidable linear ordering, there exists a sorted list m which is a permutation of l. In a formal treatment, we must give explicit representations of the notions of "sortedness" and "permutation". We use an impredicative definition of the permutation relation $\equiv$, which we do not discuss here. The predicate *Sorted* was defined by recursion:

$$\frac{}{Sorted(nil)} \quad \frac{Sorted\ (l)\ \ a \preceq l}{Sorted\ (a :: l)}\ [a{:}\alpha, l{:}list\ \alpha]$$

where

$$\frac{}{a \preceq nil}\ [a{:}\alpha] \quad \frac{a \preceq l\ \ a \leq b}{a \preceq b :: l}\ [a, b{:}\alpha, l{:}list\ \alpha]$$

As we observed in the section on dependent list recursion, *Sorted* may be seen as an instance of the ***depListof*** constructor, while the predicate $\lambda l{:}list\,\alpha.\ a \preceq l$, for $a{:}\alpha$, is just an instance of *Listof*.

A correctness proof for insert sort Here is the ML prototype for the sort that we wrote, with a view to proving correct:

```
fun listrec n c l = fold ( fn (a,m) => c a m) l n;
val swapcons = fn b => (fn p =>
                        let val (c,m) = p
                         in  (max(c,b), (min(c,b):: m))
                          end);
fun scons a l = let val (b,m) = listrec (a,nil) swapcons  l
                in b::m
                 end;
val sort = listrec nil scons;
```

and here is its translation into LEGO code:

```
[swapcons = [A|Type][b:A][p:A # (list A)]
            ((min p.1 p.2),cons (max p.1 p.2) m)];

[scons = [A|Type][a:A][l:list A]
         [p = listrec (a,nil) swapcons  l](cons p.1 p.2)];

[sort = listrec nil scons];
```

We anticipate two levels of induction in the correctness proof for this function, since sort is defined by nested recursion. We recall the derived induction principles for recursive program phrases which we used to explain recursive deliverables:

$$\text{for } \phi{:}(\mathit{list}\ \alpha) \longrightarrow \beta \longrightarrow \mathit{Prop}, \quad n{:}\beta, \quad c{:}\alpha \longrightarrow (\mathit{list}\ \alpha) \longrightarrow \beta \longrightarrow \beta$$

$$\frac{\phi\ \mathit{nil}\ n \qquad \begin{array}{c}\left[\phi\ l\ b\ \ [a{:}\alpha,\ l{:}\mathit{list}\ \alpha,\ b{:}\beta]\right] \\ \vdots \\ \phi\ (a{::}l)\ (c\ a\ l\ b)\end{array}}{\forall l{:}\mathit{list}\ \alpha.\phi\ l\ (\mathit{listrec}\ n\ c\ l)}$$

a consequence of which is the following induction principle for sorted lists, which is just the proof component of an instance of the $\mathit{depListrec}_2$ combinator we considered above:

$$\text{for } \phi{:}(\mathit{list}\ \alpha) \longrightarrow \beta \longrightarrow \mathit{Prop}, \quad n{:}\beta, \quad c{:}\alpha \longrightarrow (\mathit{list}\ \alpha) \longrightarrow \beta \longrightarrow \beta$$

$$\frac{\phi\ \mathit{nil}\ n \qquad \begin{array}{c}\left[\mathit{Sorted}\,(a{::}l),\ \ \phi\ l\ b\ \ [a{:}\alpha,\ l{:}\mathit{list}\ \alpha,\ b{:}\beta]\right] \\ \vdots \\ \phi\ (a{::}l)\ (c\ a\ l\ b)\end{array}}{\forall l{:}\mathit{list}\ \alpha.\ \mathit{Sorted}\ \ l \Longrightarrow \phi\ l\ (\mathit{listrec}\ n\ c\ l)}\ \text{Sorted List Induction}$$

This induction principle exemplifies the motivation for second-order deliverables: namely that a relation ϕ between lists holds, *only* on the condition that one of them is sorted.

The correctness proof proceeds by application of the first induction principle, for the outermost recursion: this produces as subgoals

(a) $\mathit{Sorted}(\mathit{nil}),\ \mathit{nil} \sim \mathit{nil}$
(b) $\forall a : \alpha.\forall l, m{:}\mathit{list}\ \alpha.\ (\mathit{Sorted}\,(m) \wedge l \sim m) \Longrightarrow$
$(\mathit{Sorted}\,(\text{scons}\ a\ m) \wedge a :: l \sim (\text{scons}\ a\ m))$.

Now (a) is immediate, and (b) reduces to

$$\forall a : \alpha.\forall m{:}\mathit{list}\ \alpha.\mathit{Sorted}\,(m) \Longrightarrow (\mathit{Sorted}\,(\text{scons}\ a\ m) \wedge a :: m \sim (\text{scons}\ a\ m))$$

$\forall$-introduction extends the context with the assumption $[a{:}\alpha]$, and in this extended context we define the relativised specification

$$\psi_a \equiv \lambda m, n{:}\mathit{list}\ \alpha.\mathit{Sorted}\,(n) \wedge a :: m \sim n.$$

Now we may apply Sorted List Induction to yield the following subgoals in the new context:

(c) $a :: nil \sim a :: nil \wedge Sorted\,(a :: nil)$
(d) $\forall b : \alpha.\forall m, n{:}list\ \alpha.\ Sorted\,(b :: m) \wedge \psi_a mn \Longrightarrow \psi_a(b :: m)(\text{swapcons}\ b\ m)$.

(c) is trivial, and (d) reduces to

$$\forall b, c : \alpha.\forall m, n{:}list\ \alpha.Sorted\,(b :: m) \wedge Sorted\,(c :: n) \wedge a :: m \sim c :: n \Longrightarrow$$

$$Sorted\,(\min\,(b,c) :: \max\,(b,c) :: n) \wedge a :: b :: m \sim \min\,(b,c) :: \max\,(b,c) :: n$$

This last goal rests, apart from some trivial properties of $\sim$, on the following:

Lemma 12. *Suppose $a, b, c \in \alpha$, and $m, n \in list\ \alpha$ such that:*

- $Sorted\,(c :: n)$;
- $c :: n \sim a :: m$;
- $b \preceq m$.

Then $\max\,(b,c) \preceq n$.

The proof recast in terms of deliverables We have a double recursion in the definition of our program, which translates into a nested list recursion at the level of deliverables. The base case of each induction is resolved with a simple pointwise construction. So too is the inner step case for the verification of the **swapcons** function, with some additional propositional reasoning.

However, there is a point of considerable delicacy in the course of this proof. There is a stage at which we must shift from a second-order deliverable over 1, the identically true predicate on lists, defined by the outermost recursion, to the inner recursion on second-order deliverables defined over the predicate *Sorted*. It appears that we must eliminate the *input* parameter l using the permutation relation, which moves the condition $Sorted(m)$ from being a condition on a *result* to being a condition on the *input* to *scons*. This is the reduction in case (b) above, from

$$\forall a : \alpha.\forall l, m{:}list\,\alpha.(Sorted\,(m) \wedge l \sim m) \Longrightarrow (Sorted\,(\text{scons}\ a\ m) \wedge a :: l \sim (\text{scons}\ a\ m)$$

to

$$\forall a : \alpha.\forall m{:}list\ \alpha.Sorted\,(m) \Longrightarrow (Sorted\,(\text{scons}\ a\ m) \wedge a :: m \sim (\text{scons}\ a\ m)$$

This shift allows us to apply the principle of Sorted List Induction.

This is rather inconvenient, and moreover it seems that only *ad hoc* solutions exist to resolve the difficulty. Rather than attempt a detailed exposition here, we merely remark on the rather unsatisfactory nature of this step in our derivation. The details are in [23].

We outline the significant refinement steps in the derivation.

We seek, as a top-level goal

$$1 \xrightarrow{insertsort} SortSpec \quad [1]$$

where $SortSpec\ l\ m =_{\text{def}} Sorted(m) \wedge l \sim m$.

The outermost recursion uses the $Listrec_2$ combinator, which we apply after splitting the initial goal with the composition operator.

$$\frac{1 \xrightarrow{base} ?1 \quad [1] \; ?1 \xrightarrow{step} SortSpec \quad [1]}{1 \xrightarrow{insertsort} SortSpec \quad [1]} (compose)$$

$$\frac{\Pi a{:}\alpha.\ SortSpec \xrightarrow{\mathcal{F}_{step}} (cons\ a)^{\star} SortSpec \quad [1_{list\ \alpha}]}{?1 \xrightarrow{step} SortSpec \quad [1]} (Listrec_2)$$

instantiates subgoal ?1 with the specification $nil^{\star}SortSpec$.

The base case of this recursion/induction is resolved by a pointwise construction. Moreover, we can prove that any permutation of the *nil* list must be equal to *nil*, and the *nil* list is trivially sorted. Hence we obtain the correct instantiation

$$\frac{\begin{array}{l} value = nil \\ proof : Sorted(nil) \wedge nil \sim nil \end{array}}{1 \xrightarrow{base} nil^{\star} SortSpec \quad [1]} (pointwise)$$

The step case is a little more complicated. We first introduce the parameter $a : \alpha$ in the $Listrec_2$ rule. Recall that the algorithm we considered above contains a `let` construct.

```
fun scons a l = let val (b,m) = listrec (a,nil) swapcons  l
                in b::m
                 end;
```

Now composition of deliverables explicates the `let` construct. For the second component of this composition, we just use the second-order deliverable analogue of the trivial construction of Section 3. We then fold this function into a relativised specification, with which we pursue the inner recursion/induction:

$$\frac{SortSpec \rightarrow SortSpec' \quad [1] \quad \dfrac{function = \lambda p{:}\alpha \times list\alpha.\ \pi_1(p) :: \pi_2(p)}{SortSpec' \rightarrow (cons\ a)^{\star} SortSpec \quad [1]} (func)}{SortSpec \xrightarrow{\mathcal{F}_{step}\ a} cons\ a^{\star} SortSpec \quad [1]} (compose)$$

where $SortSpec' =_{\text{def}} (\lambda p{:}\alpha \times list\alpha.\ \pi_1(p) :: \pi_2(p))^{\star}(cons\ a)^{\star} SortSpec$

Eliding the details, we massage the goal $SortSpec \twoheadrightarrow SortSpec'$ [1] into a form in which we may exploit the principle of Sorted List Induction. It turns out that the target specification remains as $SortSpec'$:

$$\frac{\dfrac{1 \twoheadrightarrow SortSpec' \;\; [Sorted]}{\cdots}}{SortSpec \twoheadrightarrow SortSpec' \;\; [1]}$$

As indicated above, we are now in a position to use the $depListrec_2$ combinator. Again, we preface its application with an appeal to composition.

$$\frac{1 \xrightarrow{base'} ? \;\; [Sorted] \; ? \xrightarrow{step'} SortSpec \;\; [Sorted]}{1 \twoheadrightarrow SortSpec \;\; [Sorted]} (compose)$$

$$\frac{\Pi b{:}\alpha.\ SortSpec' \xrightarrow{\mathcal{G}_{step'}} (cons\ b)^{\star}SortSpec' \;\; [(cons\ b)^{\star}Sorted]}{? \twoheadrightarrow SortSpec' \;\; [Sorted]} (depListrec_2)$$

The proof concludes by considering a pointwise construction in the base case of the induction,

$$\frac{\begin{array}{l} value = (a, nil) \\ proof : Sorted([a]) \wedge [a] \sim [a] \end{array}}{1 \xrightarrow{base'} nil^{\star}(SortSpec') \;\; [Sorted]} (pointwise)$$

and then an account of the verification of the **swapcons** function in the *step'* case. We omit the details, which exploit the informal argument of Lemma 12.

6 A set-theoretic model

In the foregoing discussion, our emphasis lay on using and constructing deliverables in the context of a particular type theory and its machine implementation. However, the idea of deliverables need not be restricted to this particular seeting. We observed that one could model deliverables in the framework of elementary set theory. The thesis [23] gives a detailed categorical account of this model, showing how it supports the interpretation of a subtheory of ECC. Indeed, the model is parametric in the choice of an underlying *topos*(categorical model of set theory). We sketch the main ideas. We refer the reader to [23] for a complete account.

In this interpretation, a (closed) type is given by a pair of sets (A, A'), with $A' \subseteq A$. In particular, the type *Prop* of propositions, is given by the pair (Ω, Ω), where Ω is the subobject classifier. A morphism between such objects is given by the obvious abstract notion of first-order deliverable, i.e. an arrow on the underlying sets which respects the distinguished subsets. This data gives us an interpretation of contexts and morphisms between them.

Types in a given context $\Gamma = (X_0, X_1)$ are given by the relativised specifications, i.e. pairs of sets (A, R), where now $R \subseteq X_0 \times A$. Without going into further detail, this structure gives rise to a *fibration*, whose fibres essentially consist of the second-order deliverables described above. Moreover, this fibration has sums, products and a small subfibration and thus supports an interpretation of the "*Prop*: $Type_0$: $Type_1$" fragment of ECC. In the presence of set-theoretic universes, given for example by inaccesible cardinals, we may model the whole of ECC.

7 Describing deliverables by lambda expressions

We now outline some tentative ideas which we are investigating with a student, M. Takeyama. They take us closer to Martin-Löf's subset theory approach referred to in the next section.

Deliverables, like functions, form a cartesian closed category, so just like functions they can be described in the internal language of that category, namely typed lambda terms. (For simplicity we speak here of first order deliverables.)

Thus the definition of a deliverable using categorical combinators

$$h = f \circ (\langle \pi_1, g \circ \langle \pi_2, \pi_1 \rangle \rangle)$$

(or its Lego equivalent) becomes more readable in the familiar λ-notation

$$h = \lambda xy : t.f(x, g(y, x)) \quad \text{or} \quad h[x, y : t] = f(x, g(y, x))$$

We would also like to be able to create deliverables, like functions, by refinement from a specification.

To allow λ-notation and refinement, we seek to extend ECC with a new sort, *Spec*, analogous to *Prop* and *Type*, with new operations * and $_*$. Given an object in the category of deliverables, say $(t, P : t \rightarrow Prop)$, we could form $P^* : Spec$, and given a morphism, that is a deliverable, $F : del_1 PQ$ we could form $F^* : P^* \rightarrow Q^*$. Now given also $G^* : Q^* \rightarrow R^*$ we can write $\lambda x : P^*.G^*(F^* x)$, thus describing the new deliverable using ECC λ-notation. We could also do a refinement: take $P^* \rightarrow R^*$ as a goal, then do $\rightarrow$ introduction, which adds $x : P^*$ to the context and gives R^* as the subgoal.

Finally, having developed $HH : P^* \rightarrow R^*$ it needs to be converted back to a deliverable, say $HH_* : P \rightarrow R$, using techniques familiar from categorical logic [18]. So * is a formal operation but $_*$ does some work.

The approach could be generalised. Given any cartesian closed category C, defined in ECC as a tuple, $C : CartClosedCat$, we could get a new sort, $SpecC$, and operations * and $_*$ depending on C.

These ideas and their extension to second order deliverables are still to be worked out both in theory and in practice.

8 Related work

Two approaches to a logical account of formal program development are well known. The first relies on annotating programs with logical formulae, and expressing the correctness of a program in terms of logical deductions. This has its origins in the work of Floyd, Hoare and others in the 'Sixties [12, for example]. The idea of a deliverable clearly has echoes of this idea, but brought into the functional setting. It also avoids the defect of the Floyd/Hoare style in having object language and meta-logical variables on the same footing as object variables of our chosen type theory.

The second approach, based on various intuitionistic type theories, has been to develop constructive proofs, and use realisability techniques to extract algorithmic information. Both Martin-Löf type theory [5, 27] and the Calculus of Constructions [6, 28, 29] have been used in this style. Our work uses ideas from both these schools. Most influential has been the theory of subsets in Martin-Löf type theory. This has been given an eloquent treatment in [27, Chapter 18], to which the reader is referred for a detailed discussion. The central problem in using constructive proofs as a programming discipline is that proofs contain redundant information. Dependent types allow us to express logical predicates as types, but do not permit the representation of any more recursive functions. For the Calculus of Constructions, this has a precise statement in the result of Berardi and Mohring [28, 2]:

Theorem *CC is conservative over* F_ω.

Consequently, *CC* can represent, in the standard (Church) representation of functions on inductive types, no more functions than F_ω. The proof is based on a syntactic mapping, the so-called Berardi-Mohring projection. Mohring used this mapping as an extraction function, which, coupled with the associated realisability predicate, allows a powerful and flexible approach to program development from proofs. Under this interpretation, an arbitrary type is interpreted as a type, together with a predicate (the realisability predicate) defined over it.

The approach taken in [27] is to separate computationally relevant proofs from the purely logical, via a translation of the judgments of the basic formal system into multiple judgments. This translation permits the formation of a "subset type" $\{x \in A | B(x)\}$, for which a reasonable elimination rule may be given. An attempt to equip the basic theory with such a type (which may be used to precisely hide the information of the precise nature of a proof that predicate $B(x)$ holds) yields very unsatisfactory results [32, 33, 27].

Given the basic theory of types and terms in Martin-Löf type theory, the first step is to extend the theory with a notion of ***proposition*** and a judgment P **true** for propositions. This is very straightforward, using propositions as types. A proposition is just a type in the basic theory. A proposition is true if there is some element inhabiting it, again in the basic theory. This seemingly innocent proof-irrelevance gives the subset theory its power.

The subset theory now interprets the basic judgment

$$A \textbf{ set}$$

of Martin-Löf type theory as two judgments in the underlying theory, prescribing

- a set A' in the basic theory, and
- a family of propositions $A''(x)$ **prop** $[x \in A']$, again in the basic theory.

This corresponds to our definition of specification in Section 3.

Equality of sets A, B is based on equality of the underlying sets A', B', but uses *logical* equivalence of the families A'', B''. We rejected such a choice, in favour of the decidable relation of convertibility in ECC.

The membership judgment $a \in A$ is interpreted in the obvious way: we may derive $a \in A$ if we can derive $a \in A'$ and $A''(a)$ **true** in the basic theory. This captures the essential idea, that the judgments A **set** and $a \in A$ should describe subsets of the terms in A' in the underlying theory. In this way, the proofs of the propositions $A''(a)$ are systematically suppressed. It is then relatively straightforward to see how this allows the interpretation of a subset-type constructor.

How does this compare with our approach? The resulting expressions for the various type constructors are very similar, compare for example the definition of exponential for second-order deliverables with the Π-type in the subset interpretation. We consider explicit proofs of the propositional parts of our specifications, whereas we need only know that some proof may be derived in the subset theory. However, this seems to be one of the limitations of their approach, in that we only know that certain *derivations* in the subset theory arise from certain other derivations. Our use of Σ-types, by contrast, means that we can represent the derivations of deliverables as actual *terms* within ECC, using the definable combinators which code up the explicit translation. The price we pay seems to be that we have to work with a rather clumsy language for these terms, as opposed to the conceptual elegance of reusing the basic language of types and terms in the subset theory.

The NuPrl system, developed by Constable and his co-workers [5], is based on early versions of Martin-Löf's type theory. In particular, the underlying term calculus is untyped, and the system has extensional equality types. This has the advantage of suppressing some irrelevant information in proofs. It also overcomes the limitations on the use of an explicit subset-type constructor in a theory with intensional equality, exposed in [32, 33]. The sovereign disadvantage is that the basic judgments of the theory become undecidable, coupled with a proliferation of well-formedness conditions in the application of the rules.

Recently, Hayashi has also proposed a system based on realisability, which abandons the usual type constructors Π, Σ, on which most work to date on type theory has been based, in favour of a more set-theoretic style, with union, intersection and singleton types [11]. The system he considers is, however, ingenious enough to represent dependent products and dependent sums. At the same time, the typing rules for union and intersection hide information. This allows a simple translation or extraction into a programming language with a polymorphic type discipline. Singleton types seem essential in achieving this harmony between the type system and the underlying untyped terms.

Pavlovič, in his thesis [30], elaborates in categorical terms a theory of constructions in which programs do not depend on proofs of logical propositions.

As with the models of Constructions considered by Hyland and Pitts [16], the emphasis is on extensional systems, rather than the intensional system we work with here. Proof-theoretic properties seem to be regarded as something "...an implementation would have to answer" [30, p.8].

9 Conclusions

We have described a proof development method which takes as its basic entity a program plus its correctness proof. If the programs are functional these program proof-pairs admit exactly the same combination operators as do functions. We have studied the mathematical theory of such program proof-pairs and the operations which combine them, and we have considered the case where the pre- and postconditions are contextually linked to express a relationship between input and output. The combination operations have been coded within the Lego proof system; a number of examples have been developed in Lego, and some metatheoretic results about the operations have been proved in Lego.

Acknowledgements

We would like to thank Randy Pollack for the Lego system and Zhaohui Luo for his contribution to its theoretical underpinning. We also thank our colleagues in the Lego Club for friendly comments and stimulation. We thank Per Martin-Löf, Gordon Plotkin and Susumu Hayashi for helpful comments on presentations of this work. Thanks also to Chris Owens for proof-reading the manuscript.

References

1. J.Bénabou, *Fibred categories and the foundations of naïve category theory*, JSL, 1985.
2. S. Berardi, *Type Dependence and Constructive Mathematics*, Ph.D. thesis, Dipartimento di Informatica, Torino, Italy 1990.
3. N.G. de Bruijn, *A survey of the project AUTOMATH*, in: [35].
4. R.M.Burstall and J.H.McKinna, *Deliverables: an approach to program development in Constructions*, in [15], also available as a University of Edinburgh technical report ECS-LFCS-91-133.
5. R.Constable *et al.*, *Implementing Mathematics with the NuPrl Proof Development System*, Prentice-Hall, New Jersey, 1986.
6. T.Coquand and G.Huet, *Constructions: a Higher-order Proof system for mechanizing mathematics*, in: Proceedings *EUROCAL* '85, LNCS 203, Springer-Verlag, 1985.
7. T.Coquand, *Metamathematical Investigations of a Calculus of Constructions*, in: [14].
8. P-L.Curien, *Categorical Combinators, Sequential Algorithms and Functional Programming*, Pitman Research Notes in Theoretical Computer Science, Pitman, London, 1986.

9. J-Y.Girard, *Interpretation fonctionelle et élimination des coupures dans l'arithmétique de l'ordre supérieure*, thesis, University of Paris VII, 1972.
10. S.Hayashi, *Adjunction of semifunctors: categorical structures in nonextensional lambda calculus*, in Theoretical Computer Science, Vol. 41, North-Holland, Amsterdam, 1985.
11. S.Hayashi, *Singleton, Union and Intersection Types for Program Extraction*, in: Proceedings of TACS '91, Sendai, Japan, Springer LNCS 526, Springer-Verlag, 1991.
12. C.A.R.Hoare, *An axiomatic basis for computer programming*, in: Communications of the ACM, Vol. 12, 1969.
13. W.A.Howard, *The "formulae-as-types" notion of construction*, in: [35].
14. G.Huet, T.Coquand, C.Paulin-Mohring *et al.*, *The Calculus of Constructions, Version 4.10, Documentation and user's manual*, Rapports Techniques no.110, Projet Formel, INRIA-Rocquencourt, Paris, August 1989.
15. G.Huet and G.Plotkin, eds. *Electronic Proceedings of the First Annual BRA Workshop on Logical Frameworks, Antibes, May 1990,*, distributed electronically to participating BRA sites, January 1991.
16. J.M.E.Hyland and A.M.Pitts, *The Theory of Constructions: Categorical Semantics and Topos-theoretic models*, in: Proceedings of the AMS Conference on Categories in Computer Science, Boulder, Colorado, 1986.
17. P.T.Johnstone and R.Paré, eds., *Indexed Categories and their Applications*, Springer LNM 661, Springer-Verlag, 1978.
18. J.Lambek and P.J.Scott, *An Introduction to Higher-Order Categorical Logic*, Cambridge Studies in Advanced Mathematics no. 7, Cambridge University Press, Cambridge, England, 1986.
19. Z.Luo, *ECC, an Extended Calculus of Constructions*, in: Proceedings of the Fourth IEEE Conference on Logic in Computer Science, Asilomar, California, 1989.
20. Z.Luo, *An Extended Calculus of Constructions*, Ph.D. Thesis, Department of Computer Science, University of Edinburgh, June 1990.
21. Z.Luo, *Program Specification and Data Refinement in Type Theory*, Technical Report ECS-LFCS-90-131, Department of Computer Science, University of Edinburgh, January 1991.
22. Z.Luo and R.Pollack, *LEGO Proof Development System: User's Manual*, LFCS Technical Report ECS–LFCS–92-211, 1992.
23. J.H.McKinna, *Deliverables: a categorical approach to program development in type theory*, Ph.D. thesis, University of Edinburgh, 1992.
24. P.Martin-Löf, *An Intuitionistic Theory of Types: Predicative part*, in: Logic Colloquium 73, North-Holland, Amsterdam, 1975.
25. P.Martin-Löf, *Constructive Mathematics and Computer Programming*, in: proceedings of the Conference on Logic, Philosophy and Methodology of Science VI, 1979, North-Holland, Amsterdam, 1982.
26. M.Mendler, *The Logic of Design*, Ph.D. thesis, University of Edinburgh, forthcoming, 1992.
27. B.Nordström, K.Petersson, and J.Smith, *Programming in Martin-Löf's type theory*, Oxford University Press, 1990.
28. C.Paulin-Mohring, *Extracting F_ω's programs from proofs in the Calculus of Constructions*, in: Proceedings POPL89, ACM, 1989.
29. C.Paulin-Mohring and B.Werner, *Extracting and Executing Programs developed in the Inductive Constructions System: a Progress Report*, in: [15].

30. D.Pavlovič, *Predicates and Fibrations*, proefschrift, University of Utrecht, 1990.
31. R.A.Pollack, *Implicit Syntax*, in: [15].
32. A.Salvesen and J.Smith, *On the strength of the subset type in Martin-Löf's type theory*, in: Proceedings of the Third LICS Symposium, IEEE, 1988.
33. A.Salvesen, *On Information Discharging and Retrieval in Martin-Löf's type theory*, Ph.D. thesis, Institute of Informatics, University of Oslo, 1989.
34. D.Sannella, *Formal specification of ML programs*, LFCS technical report ECS-LFCS-86-15, Dept. of Computer Science, University of Edinburgh, 1986.
35. J.P.Seldin and J.R.Hindley, eds., *To H.B.Curry, essays in Combinatory Logic, λ-calculus and Formalism*, Academic Press, 1980.
36. B.Reus and T.Streicher, *Verifying Properties of Module Constructions in Type Theory*, this volume.

COMPLEX AND COMPLEX-LIKE TRACES*

Volker Diekert

Universität Stuttgart, Institut für Informatik,
Breitwiesenstr. 20–22, D 70565 Stuttgart

Abstract. The definition and some known results on complex traces are reviewed. We also discuss some open questions concerning the Poset-property of complex traces. The main new contribution of the paper is the presentation of the notion of complex-like trace. Every complex trace is complex-like, but there are other objects such as a finite trace with some additional non-empty alphabetic information. In the sequential case this information is nothing else than explicit termination. Together with concurrency the concept leads to a rich mathematical structure. Our results show that complex-like traces form a prime algebraic and coherently complete Scott-domain. Our main theorem shows that the concatenation on this domain is continuous.

1 Introduction

The theory of Mazurkiewicz traces [17] provides us with a natural semantics for terminating concurrent processes. Highlights of the theory are (among others) the generalization of Kleene's Theorem [19, 20], Zielonka's construction of asynchronous automata [27], and the logical characterization of recognizable languages,[25]. Surveys on finite traces can be found in [1, 3, 8, 21] or in [7].
In order to cope with the behavior of infinite processes the notion of infinite trace has been introduced, [2, 10, 11, 16, 18]. The straightforward way to see an infinite trace is to look at an infinite dependence graph where every vertex has a finite past [18]. More formally, an (infinite) trace over a dependence alphabet (Σ, D) is an (infinite) node-labelled acyclic graph $[V, E, \lambda]$ where edges are between different vertices with dependent labels and where the downward closure of every vertex is a finite set. A finite or infinite Mazurkiewicz trace (dependence graph) is called a *real trace*. A first successful program in the theory of real traces has been to generalize the theory of infinite words (as exposed e.g. in [22, 24]) to real traces. In particular, recognizable real trace languages have been characterized by concurrent rational expressions [12], asynchronous automata with non-deterministic Büchi acceptance [14] and with deterministic Muller acceptance [6]. Very recently, the logical counterpart has completed the picture [9].
The main drawback of real traces is that there is no way to define any convenient notion of concatenation. This causes many problems. Think for example

* This research has been partially supported by the ESPRIT Basic Research Actions No. 6137 ASMICS II

that two non-terminating processes p and q are defined by infinite real traces r and s (or by infinitary real trace languages). For sequential processes the composition pq will just behave like p, since q will never start if p does not terminate. Hence, the equation $rs = r$ is adequate. However, dealing with concurrency, the processes p and q might be completely independent and then $pq = qp = p||q$ is the parallel execution. Hence, we need a formalism where $r \neq rs = sr \neq s$ is possible. A suitable formalism which solves this problem is given by the calculus of *complex traces*. A complex trace is simply a pair of some real trace and some finite alphabet containing the information which set of actions may not start concurrently. The resulting theory is a proper and correct generalization of the theory of finite and infinite words. It contains a rich mathematics and has a meaningful semantics. There is a topology on the set of complex traces such that this space is compact.This topology is given by the completion of some natural ultra-metrics on the set of finite traces where the concatenation is uniformly continuous [4].

We can define in a canonical way the notions of rational, concurrent-rational, and recognizable complex trace language. These languages are studied in [5] and as the main result there the generalization of the Kleene-Ochmanski-Theorem is shown. It is also shown that complex traces form a coherently complete CPO with respect to prefix ordering, [15].

The present paper is organized as follows: first we recall the notion of complex trace and we review some of the basic results. The Poset-properties with respect to prefix ordering are studied in more detail. (Poset = partially ordered set) We show that the CPO (= complete partial order) of complex traces is not algebraic as soon as the dependence alphabet contains an induced subgraph which is a line of four letters, Prop. 11. In fact we strongly conjecture that this is a characterization. We also show its relation to the question if the non-real traces form a right ideal. Dependence alphabets where non real-traces form a right-ideal are fully characterized in Prop. 10.

As in any CPO where the partial order is defined by the prefix relation in a monoid, we can not expect that the concatenation is continuous. For this reason we define another ordering $\sqsubseteq_{\mathrm{ap}}$ between complex traces, which will be called *approximation*. This new ordering leads to the notion of *complex-like trace*. A complex-like trace is a pair consisting of a real trace and some set of letters. Let us mention that a finite complex-like trace may have a non-empty imaginary part. Section 5 develops a theory of this structure. Again, in the sequential setting, we do nothing new. Our approach simply reduces to a formalism with explicit termination. However, since we are working with concurrency, some nice and non-trivial properties can be exhibited. We show that complex-like traces form a coherently complete Scott domain (which in fact is prime algebraic), see Thm. 18. The subset of finite and real traces can fully be characterized in terms of the approximation ordering $\sqsubseteq_{\mathrm{ap}}$, Cor. 19. As a main result we prove that the concatenation is continuous, Thm. 22.

2 The semantics of complex traces

We assume in this section that the reader is familiar with the basic concept of traces. All formal definitions are deferred to the next sections. A real trace can be thought as a concurrent process which has some serialization as a finite or infinite word. However, there are simple examples of concurrent processes which show that the notion of real trace can not cope with some basic phenomena of non-terminating processes. To be more concrete let us consider the following two procedures:

- **procedure** $p(x, y, z)$
  while condition(x) **do** $a(x)$ **endwhile**
  $b(x, y); c(y, z)$
  endprocedure

- **procedure** $q(x, y, z)$
  while condition(x) **do** $a(x)$ **endwhile**
  $c(y, z); b(x, y)$
  endprocedure

By a, b, c we denote some subroutines where the set of read/write variables are denoted in brackets. We obtain the following dependence alphabet:

$$(\Sigma, D) = a - b - c$$

The sequential runs of p and q are given by the sets a^*bc and a^*cb respectively. The corresponding trace sets can be depicted as follows:

$$a^* \longrightarrow b \longrightarrow c \text{ and } \begin{array}{ccc} a^* & \longrightarrow & b \\ & \nearrow & \\ c & & \end{array}$$

In both models, sequential and concurrent, we clearly see a different behavior. Now, assume that the while-condition never becomes false.
Then for a sequential model, we may describe both procedures p and q by the expression $a^\omega bc = a^\omega \in \Sigma^\omega$. The point however is that the expression a^ω gives us the important information that neither b nor c have been executed and that the values of the variables y and z are unchanged. In a concurrent execution p and q may be have differently, even if the while condition never becomes false. (Imagine a scenario of open systems where a change of the variable z may have a drastic impact on the environment.) The procedure p can neither change y nor z, the procedure q can do, since c may be executed concurrently with a. Therefore we need a semantics where the concatenation of $a^\omega, b,$ and c yields a different result from the sequential concatenation of $a^\omega, c,$ and b. Since a and c are independent it seems to be obvious that we have $a^\omega \neq a^\omega c = ca^\omega$. Less obvious is that this formalism forces $a^\omega b \neq a^\omega$, too - although a and b are dependent.

Indeed assume that we axiomatize:

$$a^\omega b^\omega = a^\omega \text{ (since } (a,b) \in D),$$
$$b^\omega c^\omega = b^\omega \text{ (since } (b,c) \in D),$$
$$a^\omega c^\omega \neq a^\omega \text{ (since } (a,c) \notin D).$$

Then these axioms can not be combined with associativity:

$$a^\omega \neq a^\omega c^\omega = (a^\omega b^\omega)c^\omega = a^\omega(b^\omega c^\omega) = a^\omega b^\omega = a^\omega$$

The calculus of complex traces solves this problem. A complex trace is a pair consisting of a real and imaginary part. The real part is a finite or infinite trace in the sense of Mazurkiewicz. The execution of every finite prefix of the real part can be seen, in principle, after some finite amount of time. The imaginary part is simply some finite alphabetic information. It describes the set of actions which never can be started concurrently. Thus, the imaginary part is never directly observable, but it has influence to the environment. In our example we will obtain the following:

$$\begin{aligned} p = a^\omega bc &= (a^\omega, D(a))(b,\emptyset)(c,\emptyset) \\ &= (a^\omega, D(a,b)) \cdot (c,\emptyset) \\ &= (a^\omega, D(a,b,c)) = (a^\omega, \Sigma) \\ q = a^\omega cb &= (a^\omega, D(a))(c,\emptyset)(b,\emptyset) \\ &= (ca^\omega, D(a))(b,\emptyset) \\ &= (ca^\omega, D(a,b)) = (ca^\omega, \Sigma) \end{aligned}$$

Furthermore these expressions show as desired $p = p^2 = pq \neq qp = q^2 = q$
A description with recognizable sets can be used to include the finite and infinite behavior over (Σ, D):

$$\begin{aligned} p = a^*bc \cup a^\omega bc &= \quad a^*bc \cup \{(a^\omega, D(a,b,c)\} \\ q = a^*cb \cup a^\omega cb &= a^*ca^*b \cup \{(a^*ca^\omega, D(a,b)\} \end{aligned}$$

3 Traces

3.1 Dependence graphs and real traces

A *dependence alphabet* (Σ, D) is a (finite) alphabet Σ together with a symmetric and reflexive *dependence relation* $D \subseteq \Sigma \times \Sigma$. The complementary relation $I = \Sigma \times \Sigma \setminus D$ is called *independence relation*. The quotient monoid $\mathbb{M} = \mathbb{M}(\Sigma, D) = \Sigma^*/\{ab = ba \mid (a,b) \in I\}$ is Mazurkiewicz' trace monoid. Every trace $t \in \mathbb{M}$ can be identified with its dependence graph which is (an isomorphism class of) a node-labelled acyclic graph $[V, E, \lambda]$ where

- V is the set of vertices,
- $E \subseteq V \times V$ is the set of edges,
- $\lambda : V \longrightarrow \Sigma$ is the labelling such that $\lambda^{-1}(D) = E \cup E^{-1} \cup id_V$,
- and the induced partial order $E^* \subseteq V \times V$ is well-founded

In order to speak of the set of all dependence graphs we need an upper bound for the cardinality of V. A convenient upper bound is *countability*. The set of all dependence graphs where the vertex set is at most countable is denoted by $\mathbb{G}(\Sigma, D)$ or simply by $\mathbb{G}$, if the reference to (Σ, D) is clear. The set $\mathbb{G}$ is a monoid by

$$[V_1, E_1, \lambda_1] \cdot [V_2, E_2, \lambda_2] = [V, E, \lambda]$$

where $V = V_1 \dot{\cup} V_2$ is the disjoint union with induced labelling $\lambda : V \longrightarrow \Sigma$ and $E = E_1 \cup E_2 \cup \{(x, y) \in V_1 \times V_2 \mid (\lambda_1(x), \lambda_2(y)) \in D\}$.
There is a canonical homomorphism $\varphi : \Sigma^* \longrightarrow \mathbb{G}$ which induces an identification of $\mathbb{M}(\Sigma, D)$ with the submonoid of all finite dependence graphs. The homomorphism φ extends to a mapping $\varphi : \Sigma^\infty = \Sigma^* \cup \Sigma^\omega \longrightarrow \mathbb{G}$. The image $\varphi(\Sigma^\infty)$ is called the set of real traces. It is denoted by $\mathbb{R} = \mathbb{R}(\Sigma, D)$. A dependence graph is real if and only if every vertex has a finite past.
A dependence graph $g \in \mathbb{G}$ splits into its real part $\mathrm{Re}(g) \in \mathbb{R}$ and its transfinite part $\mathrm{Tr}(g) \in \mathbb{G}$. A vertex of g belongs to $\mathrm{Re}(g)$ if it has a finite past and to $Tr(g)$ otherwise. Clearly, we have $g = \mathrm{Re}(g)\mathrm{Tr}(g)$.
The cardinality of a set V is denoted by $|V|$. The length of $g = [V, E, \lambda] \in \mathbb{G}$ is $|g| = |V|$ and $|g|_a = |\lambda(a)|$ is the a-length, $a \in \Sigma$. The alphabet of g is $\mathrm{alph}(g) = \{a \in \Sigma \mid |g|_a \geq 1\}$ and the alphabet at infinity is $\mathrm{alphinf}(g) = \{a \in \Sigma \mid |g|_a = \infty \text{ or } a \in \mathrm{alph}(\mathrm{Tr}(g))\}$. For $A \subseteq \Sigma$ we denote the set of dependent letters by $D(A) = \{b \in \Sigma \mid (a, b) \in D \text{ for some } a \in A\}$ and set of independent letters by $I(A) = \{b \in \Sigma \mid (a, b) \in I \text{ for all } a \in A\}$. The extension to a dependence graph $g \in \mathbb{G}$ is given by $D(g) = D(\mathrm{alph}(g)), I(g) = I(\mathrm{alph}(g))$. For $g, h \in \mathbb{G}$ such that g is a prefix of h, we denote by $g^{-1}h$ the unique dependence graph such that $g(g^{-1}h) = h$.
For $A \subseteq \Sigma$ and $g \in \mathbb{G}$ we denote by $\mu_A(g)$ the maximal real prefix of g such that $\mathrm{alph}(\mu_A(g)) \subseteq I(A)$. The μ-notation is introduced because of the following property:

$$\mathrm{Re}(gh) = \mathrm{Re}(g)\mathrm{Re}(\mu_{\mathrm{alphinf}(g)}(h)) \text{ for all } g, h \in \mathbb{G}$$

3.2 Complex traces

The monoid of complex traces $\mathbb{C} = \mathbb{C}(\Sigma, D)$ is the quotient monoid of $\mathbb{G}$ by the coarsest congruence which respects real parts, [4, Thm.5.4]. It turns out that two dependence graphs $g, h \in \mathbb{G}$ yield the same complex trace if and only if $\mathrm{Re}(g) = \mathrm{Re}(h)$ and $D(\mathrm{alphinf}(g)) = D(\mathrm{alphinf}(h))$.
Thus, a complex trace x is a pair $x = (r, D(A))$ where $r \in \mathbb{R}$ and $A \subseteq \Sigma$ such that $\mathrm{alphinf}(r) \subseteq A$ and such that every $a \in A$ is connected by a path inside of the subset A to some $b \in \mathrm{alphinf}(r)$. The subset $D(A) \subseteq \Sigma$ is called the *imaginary* part of x; it is denoted by $\mathrm{Im}(x)$. Hence, we have $x = (\mathrm{Re}(x), \mathrm{Im}(x))$.

The multiplication in $\mathbb{G}$ induces the following explicit formula for complex traces.

$$(r, D(A)) \cdot (s, D(B)) = (r\mu_A(s), D(A \cup B) \cup D(\mu_A(s)^{-1}s))$$

Example 1. Let $(\Sigma, D) = a - b - c - d$. Then we have

$$a^\omega b = (a^\omega, D(a,b)) \qquad a^\omega c = (a^\omega, D(a))$$
$$a^\omega bc = (a^\omega, D(a,b,c)) \qquad a^\omega cb = (ca^\omega, D(a,b))$$
$$abcd^\omega = (abcd^\omega, \{c,d\}) \qquad (abcd^\omega)^2 = (abacbd^\omega, \{b,c,d\}$$
$$(abcd^\omega)^n = (abacbad^\omega, \Sigma) \text{ for } 3 \le n \le \omega$$

3.3 Topology

The prefix ordering on $\mathbb{G}$ is denoted by $\le$. Thus, we have $g \le h$ if and only if $g^{-1}h$ is defined. From a topological viewpoint $\mathbf{R}$ and $\mathbb{C}$ are well understood. Consider two different ultra-metrics on $\mathbf{M}$: $d_{\mathbf{R}}(s,t) = 2^{-l_{\mathbf{R}}(s,t)}$ and $d(s,t) = 2^{-l(s,t)}$. The functions $l_{\mathbf{R}}, l : \mathbf{M} \times \mathbf{M} \longrightarrow \mathbf{N} \cup \{\infty\}$ are defined as follows:

$$l_{\mathbf{R}}(s,t) = \max\{n \mid \forall p \in \mathbf{M}, |p| \le n : p \le s \Longleftrightarrow p \le t\}$$

$$l(s,t) = \max\{n \mid \forall p \in \mathbf{M}, |p| < n : D(p^{-1}s) = D(p^{-1}t)\}$$

Both metrics induce a discrete topology on $\mathbf{M}$. The completion of $(\mathbf{M}, d_{\mathbf{R}})$ yields the compact topological space $(\mathbf{R}, \hat{d}_{\mathbf{R}})$, [16]. The drawback of the metric $d_{\mathbf{R}}$ is that the multiplication of $\mathbf{M}$ is not uniformly continuous. Contrary, the multiplication of $(\mathbf{M}, d)$ is uniformly continuous. Hence it extends to the completion of $(\mathbf{M}, d)$ and this is the monoid $\mathbb{C}$ of complex traces. The topological monoid $\mathbb{C}$ is compact, where $\mathbf{M}$ is an open, dense, and discrete subspace. All results above are shown in [4]

4 Poset properties of the prefix relation

4.1 Posets

Let $(M, \le)$ be any partially ordered set. A subset $Y \subseteq M$ is called *directed* (*coherent* respectively), if for all $x, y \in Y$ there is some $z \in Y$ ($z \in M$ respectively) such that $x \le z$ and $y \le z$. A Poset M is called complete (coherently complete respectively), if every directed (coherent respectively) subset $Y \subseteq M$ has a least upper bound $\sqcup Y$. A complete Poset is called a CPO. Thinking that the empty set is directed, we see that any CPO must have a bottom element $\bot$. Every Poset $(M, \le)$ can be embedded in its ideal completion. An ideal is simply a downward closed directed subset. The set of ideals is ordered by inclusion and clearly a CPO. The mapping $x \mapsto \{p \in M \mid p \le x\}$ defines a canonical embedding of M into this CPO. A mapping between CPO's $f : M \longrightarrow M'$ is called *continuous*, if $f(\sqcup Y)$ is the least upper bound of $f(Y)$ for all directed subsets $Y \subseteq M$.

An element $x \in M$ of a CPO $(M, \leq)$ is called *compact* (*prime* respectively) if $x \leq \sqcup Y$ for any directed set $Y \subseteq M$ ($Y \subseteq M$ where $\sqcup Y$ exists respectively) implies $x \leq y$ for some $y \in Y$. A CPO is called *algebraic* or a *Scott-domain* (prime-algebraic respectively) if every element x is the least upper bound of all compact (prime respectively) elements less than or equal to x.

4.2 The Poset of real traces

Mazurkiewicz defined an infinite trace as an infinite downwards-closed directed subset of finite traces [18]. He also observed that an infinite trace has a unique (real) dependence graph representation. This can be rephrased as follows:

Proposition 1 [18]. *The set of real traces* $\mathbb{R}$ *is the ideal completion of the set* $(\mathbb{M}, \leq)$ *of finite traces ordered by the prefix relation.*

Proof. Let $g = [V, E, \lambda] \in \mathbb{G}$ be a dependence graph. Then $\lambda^{-1}(a) \subseteq V$ is well-ordered for $a \in \Sigma$. Hence each vertex $v \in V$ with $\lambda(v) = a$ has a natural representation $v = (a, i)$ where i is an ordinal. This is called the standard representation of a trace. Let $Y \subseteq \mathbb{G}$ be any coherent set of dependence graphs in standard representation such that there is a countable ordinal α with $i \leq \alpha$ for all $(a, i) \in g \in Y, a \in \Sigma$. (For $Y \subseteq \mathbb{R}$ we may take $\alpha = \omega$).
Then it is clear that the graph theoretical union

$$h = \bigcup_{g \in Y} g$$

defines a dependence graph. Clearly, $h = \sqcup Y$ and $h \in \mathbb{R}$ if $Y \subseteq \mathbb{R}$. Hence the Poset $\mathbb{R}$ is complete and since $r = \sqcup Y$ for $Y = \{p \in \mathbb{M} \mid p \leq r\}$ for all $r \in \mathbb{R}$, the set $(\mathbb{R}, \leq)$ is the ideal completion of $(\mathbb{M}, \leq)$.
□

The reflection above also shows:

Proposition 2 [13, 15]. *The set of real traces* $\mathbb{R}$ *is a coherently complete CPO. A coherent subset* Y *of dependence graphs has a least upper bound in* $\mathbb{G}$*if and only if it is countable.*

There is a convenient characterization of compact and prime dependence graphs:

Proposition 3 [13, 15]. *A dependence graph* $g = [V, E, \lambda]$ *is compact (prime respectively) if and only if* $g = \downarrow P$ *for some finite subset (one-element subset)* $P \subseteq V$.

Corollary 4. *Every dependence graph is the least upper bound of primes. A real trace is compact if and only if it is finite. A finite trace is prime if and only if it has a single maximal element (i.e., if it is a pyramid in the sense of Viennot [26]).*

Corollary 5. *The CPO* $\mathbb{R}$ *is a prime algebraic Scott domain.*

4.3 The Poset of complex traces

The monoid of complex traces $\mathbb{C}$ is neither left- nor right-cancellative. Nevertheless, it is partially ordered by the prefix relation. This follows since $(r, D(A)) \leq (s, D(B))$ implies $r \leq s$ and $D(A) \subseteq D(B)$. The converse implication does not hold: Let $(\Sigma, D) = a - b - c$, then (a^ω, Σ) is no prefix of (ca^ω, Σ), although we have $a^\omega \leq ca^\omega$.
The Poset properties of $\mathbb{C}$ have been studied by Gastin and Petit. In particular, they showed the following result for which we will give another proof here.

Proposition 6 [15]. *The Poset* $(\mathbb{C}, \leq)$ *is coherently complete.*

Proof. Let $r \in \mathbb{R}$ be a real trace, $y \in \mathbb{C}$ be a complex trace such that $r \leq z, y \leq z$ for some $z \in \mathbb{C}$. Representing z as a dependence graph, one easily sees that $\sqcup\{r, y\} \in \mathbb{C}$ exists. Now, let $Y \subseteq \mathbb{C}$ be any coherent set of complex traces. Then $\mathrm{Re}(Y)$ is coherent and $r = \sqcup \mathrm{Re}(Y) \in \mathbb{R}$ exists by Prop. 2. By the observation above, we may now replace each $y \in Y$ by $\sqcup\{r, y\}$. Hence without restriction we have $(r, D(\mathrm{alphinf}(r)) \leq y$ for all $y \in Y$. Therefore $y = (r, D(A_y))$ for all $y \in Y$ and we obtain

$$\sqcup Y = (r, D(\bigcup_{y \in Y} A_y))$$

Hence Y has a least upper bound. □

Remark 7. The proof above shows that the natural mappings $\mathbb{R} \hookrightarrow \mathbb{C}, r \mapsto (r, D(\mathrm{alphinf}(r))$ and $\mathbb{C} \longrightarrow \mathbb{R}, x \mapsto \mathrm{Re}(x)$ are continuous.

Example 2 [15]. Let $(\Sigma, D) = a - b - c$. Then $\mathbb{C}(\Sigma, D)$ is prime algebraic.

In the example above, every non-real trace is maximal, i.e., $x \in \mathbb{C} \setminus \mathbb{R}$ and $x \leq y \in \mathbb{C}$ imply $x = y$. Slightly more general, we can state:

Remark 8. If every non-real trace is maximal, then $\mathbb{C}(\Sigma, D)$ is a prime algebraic Scott-domain.

Proof. Every real trace is least upper bound of primes. Therefore it is enough to consider a trace $x \in \mathbb{C} \setminus \mathbb{R}$. We show that x is prime. Let $x \leq \sqcup Y$. Since x is maximal, we have $x = \sqcup Y$. Since $x \notin \mathbb{R}$ there is at least one $y \in Y$ such that $y \in \mathbb{C} \setminus \mathbb{R}$. Hence y is maximal. It follows $y = \sqcup Y = x$. □

Another example where Rem. 8 applies to is given by

$$(\Sigma, D) = \begin{array}{ccc} a & - & b \\ | & & | \\ d & - & c \end{array}$$

However, in general the situation is different, since (worse) we can not expect that $\mathbb{C}(\Sigma, D)$ is algebraic. Let $x \in \mathbb{C} \setminus \mathbb{R}$ be a non-real trace such that $xy \in \mathbb{R}$ for some $y \in \mathbb{C}$, then $\mathbb{C}$ is not algebraic.

Remark 9. Let $x \in \mathbb{C}$ such that $xy \in \mathbf{R}$ for some $y \in \mathbb{C}$. Then we have

$$\sqcup\{p \in \mathbb{C} \mid p \leq x, p \text{ compact }\} = \sqcup\{p \in \mathbf{M} \mid p \leq x\} = \mathrm{Re}(x)$$

Proof. Let $p \leq x, p$ compact, and $xy \in \mathbf{R}$. Then we have $p \leq xy = \sqcup\{q \in \mathbf{M} \mid q \leq xy\}$. Since p is compact, we have $p \leq q$ for some finite trace q. □

It is clear that a situation where $\mathbb{C}\setminus\mathbf{R}$ does not form a right ideal, plays a special role. The dependence alphabets allowing this situation can be characterized using the notion of *consistent pair*. A pair (A, B) such that $A \subseteq B \subseteq \Sigma$ is called consistent, if there is some$(r, D(B)) \in \mathbb{C}$ such that $\mathrm{alphinf}(r) = A$. Obviously, a pair (A, B) with $A \subseteq B$ is consistent if and only if for all $b \in B$ there is some $a \in A$ and a path $a = a_0, a_1, \ldots, a_k = b$ such that $(a_{i-1}, a_i) \in D$ and $a_i \in B$ for $1 \leq i \leq k$.

Proposition 10. *Let (Σ, D) be a dependence alphabet. Then the following assertions are equivalent.*

i) The set $\mathbb{C} \setminus \mathbf{R}$ does not form a right ideal.
ii) There are some consistent pair (A, B) and $C \subseteq I(B)$ such that $D(A) \neq D(B)$ and $D(A \cup C) = D(B \cup C)$.
iii) There is some consistent pair (A, B) such that $D(A) \neq D(B)$ and $D(A \cup I(B)) = \Sigma$.

Proof. i) $\Longrightarrow$ ii): Let $x = (r, D(B)) \in \mathbb{C}\setminus\mathbf{R}, A = \mathrm{alphinf}(r)$, and $y = (s, \mathrm{Im}(y)) \in \mathbb{C}$ such that $xy \in \mathbf{R}$. Then we have $xy = r\mu_B(s)$. Define $C = \mathrm{alphinf}(\mu_B(s))$. It follows $C \subseteq I(B)$ and $D(B \cup C) \subseteq D(A \cup C) \subseteq D(B \cup C)$.
ii) $\Longrightarrow$ iii): Trivial.
iii) $\Longrightarrow$ i): Let $x = (r, D(B)) \in \mathbb{C} \setminus \mathbf{R}$ such that $\mathrm{alphinf}(r) = A$ and $D(A \cup I(B)) = \Sigma$. Let $I(B) = \{a_1, \ldots, a_k\}$ and $s = (a_1 \cdots a_k)^\omega$. Then we have $xs = (r, D(B))(s, D(I(B))) = (rs, \Sigma) = rs \in \mathbf{R}$. Hence $\mathbb{C} \setminus \mathbf{R}$ is not a right ideal. □

It is easy to verify from the proposition above that if $\mathbb{C} \setminus \mathbf{R}$ is not a right ideal, then (Σ, D) contains the line $a - b - c - d$ as induced subgraph. In fact, the existence of such a subgraph is enough to destroy the algebraicity of $\mathbb{C}(\Sigma, D)$.

Proposition 11. *Let $a - b - c - d$ be an induced subgraph of (Σ, D). Then $\mathbb{C}(\Sigma, D)$ is not algebraic.*

Proof. Let $D(b) \setminus D(a, c, d) = \{e_1, \ldots, e_k\} = E$ and $e = e_1 \cdots e_k$. Consider $(ae)^\omega b = ((ae)^\omega, D(a, b) \cup D(E))$. This is not a real trace since $ae \in I(c)$ and $b \in D(c)$. Since $E \cup \{a\} \subseteq D(b)$, every proper prefix of $(ae)^\omega b$ is real. Therefore $(ae)^\omega b = \sqcup Y$ implies $(ae)^\omega b \in Y$. Hence we content to show that $(ae)^\omega b$ is not compact. To see this let $y = (aed)^\omega c = ((aed)^\omega, D(a, c, d) \cup D(E))$. Since $a \in I(E) \cap I(d, c)$, we have $y = \sqcup\{a^n(ed)^\omega c \mid n \geq 0\}$. On the other hand $(ae)^\omega b \cdot d^\omega c = y$ since $D(b) \subseteq E \cup D(a, c, d)$. Hence $(ae)^\omega b \leq \sqcup\{a^n(ed)^\omega c \mid n \geq 0\}$, and $(ae)^\omega b$ is not compact. □

Example 3. Let $(\Sigma, D) =$

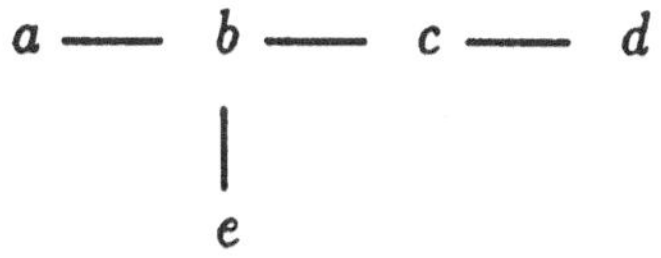

As it is shown in the proof above, the element $(ae)^\omega b$ is no least upper bound of compact elements. Note that this follows also from the fact that $(ae)^\omega b \cdot d^\omega = ((aed)^\omega, \Sigma) = (aed)^\omega \in \mathbf{R}$.

The next example is more interesting, since $a - b - c - d$ is an induced subgraph of (Σ, D) and, nevertheless, $\mathbb{C} \setminus \mathbf{R}$ is a right ideal.

Example 4. Let $(\Sigma, D) =$

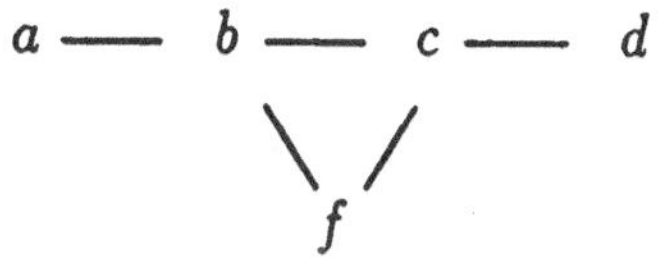

The CPO $\mathbb{C}(\Sigma, D)$ of Ex. 4 is not algebraic, since the non real trace $a^\omega b$ is no least upper bound of compact elements by the proof of Prop. 11. (Note that $D(b) \setminus D(a, c, d) = \emptyset$) The set $\mathbb{C} \setminus \mathbf{R}$ is a right ideal by Prop. 10.

Conjecture. The CPO $\mathbb{C}(\Sigma, D)$ is algebraic (i.e., a Scott domain), if and only if (Σ, D) does not contain the line $a - b - c - d$ as induced subgraph.

A verification of the conjecture above is currently under preparation in the diploma thesis [23].

5 Complex-like traces

In this section we propose a new notion of complex trace. We will give no proofs. They will appear elsewhere. Our approach is very much in the spirit of the semantics of the imaginary part of a trace, therefore we call the new objects ***complex-like traces***. The reason to introduce another notion is that the concatenation is not continuous with respect to prefix ordering. This is even true for finite traces, which has an unpleasant consequence for the semantics of a composed process. Assume p, q are processes and $p' \leq p, q' \leq q$ are prefixes. In general, $p'q'$ is no prefix of pq, hence there is not very much information of $p'q'$ about pq. The solution to this problem in the sequential case is well-known. We simply introduce a symbol for explicit termination $\surd$. Then we modify slightly the ordering $\leq$ and the rules for concatenation: $p < q$ if both p is a prefix of q and p does not stop with $\surd$, $pq = p$, if p does not stop with $\surd$, and $(p\surd)q = pq$.

A concurrent process might consist of several independent components and it makes not much sense to speak of global termination. At least this would be too restrictive and in fact we can do much better (even for the model of finite processes). The imaginary part of a complex trace is some set $D(A)$. We can think of A as the set of actions which (at any stage) will be executed in the future. Hence any $b \in D(A)$ has to be delayed and is not allowed to be executed concurrently. Now, very often the exact information about the set $D(A)$ is not available. Then we may enlarge $D(A)$. We will loose information, but we are on the safer side, since less actions can be started. Hence we content that (r, Σ) is always an approximation of $(r, D(A))$ and $(r, \emptyset)$ means explicit global termination.

Definition 12. Let (Σ, D) be a dependence alphabet. A *complex-like trace* is a pair $(r, D(A))$ where $r \in \mathbb{R}(\Sigma, D)$ is a real (finite or infinite) trace and $A \subseteq \Sigma$ is any subset such that $\text{alphinf}(r) \subseteq A$. The set of complex-like traces is denoted by $\tilde{\mathbb{C}}(\Sigma, D)$ or $\tilde{\mathbb{C}}$ if the reference to (Σ, D) is clear from the context.

The set $\tilde{\mathbb{C}}(\Sigma, D)$ is a monoid by the usual rule:

$$(r, D(A)) \cdot (s, D(B)) = (r\mu_A(s), D(A \cup B) \cup D(\mu_A(s)^{-1}s))$$

Note that every complex trace is complex-like and that $\mathbb{C}(\Sigma, D)$ is a submonoid of $\tilde{\mathbb{C}}(\Sigma, D)$. A complex-like trace $(r, D(A))$ is called *terminated*, if $D(A) = \emptyset$. Infinite traces are never terminated and finite traces can be identified with Mazurkiewicz traces, if they are terminated: $(t, \emptyset) = t \in \mathbb{M}(\Sigma, D) \subseteq \tilde{\mathbb{C}}(\Sigma, D)$.

Example 5. i) Let $D = \Sigma \times \Sigma$ be the full dependence relation, i.e., $\mathbb{M} = \Sigma^*$ and $\mathbb{R} = \mathbb{C} = \Sigma^\infty = \Sigma^* \cup \Sigma^\omega$. Then there are only three types of traces:

- finite terminated: $(t, \emptyset), t \in \Sigma^*$,
- finite non-terminated: $(t, \Sigma), t \in \Sigma^*$,
- infinite: $(t, \Sigma), t \in \Sigma^\omega$.

Write $(t, \emptyset)$ as $t\surd$ and (by abuse of language) (t, Σ) as t for $t \in \Sigma^\infty$. Then we see that our formalism is exactly the same as the formalism with explicit termination.

ii) Let (Σ, D) be a disjoint union of cliques, i.e.,

$$(\Sigma, D) = \dot{\bigcup}(\Sigma_i, \Sigma_i \times \Sigma_i), \mathbb{M} = \prod \Sigma_i^*, \text{ and } \mathbb{R} = \mathbb{C} = \prod \Sigma_i^\infty$$

Then every trace consists of independent components. Now, our formalism amounts to a formalism with an explicit termination symbol $\surd_i$ for each component.

Our goal is to define an ordering between complex-like traces such that $\tilde{\mathbb{C}}$ is a CPO where the concatenation is continuous. As already seen, we may not use the prefix ordering for this purpose. We have to define an ordering which is incomparable (but nevertheless closely related) to the usual prefix ordering.

Definition 13. Let $(r, D(A)), (s, D(B)) \in \tilde{\mathbb{C}}$ be complex like traces. We call $(r, D(A))$ an *approximation* of $(s, D(B))$, denoted by $(r, D(A)) \sqsubseteq_{\mathrm{ap}} (s, D(B))$, if $r \le s$ and $D(B) \cup D(r^{-1}s) \subseteq D(A)$.

The semantics of this ordering is clear. Assume we are interested in a finite approximation $(t, D(A))$ of a real trace $r = (r, D(\mathrm{alphinf}(r)) \in \mathbb{R} \subseteq \tilde{\mathbb{C}}$. Then t has to be a prefix of r. Every action in $t^{-1}r$ will be executed in the future. Hence, nothing in $D(t^{-1}r) \supseteq D(\mathrm{alphinf}(r))$ is allowed to start concurrently.
Any approximation $(t, D(A))$ of r should contain at least $D(t^{-1}r)$ in its imaginary part. To enlarge this part means simply some loss of information.

Example 6.

$$
\begin{gathered}
(1, \Sigma) \sqsubseteq_{\mathrm{ap}} x \text{ for all } x \in \tilde{\mathbb{C}} \\
(r, D(A)) \sqsubseteq_{\mathrm{ap}} (r, D(B)) \text{ if and only if } D(A) \supseteq D(B) \\
(r, D(A)) \sqsubseteq_{\mathrm{ap}} (s, D(A)) \text{ if and only if } r = s
\end{gathered}
$$

Remark 14. : Let $x, y \in \tilde{\mathbb{C}}$ be complex-like traces such that $x \le y$, and $x \sqsubseteq_{\mathrm{ap}} y$ i.e., x is a prefix of y and x approximates y. Then we have $x = y$.

Remark 15. In the following, we freely identify $\mathbb{M}, \mathbb{R}$, and $\mathbb{C}$ as subsets of $\tilde{\mathbb{C}}$. The set of real traces $\mathbb{R} \subseteq \tilde{\mathbb{C}}$ is the set of maximal elements. The bottom element of $\tilde{\mathbb{C}}$ is $(1, \Sigma)$.

Example 7. Let $(\Sigma, D) =$

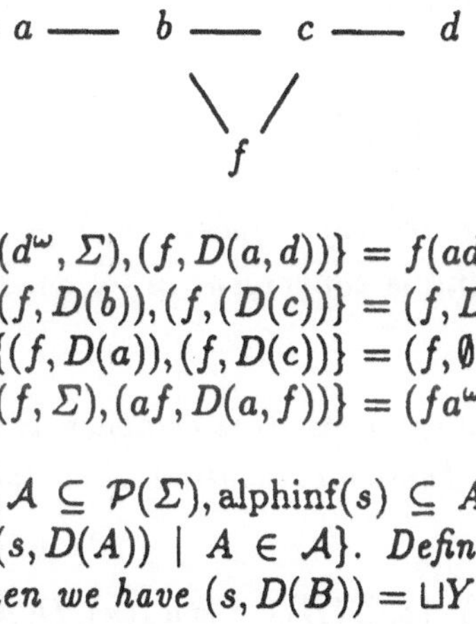

$$
\text{Then: } \begin{cases}
\sqcup\{(a^\omega, \Sigma), (d^\omega, \Sigma), (f, D(a, d))\} = f(ad)^\omega & \in \mathbb{R} \\
\sqcup\{(f, D(b)), (f, (D(c))\} = (f, D(f)) & \in \tilde{\mathbb{C}} \setminus \mathbb{C} \\
\sqcup\{(f, D(a)), (f, D(c))\} = (f, \emptyset) & \in \mathbb{M} \\
\sqcup\{(a^\omega, D(b, c)), (f, \Sigma), (af, D(a, f))\} = (fa^\omega, D(b)) & \in \mathbb{C} \setminus \mathbb{R}
\end{cases}
$$

Lemma 16. *Let* $s \in \mathbb{R}, \mathcal{A} \subseteq \mathcal{P}(\Sigma), \mathrm{alphinf}(s) \subseteq A$ *for all* $A \in \mathcal{A}$, *and* Y *be the coherent set* $Y = \{(s, D(A)) \mid A \in \mathcal{A}\}$. *Define* $B = \{b \in \Sigma \mid D(b) \subseteq D(A)$ *for all* $A \in \mathcal{A}\}$. *Then we have* $(s, D(B)) = \sqcup Y$.

Theorem 17. *The partial order* $(\tilde{\mathbb{C}}, \sqsubseteq_{\mathrm{ap}})$ *is coherently complete, in particular* $(\tilde{\mathbb{C}}, \sqsubseteq_{\mathrm{ap}})$ *is a CPO.*

Theorem 18. *: The CPO* $(\tilde{\mathbb{C}}, \sqsubseteq_{\mathrm{ap}})$ *is a Scott-domain, i.e., every element is least upper bound of compact elements. An element* $(p, D(A))$ *is compact if and only if* $p \in \mathbb{M}$ *is a finite trace.*

Corollary 19. *Let $\tilde{\mathbb{M}} \subseteq \tilde{\mathbb{C}}$ be the subset of finite complex-like traces. Then $\tilde{\mathbb{C}}$ is the ideal completion of $\tilde{\mathbb{M}}$ with respect to the ordering $\sqsubseteq_{ap}$. The set $\mathbb{M} \subseteq \tilde{\mathbb{M}}$ of Mazurkiewicz traces is the set of complex-like traces which are both compact and maximal.*

Remark 20. It is possible to characterize those $(1, D(A)), A \subseteq \Sigma$ which are prime. (Note that $(1, \Sigma)$ is prime). Then one can deduce that every $(1, D(A))$ is least upper bound of some primes and finally that $\tilde{\mathbb{C}}(\Sigma, D)$ is prime algebraic.

Since $\tilde{\mathbb{C}}(\Sigma, D)$ is a CPO we may speak of continuous mappings.
The main theorem on $\tilde{\mathbb{C}}$ is that the concatenation is continuous. This follows e.g. from an extension of the metric $d = 2^{-l}$ from Section 3.3 to a metric on $\tilde{\mathbb{C}}$ and the relation of this metric to the partial order $\sqsubseteq_{ap}$.
In particular, if two processes $x, y \in \tilde{\mathbb{C}}$ are approximated by some (finite) complex-like traces $x' = (p, D(A)) \sqsubseteq_{ap} x, y' = (q, D(B)) \sqsubseteq_{ap} y$, then we have $x'y' \sqsubseteq_{ap} xy$ and, if $D(A) \neq \Sigma$, we can expect that the approximation $x'y'$ contains more information about xy than x'. In any case, we have $\mathrm{Re}(x') \leq \mathrm{Re}(x'y') \leq \mathrm{Re}(xy)$, whether or not x' has terminated.
To be more formal, let $p \in \mathbb{M}$ be finite and $x \in \tilde{\mathbb{C}}$ be complex-like, $x = (r, D(A))$. Then we define $D(p^{-1}x)$ if and only if $p \leq r$:

$$D(p^{-1}x) = D(\mathrm{alph}(p^{-1}r) \cup A)$$

With a suitable convention about partially defined values, a metric $d = 2^{-l}$ for $\tilde{\mathbb{C}}$ can be defined by (practically) the same formula as for $\mathbb{C}$ above :

$$l(x, y) = \sup\{n \mid \forall p \in \mathbb{M}, |p| < n : D(p^{-1}x) = D(p^{-1}y)\}$$

Thus, we may view the CPO $\tilde{\mathbb{C}}$ also as a metric space $(\tilde{\mathbb{C}}, d) = (\tilde{\mathbb{C}}, 2^{-l})$. The relation to the partial order $\sqsubseteq_{ap}$ is clear from the following observation.

Proposition 21. *Every increasing sequence $(y_n)_{n \in \mathbb{N}}, (y_i \sqsubseteq_{ap} y_j$ for all $i \leq j)$ is a Cauchy-sequence in $\tilde{\mathbb{C}}$ and is convergent in $(\tilde{\mathbb{C}}, d)$. Hence $\lim_{n \to \infty} y_n = \bigsqcup\{y_n \mid n \in \mathbb{N}\}$*

Theorem 22. *The metric space $(\tilde{\mathbb{C}}, d)$ is compact, it contains the set of complex traces as a closed subset. The concatenation of complex-like traces by the usual rule*

$$\tilde{\mathbb{C}} \times \tilde{\mathbb{C}} \longrightarrow \tilde{\mathbb{C}}, (r, D(A)) \cdot (s, D(B)) = (r\mu_A(s), D(A \cup B) \cup D(\mu_A(s)^{-1}s))$$

is uniformly continuous. The concatenation is monotone. Hence, it is continuous in the CPO $(\tilde{\mathbb{C}}, \sqsubseteq_{ap})$.

Acknowledgement. I would like to thank Paul Gastin for various helpful comments and Heike Photien for her great help in producing the LaTeXversion of my handwritten manuscript.

References

1. I.J. Aalbersberg and G. Rozenberg. Theory of traces. *Theoretical Computer Science*, 60:1–82, 1988.
2. P. Bonizzoni, G. Mauri, and G. Pighizzini. About infinite traces. In V. Diekert, editor, *Proceedings of the ASMICS workshop Free Partially Commutative Monoids, Kochel am See, Oktober 1989*, Report TUM-I9002, Technical University of Munich, pages 1–10, 1990.
3. V. Diekert. *Combinatorics on Traces*. Number 454 in Lecture Notes in Computer Science. Springer, Berlin-Heidelberg-New York, 1990.
4. V. Diekert. On the concatenation of infinite traces. In Choffrut C. et al., editors, *Proceedings of the 8th Annual Symposium on Theoretical Aspects of Computer Science (STACS'91), Hamburg 1991*, number 480 in Lecture Notes in Computer Science, pages 105–117, Berlin-Heidelberg-New York, 1991. Springer. To appear 1993 in *Theoret. Comput. Sci.*
5. V. Diekert, P. Gastin, and A. Petit. Recognizable complex trace languages. In A. Tarlecki, editor, *Proceedings of the 16th Symposium on Mathematical Foundations of Computer Science (MFCS'91), Kazimierz Dolny (Poland) 1991*, number 520 in Lecture Notes in Computer Science, pages 131–140, Berlin-Heidelberg-New York, 1991. Springer. Full version: Rapport de Recherche 640 (1991), Université de Paris Sud.
6. V. Diekert and A. Muscholl. Deterministic asynchronous automata for infinite traces. In P. Enjalbert, A. Finkel, and K. W. Wagner, editors, *Proceedings of the 10th Annual Symposium on Theoretical Aspects of Computer Science (STACS'93), Würzburg 1993*, number 665 in Lecture Notes in Computer Science, pages 617–628, Berlin-Heidelberg-New York, 1993. Springer.
7. V. Diekert and G. Rozenberg, editors. *Trace Book (preliminary title)*. 1993. To appear.
8. V. Diekert, editor, editor. *Free Partially Commutative Monoids. Proceedings of a workshop of the ESPRIT Basic Research Action No 3166: Algebraic and Syntactic Methods in Computer Science (ASMICS), Kochel am See, Bavaria, FRG (1989)*, number TUM-I9002. Technical University Munich, 1990.
9. Werner Ebinger and Anca Muscholl. On logical definability of ω-trace languages. In *Proceedings of the 20th International Colloquium on Automata Languages and Programming (ICALP'93), Lund (Sweden) 1993*, Lecture Notes in Computer Science, Berlin-Heidelberg-New York, 1993. Springer. To appear.
10. M.P. Flé and G. Roucairol. Maximal serializability of iterated transactions. *Theoretical Computer Science*, 38:1–16, 1985.
11. P. Gastin. Infinite traces. In I. Guessarian, editor, *Proceedings of the Spring School of Theoretical Computer Science on Semantics of Systems of Concurrent Processes*, number 469 in Lecture Notes in Computer Science, pages 277–308, Berlin-Heidelberg-New York, 1990. Springer.
12. P. Gastin, A. Petit, and W. Zielonka. A Kleene theorem for infinite trace languages. In J. Leach Albert et al., editors, *Proceedings of the 18th International Colloquium on Automata Languages and Programming (ICALP'91), Madrid (Spain) 1991*, number 510 in Lecture Notes in Computer Science, pages 254–266, Berlin-Heidelberg-New York, 1991. Springer.
13. P. Gastin and B. Rozoy. The poset of infinitary traces. Tech. Rep. LITP 91.07, Université Paris 6 (France), 1991. To appear in *Theoret. Comp. Sci.*

14. Paul Gastin and Antoine Petit. Asynchronous automata for infinite traces. In W. Kuich, editor, *Proceedings of the 19th International Colloquium on Automata Languages and Programming (ICALP'92), Vienna (Austria) 1992*, number 623 in Lecture Notes in Computer Science, pages 583–594, Berlin-Heidelberg-New York, 1992. Springer.
15. Paul Gastin and Antoine Petit. Poset properties of complex traces. In I. M. Havel and V. Koubek, editors, *Proceedings of the 17th Symposium on Mathematical Foundations of Computer Science (MFCS'92), Prague, (Czechoslovakia), 1992*, number 629 in Lecture Notes in Computer Science, pages 255–263, Berlin-Heidelberg-New York, 1992. Springer.
16. M. Kwiatkowska. A metric for traces. *Information Processing Letters*, 35:129–135, 1990.
17. A. Mazurkiewicz. Concurrent program schemes and their interpretations. DAIMI Rep. PB 78, Aarhus University, Aarhus, 1977.
18. A. Mazurkiewicz. Trace theory. In W. Brauer et al., editors, *Petri Nets, Applications and Relationship to other Models of Concurrency*, number 255 in Lecture Notes in Computer Science, pages 279–324, Berlin-Heidelberg-New York, 1987. Springer.
19. Y. Métivier. On recognizable subsets of free partially commutative monoids. In L. Kott, editor, *Proceedings of the 13th International Colloquium on Automata Languages and Programming (ICALP'86), Rennes (France) 1986*, number 226 in Lecture Notes in Computer Science, pages 254–264, Berlin-Heidelberg-New York, 1986. Springer.
20. E. Ochmanski. Regular behaviour of concurrent systems. *Bulletin of the European Association for Theoretical Computer Science (EATCS)*, 27:56–67, Oct 1985.
21. D. Perrin. Partial commutations. In *Proceedings of the 16th International Colloquium on Automata, Languages and Programming (ICALP'89), Stresa (Italy) 1989*, number 372 in Lecture Notes in Computer Science, pages 637–651, Berlin-Heidelberg-New York, 1989. Springer.
22. D. Perrin and J.E. Pin. Mots Infinis. Tech. Rep. LITP 91.06, Université Paris 6 (France), 1991. Book to appear.
23. Dan Teodosiu. *Bereichseigenschaften komplexer Spuren*. Diplomarbeit, Universität Stuttgart, 1993.
24. Wolfgang Thomas. Automata on infinite objects. In Jan van Leeuwen, editor, *Handbook of Theoretical Computer Science*, chapter 4, pages 133–191. Elsevier Science Publishers B. V., 1990.
25. Wolfgang Thomas. On logical definability of trace languages. In V. Diekert, editor, *Proceedings of a workshop of the ESPRIT Basic Research Action No 3166: Algebraic and Syntactic Methods in Computer Science (ASMICS), Kochel am See, Bavaria, FRG (1989)*, Report TUM-I9002, Technical University of Munich, pages 172–182, 1990.
26. X.G. Viennot. Heaps of pieces I: Basic definitions and combinatorial lemmas. In G. Labelle et al., editors, *Proceedings Combinatoire énumerative, Montreal, Quebec (Canada) 1985*, number 1234 in Lecture Notes in Mathematics, pages 321–350, Berlin-Heidelberg-New York, 1986. Springer.
27. W. Zielonka. Notes on finite asynchronous automata. *R.A.I.R.O. — Informatique Théorique et Applications*, 21:99–135, 1987.

Symbolic bisimulations
(Abstract)

Matthew Hennessy

Sussex University, School of Math. & Physical Sciences
Falmer, Brighton BN1 9RH, UK
(matthewh@cogs.sussex.ac.uk)

Most of the traditional verification methods for process description languages rely on the fact that the underlying transition systems are finite branching. If processes are allowed to send and receive messages or values from an infinite data set then this is rarely the case. However if we work with "symbolic transition systems" as opposed to the usual concrete ones then many of the standard methods can still be applied. This talk will survey recent joint work with H. Lin and X. Liu on this symbolic approach to message-passing processes. This will include a theory of symbolic bisimulations, sound and complete proof systems for mesage-passing process calculi and a logical characterisation of bisimulation equivalence in terms of a first-order modal logic.

Some Results on the Full Abstraction Problem for Restricted Lambda Calculi

Furio Honsell and Marina Lenisa

Dipartimento di Matematica e Informatica,
Università di Udine, via Zanon,6 - Italy
E-mail:{honsell,lenisa}@udmi5400.cineca.it

dedicated to Corrado Böhm
on the occasion of his 70th *birthday*

Abstract. Issues in the mathematical semantics of two restrictions of the λ-calculus, i.e. λI-calculus and λ_V-calculus, are discussed. A *fully abstract* model for the natural evaluation of the former is defined using complete partial orders and strict Scott-continuous functions. A correct, albeit non-*fully abstract*, model for the SECD evaluation of the latter is defined using Girard's coherence spaces and stable functions. These results are used to illustrate the interest of the analysis of the *fine structure* of mathematical models of programming languages.

1 Introduction

D. Scott, in the late sixties, discovered a truly mathematical semantics for λ-calculus using continuous lattices. His D_∞ model was the first example of a mathematical structure, completely autonomous from syntax, which provided independent insight into the combinatorics of the evaluation of functional expressions. Since then, much energy has been devoted to develop and investigate mathematical structures which model denotations of programs. These are usually quite elaborate. It is therefore questionable if the denotational account of the semantics of programming languages is to be preferred to other forms of formal description of the semantics, e.g. operational semantics or rewrite semantics. We think that mathematical models, albeit involved, are useful to define and interesting to investigate. In the first place there is a psychological motivation. Mathematical structures have a far richer structure, e.g. topological or order theoretic, than merely that necessary to interpret the linguistic phrases. They provide independent and abstract insight in the structure of meanings, and this, being often geometric, can stimulate mental imagery, far better than pure symbol pushing. Secondly, mathematical models yield non trivial theories of *program equivalence*. This is particularly fruitful since modern methodologies of program development are often based on *program transformation*. This technique relies on powerful theories of *observational equivalence* of modules of programs, whereby one can replace a module in a program by an observationally

equivalent one, preserving the overall meaning of the main program. In practice the original modules are logically graspable but inefficient while the final ones are the opposite, being often obtained by iterated transformations of submodules. Formally, in the case of the λ-calculus, given a machine I for evaluating programs, i.e. closed λ-terms, M and N are said to be ***observationally equivalent*** if and only if:
$\forall C[\,].(\forall C[M], C[N] \in \Lambda^{\circ}.\ (I \text{ stops on input } C[M] \iff I \text{ stops on input } C[N]))$.
Of course, given an evaluation machine, the most useful models are those which provide a *fully abstract* semantics with respect to it, i.e. those which induce an equivalence on programs which coincides with the observational equivalence of the machine. This property however, is not necessary for the model to be worthwhile investigating. In [6] a denotational model for the language ISWIM of P. Landin is provided, which even if it induces a finer equivalence on terms, than the observational equivalence induced by the SECD machine, it allows nonetheless to show easily, say, that all fixed point operators are observationally equal. Of course one could prove the validity of each such equation using some involved *computation induction*, at least in principle. But, syntactic proofs are more *error-prone*. And, as it occurs in this case, we can prove easily the equations in question just by analyzing the ***fine structure*** of the model using mathematical tools, which work only because of the extra structure of the model. We can show in fact, an *ApproximationTheorem*, which allows to express the meaning of non-terminating processes in terms of limits of simpler ones. Topological and order theoretic techniques are at the core of the many results which characterize useful classes of program equivalences in denotational semantics. This motivation for building and investigating mathematical models is ultimately an instance of the general use of mathematical structures as a source of inspiration for developing interesting logical principles for reasoning about programs.

In this paper we present two constructions which exemplify the advantages that can be gained from infinite mathematical models in the study of the semantics of programming languages, of the kind discussed above. Both constructions deal with the semantics of variants of the classical λ-calculus (the λK-calculus), which axiomatize functions which *strictly* depend on all their arguments: the λI-calculus and the λ_V-calculus. The first construction has a foundational character, while the second is more in the way of applications.

The former calculus is the system originally introduced by Church in [3] in order to characterize and study the class of intuitively computable functions. The set of λI-terms is the subset of the λK-terms, consisting of those terms in which abstracted variables always occur in the body of the abstraction. This calculus has peculiar computational properties. For instance weakly normalizing terms are strongly normalizing, hence normal forms can be safely taken as values; but then, because of the lack of erasing combinators, all encodings are more cumbersome than in the λK-calculus, and one is forced to model recursive functions without fixed-point combinators. D. Scott, in [2], contrasted briefly the viewpoint of Church, with that of Curry who favored the λK-calculus. In particular he remarked: "In the λI-calculus a function is meant to strictly depend on all its arguments..." but "... what is a strict function? is it one which is undefined for undefined arguments? that seems a necessary condition, but is this all that we should say?" A complete satisfactory answer should illuminate on the general notion of *strictness* in relations. Very little has been done, up to now, even in the direction of investigating Scott's suggestion to model the λI-functions using strict functions in the context of domain semantics. In particular in [2] (p.368), an open problem concerning the mathematical semantics of the λI-calculus is raised: *Does there exist a mathematical model for the extensional theory where all terms without normal form are equated, which is known to be consistent by purely syntactic means?* The importance of this theory lies in the fact that it is the *natural* maximal theory. A positive answer to this question was announced in [5]. In this paper we will define such a mathematical model, thus providing partial evidence that Scott's original suggestion is indeed viable. To this end we shall investigate briefly the semantics of the λI-calculus.

The λ_V-calculus was introduced by Plotkin [12] in order to reason about equivalence and termination of ISWIM programs interpreted by Landin's SECD machine [10]. It is obtained from the λK-calculus by restricting the β-rule to redexes whose operand is a value (i.e. a constant, a variable or a function). The λ_V-calculus can be viewed as the abstract paradigm of functional programming languages with a lazy call-by-value evaluation mechanism. This calculus was extensively studied in [6]. It was shown there that, under very reasonable assumptions no fully abstract model exists in the category of C.P.O.'s and strict continuous function. This is due to the fact that there are too many continuous functions, some of which have a *parallel flavor* which allows to tell apart λ-terms which are observationally equal when evaluated sequentially. In this paper we investigate models for the λ_V-calculus in the category of coherence spaces [7, 8] and stable functions. Models built out of stable functions should not present the undesired phenomena of continuous functions, since parallel functions are not stable. The outcome however seems rather unsatisfactory at first. Not only it turns out that just modeling correctly the λ_V-calculus in coherence spaces is not immediate, for there is no standard way of modeling call-by-value evaluation and laziness. But moreover, even using this semantics we succeed only in defining a model which is correct but non-fully abstract with respect to the SECD machine. The model we analyze in detail turns out to be nonetheless very useful if taken together with the continuous one defined in [6]. The β_V-theories induced by the two models are in fact incomparable. Hence more sound equalities between λ-terms can be shown using the two semantics together. It comes as a pleasant and appropriate case at point, that the proof of the observational equivalence of the two terms which are used to show that the model of [6] is not fully abstract, can be readily obtained using the coherence semantics. This equivalence is proved in [6] using a lengthy and difficult syntactic computation induction, but it is immediate to check that the two terms have equal "coherent" denotations, and since the coherence model is correct they must also be observationally equal.

In conclusion, in this paper, we study the full abstraction problem for two models of different λ-calculi with respect to the observational equivalence induced by their natural evaluation machines. For simplicity we consider only calculi without constants. In particular for the λI-calculus we consider a machine that evaluates to β-normal form by reducing at each stage the leftmost redex and define a fully abstract model with respect to it. The correctness of the model will be shown using a computability argument inspired by that of [4]. To this end we define a finitary logical description of the model, i.e. an *intersection* type assignment system in which types can be seen as names for the compact elements of the semantic domain. As an interesting by-product of the correctness proof we obtain a new proof of the fact that in the λI-calculus the notion of strong normalization coincides with that of weak normalization. As far as the λ_V-calculus we follow quite closely [6] and prove an Approximation Theorem using indexed reductions. The novelty lies essentially in the construction of a correct coherence model using a particular form of stable functions as morphisms.

For lack of space we are forced to proceed directly in a terse definition/theorem style reducing proofs and comments to a minimum. We will use notation and conventions of [1] without explicit mention. In particular we will denote by $M[N/x]$ the capture avoiding substitution of all occurrences of the free variable x in M with N. We will denote with $FV(M)$ the set of free variables of the term M. Finally if $(D, \sqsubseteq)$ is a C.P.O. and $a, b \in D$, we denote by $f_{a,b}$ the step function $\lambda d \in D. if\ a \sqsubseteq d\ then\ b\ else\ \bot$.

The authors would like to thank Mariangiola Dezani Ciancaglini and Simona Ronchi Della Rocca for many helpful discussions.

2 An Adequate Semantics for the λI-calculus

In this section we define an extensional model for the λI-calculus which models the $\lambda\eta$-theory which equates all λI-terms not having a normal form. Thus we solve positively the problem raised in [2]. What we achieve can be also viewed as the construction of a fully abstract model with respect to the observational equivalence induced by a natural reduction machine.

First we define the notion of model for the λI-calculus. Then we state the fundamental I-Separability Theorem, due to Barendregt [1]. Next we define the operational equivalence induced by the reduction machine which evaluates λI-terms to β-normal form by rewriting at each stage the leftmost redex. Then we define a model for the $\lambda\eta I$-calculus, D^I, as a particular solution, in the category of Complete Partial Orders (C.P.O.'s), of the domain equation $D \simeq [D \rightarrow_\bot D]$, where $[D \rightarrow_\bot D]$ is the space of strict continuous functions from D to D. Finally we show that the model D^I is fully abstract with respect to the above operational equivalence. For this purpose we introduce a finitary logical description of the model in the style of [4] and use an argument which is based on a generalization of Krivine's presentation of Tait's computability technique for showing strong normalization of the simply typed λ-calculus.

2.1 Preliminary Definitions and Results

Definition 1. i) The set Λ_I of terms of the λI-calculus is defined inductively as follows
- $x \in \Lambda_I$
- if $M \in \Lambda_I$ and $x \in FV(M)$ then $\lambda x.M \in \Lambda_I$
- if $M, N \in \Lambda_I$ then $(MN) \in \Lambda_I$.

ii) The theory λ_I of the λI-calculus is the equational theory defined as the congruence relation induced by the following reduction schemata:
(α) $\lambda x.M \rightarrow_I \lambda y.M[y/x]$, if $y \notin FV(M)$;
(β_I) $(\lambda x.M)N \rightarrow_I M[N/x]$.
A λ_I-*theory* is a binary congruence relation which extends λ_I. A λ_I-theory is said to be *extensional* if it is also closed under the following reduction schema
(η) $\lambda x.Mx \rightarrow_I M$, if $x \notin FV(M)$.

We introduce now the definition of λI-model in terms of the notion of model for I-Combinatory Logic, cl_I (for the relations between λI-calculus and cl_I see [1]).

Definition 2. i) A model for cl_I is a structure $< D, \bullet >$, where $\bullet$ is a binary operation on D such that there are four elements $i, b, c, s \in D$ satisfying the following conditions:
- $i \bullet d = d$
- $((b \bullet d_0) \bullet d_1) \bullet d_2 = d_0 \bullet (d_1 \bullet d_2)$
- $((c \bullet d_0) \bullet d_1) \bullet d_2 = (d_0 \bullet d_2) \bullet d_1$
- $((s \bullet d_0) \bullet d_1) \bullet d_2 = (d_0 \bullet d_2) \bullet (d_1 \bullet d_2)$.

ii) A *combinatory model* for λI is a cl_I-model $< D, \bullet >$ such that there exists an element $\epsilon \in D$ with the following properties:
- $((\epsilon \bullet d_0) \bullet d_1) = d_0 \bullet d_1$
- $(\forall d \in D.(d_0 \bullet d = d_1 \bullet d)) \Rightarrow \epsilon \bullet d_0 = \epsilon \bullet d_1$
- $\epsilon \bullet \epsilon = \epsilon$.

iii) A model is said to be *trivial* if all closed terms have equal interpretation.
iv) A combinatory model for λI is *extensional* if $\epsilon = i$.

Contrary to what happens in the case of the standard λ-calculus, for a λI-model to be non trivial it is not sufficient that the model have more than one point. Moreover if any two elements among i, b, c, s are equal then i, b, c, s are all equal and the model is trivial. This can be seen also as an instance of the very important Theorem 1 below, due to H. Barendregt [1]. We need first a definition:

Definition 3. Let $\mathcal{F} = \{M_1, \ldots, M_n\} \subseteq \Lambda I$. The set $\mathcal{F}$ is said to be *I-separable* if and only if

$$\forall L_1, \ldots, L_n \in \Lambda I^\circ.(\exists F \in \Lambda I.FM_1 = L_1, \ldots, FM_n = L_n) \ .$$

Theorem 1 (I-Separability). *$\mathcal{F} \subseteq \Lambda I^\circ$ is I-separable if and only if $\mathcal{F}$ contains terms having distinct $\beta\eta$-normal forms.*

As a consequence of Theorem 1 we have also that a non trivial λI-model must be infinite. In fact, if a λI-model is finite, then there exist two distinct closed $\beta\eta$-normal forms which have the same interpretation; using the I-separability we can thus equate $\lambda x.x$ to $\lambda xyz.xz(yz)$ and get $i = s$. Another immediate consequence of Theorem 1 is the maximality of the theory that we model in this section.

We present now the operational semantics, in *S.O.S. style*, of a call-by-name sequential machine which reduces closed λI-terms to β-normal forms, by rewriting at each stage the leftmost redex.

Definition 4 (Operational Semantics). The set of programs is the set of closed λI-terms, Λ_I°. The set of values is the set of λI-terms in β-normal forms, NF_I. We define the leftmost reduction $\Downarrow_{LI}$ as the least binary relation over $\Lambda I \times NF_I$ satisfying the following rules:

$$(var)\qquad \frac{}{x \Downarrow_{LI} x}$$

$$(appl)\qquad \frac{M \Downarrow_{LI} \lambda x.P \quad P[N/x] \Downarrow_{LI} Q}{MN \Downarrow_{LI} Q}$$

$$(app2)\qquad \frac{M \Downarrow_{LI} P \quad P \text{ is not an abstraction} \quad N \Downarrow_{LI} Q}{MN \Downarrow_{LI} PQ}$$

$$(abstr)\qquad \frac{M \Downarrow_{LI} P}{\lambda x.M \Downarrow_{LI} \lambda x.P}\ .$$

We write simply $M \Downarrow_{LI}$ whenever M converges to a value, i.e. whenever the judgement $M \Downarrow_{LI} N$ can be established for some term N.

Once we have an operational semantics there is a natural operational equivalence to define on programs: the *observational equivalence*. This arises if we consider programs as *black boxes*. That is we take them to be equal if we cannot tell them apart by *observing* that for a given program context the machine halts when one is used as a subprogram but does not halt when the other is used as a subprogram. Thus we give:

Definition 5 (Observational Equivalence). i) Let $M, N \in \Lambda_I^\circ$. Then $M \equiv_{LI} N$ iff $\forall C[\,].(C[M], C[N] \in \Lambda_I^\circ \Rightarrow (C[M] \Downarrow_{LI} \Leftrightarrow C[N] \Downarrow_{LI}))$.
ii) Let T_{LI} be the λI-theory $\{(M,N) \mid M, N \in \Lambda_I^\circ \wedge M \equiv_{LI} N\}$.

2.2 λI-calculus and λ_V-calculus Models Based on Strict Continuous Functions

In this subsection we study the problem of defining models for the λI-calculus and for the λ_V-calculus in the category of C.P.O.'s, by interpreting λ-abstractions as strict continuous functions. A continuous function is said to be *strict* if it maps $\bot$ to $\bot$. In particular we consider domains D such that the space $[D \rightarrow_\bot D]$ of strict continuous functions from D to D is a retract of D. We recall that a domain D' is a retract of a domain D ($D' \rhd D$) if there exist two continuous functions $\Phi : D' \to D$ and $\Psi : D \to D'$ such that $\Psi \circ \Phi = id_{D'}$. See [6] for details concerning the λ_V-calculus.

Definition 6. Let D be such that $[D \to_\perp D] \rhd D$. We denote with $\Phi : [D \to_\perp D] \to D$ and $\Psi : D \to [D \to_\perp D]$ the continuous functions such that $\Psi \circ \Phi = id_{[D \to_\perp D]}$. Let $\bullet : D \times D \to D$ be the function defined as $e \bullet d = Ap(\Psi(e), d)$. The interpretation function $[\![\ ]\!] : \Lambda I \times (Var \to D) \to D$ is defined as follows:
- $[\![x]\!]_\rho = \rho(x)$
- $[\![MN]\!]_\rho = [\![M]\!]_\rho \bullet [\![N]\!]_\rho$
- $[\![\lambda x.M]\!]_\rho = \Phi(strict(\boldsymbol{\lambda} d \in D.[\![M]\!]_{\rho[d/x]}))$,

where $\rho : Var \to D$ and $strict : [D \to D] \to [D \to_\perp D]$ is the function defined by: $strict(f) = \boldsymbol{\lambda} d.if\ d = \perp\ then\ \perp\ else\ f(d)$.

A weak version of Proposition 2 appears in [5]. We do not have enough space to introduce properly the λN°-calculus mentioned below. We limit ourselves to remark just that it is the calculus obtained from the λK-calculus by restricting the β-rule to redexes whose operand is a closed λK-normal form. It is a calculus in the spirit of the λ_V-calculus and constitutes the natural equational theory for reasoning about pure ISWIM programs, when these are evaluated by-value eagerly (i.e. reducing bodies of abstractions before application).

Proposition 2. *Let D be a domain such that $[D \to_\perp D] \rhd D$. Then*
i) D is a model for cl_I
ii) D is a model for the λN°-calculus
iii) If $\Phi \circ \Psi \in [D \to_\perp D]$, then D is a model for the λI-calculus; otherwise, if $\Phi \circ \Psi \notin [D \to_\perp D]$, then D is a λ_V-model.
iv) If D has at least three elements then it is a non trivial λI-model.

Proof. i) Take, for any ρ, $i = [\![\lambda x.x]\!]_\rho$, $b = [\![\lambda xyz.x(yz)]\!]_\rho$, $c = [\![\lambda xyz.xzy]\!]_\rho$, $s = [\![\lambda xyz.xz(yz)]\!]_\rho$.
ii) By induction on the structure of a λK-normal form N one can show that, if the model is not trivial, then $\exists \rho.[\![N]\!]_\rho \neq \perp$. Hence closed normal forms are erasable.
iii) If $\Phi \circ \Psi \in [D \to_\perp D]$ then $\Phi(\Phi \circ \Psi)$ satisfies the requirements for being the constant ϵ of Definition 2. If $\Phi \circ \Psi \notin [D \to_\perp D]$ then the interpretation of λ-abstractions is different from $\perp$ and hence we can model the λ_V-calculus by considering $V = D \setminus \{\perp\}$ as the set of semantic values and interpreting variables in V, see [6] for more details.
iv) One can easily see that say, $i \neq c$ since, taking a_1 to be the image of the identity function and a_2 to be image of $strict(\boldsymbol{\lambda} d.a_1)$, we have $a_2 a_1 \neq a_1 a_2$. □

2.3 A Fully Abstract Model for the λI-calculus

We start by defining an extensional model for the λI-calculus as a particular solution of the domain equation $D \simeq [D \to_\perp D]$, in the category of C.P.O.'s and projection pairs.

Definition 7. The λI-model D^I is the structure $< D^I, \bullet^I >$, where:
- (D^I, Ψ, Φ) is the *inverse limit*, $lim_{\leftarrow}(D_n, i_n)$, where $D_0 = \{\bot, \phi, \top\}$ is a three point lattice with $\bot$ as the bottom element and $\top$ as the top element, and $i_0 : D_0 \rightarrow D_1$ is defined as
$i_0(\bot) = \bot$
$i_0(\phi) = f_{\top,\phi}$
$i_0(\top) = f_{\phi,\top}$;
- $\bullet^I : D^I \times D^I \rightarrow D^I$ is defined as $e \bullet^I d = (\Psi(e))(d)$, where $\Psi : D^I \rightarrow [D^I \rightarrow_\bot D^I]$;

The interpretation function in D^I will be denoted by $[\![\ ]\!]^I$.

Proposition 3. *D^I is an extensional combinatory model for the λI-calculus.*

Proof. Immediate using Proposition 2. □

Definition 8. Let T_{D^I} be the theory induced by the λI-model D^I, i.e.:

$$T_{D^I} = \{(M, N) \mid M, N \in \Lambda_I^\circ \wedge [\![M]\!]^I = [\![N]\!]^I\} .$$

We are now ready to state the *main theorem*

Theorem 4 (Full Abstraction). $T_{D^I} = T_{LI}$.

The rest of this section is devoted to the proof of the above theorem. We first prove that the model is correct, i.e. $T_{D^I} \subseteq T_{LI}$ by showing that if $M \in \Lambda I$, then: $(\exists \rho . [\![M]\!]^I_\rho \neq \bot)$ if and only if M is strongly normalizing. Continuity of contexts and the fact that M is strongly normalizing if and only if $M \Downarrow_{LI}$ then yield the result. We then show that the model is complete, i.e. $T_{D^I} \supseteq T_{LI}$, by showing that if M and N have different $\beta\eta$-normal forms then $\exists \rho . [\![M]\!]^I_\rho \neq [\![N]\!]^I_\rho$. The consistency of the model and Theorem 1 then yield the result. The proof of correctness is rather difficult. We use a generalization of the computability technique à la Krivine. To this end we need to define first a finitary logical presentation of the model D^I as intersection type assignment system, see [4] for more details.

The D^I Intersection Type Assignment System. Since the model D^I is an algebraic complete lattice we can define an *intersection type* language, T_I, out of the names of the compact elements of the model different from $\bot$. We can then introduce a *type assignment system* S_I for assigning types of this language to λI-terms in such a way that a term M has type σ if and only if the compact element corresponding to σ is in the relation $\sqsubseteq$ with $[\![M]\!]^I_\rho$, for some ρ.

Definition 9. i) The set T_I of types is the smallest set containing: the constants f and t and for all $\sigma, \tau \in T_I$ the arrow type $\sigma \rightarrow \tau$ and the intersection type $\sigma \cap \tau$.
ii) The relation $\leq$ is the smallest reflexive and transitive relation on T_I satisfying

the following conditions for all $\tau, \sigma, \tau', \sigma', \rho \in T_I$:

1. $\tau \leq f$
2. $\tau \geq t$
3. $t \rightarrow f \geq f$
4. $f \rightarrow t \leq t$
5. $(\sigma \leq \tau \wedge \sigma \leq \rho) \Rightarrow \sigma \leq (\rho \cap \tau)$
6. $\sigma \cap \tau \leq \sigma$
7. $(\sigma \rightarrow \rho) \cap (\sigma \rightarrow \tau) \leq \sigma \rightarrow (\rho \cap \tau)$
8. $\sigma' \leq \sigma, \tau \leq \tau' \Rightarrow \sigma \rightarrow \tau \leq \sigma' \rightarrow \tau'$.

One can easily check that the relation $\leq$ makes T_I an inf-semilattice. Moreover the relation $\sim$ ($\tau \sim \sigma$ if and only if $\sigma \leq \tau$ and $\tau \leq \sigma$) is an equivalence relation.

Definition 10. i) Let a *basis* be a partial function from variables to types. A basis will be denoted by $\{x : \sigma \mid B(x) = \sigma\}$.
ii) The following type assignment system S_I establishes judgements of the shape $B \vdash M : \sigma$, where B is a basis, M is a λI-term and $\sigma \in T_I$. The rules are:

$$(var)\ \frac{}{B \cup \{x : \sigma\} \vdash x : \sigma}$$

$$(\rightarrow E)\ \frac{B \vdash M : \tau \rightarrow \sigma \quad B \vdash N : \tau}{B \vdash MN : \sigma}$$

$$(\rightarrow I)\ \frac{B \cup \{x : \sigma\} \vdash M : \tau \quad x \in FV(M) \quad x \notin dom(B)}{B \vdash \lambda x.M : \sigma \rightarrow \tau}$$

$$(\cap I)\ \frac{B \vdash M : \sigma \quad B \vdash M : \tau}{B \vdash M : \sigma \cap \tau} \qquad (\leq)\ \frac{B \vdash M : \sigma \quad \sigma \leq \tau}{B \vdash M : \tau}$$

Definition 11. Let CI be the set of compact elements of $D^I \setminus \{\perp\}$. The function $* : T_I \rightarrow CI$ is inductively defined as follows:
- $f^* = \phi^*$
- $t^* = \top^*$
- $(\sigma \rightarrow \tau)^* = f_{\sigma^*, \tau^*}$
- $(\sigma \cap \tau)^* = \sigma^* \sqcup \tau^*$.

where ϕ^* and $\top^*$ are the images of ϕ and $\top$, respectively, under the canonical injection of D_0 into D^I.

Proposition 5. *i) The function $*$ is a surjection.*
ii) Let $\sigma, \tau \in T_I$. Then $\sigma \leq \tau$ iff $\sigma^* \sqsupseteq \tau^*$.

Proof. i) The set CI can be expressed as $\bigsqcup_{n \in \omega} CI_n$, where CI_n is the image under the n-th projection of CI. The proof proceeds by induction on n, taking into account that an element of CI_n is a step function or a l.u.b of step functions.
ii) The implication ($\Rightarrow$) can be shown by induction on the length of the proof $\sigma \leq \tau$; the implication ($\Leftarrow$) can be shown by induction on the number of symbols in types. □

The following theorem, which asserts that the type assignment system S_I is a sound and complete logical description of the model D^I, motivates all the above definitions:

Theorem 6 (Faithfulness). *i) If* $B \vdash M : \sigma$, *then* $\sigma^* \sqsubseteq [\![M]\!]^I_{\rho_B}$, *where* $\forall x \in dom(B).\rho_B(x) = B(x)^*$.
ii) If $\sigma^* \sqsubseteq [\![M]\!]^I_\rho$, *then there exists* B *such that* $\forall x \in dom(B).B(x)^* \sqsubseteq \rho(x)$ *and* $B \vdash M : \sigma$.

Proof. i) By induction on the length of the proof $B \vdash M : \sigma$.
ii) By induction on the structure of M. □

Using a generalization of Krivine's computability technique we show now that if $(\exists \rho.[\![M]\!]^I_\rho \neq \perp)$ then M is strongly normalizing.

Definition 12. A set X of strongly normalizing λI-terms is *saturated* if:
i) if $x \in FV(M_1)$ and $(M_1[N/x])M_2 \ldots M_n \in X$, then $(\lambda x.M_1)NM_2 \ldots M_n \in X$;
ii) if $M_1, \ldots, M_n$ are strongly normalizing λI-terms, then $xM_1M_2 \ldots M_n \in X$.

Proposition 7. *The set Sat of saturated sets is a non empty complete lattice closed under the operation* $\Rightarrow$ *defined by:*

$$(B \Rightarrow C) = \{M \in \Lambda I \mid \forall N.(N \in B \Rightarrow MN \in C)\} .$$

Proof. The set of strongly normalizing terms is saturated. Saturated sets are closed under set-theoretic intersection. The rest is routine. □

We define now saturated sets, $Comp(\sigma)$, by induction on the structure of σ making sure that type equivalence is preserved. In particular, since $f \sim t \to f$ and $t \sim f \to t$, we must make sure that $Comp(f) = Comp(t \to f)$ and $Comp(t) = Comp(f \to t)$. We achieve this by defining a monotone operator over a suitable lattice derived from Sat and taking the pair $(Comp(f), Comp(t))$ to be a fixed point of that operator.

Definition 13. i)Let $(Satp, \leq)$ be the complete lattice consisting of all pairs (A, B) of saturated sets such that $B \subseteq A$. We define

$$(A, B) \leq (A', B') \iff (A \subseteq A' \wedge B \supseteq B') .$$

ii)Let $R : (Satp, \leq) \to (Satp, \leq)$ be the operator defined as:

$$R(A, B) = ((B \Rightarrow A), (A \Rightarrow B)), \quad (A, B) \in Satp .$$

R is clearly monotonic, so let (A_0, B_0) be a fixed point of R. It is worth noticing that the least fixed point is the pair consisting of the set of strongly normalizing λI-terms, and the set of those terms which yield strongly normalizing terms no matter how many times they are applied to strongly normalizing terms.

Definition 14. The sets $Comp(\sigma)$ are inductively defined as follows:
- $Comp(f) = A_0$
- $Comp(t) = B_0$
- $Comp(\sigma \to \tau) = (Comp(\sigma) \Rightarrow Comp(\tau))$
- $Comp(\sigma \cap \tau) = Comp(\sigma) \cap Comp(\tau)$,
where (A_0, B_0) is a fixed point of the monotone operator R.

The above definition can be easily seen to be well posed using the following lemma, which is proved by induction on the structure of types:

Lemma 8. *If $\sigma \leq \tau$ then $Comp(\sigma) \subseteq Comp(\tau)$.*

Proof of the Full Abstraction of D^I.

Lemma 9. *i) Let $M, N_1, \ldots, N_n \in \Lambda I$ and $\{x_1, \ldots, x_n\} \supseteq FV(M)$. If $B \cup \{x_1 : \sigma_1, \ldots, x_n : \sigma_n\} \vdash M : \sigma$ is derivable in S_I and $\forall i. N_i \in Comp(\sigma_i)$, then $M[N_i/x_i] \in Comp(\sigma)$.*
ii)Let $M \in \Lambda I$. Then $\exists \rho. [\![M]\!]^I_\rho \sqsupseteq \phi$ if and only if M is strongly normalizing.

Proof. i) By induction on the length of the proof $B \cup \{x_1 : \sigma_1, \ldots, x_n : \sigma_n\} \vdash M : \sigma$, using Lemma 8.
ii) ($\Rightarrow$) Follows from faithfulness of S_I and from i) by considering the trivial substitution $M[x_i/x_i]$, since variables belong to every saturated set.
ii) ($\Rightarrow$) Consider the environment $\rho_\top$ mapping every variable in $\top$. By induction on the structure of normal forms M, it is easy to show that $[\![M]\!]^I_{\rho_\top} \sqsupseteq \phi$. □

The proof of the following Theorem is an interesting example of a *semantical* proof of a result which is normally proved *syntactically*.

Theorem 10 (Conservation). *Let $M \in \Lambda I$. M is weakly normalizing if and only if M is strongly normalizing.*

Proof. If M is weakly normalizing, then $\forall \rho. [\![M]\!]^I_\rho = [\![M']\!]^I_\rho$, for some normal form M'. But then by Lemma 9, M is strongly normalizing. □

Lemma 11. *Let $M \in \Lambda I$.*
i) $(\exists \rho. [\![M]\!]^I_\rho \neq \perp) \Leftrightarrow M \Downarrow_{LI}$.
ii) If M and N have different $\beta\eta$-normal forms, then $\exists \rho. [\![M]\!]^I_\rho \neq [\![N]\!]^I_\rho$.

Proof. i) From Lemma 9 and Theorem 10.
ii) Suppose that $\forall \rho. [\![M]\!]^I_\rho = [\![N]\!]^I_\rho$. We can assume that both M and N are closed, since otherwise we consider their closures, then, since M and N have different normal forms, using Theorem 1 we have that the model is trivial, but this contradicts Proposition 2. □

3 Modeling the λ_V-calculus in Coherence Spaces

Since the continuous semantics of λ_V-calculus is not entirely satisfactory *vis-à-vis* the full-abstraction problem, we investigate now the problem of modeling λ_V-calculus using coherence spaces and stable functions, see [8].

The λ_V-calculus is usually modeled in C.P.O's by taking the initial solution of the domain equation $D \simeq [D \to_\perp D]_\perp$, where $[D \to_\perp D]_\perp$ is the lifted space of strict continuous functions from D to D, see [6]. The new bottom

element, which results from lifting, is intended to denote closed terms which do not reduce to a value, i.e. undefined expressions, Thus λ_V-abstractions, which are values corresponding to *partial* functions, can be safely modeled as strict continuous functions. The immediate question which arises if we want to use coherence spaces, is then: how shall we denote undefined expressions and hence partial functions. The techniques used for Scott Domains cannot be mimicked directly in coherence spaces. In fact:
- if D, D' are coherence spaces, the natural operator *strict* defined as

$$strict(f) = \lambda d.if\ d = \emptyset\ then\ \emptyset\ else\ f(d),$$

does not yield a stable function in general even if f is stable;
- the lifting operator takes us out of the category of coherence spaces.

We could try to define a different strict operator, manipulating directly the *traces* of the stable functions, but this would allow us just to model λI-calculus. Or else we could discuss models in the wider category of *D-I domains* and stable functions, but this would take us out of the mathematically attractive universe of coherence spaces. We will show in this section that, at the price of a slight modification of the category of coherence spaces, we can nevertheless succeed in defining an adequate notion of *partial* stable function.

In the next subsection, first we recall some definitions concerning the λ_V-calculus and we present the operational semantics determined by the SECD machine. We give then the main definitions and results concerning coherence spaces. These spaces are known in the literature also as binary qualitative domains. For further details and proofs concerning coherence spaces see [7, 8]. We introduce then a category derived from that of coherence spaces, where suitable notions of *lifted* coherence space and *partial* stable function can be defined, and where reflexive domain equations can be solved. We discuss a particular λ_V-model which is correct but not complete with respect to the observational equivalence induced by the SECD machine.

3.1 Operational Semantics and Observational Equivalence for the λ_V-calculus

We will not give a formal introduction to the λ_V-calculus, the reader can see [12, 6] for more details. We just recall that the set of terms of the λ_V-calculus without constants is precisely the set of the λK-terms. The equational theory of the λ_V-calculus differs from that of the standard calculus in that the β-rule is replaced by the β_V-rule, i.e. $(\lambda x.M)N \rightarrow_V M[N/x]$ if and only if $N \in Val_V = \{\lambda x.M \mid M \in \Lambda\} \cup Var$. Where Val_V is called the set of *values*. Terms which reduce to values are called *valuable*.

The operational semantics induced by Landin's SECD machine on pure λ-terms can be defined as follows, in S.O.S. style:

Definition 15 (Operational Semantics). The set of programs is the set of closed λ-terms, Λ°. We define the lazy leftmost reduction $\Downarrow_{SECD}$ as the least binary relation over $\Lambda^{\circ} \times (Val_V \cap \Lambda^{\circ})$ satisfying the following rules:

$$(abstr)\qquad \frac{}{\lambda x.P \Downarrow_{SECD} \lambda x.P}$$

$$(app)\qquad \frac{M \Downarrow_{SECD} \lambda x.P \quad N \Downarrow_{SECD} J \quad P[J/x] \Downarrow_{SECD} N}{MN \Downarrow_{SECD} N} \;.$$

We write simply $M \Downarrow_{SECD}$ to indicate that M converges to a value.

Definition 16 (Observational Equivalence). i) $M \equiv_{SECD} N$ if and only if $\forall C[\,].(C[M], C[N] \in \Lambda^{\circ} \Rightarrow (C[M] \Downarrow_{SECD} \Leftrightarrow C[N] \Downarrow_{SECD}))$.
ii) Let T_{SECD} be the λ-theory $\{(M, N) \mid M, N \in \Lambda^{\circ} \wedge M \equiv_{SECD} N\}$.

G.Plotkin, in [12] showed that the equational theory of the λ_V-calculus is strictly included in T_{SECD}. And hence that it is an equational theory for reasoning *correctly* about the observational equivalence.

3.2 Coherence Spaces and Stable Functions

Definition 17. i) A *coherence space* X is a set of sets which satisfies:
- *binary completeness*: if $M \subseteq A$ and if $\forall a, b \in M.(a \cup b \in A)$, then $\bigcup M \in A$.
- *down closure*: if $a \in X$ and $b \subseteq a$, then $b \in X$.

Definition 18. Let X be a coherence space:
i) We denote with X_{fin} the set of finite elements of X.
ii) Two elements $a, b \in X$ are said to be *compatible* iff $a \cup b \in X$.
iii) We denote with $\mid X \mid$ the set $\{z \mid \{z\} \in X\}$. The elements of $\mid X \mid$ are called the *atoms* of X.

A coherence space X is completely determined by the set $\mid X \mid$ of its atoms and by the compatibility relation $\frown\!\!\!\!\smile$ between atoms of X defined as follows:

$$z \frown\!\!\!\!\smile z'[modX] \text{ iff } \{z, z'\} \in X \;.$$

Definition 19. Let X, Y be coherence spaces. A function $f : X \to Y$ is *stable* if and only if:

1. $\forall a, b \in X.(a \subseteq b \Rightarrow f(a) \subseteq f(b))$;
2. if $\{a_i\}_{i \in I}$ is a direct family of elements of X, then $f(\bigcup_{i \in I} a_i) = \bigcup_{i \in I} f(a_i)$;
3. $\forall a, b \in X.(a \cup b \in X \Rightarrow (f(a \cap b) = f(a) \cap f(b)))$.

The set of stable functions from X to Y will be denoted by $S(X, Y)$.

Definition 20. Let X, Y be coherence spaces.
i) Let $f : X \to Y$ be a stable function. The trace of f is the set:

$$Tr(f) = \{(a, z) \mid (a, z) \in X_{fin} \times \mid Y \mid \;,\; z \in f(a) \;,\; \forall a' \in X.(a' \subset a \Rightarrow z \notin f(a'))\}.$$

ii) The set of traces of stable functions from X to Y will be denoted by $[X \to Y]$.
iii) Stable functions can be ordered according to Berry's order relation $\sqsubseteq_s$ defined as follows: let $f, g : X \to Y$ be two stable functions, then

$$f \sqsubseteq_s g \text{ iff } \forall a, b \in X.(a \subseteq b \Rightarrow f(a) = f(b) \cap g(a)) \;.$$

It is worth noticing that the pointwise ordering of functions is different from Berry's order. In fact if $f \sqsubseteq_s g$, then $\forall a \in X.(f(a) \subseteq g(a))$, but the converse is false.

Theorem 12 (Representation Theorem). *Let X, Y be coherence spaces.*
i) A stable function $f : X \to Y$ is determined by its trace $Tr(f)$, i.e.:
$f(a) = \{z \mid \exists d \in X_{fin}.(d \subseteq a \wedge (d, z) \in Tr(f))\},\ a \in X$.
ii) $[X \to Y]$ is a coherence space isomorphic to the set $S(X, Y)$ ordered by $\sqsubseteq_s$.

3.3 A "Coherent" Version of Lifted Spaces and Strict Functions

Since one cannot build a "lifted" coherence space simply by adding a new bottom element to the space, in order to define an adequate coherence semantics we decide to take an opposite view to that which inspires the construction in C.P.O.'s. Instead of adding a point which is intended to represent undefined expressions we *add* an atom whose rôle is to mark points which denote *defined expressions*. As a consequence we have to give up the requirement that there be only one semantic nonvalue. Formally we simulate the "lifted" coherence space by adding to a coherence space a new atom $*$ compatible with all the atoms of the space and take as set of *semantic values* the set of points to which the atom $*$ belongs.

Definition 21. Let X be a coherence space and let $* \notin \mid X \mid$. We denote with X_* the coherence space defined as follows:
- $\mid X_* \mid = \mid X \mid \cup \{*\}$;
- if $z_1, z_2 \in \mid X \mid$, then $z_1 \frown\!\!\!\!\smile z_2[modX_*]$ iff $z_1 \frown\!\!\!\!\smile z_2[modX]$;
- $* \frown\!\!\!\!\smile z[modX_*]$, for all $z \in \mid X_* \mid$.

The new atom $*$ will be called the *added* atom of X_*. The space X_* has naturally associated constructor and destructor operators in_* and out_* which are stable and such that $(out_* \circ in_*)(a) = a$, for all $a \in X$:

Definition 22. Let X be a coherence space:
i) $in_* : X \to X_*$ is defined as $in_*(a) = a \cup \{*\}$, for $a \in X$
ii) $out_* : X_* \to X$ is defined as $out_*(a) = a \setminus \{*\}$, for $a \in X$.

Carrying further our idea for modeling λ_V-calculus, we try to interpret closed non valuable terms as $\emptyset$ and thus λ-abstractions as stable functions that return $\emptyset$ when the argument is a semantic nonvalue, i.e. a point to which the atom $*$ does not belong. Hence we define the analogue of the "strict" operator, as the operator which maps a stable function into a stable function which is extensionally equal to it on semantic values but which returns $\emptyset$ when the argument is a semantic nonvalue. We will call *partial stable* functions, stable functions that return $\emptyset$ when the argument is a semantic nonvalue. The results in the sequel will show that our technique is indeed adequate. Thus in order to model the λ_V-calculus, we try to solve the domain equation $D \simeq [D \to_{s*} D]_*$, where $[D \to_{s*} D]$ denotes the coherence space of traces of partial stable functions. The reason for considering partial stable functions instead of simply stable functions is motivated by the desire to keep superfluous points to a minimum. As a very interesting consequence we succeed in modeling also the η_V-rule, i.e. $\lambda x.Mx \rightarrow_V M$ if $x \notin FV(M)$ and $M \in Val_V$. In order to simplify the presentation we will solve the above domain equation in the category bqD_* of Pointed Coherence Spaces, defined below. One can also easily see that Girard's category bqD of coherence spaces and binary qualitative morphisms does not work immediately.

Proposition 13. *The category bqD_* is defined as follows:*
- the objects are the pairs $(D, *_D)$ where D is a coherence space and $*_D$ is an atom of D (the atom $*_D$ will be referred to as the base atom of the pointed space $(D, *_D)$);
- the arrows from $(D, *_D)$ to $(D', *_{D'})$ are the binary qualitative morphisms that map base atoms into base atoms, i.e. injective functions from $| D |$ to $| D' |$ which preserve compatibility and map base atoms to base atoms.

We will denote by $bqD_*((D, *_D), (D', *_{D'}))$ the set of binary qualitative morphisms from the pointed coherence space $(D, *_D)$ to the pointed coherence space $(D', *_{D'})$. Moreover we will denote by $S_{*_D}(D, D')$ the set of partial stable functions f from D to D', i.e. the set of stable functions such that, if $(a, z) \in Tr(f)$, then $*_D \in a$. We are now ready to define the function $partial_s$, which is the "coherent" analogue of the *strict* operator. One can easily check that:

Proposition 14. *Let $(D, *_D), (D', *_{D'})$ be pointed coherence spaces.*
i) The space of traces of partial stable functions from $(D, *_D)$ to $(D', *_{D'})$, denoted by $[D \rightarrow_{s*_D} D']$, is a coherence space.
ii) The function $partial_s : S(D, D') \rightarrow S_{*_D}(D, D')$ defined as follows, is stable:

$$partial_s(f) = \lambda d \in (D, *_D).(if\ *_D \in d\ then\ f(d)\ else\ \emptyset), \quad f \in S(D, D') .$$

Finally we introduce the following useful binary qualitative morphisms and endofunctors over the category bqD_*. They are instrumental for the construction of the next subsection. It is easy to check that the definitions are well posed.

Definition 23. Let $(D, *_D), (D', *_{D'}), (E, *_E), (E', *_{E'})$ be objects of the category bqD_*. Let $f \in bqD_*((D, *_D), (D', *_{D'}))$ and $g \in bqD_*((E, *_E), (E', *_{E'}))$.
i) The morphism $(f \rightarrow_* g) \in bqD_*(([D \rightarrow_{s*_D} E], (\{*_D\}, *_E)), ([D' \rightarrow_{s*_{D'}} E'], (\{*_{D'}\}, *_{E'})))$ is defined as follows:

$$(f \rightarrow_* g)((a, z)) = (f^+(a), g(z)), \quad (a, z) \in | [D \rightarrow_{s*_D} E] | ,$$

where $f^+ : D \rightarrow D'$ is the stable function defined by $f^+(a) = \{f(z) \mid z \in a\}$.
ii) The morphism $f_* \in bqD((D_*, *), (D'_*, *))$ is defined as follows:
$f_*(z) = f(z)$, if $z \in | D |$;
$f_*(*) = *$.

Definition 24. i) Let $G_1 : bqD_* \times bqD_* \to bqD_*$ be the functor defined as follows:
$G_1((D, *_D), (E, *_E)) = ([D \to_{s*_D} E], (\{*_D\}, *_E))$, for $(D, *_D), (E, *_E)$ pointed coherence spaces;
$G_1(f, g) = (f \to_* g)$, for $f \in bqD_*((D, *_D), (D', *_{D'}))$, $g \in bqD_*((E, *_E), (E', *_{E'}))$.
ii) Let $G_* : bqD_* \to bqD_*$ be the functor defined as follows:
$G_*((D, *_D)) = (D_*, *)$, for $(D, *_D)$ pointed coherence space;
$G_*(f) = f_*$, for $f \in bqD_*((D, *_D), (D', *_{D'}))$.
iii) Let $G : bqD_* \to bqD_*$ be the composition functor of G_1 and G_*:
$G = \lambda((D, *_D) \in bqD_*).(G_* \circ G_1)((D, *_D), (D, *_D))$. Then
$G((D, *_D)) = ([D \to_{s*_D} D]_*, *)$, for $(D, *_D)$ pointed coherence space;
$G(f) = (f \to_* f)_*$, for $f \in bqD_*((D, *_D), (D, *_D))$.

3.4 A Coherence Model for the λ_V-calculus

In this subsection we define a coherence model for the λ_V-calculus in the category bqD_* using the machinery introduced previously. More precisely we will take the initial solution in bqD_* of the reflexive equation $(D, *_D) \simeq ([D \to_{s*_D} D]_*, *)$, where the base atom of $([D \to_{s*_D} D]_*, *)$ is the added atom of $[D \to_{s*_D} D]_*$. Domain equations can be solved in this category with the usual "inverse limit" technique.

Proposition 15. *The domain equation $(D, *_D) \simeq ([D \to_{s*_D} D]_*, *)$ has initial solution in bqD_* equal to $lim_{\to}(D_n, i_n)$, where $(D_0, *_0)$ is a pointed coherence space with only one atom, $(D_{n+1}, *_{n+1}) = ([D_n \to_{s*_n} D_n]_*, *)$, i_0 is the morphism which maps $*_0$ to $*_1$ and $i_{n+1} = (i_n \to_* i_n)_*$.*

Proof. Follows from the general Theorem in [13] since the category bqD_* is ω-complete and the functor G is ω-continuous (see [13] for further details). □

We have achieved our goal, finally! One can easily see, in fact, that the initial solution of the domain equation $(D, *_D) \simeq ([D \to_{s*_D} D]_*, *)$ in bqD_*, can be used to model adequately the λ_V-calculus. More formally one can show that the structure defined below is a *syntactical λ_V-model* in the sense of [6].

Definition 25. The λ_V-model D^V is the structure $< D^V, \{d \in D^V \mid * \in d\}, \bullet^V, [\![\,]\!]^V >$, where:
- $((D^V, *), i_V)$ is the initial solution of the equation $(D, *_D) \simeq ([D \to_{s*_D} D]_*, *)$ in bqD_*;
- $\bullet^V : D^V \times D^V \to D^V$ is defined by:
$e \bullet^V d = if\ * \in d\ then\ Ap(out_*(i_V^+(e)), d)\ else\ \emptyset$,
where i_V^+ is the isomorphism from D^V to $[D^V \to_{s*} D^V]_*$ defined by $i_V^+(d) = \{i_V(z) \mid z \in d\}$.
- $[\![\,]\!]^V : \Lambda \times (Var \to \{d \in D^V \mid * \in d\}) \to D^V$ is defined inductively as follows:
$[\![x]\!]^V_\rho = \rho(x)$
$[\![MN]\!]^V_\rho = [\![M]\!]^V_\rho \bullet^V [\![N]\!]^V_\rho$

$[\![\lambda x.M]\!]^V_\rho = i^-_V(in_*(partial_s(\lambda d \in D^V.[\![M]\!]^V_{\rho_{ext}[d/x]})))$,
where $\rho_{ext} : Var \to D^V$ is the natural extension of ρ when the codomain is taken to be all of D^V and i^-_V is the isomorphism from $[D^V \to_{s*} D^V]_*$ to D^V defined by $i^-_V(d) = \{z \mid i_V(z) \in d\}$.

In the definition above and elsewhere we will take $[\![\;]\!]^V$ to be sometimes defined on the whole $\Lambda \times (Var \to D^V)$, when necessary and clear from the context. The reader can check that this does not introduce ambiguities. In order to show that the structure D^V is well defined, one needs to prove that in general the function $\lambda d_1 \ldots d_n.[\![M]\!]^V_{\rho_{ext}[d_1/x_1,\ldots,d_n/x_n]}$ is stable, for all $M \in \Lambda$, n-tuples of variables $x_1,\ldots,x_n$ and ρ's. This can be seen by structural induction on M. We omit this proof as well as the routine check that the model defined above is indeed a syntactical λ_V-model.

3.5 Correctness of the Coherence Semantics

In this subsection we outline the proof that the denotational semantics determined by D^V is correct with respect to the operational equivalence induced by Landin's SECD machine. In particular we show that if $T_{D^V} \subseteq T_{SECD}$, where T_{D^V} is the theory induced by D^V i.e. :

$$T_{D^V} = \{(M, N) \mid M, N \in \Lambda^\circ \wedge [\![M]\!]^V = [\![N]\!]^V\} \; .$$

First we need to recall that by the standardization theorem in [12], if $M \equiv_{SECD} N$ then $\forall C[\;].(C[M], C[N] \in \Lambda^\circ \Rightarrow (C[M] \twoheadrightarrow_V Val_V \Leftrightarrow C[N] \twoheadrightarrow_V Val_V))\}$, where $\twoheadrightarrow_V$ denotes the reflexive, transitive and contextual closure of $\to_V$.
In order to prove the correctness result we now proceed as in [6], for the canonical λ_V-model in Scott domains. We start by proving an *Approximation Theorem* for the λ_V-model D^V, whereby the interpretation of a term can be expressed as the l.u.b. of the interpretations of its *approximants*. Approximants are defined in an extension of the λ_V-calculus, the λ^*_V-calculus, obtained by adding a distinguished constant Ω, whose intended meaning is the least element of the coherence space.

Definition 26 (λ^*_V-calculus). The set of λ^*_V-terms Λ^* is inductively defined as Λ, starting from $Var \cup \{\Omega\}$. The set of values is $Val^* = Var \cup \{\lambda x.M \mid M \in \Lambda^*\}$.
The rules of the approximate reduction are:
(α^*) $(\lambda x.M) \to^*_V (\lambda y.M[y/x])$, if $y \notin FV(M)$;
(β^*_V) $(\lambda x.M)N \to^*_V M[N/x]$, if $N \in Val^*$;
(Ωl) $\Omega M \to^*_V \Omega$;
(Ωr) $M\Omega \to^*_V \Omega$.
The set NF^*_V of terms of Λ^* in $\to^*_V$-normal is defined as follows:
- $\Omega \in NF^*_V$;
- $x \in NF^*_V$;
- $\xi M_1 \ldots M_m \in NF^*_V$, if $\xi \in Var$ and $M_i \not\equiv \Omega$, $M_i \in NF^*_V$ $\forall i = 1,\ldots,m$, or $\xi \equiv \lambda x.P$, $P \in NF^*_V$, $M_i \not\equiv \Omega$, $M_i \in NF^*_V$ $\forall i = 1,\ldots,m$ and $M_1 \notin Val^*$;
- $\lambda x_1 \ldots x_n.M \in NF^*_V$, if $M \in NF^*_V$.

Definition 27. Let $M \in \Lambda^*$, the set of *approximants* of M is defined as follows: $\mathcal{A}_V(M) = \{A \mid A \in NF_V^*$ *and* A *is obtained from* M', $M' =_V M$, *by substituting* Ω *for some subterms of* $M'\}$.

In order to prove the Approximation Theorem we need to introduce the notion of indexed term and indexed $\lambda\beta_V^*$-reduction. This is a standard tool for analyzing the fine structure of models obtained as limits, in that it allows to reason about n^{th}-*approximations* of interpretations of λ-terms.

Definition 28. i) The set of indexed λ-terms, Λ_{in}^*, is the language generated by the following grammar: $M \ ::= \ \Omega^n \mid x^n \mid (\lambda x.M)^n \mid (MM)^n$, for $n \in \omega$ and $x \in Var$.
ii) We denote by $er : \Lambda_{in}^* \rightarrow \Lambda^*$ the function that maps any term $M \in \Lambda_{in}^*$ to the term which is obtained from the indexed term M by erasing indices.
iii) We denote by $\Lambda_{in,exter}^*$ the set of terms obtained from indexed terms erasing only the outermost index.
iv) Let $M, N \in \Lambda_{in,exter}^*$, $x \in Var$, we define by $M^b[N^n/x^c]$ the result of the substitution of N for any free occurrence of x in M , in such a way that, if c is the index of a free occurrence of x, c is eliminated and the index of the term N is substituted for that occurrence of x is n.

Definition 29 (Indexed $\lambda\beta_V^*$-reduction). i) The rules of the indexed $\lambda\beta_V^*$-reduction are the following:
(I_α) $(\lambda x.M^n)^m \rightarrow_{IV} (\lambda y.(M^n[y^c/x^c]))^m$, where $M \in \Lambda_{in,exter}^*$, $y \notin FV(M)$;
(I_{β_V}) $((\lambda x.M^n)^{m+1}N^p)^q \rightarrow_{IV} M^b[N^a/x^c]$, where $M, N \in \Lambda_{in,exter}^*$, $er(N^p) \in Val^*$, $p \geq 0$, $a = min(m,p,c)$, $b = min(m,n,q)$;
$(I_{\Omega l})$ $((\lambda x.M^n)^0 N^p)^q \rightarrow_{IV} \Omega^0$, where $M, N \in \Lambda_{in,exter}^*$.
ii) We denote with $\twoheadrightarrow_{IV}$ the contextual, reflexive and transitive closure of $\rightarrow_{IV}$.

Indexed λ-terms can be naturally interpreted in D^V:

Definition 30. Let $M \in \Lambda_{in}^*$. The interpretation $[\![M]\!]_\rho^V$ of M is inductively defined as follows:
- $[\![\Omega^n]\!]_\rho^V = \emptyset$
- $[\![x^n]\!]_\rho^V = (\rho(x))_n$
- $[\![(MN)^n]\!]_\rho^V = ([\![M]\!]_\rho^V \bullet^V [\![N]\!]_\rho^V)_n$
- $[\![(\lambda x.M)^n]\!]_\rho^V = (i_V^-(in_*(partial_s(\lambda d \in D^V.[\![M]\!]_{\rho_{ext}[d/x]}^V))))_n$,

where the notation $(d)_n$ used above indicate the injection into D^V of the n-th projection of $d \in D^V$ onto D_n.

We can now prove the following results:

Proposition 16. *1. Let* $M, M' \in \Lambda_{in}^*$, *if* $M \twoheadrightarrow_{IV} M'$, *then* $\forall\rho.[\![M]\!]_\rho^V = [\![M']\!]_\rho^V$.
2. Let $M \in \Lambda^*$. *Then,* $\forall\rho$, $[\![M]\!]_\rho^V = \bigcup_{M':er(M')=M} [\![M']\!]_\rho^V$.
3. If $M \in \Lambda_{in}^*$, *then there exists* $A \in \mathcal{A}_V(er(M))$ *s.t.* $\forall\rho.[\![M]\!]_\rho^V \subseteq [\![A]\!]_\rho^V$.
4. Let $M \in \Lambda^*$, *then* $\forall C[\ ]\forall\rho.([\![C[\Omega]]\!]_\rho^V \subseteq [\![C[M]]\!]_\rho^V)$.

5. *For all $M \in \Lambda$, for all $A \in \mathcal{A}_V(M)$ and all ρ, $[\![A]\!]^V_\rho \subseteq [\![M]\!]^V_\rho$.*

Proof. 1. By induction on the number of reduction steps.
2. By induction on the structure of M, since the following is a direct family

$$\{Tr(\lambda d.[\![M']\!]^V_{\rho_{ext}[d/x]})\}_{er(M')=M}$$

3. Follows form the facts that $\rightarrow_{IV}$-convertible indexed λ-terms have the same denotation and that indexed terms are normalizing. A proof of normalization for the indexed $\lambda\beta^*_V$-reduction can be achieved by showing that indexed β^*-reduction is strongly normalizing. Indexed $\lambda\beta^*_V$-reduction can be seen, in fact, as a particular reduction strategy of indexed β^*-reduction.
4. By induction on the structure of $C[\ ]$, for any n-tuple $x_1, \ldots, x_n$, we have

$$Tr(\lambda d_1 \ldots d_n [\![C[\Omega]]\!]^V_{\rho_{ext}[d_1/x_1,\ldots,d_n/x_n]}) \subseteq Tr(\lambda d_1 \ldots d_n [\![C[M]]\!]^V_{\rho_{ext}[d_1/x_1,\ldots,d_n/x_n]}).$$

5. Use the fact above and that β_V-convertible terms have the same denotation. □

Using Proposition 16 we can show immediately:

Theorem 17 (Approximation Theorem). *Let $M \in \Lambda$. Then, for all ρ,*

$$[\![M]\!]^V_\rho = \bigcup\{[\![A]\!]^V_\rho \mid A \in \mathcal{A}_V(M)\}.$$

Using the Approximation theorem we can now show that many interesting equations hold in the model. See [6] for some examples. In particular we can show the following lemma, which implies immediately the correctness of the denotational semantics w.r.t. the observational equivalence:

Lemma 18. *Let $M \in \Lambda$.*
i) If M is not valuable, then $\exists \rho . [\![M]\!]^V_\rho = \emptyset$.
ii) If M is valuable, then $\forall \rho . [\![M]\!]^V_\rho \neq \emptyset$.

Proof. i) Let M be not valuable. If $A \in \mathcal{A}_V(M)$, then $A \in NF^*_V$ and $A \notin Val^*$. By induction on the structure of $A \in NF^*_V \setminus Val^*$ one can show that, if ρ is the environment which maps all the variables to $\{*\}$, then $[\![A]\!]^V_\rho = \emptyset$. Hence using the Approximation theorem we have $[\![M]\!]^V_\rho = \emptyset$.
ii) From the definition of the λ_V-model D^V. □

Theorem 19 (Correctness Theorem). $T_{D^V} \subseteq T_{SECD}$.

A remark concerning coherence models like D^V is in order here. Contrary to what happens in Scott domains the relation $M \sqsubseteq^V N$ defined as $\forall \rho . [\![M]\!]^V_\rho \subseteq [\![N]\!]^V_\rho$ is not a congruence over Λ. In fact there exist λ-terms M, N such that $M \sqsubseteq^V N$, but $C[M] \not\sqsubseteq^V C[N]$, for some context $C[\ ]$. For instance take $S_1 \equiv (\lambda xy.\Delta\Delta)(x(\lambda x.\Delta\Delta))$ and $S_2 \equiv \lambda x.\Delta\Delta$, where $\Delta \equiv \lambda x.xx$. Then for any ρ, $[\![S_1]\!]^V_\rho \subseteq [\![S_2]\!]^V_\rho$, but if we consider the context $C[\ ] \equiv \lambda x.[\ ]$ we have that $[\![C[S_1]]\!]^V_\rho$ and $[\![C[S_2]]\!]^V_\rho$ are incomparable. The reason is that there exist pointwise comparable functions with incomparable traces i.e. incomparable under Berry's order. In any case the relation $\sqsubseteq^V$ restricted to Λ^o is a congruence relation.

3.6 The Interplay Between Different Correct but Incomplete Semantics

In this final subsection we prove first that the λ_V-model D^V is not fully-abstract with respect to the SECD operational equivalence. In particular we exhibit two operationally equivalent λ-terms, Q_1 and Q_2, which have different interpretations in the λ_V-model D^V. In order to show the operational equivalence of Q_1 and Q_2, we shall not embark in a cumbersome syntactical proof but show instead immediately that Q_1 and Q_2 have the same denotation in the canonical λ_V-model P, defined in [6] using C.P.O.'s. The counterexample to the completeness of D^V appears because there exist λ-definable stable functions which are pointwise comparable but incomparable under Berry's order.

Theorem 20 (Incompleteness Theorem). *The following λ-terms*
$Q_1 \equiv \lambda x.(\lambda xyz.\Delta\Delta)(x R_1)(x R_2)$ *and* $Q_2 \equiv \lambda x.(\lambda xy.\Delta\Delta)(x R_1)$,
where $R_1 \equiv \lambda z.((\lambda xy.\Delta\Delta)(z(\lambda x.\Delta\Delta)))$ *and* $R_2 \equiv \lambda xy.\Delta\Delta$,
are observationally equivalent but have different denotations in D^V.

Proof. The λ-terms $R_1 \equiv \lambda z.((\lambda xy.\Delta\Delta)(z(\lambda x.\Delta\Delta)))$ and $R_2 \equiv \lambda xy.\Delta\Delta$ when interpreted in D^V are incomparable under Berry's order. Hence there exists $d \in [D^V \rightarrow_{s*} D^V]_*$, say $\{*, ([\![R_1]\!]^V, *)\}$, such that $\emptyset = [\![Q_1]\!]^V \bullet^V d \neq [\![Q_2]\!]^V \bullet^V d = \{*\}$. But Q_1 and Q_2 are observationally equal, since in the canonical continuous λ_V-model P one has $[\![R_1]\!]^P \subseteq [\![R_2]\!]^P$. In fact, for all d $[\![x R_1]\!]^P_{\rho[d/x]} = \perp$ holds if and only if either $[\![x R_1]\!]^P_{\rho[d/x]} = \perp$ or $[\![x R_2]\!]^P_{\rho[d/x]} = \perp$, and hence $[\![Q_1]\!]^P = [\![Q_2]\!]^P$. □

What we have shown can be stated also as $T_{D^V} \not\supseteq T_{SECD}$ since $T_P \not\subseteq T_{D^V}$, where T_P is the theory induced by the model P. There is indeed a *chiasma* between P and D^V, both in their theories and in the technique used to show their non full abstraction. Along the same line of reasoning as above, we can in fact prove immediately that $T_P \not\supseteq T_{SECD}$, and hence the incompleteness of the λ_V-model P, by showing that the λ-terms M_1 and M_2 defined in [6], which have different denotations in P, have the same interpretation in D^V, i.e. $T_P \not\supseteq T_{D^V}$. The following semantic proof of the operational equivalence of M_1 and M_2 is certainly more easier than the intricate syntactical proof done in [6].

Theorem 21. *The following λ-terms*
$M_1 \equiv \lambda x.((\lambda xyz.\Delta\Delta)(x(\lambda x.\Delta\Delta)(\lambda xy.\Delta\Delta)))(x(\lambda xy.\Delta\Delta)(\lambda x.\Delta\Delta))$ *and*
$M_2 \equiv \lambda x.(\lambda xy.\Delta\Delta)(x(\lambda x.\Delta\Delta)(\lambda x.\Delta\Delta))$
have the same denotation in D^V *and hence are observationally equivalent.*

Proof. First, using the Approximation Theorem one computes that
$[\![\lambda x.\Delta\Delta]\!]^V_{\rho[d/x]} = \{*\}$ and $[\![\lambda xy.\Delta\Delta]\!]^V_{\rho[d/x]} = \{*, (\{*\}, *)\}$.
Then one shows that as far as the subterms of M_1 and M_2 :
$[\![x(\lambda x.\Delta\Delta)(\lambda xy.\Delta\Delta)]\!]^V_{\rho[d/x]}$ and $[\![x(\lambda xy.\Delta\Delta)(\lambda x.\Delta\Delta)]\!]^V_{\rho[d/x]}$ are semantic values if and only if $[\![x(\lambda x.\Delta\Delta)(\lambda x.\Delta\Delta)]\!]^V_{\rho[d/x]}$ is a semantic value. In fact:

- $[\![x(\lambda x.\Delta\Delta)(\lambda xy.\Delta\Delta)]\!]^V_{\rho[d/x]}$ is a semantic value iff $* \in d$ and $(\{*\}, *) \in d$ and either $(\{*\}, (\{*\}, *)) \in d$ or $(\{*\}, (\{(\{*\}, *)\}, *)) \in d$ or $(\{*\}, (\{*, (\{*\}, *)\}, *)) \in d$;
- $[\![x(\lambda xy.\Delta\Delta)(\lambda x.\Delta\Delta)]\!]^V_{\rho[d/x]}$ is a semantic value iff $* \in d$ and $((\{*\}, *) \in d \vee (\{(\{*\}, *)\}, *) \in d \vee (\{*, (\{*\}, *)\}, *) \in d)$ and $((\{*\}, (\{*\}, *)) \in d \vee (\{(\{*\}, *)\}, (\{*\}, *)) \in d \vee (\{*, (\{*\}, *)\}, (\{*\}, *)) \in d)$,

hence $(* \in [\![x(\lambda x.\Delta\Delta)(\lambda xy.\Delta\Delta)]\!]^V_{\rho[d/x]} \wedge * \in [\![x(\lambda xy.\Delta\Delta)(\lambda x.\Delta\Delta)]\!]^V_{\rho[d/x]})$ iff $(* \in d \wedge (\{*\}, *) \in d \wedge (\{*\}, (\{*\}, *)) \in d)$, that is iff $* \in [\![x(\lambda x.\Delta\Delta)(\lambda x.\Delta\Delta)]\!]^V_{\rho[d/x]}$. □

In conclusion we have that indeed the "coherent" model succeeds in equating observationally equivalent terms that the "continuous" model tells apart, as our original intuition suggested. And this is because the Scott-continuous functions which separate those terms are *indeed* parallel and hence not stable. But somewhat surprisingly Berry's order introduces perverse stable functions which tell apart observationally equivalent terms which are instead equated in the continuous model. Although neither the coherent model nor the continuous one are fully abstract, nevertheless the investigation carried out was shown to be rewarding.

References

1. Barendregt, H.: The Lambda Calculus, its Syntax and Semantics. North Holland. Amsterdam (1984)
2. Böhm, C.: Lambda Calculus and Computer Science Theory. LNCS **37** Springer-Verlag (1975)
3. Church, A.: The Calculi of Lambda Conversion. Princeton University Press Princeton (1941)
4. Coppo, M., Dezani-Ciancaglini, M., Zacchi, M.: Type Theories, Normal Forms and D_∞-Lambda-Models. Information and Computation **72(2)** (1987) 85–116
5. Dezani-Ciancaglini, M., Honsell, F., Ronchi Della Rocca, S.: Models for theories of functions depending on all their arguments (abstract). Journal of Symbolic Logic **51(3)** (1986) 399–402
6. Egidi, L., Honsell, F., Ronchi Della Rocca, S.: Operational, Denotational and Logical descriptions: a case study. Fundamenta Informaticae **16(2)** (1992) 149–169
7. Girard, J.Y.: The system F of variable types, fifteen years later. TCS **45** (1986) 159–192
8. Girard, J. Y., Lafont, Y., Taylor, P.: Proofs and Types. Cambridge University Press. Cambridge (1989)
9. Hindley, J. R., Longo, G.: Lambda Calculus Models and Extensionality. Z. Math. Logik Grundlag. Math. **26** (1980) 289–310
10. Landin, P. J.: The mechanical evaluation of expressions. Computer J. **6(4)** (1964) 308–320
11. Meyer, A.: What is a Model of the Lambda Calculus? Information and Control **52** (1982) 87–122
12. Plotkin, G. D.: Call-by-name, call-by-value and the λ-calculus. TCS **1** (1975) 125–159
13. Smith, M. B., Plotkin, G. D.: The category-theoretic solution of recursive domain equations. SIAM J. of Computing **11(5)** (1982) 761–783

Action Calculi, or Syntactic Action Structures

Robin Milner

Laboratory for Foundations of Computer Science
Computer Science Department, University of Edinburgh
The King's Buildings, Edinburgh EH9 3JZ, UK

Abstract Action structures have previously been proposed as an algebra for both the syntax and the semantics of interactive computation. Here a class of concrete action structures called action calculi is identified, which can serve as a non-linear syntax for a wide variety of models of interactive behaviour. They generalise a previously defined action structure PIC for the π-calculus. One action calculus differs from another only in its generators, called *controls*.

Several extensions to PIC are given as action calculi, giving essentially the same power as the π-calculus. An action calculus is also outlined for PT nets – a class of Petri nets – parametrized upon their places and transitions.

Finally, action calculi are characterized as the free algebras in a sub-variety of action structures, namely those which satisfy certain additional axioms.

1 Introduction

Action structures [7] have been proposed as an algebra for both the syntax and the semantics of interactive computation. In particular, the cited paper gave an example expressing the meaning of a simple language (for evaluating expressions) as a homomorphism between two action structures, one expressing the evaluation mechanism and the other containing the denoted entities. One hopes to extend this treatment to formalisms which express interaction; then the framework will apply not only to programming languages but to the wider class of communicating systems. Formidable difficulties remain in determining what the denoted entities should be in this case; one further hopes that action structures will assist in tackling this problem.

The present paper addresses a question crudely expressed thus: When is an action structure syntactical? Various authors have found it helpful to use formalism with richer structure than a free term algebra, when defining an evaluation mechanism or a communicational model for concurrent systems. Following ideas of Banâtre and Métayer [2], Berry and Boudol based their Chemical Abstract Machine [1] on the notion of a *multiset*. Prompted by them, the present author imposed a structural congruence upon the π-calculus as part of the formal language, not of the semantics [6]. Again, Meseguer and Montanari [5] have revealed the monoidal structure inherent in Petri nets. Indeed, part of the very success of Petri's notion of a net [11] is that its syntax is graphical, not linear.

By imposing an enriched monoidal structure upon action structures, we are already committed to this path; in fact action structures go further and include laws of abstraction (parametrization) over names. But the action structure axioms alone do not determine a syntactic subclass of action structures. Here we present such a class, which we call *action calculi*.

By adding six so-called *concrete* axioms to the basic ones, we arrive at action calculi as term algebras quotiented by the congruence which the axioms induce. These term algebras have an appealing concrete presentation. One action calculus differs from another only in its generators and their dynamics; the generators are best seen as *control operators*. It turns out that the action structure PIC for the synchronous π-calculus [9] is an action calculus, and has just three generators ν, **in** and **out** (restriction, input and output), while an action calculus for Petri's place-transition nets is generated by ν, **pre**, **post** and **mark** (restriction, pre-condition, post-condition and marking). Thus action calculi begin to achieve one of the main purposes of action structures – to unite different models of concurrency in a common setting.

The paper is organised as follows. Section 2 reviews the basic definition of action structures. Section 3 reviews the action structure PIC, which serves as an exemplar for the general definition of action calculus presented in Section 4. Section 5 shows how PIC can be extended to richer action calculi containing guarding, choice and replication, by adding further generators; it also outlines the treatment of Petri nets in an action calculus called NETC. Section 6 is technical; it gives the concrete axioms, substantiates the claim that each action calculus is indeed a quotient of a term algebra, and sketches the category of *concrete action structures*, in which action calculi are the initial objects. Section 7 briefly identifies some future lines of investigation.

2 Action structures reviewed

In this section we recall from [7] the basic notion of an action structure. We first give it in category-theoretic terms, then elaborate it algebraically. Those unfamiliar with categories should find no difficulty if they focus upon the algebra.

2.1 Action structure: definition An *action structure* A is a strict monoidal category, with two extra items:

- a set X_A referred to as *names*, and for each $x \in X_A$ an endo-functor upon A known as an *abstractor*;
- a preorder $\searrow_A$ over each hom-set of A, called *reaction*, which is preserved by composition, monoidal product and abstraction, and for which the units are minimal. ∎

Even more succinctly, an action structure is a preordered strict monoidal category with an indexed set of endo-functors.

2.2 Algebraic elaboration We shall work with the following algebraic characterization of action structures.

First, an action structure A has a monoid $(M_A, \otimes, \epsilon)$ of objects; these are the objects of the category and we shall call them *arities*. We shall use $k, l, m, n, \ldots$ to range over arities. An arity may be, for example, a sequence of sorts (like INT, BOOL) under concatenation; in the special case when there is only one sort, 1, the monoid M_A is just the natural numbers under addition.

Second, for each pair m, n of arities A has a family $A_{m,n}$ of *actions*; they are the morphisms of A as a category. If a is a member of this set we write $a : m \to n$ and call m and n the *source* and *target* arities of a; we even abuse terminology by calling $m \to n$ just the arity of a. We shall use a, b, c, d to range over actions.

Third, since A is a monoidal category with abstractors, there is a unit action $\mathbf{id}_m : m \to m$ for each arity m, and the operations of composition $\cdot$, tensor or monoidal product $\otimes$, and abstraction $\mathbf{ab}_x$ for each $x \in X_A$. They obey the following *arity rules*:

$$\mathbf{id}_m : m \to m \qquad \frac{a : k \to m \qquad b : m \to p}{a \cdot b : k \to p}$$

$$\frac{a : k \to n \qquad b : l \to p}{a \otimes b : k \otimes l \to n \otimes p} \qquad \frac{a : m \to n}{\mathbf{ab}_x a : \mathbf{ab}_x m \to \mathbf{ab}_x n}$$

Note that composition is written forwards; we write $a \cdot b$ where the standard in category theory is to write $b \circ a$.

Note also that the operation of $\mathbf{ab}_x$ upon arities is not specified. But throughout this paper we shall assume an arity k_x associated with each name x, and then $\mathbf{ab}_x m$ is just $k_x \otimes m$. For example if arities are sort sequences we may have $k_x = \text{INT}$, which may mean that x is an integer variable; then for $a : \text{BOOL} \otimes \text{BOOL} \to \text{INT}$ we shall have $\mathbf{ab}_x a : \text{INT} \otimes \text{BOOL} \otimes \text{BOOL} \to \text{INT} \otimes \text{INT}$.

The status of A as a strict monoidal category with functorial abstractors is directly expressed by eight basic equational axioms, which we now give. Here (and later) we imagine arities to be ascribed in any way which respects the arity rules, and which gives the same arity to each side of an equation:

BASIC AXIOMS	
$a \cdot \mathbf{id} = a = \mathbf{id} \cdot a$	$a \cdot (b \cdot c) = (a \cdot b) \cdot c$
$a \otimes \mathbf{id}_\epsilon = a = \mathbf{id}_\epsilon \otimes a$	$a \otimes (b \otimes c) = (a \otimes b) \otimes c$
$\mathbf{id} \otimes \mathbf{id} = \mathbf{id}$	$(a \cdot b) \otimes (c \cdot d) = (a \otimes c) \cdot (b \otimes d)$
$\mathbf{ab}_x \mathbf{id} = \mathbf{id}$	$\mathbf{ab}_x(a \cdot b) = (\mathbf{ab}_x a) \cdot (\mathbf{ab}_x b)$.

Finally the reaction relation $\searrow_A$, which will often be written just $\searrow$, is a preorder on each $A_{m,n}$; so $a \searrow a'$ implies that a and a' have the same source and target arities. Reaction is also preserved by every operation; that is, $a \searrow a'$ implies $b \otimes a \searrow b \otimes a'$, $\mathbf{ab}_x a \searrow \mathbf{ab}_x a'$, etc. We stipulate that if $\mathbf{id} \searrow a$ then $a = \mathbf{id}$.

2.3 Intuition We think of tensor product ($\otimes$) as parallel composition. Reactions within a and within b can occur independently within $a \otimes b$. But there may be reactions $a \otimes b \searrow c$ which do not arise from a alone or from b alone; these represent communication or interaction between a and b. In particular, a and b may "use" the same name x, and the manner of use may constitute communication via x as a port.

Abstraction $\mathbf{ab}_x$ allows parametrization upon x. In particular, when x is a means of interaction as just described, then abstraction upon x allows one to vary the interface x, i.e. to vary the partners with which communication via x may take place.

Composition $a \cdot b$ does not represent *sequential* composition. Given $a : m \rightarrow n$ and $b : n \rightarrow p$, the arity n indicates the extent to which b is parametric; to this extent, information may flow from a to b at any time during the continuing reaction within each of them or between them. (It is sometimes better to think of an action as an *activity*, since it may enjoy an unbounded number of reactions.)

The reader is invited to consider these discursive comments briefly at first, and later more thoroughly after considering the action structure for the π-calculus.

3 An action structure for the π-calculus

In this section we define an action structure PIC for the π-calculus. It is studied more fully, with some alternatives, in [9]; here we shall use it as an exemplar and motivation for our later definition of action calculi.

3.1 Particles PIC formalises two principal features of the π-calculus; the passage of names through ports which are also names, and the localisation of names by restriction. Thus the elementary constituents of an action in PIC are the *particles* π, given by

$$\pi ::= x(\vec{y}) \mid \overline{x}\langle\vec{y}\rangle \mid \nu x .$$

The first two kinds are *input* and *output* particles. In the input particle $x(\vec{y})$ the vector $\vec{y}$ consists of distinct names, which are binding occurrences. The third kind is a *restriction particle*, and its name occurrence is also binding.

3.2 Actions The essential part of an *action* of PIC is just a collection of particles. It would be a multiset, were it not for the binding discipline. We wish to allow an action to contain a sequence such as

$$\ldots\, x(y)\, \overline{y}\langle z\rangle\, \ldots$$

representing the receipt of a name y followed by its use as a port. Thus the first occurrence of y binds the second. Adopting the convention that the scope of a binding extends to the right, we therefore declare the *body* S of an action to be a *partial sequence*

$$[\,\pi_1 \ldots \pi_n\,]$$

of particles; that is, a sequence in which we allow the commutation $\pi\pi' = \pi'\pi$ of any neighbouring pair neither of which binds a name occurring in the other. The general form of an action $a : m \rightarrow n$ is then

$$a = (\vec{x})S\langle\vec{y}\rangle ,$$

where

- $\vec{x}$ is an m-vector of distinct names, the *imported* names;
- S is a partial sequence of particles, the *body*;
- $\vec{y}$ is an n-vector of names, not necessarily distinct, the *exported* names.

The imported names are bound. The scope of each bound name extends to the right of its binding occurrence, and includes the exported names. We identify actions which only differ by alpha-conversion (change of bound names). We tacitly assume in what follows that by suitable alpha-conversion, a name bound in a action is distinct from any other name in that action or in any other action under discussion.

As examples, here are five important kinds of action; in fact, it turns out that all others are expressible in terms of these:

$$\begin{aligned}
\mathbf{out}_x : m \to 0 &\stackrel{\text{def}}{=} (\vec{y})[\,\overline{x}\langle\vec{y}\rangle\,]\langle\,\rangle && (|\vec{y}| = m)\\
\mathbf{in}_x : 0 \to m &\stackrel{\text{def}}{=} (\,)[\,x(\vec{y})\,]\langle\vec{y}\rangle && (|\vec{y}| = m)\\
\langle x\rangle : 0 \to 1 &\stackrel{\text{def}}{=} (\,)[\;]\langle x\rangle\\
\nu : 0 \to 1 &\stackrel{\text{def}}{=} (\,)[\,\nu x\,]\langle x\rangle\\
\omega : 1 \to 0 &\stackrel{\text{def}}{=} (x)[\;]\langle\,\rangle\,.
\end{aligned}$$

We shall allow ourselves to omit any empty component of an action; so we may write $(\vec{y})[\,\overline{x}\langle\vec{y}\rangle\,]$ for $(\vec{y})[\,\overline{x}\langle\vec{y}\rangle\,]\langle\,\rangle$ or $(x)\langle x\rangle$ for $(x)[\;]\langle x\rangle$.

3.3 Operations First, the identities are given by

$$\mathbf{id}_m \stackrel{\text{def}}{=} (\vec{x})\langle\vec{x}\rangle \qquad (|\vec{x}| = m)\,.$$

To define the other operations we use substitutions like $\sigma = \{\vec{u}/\vec{v}\}$, where $\vec{v}$ is a vector of distinct names; $\sigma(-)$ is the result of simultaneously replacing each free occurrence of u_i in $(-)$ by the corresponding v_i, first alpha-converting $(-)$ to change any bound uses of $\vec{v}$. We shall use $^\frown$ to concatenate partial sequences. Suppose $a = (\vec{u})S\langle\vec{v}\rangle$ and $b = (\vec{x})T\langle\vec{y}\rangle$; then composition, product and abstraction are given as follows:

$$\begin{aligned}
a \cdot b &\stackrel{\text{def}}{=} (\vec{u})S^\frown\sigma T\langle\sigma\vec{y}\rangle && (\sigma = \{\vec{v}/\vec{x}\})\\
a \otimes b &\stackrel{\text{def}}{=} (\vec{u}\vec{x})S^\frown T\langle\vec{v}\vec{y}\rangle\\
\mathbf{ab}_x a &\stackrel{\text{def}}{=} (x\vec{u})S\langle x\vec{v}\rangle\,.
\end{aligned}$$

Recall our convention that a bound name differs from all names used outside its scope. Thus $a \otimes b$ is simply the juxtaposition of the two actions; no order is dictated by the concatenation of the bodies, since the convention ensures that $S^\frown T = T^\frown S$ in this case. But in $a \cdot b$ the substitution σ may replace some names free in T by names bound in S, and then $S^\frown\sigma T \neq \sigma T^\frown S$.

It is a routine matter to verify the action structure axioms. Note in particular that the export of x in $\mathbf{ab}_x a$ ensures that $\mathbf{ab}_x$ is functorial. We shall also use the following non-functorial form of abstraction:

$$(x)a \stackrel{\text{def}}{=} (x\vec{u})S\langle\vec{v}\rangle \qquad (\text{where } a = (\vec{u})S\langle\vec{v}\rangle,\ x \notin \vec{u})\,.$$

3.4 Dynamics The atomic reaction relation $\searrow^1$ of PIC is defined as follows. Whenever a contains a subsequence like $\overline{u}\langle\vec{v}\rangle\, u(\vec{w})$, i.e. a takes the form

$$(\vec{x})\, S^\frown[\,\overline{u}\langle\vec{v}\rangle\, u(\vec{w})\,]^\frown T\,\langle\vec{y}\rangle$$

(after suitable commutations), then

$$a \searrow^1 (\vec{x})\, S^\frown \sigma T\,\langle\sigma\vec{y}\rangle,$$

where σ is the substitution $\{\vec{v}/\vec{w}\}$. We call $\overline{u}\langle\vec{v}\rangle\, u(\vec{w})$ a *redex* of a.

Example Let $a = (x)[\,\overline{u}\langle x\rangle\, u(w)\, \overline{w}\langle y\rangle\, x(z)\,]\langle wz\rangle$. Then

$$\begin{array}{lll} a & \searrow^1 & (x)[\,\overline{x}\langle y\rangle\, x(z)\,]\langle xz\rangle \\ & \searrow^1 & (x)\langle xy\rangle\ . \end{array}$$

■

Note that the port of a redex (in the example, first u then x) may be free or bound.

One can check that $\searrow^1$ is indeed preserved by the operations; for example $a \searrow^1 a'$ implies $b \cdot a \searrow^1 b \cdot a'$.

The reaction relation $\searrow$ is defined to be $(\searrow^1)^*$, the transitive reflexive closure of the atomic relation.

3.5 Discussion PIC has considerable expressive power; for example it encodes the linear λ-calculus very naturally, mimicking β-reduction by reaction. The encoding is along the lines of [6]. But PIC needs extension to become a practically useful calculus of processes. This extension is done in [9], along uniform lines set out in [7]. But in that approach the resulting process calculus is not itself an action structure. By contrast, the extensions to PIC in Section 5 below given essentially the same expressive power, while remaining entirely within the action structure framework.

4 Controls and action calculi

In this section we see how to generalise PIC to a class of action structures which we shall call *action calculi*. The key to this generalisation is to observe that each particle of PIC is a special case of a richer construction which we shall call a *molecule*. This notion of molecule is not far from that of Berry and Boudol [1]. The molecules of any action calculus are formed from generators which we shall call *controls*.

4.1 Control constants Each particle π of PIC has some free and some binding occurrences of names. In 3.2 we used the symbol ν and the indexed symbols $\mathbf{in}_x, \mathbf{out}_x$ for actions containing a single particle (of each of the three kinds). But this notation is somewhat ad hoc; it conceals a regularity which will lead us to a generalization of PIC, and indeed to a generalization of the π-calculus.

Consider the action $\langle\vec{y}\rangle \cdot \mathbf{out}_x$, with arity $0 \to 0$. In our partial sequence notation, it is the singleton $[\, x\langle\vec{y}\rangle\,]$. But there is no formal reason to treat the index x differently from the name-vector $\vec{y}$. We reveal the generality of particle formation more clearly if we simply declare that there are (for PIC) three *control constants*

$$\begin{aligned} \mathbf{in} &: 1 \to m \\ \mathbf{out} &: 1{+}m \to 0 \\ \nu &: 0 \to 1\,, \end{aligned}$$

and that for any control constant $C : m \to n$ there are PIC particles of the form

$$\langle\vec{y}\rangle C(\vec{x}) \qquad (|\vec{y}| = m,\ |\vec{x}| = n)$$

where the names $\vec{x}$ are distinct and binding. Thus our three forms of PIC particle $x(\vec{y})$, $\overline{x}\langle\vec{y}\rangle$ and νx are more truly written

$$\langle x\rangle \mathbf{in}(\vec{y})\ ,\ \langle x\vec{y}\rangle \mathbf{out}(\,)\ ,\ \langle\,\rangle \nu(x)\ .$$

Moreover, let us define

$$\begin{aligned} \mathbf{out}_x : m \to 0 &\stackrel{\text{def}}{=} (\langle x\rangle \otimes \mathbf{id}_m) \cdot \mathbf{out} \\ \mathbf{in}_x : 0 \to m &\stackrel{\text{def}}{=} \langle x\rangle \cdot \mathbf{in}\ ; \end{aligned}$$

then the atomic reaction relation for PIC given in 3.4 can equally be defined by the rule

$$\mathbf{out}_x \otimes \mathbf{in}_x \searrow^{1} \mathbf{id}_m\ .$$

In a similar way we may define a wide variety of action structures, each one determined by a given set $\mathcal{K}$ of control constants and a set of reaction rules which determine their meaning. We shall call each action structure built in this way an *action calculus* and denote it by AC($\mathcal{K}$), understanding that with $\mathcal{K}$ the reaction rules are also given. Thus PIC=AC$\{\nu, \mathbf{in}, \mathbf{out}\}$ is just a special case of this uniform construction.

This class of action calculi is still too restrictive; they all lack an important control feature. In these calculi, if $a \searrow a'$ is a possible reaction, then it may occur in any context; there is no way to delay it until some other activity is complete, or to make it conditional on the outcome of that activity. We now seek to remove this weakness.

4.2 Controls We wish to generalise the notion of control constant to allow activity of subactions to be constrained, modified or delayed. An example of delay is the simple guarding construction $a.P$ of CCS; P cannot act until a has happened. Sequential composition $P;Q$ in CSP is another example; Q cannot act until P has finished acting. So it is a natural step to allow action constructors C which take actions as arguments; we shall refer to these as *control constructors*, or more briefly as *controls*.

For each control C and suitable actions $a_1, \ldots, a_r$ there is a *control construction*

$$C(a_1, \ldots, a_r)$$

which is also an action. Each C is equipped with an arity rule of the form

$$\frac{a_1 : m_1 \to n_1 \quad \cdots \quad a_r : m_r \to n_r}{C(a_1, \ldots, a_r) : m \to n} \; (\chi)$$

where the side-condition χ may constrain the values of r, m_i, n_i, m, n. For example χ may determine that r is fixed, and then we say that C has *rank* r; a control of rank 0 is just a control constant, and we then omit the brackets in the construction $C(\,)$.

We shall look at one or two examples of controls in Section 5.

4.3 Action calculi: statics We are now ready to define the statics of the *action calculus* AC($\mathcal{K}$), where $\mathcal{K}$ is a given collection of controls with associated arity rules.

First, if $C\vec{a}$ is a control construction of arity $m \to n$, then

$$\langle\vec{x}\rangle\, C\vec{a}\,(\vec{y}) \qquad (|\vec{x}| = m, |\vec{y}| = n)$$

is a *molecule*. The names $\vec{y}$ must be distinct, and are *binding* occurrences; they provide the means by which the action $C\vec{a}$ can export data, just as the names $\vec{x}$ are data imported to the action.

The rest is just as in PIC, but with molecules μ instead of particles π. An action $a : m \to n$ now takes the form

$$(\vec{x})\,[\,\mu_1 \ldots \mu_k\,]\,\langle\vec{y}\rangle \qquad (|\vec{x}| = m, |\vec{y}| = n)\,;$$

the body is again a partial sequence, i.e. we allow the commutation $\mu\mu' = \mu'\mu$ of a neighbouring pair of molecules provided that neither of them binds a name occurring in the other. The imported names $\vec{x}$ are bound. The scope of each bound name extends to the right of its binding occurrence, to the end of the smallest action containing it. Thus if a is an action argument within a molecule, the scope of any name bound within a is confined to a; however, any name free in a is also free in the molecule. As in PIC, alpha conversion is allowed.

4.4 Action calculi: dynamics As we have already suggested, the atomic reaction relation in an action calculus may be given by rules such as $\mathbf{out}_x \otimes \mathbf{in}_x \searrow^1 \mathbf{id}_m$, which we shall call *control rules*. We understand that $\searrow^1$ is the smallest relation which satisfies the control rules and is closed under product, composition and abstraction.

Arbitrary complex control mechanisms may be introduced into action calculi; by merely defining the notion of action calculus we unfortunately do not succeed in isolating a distinguished family of controls! What we can claim is that the action calculus framework saves us much repetitious work in setting up a calculus for interaction; we can also hope that it allows us more readily to compare and classify such calculi, on the basis of the complexity of their controls and associated rules.

5 Examples of action calculi

In this section we first summarise the constituents of PIC, which is just AC($\boldsymbol{\nu}$, **in**, **out**). This is the *essential* π-calculus, as it contains no more than the basic controls for restriction and name-passing. Then we enumerate a sequence of π-calculi, PIC($\mathcal{K}$), gained by adjoining extra controls $\mathcal{K}$ to PIC. We consider three controls: *boxing* (which may also be called *guarding*), *choice* and *replication*. These all come into π-calculus as defined in [10, 8], but we do not try to recover exactly their "standard" forms; our new framework presents control in a new light, and slight variations of meaning arise if we seek the simplest reaction rules.

We conclude the section by outlining an action calculus of Petri nets. Unless otherwise stated, we assume that the arities are natural numbers.

5.1 Basic π-calculus: PIC = AC(ν, in, out)

Controls :	ν, **in**, **out** (rank 0)
Arity rules :	$\nu : 0 \to 1$ $\mathbf{in} : 1 \to m \qquad \mathbf{out} : 1+m \to 0$
Derived controls :	$\mathbf{out}_x : m \to 0 \stackrel{\text{def}}{=} (\langle x \rangle \otimes \mathbf{id}_m) \cdot \mathbf{out}$ $\mathbf{in}_x : 0 \to m \stackrel{\text{def}}{=} \langle x \rangle \cdot \mathbf{in}$
Control rule :	$\mathbf{out}_x \otimes \mathbf{in}_x \searrow^1 \mathbf{id}_m$

One might have expected the control rule to be stated in the form

$$\langle \vec{y} \rangle \cdot \mathbf{out}_x \otimes \mathbf{in}_x \searrow^1 \langle \vec{y} \rangle \ ;$$

but in fact these two rules generate the same atomic reaction relation, because of the closure conditions.

5.2 π-calculus with boxing: PIC(box)

Control :	**box** (rank 1)
Arity rule :	$\dfrac{a : m \to n}{\mathbf{box}\, a : 1 \to n}$
Derived control :	$\mathbf{box}_x a : 0 \to n \stackrel{\text{def}}{=} \langle x \rangle \cdot \mathbf{box}\, a$
Control rule :	$\mathbf{out}_x \otimes \mathbf{box}_x a \searrow^1 a$

It is clear that $\mathbf{in} : 0 \to m$ is redundant in the presence of **box**, as it is essentially $\mathbf{box}\,\mathbf{id}_m$.

This calculus correspond to allowing *input* guards, but not *output* guards, in the π-calculus. From the work of Honda and Tokoro [4] we know that output guarding can in fact be defined in terms of input guarding. Written in the style of the reduction rules for π-calculus in [7], Section 2.4, the rule would appear as follows:

$$\overline{x}\langle\vec{z}\rangle \mid x(\vec{y}).P \longrightarrow \{\vec{z}/\vec{y}\}P\ .$$

5.3 π-calculus with boxing and choice: PIC(box, choose)

Control :	**choose**
Arity rule :	$\dfrac{a_1 : 0 \to n \qquad \cdots \qquad a_r : 0 \to n}{\mathbf{choose}(a_1, \ldots, a_r) : 0 \to n}$
Control rule :	$\mathbf{out}_x \otimes \mathbf{choose}(\ldots, \mathbf{box}_x a, \ldots) \searrow^1 a$

This rule, written in the style of the π-calculus COMM rule ([8], 2.4) would appear as

$$\overline{x}\langle\vec{z}\rangle \mid (\cdots + x(\vec{y}).P + \cdots) \longrightarrow \{\vec{z}/\vec{y}\}P\ ;$$

thus it corresponds to a special case of the COMM rule, allowing choice and guarding for inputs but not for outputs. The reader should have no difficulty in defining controls to match the general case of COMM.

5.4 π-calculus with replication: PIC(rep)

Control :	**rep** (rank 1)
Arity rule :	$\dfrac{a : m \to 0}{\mathbf{rep}\ a : 1 \to 0}$
Derived control :	$\mathbf{rep}_x a : 0 \to 0 \stackrel{\text{def}}{=} \langle x \rangle \cdot \mathbf{rep}\ a$
Control rule :	$\mathbf{out}_x \otimes \mathbf{rep}_x a \searrow^1 a \otimes \mathbf{rep}_x a$

This form of replication differs from that described in [8], which was written $!P$. The replication of $\mathbf{rep}_x a$ is controlled by an interaction at x, as is clear from its control rule. An even clearer (equivalent) form of the rule is

$$\langle\vec{z}\rangle \cdot \mathbf{out}_x \otimes \mathbf{rep}_x a \searrow^1 \langle\vec{z}\rangle \cdot a \otimes \mathbf{rep}_x a\ ,$$

which shows clearly that a copy of the replicated action is "spun off" with parameters $\vec{z}$ received via x. The rule corresponds to a special case of reduction as defined in [8], namely

$$\overline{x}\langle\vec{z}\rangle \mid\ !x(\vec{y}).P \longrightarrow \{\vec{z}/\vec{y}\}P \mid\ !x(\vec{y}).P\ .$$

Our use of a control rule contrasts with the way replication is treated in [8]; there, a structural congruence law $!P \equiv P \mid !P$ is imposed, so that (without any special reduction rule) $!P$ is treated as arbitrarily many copies of P already coexisting. Such a law would break our present convention that controls are given meaning only by their reactions, not by axioms. There is also a related reason to prefer the present treatment – namely that it corresponds more closely to what would happen in a machine. Since an action calculus aims to be a rather concrete model of reaction, this correspondence is appropriate.

These extensions of PIC can of course be combined into PIC(**box**, **choose**, **rep**); for practical purposes this appears to be as expressive as any existing version of the calculus.

5.5 An action calculus for Petri nets: NETC = AC(ν, mark, pre, post)

We shall present, as an action calculus NETC, one of the action structures for Petri nets which were outlined in [7]. It is essentially an algebra – with dynamics – of place-transition nets (PT nets) which are parametric in both their places and their transitions. This parametrization allows two places or transitions in different nets, or even in the same net, to be coalesced with one another.

The monoidal structure of Petri nets was first exposed by Meseguer and Montanari [5]; that idea is essential to what follows. What is extra here is the power brought to the treatment by the composition and parametrization of action structures.

To begin with, we declare that the name-set X is partitioned into *place-names* $p, q, \ldots$ and *transition-names* $t, u, \ldots$. To reflect this duality properly we require the monoid of arities to consist of *pairs* of natural numbers, so that the unit is 0,0; a place-name p has arity $1,0$ while a transition-name t has arity $0,1$. The corresponding data actions have arities as follows:

$$\langle p \rangle : 0,0 \rightarrow 1,0 \qquad \langle t \rangle : 0,0 \rightarrow 0,1\,.$$

The controls of NETC are very simple; apart from restriction they correspond to marking a place, and declaring a place to be either a pre-condition or a post-condition of a transition. We now give the constituents of NETC, and then discuss them.

Controls :	ν, **mark**, **pre**, **post** (rank 0)
Arity rules :	$\nu : 0,0 \rightarrow 1,0 \quad \nu : 0,0 \rightarrow 0,1 \quad \mathbf{mark} : 1,0 \rightarrow 0,0$ $\mathbf{pre} : 1,1 \rightarrow 0,0 \quad \mathbf{post} : 1,1 \rightarrow 0,0$
Derived controls :	$\mathbf{mark}_{\vec{p}} \stackrel{\text{def}}{=} \langle p_1 \rangle \cdot \mathbf{mark} \otimes \cdots \otimes \langle p_k \rangle \cdot \mathbf{mark} \quad (\lvert\vec{p}\rvert = k)$ $\mathbf{pre}^t_{\vec{p}} \stackrel{\text{def}}{=} \langle p_1 t \rangle \cdot \mathbf{pre} \otimes \cdots \otimes \langle p_k t \rangle \cdot \mathbf{pre} \quad (\lvert\vec{p}\rvert = k)$ $\mathbf{post}^t_{\vec{q}} \stackrel{\text{def}}{=} \langle q_1 t \rangle \cdot \mathbf{post} \otimes \cdots \otimes \langle q_k t \rangle \cdot \mathbf{post} \quad (\lvert\vec{q}\rvert = k)$ $\mathbf{trans}_{\vec{p};\vec{q}} \stackrel{\text{def}}{=} (\nu t)(\mathbf{pre}^t_{\vec{p}} \otimes \mathbf{post}^t_{\vec{q}})$
Control rule :	$\mathbf{mark}_{\vec{p}} \otimes \mathbf{trans}_{\vec{p};\vec{q}} \searrow^1 \mathbf{trans}_{\vec{p};\vec{q}} \otimes \mathbf{mark}_{\vec{q}}$

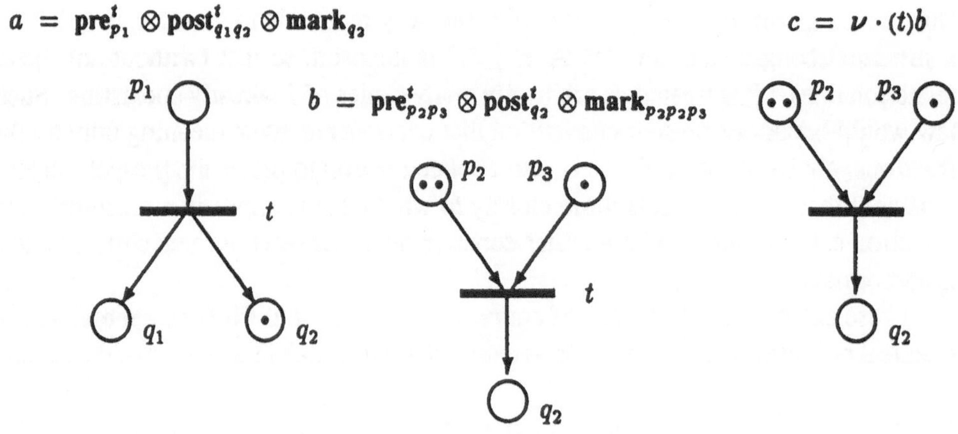

Figure 1: Three PT nets and their expression as actions

We shall first consider actions of arity $0, 0 \to 0, 0$; they are just PT nets in which some or all of the places and transitions bear names, each name occurring at most once in a net. An action has one particle for each arc of the net which it represents, and one for each token. Figure 1 shows three examples. The expressions for the first two nets, a and b, should be clear on inspection of the definition of derived controls. Note particularly that *every* place and transition is named in these two nets.

Now compare b with the third net c, whose transition is not named. The name has been hidden by composing restriction, ν, with an abstraction.

In passing, we note that abstraction may also be used for renaming. Thus the net $b' = \langle p_2 \rangle \cdot (p_3)b$ (not shown) is like b except that place p_3 is coalesced with place p_2; thus in b' the place p_2 is triply marked, and is a precondition of weight 2 to transition t.

Returning to the distinction between b and c, we observe that

$$c = \mathbf{mark}_{p_2} \otimes \mathbf{mark}_{p_2 p_3} \otimes \mathbf{trans}_{p_2 p_3 ; q_2} \cdot$$

Thus, according to the control rule, c has a reaction as follows:

$$c \searrow^1 \mathbf{mark}_{p_2 q_2} \otimes \mathbf{trans}_{p_2 p_3 ; q_2} ;$$

a token has been removed from each of p_2, p_3 and one placed on q_2, exactly as is dictated by the conventional net firing rule. On the other hand no reaction is possible for b. The reason is to do with the algebraic net constructions. In particular, the rule for forming a product requires like-named places and transitions to be coalesced. Thus a net must be seen only as a *partial* description of the pre- and post-conditions on its named transitions, since a coalescence can add further conditions. It is only when the name of a transition is restricted (i.e. removed from the net-diagram) that we can take the description to be *complete*, and thus know when it may fire. This is emphasized by the two products $a \otimes b$ and $a \otimes c$, which are illustrated in Figure 2. In the first case two transitions have been coalesced – and the resulting single transition cannot fire; but in the second case the firing condition on c's unnamed transition remains satisfied.

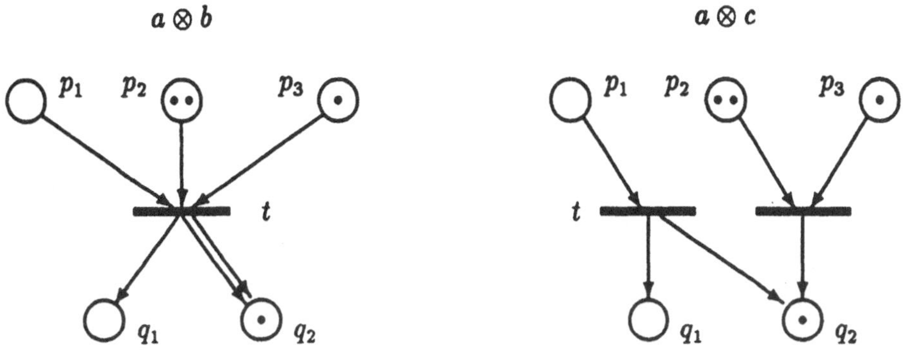

Figure 2: The products $a \otimes b$ and $a \otimes c$ of PT nets

Space prevents us from discussing the constructions of this action structure further, and we refer to [7] for more details. But several points arise even from what has been said. First, it seems clear that the notion of restriction ν is of general significance, not merely confined to the π-calculus. Second, it seems worth pursuing the theory of Petri nets enriched by parametrization; the algebraic treatment which this allows appears remarkably natural, and can be based on just four very basic control constructions. Third, it is worth considering how to combine the basic ideas of net theory with the dynamic reconfiguration suggested by the π-calculus, within the single framework of action calculi. The form of the control rule in NETC ensures that a net remains unchanged after the firing of a transition, as is standard in net theory; but in the context of action calculi it is easy to propose variants of the rule which allow some reconfiguration to occur.

Finally, it is striking that only control *constants* (controls with no action parameters) are needed to define nets. Yet, informally speaking, nets appear to lack no power in representing computational phenomena. This suggests that careful study is needed to determine in what sense parametric controls add expressive power to action calculi.

6 Action calculi as free constructions

We shall now give an alternative characterisation for the statics of action calculi, which reveals clearly their algebraic status.

The characterisation is this: For each set $\mathcal{K}$ of controls, the action calculus $\mathsf{AC}(K)$ is isomorphic to the quotient of a term algebra by a certain congruence. The terms are generated from certain constants, by the basic action structure constructions and the control constructions; the congruence is induced by the basic action structure axioms together with further axioms. The latter, which we shall call the *concrete* axioms, do not vary from one action calculus to another since they do not mention the controls $\mathcal{K}$; the latter are truly *free* constructors, and their meaning lies entirely in their dynamics.

Throughout this section we continue to assume a fixed name-set X, and to assume that the arities are the natural numbers, with $x : 1$ for each $x \in X$. We also assume a fixed but arbitrary control set $\mathcal{K}$.

6.1 Action terms We shall use $t, u, \ldots$ to stand for *action terms*; the set $\mathcal{T}(\mathcal{K})$ of action terms is the smallest closed under the following term constructions:

$\omega : 1 \rightarrow 0$	*discard*
$\langle x \rangle : 0 \rightarrow 1$	*datum* ($x \in X$)
$C\vec{t}$	control constructions ($C \in \mathcal{K}$)
$\mathbf{id}_k,\ t \cdot u,\ t \otimes u,\ \mathbf{ab}_x t$	basic constructions.

The constructions are subject to rules of arity.

We find it convenient to define the following form of abstraction, which is closer to λ-abstraction:

$$(x)t \stackrel{\text{def}}{=} \mathbf{ab}_x t \cdot (\omega \otimes \mathbf{id}_n) \qquad (t : m \rightarrow n).$$

Note that the postcomponent $\omega \otimes \mathbf{id}_n$ has arity $1+n \rightarrow n$; intuitively, it just discards its first datum. It will turn out that, in the presence of our axioms, $\mathbf{ab}_x$ is expressible in terms of (x); the axiomatization is a little smoother in terms of the latter.

6.2 Concrete axioms The concrete axioms are six in number. The first four characterize abstraction as a name-binding operation. Two of these correspond closely to the α and β rules of λ-calculus; the other two, γ and δ, give something like the η rule of λ-calculus. To state these axioms we must first define when a name x occurs *free* in a term. This is standard, once we declare for the basic constants that x is free in $\langle y \rangle$ iff $x = y$, and no name is free in ω or in $\mathbf{id}_k$. Then the substitution operation $\{y/x\}$, which changes bound names to avoid clashes, is also standard.

$$\begin{array}{lll}
\alpha: & (x)t = (y)\{y/x\}t & (y \text{ not free in } t)\\
\beta: & (\langle y \rangle \otimes \mathbf{id}_m) \cdot (x)t = \{y/x\}t & (t : m \rightarrow n)\\
\gamma: & (x)t = \omega \otimes t & (x \text{ not free in } t)\\
\delta: & (x)(t \otimes \mathbf{id}_m) = (x)t \otimes \mathbf{id}_m\,. &
\end{array}$$

We now need two further abbreviations:

$$\begin{array}{rcl}
\langle x_1 \cdots x_k \rangle & \stackrel{\text{def}}{=} & \langle x_1 \rangle \otimes \cdots \otimes \langle x_k \rangle\\
(x_1 \cdots x_k)t & \stackrel{\text{def}}{=} & (x_1)\cdots(x_k)t\,.
\end{array}$$

Using these, we define a family of permutation actions:

$$\mathbf{i}_{km} \stackrel{\text{def}}{=} (\vec{x}\vec{y})\langle \vec{y}\vec{x} \rangle \qquad (|\vec{x}| = k, |\vec{y}| = m)\,.$$

Note that $\mathbf{i}_{km} : (k \otimes m) \rightarrow (m \otimes k)$ is an isomorphism of arities; from the axioms one can show that $\mathbf{i}_{mk}$ is indeed its inverse. In fact, under composition the $\mathbf{i}_{km}$ generate all and only the action structure isomorphisms.

Of the two remaining axioms, ϵ says that the trivial permutation $\mathbf{i}_{10}$, which is just $(x)\langle x \rangle$, is the unit of arity 1; ζ says that permutation of arity agrees with permutation of the product of actions:

$$\begin{array}{ll}
\epsilon: & \mathbf{i}_{10} = \mathbf{id}_1\\
\zeta: & \mathbf{i}_{km} \cdot (u \otimes t) = (t \otimes u) \cdot \mathbf{i}_{ln} \quad (t : k \rightarrow l,\ u : m \rightarrow n)\,.
\end{array}$$

To give a feeling for the power of these axioms, together with the basic action structure axioms, we list a few of their consequences; the proofs are routine.

6.3 Lemma From the action structure axioms on terms, together with the concrete axioms, it follows that

$$\begin{array}{ll} (x)(t \cdot u) = (x)t \cdot u & (x \text{ not free in } u) \\ (x)(t \cdot u) = (\mathbf{id}_1 \otimes t) \cdot (x)u & (x \text{ not free in } t) \\ (x)(t \otimes u) = (x)t \otimes u & (x \text{ not free in } u) \\ (x)(t \otimes u) = t \otimes (x)u & (t : 0 \to n,\ x \text{ not free in } t) \\ \mathbf{i}_{k0} = \mathbf{i}_{0k} = \mathbf{id}_k & \\ \mathbf{ab}_x t = (x)(\langle x \rangle \otimes t)\,. & \end{array}$$

■

The last equation justifies our claim that $\mathbf{ab}_x$ and (x) are interdefinable.

Clearly $\mathcal{T}(\mathcal{K})/\equiv$ is an action structure, where $\equiv$ is the congruence induced by the basic and concrete axioms. Now recall that in [7] a (standard) notion of homomorphism and isomorphism was defined for action structures. With the help of the lemma we get the following significant result:

6.4 Theorem $\mathcal{T}(\mathcal{K})/\equiv$ is isomorphic to $\mathsf{AC}(\mathcal{K})$. ■

Such a result is not surprising; we would expect $\mathsf{AC}(\mathcal{K})$ to be characterized by *some* set of equational axioms. Nonetheless the axioms α–ζ have some intuitive appeal.

We have been somewhat informal in our notation. Strictly, we should have adopted different notations for the constructions of $\mathcal{T}(\mathcal{K})$, such as ω, $\langle x \rangle$ and $(x)(-)$, from those already used in $\mathsf{AC}(\mathcal{K})$. But in using notation ambiguously we have merely anticipated the isomorphism which is assured by the theorem. The isomorphism also assures us that the partial sequence notation of Section 4 can be mixed with the term language with impunity; for example if the names $\vec{x}, \vec{y}$ do not occur free in $\vec{a}$ then

$$(\vec{x})\,[\,\langle \vec{x} \rangle\, C\vec{a}\,(\vec{y})\,]\,\langle \vec{y} \rangle \;=\; C\vec{a}\,,$$

where the left-hand form is in partial sequence notation, and the right-hand term is in $\mathcal{T}(\mathcal{K})$.

6.5 Concrete action structures Let us look more closely at the concrete axioms. We have used them just to impose a congruence upon the terms $t, u, \ldots$ of $\mathcal{T}(\mathcal{K})$. But let us now regard them as axioms which may or may not be satisfied by the actions $a, b, \ldots$ of an arbitrary action structure:

CONCRETE AXIOMS		
$\alpha:$	$(x)t = (y)\{y/x\}a$	$(y \text{ not free in } a)$
$\beta:$	$(\langle y \rangle \otimes \mathbf{id}_m) \cdot (x)a = \{y/x\}a$	$(a : m \to n)$
$\gamma:$	$(x)a = \omega \otimes a$	$(x \text{ not free in } a)$
$\delta:$	$(x)(a \otimes \mathbf{id}_m) = (x)a \otimes \mathbf{id}_m$	
$\epsilon:$	$\mathbf{i}_{10} = \mathbf{id}_1$	
$\zeta:$	$\mathbf{i}_{km} \cdot (b \otimes a) = (a \otimes b) \cdot \mathbf{i}_{ln}$	$(a : k \to l,\ b : m \to n)\,.$

But we are now dealing with a richer algebraic variety than action structures, since the axioms are phrased in terms of the predicate "free in" $\subseteq X \times A$ and the renaming operation $\{^y/_x\} : A \to A$.

A variety can indeed be defined with such additional operations, which are subject to some natural laws. Let us call this variety the *concrete action structures*. Then it turns out that our action calculus AC($\mathcal{K}$) is indeed initial in the appropriate category of concrete action structures over $\mathcal{K}$; that is, action calculi are the *free* concrete action structures.

7 Future work

A first line for further investigation is to compare the way an action calculus is extended to a larger one, by adding controls, with the approach taken in [7] where it is shown how to define a process calculus (which is *not* an action calculus) on top of an arbitrary action structure. The latter approach aimed specifically at recovering the semantic congruence of bisimilarity between processes in a uniform manner. To do so, it exploited two notions which appear to be natural refinements of an action structure – the notions of *effect structure* and *incident set*. It is not yet clear how to treat bisimilarity, with or without these notions, in an action calculus; this is a vitally important task, since it is essential to divide out by a larger congruence in order to treat action calculi semantically.

More generally, the way is now open to attempt a classification of concrete models of concurrency, in terms of the controls which characterize them as action calculi. An understanding of the relative power of various control mechanisms is notably absent.

Finally, the class of concrete action structures mentioned at the end of Section 6 deserves further study.

References

[1] Berry, G. and Boudol, G., *The chemical abstract machine*, Journal of Theoretical Computer Science, Vol 96, pp217–248, 1992.

[2] Banâtre, J.P. and Métayer, D., *The GAMMA model and its discipline of programming*, Science of Computer Programming, Vol 15, pp55–77, 1990.

[3] Hoare, C.A.R., *Communicating Sequential Processes*, Communications of ACM, Vol 21, pp666–677, 1978.

[4] Honda, K. and Tokoro, M., *An object calculus for asynchronous communication*, Proc. European Conference on object-oriented programming, Lecture Notes in Computer Science, Vol 512, Springer, pp133–147, 1991.

[5] Meseguer, J. and Montanari, U., *Petri nets are monoids*, Journal of Information and Computation, Vol 88, pp105–155, 1990.

[6] Milner, R., *Functions as processes*, Math. Struct. in Comp. Science, Vol 2, pp119–141, 1992.

[7] Milner, R., *Action structures*, Research Report LFCS–92–249, Laboratory for Foundations of Computer Science, Computer Science Department, Edinburgh University, 1992.

[8] Milner, R., *The polyadic π-calculus: a tutorial*, Research Report ECS–LFCS–91–180, Laboratory for Foundations of Computer Science, Computer Science Department, Edinburgh University, 1991. (To appear in the Proceedings of the International Summer School on Logic and Algebra of Specification, Marktoberdorf, August 1991.)

[9] Milner, R., *Action structures for the π-calculus*, Research Report ECS–LFCS–93–264, Laboratory for Foundations of Computer Science, Computer Science Department, Edinburgh University, 1992.

[10] Milner, R., Parrow, J. and Walker D., *A calculus of mobile processes, Parts I and II*, Journal of Information and Computation, Vol 100, pp1–40 and pp41–77, 1992.

[11] Petri, C.A., Fundamentals of a theory of asynchronous information flow, Proc. IFIP Congress '62, North Holland, pp386–390, 1962.

Observable Properties of Higher Order Functions that Dynamically Create Local Names, or: What's *new*?

Andrew M. Pitts* and Ian D. B. Stark**

University of Cambridge Computer Laboratory,
Pembroke Street, Cambridge CB2 3QG, England

Abstract. The research reported in this paper is concerned with the problem of reasoning about properties of higher order functions involving state. It is motivated by the desire to identify what, if any, are the difficulties created purely by *locality* of state, independent of other properties such as side-effects, exceptional termination and non-termination due to recursion. We consider a simple language (equivalent to a fragment of Standard ML) of typed, higher order functions that can dynamically create fresh *names*; names are created with local scope, can be tested for equality and can be passed around via function application, but that is all. Despite the extreme simplicity of the language and its operational semantics, the observable properties of such functions are shown to be very subtle. A notion of 'logical relation' is introduced which incorporates a version of *representation independence* for local names. We show how to use it to establish observational equivalences. The method is shown to be complete (and decidable) for expressions of first order types, but incomplete at higher types.

1 Introduction

Programming languages combining higher order features with the manipulation of local state present severe problems for the traditional techniques of programming language semantics and logics of programs. For denotational semantics, the problems manifest themselves as a lack of abstraction in existing semantic models: some expressions that are observationally equivalent (i.e. that can be interchanged in any program without affecting its behaviour when executed) are assigned different denotations in the model. For operational semantics, the problems manifest themselves partly in the fact that simple techniques for analyzing observational equivalence in the case of purely functional languages (such as Milner's 'Context Lemma' [8], or more generally, notions of applicative bisimulation [1]) break down in the presence of state-based features. Furthermore, operationally based approaches to properties of programs are often inconveniently intensional, e.g. the familiar congruence properties of equational logic fail to hold. (See [6, Sect. 5(A)], for example.) These problems have been intensively studied for the case of local variables in block-structured, Algol-like languages and to a lesser extent for the case of languages

* Supported by UK SERC grant GR/G53279 and CEC ESPRIT project CLICS-II
** Supported by UK SERC studentship 91307943 and CEC SCIENCE project PL910296

involving the dynamic creation of mutable locations (such as ML-style *references*). See [17, 2, 7, 3, 18, 12, 13, 6, 4]. Our interest in this subject stems primarily from a desire to improve and deepen the techniques which are available for reasoning about program behaviour in the 'impure' functional language Standard ML [9].

Our motivation here is to try to identify what, if any, are the difficulties created purely by *locality* of state, independent of other properties such as side-effects, exceptional termination and non-termination due to recursion. Accordingly we consider higher order functions which can dynamically create fresh names of things, but which ignore completely what kind of thing (references, exceptions, etc.) is being named. Names are created with local scope, can be tested for equality, and are passed around via function application, but that is all. Because of this limited framework, there is some hope of obtaining *definitive* results—fully abstract models and complete proof techniques. As the vehicle for this study we formulate an extension of the call-by-value, simply typed lambda calculus, called the *nu-calculus* and introduced in Sect. 2. In ML terms, it contains higher order functions over ground types `bool` and `unit ref`—the latter being the type of dynamically created references to the unique element of type `unit`. This acts as a type of 'names' because only one thing can be (and is) stored in such a reference, so that its only characteristic is its name. We have purposely excluded recursion from the nu-calculus and as a result any closed expression evaluates to an essentially unique canonical form. Indeed, the nu-calculus appears at first sight to be an extremely simple system. On closer inspection, we find that nu-calculus expressions can exhibit very subtle behaviour with respect to an appropriate notion of observational equivalence. Thus our first contribution is somewhat in the spirit of Meyer and Seiber [7]: we observe that even for this extremely simple case of local state there are observationally equivalent expressions which traditional denotational techniques will fail to identify (Example 4).

In Sect. 3 we introduce a notion of 'logical relation' for the nu-calculus incorporating a version of *representation independence* for local names. Our technique is a syntactic version of the relationally parametric semantics of O'Hearn and Tennent [13]. There are also interesting similarities with Plotkin and Abadi's parametricity schema for existential types [16, Theorem 7]. We use our version of logical relations to establish the termination properties of the nu-calculus (Theorem 12) and to provide a useful notion of 'applicative' equivalence between nu-calculus expressions which implies observational equivalence (Theorem 14). Although the two notions of equivalence differ at higher order types (Example 6), they coincide for expressions of first order types (Theorem 22) and are decidable there (Corollary 23). The proof of this occupies Sect. 4 and is surprisingly hard work: although applicative equivalence provides a compositional explanation of (observational equivalence classes of) first order functions, even these can have complicated behaviour (see Example 1).

Note. This paper is an expanded version of the operationally-based results announced in [14]. That reference also contains an outline of our approach to the denotational semantics of the nu-calculus.

2 The nu-calculus

Syntactically, the nu-calculus is a kind of simply typed lambda calculus. The types, σ, are built up from a ground type o of *booleans* and a ground type ν of *names*, by forming *function types*, $\sigma \rightarrow \sigma'$. Expressions take the form

$M ::= x$	variable
$\mid\; n$	name
$\mid\; true \;\mid\; false$	truth values
$\mid\; \mathit{if}\, M \;\mathit{then}\; M \;\mathit{else}\; M$	conditional
$\mid\; M = M$	equality of names
$\mid\; \nu n \,.\, M$	local name declaration
$\mid\; \lambda x : \sigma \,.\, M$	function abstraction
$\mid\; MM$	function application

where $x \in \mathit{Var}$, an infinite set whose elements are called *variables*, and $n \in \mathit{Nme}$, an infinite set (disjoint from Var) whose elements are called *names*. Function abstraction is a variable-binding construct (occurrences of x in M are bound in $\lambda x : \sigma \,.\, M$), whereas local name declaration is a name-binding construct (occurrences of n in M are bound in $\nu n \,.\, M$). We write $\mathit{Var}(M)$ and $\mathit{Nme}(M)$ for the finite subsets of Var and Nme consisting of the free variables and the free names in an expression M. We denote by $M[M'/x]$ (respectively $M[M'/n]$) the result of substituting an expression M' for all free occurrences of x (respectively n) in M, renaming bound variables and bound names if necessary, to avoid variable and name capture.

Note. Henceforward, we implicitly identify expressions that differ up to α-conversion of bound variables and bound names. Thus when we refer to an expression M we really mean an α-equivalence class of expressions, referred to via one of its representatives M.

Expressions will be assigned types via typing assertions of the form

$$s, \Gamma \vdash M : \sigma$$

where s is a finite subset of Nme, Γ is a finite function from variables to types, σ is a type, and M is an expression (more precisely, an α-equivalence class of expressions) satisfying $\mathit{Nme}(M) \subseteq s$ and $\mathit{Var}(M) \subseteq \mathit{dom}(\Gamma)$ (the domain of definition of Γ). The rules generating the valid typing assertions are given in Table 1. In these rules $s \oplus \{n\}$ indicates the finite set of names obtained from s by adjoining $n \notin s$; and $\Gamma \oplus [x : \sigma]$ denotes the finite function obtained by extending Γ by mapping $x \notin \mathit{dom}(\Gamma)$ to σ. Clearly, if $s, \Gamma \vdash M : \sigma$ holds, then σ is uniquely determined by s, Γ and M. We write

$$\mathrm{Exp}_\sigma(s) \stackrel{\mathit{def}}{=} \{M \mid s, \emptyset \vdash M : \sigma\}$$

for the set of *closed nu-calculus expression of type σ with free names in the set s*. The subset

$$\mathrm{Can}_\sigma(s) \subseteq \mathrm{Exp}_\sigma(s)$$

Table 1. Rules for assigning types in the nu-calculus

$$\frac{}{s,\Gamma \vdash x : \Gamma(x)}\ (x \in dom(\Gamma)) \qquad \frac{}{s,\Gamma \vdash n : \nu}\ (n \in s) \qquad \frac{}{s,\Gamma \vdash b : o}\ (b = true, false)$$

$$\frac{s,\Gamma \vdash B : o \quad s,\Gamma \vdash M : \sigma \quad s,\Gamma \vdash M' : \sigma}{s,\Gamma \vdash if\ B\ then\ M\ else\ M' : \sigma} \qquad \frac{s,\Gamma \vdash N : \nu \quad s,\Gamma \vdash N' : \nu}{s,\Gamma \vdash (N = N') : o}$$

$$\frac{s \oplus \{n\},\Gamma \vdash M : \sigma}{s,\Gamma \vdash \nu n\,.\,M : \sigma} \qquad \frac{s,\Gamma \oplus [x : \sigma] \vdash M : \sigma'}{s,\Gamma \vdash \lambda x : \sigma\,.\,M : \sigma \to \sigma'} \qquad \frac{s,\Gamma \vdash F : \sigma \to \sigma' \quad s,\Gamma \vdash M : \sigma}{s,\Gamma \vdash FM : \sigma'}$$

of *canonical* nu-calculus expressions of type σ with free names in the set s consists of those closed expressions which are either names (in s), or the booleans constants *true* and *false*, or function abstractions.

We give the operational semantics of the nu-calculus in terms of an inductively defined evaluation relation which matches the computational behaviour of equivalent ML expressions. The ML equivalent of the expression $\nu n\,.\,M$ is

```
let n=ref() in M end
```

(using the ML type `unit ref` for the type of names). In other words the effect of evaluating $\nu n\,.\,M$ should be to create a fresh name n and then use it in evaluating M. In the definition of ML [9] environments are used to bind identifiers (variables) to addresses (names), whereas here we have chosen to simplify the form of the evaluation relation by using 'extended' expressions containing names explicitly. It would be possible to simplify the syntax of the nu-calculus even further by identifying the syntactic category of names with that of variables of type ν. We choose not to do so because names and variables have different semantic properties. For example, the operational semantics we give commutes with arbitrary substitutions on variables, but only with restricted forms of substitutions on names (viz. essentially just permutations of names): see Remark 2.

An appropriate notion of state for this simple language is just a finite subset of *Nme*, indicating the names which have been created so far. So we will use an evaluation relation of the form

$$s \vdash M \Downarrow_\sigma (s')C \tag{1}$$

where s and s' are *disjoint* finite sets of names, $M \in \mathrm{Exp}_\sigma(s)$ and $C \in \mathrm{Can}_\sigma(s \oplus s')$.

Note. Throughout this paper, we write $s \oplus s'$ to indicate the union of two sets s and s' that are *disjoint*.

The intended meaning of (1) is: 'in state s, expression M evaluates to canonical form C creating fresh, local names s' in the process'. The rules for generating the relation are given in Table 2. In rule (EQ) we use the notation $\delta_{nn'}$, where

$$\delta_{nn'} \stackrel{def}{=} \begin{cases} true \text{ if } n = n' \\ false \text{ if } n \neq n' \end{cases}.$$

It is important to note that the rules in Table 2 refer to the collection of judgements as in (1) that are well-formed, i.e. satisfy the conditions mentioned above. For example, in rule (LOCAL) the well-formedness of the hypothesis and the conclusion entail that n is not an element of either s or s_1.

Table 2. Rules for evaluating nu-calculus expressions

(CAN)
$$\frac{}{s \vdash C \Downarrow_\sigma C}$$

(COND1)
$$\frac{s \vdash B \Downarrow_o (s_1)\mathit{true} \quad s \oplus s_1 \vdash M \Downarrow_\sigma (s_2)C}{s \vdash \mathit{if}\ B\ \mathit{then}\ M\ \mathit{else}\ M' \Downarrow_\sigma (s_1 \oplus s_2)C}$$

(COND2)
$$\frac{s \vdash B \Downarrow_o (s_1)\mathit{false} \quad s \oplus s_1 \vdash M' \Downarrow_\sigma (s_2)C'}{s \vdash \mathit{if}\ B\ \mathit{then}\ M\ \mathit{else}\ M' \Downarrow_\sigma (s_1 \oplus s_2)C'}$$

(EQ)
$$\frac{s \vdash N \Downarrow_\nu (s_1)n \quad s \oplus s_1 \vdash N' \Downarrow_\nu (s_2)n'}{s \vdash (N = N') \Downarrow_o (s_1 \oplus s_2)\delta_{nn'}}$$

(LOCAL)
$$\frac{s \oplus \{n\} \vdash M \Downarrow_\sigma (s_1)C}{s \vdash \nu n \,.\, M \Downarrow_\sigma (\{n\} \oplus s_1)C}$$

(APP)
$$\frac{s \vdash F \Downarrow_{\sigma \to \sigma'} (s_1)\lambda x : \sigma \,.\, M' \quad s \oplus s_1 \vdash M \Downarrow_\sigma (s_2)C \quad s \oplus s_1 \oplus s_2 \vdash M'[C/x] \Downarrow_{\sigma'} (s_3)C'}{s \vdash FM \Downarrow_{\sigma'} (s_1 \oplus s_2 \oplus s_3)C'}$$

It is easy to see that evaluation is deterministic up to renaming created names, in the following sense:

Lemma 1. *If* $s \vdash M \Downarrow_\sigma (s_1)C$ *and* $s \vdash M \Downarrow_\sigma (s_2)C'$, *then there is a bijection* $R : s_1 \leftrightarrow s_2$ *so that* C' *is* α*-convertible with the expression* $C[n'/n \mid (n, n') \in R]$.

Remark 2 (States are affine linear). The initial state s in the evaluation (1) has the structural properties of an *affine linear logic context*, in the sense that derived rules of weakening and exchange are valid, but a rule of contraction is not. (Compare the use made of affine linear logic by O'Hearn in [11].) Thus

$$\text{(WEAK)}\quad \frac{s \vdash M \Downarrow_\sigma (s_1)C}{s \oplus \{n\} \vdash M \Downarrow_\sigma (s_1)C}$$

$$\text{(EXCH)}\quad \frac{s \oplus \{n\} \oplus \{n'\} \vdash M \Downarrow_\sigma (s_1)C}{s \oplus \{n'\} \oplus \{n\} \vdash M \Downarrow_\sigma (s_1)C}$$

are correct derived rules (the second trivially so, because we are using states that are sets rather than lists), but

$$\text{(CONTR)}\quad \frac{s \oplus \{n\} \oplus \{n'\} \vdash M \Downarrow_\sigma (s_1)C}{s \oplus \{n''\} \vdash M[n''/n, n''/n'] \Downarrow_\sigma (s_1)C[n''/n, n''/n']}$$

is not a correct derived rule — as can be seen, for example, by taking s and s_1 to be $\emptyset$, σ to be o, M to be $n = n'$ and C to be *false*.

More generally, given a function $f : s \to s'$ and letting $M[f]$ denote the substituted expression $M[f(n)/n \mid n \in s]$, we have that the rule (SUBST) below is a correct derived rule *provided that f is an injective function.*

$$\text{(SUBST)}\quad \frac{s \vdash M \Downarrow_\sigma (s_1)C}{s' \vdash M[f] \Downarrow_\sigma (s_1)C[f]}$$

Remark 3 (Sequentiality condition). The evaluation rules in Table 2 follow the state convention of Standard ML [9, p. 50], i.e. order of evaluation is from left to right, with state accumulating sequentially. We have formulated the operational semantics of the nu-calculus in this way to emphasize that it is (equivalent to) a fragment of ML. However, because we are dealing with state that can be created but cannot be mutated, some of this sequentiality is spurious. Table 3 gives 'de-sequentialized' versions of rules (COND1), (COND2), (EQ), and (APP). We claim that using these rules instead of the corresponding rules in Table 2 does not affect the collection of instances of evaluation that are derivable. This claim follows from the fact that a converse of the weakening rule (WEAK) is derivable:

$$\text{(STREN)}\quad \frac{s \oplus s' \vdash M \Downarrow_\sigma (s_1)C}{s \vdash M \Downarrow_\sigma (s_1)C}\ (\mathit{Nme}(M) \subseteq s \text{ and } \mathit{Nme}(C) \subseteq s \oplus s_1)\quad .$$

The evaluation relation (1) can be used to define a Morris-style contextual equivalence between nu-calculus expressions: two expressions are equivalent if they can be interchanged in any program without affecting the observable result of evaluating it. Here we will take a 'program' to be a closed expression of type o, and the possible observable results of evaluating a program to be the booleans *true* and *false*, disregarding any local names that are created in the process of evaluation. (It would not change the notion of observational equivalence given below if we also allowed programs to be of type ν and observable results to include pre-existing names.) In the following definition, as usual the 'context' $B[-]$ is an expression in which some subexpressions have been replaced by a place-holder, $-$; and then $B[M]$ denotes the result of filling the place-holder with an expression M.

Table 3. 'De-sequentialized' evaluation rules

(COND1') $$\frac{s \vdash B \Downarrow_o (s_1)\mathit{true} \quad s \vdash M \Downarrow_\sigma (s_2)C}{s \vdash \mathit{if}\ B\ \mathit{then}\ M\ \mathit{else}\ M' \Downarrow_\sigma (s_1 \oplus s_2)C}$$

(COND2') $$\frac{s \vdash B \Downarrow_o (s_1)\mathit{false} \quad s \vdash M' \Downarrow_\sigma (s_2)C'}{s \vdash \mathit{if}\ B\ \mathit{then}\ M\ \mathit{else}\ M' \Downarrow_\sigma (s_1 \oplus s_2)C'}$$

(EQ') $$\frac{s \vdash N \Downarrow_\nu (s_1)n \quad s \vdash N' \Downarrow_\nu (s_2)n'}{s \vdash (N = N') \Downarrow_o (s_1 \oplus s_2)\delta_{nn'}}$$

(APP') $$\frac{\begin{array}{c} s \vdash F \Downarrow_{\sigma \to \sigma'} (s_1)\lambda x : \sigma \,.\, M' \qquad s \vdash M \Downarrow_\sigma (s_2)C \\ s \oplus s_1 \oplus s_2 \vdash M'[C/x] \Downarrow_{\sigma'} (s_3)C' \end{array}}{s \vdash FM \Downarrow_{\sigma'} (s_1 \oplus s_2 \oplus s_3)C'}$$

Definition 4 (Observational equivalence). Given $M_1, M_2 \in \mathrm{Exp}_\sigma(s)$, we write

$$s \vdash M_1 \approx_\sigma M_2$$

to mean that for all $B[-]$ and all $b \in \{\mathit{true}, \mathit{false}\}$,

$$\exists s_1(s \vdash B[M_1] \Downarrow_o (s_1)b) \Leftrightarrow \exists s_2(s \vdash B[M_2] \Downarrow_o (s_2)b) \quad .$$

In this case we say that M_1 and M_2 are *observationally equivalent.*

The following result shows that one need only consider contexts that immediately evaluate their arguments in order to establish observational equivalence. It is the analogue of Theorem (**ciu**) in [4].

Lemma 5. *$s \vdash M_1 \approx_\sigma M_2$ if and only if for all $b \in \{\mathit{true}, \mathit{false}\}$ and all $\lambda x : \sigma \,.\, B \in \mathrm{Can}_{\sigma \to o}(s)$*

$$\exists s_1(s \vdash (\lambda x : \sigma \,.\, B)M_1 \Downarrow_o (s_1)b) \Leftrightarrow \exists s_2(s \vdash (\lambda x : \sigma \,.\, B)M_2 \Downarrow_o (s_2)b) \quad .$$

The following instances of observational equivalence are easily established using the lemma.

Corollary 6. *1. If $M \in \mathrm{Exp}_\sigma(s)$ and $n \notin s$, then $s \vdash \nu n \,.\, M \approx_\sigma M$.*

2. If $M \in \mathrm{Exp}_\sigma(s \oplus \{n\} \oplus \{n'\})$, then $s \vdash \nu n \,.\, \nu n' \,.\, M \approx_\sigma \nu n' \,.\, \nu n \,.\, M$.

3. If $s \vdash M \Downarrow_\sigma (s')C$, then $s \vdash M \approx_\sigma \nu s' \,.\, C$. Here $\nu s' \,.\, C$ stands for $\nu n_1 \ldots \nu n_k \,.\, C$ if $s' = \{n_1, \ldots, n_k\}$ for some $k > 0$, and stands for C if $s' = \emptyset$. (By part 2, up to observational equivalence, it does not matter which order we enumerate the elements of s' in $\nu s' \,.\, C$.)

4. If $s, [x : \sigma] \vdash M : \sigma'$ and $C \in \mathrm{Can}_\sigma(s)$, then $s \vdash (\lambda x : \sigma \,.\, M)C \approx_{\sigma'} M[C/x]$.

In the next section we will show that evaluation of nu-calculus expressions always terminates (Theorem 12). It follows from this and the above corollary that, up to observational equivalence, the only closed expressions of type o are *true* and *false* and the only closed expression of type ν not involving any free names is

$$new \stackrel{def}{=} \nu n \,.\, n \quad .$$

However, at higher types things become more complicated. The following example gives infinitely many expressions of type $\nu \rightarrow \nu$ which are mutually observationally inequivalent.

Example 1. For each $p \geq 1$, consider the nu-calculus expression of type $\nu \rightarrow \nu$ which first creates $p+1$ local names $n_0, \ldots, n_p$ and then acts as the function cyclically permuting these names and mapping any other name to n_0:

$$\begin{aligned} F_p \stackrel{def}{=} \nu n_0 \ldots \nu n_p \,.\, \lambda x : \nu \,.\, & \mathit{if}\; x = n_0 \;\mathit{then}\; n_1 \;\mathit{else} \\ & \mathit{if}\; x = n_1 \;\mathit{then}\; n_2 \;\mathit{else} \\ & \quad \ldots \\ & \quad \mathit{if}\; x = n_p \;\mathit{then}\; n_0 \;\mathit{else}\; n_0 \quad . \end{aligned}$$

Then $\emptyset \vdash F_p \not\approx_{\nu \rightarrow \nu} F_{p'}$ whenever $p \neq p'$, because

$$B_q \stackrel{def}{=} \lambda f : \nu \rightarrow \nu \,.\, \nu n \,.\, (f^{(q+2)}(n) = f(n))$$

has the property that for all $q \in \{1, \ldots, p\}$, $\emptyset \vdash B_q F_p \Downarrow_o (\{n_0, \ldots, n_p, n\})\mathit{true}$ if and only if $q = p$. (In B_q, $f^{(q+2)}$ indicates f iterated $q+2$ times.)

Example 2. Here is a simple example to illustrate the fact that local name declaration and function abstraction in general do not commute up to observational equivalence. The expressions

$$M \stackrel{def}{=} \nu n \,.\, \lambda x : \nu \,.\, n \quad \text{and} \quad N \stackrel{def}{=} \lambda x : \nu \,.\, \nu n \,.\, n$$

are not observationally equivalent, because $B \stackrel{def}{=} \lambda f : \nu \rightarrow \nu \,.\, (f\mathit{new} = f\mathit{new})$ has the property that $\emptyset \vdash BM \Downarrow_o (\{n, n_1, n_2\})\mathit{true}$ whereas $\emptyset \vdash BN \Downarrow_o (\{n, n_1, n_2\})\mathit{false}$.

Example 3. The rule (APP) in Table 2 embodies a form of strict, or 'call-by-value', application. Part 4 of Corollary 6 shows that the appropriate restricted form of beta-conversion (Plotkin's β_v [15]) holds up to observational equivalence. Although there is no non-termination in our simple language, the general form of beta-conversion fails for the nu-calculus, because of the dynamics of name creation. For example, the beta redex $(\lambda x : \nu \,.\, x = x)\mathit{new}$ is not observationally equivalent to the corresponding reduct $\mathit{new} = \mathit{new}$ since

$$\emptyset \vdash (\lambda x : \nu \,.\, x = x)\mathit{new} \Downarrow_o (\{n_1\})\mathit{true}$$
$$\emptyset \vdash (\mathit{new} = \mathit{new}) \Downarrow_o (\{n_1, n_2\})\mathit{false} \quad .$$

For the simple functional language PCF, Milner's context lemma [8] shows that observational equivalence may be established by testing just with ***applicative contexts***, i.e. those of the form $[-]C_1 C_2 \ldots C_k$. Not surprisingly, this fails in the nu-calculus. For example, the expressions F_p in Example 1 are in fact indistinguishable by such applicative contexts, even though they can be distinguished by more complicated contexts (like $B_q([-])$) which carry out 'anonymous' manipulation of the private names $n_0, \ldots, n_p$. It would seem that the properties of higher order functions which create and pass around private names can be quite subtle. Two contrasting examples of observational equivalence, more subtle than those in Corollary 6, are given below. The first one illustrates the fact that local names are always distinct from externally supplied names; the second illustrates the fact that any two local names are indiscernible by externally supplied boolean tests. (This second equivalence is quite delicate—it certainly would not hold in languages where evaluation of functions can have side-effects on mutable state.) The methods developed in the next section suffice to prove (2), but not (3). At the moment, the only method known to us for establishing this second equivalence is denotational, i.e. via a specific model of the nu-calculus: see [14, Sect. 4].

Example 4.

$$\emptyset \vdash \nu n \,.\, \lambda x : \nu \,.\, (x = n) \approx_{\nu \to o} \lambda x : \nu \,.\, \mathit{false} \tag{2}$$

$$\emptyset \vdash \nu n \,.\, \nu n' \,.\, \lambda f : \nu \to o \,.\, (fn = fn') \approx_{(\nu \to o) \to o} \lambda f : \nu \to o \,.\, \mathit{true} \;. \tag{3}$$

In (3), the boolean equality test $fn = fn'$ is an abbreviation for

$$\mathit{if}\ fn\ \mathit{then}\ (\mathit{if}\ fn'\ \mathit{then}\ \mathit{true}\ \mathit{else}\ \mathit{false})\ \mathit{else}\ (\mathit{if}\ fn'\ \mathit{then}\ \mathit{false}\ \mathit{else}\ \mathit{true}) \;.$$

3 Representation independence for local names

This section develops a notion of (binary) logical relation for the nu-calculus and shows how to use it to establish instances of observational equivalence between nu-calculus expressions.

Given finite subsets $s_1, s_2 \subseteq \mathit{Nme}$ of names, we write $R : s_1 \rightleftharpoons s_2$ to indicate that R is (the graph of) a *partial bijection* from s_1 to s_2. In other words, $R \subseteq s_1 \times s_2$ satisfies

$$m_1 \, R \, m_2 \wedge n_1 \, R \, n_2 \Rightarrow (m_1 = n_1 \Leftrightarrow m_2 = n_2) \;. \tag{4}$$

(We use infix notation for binary relations.) Note that $R \oplus R'$ is a partial bijection $s_1 \oplus s_1' \rightleftharpoons s_2 \oplus s_2'$ when $R : s_1 \rightleftharpoons s_2$ and $R' : s_1' \rightleftharpoons s_2'$ are disjoint partial bijections. The *identity* partial bijection, $I_s : s \rightleftharpoons s$, is given by:

$$n_1 \, I_s \, n_2 \Leftrightarrow n_1 = n_2 \;. \tag{5}$$

The domain and codomain of definition of a partial bijection $R : s_1 \rightleftharpoons s_2$ will be denoted

$$\mathit{dom}(R) \stackrel{def}{=} \{n_1 \in s_1 \mid \exists n_2 \in s_2 \,.\, n_1 \, R \, n_2\} \tag{6}$$

$$\mathit{cod}(R) \stackrel{def}{=} \{n_2 \in s_2 \mid \exists n_1 \in s_1 \,.\, n_1 \, R \, n_2\} \;. \tag{7}$$

Thus R is a bijection just in case $dom(R) = s_1$ and $cod(R) = s_2$, in which case we write $R : s_1 \leftrightarrow s_2$.

Definition 7 (Logical relations). For each type σ we define a family of binary relations between canonical expressions

$$(R_\sigma \subseteq \mathrm{Can}_\sigma(s_1) \times \mathrm{Can}_\sigma(s_2) \mid R : s_1 \rightleftharpoons s_2)$$

by induction on the structure of σ as in (9), (10) and (11) below; clause (11) makes use of associated relations between expressions, $\overline{R}_\sigma \subseteq \mathrm{Exp}_\sigma(s_1) \times \mathrm{Exp}_\sigma(s_2)$ defined by (8).

$$\begin{aligned} M_1 \; \overline{R}_\sigma \; M_2 \Leftrightarrow \exists R' : s'_1 \rightleftharpoons s'_2, C_1 \in \mathrm{Can}_\sigma(s_1 \oplus s'_1), C_2 \in \mathrm{Can}_\sigma(s_2 \oplus s'_2) \; . \\ s_1 \vdash M_1 \Downarrow_\sigma (s'_1)C_1 \wedge s_2 \vdash M_2 \Downarrow_\sigma (s'_2)C_2 \wedge C_1 \; (R \oplus R')_\sigma \; C_2 \end{aligned} \tag{8}$$

$$b_1 \; R_o \; b_2 \Leftrightarrow b_1 = b_2 \tag{9}$$

$$n_1 \; R_\nu \; n_2 \Leftrightarrow n_1 \; R \; n_2 \tag{10}$$

$$\begin{aligned} &\lambda x : \sigma \, . \, M_1 \; R_{\sigma \to \sigma'} \; \lambda x : \sigma \, . \, M_2 \Leftrightarrow \\ &\qquad \forall R' : s'_1 \rightleftharpoons s'_2, C_1 \in \mathrm{Can}_\sigma(s_1 \oplus s'_1), C_2 \in \mathrm{Can}_\sigma(s_2 \oplus s'_2) \; . \\ &\qquad C_1 \; (R \oplus R')_\sigma \; C_2 \Rightarrow M_1[C_1/x] \; (\overline{R \oplus R'})_{\sigma'} \; M_2[C_2/x] \; . \end{aligned} \tag{11}$$

(It is implicit in (8) and (11) that each s'_i is required to be disjoint from s_i.)

The family $(\overline{R}_\sigma \mid \sigma)$ is a form of binary 'logical relation' for nu-calculus expressions. Since we choose in (9) to take the logical relation to be the identity at the ground type o, the whole family is determined by what we take at the other ground type ν. We wish related expressions to be mapped to related expressions by any nu-calculus function, and we have to impose the restriction (4) on the relation R to ensure this property holds for the function testing equality of names. The following proposition expresses this fundamental property of our notion of logical relation.

Proposition 8 (Fundamental property of logical relations). *Suppose*

$$[x_1 : \sigma_1, \ldots, x_k : \sigma_k] \vdash M : \sigma \quad .$$

Then for all $R : s_1 \rightleftharpoons s_2$, $C_i \in \mathrm{Can}_{\sigma_i}(s_1)$ and $D_i \in \mathrm{Can}_{\sigma_i}(s_2)$ $(i = 1, \ldots, k)$, one has

$$\left(\bigwedge_{i=1}^{k} C_i \; R_{\sigma_i} \; D_i \right) \Rightarrow M[C_1/x_1, \ldots, C_k/x_k] \; \overline{R}_\sigma \; M[D_1/x_1, \ldots, D_k/x_k] \quad .$$

Proof. The proof proceeds by induction on the derivation of the typing assertion $[x_1 : \sigma_1, \ldots, x_k : \sigma_k] \vdash M : \sigma$, and makes use of (the only if part of) the following lemma, which is itself proved by induction on the structure of the type σ. We omit the details. □

Lemma 9. *Given $R : s_1 \rightleftharpoons s_2$ and $R' : s'_1 \rightleftharpoons s'_2$ with s_i and s'_i disjoint (for $i = 1,2$), then for all types σ and all canonical expressions $C_i \in \mathrm{Can}_\sigma(s_i)$ ($i = 1,2$), $C_1\ R_\sigma\ C_2$ if and only if $C_1\ (R \oplus R')_\sigma\ C_2$.*

Similarly, for all $M_i \in \mathrm{Exp}_\sigma(s_i)$, $M_1\ \overline{R}_\sigma\ M_2$ if and only if $M_1\ (\overline{R \oplus R'})_\sigma\ M_2$.

Remark. The main interest in Definition 7 lies in clause (8) where the relation $\overline{R}_\sigma$ on expressions is defined in terms of the relation R_σ on canonical expressions. This clause embodies a form of 'representation independence' for the dynamically created local names. (Cf. Plotkin and Abadi's parametricity schema for existential types [16, Theorem 7].) One might have expected to see not (8), but rather

$$\begin{aligned} M_1\ \overline{R}_\sigma\ M_2 \Leftrightarrow\ & (\forall s'_1, C_1 \in \mathrm{Can}_\sigma(s_1 \oplus s'_1)\ .\ s_1 \vdash M_1 \Downarrow_\sigma (s'_1)C_1 \Rightarrow \\ & \quad \exists s'_2, R' : s'_1 \rightleftharpoons s'_2, C_2 \in \mathrm{Can}_\sigma(s_2 \oplus s'_2)\ . \\ & \quad\quad s_2 \vdash M_2 \Downarrow_\sigma (s'_2)C_2 \wedge C_1\ (R \oplus R')_\sigma\ C_2) \\ \wedge\ & \\ & (\forall s'_2, C_2 \in \mathrm{Can}_\sigma(s_2 \oplus s'_2)\ .\ s_2 \vdash M_2 \Downarrow_\sigma (s'_2)C_2 \Rightarrow \\ & \quad \exists s'_1, R' : s'_1 \rightleftharpoons s'_2, C_1 \in \mathrm{Can}_\sigma(s_1 \oplus s'_1)\ . \\ & \quad\quad s_1 \vdash M_1 \Downarrow_\sigma (s'_1)C_1 \wedge C_1\ (R \oplus R')_\sigma\ C_2) \end{aligned} \tag{12}$$

This deals appropriately with the possibility of non-termination. However, the simple language we are considering here has the property (Theorem 12) that all expressions converge to canonical forms which are essentially unique (by Lemma 1), in which case (12) is equivalent to the simpler form (8).

Clause (11) of Definition 7 is a syntactic version of O'Hearn and Tennent's approach to relational parametricity in [13]. It also exhibits the characteristic feature of 'logical relations', in that two functions are defined to be related if they send related arguments to related results. To be more in keeping with the definition of applicative bisimulation in [1], one might consider an alternative definition in which two functions are related when they give related results for all arguments. For pure functional languages, such as the lazy lambda calculus, one expects the two approaches to be equivalent, and to equal observational equivalence: see [1, 5]. Here, the notion of 'applicative equivalence' we define below using Definition 7 is contained in, but not equal to observational equivalence; and we believe that replacing clause (11) by a 'related if related on all arguments' version (which we will not formulate precisely here) results in an even weaker notion of equivalence.

We will need to use Proposition 8 in the more general form given in the corollary below. Its statement makes use of the following notation for *renaming* expressions along the bijection $R : \mathit{dom}(R) \leftrightarrow \mathit{cod}(R)$ obtained from a partial bijection $R : s_1 \rightleftharpoons s_2$ by restricting it to its domain of definition (cf. definitions (6) and (7)).

Definition 10. Given a partial bijection $R : s_1 \rightleftharpoons s_2$, for any nu-calculus expression M, let $M[R]$ denote the result of simultaneously substituting for each name in $\mathit{dom}(R)$ the corresponding name in $\mathit{cod}(R)$:

$$M[R] \stackrel{def}{=} M[n'/n \mid n\ R\ n'] \ .$$

Corollary 11. *Suppose* $s_1, [x_1 : \sigma_1, \ldots, x_k : \sigma_k] \vdash M : \sigma$, *that* $R : s_1 \leftrightarrow s_2$ *is a bijection and that* $R' : s_1' \rightleftharpoons s_2'$ *is a partial bijection disjoint from* R. *Then for all* $C_i \in \mathrm{Can}_{\sigma_i}(s_1 \oplus s_1')$ *and* $D_i \in \mathrm{Can}_{\sigma_i}(s_2 \oplus s_2')$ $(i = 1, \ldots, k)$ *one has*

$$\left(\bigwedge_{i=1}^{k} C_i \ (R \oplus R')_{\sigma_i} \ D_i \right) \Rightarrow$$

$$M[C_1/x_1, \ldots, C_k/x_k] \ (\overline{R \oplus R'})_\sigma \ M[R][D_1/x_1, \ldots, D_k/x_k] \ .$$

Proof. Apply Proposition 8 to

$$[y_1 : \nu, \ldots, y_\ell : \nu, x_1 : \sigma_1, \ldots, x_k : \sigma_k] \vdash M[y_j/n_j \mid 1 \leq j \leq \ell] : \sigma$$

where $s = \{n_1, \ldots, n_\ell\}$. □

Theorem 12 (Termination). *For all closed expressions* M, *of type* σ *and with free names in the set* s *say, there is some set of names* s' *(disjoint from* s*) and some canonical expression* $C \in \mathrm{Can}_\sigma(s \oplus s')$ *such that* $s \vdash M \Downarrow_\sigma (s')C$.

Proof. The $k = 0$ case of Corollary 11 implies that $M \ (\overline{I_s})_\sigma \ M$ for all $M \in \mathrm{Exp}_\sigma(s)$. Termination follows from this, given the definition of $\overline{R}_\sigma$ in (8). □

We now show how the fundamental property of our notion of logical relation embodied in Proposition 8 can be used to establish observational equivalences.

Definition 13 (Applicative equivalence). We say that two expressions $M_1, M_2 \in \mathrm{Exp}_\sigma(s)$ are *applicatively equivalent* if $M_1 \ (\overline{I_s})_\sigma \ M_2$, where $I_s : s \rightleftharpoons s$ is the identity partial bijection on s defined in (5).

Theorem 14. *Applicative equivalence implies observational equivalence.*

Proof. Suppose $M_1 \ (\overline{I_s})_\sigma \ M_2$. We employ Lemma 5 to see that $s \vdash M_1 \approx_\sigma M_2$. By (8) there is some $R' : s_1' \rightleftharpoons s_2'$, and C_1, C_2 with $s \vdash M_i \Downarrow_\sigma (s_i')C_i$ $(i = 1, 2)$ and $C_1 \ (I_s \oplus R')_\sigma \ C_2$. Then for any $\lambda x : \sigma \, . \, B \in \mathrm{Can}_{\sigma \to o}(s)$, applying Corollary 11 with $R = I_s$ we get $B[C_1/x] \ (\overline{I_s \oplus R'})_o \ B[C_2/x]$. Hence by (8) again, there is some $R'' : s_1'' \rightleftharpoons s_2''$ and b_1, b_2 with $s \oplus s_i' \vdash B[C_i/x] \Downarrow_o (s_i'')b_i$ $(i = 1, 2)$ and $b_1 \ (I_s \oplus R' \oplus R'')_o \ b_2$, i.e. with $b_1 = b_2$ (by (9)). Applying the rules in Table 2, we deduce that $s \vdash (\lambda x : \sigma \, . \, B)M_i \Downarrow_o (s_i' \oplus s_i'')b_i$ with $b_1 = b_2$. Thus Lemma 5 and the deterministic nature of the evaluation relation (Lemma 1) imply that $M_1 \approx_\sigma M_2$. □

Example 5. Theorem 14 provides quite a powerful method for establishing some observational equivalences, since the relation $(\overline{I_s})_\sigma$ is much easier to deal with than $\approx_\sigma$. For example, the observational equivalence (2) can be established by this method. For with

$$C_1 \stackrel{def}{=} \lambda x : \nu \, . \, (x = n) \qquad \text{and} \qquad C_2 \stackrel{def}{=} \lambda x : \nu \, . \, \mathit{false}$$

it is not hard to see that $C_1 \ (I_\emptyset \oplus R)_{\nu \to o} \ C_2$ where $R : \{n\} \rightleftharpoons \emptyset$ is necessarily the empty partial bijection; hence $\nu n \, . \, C_1 \ (\overline{I_\emptyset})_{\nu \to o} \ C_2$, as required.

However, not every observational equivalence can be established via Theorem 14, as the following example shows. Thus *applicative equivalence is in general a strictly weaker relation than observational equivalence.* Nevertheless, as we shall see below (Theorem 22), the converse of Theorem 14 does hold when σ is a *first order type*, i.e. of the form $\sigma_k \rightarrow \sigma_{k-1} \rightarrow \cdots \rightarrow \sigma_0$ with each σ_i either ν or o.

Example 6. The pair of second order expressions in (3) are observationally equivalent (this can be established via the denotational methods sketched in [14, Sect. 4]), but they are not related by $(\overline{I_\emptyset})_{(\nu \rightarrow o) \rightarrow o}$. For the only possible partial bijection $R : \{n, n'\} \rightleftharpoons \emptyset$ is $R = \emptyset$; but $\lambda f : \nu \rightarrow o \,.\, (fn = fn')$ and $\lambda f : \nu \rightarrow o \,.\, \mathit{true}$ are not related by $(I_\emptyset \oplus R)_{(\nu \rightarrow o) \rightarrow o}$, because for the canonical expressions C_1 and C_2 defined in Example 5, $C_1 \; (I_\emptyset \oplus R)_{\nu \rightarrow o} \; C_2$, whereas it is not the case that $(fn = fn')[C_1/f] \; (\overline{I_\emptyset \oplus R})_o \; \mathit{true}[C_2/f]$.

4 Observational relations

To investigate further the relationship between observational and applicative equivalence, we introduce the following generalization of the notion of observational equivalence which we will see satisfies all the defining properties of applicative equivalence in Definition 7 except (11).

Definition 15. Given a partial bijection $R : s_1 \rightleftharpoons s_2$ and expressions $M_i \in \mathrm{Exp}_\sigma(s_i)$ $(i = 1, 2)$, we write

$$M_1 R^{\mathrm{obs}}_\sigma M_2$$

to mean that for all $\tau \in \{o, \nu\}$ and all $\lambda x : \sigma \,.\, P \in \mathrm{Can}_{\sigma \rightarrow \tau}(\mathit{dom}(R))$

$$(\lambda x : \sigma \,.\, P) M_1 \; \overline{R}_\tau \; (\lambda x : \sigma \,.\, P[R]) M_2 \quad .$$

In this case we say that M_1 and M_2 are *observationally R-related.* Note that because τ is a ground type, the relation $\overline{R}_\tau$, defined using (8), (9) and (10), takes a particularly simple form:

- For all $B_i \in \mathrm{Exp}_o(s_i)$, $B_1 \; \overline{R}_o \; B_2$ if and only if there is some $b \in \{\mathit{true}, \mathit{false}\}$ so that for each $i = 1, 2$ $s_i \vdash B_i \Downarrow_o (s_i')b$ for some s_i'.
- For all $N_i \in \mathrm{Exp}_\nu(s_i)$, $N_1 \; \overline{R}_\nu \; N_2$ if and only if for each $i = 1, 2$ $s_i \vdash N_i \Downarrow_\nu (s_i')n_i$ for some s_i' and some $n_i \in s_i \oplus s_i'$ satisfying

$$n_1 \; R \; n_2 \text{ or } (n_1 \in s_1' \text{ and } n_2 \in s_2') \quad .$$

The following proposition substantiates the claim that observational relations generalize the notion of observational equivalence.

Proposition 16. *Observational equivalence coincides with being observationally I_s-related. In other words, for any $M_1, M_2 \in \mathrm{Exp}_\sigma(s)$*

$$s \vdash M_1 \approx_\sigma M_2 \Leftrightarrow M_1 (I_s)^{\mathrm{obs}}_\sigma M_2 \quad .$$

Proof. Comparing Definition 15 with the characterization of observational equivalence in Lemma 5, it suffices to show that when $s \vdash M_1 \approx_\sigma M_2$ then $(\lambda x : \sigma . P)M_1 \ (\overline{I_s})_\nu \ (\lambda x : \sigma . P)M_2$, for any $\lambda x : \sigma . P \in \mathrm{Can}_{\sigma \to \nu}(s)$. Certainly $s \vdash M_1 \approx_\sigma M_2$ implies $s \vdash (\lambda x : \sigma . P)M_1 \approx_\nu (\lambda x : \sigma . P)M_2$. So in fact it suffices to show for any $N_1, N_2 \in \mathrm{Exp}_\nu(s)$ that

$$s \vdash N_1 \approx_\nu N_2 \Rightarrow N_1 \ (\overline{I_s})_\nu \ N_2 \quad . \tag{13}$$

To proof (13), first use Theorem 12 to find s_i and n_i such that $s \vdash N_i \Downarrow_\nu (s_i)n_i$. For any $n \in s$ one thus has $s \vdash (\lambda x : \nu . x = n)N_i \Downarrow_\nu (s_i)b_i$, where $b_i = \mathit{true}$ if and only if $n = n_i$. Since $s \vdash N_1 \approx_\nu N_2$, $b_1 = b_2$; hence either $n_1 = n_2 \in s$, or $n_1 \in s_1$ and $n_2 \in s_2$. Thus $N_1 \ (\overline{I_s})_\nu \ N_2$, as required. □

Lemma 17. *For any partial bijection $R : s_1 \rightleftharpoons s_2$ and any $M_i \in \mathrm{Exp}_\sigma(s_i)$ $(i = 1, 2)$*

$$M_1 \ \overline{R}_\sigma \ M_2 \Rightarrow M_1 R_\sigma^{\mathrm{obs}} M_2 \quad . \tag{14}$$

Moreover, when $\sigma \in \{o, \nu\}$ the reverse implication holds.

Proof. The implication (14) follows immediately from Corollary 11. To see that the second part of the lemma holds, note that in case $\sigma \in \{o, \nu\}$, if $M_1 R_\sigma^{\mathrm{obs}} M_2$ then in Definition 15 we can take P to be x to conclude that $(\lambda x : \sigma . x)M_1 \ \overline{R}_\sigma \ (\lambda x : \sigma . x)M_2$ and hence that $M_1 \ \overline{R}_\sigma \ M_2$ (since M_i and $(\lambda x : \sigma . x)M_i$ have the same behaviour under evaluation). □

Lemma 18. *For any $R : s_1 \rightleftharpoons s_2$, $M_i \in \mathrm{Exp}_\sigma(s_i)$ $(i = 1, 2)$ and $\lambda x : \sigma . N \in \mathrm{Can}_{\sigma \to \nu}(dom(R))$, suppose*

$$\begin{aligned} s_i \vdash M_i \Downarrow_\sigma (s_i')C_i \qquad (i = 1, 2) \\ s_1 \oplus s_1' \vdash N[C_1/x] \Downarrow_\nu (s_1'')n_1 \\ s_2 \oplus s_2' \vdash N[R][C_2/x] \Downarrow_\nu (s_2'')n_2 \quad . \end{aligned}$$

If $M_1 R_\sigma^{\mathrm{obs}} M_2$, then $n_1 \in s_1''$ if and only if $n_2 \in s_2''$.

Proof. Consider the boolean expression

$$B \stackrel{\mathit{def}}{=} (\lambda x : \sigma . N)x = (\lambda x : \sigma . N)x \quad .$$

For each $i = 1, 2$ let s_i''' be a fresh set of names in bijection with s_i'', via $R_i : s_i'' \leftrightarrow s_i'''$ say. Then

$$\begin{aligned} s_1 \vdash (\lambda x : \sigma . B)M_1 \Downarrow_o (s_1' \oplus s_1'' \oplus s_1''')b_1 \\ s_2 \vdash (\lambda x : \sigma . B[R])M_2 \Downarrow_o (s_2' \oplus s_2'' \oplus s_2''')b_2 \end{aligned}$$

where $b_i = \mathit{false}$ if and only if $n_i \neq n_i[R_i]$, i.e. if and only if $n_i \in s_i''$. If $M_1 R_\sigma^{\mathrm{obs}} M_2$ then we must have $b_1 = b_2$, from which the result follows. □

The following proposition expresses a key property of observational relations which is a precise analogue of the characteristic clause (8) in the definition of logical relation that we have been using. It shows why partial bijections between states (sets of names) play a prominent role in studying observational properties of the nu-calculus, since they can be used to explain observational equivalence (i.e. being observationally I_s-related, by Proposition 16) between general expressions in terms of observational relations between canonical expressions. The proof of the proposition is quite intricate and we give it in some detail.

Proposition 19. *For any partial bijection* $R : s_1 \rightleftharpoons s_2$ *and any* $M_i \in \mathrm{Exp}_\sigma(s_i)$ *(*$i = 1, 2$*)*

$$M_1 R^{\mathrm{obs}}_\sigma M_2 \Leftrightarrow \exists R' : s'_1 \rightleftharpoons s'_2, C_1 \in \mathrm{Can}_\sigma(s_1 \oplus s'_1), C_2 \in \mathrm{Can}_\sigma(s_2 \oplus s'_2) \,. \qquad (15)$$
$$s_1 \vdash M_1 \Downarrow_\sigma (s'_1)C_1 \wedge s_2 \vdash M_2 \Downarrow_\sigma (s'_2)C_2 \wedge C_1 (R \oplus R')^{\mathrm{obs}}_\sigma C_2 \quad .$$

Proof. Suppose that $M_1 R^{\mathrm{obs}}_\sigma M_2$. By Theorem 12, $s_i \vdash M_i \Downarrow_\sigma (s'_i)C_i$ for some $C_i \in \mathrm{Can}_\sigma(s_i \oplus s'_i)$ $(i = 1, 2)$. We begin by constructing a suitable partial bijection $R' : s'_1 \rightleftharpoons s'_2$.

Let R' consist of those pairs of names $(n, n') \in s'_1 \times s'_2$ for which there is some $\lambda x : \sigma \,.\, N \in \mathrm{Can}_{\sigma \to \nu}(dom(R))$ with

$$s_1 \oplus s'_1 \vdash (\lambda x : \sigma \,.\, N)C_1 \Downarrow_\nu (s''_1)n \qquad (16)$$
$$s_2 \oplus s'_2 \vdash (\lambda x : \sigma \,.\, N[R])C_2 \Downarrow_\nu (s''_2)n' \quad . \qquad (17)$$

To see that R' is a partial bijection, suppose $n \; R' \; n'$, witnessed by a canonical expression $\lambda x : \sigma \,.\, N$ satisfying (16) and (17), and suppose also $m \; R' \; m'$, witnessed by some $\lambda x : \sigma \,.\, M$. Applying the test $\lambda x : \sigma \,.\, (N = M) \in \mathrm{Can}_{\sigma \to o}(dom(R))$ to $M_1 R^{\mathrm{obs}}_\sigma M_2$, we have $(\lambda x : \sigma \,.\, (N = M))M_1 \; \overline{R}_o \; (\lambda x : \sigma \,.\, (N = M)[R])M_2$; from this it follows that $n = m$ if and only if $n' = m'$. Thus R' is indeed a partial bijection.

Next we show that $C_1 (R \oplus R')^{\mathrm{obs}}_\sigma C_2$. Given any $\lambda x : \sigma \,.\, P \in \mathrm{Can}_{\sigma \to \tau}(dom(R \oplus R'))$ with $\tau \in \{o, \nu\}$, we have to show that $(\lambda x : \sigma \,.\, P)C_1 \; (\overline{R \oplus R'})_\tau \; (\lambda x : \sigma \,.\, P[R \oplus R'])C_2$. Enumerate R' as $\{(n_i, n'_i) \mid 1 \le i \le k\}$ for some $k \ge 0$, and for each i let $\lambda x : \sigma \,.\, N_i \in \mathrm{Can}_{\sigma \to \nu}(dom(R))$ witness that $n_i \; R' \; n'_i$ (as in (16) and (17)). Consider

$$P' \stackrel{def}{=} (\lambda y_1 : \nu \,.\, \cdots \lambda y_k : \nu \,.\, P[y_i/n_i \mid 1 \le i \le k])N_1 \cdots N_k$$

Suppose that

$$s_1 \oplus s'_1 \vdash (\lambda x : \sigma \,.\, P)C_1 \Downarrow_\tau (s''_1)D_1 \qquad (18)$$
$$s_2 \oplus s'_2 \vdash (\lambda x : \sigma \,.\, P[R \oplus R'])C_2 \Downarrow_\tau (s''_2)D_2 \quad . \qquad (19)$$

Then by construction of P', we also have

$$s_1 \vdash (\lambda x : \sigma \,.\, P')M_1 \Downarrow_\tau (s'_1 \oplus s \oplus s''_1)D_1 \qquad (20)$$
$$s_2 \vdash (\lambda x : \sigma \,.\, P'[R])M_2 \Downarrow_\tau (s'_2 \oplus s' \oplus s''_2)D_2 \qquad (21)$$

for some s and s'. Since $\lambda x : \sigma \,.\, P' \in \mathrm{Can}_{\sigma \to \tau}(dom(R))$ and $M_1 R^{\mathrm{obs}}_\sigma M_2$, we have $(\lambda x : \sigma \,.\, P')M_1 \; \overline{R}_\tau \; (\lambda x : \sigma \,.\, P'[R])M_2$. Hence by (20) and (21),

$$D_1 \; (R \oplus S)_\tau \; D_2 \qquad (22)$$

for some $S : s_1' \oplus s \oplus s_1'' \rightleftharpoons s_2' \oplus s' \oplus s_2''$. We consider the cases $\tau = o$ and $\tau = \nu$ separately.

When $\tau = o$, (22) immediately gives $D_1 = D_2$, and hence by (18) and (19), $(\lambda x : \sigma \,.\, P)C_1 \; (\overline{R \oplus R'})_o \; (\lambda x : \sigma \,.\, P[R \oplus R'])C_2$, as required.

When $\tau = \nu$, (22) implies either $D_1 \; R \; D_2$, or $D_1 \in s_1' \oplus s \oplus s_1''$ and $D_2 \in s_2' \oplus s \oplus s_2''$. But in this second case, by Lemma 18

$$(D_1 \in s_1' \text{ and } D_2 \in s_2') \text{ or } (D_1 \in s \oplus s_1'' \text{ and } D_2 \in s' \oplus s_2'') \quad .$$

By definition of R', if $D_i \in s_i'$ $(i = 1, 2)$, then $D_1 \; R' \; D_2$. So when $\tau = \nu$ we have

$$D_1 \; R \oplus R' \; D_2 \text{ or } (D_1 \in s \oplus s_1'' \text{ and } D_2 \in s' \oplus s_2'')$$

and hence by (18) and (19), $(\lambda x : \sigma \,.\, P)C_1 \; (\overline{R \oplus R'})_\nu \; (\lambda x : \sigma \,.\, P[R \oplus R'])C_2$, as required.

This completes the proof of the implication $\Rightarrow$ in (15). The proof of the reverse implication is quite straightforward and we omit it. □

Combining Proposition 19 with Lemma 17, we have that R_σ^{obs} satisfies the defining clauses (8)–(10) of R_σ and $\overline{R}_\sigma$ in Definition 7. It cannot also satisfy clause (11) for function types, since then R_σ^{obs} and $\overline{R}_\sigma$ would coincide for all σ, and hence (by Proposition 16) observational equivalence would coincide with applicative equivalence; but by Example 6 we know that in general this is not the case. However, for function types $\sigma \to \sigma'$ with $\sigma \in \{o, \nu\}$ we can simplify clause (11) as in Proposition 21 below. To establish this proposition we need the following property of the relations $\overline{R}_\sigma$ under relabelling along a bijection; it is easily established by induction on the structure of σ, using the derived rule (SUBST) from Remark 2.

Lemma 20. *Suppose given a partial bijection $R : s_1 \rightleftharpoons s_2$, and bijections $R_1 : s_1 \leftrightarrow s_1'$ and $R_2 : s_2 \leftrightarrow s_2'$. Then for all $M_i \in \mathrm{Exp}_\sigma(s_i)$ $(i = 1, 2)$*

$$M_1 \; \overline{R}_\sigma \; M_2 \Leftrightarrow M_1[R_1] \; (\overline{R_2 \circ R \circ R_1^{-1}})_\sigma \; M_2[R_2]$$

where $R_2 \circ R \circ R_1^{-1}$ is the composed relation $\{(n_1', n_2') \mid \exists (n_1, n_2) \in R \,.\, (n_i, n_i') \in R_i (i = 1, 2)\}$.

Proposition 21. *Suppose given $R : s_1 \rightleftharpoons s_2$ and $C_i \in \mathrm{Can}_{\sigma \to \sigma'}(s_i)$ $(i = 1, 2)$.*

1. *When $\sigma = o$, $C_1 \; R_{o \to \sigma'} \; C_2$ if and only if for all $b \in \{\mathsf{true}, \mathsf{false}\}$, $C_1 b \; \overline{R}_{\sigma'} \; C_2 b$.*
2. *When $\sigma = \nu$, $C_1 \; R_{\nu \to \sigma'} \; C_2$ if and only if*
   (a) for all $(n_1, n_2) \in R$, $C_1 n_1 \; \overline{R}_{\sigma'} \; C_2 n_2$, and
   (b) $C_1 n \; (\overline{R \oplus I_{\{n\}}})_{\sigma'} \; C_2 n$
   where n is some name not in $s_1 \cup s_2$.

Proof. The 'only if' direction of each statement follows almost immediately from definition (11). For the 'if' direction, suppose given $R' : s_1 \rightleftharpoons s_2'$ and $D_i \in \mathrm{Can}_\sigma(s_i \oplus s_i')$ $(i = 1, 2)$ with

$$D_1 \; (R \oplus R')_\sigma \; D_2 \quad . \tag{23}$$

It suffices to show that

$$C_1 D_1 \; (\overline{R \oplus R'})_{\sigma'} \; C_2 D_2 \quad . \tag{24}$$

In case $\sigma = o$, (23) implies $D_1 = D_2 \in \{true, false\}$, hence $C_1 D_1 \overline{R}_{\sigma'} C_2 D_2$ holds by hypothesis, and therefore so does (24), by Lemma 9.

In case $\sigma = \nu$, (23) implies either $(D_1, D_2) \in R$ or $(D_1, D_2) \in R'$. The first possibility yields (24) much as in the case $\sigma = o$. In the second case, we can express R' as $R_1 \oplus R_2$ where $R_1 = \{(D_1, D_2)\}$ and $R_2 = R' \setminus \{(D_1, D_2)\}$. Then Lemma 20 and the assumption that $C_1 n \ (\overline{R \oplus I_{\{n\}}})_{\sigma'} \ C_2 n$ implies $C_1 D_1 \ (\overline{R \oplus R_1})_{\sigma'} \ C_2 D_2$; hence by Lemma 9, (24) holds since $R' = R_1 \oplus R_2$. □

Theorem 22. *Observational equivalence coincides with applicative equivalence for expressions of first order types. In other words, if σ is of the form $\sigma_k \to \sigma_{k-1} \to \cdots \to \sigma_0$ with each σ_i either ν or o, then for all $M_1, M_2 \in \mathrm{Exp}_\sigma(s)$*

$$s \vdash M_1 \approx_\sigma M_2 \Leftrightarrow M_1 \ (\overline{I_s})_\sigma \ M_2 \quad .$$

Proof. By Theorem 14 and Proposition 16, it suffices to prove for first order σ, and any $R : s_1 \rightleftharpoons s_2$ and $M_i \in \mathrm{Exp}_\sigma(s_i)$, that

$$M_1 R_\sigma^{\mathrm{obs}} M_2 \Rightarrow M_1 \ \overline{R}_\sigma \ M_2 \quad .$$

We do this by induction on the structure of σ. The base cases $\sigma = o, \nu$ are covered by the last part of Lemma 17. For the induction step we have to show that the property holds of $\tau \to \sigma$ ($\tau \in \{o, \nu\}$) when it does of σ. For this, by Propositions 19 and 21 it suffices to check that $R_{\tau \to \sigma}^{\mathrm{obs}}$ satisfies the analogue of the 'only if' part of the latter proposition. In other words it suffices to check that if $C_1 R_{\tau \to \sigma}^{\mathrm{obs}} C_2$, then

- when $\tau = o$, $C_1 b \ \overline{R}_\sigma \ C_2 b$ for all $b \in \{true, false\}$; and
- when $\tau = \nu$
  - for all $(n_1, n_2) \in R$, $C_1 n_1 \ \overline{R}_\sigma \ C_2 n_2$, and
  - $C_1 n \ (\overline{R \oplus I_{\{n\}}})_\sigma \ C_2 n$

  where n is any name not in $s_1 \cup s_2$.

We indicate the proof of the last of these properties (the others being straightforward to establish). So suppose $C_1 R_{\nu \to \sigma}^{\mathrm{obs}} C_2$ and $n \notin s_1 \cup s_2$. Given any $\tau \in \{o, \nu\}$ and any $\lambda x : \sigma \, . \, P \in \mathrm{Can}_{\sigma \to \tau}(dom(R \oplus I_{\{n\}}))$, we have to show

$$(\lambda x : \sigma \, . \, P)(C_1 n) \ (\overline{R \oplus I_{\{n\}}})_\tau \ (\lambda x : \sigma \, . \, P)(C_2 n) \quad . \tag{25}$$

Consider

$$P' \stackrel{def}{=} \nu n \, . \, (\lambda x : \sigma \, . \, P)(f n)$$

Since $\lambda f : \nu \to \sigma \, . \, P' \in \mathrm{Can}_{(\nu \to \sigma) \to \tau}(dom(R))$, we have

$$(\lambda f : \nu \to \sigma \, . \, P') C_1 \ \overline{R}_\tau \ (\lambda f : \nu \to \sigma \, . \, P') C_2 \quad . \tag{26}$$

So if

$$s_1 \oplus \{n\} \vdash (\lambda x : \sigma \, . \, P)(C_1 n) \Downarrow_\tau (s_1') D_1$$
$$s_2 \oplus \{n\} \vdash (\lambda x : \sigma \, . \, P[R \oplus I_{\{n\}}])(C_2 n) \Downarrow_\tau (s_2') D_2$$

then by definition of P', (26) implies $D_1\ (R \oplus R')_\tau\ D_2$ for some $R' : \{n\} \oplus s'_1 \rightleftharpoons \{n\} \oplus s'_2$. In case $\tau = o$ this immediately gives $D_1 = D_2$ and hence that (25) holds, as required. In case $\tau = \nu$, it suffices to show that

$$D_1 = n \Leftrightarrow D_2 = n \quad . \tag{27}$$

For then $D_1\ (R \oplus I_{\{n\}} \oplus R'')_\tau\ D_2$ for some R'' (namely $R'' = R' \setminus \{(n,n)\}$) and hence (25) holds, as required. To see that (27) holds, consider applying the test

$$\lambda f : \nu \to \sigma \,.\, \nu n \,.\, ((\lambda x : \sigma \,.\, P)(fn) = n) \in \mathrm{Can}_{(\nu \to \sigma) \to o}(dom(R))$$

to $C_1 R^{\mathrm{obs}}_{\tau \to \sigma} C_2$. □

Corollary 23. *The relation of observational equivalence between nu-calculus expressions of first order type is decidable.*

Proof. By the above theorem, it suffices to check that the relations $\overline{R}_\sigma$ are decidable for first order σ. For this, it is sufficient to establish the decidability of the relations R_σ (for first order σ) since Theorem 12 ensures that we can calculate s'_1 and s'_2 in clause (8), and then there are only finitely R' for which a decidable property has to be checked. The decidability of R_σ can be established by induction on the structure of the first order type σ, the base cases being trivial, and the induction step following from Proposition 21. □

5 Conclusion

The nu-calculus combines higher order functions with an extremely simple kind of dynamically created local state. Our original motivation for introducing and studying such a computationally simple language was as a vehicle for understanding what, if any, are the difficulties introduced by pure locality of state when reasoning about properties of higher order functions. Our expectation that the difficulties would not be very great has proved to be incorrect, as the results and examples in this paper show.

On a more positive note, we have developed a useful notion of logical relation which builds in a version of 'representation independence' for local names. We showed that it can be used to establish observational equivalence between expressions (Theorem 14). We expect that extensions of this logical relations approach will prove useful for studying observational equivalence in computationally more interesting languages (such as a larger fragment of ML with dynamically created references and exception names).

For the nu-calculus, this method of establishing observational equivalence is incomplete in general (Example 6), but is complete for expressions of first order type (Theorem 22). Of course, the fundamental problem is that (canonical) expressions $\lambda x : \sigma \to \sigma' \,.\, M$ of function type are not in general determined up to observational equivalence by their *extensional* behaviour, i.e. by the function on closed expressions $C \mapsto M[C/x]$ that they determine via application. Nevertheless, it may be that observational equivalence at function types, $\approx_{\sigma \to \sigma'}$, can be explained *compositionally* by applying some construction to $\approx_\sigma$ and $\approx_{\sigma'}$. Clearly this compositionality property is enjoyed by the notion of applicative equivalence (Definition 13). We leave as an open question whether observational equivalence also has this property.

This paper has taken an operationally-based approach. Section 4 of [14] outlines an approach to the denotational semantics of the nu-calculus which builds on work of Moggi [10] using categorical *monads*. The monadic approach enforces a distinction between denotations of values (expressions in canonical form) and denotations of computations (arbitrary expressions). This is helpful, since it allows us to identify explicitly and simply what structure is needed in a model to give a static meaning for the key dynamic aspect of the nu-calculus, viz. *the action of computing a new name*. Further details will appear elsewhere.

Acknowledgements We are grateful to Eugenio Moggi, Peter O'Hearn, Allen Stoughton and Robert Tennent for making their unpublished work available to us. We have benefited from many conversations with them on the topic of this paper.

References

1. S. Abramsky. The Lazy Lambda Calculus. In D. Turner (ed.), *Research Topics in Functional Programming* (Addison-Wesley, 1990), pp 65–116.
2. H.-J. Boehm. Side-effects and aliasing can have simple axiomatic descriptions, *ACM Trans. Prog. Lang. Syst.* 7(1985) 637–655.
3. M. Felleisen and D. P. Friedman. A Syntactic Theory of Sequential State, *Theoretical Computer Science* 69(1989) 243–287.
4. F. Honsell, I. A. Mason, S. Smith and C. Talcott. A Variable Typed Logic of Effects. In *Proc. Computer Science Logic 1992*, Lecture Notes in Computer Science (Springer-Verlag, Berlin, 1993), *to appear.*
5. D. J. Howe. Equality in Lazy Computation Systems. In *Proc. 4th Annual Symp. on Logic in Computer Science*, Asilomar, 1989 (IEEE Computer Society Press, Washington, 1989) pp 198–203.
6. I. A. Mason and C. Talcott. References, local variables and operational reasoning. In *Proc. 7th Annual Symp. on Logic in Computer Science*, Santa Cruz, 1992 (IEEE Computer Society Press, Washington, 1992) pp 186–197.
7. A. Meyer and K. Sieber. Towards fully abstract semantics for local variables: preliminary report. In *Conf. Record 15th Symp. on Principles of Programming Languages*, San Diego, 1988 (ACM, New York, 1988) pp 191-203.
8. R. Milner. Fully abstract models of typed λ-calculi. *Theoretical Computer Science* 4(1977) 1–22.
9. R. Milner, M. Tofte and R. Harper. *The Definition of Standard ML* (MIT Press, 1990).
10. E. Moggi. Notions of Computation and Monads, *Information and Computation* 93(1991) 55–92.
11. P. W. O'Hearn. A Model for Syntactic Control of Interference, *Mathematical Structures in Computer Science*, to appear.
12. P. W. O'Hearn and R. D. Tennent. Semantics of Local Variables. In M. P. Fourman, P. T. Johnstone and A. M. Pitts (eds), *Applications of Categories in Computer Science*, L.M.S. Lecture Note Series 177 (Cambridge University Press, 1992), pp 217–238.
13. P. W. O'Hearn and R. D. Tennent. Relational Parametricity and Local Variables. In *Conf. Record 20th Symp. on Principles of Programming Languages*, Charleston, 1993 (ACM, New York, 1993) pp 171–184.
14. A. M. Pitts and I. D. B. Stark. On the Observational Properties of Higher Order Functions that Dynamically Create Local Names (preliminary report). In *Proceedings of the ACM SIGPLAN Workshop on State in Programming Languages*, Copenhagen, 1993, Yale Univ. Dept. Computer Science Tech. Report.
15. G. D. Plotkin. Call-by-name, call-by-value and the lambda calculus. *Theoretical computer Science* 1(1975) 125–159.
16. G. D. Plotkin and M. Abadi. A Logic for Parametric Polymorphism. In *Proceedings of the Conference on Typed Lambda Calculus and its Applications*, Utrecht, 1993, Lecture Notes in Computer Science Vol. 664 (Springer-Verlag, Berlin, 1993) pp 361-375.
17. J. C. Reynolds. Syntactic Control of Interference. In *Conf. Record 5th Symp. on Principles of Programming Languages*, Tucson, 1978 (ACM, New York, 1978) pp 39–46.
18. R. D. Tennent. Semantic Analysis of Specification Logic, *Information and Computation* 85(1990) 135–162.

The Second Calculus of Binary Relations

Vaughan Pratt*
Stanford University

Abstract

We view the Chu space interpretation of linear logic as an alternative interpretation of the language of the Peirce calculus of binary relations. Chu spaces amount to K-valued binary relations, which for $K = 2^n$ we show generalize n-ary relational structures. We also exhibit a four-stage unique factorization system for Chu transforms that illuminates their operation.

1 Introduction

In 1860 A. De Morgan [DM60] introduced a calculus of binary relations equivalent in expressive power to one whose formulas, written in today's notation, are inequalities $a \leq b$ between terms $a, b, \ldots$ built up from variables with the operations of composition $a;b$, converse $a^{\smile}$, and complement a^{-}. In 1870 C.S. Peirce [Pei33] extended De Morgan's calculus with Boolean connectives $a+b$ and ab, Boolean constants 0 and 1, and an identity $1'$ for composition. In 1895 E. Schröder devoted a book [Sch95] to the calculus, and further extended it with the operations of reflexive transitive closure, a^*, and its De Morgan dual $a_\dagger$. In full this should be called the De Morgan-Peirce-Schröder-Tarski-Jónsson calculus, taking into account the further model-theoretic contributions of Tarski [Tar41] and Jónsson [JT48, JT52]. However it may reasonably be argued that Peirce did the bulk of the work of bringing the calculus to its modern form, which we recognize by calling it simply the Peirce calculus.

In 1987 J.-Y. Girard [Gir87] introduced linear logic, whose language is strikingly similar to that of the Peirce calculus. Liberally interpreted, linear logic may be regarded as subsuming the Peirce calculus, relevance logic [Dun86], quantales [Mul86], and related logics. But we feel this is too broad, since these interpretations lack the bilinear tensor product characteristic of linear algebra, present in the Chu calculus. Moreover these nonconstructive interpretations considerably predate linear logic, and are done an injustice by sweeping them all under the rubric of linear logic. At the same time the great originality and strength of linear logic are undermined by presenting it as both a nonconstructive and constructive logic, modeled by Girard with respectively phase and coherence spaces. The former is not so novel, as some in the relevance logic community have pointed out, and as others [BvN36] could also. Linear logic is seen in its best light as the realization of the Curry-Howard isomorphism for linear algebra, imaginatively moving logic into new but legitimate territory.

This paper focuses these distinctions more sharply by describing both the Peirce calculus and linear logic as calculi having the same domain, namely binary relations, and essentially the same constants and operations, but with strikingly different interpretations. Furthermore the associated equational logic consists as usual of equations for the former, but isomorphisms for the latter, characteristic of the passage from nonconstructive to constructive logic. We further sharpen this focus by eliminating all other distinctions as far as possible consistent with the substantive details of the two calculi. Key to this passage is the replacement of composition by

*This work was supported by ONR under grant number N00014-92-J-1974, and a gift from Mitsubishi.

being mathematics' basic operations of sequential and parallel composition respectively, which for some time now in our own writing about models of behavior [Pra86] we have been calling respectively *concatenation*, or sequence, and *orthocurrence*, or flow.

This juxtaposition of the calculi is achieved by translating Chu's construction of *-autonomous categories [Bar79], ordinarily given in the rarefied atmosphere of commuting diagrams, into the same elementary set-theoretic terms in which the Peirce calculus is customarily described. Although the Chu interpretation of linear logic is conventionally understood via adjunctions in terms of (co)products and tensor products, in our elementary account of this interpretation we shall not even need to define the morphisms that go with the Chu objects, which we leave to the second half of the paper.

Chu's construction was originally studied by P. Chu as a Master's thesis under the supervision of M. Barr, who supplied the basic definition. More recently Barr has commented that "At the time, the formal construction appeared not to have substantial mathematical interest, but it appears to be the most interesting part in the present context." [Bar91]. Certainly within the past four years or so Chu's construction has turned into one of the more popular constructive interpretations of linear logic [Bar91, BG90, LS91], at least relative to the still-small overall interest in the constructive aspects of linear logic.

There is still no consensus on the proper standard model for linear logic, constructive or not. Our concern is with the algebra of Chu spaces. While we feel this is what linear logic should be about, this is a decision we are happy to leave to the linear logic community, Chu spaces being of mathematical interest independently of their relevance to logic.

The second half of the paper treats Chu transforms, which lift the class of Chu spaces to a category thereof. We obtain a four-stage unique factorization property for Chu transforms that illuminates their role as structure-preserving homomorphisms. And we show that n-ary relational structures and their homomorphisms fully and concretely embed in the category of Chu spaces over the set 2^n. We will treat further aspects and applications of Chu spaces, of which there appear to be a good many even at this early stage, in sequels to this paper.

2 The Peirce and Chu Calculi of Binary Relations

2.1 The Common Language

The Peirce calculus amounts to two copies of the logical connectives *or*, *false*, *and*, *true*, *not*, and *implies*, distinguished as the logical and relative (relational) forms of those connectives. To these Schröder [Sch95] added reflexive transitive closure a_0, nowadays a^*, and its De Morgan dual a_1.

Combining the separate involutory logical and relative complements, a^- and $a^\smallsmile$, as a single involutory ($a^{\perp\perp} = a$) complement $a^{\smallsmile -} = a^\perp$ [Pra92c, p.252] weakens the Boolean structure of the Peirce calculus to that of a De Morgan lattice [Dun86, p.184,p.193], since neither $a + a^\perp = 1$ nor $aa^\perp = 0$ hold of binary relations. This seems in practice to leave the utility of the Peirce calculus largely unimpaired, whose operations are as follows.

	Logical :	$a+b$	0	ab	1
Peirce	*Relative* :	$a \dotplus b$	$0'$	$a;b$	$1'$
Language:	*Nonmonotone* :	$a^\perp$	$a\backslash b$	b/a	$a \rightarrow b$
	Closure :		a^*	a_1	

These are not independent, and a suitable basis is

Peirce	$a+b$	0
	$a;b$	$1'$
Basis:	$a^\perp$	$a^\ddagger$

where $a^{\ddagger} = a^{\perp *}$, intermediate between a^* and a_1, to go with $A^{\dagger}$ below.

We eliminate the remaining operations from consideration by reducing them to mere abbreviations, definable in terms of the basic operations as follows.

Peirce Abbreviations:

$$\begin{array}{rclcl} ab &=& (a^{\perp}+b^{\perp})^{\perp} && \\ a \dagger b &=& (b^{\perp};a^{\perp})^{\perp} && \\ a\backslash b &=& (b^{\perp};a)^{\perp} &=& a^{\perp} \dagger b \\ b/a &=& (a;b^{\perp})^{\perp} &=& b \dagger a^{\perp} \\ a^* &=& a^{\perp\ddagger} && \end{array} \qquad \begin{array}{rcl} 1 &=& 0^{\perp} \\ 0' &=& 1'^{\perp} \\ a \rightarrow b &=& a^{\perp} + b \\ && \\ a_1 &=& a^{\ddagger\perp} \end{array}$$

The language of the Chu calculus is that of linear logic, which we give as follows.[1]

Chu Language:

Additives :	$A+B$	0	$A\times B$	1
Multiplicatives :	$A\oplus B$	$\perp$	$A\otimes B$	$\top$
Nonmonotone :	$A^{\perp}$	$A \multimap B$	$A \Rightarrow B$	
Exponentials :	$!A$	$?A$		

These are intended to correspond with the Peirce connectives tabulated in the corresponding positions, with the following exceptions. The so-called *residuals* of the Peirce calculus, $a\backslash b$ and b/a, which from the table of abbreviations can be seen to behave like implications, merge in linear logic into the one "linear implication" $A \multimap B$. The implications have "currying" in common: $a\backslash(b\backslash c) = (b;a)\backslash c$ and $a\rightarrow(b\rightarrow c) = (ab)\rightarrow c$ hold in the Peirce calculus, while $A \multimap (B \multimap C) \cong (A\otimes B) \multimap C$ and $A\Rightarrow(B\Rightarrow C) \cong (A\times B)\Rightarrow C$ will be seen to obtain for the Chu calculus. And Girard's dual exponentials are loosely related to Schröder's dual closures, ideally as a sort of "cotransitive closure;" for simplicity we content ourselves below with the naive interpretation of $!A$ as the domain of A.

As with the Peirce calculus, these operations are not independent, and we choose the following basis, matching our choice of basis for the Peirce calculus. For this purpose we take as primitive not $!A$ itself but rather $A^{\dagger} = !(A^{\perp})$, explained below.

Chu Basis:

$$\begin{array}{cc} A+B & 0 \\ A\otimes B & \top \\ A^{\perp} & A^{\dagger} \end{array}$$

We can then similarly define the rest of the linear logic operations as follows.

Chu Abbreviations:

$$\begin{array}{rclcl} A\times B &=& (A^{\perp}+B^{\perp})^{\perp} && \\ A\oplus B &=& (B^{\perp}\otimes A^{\perp})^{\perp} && \\ A \multimap B &=& (B^{\perp}\otimes A)^{\perp} &=& A^{\perp}\oplus B \\ !A &=& A^{\perp\dagger} && \end{array} \qquad \begin{array}{rcl} 1 &=& 0^{\perp} \\ \perp &=& \top^{\perp} \\ A\Rightarrow B &=& !A \multimap B \\ ?A &=& A^{\dagger\perp} \end{array}$$

Except mainly for notational differences, the identification of $a\backslash b$ and b/a, and the absence of $*$ from $a \rightarrow b$, we have in this way concentrated whatever differences exist between the two calculi into the primitives, whose very different interpretations we now give.

---

[1]This is the notation now followed by Barr and Seely, and close to their earlier usage [Bar91, See89]. It replaces Girard's idiosyncratic notation $A\oplus B$, $A\&B$, and $A⅋B$ by respectively $A+B$, $A\times B$, and $A\oplus B$, and interchanges his assignments of 1 and $\top$. For vector spaces, $A \oplus B$ conventionally denotes the *biproduct* $A\times B = A+B$, but the Chu calculus distinguishes $\times$ and $+$, freeing up $\oplus$ for this other use. Actually Girard's notation goes quite tidily with our choice of primitives, which become $A\oplus B$, 0, $A\otimes B$, 1. The trouble here is that coproduct and final object have been $+$ and 1 for many decades now, and one needs a better reason than tidiness to make such a sweeping change. The tensor unit, $\top$ in Barr-Seely notation, is often written I, but almost never 1.

2.2 The Peirce and Chu Interpretations

The Peirce calculus is standardly interinterpreted for $a, b, \ldots$ ranging over subsets of X^2 where X is a fixed infinite but otherwise arbitrary set, namely as follows.

Peirce Interpretation:

$$\begin{array}{rclrcl} x(a+b)y & \Leftrightarrow & xay \vee xby & x0y & \Leftrightarrow & \mathit{false} \\ x(a;b)z & \Leftrightarrow & \exists y[xay \wedge ybz] & x1'y & \Leftrightarrow & x=y \\ x(a^{\perp})y & \Leftrightarrow & \neg(yax) & a^* & = & 1'+a+a;a+\ldots \end{array}$$

We may pass from binary relations on a single fixed set X to binary relations from a *domain* X to a *codomain* Y, which are permitted to vary from one relation to the next, provided we make the operations partial. In this extension $a+b$ is defined only when a and b have the same domain and codomain, while $a;b$ is defined just when the codomain of a is the domain of b. Furthermore every set X has its own identity $1'_X$, making $a;b$ the composition no longer of a monoid but of a category. Such structures have been called *Schröder categories* [Jón88].

The Chu calculus assumes such a variable domain and codomain at the outset. We define its connectives, acting on binary relations $A, B, \ldots, A_i, \ldots$ as subsets of $X_A \times Y_A$, $X_B \times Y_B, \ldots$, $X_i \times Y_i, \ldots$, as follows.

Chu Interpretation:

$$\begin{array}{rclrcl} A+B & = & A{\cdot}0 \bowtie B{\cdot}1 & 0 & = & \lceil\emptyset\rceil \\ A\otimes B & = & (\bowtie_{x'\in X_B} A{\cdot}x') \bowtie (\bowtie_{x\in X_A} x{\cdot}B) & \top & = & \lceil\{0\}\rceil \\ x(A^{\perp})y & \Leftrightarrow & yax & A^{\dagger} & = & \lceil Y_A\rceil \end{array}$$

We write $\lceil X\rceil$ for the membership relation from X to 2^X, which we take to be the Chu representation[2] of the set X. We write $A{\cdot}z$ for the result of renaming each x in the domain of A to (x,z) (without otherwise changing the relation); $z{\cdot}A$ renames each x to (z,x). Lastly we define the *natural join*[3] $A = \bowtie_i A_i$ of a family A_i of relations thus. Define $X = \bigcup_i X_i$, and define $Y = \{y \in \prod_i Y_i \mid \forall ij \forall x{\in}X_i{\cap}X_j[xA_iy_i = xA_jy_j]\}$. (So if the X_i's are disjoint, $Y = \prod_i Y_i$.) Define A from X to Y such that for each $x \in X_i$, $xAy = xA_iy_i$, well-defined in the event that any x appears more than once in this condition (non-disjoint X_i's) because of how we chose Y.

The Chu interpretation of 0 is the unique 0×1 relation, while $\top$ denotes the 1×2 relation $\begin{pmatrix}0 & 1\end{pmatrix}$. $A^{\dagger}$ denotes the $Y_A \times 2^{Y_A}$ relation $yA^{\dagger}Z = y{\in}Z$. $A^{\perp}$ is converse. $A+B$ is the $(X_A+X_B) \times (Y_A \times Y_B)$ relation (the 0 and 1 implement the disjoint union X_A+X_B) satisfying $(x,0)(A+B)(y,y') = xAy$ and $(x,1)(A+B)(y,y') = xBy'$, which we illustrate as follows.

$$\begin{pmatrix}1 & 0\\ 0 & 1\end{pmatrix} + \begin{pmatrix}1 & 0 & 1\\ 0 & 1 & 0\end{pmatrix} = \begin{pmatrix}1 & 0 & 1 & 0 & 1 & 0\\ 0 & 1 & 0 & 1 & 0 & 1\\ 1 & 1 & 0 & 0 & 1 & 1\\ 0 & 0 & 1 & 1 & 0 & 0\end{pmatrix}$$

The rows of $A+B$ are those of A followed by those of B, in order, while its columns are indexed in order by $(y_0,y'_0), (y_1,y'_0), (y_0,y'_1), (y_1,y'_1), (y_0,y'_2), (y_1,y'_2)$ where $y_i \in Y_A$, $y'_i \in Y_B$.

This leaves just $A\otimes B$, whose properties we summarize as follows.

Theorem 1 *$A\otimes B$ has domain $X_A \times X_B$ and codomain the set of all pairs of functions $(f : X_A \to Y_B, g : X_B \to Y_A)$ satisfying $xAg(x') = x'Bf(x)$, with $(x,x')(A\otimes B)(f,g) = xAg(x')$ $(= x'Bf(x))$.*

[2] This generalizes immediately to $\lceil(X,\leq)\rceil$ for any poset, by interpreting 2^X to consist of just the order ideals of X rather than all subsets. This is the usual open-set representation of a poset as a topological space, a set then being just a discrete or unordered poset.

[3] The join operation comes from database theory. We take X to be the attributes or columns of the relation. In database terms $A \subseteq X\times Y$ is an X-ary relation (i.e. X is the set of attributes or columns) on the domain $\{0,1\}$ consisting of a set Y (the rows) of records, each of which is an X-tuple of bits. This is the transpose of the usual view of A as an $X\times Y$ matrix.

Proof: The domain of $\bowtie_{x'\in X_B} A\cdot x'$ can be seen to be $X_A \times X_B$, while by disjointness of $A\cdot x'$ as x' varies, the codomain is $\prod_{x'\in X_B} Y_A = Y_A^{X_B}$, i.e. the set of functions $g : X_B \to Y_A$. And the resulting relation A' is defined by $(x, x')A'g = xAg(x')$. Likewise $\bowtie_{x\in X_A} x\cdot B$ has the same domain, $X_A \times X_B$, has codomain $Y_B^{X_A}$, i.e. functions $f : X_A \to Y_B$, and is the relation B' defined by $(x, x')B'f = x'Bf(x)$. Hence the join of these two joins also has domain $X_A \times X_B$, while its codomain is that subset of the product $Y_A^{X_B} \times Y_B^{X_A}$ consisting of those pairs (f, g) such that for all (x, x') in $X_A \times X_B$, $(x, x')A'g = (x, x')B'f$, that is, $xAg(x') = x'Bf(x)$, this then being the value of $(x, x')(A\otimes B)(f, g)$. ∎

Corollary 2 *The domain of $A \multimap B$ is the set of all pairs of functions $(f : X_A \to X_B, g : Y_B \to Y_A)$ such that for all $x \in X_A$, $y' \in Y_B$, $f(x)By' = xAg(y')$* (cf. the definition of Chu transform in the section of that name, also cf. continuous functions of topological spaces where the Y's are taken to consist of open sets).

As a proposition, the first join repeats A "at X_B different locations," with a fresh set of variables of A for each location, while the second repeats B at locations X_A, with a fresh set of variables of B for each location, such that the two sets of repetitions use the same set $X_A \times X_B$ of variables. The join of the two is the conjunction of these two conditions, expressing the notion of *bilinearity* characteristic of tensor product. Although there is nothing "linear" about binary relations, the "linear" in linear logic expresses the thought that the essence of linear algebra resides in this property rather than in anything to do with the structure of fields [LS91].

We may view the relation A from X to Y as denoting the Boolean proposition P whose set of variables is X and each of whose assignments $s : X \to 2$ of truth values to those variables satisfies P just when there exists $y \in Y$ such that $\forall x \in X[s(x) = xAy]$. The y's in Y then correspond to satisfying assignments, or equivalently to clauses of the DNF form of the proposition. In this view, join is exactly the notion of conjunction of Boolean propositions.

We illustrate this definition of $A\otimes B$ with the following example.

$$\begin{pmatrix} 1 & 0 \\ 0 & 1 \end{pmatrix} \otimes \begin{pmatrix} 1 & 0 & 1 \\ 0 & 1 & 0 \end{pmatrix} = \begin{pmatrix} 1 & 1 & 0 & 0 \\ 0 & 0 & 1 & 1 \\ 0 & 0 & 1 & 1 \\ 1 & 1 & 0 & 0 \end{pmatrix}$$

The rows of $A\otimes B$ in order are (x_0, x'_0), (x_0, x'_1), (x_1, x'_0) (x_1, x'_1). Its columns in order are

$$\begin{array}{l} (\{x_0 \mapsto y'_0, x_1 \mapsto y'_1\}, \ \{x'_0 \mapsto y_0, x'_1 \mapsto y_1\}), \\ (\{x_0 \mapsto y'_1, x_1 \mapsto y'_0\}, \ \{x'_0 \mapsto y_1, x'_1 \mapsto y_0\}), \\ (\{x_0 \mapsto y'_2, x_1 \mapsto y'_1\}, \ \{x'_0 \mapsto y_0, x'_1 \mapsto y_1\}), \\ (\{x_0 \mapsto y'_1, x_1 \mapsto y'_2\}, \ \{x'_0 \mapsto y_1, x'_1 \mapsto y_0\}), \end{array}$$

these four being the only compatible pairs of (f, g)'s out of the $3^2 \times 2^2 = 36$ possibilities. For example the first column is indexed by the given $(f : X \to Y', g : X' \to Y)$ specifying that the entry in the first row of that column should be $x_0Ag(x'_0) = x_0Ay_0$ $(= x'_0Bf(x_0) = x'_0By'_0) = 1$.

These operations on relations are somewhat more intricate than those of the Peirce calculus. The idea however is that one should not work directly with the interpretation but rather indirectly with its logical properties, which is also the idea behind the Peirce calculus. Since the respective logics are of comparable complexity, the gains are potentially greater with Chu logic than with Peirce logic (provided one never has to resort to the explicit interpretation), since more complex machinery is being manipulated at no additional cost in logical complexity. While more complex does not always mean more powerful, in this case a small increase in complexity turns out to lead to a considerable increase in power.

2.3 K-valued Relations

We now make a small generalization to the notion of binary relation that gives a large increase in the power of the Chu calculus. We allow K-valued binary relations where K is an arbitrary set. Thus instead of xAy either holding or not, it has a value from K. More formally, A is a triple (X, Y, a) where X and Y are sets and $a : X \times Y \to K$ is a K-valued function.

We need consider only $\lceil - \rceil$ and $\bowtie$. We generalize $\lceil X \rceil$ from membership of elements of X in elements of 2^X to application of elements of K^X (i.e. functions $f : X \to K$) to elements of X.[4] And the join operation immediately generalizes from $\{0,1\}$-valued to K-valued relations since nothing in the definition of join depended on that special case.

2.4 Equational Laws

We have $a+b = b+a$ for the Peirce calculus, and it is natural to expect this for the Chu calculus as well. However on closer inspection we notice that each x in the domain of A becomes $(x, 0)$ in $A+B$ but $(x, 1)$ in $B+A$. We may however claim the isomorphism $A+B \cong B+A$, in which $(x, 0)$ in $A+B$ is matched up with $(x, 1)$ in $B+A$. This applies to the other laws of the Chu calculus as well. (A is isomorphic to B when there exist bijections $X_A \cong X_B$, $Y_A \cong Y_B$ of their index sets making their corresponding entries equal.) The full list of isomorphisms (and equalities where possible) we know to hold for extensional T_0 Chu spaces (defined in the paragraph following Definition 3 below) is as follows.

$$\begin{array}{rclrclrcl}
A+(B+C) & \cong & (A+B)+C & A+0 & \cong & A & A+B & \cong & B+A \\
A\otimes(B\otimes C) & \cong & (A\otimes B)\otimes C & A\otimes\top & \cong & A & A\otimes B & \cong & B\otimes A \\
A\otimes(B+C) & \cong & (A\otimes B)+(A\otimes C) & A\otimes 0 & \cong & 0 & A^{\perp\perp} & = & A \\
(A+B)^{\dagger} & \cong & A^{\dagger}\otimes B^{\dagger} & A^{\dagger\perp\dagger} & = & A^{\dagger} & & &
\end{array}$$

Without attempting to be complete (though this should be pretty close to the expressible consequences of the RA axioms [JT48, JT52]), the Peirce laws include all these less $a;b = b;a$ and $(a+b)^{\ddagger} = a^{\ddagger};b^{\ddagger}$ (close), plus the Boolean properties expressible in the available language, e.g. idempotence and distributivity over each other of ab and $a+b$.

From the Chu laws and the definitions of abbreviations we can derive for example $(A\otimes B)\multimap C = (C^{\perp}\otimes(A\otimes B))^{\perp} \cong ((C^{\perp}\otimes A)\otimes B)^{\perp} = B\multimap(C^{\perp}\otimes A)^{\perp} = B\multimap(A\multimap C)$. We leave $(A\times B)\Rightarrow C \cong A\Rightarrow(B\Rightarrow C)$ as an exercise. We are not aware of any completeness results for the isomorphism theory of the Chu calculus.

An equivalent to $(A+B)^{\dagger} \cong A^{\dagger}\otimes B^{\dagger}$ is $!(A\times B) \cong\, !A\otimes !B$, but this uses the nonprimitive $\times$, our first reason for taking $A^{\dagger}$ as primitive rather than $!A$, the contravariance of $A^{\dagger}$ notwithstanding. This law just asserts the previously noted fact that the codomain of a sum is the product of the codomains of the arguments, the naturality of which is our second reason for preferring $A^{\dagger}$ over $!A$.[5] Abstracting away $A+B$, we obtain the law $\lceil X\times Y \rceil \cong \lceil X \rceil \otimes \lceil Y \rceil$ for sets X, Y. This remains valid when X and Y are generalized to posets, these forming a cartesian closed category. That is, to form the cartesian product of sets and posets when represented as Chu spaces, form their tensor product in the Chu calculus, not their direct product $A \times B$ which yields something different. The product of join-semilattices, when these are represented as Chu spaces over 2 whose rows are closed under finite bitwise OR (union), is however not formed by tensor product; in fact

[4] The generalization to posets $\lceil (X, \leq) \rceil$ is intended only for $K = 2$, though some analogous notion might be possible for larger K, in particular for power sets $K = 2^n$.

[5] It seems plausible to us that both are needed, in that $!A$ may well be more appropriately interpreted as either the symmetric (boson) or antisymmetric (fermion) tensor algebra generated by A, as contemplated in recent unpublished work of Blute, Panangaden, and Seely on " Old Foundations for Linear Logic: Holomorphic Functions in Banach Spaces as Models of Exponential Types," concerning the Fock space interpretation of $!A$, and of Blute on "Modelling linear logic with vector spaces," a forthcoming talk at the Cornell workshop on linear logic, June 1993. We hope to understand this issue better in the near future.

the tensor product of a meet-semilattice with a join-semilattice is a distributive lattice (since tensor product works by conjoining properties), details in a future paper.

Given that the linear logic primitives are all definable with $\lceil - \rceil$, $\bowtie$, and $A^\perp$, it may be worth investigating taking these as an even simpler basis for linear logic, suitably organized.

Note that $A^\dagger$, $A^{\dagger\dagger}$, $A^{\dagger\dagger\dagger}\ldots$ is $Y_A, K^{Y_A}, K^{K^{Y_A}}, \ldots$, in contrast to $!!A = !A$.

2.5 Historical Notes

In introducing linear logic, Girard proposed phase spaces and coherence spaces [Gir87] as respectively nonconstructive and constructive interpretations, the distinction being whether each sequent $\Gamma \vdash \Delta$ is considered to denote a truth value or a set of proofs of Δ from Γ. But when Girard presented his logic at a category theory conference in Boulder in 1988, M. Barr recognized the suitability for modeling linear logic of his *-autonomous categories in general and his student P. Chu's construction of such in particular [Bar79]. Chu spaces, as the objects of Chu's construction for the category of sets, seem to be a particularly attractive constructive model of linear logic.

We have been using $A+B$ and $A\otimes B$ in our concurrency work, starting with [Pra86] where they are notated respectively $A\|B$, called *concurrence* (meaning noninteractive asynchronous parallel composition) (p.47), and $A\times B$, *orthocurrence*, meaning flow or mixing of one process through or in another (p.49, also §3), an interactive form of parallel composition. Subsequently it was realized [CCMP91, p.208] that $A\times B$ should have been tensor product $A\otimes B$; the confusion with $A\times B$ occurred because the earlier work was conducted in the category **Pos**, which being cartesian closed identifies the two. The confusion was exposed when the passage from ordered time to real time broke the $A \times B$ definition. By the same token the Chu representation of the direct product of posets (conflict-free schedules) is the tensor product of the Chu representations of those posets, but this does not extend to schedules having conflicts and other forms of causal structure. Our interpretations of concurrence and orthocurrence, which have been evolving over the intervening years, appear to have been moving steadily towards the Chu interpretation. It remains to connect up the Chu interpretation with the yet more general interpretations of [CCMP91], which we expect to be only a matter of details.

With regard to orthocurrence as flow, e.g. of a sequence A of trains through a sequence B of stations, the bilinearity expressed in the equation defining tensor product corresponds to the notion that when we stand on the platform of any station $b \in X_B$ we see the same sequence A of trains, and vice versa when we watch the stations go by from any train $a \in X_A$.

We have been using $A^\perp$ only relatively recently [Pra92b, Pra92a], as the basic link between schedules and automata, and as complementarity in quantum mechanics [Pra93]. Automata express behavior as graphs with states as vertices and events as edges; schedules dualize this by interchanging them, with $A^\perp$ denoting the automaton form of the schedule A. The generalization of the event spaces of [Pra92b, Pra92a] to the Chu spaces of this paper is anticipated by the partial distributive lattices of [Pra93, §5], which are essentially Chu spaces, whose role in this application we defer to a separate paper.

3 Chu Spaces and Transforms

We now imbue a binary relation with a spatial character by taking its domain to be its point set, and its codomain to be its degrees of freedom or *states*, reflected in the following notation.

Definition 3 *A Chu space $A = (P, S, v)$ over a set K consists of a set P of points, a set S of states, and a function $v : P\times S \to K$ assigning a value $v(p,s)$ to each point p in each state s.*

We associate with $v : P\times S \to K$ the functions $v_- : P \to (S \to K)$ and $v^- : S \to (P \to K)$ satisfying $v_-(p)(s) = v^-(s)(p) = v(p,s)$. We abbreviate $v_-(p)$ to v_p, called the *extension* of p, and

$v^-(s)$ to v^s, the *extension* of s. We think of v^s as one of the permitted *paintings* of the underlying set P, with K for our palette. Dually v_p is a painting of S, understood as the varying values of one point encountered as one traverses the "possibility space" of alternative paintings of P. We write $V^* = \{v^s \mid s \in S\}$ for the set of extensions of states, through which $S \xrightarrow{v^-} K^P$ factors as $S \to V^* \to K^P$. Dually $P \xrightarrow{v_-} K^S$ factors through the set $V_* = \{v_p \mid p \in P\}$ of extensions of points as $P \to V_* \to K^S$. When all states have distinct extensions, i.e. v^- is injective ($S \cong V^*$), we call v *extensional* (shorter and more mnemonic than Barr's "right separated" [Bar91]) and say it has *enough points* (to distinguish states). (This situation is very important, allowing us to interpret (P, S) for $S \subseteq K^P$ as a Chu space.) When all points have distinct extensions ($P \cong V_*$) we call A T_0 (by analogy with the topological property of that name), and say it has *enough states* (Barr: left separated). Locales are the prototypical example of a nonextensional but T_0 space [Vic89, p.61].

Logically speaking, points are *necessary* in the sense that they are necessarily all present in the space at the one time. Dually, states are *possible*, in that the space is in one state *or* another, like the possible worlds of a Kripke structure (our conventional understanding of spaces does not permit us to imagine that the whole space is in all states simultaneously).

More states mean more degrees of freedom, corresponding to less structure. At one extreme the extensional space (P, K^P) contains all possible states and hence has the vacuous structure of a set or discrete space, which we shall identify with the set P itself. We view the omitted states, those in $K^P - V^*$, as the *atomic* properties of the space, collectively constituting the theory or *structure* of the space. At the other extreme the space $(P, \emptyset)$ omits all states, which we view as the inconsistent structure on P. A one-state space, S a singleton, is "rigid," every point having a uniquely determined or constant value. The canonical rigid space is $(K, \{0\})$, which we denote $\perp$ (not K, which a moment ago we associated, in its capacity as a set, with the discrete space (K, K^K)).

The dual of $A = (P, S, v)$ is the *state space* $A^\perp = (S, P, v^\smile)$ where $v^\smile(s, p) = v(p, s)$. Duality interchanges points and states, and hence necessity and possibility. This confers on duality one of the qualities of logical negation, another being double negation, $A^{\perp\perp} = A$.

We turn now to the notion of a transform of Chu spaces, foreshadowed in Theorem 1.

Definition 4 *A Chu transform* $(f, g) : (P, S, v) \to (Q, T, w)$ *consists of functions* $f : P \to Q$, $g : T \to S$ *satisfying* $v(p, gt) = w(fp, t)$ *for all* $p \in P$ *and* $t \in T$. *Composition of* $(f', g') : (Q, T, w) \to (R, U, X)$ *with* $(f, g) : (P, S, v) \to (Q, T, w)$ *is defined by* $(f', g')(f, g) = (f'f, gg')$, *satisfying* $v(p, gg'u) = w(fp, g'u) = x(f'fp, u)$ *and hence a Chu transform. Associativity is inherited from that of function composition. The identity transform on* (P, S, v) *is the pair* $(1_P, 1_S)$ *of identity functions on* P, S *respectively. We abbreviate* $(f, g) : (P, S, v) \to (Q, T, w)$ *to* $f : A \to B$ *when unambiguous.*

It is easy to see from the explanation of $A \otimes B$ in the previous section that defining $A \multimap B$ to be $(A \otimes B^\perp)^\perp$ makes it the Chu space whose points are the linear transformations from A to B.

We denote by **Chu**(K) the category of Chu spaces and their Chu transforms so composed.[6] $A+B$ and $A \times B$ as defined for the Chu calculus are respectively coproduct and product in this category, which $A \multimap B$ is the internal hom and $A \otimes B$ its associated tensor product.

Taking the duals $A^\perp = (S, P, v^\smile)$ and $B^\perp = (T, Q, w^\smile)$ of the domain and codomain of $(f, g) : (P, S, v) \to (Q, T, w)$ necessarily entails replacing (f, g) by (g, f). To construe this as a Chu transform we must then treat it as $(g, f) : (T, Q, w^\smile) \to (S, P, v^\smile)$. That is, duality reverses

[6] This definition of **Chu**(K) takes place in the category **Set** of sets and functions, with K as a distinguished set. By generalizing **Set** to any symmetric closed monoidal category $\mathcal{V}$ and K to any object K of $\mathcal{V}$, we may correspondingly generalize the above definition to the doubly parametrized category **Chu**$(\mathcal{V}, K)$ defined by Barr and Chu [Bar79].

the direction of transforms, much as transposing a matrix reverses the direction of the linear transformation it defines.

As mentioned in less detail earlier, Chu spaces first arose as the objects of the self-dual symmetric closed monoidal category produced by Chu's construction from a symmetric closed monoidal category $\mathcal{V}$, for the case $\mathcal{V}$ = **Set** [Bar79]. Lafont and Streicher have more recently called the objects of this case *games* [LS91]. They observed that vector spaces over a field K may be realized as games over the underlying set of K, and that topological spaces may be realized as games over 2. We observe that Chu transforms of Chu posets $\lceil(X, \leq)\rceil$ realize exactly monotone functions of posets, which generalizes to a comprehensive analysis of Stone duality viewed as a continuum from sets to complete atomic Boolean algebras, to be treated elsewhere.

4 Unique Factorization of Chu Transforms

In this section we develop a factorization yielding a useful insight into what the Chu transform accomplishes, namely the proper management of states or degrees of freedom as the complement of structure, that which is preserved by transformations.

Functions factor, uniquely up to an isomorphism at the junction of the factors, as the composition of a surjection with an injection. That is, we first perform all needed identifications, then add all needed new elements. Doing these in the other order destroys this uniqueness, since there are many nonisomorphic ways to add elements and then identify some of them. This is the basic example of an EM (for epi-mono) factorization system [FK72].

We show here a more elaborate unique factorization for Chu transforms (f, g), namely the sequence described informally as, omit states, duplicate states, identify points, add points. This draws distinctions finer than those made by epis and monos, which view both state omission and point addition as monos, and the other two as both epis. Lacking more abstract terms for these notions, we adopt the above sequence of four informally named operations and call this the *ODIA factorization* for Chu(K).

We accomplish this factorization in two steps. We refer to the OD half, omission then deletion of states, as an *erasure*, and the IA half, identification then addition of points, as a *move*. We prove unique factorization into an erasure followed by a move. The rest of the ODIA factorization is then an immediate corollary of the EM factorizations of f and g individually, and we will say nothing further about that.

Definition 5 *An erasure is a Chu transform of the form* $(1, g)$*; we let* $\mathcal{E}$ *denote the class of all erasures of* Chu(K)*. Dually a move is of the form* $(f, 1)$*, forming the class* $\mathcal{M}$.[7]

An erasure $(1, g) : (P, S, v) \to (P, T, h)$ modifies only states: every state $t \in T$ receives its extension from state $g(t) \in S$. Hence $H^{\bullet} \subseteq V^{\bullet}$, whence any state omitted from $A = (P, S, v)$ remains omitted from $H = (P, T, h)$, i.e. *erasure preserves structure*. (Thus we may identify the *omitted* states, namely $K^P - V^{\bullet}$, with the *structure* of the Chu space (P, S, v).) A state of A may be *duplicated* in H, but we distinguish duplication of existing states from creation of new states.

A move $(f, 1) : (P, T, h) \to (Q, T, w)$ modifies only points, mapping each point $p \in P$ to a point $f(p) \in Q$ having the same extension as p. Since the image $f(H)$ of a move "receives" its states from H, states are neither created nor destroyed in $f(H)$. The "shape" of a state changes concomitantly with that of P under f, but the values of points of P in a given state are not changed as they move to their new locations in Q, all necessary changes in value having been previously accomplished by the erasure.

[7]Thus $\mathcal{E} \cap \mathcal{M}$ consists of the identities of Chu(K); when it is necessary that it consist of the isomorphisms, here bijections, it suffices to close $\mathcal{E}$ and $\mathcal{M}$ under composition with bijections. Normally $\mathcal{E}$ and $\mathcal{M}$ consist of respectively epis and monics, our little pun receives some legitimacy from the observation that $(1, g)$ is an epi of Chu(K), and $(f, 1)$ a monic, just when both f and g are injective.

Bear in mind that the transform $f : A \to B$ maps A *into* B, and that the noncreation of states refers only to A thus transformed, not to the whole target. While the image $f(\mathbb{R})$ of the linear transformation $f: \mathbf{R} \to \mathbb{R}^2$ has at most the degrees of freedom of $\mathbf{R}$, $\mathbb{R}^2$ has visibly more degrees of freedom than $\mathbb{R}$.

Theorem 6 *Every transform factors uniquely and functorially as the composition me of a move m with an erasure e.*

Proof: Uniqueness is immediate: $(f,g) : (P,S,v) \to (Q,T,w)$ must factorize as $(f,1)(1,g)$. For existence define the intermediate object (P,T,h) as $h(p,t) = v(p,gt)$, making $(1,g)$ a transform. But $v(p,gt) = w(fp,t)$, whence $(f,1)$ is also a transform.

Functoriality means that any transform of h to h', as a commuting square with sides u, w (Figure 1(a)), factors uniquely as the composition of transforms m to m' with e to e' (Figure 1(b)), mediated by a unique transform v.

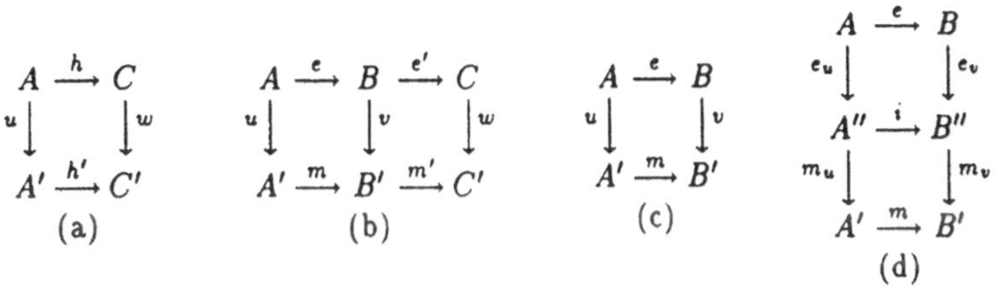

Figure 1

This can be seen to be equivalent to the requirement that all squares of the form shown in Figure 1(c) have a unique *diagonal fill-in* from B to A'. This is seen by factoring u and v as $m_u e_u$ and $m_v e_v$ respectively as in Figure 1(d). Each of the equal sides $AA'B'$, ABB' of the commuting square has now been $\mathcal{EM}$-factored, as $mm_u e_u$ and $m_v e_v e$ respectively. But $\mathcal{EM}$ factorization is unique (up to a bijection if we have closed $\mathcal{E}$ and $\mathcal{M}$ under composition with bijections), yielding the identity (or a bijection) i from A'' to B''. The sides ABB', $AA'B$ of the square determine $m_v i e_u : A \to B'$, hence $i : A'' \to B''$, hence $i^{-1} : B'' \to A''$, making the diagonal fill-in $m_u i^{-1} e_v : B \to A'$. ∎

Returning to the notion of S as the possible states of (P,S,v), the geometric significance of this unique factorization is that every transform can be viewed as taking place in two stages. First the theory of the space being transformed is strengthened in preparation for the coming move, by erasing suitable states (and permuting and duplicating some of the surviving states), without however moving the points themselves, and without introducing any new states. Then the points are moved in a way that does not modify their assignments in each of the new states.

The $\mathcal{EM}$ factorization thus separates transforms into a purely structure-preserving part followed by a purely point-moving part. This constitutes the dynamic confirmation of our previous static analysis of P and S as the sets of respectively necessary points and possible states. We may think of the omitted states dually as necessary facts. From this perspective, transforms preserve that which is necessary, namely points and facts.

It should be clear from this analysis of the factorization of (f,g) through H that f, g, and h are by no means independent. Indeed either of f or g suffice to determine h. Further, if A is extensional then h (and hence f) determines g, while if B is T_0, h (and hence g) determines f. It follows that our convention of abbreviating (f,g) to $f : A \to B$ is unambiguous when A is extensional.

5 Power of Chu Spaces

We have already mentioned Lafont and Streicher's observation [LS91, p.45] that the category of vector spaces over a field **K** is a full subcategory of **Chu**(K), and that the category **Top** of topological spaces is a full subcategory of **Chu**(2). We improve on these observations by showing that *every* n-ary relational structure is realizable as an object of **Chu**(2^n), giving a very strong sense in which Chu spaces form a universal category.

The earliest instance of a universal category is due to Trnková [Trn66]. The universality of the category of semigroups was established by Hedrlín and Lambek [HL69]. These and a number of other such embeddings all took the form of a full and faithful functor that did not preserve underlying sets, for example representing some finite objects as infinite ones. The advantages accruing from the unifying framework of say semigroups are then more than offset by the radically different discipline required to do mathematics in the absence of the expected underlying set.

Pultr and Trnková [PT80] call the kind of *concrete* full embedding we aim for here a *realization*: the functor $F : C \to D$ realizes object A of C when not only is F full and faithful, but $U_D(F(A)) = U_C(A)$, where $U_C : C \to \mathbf{Set}$, $U_D : D \to \mathbf{Set}$ are the respective underlying-set functors. Pultr and Trnková give hardly any realizations, concentrating on weaker forms of full embeddings. In contrast the embedding here is a realization, and a simple one at that.

Here by "A represented as B" we shall mean throughout that the category C_A of all A's *fully* embeds[8] in the category C_B of all B's.

Definition 7 *For any ordinal n, an n-ary relational structure (X, ρ) consists of a set X, the carrier, and an n-ary relation $\rho \subseteq X^n$ on X. A homomorphism $f : (X,\rho) \to (Y,\sigma)$ between two such structures is a function $f : X \to Y$ between their underlying sets for which $f\rho \subseteq \sigma$. Here $f\rho$ denotes $\{f\mathbf{a} \mid \mathbf{a} \in \rho\}$, where $\mathbf{a}$ denotes $(a_0, \ldots, a_{n-1})$ and $f\mathbf{a}$ denotes $(fa_0, \ldots, fa_{n-1})$. We denote by $\mathbf{Str}_n$ the category formed by the n-ary relational structures and their homomorphisms.*

It suffices to treat structures with a single carrier and relation, since k carriers can be combined as their disjoint union, kept track of with k unary relations ($\lceil \log_2 k \rceil$ is enough information-theoretically, but not enough to ensure that homomorphisms respect type). Multiple nonempty relations on a set can be joined to form a single relation on the same set, of arity at most the sum of the arities of its constituent relations. For algebras, structures all of whose $(n+1)$-ary relations are n-ary operations, the join may share the input coordinates of the operations, reducing the total arity to the maximum of the input arities plus the number of operations (including constants).

This notion of homomorphism is standard in the strong sense that *any* class of n-ary relational structures and their homomorphisms constitutes a full subcategory of $\mathbf{Str}_n$. Familiar examples of such categories and their arities include those of semigroups (3), monoids (4), groups (3), rings (4), rings with a multiplicative unit (5), fields (4), lattices (3), lattices with top and bottom (5), Boolean algebras (3), vector spaces (4),[9] directed graphs or binary relations (2), multigraphs (4), posets (2), and categories (4).

Many of these numbers benefit from group structure, for which homomorphisms preserve inverses and identities even when these operations are not given explicitly as part of the relation. Units of monoids, including tops and bottoms of lattices, are not so fortunate and each requires its own unary relation in order to be recognized and preserved by homomorphisms.

The universality achieved here is of a different kind from that achieved by say ZF set theory. Externally a model of ZF is a single object of $\mathbf{Str}_2$ of some cardinality, with membership as

[8] An embedding is a faithful functor $F : C_A \to C_B$, i.e. for distinct morphisms $f \neq g$ of C_A, $F(f) \neq F(g)$, and is *full* when for all pairs a, b of objects of C_A and all morphisms $g : F(a) \to F(b)$ of C_B, there exists $f : a \to b$ in C_A such that $g = F(f)$.

[9] Treat as partial rings, with uv defined just when u is on a specified axis. This works equally well for homogeneous vector spaces (all over the one field) and heterogeneous, the only nontrivial field endomorphisms being automorphisms.

its only relation, "internally" coding objects larger than any fixed cardinal including its own. Our universality has no separate notion of an internal world; instead we code our objects purely externally.

We now define the promised functor $F : \mathbf{Str}_n \to \mathbf{Chu}(\mathbf{2}^n)$, namely in definitions 9 and 13, and prove that it is full, faithful, *and concrete.*

The complementarity of constraints and states indicates ρ and $\overline{\rho}$ as the appropriate respective sources of each. We shall define a state to be essentially a subset of $\overline{\rho}$, with however a small but essential refinement. The following lemma obtains from the standard constraint-based definition of homomorphism an equivalent state-based characterization.

Lemma 8 $f\rho \subseteq \sigma \;\Leftrightarrow\; f^{-1}\overline{\sigma} \subseteq \overline{\rho}$. Here $\overline{\rho} = A^n - \rho$ and $\overline{\sigma} = B^n - \sigma$.

Proof:

$$\begin{array}{rlll} f\rho \subseteq \sigma & \Leftrightarrow & \rho \subseteq f^{-1}\sigma & \text{(Definition of } f^{-1}) \\ & \Leftrightarrow & \overline{f^{-1}\sigma} \subseteq \overline{\rho} & \text{(Complement)} \\ & \Leftrightarrow & f^{-1}\overline{\sigma} \subseteq \overline{\rho} & (f^{-1} \text{ preserves Boolean operations}) \end{array}$$

∎

Definition 9 (F on objects). *Let $\mathbf{2}^n$ denote the set of n-bit bit vectors, that is, n-tuples over $\mathbf{2}$. We define the object part of the functor $F : \mathbf{Str}_n \to \mathbf{Chu}(\mathbf{2}^n)$ as taking the n-ary relational structure (A, ρ) to the Chu space (A, R, v) defined as follows. Take R to consist of those n-tuples $r \in (2^A)^n$ of subsets of A for which $\prod_i r_i \subseteq \overline{\rho}$. Let $v : A \times R \to 2^n$ satisfy $v(a, r)_i = 1$ if $a \in r_i$, and 0 otherwise.*

It might seem that R could be represented more naturally and conveniently as just the power set of $\overline{\rho}$. But observe that a state r as defined here can be recovered from the set $\prod_i r_i$ of its n-tuples just when no component r_i is empty. The definition of $f^{-1} : S \to R$ in Definition 13 below requires each r_i to be available independently even when some are empty.

The crucial test of whether (A, R, v) faithfully represents (A, ρ) is whether ρ can be recovered from (A, R). We show this constructively as follows.

Lemma 10 *For all* $\mathbf{a} \in A^n$, $\mathbf{a} \in \rho \;\Leftrightarrow\; \forall r \in R\; \exists i < n : v(a_i, r)_i = 0$.

Proof:

$$\begin{array}{rlll} \mathbf{a} \in \rho & \Leftrightarrow & \forall r \in R : \mathbf{a} \notin \prod_i r_i & \text{(Construction of } R) \\ & \Leftrightarrow & \forall r \in R\; \exists i < n : a_i \notin r_i & \text{(Definition of product)} \\ & \Leftrightarrow & \forall r \in R\; \exists i < n : v(a_i, r)_i = 0 & \text{(Construction of } v) \end{array}$$

∎

Corollary 11 *F is injective on objects.*

Lemma 12 *(A, R, v) is extensional.*

Proof: If $v^r = v^{r'}$ then $\forall i [a \in r_i \Leftrightarrow a \in r'_i]$, so $\forall i : r_i = r'_i$, whence $r = r'$. ∎

If we regard R as a subset not of $(2^A)^n$ but of the isomorphic $(2^n)^A$, this makes Lemma 12 clear by qualifying (A, R) as an extensional object of $\mathbf{Chu}(2^n)$. We may view (A, R) for arbitrary $R \subseteq (2^n)^A$ as a generalization of $(A, \overline{\rho})$, which in the case $n = 1$ reduces to ordinary binary relations, which as previously noted capture topological spaces along with other similar structures such as complete lattices etc.

Definition 13 (F on maps). *Let $f : (A, \rho) \to (B, \sigma)$ be a homomorphism, with $F(A, \rho) = (A, R, v)$ and $F(B, \sigma) = (B, S, w)$ as per Definition 9. Define $f^{-1} : (2^n)^B \to (2^n)^A$ to take $g : B \to 2^n$ to $gf : A \to 2^n$. Now for all $s \in S$, $\prod_i s_i \subseteq \overline{\sigma}$ by construction of S. Hence $\prod_i f^{-1}s_i \subseteq \overline{\rho}$, by Lemma 8. Thus $f^{-1}s \in R$ by construction of R. We may therefore define $F(f)$ as (f, f^{-1}) where $f^{-1} : S \to R$.*

Theorem 14 *The functor F of Definitions 9 and 13 is concrete, faithful, and full.*

Proof: F is concrete by construction, and *a fortiori* faithful.

For fullness consider any Chu transform $(f, g) : F(A, \rho) \to F(B, \sigma)$ where $F(A, \rho) = (A, R)$ and $F(B, \sigma) = (B, S)$. If $\mathbf{a} \in \rho$, then for every $s \in S$ there exists $i < n$ such that

$$\begin{array}{rcll} v(a_i, gs)_i & = & 0 & \text{(Lemma 10 with } r = gs\text{)}, \\ \text{whence } w(fa_i, s)_i & = & 0 & \text{((}f, g\text{) is a Chu transform).} \end{array}$$

Hence by Lemma 10, $f\mathbf{a} \in \sigma$, establishing that f is a homomorphism. And since (A, R, v) is extensional, by Lemma 12, g is determined by f. Hence $F(f) = (f, g)$. ∎

Remarks. (i) Where size matters, R need contain only those states representable as the inverse image of a tuple of singletons. These can be characterized explicitly as those states r with the property that either $r_i = r_j$ or $r_i \cap r_j = \emptyset$ for all $i, j < n$, observing that f^{-1} preserves this property. (ii) Lemma 12 is an inessential bonus. Had Definition 9 produced a nonextensional (A, R, v), we would simply have enforced extensionality, needed for fullness, by identifying those states having the same extension.

Acknowledgements. I am deeply indebted to Michael Barr for many generously shared insights about Chu spaces, including the suggestion that semigroups might be realizable in **Chu**(2^3) (in the sense of a concrete full embedding [PT80, p.49]).

References

[Bar79] M. Barr. **-Autonomous categories, LNM 752.* Springer-Verlag, 1979.

[Bar91] M. Barr. *-Autonomous categories and linear logic. *Math Structures in Comp. Sci.*, 1(2), 1991.

[BG90] C. Brown and D. Gurr. A categorical linear framework for Petri nets. In J. Mitchell, editor, *Logic in Computer Science*, pages 208–218. IEEE Computer Society, June 1990.

[BvN36] G. Birkhoff and J. von Neumann. The logic of quantum mechanics. *Annals of Mathematics*, 37:823–843, 1936.

[CCMP91] R.T Casley, R.F. Crew, J. Meseguer, and V.R. Pratt. Temporal structures. *Math. Structures in Comp. Sci.*, 1(2):179–213, July 1991.

[DM60] A. De Morgan. On the syllogism, no. IV, and on the logic of relations. *Trans. Cambridge Phil. Soc.*, 10:331–358, 1860.

[Dun86] J.M. Dunn. Relevant logic and entailment. In D. Gabbay and F. Guenthner, editors, *Handbook of Philosophical Logic*, volume III, pages 117–224. Reidel, Dordrecht, 1986.

[FK72] P. Freyd and G. M. Kelly. Categories of continuous functors I. *Journal of Pure and Applied Algebra*, 2(3):169–191, 1972.

[Gir87] J.-Y. Girard. Linear logic. *Theoretical Computer Science*, 50:1–102, 1987.

[HL69] Z. Hedrlín and J. Lambek. How comprehensive is the category of semigroups. *J. Algebra*, 11:195–212, 1969.

[Jón88] B. Jónsson. Relation algebras and Schröder categories. *Discrete Mathematics*, 70:27–45, 1988.

[JT48] B. Jónsson and A. Tarski. Representation problems for relation algebras. *Bull. Amer. Math. Soc.*, 54:80,1192, 1948.

[JT52] B. Jónsson and A. Tarski. Boolean algebras with operators. Part II. *Amer. J. Math.*, 74:127–162, 1952.

[LS91] Y. Lafont and T. Streicher. Games semantics for linear logic. In *Proc. 6th Annual IEEE Symp. on Logic in Computer Science*, pages 43–49, Amsterdam, July 1991.

[Mul86] C.J. Mulvey. &. In *Second Topology Conference*, Rendiconti del Circolo Matematico di Palermo, ser.2, supplement no. 12, pages 99–104, 1986.

[Pei33] C.S. Peirce. Description of a notation for the logic of relatives, resulting from an amplification of the conceptions of Boole's calculus of logic. In *Collected Papers of Charles Sanders Peirce. III. Exact Logic.* Harvard University Press, 1933.

[Pra86] V.R. Pratt. Modeling concurrency with partial orders. *Int. J. of Parallel Programming*, 15(1):33–71, February 1986.

[Pra92a] V.R. Pratt. The duality of time and information. In *Proc. of CONCUR'92, LNCS 630*, pages 237–253, Stonybrook, New York, August 1992. Springer-Verlag.

[Pra92b] V.R. Pratt. Event spaces and their linear logic. In *AMAST'91: Algebraic Methodology and Software Technology*, Workshops in Computing, pages 1–23, Iowa City, 1992. Springer-Verlag.

[Pra92c] V.R. Pratt. Origins of the calculus of binary relations. In *Proc. 7th Annual IEEE Symp. on Logic in Computer Science*, pages 248–254, Santa Cruz, CA, June 1992.

[Pra93] V.R. Pratt. Linear logic for generalized quantum mechanics. In *Proc. Workshop on Physics and Computation (PhysComp'92)*, Dallas, 1993. IEEE.

[PT80] A. Pultr and V. Trnková. *Combinatorial, Algebraic and Topological Representations of Groups, Semigroups, and Categories.* North-Holland, 1980.

[Sch95] E. Schröder. *Vorlesungen über die Algebra der Logik (Exakte Logik). Dritter Band: Algebra und Logik der Relative.* B.G. Teubner, Leipzig, 1895.

[See89] R.A.G Seely. Linear logic, $*$-autonomous categories and cofree algebras. In *Categories in Computer Science and Logic*, volume 92 of *Contemporary Mathematics*, pages 371–382, held June 1987, Boulder, Colorado, 1989.

[Tar41] A. Tarski. On the calculus of relations. *J. Symbolic Logic*, 6:73–89, 1941.

[Trn66] V. Trnková. Universal categories. *Comment. Math. Univ.Carolinae*, 7:143–206, 1966.

[Vic89] S. Vickers. *Topology via Logic.* Cambridge University Press, 1989.

An Introduction to Dynamic Labeled 2-structures

A. Ehrenfeucht

Department of Computer Science
University of Colorado at Boulder
Boulder, CO 80309, U.S.A.

and

G. Rozenberg

Department of Computer Science
Leiden University
P.O. Box 9512
2300 RA Leiden, The Netherlands

Department of Computer Science
University of Colorado at Boulder
Boulder, CO 80309, U.S.A.

Abstract. The notion of a *dynamic labeled 2-structure* is introduced and investigated. It generalizes the notion of a labeled 2-structure (ℓ2s), see [ER1], by making it possible to change the (label) relationships between the nodes. This is achieved by storing in the nodes of a ℓ2s output and input functions which can change the outgoing and incoming labels, respectively. The notion of a clan which is central in the theory of ℓ2s's is transferred to the framework of dℓ2s's, and the basic properties of clans of dℓ2s's are investigated.

Introduction

The theory of *2-structures* forms a convenient framework for considering various kinds of formal structures encountered in mathematics and computer science (see, e.g., [ER1], [ER2], and [ER4]). A *labeled 2-structure*, abbreviated ℓ2s, is a finite domain D together with a (labeling) function λ from the set of all 2-edges over D into some alphabet Δ (a *2-edge* over D is an ordered pair of different elements of D). Hence a ℓ2s $g = (D,\Delta,\lambda)$ may be seen as representing a set of nodes D together with the relationships between all pairs of different nodes, those relationships are given by λ. Such a ℓ2s is *static* in the sense that the relationships given by λ are given once and forever - they cannot change. This may be a disadvantage in modeling systems that are dynamic, where the relationships between elements of a system may change during the evolution of a system; graph grammars and computer networks are examples of systems of such a nature.

In this paper we introduce the notion of a *dynamic labeled 2-structure*, abbreviated dℓ2s. It consists of two components:

(i) a set of nodes D where in each node x a set of output functions O_x and a set of input functions I_x is stored; this component is called a *mutating scheme*,

and

(ii) a set G of ℓ2s's with D as the common domain, where G is closed w.r.t. transformations of the 2-edges induced by output and input functions in the nodes of D.

Applying an output function φ in a node x changes the current label $\lambda(x,y)$ of each outgoing 2-edge (x,y) into $\varphi(\lambda(x,y))$, and similarly, applying an input function γ in a node x changes the current label $\lambda(y,x)$ of each incoming 2-edge (y,x) into $\gamma(\lambda(y,x))$.

Clearly, in order to make the above setup workable (e.g., what happens to the label $\lambda(x,y)$ if "simultaneously" an output function is applied in x and an input function is applied in y?) one needs to make some basic assumptions about the underlying mutating scheme. To this aim we will give a set of four axioms and consider only mutating schemes satisfying this set of axioms - such mutating schemes are called *simply transitive*.

It turns out that (simply transitive) dℓ2s's have an elegant mathematical structure. First of all we prove that the sets of output (input) functions in all nodes are equal, and moreover each of them is a group. Then we prove that also the set of labels with a suitably chosen operation forms a group $\underline{\Delta}$, and applying an output function φ in a node x amounts to the left multiplication in $\underline{\Delta}$ of labels $\lambda(x,y)$ by a specific symbol a associated with φ, while applying an input function γ in a node x amounts to the right multiplication in Δ of labels $\lambda(y,x)$ by a specfic symbol b associated with γ. Thus the investigation of dℓ2s's happens to large extent within the framework of the theory of groups.

In this paper we demonstrate how to transfer some of the central notions of the theory of ℓ2s's such as *reversibility* and *clans* into the framework of dℓ2s's. In particular we show that the notion of reversibility carries over quite naturally to dℓ2s's using involutions on the group of labels. Also, the notion of a clan carries over naturally to the framework of dℓ2s's. Using a basic technique of *transformations w.r.t. horizons* we prove some basic properties of clans, and in particular we prove that the set of clans of a dℓ2s has strong closure properties.

This paper should be seen as an exposition of [ER5] which is a full paper and contains the proofs of all the results given here, as well as many additional comments, examples, technical results, etc., concerning the properties of d*ℓ*2s's.

0. Preliminaries

In this section we settle some basic terminology and notation concerning sets and semigroups and we recall some notions concerning labeled 2-structures.

We assume the reader to be familiar with basic theory of semigroups and groups (see, e.g., [BB]). The familiarity with the theory of 2-structures (see, e.g., [ER1]) will give the reader more intuition concerning the notions and results of this paper, however, in order to make the paper self contained in this section we recall the basic notions of labeled 2-structures.

For a set X, $|X|$ denotes the cardinality of X, and $E_2(X)$ denotes the set of all *2-edges over* X, i.e., $E_2(X) = \{(u,v): u,v \in X \text{ and } u \neq v\}$. The empty set is denoted by $\emptyset$.

For sets X and Y, we use Y^X to denote the set of all functions from X to Y. The identity mapping on X is denoted by id_X . A bijection $\varphi : X \to X$ is an *involution* iff $\varphi^2 = id_X$.

For monoids $A = (X,\circ)$ and $B = (Y,\Box)$ a bijection $\varphi: X \to Y$ is an *antiisomorphism of* A *onto* B, iff for all $u,v \in X$, $\varphi(u \circ v) = \varphi(v) \Box \varphi(u)$. If $A = B$, then φ is an *antiautomorphism of* A.

A *labeled 2-structure* (abbreviated *ℓ2s*) is a 3-tuple $g = (D,\Delta,\lambda)$, where D is a finite nonempty set, Δ is a finite alphabet, and λ is a function from $E_2(D)$ into Δ. The set of *nodes* D is called the *domain of* g, denoted *dom*(g), λ is called the *labeling function of* g, denoted lab_g , and Δ is called the *alphabet* of g, denoted *alph*(g). It is usually assumed that Δ is the *useful* alphabet, i.e., $\Delta = \{ \lambda(x,y) : (x,y) \in E_2(D) \}$. Then g may be given in the form (D,λ).

A *ℓ*2s g may be seen as a complete directed graph on D with edges labeled by elements of Δ and so we may use the usual pictorial representation of directed graphs to give *ℓ*2s's.

The basic technical notion concerning *ℓ*2s's is the notion of a clan. A set $X \subseteq D$ is a *clan* of g iff for all $x,y \in X$ and all $z \in D-X$, $\lambda(z,x) = \lambda(z,y)$ and $\lambda(x,z) = \lambda(y,z)$. Hence a clan of g is a subset of D such that each element z of D outside X "sees" all elements of X in the same way, and all elements of X "see" z in the same way.

The set of all clans of g is denoted by $C(g)$.

It follows directly from the definition of a clan that $\emptyset \in C(g)$,

$D \in C(g)$, and for each $x \in D$, $\{x\} \in C(g)$, these clans are called the *trivial clans* - the set of trivial clans of g is denoted by $TC(g)$. We call g *primitive* iff it has only trivial clans.

Example 0.1.

Let g be the following *ℓ2s*:

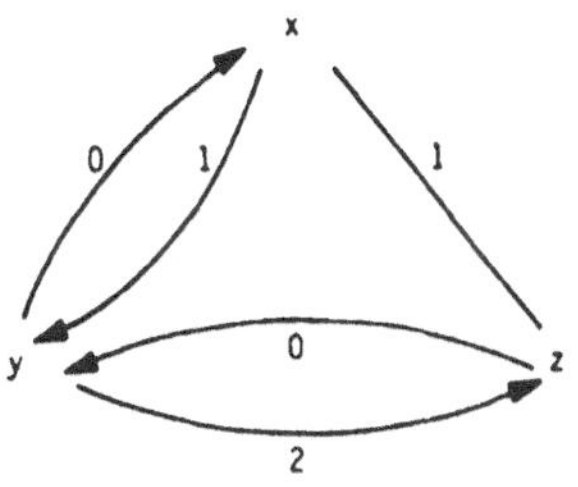

Then $C(g) = TC(g) \cup \{ \{x,v\}, \{y,u,z\}, \{y,u\}, \{y,z\}, \{u,z\} \}$, and so g is not primitive. □

A *ℓ2s* $g = (D,\lambda)$ is called *reversible* iff for all $(x,y), (u,z) \in E_2(D)$, $\lambda(x,y) = \lambda(u,z)$ implies $\lambda(y,x) = \lambda(z,u)$. Technically it is easier to deal with reversible rather than arbitrary *ℓ2s*'s (e.g., to prove that $X \subseteq D$ is a clan of g it suffices to prove only that $\lambda(z,x) = \lambda(z,y)$ for each $z \in D-X$ and all $x,y \in X$). Moreover, in the research centered around the properties of clans, it suffices to consider only reversible *ℓ2s*'s because (as proved in [ER1]) for each *ℓ2s* g there exists a reversible *ℓ2s* h such that $C(g) = C(h)$.

We will use the following terminology concerning transformation semigroups.

Let Δ be a set (possibly infinite) and let T_Δ be the full transformation semigroup on Δ. A subsemigroup Q of T_Δ is *transitive* iff for all $a,b \in \Delta$ there exists a $\psi \in Q$ such that $\psi(a) = b$. The *centralizer* of Q, denoted CN(Q), is the set $\{ \varphi \in T_\Delta : \psi\varphi = \varphi\psi \text{ for all } \psi \in Q \}$. Two transformation semigroups Q and R *permute* iff $\psi\varphi = \varphi\psi$ for all $\varphi \in Q$, $\psi \in R$.

A permutation group Q is *simply transitive* iff Q is transitive and, for each $a \in \Delta$, $\{ \psi \in Q : \psi(a) = a \} = \{ id_\Delta \}$.

1. Mutating Schemes and Dynamic Labeled 2-structures

In this section the basic notion of this paper - a *dynamic labeled 2-structure* - is introduced. It is based on the notion of a mutating scheme.

A possible intuition for the notion of a mutating scheme is that of the set D of nodes of a network. Each such node x is a processor of a network and the actions of such a processor consist of *output* actions O_x and *input* actions I_x . In a given global state of a network all kinds of relationships hold between the nodes of the network. Hence a global state is represented by a labeled 2-structure g with *dom*(g) = D where for the 2-edge (x,y), $lab_g(x,y) = b$ says that b is the relationship between x and y in this global state (g). Then, when an output action $\varphi \in O_x$ takes place in x it will affect the relationships between x and the other nodes by changing the label $lab_g(x,y)$ of each outgoing 2-edge (x,y) to $\varphi(lab_g(x,y))$. Analogously, an input action $\eta \in I_x$ will change the label $lab_g(y,x)$ of each incoming 2-edge (y,x) to $\gamma(lab_g(y,x))$.

Formally a mutating scheme is defined as follows.

Definition 1.1. A *mutating scheme* is a 4-tuple $K = (D,\Delta,O,I)$, where D and Δ are nonempty sets,
$O = \{ O_x : x \in D \}$,
$I = \{ I_x : x \in D \}$, and
for each $x \in D$, $O_x \subseteq \Delta^\Delta$ and $I_x \subseteq \Delta^\Delta$. □

We also say that K is a *mutating scheme on* D or more specifically that K is a Δ-*mutating scheme on* D. Here D is called the *domain of* K, denoted by *dom*(K), and Δ is called the *alphabet of* K, denoted by *alph*(K). For each $x \in D$, O_x are the *output functions in* x, and I_x are the *input functions in* x.

We will use $out_K(x)$ to denote O_x and $inp_K(x)$ to denote I_x. Also, throughout the paper we will use the symbol φ (possibly with an index) to denote an output function (for some O_x) and we will use the symbol γ (possibly with an index) to denote an input function (for some I_x).

Clearly in specifying K we may omit Δ, since it is given by the functions from O (and I); however it is often convenient to keep Δ in the specification of K.

We will formalize now how to use a function stored in a node of a $\ell2s$ to change the labels of the 2-edges outgoing from and incoming to this node.

Definition 1.2. Let $g = (D,\lambda)$ be a $\ell2s$. Let $x \in D$ and let $\varphi,\psi \in \Delta^\Delta$, where $alph(g) \subseteq \Delta$.

(1) The *output* ψ*-renaming of* g *in* x, denoted $oren_\psi(g,x)$, is the $\ell2s$ (D,λ'), where, for each $(u,v) \in E_2(D)$,

$$\lambda'(u,v) = \begin{cases} \lambda(u,v) & \text{if } u \neq x, \\ \psi(\lambda(u,v)) & \text{if } u = x. \end{cases}$$

(2) The *input* ψ*-renaming of* g *in* x, denoted $iren_\psi(g,x)$, is the $\ell2s(D,\lambda')$ where, for each $(u,v) \in E_2(D)$,

$$\lambda'(u,v) = \begin{cases} \lambda(u,v) & \text{if } v \neq x, \\ \psi(\lambda(u,v)) & \text{if } v = x. \end{cases}$$

(3) The (φ,ψ)*-renaming of* g *in* x, denoted $ren_{(\varphi,\psi)}(g,x)$, is the $\ell2s$ $oren_\varphi(iren_\psi(g,x),x)$. □

Clearly, in (3) above, $oren_\varphi(iren_\psi(g,x),x) = iren_\psi(oren_\varphi(g,x),x)$.

Choosing an output and an input function in each node of a $\ell 2s$ is formalized through the notion of a selector.

Definition 1.3. Let K be a mutating scheme. A *selector for* K is a function $S : D \to \bigcup_{x\in D} O_x \times I_x$ such that for each $x \in D$, $S(x) = (\varphi,\gamma)$, where $\varphi \in O_x$ and $\gamma \in I_x$. □

As pointed out already, a mutating scheme K may be seen as a set of nodes D, where each node $x \in D$ contains output functions O_x and input functions I_x. If we span a $\ell 2s$ g on D, then as the result of applying the output and input functions in the nodes of D, g will be transformed into a family of $\ell 2s$'s. Now if we have a set G of $\ell 2s$'s such that for each $g \in G$ the family of $\ell 2s$'s resulting from transforming g within K will be contained in G, then G may be seen as a possible "implementation" of K. Formally we get the following definition.

Definition 1.4. Let K be a mutating scheme.

(1) A $\ell 2s$ g is *compatible with* K iff $dom(g) = dom(K)$ and $alph(g) \subseteq alph(K)$.

(2) A set G of $\ell 2s$'s *implements* K iff for each $g \in G$, g is compatible with K, $oren_\varphi(g,x) \in G$ and $iren_\gamma(g,x) \in G$ for all $x \in dom(g)$, $\varphi \in out_K(x)$, and $\gamma \in inp_K(x)$. □

We use $comp(K)$ to denote the family of all $\ell 2s$'s compatible with K.

We will consider now a set of axioms for mutating schemes.

Definition 1.5. A mutating scheme $K = (D,\Delta,\mathcal{O},\mathcal{I})$ is called *simply transitive* iff it satisfies the following four axioms.

A1. For each $x \in D$, $out_K(x)$ and $inp_K(x)$ are closed under compositions.

A2. For all $a,b \in \Delta$ and each $x \in D$ there exist $\varphi \in out_K(x)$ and $\psi \in inp_K(x)$ such that $\varphi(a) = b$ and $\psi(a) = b$.

A3. For all $(x,y) \in E_2(D)$, $\varphi \in out_K(x)$, $\gamma \in inp_K(y)$ and $a \in \Delta$, $\varphi(\gamma(a)) = \gamma(\varphi(a))$.

A4. $|D| \geq 3$. □

In this paper we will consider only simply transitive mutating schemes.

We have already used the term "simply transitive" for a subfamily of permutation groups. The technical reason for using this term again will become apparent in Section 2 (see Remark 2.1).

The Axioms A1 through A3 have a rather clear intuition.

Axiom A1 says that each change of labels outgoing from x (incoming to x) that can be achieved by applying a composition of functions from $O_x(I_x)$ can be achieved in "one stroke" by using one output (input) function in x.

Axiom A2 says that if one wishes to change a specific label a outgoing from x (incoming to x) into a specific label b outgoing from x (incoming to x) then it is possible to do it by appropriately choosing an output (an input) function in x.

Axiom A3 allows applying output and input functions concurrently throughout any *ℓ2s* spanned on D.

Axiom A4 is rather technical - its usefulness will be discussed in the next section.

Given a simply transitive mutating scheme K, each choice of a selector for K leads to a specific transformation of the class of *ℓ2s*'s spanned on the domain of K.

Definition 1.6. Let K be a simply transitive mutating scheme and let S be a selector for K such that, for each $x \in D = dom(K)$, $S(x) = (\varphi_x , \gamma_x)$. The *transformation induced by* S, denoted by tr_S , is the function from *comp*(K) into *comp*(K) defined by: for $g = (D,\lambda) \in comp(K)$, $tr_S(g) = (D,\lambda')$, where for each $(x,y) \in E_2(D)$, $\lambda'(x,y) = \gamma_y(\varphi_x (\lambda(x,y)))$. □

Also the following result follows easily from Axiom A3.

Theorem 1.1. Let K be a simply transitive mutating scheme and let S be a selector for K. For each $g \in comp(K)$,
$tr_S(g) = ren_{S(x_m)}(ren_{S(x_{m-1})}(\ldots ren_{S(x_2)}(ren_{S(x_1)}(g,x_1),x_2),\ldots),x_m)$,
where x_1 , x_2 , ..., x_m is an arbitrary ordering of D. □

The above theorem is a typical "sequentialization lemma" in the theory of concurrent systems (see, e.g., [R]) or in the theory of graph grammars (see, e.g., [EKR]). The result of applying renamings concurrently in all the nodes of g is equivalent to applying renamings sequentially, i.e., node-by-node, and the order in which nodes are chosen (for applying renamings) is irrelevant, i.e., the resulting *ℓ2s* will always be the same.

We are ready now to define the main notion of this paper.

Definition 1.7.

(1) A *dynamic labeled 2-structure*, abbreviated *dℓ2s*, is an ordered pair (K,G) such that K is a simply transitive mutation scheme and G is a set of *ℓ2s*'s implementing K.

(2) A *dℓ2s* (K,G) is a *single axiom dℓ2s* iff there exists a $g \in G$ such that for each $h \in G$ there exists a selector S for K such that $h = tr_S(g)$. □

A g satisfying (2) above is called an *axiom* of (K,G).

Given a *dℓ2s* $\mathcal{A} = (K,G)$, we can specify G as an ordered pair (D,Λ), where D is the common domain of all *ℓ2s*'s of G (i.e, $D = dom(K)$) and Λ is the set of labeling functions of $E_2(D)$ such that $G = \{ (D,\lambda) : \lambda \in \Lambda \}$. If $\mathcal{A}$ is a single axiom *dℓ2s* and g_0 is an axiom of $\mathcal{A}$, then we can specify $\mathcal{A}$ in the form (K,g_0).

2. The Group of Input and Output Functions

In this section we will investigate the structure of the set of output and input functions stored in a node of a simply transitive mutating scheme K as well as the relationship between the sets of transformations stored in different nodes of K.

In order to simplify the statements of results, unless explicitely clear otherwise, *in the rest of this paper we will consider an arbitrary but fixed simply transitive mutating scheme* $K = (D,\Delta,O,I)$. *Hence whenever in the sequel we write that* $G = (D,\Delta)$ *is a dℓ2s, we mean that* (K,G) *is a dℓ2s.*

Note that in terms of semigroups Axiom A1 says that all O_x and I_x are transformation semigroups, Axiom A2 says that all O_x and I_x are transitive semigroups, and Axiom A3 says that O_x and I_y permute for all $x \neq y$.

The following three results give the basic properties of the sets of input and output functions.

Theorem 2.1. For each $(x,y) \in E_2(D)$, O_x and I_y are isomorphic transitive groups of permutations. □

Theorem 2.2. Let $x \in D$ and let $\Psi \in \{ out_K(x), inp_K(x) \}$.
(1) For all $\psi \in \Psi$, if $\psi(a) = a$ for some $a \in \Delta$, then $\psi = id_\Delta$.
(2) For all $\psi_1 , \psi_2 \in \Psi$, if $\psi_1(a) = \psi_2(a)$ for some $a \in \Delta$, then $\psi_1 = \psi_2$. □

Theorem 2.3. For all $x,y \in D$, $O_x = O_y$ and $I_x = I_y$. □

Thus we have two sets of functions only: the set of *output functions* O_K such that, for each $x \in D$, $out_K(x) = O_K$, and the set of *input functions* I_K such that, for each $x \in D$, $inp_K(x) = I_K$. The set of output functions O_K with the operation of composition forms the group $\underline{O}_K$ and the set of input functions I_K with the operation of composition forms the group $\underline{I}_K$.

Remark 2.1. Note that Theorems 2.1 and 2.2(1) imply that $\underline{O}_K$ and $\underline{I}_K$ are isomorphic *simply transitive* groups of permutations. This justifies the name "simply transitive" given to mutating schemes satisfying Axioms A1 through A4. □

Let a_0 be a fixed but arbitrary element of Δ.

Then for each $b \in \Delta$ let φ_b be the element of O_K such that $\varphi_b(a_0) = b$, and let γ_b be the element of I_K such that $\gamma_b(a_0) = b$. The existence and the uniqueness of φ_b and γ_b follows from Axiom A2 and Theorem 2.2(2). Hence to specify a selector S for K, we may also write $S(x) = (a,b)$ for $x \in D$, where $a,b \in \Delta$ are such that $S(x) = (\varphi_a , \gamma_b)$.

Our next result is a refinement of Theorem 2.1.

Theorem 2.4. Let $\beta : O_K \to I_K$ be such that, for each $a \in \Delta$, $\beta(\varphi_a) = \gamma_a$. Then β is an antiisomorphism from $\underline{O}_K$ onto $\underline{I}_K$. □

We will investigate now the structure of the set of selectors for K.

The following lemma follows directly from the definitions.

Lemma 2.1. Let K be a simply transitive mutating scheme.
(1) Let S be a selector for K, and let $T : D \to O_K \times I_K$ be defined by: for all $x \in D$, $T(x) = (\varphi^{-1}, \gamma^{-1})$, where $S(x) = (\varphi, \gamma)$. Then T is also a selector for K. Moreover, for each *ℓ2s* g compatible with K, $tr_T(tr_S(g)) = tr_S(tr_T(g)) = g$.
(2) Let S_1, S_2 be selectors for K and let $T : D \to O_K \times I_K$ be defined by: for all $x \in D$, $T(x) = (\varphi_2\varphi_1, \gamma_2\gamma_1)$, where $S_1(x) = (\varphi_1, \gamma_1)$ and $S_2(x) = (\varphi_2, \gamma_2)$. Then T is also a selector for K.
Moreover, for each *ℓ2s* g compatible with K, $tr_T(g) = tr_{S_2}(tr_{S_1}(g))$. □

The selector T from (1) above is thus the *inverse of* S, denoted S^{-1}, and the selector T from (2) above is thus the *composition of* S_1 *with* S_2, denoted S_1 *com* S_2.

Let TR(K) denote the set of all transformations tr_S induced by all selectors S for K, and let $\cdot$ denote the composition of functions. Lemma 2.1 implies the following result.

Theorem 2.5. (TR(K),·) is a group. □

If K is a simply transitive mutating scheme, $g \in comp(K)$ and S is a selector for K, then applying S to g (i.e., computing $tr_S(g)$) is a *concurrent* operation: in *each* node x of g the (φ_x, γ_x)-renaming of g takes place, where $S(x) = (\varphi_x, \gamma_x)$. Since we know now that $\underline{O}_K$ and $\underline{I}_K$ are monoids (even groups), the *sequential* operations of renaming of g in a single node can be also performed by a selector. It is clear that for a node x and $\varphi \in O_K$, $\gamma \in I_K$, $ren_{(\varphi,\gamma)}(g,x) = tr_S(g)$, where S is such that $S(x) = (\varphi,\gamma)$ and $S(y) = (id_\Delta, id_\Delta)$ for $y \neq x$ (where $\Delta = alph(K)$). Hence, by Theorem 1.1, (TR(K),·) is generated by renamings. As a matter of fact (TR(K),·) is a subsemigroup of the transformation semigroup $T_{comp(K)}$, and (TR(K),·) is a subgroup of the group of permutations on *comp*(K).

It is instructive to notice that in a single axiom *dℓ2s* G one can move between arbitrary elements of G by choosing an appropriate selector.

Lemma 2.2. Let G be a single axiom *dℓ2s*. For each ordered pair (g_1, g_2) of elements of G there exists a selector S for g_1 such that $tr_S(g_1) = g_2$. □

3. The Group of Labels

In this section we demonstrate how the group of input and output functions induces a group on labels and consequently the relabeling operations in K become simply group operations in the group of labels.

Definition 3.1. Let $\circ$ be the binary operation on Δ defined by: for all $a,b,c \in \Delta$, $a \circ b = c$ iff $\varphi_a\varphi_b = \varphi_c$. □

Obviously $\underline{\Delta} = (\Delta,\circ)$ is a well-defined group, and $\alpha: \Delta \to O_K$ be defined by: for all $a \in \Delta$, $\alpha(a) = \varphi_a$, is an isomorphism of $\underline{\Delta}$ onto $\underline{O}_K$. Hence by Theorem 2.4 the mapping $\beta: \Delta \to I_K$ such that $\beta(a) = \gamma_a$ for all $a \in \Delta$, is an antiisomorphism of $\underline{\Delta}$ onto $\underline{I}_K$.

The basic properties of the operation $\circ$ is given by the following results.

Theorem 3.1. For all $a,b \in \Delta$, $a \circ b = \varphi_a(b)$. □

Theorem 3.2. For each $b \in \Delta$, each $\varphi \in O_K$, and each $\gamma \in I_K$, $\varphi(b) = \varphi(a_0) \circ b$ and $\gamma(b) = b \circ \gamma(a_0)$. □

Hence computing $\varphi(b)$ amounts to multiplying on the left (in $\underline{\Delta}$) by $\varphi(a_0)$, and computing $\gamma(b)$ amounts to multiplying on the right (in $\underline{\Delta}$) by $\gamma(a_0)$. Consequently to specify K it suffices to give D and $\underline{\Delta}$; we may write $K = (D,\underline{\Delta})$.

In the above we have investigated the basic properties of simply transitive mutating schemes, i.e., mutating schemes satisfying Axioms A1 through A4. It is not difficult to give examples of simply transitive mutating schemes - hence Axioms A1 through A4 are consistent. As a matter of fact we will demonstrate now that any group induces a simply transitive mutating scheme.

Definition 3.2. Let $\underline{\Delta} = (\Delta,\square)$ be a group and let D be a nonempty set. Let $K(D,\underline{\Delta}) = (D,\Delta,O,I)$ be the Δ-mutating scheme on D such that, for each $x \in D$,
$\varphi \in O_x$ iff there exists $a \in \Delta$ such that, for all $b \in \Delta$, $\varphi(b) = a \square b$, and
$\gamma \in I_x$ iff there exists $a \in \Delta$ such that, for all $b \in \Delta$, $\gamma(b) = b \square a$. □

Theorem 3.3. $K(D,\underline{\Delta})$ satisfies Axioms A1, A2, A3. □

This if $|D| \geq 3$, then $K(D,\underline{\Delta})$ is a simply transitive mutating scheme.

Axiom A4 requiring that $|D| \geq 3$ has been used only in the proof of Theorem 2.3. Hence requiring that $|D| \geq 3$ guarantees that we have only one set of output functions and one set of input functions in K. Consequently we deal with only one group $(\Delta,\circ)$. If $|D| = 2$ (and A1 through A3 are satisfied), then we *may* have two sets of output functions O_x , O_y and two sets of input functions I_x , I_y (where $D = \{x,y\}$). Then we deal with two groups $\underline{\Delta}_1$ and $\underline{\Delta}_2$ where $\underline{\Delta}_1$ "governs" the 2-edge (x,y) and $\underline{\Delta}_2$ "governs" the 2-edge (y,x). This means that applying a $\varphi \in O_x$ in x means multiplying the label of (x,y) by some b on the left in $\underline{\Delta}_1$, while applying a $\gamma \in I_y$ in y means multiplying the label of (x,y) by some b on the right in $\underline{\Delta}_1$. Similarly, applying a $\varphi \in O_y$ in y means multiplying the label of (y,x) by some b on the left in $\underline{\Delta}_2$, while applying a $\gamma \in I_x$ in x means multiplying the label of (y,x) by some b on the right in $\underline{\Delta}_2$.

4. Reversibility

Reversibility is a very important and convenient feature of 2-structures. In this section we discuss how to transfer it into the framework of dynamic 2-structures.

We begin with the following definition.

Definition 4.1. Let $\delta : \Delta \to \Delta$ be an involution. A ℓ2s (D,λ) is *δ-reversible* iff for all $(x,y) \in E_2(D)$, $\lambda(x,y) = \delta(\lambda(y,x))$. □

Note that for an involution δ on Δ we have $\delta\delta(a) = a$ for each $a \in \Delta$. Consequently, for each $a \in \Delta$, $\delta^{-1}\delta\delta(a) = \delta^{-1}(a)$ and so $\delta(a) = \delta^{-1}(a)$. Hence $\delta = \delta^{-1}$.

In order to simplify the statements of results, *unless explicitely clear otherwise, for the rest of this paper we will fix an arbitrary involutive antiautomorphism of $\underline{\Delta}$ (i.e., an antiautomorphism δ of $\underline{\Delta}$ that is an involution).*

Definition 4.2. $\varphi \in O_K$ and $\gamma \in I_K$ are *δ-conjugated* iff $\varphi = \delta\gamma\delta (= \delta^{-1}\gamma\delta)$.□

We also say that the ordered pair (φ,γ) is *δ-conjugated.*

The following result provides an alternative definition of δ-conjugated pairs.

Theorem 4.1. For $\varphi \in O_K$ and $\gamma \in I_K$, (φ,γ) is δ-conjugated iff for each δ-reversible ℓ2s $g = (D,\lambda)$ and each $x \in D$, $ren_{(\varphi,\gamma)}(g,x)$ is δ-reversible. □

δ-conjugated pairs can be characterized as follows.

Theorem 4.2. For $a,b \in \Delta$, (φ_a , γ_b) is δ-conjugated iff $b = \delta(a)$. □

By Theorem 4.1 if we deal with δ-reversible ℓ2s's only, then we have to use selectors that use δ-conjugated pairs only.

Definition 4.3. A selector S for K is a *δ-selector* iff for each $x \in D$, $S(x)$ is δ-conjugated. □

Note that in specifying a δ-selector S it suffices to specify for each $x \in D$ only the first component of $S(x)$, i.e., rather to write that $S(x) = (\varphi_a , \gamma_b)$ (or simply $S(x) = (a,b)$) we may write $S(x) = \varphi_a$ (or simply $S(x) = a$) because γ_b is uniquely determined, viz. $\gamma_b = \gamma_{\delta(a)}$ (resp., $b = \delta(a)$).

Since we want to deal with δ-reversible ℓ2s's, the definition of an implementation of K has to be modified as follows.

Definition 4.4.

(1) A set G of ℓ2s's is a *δ-implementation of* K iff

(i) each $g \in G$ is compatible with K,

(ii) each $g \in G$ is δ-reversible, and

(iii) for each $g \in G$, and each δ-selector S, $tr_S(g) \in G$.

(2) A dℓ2s $\mathcal{A} = (K,G)$ is *δ-reversible* (or simply *reversible*) iff G is a δ-implementation of K. □

Note that to specify a single axiom δ-reversible dℓ2s it suffices to specify the group $\underline{\Delta}$ of labels, δ, and an axiom of G.

5. Clans, Horizons and Translations

The notion of a clan is central in the theory of ℓ2s's. In this section we transfer this notion into the framework of dℓ2s's. We also introduce and investigate the technical notions of a horizon and of a translation (w.r.t. a horizon) and demonstrate that usefulness for the investigation of clans in dℓ2s's.

We begin by defining the notion of a clan for a dℓ2s.

Definition 5.1. Let $\mathcal{G} = (D,\Lambda)$ be a dℓ2s and let $X \subseteq D$. X is a *clan of* $\mathcal{G}$ iff there exists $g \in \mathcal{G}$ such that $X \in C(g)$. □

We will use $C(\mathcal{G})$ to denote the set of clans of $\mathcal{G}$.

Let $Q = (Z,\circ)$ be an arbitrary group.

Then let $\hat{Q} = (Z \times Z, \square)$ be the group such that for all $(a,b),(c,d) \in Z \times Z$, $(a,b) \square (c,d) = (a \circ c, d \circ b)$, and let $\delta : Z \times Z \to Z \times Z$ be defined by: $\delta(a,b) = (b,a)$ for all $(a,b) \in Z \times Z$.

For $(a,b),(c,d) \in Z \times Z$ we have

$\delta((a,b) \square (c,d)) = \delta(a \circ c, d \circ b) = (d \circ b, a \circ c) = (d,c) \square (b,a) = \delta(c,d) \square \delta(a,b)$.

Hence δ is an involutive antiautomorphism of $\hat{Q}$.

Now let $g = (D,\lambda)$ be a $\ell 2s$ such that *alph*$(g) \subseteq Z$. Then let $\hat{g} = (D,\hat{\lambda})$ where, for each $(x,y) \in E_2(D)$, $\hat{\lambda}(x,y) = (\lambda(x,y), \lambda(y,x))$. It is easily seen that $\hat{g}$ is δ-reversible; $\hat{g}$ is the *δ-reversible version of* g. As a matter of fact the construction of $\hat{g}$ is the standard construction from [ER1] to obtain the reversible version of a $\ell 2s$, where it was also proved that $C(g) = C(\hat{g})$.

Consider $(x,y) \in E_2(D)$.

Let S be a selector in the simply transitive mutating scheme K determined by D and Q. Let $S(x) = (a,b)$ and for the clarity of considerations assume that, for all $z \in D - \{x\}$, $S(z) = (e,e)$ where e is the identity of Q.

Assume that in g, $\lambda(x,y) = p$ and $\lambda(y,x) = q$:

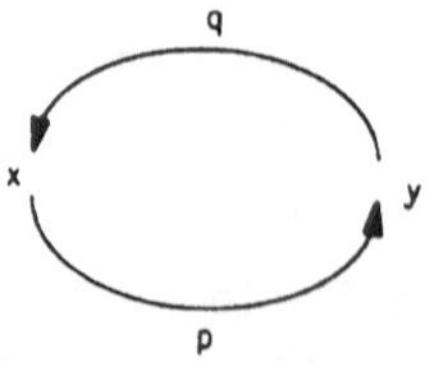

Then in $\hat{g}$ we have:

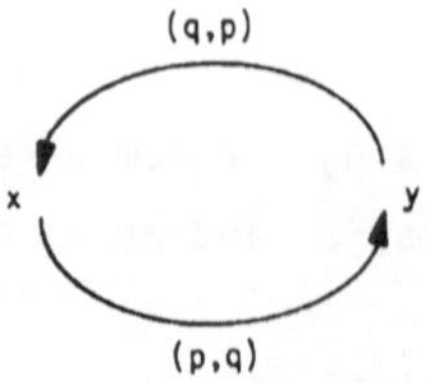

Consider the selector $\hat{S}$ in the simply transitive mutating scheme $\hat{K}$ determined by D and $\hat{Q}$ such that, for each $z \in D$, $\hat{S}(z) = ((c,d),(d,c))$ if $S(z) = (c,d)$. Hence $\hat{S}(x) = ((a,b),(b,a))$ and $\hat{S}(t) = ((e,e),(e,e))$ for all $t \in D - \{x\}$.

Hence in $tr_S(g)$ we have:

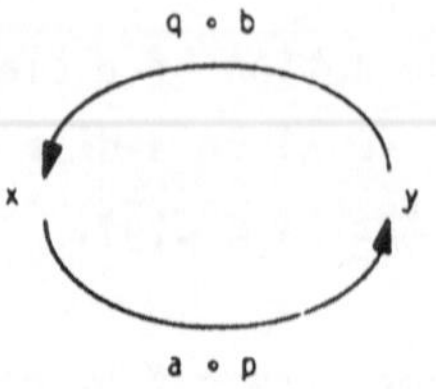

and in $tr_{\hat{S}}(\hat{g})$ we have:

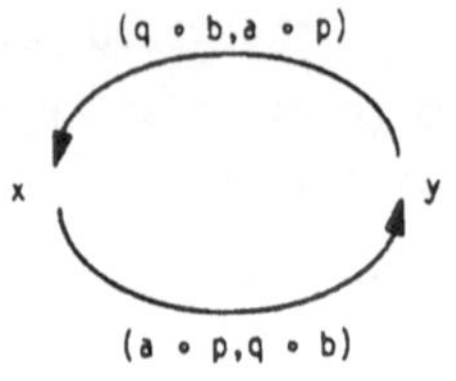

Note that $\hat{S}$ is a δ-selector and that the labels of (x,y) and (y,x) in $tr_{\hat{S}}(\hat{g})$ are δ-conjugated.

Now based on the above it is easy to prove that for an arbitrary *ℓ2s* g and an arbitrary selector S the following diagram commutes:

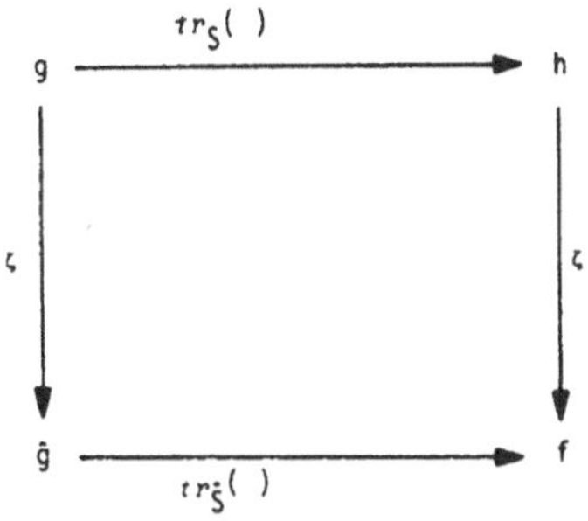

where ς is the transformation leading a *ℓ2s* to its δ-reversible version as explained above, and $\hat{S}$ is obtained from S as explained above.

It is easy to see that if G is a *dℓ2s*, then $\hat{G} = \{ \hat{g} : g \in G \}$ is a δ-implementation of $\hat{K}$ where $C(G) = C(\hat{G})$. Hence $\hat{\mathcal{A}} = (\hat{K},\hat{G})$ is a δ-reversible *dℓ2s*.

Consequently, as is the case in the theory of labeled 2-structures, it suffices to consider reversible *dℓ2s's* only.

As a matter of fact, *unless explicitely clear otherwise, in the rest of the paper we will consider an arbitrary but fixed* dℓ2s $G = (D,\Lambda)$ *which is a* δ-*reversible implementation of* K.

Definition 5.2. Let $g \in G$ and let $x \in D$. x is a *horizon of* g iff for each $y \in D - \{x\}$, $\lambda(x,y) = e$. □

We will use $isol(g)$ to denote the set of all horizons of g, and HOR(G) to denote the set of all $g \in G$ such that $isol(g) \neq \emptyset$.

Using the graph-theoretic intuition, one can interpret the label $\lambda(x,y) = e$ as the "absence of a 2-edge (x,y)" in g. In this way a horizon x of g is a *disconnected element* of g; hence we can talk about *connected components* and *maximal connected components* of g. This intuition also explains the notation $isol(g)$, where *isol* stands for "isolated elements of".

Definition 5.3. Let $g = (D,\lambda) \in G$ and let $x \in D$.

(1) For $y \in D - \{x\}$, the x-*pair for* y *in* g is the conjugated pair $(\varphi_a , \gamma_{\delta(a)})$, where $a = (\lambda(y,x))^{-1}$. The x-*pair for* x *in* g is the pair (φ_e , γ_e).

(2) The x-*selector for* g, denoted sel_x , is the selector S such that, for each $z \in D$, $S(z)$ is the x-pair for z in g.

(3) The x-*translation of* g, denoted $trans_x(g)$, is the *ℓ2s* $tr_{sel_x}(g)$. □

Theorem 5.1. For each $g \in G$ and each $x \in D$, $x \in isol(trans_x(g))$. □

Corollary 5.1. For each $x \in D$ there is a $g \in G$ such that $x \in isol(g)$. □

Corollary 5.2. For each $x \in D$, $D - \{x\} \in C(G)$. □

Following the above corollary, also complements of singletons in D are called *trivial* clans. Hence for a dℓ2s trivial clans are: the empty set, the whole domain, singletons, and complements of singletons. Consequently the phrase "X is a trivial clan in g" where $g \in G$ and the phrase" X is a trivial clan in G" have different meanings. The set of trivial clans of G will be denoted by $TC(G)$.

The next result gives a useful sufficient condition for preserving a clan when applying a selector to a ℓ2s.

Theorem 5.2. Let $g \in G$, let $X \in C(g)$, and let S be a selector. $S(x_1) = S(x_2)$ for all x_1 , $x_2 \in X$ iff $X \in C(tr_S(g))$. □

Corollary 5.3. Let $g \in G$ and let $x \in D$. If $X \in C(g)$ is such that $x \notin X$, then $X \in C(trans_x(g))$.

Proof.

By Remark 5.1, $trans_x(g)$ is a selector S that satisfies the statement of Theorem 5.2. □

Hence, for each $g \in G$, each clan X of g is a clan of $trans_x(g)$ for some $x \in D$. Thus from the point of view of clans of G it suffices to consider x-translations (where $x \in D$) of elements of G.

Definition 5.4. Let $g \in G$ and let $X \subseteq D$. X is *isolated in* g iff for each $x \in X$ and each $y \in D - X$, $lab_g(x,y) = e$. □

Thus, using the intuitions explained at the beginning of this section, if X is isolated in g, then X is a union of connected components of g. Our next result says that each clan of G is a union of connected components in some $h \in G$.

Theorem 5.3. For each $g \in G$ and each $X \in C(g)$ there exists $h \in G$ such that X is isolated in h. □

Corollary 5.4. For each $X \in C(G)$, $\bar{X} \in C(G)$. □

Note that from the definition of a horizon it easily follows that if x is a horizon of $g \in G$, then each clan of g containing x is a union of connected components of g.

The following result concerns the transfer of clans between elements of G using common horizons.

Theorem 5.4. Let G be a single axiom dℓ2s. Let g_1 , $g_2 \in$ G be such that $isol(g_1) \cap isol(g_2) \neq \emptyset$, and let $x \in isol(g_1) \cap isol(g_2)$. If $X \in C(g_1)$ is such that $x \notin X$, then $X \in C(g_2)$. □

6. Closure Properties of Clans

In this section we investigate closure properties of $C(G)$, where *it is assumed that* G *is a single axiom* dℓ2s.

Our first result says that for each $g \in HOR(G)$ the clans of g represent all clans of G.

Theorem 6.1. For each $g \in HOR(G)$, $C(G) = C(g) \cup \{ X : \overline{X} \in C(g) \}$. □

Theorem 6.1 is very useful in computing clans of G.

Example 6.1.

Consider the single axiom δ-reversible dℓ2s G given by $\underline{\Delta} = (\Delta, \circledcirc)$ from Example 3.1, $\delta = id_\Delta$, and the following axiom g_0 on the domain $D = \{ 1,...,12 \}$:

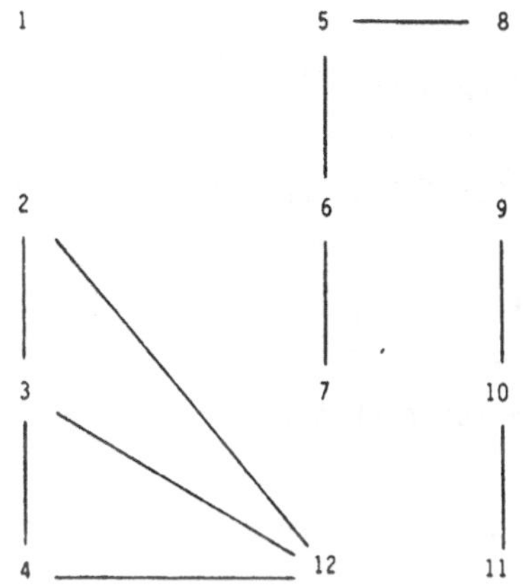

Here we have given all symmetric 2-edges labeled 1 and omitted all symmetric 2-edges labeled by 0 (the identity of $\underline{\Delta}$).

G has enormous number of elements and so computing $C(G)$ by going through all elements of G is certainly not feasible. However, since g_0 has a horizon (viz. 1), by Theorem 6.1 the clans of G are exactly the clans of g_0 and their complements.

Hence the clans of G consist of:

(1) all trivial clans of g_0 and their compelements:

(2) all unions of connected components of g_0 ,

(3) nontrivial clans within connected components and their complements ; these are: {9,11}, D - {9,11}, {2,3,4}, D - {2,3,4}, {2,4}, D - {2,4}, {3,12}, D - {3,12}.

Since the clans from (1) and (2) above are easy (but tedious) to write down, it is really easy to find the clans of G. □

We proceed now to establish the closure properties of $C(G)$.

We have already established (Corollary 5.4) that for each $X \in C(G)$, $\overline{X} \in C(G)$.

Then we recall the following result (from [ER1]) concerning the closure properties of the set of clans of a $\ell 2s$.

Proposition 6.1. Let g be a $\ell 2s$ and let $X,Y \in C(g)$.

(1) $X \cap Y \in C(g)$,

(2) if $X \cap Y \neq \emptyset$, then $X \cup Y \in C(g)$,

(3) if $Y - X \neq \emptyset$, then $X - Y \in C(g)$. □

The notion of an independent Boolean algebra has turned out to be a useful tool for investigating $\ell 2s$'s (see, e.g., [ER3]).

Definition 6.1.

(1) Sets $X,Y \subseteq D$ are *independent* iff $X \cap Y \neq \emptyset$, $\overline{X} \cap Y \neq \emptyset$, $X \cap \overline{Y} \neq \emptyset$, and $\overline{X} \cap \overline{Y} \neq \emptyset$.

(2) A family F of subsets of D is an *independent Boolean algebra* (*over* D), abbreviated *iba*, iff

(i) for all independent $X,Y \in F$, $X \cup Y \in F$, $X \cap Y \in F$ and $X - Y \in F$,

(ii) for each $X \in F$, $\overline{X} \in F$,

(iii) $\emptyset \in F$ and $\{x\} \in F$ for each $x \in D$. □

Here is the main result concerning the closure properties of $C(G)$.

Theorem 6.2. $C(G)$ is an iba.

Acknowledgements

The authors are indebted to T. Harju and P. ten Pas for useful discussions on dynamic labeled 2-structures. The research presented here has been carried out within the framework of the ESPRIT Basic Research Actions "COMPUGRAPH" and "CALIBAN".

References

[BB] G. Birkhoff and T.C. Bartee, *Modern Applied Algebra*, McGraw Hill, New York, 1970.

[ER1] A. Ehrenfeucht and G. Rozenberg, Theory of 2-structures, parts I and II, *Theoretical Computer Science*, v. 70, 277-342, 1990.

[ER2] A. Ehrenfeucht and G. Rozenberg, Angular 2-structures, *Theoretical Computer Science*, v. 92, 227-248, 1992.

[ER3] A. Ehrenfeucht and G. Rozenberg, Square systems, Dept. of Computer Science, Leiden University, Technical Report No. 31, 1992.

[ER4] A. Ehrenfeucht and G. Rozenberg, T-structures, T-functions and texts, *Theoretical Computer Science*, 1993, to appear.

[ER5] A. Ehrenfeucht and G. Rozenberg, Dynamic labeled 2-structures, Dept. of Computer Science, Leiden University, Techn. Rep. No. 1, 1993.

[EHR] A. Ehrenfeucht, T. Harju and G. Rozenberg, Permutable transformation semigroups, *Semigroup Forum*, 1993, to appear.

[EKR] H. Ehrig, H.-J. Kreowski and G. Rozenberg, (Eds.), Graph grammars and their application to computer science, Lecture Notes in Computer Science 532, Springer Verlag, Heidelberg, 1991.

[R] W. Reisig, *Petri Nets, An Introduction*, Springer Verlag, Berlin, Heidelberg, 1985.

POST CORRESPONDENCE PROBLEM : PRIMITIVITY AND INTERRELATIONS WITH COMPLEXITY CLASSES

ALEXANDRU MATEESCU

University of Bucharest,
Faculty of Mathematics,
Academiei 14, sector 1,
70109 Bucharest, Romania

ARTO SALOMAA

Academy of Finland and
Mathematics Department,
University of Turku,
20500 Turku, Finland

Abstract. We first introduce the notion of a general primality type, for a systematic study of "simple", "primitive" or "prime" solutions of the Post Correspondence Problem. We give an exhaustive charcterization of general primality types. We then introduce PCP-related complexity classes, for instance, the time complexity classes PCP-P and PCP-NP. We obtain the chain of inclusions

$$P \subsetneq \text{PCP-P} \subsetneq \text{PCP-NP} = NP \subsetneq \text{P-SPACE} = \text{PCP-P-SPACE} = \text{PCP-NP-SPACE}.$$

1. Background

Apart from being a basic reference point in undecidability proofs, Post Correspondence Problem constitutes a way of representing successful computations. The idea in such a representation is that the target word in a solution encodes the sequence of instantaneous descriptions, whereas the source word being a solution guarantees that the sequence indeed comes from a successful computation. This idea is used in most of the standard references and is present also in Post's original [5].

Whenever w_1 and w_2 are solutions for an instance of the Post Correspondence Problem (PCP), then so is w_1w_2 . It is natural to consider w_1 and w_2 to be "simpler" solutions than w_1w_2 . The same holds for the represented computations as well. Simplifications may be done also in a more sophisticated way, for instance, by removing a subword from a solution w , or from the target word corresponding to w. In each case a simpler computation results.

For a systematic study of "simple" , "primitive" or "prime" solutions, the notion of a *primality type* was introduced and investigated in [7]. This investigation was extended in [2] to concern certain problems in the combinatorics of words. Such considerations had already earlier turned out to be important in characterization and representation results, [1], as well as in approaching the borderline between decidable and undecidable, [6].

In this paper we first introduce the notion of a *general primality type*, where both source and target words are taken into consideration. We give an exhaustive characterization of general primality types. The basic tool is our fundamental lemma that establishes a rather surprising link between components corresponding to source and target words. Results such as this link express certain fundamental properties of computations. The properties have not yet been described in more direct terms. We then define a variant of the Post Correspondence Problem, the *machine - oriented* Post Correspondence Problem or MOPCP, especially suitable for complexity considerations. All recursively enumerable sets can be represented in terms of instances of MOPCP and, moreover, deterministic and nondeterministic time and space complexities have their natural counterpart in the representation.

2. General primality types

The reader is referred to [6] for all unexplained notions.

Let g and h be nonerasing morphisms of Σ^* into Δ^* , where Σ and Δ are finite alphabets. The *equality* set between g and h is defined by :

$$E(g,h) = \{ w \in \Sigma^+ \mid g(w) = h(w) \} .$$

(Observe that the empty word λ is not considered to be a member of the equality set.)

The pair (g , h) = PCP is also referred to as an instance of the Post Correspondence Problem. Words in E(g , h) , if any, are called *solutions* of PCP.

For a word w over Σ, we consider sets of words obtained from w by removing a final subword, a subword, or a scattered subword, respectively. By definition,

$\mathit{fin}\ (w) = \{ v_1 \mid w = v_1v_2 \text{, for some } v_2 \in \Sigma^* \}$,

$\mathit{sub}\ (w) = \{ v_1v_2 \mid w = v_1xv_2 \text{, for some } v_1, v_2, x \in \Sigma^* \}$,

$\mathit{scatsub}\ (w) = \{ v_1 \ldots v_k \mid w = x_1v_1 \ldots x_kv_kx_{k+1} \text{ for some } x_i, v_i \in \Sigma^* \}$.

Three further sets determined by the pair (g , h) are now defined as follows :

$F(g , h) = \{ w \in E(g , h) \mid \mathit{fin}\ (w) \cap E(g , h) = \{w\} \}$,

$S(g , h) = \{ w \in E(g , h) \mid \mathit{sub}\ (w) \cap E(g , h) = \{w\} \}$,

$P(g , h) = \{ w \in E(g , h) \mid \mathit{scatsub}\ (w) \cap E(g , h) = \{w\} \}$.

Words in the three sets are called F - *prime* , S - *prime* , and P - *prime solutions* for the instance PCP = (g , h) , respectively.

The triple (p , s , f), where p, s and f are the cardinalities of the sets P(g , h), S(g , h) and F(g , h), respectively, is defined to be the *primality type* of the instance PCP = (g , h). Thus, p, s and f are nonnegative integers or ∞ .

The following two results were established in [7].

Lemma 1. *Assume that the word xyz is in F(g , h), where x, y, z are nonempty, and that xz is in E(g , h). Then* xy^nz *is in F(g , h), for all* $n \geq 0$.

Theorem 1. *A triple (p , s , f) is a primality type iff either* $p = s = f = 0$ *, or else (i)* $1 \leq p \leq s \leq f$*, (ii) p is finite, (iii) if* $s < f$ *then* $f = \infty$ *. An example for each possible type can be effectively constructed.*

We now consider the target alphabet Δ, and introduce the co - sets corresponding to the sets E, F, S, P :

$\text{co - }E(g , h) = \{ \alpha \in \Delta^+ \mid g(w) = h(w) = \alpha, \text{ for some } w \in \Sigma^+ \}$,

$\text{co - }F(g , h) = \{ \alpha \in \text{co - }E(g , h) \mid \text{fin}\ (\alpha) \cap \text{co - }E(g , h) = \{ \alpha \} \}$,

$\text{co - }S(g , h) = \{ \alpha \in \text{co - }E(g , h) \mid \text{sub}\ (\alpha) \cap \text{co - }E(g , h) = \{ \alpha \} \}$,

$\text{co - }P(g , h) = \{ \alpha \in \text{co - }E(g , h) \mid \text{scatsub}\ (\alpha) \cap \text{co - }E(g , h) = \{ \alpha \} \}$.

Words in the latter three sets are called F_c *- prime* , S_c *- prime* and P_c *- prime* with respect to the instance PCP = (g , h).

Let p_c , s_c and f_c be the cardinalities of the sets P(g , h), S(g , h) and F(g , h), respectively. Then the six - tuple

$$(p_c , p , s_c , s , f_c , f)$$

is defined to be the *general primality type* of the instance PCP = (g , h).

The fairly easy proofs of the following two lemmas are omitted here.

Lemma 2. *$p_c \le p$, $s_c \le s$, $f_c \le f$, where strict inequality is possible in each case.*

Lemma 3. *For any instance (g , h), we have $p_c \le s_c \le f_c$. Moreover, if one of the six numbers p , s , f , p_c , s_c , f_c equals 0 , they all are equal 0 .*

3. Fundamental lemma and characterization

We are now ready to present the most important technical tool in the characterzation of general primality types. Apart from the inequalities

$$p_c \le p, \quad s_c \le s, \quad f_c \le f$$

we have so far no results that would tie up conditions involving target words (computations) with conditions involving source words . Our fundamental lemma below gives such a result . In view of the characterization presented below, the result of the lemma is exhaustive.

Lemma 4. *If $s = 1$, then $f_c = 1$ or $f_c = \infty$.*

Proof. The proof is indirect. We show that the conditions

(1) $$s = 1 \text{ and } 1 < f_c < \infty$$

lead to a contradiction. For notational convenience, we write x_g and x_h instead of g(x) and h(x).

Let w be the only element of S. (1) implies that $f \ge f_c > 1$. Hence, there is a word $u = xyz$ in F(g , h) , where x, y, z are nonempty, such that $xz = w$ and u_g is in co - F(g , h) . Without loss of generality, we assume that the decomposition of w is chosen in such a way that z_g is the shortest possible . (In other words, whenever $u' = x'y'z'$ is in F(g , h) - S(g , h), u'_g is in co - F(g , h) and $w = x'z'$, then $|z'_g| \ge |z_g|$.)

By Lemma 1, all words xy^nz , $n \ge 0$, are in F(g , h) . Because f_c is finite , some word $x_g y_g^k z_g$, $k \geq 2$, is not in co - F(g , h) . We choose again k to be the smallest possible.

We know that a proper prefix of $x_g y_g^k z_g$ is in co - F(g , h) . We consider separately the cases that this proper prefix is w_g or some other word. We assume first that w_g is a proper prefix of $x_g y_g^k z_g$, that is ,

$$x_g y_g^k z_g = w_g \alpha ,$$

where α is nonempty. On the other hand , $w_g = x_g z_g$ and , consequently ,

$$y_g^k z_g = z_g \alpha .$$

By a result well - known in combinatorics of words, there are words β_1 , β_2 and an integer $p \geq 0$ such that

$$z_g = (\beta_1\beta_2)^p\beta_1 , \ \alpha = \beta_2\beta_1 , \ y_g^k = \beta_1\beta_2 .$$

(This result is often referred to as Lyndon's Theorem. A formulation of it suitable for us appears as Lemma 3.2 in [7] .)

We now write y_g in the form $y_g = \eta_1\eta_2$, where

$$\beta_1 = y_g^i \eta_1 , \ \beta_2 = \eta_2 y_g^j , \ i + j + 1 = k .$$

A direct computation now gives the results

$$x_g z_g = x_g y_g^{kp} y_g^i \eta_1 = x_g y_g^{kp+i} \eta_1 ,$$

$$x_g y_g z_g = x_g y_g^{kp+i+1} \eta_1 = x_g y_g^{kp+i} \eta_1 \eta_2 \eta_1 = x_g z_g \eta_2 \eta_1 .$$

Hence, $w_g = x_g z_g$ is a proper prefix of $u_g = x_g y_g z_g$, which contradicts the fact that u_g is in co - F(g , h) .

Consequently, the remaining possibility is that $x_g y_g^k z_g$ has a proper prefix in co - F(g , h) different from w_g . This implies that a word x'vz' is in F(g , h) - S(g , h) such that x'z' = w and $x_g' v_g z_g'$ is a proper prefix of $x_g y_g^k z_g$. Thus, for some nonempty word δ ,

(2) $$x_g' v_g z_g' \delta = x_g y_g^k z_g .$$

Moreover, $x_g' z_g' = w_g = x_g z_g$ and $| z'_g | \geq | z_g |$, by our assumption concerning the decomposition w = xz . This implies the existence of a (possible empty) word t such that

$$z_g' = t z_g \quad \text{and} \quad x_g = x_g' t .$$

We now infer, by (2), that

(3) $$v_g t z_g \delta = t y_g^k z_g \, .$$

Since δ is nonempty, we must have

$$|v_g| < |y_g^k| \, .$$

Consequently, there are nonempty words y_g' and y_g'' with the properties

(4) $$y_g = y_g' y_g'' \quad , \; v_g t = t y_g^d y_g' \, , \quad z_g \delta = y_g'' y_g^e z_g \quad ,$$

where the integers d and e satisfy $d + e + 1 = k$.

(Indeed, this decomposition of y_g results from (3). If $y_g' = \lambda$ or $y_g'' = \lambda$, we must have $d + e = k$ with $d > 0$ and $e > 0$, which contradicts the fact that k is the smallest number satisfying $x_g y_g^k z_g \notin$ co - F(g , h).)

We apply again Lyndon's Theorem, now to the last equation (4), and conclude the existence of words η, ν and an integer $p \geq 0$ with the properties

$$y_g'' y_g^e = \eta\nu \quad , \quad \delta = \nu\eta \quad , \quad z_g = (\eta\nu)^p \eta \quad .$$

Hence,

(5) $$z_g = (y_g'' y_g^e)^p \eta \quad .$$

From the equation $x_g z_g = x_h z_h$ we infer that one of the words z_g and z_h is a proper suffix of the other. We may assume that

(6) $$z_g = \gamma z_h \quad , \quad \gamma \neq \lambda \quad ,$$

because otherwise we interchange the roles of g and h .

On the other hand, the equation $x_g y_g z_g = x_h y_h z_h$ implies the equation $y_g \gamma = \gamma y_h$. (Observe that $x_g z_g = x_h z_h$ implies also the equation $x_g \gamma = x_h$.) Another application of Lyndon's Theorem gives the relations

$$y_g = \xi\tau \quad , \quad y_h = \tau\xi \quad , \quad \gamma = (\xi\tau)^q \xi \quad .$$

Hence,

(7) $$\gamma = y_g^q \xi \quad .$$

There still remain some straightforward computations, based on the results already obtained. Comparing the two expressions for z_g in (5) and (6), we write γ in the form

(8) $$\gamma = (y_g'' y_g^e)^r \mu \quad ,$$

where μ is a prefix of $(y_g"y_g^e)^{p-r}\eta$ .. The equations (7) and (8) give first

$$(y_g"y_g^e)^r\mu = y_g^q\xi$$

and then, by the first equation in (4),

$$(y_g"(y_g'y_g")^e)^r\mu = (y_g'y_g")^q\xi \quad .$$

Taking from both sides the prefix of length $|y_g|$, we see that y_g' and $y_g"$ commute and, hence, must be powers of the same word :

$$y_g' = f^{i_1} \quad , \qquad y_g" = f^{i_2} \quad , \qquad y_g = f^{i_1+i_2} \quad .$$

(Observe that all of the integers e, r, q are positive.) By (5),

$$z_g = f^{pe(i_1+i_2)+pi_2}\eta \quad .$$

Moreover,

$$\eta\nu = y_g"y_g^e = f^{e(i_1+i_2)+i_2} \quad .$$

Consequently, η is a prefix of $f^{e(i_1+i_2)+i_2}$ and z_g is a proper prefix of the word

$$y_gz_g = f^{pe(i_1+i_2)+pi_2+i_1+i_2}\eta \quad .$$

Therefore, $w_g = x_gz_g$ is a proper prefix of $u_g = x_gy_gz_g$. Hence, u_g is not in co - F(g , h). Thus, we have reached a contradiction in both cases, and Lemma 4 follows.

We now summarize the restrictions presented in Theorem 1 and Lemmas 2 - 4 :

(1) $p \leq s \leq f$,

(2) $p_c \leq s_c \leq f_c$,

(3) $p_c \leq p$, $s_c \leq s$, $f_c \leq f$,

(4) p is finite,

(5) if $s < f$, then $f = \infty$,

(6) if $s = 1$, then $f_c = 1$ or $f_c = \infty$.

Thus, we established the following result.

Theorem 2. *Assume that the six-tuple* $(p_c , p , s_c , s , f_c , f)$ *is the general primality type of some instance PCP = (g , h). Then the conditions (1) - (6) are satisfied.*

We have shown what general primality types *possibly can occur*. There still remains the problem about what types *actually do occur*. This problem was solved for primality types in [7]. For general primality types it is solved in the next theorem.

Theorem 3. *For every six - tuple* (p_c, p, s_c, s, f_c, f), *with nonzero components satisfying (1)-(6), an instance PCP = (g , h) of the Post Correspondence Problem having the general primality type* (p_c, p, s_c, s, f_c, f) *can be effectively constructed.*

The space does not permit us to present a detailed proof. The main case division is based on the number of components ∞ in the given six - tuple. We refer the reader to [3] for all details.

4. Machine - oriented Post Correspondence Problem

We will now show how the customary time and space complexity classes have their natural counterparts in terms of PCP. In this way *Post complexity classes* are introduced. We can show that they coincide with the ordinary complexity classes, except that in the case of P we can only show the inclusion in the corresponding Post class, P $\subseteq$ PCP-P. For all proofs we refer the reader to [4] . We first give the necessary definitions.

An injective morphism is termed a *code* . Clearly, a code is always nonerasing. A code h is of *bounded delay* iff, for some integer $k \geq 1$, whenever h(u) is a prefix of h(w) and $|u| \geq k$, then u and v begin with the same letter.

Bounded delay means that only a bounded amount of lookahead from left to right is needed in parsing. If we want to find w from h(w) , we always find the first letter of w by inspecting a sufficiently long prefix of w. (The length is independent of w. In fact the length km will always suffice, where m is the maximal length of the h - images of single letters.)

Let Δ be a subalphabet of Σ . The morphism π_Δ defined by : $\pi_\Delta(a) = a$ for $a \in \Delta$ and $\pi_\Delta(a) = \lambda$ for $a \in \Sigma - \Delta$ is termed a *projection* .

Let M = (g,h) be an instance of the Post Correspondence Problem. As before, Σ and Δ are the domain and range alphabets of the morphisms. The instance M is termed an instance of the *machine - oriented Post Correspondence Problem, MOPCP,* iff

$\Sigma = \Sigma_1 \cup \Sigma_2 \cup \Sigma_3 \cup \Sigma_4$, where the subalphabets Σ_i, are nonempty and pairwise disjoint and, moreover,

$$|g(a)| > |h(a)| \text{ if } a \in \Sigma_1 ,$$

$$|g(a)| = |h(a)| \text{ if } a \in \Sigma_2 \cup \Sigma_3 ,$$

$$|g(a)| < |h(a)| \text{ if } a \in \Sigma_4 .$$

In the cases, in which we will be particularly interested and which give the background for the subsequent definitions concerning the generated language and complexity, the solutions will be of the form $\alpha_1\alpha_2\alpha_3\alpha_4$, with $\alpha_i \in \Sigma_i^+$, for i = 1,2,3,4. Here α_1 is used to make the g - image sufficiently much longer than the h - image. The part α_2 actually creates the input for the computation. The part α_3 serves the double purpose of first producing enough workspace for the computation (for this purpose α_1 has to be sufficiently long) and then performing the computation itself. The final part α_4 makes it possible for h to catch up if the computation is a correct one.

The details will become explicit below. Considering the role of α_2 , the following definition is natural.

The *language generated* by an instance M = (g,h) of the MOPCP is defined by

$$L(M) = \pi_{\Sigma_2}(E(g,h) \cap \Sigma_1^+\Sigma_2^+\Sigma_3^+\Sigma_4^+).$$

Thus, we take the α_2 - part of solutions of the form $\alpha_1\alpha_2\alpha_3\alpha_4$. It will be seen below that all recursively enumerable languages are of this form.

Let M = (g,h) be an instance of the MOPCP and L = L(M) the generated language. The *nondeterministic time complexity* of M is the function t(n) from nonegative integers into nonegative integers defined by

$$t(n) = \max \{|\alpha| \mid \alpha = \alpha_1\alpha_2\alpha_3\alpha_4 \text{ is a solution of M}, \alpha_i \in \Sigma_i^+ , 1 \le i \le 4, |\alpha_2| = n ,$$
$$\text{and there is no solution } \alpha_1'\alpha_2\alpha_3'\alpha_4' \text{ with } |\alpha_1'\alpha_2\alpha_3'\alpha_4'| < |\alpha| \}.$$

By definition, max $\emptyset$ = 1.

Thus, for each word $\alpha_2 \in L$ of length n, we take the shortest possible solution "producing" α_2 and then maximize the length of these shortest solutions. If there are no words of length n in L, the time complexity t(n) is defined to be 1.

If p(n) is a polynomial and M an instance whose time complexity function t(n) satisfies $t(n) \leq p(n)$, for all n, then the language L = L(M) is said to be of *polynomial nondeterministic PCP - time complexity*. The resulting class of languages is denoted by *PCP - NP*.

The *nondeterministic space complexity* of the instance M is defined similarly :

$$s(n) = \max \{|\alpha_1| \mid \alpha = \alpha_1\alpha_2\alpha_3\alpha_4 \text{ is a solution of M}, \alpha_i \in \Sigma_i^+, 1 \leq i \leq 4, |\alpha_2| = n, \text{ and there is no solution } \alpha_1'\alpha_2\alpha_3'\alpha_4' \text{ with } |\alpha_1'| < |\alpha_1| \}.$$

The notion of *nondeterministic polynomial PCP - space complexity* and the class *PCP - NP - SPACE* are defined analogously.

Finally, we say that the instance M is *deterministic* iff the restriction of the morphism h to Σ_3 is of bounded delay. The specializations of the above definitions to the deterministic case are now straightforward. This gives rise to the complexity classes PCP - P and PCP - P - SPACE. Thus, a language L is in PCP - P iff there are a deterministic instance M of the MOPCP and a polynomial p(n) such that L = L(M) and the time complexity function t(n) of M satisfies $t(n) \leq p(n)$, for all n.

We are now ready to state our main results .

Theorem 4. *For every recursively enumerable language L, an instance M of the MOPCP can be effectively constructed such that L = L(M).*

Theorem 5. *The following inclusions hold :*

$$P \subsetneq PCP\text{-}P \subsetneq PCP\text{-}NP = NP \subsetneq P\text{-}SPACE = PCP\text{-}P\text{-}SPACE = PCP\text{-}NP\text{-}SPACE.$$

References

[1] K. Culik. A purely homomorphic characterization of recursively enumerable languages . *Journal of the Association for Computing Machinery* 26 (1979) 345 - 350.

[2] A. Mateescu and A. Salomaa . PCP - prime words and primality types . *RAIRO / Theoretical Informatics and Applications*, 27, 1 (1993) 57 - 70 .

[3] A. Mateescu and A. Salomaa . On simplest possible solutions for Post Correspondence Problems . *Acta Informatica*, to appear.

[4] A. Mateescu , A. Salomaa and Sheng Yu. P , NP and Post Correspondence Problem. Submitted for publication.

[5] E. Post . A variant of a recursively unsolvable problem . *Bulletin of the American Mathematical Society* 52 (1946) 264 - 268 .

[6] A. Salomaa . *Jewels of Formal Language Theory* . Computer Science Press (1981) .

[7] A. Salomaa , K. Salomaa and Sheng Yu . Primality types of instances of the Post Correspondence Problem . *EATCS Bulletin* 44 (1991) 226 - 241 .

A taste of linear logic

Philip Wadler

Department of Computing Science, University of Glasgow, G12 8QQ, Scotland
(wadler@dcs.glasgow.ac.uk)

Abstract. This tutorial paper provides an introduction to intuitionistic logic and linear logic, and shows how they correspond to type systems for functional languages via the notion of 'Propositions as Types'. The presentation of linear logic is simplified by basing it on the Logic of Unity. An application to the array update problem is briefly discussed.

1 Introduction

Some of the best things in life are free; and some are not.

Truth is free. Having proved a theorem, you may use this proof as many times as you wish, at no extra cost. Food, on the other hand, has a cost. Having baked a cake, you may eat it only once. If traditional logic is about truth, then linear logic is about food.

In traditional logic, if a fact is used to conclude another fact, the first fact is still available. For instance, given that A implies B and given A, one may deduce both A and B. In symbols, this is written as the judgement

(i) $$A \rightarrow B,\ A \vdash A \times B$$

where $A \rightarrow B$ is read 'A implies B', and $A \times B$ is read 'A and B'. (It will shortly become apparent why $\times$ is written for 'and'.) The assumption A is used twice in the proof of this judgement, and that is reasonable because truth is free.

Traditional logic comes in many forms: the one we shall concentrate on is *intuitionistic* logic. This logic is of special interest because its terms are in one-to-one correspondence with a typed functional programming language. This correspondence goes by the name 'Curry-Howard isomorphism', after the logicians who discovered it, or by 'Propositions as types', because it views propositions of logic as types in a functional program. Under this correspondence, one proof of judgement (i) may be written as the program

(ii) $$f : A \rightarrow B,\ x : A \vdash (x, f\,(x)) : A \times B$$

where f is a free variable denoting a function of type $A \rightarrow B$, and x is a free variable denoting a value of type A, and $(x, f\,(x))$ is a term denoting a pair of type $A \times B$. The first component of the pair is the value x, the second component is the result of applying function f to value x. The free variable x appears twice in the term, just as assumption A is used twice in the proof of the corresponding judgement. This is reasonable if the value in x costs little to copy, so the name 'free' variable is doubly appropriate.

But not all things in life are free. To capture this notion, the logician Jean-Yves Girard devised linear logic, a 'resource concious' logic. In linear logic, assumptions may not be freely copied, nor may they be foolishly discarded: each assumption is used exactly once. It is no longer the case that given A implies B and given A one can deduce both A and B. In symbols, this is written as the non-judgement

(iii) $$\langle A \multimap B\rangle,\ \langle A\rangle \nvdash A \otimes B$$

where $A \multimap B$ is read 'consuming A yields B' and $A \otimes B$ is read 'both A and B', and angle brackets are written around each assumption to indicate that it must be used exactly once. Taking A to be the proposition 'I have a cake', and B to be the proposition 'I am full', we discover that one cannot have one's cake and eat it too.

Just as lambda calculus corresponds to intuitionistic logic, there is a programming calculus that correspond to linear logic. Corresponding to (iii) is the program

(iv) $$\langle f : A \multimap B\rangle,\ \langle x : A\rangle \nvdash \langle x, f\,\langle x\rangle\rangle : A \otimes B.$$

If f were a function that takes a cake into a feeling of fullness, and x was a cake, then $\langle x, f\,\langle x\rangle\rangle$ is a pair consisting of the original cake and the fullness resulting from eating it. Just as judgement (iii) is invalid, program (iv) is ill-typed: the program that keeps a cake and eats it too has a type error.

Such a type system has less fanciful uses. It can allow for fine control over the efficient use of storage, and in particular for a new answer to the old question of how to handle arrays efficiently in a functional language.

This paper is intended as a general introduction to intutionistic logic and the Curry-Howard isomorphism, and to linear logic and some of its applications in functional programming. The presentation of linear logic is simplified by basing it on Girard's Logic of Unity, a refinement of the concept of linear logic.

The organisation follows the development above. Section 1 reviews intuitionistic logic. Section 2 describes how the Curry-Howard correspondence relates this logic to lambda calculus. Section 3 introduces linear logic. Section 4 uses the Curry-Howard correspondence to derive a linear lambda calculus, and outlines its applications. Section 5 discusses related work and concludes.

It is worth noting what is *not* covered. First, the paper concentrates on the application of linear logic to functional programming, and ignores other intriguing applications such as logic programming or concurrent programming. Second, the paper describes only *intuitionistic* linear logic, and ignores the related theory of *classical* linear logic. However, it is hoped that this paper will put the reader in a better position to appreciate the literature on the subject.

As this paper is mainly tutorial in nature, it is to a large extent a gloss on the work of others. It is a pleasure to acknowledge here some of the people who have taught me these ideas: Jean-Yves Girard, Samson Abramsky, Martin Hyland, Yves Lafont, Gordon Plotkin, Vaughn Pratt, and Robert Seely.

2 Intuitionistic logic

Most computer scientists have seen one or another of the many ways of formulating logic. The version presented here is based on natural deduction, and specially emphasises the Contraction and Weakening rules, which are the key to understanding linear logic.

A *proposition* is built up from propositional constants using the combining forms *implies*, written $\rightarrow$, *and*, written $\times$, and *or*, written $+$. Let A, B, C range over propositions, and X range over propositional constants. Then the grammar of propositions is as follows.

$$A, B, C ::= X \mid A \rightarrow B \mid A \times B \mid A + B$$

An *assumption* is a sequence of zero or more propositions. Let Γ, Δ, Θ range over assumptions. A *judgement* has the form $\Gamma \vdash A$, meaning that from assumptions Γ one can conclude proposition A.

A *rule* consists of zero or more judgements written above a line, and one judgement written below. If all the judgements above the line are derivable, then the judgement below is derivable also. Note that there are three different levels of implication: $\rightarrow$ in a proposition, $\vdash$ in a judgement, and the line in a rule.

$$\frac{}{A \vdash A}\ \text{Id} \qquad \frac{\Gamma, \Delta \vdash A}{\Delta, \Gamma \vdash A}\ \text{Exchange}$$

$$\frac{\Gamma, A, A \vdash B}{\Gamma, A \vdash B}\ \text{Contraction} \qquad \frac{\Gamma \vdash B}{\Gamma, A \vdash B}\ \text{Weakening}$$

$$\frac{\Gamma, A \vdash B}{\Gamma \vdash A \rightarrow B}\ \rightarrow\text{-I} \qquad \frac{\Gamma \vdash A \rightarrow B \qquad \Delta \vdash A}{\Gamma, \Delta \vdash B}\ \rightarrow\text{-E}$$

$$\frac{\Gamma \vdash A \qquad \Delta \vdash B}{\Gamma, \Delta \vdash A \times B}\ \times\text{-I} \qquad \frac{\Gamma \vdash A \times B \qquad \Delta, A, B \vdash C}{\Gamma, \Delta \vdash C}\ \times\text{-E}$$

$$\frac{\Gamma \vdash A}{\Gamma \vdash A + B}\ +\text{-I}_1 \qquad \frac{\Gamma \vdash B}{\Gamma \vdash A + B}\ +\text{-I}_2 \qquad \frac{\Gamma \vdash A + B \qquad \Delta, A \vdash C \qquad \Delta, B \vdash C}{\Gamma, \Delta \vdash C}\ +\text{-E}$$

Fig. 1. Intuitionistic logic

The rules for intuitionistic logic are shown in Figure 1. They come in three groups.

First is the lone *axiom*, Id, the only rule with no judgements above the line. This rule expresses tautology: from hypotheses A one can deduce conclusion A.

Second are the three *structural* rules, Exchange, Contraction, and Weakening. Exchange expresses that the order of hypotheses is irrelevant, Contraction expresses that any hypothesis may be duplicated, and Weakening expresses that any hypothesis may be discarded.

Third are the *logical* rules. These come in pairs. In the *introduction* rule $\to$-I a judgement ending in a proposition formed with $\to$ appears below the line, while in the *elimination* rule $\to$-E a judgment ending in a proposition formed with $\to$ appears above the line. Similarly for the other logical connectives, $\times$ and $+$.

Each rule has a straightforward logical reading. For instance, $\to$-I expresses the 'deduction theorem': if from assumptions Γ and A one can deduce B, then from assumption Γ alone one can deduce A implies B. Similarly, $\to$-E expresses 'modus ponens': if from assumptions Γ one can deduce that A implies B, and from assumptions Δ one can deduce A, then from assumptions Γ and Δ one can deduce B.

Derivations can be written as proof trees. The judgement derived is at the root, each branch represents a use of a rule, and the leaves are axioms. Here is a tree deriving the judgement $A \to B, A \vdash A \times B$.

$$\dfrac{\dfrac{\dfrac{\dfrac{}{A \vdash A.}\,\mathrm{Id} \qquad \dfrac{\dfrac{}{A \to B \vdash A \to B}\,\mathrm{Id} \qquad \dfrac{}{A \vdash A}\,\mathrm{Id}}{A \to B,\, A \vdash B}\,{\to}\text{-E}}{A,\, A \to B,\, A \vdash A \times B}\,\times\text{-I}}{A \to B,\, A,\, A \vdash A \times B}\,\text{Exchange}}{A \to B,\, A \vdash A \times B}\,\text{Contraction}$$

In this proof, the assumption A is used twice, once to conclude A itself, and once to conclude B. The double usage of A is justified by Contraction.

2.1 Alternative rules for conjunction

The rules for conjunction may also be expressed in an alternative form.

$$\dfrac{\Gamma \vdash A \qquad \Gamma \vdash B}{\Gamma \vdash A \times B}\,\times\text{-I}' \qquad \dfrac{\Gamma \vdash A \times B}{\Gamma \vdash A}\,\times\text{-E}'_1 \qquad \dfrac{\Gamma \vdash A \times B}{\Gamma \vdash B}\,\times\text{-E}'_2$$

These are equivalent to the previous rules. The new rules may be derived from the old, plus Weakening and Contraction. Furthermore, the old rules may be derived from the new, plus Weakening, Contraction, $\to$-I, and $\to$-E.

The new $\times$-I$'$ is derived from the old $\times$-I together with Contraction.

$$\dfrac{\dfrac{\Gamma \vdash A \qquad \Gamma \vdash B}{\Gamma, \Gamma \vdash A \times B}\,\times\text{-I}}{\Gamma \vdash A \times B}\,\text{Contraction}$$

The double line stands for one use of Contraction for each assumption in Γ, and also hides some uses of Exchange to bring matching hypotheses together. The Exchange rule is so dull that its uses will not usually be mentioned.

The new $\times$-E_1' is derived from the old $\times$-E plus Weakening.

$$\dfrac{\Gamma \vdash A \times B \qquad \dfrac{\dfrac{}{A \vdash A}\,\text{Id}}{A,\, B \vdash A}\,\text{Weakening}}{\Gamma \vdash A}\,\times\text{-E}$$

One might think that the Weakening rule is a little silly - why bother to make an extra assumption? - but this proof demonstrates its utility. The rule $\times$-E_2' is derived similarly.

The other way around, the rule $\times$-I can be derived from $\times$-I$'$ and Weakening.

$$\dfrac{\dfrac{\Gamma \vdash A}{\Gamma,\, \Delta \vdash A}\,\text{Weakening} \qquad \dfrac{\Delta \vdash B}{\Gamma,\, \Delta \vdash B}\,\text{Weakening}}{\Gamma,\, \Delta \vdash A \times B}\,\times\text{-I}'$$

The double lines stand for one use of Weakening for each assumption in Γ and Δ, and also hide some uses of Exchange.

Finally, the rule $\times$-E can be derived from $\times$-E_1' and $\times$-E_2' togther with Contraction, $\multimap$-I, and $\multimap$-E.

$$\dfrac{\dfrac{\dfrac{\dfrac{\dfrac{\Delta,\, A,\, B \vdash C}{\Delta,\, A \vdash B \multimap C}\,\multimap\text{-I}}{\Delta \vdash A \multimap B \multimap C}\,\multimap\text{-I} \qquad \dfrac{\Gamma \vdash A \times B}{\Gamma \vdash A}\,\times\text{-E}_1'}{\Delta,\, \Gamma \vdash B \multimap C}\,\multimap\text{-E} \qquad \dfrac{\Gamma \vdash A \times B}{\Gamma \vdash B}\,\times\text{-E}_2'}{\Delta,\, \Gamma,\, \Gamma \vdash C}\,\multimap\text{-E}}{\Gamma,\, \Delta \vdash C}\,\text{Contraction}$$

Note that the hypothesis $\Gamma \vdash A \times B$ is used *twice* in this proof, the multiple uses of Γ being reduced to a single use by Contraction.

The key difference of linear logic is that it restricts the use of Contraction and Weakening, so the two different rule sets are no longer equivalent. Instead of a single connective $\times$, there will be two connectives, $\otimes$ and $\&$, one with rules analogous to $\times$-I and $\times$-E, and the other with rules analogous to $\times$-I$'$, $\times$-E_1', and $\times$-E_2'.

2.2 An alternative to structural rules

The rules for Exchange, Weakening, and Contraction may appear unfamiliar to some readers, because logic is often presented in a different form that hides these rules.

$$\frac{}{\Gamma, A, \Delta \vdash A}\ \mathrm{Id}'$$

$$\frac{\Gamma, A \vdash B}{\Gamma \vdash A \rightarrow B}\ {\rightarrow}\text{-I} \qquad \frac{\Gamma \vdash A \rightarrow B \quad \Gamma \vdash A}{\Gamma \vdash B}\ {\rightarrow}\text{-E}'$$

The rule Id′ may be derived from Id, Weakening, and Exchange, and the rule →-E′ may be derived from →-E, Contraction, and Exchange. The system consisting of Id′, →-I, and →-E′ derives exactly the same judgements as the system consisting of Id, Exchange, Contraction, Weakening, →-I, and →-E. The other logical rules can be similarly modified, yielding an equivalent of the full system.

This system is convenient in that it eliminates all concern with the structural rules; but it has the disadvantage that it obscures the role of Contraction and Weakening.

2.3 Commuting conversions for structural rules

One lesson of the alternative system is that the order in which the structural rules are applied is irrelevant. This can be captured in our system by *commuting conversions*, which define an equivalence relation on proofs, written $\Longleftrightarrow$. Here are the commuting conversions for Contraction with the rule →-E.

$$\frac{\dfrac{\Gamma, C, C \vdash A \rightarrow B}{\Gamma, C \vdash A \rightarrow B}\ \text{Cont'n} \qquad \Delta \vdash A}{\Gamma, C, \Delta \vdash B}\ {\rightarrow}\text{-E} \quad \Longleftrightarrow \quad \frac{\dfrac{\Gamma, C, C \vdash A \rightarrow B \quad \Delta \vdash A}{\Gamma, C, C, \Delta \vdash B}\ {\rightarrow}\text{-E}}{\Gamma, C, \Delta \vdash B}\ \text{Cont'n}$$

$$\frac{\Gamma \vdash A \rightarrow B \qquad \dfrac{\Delta, C, C \vdash A}{\Delta, C \vdash A}\ \text{Cont'n}}{\Gamma, \Delta, C \vdash B}\ {\rightarrow}\text{-E} \quad \Longleftrightarrow \quad \frac{\dfrac{\Gamma \vdash A \rightarrow B \quad \Delta, C, C \vdash A}{\Gamma, \Delta, C, C \vdash B}\ {\rightarrow}\text{-E}}{\Gamma, \Delta, C \vdash B}\ \text{Cont'n}$$

There are similar commuting conversions for Exchange, Contraction, and Weakening with each of the structural and logical rules.

As was already noted, a judgement is derivable in the system with explicit structural rules if and only if it is derivable in the system with structural rules 'built in'. With the commuting conversions, one can show not merely that the two systems derive the same judgements, but that proofs in the two systems are equivalent.

2.4 Proof reduction

A judgement may have several distinct derivations. Some of these may represent profoundly different proofs, while others may represent trivial variations. In particular, there is little point in having an introduction rule for a connective followed immediately by an elimination rule for the same connective.

Here is a sketch of a proof that introduces the proposition $A \to B$, only to immediately eliminate it.

$$\dfrac{\dfrac{\dfrac{\dfrac{A \vdash A \cdots}{\vdots\, u}}{\Gamma, A \cdots \vdash B}}{\dfrac{\Gamma, A \vdash B}{\Gamma \vdash A \to B}\to\text{-I}} \qquad \dfrac{\vdots\, t}{\Delta \vdash A}}{\Gamma, \Delta \vdash B}\to\text{-E}$$

Here t labels a subtree of the proof, ending in the judgement $\Delta \vdash A$, and u labels another subtree, ending in the judgement $\Gamma, A \vdash B$. Thanks to Weakening and Contraction, the assumption A may be used zero or more times in proof u, where each use corresponds to one appearance of the leaf $A \vdash A$. The commuting conversions allow all uses of Weakening and Contraction on A in u to be moved to the end of the subtree, where they are indicated by the double line.

The unnecessary proposition $A \to B$ can be eliminated by replacing each occurrence of the leaf $A \vdash A$ in the proof u by a copy of the proof t of the judgement $\Gamma \vdash A$.

$$\begin{array}{c} \vdots\, t \\ \Delta \vdash A \cdots \\ \vdots\, u \\ \Gamma, \Delta \cdots \vdash B \\ \hline\hline \Gamma, \Delta \vdash B \end{array}$$

The Weakenings and Contractions previously performed on assumption A now need to be performed once for each assumption in Δ.

The replacement of the first proof by the second proof is called a *proof reduction*. If the assumption A is used more than once, then the proof t will be copied many times; so reduction does not always make a proof smaller. However, reduction is guaranteed to eliminate the unnecessary occurrence of $A \to B$ from the proof.

Similar proof reductions apply for the other connectives as well. Figure 2 shows the reduction rules for $\to$ and $\times$ and one of the two reduction rules for $+$ (the other is almost identical). Reductions form a partial order, denoted by $\Longrightarrow$. A proof to which no reductions apply is said to be in *normal form*. Remarkably, every proof has a normal form, and this normal form is unique modulo the commuting conversions.

A proof in normal form contains no extraneous propositions. More precisely, if a proof of a judgement $\Gamma \vdash A$ is in normal form, every proposition in the proof is a subformula of Γ or A. (The subformulas are defined in the obvious way; for instance, the subformulas of $A \to B$ are $A \to B$ itself plus the subformulas of A and B.)

For example, consider the judgement $A \to B, A \vdash B \times B$. One way to prove this is to use $\to$-I and Contraction to prove $\vdash B \to B \times B$, and then use $\to$-E

$$\dfrac{\dfrac{\dfrac{\begin{array}{c}A\vdash A\cdots\\ \vdots\, u\\ \Gamma, A\cdots\vdash B\end{array}}{\Gamma, A\vdash B}}{\Gamma\vdash A\to B}\,\to\text{-I}\qquad \begin{array}{c}\vdots\, t\\ \Delta\vdash A\end{array}}{\Gamma, \Delta\vdash B}\,\to\text{-E}\quad\Longrightarrow\quad \dfrac{\begin{array}{c}\vdots\, t\\ \Delta\vdash A\cdots\\ \vdots\, u\\ \Gamma, \Delta\cdots\vdash B\end{array}}{\Gamma, \Delta\vdash B}$$

$$\dfrac{\dfrac{\begin{array}{c}\vdots\, t\\ \Gamma\vdash A\end{array}\qquad\begin{array}{c}\vdots\, u\\ \Delta\vdash B\end{array}}{\Gamma, \Delta\vdash A\times B}\,\times\text{-I}\qquad \dfrac{\begin{array}{c}A\vdash A\cdots\quad B\vdash B\cdots\\ \vdots\, v\\ \Theta, A\cdots, B\cdots\vdash C\end{array}}{\Theta, A, B\vdash C}}{\Gamma, \Delta, \Theta\vdash C}\,\times\text{-E}\quad\Longrightarrow\quad \dfrac{\begin{array}{c}\vdots\, t\qquad\quad \vdots\, u\\ \Gamma\vdash A\cdots\quad \Delta\vdash B\cdots\\ \vdots\, v\\ \Gamma\cdots, \Delta\cdots, \Theta\vdash C\end{array}}{\Gamma, \Delta, \Theta\vdash C}$$

$$\dfrac{\dfrac{\begin{array}{c}\vdots\, t\\ \Gamma\vdash A\end{array}}{\Gamma\vdash A+B}\,+\text{-I}_1\qquad \dfrac{\begin{array}{c}A\vdash A\cdots\\ \vdots\, v\\ \Delta, A\cdots\vdash C\end{array}}{\Delta, A\vdash C}\qquad \dfrac{\begin{array}{c}B\vdash B\cdots\\ \vdots\, w\\ \Delta, B\cdots\vdash C\end{array}}{\Delta, B\vdash C}}{\Gamma, \Delta\vdash C}\,+\text{-E}\quad\Longrightarrow\quad \dfrac{\begin{array}{c}\vdots\, t\\ \Gamma\vdash A\cdots\\ \vdots\, v\\ \Gamma\cdots, \Delta\vdash C\end{array}}{\Gamma, \Delta\vdash C}$$

Fig. 2. Proof reduction for intuitionistic logic

to discharge the antecedent B with a proof of $A\to B,\ A\vdash B$. Here is the proof in full.

$$\dfrac{\dfrac{\dfrac{\dfrac{\dfrac{}{B\vdash B}\,\text{Id}\qquad \dfrac{}{B\vdash B}\,\text{Id}}{B, B\vdash B\times B}\,\times\text{-I}}{B\vdash B\times B}\,\text{Contraction}}{\vdash B\to(B\times B)}\,\to\text{-I}\ (\dagger)\qquad \dfrac{\dfrac{}{A\to B\vdash A\to B}\,\text{Id}\qquad \dfrac{}{A\vdash A}\,\text{Id}}{A\to B,\ A\vdash B}\,\to\text{-E}}{A\to B,\ A\vdash B\times B}\,\to\text{-E}\ (\dagger)$$

This proof contains a proposition, $B\to B\times B$, that is not a subformula of the judgement proved. It also contains an introduction rule immediately followed by an elimination rule, each marked with (†). Applying the reduction rule for $\to$ results in the following proof.

$$\dfrac{\dfrac{\dfrac{}{A \to B \vdash A \to B}\,\mathrm{Id} \qquad \dfrac{}{A \vdash A}\,\mathrm{Id}}{A \to B,\, A \vdash B}\,{\to}\text{-E} \qquad \dfrac{\dfrac{}{A \to B \vdash A \to B}\,\mathrm{Id} \qquad \dfrac{}{A \vdash A}\,\mathrm{Id}}{A \to B,\, A \vdash B}\,{\to}\text{-E}}{\dfrac{A \to B,\, A,\, A \to B,\, A \vdash B \times B}{A \to B,\, A \vdash B \times B}\,\text{Contraction}}\,\times\text{-I}$$

Contraction on B in the first proof has been replaced in the second by copying the proof of $A \to B,\, A \vdash B$ twice, and applying Contraction to the assumptions $A \to B$ and A. The second proof is in normal form, and does indeed satisfy the subformula property.

3 Intuitionistic terms

We now augment our judgements to contain terms. From the logicians' point of view, terms encode proofs. From the computer scientists' point of view, terms are a programming language for which the corresponding logic provides a type system.

In a constructive logic, propositions can be read as types, and proofs read as terms of the corresponding type. Implication is read as function space: a proof of $A \to B$ is a function that takes any proof of A into a proof of B. Conjunction is read as cartesian product: a proof of $A \times B$ is a pair consisting of a proof of A and a proof of B. Disjunction is read as disjoint sum: a proof of $A + B$ is either a proof of A or a proof of B. Henceforth *type* will be a synonym for proposition.

Terms encode the rules of the logic. There is one term form for the axiom Id, namely variables, and one term form for each of the logical rules. An introduction rule corresponds to a term that *constructs* a value of the corresponding type, while an elimination rule corresponds to a term that *deconstructs* a value of the corresponding type. There are no term forms for the structural rules, Exchange, Contraction, and Weakening. Let x, y, z range over variables, and s, t, u, v, w range over terms. The grammar of terms is as follows.

$$\begin{array}{rl} s, t, u, v, w ::= & x \\ | & \lambda x.\, u \mid s\,(t) \\ | & (t, u) \mid \text{case } s \text{ of } (x, y) \to v \\ | & \text{inl}\,(t) \mid \text{inr}\,(u) \mid \text{case } s \text{ of inl}\,(x) \to v;\ \text{inr}\,(y) \to w \end{array}$$

Assumptions are now written in the form

$$x_1 : A_1, \ldots, x_n : A_n$$

where $x_1, \ldots, x_n$ are distinct variables, $A_1, \ldots, A_n$ are types, and $n \geq 0$. As before, let Γ and Δ range over assumptions. Write Γ, Δ to denote the concatenation of assumptions; an implicit side condition of such a concatenation is that Γ and Δ contain distinct variables.

Judgements are now written in the form $\Gamma \vdash t : A$. One can think of Γ as a declaration specifying the types of the free variables of term t, which itself has type A.

Each rule is decorated with variables and terms, as shown in Figure 3.

$$\frac{}{x : A \vdash x : A}\ \text{Id} \qquad \frac{\Gamma, \Delta \vdash t : A}{\Delta, \Gamma \vdash t : A}\ \text{Exchange}$$

$$\frac{\Gamma, y : A, z : A \vdash u : B}{\Gamma, x : A \vdash u[x/y, x/z] : B}\ \text{Contraction} \qquad \frac{\Gamma \vdash u : B}{\Gamma, x : A \vdash u : B}\ \text{Weakening}$$

$$\frac{\Gamma, x : A \vdash u : B}{\Gamma \vdash \lambda x.t : A \to B}\ \to\text{-I} \qquad \frac{\Gamma \vdash s : A \to B \qquad \Delta \vdash t : A}{\Gamma, \Delta \vdash s(t) : B}\ \to\text{-E}$$

$$\frac{\Gamma \vdash t : A \qquad \Delta \vdash u : B}{\Gamma, \Delta \vdash (t, u) : A \times B}\ \times\text{-I} \qquad \frac{\Gamma \vdash s : A \times B \qquad \Delta, x : A, y : B \vdash v : C}{\Gamma, \Delta \vdash \text{case } s \text{ of } (x, y) \to v : C}\ \times\text{-E}$$

$$\frac{\Gamma \vdash t : A}{\Gamma \vdash \text{inl}\,(t) : A + B}\ +\text{-I}_1 \qquad \frac{\Gamma \vdash u : B}{\Gamma \vdash \text{inr}\,(u) : A + B}\ +\text{-I}_2$$

$$\frac{\Gamma \vdash s : A + B \qquad \Delta, x : A \vdash v : C \qquad \Delta, y : B \vdash w : C}{\Gamma, \Delta \vdash \text{case } t \text{ of inl}\,(x) \to v;\ \text{inr}\,(y) \to w : C}\ +\text{-E}$$

Fig. 3. Intuitionistic types

If a concatenation of the form Γ, Δ appears in a rule, then Γ and Δ must contain distinct variables. Thus in the $\to$-I rule, the variable x may not appear in Γ, so there is no shadowing of variables.

A variable in the assumption can appear more than once in a term only if Contraction is used. In the Contraction rule, the notation $t[x/y, x/z]$ stands for term t with variable x replacing all occurrences of variables y and z. Similarly, a variable in the assumption must appear in the corresponding term unless Weakening is used.

Here is the proof of the judgement $A \to B, A \vdash A \times B$, decorated with terms.

$$\dfrac{\dfrac{\dfrac{\dfrac{}{y : A \vdash y : A}\ \text{Id} \qquad \dfrac{\dfrac{}{f : A \to B \vdash f : A \to B}\ \text{Id} \qquad \dfrac{}{z : A \vdash z : A}\ \text{Id}}{f : A \to B, z : A \vdash f(z) : B}\ \to\text{-E}}{y : A, f : A \to B, z : A \vdash (y, f(z)) : A \times B}\ \times\text{-I}}{f : A \to B, y : A, z : A \vdash (y, f(z)) : A \times B}\ \text{Exchange}}{f : A \to B, x : A \vdash (x, f(x)) : A \times B}\ \text{Contraction}$$

The judgement at the root of this tree encodes the entire proof tree. In general, the judgement at the bottom of a labeled proof tree uniquely encodes the entire tree, modulo the commuting conversions for the structural rules.

3.1 Alternate terms for conjunction

The alternate rules for conjunction elimination correspond to alternate term forms for deconstructing pairs, fst (s) and snd (s).

$$\frac{\Gamma \vdash s : A \times B}{\Gamma \vdash \mathrm{fst}\,(s) : A}\ \times\text{-E}_1' \qquad \frac{\Gamma \vdash s : A \times B}{\Gamma \vdash \mathrm{snd}\,(s) : B}\ \times\text{-E}_2'$$

Just as the various rules can be expressed in terms of each other, the various terms can be defined in terms of each other. The terms 'fst' and 'snd' can be defined using 'case'.

$$\begin{aligned} \mathrm{fst}\,(s) &= \mathrm{case}\ s\ \mathrm{of}\ (x, y) \to x \\ \mathrm{snd}\,(s) &= \mathrm{case}\ s\ \mathrm{of}\ (x, y) \to y \end{aligned}$$

Conversely, the term 'case' can be defined using 'fst' and 'snd'.

$$\mathrm{case}\ s\ \mathrm{of}\ (x, y) \to v = (\lambda x.\, \lambda y.\, v)\,(\mathrm{fst}\,(s))\,(\mathrm{snd}\,(s))$$

These definitions can be read off from the corresponding proof trees seen previously.

3.2 Term reduction

Since terms encode proofs, the proof reductions seen previously can be encoded more compactly as term reductions. The operation of substituting a proof subtree for a leaf becomes the operation of substituting a term for a variable.

$$\begin{aligned} (\lambda x.\, u)\,(t) &\Longrightarrow u[t/x] \\ \mathrm{case}\ (t, u)\ \mathrm{of}\ (x, y) \to v &\Longrightarrow v[t/x, u/y] \\ \mathrm{case}\ \mathrm{inl}\,(t)\ \mathrm{of}\ \mathrm{inl}\,(x) \to v;\ \mathrm{inr}\,(y) \to w &\Longrightarrow v[t/x] \\ \mathrm{case}\ \mathrm{inr}\,(u)\ \mathrm{of}\ \mathrm{inl}\,(x) \to v;\ \mathrm{inr}\,(y) \to w &\Longrightarrow w[u/y] \end{aligned}$$

These rules are familiar to any programmer; in particular, the first is the beta reduction rule of lambda calculus.

The properties of proof reduction carry over into properties of term reduction. The correspondence of proof reductions with term reductions guarantees that reducing a well-typed term yields a well-typed term; this is called the *Subject Reduction* property. The uniqueness of normal forms for proofs corresponds to the uniqueness of normal forms for terms, which is analogous to the Church-Rosser theorem for untyped lambda calculus. The existence of normal forms for proofs means that every reduction sequence starting with a well-typed term eventually terminates; a property which certainly does *not* hold for untyped lambda calculus.

It is fascinating to observe that the notion of proof reduction, which was formulated to demonstrate the properties of proof systems, corresponds precisely to term reduction, which was formulated as a model of computation. This is the essence of the Curry-Howard isomorphism.

3.3 Fixpoints

Since every typed term has a normal form, the typed term calculus is strictly less expressive than untyped lambda calculus. To regain the power to express every computable function, one adds for each type A a constant

$$\mathit{fix} : (A \rightarrow A) \rightarrow A$$

with the reduction rule

$$\mathit{fix}(f) \Longrightarrow f\,(\mathit{fix}(f)).$$

A term containing this constant may not have a normal form, even if it is well typed. In particular, for every type A, the judgement $\vdash \mathit{fix}(\lambda x.\, x) : A$ is derivable, but this term has no normal form. Perhaps that is just as well, since this judgement corresponds to a proof that *any* proposition A is true!

4 Linear logic

Traditional logic encourages reckless use of resources. Contraction profligately duplicates assumptions, Weakening foolishly discards assumptions. This makes sense for logic, where truth is free; and it makes sense for some programming languages, where copying a value is as cheap as copying a pointer. But it is not always sensible. Linear logic encourages more careful use of resources. It is a logic for the 90s. If you lean to the right, view it as a logic of realistic accounting: no more free assumptions. If you lean to the left, view it as an eco-logic: resources must be conserved.

The obvious thing to do is to simply get rid of Contraction and Weakening entirely, but that is perhaps too severe. Our desire is to find a logic that allows control over Contraction and Weakening, but is still powerful enough that traditional intuitionistic logic may be embedded within it. In programming language terms, this corresponds to a language that allows control over duplication and discarding for some variables, but is still powerful enough that all traditional programs may be expressed within it.

So rather than getting rid of Contraction and Weakening entirely, we shall 'bell the cat'. If Contraction or Weakening are used in a proof, then this will be explicitly visible in the proposition proved.

The absence of Contraction and Weakening profoundly changes the nature of the logical connectives. Implication is now written $A \multimap B$ and pronounced 'consume A yielding B'. (The symbol $\multimap$ on its own is pronounced 'lollipop'.) As noted previously, in the absence of Contraction and Weakening, there are two distinct ways to formulate conjunction, corresponding to two distinct connectives in linear logic. These are written $A \otimes B$, pronounced 'both A and B', and $A \,\&\, B$, pronounced 'choose from A and B'. (The symbols $\otimes$ and $\&$ are pronounced 'tensor' and 'with'.) Disjunction is written $A \oplus B$ and still pronounced 'either A or B'. Finally, a new form of proposition is introduced to indicate where Contraction or Weakening may be used. It is written $!A$ and pronounced 'of course A'. (The symbol ! is pronounced 'pling' or 'bang!'.)

As before, let A, B, C range over propositions, and X range over propositional constants. The grammar of linear propositions is as follows.

$$A, B, C ::= X \mid A \multimap B \mid A \otimes B \mid A \,\&\, B \mid A \oplus B \mid !A$$

The particular logical system that we will study is based on Girard's Logic of Unity, which is a refinement of linear logic. This achieves some significant technical simplifications by using *two* forms of assumption. One form of assumption, called *linear*, does not allow Contraction or Weakening, and is written in angle brackets, $\langle A \rangle$. The other form of assumption, called *intuitionistic*, does allow Contraction and Weakening, and is written in square brackets, $[A]$.

As we shall see, an assumption of the form $\langle !A \rangle$ is in a sense equivalent to an assumption of the form $[A]$. However, they differ in that a linear assumption – even one of the form $\langle !A \rangle$ – must be used *exactly once* in a proof, while an intuitionistic assumption may be used any number of times.

As before, let Γ, Δ range over sequences of zero or more assumptions, which may be of either sort. Write $[\Gamma]$ for a sequence of zero or more intuitionistic assumptions.

Judgements, as before, are written in the form $\Gamma \vdash A$. It is only assumptions that are labeled with angle or square brackets – these brackets only appear to the left of $\vdash$, never to the right.

The rules of linear logic are shown in Figure 4. There are now two axioms, $\langle$Id$\rangle$ for a linear assumption and [Id] for an intuitionistic assumption. As one might expect, the rules Contraction, Weakening, and the logical rules for ! use intuitionistic assumptions. The logical rules for the other conectives, $\multimap$, $\otimes$, $\&$, and $\oplus$, all use linear assumptions.

Apart from the switch to linear assumptions, the logical rules for $\multimap$, $\otimes$, and $\oplus$ are identical to the logical rules for $\rightarrow$, $\times$, and $+$; and the logical rules for $\&$ are identical to the alternate logical rules for $\times$. Note that $\otimes$-I uses two distinct assumption lists, while $\&$-I uses the same list twice.

Before looking in detail at the rules for !, let's take a moment to explore the meaning other logical connectives.

4.1 A logic of resources

We can read the new rules in terms of combinations of resources. Take A to be the proposition 'I have ten zlotys', B to be the proposition 'I have a pizza', and C to be the proposition 'I have a cake'. Adding to our logic the axioms

$$\langle A \rangle \vdash B \qquad \langle A \rangle \vdash C$$

expresses that for ten zlotys I may buy a pizza, and for ten zlotys I may buy a cake. Instantiating $\otimes$-I yields

$$\frac{\langle A \rangle \vdash B \qquad \langle A \rangle \vdash C}{\langle A \rangle, \langle A \rangle \vdash B \otimes C}\ \otimes\text{-I}$$

$$\frac{}{\langle A\rangle \vdash A}\langle\mathrm{Id}\rangle \qquad \frac{}{[A] \vdash A}[\mathrm{Id}] \qquad \frac{\Gamma, \Delta \vdash A}{\Delta, \Gamma \vdash A}\text{Exchange}$$

$$\frac{\Gamma, [A], [A] \vdash B}{\Gamma, [A] \vdash B}\text{Contraction} \qquad \frac{\Gamma \vdash B}{\Gamma, [A] \vdash B}\text{Weakening}$$

$$\frac{[\Gamma] \vdash A}{[\Gamma] \vdash !A}\text{!-I} \qquad \frac{\Gamma \vdash !A \qquad \Delta, [A] \vdash B}{\Gamma, \Delta \vdash B}\text{!-E}$$

$$\frac{\Gamma, \langle A\rangle \vdash B}{\Gamma \vdash A \multimap B}\multimap\text{-I} \qquad \frac{\Gamma \vdash A \multimap B \qquad \Delta \vdash A}{\Gamma, \Delta \vdash B}\multimap\text{-E}$$

$$\frac{\Gamma \vdash A \qquad \Delta \vdash B}{\Gamma, \Delta \vdash A \otimes B}\otimes\text{-I} \qquad \frac{\Gamma \vdash A \otimes B \qquad \Delta, \langle A\rangle, \langle B\rangle \vdash C}{\Gamma, \Delta \vdash C}\otimes\text{-E}$$

$$\frac{\Gamma \vdash A \qquad \Gamma \vdash B}{\Gamma \vdash A \,\&\, B}\&\text{-I} \qquad \frac{\Gamma \vdash A \,\&\, B}{\Gamma \vdash A}\&\text{-E}_1 \qquad \frac{\Gamma \vdash A \,\&\, B}{\Gamma \vdash B}\&\text{-E}_2$$

$$\frac{\Gamma \vdash A}{\Gamma \vdash A \oplus B}\oplus\text{-I}_1 \qquad \frac{\Gamma \vdash B}{\Gamma \vdash A \oplus B}\oplus\text{-I}_2 \qquad \frac{\Gamma \vdash A \oplus B \qquad \Delta, \langle A\rangle \vdash C \qquad \Delta, \langle B\rangle \vdash C}{\Gamma, \Delta \vdash C}\oplus\text{-E}$$

Fig. 4. Linear logic

meaning that for twenty zlotys I can buy both a pizza and a cake. Instantiating &-I yields

$$\frac{\langle A\rangle \vdash B \qquad \langle A\rangle \vdash C}{\langle A\rangle \vdash B \,\&\, C}\&\text{-I}$$

meaning that for ten zlotys I can buy whichever I choose from a cake and a pizza. Instantiating $\oplus$-I_1 yields

$$\frac{\langle A\rangle \vdash B}{\langle A\rangle \vdash B \oplus C}\oplus\text{-I}_1$$

meaning that for ten zlotys I can buy either a pizza or a cake. But I no longer have a choice: the proof implies that I must buy a pizza. If the bakery closes, so that $\langle A\rangle \vdash C$ is no longer an axiom, then $\langle A\rangle \vdash B \,\&\, C$ is no longer provable, but $\langle A\rangle \vdash B \oplus C$ remains provable as long as the pizzeria remains open.

Take D to be the proposition 'I am happy'. Then $\langle B \otimes C\rangle \vdash D$ expresses that I will be happy given *both* a pizza and a cake; and $\langle B \,\&\, C\rangle \vdash D$ expresses that I will be happy given *my choice* from a pizza and a cake; and $\langle B \oplus C\rangle \vdash D$ expresses that I will be happy given *either* a pizza or a cake, I don't care which.

For any A and B, one can prove $\langle A\&B\rangle \vdash A\oplus B$. Indeed, there are two proofs, one choosing A and one choosing B. But there is no way to prove the converse,

$\langle A \oplus B \rangle \vdash A \mathbin{\&} B$. Furthermore, neither $\langle A \otimes B \rangle \vdash A \mathbin{\&} B$ nor $A \mathbin{\&} B \vdash A \otimes B$ is provable.

Thanks to the absence of Contraction for linear assumptions, it is impossible to prove $\langle A \rangle \vdash A \otimes A$; disappointingly, ten zlotys cannot magically become twenty. Thanks to the absence of Weakening, it is impossible to prove $\langle A \otimes A \rangle \vdash A$; reassuringly, twenty zlotys will not mysteriously become ten.

4.2 Unlimited resources

A linear assumption, $\langle A \rangle$, can be thought of as supplying exactly one occurrence of A, while an intuitionistic assumption, $[A]$, can be thought of as supplying an unlimited number of occurrences of A: multiple occurrences of A may be supplied if Contraction is used, and no occurrences of A may be supplied if Weakening is used. Taking A to be 'I have ten zlotys', then the linear assumption $\langle A \rangle$ holds if I have a ten zloty note in my pocket, while the intuitionistic assumption $[A]$ holds if I have an unlimited supply of ten zloty notes.

It is instructive to compare the two axioms.

$$\frac{}{\langle A \rangle \vdash A}\langle \mathrm{Id} \rangle \qquad \frac{}{[A] \vdash A}[\mathrm{Id}]$$

The axiom on the left says that if I have ten zlotys in my pocket, I can reach in and extract ten zlotys. The axiom on the right says that if I have an indefinitely large supply of zloty notes, then from it I may withdraw a single ten zloty note.

If $\Gamma,\ \langle A \rangle \vdash B$ is provable, then $\Gamma,\ [A] \vdash B$ is also provable.

$$\frac{\dfrac{\Gamma,\ \langle A \rangle \vdash B}{\Gamma \vdash A \multimap B}\multimap\text{-I} \qquad \dfrac{}{[A] \vdash A}[\mathrm{Id}]}{\Gamma,\ [A] \vdash B}\multimap\text{-E}$$

Taking $\langle A \rangle \vdash B$ to mean that for ten zlotys I can buy a pizza, it follows that $[A] \vdash B$, so if I have an indefinitely large supply of ten zloty notes then I can still buy a pizza.

Since the logical connectives are defined in terms of linear assumptions, some method is required to turn an intuitionistic assumption into a linear one. This is supplied by the connective !. An intuitionistic assumption $[A]$ is equivalent to a linear assumption $\langle !A \rangle$. Thus, the proposition $!A$ holds when there is an unlimited supply of A.

In the !-I rule, $[\Gamma]$ is written to indicate that all the assumptions must be intuitionistic. The rule states that if from such an assumption list one can prove A, then from the same assumption list one can prove $!A$. In other words, if given unlimited assumptions one can derive a single A, then from the same assumptions one can also derive an unlimited number of A. Instantiating !-I yields

$$\frac{[A] \vdash B}{[A] \vdash !B}\text{!-I}.$$

If I have an unlimited supply of ten zloty notes, then not only can I buy one pizza, there is no limit to the number I can buy.

The rule !-E states that an intuitionistic assumption $[A]$ may be satisfied by a proof of the proposition $!A$. Adding to our logic the axiom

$$[B] \vdash D$$

expresses that an unlimited supply of pizza leads to happiness. Instantiating !-E yields

$$\frac{[A] \vdash !B \qquad [B] \vdash D}{[A] \vdash D}\ \text{!-E}.$$

Given an unlimited supply of zlotys, I can produce an unlimited supply of pizza, and this in turn produces happiness.

A linear assumption $\langle !A \rangle$ is equivalent to an intuitionistic assumption $[A]$. That is, $\Gamma, \langle !A \rangle \vdash B$ is provable if and only if $\Gamma, [A] \vdash B$ is provable. In the forward direction, this is shown with the aid of !-I.

$$\frac{\dfrac{\Gamma, \langle !A \rangle \vdash B}{\Gamma \vdash !A \multimap B}\ \multimap\text{-I} \qquad \dfrac{\dfrac{}{[A] \vdash A}\ [\text{Id}]}{[A] \vdash !A}\ \text{!-I}}{\Gamma, [A] \vdash B}\ \multimap\text{-E}$$

In the reverse direction, it is shown with the aid of !-E.

$$\frac{\dfrac{}{\langle !A \rangle \vdash !A}\ \langle\text{Id}\rangle \qquad \Gamma, [A] \vdash B}{\Gamma, \langle !A \rangle \vdash B}\ \text{!-E}$$

One might therefore ask: Why bother with assumptions of the form $[A]$? Why not just use $\langle !A \rangle$ instead? There is a good reason for distinguishing the two, which will be explained after proof reductions are discussed.

The propositions $!(A \,\&\, B)$ and $!A \otimes !B$ are equivalent in linear logic. That is, both of the following are provable:

$$\langle !(A \,\&\, B) \rangle \vdash !A \otimes !B,$$
$$\langle !A \otimes !B \rangle \vdash !(A \,\&\, B).$$

Consider the following two airlines. On one, there is a single steward, who can be called an unlimited number of times. Whenever called, he will give you your choice of either a cup of coffee or a cup of tea. On the other, there are separate coffee and tea stewards, each of whom can be called an unlimited number of times, and each of whom provides a cup of the appropriate beverage when called. Thanks to the equivalence displayed above, you realize that you should not let this distinction determine your choice of airline.

4.3 Embedding intuitionistic logic in linear logic

One reason for incorporating ! into our logic is so that it has sufficient power that intuitionistic logic can be embedded within it. The idea is to find a translation of the connectives $\rightarrow$, $\times$ and $+$ into the connectives !, $\multimap$, &, and $\oplus$, such that a judgement $\Gamma \vdash A$ is provable in intuitionistic logic if and only if the corresponding judgement $[\Gamma] \vdash A$ is provable in linear logic. As one might expect, the embedding takes (unlabeled) assumptions of the intuitionistic judgement into intuitionistic assumptions in the linear judgement. This is necessary because Contraction and Weakening may be applied to any assumption in an intuitionistic proof.

To find the translation for $\rightarrow$, compare the rules for $\rightarrow$ in Figure 1 with the rules for $\multimap$ in Figure 4. The key difference is that A appears as an intuitionistic assumption in $\rightarrow$-I, but as a linear assumption in $\multimap$-I. This makes it reasonable to define

$$A \rightarrow B = {!A} \multimap B.$$

With this definition, the intuitionistic rules can be defined from the corresponding rules of linear logic, together with the rules for !-I and !-E.

$$\dfrac{\dfrac{\dfrac{}{\langle !A\rangle \vdash {!A}}\,\langle\mathrm{Id}\rangle \qquad \Gamma, [A] \vdash B}{\Gamma, \langle !A\rangle \vdash B}\,\text{!-E}}{\Gamma \vdash {!A} \multimap B}\,{\multimap}\text{-I} \qquad\qquad \dfrac{\Gamma \vdash {!A} \multimap B \qquad \dfrac{[\Delta] \vdash B}{[\Delta] \vdash {!B}}\,\text{!-I}}{\Gamma, [\Delta] \vdash B}\,{\multimap}\text{-E}$$

The derived rules are slightly more general than required, in that Γ may contain assumptions of either type, though $[\Delta]$ is indeed restricted to contain intuitionistic assumptions.

A similar comparison of the rules for $\times$ with the rules for &, and of the rules for $+$ with the rules for $\otimes$, yields the complete embedding.

$$\begin{aligned} A \rightarrow B &= {!A} \multimap B \\ A \times B &= A \mathbin{\&} B \\ A + B &= {!A} \oplus {!B} \end{aligned}$$

It is an easy exercise to work out the corresponding derived rules.

Comparing the rules for $\times$ with the rules for $\otimes$ yields an alternative embedding.

$$A \times B = {!A} \otimes {!B}$$

Observe that $\langle !A \otimes !B\rangle \vdash A \mathbin{\&} B$ but that $\langle A \mathbin{\&} B\rangle \nvdash {!A} \otimes {!B}$. So in some sense the embedding with & is tighter than the embedding with $\otimes$.

4.4 Proof reductions

The new rules for proof reduction are show in Figure 5. Whereas before the double bars standing for potential occurrences of Contraction and Weakening were ubiquitous, now they appear only in the reduction rule for !. Furthermore, rule ! is now the only rule which may copy (or discard) a proof sub-tree. The rules for $\multimap$, $\otimes$, $\oplus$ are simpler than the corresponding rules for $\rightarrow$, $\times$, and $+$, because the linear connectives involve only linear assumptions, which are guaranteed to be used exactly once in the proof tree.

$$\cfrac{\cfrac{\begin{array}{c}\vdots\, t\\ [\Gamma]\vdash A\end{array}}{}\;!\text{-I}\qquad \cfrac{\begin{array}{c}[A]\vdash A\cdots\\ \vdots\, u\\ \Delta,[A]\cdots\vdash B\end{array}}{\Delta,[A]\vdash B}}{[\Gamma],\Delta\vdash B}\;!\text{-E} \quad\Longrightarrow\quad \cfrac{\begin{array}{c}\vdots\, t\\ [\Gamma]\vdash A\cdots\\ \vdots\, u\\ [\Gamma]\cdots,\Delta\vdash B\end{array}}{[\Gamma],\Delta\vdash B}$$

$$\cfrac{\cfrac{\begin{array}{c}\langle A\rangle\vdash A\\ \vdots\, u\\ \Gamma, A\vdash B\end{array}}{\Gamma\vdash A\multimap B}\;\multimap\text{-I}\qquad \begin{array}{c}\vdots\, t\\ \Delta\vdash A\end{array}}{\Gamma,\Delta\vdash B}\;\multimap\text{-E} \quad\Longrightarrow\quad \begin{array}{c}\vdots\, t\\ \Delta\vdash A\\ \vdots\, u\\ \Gamma,\Delta\vdash B\end{array}$$

$$\cfrac{\cfrac{\begin{array}{c}\vdots\, t\\ \Gamma\vdash A\end{array}\qquad \begin{array}{c}\vdots\, u\\ \Delta\vdash B\end{array}}{\Gamma,\Delta\vdash A\otimes B}\;\otimes\text{-I}\qquad \begin{array}{c}\langle A\rangle\vdash A\quad \langle B\rangle\vdash B\\ \vdots\, v\\ \Theta, A, B\vdash C\end{array}}{\Gamma,\Delta,\Theta\vdash C}\;\otimes\text{-E} \quad\Longrightarrow\quad \begin{array}{c}\begin{array}{cc}\vdots\, t & \vdots\, u\\ \Gamma\vdash A & \Delta\vdash B\end{array}\\ \vdots\, v\\ \Gamma,\Delta,\Theta\vdash C\end{array}$$

$$\cfrac{\cfrac{\begin{array}{c}\vdots\, t\\ \Gamma\vdash A\end{array}\qquad \begin{array}{c}\vdots\, u\\ \Gamma\vdash B\end{array}}{\Gamma\vdash A\,\&\,B}\;\&\text{-I}}{\Gamma\vdash A}\;\&\text{-E}_1 \quad\Longrightarrow\quad \begin{array}{c}\vdots\, t\\ \Gamma\vdash A\end{array}$$

$$\cfrac{\cfrac{\begin{array}{c}\vdots\, t\\ \Gamma\vdash A\end{array}}{\Gamma\vdash A\oplus B}\;\oplus\text{-I}_1\qquad \begin{array}{c}\langle A\rangle\vdash A\\ \vdots\, v\\ \Delta, A\vdash C\end{array}\qquad \begin{array}{c}\langle B\rangle\vdash B\\ \vdots\, w\\ \Delta, B\vdash C\end{array}}{\Gamma,\Delta\vdash C}\;\oplus\text{-E} \quad\Longrightarrow\quad \begin{array}{c}\vdots\, t\\ \Gamma\vdash A\\ \vdots\, v\\ \Gamma,\Delta\vdash C\end{array}$$

Fig. 5. Proof reduction for linear logic

4.5 The need for intuitionistic assumptions

As noted previously, an intuitionistic assumption $[A]$ is equivalent to a linear assumption $\langle !A\rangle$. So why bother with intuitionistic assumptions? One is tempted to define a simpler version of linear logic by replacing $[A]$ by $\langle !A\rangle$ in the rules Contraction and Weakening, and adopting some variant of the [Id], !-I, and !-E rules. Surely such a system would be better?

Alas, this system is *too* simple. The problem shows up when one considers proof reduction. Observe that a judgement can prove a proposition begining with ! even if none of the assumptions begin with !.

$$\dfrac{\dfrac{}{\langle A \multimap !B\rangle \vdash A \multimap !B}\,\langle\mathrm{Id}\rangle \qquad \dfrac{}{\langle A\rangle \vdash A}\,\langle\mathrm{Id}\rangle}{\langle A \multimap !B\rangle,\ \langle A\rangle \vdash !B}\,\multimap\text{-E}$$

Now consider the following proof.

$$\dfrac{\dfrac{\dfrac{\dfrac{\dfrac{}{\langle !B\rangle \vdash !B}\,\langle\mathrm{Id}\rangle \qquad \dfrac{}{\langle !B\rangle \vdash !B}\,\langle\mathrm{Id}\rangle}{\langle !B\rangle,\ \langle !B\rangle \vdash !B \otimes !B}\,\otimes\text{-I}}{\langle !B\rangle \vdash !B \otimes !B}\,\text{Cont'n}}{\vdash !B \multimap (!B \otimes !B)}\,\multimap\text{-I }(\dagger) \qquad \begin{array}{c}\vdots\\ \langle A \multimap !B\rangle,\ \langle A\rangle \vdash !B\end{array}}{\langle A \multimap !B\rangle,\ \langle A\rangle \vdash !B \otimes !B}\,\multimap\text{-E }(\dagger)$$

This contains a $\multimap$-I rule followed immediately by a $\multimap$-E rule, both marked with $(\dagger)$. Applying proof reduction yields the following proof.

$$\dfrac{\dfrac{\begin{array}{c}\vdots\\ \langle A \multimap !B\rangle,\ \langle A\rangle \vdash !B\end{array} \qquad \begin{array}{c}\vdots\\ \langle A \multimap !B\rangle,\ \langle A\rangle \vdash !B\end{array}}{\langle A \multimap !B\rangle,\ \langle A\rangle,\ \langle A \multimap !B\rangle,\ \langle A\rangle \vdash !B \otimes !B}\,\otimes\text{-I}}{\langle A \multimap !B\rangle,\ \langle A\rangle \vdash !B \otimes !B}\,\text{Cont'n}$$

But this proof is ***illegal***! Contraction has been applied to assumptions $\langle A \multimap !B\rangle$ and $\langle A\rangle$, which is not allowed.

The problem is that the soundness of the proof reduction rule for $\multimap$ depends on the fact that in a proof of $\Gamma,\ \langle A\rangle \vdash B$, the assumption $\langle A\rangle$ is used exactly once. But this is no longer true with the modified Contraction and Weakening rules.

This problem does not arise for the system given here. The closest one can come to the bad proof is the following.

$$
\dfrac{
\dfrac{
\dfrac{}{\langle !B\rangle \vdash !B}\langle \mathrm{Id}\rangle
\qquad
\dfrac{
\dfrac{
\dfrac{\dfrac{}{[B]\vdash B}[\mathrm{Id}]}{[B]\vdash !B}\,!\text{-I}
\qquad
\dfrac{\dfrac{}{[B]\vdash B}[\mathrm{Id}]}{[B]\vdash !B}\,!\text{-I}
}{[B],\,[B]\vdash !B\otimes !B}\otimes\text{-I}
}{[B]\vdash !B\otimes !B}\,\text{Cont'n}
}{
\dfrac{\langle !B\rangle \vdash !B\otimes !B}{\vdash !B\multimap(!B\otimes !B)}\multimap\text{-I }(\dagger)
}\,!\text{-E}
\qquad
\begin{array}{c}\vdots\\ \langle A\multimap !B\rangle,\,\langle A\rangle\vdash !B\end{array}
}{\langle A\multimap !B\rangle,\,\langle A\rangle\vdash !B\otimes !B}\multimap\text{-E }(\dagger)
$$

Applying proof reduction yields the following proof.

$$
\dfrac{
\begin{array}{c}\vdots\\ \langle A\multimap !B\rangle,\,\langle A\rangle\vdash !B\end{array}
\qquad
\dfrac{
\dfrac{
\dfrac{\dfrac{}{[B]\vdash B}[\mathrm{Id}]}{[B]\vdash !B}\,!\text{-I}
\qquad
\dfrac{\dfrac{}{[B]\vdash B}[\mathrm{Id}]}{[B]\vdash !B}\,!\text{-I}
}{[B],\,[B]\vdash !B\otimes !B}\otimes\text{-I}
}{[B]\vdash !B\otimes !B}\,\text{Cont'n}
}{\langle A\multimap !B\rangle,\,\langle A\rangle\vdash !B\otimes !B}\,!\text{-E}
$$

Thanks to !-E, the linear assumption $\langle !B\rangle$ is used exactly once in the proof of $\langle !B\rangle \vdash !B\otimes !B$, and so proof reduction poses no problem.

It is indeed possible to formulate a system of linear logic. One way to do so is to replace the rules [Id], Contraction, Weakening, !-I, and !-E of Figure 4 by the rules in Figure 6. The rule !-E$'_1$ combines [Id] and !-E, and the rules !-E$'_2$ and !-E$'_3$ correspond to Contraction and Weakening. This system is considerably more complex, especially when one compares !-I with !-I$'$.

$$
\dfrac{\Gamma_1\vdash !A_1 \quad \cdots \quad \Gamma_n\vdash !A_n \quad \langle !A_1\rangle,\ldots,\langle !A_n\rangle\vdash B}{\Gamma_1,\ldots,\Gamma_n\vdash !B}\,!\text{-I}'
$$

$$
\dfrac{\Gamma\vdash !A \quad \Delta,\langle A\rangle\vdash B}{\Gamma,\Delta\vdash B}\,!\text{-E}'_1
\qquad
\dfrac{\Gamma\vdash !A \quad \Delta,\langle !A\rangle,\langle !A\rangle\vdash B}{\Gamma,\Delta\vdash B}\,!\text{-E}'_2
\qquad
\dfrac{\Gamma\vdash !A \quad \Delta\vdash B}{\Gamma,\Delta\vdash B}\,!\text{-E}'_3
$$

Fig. 6. Alternate rules for linear logic

5 Linear types

At last, all the pieces are in place to carry through our plan. We have seen intuitionistic logic and how it induces a typed functional language, and we have seen linear logic. Having set everything up carefully, it is straightforward to induce a typed language based on linear logic.

As before, there is one term form for variables, and one for each of the logical introduction and elimination rules. The term forms follow closely what has already been seen for intuitionistic logic. To distinguish the new term forms, they are generally written with angle brackets rather than round brackets. There are two pair constructors, $\langle t, u\rangle$ for types of the form $A \otimes B$, and $\langle\!\langle t, u\rangle\!\rangle$ for types of the form $A \mathbin{\&} B$. The grammar of terms is as follows.

$$\begin{aligned}
s, t, u, v, w ::= {} & x \\
\mid {} & \lambda\langle x\rangle.\, u \mid s\,\langle t\rangle \\
\mid {} & !t \mid \mathrm{case}\ s\ \mathrm{of}\ !x \rightarrow u \\
\mid {} & \langle t, u\rangle \mid \mathrm{case}\ s\ \mathrm{of}\ \langle x, y\rangle \rightarrow v \\
\mid {} & \langle\!\langle t, u\rangle\!\rangle \mid \mathrm{fst}\,\langle s\rangle \mid \mathrm{snd}\,\langle s\rangle \\
\mid {} & \mathrm{inl}\,\langle t\rangle \mid \mathrm{inr}\,\langle u\rangle \mid \mathrm{case}\ s\ \mathrm{of}\ \mathrm{inl}\,\langle x\rangle \rightarrow v;\ \mathrm{inr}\,\langle y\rangle \rightarrow w
\end{aligned}$$

Assumptions now take two forms: linear, written $\langle x : A\rangle$, and intuitionistic, written $[x : A]$. As before, let Γ, Δ range over lists of zero or more assumptions, where all of the variables are distinct, and let $[\Gamma]$ denote a list containing only intuitionistic assumptions. Judgements have the form $\Gamma \vdash t : A$.

The rules for linear terms are shown in Figure 7.

Here is a sample proof augmented with terms.

$$\dfrac{\dfrac{}{\langle x : !B\rangle x : !B}\langle\mathrm{Id}\rangle \qquad \dfrac{\dfrac{\dfrac{\dfrac{}{[y : B] \vdash y : B}[\mathrm{Id}]}{[y : B] \vdash !y : !B}\,!\text{-I} \qquad \dfrac{\dfrac{}{[y : B] \vdash y : B}[\mathrm{Id}]}{[y : B] \vdash !y : !B}\,!\text{-I}}{[y : B], [y : B] \vdash \langle !y, !y\rangle : !B \otimes !B}\otimes\text{-I}}{[y : B] \vdash \langle !y, !y\rangle : !B \otimes !B}\,\text{Cont'n}}{\dfrac{\langle x : !B\rangle \vdash \mathrm{case}\ x\ \mathrm{of}\ !y \rightarrow \langle !y, !y\rangle : !B \otimes !B}{\vdash \lambda\langle x\rangle.\, \mathrm{case}\ x\ \mathrm{of}\ !y \rightarrow \langle !y, !y\rangle : !B \multimap !B \otimes !B}\multimap\text{-I}}\,!\text{-E}$$

As before, the judgement at the root uniquely encodes the entire proof tree, modulo the commuting conversions.

5.1 Term reductions

Proof reductions may be read off as term reductions. As before, the properties of proof reductions guarantee that every term has a unique normal form, and that term reduction preserves well typing.

$$\begin{aligned}
\mathrm{case}\ !t\ \mathrm{of}\ !x \rightarrow u &\Longrightarrow u[t/x] \\
(\lambda\langle x\rangle.\, u)\,\langle t\rangle &\Longrightarrow u[t/x] \\
\mathrm{case}\ \langle t, u\rangle\ \mathrm{of}\ \langle x, y\rangle \rightarrow v &\Longrightarrow v[t/x, u/y] \\
\mathrm{fst}\,\langle\ \langle\!\langle t, u\rangle\!\rangle\ \rangle &\Longrightarrow t \\
\mathrm{snd}\,\langle\ \langle\!\langle t, u\rangle\!\rangle\ \rangle &\Longrightarrow u \\
\mathrm{case}\ \mathrm{inl}\,\langle t\rangle\ \mathrm{of}\ \mathrm{inl}\,\langle x\rangle \rightarrow v;\ \mathrm{inr}\,\langle y\rangle \rightarrow w &\Longrightarrow v[t/x] \\
\mathrm{case}\ \mathrm{inr}\,\langle u\rangle\ \mathrm{of}\ \mathrm{inl}\,\langle x\rangle \rightarrow v;\ \mathrm{inr}\,\langle y\rangle \rightarrow w &\Longrightarrow w[u/y]
\end{aligned}$$

$$\frac{}{\langle x : A\rangle \vdash x : A}\,\langle\mathrm{Id}\rangle \qquad \frac{}{[x : A] \vdash x : A}\,[\mathrm{Id}] \qquad \frac{\Gamma,\ \Delta \vdash t : A}{\Delta,\ \Gamma \vdash t : A}\,\text{Exchange}$$

$$\frac{\Gamma,\ [y : A],\ [z : A] \vdash u : B}{\Gamma,\ [x : A] \vdash u[x/z, y/z] : B}\,\text{Contraction} \qquad \frac{\Gamma \vdash u : B}{\Gamma,\ [x : A] \vdash u : B}\,\text{Weakening}$$

$$\frac{[\Gamma] \vdash t : A}{[\Gamma] \vdash\ !t : !A}\,!\text{-I} \qquad \frac{\Gamma \vdash s : !A \qquad \Delta,\ [x : A] \vdash u : B}{\Gamma,\ \Delta \vdash \mathsf{case}\ s\ \mathsf{of}\ !x \rightarrow u}\,!\text{-E}$$

$$\frac{\Gamma \vdash t : A \qquad \Delta \vdash u : B}{\Gamma,\ \Delta \vdash \langle t, u\rangle : A \otimes B}\,\otimes\text{-I} \qquad \frac{\Gamma \vdash s : A \otimes B \qquad \Delta,\ \langle x : A\rangle,\ \langle y : B\rangle \vdash v : C}{\Gamma,\ \Delta \vdash \mathsf{case}\ s\ \mathsf{of}\ \langle x, y\rangle \rightarrow v : C}\,\otimes\text{-E}$$

$$\frac{\Gamma \vdash t : A \qquad \Gamma \vdash u : B}{\Gamma \vdash \langle\langle t, u\rangle\rangle : A \,\&\, B}\,\&\text{-I} \qquad \frac{\Gamma \vdash s : A \,\&\, B}{\Gamma \vdash \mathsf{fst}\,\langle s\rangle : A}\,\&\text{-E}_1 \qquad \frac{\Gamma \vdash s : A \,\&\, B}{\Gamma \vdash \mathsf{snd}\,\langle s\rangle : B}\,\&\text{-E}_2$$

$$\frac{\Gamma,\ x : A \vdash u : B}{\Gamma \vdash \lambda\langle x\rangle.\,t : A \multimap B}\,\multimap\text{-I} \qquad \frac{\Gamma \vdash s : A \multimap B \qquad \Delta \vdash t : A}{\Gamma,\ \Delta \vdash s\langle t\rangle : B}\,\multimap\text{-E}$$

$$\frac{\Gamma \vdash t : A}{\Gamma \vdash \mathsf{inl}\,\langle t\rangle : A \oplus B}\,\oplus\text{-I}_1 \qquad \frac{\Gamma \vdash u : B}{\Gamma \vdash \mathsf{inr}\,\langle u\rangle : A \oplus B}\,\oplus\text{-I}_2$$

$$\frac{\Gamma \vdash s : A + B \qquad \Delta,\ \langle x : A\rangle \vdash v : C \qquad \Delta,\ \langle y : B\rangle \vdash w : C}{\Gamma,\ \Delta \vdash \mathsf{case}\ t\ \mathsf{of}\ \mathsf{inl}\,\langle x\rangle \rightarrow v;\ \mathsf{inr}\,\langle y\rangle \rightarrow w : C}\,\oplus\text{-E}$$

Fig. 7. Linear types

Thanks to linearity, the only substitution in the above that may duplicate a variable is that for !. In a lazy language, this means that only evaluation of ! requires overwriting.

It is possible to arrange an implementation so that a variable corresponding to a linear assumption contains the sole pointer to a value. However, this requires that a term of the form $!t$ be re-evaluated *each* time it is examined, which is prohibitively expensive for most purposes.

More commonly, a term of the form $!t$ will be overwritten with its value the first time it is accessed, and future accesses will copy a pointer to that value. In this case, a variable corresponding to a linear assumption may not contain the sole pointer to a value. However, the absence of Contraction and Weakening on linear assumptions makes it possible to guarantee the following useful property: if a variable corresponding to a linear assumption ever contains the sole pointer to a variable, then it will continue to do so. Some applications of this property will be discussed later.

5.2 Embedding intuitionistic logic into linear logic

As we have seen, the connectives of intuitionistic logic can be regarded as abbreviations for combinations of connectives in linear logic.

$$
\begin{aligned}
A \to B &= !A \multimap B \\
A \times B &= A \mathbin{\&} B \\
A + B &= !A \oplus !B
\end{aligned}
$$

Similarly, the term calculus of intuitionistic logic can be regarded as abbreviations for combinations of linear terms.

$$
\begin{aligned}
\lambda x.\, u &= \lambda\langle x'\rangle.\, \text{case } x' \text{ of } !x \to u \\
t\,(u) &= t\,\langle !u\rangle \\
(t, u) &= \langle\!\langle t, u\rangle\!\rangle \\
\mathsf{fst}\,(s) &= \mathsf{fst}\,\langle s\rangle \\
\mathsf{snd}\,(s) &= \mathsf{snd}\,\langle s\rangle \\
\mathsf{inl}\,(t) &= \mathsf{inl}\,\langle !t\rangle \\
\mathsf{inr}\,(u) &= \mathsf{inr}\,\langle !u\rangle \\
\text{case } s \text{ of } \mathsf{inl}\,(x) \to v;\ \mathsf{inr}\,(y) \to w &= \text{case } s \text{ of} \\
&\qquad \mathsf{inl}\,\langle x'\rangle \to \text{case } x' \text{ of } !x \to v; \\
&\qquad \mathsf{inr}\,\langle y'\rangle \to \text{case } y' \text{ of } !y \to w
\end{aligned}
$$

An intuitionistic judgement $\Gamma \vdash t : A$ is provable if and only if the corresponding linear judgement $[\Gamma] \vdash t : A$ is provable. In this way, the ordinary lambda calculus can be regarded as a subset of the linear lambda calculus.

There is an alternate translation for the product types.

$$A \times B = !A \otimes !B$$

This gives rise to an alternate translation for terms.

$$
\begin{aligned}
(t, u) &= \langle !t, !u\rangle \\
\text{case } s \text{ of } (x, y) \to v &= \text{case } s \text{ of } \langle x', y'\rangle \to \\
&\qquad \text{case } x' \text{ of } !x \to \\
&\qquad\quad \text{case } y' \text{ of } !y \to v
\end{aligned}
$$

In some circumstances, this alternate translation may be more efficient.

The embedding of intutionistic logic into linear logic also works for the fixpoint constant.

$$\mathit{fix} : !(!A \multimap A) \multimap A$$

This has the following reduction rule.

$$\mathit{fix}\,\langle !f\rangle \Longrightarrow f\,\langle !\mathit{fix}\,\langle !f\rangle\rangle$$

5.3 Array update

As mentioned, if a variable corresponding to a linear assumption ever contains the sole pointer to a value, then it will continue to do so. This can be exploited to provide a solution to the old problem of in-place update for arrays.

An array (of type *Arr*) is a mapping of indices (of type *Ix*) to values (of type *Val*). The usual operations provided on arrays are as follows.

$$\begin{aligned} \mathit{new} &: \mathit{Val} \rightarrow \mathit{Arr} \\ \mathit{lookup} &: \mathit{Ix} \rightarrow \mathit{Arr} \rightarrow \mathit{Val} \\ \mathit{update} &: \mathit{Ix} \rightarrow \mathit{Val} \rightarrow \mathit{Arr} \rightarrow \mathit{Arr} \end{aligned}$$

Here $\mathit{new}\,(v)$ returns an array with each location initialised to v, and $\mathit{lookup}\,(i)\,(a)$ returns the value at location i in array a, and $\mathit{update}\,(i)\,(v)\,(a)$ returns an array identical to a except that location i contains value v. The trouble with this is that the update operation can be prohibitively expensive, as it may need to copy the entire array.

A version of these operators may be devised for the linear type calculus that places a newly allocated array in a variable corresponding to a linear assumption. Since the variable contains the sole pointer to the array, one can guarantee that it will continue to do so. Hence the update operation may safely be performed in place. Here are the new versions of the operations.

$$\begin{aligned} \mathit{new} &: !\,\mathit{Val} \multimap (\mathit{Arr} \multimap \mathit{Arr} \otimes X) \multimap X \\ \mathit{lookup} &: !\mathit{Ix} \multimap \mathit{Arr} \multimap \mathit{Arr} \otimes !\,\mathit{Val} \\ \mathit{update} &: !\mathit{Ix} \multimap !\,\mathit{Val} \multimap \mathit{Arr} \multimap \mathit{Arr} \end{aligned}$$

Here $\mathit{new}\,\langle v\rangle\,\langle f\rangle$ allocates a new array a with each location initialised to v, and then computes $f\,\langle a\rangle$ which will return a pair $\langle a', x\rangle$, and then deallocates the final array a' and returns the value x; thus *new* acts very much like a block, in that it allocates an array, processes it, and deallocates it. The call $\mathit{lookup}\,\langle a\rangle\,\langle i\rangle$ returns the pair $\langle a, v\rangle$ where v is the value in location i of a. Note that since *lookup* is passed the sole pointer to the array, it must return the array it is passed. As before, the call $\mathit{update}\,\langle i\rangle\,\langle v\rangle\,\langle a\rangle$ returns an array identical to a except that location i contains value v; but since this call is passed the sole pointer to array a, it may be safely implemented by updating the array in place.

For example, evaluating

$$\begin{aligned} \mathit{new}\,\langle 6\rangle\,(\lambda\langle a_0\rangle.\ &\text{case } \mathit{lookup}\,\langle 1\rangle\,\langle a_0\rangle \text{ of } \langle a_1, x\rangle \rightarrow \\ &\text{case } \mathit{update}\,\langle 2\rangle\,\langle x+1\rangle\,\langle a_1\rangle \text{ of } a_2 \rightarrow \\ &\text{case } \mathit{lookup}\,\langle 2\rangle\,\langle a_2\rangle \text{ of } \langle a_3, y\rangle \rightarrow \\ &\langle a_3, x \times y\rangle) \end{aligned}$$

returns 42. (In the second line, case t of $x \rightarrow u$ is a convenient abbreviation for $(\lambda\langle x\rangle.\,u)\,(t)$.)

This approach requires further refinement. The form given here is too unwieldy for convenient use. But it should give a hint as to the practical applications of linear logic.

6 Conclusions and related work

Traditional logic has close ties to computing in general and functional languages in particular. The Curry-Howard isomorphism specifies a precise correspondence between intuitionistic logic, on the one hand, and typed lambda calculus, on the other [4, 11].

As a result, logicians and computer scientists sometimes discover the same system independently, usually with the logician getting there first. The type inference algorithm published by Milner in 1978, is at the heart of the type system used in ML, Miranda, and Haskell [14]. Unbeknown to Milner, the same idea was published by Hindley in 1969 [9]. Reynolds described a polymorphic lambda calculus in 1974 that generalised Milner's type system, and also generalised the power of generic type variables in languages such as Ada [17, 18]. Unbeknown to Reynolds, the same generalisation was described by Girard in 1972 [5].

In 1987, Girard published the first description of linear logic [6]. By now, the computer scientists and the logicians had realised that they had something to say to one another: the seminal paper appeared in *Theoretical Computer Science.* Computer scientists have been active ever since, exploring the applications of linear logic to computing.

Early computational models were discussed by Lafont [12] and Holström [10]. Abramsky wrote a highly influential paper that explored computing applications of both intuitionistic and classical linear logic [1]. Other models have been discussed by Chirimar, Gunter, and Riecke [3], Lincoln and Mitchell [13], Reddy [16], and Wadler [21, 22].

The particular formulation of linear logic presented here is based on Girard's Logic of Unity, a refinement of linear logic [7]. This overcomes some technical problems with other presentations of linear logic, some of which are discussed by Benton, Bierman, de Paiva, and Hyland [2], and Wadler [23, 24]. Much of the insight for this work comes from categorical models of linear logic [19, 15]. The particular system presented here was suggested to the author by Girard, and a similar system has been suggested by Plotkin.

For further background on traditional logic see the wonderful introduction by Girard, Lafont, and Taylor [8], and for further details on linear logic see the helpful textbook by Troelstra [20].

Acknowledgements. I thank Cordy Hall for her detailed and timely comments. The paper was produced using Knuth's Tex, Lamport's Latex, Taylor's tree macros, and style macros from Springer-Verlag.

References

1. S. Abramsky, Computational interpretations of linear logic. Presented at *Workshop on Mathematical Foundations of Programming Language Semantics*, 1990. To appear in *Theoretical Computer Science.*
2. N. Benton, G. Bierman, V. de Paiva, and M. Hyland, Type assignment for intuitionistic linear logic. Draft paper, August 1992.

3. J. Chirimar, C. A. Gunter, and J. G. Riecke. Linear ML. In *Symposium on Lisp and Functional Programming*, ACM Press, San Francisco, June 1992.
4. H. B. Curry and R. Feys, *Combinatory Logic*, North Holland, 1958.
5. J.-Y. Girard, *Interprétation functionelle et élimination des coupures dans l'arithmétique d'ordre supérieure*. Ph.D. thesis, Université Paris VII, 1972.
6. J.-Y. Girard, Linear logic. *Theoretical Computer Science*, 50:1–102, 1987.
7. J.-Y. Girard, On the unity of logic. Manuscript, 1991.
8. J.-Y. Girard, Y. Lafont, and P. Taylor, *Proofs and types*, Cambridge University Press, 1989.
9. R. Hindley, The principal type scheme of an object in combinatory logic. *Trans. Am. Math. Soc.*, 146:29–60, December 1969.
10. S. Holmström, A linear functional language. Draft paper, Chalmers University of Technology, 1988.
11. W. A. Howard, The formulae-as-types notion of contstruction. In J. P. Seldin and J. R. Hindley, editors, *To H. B. Curry: Essays on Combinatory Logic, Lambda Calculus, and Formalism*, Academic Press, 1980. (The original version was circulated privately in 1969.)
12. Y. Lafont, The linear abstract machine. *Theoretical Computer Science*, 59:157–180, 1988.
13. P. Lincoln and J. Mitchell, Operational aspects of linear lambda calculus. In *7'th Symposium on Logic in Computer Science*, IEEE Press, Santa Cruz, California, June 1992.
14. R. Milner, A theory of type polymorphism in programming. *J. Comput. Syst. Sci.*, 17:348–375, 1978.
15. V. Pratt, Event spaces and their linear logic. In *AMAST '91: Algebraic Methodology And Software Technology*, Iowa City, Springer Verlag LNCS, 1992.
16. U. S. Reddy, A typed foundation for directional logic programming. In E. Lamma and P. Mello, editors, *Extensions of logic programming*, Lecture Notes in Artificial Intelligence 660, Springer-Verlag, 1993.
17. J. C. Reynolds, Towards a theory of type structure. In B. Robinet, editor, *Proc. Colloque sur la Programmation*, LNCS 19, Springer-Verlag.
18. J. C. Reynolds, Three approaches to type structure. In *Mathematical Foundations of Software Development*, LNCS 185, Springer-Verlag, 1985.
19. R. A. G. Seely, Linear logic, $*$-autonomous categories, and cofree coalgebras. In *Categories in Computer Science and Logic*, June 1989. AMS Contemporary Mathematics 92.
20. A. S. Troelstra, *Lectures on Linear Logic*. CSLI Lecture Notes, 1992.
21. P. Wadler, Linear types can change the world! In *IFIP TC 2 Working Conference on Programming Concepts and Methods*, Sea of Galilee, Israel, April 1990. Published as M. Broy and C. Jones, editors, *Programming Concepts and Methods*, North Holland, 1990.
22. P. Wadler, Is there a use for linear logic? In *Conference on Partial Evaluation and Semantics-Based Program Manipulation (PEPM)*, ACM Press, New Haven, Connecticut, June 1991.
23. P. Wadler, There's no substitute for linear logic. Presented at *Workshop on Mathematical Foundations of Programming Language Semantics*, Oxford, April 1992.
24. P. Wadler, A syntax for linear logic. Presented at *Conference on Mathematical Foundations of Programming Language Semantics*, New Orleans, April 1993.

On the tree inclusion problem

Laurent Alonso* René Schott **

Abstract. We consider the following problem: Given ordered labeled trees S and T can S be obtained from T by deleting nodes ? Deletion of the root node u of a subtree with children $(T_1, \ldots, T_n)$ means replacing the subtree by the trees $T_1, \ldots, T_n$. For the tree inclusion problem,there can generally be exponentially many ways to obtain the included tree. Recently, P.Kilpeläinen and H.Mannila [KM] gave an algorithm based on dynamic programming requiring $O(| S | . | T |)$ time and space in the worst case and also on the average. We give a new algorithm which improves the previous one on the average and breaks the $| S | . | T |$ barrier.

1 Introduction

Extracting a node is a natural operation on trees. This operation is defined as follows: let T be a tree, v one of his nodes, T_1, T_2, ..., T_k the subtrees which are sons of v. Extracting a node v of T corresponds to the replacement of T by :

- the forest T_1, ..., T_k if v is the root of T.
- the tree T where the node v has been removed and the edges connecting v to its sons replaced by the edges connecting the father of v to the sons of v (the ordering of successors of nodes is not modified).

Example 1. Extraction of the left tree's root : | Extraction of the internal node circled in the left tree :

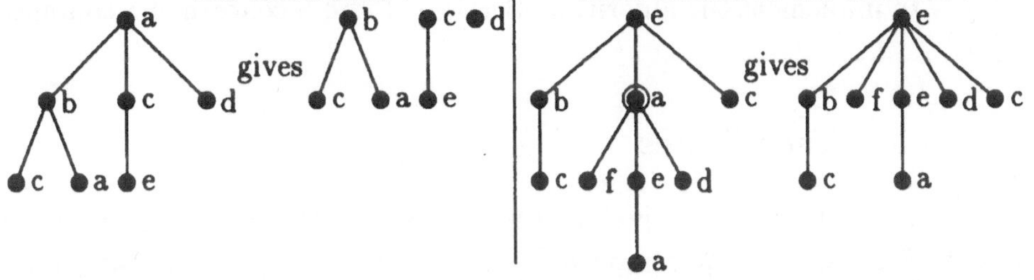

Now we must extend this definition to forests. Extracting a node v from a forest T_1, T_2, ..., T_k is done by replacing the tree T_i which contains the node v by the tree T_i where the node v has been extracted. We will say that a tree S is included or embedded in a tree T if and only if we can obtain S by extracting iteratively a finite number of nodes from T.

* E.N.S., 45 rue d'Ulm, 75005 Paris, France and CRIN, INRIA-Lorraine, Université de Nancy 1, 54506 Vandoeuvre-lès-Nancy, France
** CRIN, INRIA-Lorraine, Université de Nancy 1, 54506 Vandoeuvre-lès-Nancy, France

Example 2. The tree S is included in the tree T.

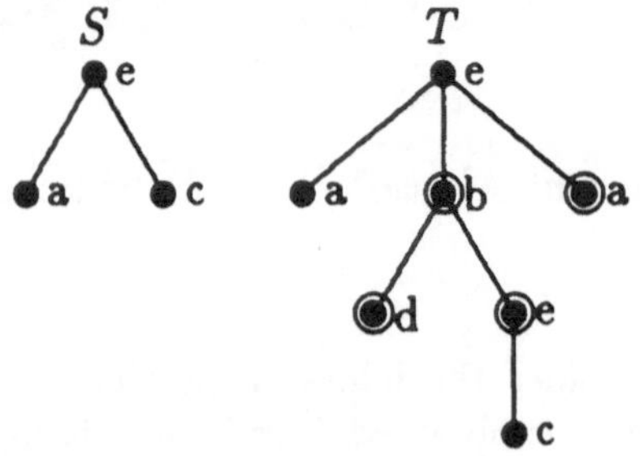

In fact, the tree S can be obtained from the tree T by extracting the nodes circled above.

Pekka Kilpeläinen and Heikki Mannila gave an algorithm which uses arrays and whose complexity is $O(|S||T|)$ in the worst, best and average cases. This complexity is reasonable for trees of (approximatively) equal sizes but is too much high if $|S|$ is small with respect to $|T|$. This corresponds to the case where the probability of finding a possible inclusion by examining a small number of nodes is high.

We give an algorithm with worst case complexity in $O(|S||T|)$ but which behaves on the average like $O(|T| + f(|S|, |T|))$ where f is a function satisfying the following properties :

- $\forall x, \forall y, f(x, y) \leq x\ y$,
- if x is fixed, the function $y \mapsto f(x, y)$ has a constant limit as y goes to infinity.

The first property follows from the worst case complexity and is of limited interest. The second one is more important and shows that as larger the size of T is as smaller is the number of nodes of T to be examined until the first inclusion is found.
In Section 2, we explain the inclusion search strategies while in Section 3 we show how to implement the algorithm. Sections 4 and 5 concern the complexity analysis.

2 Applied Techniques

We try to know if a tree S is included in a tree T. If the answer is yes, we will give a set $\mathcal{E}$ of nodes to be removed from T in order to get S. In fact, we will provide the set of remaining nodes and a function which associates to each node p of S, the node q of $\mathcal{E}$ which will be in the place of p after the transformation of T in S. This function will be called embedding function.
In this section, we study the properties of the embedding function and give the principles of the inclusion search.

Proposition 1. *If the tree S is included in T and if the labels of the roots of S and T are the same, then there exists an embedding which transforms the root of S in the root of T.*

Proof. Let f be the embedding function of S in T. This function transforms the root s of S in a node v of T. If v is equal to the root t of T then we stop, otherwise we define a function g as follows :

- $\forall u \neq s, g(u) = f(u)$
- $g(s) = t$

We can then show that g is an embedding function of S in T (see [AS]).

We will need later the two following proprositions which are proved in [AS]

Proposition 2. *If the following assumptions are fulfilled :*

- *$\prec$ represents the preorder or postorder traversal of a tree,*
- *f is an embedding function of S in T,*

then $\forall u, v$ nodes of S, $f(u) \prec f(v)$

Proposition 3. *Let f be the injective mapping from the nodes of S to the nodes of T such that :*

- *$\forall u \in S, \forall v \in S, u \prec v \Leftrightarrow f(u) \prec f(v)$ for the preorder traversal,*
- *$\forall u \in S$, $label(u) = label(f(u))$*

then f is an embedding of S in T.

Now we can indicate two new principles of the inclusion search.

Proposition 4. *If S is included in T and if*

- *the labels of the roots s of S and t of T are the same,*
- *the sons of s are the nodes s_1, s_2, ..., s_k,*
- *u is the smallest node of T (for the postorder) for which the subtree defined by s_1 is included in the subtree defined by u,*

then there exists an embedding function f which transforms s_1 in u.

Proof. We choose for :

- f one of the embeddings of S in T which transforms the root s of S in the root t of T
- g an embedding of the subtree s_1 in the subtree of u.

Proposition 1 and the first assumption prove the existence of f. Let:

- $t_1 = f(u)$
- C_1 : the set of nodes belonging to the subtree of s_1,
- C_2 : the set of nodes belonging to one of the subtrees of s_2, ..., s_k.

The case $t_1 = u$ permits to conclude directly, therefore we will assume that $t_1 \neq u$. Let h be the function which associates :

- s to t
- the nodes v of C_1 to $g(v)$
- the nodes v of C_2 to $f(v)$

We then have

$$\forall v, w \text{ nodes of } S,\ v \prec w \Leftrightarrow g(u) \prec g(w)$$

for the preorder and postorder traversals.

The result follows from Proposition 3 applied to the function h.

This proposition is very important. It allows, in fact, to test easily if an embedding which transforms s in the node v of the tree T is possible. It's sufficient to search an inclusion of s_1 in v . If this test fails, the answer is no. But if we succeed in associating the node s_1 to a node u and if u is the smallest possible, we search an inclusion of s_2, s_3, ..., s_k in the subtree of v composed by nodes whose preorder is greater than that of u.

Proposition 5. *If S is embedded in T and if :*

- *the labels of the roots s of S and t of T are the same,*
- *the sons of s are the nodes s_1, s_2, ..., s_k*
- *$i < k$ and v_i is the smallest node for the postorder traversal which permits*
  - *the embedding of the tree composed by s and the subtrees defined by s_1, ... s_i in T*
  - *to associate s_i with v_i*
- *v_{i+1} the smallest node of T for the postorder*
  - *is greater than v_i for the postorder and preorder,*
  - *permits the association of a subtree of s_{i+1} to the subtree of v_{i+1}*

then there exists an embedding f which transforms s_{i+1} in v_{i+1}.

Proof. The proof is similar to the proof of Proposition 4.

This proposition is, in fact, a main result: It shows if a node u of S can be associated with a node v of T by searching successively for the best possible embeddings of the sons of u in v. More precisely, we search first for the best possible embedding of the sons of u in the subtree v, then we search for a place for the second son s_2 of s in the subtree of v composed of the nodes whose preorder is greater than the preorder of v_1. We continue until we find a possible association for all the sons of s or fail.

3 Algorithm

The algorithm is based on the propositions of the preceding section. These propositions suggest the definition of three recursive functions :

- the function *embed_root* which searches an embedding of a subtree P of S in a subtree Q of T, by associating the root p of P with the root q of Q

- the function *embed* which searches for an embedding of a subtree P of S in a subtree Q of T
- the function *include* which searches for an embedding of a subtree P of S in a tree composed by nodes of T belonging to a subtree Q and whose preorder and postorder are greater than r.

The call to the function *embed* with the parameters S and T solves this problem. The most natural way for defining these functions is the following : for the function *embed* P in Q

- classify the nodes of Q with respect to the postorder,
- visit the nodes in increasing order and use the procedure *embed_root* as soon as you find a node v with the same label as the root p of P. Stop as soon as the first embedding has been discovered.

For the function *embed_root* P in Q :

- if p is a leaf, return the embedding which transforms p in q
- try to embed the right son p_1 of p in the subtree v
- if no embedding has been found, return a negative answer.
- $i = 1$,
- while i is less than the number of sons of p do
  - denote by E the set of nodes of Q which are greater than the embedding of q_i in Q,
  - $i = i + 1$,
  - look for an embedding of p_i, the ith son of p, in E,
  - if no embedding has been found, return a negative result
- return the embedding which has been found

The definition of the function *include* is very close to the definition of *embed*. In fact, it is enough to replace the occurrences of Q by $\mathcal{E}$, where $\mathcal{E}$ is described by two parameters :

- Q a subtree to which it belongs,
- r a node of the tree T

$\mathcal{E}$ stands for the set of nodes of T which are greater than v for the preorder and postorder. Using directly these definitions is possible but gives rise to an algorithm whose maximal complexity is high : the complexity grows like $exp(|T|)$ since some subtree embedding operations can be repeated a lot of times. In order to avoid this problem, we store the results of the calls to the functions *embed*, *include* and *embed_root*. This is possible using arrays attached to the leaves of T.

3.1 Data structures used in the algorithm

To each node of the tree T, we associate three arrays (for the sons, the inclusion and the embedding). These arrays have a similar structure. They are connected to a node q of T and associate with each node p of S a value v which is generally a node of T but which can also be Ω, *yes* or *no* (Ω is an undefined state, the initial value we put in each box of the array). The values *yes* and *no* appear only in the array of sons and have the following meaning :

- p is a leaf of S and the labels of p and q are the same,
- there exists no embedding which transforms p in q.

The presence of a node v is interpreted differently for each array.

- For the array of sons : this means that there exists an embedding of the subtree p in the subtree q which transforms the left son of p in v.
- For the embedding array : this means that each embedding which transforms the node p in r, where r has a postorder greater than that of q, has to be searched by starting on the node v.
- For the inclusion array : this means that each embedding which transforms p in r, where r has postorder and preorder greater than that of q, has to be searched by starting on the node v.

This will be done by storing the value v in the first box of the array while the other boxes of the array will be initialized with Ω's.

3.2 Implementation of the algorithm

We give below some details concerning the implementation of our algorithm. This is done using three procedures :

- for sharing the arrays,
- recursive procedures,
- the inclusion search procedure (this procedure is called to start the search)

We assume that going from a node to the next can be done in time $O(1)$ (for the preorder and postorder). If this is not the case, it is enough to create in the initialization step the corresponding arrays of size $|S|$ and $|T|$. The corresponding complexity is in $O(|S| + |T|)$.

Sharing the arrays. We need two procedures ; the first one stores a value in an array, the second one shows if a value has been stored and if yes which one.

```
add_value (p, q, kind_of_array, value)
        if q has no array of type kind_of_array
                allocate an array of size |S|
                initialize the boxes with the value Ω
                link this array to the node q
        call t the array of type kind_of_array of q
        t[p] = value
```

To store a value, we look if the array in which we will store it still exists. If the answer is no, we create it and then store the value in the corresponding box.

Remark. kind_of_array can take three values which indicate if we must use the array of the sons, of the embeddings or of the inclusions.

To search a value in an array, we have similarly :

```
search_value (p, q, kind_of_array)
        if q has no array of the type kind_of_array
                result = Ω
        otherwise
                let t be the array of type kind_of_array of q
                result = t[p]
        return result
```

The recursive procedures. We will redefine the functions *embed_root*, *embed*, *include* in order to avoid the repetition of the same computations. We begin with the function *embed_root* which does the main job.

```
embed_root (p,q)
        result = search_value (p,q, son_array)
        if result ≠ Ω
                return result
        if p is a leaf of S
                add_value (p,q, son_array, 'yes")
                return "yes"
        denote by p1, p2, ..., pk the sons of p
        place1 = embed (p1,q)
        if place1 = "no" or place1 = q
                add_value (p,q, son_array, 'no")
                return "no"
        i = 2
        while i ≤ k do
                placei = embed(pi,q,placei-1)
                if placei = "no"
                        add_value (p,q, son_array, "no")
                        return "no"
                i=i+1
        add_value (p,q, son_array, place1)
        return "yes"
```

Then the two functions which are called by *embed_root*:

```
embed (p,q)
        e = a label of p
        first_node = the smallest node (for the postorder traversal which belongs to q
        v = first_node
        while v ⪯ q for the postorder traversal
                if e = a label of v
                        result = embed_root (p, v)
                        if result ≠ "no"
                                add_search (p, first_node, embedding_array, v)
                                return v
```

```
        result = search_value (p,v, embedding_array)
        if result ≠ Ω
                v = result
        else
                v = v+1
    add_search (p, first_node, embedding_array, v)
    return "no"
```

and

```
include (p,q,r)
    e = a label of p
    result = search_value (p,r, inclusion_array)
    if result = Ω
            result = search_value (p,r, embedding_array)
    if result ≠ Ω
            v = result
    else
            v = r+1
    while v ≺ q for the postorder
        if e = a label of v
            result = embed_root (p, v)
            if result ≠ "no"
                if r ≺ v for the preorder
                    add_search (p, r, inclusion_array, v)
                    add_search (p, r, embedding_array, v)
                    return v
                else
                    add_search (p, r, embedding_array, v)
                    result = include (p, q, v)
                    v = search_value (p,v, inclusion_array)
                    add_search (p, r, embed_array, v)
                    return result
        result = search_value (p,v, embed_array)
        if result ≠ Ω
                v = result
        else
                v = v+1
    add_search (p, r, inclusion_array,v)
    add_search (p, r, embed_array,v)
    return "no"
```

The inclusion search procedure for S in T.

```
search(S,T)
    size = 2|S|
    search in T all the subtrees of size SIZE, and denote it T_1, ..., T_k
    i = 1
    f_continue = true // a small flag
    while i ≤ k do
```

```
        result = embed(s, Ti)
        if result not equal to "no" then
                v = search_value (s,Ti,son_array)
                f_continue = false
        else
                i = i+1
    if f_continue = true
            if |T| = size
                    return "no"
            else
                    result = embed (s, T)
                    if result not equal to "no" then
                            return "no"
                    else
                            v = search_value (s,Ti,son_array)
    recreate the embedding of S in T which transforms s in v
```

In case of success, we must extract from our datas an embedding of S in T. We know now that there exists an embedding which transforms s in v. Theferore, we search in the array of sons of v the node v_1 which corresponds to the leftson of s, then in the array of inclusions of v_1, the node v_2 which corresponds to the right brother of s_1, ... In so doing, we prove the existence of an embedding of S in T.

4 The worst case complexity

In this section we show that the worst case complexity of our algorithm is upper bounded by $O(|S||T|)$. For this purpose we have to analyse the calls to the function *embed_root*.

4.1 The calls to the function *embed_root*

We choose a node p of S and show how the procedure *embed_root* is used. Let q_i' be the value of q's second parameter at the $i^{ième}$th call, p being the argument. Since the result of each call is memorized, only the values of the second parameter which are different are interesting. Let q_i be the list of elements $q_1', \ldots, q_{k'}'$ ordered with respect to their first appearance, in other words $q_1, \ldots, q_k$ is the list q_1', $\ldots, q_{k'}'$ in which we deleted the elements q_i' so that :

$$\exists j < i, q_j' = q_i'.$$

Then we have the following property :

Property 1 *If $j < i$ and if the intersection of the subtrees q_i and q_j is not empty, then the subtree q_i is included in the subtree q_j.*

Proof. A proof of this property by induction on the distance between a node p and the root s of S can be found in [AS].

4.2 The worst case complexity

Let $M(S,T)$ be the number of T's nodes which have been examined during the inclusion search.

Theorem 6. *The worst-case complexity of our algorithm is in* $O(|T| + M(S,T)|S|)$.

Proof. see [AS].

Remark. Since $M(S,T) \leq |S|$, the worst case complexity is in $O(|S||T|)$ as announced.

5 An upper bound for the average complexity

We will prove in this section that the average value of $M(S,T)$ tends to a constant as $|T|$ goes to infinity ($|S|$ being fixed)

Let us first define the notion of complexity which we will use. For a tree S, we will denote by v_S the number of its distinct labels. The average complexity $C(n)$ is over the set $T(n)$ of trees of size n whose labels belong to v_S. The cardinality of $T(n)$ is

$$v_S^{|T|}\frac{2n-2!}{n!(n-1)!}$$

If the tree T has labels which do not appear in S, we can easily come back to the previous case by removing the nodes corresponding to these label in time $O(n\ log(v_S))$. We get then a tree whose size is less than the size of the initial tree, the search is therefore more easier. The value of $C(n)$ is then less than

$$n + |S|\frac{n!(n-1)!}{v_S^n(2n-2)!}\sum_{T\in T(n)} M(T,S)$$

Let ρ be the probability to find an inclusion of S in a tree of size $2|S|$, we have, of course, $0 < \rho \leq 1$ .The case $\rho = 1$ appears if S is reduced to a leaf. The number of subtrees of size $2|S|$ which appear in a tree T is highly important. In fact, assume that the initial tree T has k subtrees of size $2|S|$. We first start a search in the first subtree of size $2|S|$: this requires a time complexity in $O(|S|^2)$. This search succeeds with probability ρ and fails with probability $1-\rho$. In case of failure, we start a search in the second subtree of size $2|S|$, ...

The average complexity of the examination of the tree T: the average complexity over all trees of size n with k subtrees of size $2|S|$ is therefore upper bounded by

$$|T| + |S|^2\left(1 + (1-\rho) + \cdots + (1-\rho)^{k-1}\right) + (1-\rho)^k|T||S|$$

i.e. by

$$|T| + \frac{1}{\rho}|S|^2 + (1-\rho)^k|T||S|.$$

Let $N(q,n)$ be the percentage of trees of size n with less than $n^{\frac{1}{4}}$ subtrees of size $2q$.

Since the average complexity for the trees with more than $n^{\frac{1}{4}}$ subtrees of size $2q$ is in $O(|S|^2)$, the average complexity is upper bounded by

$$O(n + N(2|S|,n)n|S| + \frac{1}{\rho}|S|^2)$$

And since $N(2|S|,n) = O(\frac{1}{n})$ (see [AS]), we finally get an average complexity in $O(|T| + M(S,T)|S|)$.

6 Conclusion

We have designed an algorithm for the tree inclusion problem with worst case complexity in $O(|S||T|)$. The average complexity is less than $O(|T|+M(S,T)|S|)$ where $M(S,T)$ is an upper bounded function when S is fixed. If the size of T is much greater than the size of S, only a small part of T is examined.

References

[AS] L.Alonso, R.Schott, On the tree inclusion, preprint.

[CPT] J.Cai, R.Paige, R.Tarjan, More efficient bottom-up tree pattern matching, Proc. CAAP'90, LNCS 431, 72-86.

[HD] C.M.Hoffman, M.J.O'Donnell, Pattern matching in trees, J.A.C.M., 29, 1, 68-95, 1992.

[KM] P.Kilpeläinen, H.Mannila, The tree inclusion problem, CAAP 91, LNCS 493, 202-214, 1991.

[KNU] D.E.Knuth, The art of computer programming, Fundamental Algorithms, 1, Addison Wesley, 1973.

[ZS1] K.Zhang, D.Shasha, Simple fast algorithms for editing distance between trees and related problems, SIAM Journal of Computing, 18, 1245-1262, 1989

[ZS2] K.Zhang, D.Shasha, Fast Algorithms for the Unit Cost Editing Distance between Trees, Journal of Algorithms 11, 581-621, 1990

On the Adequacy of Per Models [1]

Roberto M. Amadio
CRIN-CNRS & Inria-Lorraine, Nancy[2]

Abstract

We consider a fixed point extension of the second order lambda calculus equipped with a call by value evaluation mechanism. We interpret the language in a partial cartesian closed category of "directed complete" partial equivalence relations (pers) over a domain theoretic model of a type-free, call-by-value, lambda calculus. Our main result is that the notions of "syntactic" and "semantic" convergence coincide.

Keywords: Second order lambda-calculus, Per models, Denotational vs. Operational Semantics.

Introduction

Recently the research in the area of "synthetic domain theory" (see, e.g., [10]) has especially addressed the problem of discovering subcategories of partial equivalence relations (pers) over a partial combinatory algebra (pca) that enjoy good completeness properties, and that admit certain constructions typical of domain theory, such as the solution of recursive function and domain equations.

Our concern here does not lie in the construction of models, or in the categorical abstraction of such construction, but in an attempt at connecting such models to issues arising in the design and semantics of, say, typed functional languages. In particular we connect a certain per-interpretation to a call-by-value evaluation discipline that corresponds to current implementations of higher order typed functional languages with a static type checker (see, e.g., [6]).

The *main result* establishes the equivalence of the syntactic and semantic notions of convergence. This is a classical "adequacy" result for domain theoretic interpretations, as sketched for example in [13], after [11]. However, as far as we know, no results were available, in the case of per-interpretations.

There are two additional points we wish to emphasize:

• In order to prove such an adequacy theorem *very little structure* on the per model is needed, in particular we do not require working with an O-category.

• The adequacy of the per interpretation is largely *independent* from the adequacy of the underlying pca.

Section 5 is the technical core of this paper. The proof of adequacy for standard domain theoretic interpretations requires the combination of "admissible predicates" techniques and "reducibility candidates" techniques. In the proof we propose here, there are two additional twists that are due to the presence of second order types, and to the interplay between the typed and the type-free structure. In particular the key of the result lies in the definition of *adequacy relation* (5.1), and *in the way one associates an adequacy relation to a type* (5.2).

For lack of space we have omitted the more or less standard proofs of the first four sections which can be found in [3]. Such sections contain respectively: (1) the definition of a fixed point extension of the second order lambda calculus; (2) the definition of the evaluation mechanism of such language; (3) the description of the basic properties of the partial cartesian closed category of "directed complete" pers over a cpo model of call-by-value, type-free lambda calculus; (4) the interpretation of the language in the semantic structure.

---

[1] This paper is a short version of [3] and was first presented at a CLICS meeting in Aarhus (April 92).
[2] CRIN, B.P. 239, 54506, Vandoeuvre-lès-Nancy, FRANCE. E-mail: amadio@loria.loria.fr.

1. Language

Types and raw terms are defined by the following BNFs:

Type Variables:	tv::= t|s|...	*Types:*	α ::= 1|tv|($\alpha \rightharpoonup \alpha$)|($\forall$tv.$\alpha$)
Term Variables:	v::= x|y|...	*Terms:*	M ::= $*$|v|(λv:α.M)|(MM)|(λtv.M)|(Mα)|(Y_αM)

In the following we will feel free to spare on parentheses and to omit the type label in the Y combinator. All types and terms are considered up to α-redenomination. A *well formed context* Γ is given by a list of pairs, v: α, in which all variables are distinct. We write Γ, v: α, to point out the last element of the list, we write v:$\alpha \in \Gamma$ to state that v:α occurs in Γ, and we denote with ftv(Γ) the collection of type variables free in types occurring in Γ. Note that in the calculus presented here type variables contexts are left implicit.

As usual a *substitution*, say σ, is a function that associates variables, say v, v_1, ..., to formal expressions, say exp, exp_1, ... The domain of a substitution is $\{v \mid \sigma(v) \neq v\}$. We assume that such domain is always finite. We denote with $[exp_1/v_1,..., exp_n/v_n]$ a substitution whose domain is contained in $\{v_1,..., v_n\}$, and that associates exp_i to v_i, for i=1,...,n. If σ is a substitution and exp is an expression then σexp denotes the expression resulting from the application of the substitution to the expression, according to the standard rules that take care of bound variables. We abbreviate an iteration of substitutions, say $\sigma_1(..(\sigma_n\ exp)..)$, with $\sigma_1..\sigma_n$ exp, so, for example, $[r/s][s/t]\ (t \rightharpoonup s) = (r \rightharpoonup r)$.

In a formal system the symbol "$\Rightarrow$" separates, in an inference rule, the premises from the conclusion. If "J" is a judgment of the formal system then we write "$\vdash$J", if such judgment is derivable. A *typing judgment* is of the shape $\Gamma \supset M:\alpha$ where Γ is always a well formed context. Derivable typing judgments are specified by the following formal system:

(*)		$\Rightarrow \Gamma \supset *: 1$
(asmp)	x: $\alpha \in \Gamma$	$\Rightarrow \Gamma \supset x: \alpha$
($\rightharpoonup$I)	$\Gamma, x:\alpha \supset M:\beta$	$\Rightarrow \Gamma \supset (\lambda x:\alpha.M): (\alpha \rightharpoonup \beta)$
($\rightharpoonup$E)	$\Gamma \supset M: (\alpha \rightharpoonup \beta)$, $\Gamma \supset N:\alpha$	$\Rightarrow \Gamma \supset (MN): \beta$
($\forall$I)	$\Gamma \supset M: \alpha$ $t \notin ftv(\Gamma)$	$\Rightarrow \Gamma \supset (\lambda t.M): (\forall t.\alpha)$
($\forall$E)	$\Gamma \supset M: (\forall t.\alpha)$	$\Rightarrow \Gamma \supset (M\beta): [\beta/t]\alpha$
(Y)	$\Gamma \supset M: (1 \rightharpoonup \alpha) \rightharpoonup \alpha$	$\Rightarrow \Gamma \supset Y_\alpha M: \alpha$

This language is intended to represent a second order lambda calculus with a fixed point combinators over lifted types. One can think of the constant Y_α as having type $((\alpha)_\perp \to (\alpha)_\perp) \to (\alpha)_\perp$, where: $(\alpha)_\perp \equiv (1 \rightharpoonup \alpha)$, and $(\alpha \rightharpoonup \beta) \equiv (\alpha \to (\beta)_\perp)$. The type $(\alpha \rightharpoonup \beta)$ should be thought as the type of the *partial* functions from α to β.

2. Evaluation

We denote with $\Lambda^\circ\alpha$ the collection of terms of type α without free variables, but possibly with free type variables. The *canonical forms* are the terms in $\Lambda^\circ\alpha$, for some α, which are generated by the following grammar:

C::= * | (λv: α. M) | (λtv. C)

The evaluation "$\mapsto$" is specified as a relation between terms without free variables and canonical forms. If M$\mapsto$C then M and C have the same type, so one may also think of "$\mapsto$" as a family of relations indexed over types. The definition of the evaluation relation proceeds by induction on the structure of a well-typed closed term.

(*)		$\Rightarrow$ $*\mapsto*$
(asmp)	"we never evaluate a free variable"	
($\rightharpoonup$I)		$\Rightarrow$ $\lambda x{:}\alpha.M \mapsto \lambda x{:}\alpha.M$
($\rightharpoonup$E)	$M \mapsto \lambda x{:}\alpha.M'$, $N \mapsto C'$, $[C'/x]M' \mapsto C$	$\Rightarrow$ $MN \mapsto C$
($\forall$I)	$M \mapsto C$	$\Rightarrow$ $\lambda t.M \mapsto \lambda t.C$
($\forall$E)	$M \mapsto \lambda t.C$	$\Rightarrow$ $M\alpha \mapsto [\alpha/t]C$
(Y)	$M(\lambda x{:}1.YM) \mapsto C$	$\Rightarrow$ $YM \mapsto C$ (x fresh variable)

We write $M\downarrow$ if $\vdash M \mapsto C$, for some canonical form C, and $M\uparrow$ otherwise. Note that the definition of "$\mapsto$" gives directly a deterministic procedure to reduce, if possible, a closed term to a canonical form. Hence each term can reduce at most to a canonical form. Canonical forms always reduce to themselves.

Observe that we *evaluate under type abstraction*. On one hand this corresponds to the fact that in actual implementations of the language the type-checker is static, hence no information about type abstraction and type application appears at run time. On the other hand, as it will become clear in section 4, this choice is important in capturing the behavior of the interpretation of second order types as intersection.

3. Semantic Structure

In the presentation of the per-model we take a minimalist approach, by presenting only those properties and constructions that are needed in the proof. We refer to [2] for more information about the relevance and the context of the structures discussed below.

3.1 Realizability Structure [3]

We assume to have an object D in the category of **dcpo** that has its partial functional space as a retract, i.e. there exists i: $(D \rightharpoonup D) \rightharpoonup D$, j: $D \rightharpoonup (D \rightharpoonup D)$, $j \circ i = id_{D \rightharpoonup D}$ in **dcpo.** We define a partial operation of application over D as: $de \triangleq j(d)(e)$. From this operation one can define, as usual, continuous operations of pairing, $<\,,\,>{:}D \times D \to D$, and projection $\pi_i{:}D \to D$, (i=1,2) such that $\pi_1<d,d'> = d$, and $\pi_2<d,d'> = d'$.

Category of ppers A partial equivalence relation over D (per) is a binary relation over D that is symmetric and transitive. We denote with A, B,... pers over D. We write: dAe for $(d,e) \in A$, $[d]_A$ for $\{e \in D \mid dAe\}$, $|A|$ for $\{d \in D \mid dAd\}$, [A] for $\{[d]_A \mid d \in |A|\}$. A partial morphism between pers, f: $A \rightharpoonup B$, is a map f: $[A] \rightharpoonup [B]$ such that

$$\exists h \in D.\ \forall d \in |A|.\ (\ f([d]_A)\Downarrow\ \wedge\ hd \in f([d]_A))\ \vee\ (f([d]_A)\Uparrow\ \wedge\ hd\Uparrow)).$$

We denote with **pper** the category of pers and partial morphisms of pers. Such category is (equivalent to) a pccc where terminal, product (in the related category of total morphisms), and partial exponent pers are defined as follows:

3 *Doman-Theoretic Conventions.* A set X is *directed* in the poset (P, $\leq$) if $\emptyset \neq X \subseteq P$, and $\forall x,y \in X. \exists z \in X.(x \leq z\ \wedge\ y \leq z)$. A poset is *directed complete* (dcpo) if it has joins of its directed subsets. Two mathematical expressions including partial operations, say e_1, e_2, are *Kleene equivalent*, written $e_1 \simeq e_2$, if either they are both defined and they are equal, or they are both undefined. We also write $e\Downarrow$ ($e\Uparrow$) if a mathematical expression is defined (undefined). A *partial (Scott-)continuous function* between two dcpos, $h{:}(D,\leq_D) \rightharpoonup (E, \leq_E)$, is a partial function between the dcpos D and E such that for any directed set X in D, $f(\sqcup X) \simeq \sqcup f(X)$ (whenever we take the join of an indexed set of mathematical expressions, such join is defined if the join of the *defined* mathematical expressions is defined). We denote with **dcpo** the category of dcpos and partial continuous functions. This category is (equivalent to) a *partial cartesian closed category* in the sense of, e.g., [12]. Given two dcpos, D, E, we denote with $D \rightharpoonup E$, the partial exponent, that is the collection of partial continuous functions pointwise ordered.

$$\mathbf{1} \triangleq D\times D, \qquad d(A\times B)e \Leftrightarrow (\pi_1 d)A(\pi_1 e) \wedge (\pi_2 d)B(\pi_2 e),$$
$$h\ pexp(A, B)\ k \Leftrightarrow \forall d,e.\ dAe \Rightarrow (hdBke \vee (hd\Uparrow \wedge ke\Uparrow)).$$

The interpretation of the language is based on the following category of directed complete pers and partial maps. One may think of this category as a loose analogous of the category of dcpos and partial continuous maps. We will see that it retains the basic properties of the category of ppers, and moreover it has a fixed point combinator over "lifted" objects. The proofs of these results follow [1].

3.2 Definition *(directed complete pers)*

A per A is directed complete (dcper) if for any directed set X in D×D, if $X\subseteq A$ then $\bigsqcup_{D\times D}X\in A$. We define **dcpper** as the full subcategory of **pper** whose objects are dcpers.

3.3 Proposition *(basic properties of the semantic structures)*

(1) **dcpper** is a pccc. (2) dcpers are closed under arbitrary intersections. (3) **dcpper** is reflective in **pper**, that is the inclusion functor from **dcpper** to **pper** has a left adjoint. (4) **dcpper** has fixpoints over objects of the shape pexp(**1**,A).

4. Interpretation

In this section we define an interpretation of the language based on the semantic structures just introduced. By convention if $\tau: V \rightharpoonup W$ is a partial function from a collection of variables, V, to a set of values, W, then for $v\in V$, and $a\in W$ we define: $\tau[a/v](v') \simeq$ if $v' \equiv v$ then a else $\tau(v')$.

Types. The interpretation of a type, given a type assignment η: tv→dcper, is a dcper, defined by induction as follows:

$$[\![1]\!]\eta = \mathbf{1} \qquad [\![t]\!]\eta = \eta(t)$$
$$[\![\alpha\rightharpoonup\beta]\!]\eta = pexp([\![\alpha]\!]\eta, [\![\beta]\!]\eta) \qquad [\![\forall t.\alpha]\!]\eta = \bigcap_{A\ dcper}[\![\alpha]\!]\eta[A/t]$$

Terms. An assignment is a partial function $\rho: v\rightharpoonup\bigcup_{A\ dcper}[A]$. A type assignment η is compatible with an assignment ρ, w.r.t. a well-formed context Γ, if for any $x{:}\alpha\in\Gamma$, $([\![\alpha]\!]\eta = \emptyset \Rightarrow \rho(x)\Uparrow) \wedge ([\![\alpha]\!]\eta \neq \emptyset \Rightarrow \rho(x)\in[[\![\alpha]\!]\eta])$; we shortly write this as $\eta\uparrow_\Gamma\rho$. The interpretation of a judgment $\vdash\Gamma\supset M{:}\ \alpha$, given η, ρ, such that $\eta\uparrow_\Gamma\rho$, is either undefined or an element in $[[\![\alpha]\!]\eta]$ (equivalently it is a partial map from the terminal object to $[\![\alpha]\!]\eta$). Such interpretation is defined by induction on the length of the typing judgment. Observe that some clause may fail to be defined, hence the use of Kleene equivalence. Suppose $\vdash\emptyset\supset M{:}\ \alpha$, we write $M\Downarrow$ if for any type assignment η, and any ρ, $[\![\emptyset\supset M{:}\alpha]\!]\eta\rho\Downarrow$, and $M\Uparrow$ otherwise.

(*) $\quad [\![\Gamma\supset *{:}\ 1]\!]\eta\rho = [d]_1$, for $d\in D$.

(asmp) $\quad [\![\Gamma\supset x_i{:}\ \alpha_i]\!]\eta\rho \simeq \rho(x_i)$

($\rightharpoonup$I) $\quad [\![\Gamma\supset\lambda x{:}\alpha.M{:}\alpha\rightharpoonup\beta]\!]\eta\rho = \{h\in D \mid \forall d\in|A|.\ (f(d)\Downarrow \Rightarrow hd\in f(d)) \wedge (f(d)\Uparrow \Rightarrow hd\Uparrow)\}$
where: $A = [\![\alpha]\!]\eta$, $f(d) \simeq [\![\Gamma,x{:}\alpha\supset M{:}\beta]\!]\eta\rho[[d]_A/x]$.

($\rightharpoonup$E) $\quad [\![\Gamma\supset MN{:}\beta]\!]\eta\rho \simeq [\![\Gamma\supset M{:}\alpha\rightharpoonup\beta]\!]\eta\rho \cdot [\![\Gamma\supset N{:}\alpha]\!]\eta\rho$
where: $A \triangleq [\![\alpha]\!]\eta$, $B = [\![\beta]\!]\eta$, $[h]_{pexp(A,B)}\cdot[d]_A \simeq [hd]_B$.

($\forall$I) $\quad [\![\Gamma\supset\lambda t.M{:}\forall t.\alpha]\!]\eta\rho \simeq$ *if* $\exists A.(f(A)\Uparrow)$ *then* $\Uparrow$ *else* $\{h\in D \mid \forall A\ \text{dcper}.\ h\in f(A)\}$
where: $f(A) \simeq [\![\Gamma\supset M{:}\alpha]\!]\eta[A/t]\rho$.

(∀E) $[\![\Gamma \supset M\beta:[\beta/t]\alpha]\!]\eta\rho \cong [\![\Gamma \supset M:\forall t.\alpha]\!]\eta\rho \cdot [\![\beta]\!]\eta$
where: $F = \lambda A.[\![\alpha]\!]\eta[A/t]$, $\cap F = \cap_{A\ \mathrm{dcper}} F(A)$, $[h]_{\cap F} \cdot B \cong [h]_{F(B)}$.

(Y) $[\![\Gamma \supset YM:\alpha]\!]\eta\rho \cong [\ \sqcup_{n<\omega} k(n)\ ]_A$
where: $B = [\![(1 \rightarrow \alpha) \rightarrow \alpha]\!]\eta$, $[k]_B \cong [\![\Gamma \supset M:(1 \rightarrow \alpha) \rightarrow \alpha]\!]\eta\rho$,
$k(0) \cong \Uparrow$, $k(n+1) \cong k\ i(\lambda d \in D.k(n))$.

Notes We retain the attention of the reader on three points: (1) Something has to be done in order to show in the clauses (→I) and (∀I) that certain collections of realizers are not empty. (2) When we apply a term to a type (clause (∀E)), we keep the same realizer, this connects with the choice of evaluating under type abstraction. (3) In the (Y) clause the existence of the fixed point combinator, which was announced in section 3, takes a concrete shape. Its construction takes advantage of an iterator one can build in the realizability structure.

The following is proved by connecting the interpretation of a typed term in the per-model to the interpretation of its underlying type free-term in the realizability structure.

4.1 Proposition *(Typing Soundness)*
If $\vdash \Gamma \supset M:\alpha$ then, for any η and ρ such that $\eta \uparrow_\Gamma \rho$, we have:
$[\![\Gamma \supset M:\alpha]\!]\eta\rho \Downarrow \ \Rightarrow\ [\![\Gamma \supset M:\alpha]\!]\eta\rho \in [\ [\![\alpha]\!]\eta\ ]$.

In the following σ, τ,..., denote substitutions of types for type variables. There is an obvious inductive definition of what is the result of applying a substitution σ to a type, to a context, to a term, and to a typing judgment. Having defined such notion the next thing to verify is that provability is invariant under type substitution, and that type and term substitutions commute with the respective semantic substitutions. The following lemmas are proved by induction on the length of the typing.

4.2 Lemma *(Type Substitution)*
Suppose $\vdash \Gamma \supset M:\alpha$. Then: (1) If σ is a type substitution then $\vdash \sigma(\Gamma \supset M:\alpha)$. (2) For any type-assignment η, for any type substitution σ, and for any assignment ρ such that $\eta \uparrow_{\sigma\Gamma} \rho$, we have: $[\![\sigma(\Gamma \supset M:\alpha)]\!]\eta\rho \cong [\![\Gamma \supset M:\alpha]\!]\eta'\rho$ where: $\eta'(t) \triangleq [\![\sigma t]\!]\eta$.

4.3 Lemma *(Term Substitution)*
If $\vdash \Gamma, x:\alpha \supset M:\beta$, and $\vdash \Gamma \supset N:\alpha$ then (1) $\vdash \Gamma \supset [N/x]M:\beta$.
(2) For any type-assignment η, for any assignment ρ such that $\eta \uparrow_\Gamma \rho$,
$[\![\Gamma \supset N:\alpha]\!]\eta\rho\Downarrow \ \Rightarrow\ [\![\Gamma \supset [N/x]M:\beta]\!]\eta\rho \cong [\![\Gamma, x:\alpha \supset M:\beta]\!]\eta\rho'$, where: $\rho' \triangleq \rho[\ [\![\Gamma \supset N:\alpha]\!]\eta\rho\ /x\ ]$.

The following is proved by induction on the structure of C.

4.4 Lemma *(Canonical Forms are Defined)*
If $\vdash \emptyset \supset C:\alpha$, where C is a canonical form, then for any η and ρ, $[\![\emptyset \supset C:\alpha]\!]\eta\rho \Downarrow$.

The following is proved by induction on the deduction of the evaluation judgment.

4.5 Lemma *(Invariance under Evaluation)*
If $\vdash M \mapsto C$ then, for any η and ρ, $[\![\emptyset \supset M:\alpha]\!]\eta\rho = [\![\emptyset \supset C:\alpha]\!]\eta\rho$.

5. Adequacy

We want to prove that given a well typed closed term M, $M\Downarrow$ iff $M\downarrow$. It is easy to show that if $M\downarrow$ then $M\Downarrow$, as the interpretation is invariant under evaluation (4.5), and the interpretation of a canonical form is defined (4.4). In the other direction an iterated attempt of generalizing the induction hypothesis leads to the following

5.1 Definition *(adequacy relation)*

Let η be a type assignment. A relation $S \subseteq |[\![\alpha]\!]\eta| \times \Lambda^o{}_\alpha$ is an adequacy relation of type α, w.r.t. the type assignment η, if it satisfies the following conditions:

(C.1)	$h\, S\, M$	$\Rightarrow M\downarrow$.
(C.2)	$\{h_n\}_{n<\omega}$ directed in D $\wedge\ \forall n.\ h_n S\, M$	$\Rightarrow (\sqcup_{n<\omega} h_n)\, S\, M$.
(C.3)	$(h\, S\, M \wedge \vdash M \mapsto C \wedge \vdash M' \mapsto C)$	$\Rightarrow h\, S\, M'$.
(C.4)	$(h\, S\, M \wedge h\, [\![\alpha]\!]\eta\, h')$	$\Rightarrow h'\, S\, M$.

We denote with $AR(\alpha, \eta)$ the collection of adequacy relations of type α, w.r.t. the type assignment η. Observe that, for any type α, the empty set is an adequacy relation of type α.

If one thinks of h as an element in the equivalence class corresponding to the interpretation of the term M, then condition (C.1) corresponds to what we need to prove. Condition (C.2) says that an adequacy relation is a kind of admissible predicate, in that it is closed under directed sets. Condition (C.3) says that an adequacy relation is invariant w.r.t terms that reduce to the same canonical form. The formulation of this condition seems to be new; it has the advantage of being simple and of not requiring a finer analysis of the evaluation relation as the closure of a "one-step" reduction relation. Condition (C.4) says that an adequacy relation is invariant w.r.t. the equivalence induced by the corresponding per. This condition comes from the choice of representing adequacy relations as relations over the field of a per rather than over the collection of equivalences classes.

We wish to assign to each type α an adequacy relation, parametrically in a type assignment η. In order to do this correctly we need to introduce two further parameters:[4]

(i) A *type substitution* σ.

(ii) An *adequacy relation assignment,* θ: $tv \rightarrow \cup_{\alpha\ type} AR(\alpha, \eta)$ (henceforth ar-assignment) that depends on η.

Let η be a type assignment. A substitution σ and an ar-assignment θ are *compatible*, w.r.t η, if $\theta(t) \in AR(\sigma t, \eta)$, for any t. We write this as: $\sigma \uparrow_\eta \theta$.[5]

5.2 Definition *(associating adequacy relations to types)*

Let α be any type. For any type assignment η, any substitution σ, and any ar-assignment θ, such that $\sigma \uparrow_\eta \theta$, we define a relation

$R(\alpha, \sigma, \theta) \subseteq |[\![\sigma\alpha]\!]\eta| \times \Lambda^o{}_{\sigma\alpha}$

by induction on the structure of α, as follows:

(1) $R(1, \sigma, \theta) \quad \triangleq \{(h, M) \in D \times \Lambda^o{}_1 \mid M\downarrow\}$

(tv) $R(t, \sigma, \theta) \quad \triangleq \theta(t)$

---

4 A similar problem arises in the proof of (strong) normalization of system F, see [9], chpt. 14.

5 In the following whenever we bring a substitution under a bound variable we assume that the bound variable has been suitably renamed so that it does not interact with the substitution.

$(\rightharpoonup)$ $\quad R(\alpha\rightharpoonup\beta, \sigma, \theta) \quad \triangleq \{(h, M)\in |[\![\sigma(\alpha\rightharpoonup\beta)]\!]\eta| \times \Lambda^o{}_{\sigma(\alpha\rightharpoonup\beta)} \mid$
$M\downarrow \wedge (d\, R(\alpha, \sigma, \theta)\, N \Rightarrow (hd\, R(\beta, \sigma, \theta)\, MN \vee hd\Uparrow))\}$

$(\forall)$ $\quad R(\forall t.\alpha, \sigma, \theta) \quad \triangleq \{(h, M)\in |[\![\sigma(\forall t.\alpha)]\!]\eta| \times \Lambda^o{}_{\sigma(\forall t.\alpha)} \mid$
$M\downarrow \wedge \forall\beta. \forall S\in AR(\beta, \eta).\ h\, R(\alpha, [\beta/t]\sigma, \theta[S/t])\, M\beta\}$

Example Let id be the identity over the realizability structure D, let Id $\equiv \lambda t.\lambda x{:}t.\ x$ be the polymorphic identity, and let σ, θ, η be given so that $\sigma\!\uparrow_\eta\!\theta$. Then:
id $R(\forall t.t\rightharpoonup t, \sigma, \theta)$ Id , as
Id$\downarrow$ and $\forall\beta. \forall S\in AR(\beta, \eta)$. id $R(t\rightharpoonup t, [\beta/t]\sigma, \theta[S/t])$ Idβ , as
Id$\beta\downarrow$ and $\forall\beta. \forall S\in AR(\beta, \eta)$. d S' N $\Rightarrow$ d S' IdβN, where $S'\equiv R(t, [\beta/t]\sigma, \theta[S/t])$, as d S' N and (C.1) implies N$\downarrow$. Moreover N$\downarrow$ and (C.3) implies d S' IdβN. □

5.3 Lemma *(coherence of the definition)*

Let α be any type. For any type assignment η, any substitution σ, and any ar-assignment θ, such that $\sigma\!\uparrow_\eta\!\theta$, we have: $R(\alpha, \sigma, \theta)\in AR(\sigma\alpha, \eta)$.

Proof

By induction on the structure of the type α we verify that $R(\alpha, \sigma, \theta)$ is well defined and belongs to $AR(\sigma\alpha, \eta)$ as it satisfies (C.1-4).

(1) $R(1, \sigma, \theta)\in AR(1, \eta)$, as one checks that $\{(h, M)\in D\times\Lambda^o{}_1 \mid M\downarrow\}$ satisfies (C.1-4).

(tv) $R(t, \sigma, \theta)=\theta(t)\in AR(\sigma t, \eta)$, by the assumption: $\sigma\!\uparrow_\eta\!\theta$.

$(\rightharpoonup)$ By definition $R(\alpha\rightharpoonup\beta, \sigma, \theta) \subseteq |[\![\sigma(\alpha\rightharpoonup\beta)]\!]\eta| \times \Lambda^o{}_{\sigma(\alpha\rightharpoonup\beta)}$, and satisfies (C.1).

(C.2): Suppose $\{h_n\}_{n<\omega}$ directed, and $\forall n.\ h_n R(\alpha\rightharpoonup\beta, \sigma, \theta)$ M. Then:
(i) $\sqcup_{n<\omega} h_n \in |[\![\sigma(\alpha\rightharpoonup\beta)]\!]\eta|$, because $[\![\sigma(\alpha\rightharpoonup\beta)]\!]\eta\in$ dcper.
(ii) For any $d\in D$, $(\sqcup_{n<\omega} h_n)d \equiv \sqcup_{n<\omega} h_n d$, and $\{h_n d \mid h_n d\Downarrow\}_{n<\omega}$ is directed, if not empty.
(iii) Suppose d $R(\alpha, \sigma, \theta)$ N. There are two possibilities:
(a) $(\sqcup_{n<\omega} h_n)d\Uparrow$, and we are done. (b) $(\sqcup_{n<\omega} h_n)d\Downarrow$, that implies $h_m d\Downarrow$, for some m, and therefore $h_n d\ R(\beta, \sigma, \theta)$ MN, for all $n\geq m$. We can then apply the ind. hyp. on β to conclude $(\sqcup_{n<\omega} h_n d)\ R(\beta, \sigma, \theta)$ MN, i.e. by (i, ii) $(\sqcup_{n<\omega} h_n)d\ R(\beta, \sigma, \theta)$ MN.
(iv) By combining cases (iii.a-b) we have: $\sqcup_{n<\omega} h_n R(\alpha\rightharpoonup\beta, \sigma, \theta)$ M.

(C.3): Suppose h $R(\alpha\rightharpoonup\beta, \sigma, \theta)$ M $\wedge \vdash M\mapsto C \wedge \vdash M'\mapsto C$. Then h $R(\alpha\rightharpoonup\beta, \sigma, \theta)$ M' as:
(i) $M'\in\Lambda^o{}_{\sigma(\alpha\rightharpoonup\beta)}$ and M'$\downarrow$ (as the evaluation preserves the type).
(ii) Given N, $\vdash MN\mapsto C'$ iff $\vdash M'N\mapsto C'$.
(iii) Suppose d $R(\alpha, \sigma, \theta)$ N. By (ii): $hd\Uparrow \vee (hd\Downarrow \wedge MN\downarrow \wedge M'N\downarrow)$. In the first case we are done, in the latter one shows: hd $R(\beta, \sigma, \theta)$ M'N by applying the inductive hypothesis on β, and (C.3) with (ii).

(C.4): Suppose (h $R(\alpha\rightharpoonup\beta, \sigma, \theta)$ M $\wedge$ h $[\![\sigma(\alpha\rightharpoonup\beta)]\!]\eta$ h').
Then, by definition of pexp: d $[\![\sigma\alpha]\!]\eta$ d' $\Rightarrow$ (hd $[\![\sigma\beta]\!]\eta$ h'd' $\vee$ (hd $\Uparrow \wedge$ h'd $\Uparrow$)).
To show h' $R(\alpha\rightharpoonup\beta, \sigma, \theta)$ M we have to verify:

d $R(\alpha, \sigma, \theta)$ N $\Rightarrow$ (h'd $R(\beta, \sigma, \theta)$ MN $\vee$ h'd$\Uparrow$).

But: h'd$\Uparrow \Leftrightarrow$ hd$\Uparrow$, and h'd$\Downarrow \Rightarrow$ hd$\Downarrow \Rightarrow$ (hd $R(\beta, \sigma, \theta)$ MN)
$\Rightarrow$ (h'd $R(\beta, \sigma, \theta)$ MN), the last implication by ind. hyp. on β and (C.4).

$(\forall)$ By definition $R(\forall t.\alpha, \sigma, \theta) \subseteq |[\![\sigma(\forall t.\alpha)]\!]\eta| \times \Lambda^o{}_{\sigma(\forall t.\alpha)}$ and satisfies (C.1).
Observe that for any type β (see proviso): $(\sigma\!\uparrow_\eta\!\theta \wedge S\in AR(\beta, \eta)) \Rightarrow [\beta/t]\sigma\!\uparrow_\eta\!\theta[S/t]$.

(C.2) Suppose $\{h_n\}_{n<\omega}$ directed, and $\forall n.\ h_n R(\forall t.\alpha, \sigma, \theta)$ M. Then,

$$\forall\beta.\ \forall S\in AR(\beta, \eta).\ h_n\ R(\alpha, [\beta/t]\sigma, \theta[S/t])\ M\beta$$

Hence, by induction hypothesis on α, and (C.2):

$$\forall\beta.\ \forall S\in AR(\beta, \eta).\ \sqcup_{n<\omega} h_n\ R(\alpha, [\beta/t]\sigma, \theta[S/t])\ M\beta$$

and this implies by definition of $R(\forall t.\alpha, \sigma, \theta)$:

$$\sqcup_{n<\omega} h_n\ R(\forall t.\alpha, \sigma, \theta)\ M$$

(C.3): Suppose $(h\ R(\forall t.\alpha, \sigma, \theta)\ M\ \wedge\ \vdash M\mapsto C\ \wedge\ \vdash M'\mapsto C)$.
Then $h\ R(\forall t.\alpha, \sigma, \theta)\ M'$ because:
(i) $M'\in\Lambda^o{}_{\sigma(\forall t.\alpha)}$ and $M'\downarrow$.
(ii) For any β, $\vdash M\beta\mapsto C'$ iff $\vdash M'\beta\mapsto C'$.
(iii) For any β, for any $S\in AR(\beta, \eta)$, we can apply (C.3) on α (using (ii) and ind. hyp.), with substitution $[\beta/t]\sigma$, and ar-assignment $\theta[S/t]$ to conclude:
$h\ R(\alpha, [\beta/t]\sigma, \theta[S/t])\ M'\beta$. Hence: $h\ R(\forall t.\alpha, \sigma, \theta)\ M'$.

(C.4): Suppose $(h\ R(\forall t.\alpha, \sigma, \theta)\ M\ \wedge\ h\ [\![\sigma(\forall t.\alpha)]\!]\eta\ h')$.
Then, by definition: $\forall B$ dcper. $h\ [\![\sigma\alpha]\!]\eta[B/t]\ h'$.
To show $h'\ R(\forall t.\alpha, \sigma, \theta)$ M we have to verify:

$$\forall\beta.\ \forall S\in AR(\beta, \eta).\ h'\ R(\alpha, [\beta/t]\sigma, \theta[S/t])\ M\beta\ .$$

But for $B = [\![\beta]\!]\eta$, we have $[\![\sigma\alpha]\!]\eta[B/t] = [\![[\beta/t]\sigma\alpha]\!]\eta$, by the type substitution lemma. Hence $h\ [\![[\beta/t]\sigma\alpha]\!]\eta\ h'$, and we can apply ind. hyp. on α, and (C.4). □

5.4 Theorem *(semantic and syntactic convergence coincide)*

Suppose $\vdash (x_1: \alpha_1),...,(x_n: \alpha_n) \supset M: \alpha$. Then for any type assignment η, any substitution σ, and any ar-assignment θ, such that $\sigma\uparrow_\eta\theta$, we have:

if $d_i\ R(\alpha_i, \sigma, \theta)\ C_i$, for i=1,..,n, and $[\![\sigma(\Gamma\supset M: \alpha)]\!]\eta[d/x] = [h]_A$
then $h\ R(\alpha, \sigma, \theta)\ [C/x]\sigma M$.

where: $\Gamma \equiv (x_1: \alpha_1),...,(x_n: \alpha_n)$; $[d/x] \equiv [\ [d_1]_{A_1}/x_1,...,[d_n]_{A_n}/x_n]$; $A_i \equiv [\![\sigma\alpha_i]\!]\eta$ for i=1,..,n; $A \equiv [\![\sigma\alpha]\!]\eta$; $[C/x] \equiv [C_1/x_1,..., C_n/x_n]$.

Proof

In order to have a quick understanding of the statement observe, as an instance: if $\vdash \varnothing \supset$ M: α where M and α have no free type variables, and $[\![\varnothing\supset M: \alpha]\!]\eta\rho = [h]_A$ then $hR(\alpha, id, \theta)M$, for arbitrary η, ρ, θ. Hence, by (C.1), we have: $M\Downarrow\ \Rightarrow\ M\downarrow$.

We now proceed with the proof, by induction on the length of the typing (we have emphasized in italic the points of the proof where one discovers the need to generalize the inductive hypothesis, and to adapt the definition of R).

(*) Then $[\![\sigma\Gamma\supset *: 1]\!]\eta[d/x] = [d]_1$, and $d\ R(1, \sigma, \theta)$ *, by def. of $R(1, \sigma, \theta)$.

(asmp) Then $[\![\sigma\Gamma\supset x_i: \sigma\alpha_i]\!]\eta[d/x] = [d_i]_{A_i}$, and $d_i\ R(\alpha_i, \sigma, \theta)\ C_i$, by assumption.

(→I) Suppose $h\in[\![\sigma(\Gamma\supset\lambda x{:}\alpha.M{:}\alpha\to\beta)]\!]\eta[d/x]$. Then by the interpretation definition: $\forall d\in|A|.\ (f(d)\Downarrow\ \Rightarrow hd\in f(d)) \wedge (f(d)\Uparrow\ \Rightarrow hd\Uparrow)\}$ where: $A = [\![\sigma\alpha]\!]\eta$,
$f(d) = [\![\sigma(\Gamma,x{:}\alpha\supset M{:}\beta)]\!]\eta[d/x][[d]_A/x]$.
We have to show: $h\ R(\alpha\to\beta, \sigma, \theta)\ [C/x]\sigma(\lambda x{:}\alpha.M)$, that is:
(i) $[C/x]\sigma(\lambda x{:}\alpha.M)\downarrow$, that holds because an abstraction always converges.
(ii) $d\ R(\alpha, \sigma, \theta)\ N\ \Rightarrow\ (hd\ R(\beta, \sigma, \theta)\ [C/x]\sigma(\lambda x{:}\alpha.M)\ N) \vee (hd\Uparrow)$.
Observe: $[C/x]\sigma(\lambda x{:}\alpha.M) \equiv \lambda x{:}\sigma\alpha.[C/x]\sigma M$.
Also: $\vdash (\lambda x{:}\sigma\alpha.[C/x]\sigma M)N\mapsto C$ iff $\vdash [C'/x][C/x]\sigma M\mapsto C$ and $\vdash N\mapsto C'$.

Suppose hd$\Downarrow$, we can apply the inductive hypothesis on the provable judgment $\Gamma, x{:}\alpha \supset M{:}\beta$ to get hd $R(\beta, \sigma, \theta)$ $[C'/x][C/x]\sigma M$. By (C.1) $[C'/x][C/x]\sigma M\downarrow$, *and by (C.3)* we get (ii).

($\rightarrow$E) Suppose $d \in [\![\sigma(\Gamma\supset MN{:}\beta)]\!]\eta[d/x]$ then we have to show d $R(\beta, \sigma, \theta)$ $[C/x]\sigma(MN)$. By the definition of the interpretation we have $h \in [\![\sigma(\Gamma\supset M{:}\alpha\rightarrow\beta)]\!]\eta[d/x]$, and $d' \in [\![\sigma(\Gamma\supset N{:}\alpha)]\!]\eta[d/x]$, such that $hd' [\![\sigma\beta]\!]\eta$ d.
By induction hypothesis we have: h $R(\alpha\rightarrow\beta, \sigma, \theta)$ $[C/x]\sigma(M)$, and
d' $R(\alpha, \sigma, \theta)[C/x]\sigma(N)$. *By definition of $R(\alpha\rightarrow\beta, \sigma, \theta)$* we can conclude:
hd' $R(\beta, \sigma, \theta)$ $[C/x]\sigma(MN)$, and by (C.4) d $R(\beta, \sigma, \theta)$ $[C/x]\sigma(MN)$.

($\forall$I) Suppose $h \in [\![\sigma(\Gamma\supset\lambda t.M{:}\forall t.\alpha)]\!]\eta[d/x]$ then we have to show:
(i) $[C/x]\sigma(\lambda t.M)\downarrow$. (ii) $\forall\beta.\ \forall S\in AR(\beta, \eta)$. h $R(\alpha, [\beta/t]\sigma, \theta[S/t])$ $[C/x]\sigma(\lambda t.M)\beta$.
Observe: $[C/x]\sigma(\lambda t.M) \equiv \lambda t.[C/x]\sigma M$. By induction hypothesis applied to $\Gamma\supset M{:}\ \alpha$, we have, $\forall\beta.\ \forall S\in AR(\beta, \eta)$: h $R(\alpha, [\beta/t]\sigma, \theta[S/t])$ $[C/x][\beta/t]\sigma M$.
Observe: $\vdash [C/x]\sigma(\lambda t.M)\beta \mapsto C'$ iff $\vdash [C/x][\beta/t]\sigma M \mapsto C'$, *and apply (C.3).*

($\forall$E) Suppose $h \in [\![\sigma(\Gamma\supset M\beta{:}[\beta/t]\alpha)]\!]\eta[d/x]$, then we have to show:
h $R([\beta/t]\alpha, \sigma, \theta)$ $[C/x]\sigma(M\beta)$.
By induction hypothesis applied to $\Gamma\supset M{:}\ \forall t.\alpha$, and the *definition of $R(\forall t.\alpha, \sigma, \theta)$* we have: h $R(\alpha, [\sigma\beta/t]\sigma, \theta[S/t])$ $[C/x]\sigma M\sigma\beta$, and observe $\sigma[\beta/t]\alpha = [\sigma\beta/t]\sigma\alpha$.

(Y) Suppose $[\ \sqcup_{n<\omega} k(n)\ ]_A = [\![\sigma(\Gamma\supset YM{:}\alpha)]\!]\eta[d/x]$, where:
$k \in [\![\sigma(\Gamma\supset M{:}(1\rightarrow\alpha)\rightarrow\alpha)]\!]\eta[d/x]$ for some k. We show that for any n big enough
$k(n)$ $R(\alpha, \sigma, \theta)$ $[C/x]\sigma YM$. We conclude *by (C.2)*: $\sqcup_{n<\omega} k(n)$ $R(\alpha, \sigma, \theta)$ $[C/x]\sigma YM$. By induction hypothesis on
$\Gamma\supset M{:}\ (1\rightarrow\alpha)\rightarrow\alpha$ we know: k $R((1\rightarrow\alpha)\rightarrow\alpha, \sigma, \theta)$ $[C/x]\sigma M$. Observe:

(i) $i(\lambda d\in D.\Uparrow)$ $R((1\rightarrow\alpha), \sigma, \theta)$ $\lambda x{:}1.[C/x]\sigma YM$.

(ii) $i(\lambda d\in D.k(n))$ $R((1\rightarrow\alpha), \sigma, \theta)$ $\lambda x{:}1.[C/x]\sigma YM \Rightarrow$

$$k\ i(\lambda d\in D.k(n))\ R(\alpha, \sigma, \theta)\ [C/x]\sigma M\ (\lambda x{:}1.[C/x]\sigma YM)\ \vee\ k(n+1)\Uparrow .$$

Use (C.3) to conclude $k(n)$ $R(\alpha, \sigma, \theta)$ $[C/x]\sigma YM$, for n big enough. □

6. Conclusion

(1) *Relating model-theoretic and operational pre-orders.* We write $\Gamma\supset M\leq_{obs}N{:}\ \alpha$, if $\vdash\Gamma\supset M{:}\ \alpha$, $\vdash\Gamma\supset N{:}\ \alpha$, and for any "context" C[ ] such that $\vdash\emptyset\supset C[M]{:}\ \beta$, and $\vdash\emptyset\supset C[N]{:}\ \beta$, we have: $C[M]\downarrow\ \Rightarrow\ C[N]\downarrow$. This defines an *operational* preorder on terms. In the standard domain theoretic case it is easy to prove, as a corollary of the correspondence between syntactic and semantic convergence, that the pre-order induced by the model (definition left to the reader) is contained in the operational preorder defined above. When considering per-models there is a natural *intrinsic* way [14] to order the equivalence classes. Let A be a per over D. The intrinsic preorder $\leq_A$ over [A] is defined as follows:

$$x \leq_A y \text{ if } \forall h{:}\ A\rightarrow\mathbf{1}.\ (h(x)\Downarrow\ \Rightarrow\ h(y)\Downarrow)$$

How does the intrinsic pre-order relate to the operational pre-order ? We expect that the former is included in the latter, but to have a proof as simple as in the domain-theoretic case some additional model-theoretic information seems needed. In particular one needs to show that contexts induce monotone operators w.r.t. the intrinsic pre-order, and this seems to depend on the fact that products, used in the interpretation of second order types, have a pointwise intrinsic preordering. We do not have such a property available for the model described here. There are other per models that enjoy this property, e.g. the model of complete extensional pers over Kleene partial combinatory algebra [8]. We expect that the "architecture" of the proof of our main result (5.4) can be applied to such per-models. Hence we suspect that for such models a standard proof of the theorem relating intrinsic and operational pre-ordering will go through.

(2) *Call-by-name, Subtyping, and Recursive Types.* It seems worthwhile to recall that a call-by-name version of the calculus can be easily coded in the calculus presented here with the standard idea that the call-by-name functional space, say $\alpha \rightharpoonup_n \beta$, is coded as $(1 \rightharpoonup \alpha) \rightharpoonup \beta$. We also recall that per-models have been used as a semantic foundation for typed functional languages with a notion of *subtyping* [7]. Our result suggests that this is an *adequate* approach. Moreover we expect that the adequacy result extends to recursive types once we take as semantic structure the collection of complete uniform pers over a D-infinity model, as presented in [1]. The basic idea is to "stratify" the definition of the adequacy relation associated to a type. Formally one introduces an intermediary family of adequacy relation $R(n, \alpha, \sigma, \theta)$, where $n \in \omega$. The adequacy relation $R(\alpha, \sigma, \theta)$ is obtained by a process of completion of the sequence $\{R(n, \alpha, \sigma, \theta) \mid n \in \omega\}$. Roughly $R(n, \alpha, \sigma, \theta)$ represents $R(\alpha, \sigma, \theta)$ cut at the n-th level of the construction of the underlying D_∞ structure. We refrain from going into this point since the development of the model requires a certain number of rather ad hoc conditions that would only obscure the main ideas we have discussed.

(3) *Independence from the Adequacy of the Realizability Structure.* It is well know that to every, say closed, term M of type α one can associate its "erasure", i.e. a type free term er(M) such that: $[\![M]\!]^{per} \cong [\, [\![er(M)]\!]^D]_{[A]}$ (a). In [2] we suggested that a "cheap" adequacy theorem could be obtained by the following schema of implications:

$$[\![M]\!]^{per} \Downarrow \;\Rightarrow_{(1)}\; [\![er(M)]\!]^D \Downarrow \;\Rightarrow_{(2)}\; er(M)\downarrow \;\Rightarrow_{(3)}\; M\downarrow$$

where (1) follows by a result of type (a), (2) follows by the adequacy of the realizability structure, and (3) follows by a comparison of the evaluations. The weak point of this chain of implications is (2). As a matter of fact we have shown that the adequacy of the per model is *independent* from the adequacy of the realizability structure w.r.t. the related type-free language (following [5], we can build a realizability structure D that is *not* adequate in that $[\![\Omega]\!]^D \Downarrow$, where $\Omega \equiv (\lambda x.xx)(\lambda x.xx)$).

References

[1] Amadio R. [1989] "Recursion over realizability structures", Info.&Comp., 91, 1, (55-85), 1991. Also appeared as TR 1/89 Dipartimento di Informatica, Università di Pisa.

[2] Amadio R. [1990] "Domains in a Realizability Framework", in Proc. CAAP91, Brighton, SLNCS 493, Abramsky S., Maibaum T. (eds.), (241-263), full version appeared as Liens TR 19-90.

[3] Amadio R. [1992] "On the adequacy of per models", RR 1579 Inria-Lorraine, January 1992 (20 pages).

[4] Amadio R., Cardelli L. [1990] "Subtyping Recursive Types", in Proc. ACM-POPL91, Orlando. Full version appeared as DEC-SRC TR #62 and TR 133 Inria-Lorraine, in press on ACM-TOPLAS.

[5] Baeten J, Boerboom B. [1979] "Ω can be anything it shouldn't be", Indag. Math. 41, (111-120).

[6] Cardelli L. [1989] "The Quest language and system", internal note, DEC-SRC, Palo Alto.

[7] Cardelli L., Longo G.[1990] "A semantic basis for Quest", ACM-Lisp and F. P. 90, Nice.

[8] Freyd P., Mulry P., Rosolini G., Scott D. [1990] "Extensional Pers", 5th IEEE-LICS, Philadelphia.

[9] Girard J.Y., Lafont Y., Taylor P. [1989] "Proofs and Types", Cambridge University Press.

[10] Hyland M. [1991] "First steps in synthetic domain theory", in Proc. Category Theory 90, Como, Carboni&al. (eds.), Springer-Verlag.

[11] Martin-Löf [1983] "The domain interpretation of type theory", in Proc. Workshop on the Semantics of Programming Languages, Dybier&al (eds.), Chalmers University, Göteborg.

[12] Moggi E. [1988] "Partial morphisms in categories of effective objects", Info.&Comp., 76, (250-277).

[13] Plotkin G. [1985] "Denotational semantics with partial functions", lecture notes, CSLI, Stanford 1985.

[14] Rosolini G. [1986] "Continuity and effectiveness in Topoi", PhD Thesis, Oxford University.

Hausdorff Reductions to Sparse Sets and to Sets of High Information Content

V. Arvind*[1] and J. Köbler**[2] and M. Mundhenk**[2]

[1] Department of Computer Science and Engineering, Indian Institute of Technology, Delhi, New Delhi 110016, India,
[2] Universität Ulm, Oberer Eselsberg, D-89069 Ulm, Germany

Abstract. We investigate the complexity of sets that have a rich internal structure and at the same time are reducible to sets of either low or very high information content. In particular, we show that every length-decreasing or word-decreasing self-reducible set that reduces to some sparse set via a non-monotone variant of the Hausdorff reducibility is low for Δ_2^p.
Measuring the information content of a set by the space-bounded Kolmogorov complexity of its characteristic sequence, we further investigate the (non-uniform) complexity of sets A in EXPSPACE/poly that reduce to some set having very high information content. Specifically, we show that if the reducibility used has a certain property, called "reliability," then A in fact is reducible to a sparse set (under the same reducibility). As a consequence of our results, the existence of hard sets (under "reliable" reducibilities) of very high information content is unlikely for various complexity classes as for example NP, PP, and PSPACE.

1 Introduction

An important subject of research in structural complexity theory is the study of reductions to sets of low information content (as for example sparse sets). Historically, this study originated in the Berman-Hartmanis conjecture that all NP-complete sets are polynomial-time isomorphic, a hypothesis that is empirically supported by all natural examples of NP-complete sets. An intuitive interpretation of the conjecture is that all NP-complete sets are only different encodings of the same information. Since polynomial-time isomorphic sets must have similar densities, and since there are NP-complete sets of exponential density, the Berman-Hartmanis conjecture implies that sparse sets cannot be NP-complete. Motivated by the Berman-Hartmanis conjecture, several results concerning reductions of NP-complete sets to sparse sets have been established. The basic results along this line of research were Mahaney's theorem that NP-complete sets are not many-one reducible to a sparse set unless P = NP [13], and the

* Work done while visiting Universität Ulm. Supported in part by an Alexander von Humboldt research fellowship.

** Work supported in part by the DAAD through Acciones Integradas 1992, 313-AI-e-es/zk

result of Karp, Lipton, and Sipser, that a sparse set cannot be Turing-hard for NP unless the polynomial hierarchy collapses to Σ_2^p [8]. The more recent result of Ogiwara and Watanabe [14], that the existence of sparse hard sets for NP under polynomial time bounded truth-table reductions implies P = NP, has been followed up by analogous results for more general reductions in [1, 15, 2].

Recently, Book and Lutz [7] have considered reductions to sets whose characteristic sequences are of very high space-bounded Kolmogorov complexity (they call the class of such sets HIGH). Book and Lutz obtained the surprising result that every set in ESPACE that is (polynomial-time) bounded truth-table reducible to a set in HIGH is actually bounded truth-table reducible to some sparse set. Thus, oracles of very high space-bounded Kolmogorov complexity are only as useful as sparse sets provided that the oracle access is restricted to a constant number of queries to retrieve the information from the oracle set. As a consequence, if SAT bounded truth-table reduces to a set in HIGH, then SAT bounded truth-table reduces to a sparse set, and thus, using the result of Ogiwara and Watanabe, it follows that P = NP.

Of course, one would expect similar results for a wider class of reducibility types, not only for the bounded truth-table case. The aim of the present paper is to identify the properties (of reducibilities) that are responsible for their inability to extract a substantial part of the information hidden in the sets in HIGH. We prove results analogous to the one of Book and Lutz for a wide range of reducibilities which we call "reliable reducibilities". Intuitively, A reduces to B via a reliable reducibility if not only A is determined by B (and the specific reduction used), but information about B can also be retrieved from the knowledge of A. More precisely, in the case of a reliable reduction, given A and the reduction procedure, it is possible to compute a *minimum* set B' of strings to which A reduces via the given reduction, that is, B' is a subset of every set to which A reduces.

For example, deterministic many-one reductions f are reliable, since for every $x \in A$ the membership of the image $f(x)$ in B is guaranteed, and A correctly reduces to the set $B' = \{f(x) \mid x \in A\}$ of all these images (a similar consideration applies for conjunctive reductions). In contrast, Turing reductions are not considered to be reliable since in general the information about B that can be derived from the knowledge of A and the oracle machine performing the reduction is rather vague. On the other hand, as we show in this paper, the Hausdorff reducibility [17] is reliable. In fact, it turns out that even an extension of the Hausdorff reducibility (called non-monotone Hausdorff reducibility) is reliable.

In spite of its reliability, the Hausdorff reducibility is much more powerful than the many-one reducibility, and in some cases it has been shown that the Hausdorff reducibility has the same power as the truth-table reducibility. For example, Wagner [17] proved that the closure of NP under the Hausdorff reducibility coincides with the truth-table closure of NP. Furthermore, for every set ring $\mathcal{C}$ it is known (cf. [18]) that the closure of $\mathcal{C}$ under the bounded Hausdorff reducibility is the same as its closure under the bounded truth-table reducibility (as an example we mention the equality $R^p_{bhd}(R^p_c(\text{SPARSE})) = R^p_{btt}(R^p_c(\text{SPARSE}))$ [3],

where R_c^p(SPARSE) is the closure of sparse sets under the conjunctive reducibility).
In the first part of this paper we consider sets that have a rich internal structure (self-reducibility) and at the same time are Hausdorff reducible to sets of low information content (i.e. sparse sets). Exploiting properties of the Hausdorff reducibility we show that every length-decreasing or word-decreasing self-reducible set that non-monotone Hausdorff reduces to some sparse set is low for P^{NP}, i.e. it does not provide this class with any additional power when used as an oracle. This generalizes and improves results of Lozano and Torán [12] stating that any length-decreasing or word-decreasing self-reducible set that many-one reduces to a sparse set is low for Δ_2^p.
In the second part we show that if a set A in EXPSPACE/poly reduces via a reliable reducibility to some set in HIGH, then A reduces (via the same reducibility) to a sparse set. As a consequence of this result, it is unlikely that there exist hard sets (under the non-monotone Hausdorff and other reliable reducibilities) for complexity classes like UP, NP, PP, or PSPACE.

2 Preliminaries

Let $\Sigma = \{0, 1\}$ be the standard alphabet, and let $A \subseteq \Sigma^*$ be a set. The length of a string x is denoted by $|x|$. $A^{=n}$ ($A^{\leq n}$) denotes the set of all strings in A of length n (up to length n, respectively). χ_A denotes the characteristic function of A. $\chi_{\leq n}^A$ denotes the characteristic sequence of A for all strings up to length n, i.e., $|\chi_{\leq n}^A| = 2^{n+1} - 1$, and the i-th bit of $\chi_{\leq n}^A$ equals $\chi_A(s_i)$ where s_i is the i-th string in Σ^* in lexicographical order. The cardinality of A is denoted by $|A|$. The census function of a set A is $census_A(1^n) = |A^{\leq n}|$. A set S is called sparse if its census function is bounded above by a polynomial. A set T is called a tally set if $T \subseteq 0^*$. We use TALLY and SPARSE to represent the classes of tally and sparse sets, respectively. The empty string is denoted by ε. $\langle \cdot, \cdot \rangle$ denotes a standard polynomial time computable pairing function whose inverses are also computable in polynomial time. For any reducibility type α, let $R_\alpha(\mathcal{C}) = \{A \mid \exists B \in \mathcal{C} : A \leq_\alpha B\}$ denote the class of sets $\leq_\alpha$-reducible to some set in $\mathcal{C}$. The reducibilities discussed in this paper are the standard polynomial-time reducibilities defined by Ladner, Lynch, and Selman [10] and the following Hausdorff reducibility introduced by Wagner [16].

Definition 1. A is Hausdorff reducible to B (in symbols: $A \leq_{hd}^p B$), if there exists a polynomial-time computable function f mapping every string x to a sequence of queries, such that for all $x \in \Sigma^*$, if $f(x) = \langle y_1, \ldots, y_k \rangle$ then

- $y_{i+1} \in B$ implies $y_i \in B$ for all $i = 1, \ldots, k-1$, and
- $x \in A \iff \max\{j \mid 0 \leq j \leq k$ and for all $i = 1, \ldots, j : y_i \in B\}$ is odd.

We call f a bounded Hausdorff reduction ($A \leq_{bhd}^p B$) if the number $k(x)$ of queries produced by f on x is bounded by a constant for all x.

In this context, the i-th query y_i computed by $f(x)$, $1 \leq i \leq k(x)$, is also denoted by $f(x,i)$. By removing in the definition of the Hausdorff reducibility the monotony condition we obtain the following generalization of the Hausdorff reducibility which, as we will see in Proposition 3, is equivalent in power to the composition of the Hausdorff and the conjunctive reducibilities.

Definition 2. A is non-monotone Hausdorff reducible to B ($A \leq^p_{nhd} B$), if there exists a polynomial-time computable function f mapping every string x to a sequence of queries, such that for all $x \in \Sigma^*$, if $f(x) = \langle y_1, \ldots, y_k \rangle$ then

$$x \in A \iff \max\{j \mid 0 \leq j \leq k \text{ and for all } i = 1, \ldots, j : y_i \in B\} \text{ is odd.}$$

The following proposition gives examples for equalities of reduction classes defined by (bounded) truth-table reducibilities and Hausdorff reducibilities.

Proposition 3.

1. *For every class $\mathcal{C}$ of languages, $R^p_{nhd}(\mathcal{C}) = R^p_{hd}(R^p_c(\mathcal{C}))$.*
2. [17]$R^p_{hd}(\mathrm{NP}) = R^p_{nhd}(\mathrm{NP}) = R^p_{tt}(\mathrm{NP})$ $(= \Theta^p_2)$.
3. [3] $R^p_{bhd}(R^p_c(\mathrm{SPARSE})) = R^p_{btt}(R^p_c(\mathrm{SPARSE}))$.

Let M be a Turing machine, z be a string and let d, s be natural numbers. We say that $z \in KS_M[d,s]$, if M on some input of length at most d outputs z using space at most s. In other words, $KS_M[d,s]$ is the set of strings whose *s-space-bounded Kolmogorov complexity* relative to M is bounded by d. Similarly, for a string y, $KS_M[d, s|y]$ is the set of all strings z for which there is a string x of length at most d such that M on input $\langle x, y\rangle$ outputs z using space at most s. Well known simulation-techniques (see [11]) show that there is a Universal Turing machine U such that for every machine M there is a constant c such that for all d, s: $KS_M[d,s] \subseteq KS_U[d+c, cs+c]$. Henceforth, we fix such a Universal Turing machine and omit the subscript. Note that there is a constant c such that for every set A and for all n, the characteristic sequence $\chi^A_{\leq n}$ of A restricted to the set of strings up to length n is in $KS[2^{n+1}+c, 2^{cn}]$.
Next we give a definition for the class HIGH containing only sets of very high information content. (We say that a property holds "for almost every n" if it holds for all but finitely many n.)

Definition 4. A set B is in HIGH, if for every constant $c > 0$ there exists a polynomial q such that for almost every n, $\chi^B_{\leq n} \notin KS[2^{n+1} - q(n), 2^{cn}]$.

Observe that the definition above for the class HIGH is an extension of the original one given by Book and Lutz [7] who required for q the fixed polynomial $2n$. It is easy to see that for every set A in the class EXPSPACE/poly there is a polynomial p such that for every n, $\chi^A_{\leq n} \in KS[p(n), 2^{p(n)}]$, where EXPSPACE = DSPACE($2^{n^{O(1)}}$).
Self-reducibility is an important structural property. The types of self-reducibility we will use are the length-decreasing (as defined by Ko [9]) and the word-decreasing self-reducibility (see [5]).

Definition 5. [9] A partial order $\prec$ on Σ^* is said to be polynomially well-founded and length-related (abbr. *polynomially related*) if there is a polynomial p such that

1. for all $x, y \in \Sigma^*$: if $x \prec y$, then $|x| \leq p(|y|)$,
2. $x \prec y$ is decidable in time polynomial in $|x| + |y|$, and
3. for all $x_1, x_2, \ldots, x_k \in \Sigma^*$: if $x_1 \prec x_2 \prec \cdots \prec x_k$, then $k \leq p(|x_k|)$.

Definition 6. [9, 5] A set A is length-decreasing self-reducible, if there is some polynomial-time oracle machine M and a polynomially related ordering $\prec$ such that $A = L(M, A)$ and, on any input x, M queries the oracle only about words $q \prec x$. [3] A set A is word-decreasing self-reducible, if there is some polynomial-time oracle machine M such that $A = L(M, A)$ and, on any input x, M queries the oracle only about words lexicographically smaller than x.

A set A is called $\mathrm{FP}^{\mathrm{NP}}$*-printable*, if there exists a polynomial-time algorithm that uses an oracle from NP and computes on input 1^n (an encoding of) the finite set $A^{\leq n}$. For further definitions used in this paper we refer the reader to some standard book on structural complexity theory (as for example [6]).

3 On Hausdorff reductions of self-reducible sets to sparse sets

In this section we consider consequences of the assumption that a self-reducible set A $\leq^p_{nhd}$-reduces to a sparse set S. Exploiting the self-reducibility structure of the set A we are able to construct an $\mathrm{FP}^{\mathrm{NP}}$-printable subset of S to which A reduces. We only sketch the proof for the case of length-decreasing self-reducible sets, the word-decreasing case can be proved similarly.

Theorem 7. *If a length-decreasing self-reducible set A $\leq^p_{nhd}$-reduces to a sparse set S, then there exists an $\mathrm{FP}^{\mathrm{NP}}$-printable subset $S' \subseteq S$ such that $A \leq^p_{nhd} S'$.*

Proof Sketch. Let A be a length-decreasing self-reducible set and let S be a sparse set such that $A \leq^p_{nhd} S$ via some reduction function f. Let k be a polynomial such that f on input x produces $k(|x|)$ many queries $f(x, i)$, $1 \leq i \leq k(|x|)$, all of length $r(|x|)$ for some strictly increasing polynomial r. Let M be an oracle Turing machine and let $\prec$ be a polynomially related ordering witnessing the self-reducibility of A.
Let $j(x, B) = \max\{0 \leq j \leq k(|x|) \mid \text{for all } i = 1, \ldots, j : f(x, i) \in B\}$, i.e., $x \in A \iff j(x, S)$ is odd. Let M' be an oracle Turing machine that works as follows: M' simulates M; if M asks an oracle query then M uses its oracle (say O) to compute $j(q, O)$, and continues the simulation with answer "yes" if $j(q, O)$ is odd, and with answer "no" otherwise. It is clear that $A = L(M', S)$.
Next we show how to construct for $i = 0, 1, 2, \ldots$ a subset S_i of $S^{=r(i)}$ such that the reduction from A to S_i via f is correct for all instances in Σ^i (i.e.

[3] Observe that by part 3 of the preceding definition, $x \not\prec x$.

$x \in \Sigma^i \Rightarrow [x \in A \iff j(x, S_i)$ is odd]). Using the modified self-reduction machine M', the sets S_i can be constructed by a polynomial-time oracle machine using an NP-oracle. Define the *distance* (cf. [12]) between two strings y and x as the maximum length of an increasing chain from y to x w.r.t. $\prec$ (if such a chain exists): $dist(y, x) = \max\{k \mid \exists x_0, x_1, \ldots, x_k : y = x_0 \prec x_1 \prec \cdots \prec x_k = x\}$. Since $\prec$ is polynomially related, there exists a polynomial p such that $dist(y, x) \leq p(|x|)$ for every pair (y, x) on wich $dist$ is defined. Clearly, $dist(x, x) = 0$. By the "inductive structure" of self-reducible sets it follows for all $x \in \Sigma^*$ and $i \in \mathbb{N}$: if for all y with $dist(y, x) > i$ it holds that $y \in L(M', Q) \iff j(y, Q)$ is odd, then for all y with $dist(y, x) \geq i$ it holds that $y \in L(M', Q) \iff y \in A$. The following function uses these properties of M', j, and A to construct on input 0^i a set Q, which is a finite subset of S to which all instances of A having length at most i are correctly reduced by f. Define $E = \{\langle y, 0^m, 0^i, Q\rangle \mid \exists x \in \Sigma^i : dist(y, x) \geq m \wedge y \in L(M', Q) \not\Leftrightarrow j(y, Q)$ is odd $\}$. Note that E is in NP.

CONSTRUCT(0^i)
  $Q := \emptyset$
  while $\exists j \leq p(i), y \in \Sigma^{\leq p(i)} : \langle y, 0^j, 0^i, Q\rangle \in E$ **do**
    $k := \max\{m \mid \exists z \in \Sigma^{\leq p(i)} : \langle z, 0^m, 0^i, Q\rangle \in E\}$
    $x := \min\{y \mid \langle y, 0^k, 0^i, Q\rangle \in E\}$ w.r.t. lexicographical order
    $Q := Q \cup \{f(x, j(x, Q) + 1)\}$
  end
  return $Q \cap \Sigma^{r(i)}$

Define $S' = \bigcup_{i \geq 0}$ CONSTRUCT(0^i). It follows that S' is sparse, that $A \leq^p_{nhd} S'$ via f, and finally that S' is $\mathrm{FP}^{\mathrm{NP}}$-printable. ∎

Theorem 8. *If a word-decreasing self-reducible set A $\leq^p_{nhd}$-reduces to a sparse set S, then there exists an $\mathrm{FP}^{\mathrm{NP}}$-printable subset $S' \subseteq S$ such that $A \leq^p_{nhd} S'$.*

Corollary 9. *If a length-decreasing or word-decreasing self-reducible set A $\leq^p_{nhd}$-reduces to a sparse set, then A is low for Δ^p_2, i.e. $\Delta^p_2(A) = \Delta^p_2$.*

Since the PSPACE-complete set QBF and the NP-complete set SAT are word-decreasing self-reducible we can state the following theorem.

Theorem 10.
1. *If* SAT *is $\leq^p_{nhd}$-reducible to a sparse set, then* $\mathrm{PH} = \Delta^p_2$.
2. *If QBF is $\leq^p_{nhd}$-reducible to a sparse set, then* $\mathrm{PSPACE} = \Delta^p_2$.

4 Reliable reducibilities

In this section, we introduce the concept of a reliable reducibility and prove the reliability of various reducibilities as for example the (non-monotone) Hausdorff reducibility, the composition of the bounded Hausdorff and the conjunctive reducibilities (henceforth called *bhd-c* reducibility), and the *co-np* as well as the *co-rp* many-one reducibilities.

Definition 11. A reducibility $\leq_\alpha$ is reliable if for all sets A, B such that $A \leq_\alpha B$ there exist a polynomial p and a sequence $B_n \subseteq B$, $n \geq 0$, of sets fulfilling the following properties:

- all instances for A up to length n are correctly reduced (via the given reduction) to B_n,
- for every $n \geq 0$, $\chi_{\leq n}^{B_n} \in KS\left[p(n), 2^{p(n)} \,\middle|\, \chi_{\leq n}^{A}\right]$.

Theorem 12. *The composition of the Hausdorff reducibility and the conjunctive reducibility is reliable.*

Proof Sketch. Let $A \leq_{hd}^{p} B$ via a polynomial time computable function f, and let B conjunctively reduce to C via a polynomial time computable function g. Let $k(x)$ be the number of queries in the list $f(x)$. The following algorithm computes on input $\chi_{\leq n}^{A}$ the characteristic string $\chi_{\leq n}^{C_n}$ of a finite set $C_n \subseteq C$ such that A reduces to C_n via f and g for all instances up to length n. In fact, C_n is the smallest set with this property.

```
input χ^A_{≤n}
  C_n := ∅; x := ε;
repeat
  l := max({0} ∪ {j | 1 ≤ j ≤ k(|x|) and g(f(x,j)) ⊆ C_n})
  if ⋃{g(f(x,j)) | 1 ≤ j ≤ l} ⊄ C_n then
    C_n := C_n ∪ ⋃{g(f(x,j)) | 1 ≤ j ≤ l}
    x := ε
  elsif x ∈ A ⟺ l is even then
    C_n := C_n ∪ g(f(x, l+1))
    x := ε
  else x := succ(x) (* in lexicographical order *)
  end
until x = 0^{n+1}
output χ^{C_n}_{≤n}
```

Claim 13.
1. *C_n is a subset of C for all $n \geq 0$.*
2. *For all $n \geq 0$, A reduces to C_n via f and g for all instances $x \in \Sigma^{\leq n}$.*

Observe that $|C_n| = 2^{O(n)}$, and therefore the algorithm can be performed in time $2^{O(n)}$. As a consequence, it is easy to see that there exists a constant c such that for every n, $\chi_{\leq n}^{C_n} \in KS\left[c, 2^{cn} \,\middle|\, \chi_{\leq n}^{A}\right]$. ∎

Since the many-one, conjunctive, (bounded) Hausdorff, and *bhd-c* reducibilities are all special cases of the composed Hausdorff and conjunctive reducibilities, we can state the following corollary.

Corollary 14. *The many-one, conjunctive, (bounded) Hausdorff, and bhd-c reducibilities as well as the non-monotone Hausdorff reducibility are reliable.*

As shown in [4], also the *co-np* many-one reducibility, the *co-rp* many-one reducibility, and the reducibility obtained by composing the Hausdorff, the *co-rp* many-one, and the conjunctive reducibilities are reliable.

5 On reliable reductions to HIGH sets

In this section we prove that no set A in EXPSPACE/poly reduces via a reliable reduction to a set B in HIGH unless A is reducible via a reduction of the same type to a sparse set. In order to give an intuitive explanation how the proof works, consider the case that A reduces to B via a many-one reduction function f. Since the set A is of relatively low space-bounded Kolmogorov complexity, membership in B can be decided for all the instances in the range of f using only a relatively small amount of resources (assuming that f is honest). Therefore, the range of f cannot be too large since otherwise a substantial part of the characteristic sequence of B would contain only little information, contradicting the fact that B is in HIGH.

For the proof we need two lemmas that are of independent interest. Intuitively spoken, the following lemma shows that the Kolmogorov complexity of a string cannot be very high, if "many" 1's of the string are easily computable. For $b \in \{0,1\}$ let $\#_b(y)$ be the number of bits equal to b in the string y. Further, let $\preceq$ be the partial ordering on Σ^* defined by $a_1 \ldots a_k \preceq b_1 \ldots b_l$, if $k \leq l$ and $a_i \leq b_i$, for $i = 1, \ldots, k$.

Lemma 15. *There exists a constant c^* such that for all $c, d \in \mathbb{N}$ and for all $x, y \in \Sigma^*$, if $x \preceq y$ and $x \in KS[d, c]$, then $y \in KS[2\log|d| + d + |y| - \#_1(x) + c^*, \max(c, c^* \cdot |y|)]$.*

The second lemma that we need states that if in the reduction of a set A to some set B the number of positively answered queries on instances of length n is polynomially bounded, then the given reduction can be modified (by padding the queries) to yield a reduction from A to a sparse set.

Lemma 16. *Assume that a set A reduces via a given reduction to some set. If there is a polynomial r such that for every $n \geq 0$ there is a set S_n of cardinality $|S_n| \leq r(n)$ such that all length n instances for A are correctly reduced to S_n, then A reduces via the same type of reduction (that is, only the queries need to be padded) to a sparse set $\hat{S}$.*

Now we are ready to prove the main result of this section. It states that in order to decide a set in EXPSPACE/poly, all the information that a reliable reduction is able to extract out of a set containing a lot of information can be provided just as well by a sparse set (which is of very low information content).

Theorem 17. *If a set $A \in$ EXPSPACE/poly reduces via a reliable reduction to a set $B \in$ HIGH, then A reduces via the same type of reduction (that is, only the queries need to be padded) to a sparse set.*

Proof Sketch. Let $A \in$ EXPSPACE/poly, $B \in$ HIGH and assume that A reduces to B via a reliable reduction. Intuitively, using the reliability of the reduction and the fact that $A \in$ EXPSPACE/poly, it is possible to compute within the resource bounds provided by the complexity class EXPSPACE/poly a sequence of sets $B_n \subseteq B$ such that for all instances up to length n, A reduces correctly to B_n. Since $B \in$ HIGH, the number of 1's in the characteristic sequences of the sets B_n cannot be large, and therefore, as we will see, A reduces to a sparse set. In the sequel we give the formal proof.

Since $A \in$ EXPSPACE/poly, there exists a polynomial s such that for every n, $\chi^A_{\leq n} \in KS[s(n), 2^{s(n)}]$. Also, since A reduces to B via a reliable reduction, there are a polynomial s' and a sequence $B_n \subseteq B$, $n \geq 0$, of sets such that for every $n \geq 0$, all instances for A up to length n are correctly reduced (via the given reduction) to B_n, and $\chi^{B_n}_{\leq n} \in KS\left[s'(n), 2^{s'(n)} \,|\, \chi^A_{\leq n}\right]$. As an immediate consequence of the reliability of the reduction from A to B and the fact that $A \in$ EXPSPACE/poly we have (1) $\chi^{B_n}_{\leq n} \in KS[p(n), 2^{p(n)}]$ for some polynomial p and every $n \geq 0$. Next we show that there is a polynomial q such that (2) $census_{B_n}(1^n) \leq q(n)$ for all n. Assume otherwise, then for every polynomial q there are infinitely many n such that $census_{B_n}(1^n) > q(n)$. Since $B_n \subseteq B$ it follows by (1) above and by Lemma 15 (letting $x = \chi^{B_n}_{\leq n}$ and $y = \chi^B_{\leq p(n)}$) that there is a constant c^* such that for every polynomial q there are infinitely many n for which $\chi^B_{\leq p(n)} \in KS[2\log|p(n)| + p(n) + 2^{p(n)} - q(n) + c^*, c^* \cdot 2^{p(n)}]$, contradicting the fact that $B \in$ HIGH.

To complete the proof of the theorem let r be a (non-decreasing) polynomial bounding the length of the queries of the given reliable reduction from A to B. Then for all length n instances, A reduces correctly to the set $B^{\leq r(n)}_{r(n)}$. By (2) above it follows that $census_{B_{r(n)}}(1^{r(n)}) \leq q(r(n))$ for all n, and thus the theorem follows by Lemma 16. ∎

As a consequence of the preceding theorem and the reliability results of Section 4 we obtain the following theorem.

Theorem 18. *Let A be in* EXPSPACE/poly *and let α be one of the following reducibility types:*

non-monotone Hausdorff reducibility, bhd-c reducibility, co-np many-one reducibility, or the composition of the bhd, co-rp many-one, and conjunctive reducibilities.

Then A is in R_α(HIGH) if and only if A is in R_α(SPARSE).

By Theorem 17 we know that the existence of hard sets in HIGH for any complexity class $\mathcal{C} \subseteq$ EXPSPACE/poly with respect to a reliable reducibility implies the existence of a sparse hard set for $\mathcal{C}$ with respect to that reducibility. Thus the existence of hard sets in HIGH leads to the same consequences as the existence of sparse hard sets. Using Theorems 10 and 18, and results of [2] we get

Corollary 19.

1. *Let $\mathcal{C}$ be a class from* {UP, NP, $C_=$P, PP}. *If $\mathcal{C} \subseteq R^p_{bhd}(R^p_c(\text{HIGH}))$, then $\mathcal{C} =$* P.
2. *For $\mathcal{C} \in$* {NP, PSPACE}, *if $\mathcal{C} \subseteq R^p_{nhd}(\text{HIGH})$, then $\mathcal{C}$ is low for Δ^p_2.*
3. *If* NP *is contained in $R^p_{bhd}(R^{co\text{-}rp}_m(R^p_c(\text{HIGH})))$, then* NP = RP.

Acknowledgements. The authors thank Ricard Gavaldà, Montse Hermo and Elvira Mayordomo for helpful comments.

References

1. V. Arvind, Y. Han, L.A. Hemachandra, J. Köbler, A. Lozano, M. Mundhenk, M. Ogiwara, U. Schöning, R. Silvestri, and T. Thierauf. Reductions to sets of low information content. *Proceedings of the 19th ICALP*, Lecture Notes in Computer Science, #623:162-173, Springer Verlag, 1992.
2. V. Arvind, J. Köbler, and M. Mundhenk. On bounded truth-table, conjunctive, and randomized reductions to sparse sets. In *Proceedings 12th Conference on the Foundations of Software Technology & Theoretical Computer Science,* Lecture Notes in Computer Science, #652:140–151, Springer Verlag, 1992.
3. V. Arvind, J. Köbler, and M. Mundhenk. Lowness and the complexity of sparse and tally descriptions. In *Proceedings Third International Symposium on Algorithms and Computation,* Lecture Notes in Computer Science, #650:249–258, Springer Verlag, 1992.
4. V. Arvind, J. Köbler, and M. Mundhenk. Reliable reductions, high sets and low sets. Technical Report, Universität Ulm, *Ulmer Informatik-Bericht 92-19*, 1992.
5. J. Balcazár. Self-reducibility. *Journal of Computer and System Sciences*, 41:367-388, 1990.
6. J.L. Balcázar, J. Díaz, and J. Gabarró. *Structural Complexity I.* EATCS Monographs on Theoretical Computer Science, Springer-Verlag, 1988.
7. R. Book and J. Lutz. On languages with very high information content. *Proceedings of the 7th Structure in Complexity Theory Conference*, 255-259, IEEE Computer Society Press, 1992.
8. R. Karp and R. Lipton. Some connections between nonuniform and uniform complexity classes. *Proceedings of the 12th ACM Symposium on Theory of Computing,* 302-309, April 1980.
9. K. Ko. On self-reducibility and weak p-selectivity. *Journal of Computer and System Sciences*, 26:209-221, 1983.
10. R. Ladner, N. Lynch, and A. Selman. A comparison of polynomial time reducibilities. *Theoretical Computer Science*, 1(2):103-124, 1975.
11. M. Li and P. Vitanyi. *An introduction to Kolmogorov complexity and its application.* Addison-Wesley, 1992.
12. A. Lozano and J. Torán. Self-reducible sets of small density. *Mathematical Systems Theory*, 24:83-100, 1991.
13. S. Mahaney. Sparse complete sets for NP: solution of a conjecture of Berman and Hartmanis. *Journal of Computer and System Sciences*, 25(2):130-143, 1982.
14. M. Ogiwara and O. Watanabe. On polynomial-time bounded truth-table reducibility of NP sets to sparse sets. *SIAM Journal on Computing*, 20(3):471-483, 1991.
15. D. Ranjan and P. Rohatgi. Randomized reductions to sparse sets. *Proceedings of the 7th Structure in Complexity Theory Conference*, IEEE Computer Society Press, 239-242, 1992.
16. K.W. Wagner. More complicated questions about maxima and minima, and some closures of NP. *Theoretical Computer Science*, 51:53-80, 1987.
17. K.W. Wagner. Bounded query classes. *SIAM Journal on Computing*, 19(5):83-846, 1990.
18. G. Wechsung and K.W. Wagner. On the boolean closure of NP. Manuscript. (Extended abstract by: G. Wechsung, On the boolean closure of NP, in *Proc. International Conference on Fundamentals of Computation Theory*, Lecture Notes in Computer Science, #199:485–493, Springer-Verlag, 1985.)

Stores as Homomorphisms and their Transformations*

Egidio Astesiano – Gianna Reggio – Elena Zucca

Università di Genova – Dipartimento di Informatica e Scienze dell'Informazione
Viale Benedetto XV,3 16132, Genova, Italy
astes, reggio, zucca @ disi.unige.it

Introduction

In the classical denotational model of imperative languages (see e.g. [7], Chap. 7.3) handling structured types, like arrays, requires an ad-hoc treatment for each data type, including e.g. an ad-hoc allocation and deallocation mechanism. Our aim is to give a homogeneous approach that can be followed whichever is the data structure of the language.

We start from the traditional model for Pascal-like languages, which uses a notion of store as a mapping from left values (containers for values, usually called *locations*), into right values; we combine this idea with the well-known algebraic approach for modelling data types. More precisely, we consider an algebraic structure both for the right and the left values; consequently, the store becomes a homomorphic mapping of the left into the right structure.

Seeing a store as a homomorphism has a number of interesting consequences. First of all, the transformations over a store can be uniformly and rigorously defined on the basis of the principle that they are minimal variations compatible with some basic intended effect (e.g., some elementary substitution). Thus semantic clauses too, which rely on these transformations as auxiliary functions, can be given uniformly; for example, we can give a unique clause for assignment for any data type in Pascal and Ada-like languages.

In Sect. 1 we present the problem and outline the solution, while in Sects. 2 and 3 we give the formal model; in the conclusion we mention some related work.

An extended presentation of the ideas of this paper, including proofs, together with many examples of application, is given in [2].

1 A Motivating Example

As stated in [7], Chap. 5, *imperative languages* can be informally defined as languages that utilize the *store*, which is an abstraction of the computer's memory, and include syntactic constructs (usually called *commands*) whose semantics is, roughly speaking, a store transformation. The most simple example is an assignment like e.g. `x:=y+1` .

* This work has been supported by IS-CORE-Esprit-BRA-W.G. n. 6071 and the project MURST 40% "Metodi e specifiche per la concorrenza".

In the classical denotational model (here and in the sequel we follow in the essence [7] Chap. 7) the store is formalized as a mapping from containers for values, usually called *locations*, into values (like integers). Formally

$Store = [Loc \rightarrow Val]_{fin}$.[2]

This definition is slightly different from the traditional one for distinguishing the case of a location which is unused in the current store ($l \notin Dom(\sigma)$) from the case of a location which is in use, but not yet initialized ($l \in Dom(\sigma)$, $\sigma(l)$ undefined).

The effect of an assignment like above is, roughly speaking, to add to the current store an association from the location corresponding to x to the value obtained evaluating y+1. The formalization depends on the overall semantics of the language. For example, the typical model based on environment and store formalizes the effect of a command as a function which, for a given environment, returns a store transformation.

$Env = [Id \rightarrow Den], \qquad Den = Loc + \ldots$

$\mathcal{C}\colon Com \rightarrow [Env \rightarrow [Store \rightarrow Store]]$.

The semantics of the above assignment is as follows:

$\mathcal{C}[\![\mathtt{x} := \mathtt{y} + 1]\!] \rho\sigma = \sigma[\sigma(\rho(\mathtt{y})) +^{\mathbb{Z}} 1/\rho(\mathtt{x})]$.

Introducing two different semantic functions for left and right expressions, i.e. expressions which may appear in the left-hand (resp. right-hand) side of an assignment, the above clause can be obtained as an instance of the following general clause (+) for assignment:

$$(+)\ \mathcal{C}[\![lexpr := expr]\!] \rho\sigma = \sigma[\mathcal{E}[\![expr]\!] \rho\sigma / \mathcal{LE}[\![lexpr]\!] \rho\sigma]$$

where

$\mathcal{LE}\colon LExpr \rightarrow [Env \rightarrow [Store \rightarrow Loc]], \qquad \mathcal{E}\colon Expr \rightarrow [Env \rightarrow [Store \rightarrow Val]]$;

of course in the case of an identifier we have

$\mathcal{LE}[\![id]\!] \rho\sigma = \rho(id), \qquad \mathcal{E}[\![id]\!] \rho\sigma = \sigma(\rho(id))$.

Let us consider the case of compound data structures ([7] 7.3), for example an array declaration like

```
type a = array [1..10] of int.
```

A variable identifier of this type, say `arr`, has a denotation in the environment which is, accordingly with the intuition, a partial function from indexes into integer locations. However, the store remains a mapping from locations of basic types (like `int`) into basic values; no associations are introduced in the store for compound types.

Hence, an assignment to `arr` cannot be modelled using the general schema (+), but is actually expanded to ten assignments, one for each of the components:

$$\mathcal{C}[\![\mathtt{arr} := \mathrm{expr}]\!] \rho\sigma = \sigma[z_1/\rho(\mathtt{arr})[1]] \ldots [z_{10}/\rho(\mathtt{arr})[10]],$$
$$\text{if } (\mathcal{E}[\![\mathrm{expr}]\!] \rho\sigma)(i) = z_i, \text{ for } i = 1, \ldots, 10.$$

Analogously, allocation and deallocation for a variable identifier of an array type cannot be modelled following a general schema; allocation for `arr` is expanded to ten allocations of integer locations (see [7] for the details).

[2] Here and in the following $[A \rightarrow B]_{fin}$ denotes the set of the partial functions from a finite subset of A into B.

What we look for in this paper is a general and more abstract model, which allows to treat assignment, allocation and deallocation in a uniform way for any data structure, thus providing a basis for a systematic approach to proving semantic properties.

We start from the idea of modelling a data type as a (many-sorted partial) algebra. That means that, for example, a language operator like the array selector $\_[\_]$, which combines an expression of type a and an expression of type `[1..10]`, giving an expression of type `int`, has two different semantic counterparts: an operation which takes a location of type a, an index in $\{1,\ldots,10\}$ and gives an integer location, and an operation which takes a value of type a, an index in $\{1,\ldots,10\}$ and gives an integer. That is formalized by giving a signature Σ_a in which we distinguish left and right sorts, and a corresponding algebra A, as shown in Fig. 1.

sig $\Sigma_a =$
  sorts *L-int, R-int, L-a, R-a, ind*
  opns
    $\_+\_, \_\times\_$: *R-int R-int* $\rightarrow$ *R-int*
    $\_[\_]_L$: *L-a ind* $\rightarrow$ *L-int*
    $\_[\_]_R$: *R-a ind* $\rightarrow$ *R-int*

$A_{L\text{-}int} = Loc_{int} \cup \{l[i] \mid l \in Loc_a, i \in A_{ind}\}$, $\quad A_{R\text{-}int} = \mathbb{Z}$
$A_{L\text{-}a} = Loc_a$, $\quad A_{R\text{-}a} = [A_{ind} \rightarrow A_{R\text{-}int}]$
$A_{ind} = \{1,\ldots,10\}$
where Loc_a, Loc_{int} are two infinite denumerable sets
$\_+\_^A$, $\_\times\_^A$ are the usual sum and product in $\mathbb{Z}$
$\_[\_]_L^A(l,i) =_\perp l[i]$, for each $l \in A_{L\text{-}a}$, $i \in A_{ind}$
$\_[\_]_R^A$ is the usual function application.

Fig. 1. Left-right algebra for the type a

Consider now the store σ. Having different sorts, σ is a sort-indexed family of maps $\{\sigma_a, \sigma_{int}\}$, from left to right values of the corresponding sorts; moreover we have some consistency requirements. First, whenever a location of type a is in use, all its subcomponents of type `int` are in use, too, and conversely. Formally:

(*) for each $l \in Loc_a$, $i \in \{1,\ldots,10\}$, $l[i] \in Dom(\sigma)$ iff $l \in Dom(\sigma)$.

Moreover, it is easy to see that each store σ must satisfy the condition

(**) $\sigma_{int}(l[i]) =_\perp \sigma_a(l)[i]$, for each $i \in \{1,\ldots,10\}$,

where $=_\perp$ denotes strong equality (either the two sides are defined and equal or both are undefined).

The properties (*) and (**) can be expressed in a general way for any data type, by structuring the domain of the store $Dom(\sigma)$ as an algebra, and seeing σ as a structure preserving mapping, i.e. a homomorphism.

In this way, the property (*) above can be generalized requiring that $Dom(\sigma)$ is a strong subalgebra of A; the property (**) corresponds to requiring that by extending the store by the identity over right values, we get a partial homomorphism from $Dom(\sigma)$ into the restriction of A to only right sorts and operations (refer to Sect. 2 for the detailed definitions).

A major consequence of seeing the store as homomorphism is the possibility of qualifying the store transformations that can occur in a program execution in a way that it is independent of the particular data structure. Considering for example substitution: the basic intended effect is that a new association is added from a used location, say l of type a, into a right value, removing any preceding association with l. As a consequence, in order to keep the homomorphic structure of the store, a new association is added also for each location which is a subcomponent of l, say $l[1]$, ..., $l[10]$. Now we can define substitution essentially as the minimal variation of the store which has the above intended effect and is compatible with its homomorphic structure, i.e. gives a new store which is still a homomorphism. Analogously for allocation, deallocation and alike (the definitions are in Sect. 3).

Then we get immediately two important applications.

- We can provide uniform semantic clauses, independently of the data type, since we can use the global definition of the store transformations as auxiliary functions. For example the assignment clause takes the general form
  $$\mathcal{C}[\![\mathit{lexpr} := \mathit{expr}]\!] \rho\sigma = \sigma[\mathcal{E}[\![\mathit{expr}]\!] \rho\sigma // \mathcal{LE}[\![\mathit{lexpr}]\!] \rho\sigma]$$
  where now $\_[\_//\_]$ denotes substitution (in the sense described before and formally given in Sect. 3).
- For every data type we can check whether the explicit definition of the store transformations is correct, in the sense that the resulting transformations, usually given by a series of detailed clauses, are the same as that given by general definition.

2 Stores as Homomorphisms

Before formally introducing stores, we have to define the overall algebraic structure for left and right values, that we call a *left-right algebra*. That is an algebra over a particular kind of signature, that we call *left-right static signature*, which, informally, provides two different kinds of value types: the types of the values which can be stored, called *left-right types* (modelled by two sorts, one for the actual values, called right, and one for the corresponding locations, called left), and the types of the values which cannot be stored, called *right types* (modelled by only the right sort).

Correspondingly, there are three different kinds of operations: pairs of operations returning a left value and a right value in a "corresponding" way (e.g. array selectors); operations returning left values for which there is no right analogous (e.g. an operation which, given a location, returns the next location in the store); operations returning right values for which there is no left analogous (e.g. integer sum, product and so on).

Definition 1. A *(left-right) static signature* is a 5-tuple $ST\Sigma = (T, RT, OP, LOP, ROP)$ where:

- T (left-right types), RT (right types) are two sets of symbols; let
  $LSorts(ST\Sigma) = \{L\text{-}t \mid t \in T\}$,
  $RSorts(ST\Sigma) = RT \cup \{R\text{-}t \mid t \in T\}$,
  $Sorts(ST\Sigma) = LSorts(ST\Sigma) \cup RSorts(ST\Sigma)$
  be the sets of the *left sorts*, *right sorts* and *sorts* of $ST\Sigma$, respectively;
- $OP = \{OP_{w,L\text{-}t}\}_{w \in Sorts(ST\Sigma)^*, t \in T}$;
- $LOP = \{LOP_{w,L\text{-}t}\}_{w \in Sorts(ST\Sigma)^*, t \in T}$;
- $ROP = \{ROP_{w,rs}\}_{w \in RSorts(ST\Sigma)^*, rs \in RSorts(ST\Sigma)}$.

Let in what follows $ST\Sigma = (T, RT, OP, LOP, ROP)$ be a (left-right) static signature; then
  $OP_L = \{op_L : s_1 \ldots s_n \to L\text{-}t \mid op \in OP_{s_1 \ldots s_n, L\text{-}t}\}$,
  $OP_R = \{op_R : rs_1 \ldots rs_n \to R\text{-}t \mid op \in OP_{s_1 \ldots s_n, L\text{-}t}\}$,
where $rs_i = s_i$ if $s_i \in RSorts(ST\Sigma)$, $R\text{-}t_i$ if $s_i = L\text{-}t_i$.

Note that LOP models left operations with no right analogous, while OP_L models the left version of operations having also the right one; the same difference holds between ROP and OP_R.

It is easy to see that $ST\Sigma$ uniquely determines a usual many-sorted signature, denoted by $WholeSig(ST\Sigma)$, defined as the pair

$$<Sorts(ST\Sigma), LOP \cup OP_L \cup OP_R \cup ROP>.$$

Moreover, we associate with $ST\Sigma$ other two signatures (which are subsignatures of $ST\Sigma$) keeping only the operations which must be preserved by the homomorphic structure of the stores, in the left and right version respectively:
  $LSig(ST\Sigma) = <Sorts(ST\Sigma), OP_L>$, the *left part* of $ST\Sigma$,
  $RSig(ST\Sigma) = <RSorts(ST\Sigma), OP_R>$, the *right part* of $ST\Sigma$.
Finally, we denote by $\phi_{ST\Sigma}$ the signature morphism from $LSig(ST\Sigma)$ into $RSig(ST\Sigma)$ which maps right sorts into themselves, left sorts and operations into corresponding right sorts and operations.

Definition 2. A *left-right $ST\Sigma$-algebra* is a partial algebra A over $WholeSig(ST\Sigma)$.

If A is a left-right $ST\Sigma$-algebra, then the *left* (resp. *right*) part of A, denoted by A^L (resp. A^R), is the restriction of A to $LSig(ST\Sigma)$ (resp. $RSig(ST\Sigma)$).

An element l belonging to $A_{L\text{-}t}$ is called a ***principal left value*** iff there exist no $op_L : s_1 \ldots s_n \to L\text{-}t$ in OP, $a_1 \in A_{s_1}, \ldots, a_n \in A_{s_n}$ s.t. $l = op_L{}^A(a_1, \ldots, a_n)$. Intuitively, principal left values are left values which are not subcomponents of other left values. For each left-right type t, let $Loc^A{}_t$ denote the set of the principal left values in $A_{L\text{-}t}$.

We define now ***stores***. Roughly speaking a store is a mapping from (currently existing) locations into right values, satisfying some consistency requirements. More precisely: existing locations consist in a finite family of principal locations together with all their subcomponents (compare Sect. 1 (*)); the associations from left into right values respect the operations (compare Sect. 1 (**)). These requirements can be formally expressed as below.

Definition 3. If A is a left-right $ST\Sigma$-algebra, $\Sigma = WholeSig(ST\Sigma)$, then a *store* of A is a Σ-homomorphism $\sigma: D \rightarrow A^R|_{\phi_{ST\Sigma}}$ which satisfies the following assumptions. Set $Loc^D{}_t = D_{L\text{-}t} \cap Loc^A{}_t$, for all $t \in T$.

1. $Loc^D{}_t$ is finite, for all $t \in T$;
2. D is the subalgebra of A^L generated by $Loc^D \cup \{A_{rs}\}_{rs \in RSorts(ST\Sigma)}$;
3. $\sigma_{rs} = Id_{A_{rs}}$ (the identity of A_{rs}), for all $rs \in RSorts(ST\Sigma)$.

We denote by A_{store} the set of the stores of A.

Here above $A^R|_{\phi_{ST\Sigma}}$ denotes the reduct of A^R w.r.t. the signature morphism $\phi_{ST\Sigma}$.

Due to the requirements that a store must satisfy, it turns out that it is uniquely determined by fixing which are the currently existing principal locations and their associated right values. Hence it is more convenient to introduce a notion of *store kernel* which consists in a mapping from a finite set of principal locations into right values. In this way a store can be defined as the minimal mapping which extends a store kernel and moreover satisfies the consistency requirements, i.e. conditions 1, 2, 3 above.

Definition 4. If A is a left-right $ST\Sigma$-algebra, then a *(store) kernel of* A is a T-family of partial functions $\kappa = \{\kappa_t\}_{t \in T}$ s.t., for all $t \in T$, $\kappa_t \in [Loc^A{}_t \rightarrow A_{R\text{-}t}]_{\mathsf{fin}}$. We denote by A_{kernel} the set of the kernels of A.

Proposition 5. *There exists a bijective correspondence* $\overline{}: A_{\mathsf{kernel}} \rightarrow A_{store}$ *which associates with each kernel* κ *the store* $\overline{\kappa}$ *generated by* κ.

3 Store Transformations

The purpose of a left-right algebra is to give the algebraic structure of all intermediate configurations in the execution of an imperative program; each store models one configuration. In order to have a complete model of the execution, we must add *dynamic operations*, i.e. operations which model store transformations.

Definition 6. A *left-right signature* is a pair $\Sigma = (ST\Sigma, DOP)$ where:

- $ST\Sigma$ is a left-right static signature;
- DOP is an $S^* \times S$-family of symbols called *dynamic operation symbols*, where $S = Sorts(ST\Sigma)$; if $dop \in DOP_{s_1 \ldots s_n, s}$, then we write $dop: s_1 \ldots s_n \Rightarrow s$.

Let in what follows $\Sigma = (ST\Sigma, DOP)$ be a left-right signature.

Definition 7. A *left-right structure* over Σ is a pair $LRS = (A, \{dop^{LRS}\}_{dop \in DOP})$ where:

- A is a left-right $ST\Sigma$-algebra;
- for each $dop: s_1 \ldots s_n \Rightarrow s$,
  $dop^{LRS}: A_{store} \times A_{s_1} \times \ldots \times A_{s_n} \rightarrow A_{store} \times A_s$.

Dynamic operations returning just a store can be obtained by adding a dummy sort whose carrier is a singleton and are denoted by $dop: s_1 \dots s_n \Rightarrow$.

Note that, due to Prop. 5 above, in order to define a dynamic operation, it is sufficient to define a corresponding operation which acts on kernels. Formally, let $f: A_{\text{kernel}} \times A_{s_1} \times \dots \times A_{s_n} \to A_{\text{kernel}} \times A_s$ be an operation which acts on kernels; then we define $\overline{f}: A_{\text{store}} \times A_{s_1} \times \dots \times A_{s_n} \to A_{\text{store}} \times A_s$ by $\overline{f}(\overline{\kappa}) = \overline{f(\kappa)}$.

We define now a family of dynamic operations sufficient for modelling store transformations in Pascal-like languages, by giving the corresponding ones on the kernels.

Let in what follows $LRS = (A, \{dop^{LRS}\}_{dop \in DOP})$ be a left-right structure.

Definition 8. The *empty kernel*, denoted by $\emptyset$, is the kernel with empty domain.

Fact 9. *The store generated by* $\emptyset$ *is the store with empty domain.*

Definition 10. For each $t \in T$, let new_t denote the predicate over $A_{\text{kernel}} \times Loc^A{}_t$ defined by: $\text{new}_t(\kappa, l)$ holds iff $l \notin Dom(\kappa)$.

Definition 11. For each $t \in T$, the *extension* (or *allocation*) operation of type t is the function

$\_ +_t \_ : A_{\text{kernel}} \times Loc^A{}_t \to A_{\text{kernel}}$

defined as follows:

$\kappa +_t l = \kappa'$ if $\text{new}_t(\kappa, l)$ holds, undefined otherwise where:

$Dom(\kappa') = Dom(\kappa) \cup \{l\}$; $\qquad Graph(\kappa') = Graph(\kappa)$.

Definition 12. For each $t \in T$, the *restriction* (or *deallocation*) operation (of type t) is the function

$\_ \backslash_t \_ : A_{\text{kernel}} \times Loc^A{}_t \to A_{\text{kernel}}$

defined as follows:

$\kappa \backslash_t l = \kappa'$, where

$Dom(\kappa') = Dom(\kappa) - \{l\}$; $\qquad \kappa'(l') =_\perp \kappa(l')$, for each $l' \in Dom(\kappa')$.

For defining in a general way the substitution operation, we need to assume that, for each non principal location, say l, l can be obtained in a unique way as a subcomponent of a unique principal location, say l' (that implies in particular that each operation in OP has only an argument of left sort); moreover, changing the right value associated with l uniquely determines a corresponding change of the right value associated with l'.

These assumptions allow to uniquely define the store transformation induced by updating whatever location; actually less restrictive assumptions would be sufficient (allowing more arguments of left sorts in operations in OP), but the above version allows a simpler formalization, and is satisfied by all usual imperative languages, as Pascal, Algol, Ada, Common Lisp and so on.

Assumption Upd1. The operations in OP are all of the form

$op: L\text{-}t\ rs_1 \dots rs_n \to L\text{-}t'$, with $t, t' \in T, rs_i \in RSorts(ST\Sigma)$ for $i = 1, \dots, n$.

For formally expressing the second assumption, we need some technical definitions.

Definition 13. Let A be a left-right $ST\Sigma$-algebra.

For each operation $op: L\text{-}t\ rs_1 \dots rs_m \to L\text{-}t' \in OP$, $r_1 \in A_{rs_1}, \dots, r_m \in A_{rs_m}$, we say that $sel = op[r_1, \dots, r_m]$ is a *selector* from t into t' and define:

- $sel_L: A_{L\text{-}t} \to A_{L\text{-}t'}$, $sel_L(l) = op_L(l, r_1, \ldots, r_m)$;
- $sel_R: A_{R\text{-}t} \to A_{R\text{-}t'}$, $sel_R(r) = op_R(r, r_1, \ldots, r_m)$.
- $l \xrightarrow{sel} l'$ iff $l' = sel_L(l)$;
- $l \xrightarrow{sel_1 \cdot \ldots \cdot sel_n} l'$ iff there exist $l_1, \ldots, l_n$ s.t. $l \xrightarrow{sel_1} l_1 \to \ldots \to l_{n-2} \xrightarrow{sel_{n-1}} l_{n-1} \xrightarrow{sel_n} l_n = l'$, $(n \geq 0)$.

Assumption Upd2. For each $t \in T$, $l \in A_{L\text{-}t}$, there exist unique $t' \in T$, $l' \in Loc^A{}_{t'}$, $sel_1 \cdot \ldots \cdot sel_n$ selector list s.t. $l' \xrightarrow{sel_1 \cdot \ldots \cdot sel_n} l$.

Assumption Upd3. For each selector sel from t into t', we assume that there exist two functions:

$\text{upd}(sel): A_{R\text{-}t} \times A_{R\text{-}t'} \to A_{R\text{-}t}$ $\quad$ $\text{upd}^{\perp}(sel): A_{R\text{-}t'} \to A_{R\text{-}t}$ such that:

- $sel_R(\text{upd}(sel)(r, r')) = r'$; $\quad$ $sel_R(\text{upd}^{\perp}(sel)(r')) = r'$;
- for each sel' selector from t into t'', $sel' \neq sel$,
  $sel'_R(\text{upd}(sel)(r, r')) =_{\perp} sel'_R(r)$ $\quad$ $sel'_R(\text{upd}^{\perp}(sel)(r'))$ undefined.

The assumption above informally means that for each selector sel from t into t' (e.g. _[1] which takes the first element of an array), there exist two corresponding operations:

- upd(sel) which models the effect of changing the sel-component of a value of type t (e.g. _[_/1] which returns the array obtained updating the first element);
- upd$^{\perp}$(sel) which constructs a value of type t from a value of type t' (e.g. $\emptyset$[_/1] which returns an array having only the first element).

These two functions can be naturally extended to lists of selectors (we omit the formal definition).

Definition 14. For each $t \in T$, the ***substitution operation*** (of type t) is the function

$\_[\_//\_]_t: A_{\text{kernel}} \times A_{R\text{-}t} \times A_{L\text{-}t} \to A_{\text{kernel}}$

defined as follows:

$\kappa[r//l]_t = \kappa'$ if there exist $l' \in Dom(\kappa)$, $sel\text{-}list$ list of selectors s.t. $l' \xrightarrow{sel\text{-}list} l$, undefined otherwise, where

$Dom(\kappa') = Dom(\kappa)$;
$\kappa'(l') = \text{upd}(sel\text{-}list)(\kappa(l'), r)$, if $\kappa(l')$ is defined;
$\kappa'(l') = \text{upd}^{\perp}(sel\text{-}list)(r)$, if $\kappa(l')$ is undefined;
$\kappa'(l'') = \kappa(l'')$, for each $l'' \neq l'$.

Note that if there exists l' and $sel\text{-}list$, then they are unique by assumption **Upd2**.

Below we give an explicit definition of the three basic dynamic operations as acting on the stores, in the case that assumptions **Upd1**, **Upd2** and **Upd3** hold.

In the following we write $l \to l'$ ($l \not\to l'$) iff there exists (resp. there does not exist) a list of selectors $sel\text{-}list$ s.t. $l \xrightarrow{sel\text{-}list} l'$.

Fact 15. *For each $t \in T$, $\sigma \in A_{\text{store}}$ and $l \in Loc^A{}_t$ s.t.* $\text{new}_t(ker(\sigma), l)$ *holds,* $ker(\sigma) +_t l = \sigma'$, *where:*
$Dom(\sigma') = Dom(\sigma) \cup \{l' \mid l \to l'\}$; $Graph(\sigma') = Graph(\sigma)$.

In other words, the store obtained by an allocation includes the new location with all its subcomponents and is unchanged elsewhere.

Fact 16. *For each $t \in T$, $\sigma \in A_{store}$, $l \in Loc^A{}_t$,*
$\overline{ker(\sigma)}\backslash_t l = \sigma'$, *where*
$Dom(\sigma') = Dom(\sigma) - \{l' \mid l \to l'\}$; $\sigma'(l') =_\perp \sigma(l')$, *for each* $l' \in Dom(\sigma')$.

In other words, the store obtained by a deallocation keeps only the locations which are not subcomponents of the deleted location with their associated right values.

Proposition 17. *For each $t \in T$, $\sigma \in A_{store}$, $r \in A_{R\text{-}t}$, $l \in A_{L\text{-}t}$ s.t. $l \in Dom(\sigma)$,*
$\overline{ker(\sigma)}[r//l]_t = \sigma'$, *where σ' is inductively defined by*

$Dom(\sigma') = Dom(\sigma)$;
$\sigma'(l) = r$;
for each $l' \in Dom(\sigma)$ *s.t.* $l \not\to l'$, $l' \not\to l$, $\sigma'(l') =_\perp \sigma(l')$;
for each l' *s.t.* $l \to l'' \xrightarrow{sel} l'$, *if* $\sigma'(l'') = r''$, *then* $\sigma'(l') =_\perp sel_R(r'')$;
for each l' *s.t.* $l' \xrightarrow{sel} l'' \to l$;
if $\sigma'(l'') = r''$, $\sigma(l') = r'$, *then* $\sigma'(l') = \mathrm{upd}(sel)(r', r'')$;
if $\sigma'(l'') = r''$, $\sigma(l')$ *is undefined, then* $\sigma'(l') = \mathrm{upd}(sel)^\perp(r'')$.

In other words, the store obtained by a substitution is the minimal store which contains the new association and leaves unchanged all the unrelated locations (i.e., which are neither subcomponents of the updated location, nor conversely).

Allocation, deallocation and substitution can be considered the basic store transformations in usual imperative languages, in the sense that the final store obtained as the result of a program can be always obtained starting from the initial store (empty) by applying a finite sequence of these operations. That property is formally expressed here below.

Proposition 18. *Assume that LRS satisfies assumptions* **Upd1, Upd2** *and* **Upd3.** *Then the operations empty store, extension and substitution are a generating family for A_{store}, in the sense that the stores are the family inductively defined by:*

(1) $\emptyset \in A_{store}$;
(2) if $\sigma \in A_{store}$, *then* $\sigma + l \in A_{store}$, *for each principal left value l s.t.* $\mathrm{new}(\sigma, l)$ *holds;*
(3) if $\sigma \in A_{store}$, *then* $\sigma[r//l] \in A_{store}$, *for each* $l \in A_{L\text{-}t}$, $r \in A_{R\text{-}t}$, *s.t.* $l \in Dom(\sigma)$.

4 Conclusion

We have defined a mathematical structure, the left-right structure, which extends to the imperative case the usual algebraic framework for data types. That means fixing a structure for right values and for locations (left values); moreover the association from locations into values must respect this structure, i.e. the store is a homomorphism. For what concerns dynamics, the underlying data structure determines also which are the possible basic transformations of the store: these are modelled in turn as operations which involve the store and (either left or right) values, called dynamic operations.

The main result of the paper is to provide an abstract uniform setting for the semantics of programming languages, which in a sense rounds up and completes the well-known denotational approach. Our framework can be used also in the context of the wider approach of inductive semantics, advocated by the first of the authors; actually the first application of left-right structures has been shown in that context (see [1]).

There is an interesting issue that we have not treated here and will be exposed in some further paper, namely the relationship to the approaches whose early representative is the "evolving algebra" framework (see [4]). The distinguishing feature in our approach is the concept of store as homomorphism; however it should be possible to pass from one formalism to the other in some canonical way.

An important paper also dealing with L-values and R-values is [6], where however the main aim is the (functorial) treatment of the locality of variables and the issue of structured data types is not tackled.

We mention also a very recent paper on "mutation algebras" (see [5]), which has some similarity in the aims, but a completely different technique, without reference to the concept of a homomorphic store.

Finally, it is possible to describe also object based languages using left/right structures modelling objects in the same way of values of pointer types; nevertheless, this kind of model looks a too low level for objects. A better model which is a generalization of left/right structures is proposed in [3].

Acknowledgment We thanks the referee for various helpful comments and for reminding us the work in [6].

References

1. E. Astesiano. Inductive semantics. In *Formal Descriptions of Programming Concepts*, Berlin, 1991. Springer Verlag.
2. E. Astesiano, G. Reggio, and E. Zucca. Stores as homomorphisms and their transformations - A uniform approach to structured data types in denotational semantics. Technical report, 1993. Submitted.
3. E. Astesiano and E. Zucca. A semantic model for dynamic systems. In U.W. Lipeck and B. Thalheim, editors, *Fourth International Workshop on Foundations of Models and Languages for Data and Objects - Modelling Database Dynamics*, Volkse (Germany), October 1992.
4. Y. Gurevich. Evolving algebra, a tutorial introduction. *Bulletin of the EATCS*, (43):264–284, 1991.
5. G.T. Leavens and K.K. Dhara. A foundation for the model theory of abstract data types with mutation and aliasing. Technical Report 92-35, Department of Computer Science, Iowa State University, Amsterdam, November 1992.
6. J.C. Reynolds. The essence of Algol. In *Intl. Symp. on Algorithmic Languages*, Amsterdam, 1981. North-Holland.
7. D.A. Schmidt. *Denotational Semantics: a methodology for language development.* Wm. C. Brown Publishers, Duboque, Iowa, 1986.

Comparative Semantics for Linear Arrays of Communicating Processes

a study of the UNIX* fork and pipe commands

J.W. de Bakker[1,2], F. van Breugel[1,2,**] and A. de Bruin[3]

[1] Department of Software Technology, CWI, P.O. Box 4079, 1009 AB Amsterdam
[2] Department of Mathematics and Computer Science, Vrije Universiteit, P.O. Box 7161, 1007 MC Amsterdam
[3] Faculty of Economics, Erasmus Universiteit, P.O. Box 1738, 3000 DR Rotterdam

Abstract. Operational ($\mathcal{O}$) and denotational ($\mathcal{D}$) semantic models are designed for a language incorporating a version of the UNIX fork and pipe commands. Taking a simple while language as starting point, a number of programming constructs are added which achieve that a program can generate a dynamically evolving linear array of processes connected by channels. Over these channels sequences of values ('streams') are transmitted. Both $\mathcal{O}$ and $\mathcal{D}$ are defined as (unique) fixed point of a contractive higher order operator. This allows a smooth proof that $\mathcal{O}$ and $\mathcal{D}$ are equivalent. Additional features are the use of hiatons, and of the closely related syntactic resumptions and semantic continuations.

1 Introduction

We present a comparative semantic study of a simple imperative language $\mathcal{L}$ which features the construction of dynamically evolving linear arrays of communicating processes. Our investigation was in particular motivated by the UNIX fork and pipe commands which return in somewhat adapted form in $\mathcal{L}$.

Both operational ($\mathcal{O}$) - based on an SOS style transition system ([Plo81]) - and denotational ($\mathcal{D}$) semantics for $\mathcal{L}$ will be presented, and their equivalence will be established. Simple topological techniques will suffice for the mathematical underpinning of both models. In fact, Banach's fixed point theorem ([Ban22]) is all we need. ([BR92] gives an overview of more advanced uses of topological modelling.)

Forks and pipes occur in several papers on programming language design and application (forks, e.g., in [HSS91], pipes in [KK92]). *Semantic* studies focusing on these topics are scarce (e.g. [AW85, Ben82, Bru86, MA89, RS83, RS92]), and none of them develops *both* operational and denotational models. Accordingly, we see the *comparative* result as the main contribution of our paper.

In the remainder of this introduction we informally introduce $\mathcal{L}$, and present three simple examples of its use culminating in a version of the sieve of Eratosthenes. Sections 2 and 3 present the operational and denotational semantics, respectively. In the design of $\mathcal{O}$, arrays of processes are modelled using the concept of (nested) *resumptions*. For $\mathcal{D}$, *continuations* are an essential tool. In Section 4, we prove the equivalence of $\mathcal{O}$ and $\mathcal{D}$ using the unique fixed point proof principle from [KR90]. Let us mention one subtlety in the semantic models: in order to apply Banach's theorem, we require contractiveness at various instances. At appropriate points a version of Park's *hiaton* ([Par83]) is used to enforce contractiveness if this would not arise naturally.

We now present the syntax of $\mathcal{L}$. It is a simple imperative language with assignment, while statements and the like, to which three further constructs are added: **write** (e), **read** (v), and **fork** (v). The syntax for $\mathcal{L}$ follows

$$s ::= v := e \mid \textbf{skip} \mid \textbf{write}\,(e) \mid \textbf{read}\,(v) \mid \textbf{fork}\,(v) \mid s\,;s \mid \textbf{if}\ b\ \textbf{then}\ s\ \textbf{else}\ s\ \textbf{fi} \mid \textbf{while}\ b\ \textbf{do}\ s\ \textbf{od}.$$

In the sequel, a program in execution will be called a *process*. Each process has exactly one input channel and one output channel connected to it (see Figure 1). Execution of the *write statement* **write** (e) has the effect that the value of the expression e is written on the output channel, the effect of the *read statement*

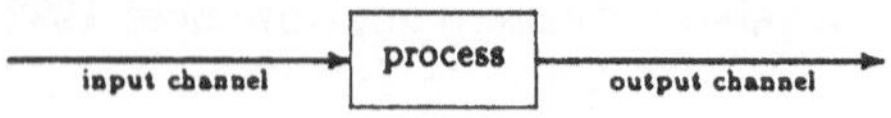

Figure 1

* UNIX is a trademark of Bell Laboratories.

** This work was partially supported by the Netherlands *Nationale Faciliteit Informatica* programme, project Research and Education in Concurrent Systems (REX).

read (v) is that a new value is read from the input channel which is then assigned to the variable v. If there are no more values on the input channel then the process blocks (terminates).

A process can be modelled by a function which takes an input stream as an argument and yields an output stream as a result. The input stream is the sequence of all values assumed to be preloaded on the input channel, and the output stream is the sequence of all values to be written by the process on the output channel. Both streams can very well be infinite, and this means that nonterminating processes are meaningful in this setting. We give as first example, a '2-filter' described by the program

```
while true
do read (v);
    if v mod 2 ≠ 0 then write (v) else skip fi
od.
```

This program filters all even numbers, passing only the odd numbers from its input channel to its output channel.

The other new concept in the language is the *fork statement*, described by a statement of the form **fork** (v). This statement can be regarded as a combination of the UNIX fork and the UNIX pipe. When a process executes the statement **fork** (v), the effect is that an almost identical copy of the process is constructed. We call the original process the *parent* and the new process the *child*. After the fork statement has been evaluated both processes continue execution with the statement following the fork statement. There is no sharing of variables, each process has its own set of variables all (but for the variable v, see below) having the values they had in the parent process when the fork statement was executed.

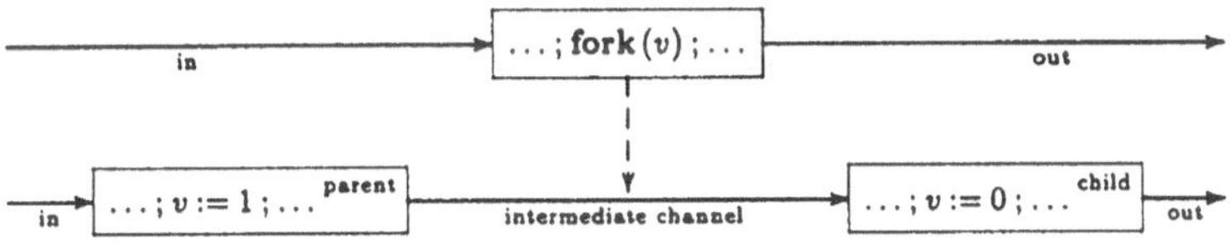

Figure 2

There are two differences between the two processes. The first one has to do with the fact that executing **fork** (v) has as a side effect that a value is assigned to v. In the parent process the assignment $v := 1$ is performed, in the child process the value 0 is assigned to v. The other difference has to do with the input and output channels of the original process. On execution of the fork statement a new intermediate channel is constructed which behaves like a UNIX pipe. The parent process remains connected to the original input channel, but from now on writes on the new intermediate channel. The child will write on the original output channel, but reads from the intermediate channel. The effect of a fork statement is depicted in Figure 2. The second example is the program

```
read (v);
write (v);
fork (w);
if w = 1
then while true
        do read (v);
            if v mod 2 ≠ 0 then write (v) else skip fi
        od
else  while true
        do read (v);
            if v mod 3 ≠ 0 then write (v) else skip fi
        od
fi.
```

The original process passes one value from the input to the output unaltered, and then splits into two filters: the parent filters out all even numbers, passing only the odd input numbers to the child. The child filters out all the numbers which are a multiple of 3. The effect is a filter that passes its first input number unaltered, and then passes only those inputs values that are not multiples of 2 or 3.

The final example is a version of the sieve of Eratosthenes:

```
read (v);
while true
do read (v);
   write (v);
   fork (w);
   if w = 1
   then while true
        do read (x);
           if x mod v ≠ 0 then write (x) else skip fi
        od
   else skip
   fi
od
```

If on the input channel for the original process the stream of the positive natural numbers is inserted, then execution of this program will result in an expanding array of processes which in cooperation yield an output stream consisting of all prime numbers. The original process can be called an 'expander' (e in Figure 3), it reads a number n and expands into a filter process (the parent) which blocks all multiples of n (the parent process is denoted by n in Figure 3), and a new expander process (the child) which behaves like the original process. How this array evolves is shown in Figure 3.

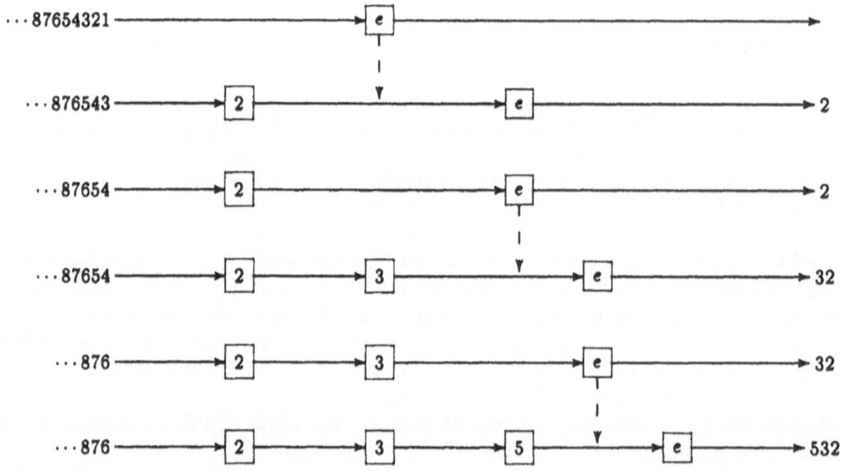

Figure 3

2 Operational Semantics

Before we come to the operational semantics, we first repeat the definition of the syntax for $\mathcal{L}$. Let $(v \in)$ *Var* be the syntactic class of *variables*.

Definition 1. The language $(s \in) \mathcal{L}$ is defined by

$$s ::= v := e \mid \textbf{skip} \mid \textbf{write}\,(e) \mid \textbf{read}\,(v) \mid \textbf{fork}\,(v) \mid s\,;s \mid \textbf{if}\ b\ \textbf{then}\ s\ \textbf{else}\ s\ \textbf{fi} \mid \textbf{while}\ b\ \textbf{do}\ s\ \textbf{od}.$$

Here e and b range over the syntactic classes of *expressions* and *boolean expressions*, respectively. We assume a simple syntax for these which we do not bother to specify. Programs in $\mathcal{L}$ operate on streams of input values, delivering streams of output values. Let us use $(\alpha \in)$ *Val* to denote the set of these (input and output) *values*. In addition, we shall have occasion to use the 'silent' value τ. We write $(\beta \in)\ Val_\tau = Val \cup \{\tau\}$. The role of the τ-value - sometimes also called *hiaton* - will be, in the transition system to be introduced in a moment, to signal a 'silent' transition. Such a transition does not correspond to delivering a 'normal' value (from *Val*); it is employed in a situation where the metric framework requires a step to achieve contractiveness.

The operational semantics for $\mathcal{L}$ will be based on a transition system in the familiar SOS style. In this system we encounter

* The set $(\sigma \in)$ $State = Var \rightarrow Val$ of *states*. The notation $\sigma\{\alpha/v\}$ is used for a state which is like σ, but for its value in v which equals α.
* The set $(\varsigma \in)$ Val_τ^∞, consisting of all finite and infinite sequences (the 'streams' mentioned earlier) of elements from Val_τ.
* The special symbol E standing for *termination*.
* Auxiliary syntactic categories of so-called *resumptions* and *nested resumptions*. These are introduced in

Definition 2. The class of resumptions $(r \in)$ *Res* is defined by

$$r ::= \mathrm{E} \mid s : r.$$

The class of nested resumptions $(\rho \in)$ *NRes* is defined by

$$\rho ::= \varsigma \mid < r, \sigma, \rho > .$$

Resumptions are sequences of statements ending in E. Nested resumptions have a structure of the form

$$\rho = < r_1, \sigma_1, < r_2, \sigma_2, \ldots < r_n, \sigma_n, \varsigma > \ldots >> .$$

Nested resumptions correspond to process arrays as described in the introduction in the following way:

* For $n = 0$, we have that $\rho = \varsigma$. In this case ρ consists of no more than the input stream ς.
* If $n = 1$, then $\rho = < r_1, \sigma_1, \varsigma >$. The process ρ executes the (sequence of) statements specified by r_1, for state σ_1 and input stream ς.
* For $n = 2$, we obtain $\rho = < r_1, \sigma_1, < r_2, \sigma_2, \varsigma >>$. In this case, ρ consists of a parent process $\rho_p = < r_2, \sigma_2, \varsigma >$ - interpreted as just described - the output of which acts as input for the child process $(\rho =) \rho_c = < r_1, \sigma_1, \rho_p >$.
* For $n > 2$, we obtain a process array of length n as described above.

In the transition system $\mathcal{T}$ to be presented in the next definition, we use $\mathcal{V}(e)(\sigma)$ (yielding an element in *Val*) and $\mathcal{B}(b)(\sigma)$ to denote the values of e and b in state σ.

Definition 3. The transition system $\mathcal{T} = (NRes, Val_\tau, \rightarrow)$ has *NRes* as the set of its configurations and Val_τ as its set of labels. The transition relation $\rightarrow$ is the smallest subset of $NRes \times Val_\tau \times NRes$ satisfying the rules given below. We use the notation

$$\rho \xrightarrow{\beta} \rho'$$

as short hand for $(\rho, \beta, \rho') \in \rightarrow$. A rule of the form

$$\text{if } \rho_1 \xrightarrow{\beta} \rho \text{ then } \rho_2 \xrightarrow{\beta} \rho$$

will be abbreviated to $\rho_2 \rightarrow_0 \rho_1$; the 0-subscript indicates that we have here a *zero-step* transition.

(1) $< (v := e) : r, \sigma, \rho > \rightarrow_0 < r, \sigma\{\alpha/v\}, \rho >$, where $\alpha = \mathcal{V}(e)(\sigma)$

(2) $< \mathbf{skip} : r, \sigma, \rho > \rightarrow_0 < r, \sigma, \rho >$

(3) $< \mathbf{write}(e) : r, \sigma, \rho > \xrightarrow{\alpha} < r, \sigma, \rho >$, where $\alpha = \mathcal{V}(e)(\sigma)$

(4) if $\rho \xrightarrow{\alpha} \rho'$ then $< \mathbf{read}(v) : r, \sigma, \rho > \xrightarrow{\tau} < r, \sigma\{\alpha/v\}, \rho' >$

(5) if $\rho \xrightarrow{\tau} \rho'$ then $< \mathbf{read}(v) : r, \sigma, \rho > \xrightarrow{\tau} < \mathbf{read}(v) : r, \sigma, \rho' >$

(6) $< \mathbf{fork}(v) : r, \sigma, \rho > \rightarrow_0 < r, \sigma\{0/v\}, < r, \sigma\{1/v\}, \rho >>$

(7) $< (s_1 ; s_2) : r, \sigma, \rho > \rightarrow_0 < s_1 : (s_2 : r), \sigma, \rho >$

(8) if $\mathcal{B}(b)(\sigma)$ then $< \mathbf{if}\ b\ \mathbf{then}\ s_1\ \mathbf{else}\ s_2\ \mathbf{fi} : r, \sigma, \rho > \rightarrow_0 < s_1 : r, \sigma, \rho >$

(9) if $\neg \mathcal{B}(b)(\sigma)$ then $< \mathbf{if}\ b\ \mathbf{then}\ s_1\ \mathbf{else}\ s_2\ \mathbf{fi} : r, \sigma, \rho > \rightarrow_0 < s_2 : r, \sigma, \rho >$

(10) $< \mathbf{while}\ b\ \mathbf{do}\ s\ \mathbf{od} : r, \sigma, \rho > \xrightarrow{\tau} < \mathbf{if}\ b\ \mathbf{then}\ s ; \mathbf{while}\ b\ \mathbf{do}\ s\ \mathbf{od}\ \mathbf{else}\ \mathbf{skip}\ \mathbf{fi} : r, \sigma, \rho >$

(11) $\beta \cdot \varsigma \xrightarrow{\beta} \varsigma$

We add some explanations:

* A transition $\rho \xrightarrow{\beta} \rho'$ expresses that (the process corresponding to) ρ performs a one-step transition to process ρ', while producing a value β (either a normal or a silent value) which is appended to the current output stream.
* Note that there is no transition defined for a configuration $< \mathrm{E}, \sigma, \rho >$. As a consequence, neither is there a transition possible for, e.g., $< (v := e) : \mathrm{E}, \sigma, \rho >$, $< \mathbf{read}(v) : \mathrm{E}, \sigma, < \mathrm{E}, \sigma, \rho >>$, etc. We emphasize that transitions become observable only by delivering output values (including an occasional silent value); note that this is quite different from more customary models where state changes - from σ to some σ' - are observable.
* The rules for $v{:=}e$, **skip**, $s_1 ; s_2$, and **if** b **then** s_1 **else** s_2 **fi** should be clear. The while statement always induces a silent step. (A zero-step transition would not work in this case, this being incompatible with a subsequent crucial property of zero-step transitions, cf. Lemma 7.)
* The effect of $\mathbf{write}(e) : r$ is to append $\alpha (= \mathcal{V}(e)(\sigma))$ to the output stream, and continue with r.
* For a **read** (v) statement - with respect to current r, σ, and ρ - we distinguish two cases. In the 'normal' situation, an input α is available, produced (as output) by ρ when it turns itself into ρ'. We then assign α to v, and continue with r, the updated state $\sigma\{\alpha/v\}$, and the new parent process ρ'. Otherwise, i.e. when ρ produces a silent step τ, we reject this as possible value for v - recall that the codomain of any state equals *Val* rather than Val_τ -, maintain the requirement for an input **read** (v), and continue with r, σ, and parent process ρ'. (As for the while statement also in this case a zero-step transition would not work.)
* The fork statement **fork** (v) - with respect to current r, σ, and ρ - creates two processes, the parent process

  $$\rho_p = < r, \sigma\{1/v\}, \rho >$$

  and the child process

  $$\rho_c = < r, \sigma\{0/v\}, \rho_p > .$$

  We observe that
  - The forking process performs a zero-step transition to ρ_c.
  - Both ρ_p and ρ_c execute the resumption r.
  - In ρ_p, the fork variable is set to 1, in ρ_c it is set to 0. This offers the possibility to 'program' in r so as to have different executions in ρ_p and ρ_c, respectively (cf. the examples in the introduction).
  - Since ρ_p occurs as part of ρ_c, the net effect of this is that the output of ρ_p acts as input for ρ_c, cf. also the way the read and write rules are defined.
* The final rule simply describes how an input stream $\beta \cdot \varsigma$ performs a one step transition delivering the output β, and turns itself into ς.
* The transition system $\mathcal{T}$ specifies deterministic behaviour (see Lemma 8) and synchronous communication. Concerning the former phenomenon, adding the metarule

  $$\text{if } \rho \rightarrow_0 \rho' \text{ then } < r, \sigma, \rho > \rightarrow_0 < r, \sigma, \rho' >$$

  would allow some form of parallelism in the execution of processes. As a consequence of the latter phenomenon, a parent process can only write when its child is willing to read. As we will see, a communication between a parent and its child will not be visible in the operational semantics (apart from a silent transition). Asynchronous communication could be handled by adding an output sequence to the nested resumptions which then take the form $< \varsigma, r, \sigma, \rho >$. A study of these variations is outside the scope of the present paper.

We now describe how to obtain the operational semantics $\mathcal{O} : \mathcal{L} \rightarrow Proc$, where $Proc = State \rightarrow Val_\tau^\infty \rightarrow Val_\tau^\infty$. We see that $\mathcal{O}[\![s]\!](\sigma)$ yields a function transforming streams to streams, in accordance with the intended model for $\mathcal{L}$. We shall employ an intermediate mapping $\mathcal{O} : NRes \rightarrow Val_\tau^\infty$; $\mathcal{O}$ is the function which, for argument ρ, collects the sequence of labels produced successively by the transitions as specified by $\mathcal{T}$, starting from ρ. Thus $\mathcal{O}(\rho) = \varsigma$ states that the process ρ yields output stream ς. (Recall that the *input* to ρ is included in its own description.) Let us use the terminology ρ *blocks* in case ρ cannot make any transitions, that is

$$\neg \exists \beta, \rho' : \rho \xrightarrow{\beta} \rho'.$$

As defining properties for $\mathcal{O}$ we want the following to be satisfied:

$$\mathcal{O}(\rho) = \begin{cases} \varepsilon & \text{if } \rho \text{ blocks} \\ \beta \cdot \mathcal{O}(\rho') & \text{if } \rho \xrightarrow{\beta} \rho' \end{cases}$$

Note that ρ' is not necessarily of smaller syntactic complexity than ρ, so this 'definition' cannot be shown to be well-formed simply by structural induction on ρ. Instead, we use a familiar technique for dealing with recursive definitions, viz. through the use of fixed points of some higher-order operator. Let Φ be an operator which maps meanings ϕ to meanings ϕ' in the following way:

Definition 4. Let $(\phi \in)\, Sem_{\mathcal{O}} = NRes \rightarrow Val_{\tau}^{\infty}$, and let $\Phi : Sem_{\mathcal{O}} \rightarrow Sem_{\mathcal{O}}$ be defined as follows:

$$\Phi(\phi)(\rho) = \begin{cases} \varepsilon & \text{if } \rho \text{ blocks} \\ \beta \cdot \phi(\rho') & \text{if } \rho \xrightarrow{\beta} \rho' \end{cases}$$

Well-definedness of this definition requires that $\mathcal{T}$ is deterministic, i.e. that each ρ can make at most one transition. Lemma 8 below states this result.

By the definition of Φ, it is immediate that it is contractive[4] in ϕ. Since $Sem_{\mathcal{O}}$ is a complete metric space[5], we have, by Banach's theorem[6], that Φ has a unique fixed point, and we have justified

Definition 5. The operational semantics $\mathcal{O} : Sem_{\mathcal{O}}$ is defined by

$$\mathcal{O} = \mathit{fix}\,(\Phi).$$

In addition to its serving as a means to define $\mathcal{O}$, Φ will play a crucial role (in Section 4) in the proof that $(*)$ $\mathcal{O} = \mathcal{D}$ (the denotational semantics to be introduced in Section 3). In fact, $(*)$ follows as an immediate corollary of an argument exploiting the unique fixed point property of Φ.

The next step in the technical development is the introduction of the complexity measure $c : NRes \rightarrow \mathbb{N}$ in

Definition 6. The complexity measure $c : NRes \rightarrow \mathbb{N}$ is defined by

$$c(\varsigma) = 1 \qquad c(< r, \sigma, \rho >) = c(r) + c(\rho)$$

where

$$c(\mathrm{E}) = 1 \qquad c(s : r) = c(s) * c(r)$$

where

$$\begin{array}{llll} c(v := e) & = 2 & c(\mathbf{fork}\,(v)) & = 3 \\ c(\mathbf{skip}) & = 2 & c(s_1\,;\, s_2) & = c(s_1) * c(s_2) + 1 \\ c(\mathbf{write}\,(e)) & = 1 & c(\mathbf{if}\ b\ \mathbf{then}\ s_1\ \mathbf{else}\ s_2\ \mathbf{fi}) & = c(s_1) + c(s_2) \\ c(\mathbf{read}\,(v)) & = 1 & c(\mathbf{while}\ b\ \mathbf{do}\ s\ \mathbf{od}) & = 2 \end{array}$$

The measure c is used in the proof of the following two lemmas.

Lemma 7. *For all ρ and ρ', if $\rho \rightarrow_{\mathcal{O}} \rho'$ then $c(\rho) > c(\rho')$.*

Proof. Only a few cases of the proof of this lemma are elaborated on.

---

[4] Let (X, d_X) and $(X', d_{X'})$ be metric spaces. A function $f : X \rightarrow X'$ is called contractive if there exists an δ, with $0 \leq \delta < 1$, such that, for all x and x',

$$d_{X'}(f(x), f(x')) \leq \delta \cdot d_X(x, x').$$

[5] The set Val_{τ}^{∞} is endowed with the metric

$$d(\varsigma, \varsigma') = \begin{cases} 0 & \text{if } \varsigma = \varsigma' \\ 2^{-n} & \text{otherwise} \end{cases}$$

where n is the longest common prefix of the sequences ς and ς'. By means of this metric we can endow $Sem_{\mathcal{O}}$ with the metric

$$d(\phi, \phi') = \sup\{d(\phi(\rho), \phi'(\rho)) \mid \rho \in NRes\}.$$

These metrics are ultrametrics, i.e., for all x, x', and x'',

$$d(x, x'') \leq \max\{d(x, x'), d(x', x'')\}.$$

[6] Let (X, d_X) be a complete metric space. If $f : X \rightarrow X$ is contractive then f has a unique fixed point $\mathit{fix}\,(f)$.

1. Let $\rho \equiv <(v := e) : r, \sigma, \bar{\rho}>$. Then

$$\begin{aligned}&c(<(v := e) : r, \sigma, \bar{\rho}>)\\&= 2 * c(r) + c(\bar{\rho})\\&> c(r) + c(\bar{\rho})\\&= c(<r, \sigma\{\alpha/v\}, \bar{\rho}>).\end{aligned}$$

2. Let $\rho \equiv <\textbf{fork}(v) : r, \sigma, \bar{\rho}>$. Then

$$\begin{aligned}&c(<\textbf{fork}(v) : r, \sigma, \bar{\rho}>)\\&= 3 * c(r) + c(\bar{\rho})\\&> 2 * c(r) + c(\bar{\rho})\\&= c(<r, \sigma\{0/v\}, <r, \sigma\{1/v\}, \bar{\rho}>>).\end{aligned}$$

□

Lemma 8. *The transition system $\mathcal{T}$ is deterministic.*

Proof. We can show that, for all ρ, $|\{(\beta, \rho') \mid \rho \xrightarrow{\beta} \rho'\}| \leq 1$ by induction on the complexity of ρ. □

We are now ready for the key definition of this section.

Definition 9. The operational semantics $\mathcal{O} : \mathcal{L} \to \mathit{Proc}$ is defined by

$$\mathcal{O}[\![s]\!] = \lambda\sigma . \lambda\varsigma . \mathcal{O}(<s : \text{E}, \sigma, \varsigma>).$$

The final program of the introduction with an arbitrary initial state and the input stream 12345678 will produce the output stream $\tau^3 2\tau^4 3\tau^8 5\tau^{10} 7\tau^7$ and terminate as the reader may verify.

3 Denotational Semantics

The denotational semantics for $\mathcal{L}$ uses the set of *continuations* $(\theta \in)\mathit{Cont} = \mathit{State} \to \mathit{Val}_\tau^\infty \to^1 \mathit{Val}_\tau^\infty$. Note that, but for the specialization to the nonexpansive[7] function space $\to^1$, *Cont* equals *Proc* as introduced earlier. Continuations correspond to resumptions in the sense that, as we shall see in Definition 14, meanings of *Res* reside in *Cont*.

We shall use *first* (ς) to denote the first element of the nonempty sequence ς, and *rest* (ς) to denote the result of omitting the first element from the nonempty sequence ς.

The denotational semantics $\mathcal{D}$ for $\mathcal{L}$ is presented in

Definition 10. Let $(\psi \in) \mathit{Sem}_\mathcal{D} = \mathcal{L} \to \mathit{Cont} \to^1 \mathit{Cont}$. Let $\Psi : \mathit{Sem}_\mathcal{D} \to \mathit{Sem}_\mathcal{D}$ be defined by

$$\begin{array}{lll}
\Psi(\psi)(v := e)(\theta) & = \lambda\sigma . \theta(\sigma\{\alpha/v\}) & \text{where } \alpha = \mathcal{V}(e)(\sigma)\\
\Psi(\psi)(\textbf{skip})(\theta) & = \theta & \\
\Psi(\psi)(\textbf{write}(e))(\theta) & = \lambda\sigma . \lambda\varsigma . \alpha \cdot \theta(\sigma)(\varsigma) & \text{where } \alpha = \mathcal{V}(e)(\sigma)\\
\Psi(\psi)(\textbf{read}(v))(\theta) & = \lambda\sigma . \lambda\varsigma . \begin{cases} \varepsilon & \text{(a)}\\ \tau \cdot \theta(\sigma\{\mathit{first}(\varsigma)/v\})(\mathit{rest}(\varsigma)) & \text{(b)}\\ \tau \cdot \psi(\textbf{read}(v))(\theta)(\sigma)(\mathit{rest}(\varsigma)) & \text{(c)}\end{cases} & \\
\Psi(\psi)(\textbf{fork}(v))(\theta) & = \lambda\sigma . \lambda\varsigma . (\theta(\sigma\{0/v\}))(\theta(\sigma\{1/v\})(\varsigma)) & \\
\Psi(\psi)(s_1 ; s_2)(\theta) & = \Psi(\psi)(s_1)(\Psi(\psi)(s_2)(\theta)) & \\
\Psi(\psi)(\textbf{if } b \textbf{ then } s_1 \textbf{ else } s_2 \textbf{ fi})(\theta) & = \lambda\sigma . \begin{cases} \Psi(\psi)(s_1)(\theta)(\sigma) & \text{(d)}\\ \Psi(\psi)(s_2)(\theta)(\sigma) & \text{(e)}\end{cases} & \\
\Psi(\psi)(\textbf{while } b \textbf{ do } s \textbf{ od})(\theta) & = \lambda\sigma . \lambda\varsigma . \tau \cdot \psi(\textbf{if } b \textbf{ then } s ; \textbf{while } b \textbf{ do } s \textbf{ od else skip fi})(\theta)(\sigma)(\varsigma) &
\end{array}$$

where

(a) if $\varsigma = \varepsilon$
(b) if $\varsigma \neq \varepsilon$ and *first* $(\varsigma) \neq \tau$
(c) if $\varsigma \neq \varepsilon$ and *first* $(\varsigma) = \tau$
(d) if $\mathcal{B}(b)(\sigma)$
(e) if $\neg\mathcal{B}(b)(\sigma)$

[7] Let (X, d_X) and $(X', d_{X'})$ be metric spaces. A function $f : X \to X'$ is called nonexpansive if, for all x and x', $d_{X'}(f(x), f(x')) \leq d_X(x, x')$.

The denotational semantics $\mathcal{D} : Sem_{\mathcal{D}}$ is defined by

$\mathcal{D} = \mathit{fix}\,(\Psi)$.

Some remarks:

* Much of the structure of the above clauses may be understood by consulting $\mathcal{T}$. For example, the clause for the fork statement amounts to

$$\mathcal{D}\,(\mathbf{fork}\,(v))(\theta)(\sigma)(\varsigma) = (\theta\,(\sigma\{0/v\}))(\theta\,(\sigma\{1/v\})(\varsigma)).$$

Now using the correspondence between the semantic continuation θ and the syntactic resumption r, we see that this is an immediate counterpart of the transition

$$< \mathbf{fork}\,(v) : r,\ \sigma,\ \varsigma > \rightarrow_0 < r,\ \sigma\{0/v\},\ < r,\ \sigma\{1/v\},\ \varsigma >> .$$

* Similar to what we did for $\mathcal{O}$, we have defined $\mathcal{D}$ here as (unique) fixed point of a higher-order mapping. Such a 'global' fixed point approach is attractive, were it only for symmetry reasons. However, a more traditional ('local') approach, where the taking of fixed points is restricted to the clauses for the read and while statement, would also serve our purposes.

Definition 10 is justified in

Lemma 11. *For all ψ, s, θ, and σ,*

the mapping $\Psi\,(\psi)(s)(\theta)(\sigma)$ is nonexpansive (in ς),
the mapping $\Psi\,(\psi)(s)$ is nonexpansive (in θ), and
the mapping Ψ is contractive (in ψ).

Proof. We only consider the second property. It can be shown that, for all ψ, s, θ_1, θ_2, σ, and ς,

$$d\,(\Psi\,(\psi)(s)(\theta_1)(\sigma)(\varsigma), \Psi\,(\psi)(s)(\theta_2)(\sigma)(\varsigma)) \leq d\,(\theta_1, \theta_2)$$

by structural induction on s. Only a few cases are elaborated on.

1. Let $s \equiv \mathbf{read}\,(v)$. We distinguish three cases.
   (a) If $\varsigma = \varepsilon$, then
   $$\begin{aligned} & d\,(\Psi\,(\psi)(\mathbf{read}\,(v))(\theta_1)(\sigma)(\varsigma), \Psi\,(\psi)(\mathbf{read}\,(v))(\theta_2)(\sigma)(\varsigma)) \\ &= d\,(\varepsilon, \varepsilon) \\ &\leq d\,(\theta_1, \theta_2). \end{aligned}$$
   (b) If $\varsigma \neq \varepsilon$ and $\mathit{first}\,(\varsigma) \neq \tau$, then
   $$\begin{aligned} & d\,(\Psi\,(\psi)(\mathbf{read}\,(v))(\theta_1)(\sigma)(\varsigma), \Psi\,(\psi)(\mathbf{read}\,(v))(\theta_2)(\sigma)(\varsigma)) \\ &= d\,(\tau \cdot \theta_1\,(\sigma\{\mathit{first}\,(\varsigma)/v\})(\mathit{rest}\,(\varsigma)), \tau \cdot \theta_2\,(\sigma\{\mathit{first}\,(\varsigma)/v\})(\mathit{rest}\,(\varsigma))) \\ &= \tfrac{1}{2} \cdot d\,(\theta_1\,(\sigma\{\mathit{first}\,(\varsigma)/v\})(\mathit{rest}\,(\varsigma)), \theta_2\,(\sigma\{\mathit{first}\,(\varsigma)/v\})(\mathit{rest}\,(\varsigma))) \\ &\leq \tfrac{1}{2} \cdot d\,(\theta_1, \theta_2). \end{aligned}$$
   (c) If $\varsigma \neq \varepsilon$ and $\mathit{first}\,(\varsigma) = \tau$, then
   $$\begin{aligned} & d\,(\Psi\,(\psi)(\mathbf{read}\,(v))(\theta_1)(\sigma)(\varsigma), \Psi\,(\psi)(\mathbf{read}\,(v))(\theta_2)(\sigma)(\varsigma)) \\ &= d\,(\tau \cdot \psi\,(\mathbf{read}\,(v))(\theta_1)(\sigma)(\mathit{rest}\,(\varsigma)), \tau \cdot \psi\,(\mathbf{read}\,(v))(\theta_2)(\sigma)(\mathit{rest}\,(\varsigma))) \\ &= \tfrac{1}{2} \cdot d\,(\psi\,(\mathbf{read}\,(v))(\theta_1)(\sigma)(\mathit{rest}\,(\varsigma)), \psi\,(\mathbf{read}\,(v))(\theta_2)(\sigma)(\mathit{rest}\,(\varsigma))) \\ &\leq \tfrac{1}{2} \cdot d\,(\psi\,(\mathbf{read}\,(v))(\theta_1), \psi\,(\mathbf{read}\,(v))(\theta_2)) \\ &\leq \tfrac{1}{2} \cdot d\,(\theta_1, \theta_2). && [\psi\,(\mathbf{read}\,(v)) \text{ is nonexpansive}] \end{aligned}$$
2. Let $s \equiv \mathbf{fork}\,(v)$. Then
$$\begin{aligned} & d\,(\Psi\,(\psi)(\mathbf{fork}\,(v))(\theta_1)(\sigma)(\varsigma), \Psi\,(\psi)(\mathbf{fork}\,(v))(\theta_2)(\sigma)(\varsigma)) \\ &= d\,((\theta_1\,(\sigma\{0/v\}))(\theta_1\,(\sigma\{1/v\})(\varsigma)), (\theta_2\,(\sigma\{0/v\}))(\theta_2\,(\sigma\{1/v\})(\varsigma))) \\ &\leq \max\{d\,((\theta_1\,(\sigma\{0/v\}))(\theta_1\,(\sigma\{1/v\})(\varsigma)), (\theta_1\,(\sigma\{0/v\}))(\theta_2\,(\sigma\{1/v\})(\varsigma))), \\ &\qquad d\,((\theta_1\,(\sigma\{0/v\}))(\theta_2\,(\sigma\{1/v\})(\varsigma)), (\theta_2\,(\sigma\{0/v\}))(\theta_2\,(\sigma\{1/v\})(\varsigma)))\} && [\text{ultrametricity}] \\ &\leq \max\{d\,(\theta_1\,(\sigma\{1/v\})(\varsigma), \theta_2\,(\sigma\{1/v\})(\varsigma)), d\,(\theta_1\,(\sigma\{0/v\}), \theta_2\,(\sigma\{0/v\}))\} \\ & && [\theta_1\,(\sigma\{0/v\}) \text{ is nonexpansive}] \\ &\leq d\,(\theta_1, \theta_2). \end{aligned}$$

□

We conclude this section with

Definition 12. The denotational semantics $\mathcal{D} : \mathcal{L} \rightarrow \mathit{Cont}$ is defined by

$\mathcal{D}\,[\![s]\!] = \mathcal{D}\,(s)(\lambda\sigma\,.\,\lambda\varsigma\,.\,\varepsilon)$.

4 Equivalence Theorem

Theorem 13. *For all $s \in \mathcal{L}$, $\mathcal{O}[\![s]\!] = \mathcal{D}[\![s]\!]$.*

On the way to the proof of this theorem, we first introduce two intermediate semantics.

Definition 14. The mapping $\mathcal{H}: Res \to Cont$ is defined by

$$\begin{aligned}\mathcal{H}(\mathrm{E}) &= \lambda\sigma.\lambda\varsigma.\epsilon\\ \mathcal{H}(s:r) &= \mathcal{D}(s)(\mathcal{H}(r))\end{aligned}$$

The mapping $\mathcal{I}: NRes \to Val_\tau^\infty$ is defined by

$$\begin{aligned}\mathcal{I}(\varsigma) &= \varsigma\\ \mathcal{I}(< r, \sigma, \rho >) &= \mathcal{H}(r)(\sigma)(\mathcal{I}(\rho))\end{aligned}$$

The following properties of $\mathcal{I}$ are furthermore of importance.

Lemma 15. *For all ρ, ρ', and β,*

if $\rho \to_0 \rho'$ then $\mathcal{I}(\rho) = \mathcal{I}(\rho')$, and
if $\rho \xrightarrow{\beta} \rho'$ then $\mathcal{I}(\rho) = \beta \cdot \mathcal{I}(\rho')$.

Proof. We only consider a few cases of the proof of the first property.

1. Let $\rho \equiv < (v := e) : r, \sigma, \bar{\rho} >$. Then

$$\begin{aligned}&\mathcal{I}(< (v := e) : r, \sigma, \bar{\rho} >)\\ &= \mathcal{H}((v := e) : r)(\sigma)(\mathcal{I}(\bar{\rho}))\\ &= \mathcal{D}(v := e)(\mathcal{H}(r))(\sigma)(\mathcal{I}(\bar{\rho}))\\ &= \mathcal{H}(r)(\sigma\{\alpha/v\})(\mathcal{I}(\bar{\rho}))\\ &= \mathcal{I}(< r, \sigma\{\alpha/v\}, \bar{\rho} >).\end{aligned}$$

2. Let $\rho \equiv < \mathbf{fork}(v) : r, \sigma, \bar{\rho} >$. Then

$$\begin{aligned}&\mathcal{I}(< \mathbf{fork}(v) : r, \sigma, \bar{\rho} >)\\ &= \mathcal{H}(\mathbf{fork}(v) : r)(\sigma)(\mathcal{I}(\bar{\rho}))\\ &= \mathcal{D}(\mathbf{fork}(v))(\mathcal{H}(r))(\sigma)(\mathcal{I}(\bar{\rho}))\\ &= (\mathcal{H}(r)(\sigma\{0/v\}))(\mathcal{H}(r)(\sigma\{1/v\})(\mathcal{I}(\bar{\rho})))\\ &= (\mathcal{H}(r)(\sigma\{0/v\}))(\mathcal{I}(< r, \sigma\{1/v\}, \bar{\rho} >))\\ &= \mathcal{I}(< r, \sigma\{0/v\}, < r, \sigma\{1/v\}, \bar{\rho} >>).\end{aligned}$$

□

The main step in the proof of Theorem 13 now follows. Recall that Φ is the higher-order operator used in the definition of $\mathcal{O}$.

Lemma 16. $\Phi(\mathcal{I}) = \mathcal{I}$.

Proof. We can show that, for all ρ,

$$\Phi(\mathcal{I})(\rho) = \mathcal{I}(\rho)$$

by induction on the complexity of ρ (cf. Definition 6). Only a few cases are elaborated on.

1. Let $\rho \equiv < (v := e) : r, \sigma, \rho' >$. Then

$$\begin{aligned}&\Phi(\mathcal{I})(< (v := e) : r, \sigma, \rho' >)\\ &= \Phi(\mathcal{I})(< r, \sigma\{\alpha/v\}, \rho' >) && [< (v := e) : r, \sigma, \rho' > \to_0 < r, \sigma\{\alpha/v\}, \rho' >]\\ &= \mathcal{I}(< r, \sigma\{\alpha/v\}, \rho' >) && [\text{Lemma 7, induction}]\\ &= \mathcal{I}(< (v := e) : r, \sigma, \rho' >). && [\text{Lemma 15}]\end{aligned}$$

2. Let $\rho \equiv < \mathbf{read}(v) : r, \sigma, \rho' >$. We distinguish three cases.

(a) Assume $\rho' \xrightarrow{\alpha} \rho''$. Then

$\Phi(\mathcal{I})(<\mathbf{read}(v) : r, \sigma, \rho'>)$

$= \tau \cdot \mathcal{I}(<r, \sigma\{\alpha/v\}, \rho''>)$ $\quad [<\mathbf{read}(v) : r, \sigma, \rho'> \xrightarrow{\tau} <r, \sigma\{\alpha/v\}, \rho''>]$

$= \mathcal{I}(<\mathbf{read}(v) : r, \sigma, \rho'>)$. $\quad$ [Lemma 15]

(b) Assume $\rho' \xrightarrow{\tau} \rho''$. Then

$\Phi(\mathcal{I})(<\mathbf{read}(v) : r, \sigma, \rho'>)$

$= \tau \cdot \mathcal{I}(<\mathbf{read}(v) : r, \sigma, \rho''>)$ $\quad [<\mathbf{read}(v) : r, \sigma, \rho'> \xrightarrow{\tau} <\mathbf{read}(v) : r, \sigma, \rho''>]$

$= \mathcal{I}(<\mathbf{read}(v) : r, \sigma, \rho'>)$. $\quad$ [Lemma 15]

(c) Assume ρ' blocks. Then $<\mathbf{read}(v) : r, \sigma, \rho'>$ blocks and hence

$\Phi(\mathcal{I})(<\mathbf{read}(v) : r, \sigma, \rho'>) = \varepsilon$.

Since ρ' blocks, $\Phi(\mathcal{I})(\rho') = \varepsilon$. By induction, $\mathcal{I}(\rho') = \varepsilon$. Consequently,

$\mathcal{I}(<\mathbf{read}(v) : r, \sigma, \rho'>) = \varepsilon$.

□

We have arrived at the proof of Theorem 13:

Proof. Because both $\mathcal{O}$ and $\mathcal{I}$ are fixed point of Φ (Definition 5 and Lemma 16) and Φ has a unique fixed point, $\mathcal{O}$ and $\mathcal{I}$ are equal. Consequently,

$$\begin{aligned}
&\mathcal{O}[\![s]\!](\sigma)(\varsigma)\\
&= \mathcal{O}(<s : \mathrm{E}, \sigma, \varsigma>)\\
&= \mathcal{I}(<s : \mathrm{E}, \sigma, \varsigma>)\\
&= \mathcal{H}(s : \mathrm{E})(\sigma)(\mathcal{I}(\varsigma))\\
&= \mathcal{D}(s)(\mathcal{H}(\mathrm{E}))(\sigma)(\varsigma)\\
&= \mathcal{D}(s)(\lambda\sigma . \lambda\varsigma . \varepsilon)(\sigma)(\varsigma)\\
&= \mathcal{D}[\![s]\!](\sigma)(\varsigma).
\end{aligned}$$

□

References

[AW85] S.K. Abdali and D.S. Wise. Standard, Storeless Semantics for ALGOL-style Block Structure and Call-by-Name. In A. Melton, editor, *Proceedings of the 1st International Conference on Mathematical Foundations of Programming Semantics*, volume 239 of *Lecture Notes in Computer Science*, pages 1–19, Manhattan, April 1985. Springer-Verlag.

[Ban22] S. Banach. Sur les Opérations dans les Ensembles Abstraits et leurs Applications aux Equations Intégrales. *Fundamenta Mathematicae*, 3:133–181, 1922.

[Ben82] D.B. Benson. Machine-Level Semantics for Nondeterministic, Parallel Programs. In M. Dezani-Ciancaglini and U. Montanari, editors, *Proceedings of the 5th International Symposium on Programming*, volume 137 of *Lecture Notes in Computer Science*, pages 15–25, Turin, April 1982. Springer-Verlag.

[BR92] J.W. de Bakker and J.J.M.M. Rutten, editors. *Ten Years of Concurrency Semantics, selected papers of the Amsterdam Concurrency Group.* World Scientific, Singapore, 1992.

[Bru86] A. de Bruin. *Experiments with Continuation Semantics: jumps, backtracking, dynamic networks.* PhD thesis, Vrije Universiteit, Amsterdam, May 1986.

[HSS91] T. Hagerup, A. Schmitt, and H. Seidl. FORK: A High-Level Language for PRAMs. In E.H.L. Aarts, J. van Leeuwen, and M. Rem, editors, *Proceedings of the 3rd International PARLE Conference*, volume 505 of *Lecture Notes in Computer Science*, pages 304–320, Eindhoven, June 1991. Springer-Verlag.

[KK92] E. Klein and K. Koskimies. How to Pipeline Parsing with Parallel Semantic Analysis. *Structured Programming*, 13(3):99–107, 1992.

[KR90] J.N. Kok and J.J.M.M. Rutten. Contractions in Comparing Concurrency Semantics. *Theoretical Computer Science*, 76(2/3):179–222, 1990.

[MA89] C. McDonald and L. Allison. Denotational Semantics of a Command Interpreter and their Implementation in Standard ML. *The Computer Journal*, 32(5):422–431, October 1989.

[Par83] D. Park. The "Fairness" Problem and Nondeterministic Computing Networks. In J.W. de Bakker and J. van Leeuwen, editors, *Foundations of Computer Science IV, Distributed Systems*, part 2: Semantics and Logic, volume 159 of *Mathematical Centre Tracts*, pages 133–161. Mathematical Centre, Amsterdam, 1983.

[Plo81] G.D. Plotkin. A Structural Approach to Operational Semantics. Report DAIMI FN-19, Aarhus University, Aarhus, September 1981.

[RS83] J.-C. Raoult and R. Sethi. Properties of a Notation for Combining Functions. *Journal of the ACM*, 30(3):595–611, July 1983.

[RS92] G. Rünger and K. Sieber. A Trace-Based Denotational Semantics for the PRAM-Language FORK. Report 1/1992, Universität des Saarlandes, Saarbrücken, 1992.

Rabin Tree Automata and Finite Monoids

Danièle Beauquier and Andreas Podelski

LITP, Institut Blaise Pascal (IBP)
4 Place Jussieu, 75252 Paris Cedex 05, France
dab@litp.ibp.fr
Digital Equipment Corporation, Paris Research Laboratory (PRL)
85, Avenue Victor Hugo, 92563 Rueil-Malmaison, France
podelski@prl.dec.com

Abstract. We incorporate finite monoids into the theory of Rabin recognizability of infinite tree languages. We define a free monoid of infinite trees and associate with each infinite tree language L a language $\widehat{L}$ of infinite words over this monoid. Using this correspondence we introduce *strong* monoid recognizability of infinite tree languages (strengthening the standard notion for infinite words) and show that it is equivalent to Rabin recognizability. We also show that there exists an infinite tree language L which is not Rabin recognizable, but its associated language $\widehat{L}$ is monoid recognizable (in the standard sense). Our positive result opens the theory of varieties of infinite tree languages, extending those for finite and infinite words and finite trees.

1 Introduction

The importance of Rabin automata on infinite trees comes from the fact that they yield a powerful decision tool, namely for all those problems which are reducible to the monadic second-order logic over the infinite binary tree [Rab69, Rab77]. These include, for example, many decidability problems for properties of sequential and parallel programs (*cf.*, [KT90]). The theory of Rabin automata is by now well established; for a survey, *cf.*, [Tho90], for a collection of recent results, *cf.*, [NP92]. It can be viewed as an extension of the theory of automata on infinite words [Bu62], words being the special case of unary trees. This extension is, however, not at all a straightforward one. This is indicated by the complexity bounds for the corresponding algorithms as well as the difficulty of the proofs of the corresponding results. It is confirmed also by the results of this paper.

The characterization of the recognizability of languages of infinite words in terms of finite monoids is one of the cornerstones of the theory of automata on infinite words. This paper deals with the extension of this characterization. We introduce the notion of *strong monoid recognizability* of a language of infinite trees. We show that it is equivalent to recognizability by a Rabin automaton.

This result extends a corresponding one for the case of finite trees respectively words [NP89]: A language L of finite trees is recognizable if and only if it is monoid-recognizable. The latter means that the language $\widehat{L}$ of *pointed trees* associated with L is recognizable as a set of finite words over an alphabet (of pointed trees of height 1). In terms of the theory of varieties (of monoids and formal languages, *cf.*, [Eil76, Pi84]), the result states the correspondence between the one of finite monoids and the one of recognizable languages. For results on other varieties for finite trees, *cf.*, [Tho84, NP89, Stei92, PP92], for infinite words, *cf.*, [Per84, Pec87]. Now, our result, establishing a connection between recognizable sets of infinite trees and finite monoids, indicates the possibility to open the theory of varieties of languages of infinite trees.

We also show that the straightforward extension of the characterization of the recognizability of languages of infinite words in terms of finite monoids is not possible. Namely, if *monoid recognizable* is the straightforward extension of the notion for infinite words (and weaker than strong monoid recognizable), then there exists a language of infinite trees which is monoid recognizable, but not recognizable by a Rabin automaton.

The following section provides the background material which is necessary to make this paper self-contained. Section 3 introduces the *free monoid* of infinite *pointed trees*. Infinite words over this free monoid correspond to infinite *marked trees*. We go from infinite trees to infinite words by associating with a language L of infinite trees a set $\widehat{L}$ of marked trees. In this setting, the notion of the *transition monoid* of a Rabin automaton is readily obtained, as well as the fact that it is always finite.

In Section 4, we define *strong morphisms* from the free monoid of infinite pointed trees into a finite monoid. In order to show that Rabin recognizability implies strong monoid recognizability (*i.e.*, by a strong morphism) we use the Boolean closure properties of the family of Rabin recognizable languages. For the other direction, we transform a sequential Rabin automaton recognizing $\widehat{L}$ (as a language of infinite words) into a Rabin automaton recognizing the language L of infinite trees with which $\widehat{L}$ is associated. The two directions together form the main result of this paper. The requirement on the morphism to be a strong one in this result cannot be dropped. This is demonstrated by a counter example in Section 5. Namely, there exists an infinite tree language L which is not Rabin recognizable, but its associated language $\widehat{L}$ is monoid recognizable (*i.e.*, as a language of infinite words). Interestingly, we were not able to construct such a counter example; we use a countability argument instead. Finally, we conclude with a discussion of further work.

2 Preliminaries

Given a set X, we denote X^* the set of finite words over X and X^ω the set of infinite words over X. The empty word is denoted by λ.

The *initial segment* or *prefix* relation is denoted $\leq$ (the proper one by $<$). It is defined by: $u \leq uv$ for all $u \in X^*$, $v \in X^* \cup X^\omega$. In particular, the empty word is prefix of any word.

A non-empty (possibly infinite) subset T of X^* closed under initial segments is called a *tree domain*. The elements of T are called *nodes*, the $\leq$-maximal nodes *leaves*, and λ the *root* of T. If $u \in T$, $x \in X$ and $ux \in T$ then ux is an *immediate successor* of u in T. The *boundary* $b(T)$ of T is the set of "occurrences just outside of T," *i.e.*,

$$b(t) = \{ux \in X^* \mid u \in T,\ x \in X,\ ux \notin T\}.$$

A *path* P in T is an infinite sequence $P = (w_0, w_1, \ldots)$ of successive nodes of T, starting in the root of T. That is, $w_0 = \lambda$ and w_{m+1} is an immediate successor of w_m; *i.e.*, $w_{m+1} = w_m1$ or $w_{m+1} = w_m2$ for each m. Note that a finite tree T does not have any paths.

The infinite word $f \in \{1, 2\}^\omega$ is also called a *path* of T if each of its prefixes $u_i \in \{1, 2\}^*$ is a node of T. That is, if the infinite sequence $(u_1, u_2, \ldots)$ is a path in T in the other sense.

Given a set S, an *S-valued* tree, or shortly, an *S-tree* t is a mapping $t : T \mapsto S$ where T is a tree doamin. Then, T is called the *domain* of t, $T = dom(t)$. For $u \in T$, $t(u)$ is the *label* of the node u in t. We use root of t, path in t, boundary of t *etc.*, in order to refer to the corresponding objects of T.

If $P = (w_0, w_1, \ldots)$ is a path in t, let:

$$Inf(t, P) = \{s \in S \mid t(w_m) = s \text{ for infinitely many } m\}.$$

Observe that if S is finite then $Inf(t, P)$ is always nonempty, and there exists some index m_0 such that $t(w_m) \in Inf(t, P)$ for all $m > m_0$.

For an S-tree $t : dom(t) \mapsto S$ and a node $v \in dom(t)$, the *subtree* $t.v$ of t rooted in v is the S-tree defined by: $dom(t.v) = \{w \mid vw \in dom(t)\}$ and $t.v(w) = t(vw)$ for $w \in dom(t.v)$.

Finally, we define the *label of a path* $f \in \{1, 2\}^\omega$ as the infinite word $t(f) = t(u_1)t(u_2)\ldots \in \Sigma^\omega$ constituted by the labels of the nodes u_i on this path (*i.e.*, the finite words u_i are the prefixes of f).

From now on, for notational convenience, we will focus on full binary trees over a given fixed alphabet Σ, *i.e.*, on Σ-trees with $dom(t) = \{1,2\}^*$. Thus, any node $w \in dom(t)$ has exactly two immediate successors $w1$ and $w2$. Let T_Σ^ω be the collection of all full binary Σ−trees, *i.e.*, trees of the form $t : \{1,2\}^* \mapsto \Sigma$. We will refer to them simply by (infinite) trees. The extension of our results to n-ary trees (with $n \geq 2$) or ranked trees (where the number of successors of a node may vary with its label) is straightforward.

We now give the classical definition of an automaton on infinite trees with the Rabin acceptance condition.

A *Rabin automaton* on Σ−trees is a tuple $\mathcal{A} = (Q, q_0, \delta, \mathcal{T})$ where Q is a nonempty finite set of states, $q_0 \in Q$ (the initial state), $\delta \subseteq Q \times \Sigma \times Q \times Q$ (the set of transitions), and $\mathcal{T} = \{(L_1, U_1), \ldots, (L_N, U_N)\}$ where L_i, $U_i \subseteq Q$ (the collection of accepting pairs of states).

A q-run of the automaton $\mathcal{A}$ on a tree t is a Q-tree r, $r : dom(t) \mapsto Q$, such that: $r(\lambda) = q$, and $(r(w), t(w), r(w1), r(w2)) \in \delta$ for each node $w \in dom(t)$. A q_0-run is just called a *run*.

A path $P = (w_0, w_1, \ldots)$ of a given run r is called an *accepting path* if there exists some $i \in \{1, \ldots, N\}$ such that $Inf(r, P) \cap L_i \neq \emptyset$ and $Inf(r, P) \cap U_i = \emptyset$. If all its paths are accepting, r is called an *accepting run*. A tree t is *accepted* by an automaton $\mathcal{A}$ if there exists an accepting run of $\mathcal{A}$ on t. The set of trees accepted by $\mathcal{A}$ is denoted by $L(\mathcal{A})$.

For a state q of $\mathcal{A}$, we write $L_q(\mathcal{A})$ for the set of trees for which there exists an accepting q-run of $\mathcal{A}$ on t. Or, which are recognized by the automaton obtained from $\mathcal{A}$ by setting its initial state to be q.

In order to obtain the notions above for infinite words instead of trees, we note that words can be viewed as unary trees, *i.e.*, with domain $\{1\}^*$. A *sequential Rabin automaton* on infinite words over Σ can be defined just as a Rabin automaton $\mathcal{A} = (Q, q_0, \delta, \mathcal{T})$ on *unary* Σ−trees; that is, the set of transitions is now $\delta \subseteq Q \times \Sigma \times Q$. The notions of run, acceptance and the sets $L_q(\mathcal{A})$ are defined accordingly.

3 Transition Monoids

We next introduce the objects which, roughly, will allow us to go from infinite trees to infinite words. Namely, as we will see later, they correspond to infinite words.

Definition 1 (Marked Trees of a tree language). A *marked tree* is a pair (t, f) where $t \in T_\Sigma^\omega$ is a (full binary Σ−) tree, and $p \in \{1,2\}^\omega$ is a path (in t). If $L \subseteq T_\Sigma^\omega$ is a tree language, we associate with L the *set of marked trees of* L:

$$\widehat{L} = \{(t, f) \mid t \in L, f \in \{1,2\}^\omega\}.$$

The objects in the following definition are, as we will see shortly, finite words which are the prefixes to the infinite words above.

Definition 2 (Pointed Trees). A *pointed tree* t is a (binary) Σ−tree whose boundary $b(t)$ is a singleton.

That is, if t is a pointed tree with boundary $\{w\}$, then $dom(t) = \{1,2\}^* - w\{1,2\}^*$.

The set of pointed trees has a monoid structure in a natural way. Namely, the concatenation t_1t_2 of two pointed trees t_1 (with boundary $\{w_1\}$) and t_2 is obtained by sticking the root of t_2 into w_1. Formally, the pointed tree t_1t_2 is given by: $dom(t_1t_2) = dom(t_1) \cup w_1 dom(t_2)$, and $t_1t_2(v) = t_1(v)$ for $v \in dom(t_1)$ and $t_1t_2(w_1v) = t_2(v)$ for $v \in dom(t_2)$. Let us note that $b(t_1t_2) = b(t_1)b(t_2)$. The unity is the empty tree (the pointed tree with boundary $\{\lambda\}$).

We introduce the infinite set of pointed "base" trees:

$$\Gamma_\Sigma = \{t \mid b(t) \subset \{1,2\}\}.$$

Each element t of Γ_Σ can be represented as a triple (t', a, x) where $a = t(\lambda)$ is the label of the root, $x \in b(t)$ is the boundary element, and $t' = t.x'$ is the subtree of t rooted in the (only) immediate successor x' of the root. That is, $x' = 1$ if $x = 2$, and $x' = 2$ if $x = 1$. The monoid of pointed trees is a free monoid over Γ_Σ, and, hence, denoted Γ_Σ^*. This formalizes our viewing pointed trees as finite words.

If we write a pointed tree t with border node w as the pair (t, w), then we see that a marked tree is a degenerate case of a pointed tree, namely where $w \in \{1,2\}^\omega$. Clearly, a marked tree (t, w) corresponds uniquely to an infinite sequence $((t_1, x_1), (t_2, x_2), \ldots)$ of pointed base trees $(t_i, x_i) \in \Gamma_\Sigma$ (where, of course, $w = x_1x_2\ldots$), hence, to an infinite word.

For $a \in \Sigma$, $\alpha \in \{1,2\}$, we define the subset $\Gamma_\Sigma(a,\alpha) \subset \Gamma_\Sigma$ of pointed base trees with a as root label and α as border element,

$$\Gamma_\Sigma(a,\alpha) = \{t \in \Gamma_\Sigma \mid b(t) = \alpha,\ t(\lambda) = a\}.$$

The collection of these sets form a finite partition of Γ_Σ.

Definition 3 (Runs on pointed trees). A *(q,q')–run* of the automaton $\mathcal{A}$ over a pointed tree t with boundary $\{w\}$ is a Q-tree r with domain $dom(r) = dom(t) \cup \{w\}$, such that $r(\lambda) = q$, $r(w) = q'$, and $(r(w), t(w), r(w1), r(w2)) \in \delta$ for each node $w \in dom(t)$ of t.

The (q,q')-run $r : dom(t) \cup \{w\} \mapsto Q$ on the pointed tree t with border element w is *accepting* (and then called an accepting (q,q')-run on t) if all its (infinite!) paths are accepting. Thus, all paths in the set $\{1,2\}^\omega - w\{1,2\}^\omega$.

Finally, given a (q,q')-run r of $\mathcal{A}$ on t with border element w, we define $States(r) = \{r(w) \mid \lambda \leq w < b(t)\}$ as the set of all states encountered on the nodes from the root to the "hole" of the pointed tree.

Let $\mathcal{A} = (Q, q_0, \delta, T)$ be a Rabin tree automaton, where $\delta \subseteq Q \times \Sigma \times Q \times Q$, $T = ((L_i, U_i))_{1 \leq i \leq N}$, and $q_0 \in Q$. We define the equivalence relation $\sim_\mathcal{A}$ over Γ_Σ^* by saying that $t \sim_\mathcal{A} t'$ iff for all states $q, q' \in Q$ the following equivalences hold:

(1) There exists an accepting (q,q')-run r of $\mathcal{A}$ on t iff there exists an accepting (q,q')-run r of $\mathcal{A}$ on t'.

(2) There exists an accepting (q,q')-run r of $\mathcal{A}$ on t with $States(r) \cap L_i \neq \emptyset$ iff there exists an accepting (q,q')-run r of $\mathcal{A}$ on t' with $States(r) \cap L_i \neq \emptyset$, for all $i = 1,\ldots,n$.

(3) There exists an accepting (q,q')-run r of $\mathcal{A}$ on t with $States(r) \cap U_i = \emptyset$ iff there exists an accepting (q,q')-run r of $\mathcal{A}$ on t' with $States(r) \cap U_i = \emptyset$, for all $i = 1,\ldots,n$.

Lemma 4. *The relation $\sim_\mathcal{A}$ is a congruence of finite index.*

Proof The proof without any difficulty is left to the reader. □

Definition 5 (Transition Monoid of a Rabin Tree Automaton). Given the Rabin tree automaton $\mathcal{A}$, its transition monoid is the quotient of the free monoid of pointed trees with the equivalence relation $\sim_{\mathcal{A}}$,

$$\mathcal{M}(\mathcal{A}) = \Gamma^*_{\Sigma/\sim_{\mathcal{A}}}.$$

The lemma above implies that $\mathcal{M}(\mathcal{A})$ is always finite.

4 Strong-Morphism Recognizability

Given a Rabin tree automaton $\mathcal{A}$, the canonical morphism $\theta : \Gamma^*_\Sigma \mapsto \mathcal{M}(\mathcal{A}),\ t \mapsto [t]_{\sim_{\mathcal{A}}}$ yields a partition of Γ_Σ which is finite. We will see that it has an additional property, which is important in the following. Hence, we classify morphisms with this property in the following definition.

Definition 6 (Strong Morphisms). Let ψ be a morphism $\psi : \Gamma^*_\Sigma \longrightarrow M$ into a monoid M. The morphism ψ is called a strong morphism if, for all $a \in \Sigma$, $\alpha \in \{1, 2\}$, $m \in M$, the sets:

$$\{t' \in T^\omega_\Sigma \mid \psi(t', a, \alpha) = m\}$$

are Rabin recognizable.

These are the sets of the subtrees of the pointed base trees $t \in \Gamma_\Sigma$ with the same root label a, the same border element α and the same image m under ψ. Thus, of $t \in \Gamma_\Sigma(a, \alpha) \cap \psi^{-1}(m)$.

Lemma 7. *The canonical mapping $\theta : \Gamma^*_\Sigma \longrightarrow \mathcal{M}(\mathcal{A}),\ t \mapsto [t]_{\sim_{\mathcal{A}}}$ is a strong morphism.*

Proof Let $m \in \mathcal{M}(\mathcal{A})$, $a \in \Sigma$, and $\alpha \in \{1, 2\}$ be given. The two cases $\alpha = 1$ and $\alpha = 2$ being symmetrical, we will prove the statement for the first one; *i.e.*, that $\{t' \in T^\omega_\Sigma \mid \psi(t', a, 1) = m\}$ is Rabin recognizable.

If, for some pointed tree $t \in \Gamma^*_\Sigma$, $m = [t]_{\sim_{\mathcal{A}}}$, then $\psi(t', a, 1) = m$ iff $(t', a, 1) \sim_{\mathcal{A}} t$. An equivalence class can be represented as a Boolean combination of sets $L_{q''}(\mathcal{A})$, which, here, expresses the equivalences in the definition of $\sim_{\mathcal{A}}$. Clearly, r is a (q, q')-run of $\mathcal{A}$ on the pointed tree $(t', a, 1)$ iff there exists a state q'' such that $(q, a, q', q'') \in \delta$ and $t' \in L_{q''}(\mathcal{A})$. Also, if r is a (q, q')-run on the pointed base tree $(t', a, 1)$, then $States(r) = \{q\}$. That is, $\psi^{-1}(m)$ is the set:

$$\Big(\bigcap^{q,q',q'' \in Q,\ (q,a,q',q'') \in \delta}_{\text{ex. } (q,q')-\text{run on } t} \{(t', a, 1) \in \Gamma_\Sigma \mid t' \in L_{q''}(\mathcal{A})\}$$

$$- \bigcup^{q,q',q'' \in Q\ (q,a,q',q'') \in \delta}_{\text{ex. no } (q,q')-\text{run on } t} \{(t', a, 1) \in \Gamma_\Sigma \mid t' \in L_{q''}(\mathcal{A})\} \Big)$$

$$\cap \bigcap_{i=1,\ldots,N} \Big(\bigcap^{q \in L_i,\ q',q'' \in Q,\ (q,a,q',q'') \in \delta}_{\text{ex. } (q,q')-\text{run on } t,\ States(r) \cap L_i \neq \emptyset} \{(t', a, 1) \in \Gamma_\Sigma \mid t' \in L_{q''}(\mathcal{A})\}$$

$$- \bigcup^{q \in L_i,\ q',q'' \in Q\ (q,a,q',q'') \in \delta}_{\text{ex. no } (q,q')-\text{run on } t,\ States(r) \cap L_i \neq \emptyset} \{(t', a, 1) \in \Gamma_\Sigma \mid t' \in L_{q''}(\mathcal{A})\} \Big)$$

$$\cap \bigcap_{i=1,\ldots,N} \Big(\bigcap^{q \in U_i,\ q',q'' \in Q,\ (q,a,q',q'') \in \delta}_{\text{ex. } (q,q')-\text{run on } t,\ States(r) \cap U_i = \emptyset} \{(t', a, 1) \in \Gamma_\Sigma \mid t' \in L_{q''}(\mathcal{A})\}$$

$$- \bigcup^{q \in U_i,\ q',q'' \in Q\ (q,a,q',q'') \in \delta}_{\text{ex. no } (q,q')-\text{run on } t,\ States(r) \cap U_i = \emptyset} \{(t', a, 1) \in \Gamma_\Sigma \mid t' \in L_{q''}(\mathcal{A})\} \Big).$$

Thus, $\{t' \in T_\Sigma^\omega \mid \psi(t', a, 1) = m\}$ can be represented as the Boolean combination of sets $L_{q''}(\mathcal{A})$; this Boolean combination is obtained from the above one by replacing $\{(t', a, 1) \in \Gamma_\Sigma \mid t' \in L_{q''}(\mathcal{A})\}$ with $L_{q''}(\mathcal{A})$. Thus, it is Rabin recognizable, thanks to the Boolean closure of Rabin recognizable sets. □

Definition 8 (Strong Monoid Recognizability). The language $L \subseteq T_\Sigma^\omega$ of infinite trees is called *strong monoid recognizable* if there exists a strong morphism $\psi : \Gamma_\Sigma^* \longrightarrow M$ into a *finite* monoid M and the set $\widehat{L}$ is recognized by ψ, as a language of infinite words over the alphabet Γ_Σ, *i.e.*, $\widehat{L} \subseteq \Gamma_\Sigma{}^\omega$.

According to the usual definition, a morphism $\psi : A^* \mapsto M$ recognizes a language L of infinite words over A, $L \subseteq A^\omega$, if, for some set $P \subseteq M \times M$,

$$L = \bigcup_{(m,e)\in P} \psi^{-1}(m)(\psi^{-1}(e))^\omega.$$

A *linked pair* of elements of M is a pair (m, e) such that $me = m$, $e^2 = e$. The language $L \subseteq A^\omega$ is *saturated* by the morphism ψ if, for every linked pair $(m, e) \in M$:

$$L \cap \psi^{-1}(m)(\psi^{-1}(e))^\omega \neq \emptyset \Rightarrow \psi^{-1}(m)(\psi^{-1}(e))^\omega \subseteq L.$$

Clearly, a morphism saturating the language L also recognizes it. A partial converse to this property can be formulated. We need the following notion. The *Schützenberger product* of two monoids M and N is the set:

$$M \diamond N = \{(m, \rho, n) \mid m \in M,\ \rho \in M \times N,\ n \in N\}$$

equipped with the product:

$$(m, \rho, n)(m', \rho', n') = (mm', m\rho' \cup \rho n', nn'),$$

where:

$$m\rho' = \{(mr', s') \mid (r', s') \in \rho'\},$$
$$\rho n' = \{(r, sn') \mid (r, s) \in \rho\}.$$

Let $\phi : A^* \mapsto M$ and $\psi : A^* \mapsto N$ be two morphisms from the free monoid over A into the monoids M and N, respectively. Then one denotes by $\phi \diamond \psi$ the morphism:

$$\phi \diamond \psi : A^* \mapsto M \diamond N$$

from the free monoid over A into the monoid $M \diamond N$ which is defined by (on $w \in A^*$):

$$\phi \diamond \psi(w) = (\phi(w), \rho(w), \psi(w))$$

where $\rho(w) = (\{\phi(u), \psi(v)) \mid w = uv\}$.

Fact 9 *Let $\phi : A^* \mapsto M$ be a morphism from A^* into a finite monoid M and let L be a subset of A^ω. If ϕ recognizes L then $\phi \diamond \phi$ saturates L.*

For our notion of recognizability as in Definition 8, we need the following property.

Proposition 10. *If $\phi : \Gamma_\Sigma^* \mapsto M$ is a strong morphism, then $\phi \diamond \phi : \Gamma_\Sigma^* \mapsto M \diamond M$ is also a strong morphism.*

Proof The sets:

$$\{t' \in T_\Sigma^\omega \mid (t', a, \alpha) \in \Gamma_\Sigma(a, \alpha) \cap (\phi \circ \phi)^{-1}(m, \rho, n)\}$$

are Boolean combinations of Rabin recognizable sets. □

Lemma 11. *Let $L \subseteq T_\Sigma^\omega$ be a tree language recognized by a Rabin tree automaton $\mathcal{A}$. The canonical morphism $\theta : \Gamma_\Sigma^* \mapsto \mathcal{M}(\mathcal{A})$ saturates $\widehat{L}$.*

Proof Let (m, e) be a linked pair such that $\theta^{-1}(m)(\theta^{-1}(e))^\omega \cap \widehat{L} \neq \emptyset$. Let (t, f) be a marked tree in this intersection. Since $t \in L$, there exists an accepting run r of $\mathcal{A}$ over t. We can represent (t, f) as the infinite product $(t, f) = t_0 t_1 t_2 \ldots$ of pointed trees, $t_n \in \Gamma_\Sigma^*$ for $n = 0, 1, 2, \ldots$, such that:

- $\theta(t_0) = m$;
- $\theta(t_n) = e$, for $n = 1, 2, \ldots$;
- there exists an accepting (q_0, q)-run on t_0;
- there exists an accepting (q, q)-run r_i on t_n such that $States(r_n) \cap L_i \neq \emptyset$ and $States(r_n) \cap U_i = \emptyset$ for $n = 1, 2, \ldots$, for some i $(i = 1, \ldots, N)$.

Let $(t', f') \in \theta^{-1}(m)\theta^{-1}(e)^\omega$. Then (t', f') can be represented as $(t', f') = t'_0 t'_1 t'_2 \ldots$ with $\theta(t'_0) = m$, and for $i = 1, 2, \ldots$, $\theta(t'_n) = e$. This means there exists an accepting (q_0, q)-run r'_0 on t_0. For $i = 1, 2, \ldots$, there exists an accepting run r'_n on t'_n such that $States(r'_n) \cap L_i \neq \emptyset$ and $States(r'_n) \cap U_i = \emptyset$. Thus, by composing the runs $r'_0, r'_1, r'_2 \ldots$ we obtain an accepting run r' of $\mathcal{A}$ over the tree t'. Hence, $(t', f') \in \widehat{L}$. □

Corollary 12. *Let L be a tree language recognized by a Rabin tree automaton $\mathcal{A}$. The canonical morphism $\theta : \Gamma_\Sigma^* \mapsto \mathcal{M}(\mathcal{A})$ is a strong morphism recognizing L.*

Proof The statement is a consequence of Lemma 7 and Lemma 11. □

We will now prove the reverse of the statement above.

Proposition 13. *Let $\psi : \Gamma_\Sigma^* \mapsto M$ be a strong morphism into a finite monoid M, and $L \subseteq T_\Sigma^\omega$ a tree language such that ψ recognizes L. Then L is Rabin recognizable.*

Proof Since $\widehat{L}$ is recognized by the morphism ψ, it is recognized by a deterministic sequential Rabin automaton $\mathcal{B} = (Q, \Gamma_\Sigma, \delta, q_0, \mathcal{F})$, where $\delta \subseteq Q \times \Gamma_\Sigma \times Q \times Q$ and $\mathcal{F} = ((L_i, U_i))_{1 \leq i \leq N}$.

The monoid M acts on Q in the following way: $q \cdot_s m = q'$ if there is a pointed base tree (t, a, α) in (the alphabet) Γ_Σ such that $(q, (t, a, \alpha), q') \in \delta$ and $\psi(t, a, \alpha) = m$.

Moreover, since ψ is a strong morphism, for each $a \in \Sigma$, $\alpha \in \{1, 2\}$, $m \in M$ the set $\{t' \in T_\Sigma^\omega \mid \psi(t', a, \alpha) = m\}$ is recognizable by a Rabin tree automaton $\mathcal{A}_{(m,a,\alpha)} = (Q_{(m,a,\alpha)}, \Sigma, \delta_{(m,a,\alpha)}, q_{0_{(m,a,\alpha)}}, \mathcal{F}_{(m,a,\alpha)})$.

We will build a Rabin tree automaton $\mathcal{A}$ which, roughly, simulates the sequential Rabin automaton $\mathcal{B}$ on a path $f \in \{1, 2\}^\omega$ of a tree $t \in T_\Sigma^\omega$ in the following way. At each node w along the path f (*i.e.*, $w \in \{1, 2\}^*$, $w < f$), if a is its label, $\alpha \in \{1, 2\}$ the next direction on f (*i.e.*, $w\alpha < f$), and t' the subtree of t rooted in the other successor node (*i.e.*, in $w\alpha'$ where $\alpha' \neq \alpha$), it "guesses" the image m in M of the pointed base tree $(t', a, \alpha) \in \Gamma_\Sigma$; then it "checks" its guess by using the appropriate automaton $\mathcal{A}_{(m,a,\alpha)}$. More precisely, we define:

$$\mathcal{A} = (Q \cup \bigcup_{\alpha \in \{1,2\}}^{m \in M, a \in \Sigma} Q_{(m,a,\alpha)},\ \Sigma,\ \bigcup_{\alpha \in \{1,2\}}^{m \in M, a \in \Sigma} \delta_{(m,a,\alpha)} \cup \delta',\ q_0,\ \mathcal{F} \cup \bigcup_{\alpha \in \{1,2\}}^{m \in M, a \in \Sigma} \mathcal{F}_{(m,a,\alpha)})$$

where:

- $(q, a, q', q0_{(m,a,\alpha)}) \in \delta'$ iff there exists some $(t, a, 1) \in \Gamma_\Sigma$ such that $\psi(t, a, 1) = m$ and $q \cdot_{\mathcal{B}} m = q'$;
- $(q, a, q0_{(m,a,\alpha)}, q') \in \delta'$ iff there exists some $(t, a, 2) \in \Gamma_\Sigma$ such that $\psi(t, a, 2) = m$ and $q \cdot_{\mathcal{B}} m = q'$.

We show that the language recognized by $\mathcal{A}$ is exactly L.

First, let $t \in L$. We choose an arbitrary path $f \in \{1,2\}^\omega$. Since $(t,f) \in \widehat{L}$, the marked tree, as an infinite word over Γ_Σ, $(t,f) = t_0 t_1 t_2 \ldots$ where $t_0, t_1, t_2 \in \Gamma_\Sigma$, admits an accepting run $r = (q_0, q_1, q_2, \ldots)$ by the sequential Rabin automaton $\mathcal{B}$.

Thus, there exists a state q, an integer i and an increasing sequence $(n_p)_{p \in \mathbb{N}}$ such that r can be decomposed into an accepting (q_0, q)-run r_{n_0} on the pointed tree $t_{n_0+1} t_{n_0+2} \ldots t_{n_1} \in \Gamma_\Sigma^*$, followed by the accepting (q, q)-runs r_{n_p} on the pointed tree $t_{n_p+1} t_{n_p+2} \ldots t_{n_{p+1}} \in \Gamma_\Sigma^*$ such that $States(r'_{n_p}) \cap L_i \neq \emptyset$ and $States(r'_{n_p}) \cap U_i = \emptyset$.

We choose $\alpha_n \in \{1,2\}$ for $n = 0, 1, 2, \ldots$ such that $f = \alpha_0\alpha_1\alpha_2 \ldots$; we set $f_n = \alpha_0\alpha_1\alpha_2 \ldots$, $a_n = t(f_n) \in \Sigma$, and $\psi(t_n) = m_n$.

We denote by $\bar{\alpha}_n$ the element in $\{1,2\}$ different from α_n. Now we can give an accepting run r of $\mathcal{A}$ over t, by setting, for $n \geq 0$,

- $r(f_n) = q_0 \cdot_{\mathcal{B}} (t_0 \ldots t_n)$;
- $r(f_n\bar{\alpha}_n) = q0_{(m_n,a_n,\alpha_n)}$;
- r is defined on the subtree $t.f_n\bar{\alpha}_n$ identical to an accepting run of the automaton $\mathcal{A}_{(m_n,a_n,\alpha_n)}$; *i.e.*, starting in its initial state $q0_{(m_n,a_n,\alpha_n)}$.

Now, it is clear that r is an accepting run on f since, restricted to f, r is the behaviour of $\mathcal{B}$ on (t,f). For each subtree $t.f_n\bar{\alpha}_n$, r is the behaviour of a Rabin automaton which recognizes this subtree. So, r is an accepting run on the whole tree t.

Reversely, let r be an accepting run of the automaton $\mathcal{A}$ over a tree $t \in T_\Sigma^\omega$. On some path f such that $(t,f) = t'_0 t'_1 t'_2 \ldots$, where $t_0, t_1, t_2, \ldots \in \Gamma_\Sigma$, r is the run of the automaton $\mathcal{B}$ over the infinite word $t'_0 t'_1 t'_2 \ldots$.

Again, we choose $\alpha_n \in \{1,2\}$ for $n = 0, 1, 2, \ldots$ such that $f = \alpha_0\alpha_1\alpha_2 \ldots$; we set $f_n = \alpha_0\alpha_1\alpha_2 \ldots$, and $a_n = t(f_n) \in \Sigma$.

There exists an element $m_n = \psi(t'_n)$ such that $r(f_n\bar{\alpha}_n) = q0_{(m_n,a_n,\alpha_n)}$ and since the run is an accepting one, it proves that $\psi(t_n) = \psi(t'_n)$. Since $t'_0 \ldots t'_n \ldots \in \widehat{L}$ then $(t,f) \in \widehat{L}$ and therefore $t \in L$. □

We can summarize the previous facts as follows.

Theorem 14. *A language $L \subseteq T_\Sigma^\omega$ of infinite trees is Rabin recognizable if and only if it is strong monoid recognizable. Moreover, we may assume that the morphism saturates the language $\widehat{L}$.*

That is, if and only if $L \subseteq T_\Sigma^\omega$ is Rabin recognizable then $\widehat{L} \subseteq {\Gamma_\Sigma}^\omega$ is recognized by a strong morphism $\psi : \Gamma_\Sigma^* \mapsto M$ into a finite monoid M.

Proof The "if" direction is exactly Proposition 13. The "only if" direction is Corollary 12. By Fact 9 and Proposition 10 we may assume that the morphism saturates $\widehat{L}$. □

As an immediate consequence of this theorem, we obtain the following result:

Theorem 15. *Let $\psi : \Gamma_\Sigma^* \mapsto M$ be a morphism into a finite monoid, and L a tree language such that ψ saturates $\widehat{L}$. Then, ψ recognizes the complement of L.*

5 A Counter Example

We will show that in Theorem 14 we need the condition that the morphism be a strong one.

Proposition 16. *Let Σ be an alphabet with at least two letters. There exists a morphism $\psi : \Gamma_\Sigma^* \mapsto M$ into a finite monoid M and a tree language L such that $\widehat{L}$ is recognized by ψ and L is not Rabin recognizable.*

Proof Let $\Sigma \subseteq \{a, b\}$. We shorten the quantification "there exists at most countably many" by $\exists^{\leq\omega}$. Thus, $\exists^{\leq\omega}. x \in S$ holds iff S is finite. For $u, v \in \Sigma^\omega$, let:

$$u \sim_{\text{suffix}} v \text{ iff } \exists w \in \Sigma^\omega \ \exists u', v' \in \Sigma^*.\ u = u'w,\ v = v'w;$$

i.e., iff u and v share a common suffix w. Clearly, $\sim_{\text{suffix}}$ defines an equivalence relation on Σ^ω. The equivalence class of u is $[u]_{\sim_{\text{suffix}}} = \Sigma^* \cdot ((\Sigma^*)^{-1}u)$. Thus, $[u]_{\sim_{\text{suffix}}}$ is an enumerable set. Since the cardinality of $\{1, 2\}^\omega$ is the continuum, the number of different sets $[u]_{\sim_{\text{suffix}}}$, $u \in \Sigma^\omega$, is the continuum.

Given an infinite word $u \in \Sigma^\omega$, we define the language L_u of all infinite trees which have at most countably many paths f with a label $t(f)$ which has a suffix in common with u,

$$L_u = \{t \in T_\Sigma^\omega \mid \exists^{\leq\omega} f \in \{1,2\}^\omega \mid t(f) \sim_{\text{suffix}} u\}.$$

We next show that if $u \not\sim_{\text{suffix}} v$ then $L_u \neq L_v$; this yields that the number of different languages L_u, $u \in \Sigma^\omega$ is the continuum. Let $t_u \in T_\Sigma^\omega$ be the infinite tree such that all its paths are labeled by u. Then $t_u \in L_v - L_u$ for any $v \not\sim_{\text{suffix}} u$.

Let us consider the finite monoid $M = \{1, 0, e\}$ which is the two-element lattice together with the extra neutral element e. That is, the monoid product is $1 \cdot 1 = 1, 1 \cdot 0 = 0 \cdot 1 = 0 \cdot 0 = 0$. The map $\psi_u : \Gamma_\Sigma^* \mapsto M$ defined by:

$$\psi_u(t) = \begin{cases} 1 \text{ if } \exists^{\leq\omega} f \in \{1,2\}^\omega.\ t(f) \sim_{\text{suffix}} u; \\ 0 \text{ otherwise.} \end{cases}$$

is clearly a morphism. And $\widehat{L}$ is recognized by ψ since $\widehat{L}_u = (\psi_u^{-1}(1))^\omega$. Since the set of Rabin languages over the alphabet Σ is enumerable, there exists an infinite word u such that $\widehat{L}_u$ is recognized by ψ_u and L_u is not Rabin recognizable. □

6 Conclusion and Further Work

We have shown how one can incorporate finite monoids into the theory of recognizability of infinite tree languages by Rabin automata. The notions of syntactic congruence, syntactic monoid and (star-free) regular expression for a language L of infinite trees can now be readily obtained by taking those of the associated language $\widehat{L}$ of pointed trees, viewed as infinite strings over the free monoid Γ_Σ^*. This way, we can also obtain the extension of the Theorem of Kleene to infinite trees, *i.e.*, the characterization of recognizability by Rabin tree automata through regular expressions.

Following the lines of [Per84, Pec87] and of [Tho84, NP89, PP92] for the cases of infinite strings and finite trees, respectively, one can now start the investigation of a theory of varieties of languages of infinite trees.

We note that our notion of strong monoid recognizability (and the characterization in Theorem 14) can not be used to *define* Rabin recognizability. This does not, however, concern the research proposed above, since the classification theory relies on the fact that the automata recognizability implies the finiteness of the recognizing monoid; hence, in our case, on the other direction of Theorem 14.

If, in Definition 6, one replaces Rabin by Büchi automata over infinite trees, one obtains a new notion of strong monoid recognizability. In a forthcoming paper we will show that one can obtain a characterization of the *weak* monadic second-order logic in this way.

One motivation of the direction of this work is to circumvent the complementation of Rabin automata by using the Boolean invariance of monoid morphisms. The previous section shows that this is not possible by taking the straightforward notion of monoid recognizability. Is there, however, a condition on the sets coming from the partition of Γ_Σ in Definition 6 (of strong monoid recognizability) which is weaker than the present one and such that: Theorem 14 still holds, but its proof does not use the complementation of Rabin automata?

Acknowledgements. We would like to thank Maurice Nivat and Damian Niwiński for fruitful discussions and an anonymous referee for his insightful comments.

References

[Bu62] John R. Büchi "On a decision method in restricted second order arithmetic". In Erich Nagel et al., editors, *Logic, Methodology, and Philosophy of Science*, pages 1–11. Stanford Univ. Press, 1960.

[Eil76] Samuel Eilenberg. *Automata, Language and Machine*, volume B of *Applied and Pure Mathematics*. Academic Press, 1976.

[KT90] Dexter Kozen and Jerzy Tiuryn. "Logics of programs". In Jan van Leeuwen, editor, *Handbook of Theoretical Computer Sience*, volume B, chapter 4, pages 789–840. Elsevier, 1990.

[NP89] Maurice Nivat and Andreas Podelski. "Tree Monoids and Recognizable Sets of Trees". In Hassan Aït-Kaci and Maurice Nivat, editors, ***Resolution of Equations in Algebraic Structures***, Vol. 1 (Academic Press, London, 1989).

[NP92] Maurice Nivat and Andreas Podelski, editors, *Tree Automata, Advances and Open Problems*. Elsevier Science, 1992.

[Pec87] Jean-Pierre Pécuchet. "Etude syntaxique des parties reconnaissables de mots infinis". In Laurent Kott, editor, *Proc. 13th ICALP*, LNCS 226, pages 294–303, 1987.

[PP92] Pierre Peladeau and Andreas Podelski. "On Reverse and General Definite Tree Languages". In *Proc. 18th ICALP*, LNCS. 1992.

[Per84] Dominique Perrin. "Recent results on automata and infinite words". In M.P. Chytil and V. Koubek, editors, *Math. Found. of Comp. Sci.*, LNCS 176, pages 134–148. 1984.

[Pi84] Jean-Eric Pin, *Variétés de langages formels*, Masson, Paris, 1984, and *Varieties of Formal Langages*, Plenum, London, 1986.

[Rab69] Michael O. Rabin. "Decidability of second-order theories and automata on infinite trees". *Trans. Amer. Math. Soc.* 141, pages 1–35. 1969.

[Rab77] Michael O. Rabin. "Decidable theories". In John Barwise, editor, *Handbook of Mathematical Logic*, pages 595–630. North-Holland, 1977.

[Stei92] Magnus Steinby, "A Theory of Tree Language Varieties". In Maurice Nivat and Andreas Podelski, editors, *Tree Automata, Advances and Open Problems*. Elsevier Science, 1992.

[Tho84] Wolfgang Thomas. "Logical aspects in the study of tree languages", In Bruno Courcelle, editor, ***Proc. Ninth Coll. Trees in Algebra and in Programming***, pages 31-49. Cambridge Univers. Press, Cambridge, 1984.

[Tho90] Wolfgang Thomas. "Automata on infinite objects". In Jan van Leeuwen, editor, *Handbook of Theoretical Computer Sience*, volume B, chapter 4, pages 133–161. Elsevier, 1990.

Efficient type reconstruction in the presence of inheritance

(Extended abstract)

Marcin Benke

Institute of Informatics
Warsaw University
ul. Banacha 2
02-097 Warsaw, POLAND
e-mail: benke@mimuw.edu.pl

Abstract. The complexity of type reconstruction for simply-typed lambda calculus with subtype relation resulting from single inheritance (i.e. being a disjoint union of tree-like posets) is analyzed. As a result a class of posets including (but not restricted to) trees is defined, for which the said problem is solvable in polynomial time.

Introduction

In 1984, John Mitchell [4], stated the problem of type inference for the simply-typed lambda calculus with coercions between base types (we shall call this problem TRS). He proved that it is reducible to the problem of satisfiability of systems of inequalities between type variables and base type constants (called PO-SAT or FLAT-SSI), and therefore decidable. He did not though consider in depth the question of complexity.

In 1989 Wand and O'Keefe [9] claimed that PO-SAT is NP-complete and that TRS reduces to it in polynomial time. Regretfully, the latter proof is incorrect. [3] Lincoln and Mitchell observe that algorithm for trees works in exponential rather than polynomial time.

TRS reduces to the problem of solving system of inequalties involving not only constants and variables but also an arrow — a binary functor (this problem, called PO-SAT will be specified in detail later). Tiuryn [7] proved that SSI is PSPACE-hard.

Further Wand and O'Keefe observe that if partial order is known to be a tree, then the satisfiability problem is solvable in polynomial time. In fact, their algorithm for trees works in exponential rather than polynomial time.

Main result In this paper we prove a theorem from which it follows that the SSI problem for trees is indeed satisfiable in polynomial time. Therefore the type reconstruction problem for such posets is in PTIME.

Why trees are important The original motivation for studying orders being trees was that trees correspond to the coercion structure of single-inheritance object systems. Although traditionally, recursive types are used for inheritance mechanism, recent works show that it is possible to construct type systems for inheritance which do not rely on recursive types (cf. e.g. [6]).

On the other hand trees are present even if we consider simple coercion. Perhaps the simplest possible example is a numeric or enumeration type and a couple of its subrange types.

Example One may want consider types (like in the language C) *signed int* $= -2^{15}..2^{15}-1$, *unsigned int* $= 0..2^{16}-1$, *long int* $= -2^{31}..2^{31}-1$ and coercions between them based on respective inclusions. Then the poset would look like

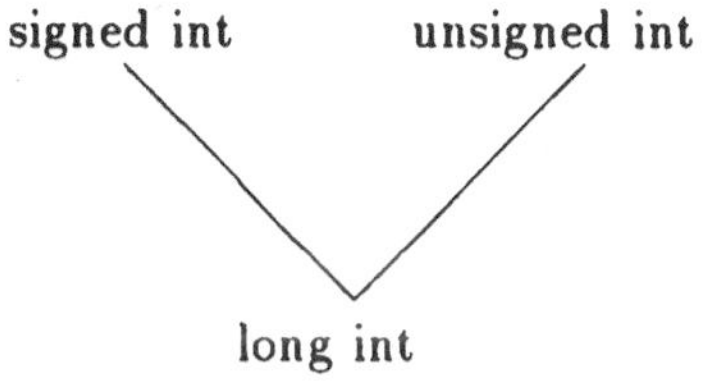

Fig. 1. A tree like coercion structure

Why trees aren't trivial The main problem is that since we consider non-flat inequalities, any result for trees actually has to cover a wider range of posets, since any algorithm working for given poset, must obviously work for poset formed of terms of given shape over this poset. Therefore the problem is in finding a class of posets wide enough to facilitate inductive proof, and narrow enough to fulfill satisfactory thesis.

Example Consider set of types from previous exmaple

$$C = \{\textit{signed int}, \textit{unsigned int}, \textit{long}\}$$

and let

$$C_{\star\to\star} = \{\rho \to \tau : \rho, \tau \in C\}$$

then coercions between elements of $C_{\star\to\star}$ form a poset shown in Fig. 2

This poset seems all but trivial. In fact it bear some resemblance to the posets for which the type reconstruction is NP-hard [3] and the SSI problem is PSPACE-hard [7] — crowns

In fact, because of this resemblance it had been for some time conjectured that the SSI problem for trees is NP-hard. However, as this work proves, said resemblance was deceiving.

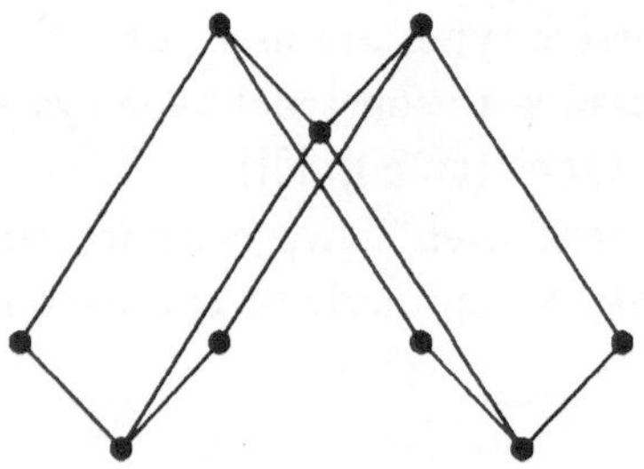

Fig. 2. An example of function space ordering

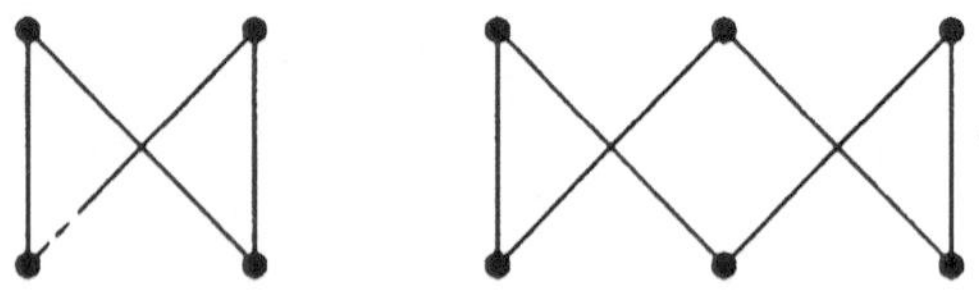

Fig. 3. Crowns

Related work A lot of effort has been devoted to studying type reconstruction recently. Bruce and Mitchell [1] present models of subtyping, recursive types and higher order polymorphism with types interpreted as partial equivalence relations over a D_∞ lambda model. Tiuryn and Wand [8] study complexity of type reconstruction with subtyping and recursive types. Kozen, Palsberg and Schwartzbach [2] analyze subtype relations for inferring partial types, where all types regardless of their rank are in relation with Ω, representing the 'unknown' type.

1 Definitions

1.1 The problem

Let $\langle Q, \leq \rangle$ be a finite poset. Let $\rightarrow$ be a binary infix operation symbol, and let $\mathcal{T}_Q$ be the term algebra over $(Q, \rightarrow)$. Its carrier consists of finite terms without variables and is partially ordered by extending the order from Q to all terms by the following rule

$$(t_1 \rightarrow t_2) \leq (r_1 \rightarrow r_2) \quad \text{iff} \quad r_1 \leq t_1 \text{ and } t_2 \leq r_2$$

That means that we can think of $\rightarrow$ as of operation which is antimonotonic in the first argument and monotonic in second.

A system Σ of inequalities is a finite set of formulas of the form

$$\Sigma = \{\tau_1 \leq \rho_1, \ldots, \tau_n \leq \rho_n\}$$

where τ's and ρ's are terms over the signature $(Q, \rightarrow)$ with variables from some set V.

Σ is said to be *flat* if every term in it is of size 1 i.e. is either a constant or a variable. Note that if Σ is not flat then by its size we mean the total size of terms it contains, rather than the number of inequalities.

Σ is said to be *satisfiable* in $\mathcal{T}_Q$ if there is a valuation $v : V \rightarrow \mathcal{T}_Q$ such that $\tau_i[v] \leq \rho_i[v]$ holds in $\mathcal{T}_Q$ for all i.

There are actually two problems which are addressed in this paper:

(Q-SSI) *Given a system Σ of inequalities, is it satisfiable in $\mathcal{T}_Q$?*

(Q-FLAT-SSI) *Given a flat system Σ of inequalities, is it satisfiable in $\mathcal{T}_Q$?*

1.2 Weak satisfiability

Let $\mathcal{T}_\star$ be the set of terms without variables over single constant $\star$ and binary operation $\rightarrow$. The elements of $\mathcal{T}_\star$ we call *shapes*. We use a canonical map $(\cdot)_\star : \mathcal{T}_Q(V) \rightarrow \mathcal{T}_\star(V)$ which maps all constants to $\star$ but leaves the variables untouched. We call $(\tau)_\star$ the shape of τ. It is defined as follows:

$$(c)_\star = \star \quad \text{if } c \in Q$$

$$(v)_\star = v \quad \text{if } v \in V$$

$$(\tau \rightarrow \rho)_\star = (\tau)_\star \rightarrow (\rho)_\star$$

For $\sigma \in \mathcal{T}_\star$ let

$$\mathcal{T}_\sigma = \{\rho \in \mathcal{T}_Q : (\rho)_\star = \sigma\}$$

A system $\Sigma = \{\tau_1 \leq \rho_1, \ldots, \tau_n \leq \rho_n\}$ is said to be weakly satisfiable if $\Sigma_\star = \{(\tau_1)_\star = (\rho_1)_\star, \ldots, (\tau_n)_\star = (\rho_n)_\star\}$ is satisfiable in $\mathcal{T}_\star$. Weak satisfiability is clearly a necessary condition for satisfiability. It is decidable in polynomial time since it is an instance of the unification problem.

1.3 Distances and disks in a poset

We say that the *up-distance from x to y in Q doesn't exceed n*, in symbols $Q \models d^{+1}(x, y) \leq n$, if there exist $x_0, \ldots, x_n \in Q$, such that $x_0 = x, x_n = y$, and $x_{2i} \leq x_{2i+1} \geq x_{2i+2}$. (Such a set $\{x_0, \ldots, x_n\}$ is called an up-fence of length n)

We say that the *down-distance from x and y in Q doesn't exceed n*, in symbols $Q \models d^{-1}(x, y) \leq n$, if there exist $x_0, \ldots, x_n \in Q$, such that $x_0 = x, x_n = y$ and $x_{2i} \geq x_{2i+1} \leq x_{2i+2}$. (Such a set $\{x_0, \ldots, x_n\}$ is called a down-fence of length n)

By a *disk in Q* with the center a and radius (r^{+1}, r^{-1}) we shall mean the set

$$D_Q(a, (r^{+1}, r^{-1})) = \{x \in Q : Q \models d^\varepsilon(a, x) \leq r^\varepsilon, \varepsilon = \pm 1\}$$

An ordered set satisfies the two disk property (alias Helly property) if, for each family $\mathcal{D}$ of disks

$$\bigcap \mathcal{D} \neq \emptyset$$

whenever

$$D_1 \cap D_2 \neq \emptyset$$

for each D_1, D_2 in $\mathcal{D}$.

The two disk property is preserved by products. Hence, if Q satisfies it, so do sets of terms of any fixed shape.

1.4 Retractions and absolute retracts

We say that $P \supseteq Q$ retracts to Q ($P \triangleright Q$) if there exists an order preserving and idempotent map $f : P \rightarrow Q$. Nevermann and Rival [5] show a class of posets for which exists a simple condition which is necessary and sufficient for retractibility.

A set

$$\mathcal{H} = \{(v_t, \delta_t^{+1}, \delta_t^{-1}) : t \in T\}$$

where T is an index set, $v_t \in Q$, is a hole in Q if

$$\bigcap_{t \in T} D_Q(v_t, \delta_t^{+1}, \delta_t^{-1}) = \emptyset$$

and $\mathcal{H}$ contains no proper and nonempty subfamily of disks with empty intersection.

We say that this hole is separated in $P \supseteq Q$ if

$$\bigcap_{t \in T} D_P(v_t, \delta_t^{+1}, \delta_t^{-1}) = \emptyset$$

A set Q is called an absolute retract if it is a retract of every ordered set in which all its holes are separated. An ordered set that satisfies the two disk property is an absolute retract [5]. Since in a poset that satisfies the two disk property, every hole consists of two elements, we shall henceforth restrict discussion to such holes.

It is not difficult to see that trees (and disjoint unions of them) do satisfy the two disk property and are therefore absolute retracts. On the other hand, crowns, which are known to be intractable [7] are not absolute retracts.

1.5 Inference system $\vdash_s$

Given system of ineqaualities we introduce a deduction system which allows to infer inequalities that must be satisfied by any solution of Σ. [7]

$$[\text{s-Ax}]\ \Sigma \vdash_s \sigma \leq \tau \qquad \text{for } \sigma \leq \tau \in \Sigma \tag{1}$$

$$[\text{s-Trans}]\ \frac{\Sigma \vdash_s \sigma \leq \tau, \quad \Sigma \vdash_s \tau \leq \rho}{\Sigma \vdash_s \sigma \leq \rho} \tag{2}$$

$$[\text{s-Subterm}]\ \frac{\Sigma \vdash_s \sigma_1 \rightarrow \sigma_2 \leq \tau_1 \rightarrow \tau_2}{\Sigma \vdash_s \tau_1 \leq \sigma_1, \quad \sigma_2 \leq \tau_2} \tag{3}$$

Call Σ *ground consistent* whenever for each $c_1, c_2 \in Q$ if $\Sigma \vdash_s c_1 \leq c_2$, then $Q \models c_1 \leq c_2$.

1.6 Inference system $\vdash_d$

Given system of inequalities we introduce a deduction system which allows to infer facts about distances between elements that must be satisfied by any solution of Σ.

$$\text{[d-Ax]} \qquad \Sigma \vdash_d d^{+1}(\sigma,\tau) \leq 1 \quad \text{dla } \sigma \leq \tau \in \Sigma \tag{4}$$

$$\text{[d-Symm]} \qquad \frac{\Sigma \vdash_d d^{\varepsilon}(\sigma,\tau) \leq n}{\Sigma \vdash_d d^{(\varepsilon(-1)^n)}(\tau,\sigma) \leq n} \tag{5}$$

$$\text{[d-Dual]} \qquad \frac{\Sigma \vdash_d d^{\varepsilon}(\sigma,\tau) \leq n}{\Sigma \vdash_d d^{(-\varepsilon)}(\sigma,\tau) \leq n+1} \tag{6}$$

$$\text{[d-Trans]} \qquad \frac{\Sigma \vdash_d d^{\varepsilon}(\sigma,\tau) \leq n,\ d^{\varepsilon(-1)^{n-1}}(\tau,\rho) \leq k}{\Sigma \vdash_d d^{\varepsilon}(\sigma,\rho) \leq n-1+k} \tag{7}$$

$$\text{[d-Subterm]} \ \frac{\Sigma \vdash_d d^{\varepsilon}(\sigma_1 \to \sigma_2, \tau_1 \to \tau_2) \leq n}{\Sigma \vdash_d d^{(-\varepsilon)}(\sigma_1,\tau_1) \leq n, \Sigma \vdash_d d^{\varepsilon}(\sigma_2,\tau_2) \leq n} \tag{8}$$

Call Σ *distance consistent* whenever for each $c_1, c_2 \in Q$ and $n \in N$ if $\Sigma \vdash_d d^{\varepsilon}(c_1,c_2) \leq n$, then $Q \models d^{\varepsilon}(c_1,c_2) \leq n$.

Distance consistency is decidable in polynomial time. We can construct two $O(|\Sigma|) \times O(|\Sigma|)$ arrays, for d^{+1} and d^{-1} respectively in such a way that in the array for d^{ε}, in the cell indexed by τ and ρ there is the least number n such that $\Sigma \vdash_d d^{\varepsilon}(\tau,\rho) \leq n$ holds.

2 The result

The main result presented in this paper is that a system of inequalities over a poset satisfying the Helly property is satisfiable iff it is weakly satisfiable and distance consistent. Thus satisfiability of such systems is decidable in polynomial time.

Lemma 1. *$\Sigma \vdash_s x \leq y$ iff $\Sigma \vdash_d d^{+1}(x,y) \leq 1$*

Proof. The left-to-right implication follows from axiom 4. The opposite one can be proved by simple induction on inference of $\Sigma \vdash_d d^{+1}(x,y) \leq 1$. □

Corollary 2. *If Σ is distance consistent then it is also ground consistent.*

Let Σ be ground consistent. Let us define the following equivalence relation $\simeq$ on $var(\Sigma)$:

$$x \simeq y \iff \Sigma \vdash_s x \leq y \wedge \Sigma \vdash_s y \leq x$$

By an *extension of Q generated by Σ* we shall mean the set

$$Q_{\Sigma} = Q \cup var(\Sigma)/\simeq$$

with the order defined as a transitive closure of the sum of the order on Q and the order on $Q \cup var(\Sigma)/\simeq$, induced by $\Sigma \vdash_s$.

Lemma 3. *If Σ is a flat system of inequalities then it is satisfiable if and only if Σ is ground consistent and $Q_\Sigma \triangleright Q$*

Proof. Each retraction $f : Q_\Sigma \rightarrow Q$ determines a solution of Σ:

$$\sigma(x) = f([x]_\simeq)$$

On the other hand, if σ is a solution of Σ, then

$$x \simeq y \Rightarrow \sigma(x) = \sigma(y)$$

□

Lemma 4. *Let Σ — ground consistent. Then:*

(a) *If $Q_\Sigma \models [x] \leq c$ for some $x \in var(\Sigma)$, $c \in Q$ then there exists $c' \in Q$ such that*

$$\Sigma \vdash_s x \leq c' \quad Q \models c' \leq c$$

(b) *If $Q_\Sigma \models c \leq [x]$ for some $x \in var(\Sigma)$, $c \in Q$ then there exists $c' \in Q$ such that*

$$\Sigma \vdash_s c' \leq x \quad Q \models c \leq c'$$

(c) *If $Q_\Sigma \models [x_1] \leq [x_2]$ for some $x_1, x_2 \in var(\Sigma)$ then either $\Sigma \vdash_s x_1 \leq x_2$ or there exist $c_1, c_2 \in Q$ such that*

$$\Sigma \vdash_s x_1 \leq c_1 \quad Q \models c_1 \leq c_2 \quad \Sigma \vdash_s c_2 \leq x_2$$

(d) *If $Q_\Sigma \models c_1 \leq c_2$ for some $c_1, c_2 \in Q$ then also $Q \models c_1 \leq c_2$*

Proof. Easy and technical (see the full paper).

□

Lemma 5. *If Σ is distance consistent than we have for any $c_1, c_2 \in Q$*

$$Q_\Sigma \models d^\varepsilon(c_1, c_2) \leq n \iff Q \models d^\varepsilon(c_1, c_2) \leq n$$

Proof. The implication ($\Leftarrow$) is obvious. To prove ($\Rightarrow$) assume the contrary, i.e. that

$$Q_\Sigma \models d^\varepsilon(c_1, c_2) \leq n$$

and

$$\neg(Q \models d^\varepsilon(c_1, c_2) \leq n).$$

We show that there exist $d_1, d_2 \in Q$, $m \leq 0$ such that

$$\Sigma \vdash_d d^\varepsilon(d_1, d_2) \leq m$$

and

$$\neg Q \models d^\varepsilon(d_1, d_2) \leq m.$$

The proof is by induction on n. The basis of induction follows from the lemma 4. The induction step goes as follows.

Let $c_1 = x_0, x_1, \ldots, x_n = c_2$ be a fence in Q_Σ, connecting c_1 with c_2. If $x_1 \in Q$, then we may use the induction hypothesis since

$$Q_\Sigma \models d^{-\epsilon}(x_1, c_2) \leq n - 1$$

$$\neg(Q \models d^{-\epsilon}(x_1, c_2) \leq n - 1)$$

On the other hand, if $x_1 = [v]$, $v \in var(\Sigma)$, then choose largest i such that $\Sigma \vdash_d d^{-\epsilon}(x_1, x_i) \leq i - 1$ (or $i = 1$, if no such exists). From the previous lemma it follows that there exists $d_1 \in Q$ such that $Q \models d^\epsilon(c_1, d_1) \leq 1$ and $\Sigma \vdash_d d^\epsilon(d_1, x_1) \leq 1$. Similarly, if $x_1 \notin Q$, then there exists $d_2 \in Q$ such that $\Sigma \vdash_d d^{\epsilon(-1)^i}(x_i, d_2) \leq 1$.

Now, one of the following must hold:

$$\neg(Q \models d^\epsilon(d_1, d_2) \leq i)$$

or

$$Q_\Sigma \models d^{\epsilon(-1)^i}(d_2, c_2) \leq n - i - 1,$$

$$\neg Q \models d^{\epsilon(-1)^i}(d_2, c_2) \leq n - i - 1$$

The first case obviously contradicts the distance consistency of Σ and the second one allows us to use the induction hypothesis □

Theorem 6. *A flat system of inequalities over a poset satisfying the Helly property is satisfiable if and only if it is distance consistent.*

Proof. Again, the left-to-right implication is pretty obvious. By the lemma 3, in order to prove the opposite one, it suffices to prove that $Q_\Sigma \triangleright Q$. Furthermore, since Q satisfies the two-disk property and thus is an absolute retract, all we have to prove that all holes in Q of cardinality 2 are separated in Q_Σ. It is shown (see the full paper) that assuming the contrary, i.e. the existence of a pair of disks which is a hole in Q that is not separated in Q_Σ, leads to a contradiction with distance consistency of Σ

□

This result can be lifted to arbitrary systems of inequalities leading us to the following

Theorem 7. *A system of inequalities Σ over a poset satisfying the Helly property is satisfiable iff it is weakly satisfiable and distance consistent.*

Proof. The ($\Rightarrow$) implication is obvious. The opposite implication is proved by induction on the number of equivalence classes of $\sim$ defined on $var(\Sigma)$ as follows

$$x \sim y \quad \text{iff} \quad \Sigma_* \models x = y$$

The proof is indeed nearly identical to the one in [7], only the notion of ground consistency has to be replaced with distance consistency. Note that the latter proof does not depend on poset being disjoint union of lattices. In fact we only need to extend the result for the flat case to cover the general case. The crucial point that the poset representing the 'function space' shares properties of its components. More precisely, let σ be a shape as defined in section 1.2, and let

$$\mathcal{T}_\sigma = \{\rho \in \mathcal{T}_Q : (\rho)_\star = \sigma\}$$

The proof in [7] depended on the obvious fact that if Q is a disjoint union of lattices, then $\mathcal{T}_\sigma$ is a lattice for every shape σ[1]

Trees differ from lattices in that every set have a g.l.b. but not necessarily the l.u.b. (or conversely, which does not matter). But this property is not shared by $\mathcal{T}_\sigma$ as easily seen for exemaple on the figures in the introduction.

The proof of this theorem (see the full paper) depends by the fact (following from [5]) that if Q satisfies the two-disk property then so does $\mathcal{T}_\sigma$ for every shape σ.

References

1. Kim Bruce and John C. Mitchell. Per models of subtyping, recursive types and higher-order polymorphism. In *Conf. Rec. ACM Symp. Principles of Programming Languages*, 1992.
2. Dexter Kozen, Jens Palsberg, and Michael I. Schwartzbach. Efficient inference of partial types. Technical Report DAIMI PB-394, Computer science Dept., Aarhus University, April 1992.
3. Patrick Lincoln and John C. Mitchell. Algorithmic aspects of type inference with subtypes. In *Conf. Rec. ACM Symp. Principles of Programming Languages*, pages 293–304, 1992.
4. John C. Mitchell. Coercion and type inference. In *Conf. Rec. ACM Symp. Principles of Programming Languages*, pages 175–185, 1984.
5. P. Nevermann and I. Rival. Holes in ordered sets. *Graphs and Combinatorics*, (1):339–350, 1985.
6. Benjamin C. Pierce and David N. Turner. Object-oriented programming without recursive types. Technical Report ECS-LFCS-92-225, LFCS, University of Edinburgh, August 1992.
7. Jerzy Tiuryn. Subtype inequalities. In *Proc. 7th IEEE Symp. Logic in Computer Science*, pages 308–315, 1992.
8. Jerzy Tiuryn and Mitchell Wand. Type reconstruction with recursive types and atomic subtyping. In M.-C. Gaudel and J.-P. Jouannaud, editors, *TAPSOFT'93: Theory and Practice of Software Development, LNCS 668*, pages 686–701, 1993.
9. M. Wand and Patrick O'Keefe. On the complexity of type inference with coercion. In *Proc. ACM Conf. Functional Programming and Computer Architecture*, 1989.

[1] [7] used a slightly different definition of shape which distinguished different connected components. For our purposes the simpler definition given in section 1.2 is sufficient, because connected components are distinguished with distance.

A Characterization of Sturmian Morphisms *

Jean Berstel[1] and Patrice Séébold[2]

[1] LITP, Institut Blaise Pascal, Paris
[2] LAMIFA, Amiens
France

Abstract. A morphism is called *Sturmian* if it preserves all Sturmian (infinite) words. It is *weakly Sturmian* if it preserves at least one Sturmian word. We prove that a morphism is Sturmian if and only if it keeps the word $ba^2ba^2baba^2bab$ balanced. As a consequence, weakly Sturmian morphisms are Sturmian. An application to infinite words associated to irrational numbers is given.

1 Introduction

A one-sided infinite word is *balanced* if the difference of the number of occurrences of a letter in two factors of the same length never exceeds one. It is *Sturmian* if it is balanced and not ultimately periodic.

Sturmian words have a long history. A clear exposition of early work by J. Bernoulli, Christoffel, and A. A. Markov is given in the book by Venkov [22]. The term "Sturmian" has been used by Hedlund and Morse in their development of symbolic dynamics [9, 10, 11]. These words are also known as Beatty sequences, cutting sequences, or characteristic sequences. There is a large literature about properties of these sequences (see for example Coven, Hedlund [6], Series [20], Fraenkel *et al.* [8], Stolarsky [21]). From a combinatorial point of view, they have been considered by S. Dulucq and D. Gouyou-Beauchamps [7], Rauzy [16, 17, 18], Brown [3], Ito, Yasutomi [12], Crisp *et al.* [5] in particular in relation with iterated morphisms, and by Séébold [19], Mignosi [13]. Sturmian words appear in ergodic theory [15], in computer graphics [2], in crystallography [1], and in pattern recognition.

A morphism is *Sturmian* if the image of every Sturmian word is a Sturmian word. Sturmian morphisms appear in number theory in connection with the so-called substitutions of characteristic sequences. A recent account of results in this direction is given by T. C. Brown in [4]. In this paper, we show that in order to test whether a morphism f is Sturmian, it suffices to check whether the single word $f(ba^2ba^2baba^2bab)$ is balanced. This is in fact a strengthening of a result by Mignosi, Séébold [14]. The decidability is an immediate consequence. We also get a simpler proof of a theorem by Crisp *et al.* [5] characterizing those irrational numbers for which the characteristic sequence is a fixed point of a (Sturmian) morphism.

* Partially supported by the PRC "Mathématiques et Informatique".

2 Definitions

Let $A = \{a, b\}$ be a two letter ***alphabet***. A^* is the set of (finite) ***words*** on A and ε is the ***empty word***. A^ω is the set of ***infinite words*** on A and $A^\infty = A^* \cup A^\omega$.

A word $w \in A^*$ is ***primitive*** if it is not a power of another word, i.e. if $w = u^p$ for $u \in A^*$ and $p \in \mathbb{N}$ implies $w = u$.

For any $u \in A^*$, $|u|$ denotes the length of u and $|u|_x$ denotes the number of occurrences of the letter x in the word u.

A ***morphism*** h is a mapping from A^* into itself such that $h(uv) = h(u)h(v)$ for all words u, v. A morphism is ***nonerasing*** if neither $h(a)$ nor $h(b)$ is the empty word. For any morphism f, $||f||$ denotes the ***length*** of f which is $|f(a)| + |f(b)|$. In the sequel, all morphisms f will be supposed to be distinct from the ***null*** morphism which maps all letters into the empty word (thus $||f|| \geq 1$). Consider the morphism ϕ defined by

$$\phi(a) = ab, \qquad \phi(b) = a$$

Setting, for $n \geq 1$,

$$u_n = \phi^n(a), \qquad v_n = \phi^n(b)$$

it is easily seen that $u_{n+1} = u_n u_{n-1}$, $v_{n+1} = u_n$. The morphism ϕ can be extended to infinite words ; it has a unique fixed point

$$\mathbf{F} = abaababaabaababaababa\ldots = \phi(\mathbf{F})$$

For any $w \in A^\infty$, $Fact(w)$ denotes the set of ***finite factors*** of w. Setting, for any $u, v \in A^*$ such that $|u| = |v|$, $\delta(u, v) = \Big||u|_a - |v|_a\Big|$, we call ***balanced*** a word $w \in A^\infty$ such that $\delta(u, v) \leq 1$ for any $u, v \in Fact(w)$ with $|u| = |v|$.

A word $\mathbf{x} \in A^\omega$ is ***Sturmian*** if it is a non ultimately periodic balanced word. It is a well-known property that

Property 1. *The word* $\mathbf{F}$ *is Sturmian.*

Sturmian words are intimately related to cutting sequences in the plane (also known as Beatty sequences). Let α, ρ be real numbers with $0 \leq \alpha < 1$. Then the infinite word $\mathbf{f}_{\alpha,\rho} = a_0 a_1 \cdots a_n \cdots$ defined by

$$a_n = \begin{cases} a & \text{if } \lfloor \alpha(n+1) + \rho \rfloor = \lfloor \alpha n + \rho \rfloor \\ b & \text{otherwise} \end{cases}$$

is Sturmian. The special case $\rho = \alpha$ has additional properties. In this case, we write $\mathbf{s}_\alpha$ for $\mathbf{f}_{\alpha,\alpha}$. The word $\mathbf{s}_\alpha$ is the ***characteristic sequence*** of α. Those words $\mathbf{s}_\alpha$ that are fixed points of morphisms have been characterized by Crisp *et al.* [5].

A morphism h is called ***Sturmian*** if $h(\mathbf{x})$ is Sturmian for every Sturmian word $\mathbf{x}$. The morphism Id_A and the morphism E that exchanges the letters a and b are obviously Sturmian. Let $\tilde{\phi}$ be the morphism defined by

$$\tilde{\phi}(a) = ba \qquad \tilde{\phi}(b) = a$$

It is well-known (see e.g. Séébold [19]) that

Property 2. *The morphisms ϕ and $\tilde{\phi}$ are Sturmian.*

A morphism h is called *weakly Sturmian* if there exists at least one Sturmian word $\mathbf{x} \in A^\omega$ such that $h(\mathbf{x})$ is Sturmian. Obviously every Sturmian morphism is weakly Sturmian. As we shall see, in fact the converse also holds.

3 Results

Notation

Let $m \geq 1$ and $r \geq 1$ be two integers. In the rest of this paper, the following notation will be used:

$$w_{m,r} = b(a^{m+1}b)^{r+1}a^m b(a^{m+1}b)^r a^m b$$
$$w'_{m,r} = ab(a^m b)^{r+1}a^{m+1}b(a^m b)^r a^{m+1}b$$

These words are balanced and primitive. Conversely, every Sturmian word contains as a factor a word $w_{m,r}$ or $w'_{m,r}$ (resp. $E(w_{m,r})$ or $E(w'_{m,r})$) for some $m, r \geq 1$.

The main result of this paper is the following theorem:

Theorem 3. *Let f be a morphism. For every integers m and r with $m, r \geq 1$, the following three conditions are equivalent:*
(i) *f is a composition of the morphisms E, ϕ and $\tilde{\phi}$;*
(ii) *$f(w_{m,r})$ is a primitive balanced word;*
(iii) *$f(w'_{m,r})$ is a primitive balanced word.*

This result shows that in order to test whether a morphism is Sturmian, it suffices to check the image of $w_{m,r}$ for any arbitrary m and r, the shortest being $w_{1,1} = ba^2ba^2baba^2bab$. Thus, we obtain

Corollary 4. *A morphism f is Sturmian iff the word $f(ba^2ba^2baba^2bab)$ is primitive and balanced. In particular, it is decidable whether a morphism is Sturmian.*

Another direct consequence of this result is the following

Theorem 5. *Let f be a morphism. The following conditions are equivalent:*
(i) *f is a composition of the morphisms E, ϕ and $\tilde{\phi}$;*
(ii) *f is Sturmian;*
(iii) *f is weakly Sturmian.*

This result plays a major role in the characterization of morphisms of characteristic sequences associated to irrational numbers.

Proposition 6. *Let f be a morphism, and let α, β be two irrational numbers with $0 < \alpha, \beta < 1$ such that*

$$\mathbf{s}_\alpha = f(\mathbf{s}_\beta).$$

Then f is a product of E and ϕ.

Observe that there is no occurrence of the morphism $\tilde{\phi}$ in the factorization given by this proposition. This is due to the following property of the words s_α:

Property 7. *Let $0 < \alpha < 1$ be an irrational number. Then the word as_α is lexicographically less than all its proper suffixes. Symmetrically, the word bs_α is lexicographically greater than all its proper suffixes.*

From these results, one can obtain rather easily the following characterization of those irrational numbers α whose characteristic sequence s_α is a fixed point of a morphism (which necessarily is Sturmian). This characterization is due to Crisp *et al.* [5]:

Theorem 8. *Let $0 < \alpha < 1$ be an irrational number. The word s_α is a fixed point of a morphism which is not the identity iff the continued fraction development of α has one of the following three forms:*

(*i*) $[0; r_0, \overline{r_1, \ldots r_n}]$, $r_n \geq r_0 \geq 1$;

(*ii*) $[0; 1 + r_0, \overline{r_1, \ldots r_n}]$, $r_n = r_0 \geq 1$;

(*iii*) $[0; 1, r_0, \overline{r_1, \ldots r_n}]$, $r_n > r_0 \geq 1$.

4 Proofs

The most involved part of the paper is the proof of theorem 3. The proof is through three lemmas. We start with a definition. A morphism f is called (m, r)-*balanced* if $f(w_{m,r})$ is balanced or $f(w'_{m,r})$ is balanced. By Theorem 3, these two words are either both balanced or not. The morphism f is *balanced* if it is (m, r)-balanced for some integers $m, r \geq 1$.

Lemma 9. *Let f be a balanced morphism. If $f(a) = a$ and $f(b) \in bA^* \cap A^*b$, then $f(b) = b$.*

Lemma 10. *Let f be a balanced morphism. If $f(a) \in aA^*a$, then $f(b) \in aA^* \cup A^*a$.*

Lemma 11. *Let f and g be two morphism such that $f = \phi \circ g$ or $f = \tilde{\phi} \circ g$, and let $m, r \geq 1$. Then f is (m, r)-balanced iff g is (m, r)-balanced.*

Proof of Theorem 3: Let f be a morphism. Since E, ϕ and $\tilde{\phi}$ are Sturmian (see e.g. Séébold [19]), it is easily seen that (i) $\Rightarrow$ (ii) and (i) $\Rightarrow$ (iii). By symmetry, it is enough to prove the implication (ii) $\Rightarrow$ (i).

Let f be a morphism such that the word $f(w_{m,r})$ is a primitive balanced word. Since $f(w_{m,r})$ is primitive, $f(a)$ and $f(b)$ are not the empty word ε thus $||f|| \geq 2$ and the result holds for $f = Id_A$ and $f = E$.

Consequently, let us suppose $||f|| \geq 3$. We observe first that $f(a)$ and $f(b)$ start or end with the same letter. Assume indeed that $f(a)$ starts with a and $f(b)$ starts with b (if $f(a)$ starts with b and $f(b)$ starts with a, then consider $E \circ f$). From Lemmas 9 and 10 it follows that if $f(a)$ and $f(b)$ do not end with the same letter, then $f(a)$ ends with b and $f(b)$ ends with a. But in this case,

$f(ab)$ contains the factor bb and $f(ba)$ contains the factor aa, which contradicts the hypothesis $f(w_{m,r})$ balanced. Consequently, let us suppose that $f(a)$ and $f(b)$ both start with the letter a (if it is the letter b then consider the morphism $E \circ f$ and if $f(a)$ and $f(b)$ end with the same letter, ϕ is replaced in what follows by $\tilde{\phi}$). Furthermore, let us suppose that $f(a)$ and $f(b)$ do not contain the factor bb. Then $f(a), f(b) \in \{a,\ ab\}^*$, thus there exist two words x and y such that $f(a) = \phi(x)$, $f(b) = \phi(y)$. Denoting by g the morphism defined by

$$g(a) = x \qquad g(b) = y$$

one obtains $f = \phi \circ g$ (if $f(a)$ or $f(b)$ contains the factor bb then $f = E \circ \tilde{\phi} \circ g$). Now, Lemma 11 implies that $g(w_{m,r})$ is balanced. Furthermore, $g(w_{m,r})$ is primitive (since $f(w_{m,r})$ is so) thus $|g(ab)|_a \neq 0$. Consequently, $||f|| > ||g||$ and the result follows by induction. □

Proof of theorem 5: The implications (i) ⇒ (ii) and (ii) ⇒ (iii) are clear and we have only to prove (iii) ⇒ (i). So let f be a weakly Sturmian morphism and $\mathbf{x}$ a Sturmian word such that $f(\mathbf{x})$ is Sturmian. Furthermore, let us suppose that $\mathbf{x}$ contains the factor aa. In this case, there exist some integers m and r, $m, r \geq 1$, such that $\mathbf{x}$ contains $w_{m,r}$ or $w'_{m,r}$ as a factor. It follows, from theorem 3, that f can be obtained by composition of E, ϕ and $\tilde{\phi}$. If $\mathbf{x}$ contains the factor bb, then the above proof holds for $g = f \circ E$ and, consequently, $f = g \circ E$ has the required property. □

Proof of Lemma 9: Let f be a morphism such that $f(a) = a$ and $f(b) \in bA^* \cap A^*b$, and let m and r be two integers, $m, r \geq 1$ such that $f(w_{m,r})$ or $f(w'_{m,r})$ is a balanced word. Since $f(a) = a$, both $f(w_{m,r})$ and $f(w'_{m,r})$ contain the factors a^m and a^{m+1}, $m \geq 1$, thus $f(b)$ does not contain bb and if $f(b) \neq b$ then $f(b)$ starts (resp. ends) with ba^mb or $ba^{m+1}b$.

For all integers $p \geq 0, p' \geq 1$, define

$$u_{p,p'} = b(a^{m+1}b)^{p+1}(a^mb)^{p'}a^{m+1}, \quad v_{p,p'} = b(a^mb)^{p'}(a^{m+1}b)^pa^mb$$

The word $w_{m,r}$ contains both $u_{r,1}$ and $v_{r,1}$, and $w'_{m,r}$ contains both $u_{0,r}$ and $v_{0,r}$. If $f(b)$ starts with ba^mb then $f(u_{p,p'})$ contains the factor

$$z = a^{m+1}f(b(a^{m+1}b)^p(a^mb)^{p'})a^{m+1}$$

and $f(v_{p,p'})$ contains the factor $z' = f(b(a^mb)^{p'}(a^{m+1}b)^p)a^mba^mb$.

Otherwise, $f(b) = ba^{m+1}bv$ for some word v. But in this case $f(u_{p,p'})$ contains the factor $z = a^{m+1}bvf((a^{m+1}b)^p)a^{m+1}ba^{m+1}$ and, since $p' \geq 1$, $f(v_{p,p'})$ contains the factor $z' = ba^mba^{m+1}bvf((a^{m+1}b)^p)a^mb$.

In both cases, $\delta(z, z') = 2$ which contradicts the hypothesis that f is (m,r)-balanced. Thus $f(b) = b$ and the lemma is proved. □

Proof of Lemma 10 : Let f be a morphism such that $f(a) \in aA^*a$ and let m and r be two integers, $m, r \geq 1$ such that $f(w_{m,r})$ or $f(w'_{m,r})$ is a balanced word. We set, for $k \in \mathbb{N}$,

$$u_k = a^{m+1}b(a^mb)^ka^{m+1}, \quad v_k = ab(a^mb)^ka^mba$$

$$x_k = b(a^{m+1}b)^{k+1}, \quad y_k = ba^m b(a^{m+1}b)^k a^m b.$$

By construction, $w_{m,r}$ both contains u_0, v_0, and x_r, y_r. Symmetrically, $w'_{m,r}$ both contains u_r, v_r, and x_0, y_0.

Assume, by way of contradiction, that $f(b) \in bA^* \cap A^*b$. If $f(a) = uaav$ then $f(a^{m+1})$ contains the factor $z = aavf(a^{m-1})uaa$ and $f(ba^m b)$ contains the factor $z' = buaavf(a^{m-1})b$. Then $\delta(z, z') = 2$ which contradicts the hypothesis. Thus $f(a)$ does not contain aa and, since $f(a) \neq a$, there exists an integer $n \geq 0$ such that

$$f(a) = (ab)^n aba$$

But, in this case, $f(ab)$ contains bab and $f(aa)$ contains $baab$. Thus $f(b)$ does not contain bb nor a^3, which implies that either $f(b) = b$ or $f(b)$ starts with bab or $baab$.

We shall now prove that $f(b) = (ba)^{n'}b$, for some $n' \geq 0$. This holds if $f(b) = b$. Thus, assume by way of contradiction that $f(b) = uaav$ for some u and v.

Observe first that $f(b)$ starts and ends with $baab$. Indeed, if $f(b)$ starts with bab then $f(ba^{m+1})$ contains the factor $z = aavf(a^m)a$ and $f(ba^m b)$ contains the factor $z' = vf(a^m)bab$. If $f(b)$ ends with bab then $f(a^{m+1}b)$ contains the factor $z = af(a^m)uaa$ and $f(ba^m b)$ contains the factor $z' = babf(a^m)u$. In both cases, $\delta(z, z') = 2$ which contradicts the assumption.

Thus $f(b)$ starts and ends with $baab$. If $m = 1$ then let $f(b) = baabv'$ for some v'. In this case, $f(x_k)$ contains the factor

$$z = a^2bv'f((a^{m+1}b)^k a^{m+1})ba^2$$

and $f(y_k)$ contains the factor

$$z' = bf(a^m)ba^2bv'f((a^{m+1}b)^k a^m)b.$$

Then $\delta(z, z') = 2$ which contradicts the hypothesis.

Otherwise $m > 1$. In this case, both $f(w_{m,r})$ and $f(w'_{m,r})$ contain both of the words $f(a^3)$ and $f(a^2b)$. But $f(a^3)$ contains the factor $baa(ba)^n baab$ and $f(a^2b)$ contains the factor $baa(ba)^{n+1}baab$. Consequently, if $f(b)$ contains as a factor a power of ba, then this power is $(ba)^n$ or $(ba)^{n+1}$. Thus there exist two integers $p, p' \geq 0$, such that $f(b)$ starts with $(baa(ba)^n)^p(baa(ba)^{n+1})^{p'}baab$ (remark that, if $p' = 0$ then $f(b) = (baa(ba)^n)^p baab$).

If $p < m - 1$ then $f(u_k)$ contains the factor

$$z = a(ab)^n aba(ab)^n aba((ab)^n aba)^p f(b(a^m b)^k a^m)a$$

and $f(v_k)$ contains the factor

$$z' = ba(ba)^n f(b(a^m b)^k a^m)(baa(ba)^n)^p baa(ba)^n bab.$$

If $p \geq m - 1$ then $f(b)$ starts with $(baa(ba)^n)^{m-1}baa$ and we denote by v' the word such that $f(b) = bv'$. In this case, $f(x_k)$ contains the factor

$$z = v'f((a^{m+1}b)^k a^m)f(a)(baa(ba)^n)^{m-1}baa$$

and $f(y_k)$ contains the factor

$$z' = bf(a^m(ba^{m+1})^k)bv'f(a^m)b$$

and $\delta(f(a^m), f(a)(baa(ba)^n)^{m-1}) = 0$. Thus, in both cases, $\delta(z,z') = 2$ which contradicts the hypothesis.

Consequently $f(b)$ does not contain aa, thus $f(b) = (ba)^{n'}b$, for some $n' \geq 0$. But in this case, since $m \neq 0$, the word $f(u_k)$ contains the factor

$$z = af((a^m b)^{k+1})(ab)^n abaa$$

and $f(v_k)$ contains the factor

$$z' = babf((a^m b)^{k+1})(ab)^n ab.$$

Again, $\delta(z,z') = 2$ which contradicts the hypothesis, thus $f(b) \in aA^* \cup A^*a$ and the lemma is proved. □

Proof of Lemma 11: The "only if" part is straightforward. For the "if" part, assume $f = \phi \circ g$ (the case $f = \tilde{\phi} \circ g$ could be done exactly in the same way). If $g(w_{m,r})$ is not balanced then there exist two words u and v such that $g(w_{m,r}) = u_1uu_2 = v_1vv_2$ with $|u| = |v|$ and $\delta(u,v) = 2$. Furthermore, u can be choosen of minimal length, which implies that there exist $x, y \in A$, $x \neq y$ and $t, t' \in A^*$ such that $u = xtx$, $v = yt'y$ and $\delta(t,t') = 0$. Let us assume $x = a$ and $y = b$ (the other case is exactly the same). Then $g(w_{m,r}) = u_1atau_2 = v_1bt'bv_2$ and $f(w_{m,r}) = \phi(u_1)ab\phi(t)ab\phi(u_2) = \phi(v_1)a\phi(t')a\phi(v_2)$.

If $v_2 \neq \varepsilon$, then $\phi(v_2)$ starts with a and $f(w_{m,r})$ is not balanced, contradiction.

If $v_2 = \varepsilon$, then $bt'b$ is a suffix of $g(w_{m,r})$. Two cases arise:

- If $|bt'b| \leq |g(a^m b(a^{m+1}b)^r a^m b)|$ then $g(a^m b(a^{m+1}b)^r a^m b) = v_1'bt'b$ and $g(w_{m,r}) = g(ba)v_1'bt'bg((a^{m+1}b)^r a^m b)$ which is the same case as $v_2 \neq \varepsilon$.

- If $|bt'b| > |g(a^m b(a^{m+1}b)^r a^m b)|$ then, since $m \geq 1$, one has $|ata| > |g(b(a^{m+1}b)^{r+1})|$. Consequently, there exist three words z, z', z'' with $z \neq \varepsilon$ and $|z'| = |z''|$ such that $ata = z'z$ and $bt'b = zz''$. But $\delta(z',z'') = \delta(u,v) = 2$, and since $z \neq \varepsilon$, one has $|z'| < |u|$ which contradicts the minimality of $|u|$.

If $g(w'_{m,r})$ is not balanced then the same contradiction holds when we compare $|bt'b|$ and $|g(a^m b(a^m b)^r a^{m+1} b)|$, thus the lemma is proved. □

We now turn to the proofs of the number-theoretic applications. Given two infinite words $\mathbf{x} = a_0a_1\cdots$ and $\mathbf{y} = b_0b_1\cdots$ over the alphabet $A = \{a,b\}$, ordered by $a < b$, we write $\mathbf{x} < \mathbf{y}$ when $\mathbf{x}$ is lexicographically less then $\mathbf{y}$, that is when there exists an integer n such that $a_n < b_n$ and $a_i = b_i$ for $0 \leq i < n$. Property 7 is a consequence of the more general

Lemma 12. *Let* $0 \leq \rho, \rho' < 1$ *and* $0 < \alpha < 1$, *with* α *irrational. Then*

$$\mathbf{f}_{\alpha,\rho} < \mathbf{f}_{\alpha,\rho'} \iff \rho < \rho'.$$

Proof. Since α is irrational, one has $\rho < \rho'$ if and only if there exists an integer n such that $\lfloor \alpha n+\rho' \rfloor = 1+\lfloor \alpha n+\rho \rfloor$ Let m be the smallest integer n satisfying this relation, and set $k = m-1$. Then, setting $\mathbf{f}_{\alpha,\rho} = a_0 a_1 \cdots$ and $\mathbf{f}_{\alpha,\rho'} = a'_0 a'_1 \cdots$, one gets $a_j = a'_j$ for $0 \leq j < k$ and $a_k < a'_k$. This proves the lemma. □

Proof of Property 7. Let us prove the first inequality, namely that $a\mathbf{s}_\alpha < \mathbf{x}$ for any proper suffix $\mathbf{x}$ of $a\mathbf{s}_\alpha$. For this, observe that $a\mathbf{s}_\alpha = \mathbf{f}_{\alpha,0}$, and that $\mathbf{x} = \mathbf{f}_{\alpha,n\alpha-\lfloor n\alpha \rfloor}$ for some integer $n > 0$. Since α is irrational, the conclusion follows from the preceding lemma. The other inequality is shown symmetrically. □

Before proceeding to the proof of Proposition 6, recall the following relations which are well-known (see e.g. [3]) :

$$E(\mathbf{s}_\beta) = \mathbf{s}_{1-\beta}, \quad \phi(\mathbf{s}_\beta) = \mathbf{s}_{(1-\beta)/(2-\beta)}$$

Proof of Proposition 6. By induction on the length $||f||$ of f. We may assume that $\mathbf{s}_\beta$ contains the factor aa. Otherwise we replace $\mathbf{s}_\beta$ by $\mathbf{s}_{1-\beta} = E(\mathbf{s}_\beta)$ and f by $f \circ E$. We also can suppose that $\mathbf{s}_\alpha$ contains the factor aa. Otherwise, we replace f by $E \circ f$ and $\mathbf{s}_\alpha$ by $\mathbf{s}_{1-\alpha}$. These normalisations do not increase the length of f. Since $b\mathbf{s}_\beta$ is Sturmian, the word $\mathbf{s}_\beta$ starts with the letter a. Similarly, $\mathbf{s}_\alpha$ starts with the letter a. In particular, $f(a)$ starts with an a, and neither $f(a)$ nor $f(b)$ contain a factor bb.

If the word $f(b)$ also starts with the letter a, then both $f(a)$ and $f(b)$ are products of words in $\{ab, a\}$. Thus $f = \phi \circ g$ for some shorter morphism g, and an appropriate word $\mathbf{s}_\gamma$. To conclude, it suffices to prove that $f(b)$ cannot start with a letter b. Indeed, otherwise $f(a)$ and $f(b)$ finish with the same letter. If this letter is a b, then $\mathbf{s}_\alpha$ contains the factor bb. Thus $f(a)$ and $f(b)$ finish by an a. Now let $r \geq 1$ by the integer such that $a^r b$ is a prefix of $\mathbf{s}_\beta$. Then $a^{r+1}b$ is a factor of $\mathbf{s}_\beta$. The word $af(a^r)b$ is a prefix of $a\mathbf{s}_\alpha$, and $af(a^r)a$ is a factor of $\mathbf{s}_\alpha$. But this shows that $a\mathbf{s}_\alpha$ is lexicographically greater than one of its suffixes. Contradiction. □

We conclude with a proof of Theorem 8. Our proof is shorter than, though not very different from [5]. It will be convenient to introduce the morphism $\gamma = \phi \circ E$. Thus $\gamma(a) = a$, $\gamma(b) = ab$. Clearly, a morphism is a composition of E and ϕ iff it is a composition of E and γ. The morphism γ is used in conjonction with the morphism $\theta_m = \gamma^m \circ E$. We observe that ([3, 22])

$$\mathbf{s}_{\beta/(1+\beta)} = \gamma(\mathbf{s}_\beta), \quad \mathbf{s}_{1/(m+\beta)} = \theta_m(\mathbf{s}_\beta)$$

Proof of Theorem 8. Let

$$\alpha = [0; r_1, r_2, \ldots]$$

be the development into continued fraction of α. Let f be such that $\mathbf{s}_\alpha = f(\mathbf{s}_\alpha)$. Then f is a product of the morphisms γ and E. Clearly $f \neq E$ and f is not a product of γ only. Consequently,

$$f = \gamma^{n_1} E \gamma^{n_2} \ldots E \gamma^{n_k} E \gamma^{n_{k+1}}$$

for some $k \geq 1$ and $n_1 \geq 0$, $n_2, \ldots, n_k \geq 1$, $n_{k+1} \geq 0$. We distinguish two cases.

First case: $n_{k+1} \geq 1$. Then

$$f = \theta_{n_1+1}\theta_{n_2}\cdots\theta_{n_k}\gamma^{n_{k+1}-1}$$

Since s_α is a fixed point,

$$[0; r_1, r_2, \ldots] = [0; 1+n_1, n_2, \ldots, n_k, n_{k+1} - 1 + r_1, r_2, \ldots]$$

whence $r_1 = 1 + n_1$, $r_2 = n_2, \ldots, r_k = n_k$, $r_{k+1} = n_{k+1} + n_1$ and $r_j = r_{j+k}$ for $j \geq 2$. Thus

$$\alpha = [0; r_1, \overline{r_2, \ldots, r_{k+1}}], \qquad r_{k+1} \geq r_1$$

which is case (i) of the theorem.

Second case: $n_{k+1} = 0$. Set $f' = EfE$. Since $s_\alpha = f(s_\alpha)$, one has $f'(Es_\alpha) = Es_\alpha$ and $f(s_\beta) = s_\beta$ where $\beta = 1 - \alpha$. Now

$$f' = E\gamma^{n_1}E\gamma^{n_2}\cdots E\gamma^{n_k} \quad \text{and} \quad n_k \geq 1.$$

This has the same form as above, excepted when $n_1 = 0$. Again, we consider two cases:

First, assume $n_1 = 0$. Then $k \geq 3$ and $f' = \theta_{n_2+1}\theta_{n_3}\cdots\theta_{n_{k-1}}\gamma^{n_k-1}$ whence, using the first case, $\beta = [1 + n_2, \overline{n_3, \ldots, n_{k-1}, n_2 + n_k}]$ and since $n_2 \geq 1$, one gets for $\alpha = 1 - \beta$ the development

$$\alpha = [0; 1, n_2, \overline{n_3, \ldots, n_{k-1}, n_2 + n_k}]$$

This is case (iii) of the theorem.

Finally, assume $n_1 > 0$. Then $f' = \theta_{n_0+1}\theta_{n_1}\cdots\theta_{n_{k-1}}\gamma^{n_k-1}$ with $n_0 = 0$. Applying the first case, we get $\beta = [0; 1, \overline{n_1, \ldots, n_k}]$ and consequently

$$\alpha = [0; 1 + n_1, \overline{n_2, \ldots, n_k, n_1}]$$

which is precisely case (ii) of the statement. □

References

1. E. Bombieri, J. E. Taylor, Which distributions of matter diffract? An initial investigation, *J. Phys.* **47** (1986), Colloque C3, 19–28.
2. J. E. Bresenham, Algorithm for computer control of a digital plotter, *IBM Systems J.* **4** (1965), 25–30.
3. T. C. Brown, A characterization of the quadratic irrationals, *Canad. Math. Bull.* **34** (1991), 36–41.
4. T. C. Brown, Descriptions of the characteristic sequence of an irrational, *Canad. Math. Bull.* **36** (1993), 15–21.
5. D. Crisp, W. Moran, A. Pollington, P. Shiue, Substitution invariant cutting sequences, Sémin. Théorie des Nombres, Bordeaux, 1993, to appear.
6. E. Coven, G. Hedlund, Sequences with minimal block growth, *Math. Systems Theory* **7** (1973), 138–153.

7. S. Dulucq, D. Gouyou-Beauchamps, Sur les facteurs des suites de Sturm, *Theoret. Comput. Sci.* **71** (1990), 381–400.
8. A. S. Fraenkel, M. Mushkin, U. Tassa, Determination of $\lfloor n\theta \rfloor$ by its sequence of differences, *Canad. Math. Bull.* **21** (1978), 441–446.
9. G.A. Hedlund, Sturmian minimal sets, *Amer. J. Math* **66** (1944), 605–620.
10. G.A. Hedlund, M. Morse, Symbolic dynamics, *Amer. J. Math* **60** (1938), 815–866.
11. G.A. Hedlund, M. Morse, Sturmian sequences, *Amer. J. Math* **61** (1940), 1–42.
12. S. Ito, S. Yasutomi, On continued fractions, substitutions and characteristic sequences, *Japan. J. Math.* **16** (1990), 287–306.
13. F. Mignosi, On the number of factors of Sturmian words, *Theoret. Comput. Sci.* **82** (1991), 71–84.
14. F. Mignosi, P. Séébold, Morphismes sturmiens et règles de Rauzy, Techn. Rep. LITP-91-74, Paris, France.
15. M. Queffélec, *Substitution Dynamical Systems – Spectral Analysis*, Lecture Notes Math.,vol. 1294, Springer-Verlag, 1987.
16. G. Rauzy, Suites à termes dans un alphabet fini, *Sémin. Théorie des Nombres* (1982–1983), 25-01,25-16, Bordeaux.
17. G. Rauzy, Mots infinis en arithmétique, in: *Automata on infinite words* (D. Perrin ed.), *Lect. Notes Comp. Sci.* **192** (1985), 165–171.
18. G. Rauzy, Sequences defined by iterated morphisms, in: *Workshop on Sequences* (R. Capocelli ed.), *Lecture Notes Comput. Sci.*, to appear.
19. P. Séébold, Fibonacci morphisms and Sturmian words, *Theoret. Comput. Sci.* **88** (1991), 367–384.
20. C. Series, The geometry of Markoff numbers, *The Mathematical Intelligencer* **7** (1985), 20–29.
21. K. B. Stolarsky, Beatty sequences, continued fractions, and certain shift operators, *Cand. Math. Bull.* **19** (1976), 473–482.
22. B. A. Venkov, *Elementary Number Theory*, Wolters-Noordhoff, Groningen, 1970.

On the complexity of scheduling incompatible jobs with unit-times

Hans L. Bodlaender[1] and Klaus Jansen[2]

[1] Department of Computer Science, Utrecht University, P.O.Box 80.089, 3508 TB Utrecht, The Netherlands
[2] Fachbereich IV, Mathematik und Informatik, Universität Trier, Postfach 3825, W-5500 Trier, Germany

Abstract. We consider scheduling problems in a multiprocessor system with incompatibile jobs of unit-time length where two incompatible jobs can not be processed on the same machine. Given a deadline k' and a number of k machines, the problem is to find a feasible assignment of the jobs to the machines. We prove the computational complexity of this scheduling problem restricted to different graph classes, arbitary and constant numbers k and k'.

1 Introduction

Let J be a set of n jobs, all with job processing time or job length one, and let M be a set of k machines. Incompatibilities between jobs are described by an undirected graph $G = (J, E)$ with vertex set J. If G contains an edge between two jobs j and j' ($j \neq j'$), we demand that these two jobs can not be executed by the same machine. An assignment of the jobs to the machines, which satisfies the incompatibility relation, is called a schedule. The processing time of a machine is given by the number of assigned jobs, since each job has processing time one. Given a deadline k', we consider the problem to find a feasible assignment of the jobs to k machines such that the number of assigned jobs to each machine is less than or equal to k'.

An important combinatorial problem which is related to the scheduling problem is the coloring problem of an undirected graph $G = (V, E)$. A k-coloring is a mapping $f : V \rightarrow \{1, \ldots, k\}$ with for all edges $(v, w) \in E$, $f(v) \neq f(w)$. A set U is called independent if each pair $v, w \in U$ with $v \neq w$ is not connected by an edge. The k-coloring problem corresponds to the problem of finding a partition of the vertices into k independent sets. It turns out, that the scheduling problem is equal to the problem to find a partition of the incompatibility graph into k independent sets, each of size at most k'. Since a partition into independent sets forms a coloring of the incompatibility graph, the problem is, given a graph G, and two integers k, k', to search for a coloring of G with at most k colors, such that for each color, there are at most k' vertices with that color. We call this problem PARTITION INTO BOUNDED INDEPENDENT SETS and give in the following the complete definition.

Problem: PARTITION INTO BOUNDED INDEPENDENT SETS
Instance: Undirected graph $G = (V, E)$, integers $k, k' \in \mathbb{N}$.
Question: Is there a partition of V into independent sets $U_1, \ldots, U_k$ with $|U_i| \leq k'$ for $1 \leq i \leq k$?

Karp [5] has shown that the coloring problem is NP-complete for undirected graphs and up to this time no polynomial-time algorithm is known for this problem. Hence, the problem PARTITION INTO BOUNDED INDEPENDENT SETS is also NP-complete for general undirected graphs.

We consider also the scheduling problem on the complement of G. Then it becomes the following problem:

Problem: PARTITION INTO BOUNDED CLIQUES
Instance: Undirected graph $G = (V, E)$, integers $k, k' \in \mathbb{N}$.
Question: Is there a partition of V into cliques $C_1, \ldots, C_k$ with $|C_i| \leq k'$ for $1 \leq i \leq k$?

This problem is NP-complete, because it contains the problem PARTITION INTO CLIQUES [5]. We denote with $\chi(G, k')$ the minimum number of independent sets of size at most k' that cover G and we denote with $\kappa(G, k')$ the minimum number of cliques of size k' that cover G.

However, the coloring problem and the problem partition into cliques become much easier, when we restrict the inputs to certain special graph classes. For example, there are efficient algorithms for interval graphs [4], cographs [8] and bipartite graphs. Therefore we have analysed the PARTITION INTO BOUNDED CLIQUES (INDEPENDENT SETS) problems for these graph classes. For split graphs both problem are solvable in polynomial time, proved by Lonc [6].

But we have also considered the cases that the numbers k or k' are constants. This is related to the case of a constant number of machines or a constant deadline.

2 Cographs

Cographs are graphs without a path with four vertices as induced subgraph [8]. These graphs can be generated by disjoint union and join operations on graphs starting with single-vertex graphs and can be represented about these operations. For graphs $G_i = (V_i, E_i)$ with $V_1 \cap V_2 = \emptyset$ the union of G_1 and G_2, $\cup(G_1, G_2)$ is given by $(V_1 \cup V_2, E_1 \cup E_2)$. The join of G_1 and G_2, denoted by $+(G_1, G_2)$ is obtained by first taking the union of G_1 and G_2, and then adding all edges $\{v_1, v_2\}$ with $v_i \in V_i$. The join of three or more graphs $G_1, \ldots, G_r$ is obtained similary: take the disjoint union, and add all edges between vertices in different graphs G_i.

To each cograph G one can associate a corresponding rooted binary tree T, called a *cotree* of G, in the following way. Each non-leaf node in the tree is labeled with either $\cup$ (union-nodes) or $+$ (join nodes). Each non-leaf node has exactly two children. Each node of the cotree corresponds to a cograph and a leaf node to a single-vertex graph. We remark that the usual definition of cotrees allows for arbitary degree of internal nodes. However, it is easy to see that both definitions has the same power and that arbitrary cotrees can be transformed to cotrees with two children per internal node. In [1] it is shown that one can decide in $O(n + e)$ time, whether a graph is a cograph, and build a corresponding cotree.

Since the complement of a cograph is again a cograph, the same results hold for the complexity of PARTITION INTO BOUNDED CLIQUES and PARTITION INTO BOUNDED INDEPENDENT SETS when restricted to cographs.

Theorem 1. *The problems* PARTITION INTO BOUNDED CLIQUES (INDEPENDENT SETS) *remain NP-complete for cographs.*

Proof. We proof the result for PARTITION INTO BOUNDED CLIQUES. Clearly the problem is in NP. To prove NP-hardness, use a transformation from the BIN-PACKING problem to the PARTITION INTO BOUNDED CLIQUES problem on cographs. An instance of the bin-packing problem is given by numbers $a_1, \ldots, a_n \in \mathbb{N}$, by $K \in \mathbb{N}$ bins and by a bin-capacity $B \in \mathbb{N}$ with $B > a_i$ and $K > 1$. The question is to decide whether there exists a partition of the set $\{1, \ldots, n\}$ in sets $I_1, \ldots, I_K$ with $\sum_{i \in I_j} a_i < B$ for each $1 \leq j \leq K$. This problem is NP-complete, see [2]. We may assume that the bin-capacity B is greater than the number n. (Otherwise, multiply B and all a_i by n.)

For every number a_i, construct a graph G_{a_i} in the following way. Take a complete graph with $B \cdot a_i$ vertices, $C_{B \cdot a_i}$, and take the union of this complete graph with $K-1$ independent vertices. Let $G_{a_1,\ldots,a_n}$ be the graph, obtained by taking the join of all G_{a_i} $(1 \leq i \leq n)$. A maximal clique in this graph can be represented by an index set $I \subset \{1, \ldots, n\}$ such that the vertices of this clique are given by the $B \cdot a_i$ vertices in G_{a_i} for each $i \in I$ and by one of the $K-1$ independent vertices in G_{a_i} for each $i \notin I$.

Then, we can prove the equivalence that there is a partition $I_1, \ldots, I_K$ of $\{1, \ldots, n\}$ with $\sum_{i \in I_j} a_i < B$ for $1 \leq j \leq K$ iff the graph $G_{a_1,\ldots,a_n}$ has a partition into K cliques each of size at most $B^2 - B + n$. □

Since the problems to partition into k cliques or independent sets of size at most k' are NP-complete, we look at instances with constant k or k'. When $k' = 3$ and $k = |V|/3$, we get the problem PARTITION INTO TRIANGLES. For constant k' we can use the recursive structure of the cograph.

We consider vectors $x = (x_1, \ldots, x_{k'})$ where x_i gives the number of cliques of size i used for the cograph. We call such a vector $x = (x_1, \ldots, x_{k'})$ *feasible with respect to* G, if there is a partition of G into cliques where x_i cliques have size i for $1 \leq i \leq k'$. For every graph G, let $L(G)$ be the set of feasible vectors. We compute a partition into a minimum number of cliques of size at most k' by computing sets $L(G)$. If $V = \{v\}$ then the set $L(G)$ consists of one vector $(1, 0, \ldots, 0)$. The generation of these vector sets can be done recursively on the cograph.

Lemma 2. *Let $G = (V, E)$ be a cograph. If $G = \cup(G_1, G_2)$ then*

$$L(G) = \{x + y | x \in L(G_1), y \in L(G_2)\}.$$

For $G = +(G_1, G_2)$, $z \in L(G)$, if and only if there exist $x \in L(G_1)$, $y \in L(G_2)$ and a mapping $f : \{(i,j) \mid i,j \geq 0,\ 1 \leq i+j \leq k'\} \to N_0$, such that:

$$(1) \sum_{j | i+j \leq k'} f(i,j) = x_i \quad 1 \leq i \leq k'$$
$$(2) \sum_{i | i+j \leq k'} f(i,j) = y_j \quad 1 \leq j \leq k'$$
$$(3) \sum_{i,j | i+j = h} f(i,j) = z_h \quad 1 \leq h \leq k'.$$

Proof. For the union the assertion is clear. For the join let $x \in L(G_1)$ and $y \in L(G_2)$ and let $C_1, \ldots, C_k$ be a partition into cliques of size at most k'. A clique in G of size k' is given either by a clique of size k' in G_1 or G_2 or is given by a clique of size i with $1 \leq i < k'$ in one of both graphs and of size $k' - i$ in the other. For a clique of size less than k' we have a similar representation. Using x_i for the numbers of cliques of size i we can describe an assignment of cliques in G_1, G_2 by such a mapping f. □

We note that we must only store $k' - 1$ of the components of the vector, because the last component is given by the number of vertices and the other components. Therefore only $O(n^{k'-1})$ vectors are possible for each cograph.

Theorem 3. *Given a constant k', the problems* PARTITION INTO BOUNDED CLIQUES (INDEPENDENT SETS) *with each set of size at most k' can be computed in polynomial time for cographs.*

Proof. We consider PARTITION INTO BOUNDED CLIQUES; for the other problem PARTITION INTO BOUNDED INDEPENDENT SETS, consider the complement of the input graph. For every node of the cotree, we compute the set $L(H)$, where H is the cograph, associated to the node. Note that the number of feasible vectors in a set $L(G)$ is polynomial, for constant k'. The computation for the set $L(G)$, when G is obtained by the union of G_1 and G_2, can be done in $O(n^{2(k'-1)})$ steps, given the sets $L(G_1)$ and $L(G_2)$. Since the set $\{(i,j)|i,j \in \mathbb{N}_0, 1 \leq i+j \leq k'\}$ has only a constant number $\frac{k' \cdot k' + 3k'}{2} \leq k' \cdot k' + 1$ of elements, for each vector pair x, y there are at most polynomial $O(n^{k' \cdot k' + 1})$ many feasible mappings f. Therefore, if G is obtained by the join of G_1 and G_2, $L(G)$ can be determined in polynomial, namely $O(n^{k'(k'+2)-1})$ many steps. In this way, we can compute $L(H)$ for every cograph H, associated with a node of the cotree. At the end, we can choose a feasible vector $x \in L(G)$ with a minimum number of cliques, i.e. with $\sum_{i=1}^{k} x_i$ minimum over all $x \in L(G)$. □

A similar approach we can use for the problems where the number of cliques k is constant. For these problems we describe partitions into cliques $C_1, \ldots, C_k$ by the sizes $|C_1|, \ldots, |C_k|$ and consider sequences of sizes instead of subsets of the vertices V. We call a sequence of sizes *feasible*, if it corresponds to a partition. Since k is constant the number of different feasible sequences for a graph can be bounded by a polynomial, $O(n^k)$. Using that one component is given by the size $|V|$ and the other components, we must store only $O(n^{k-1})$ sequences. Now we give a recursive formula for the sets $S(G)$ of feasible sequences and a cograph G. If $V = \{v\}$ then $S(G) = \{1\}$. Let $\circ$ denote the concatenation of two sequences and let $\ell(s)$ denote the length of a sequence s. Given a sequence s with $\ell(s) \leq k$ we denote with $\bar{s}$ the sequence of length k, obtained from s by adding zero or more zeros at the end of the sequence s.

Lemma 4. *Let $G = (V, E)$ be a cograph and let k be a positive integer. If $G = \cup(G_1, G_2)$ then $S(G) = \{t = s \circ s' | s \in S(G_1), s' \in S(G_2), \ell(t) \leq k\}$. If $G = +(G_1, G_2)$ then $S(G) = \{w_1 + u_1 \ldots w_k + u_k \mid w = \bar{s}, s \in S(G_1), t \in S(G_2), w_i + u_i \leq k'$ for $1 \leq i \leq k$, and u is a permutation of $\bar{t}\}$.*

Proof. For the union the assertion is clear. Let us consider the disjoint join of two graphs $G_i = (V_i, E_i)$. Let $C_1, \ldots, C_{k_1}$ and $C'_1, \ldots, C'_{k_2}$ be partitions of G_1 and G_2

into cliques with $|C_i|, |C'_i| \leq k'$ and with $k_1, k_2 \leq k$. By adding to the collections $C_1, \ldots, C_{k_1}$, and $C'_1, \ldots, C'_{k_2}$ a number of empty sets, we can assume that $k_1 = k_2 = k$. Now for every permuation π of $\{1, \ldots, k\}$, we have that $C_1 \cup C'_{\pi(1)}, \ldots, C_k \cup C'_{\pi(k)}$ is a partition of G into cliques. If the sizes are bounded by k' we get a solution of $S(G)$.

For the other direction let $C_1, \ldots, C_h$ be a partition into $h \leq k$ cliques of size $|C_i| \leq k'$. By intersection of these cliques with vertices V_1 and V_2 we get partitions of G_1 and G_2 with the properties described above. □

Theorem 5. *Give a constant k, the problems* PARTITION INTO K BOUNDED CLIQUES (INDEPENDENT SETS) *can be solved in polynomial time for cographs.*

Proof. We use that each set $S(G)$ contains at most $O(n^{k-1})$ sequences. Compute for each node of the cotree the set $S(H)$ with H the cograph corresponding to that node, after these sets have been computed for the children of the node. From lemma 4 it follows that these computations cost at most $O(k \cdot k! \cdot n^{2(k-1)})$ time. Hence, the total time for the algorithm is bounded by $O(n^{2(k-1)+1})$, given a constant k. □

3 Bipartite graphs

A graph $G = (V, E)$ is called bipartite if there is a partition of the vertices V into two disjoint set V_1, V_2 where the set of edges E forms a subset of $\{\{v, w\} | v \in V_1, w \in V_2\}$.

Cliques in a bipartite graph have at most the size two and therefore the problem PARTITION INTO BOUNDED CLIQUES can be solved in polynomial time using a matching algorithm. But for independent sets the situation is more difficult.

Theorem 6. *The problem* PARTITION INTO THREE BOUNDED INDEPENDENT SETS *remains NP-complete for bipartite graphs.*

Proof. We give a transformation from CLIQUE to this partition problem on bipartite graphs. Let $G = (V, E)$ be a graph and $k > 5$. It is not hard to see that CLIQUE remains NP-complete under the restriction, that $|E| < |V| + k(k-2)$.

Suppose we have a graph $G = (V, E)$ with $|E| < |V| + k(k-2)$. Define a bipartite graph $G_B = (V_1 \cup V_2, E_B)$ with $V_1 = V \cup A$, $V_2 = E \cup B$. The number of vertices in A is $|V| - 2k + \frac{k(k-1)}{2} > 0$ and the number of vertices in B is $|V| + k(k-2) - |E| > 0$. As edges we take all pairs $\{a, v_2\}, \{v_1, b\}$ with $a \in A, b \in B$ and $v_i \in V_i$ and for an edge $e = \{v, w\} \in E$ we take $\{v, e\}$ and $\{w, e\}$.

For this graph G, the following equivalence holds: G has a clique of size k iff G_B has a partition into three independent sets of size $k' = |V| - k + \frac{k(k-1)}{2}$. We omit the proof of this fact. The theorem now follows. □

Now we analyse the problem PARTITION INTO BOUNDED INDEPENDENT SETS of at most constant size k'.

Lemma 7. *Let $G = (V, E)$ be a bipartite graph with partition $V = V_1 \cup V_2$, $E \subset \{\{v_1, v_2\} | v_i \in V_i\}$ and let $a = |V_1| mod(k') + |V_2| mod(k')$ and $b = \frac{|V_1| + |V_2| - a}{k'}$. Then* *we get*

$$\chi(G,k')\begin{cases} = b & \text{for } a = 0 \\ \in \{b+1, b+2\} & \text{for } 0 < a \leq k' \\ = b+2 & \text{otherwise.} \end{cases}$$

If exactly one of the terms $|V_i| mod(k') = 0$, we get $\chi(G,k') = b+1$.

Proof. Note that for every bipartite graph, one has: $\lceil \frac{|V_1|+|V_2|}{k'} \rceil \leq \chi(G,k') \leq \lceil \frac{|V_1|}{k'} \rceil + \lceil \frac{|V_2|}{k'} \rceil$. □

Now we consider the case that $a = |V_1| mod(k') + |V_2| mod(k') = k'$. In this case we need either $\frac{|V_1|+|V_2|}{k'}$ independent sets of size k' or one set more. The other cases with $0 < a \leq k'$ can be transformed to this problem by adding some isolated vertices.

We must decide whether $\chi(G,k') = b+1$ or $\chi(G,k') = b+2$, (with b as in lemma 7. We transform this decision problem to a problem of finding a sequence of independent sets in G.

Lemma 8. *Let $G = (V,E)$ be a bipartite graph with partition $V = V_1 \cup V_2$, $E \subset \{\{v_1,v_2\} | v_i \in V_i\}$ and let $a = |V_1| mod(k') + |V_2| mod(k') = k'$.*

Then $\chi(G,k') = \frac{|V_1|+|V_2|}{k'}$ if and only if there is a sequence of $t > 0$ pairwise disjoint independent sets $U_{a_i,b_i} = U_{a_i} \cup U_{b_i}$ of size k' which satisfies the following conditions:

$$(1)\ U_{a_i} \subset V_1, 0 < |U_{a_i}| = a_i < k'$$
$$(2)\ U_{b_i} \subset V_2, 0 < |U_{b_i}| = b_i < k'$$
$$(3)\ (\textstyle\sum_{i=1}^{t} a_i) mod(k') = |V_1| mod(k')$$
$$(4)\ (\textstyle\sum_{i=1}^{t} b_i) mod(k') = |V_2| mod(k').$$

We can omit condition (4), as it follows from conditions (1) - (3).

Theorem 9. *Given a constant k', the problem* PARTITION INTO BOUNDED INDEPENDENT SETS *with each set of size at most k' can be solved in polynomial time for bipartite graphs.*

Proof. By using of Lemma 8. We search for sequences of independent sets of size k' with the given conditions. We can show at first that the length t of these sequences can be bounded by $k'-1$.

Consider $c = |V_1| mod(k') > 0$ and sequences of t pairwise disjoint independent sets U_{a_i,b_i} of size k' with $(\sum_{i=1}^{t} a_i) mod(k') = c$. We must only consider sequences where no subsequence satisfies these conditions. Using this fact the values $(\sum_{i=1}^{t'} a_i) mod(k')$ with $1 \leq t' \leq t$ must all be different and must lie between 1 and $k'-1$. Therefore we can bound t by $k'-1$. Since for all i: $a_i \in \{1,\ldots,k'-1\}$ we have only a constant number of these sequences which must be considered. For each such sequence $a_1 \ldots a_t$ we can test in polynomial time whether G has such a sequence of disjoint independent sets, because the size of these sets can be bounded by $(k'-1)^{k'-1}$. □

4 Interval graphs

A graph $G = (V, E)$ is an interval graph, iff to each vertex $v \in V$, a closed interval I_v in the real line can be associated, such that for each pair of vertices $u, v \in V$, $u \neq v$, $\{u, v\} \in E$, if and only if $I_u \cap I_v \neq \emptyset$. The complement G^c of an interval graph G can be transitively oriented with $(u, v) \in A$ iff $(x \in I_u, y \in I_v \rightarrow x < y)$. This orientation A induces a partial order $P = (V, A)$, called interval order.

In this section we show that the problem PARTITION INTO BOUNDED CLIQUES can be solved in linear time. For the other problem we can show the NP-completeness even if the sizes of independent sets are bounded by a constant.

Lemma 10. *Let $G = (V, E)$ be an interval graph, let $P = (V, A)$ be the corresponding interval order and let $k, k' \in \mathbb{N}$. Then there is a partition of G into k cliques of size at most k' iff there is a feasible schedule of P with unit-times which needs at most k' machines and k time steps.*

Proof. Let $C_1, \ldots, C_k$ be a partition of G into cliques with $|C_i| \leq k'$. For each clique C_i there is at least one point x on the real line with $x \in \bigcap_{v \in C_i} I_v$. We assume that the cliques are ordered according to these points on the real line. Define $T(v) = i$ if $v \in C_i$. We now prove that T gives a feasible schedule. Let $v, w \in V$ with $(v, w) \in A$. Since the interval I_v lies on the left side of I_w the corresponding cliques C_i with $v \in C_i$ and C_j with $w \in C_j$ satisfy $i < j$ and therefore we have $T(v) = i < j = T(w)$. The number of vertices at each time step is less or equal k' and the number of steps is k.

Now let $T : V \rightarrow \{1, \ldots, k\}$ be a feasible schedule where for each $1 \leq i \leq k$ we have $|\{v|T(v) = i\}| \leq k'$. Define $C_i = \{v|T(v) = i\}$. If C_i is not a clique there are vertices $v, w \in C_i$ with $\{v, w\} \notin E$. This means that the intervals $I_v \cap I_w = \emptyset$. Therefore I_v lies to the left or to the right side of I_w. In both cases we have an arc in P and hence, we have not a feasible schedule. Therefore each set C_i is a clique and we get a partition into k cliques with $|C_i| \leq k'$. □

Theorem 11. *The problem* PARTITION INTO BOUNDED CLIQUES *can be solved in linear time for interval graphs.*

Proof. Apply lemma 10 and solve the scheduling problem for interval orders using a linear time algorithm by Papadimitriou and Yannakakis [7]. □

Theorem 12. *The problem* PARTITION INTO BOUNDED INDEPENDENT SETS *remains NP-complete for interval graphs.*

Proof. We can use basically the same transformation as in the proof of theorem 1. (Use that the complement of $G_{a_1, \ldots, a_n}$ is an interval graph.) □

For the problem with constant number k of independent sets we can use the same approach as for the cographs. A partition $U_1, \ldots, U_h$ with $h \leq k$ for an interval graph can be identified with a sequence of sizes $|U_1|, \ldots, |U_h|$ and the last endpoint $x_i = max_{v \in U_i} max(I_v)$ on the real line for each set U_i. Using dynamic programming we can generate all feasible sequences.

Theorem 13. *Given a constant k, the problem* PARTITION INTO K BOUNDED INDEPENDENT SETS *each set of size at most k' can be solved in polynomial time for interval graphs.*

The complexity of PARTITION INTO BOUNDED INDEPENDENT SETS each of size at most k' is open for $k' = 3$. This problem (for $k' = 3$) contains the problem PARTITION INTO TRIANGLES for the complement of interval graphs (namely when $k = |V|/3$).

Theorem 14. *The problem* PARTITION INTO BOUNDED INDEPENDENT SETS *each of size at most four remains NP-complete for interval graphs.*

Proof. We give a transformation from NUMERICAL 3-DIMENSIONAL MATCHING [2] to the partition problem. An instance of numerical 3-dimensional matching is given by disjoint sets W, X and Y each containing m elements, a size $s(a) \in \mathbb{N}$ for each element $a \in W \cup X \cup Y$ and a bound Z such that $\sum_{a \in W \cup X \cup Y} s(a) = mZ$. The question is to decide whether $W \cup X \cup Y$ can be partitioned into m disjoint sets A_i such that each A_i contains exactly one element from each of W, X and Y and such that for $1 \leq i \leq m$, $\sum_{a \in A_i} s(a) = Z$. This problem remains NP-complete if we require that $s(a) < \frac{Z}{2}$ for all $a \in W \cup X \cup Y$. This can be proved by transforming the original problem in one where this assumption holds, by adding the value Z to each $a \in W \cup X \cup Y$ and by setting $Z' = 4Z$.

Now we give the construction of a set of intervals. The interval graph that is modeled by this set of intervals forms the instance for the partition problem. Write $W = \{w_1, \ldots, w_m\}$, $X = \{x_1, \ldots, x_m\}$ and $Y = \{y_1, \ldots, y_m\}$.

1. take for each $w_i \in W$ an interval $a_i = [0, w_i]$.
2. take for each $w_i \in W, x_j \in X$ an interval $b_{i,j} = [w_i + 1, w_i + x_j + (j\,Z)]$.
3. take for each $x_j \in X, y_k \in Y$ an interval $c_{j,k} = [(j+1)Z - y_k + 1, (m+1)Z + k]$.
4. take for each $1 \leq k \leq m$ an interval $d_k = [(m+1)Z + k + 1, (m+3)Z + 1]$.
5. take for each $w_i \in W$ $(m-1)$ intervals $e_{i,\ell} = [1, w_i]$ and $(m-1)$ intervals $f_{i,\ell} = [0, 0]$.
6. take for each $1 \leq j \leq m$ $(m-1)$ intervals $g_{j,\ell} = [(j+1)Z, (m+3)Z + 1]$ and $(m-1)$ intervals $h_{j,\ell} = [0, jZ]$.
7. take for each $1 \leq k \leq m$ $(m-1)$ intervals $p_{k,\ell} = [(m+1)Z + k + 1, (m+3)Z]$ and $q_{k,\ell} = [(m+3)Z + 1, (m+3)Z + 1]$.

Denote the set of all intervals a_i $(1 \leq i \leq m)$ by A. In a similar way, define sets B, C, D, E, F, G, H, P, Q; each of these sets contains all intervals denoted with the same letter. At first let us consider which sets of vertices form a clique. These are $A \cup F \cup H$, $A \cup E \cup H$, $B \cup H$, $C \cup G$, $P \cup G \cup D$ and $P \cup Q \cup D$ and some other which depend on the instance. We note that each independent set in the interval graph has size at most five.

The sizes of the sets are $|A| = |D| = m$, $|B| = |C| = m^2$ and the other sets E, F, G, H, P, Q have size $m(m-1)$. In total, this are $8m^2 - 4m$ vertices. We consider the problem to partition the interval graph, corresponding to the set of intervals into $2m^2 - m$ independent sets of size at most four. Note that each independent set must have size exacly four.

Let $h \in H$ and consider an independent set U of size four which contains h. Then the only possibility is to choose one vertex $c \in C$, one vertex $p \in P$ and one vertex $q \in Q$ for the set U. For a vertex $g \in G$ and an independent U which contains g we can only take one vertex $b \in B$, one vertex $e \in E$ and one vertex $f \in F$. If we delete these vertices, we have only m elements of A, B, C and D.

We now study 'cuts' between two sets of vertices in the interval graph. Consider the following cuts: (1) $A \cup E$ and B; (2) $B \cup H$ and $C \cup G$; (3) C and $P \cup D$.

We consider first the last of these three cuts. We see that the sizes $|C|$ and $|P \cup D|$ are equal to m^2. Since G has $m^2 - m$ vertices and since $C \cup G$ and $P \cup D \cup G$ are cliques, we must choose for the m^2 independent sets one vertex of C and one of $P \cup D$.

Now we prove that for each vertex $c_{j,k}$ there is a vertex $p_{k,\ell} \in P$ or a vertex $d_k \in D$ such that both are together in one of the independent sets. This means that there is no independent set U with $\{c_{j,k}, p_{k',\ell}\} \subset U$ or $\{c_{j,k}, d_{k'}\} \subset U$ if $k \neq k'$. Assume that this is not the case. We have m^2 independent sets where each contains exactly one element of C and one of P or D. Let $c_{j,k}$ a vertex with minimum k which lies in an independent set with a vertex $p_{k',\ell}$ or $d_{k'}$ for $k \neq k'$. If $k > k'$ the intervals overlap and therefore this case is not possible. But if $k < k'$, the vertices with second index less than k are correctly connected. Therefore at least one of the vertices in $\{p_{k,\ell} | 1 \leq \ell \leq m-1\} \cup \{d_k\}$ must be connected to a vertex $c_{j',k''}$ with $k'' > k$. Since the corresponding intervals overlap, we get a contradiction.

Let us consider the second cut with $B \cup H$ on the left and with $C \cup G$ on the right side. Since we have $2m^2 - m$ vertices in both sets, and since both sets are cliques, each independent sets must have one element from $B \cup H$ and one from $C \cup G$. We see that $w_i + x_j + jZ < (j+1)Z$ and that the interval $c_{j,k}$ with left endpoint $(j+1)Z - y_k + 1$ intersect with $h_{j+1,\ell}$. Therefore we can prove in similar way as above, that there is no independent set U with $\{b_{i,j}, c_{j',k}\} \subset U$ or with $\{b_{i,j}, g_{j',\ell}\} \subset U$ or with $\{h_{j,\ell}, c_{j',k}\} \subset U$ for $j \neq j'$.

For the first cut $A \cup E$ and B we get with the same argument that only vertices of $a_i, e_{i,\ell}$ are together with vertices $b_{i',j}$ if $w_i = w_{i'}$. It is possible to swap elements between independent sets, such that each a_i and each $e_{i,\ell}$ lies in an independent set together with one $b_{i,j}$.

From this analyse of cuts, it follows that we may assume that the independent sets contain pairs of intervals, illustrated in the following table. In other words, there is for example no independent set which contains $\{a_i, b_{i',\ell}\}$ or $\{e_{i,\ell}, b_{i',\ell'}\}$ for $i \neq i'$.

first interval	second interval
a_i or $e_{i,-}$	$b_{i,-}$
$b_{-,j}$ or $h_{j,-}$	$c_{j,-}$ or $g_{j,-}$
$c_{-,k}$	d_k or $p_{k,-}$.

Now consider the interval graph after deleting all independent sets U which contain $h \in H$ or $g \in G$. We now have m independent sets U_i which cover the vertices in A, D and the remaining vertices in B, C. Using that each $g_{j,-}$ is connected to one $b_{-,j}$ and that each $h_{j,-}$ is connected to one $c_{j,-}$ we have for each j exactly one vertex $b_{-,j}$ in the rest of B and one $c_{j,-}$ in the rest of C. Therefore each independent set U_i has the form $U_i = \{a_i, b_{i,j}, c_{j,k}, d_k\}$.

Now we can prove that there is a partition of $W \cup X \cup Y$ into sets A_i with exactly one element of W, X and Y and with $\sum_{a \in A_i} s(a) = Z$ iff the constructed interval graph has a partition into $2m^2 - m$ independent sets of size at most four.

Let $U_1, \ldots, U_{2m^2-m}$ be such a partition. From the analyse above, it follows that we may assume w.l.o.g. the first m independent sets have the form $U_i = \{a_i, b_{i,j}, c_{j,k}, d_k\}$ such that the sets $\{j | b_{i,j} \in U_i, 1 \leq i \leq m\}$, $\{k | c_{j,k} \in U_i, 1 \leq i \leq m\}$ are equal to $\{1, \ldots, m\}$. Using that U_i is an independent set, we have $w_i + x_j + jZ < (j+1)Z - y_k + 1$ and therefore we get $w_i + x_j + y_k \leq Z$. Since each index appears exacly once, we have $\sum_{i=1}^{m} w_i + \sum_{j=1}^{m} x_j + \sum_{k=1}^{m} y_k = mZ$. Therefore $w_i + x_j + y_k = Z$ and the sets $A_i = \{w_i, x_j, y_k\}$ given by the intervals U_i solve the matching problem.

To prove the equivalence in other direction, let $A_i = \{w_i, x_j, y_k\}$ be the sets with $\sum_{a \in A_i} s(a) = Z$. As the first m sets we choose $U_i = \{a_i, b_{i,j}, c_{j,k}, d_k\}$.

The interval a_i lies on the left side to $b_{i,j}$ and the interval d_k lies on the right side to $c_{j,k}$. To prove that $b_{i,j}$ lies on the left side to $c_{j,k}$ we compare the right endpoint of $b_{i,j}$ and the left endpoint of $c_{j,k}$. Using that $w_i + x_j + y_k = Z$, we get $w_i + x_j + (jZ) < (j+1)Z - y_k + 1$. Therefore the set U_i is independent.

Let $B' \subset B$ be the set of vertices which are not covered and construct iteratively independent sets. Let $b_{i,j} \in B'$. Then take vertices $e_{i,\ell}, f_{i,\ell}, g_{j,\ell'}$ which are not covered and put them together in one set U. Clearly, this set is independent. The construction is correct, because each index i and j appeares only $(m-1)$ times in B'. Now consider the set $C' \subset C$ of vertices which are not covered and construct in a similar way independent sets. For these take for each $c_{j,k} \in C'$ vertices $h_{j,\ell}, p_{k,\ell'}, q_{k,\ell'}$ which are not covered. After these steps all vertices are covered and we have $2m^2 - m$ independent sets. □

References

1. Corneil, D.G., Perl, Y., Stewart, L.K.: A linear recognition algorithm for cographs. SIAM Journal of Computing **4** (1985) 926–934
2. Garey, M.R., Johnson, D.S.: Computers and Intractability: A Guide to the Theory of NP-Completeness. Freeman, San Francisco (1979)
3. Golumbic, M.C.: Algorithmic graph theory and perfect graphs. Academic Press, New York (1980)
4. Gupta, U.I., Lee, D.T., Leung, J.Y.-T.: Efficient algorithms for interval graphs and circular arc graphs. Networks **12** (1982) 459–467
5. Karp, R.M.: Reducibility among combinatorial problems. In: Miller, Thatcher (eds.): Complexity of computer computations, Plenum Press (1972) 85–104
6. Lonc, Z.: On complexity of some chain and antichain partition problems. WG Conference, LNCS (1991) 97–104
7. Papadimitriou, C.H., Yannakakis, M.: Scheduling interval-ordered tasks. SIAM Journal of Computing **8** (1979) 405–409
8. Seinsche, D.: On a property of the class of n-colorable graphs. Journal of Combinatoral Theory **B 16** (1974) 191–193

Isomorphisms between Predicate and State Transformers

Marcello Bonsangue[1]* and Joost N. Kok[2]

[1] *CWI*, P.O. Box 4079, 1009 AB Amsterdam, The Netherlands. marcello@cwi.nl
[2] Utrecht University, Department of Computer Science, P.O. Box 80.089, 3508 TB Utrecht, The Netherlands. joost@cs.ruu.nl

Abstract. We study the relation between state transformers based on directed complete partial orders and predicate transformers. Concepts like 'predicate', 'liveness', 'safety' and 'predicate transformers' are formulated in a topological setting. We treat state transformers based on the Hoare, Smyth and Plotkin power domains and consider continuous, monotonic and unrestricted functions. We relate the transformers by isomorphisms thereby extending and completing earlier results and giving a complete picture of all the relationships.

1 Introduction

In this paper we give a full picture of the relationship between state transformers and predicate transformers. For the state transformers we consider the Hoare, Smyth and Plotkin power domains. We give a full picture in the sense that we consider algebraic directed complete partial orders (with a bottom element) (and not only flat domains), we consider not only continuous state transformers, but also the monotonic ones and the full function space, we do not restrict to bounded nondeterminism, and we treat all the three power domains with or without empty set. The first item is important when we want to use domains for concurrency semantics. The second and third item give more freedom in the sense that we can use these transformations also for specification purposes without constraints on computability. Having the empty set in a power domain can be important to treat deadlock. Our treatment includes the Plotkin power domain.

For state transformers we use an extension of the standard power domains. For predicate transformers we start from the (informal) classification of predicates in liveness and safety predicates of Lamport [Lam77]. Later Smyth [Smy83] followed by [AS85, Kwi91] used topology to formalize this classification. Also we use topology for defining predicates and safety and liveness predicate transformers. We consider predicate transformers with predicates that are the intersection of safety and liveness predicates.

* The research of this author was supported by a grant of the Consiglio Nazionale delle Ricerche (CNR), Italy, announcement no. 203.15.3 of 15/2/90.

We prove that the Hoare state transformers are isomorphic to safety predicate transformers, the Smyth state transformers are isomorphic to the liveness predicate transformers, and that the Plotkin state transformers are isomorphic to the "intersection" predicate transformers. So for the first time we are able to give a full picture of all the relationships filling several gaps that were present in the literature.

Next we discuss how this paper is related to previous work. Power domains for dcpo's were introduced in [Plo76], [Smy78] and [Plo81]. Our power domains are slightly more general in the sense that we do no restrict to non-empty (Scott-) compact sets. Besides the standard ways of adding the empty set to the Smyth [Smy83] and to the Plotkin [MM79, Plo81, Abr91] power domains, we also add the empty set in all the three power domain as a separate element, comparable only with itself and with the bottom.

Predicate transformers were introduced in [Dij76] with a series of healthness conditions. Back and von Wright [Bac80, vW90] use only the monotonicity restriction on predicate transformers. They use predicate transformers for refinement and provide a nice lattice theoretical framework. Nelson [Nel89] has (for the flat case) used "compatible" pairs of predicate transformers for giving semantics to a language with backtracking. Smyth [Smy83] introduced predicate transformers (with the Dijkstra healthiness conditions) for non-flat domains. Our definition of predicate transformers is parametric with respect to the collection of predicates (observable, liveness and safety). Furthermore, a new (generalization) of the multiplicativity restriction is introduced and a generalization of "compatible" pairs of predicate transformers is given.

Isomorphisms between state and predicate transformers have been given for the flat case of the Smyth power domain in [Plo79] (and for countable nondeterminism in [AP86]), and for the flat case of the Hoare power domain in [Plo81]. Also De Bakker and De Roever [Bak80, Roe76] studied (from a semantical point of view) for the flat case the relation between state transformer and predicate transformer semantics. Moreover, for the flat case of the Plotkin power domain we have proposed an isomorphism in [BK92].

For the general case of the compact Smyth power domain in the paper [Smy83] an isomorphism is given for continuous state transformers. He uses a topological technique which Plotkin later used in [Plo81] for the continuous Hoare state transformers. A recent work includes an operational point of view in Van Breugel [Bre93]. In the present paper we give some new isomorphisms for the Hoare and the Smyth power domains, showing also how the previous ones can be obtained as combinations of the new isomorphisms. Our definition of multiplicativity for predicate transformers permits us to use a technique similar to that used for the flat case. Furthermore we give isomorphisms for the Plotkin power domain. As far as we know no isomorphism was known for the non-flat Plotkin power domain (as for example is remarked in [Plo81] and in [Smy83]).

For reasons of space, proofs are not given in this paper. They can be found in [BK93].

2 Mathematical Preliminaries

We introduce some basic notions on domain theory and topology. For a more detailed discussion on domain theory consult for example [Plo81], and for topology we refer to [Eng77].

Let P be a set ordered by $\sqsubseteq_P$, $x \in P$ and let A be a subset of P. Define $x \uparrow = \{y | y \in P \wedge x \sqsubseteq y\}$ and $A \uparrow = \bigcup\{x \uparrow | x \in A\}$. A set A is called upper-closed if $A = A \uparrow$. A subset A of a partially ordered set P is said to be *directed* if it is non empty and every finite subset of A has an upper bound in A. P is a (pointed) *directed complete partially order set* (dcpo) if there exists a least element $\bot_P$ and every directed subset A of P has least upper bound (lub) $\bigsqcup A$. A directed set A is *eventually constant* if $\bigsqcup A \in A$.

An element b of a dcpo P is *finite* if for every directed set $A \subseteq P$, $b \sqsubseteq \bigsqcup A$ implies $b \sqsubseteq x$ for some $x \in A$. The set of all finite elements of P is denoted by B_P and is called *base*. A dcpo P is *algebraic* if for every element $x \in P$ the set $\{b | b \in B_P \wedge b \sqsubseteq x\}$ is directed and has least upper bound x; it is ω*-algebraic* if it is algebraic and its base is denumerable.

Let P, Q be two partially ordered sets. A function $f : P \to Q$ is *monotone* (denoted by $f : P \to_m Q$) if for all $x, y \in P$ with $x \sqsubseteq_P y$ we have $f(x) \sqsubseteq_Q f(y)$. If P ia a dcpo we say f is *continuous* (denoted by $f : P \to_c Q$) if $f(\bigsqcup A) = \bigsqcup f(A)$ for each directed set $A \subseteq P$; moreover f is *stabilizing* (denoted by $f : P \to_{c_s} Q$) if it is continuous and for every directed set $A \subseteq P$ $f(A)$ is an eventually constant directed set in Q. If $f : P \to_c Q$ is continuous then f is monotone. Given a set $\beta \subseteq Q$, a continuous function $f : P \to_c Q$ is said β*-algebraic* (denoted by $f : P \to_{a(\beta)} Q$) if for every directed set $S \subseteq P$ and for every $q \in \beta$ such that $q \sqsubseteq f(\bigsqcup S)$ there exists an $x \in S$ such that $q \sqsubseteq f(x)$. Clearly, for Q an algebraic dcpo Q with base B_Q, a function $f : P \to Q$ is continuous if and only if f is B_Q-algebraic. A function f is *strict* (denoted by $f : P \to_s Q$) if $f(\bot_P) = \bot_Q$; dually f is *top preserving* (denoted by $f : P \to_t Q$) if and only if $f(\top_P) = \top_Q$.

Let P be a dcpo and $f : P \to P$. We denote by $\mu.f$ the *least fixed point* of f, that is, $f(\mu.f) = \mu.f$ and for every other $x \in P$ such that $f(x) = x$ then $\mu.f \sqsubseteq x$. For a monotone function $f : P \to_m P$, where P is a dcpo, the least fixed point of f always exists and can be calculated by iteration, that is, there exists an ordinal λ such that $\mu.f = f^\lambda$, where the α-iteration of f is defined by $f^\alpha = f(\bigsqcup_{k<\alpha} f^k)$ for every ordinal α [HP72]. If f is also continuous then $\lambda \leq \omega$. It can be of interest to consider also non-monotonic functions, at least when they are representation as quotient of some monotonic functions between dcpo, as shown in the following transfer lemma [BK92]: let P be a dcpo and Q be a partially ordered set, $f : P \to_m P$ be a monotone function, $h : P \to_c Q$ be an onto and continuous function and $g : Q \to Q$ be a (possibly non monotone) function such that $g \circ h = h \circ f$. Then for every ordinal α the α-iteration from the bottom element g^α exists. Moreover, if for each $y \in Q$ the partially ordered set $h^{-1}(y) \subseteq P$ is finite or has either the bottom or the top element then the smallest fixed point $\mu.g$ exists and $\mu.g = h(\mu.f)$.

We now introduce some basic topological notions. A *topology* $\mathbf{O}(X)$ on a set X is a collection of subsets of X that is closed under finite intersections and arbitrary unions. The pair $(X, \mathbf{O}(X))$ is called *topological space* and the elements of $\mathbf{O}(X)$ are the *open sets* of the space X. A *base* of a topology $\mathbf{O}(X)$ on X is a subset $\mathbf{B} \subseteq \mathbf{O}(X)$ such that every open set is the union of elements of $\mathbf{B}$. A set $S \subseteq X$ is *dense* if and only if $X \setminus S$ contains no non-empty open sets. A G_δ-*set* is a countable intersection of open sets.

For example, given a partially ordered set X, its *discrete* topology is $\mathbf{O}_d(X) = \mathcal{P}(X)$ while its *Alexandroff topology* $\mathbf{O}_{Al}(X)$ consists of all the upper-closed subsets of X. If X is a dcpo, a finer topology of X is the Scott topology $\mathbf{O}_{Sc}(X)$, where $o \in \mathbf{O}_{Sc}(X)$ if and only if o is upper-closed and for any directed set $S \subseteq X$ if $\bigsqcup S \in o$ then $S \cap o \neq \emptyset$. Let A be a collection of subsets of X; the closure under arbitrary intersection of A is denoted by $A^{\cap} = \{Q | Q = \bigcap A' \wedge A' \subseteq A\}$.

We can describe a topology by its closed sets instead of its open sets. A subset of a set X is *closed* if and only if it is the complement of an open set of a given topology on X. The collection of closed sets of a topological space is denoted by $\mathbf{C}(X)$ and, dually to the case of open sets, is closed under finite unions and arbitrary intersections. For every $A \subseteq X$ there exists a closed set c and a dense set d such that $A = c \cap d$.

For example, given a dcpo P the closed sets of the Alexandroff topology are all the lower closed sets, while a set $c \subseteq P$ is closed with respect to the Scott topology if c is lower closed and for every directed set $S \subseteq P$ if $S \subseteq c$ then $\bigsqcup S \in c$.

Let $\mathbf{O}(X)$ be a topology on a set X. A subset $A \subseteq X$ is *compact* in $\mathbf{O}(X)$ if and only if for every collection of open sets $o_i \in \mathbf{O}(X)$ with $i \in I$ such that $A \subseteq \bigcup_I o_i$ there exists a finite subcollection o_j such that $A \subseteq \bigcup_J o_j$. For example $A \subseteq X$ is compact in $\mathbf{O}_d(X)$ if and only if it is a finite set. The intersection of a closed set with a compact one is always compact.

3 Predicates and Predicate Transformers

A *predicate* P is a function from a set X to the boolean set $\mathbf{Bool} = \{tt, ff\}$ or, equivalently, is a subset of X. Topology provides an elegant way of expressing predicates of programs (see [Smy83], [Kwi91]) in which the open sets of a topological space X are seen as the *computable predicates.* Taking different topologies corresponds to different restrictions on the function space. For example, with $ff \sqsubseteq tt$ we have that $\mathbf{O}_d(X)$ is isomorphic to the set of all the predicates from X to **Bool**, $\mathbf{O}_{Al}(X)$ is isomorphic to the set of all the monotone predicates from X to **Bool** and $\mathbf{O}_{Sc}(X)$ is isomorphic to the set of all the continuous predicates from X to **Bool**.

In [Lam77] two classes of predicates were introduced: *safety* and *liveness predicates.* In the topological view of Smyth [Smy83, Smy92] closed sets represent *safety predicates* while *liveness predicates* are G_δ sets. We take arbitrary intersections of open sets as liveness predicates (saturated sets). This differs also from [AS85] where liveness predicates are dense sets (the complement does not contain non-empty open sets). In [Kwi91] liveness predicates are also G_δ-sets.

In this paper we consider algebraic dcpo's together with the Scott topology. The Scott-closed sets are the safety predicates and are ordered by $\supseteq$. The Scott-open sets of a dcpo Y represent the computable (observable) predicates and are ordered by $\subseteq$. They are *finitary* in the sense that $y \in o$ if and only if there exists a $b \in B_Y$ such that $b \in o$ and $b \sqsubseteq y$. In other words, a predicate P is finitary if we can test whether P holds for y by testing only the finite elements smaller than y. Liveness predicates are the arbitrary intersection of Scott-open sets (that is Alexandroff open sets) and are ordered by $\subseteq$.

Consider for example the set of sequences (finite and infinite) over an alphabet Σ ordered by the prefix ordering (for more examples see [Kwi91]).

- Safety predicate: *always* $a = \{x | x = a^* \vee x = a^\omega\}$ (Scott closed),
- Observable predicate: *eventually* $a = \{xa | x \in \Sigma^*\} \uparrow$ (Scott open),
- Observable predicate: *start with* $x = x \uparrow$ (Scott open), where $x \in \Sigma^*$,
- Liveness predicate: *infinitely often* $a = \bigcap_{n \in N}\{x | x \in \Sigma^* \ \wedge |x|_a = n\} \uparrow$ (Alexandroff open but not Scott open),
- Neither safety nor liveness predicate: *always a but starting with* $x =$ *always* $a \wedge$ *start with* x.

Let $P(Y)$ and $P(X)$ be two collections of predicates on the space Y and X, respectively. We define *predicate transformers* as the monotone functions from $P(Y)$ to $P(X)$. Another natural restriction (besides monotonicity) in the case that $Y \in P(Y)$ is to require that a predicate transformer must be top-preserving. Predicate transformers mapping observable predicates to observable predicates are denoted by $P(Y) \to_O P(X)$.

A predicate transformer $\pi : P(Y) \to P(X)$ is *multiplicative* (denoted by $\pi : P(Y) \to_M P(X)$) if and only if for collections of predicates $P, Q \subseteq P(Y)$ if $\bigcap P \subseteq \bigcap Q$ then $\bigcap_{p \in P} \pi(p) \subseteq \bigcap_{q \in Q} \pi(q)$. If $P(Y)$ is closed under arbitrary intersection then π is multiplicative if and only if $\bigcap_{p \in P} \pi(p) = \pi(\bigcap P)$ for every collection of predicates $P \subseteq P(Y)$. If $Y \in P(Y)$ then π multiplicative implies π top preserving. Given a predicate transformer $\pi : P(Y) \to P(X)$ its *dual* $\pi^\circ : P(Y)^\circ \to P(X)^\circ$ is given by $\pi^\circ(p) = X \setminus \pi(Y \setminus p)$, for every $p \in P(Y)^\circ = \{Y \setminus q | q \in P(Y)\}$. A predicate transformer is *additive* (denoted by $\pi : P(Y) \to_A P(X)$) if and only if its dual is multiplicative.

A predicate transformer $\pi : P(Y) \to P(X)$ is *intersection extensible* (denoted by $\pi : P(Y) \to_I P(X)$) if $\bigcap_{p \in P} \pi(p) \in P(X)$ for every collection of predicates $P \subseteq P(Y)$. Dually, if $\bigcup_{p \in P} \pi(p) \in P(Y)$ then π is called *union extensible* (denoted by $\pi : P(Y) \to_U P(X)$).

Intuitively, multiplicative predicate transformers $\pi : P(Y) \to P(X)$ preserve the logical '$\forall$' on predicates on Y, while the additive ones preserve the logical '$\exists$' (even if they are not a predicates in $P(Y)$ or $P(X)$). If π is also intersection (union) extensible then the logical '$\forall$' ('$\exists$') of π of an arbitrary collection of predicates on Y is always a predicate in $P(X)$.

We now define a restricted version of the Cartesian product on (multiplicative) predicate transformers by requiring (multiplicativity) monotonicity on the intersection.

Definition 1. Let $P_1(Y)$, $P_2(Y)$ be two collections of predicates on Y and $Q_1(X), Q_2(X)$ be two collections of predicates on X. Define $(P_1(Y) \to_m Q_1(X)) \otimes (P_2(Y) \to_m Q_2(X))$ as the subset of $(P_1(Y) \to_m Q_1(X)) \times (P_2(Y) \to_m Q_2(X))$ given by:

$$\{(\pi, \rho) | \forall\, p, p' \in P_1(Y), q, q' \in P_2(Y) : p \cap q \subseteq p' \cap q' \Rightarrow \pi(p) \cap \rho(q) \subseteq \pi(p') \cap \rho(q')\}.$$

Similarly, define $(P_1(Y) \to_M Q_1(X)) \otimes_M (P_2(Y) \to_M Q_2(X))$ as

$$\{(\pi, \rho) |\ \forall\ P, P' \subseteq P_1(Y), Q, Q' \subseteq P_2(Y) : \bigcap P \cap \bigcap Q \subseteq \bigcap P' \cap \bigcap Q' \\ \Rightarrow \bigcap_{p \in P} \pi(p) \cap \bigcap_{q \in Q} \rho(q) \subseteq \bigcap_{p' \in P'} \pi(p') \cap \bigcap_{q' \in Q'} \rho(q')\}.$$

They are ordered pointwise.

Now we come to the definition of safety and liveness predicate transformers used in this paper. Let X and Y be algebraic dcpo's. The liveness predicate transformers are functions in $\mathbf{O}_{Sc}(Y)^{\cap} \to_{OM} \mathbf{O}(X)^{\cap}$ for some topology $\mathbf{O}(X)$ on X. They are ordered pointwise by subset inclusion. The safety predicate transformers are functions in $\mathbf{C}_{Sc}(Y) \to_M \mathbf{C}(X)$ for some collection of closed set $\mathbf{C}(X)$ on X. They are ordered pointwise by superset inclusion.

We can define liveness and safety predicate transformers in terms of observable predicate transformers since $(\mathbf{O}_{Sc}(Y)^{\cap} \to_{OM} \mathbf{O}(X)^{\cap}) \cong (\mathbf{O}_{Sc}(Y) \to_M \mathbf{O}(X))$ and $(\mathbf{C}_{Sc}(Y) \to_M \mathbf{C}(X)) \cong (\mathbf{O}_{Sc}(Y) \to_A \mathbf{O}(X))$. We can also change the subset order of the liveness predicate transformers to the deadlock order of [BK92], where for every $\pi_1, \pi_2 \in \mathbf{O}_{Sc}(Y) \to_M \mathbf{O}(X)$, $\pi_1 \sqsubseteq_{LD} \pi_2$ if

$$(\pi_1(\emptyset) \subseteq \pi_2(\emptyset)) \ \wedge\ (\forall o \in \mathbf{O}_{Sc}(Y) : o \neq Y \Rightarrow \pi_1(o) \setminus \pi_1(\emptyset) \subseteq \pi_2(o) \setminus \pi_2(\emptyset)).$$

and for safety predicate transformers ρ_1, ρ_2, $\rho_1 \sqsubseteq_{SD} \rho_2$ if

$$(\rho_1(\emptyset) \subseteq \rho_2(\emptyset)) \ \wedge\ (\forall c \in \mathbf{C}_{Sc}(Y) : c \neq \emptyset \Rightarrow \rho_1(c) \supseteq \rho_2(c) \setminus \rho_1(\emptyset)).$$

4 State Transformers

In this section we give generalizations of the three 'classical' power domains on ω-algebraic dcpo's, the so-called Hoare, Smyth and Plotkin power domains ([Plo76], [Smy78] and [Plo81]).

Let X be an algebraic dcpo and $A \subseteq X$. Define $\overline{A} = \{x | \forall b \in B_X : b \sqsubseteq x \Rightarrow \exists x_b \in A : b \sqsubseteq x_b\}$ and $A^* = \{x | (\exists x' \in A : x' \sqsubseteq x) \ \wedge\ (\forall b \in B_X : b \sqsubseteq x \Rightarrow \exists x_b \in A : b \sqsubseteq x_b)\}$. We have $\overline{A} = A$ if and only if $A \in \mathbf{C}_{Sc}(X)$. Further $A \subseteq A^*$, $(A^*)^* = A^*$, and $A = A^* \ \Leftrightarrow\ A = A \uparrow \cap \overline{A}$. Next we define the power domains:

Definition 2. Let X be an algebraic dcpo. Define

- the Hoare power domain $\mathcal{H}(X) = \langle \{A | A \subseteq X \ \wedge\ A = \overline{A}\}, \sqsubseteq_H \rangle$, where $A \sqsubseteq_H B$ if $A \subseteq B$,
- the Smyth power domain $\mathcal{S}(X) = \langle \{A | A \subseteq X \ \wedge\ A = A \uparrow\}, \sqsubseteq_S \rangle$, where $A \sqsubseteq_S B$ if $A \supseteq B$,

- the Plotkin power domain $\mathcal{P}(X) = \langle\{A | A \subseteq X \ \wedge \ A = A^*\}, \sqsubseteq_P\rangle$, where $A \sqsubseteq_P B$ if $A\uparrow \sqsubseteq_S B\uparrow$ and $\overline{A} \sqsubseteq_H \overline{B}$.

If we consider only bounded nondeterminism then we can restrict the power domains to those sets that are compact in the Scott topology (denoted by the subscript *co*). Since every Scott closed set of a dcpo is compact in the Scott topology we have $\mathcal{H}_{co}(X) = \mathcal{H}(X)$. The standard definitions of the Hoare, Smyth and Plotkin power domains are $\mathcal{H}^+(X), \mathcal{S}^+_{co}(X)$ and $\mathcal{P}^+_{co}(X)$, where the superscript $^+$ denotes that the power domains should be taken without the empty set. The same domains but with a slight different order are denoted by $\mathcal{H}\delta(X), \mathcal{S}\delta(X)$ and $\mathcal{P}\delta(X)$. Their ordering are respectively given by

- $A \sqsubseteq_{H\delta} B$ if $(A = \{\bot\}) \vee (A = \emptyset \Rightarrow B = \emptyset) \vee (A \neq \emptyset \wedge A \sqsubseteq_H B)$,
- $A \sqsubseteq_{S\delta} B$ if $(A = X) \vee (A = \emptyset \Rightarrow B = \emptyset) \vee (B \neq \emptyset \wedge A \sqsubseteq_S B)$,
- $A \sqsubseteq_{P\delta} B$ if $(A = \{\bot\}) \vee (A = \emptyset \Rightarrow B = \emptyset) \vee (A \neq \emptyset \wedge A \sqsubseteq_P B)$.

Most of the usual operations (such as the sequential composition) are not monotone w.r.t. $\sqsubseteq_{S\delta}$. However, we can use the transfer lemma given in the Mathematical Preliminaries for calculating least fixed points [BK92]. For an algebraic dcpo X, the Hoare power domain is a complete lattice. Furthermore it is also an algebraic dcpo with as finite elements the Scott closure of finite subsets of B_X. Also the Smyth power domain is a complete lattice, but in general not an algebraic dcpo. However its restriction $\mathcal{S}_{co}(X)$ is an algebraic dcpo with finite elements the upper closure of finite subsets of B_X [Smy78]. In general, the Plotkin power domain is neither a complete lattice nor a dcpo (there is no bottom element because the empty set is related only with itself and $\{\bot\}$ is less than any other set different from empty set). However if X is algebraic then $\mathcal{P}^+_{co}(X)$ is an algebraic dcpo with finite elements the *-closure of finite subsets of B_X (see [Plo76] and extensions in [Hrb87],[Hrb89]). Furthermore, $\mathcal{P}\delta_{co}(X)$ coincides with the standard way of adding the empty set to the Plotkin power domain [MM79], [Plo81] and also [Abr91].

State transformers are functions (ordered pointwise) from an algebraic dcpo X to one of the power domains over an algebraic dcpo Y.

5 Relations

In this section we give the isomorphisms between the state transformers and predicate transformers domains. We start with the relation between safety predicate transformers and the Hoare state transformers:

Theorem 3. *Let X and Y be two algebraic dcpo's. We have the following order-isomorphisms:*

1. $X \to \mathcal{H}(Y) \cong \mathbf{C}_{Sc}(Y) \to_M \mathbf{C}_d(X)$,
2. $X \to \mathcal{H}^+(Y) \cong \mathbf{C}_{Sc}(Y) \to_{sM} \mathbf{C}_d(X)$,
3. $X \to_m \mathcal{H}(Y) \cong \mathbf{C}_{Sc}(Y) \to_M \mathbf{C}_{Al}(X)$,
4. $X \to_c \mathcal{H}(Y) \cong \mathbf{C}_{Sc}(Y) \to_M \mathbf{C}_{Sc}(X)$,

5. $X \to_{c_s} \mathcal{H}(Y) \cong \mathbf{C}_{Sc}(Y) \to_{UM} \mathbf{C}_{Sc}(X)$.

In all cases the isomorphism is given by the function γ: $\gamma(m)(c) = \{x | m(x) \subseteq c\}$

The function γ is the generalization of the weakest liberal precondition and its inverse is given by $\gamma^{-1}(\rho)(x) = \bigcap\{c | x \in \rho(c)\}$. Because the isomorphism is always the same we can combine cases of the theorem (for example combining 2. and 4. we get the result of [Plo81]: $X \to_c \mathcal{H}^+(Y) \cong \mathbf{C}_{Sc}(Y) \to_{sM} \mathbf{C}_{Sc}(X)$). Theorem 3 holds also if we substitute $\mathcal{H}\delta(Y)$ for $\mathcal{H}(Y)$ and we take $\sqsubseteq_{SD}$ as order for the safety predicate transformers.

Now we relate liveness predicate transformers and Smyth state transformers:

Theorem 4. *Let X and Y be two algebraic dcpo's. We have the following order-isomorphisms:*

1. $X \to \mathcal{S}(Y) \cong \mathbf{O}_{Sc}(Y) \to_M \mathbf{O}_d(X)$,
2. $X \to \mathcal{S}^+(Y) \cong \mathbf{O}_{Sc}(Y) \to_{sM} \mathbf{O}_d(X)$,
3. $X \to \mathcal{S}_{co}(Y) \cong \mathbf{O}_{Sc}(Y) \to_{cM} \mathbf{O}_d(X)$,
4. $X \to_m \mathcal{S}(Y) \cong \mathbf{O}_{Sc}(Y) \to_M \mathbf{O}_{Al}(X)$,
5. $X \to_{a(\beta)} \mathcal{S}(Y) \cong \mathbf{O}_{Sc}(Y) \to_M \mathbf{O}_{Sc}(X)$,
6. $X \to_{c_s} \mathcal{S}(Y) \cong \mathbf{O}_{Sc}(Y) \to_{IM} \mathbf{O}_{Sc}(X)$,

where $\beta = \{B \uparrow\in \mathcal{S}(Y) | B \subseteq B_Y\}$. *In all cases the isomorphism is given by the function* ω: $\omega(m)(o) = \{x | m(x) \subseteq o\}$

The function ω is a generalization of the weakest precondition and its inverse is given by $\omega^{-1}(\pi)(x) = \bigcap\{o | x \in \pi(o)\}$. Also in this case we can combine 2., 3. and 5. to obtain the result of [Smy83], because $\mathcal{S}_{co}(Y)$ is an algebraic dcpo with finite elements the upper closure of finite subset of B_Y, thus every continuous function from X to $\mathcal{S}_{co}(Y)$ is also β-algebraic. Theorem 4 holds also if we substitute $\mathcal{S}\delta(Y)$ for $\mathcal{S}(Y)$ and we order of the liveness predicate transformer by $\sqsubseteq_{LD}$.

To prove these theorems we need the following extension of the *stability lemma* of Plotkin [Plo79, AP86]:

Lemma 5. *Let* $\pi : P(Y) \to_M P(X)$ *be a multiplicative predicate transformer. Then* $x \in \pi(\hat{p}) \Leftrightarrow \bigcap\{p | x \in \pi(p)\} \subseteq \hat{p}$ *for every* $\hat{p} \in P(Y)$ *and* $x \in X$.

Finally we relate the Plotkin state transformers with pairs of safety and liveness predicate transformers:

Theorem 6. *Let X and Y be two algebraic dcpo's. We have the following order-isomorphisms:*

1. $X \to \mathcal{P}(Y) \cong (\mathbf{O}_{Sc}(Y) \to_M \mathbf{O}_d(X)) \otimes_M (\mathbf{C}_{Sc}(Y) \to_M \mathbf{C}_d(X))$
2. $X \to \mathcal{P}^+(Y) \cong (\mathbf{O}_{Sc}(Y) \to_{sM} \mathbf{O}_d(X)) \otimes_M (\mathbf{C}_{Sc}(Y) \to_{sM} \mathbf{C}_d(X))$,
3. $X \to \mathcal{P}_{co}(Y) \cong (\mathbf{O}_{Sc}(Y) \to_{cM} \mathbf{O}_d(X)) \otimes_M (\mathbf{C}_{Sc}(Y) \to_M \mathbf{C}_d(X))$,
4. $X \to_m \mathcal{P}(Y) \cong (\mathbf{O}_{Sc}(Y) \to_M \mathbf{O}_{Al}(X)) \otimes_M (\mathbf{C}_{Sc}(Y) \to_M \mathbf{C}_{Al}(X))$,

5. $X \to_c \mathcal{P}_{co}(Y) \cong (\mathbf{O}_{Sc}(Y) \to_{cM} \mathbf{O}_{Sc}(X)) \otimes_M (\mathbf{C}_{Sc}(Y) \to_M \mathbf{C}_{Sc}(X))$,
6. $X \to_{c_s} \mathcal{P}(Y) \cong (\mathbf{O}_{Sc}(Y) \to_{IM} \mathbf{O}_{Sc}(X)) \otimes_M (\mathbf{C}_{Sc}(Y) \to_{UM} \mathbf{C}_{Sc}(X))$.

In all cases the isomorphism is given by the function η*:*

$$\eta(m)(o,c) = (\{x|m(x) \uparrow\subseteq o\}, \{x|\overline{m(x)} \subseteq c\})$$

The inverse of η is given by $\eta^{-1}((\pi,\rho))(x) = \bigcap\{o|x \in \pi(o)\} \cap \bigcap\{c|x \in \rho(c)\}$. Theorem 6 holds also if we substitute $\mathcal{P}\delta(Y)$ for $\mathcal{P}(Y)$ and we order the safety and the liveness predicate transformer respectively by $\sqsubseteq_{SD}$ and $\sqsubseteq_{LD}$.

To prove this theorem we need a different *stability lemma*:

Lemma 7. *Let* $(\pi,\rho) : (P_1(Y) \to_M Q_1(X)) \otimes_M (P_2(Y) \to_M Q_2(X))$. *Then for every* $x \in X$, $\hat{p} \in P_1(Y)$, *and* $\hat{q} \in P_2(Y)$ *we have:*

1. $x \in \pi(\hat{p}) \Leftrightarrow \bigcap\{p|x \in \pi(p)\} \cap \bigcap\{q|x \in \rho(q)\} \subseteq \hat{p}$,
2. $x \in \rho(\hat{q}) \Leftrightarrow \bigcap\{p|x \in \pi(p)\} \cap \bigcap\{q|x \in \rho(q)\} \subseteq \hat{q}$.

6 Conclusions and Future Work

We have proposed a formal definition of safety and liveness predicates and of predicate transformers following the line of [Smy83, Kwi91]. Furthermore we have give generalizations of the standard definitions of power domains and of state transformers, and which give us a complete series of isomorphisms between predicate and state transformers (including the Plotkin state transformers).

Future work includes: a generalization of the results to arbitrary topological spaces and applications of predicate transformers to non-flat domains for concurrency and communication.

Acknowledgements: We wish to thanks all the members of the Amsterdam Concurrency Group especially Jaco de Bakker, Jan Rutten, Daniele Turi, and Franck van Breugel, for discussions and suggestions about the contents of this paper. Thanks also to Peter Knijnenburg of Utrecht University and the anonymous referees for their helpful comments.

References

[Abr91] S. Abramsky. A domain equation for bisimulation. *Information and Computation*, 92:161–218, 1991.

[AP86] K. R. Apt and G. Plotkin. Countable nondeterminism and random assignment. *Journal of the ACM*, 33(4):724–767, October 1986.

[AS85] B. Alpern and F.B. Schneider. Defining liveness. *Information Processing Letters*, 21:181–185, 1985.

[Bac80] R.-J.R. Back. *Correctness Preserving Program Refinements: Proof Theory and Applications*, volume 131 of *Mathematical Centre Tracts*. Mathematical Centre, Amsterdam, 1980.

[Bak80] J. W. de Bakker. *Mathematical Theory of Program Corretness*. Prentice-Hall, 1980.

[BK92] M. Bonsangue and J. N. Kok. Semantics, orderings and recursion in the weakest precondition calculus. Technical Report CS-R9267, Centre for Mathematics and Computer Science, Amsterdam, 1992. Extended abstract to appear in the proceedings of the Rex Workshop '92 'Semantics: Foundations and Applications' LNCS 666, 1993.

[BK93] M. Bonsangue and J. N. Kok. Isomorphisms between state and predicate transformers. Technical report, Centre for Mathematics and Computer Science, Amsterdam, 1993. Available through anonymous ftp from ftp.cwi.nl.

[Bre93] F. van Breugel. Relating state transformation semantics and predicate transformer semantics for parallel programs. To appear, 1993.

[Dij76] E.W. Dijkstra. *A Discipline of Programming.* Prentice-Hall, 1976.

[Eng77] R. Engelking. *General Topology.* Polish Scientific Publishers, 1977.

[HP72] P. Hitchcock and D. Park. Induction rules and termination proofs. In *International Conference on Automata, Languages and Programming*, 1972.

[Hrb87] K. Hrbacek. Convex Powerdomains I. *Information and computation*, 74:198–225, 1987.

[Hrb89] K. Hrbacek. Convex Powerdomains II. *Information and computation*, 81:290–317, 1989.

[Kwi91] M.Z. Kwiatowska. On topological characterization of behavioural properties. In G.M. Reed, A.W. Roscoe, and R.F. Wachter, editors, *Topology and Category Theory in Computer Sciences - Proc. Oxford Topology Symposioum, June 1989*, pages 153–177. Oxford Science Pubblications, 1991.

[Lam77] L. Lamport. Proving the correctness of a multiprocess programs. *IEEE Transaction on Software Eng.*, SE-3:125–143, 1977.

[MM79] G. Milne and R. Milner. Concurrent processes and their syntax. *J. ACM*, 26, 2:302–321, 1979.

[Nel89] G. Nelson. A generalization of Dijkstra's calculus. *ACM Transaction on Programming Languages and Systems*, 11 - 4:517–561, 1989.

[Plo76] G. D. Plotkin. A powerdomain construction. *SIAM J. Comput.*, 5:452–487, 1976.

[Plo79] G. D. Plotkin. Dijkstra's predicate transformer and Smyth's powerdomain. In *Proceedings of the Winter School on Abstract Software Specification*, volume 86 of *Lecture Notes in Computer Science*, pages 527–553. Springer-Verlag, Berlin, 1979.

[Plo81] G.D. Plotkin. Post-graduate lecture notes in advanced domain theory (incorporating the "Pisa Notes"). Department of Computer Science, Univ. of Edinburgh, 1981.

[Roe76] W.P. de Roever. Dijkstra's predicate transformer, non-determinism, recursion, and terminations. In *Mathematical foundations of computer science*, volume 45 of *Lecture Notes in Computer Science*, pages 472–481. Springer-Verlag, Berlin, 1976.

[Smy78] M.B. Smyth. Power domains. *J. Comput. Syst. Sci.*, 16,1:23–36, 1978.

[Smy83] M.B. Smyth. Power domains and predicate transformers: A topological view. In *Proceedings of ICALP '83 (Barcelona)*, volume 154 of *Lecture Notes in Computer Science*, pages 662–675. Springer-Verlag, Berlin, 1983.

[Smy92] M.B. Smyth. Topology. In S. Abramsky, D.M. Gabbay, and T.S.E. Malbaum, editors, *Handbook of Logic in Computer Science*, volume I - Background: Mathematical Structures. Oxford University Press, 1992.

[vW90] J. von Wright. *A Lattice-theoretical Basis for Program Refinement.* PhD thesis, Abo Akademi, 1990.

On the Amount of Nondeterminism and the Power of Verifying
(Extended Abstract)

Liming Cai * and Jianer Chen **

Texas A&M University, College Station TX 77843, USA

Abstract. The relationship between nondeterminism and other computational resources is studied based on a special interactive-proof system model GC. Let $s(n)$ be a function and $\mathcal{C}$ be a complexity class. Define $GC(s(n), \mathcal{C})$ to be the class of languages that are accepted by verifiers in $\mathcal{C}$ that can make an extra $O(s(n))$ amount of nondeterminism. Our main results are (1) A systematic technique is developed to show that for many functions $s(n)$ and for many complexity classes $\mathcal{C}$, the class $GC(s(n), \mathcal{C})$ has natural complete languages; (2) The class Π_h^0 of languages accepted by log-time alternating Turing machines making h alternations is precisely the class of languages accepted by uniform families of circuits of depth h; (3) The classes $GC(s(n), \Pi_h^0)$, $h \geq 1$, characterize precisely the fixed-parameter intractability of NP-hard optimization problems. In particular, the $(2h)$th level $W[2h]$ of W-hierarchy introduced by Downey and Fellows collapses if and only if $GC(s(n), \Pi_{2h}^0) \subseteq P$ for some $s(n) = \omega(\log n)$.

1 Introduction

The study of the power of nondeterminism is central to Complexity Theory. The relationship between nondeterminism and other computational resources still remains unclear. Two fundamental questions are how much computational resource we should pay in order to eliminate nondeterminism, and how much computational resource we can save if we are granted nondeterminism. A computation with nondeterminism can be basically decomposed into the phase of guessing (nondeterministically) and the phase of verifying (using other computational resources). In general, the phase of guessing and the phase of verifying work interactively.

The notion of classifying problems according to the amount of nondeterminism and the power of verifying in a computation has appeared in recent research. Díaz and Torán [9] have studied the classes β_k, for $k \geq 1$, by allowing a deterministic polynomial time computation to make an $O(\log^k n)$ amount of nondeterminism.[3] Buss and Goldsmith [5] considered the classes $N^k P_h$, for

* Supported by Engineering Excellence Award from Texas A&M University.

** Supported by the National Science Foundation under Grant CCR-9110824.

[3] These classes were originally introduced in Kintala and Fisher [12].

$k, h \geq 1$, in which languages can be recognized by an $O(n^h \log^{O(1)} n)$-time multitape Turing machine making at most $k \log n$ binary nondeterministic choices. Wolf [15] has studied the models that are NC circuits with nondeterministic gates.

We generalize the above ideas by introducing a computation model GC (Guess-then-Check). Let $s(n)$ be a function and let C be a complexity class, then $GC(s(n), C)$ is the class of languages that can be recognized by first nondeterministically guessing $O(s(n))$ binary bits then using the power of C to verify. The reader should realize that this GC model is a special version of the interactive-proof system that has received considerable attention recently (for a survey, see [11]).

We develop a systematic technique to show that for a large class of functions $s(n)$ and for many complexity classes C, the class $GC(s(n), C)$ has natural complete languages. The technique involves characterizing the computation of a verifier by a circuit and encoding a nondeterministic string of length $s(n)$ as a length $(n \cdot s(n))/\log n$ input for the circuit. Our technique also provides new complete languages for a number of existing complexity classes. In particular, we prove that one can obtain complete languages for the class β_k by restricting the amount of nondeterminism in NP-complete problems. This result is opposite to a conjecture made by Díaz and Torán [9]. We show that the Bounded-Weight Circuit-Satisfiability problem with weight bound k is a new, natural complete language for the class $N^k P_h$ proposed by Buss and Goldsmith [5]. We also give complete languages for the classes $NNC^k(\log^i n)$ studied by Wolf [15], for which it was unknown whether there exist complete languages.

An important property of the GC model is its connections to computational optimization problems. Many computational optimization problems can be characterized by GC classes in which the length of the proof is slightly longer than $\Theta(\log n)$ and the verifier is a bounded depth circuit, which is a computation model provably strictly weaker than polynomial-time Turing machines[16]. For this, we study in detail the class $GC(s(n), \Pi_k^0)$, where Π_k^0 is the class of languages accepted by log-time alternating Turing machines making at most k alternations. New techniques are developed for this kind of alternating Turing machines and for circuits of bounded depth so that they can simulate each other efficiently. A new uniformity of circuit families is proposed. With this new uniformity, we show that the class of languages accepted by log-time alternating Turing machines making at most k alternations is precisely the class of languages accepted by uniform families of circuits of depth k. To the best of our knowledge, this is the first exact circuit characterization of the log-time alternating Turing machines making at most k alternations, improving a result of [6].

This exact correspondence between alternating Turing machines and uniform circuit families leads to important characterizations of the fixed parameter intractability [1, 2, 10] of NP optimization problems. We prove that the $(2k)$th level of the fixed parameter hierarchy (known as the W-hierarchy) introduced by Downey and Fellows collapses if and only if the class $GC(s(n), \Pi_{2k}^0)$ is a subclass of P for some function $s(n) = \omega(\log n)$. In particular, the class

$GC(s(n), \Pi_2^0)$ is a subclass of P for some function $s(n) = \omega(\log n)$ if and only if a large class of NP-hard optimization problems, including Dominating-Set and Zero-One Integer-Programming, are fixed parameter tractable.

We assume that the reader is sufficiently familiar with the theory of alternating Turing machines and uniform circuit families (see [3, 4, 13]). Some terminologies are explained as follows.

A circuit α_n is *well-leveled* if the gates of α_n can be partitioned into levels such that the first level consists of the output gate, the last level consists of all the input gates, a gate in the ith level receives inputs only from the gates in the $(i+1)$st level, and gate type alternates level by level (this implies that all gates in the same level have the same type). In order to well-level a circuit, we will allow a gate of type **AND** or **OR** to have fan-in 1. In this case, the gate simply passes the input value to the output. This kind of gates will be called *identity gates*. It is known that given a circuit of size s and depth d, an equivalent well-leveled circuit of size $O(sd)$ and depth d can be constructed in deterministic $O(\log s)$-space [8].

A circuit α_n is a *Π-circuit* (*Σ-circuit*) if the output gate of α_n is of type **AND** (**OR**). Similarly, an alternating Turing machine is a *Π-ATM* (*Σ-ATM*) if its first alternation is a universal branch (an existential branch).

Let $b(n) \leq n$ be a function. A binary string z of length n can be partitioned into $\lceil n/b(n) \rceil$ segments of length $b(n)$. These segments will be called the *$b(n)$-blocks* of z. The ith $b(n)$-block of the string z consists of the substring from the $((i-1)b(n)+1)$st bit to the $(ib(n))$th bit.[4] Similarly, a string of Boolean variables $x = x_1x_2\cdots x_n$, which in general denotes the input of a circuit or of a Turing machine, can be partitioned into $b(n)$-blocks.

The *weight* of a binary string is the number of 1's in the string.

2 The GC model and complete languages

In this section, we introduce a model GC (Guess-then-Check). Roughly speaking, $GC(s(n), C)$ is a restricted interactive-proof system in which the prover passes a proof of length $O(s(n))$ to the verifier, which has the power of C, at beginning of the computation, and no more communication between the prover and the verifier is allowed after that. A more formal definition is given as follows.

Definition 1. Let $s(n)$ be a function and let C be a complexity class. A language L is in the class $GC(s(n), C)$ if there is a language $A \in C$ and an integer $c > 0$ such that for all $x \in \{0,1\}^*$, $x \in L$ if and only if $\exists y \in \{0,1\}^*$, $|y| \leq c \cdot s(|x|)$, and $\langle x, y \rangle \in A$.

Note that in the above definition, the condition $|y| \leq c \cdot s(|x|)$ can be replaced by the equality $|y| = c \cdot s(|x|)$ so that the length of the guessed string y only depends on the length of x. In fact, we can encode the symbol 0 by 00 and the symbol 1 by 11, and let 01 and 10 denote "useless symbols" so that every string

[4] The last $b(n)$-block may contain less than $b(n)$ bits.

y in $\{0,1\}^*$ of length at most $c \cdot s(|x|)$ can be encoded into a string in $\{0,1\}^*$ of length $2c \cdot s(|x|)$ in which the first $2|y|$ bits encode the string y and the remaining bits encode useless symbols. A straightforward modification on the verifier will enable it to accept $\langle x, y' \rangle$ for some y' of length $2|y| = 2c \cdot s(|x|)$ if and only if $x \in L$.

If we require that the length $|y|$ of y is strictly bounded by $s(|x|)$, we call the model a "strict-GC" model. More formally,

Definition 2. Let $s(n)$ be a function and let C be a complexity class. A language L is in the class *strict-GC*$(s(n), C)$ if there is a language $A \in C$ such that for all $x \in \{0,1\}^*$, $x \in L$ if and only if $\exists y \in \{0,1\}^*$, $|y| \leq s(|x|)$, and $\langle x, y \rangle \in A$.

Many complexity classes can be characterized by this model. For example, the class NP can be characterized by $GC(n^{O(1)}, P)$. The class NL is precisely the class $GC(n^{O(1)}, DL)$. We will develop a general technique to show that many GC classes have natural complete languages.

Following Buss and Goldsmith [5], let P_h be the class of languages accepted by deterministic multitape Turing machines of running time $O(n^h \log^{O(1)} n)$. A circuit family $F = \{\alpha_n \mid n \geq 1\}$ is P_h*-uniform* if there is a deterministic multitape Turing machine M that, on input 1^n, generates the circuit α_n in time $O(n^h \log^{O(1)} n)$.

Lemma 3. *A language L is in P_h if and only if L is accepted by a P_h-uniform family of circuits.*

Let $N^k P_h$ denote the class of languages that is accepted by a deterministic multitape Turing machine of running time $O(n^h \log^{O(1)} n)$ that can make at most $k \log n$ binary guesses [5]. It is easy to see that $N^k P_h$ is identical to the class *strict-GC*$(k \log n, P_h)$.

A language L is *complete* for the class $N^k P_h$ under P_h*-reduction* if L is in $N^k P_h$, and for every language L' in $N^k P_h$, there is a function f such that $x \in L'$ if and only if $f(x) \in L$, and the function $f(x)$ is computable by a deterministic multitape Turing machine in $O(n^h \log^{O(1)} n)$ time.

Consider the following language, where f is a function.

BWCS[f] (Bounded-Weight Circuit-Satisfiability)
Instance: A circuit α of m inputs
Question: Does α accept an input vector of weight $\leq f(m)$?

In particular, for any fixed constant k, we denote by $BWCS[k]$ the language $BWCS[c_k]$, where c_k is the constant function k, i.e., $c_k(m) = k$ for all $m \geq 1$.

Theorem 4. *$BWCS[k]$ is complete for the class strict-$GC(k \log n, P_h)$ $= N^k P_h$ under P_h-reduction, for all $k, h \geq 1$.*

Proof. (Sketch) We sketch the proof that under the P_h-reduction, $BWCS[k]$ is hard for *strict*-$GC(k \log n, P_h)$. Let L be a language in *strict*-$GC(k \log n, P_h)$. Then there is a language $A \in P_h$ such that $x \in L$ if and only if there is a y, $|y| = k \log n$ such that $\langle x, y \rangle \in A$. By Lemma 3, A is accepted by a P_h-uniform family $\{\alpha_m | m \geq 1\}$ of circuits. Thus, given an input x of length n, we assign the first n input bits of the circuit α_m by x, where $m = n + k \log n$, resulting in a circuit $\alpha_m(x)$ with input length $k \log n$. Now x is in L if and only if the circuit $\alpha_m(x)$ accepts some input y of length $k \log n$. Each $(\log n)$-*block* of y can be regarded as a binary number of value between 0 and $n - 1$. Based on the circuit $\alpha_m(x)$, we construct another circuit $\alpha'_m(x)$ with input length kn such that the ith bit of the jth n-*block* in the input of $\alpha'_m(x)$ is 1 if and only if the jth $(\log n)$-*block* in the input of $\alpha_m(x)$ encodes the number $i - 1$. Therefore, $x \in L$ if and only if the circuit $\alpha'_m(x)$ accepts a string of weight at most k. It can be verified that the circuit $\alpha'_m(x)$ can be constructed in time $O(n^h \log^{O(1)} n)$. □

The P_1-reduction is also called ***quasilinear-time reduction***. Theorem 4 in particular presents a new (very) natural complete language for the class $N^k P_1$ under the quasilinear-time reduction. (see [5] for the discussion of $N^k P_1$-complete languages under quasilinear-time reduction.)

The proof technique for Theorem 4 can be used systematically to construct complete languages for many other GC classes. We list some of them below.

The class β_f, where f is a function, was introduced by Kintala and Fisher [12] and studied in great detail by Díaz and Torán [9]. By our GC model, the class β_f is identical to the class $GC(f(n), P)$.

Theorem 5. *Let f be a function constructible in deterministic $O(\log n)$ space. Then $BWCS[f]$ is complete for the class $\beta_{f \log n} = GC(f(n) \log n, P)$ under* log-*space reduction.*

In particular, the language $BWCS[\log^{i-1} n]$ is complete for the class $\beta_i = \beta_{\log^i n}$, for all integer $i \geq 2$. Note that the language $BWCS[\log^{i-1} n]$ is a restricted version of the Circuit-Satisfiability problem, which is complete for the class NP. This answers a question posed by Díaz and Torán [9], who were able to construct complete languages for β_i from certain complete problems for P by adding nondeterminism, and conjectured that complete languages for β_i may not be constructed from complete problems for NP by restricting the nondeterminism. (The complete problems for the class β_i are also discussed in [14].)

Let f be a function. Now we consider the class $GC(f(n), NC^k)$ for all $k \geq 1$. In particular, the classes $GC(\log^i n, NC^k)$ have been investigated by Wolf [15], using his notation $NNC^k(\log^i n)$.

Consider the following language.

BWCS[f, k]
Instance: A circuit α of input length m and depth $\log^k m$
Question: Does α accept an input vector of weight $\leq f(m)$?

Theorem 6. *Let f be a function constructible in deterministic $O(\log n)$ space and let $k \geq 1$ be an integer. Then the language $BWCS[f, k]$ is complete for the class $GC(f(n) \log n, NC^k)$ under* log-*space reduction.*

In particular, the language $BWCS[\log^{i-1} n, k]$ is complete for $NNC^k(\log^i n)$ under log-space reduction, for all $i \geq 2$, and $k \geq 1$. This is the first language that is known to be complete for the class $NNC^k(\log^i n)$.

3 A precise circuit characterization of alternating Turing machines making k alternations

The most interesting property of the GC model is its close connections to computational optimization problems. Many computational optimization problems can be characterized by GC classes with a very weak verifier that is a subclass of the class AC^0. For this reason, we first investigate these weak classes.

Buss et al. [6] have shown that $O(\log n)$-time alternating Turing machines making constant number of alternations accept precisely the class AC^0. We exhibit a more precise relationship between these two models by showing that the $O(\log n)$-time alternating Turing machines making k alternations are basically identical to the uniform families of circuits of depth k.

Because these are two very weak computation models, we must be very careful in the definitions. We define them in a way slightly different from the standard definitions. These modifications have no influence on any existing complexity class based on alternating Turing machines, such as the classes AC^k for $k \geq 0$, and the classes NC^h for $h \geq 1$. Roughly speaking, we slightly weaken the power of an alternating Turing machine and slightly strengthen the power of a circuit.

Definition 7. A *Π_k^0-ATM (Σ_k^0-ATM) M* is an $O(\log n)$-time Π-ATM (Σ-ATM) making at most k alternations. Moreover, there is a constant c such that for every input x of length n, the last phase of any computation path of M on x reads input bits from at most c $(\log n)$-blocks of the input x.

To relate a Π_k^0-ATM or a Σ_k^0-ATM to a family of circuits of depth k, we introduce a new circuit family uniformity that is slightly different from the log-time uniformity studied in [3, 4].

Definition 8. A family $F = \{\alpha_n \mid n \geq 1\}$ of circuits of polynomial size is *D^*-uniform* if the function $f(n) = \overline{\alpha}_n$ is computable by a linear time deterministic Turing machine M in the following sense: given a pair $\langle \overline{n}, \overline{i} \rangle$, M prints out the length $\log n$ substring of $\overline{\alpha}_n$, starting from the ith bit, here $\overline{\alpha}_n$ is an encoding of the circuit α_n, and $\overline{n}$ and $\overline{i}$ are the binary representations of the numbers n and i, respectively.

It is easy to see that the D^*-uniformity is equivalent to the log-time uniformity [4] for all existing circuit classes, such as the classes AC^k, $k \geq 0$, and the classes NC^h, $h \geq 1$.

Let $x = x_1x_2\cdots x_n$ be a string of Boolean variables and let $b(n) \leq n$ be a function. A *complete instance* of a $b(n)$-block $x_{h+1}\cdots x_{h+b(n)}$ of x, where $h = ib(n)$ for some i, is a string $z_{h+1}\cdots z_{h+b(n)}$, where z_j is either x_j or $\overline{x}_j$, $h+1 \leq j \leq h+b(n)$.

Let α_n be a circuit and g a gate in α_n. Let $b(n) \leq n$ be a function and c an integer. We say that the inputs of the gate g are *c $b(n)$-blocks of the input $I(\alpha_n)$ of α_n* if the inputs of g can be partitioned into c groups such that each group is a complete instance of a $b(n)$-block of $I(\alpha_n)$.

Let $Log(n)$ be the function defined as follows:

$$Log(n) = \min\{2^h \mid h \text{ is an integer and } \log n \leq 2^h\}$$

Definition 9. A family $F = \{\alpha_n \mid n \geq 1\}$ of circuits is called a *family of Π_k^0-circuits* (a *family of Σ_k^0-circuits*) if there exist two integers c and d such that for all $n \geq 1$, α_n is a well-leveled Π-circuit (Σ-circuit) of depth $k+1$ in which the inputs of each gate in level $k+1$ are c' $(dLog(n))$-blocks of the input $I(\alpha_n)$ of α_n for some $c' \leq c$.

We remark that in a Π_k^0-circuit or a Σ_k^0-circuit α_n, we assume a special encoding for the gates in level $k+1$. Each gate g in level $k+1$ is encoded as a sequence $\langle t; a_1, s_1; \cdots, a_c, s_c\rangle$, where t is the gate type of g. For $1 \leq i \leq c$, a_i and s_i specify a complete instance of a $(dLog(n))$-block of $I(\alpha_n)$, where a_i is the starting address of the $(dLog(n))$-block, and s_i is a binary string of length $dLog(n)$ indicating the corresponding complete instance for the $(dLog(n))$-block.

Lemma 10. *If a language L is accepted by a D^*-uniform family of Π_k^0-circuits, then L is accepted by a Π_k^0-ATM.*

The converse of Lemma 10 is less trivial. Let M be a Π_k^0-ATM with universal states $u_1, \cdots, u_{c_1}$, existential states $e_1, \cdots, e_{c_2}$, and deterministic states $d_1, \cdots, d_{c_3}$. Consider the set

$$\Lambda_n = \bigcup_{i=1}^{c_1}\{u_i^{(l)}, u_i^{(r)}\} \cup \bigcup_{i=1}^{c_2}\{e_i^{(l)}, e_i^{(r)}\} \cup \bigcup_{i=1}^{c_3}\{d_i\} \cup \bigcup_{i=1}^{n}\{x_i, \overline{x}_i\}$$

where the symbol $u_i^{(l)}$ ($u_i^{(r)}$) means "take the left (right) branch on the universal state u_i", the symbols $e_i^{(l)}$ and $e_i^{(r)}$ are similarly defined for existential states, the symbol d_i means "make a deterministic move d_i", the symbol x_i ($\overline{x}_i$) for $i = 1, \cdots, n$, denotes "read the ith input bit and get value 1 (0)" [5].

Given an input x of length n, a (partial) computation path P from the root to a node s in the computation tree of M on x can be represented by a sequence ρ in Λ_n^* of length at most $c\log n$, where c is a constant, such that the qth symbol in ρ specifies the action of the qth step of M along the computation path P. The sequence ρ will be called the *moving sequence* corresponding to the node s

[5] Without loss of generality, we will assume that the input bits are always read deterministically.

in the computation tree of M on x. A sequence ρ in Λ_n^* is a *moving sequence of M on input length n* if it is a moving sequence corresponding to a node in the computation tree of M on some input of length n.

Lemma 11. *The number of moving sequences of a Π_k^0-ATM on input length n is bounded by a polynomial of n.*

Lemma 12. *For all fixed $k \geq 1$, if a language L is accepted by a Π_k^0-ATM, then L is accepted by a D^*-uniform family of Π_k^0-circuits.*

Buss et al.[6] have shown that an alternating Turing machine of $O(\log n)$-time and k alternations can be simulated by a uniform family of circuits of depth k. However, their alternating Turing machine model is much weaker than ours. In their model, each computation path of an alternating Turing machine can read at most 1 input bit, and the input reading must be done at the end of the computation path.

Theorem 13. *A language L is accepted by a Π_k^0-ATM if and only if it is accepted by a D^*-uniform family of Π_k^0-circuits.*

Corollary 14. *A language L is accepted by a Σ_k^0-ATM if and only if it is accepted by a D^*-uniform family of Σ_k^0-circuits.*

4 *GC* classes and optimization problems

The circuit characterization of $O(\log n)$-time alternating Turing machines making k alternations presented in the last section enables us to construct natural complete languages for GC classes with very weak verifiers.

A circuit α_n has *weft k* if it is a well-leveled unbounded fan-in circuit of depth $k+1$ such that all gates in level $k+1$ are bounded fan-in [10]. A weft k circuit is of *bounded negation* if for each gate in level k, at most two negative-input gates can affect it.

Definition 15. $BNC(s(n), k)$ is the set of pairs $\langle \alpha, w \rangle$ where α is a bounded negation Π-circuit of weft k, and α accepts an input of weight w, $w \leq s(n)$.

Let Π_k^0 be the class of languages that are accepted by Π_k^0-ATMs.

Theorem 16. *Let $s(n) \leq n$ be a function computable in deterministic $O(\log n)$ space. Then the language $BNC(s(n), 2k)$ is complete for $GC(s(n)Log(n), \Pi_{2k}^0)$ under $O(\log n)$-space reduction, for all $k \geq 1$.*

The GC classes with very weak verifiers have a close connection to computational optimization problems. Here we briefly introduce the connection of GC to fixed parameter intractability of NP optimization problems. For further discussions, we refer our readers to [7]

The study of fixed parameter tractability has been initialized recently in [1, 2, 10], with the aim of refining the class of NP-hard optimization problems and of solving NP-hard optimization problems in practice. It has been observed that many NP-hard optimization problems can be parameterized, while their time complexity behaves very differently with respect to the parameter [10]. For example, with a fixed parameter q, the problem Vertex-Cover can be solved in time $O(n^c)$, where c is a constant independent of q, while the problem Dominating-Set has the contrasting situation where essentially no better algorithm is known than the "trivial" one which just exhaustively tries all possible solutions. For each fixed q, Dominating-Set is solvable in this way in time $\Theta(n^{q+1})$.

Downey and Fellows [10] have introduced a new framework to discuss the *fixed parameter tractability* of NP-hard problems. A hierarchy, called the *W-hierarchy*, has been established to classify the fixed parameter intractability of NP-hard optimization problems. At the bottom of the hierarchy (the zeroth level) is the class of NP-hard problems whose parameterized version can be solved in time $O(n^c)$, where c is a constant independent of the parameter q. Vertex-Cover problem, for example, is in the bottom level of the hierarchy. Independent-Set, Clique, and many others are in the first level of the hierarchy; and in the second level, there are Dominating-Set, Zero-One Integer-Programming, and many others. A fundamental question is whether this hierarchy collapses.

Based on Theorem 16, we give in the following a sufficient and necessary condition for the collapsing of W-hierarchy to the zeroth level.

Theorem 17. *The $(2k)$th level of the W-hierarchy collapses to the zeroth level if and only if $GC(t(n)Log(n), \Pi^0_{2k}) \subseteq P$ for some unbounded, non-decreasing function $t(n)$ constructible in deterministic $O(\log n)$ space.*

Corollary 18. *Dominating Set and Zero-One Integer-Programming are fixed parameter tractable if and only if $GC(t(n)Log(n), \Pi^0_2) \subseteq P$ for some unbounded, non-decreasing function $t(n)$ constructible in deterministic $O(\log n)$ space.*

The relationship between GC model with weak verifier and polynomial-time oracle machines gives an equivalent characterization of the upward collapsing property of the W-hierarchy.

Theorem 19. *For any $j > i$, the $(2j)$th level of the W-hierarchy collapses to the $(2i)$th level if and only if $GC(t(n)Log(n), \Pi^0_{2j}) \subseteq P^{GC(g(t(n))Log(n), \Pi^0_{2i})}$ for some unbounded, non-decreasing function $t(n)$ and $g(n)$ constructible in deterministic $O(\log n)$ space.*

Recall that β_k is the class of languages accepted by polynomial-time Turing machines that can make at most $O(\log^k n)$ nondeterminism [9]. An open problem is whether $P = \beta_k$ (note that β_k is a subclass of NP). The following corollary, however, shows that $P = \beta_k$ is very unlikely.

Theorem 20. *The class P is a proper subclass of the class β_k for all $k > 1$ unless all optimization problems in the W-hierarchy are fixed-parameter tractable.*

The above results show that questions in computational optimization, such as the fixed-parameter tractability of *NP*-hard problems, are actually asking if a deterministic polynomial time computation could guess a string of length larger than $\Theta(\log n)$.

Acknowledgements: We would like to thank Karl Abrahamson, Rod Downey, Mike Fellows, Don Friesen, Arkady Kanevsky, Robert Szelepcsényi, and Chee Yap for useful comments on this work.

References

1. K. ABRAHAMSON, R. G. DOWNEY AND M. R. FELLOWS, Fixed-parameter intractability II, To appear in *Proc. 10th Symposium on Theoretical Aspects of Computer Science*, Wurzburg, Germany, (1993).
2. K. R. ABRAHAMSON, J. A. ELLIS, M. R. FELLOWS, AND M. E. MATA, On the complexity of fixed parameter problems, *Proc. 30th Annual Symposium on Foundations of Computer Science*, (1989), pp. 210-215.
3. D. A. BARRINGTON, N. IMMERMAN, AND H. STRAUBING, On uniformity within NC^1, *Journal of Computer and System Sciences 41*, (1990), pp. 274-306.
4. S. R. BUSS, The Boolean formula value problem is in ALOGTIME, *Proc. 19th Annual ACM Symposium on Theory of Computing*, (1987), pp. 123-131.
5. J. F. BUSS AND J. GOLDSMITH, Nondeterminism within *P*, *Lecture Notes in Computer Science 480*, (1991), pp. 348-359.
6. S. BUSS, S. COOK, A. GUPTA, AND V. RAMACHANDRAN, An optimal parallel algorithm for formula evaluation, *SIAM J. Comput. 21*, (1992), pp. 755-780.
7. L. CAI AND J. CHEN, Fixed parameter tractability and approximability of *NP*-hard optimization problems, *Proc. 2rd Israel Symposium on Theory of Computing and Systems*, (1993), to appear.
8. J. CHEN, Characterizing parallel hierarchies by reducibilities, *Information Processing Letters 39*, (1991), pp. 303-307.
9. J. DÍAZ, AND J. TORÁN, Classes of bounded nondeterminism, *Math. System Theory 23*, (1990), pp. 21-32.
10. R. G. DOWNEY AND M. R. FELLOWS, Fixed-parameter intractability, *Proc. 7th Structure in Complexity Theory Conference*, (1992), pp. 36-49.
11. D. S. JOHNSON, The *NP*-completeness column: an ongoing guide, *Journal of Algorithms 13*, (1992), pp. 502-524.
12. C. KINTALA AND P. FISHER, Refining nondeterminism in relativized complexity classes, *SIAM J. Comput. 13*, (1984), pp. 329-337.
13. W. L. RUZZO, On uniform circuit complexity, *J. Comput. System Sci. 21*, (1981), pp. 365-383.
14. ROBERT SZELEPCSÉNYI, β_k-complete problems and greediness, *Proc. 9th British Colloquium for Theoretical Computer Science*, (1993).
15. M. J. WOLF, Nondeterministic circuits, space complexity and quasigroups, *Manuscript*, (1992).
16. A. YAO, Separating the polynomial-time hierarchy by oracles, *Proc. 26th Annual Symposium on Foundations of Computer Science*, (1985), pp. 1-10.

Observing distribution in processes

Ilaria Castellani*
INRIA Sophia-Antipolis
06565 Valbonne, France
email: ic@sophia.inria.fr

Abstract. The distributed structure of CCS processes can be made explicit by assigning different *locations* to their parallel components. These locations then become part of what is observed of a process. The assignment of locations may be done statically, or dynamically as the execution proceeds. The dynamic approach was developed first, by Boudol et al. in [BCHK91a], [BCHK91b], as it seemed more convenient for defining notions of *location equivalence* and *preorder*. However, it has the drawback of yielding infinite transition system representations. The static approach, which is more intuitive but technically more elaborate, was later developed by L. Aceto [Ace91] for *nets of automata*, a subset of CCS where parallelism is only allowed at the top level. In this approach each net of automata has a finite representation, and one may derive notions of equivalence and preorder which coincide with the dynamic ones. The present work generalizes the static treatment of Aceto to full CCS. The result is a distributed semantics which yields finite transition systems for all CCS processes with a regular behaviour and a finite degree of parallelism.

1 Introduction

This work is concerned with *distributed semantics* for CCS, accounting for the spatial distribution of processes. Such semantics focus on different aspects of behaviour than most non-interleaving semantics for CCS considered so far in the literature, which are based on the notion of causality. Roughly speaking, a distributed semantics keeps track of the behaviour of the local components of a system, and thus is appropriate for describing phenomena like a local deadlock. On the other hand a causal semantics, such as those described in [DDNM87], [GG89], [DD90] [BC91], is concerned with the flow of causality among activities and thus is better suited to model the interaction of processes and the global control structure of a system.

The distributed structure of CCS processes can be made explicit by assigning different *locations* to their parallel components. To this end we use the *location prefixing* construct $l :: p$ of [BCHK91a,b], which represents process p residing at location l. The actions of such a process are observed together with their location. We have for instance:

$$(l :: a \mid k :: b) \xrightarrow[l]{a} (l :: nil \mid k :: b) \xrightarrow[k]{b} (l :: nil \mid k :: nil)$$

In general, because of the nesting of parallelism, the locations of actions will not be simple letters $l, k, \ldots$ but rather words $u = l_1 \cdots l_n$. Then a "distributed process" will perform transitions of the form $p \xrightarrow[u]{a} p'$.

The idea is now to compare CCS terms by comparing their possible *distributions*, which are obtained by transforming each subprocess $(p \mid q)$ into $(l :: p \mid k :: q)$, where l and k are distinct locations. Intuitively, such an assignment of locations should be done statically, and

*This work has been partly supported by the Project 502-1 of the Indo-French Centre for the Promotion of Advanced Research.

then become part of what is observed of a process. This will allow us to distinguish for example $(a \mid b)$ from $(a.b + b.a)$, since any distribution of the first process will perform actions a and b at different locations. For more interesting examples we refer the reader to the introductions of [BCHK91a,b].

We thus want to define a notion of (weak) bisimulation on distributed processes, which equates processes exhibiting the same "location transitions". However, it would be too strong a requirement to ask for the identity of locations in corresponding transitions. In fact, if we want to observe distribution, we still aim, to some extent, at an extensional semantics. For instance, we do not want to observe the order in which parallel components have been assembled in a system, nor indeed the number of these components. We are only interested in the components which are active in each computation. We would like e.g. to identify the distributions of the CCS processes:

$$a \mid (b \mid c) \quad \text{and} \quad (a \mid b) \mid c$$
$$a \quad \text{and} \quad a \mid nil$$

Then transitions must be compared modulo an *association* between their locations. For instance to relate the distributions $l :: a \mid k :: (l' :: b \mid k' :: c)$ and $l :: (l' :: a \mid k' :: b) \mid k :: c$, we need to "identify" the locations l, kl', kk' of the first respectively with ll', lk', k in the second. However, it appears that this association cannot in general be fixed statically. For consider the two CCS processes:

$$p = [(\alpha + b) \mid \bar{\alpha}.b]\backslash\alpha \quad \text{and} \quad q = b$$

Intuitively, we would like to equate p and q because the observable behaviour of any distributions of these processes consists in just one action b at some location l. But here the required association of locations will depend on which run is chosen in the first process. Hence it is not obvious how to define a notion of equivalence formalising our intuition about abstract distributed behaviours.

Because of this difficulty, the static approach was initially abandoned in favour of a different one, where locations are introduced dynamically as the execution proceeds. This *dynamic approach*, where locations are associated with actions rather than with parallel components, has been presented in [BCHK91a,b]. In this setting, the notion of *location equivalence* is particularly simple: it is just the standard notion of bisimulation, applied to the transitions $p \xrightarrow[u]{a} p'$. Moreover, by weakening a little the definition of the equivalence, we obtain a notion of *location preorder*, which formalises the idea that one process is more sequential or less distributed than another. Such a notion is particularly useful when dealing with truly concurrent semantics, where an implementation is often not equivalent to its - generally more sequential - specification. Since location equivalence and preorder are essentially bisimulation relations, many proof techniques familiar from the theory of standard bisimulation may be applied to them: for example both these relations have a complete axiomatisation and a logical characterisation in the style of Hennessy and Milner, see [BCHK91a,b].

However, the dynamic approach has the drawback of yielding infinite transition systems even for regular processes, and thus cannot be directly used for verification purposes. Moreover in this approach locations represent access paths for actions rather than sites in a system, and thus are somehow remote from the original intuition. For these reasons, it was interesting to resume the initial attempt at a static approach. The problem of finding the appropriate notion of bisimulation was solved by L. Aceto in [Ace91] for *nets of automata*, a subset of CCS where parallelism is only allowed at the top level. The key idea here is to replace the usual notion of a bisimulation relation by that of a family of relations indexed by location associations. Aceto shows that the notions of static location equivalence and preorder thus obtained coincide with the dynamic ones, and thus may be used as "effective" versions of the latter.

The purpose of the present work is to generalize the static treatment of Aceto to full CCS. Having established the notion of distribution for general CCS processes, the main point is to adapt Aceto's definitions of static location equivalence and preorder. Because of the arbitrary nesting of parallelism and prefixing in CCS terms, and of the interplay between sum and parallelism, this is not completely straightforward. A step in this direction was done recently by Mukund and Nielsen in [MN92], where a notion of bisimulation equivalence based on static locations is proposed for a class of asynchronous transition systems modelling CCS with guarded sums. The notion of equivalence we present here is essentially the same (extended to all CCS), and our main result is that it coincides with the dynamic location equivalence of [BCHK91b]. We also show that a similar result holds for the location preorders.

A transition system for CCS labelled with static locations, called "spatial transition system", has been also presented in [MY92], [Yan93]. Here locations are essentially used to build a second transition system, labelled by partial orders representing *local causality*, on which is based the theory of equivalence (as well as of preorders, in [Yan93]). Again, this partial order transition system gives finite representations only for finite behaviours. This work also confirms what had been previously shown by A.Kiehn in [Kie91], namely that observing dynamic locations amounts to observe *local causality* in computations. In [Kie91] one may also find a detailed comparison of distributed and causality-based semantics.

In this extended abstract all proofs are omitted. For a full account we refer to the complete version of the paper [Cas93].

2 A language for processes with locations

We introduce here a language for specifying processes with locations, called LCCS. This language is a simple extension of CCS, including a new construct to deal with locations.

We start by recalling some conventions of CCS [Mil80]. One assumes a set of names Λ, ranged over by $\alpha, \beta, \ldots$, and a corresponding set of co-names $\bar{\Lambda} = \{\bar{\alpha} \mid \alpha \in \Lambda\}$, where $^-$ is a bijection such that $\bar{\bar{\alpha}} = \alpha$ for all $\alpha \in \Lambda$. The set of visible actions is given by $Act = \Lambda \cup \bar{\Lambda}$. Invisible actions - representing internal communications - are denoted by the symbol $\tau \notin Act$. The set of all actions is then $Act_\tau =_{\text{def}} Act \cup \{\tau\}$. We use $a, b, c, \ldots$ to range over Act and $\mu, \nu, \ldots$ to range over Act_τ. We also assume a set V of process variables, ranged over by $x, y \ldots$.

In addition to the operators of CCS, which we suppose the reader to be familiar with, LCCS includes a construct for building processes with explicit locations. Let *Loc*, ranged over by $l, k, \ldots$, be an infinite set of atomic locations. The new construct of *location prefixing*, noted $l :: p$, is used to represent process p residing at location l. Intuitively, the actions of such a process will be observed "within location l". The syntax of LCCS is as follows:

$$p ::= nil \;\mid\; \mu.p \;\mid\; (p \mid q) \;\mid\; (p+q) \;\mid\; p\backslash\alpha \;\mid\; p\langle f\rangle \;\mid\; x \;\mid\; rec\,x.\,p \;\mid\; l :: p$$

We use the slightly nonstandard notation $p\langle f\rangle$ to represent the relabelling operator of CCS. In a previous paper [BCHK91b], this language has been given a location semantics based on a dynamic assignment of locations to processes. Here we shall present a location semantics based on a static notion of location, and show that the two approaches, dynamic and static, give rise to the same notions of equivalence and preorder on CCS processes. The basic idea, common to both approaches, is that the actions of processes are observed together with the locations at which they occur. In general, because of the nesting of parallelism and prefixing in terms, the locations of actions will not be atomic locations of *Loc*, but rather *words* over these locations.

Thus general locations will be elements $u, v \ldots$ of Loc^*, and processes will be interpreted as performing transitions

$$p \xrightarrow[u]{\mu} p'$$

where μ is an action and u is the location at which it occurs.

However, locations do not have the same intuitive meaning in the two approaches. In the static approach locations represent sites - or parallel components - in a distributed system, much as one would expect. In the dynamic approach, on the other hand, the location of an action represents the sequence of actions which are locally necessary to enable it, and thus is more properly viewed as an access path to that action within the component where it occurs. Because of this difference in intuition, it is not immediately obvious that the two approaches should yield the same semantic notions. The fact that they do means that observing distribution is essentially the same as observing local causality.

3 Static approach

We start by presenting an operational semantics for LCCS based on the static notion of location. The idea of this semantics is very simple. Processes of LCCS have some components of the form $l :: p$, and the actions arising from these components are observed together with their location. The distribution of locations in a term remains fixed through execution. Location prefixing is a static construct and the operational rules do not create new locations; they simply exhibit the locations which are already present in terms. Formally, this is expressed by the operational rules for action prefixing and location prefixing. Recall that locations are words $u, v, \ldots \in Loc^*$. The empty word ε represents the location of the overall system. The rules for $\mu.p$ and $l :: p$ are respectively:

$$\text{(S1)} \qquad \mu.p \xrightarrow[\varepsilon]{\mu}_s p$$

$$\text{(S2)} \qquad p \xrightarrow[u]{\mu}_s p' \quad \Rightarrow \quad l :: p \xrightarrow[lu]{\mu}_s l :: p'$$

Rule (S1) says that an action which is not in the scope of any location l is observed as a global action of the system. Rule (S2) shows how locations are transferred from processes to actions. The rules for the remaining operators, apart from the communication rule, are similar to the standard interleaving rules for CCS, with transitions $\xrightarrow[u]{\mu}_s$ replacing the usual transitions $\xrightarrow{\mu}$. The set of all rules specifying the operational semantics of LCCS is given in Figure 1. The rule for communication (S4) requires some explanation. In the strong location transition system we take the location of a communication to be that of the smallest component which includes the two communicating subprocesses: the notation $u \sqcap v$ in rule (S4) stands for the longest common prefix of u and v. For instance we have:

Example 3.1

$$l :: \alpha \mid k :: \bar{\alpha}.(l' :: \beta \mid k' :: \bar{\beta}) \xrightarrow[\varepsilon]{\tau}_s l :: nil \mid k :: (l' :: \beta \mid k' :: \bar{\beta}) \xrightarrow[k]{\tau}_s l :: nil \mid k :: (l' :: nil \mid k' :: nil)$$

However, we shall mostly be interested here in the *weak location transition system*, where τ-transitions will have no explicit location: since the transitions themselves are not observable, it would not make much sense to attribute a location to them. The weak location transitions $\overset{a}{\underset{u}{\Longrightarrow}}_s$ and $\overset{\tau}{\Longrightarrow}_s$ are thus defined by:

$$p \overset{\tau}{\Longrightarrow}_s q \quad \Leftrightarrow \quad \exists u_1, \ldots, u_n, p_0, \ldots, p_n \;\; s.t. \;\; p = p_0 \xrightarrow[u_1]{\tau}_s p_1 \cdots \xrightarrow[u_n]{\tau}_s p_n = q$$

$$p \overset{a}{\underset{u}{\Longrightarrow}}_s q \quad \Leftrightarrow \quad \exists p_1, p_2 \;\; s.t. \;\; p \overset{\tau}{\Longrightarrow}_s p_1 \xrightarrow[u]{a}_s p_2 \overset{\tau}{\Longrightarrow}_s q$$

We shall use the weak location transition system as the basis for defining a new semantic theory for CCS, and in particular notions of equivalence and preorder which account for the degree of distribution of processes. The reader may have noticed, however, that applying the rules of Figure 1 to CCS terms just yields a transition $p \overset{\mu}{\underset{\varepsilon}{\longrightarrow}} p'$ whenever the standard semantics yields a transition $p \overset{\mu}{\longrightarrow} p'$. In fact, we shall not apply these rules directly to CCS terms. Instead, the idea is to first bring out the parallel structure of CCS terms by assigning locations to their parallel components, thus transforming them into particular LCCS terms which we call "distributed processes", and then execute these according to the given operational rules. The set DIS $\subseteq$ LCCS of *distributed processes* is given by the grammar:

$$p ::= nil \quad | \quad \mu.p \quad | \quad \underbrace{(l :: p \,|\, k :: q)}_{l \neq k} \quad | \quad (p+q) \quad | \quad p\backslash\alpha \quad | \quad p\langle f\rangle \quad | \quad x \quad | \quad rec\, x.\, p$$

Essentially, a distributed process is obtained by inserting a pair of distinct locations in a CCS term wherever there occurs a parallel operator. This is formalised by the notion of *distribution*.

Definition 3.2 The *distribution relation* is the least relation $\mathcal{D} \subseteq (\text{CCS} \times \text{DIS})$ satisfying:

- $nil\ \mathcal{D}\ nil$ and $x\,\mathcal{D}\,x$
- $p\mathcal{D}r \;\Rightarrow\; \mu.p\,\mathcal{D}\,\mu.r$
  $p\backslash\alpha\ \mathcal{D}\ r\backslash\alpha$
  $p\langle f\rangle\ \mathcal{D}\ r\langle f\rangle$
  $(rec\,x.\,p)\ \mathcal{D}\ (rec\,x.\,r)$
- $p\mathcal{D}r\ \&\ q\mathcal{D}s \;\Rightarrow\; (p\,|\,q)\ \mathcal{D}\ (l :: r \,|\, k :: s),\ \ \forall l,k$ s.t. $l \neq k$
  $(p+q)\ \mathcal{D}\ (r+s)$

If $p\mathcal{D}r$ we say that r is a *distribution* of p.

Note that the same pair of locations may be used more than once in a distribution. We shall see in fact, at the end of this section, that distributions involving just *two* atomic locations are sufficient for describing the distributed behaviour of CCS processes.

3.1 Static location equivalence

We want to define an equivalence relation $\approx_\ell^s$ on CCS processes, based on a bisimulation-like relation between their distributions. The intuition for two CCS processes p, q to be equivalent is that there exist two distributions of them, say $\tilde{p}$ and $\tilde{q}$, which perform "the same" location transitions at each step. However, as we argued already in the introduction, we cannot require the identity of locations in corresponding transitions. If we want to identify the following processes:

$$a \,|\, (b \,|\, c) \quad \text{and} \quad (a \,|\, b) \,|\, c$$
$$a \quad \text{and} \quad a \,|\, nil$$

it is clear that, whatever distributions we choose, we must allow corresponding transitions to have different – although somehow related – static locations. In general transitions will be compared modulo an *association* between their locations. The idea is directly inspired from that used by Aceto for nets of automata [Ace91]; however in our case the association will not be a bijection as in [Ace91], nor even a function. For example, in order to equate the two processes:

$$a.(b.c \,|\, nil) \quad \text{and} \quad a.b.(c \,|\, nil)$$

we need an association containing the three pairs $(\varepsilon,\varepsilon),(l,\varepsilon),(l,l')$, for some $l,l' \in Loc$.

In fact, the only property we will require of location associations is that they respect independence of locations. To make this precise, let $\ll$ denote the prefix ordering on Loc^*. If $u \ll v$ we say that v is an extension or a *sublocation* of u. If $u \not\ll v$ and $v \not\ll u$, what we indicate by $u \diamond v$, we say that u and v are *independent*.

Definition 3.3 A relation $\varphi \subseteq (Loc^* \times Loc^*)$ is a *consistent location association (cla)* if:

$$(u,v) \in \varphi \;\;\&\;\; (u',v') \in \varphi \quad \Rightarrow \quad (u \diamond u' \;\Leftrightarrow\; v \diamond v')$$

Essentially the same notion of consistent association has been proposed in [MN92] for a class of asynchronous transition systems modelling CCS with guarded sums.

Now Aceto showed in [Ace91] that, for a given pair of distributed processes we want to equate, the required *cla* cannot in general be fixed statically, but has to be built incrementally. For consider the two distributed processes, which are intuitively equivalent since both perform actions a and b in either order at different locations:

$$(\,l :: (a.\gamma + b.\bar{\gamma}) \mid k :: (\bar{\gamma}.b + \gamma.a)\,)\backslash\gamma \qquad \text{and} \qquad (l :: a \mid k :: b)$$

Here, depending on which summand is chosen in the left component of the first process, one needs to use the association $\varphi = \{(l,l),(k,k)\}$ or the association $\varphi' = \{(l,k),(k,l)\}$ (note that $\varphi \cup \varphi'$ is not consistent). Another example is given in the introduction.

To dynamically build up associations, we use the same technique as in [Ace91]. Let Φ be the set of consistent location associations. We define particular Φ-indexed families of relations S_φ over distributed processes, which we call *progressive bisimulation families* (although the relations that constitute a family are not themselves bisimulations). The idea is to start with the empty association of locations and extend it consistently as the bisimulation proceeds.

Definition 3.4 A *progressive bisimulation family (pbf)* is a Φ-indexed family $\mathbf{S} = \{S_\varphi \mid \varphi \in \Phi\}$ of relations over DIS, such that if $pS_\varphi q$ then for all $a \in Act, u \in Loc^*$:

(1) $p \overset{a}{\underset{u}{\Longrightarrow}}_s p' \;\Rightarrow\; \exists q', v$ such that $q \overset{a}{\underset{v}{\Longrightarrow}}_s q'$, $\varphi \cup \{(u,v)\} \in \Phi$ and $p' \, S_{\varphi\cup\{(u,v)\}} \, q'$

(2) $q \overset{a}{\underset{v}{\Longrightarrow}}_s q' \;\Rightarrow\; \exists p', u$ such that $p \overset{a}{\underset{u}{\Longrightarrow}}_s p'$, $\varphi \cup \{(u,v)\} \in \Phi$ and $p' \, S_{\varphi\cup\{(u,v)\}} \, q'$

(3) $p \overset{\tau}{\Longrightarrow}_s p' \;\Rightarrow\; \exists q'$ such that $q \overset{\tau}{\Longrightarrow}_s q'$ and $p' S_\varphi q'$

(4) $q \overset{\tau}{\Longrightarrow}_s q' \;\Rightarrow\; \exists p'$ such that $p \overset{\tau}{\Longrightarrow}_s p'$ and $p' S_\varphi q'$

Using these progressive bisimulation families, we may now define the *location equivalence* $\approx_\ell^s$ on CCS terms as follows:

Definition 3.5 (Static location equivalence) For $p, q \in$ CCS, we let $p \approx_\ell^s q$ if and only if for some $\bar{p}, \bar{q} \in$ DIS such that $p\,\mathcal{D}\,\bar{p}$ and $q\,\mathcal{D}\,\bar{q}$, there exists a progressive bisimulation family $\mathbf{S} = \{S_\varphi \mid \varphi \in \Phi\}$ such that $\bar{p}\, S_\emptyset\, \bar{q}$.

The reader may have noticed that the inverse $\mathcal{D}^{-1}$ of the distribution relation is a function. If we let $\pi =_{\text{def}} \mathcal{D}^{-1}$, then $\pi(p)$ gives the CCS process underlying the distributed process p. It may be easily shown that all distributions of the same process are in the relation $S_\emptyset$ for some progressive bisimulation family S:

Proposition 3.6 *Let* $p_1, p_2 \in$ DIS. *Then* $\pi(p_1) = \pi(p_2) \Rightarrow \exists$ *pbf* S *s.t.* $p_1 S_\emptyset p_2$.

Using this fact, we may prove that $\approx_\ell^s$ is indeed an equivalence relation and that, moreover, it is independent from the particular distributions that are chosen.

Proposition 3.7 (Properties of $\approx_\ell^s$)

1. *The relation* $\approx_\ell^s$ *is an equivalence on CCS processes.*
2. *For any* $p, q \in$ CCS: $p \approx_\ell^s q \Leftrightarrow$ *for all* $\bar{p}, \bar{q} \in$ DIS *such that* $p \mathcal{D} \bar{p}$ *and* $q \mathcal{D} \bar{q}$ *there exists a progressive bisimulation family* $\mathrm{S} = \{S_\varphi \mid \varphi \in \Phi\}$ *such that* $\bar{p} S_\emptyset \bar{q}$.

Thus to check the equivalence of CCS processes we may pick arbitrary distributions of them. By virtue of this result, we can restrict our attention to particular "binary" distributions, systematically associating location 0 to the left operand and location 1 to the right operand of a parallel composition. A distribution of this kind will be called *canonical*, and elements of $\{0,1\}^*$ will be called *canonical locations*. These are exactly the locations used in [MN92] and, with a slightly different notation, in [MY92],[Yan93].

Let us see now a simple example, which shows the difference between location equivalence and causality-based equivalences, such as the causal bisimulation of [DD90]:

Example 3.8 $\quad a.b + b.a \quad \not\approx_\ell^s \quad (a.\gamma \mid \bar{\gamma}.b)\backslash\gamma \; + \; (b.\gamma \mid \bar{\gamma}.a)\backslash\gamma \quad \approx_\ell^s \quad a \mid b$

As we announced earlier, $\approx_\ell^s$ will be shown to coincide with the dynamic equivalence $\approx_\ell^d$ of [BCHK91b]. Therefore all the examples given there for $\approx_\ell^d$ apply to $\approx_\ell^s$ as well.

3.2 Static location preorder

We define now a preorder $\sqsubseteq_\ell^s$ on CCS processes, which formalises the idea that one process is more sequential or *less distributed* than another. This preorder is obtained by slightly relaxing the notion of consistent association. The intuition for $p \sqsubseteq_\ell^s q$ is that there exist two distributions $\bar{p}$ and $\bar{q}$ of them such that whenever $\bar{p}$ can perform two transitions at independent locations, then $\bar{q}$ performs corresponding transitions at locations which are also independent, while the reverse is not necessarily true. This is expressed by the following notion of left-consistency:

Definition 3.9 A relation $\varphi \subseteq (Loc^* \times Loc^*)$ is a *left-consistent location association* if:

$$(u, v) \in \varphi \;\; \& \;\; (u', v') \in \varphi \quad \Rightarrow \quad (u \diamond u' \;\; \Rightarrow \;\; v \diamond v')$$

Now, if Ψ is the set of left-consistent location associations, we may obtain a notion of *progressive pre-bisimulation family (ppbf)* on distributed processes of DIS by simply replacing Φ by Ψ in Definition 3.4. Again, this gives rise to a relation on CCS processes:

Definition 3.10 (Static location preorder) If $p, q \in$ CCS, let $p \sqsubseteq_\ell^s q$ if and only if for some $\bar{p}, \bar{q} \in$ DIS such that $p \mathcal{D} \bar{p}$ and $q \mathcal{D} \bar{q}$, there exists a progressive pre-bisimulation family $\mathrm{S} = \{S_\psi \mid \psi \in \Psi\}$ such that $\bar{p} S_\emptyset \bar{q}$.

It is easy to see that $p \approx_\ell^s q \Rightarrow p \sqsubseteq_\ell^s q$. As may be expected the reverse is not true. We have for instance, for the processes of Example 3.8 above:

Example 3.11 $\quad a.b + b.a \quad \sqsubseteq_\ell^s \quad (a.\gamma \mid \bar{\gamma}.b)\backslash\gamma \; + \; (b.\gamma \mid \bar{\gamma}.a)\backslash\gamma$

We shall show that this static preorder coincides with the dynamic location preorder $\sqsubseteq_\ell^d$ of [BCHK91b], and thus inherits the theory of the latter.

For each $\mu \in Act_\tau$, $u \in Loc^*$, let $\xrightarrow[u]{\mu}_s$ be the least relation $\xrightarrow[u]{\mu}$ on LCCS processes satisfying the following axiom and rules.

(S1)	$\mu.p \xrightarrow[\varepsilon]{\mu} p$		
(S2)	$p \xrightarrow[u]{\mu} p'$	$\Rightarrow$	$l :: p \xrightarrow[lu]{\mu} l :: p'$
(S3)	$p \xrightarrow[u]{\mu} p'$	$\Rightarrow$	$p \mid q \xrightarrow[u]{\mu} p' \mid q$ $q \mid p \xrightarrow[u]{\mu} q \mid p'$
(S4)	$p \xrightarrow[u]{\alpha} p',\ q \xrightarrow[v]{\bar{\alpha}} q'$	$\Rightarrow$	$p \mid q \xrightarrow[u \sqcap v]{\tau} p' \mid q'$
(S5)	$p \xrightarrow[u]{\mu} p'$	$\Rightarrow$	$p + q \xrightarrow[u]{\mu} p'$ $q + p \xrightarrow[u]{\mu} p'$
(S6)	$p \xrightarrow[u]{\mu} p',\ \mu \notin \{\alpha, \bar{\alpha}\}$	$\Rightarrow$	$p\backslash\alpha \xrightarrow[u]{\mu} p'\backslash\alpha$
(S7)	$p \xrightarrow[u]{\mu} p'$	$\Rightarrow$	$p\langle f\rangle \xrightarrow[u]{f(\mu)} p'\langle f\rangle$
(S8)	$p[rec\,x.\,p/x] \xrightarrow[u]{\mu} p'$	$\Rightarrow$	$rec\,x.\,p \xrightarrow[u]{\mu} p'$

Figure 1: Static location transitions

Let $p \xrightarrow[u]{\tau}_d q \;\Leftrightarrow_{\text{def}}\; p \xrightarrow[u]{\tau}_s q$, and for each $a \in Act$, $u \in L^*$, let $\xrightarrow[u]{a}_d$ be the least relation $\xrightarrow[u]{a}$ on LCCS processes satisfying rules (S2), (S3), (S5), (S6), (S7), (S8) and the axiom:

(D1) $\quad a.p \xrightarrow[l]{a} l :: p \quad$ for any $l \in Loc$

Figure 2: Dynamic location transitions

---

4 Dynamic approach

We briefly recall here the dynamic approach of [BCHK91b], and in particular the definitions of $\approx_\ell^d$ and $\sqsubseteq_\ell^d$. In the dynamic approach, locations are associated with actions rather than with parallel components. This association is built dynamically, according to the rule:

$$\text{(D1)} \qquad a.p \xrightarrow[l]{a}_d l :: p \qquad \text{for any } l \in Loc$$

In some sense locations are transmitted from transitions to processes, whereas in the static case we had the inverse situation. Rule (D1) is the essence of the dynamic location semantics. The remaining rules are just as in the static semantics, see Figure 2. We refer to [BCHK91b] for more intuition on the dynamic notion of location: let us just observe that these locations increase at each step, even if the execution goes on within the same parallel component. In fact the location l which appears in rule (D1) may be seen as an identifier for the action a. Then the location u of a generic transition $p \xrightarrow[u]{a}_d p'$ is a record of all the actions which causally precede a, what we shall call also the *access path* to a.

Because of rule (D1), the dynamic location transition system is both infinitely branching and infinitely progressing: it gives infinite representations for all regular processes. Indeed, this has been the main criticism to this dynamic approach, see [Ace91],[MY92],[MN92]. In fact, while the infinite branching may be overcome easily (through a *canonical* choice of dynamic locations, see [Cas93]) the infinite progression is really intrinsic to this semantics.

Note that for τ-transitions, for which we do not want to introduce additional locations, we simply use the static transition rules. Although this last point differentiates our strong dynamic location transition system from that originally introduced in [BCHK91b], the resulting *weak (dynamic) location transition system* is the same. The definition of the weak transitions $\overset{a}{\underset{u}{\Longrightarrow}}_d$ and $\overset{\tau}{\Longrightarrow}_d$ is similar to that of the $\overset{a}{\underset{u}{\Longrightarrow}}_s$ and $\overset{\tau}{\Longrightarrow}_s$.

We define now the *dynamic location equivalence* $\approx_\ell^d$ and the *dynamic location preorder* $\sqsubseteq_\ell^d$. Because of the flexibility in the choice of locations, these definitions are much simpler than in the static case. In [BCHK91b] the relations $\approx_\ell^d$ and $\sqsubseteq_\ell^d$ are obtained as instances of a general notion of *parameterized location bisimulation*. We shall use here directly the instantiated definitions.

Definition 4.1 (Dynamic location equivalence) A symmetric relation $R \subseteq$ LCCS $\times$ LCCS is called a *dynamic location bisimulation (dlb)* iff for all $(p,q) \in R$ and for all $a \in Act, u \in Loc^+$:

$$(1)\ p \overset{a}{\underset{u}{\Longrightarrow}}_d p' \ \Rightarrow\ \exists q' \in \text{LCCS such that } q \overset{a}{\underset{u}{\Longrightarrow}}_d q' \text{ and } (p',q') \in R$$

$$(2)\ p \overset{\tau}{\Longrightarrow}_d p' \ \Rightarrow\ \exists q' \in \text{LCCS such that } q \overset{\tau}{\Longrightarrow}_d q' \text{ and } (p',q') \in R$$

The largest *dlb* is called *dynamic location equivalence* and denoted $\approx_\ell^d$.

We refer to [BCHK91b] for examples and results concerning $\approx_\ell^d$. Consider now the location preorder $\sqsubseteq_\ell^d$. Here, instead of requiring the identity of locations in corresponding transitions, we demand that the locations in the second (more distributed) process be subwords of the locations in the first (more sequential) process. Formally, the *subword* relation $\leq_{\text{sub}}$ on Loc^* is defined by: $v \leq_{\text{sub}} u \Leftrightarrow \exists v_1, \ldots, v_k,\ \exists w_1, \ldots, w_{k+1}$ s.t. $v = v_1 \cdots v_k$ and $u = w_1 v_1 \cdots w_k v_k w_{k+1}$.

Definition 4.2 (Dynamic location preorder) A relation $R \subseteq \text{LCCS} \times \text{LCCS}$ is called a *dynamic location pre-bisimulation (dlpb)* iff for all $(p,q) \in R$ and for all $a \in Act, u \in Loc^+$:

$$(1)\quad p \overset{a}{\underset{u}{\Longrightarrow}}_d p' \;\Rightarrow\; \exists v \leq_{sub} u, \exists q' \in \text{LCCS} \text{ such that } q \overset{a}{\underset{v}{\Longrightarrow}}_d q' \text{ and } (p',q') \in R$$

$$(2)\quad q \overset{a}{\underset{v}{\Longrightarrow}}_d q' \;\Rightarrow\; \exists u.\, v \leq_{sub} u, \exists p' \in \text{LCCS} \text{ such that } p \overset{a}{\underset{u}{\Longrightarrow}}_d p' \text{ and } (p',q') \in R$$

$$(3)\quad p \overset{\tau}{\Longrightarrow}_d p \;\Rightarrow\; \exists q' \in \text{LCCS} \text{ such that } q \overset{\tau}{\Longrightarrow}_d q' \text{ and } (p',q') \in R$$

$$(4)\quad q \overset{\tau}{\Longrightarrow}_d q' \;\Rightarrow\; \exists p' \in \text{LCCS} \text{ such that } p \overset{\tau}{\Longrightarrow}_d p' \text{ and } (p',q') \in R$$

The largest *dlpb* is called *dynamic location preorder* and denoted $\sqsubseteq_\ell^d$.

The intuition is as follows. If p is a sequentialized version of q, then each component of p corresponds to a group of parallel components in q. Then the local causes of any action of q will correspond to a subset of local causes of the corresponding action of p. This may be easily verified for the following examples:

Example 4.3 $\quad a.a.a \;\sqsubseteq_\ell^d\; a.a \mid a \quad$ and $\quad a.b + b.a \;\sqsubseteq_\ell^d\; a \mid b$

We shall not comment further here on the relations $\approx_\ell^d$ and $\sqsubseteq_\ell^d$, referring again the reader to [BCHK91b] for more examples and for results concerning these relations. We proceed now to state our main result, namely that the dynamic relations $\approx_\ell^d$ and $\sqsubseteq_\ell^d$ coincide with the static relations $\approx_\ell^s$ and $\sqsubseteq_\ell^s$ introduced in the previous section.

Theorem 4.4 *Let $p,q \in CCS$. Then:* $\; p \approx_\ell^s q \Leftrightarrow p \approx_\ell^d q \;$ *and* $\; p \sqsubseteq_\ell^s q \Leftrightarrow p \sqsubseteq_\ell^d q.$

To prove this results, we use a new transition system on CCS, called *occurrence system*, which is essentially a simplification of the *event system* introduced in [BC91] to compare different models of CCS. This transition system, which incorporates the information of both location transition systems, is used as an intermediate between the static and the dynamic semantics. The main point is to prove that starting from a static or a dynamic location computation, one may always reconstruct a corresponding occurrence computation. This means, essentially, that all the information about distribution and local causality is already present in both location transition systems. The proof may be found in [Cas93].

Acknowledgements

The idea of a static assignment of locations was originally put forward by G. Boudol in the course of a CEDISYS meeting in Brighton, in September 1990. I would like to thank him for inspiration and for innumerable comments and advices. I also benefitted from discussions with L. Aceto, who was the first to formalise the "static view of locations" for a subset of CCS. I am grateful to P.S. Thiagarajan, for commenting on an earlier draft of this paper and for raising several interesting questions. Part of this work was done while visiting Thiagarajan and his colleagues in Madras. I would like to thank all of them for their interest and comments.

References

[Ace91] L. Aceto. A static view of localities. Report 1483, INRIA, 1991. To appear in *Formal Aspects of Computing*.

[BC91] G. Boudol and I. Castellani. Flow models of distributed computations: three equivalent semantics for CCS. Report 1484, INRIA, 1991. To appear in *Information and Computation*. Previous version in Proc. La Roche-Posay, LNCS 469, 1990.

[BCHK91a] G. Boudol, I. Castellani, M. Hennessy, and A. Kiehn. Observing localities. Report 4/91, Sussex University, and INRIA Res. Rep. 1485, 1991. To appear in *Theoretical Computer Science*. Extended abstract in Proc. MFCS 91, LNCS 520, 1991.

[BCHK91b] G. Boudol, I. Castellani, M. Hennessy, and A. Kiehn. A theory of processes with localities. Report 1632, INRIA, 1991. To appear in *Formal Aspects of Computing*. Extended abstract in Proc. CONCUR92, LNCS 630, 1992.

[Cas93] I. Castellani. Full version of this paper. Report, INRIA, 1993. To appear.

[DD90] Ph. Darondeau and P. Degano. Causal trees: interleaving + causality. In *Proceedings LITP Spring School, La Roche-Posay*, number 469 in LNCS, 1990.

[DDNM87] P. Degano, R. De Nicola, and U. Montanari. Observational equivalences for concurrency models. In M. Wirsing, editor, *Formal Description of Programming Concepts-III, Proceedings of the 3th IFIP WG 2.2 working conference*, Ebberup 1986, pages 105–129. North-Holland, 1987.

[GG89] R.J. van Glabbeek and U. Goltz. Equivalence notions for concurrent systems and refinement of actions. Arbeitspapiere der GMD 366, Gesellschaft für Mathematik und Datenverarbeitung, Sankt Augustin, 1989. Extended abstract in Proc. MFCS 89, LNCS 379, 1989.

[Kie91] A. Kiehn. Local and global causes. Report 342/23/91, Technische Universität München, 1991. Submitted for publication.

[Mil80] R. Milner. *A Calculus of Communicating Systems*, volume 92 of *Lecture Notes in Computer Science*. Springer–Verlag, 1980.

[MN92] M. Mukund and M. Nielsen. CCS, locations and asynchronous transition systems. In *Proceedings FST-TCS 92*, 1992.

[MY92] U. Montanari and D. Yankelevich. A parametric approach to localities. In *Proceedings ICALP 92*, number 623 in LNCS, 1992.

[Yan93] D. Yankelevich. *Parametric Views of Process Description Languages*. Ph.d. thesis, University of Pisa, 1993.

Speedup of Recognizable Trace Languages

Christophe Cérin[1,2] * and Antoine Petit[2] **

[1] Université de Picardie Jules Verne,
LAMIFA, UFR de mathématiques et d'informatique, 33 rue St Leu,
F-80039 Amiens Cedex
email: cerin@lri.lri.fr, fax:(33)-1-22827502
[2] Université Paris Sud,
LRI, URA CNRS 410, Bât. 490,
F-91405 Orsay Cedex
email: petit@lri.lri.fr, fax:(33)-1-69416586

Abstract. Traces have been defined by A.Mazurkiewicz in order to modelize concurrent processes. The decomposition of a trace in Foata normal form gives the "best" parallel execution of a trace. We define naturally the speedup of a trace as the quotient of its sequential execution time by its parallel execution time. We generalize this definition to trace languages and we prove that this speedup can be computed in a modular way for any recognizable trace language.

Keywords: speedup, trace, recognizable language.

1 Introduction

The notion of traces has been introduced by A.Mazurkiewicz [Maz77] in order to describe non-sequential processes. A trace can be thought as the set of all sequential observations of a concurrent process. Each sequential observation is described, in a usual way, by a word and a trace is an equivalence class of words. In fact, the set of traces is the quotient, by a suitable congruence, of the free monoid. It has thus a structure of monoid himself which has been introduced and studied by combinatorists [CF69]. Since the original work of A.Mazurkiewicz, it has been systematically studied as a model of concurrent systems (see surveys [Maz87], [AR88], [Per89] or the monograph [Die90]).

The decomposition in Foata normal form of a trace [CF69] provides a way to execute a trace as a sequence of steps, each step being the execution in parallel of independent letters. Moreover a trace can not be decomposed in a smaller number of such steps. Hence the decomposition in Foata normal form gives the "best" parallel execution of a trace.

* This work was partially supported by "Conseil Régional de Picardie, Pôle Modélisation"

** This work was partially supported by ESPRIT WG 6317 (ASMICS2) and WG 6067 (CALIBAN)

This paper deals with the important problem to compare these "best" parallel executions with the sequential ones for a recognizable set of traces. We show that we can assume without loss of generality that each letter has the same (sequential) execution time. Therefore, the sequential execution time of a trace is reduced to its length and its parallel execution time to the number of steps in its Foata normal form decomposition. Then we define naturally the speedup of a trace as the quotient of its sequential execution time by its parallel execution time. Note that this quantity has been already defined by N.Saheb [Sah89] who studied its limit when a trace of arbitrarily great length is randomly chosen.

We extend the notion of speedup to a trace language as being the greatest lower bound of the speedup of its elements. Our main result consists in proving that this speedup can be effectively computed in a modular way for any recognizable trace language. In this way, we can evaluate the gain between the sequential execution of a recognizable trace language and its fastest parallel execution.

2 Preliminaries

A dependence alphabet is an ordered pair (X, D) where X is a finite alphabet and D the dependence relation, a reflexive and symmetric relation on X. The complement relation $I = (X \times X)\backslash D$ of D is called the independence relation. For any subset $Y \subseteq X$, $D(Y)$ is the set $\{x \in X \mid \exists y \in Y, (x,y) \in D\}$ whereas $I(Y) = X\backslash D(Y) = \{x \in X \mid \forall y \in Y, (x,y) \in I\}$.

In a classical way, X^* is the free monoid generated by X. The relation $\sim_I$ or simply $\sim$, is defined on X^* as the reflexive and transitive closure of the relation $\{(uabv, ubav) \mid u, v \in X^*, (a,b) \in I\}$. In fact, it turns out that this relation is a congruence and the quotient $X^*/_{\sim_I}$ is the trace monoid [Maz77] generated by X and I (or D). This monoid is denoted $\mathbb{M}(X, D)$, the subset of non empty traces by $\mathbb{M}^+(X, D)$ and $\varphi : X^* \to \mathbb{M}(X, D)$ is the canonical surjection from X^* onto $\mathbb{M}(X, D)$.

The length of a word $u \in X^*$ is denoted by $|u|$. Since two equivalent words have clearly the same length, we define for a trace $t \in \mathbb{M}(X, D)$, the length of t as $|t| = |u|$ for $u \in \varphi^{-1}(t)$.

A clique of the graph (X, I) is a subset $Y \subseteq X$ such that $(Y, I \cap (Y \times Y))$ is a complete graph. The set of cliques of (X, I) is denoted by $\mathcal{C}\ell$. For any clique $C \in \mathcal{C}\ell$, $|C|$ is the number of elements of C.

We consider the morphism $\mathcal{X} : \mathcal{C}\ell^* \to \mathbb{M}(X, D)$ which associates, with each clique $C = \{c_1, \cdots, c_k\} \in \mathcal{C}\ell$, the trace $\varphi(c_1 \cdots c_k)$. Note that since the letters of C are pairwise independent, it holds $\varphi(c_1 \cdots c_k) = \varphi(c_{\sigma(1)} \cdots c_{\sigma(k)})$ for any permutation σ of $\{1, \cdots, k\}$

The set FNF of Foata Normal Forms is the rational subset of the free monoid $\mathcal{C}\ell^*$ given by a rational expression of its complementary:

$$\mathcal{C}\ell^* \backslash FNF = \bigcup_{(C,C') \in K} \mathcal{C}\ell^* CC' \mathcal{C}\ell^*$$

where $K = \{(C, C') \in \mathcal{C}\ell \times \mathcal{C}\ell \mid \exists a \in C', \forall b \in C, (a,b) \in I\}$.

A sequence $(t_1, \cdots, t_r)$ of traces is a decomposition in Foata normal form of a trace t if $t = t_1 \cdots t_r$, $\exists C_1, \cdots, C_r \in \mathcal{C}\ell$ such that $\mathcal{X}(C_i) = t_i$ for $1 \le i \le r$ and $C_1 \cdots C_r \in FNF$. From Cartier and Foata's theorem [CF69], any trace $t \in \mathbb{M}^+(X, D)$ admits a unique decomposition in Foata normal form. In the sequel, this decomposition is denoted by $fnf(t)$ and the length of the decomposition, i.e., the number of traces constituting the decomposition is denoted by $lfnf(t)$. Recall that the decomposition in Foata normal form of a trace over a fixed dependence alphabet (X, D) can be obtained in linear time in the length of t (see [Die90]).

Example 1. Let $(X, D) = a — b — c$. The words *acbcaab* and *cabacab* are $\sim_I$-equivalent. The cliques of (X, I) are $C_a = \{a\}, C_b = \{b\}, C_c = \{c\}, C_{a,c} = \{a, c\}$. The set K used to define FNF is thus $K = \{(C_a, C_c), (C_a, C_{a,c}), (C_c, C_a), (C_c, C_{a,c})\}$. Hence the word $C_{a,c}C_bC_{a,c}C_aC_b$ belongs to FNF and $(\mathcal{X}(C_{a,c}), \mathcal{X}(C_b), \mathcal{X}(C_{a,c}), \mathcal{X}(C_a), \mathcal{X}(C_b)) = (\varphi(ac), \varphi(b), \varphi(ac), \varphi(a), \varphi(b))$ is the decomposition in Foata normal form of the trace $t = \varphi(acbcaab)$. Therefore, it holds that $lfnf(t) = 5$.

As in any classical monoid (cf [Ber79]) the notion of recognizable language is defined through the notion of recognizable morphism or by finite deterministic automata. In this paper we only use the following characterization: a trace language $T \subseteq \mathbb{M}(X, D)$ is recognizable if and only if $\varphi^{-1}(T) \subseteq X^*$ is a recognizable word language.

3 Foata Normal Form and Parallelization

In the modelization of a sequential process by a word, we subtend that there is a unique sequential processor which performs the tasks corresponding to the letters of the word to execute one after each other. In the case of a distributed process depicted by a trace, a notion of the best parallel execution of tasks is given by the decomposition in Foata normal form. More precisely, let $t \in \mathbb{M}^+(X, D)$ be a trace and let $fnf(t) = (t_1, \cdots, t_r)$ be its decomposition in Foata normal form. For any $1 \le i \le r$, the letters of t_i are pairwise independent, hence the trace t_i can be executed in parallel by $|t_i|$ distinct processors. Therefore, if we have enough sequential processors (at least the maximal length of a clique of $\mathcal{C}\ell$), a parallel execution of the trace t is given by executing in parallel the tasks of t_1, then the tasks of t_2 and so on until the parallel execution of the tasks of t_r.

Thus, the decomposition in Foata normal form of a trace t gives a way to execute the trace t in $lfnf(t)$ steps, each step being the parallel execution of independent letters.

The following proposition proves that the number of such steps cannot be less than the one given by the decomposition in Foata normal form. For the sake of completeness, we give below a proof of this result which is currently informally stated as "the Foata normal form gives the best parallel execution of a trace". A nive and elegant proof can also be achieved using the framework of dependence graphs by which Foata normal form has a very natural and intuitive interpretation (see e.g [Die90]). Note that we do not assume any condition on the execution time of the tasks of the alphabet X.

Proposition 1. *Let $t \in \mathbb{M}^+(X, D)$ be a non empty trace and let $C_1, \cdots, C_k \in \mathcal{C}\ell$ be such that $t = \mathcal{X}(C_1)\cdots\mathcal{X}(C_k)$. Then $lfnf(t) \leq k$.*

Proof. The proof is over an induction on the length of t. The result is trivially true for any trace of length 1. Assume that the proposition holds for traces of length at most $n-1$ and let t be a trace of length n. Let $(\mathcal{X}(C'_1), \cdots, \mathcal{X}(C'_r))$ be the decomposition of t in Foata normal form.

Recall [CP85] that two traces r and s are equal if and only if for any letters $(a, b) \in D$, $\prod_{\{a,b\}}(r) = \prod_{\{a,b\}}(s)$ where $\prod_{\{a,b\}}(r)$ is the trace obtained from r by erasing all letters distinct from a and b.

Now, let $a \in C_1$ and let C'_i be the first clique of the sequence $(C'_1, C'_2, \cdots, C'_r)$ to which a belongs. Assume that $i \geq 2$, from the definition of FNF, there exists some $b \in C'_{i-1}$ such that $(a, b) \in D$. Therefore $\prod_{\{a,b\}}(\mathcal{X}(C'_1)\cdots\mathcal{X}(C'_r))$ has b as first letter whereas $\prod_{\{a,b\}}(\mathcal{X}(C_1)\cdots\mathcal{X}(C_k))$ admits a as first letter. From the characterization recalled just above, this implies $\mathcal{X}(C'_1)\cdots\mathcal{X}(C'_r) \neq \mathcal{X}(C_1)\cdots\mathcal{X}(C_k)$ which leads to a contradiction with the hypothesis. Hence it holds $a \in C'_1$ and therefore $C_1 \subseteq C'_1$.

Let t' be the trace obtained from t by erasing the left factor $\mathcal{X}(C'_1)$. Since $\mathbb{M}(X, D)$ is left cancellative [CP85] we obtain $t' = \mathcal{X}(C'_2)\cdots\mathcal{X}(C'_r) = \mathcal{X}(C''_1)\cdots \mathcal{X}(C''_k)$ where the sequences $(C''_i) \subseteq \mathcal{C}\ell \cup \{1\}$ (1 is the empty word of $\mathcal{C}\ell$) and (R_i) are defined inductively by $R_0 = C'_1$ and for any $1 \leq i \leq k$, $C''_i = C_i \backslash R_i$ and $R_i = R_{i-1} \backslash C_i$.

The trace t' admits clearly $(\mathcal{X}(C'_2), \cdots, \mathcal{X}(C'_r))$ as decomposition in Foata normal form. Hence by induction hypothesis, we obtain $lfnf(t') \leq |\{i/1 \leq i \leq k \text{ and } C''_i \neq 1\}|$. Since $C_1 \subseteq C'_1$, we have $C''_1 = 1$. Therefore $lfnf(t') \leq k-1$ and thus $lfnf(t) \leq k$, the proposition is proved. □

In order to compare, from a time point of view, the sequential execution of a trace t and its parallel execution given by its decomposition in Foata normal form, it would be more convenient if any $a, b \in X$ have the same (sequential) execution time. This can be done without loss of generality by replacing any letter a by a sequence $a_1 \cdots a_p$ where p is the common divisor of the sequential execution times (assume to be integers) of the tasks $x \in X$. Moreover such a transformation preserves the recognizability of a language as it can be immediately deduced by induction from Proposition 2 below.

Let Y be a finite alphabet, let x, x_1, x_2 be three letters not in Y and let $(Y \cup \{x\}, D)$ be a dependence alphabet. We consider the morphism $\eta : Y \cup \{x_1, x_2\} \to Y \cup \{x\}$ defined by $\eta(x_1) = \eta(x_2) = x$ and $\eta(y) = y$ for $y \in Y$. Then we define the dependence relation D' on $Y \cup \{x_1, x_2\}$ by $(y, z) \in D'$ if $(\eta(y), \eta(z)) \in D$. The application $\phi : Y \cup \{x\} \to Y \cup \{x_1, x_2\}$ defined by $\phi(x) = x_1x_2$ and $\phi(y) = y$ for $y \in Y$ induces clearly a morphism, still denoted by ϕ, from $\mathbb{M}(Y \cup \{x\}, D)$ in $\mathbb{M}(Y \cup \{x_1, x_2\}, D')$. As it can be deduced from a more general result of Ochmanski [Och88], the morphism ϕ preserves recognizable languages. For the sake of completeness, we propose here a simple proof in this particular case.

Proposition 2. *Let $T \subseteq \mathbb{M}(Y \cup \{x\}, D)$ be a recognizable trace language, then $\phi(T) \subseteq \mathbb{M}(Y \cup \{x_1, x_2\}, D')$ is a recognizable trace language.*

Proof. Let $\varphi : (Y \cup \{x\})^* \to \mathbb{M}(Y \cup \{x\}, D)$ and $\varphi' : (Y \cup \{x_1, x_2\})^* \to \mathbb{M}(Y \cup \{x_1, x_2\}, D')$ be the canonical surjections. From the characterization of recognizable trace languages recalled in Section 2, we have to prove that $\varphi'^{-1}(\phi(T)) \subseteq (Y \cup \{x_1, x_2\})^*$ is a recognizable word language under the hypothesis that $\varphi^{-1}(T) \subseteq (Y \cup \{x\})^*$ is a recognizable word language. Let $\mu : (Y \cup \{x_1, x_2\})^* \to (Y \cup \{x\})^*$ be the morphism defined by $\mu(y) = y$ if $y \in Y$, $\mu(x_1) = x$, $\mu(x_2) = 1$ (where 1 is the empty word of $(Y \cup \{x\})^*$). From the definition of ϕ, it is easy to verify that $\varphi'^{-1}(\phi(T)) = \mu^{-1}(\varphi^{-1}(T)) \cap Y^*(x_1 I(\{x\})^* x_2 Y^*)^*$. Since recognizable word languages are closed under inverse morphisms and intersection by recognizable sets, we deduce that $\varphi'^{-1}(\phi(T))$ is recognizable and the proposition is proved. □

Now, under the hypothesis that each letter $x \in Y$ has the same (sequential) execution time, we may assume that this execution time is equal to one time unit. Hence any clique $C \in \mathcal{Cl}$ can be computed in parallel by $|C|$ processes in one time unit. This leads naturally to the following definitions:

Definition 3. Let $t \in \mathbb{M}^+(X, D)$ be a non-empty trace.
The sequential execution time of t is $Ts(t) = |t|$.
The parallel execution time of t is $Tp(t) = \min\{r \mid \exists C_1, \cdots, C_r \in \mathcal{Cl} \text{ st } \mathcal{X}(C_1) \cdots \mathcal{X}(C_r) = t\}$.

From Proposition 1, we get immediately:

Corollary 4. *Let $t \in \mathbb{M}^+(X, D)$. Then $Tp(t) = lfnf(t)$.*

For any non-empty trace $t \in \mathbb{M}^+(X, D)$, we define the speedup of t as the ratio $\mathfrak{T}(t) = \frac{Ts(t)}{Tp(t)}$ and, for technical convenience, we set $\mathfrak{T}(1) = +\infty$.

Note that we have thus $1 \le \mathfrak{T}(t) \le \max\{|C|, C \in \mathcal{Cl}\}$ for any non-empty trace t. We extend the definition of the speedup to trace language by setting, for any non-empty trace language $T \subseteq \mathbb{M}(X, D)$,

$$\mathfrak{T}_{min}(T) = \inf\{\mathfrak{T}(t), t \in T\}.$$

Thus $\mathfrak{T}_{min}(T)$ represents the worst speedup of the traces of T. Note that $\mathfrak{T}_{min}(T)$ is not always realized by a trace of T. For instance, let $(X, D) = a \quad b$ and let $T = \{\varphi[(ab)a^n], n \in \mathbb{N}\}$. For any $n \in \mathbb{N}$, $\mathfrak{T}(\varphi((ab)a^n)) = \frac{n+2}{n+1} \ne 1$ whereas $\mathfrak{T}_{min}(T) = 1$.

The principal aim of this paper is to show that the speedup $\mathfrak{T}_{min}(T)$ can be computed, in an effective way, for any recognizable trace language T.

Remark: we could also define $\mathfrak{T}_{max}(T) = \sup\{\mathfrak{T}(t), t \in T\}$. Results we proved below about $\mathfrak{T}_{min}$ can be translated without difficulty to $\mathfrak{T}_{max}$.

4 Modular Calculus of the Speedup

One of the difficulties with a modular calculus of $\mathfrak{T}_{min}$ is the behaviour of the parallel execution time Tp with respect to the concatenation. For instance, assume that $(a,b) \in I$, then $Tp(a^n) = Tp(b^n) = n$ for any $n \in \mathbb{N}$ whereas $Tp(a^n a^n) = 2n$ and $Tp(a^n b^n) = n$. In order to bypass this problem, we will carry out calculus in the free monoid $\mathcal{C}\ell^*$ rather than in the trace monoid $\mathbb{M}(X,D)$. To this purpose, we extend the definitions of the previous section by setting $\mathfrak{T}(1) = +\infty$ and for any word $f = C_1 \cdots C_r \in \mathcal{C}\ell^+$, $Ts(f) = |C_1| + \cdots + |C_r|$, $Tp(f) = r$, $\mathfrak{T}(f) = \frac{Ts(f)}{Tp(f)}$ and for any non-empty language $L \subseteq \mathcal{C}\ell^*$, $\mathfrak{T}_{min}(L) = \inf\{\mathfrak{T}(f), f \in L\}$. The link between the speedup $\mathfrak{T}_{min}$ in the free monoid $\mathcal{C}\ell^*$ and in the trace monoid $\mathbb{M}(X,D)$ is given by the following result:

Proposition 5. *Let $T \subseteq \mathbb{M}^+(X,D)$ be a non-empty trace language, then*

$$\mathfrak{T}_{min}(T) = \mathfrak{T}_{min}(\mathcal{X}^{-1}(T) \cap FNF)$$

Proof. From the definition of $\mathfrak{T}$ in $\mathcal{C}\ell^+$ and $\mathbb{M}^+(X,D)$ and from the unicity of the decomposition of a trace in Foata normal form, it holds $\mathfrak{T}(t) = \mathfrak{T}(\mathcal{X}^{-1}(t) \cap FNF)$ for any trace $t \in \mathbb{M}^+(X,D)$. The proposition follows trivially. □

If T is a recognizable trace language, then $\mathcal{X}^{-1}(T)$ and hence $\mathcal{X}^{-1}(T) \cap FNF$ are recognizable languages of $\mathcal{C}\ell^*$. A first natural idea to calculate the measure $\mathfrak{T}_{min}(L)$ for a recognizable language L of $\mathcal{C}\ell^*$ is to resort to an induction on rational expressions and to calculate $\mathfrak{T}_{min}(L^*)$, $\mathfrak{T}_{min}(L \cup L')$, $\mathfrak{T}_{min}(LL')$ in function of $\mathfrak{T}_{min}(L)$ and $\mathfrak{T}_{min}(L')$. The case of the union operation is straightforward:

Lemma 6. *Let $L, L' \subseteq \mathcal{C}\ell^+$ be non-empty languages. Then*

$$\mathfrak{T}_{min}(L \cup L') = \min\{\mathfrak{T}_{min}(L), \mathfrak{T}_{min}(L')\}$$

More surprisingly, the speedup of the iteration of a language is the same than the speedup of this language:

Lemma 7. *Let $L \subseteq \mathcal{C}\ell^+$ be a non-empty language. Then*

$$\mathfrak{T}_{min}(L^*) = \mathfrak{T}_{min}(L)$$

Proof. Since $L \subseteq L^*$ it holds $\mathfrak{T}_{min}(L) \geq \mathfrak{T}_{min}(L^*)$.

Conversely, let $u \in L^*$, $u = u_1 u_2 \ldots u_k$, $u_i \in L$ for $1 \leq i \leq k$. Then

$$\mathfrak{T}(u) = \frac{Ts(u)}{Tp(u)} = \frac{Ts(u_1) + Ts(u_2) + \cdots + Ts(u_k)}{Tp(u_1) + Tp(u_2) + \cdots + Tp(u_k)}$$

For any positive integers (or reals), $a_1, \cdots, a_n, b_1, \cdots, b_n$ it is easy to verify, for instance by an induction on n, that:

$$\min\left\{\frac{a_1}{b_1}, \frac{a_2}{b_2}, \ldots, \frac{a_n}{b_n}\right\} \leq \frac{a_1 + a_2 + \cdots + a_n}{b_1 + b_2 + \cdots + b_n} \leq \max\left\{\frac{a_1}{b_1}, \frac{a_2}{b_2}, \ldots, \frac{a_n}{b_n}\right\} \quad (1)$$

Therefore, we deduce that $\mathfrak{T}(u) \geq \min\left\{\frac{Ts(u_i)}{Tp(u_i)}, 1 \leq i \leq k\right\}$. Hence by definition of $\mathfrak{T}_{min}(L)$, $\mathfrak{T}(u) \geq \mathfrak{T}_{min}(L)$. Since this inequality is verified for any word $u \in L^*$, we obtain $\mathfrak{T}_{min}(L^*) \geq \mathfrak{T}_{min}(L)$ and the lemma is proved. □

Unfortunately, whereas for any words $f, g \in \mathcal{C}\ell^+$, $Ts(f \cdot g) = Ts(f) + Ts(g)$ and $Tp(f \cdot g) = Tp(f) + Tp(g)$, the following example shows that $\mathfrak{T}_{min}(L \cdot L')$ cannot be deduced from $\mathfrak{T}_{min}(L)$ and $\mathfrak{T}_{min}(L')$.

Example 1 (continued): Let $L_1 = (C_bC_{a,c})^*C_bC_{a,c}C_b$ and $L_1' = C_bC_{a,c}C_b$. $\mathfrak{T}_{min}(L_1) = \inf\{\frac{3n+4}{2n+3}, n \in \mathbb{N}\} = \frac{4}{3}$, $\mathfrak{T}_{min}(L_1') = \frac{4}{3}$. Let $L_2 = C_{a,c}C_{a,c}$, $\mathfrak{T}_{min}(L_1' \cdot L_2) = \inf\{\frac{3n+8}{2n+5}, n \in \mathbb{N}\} = \frac{3}{2}$ and $\mathfrak{T}_{min}(L_1' \cdot L_2) = \frac{8}{5}$. Note that $L_1 \cdot L_2$ and $L_1' \cdot L_2$ are included in FNF.

Therefore the speedup of a language cannot be calculated by a rational induction. Nevertheless, the following lemma permits us to bypass the problem of concatenation and to compute the speedup by induction on the star height of a rational expression defining the language as it will be explained in details below.

Lemma 8. *Let $L_1, L_2, L_3 \subseteq \mathcal{C}\ell^*$ be non-empty languages. Then*

$$\mathfrak{T}_{min}(L_1 L_2^* L_3) = \min\{\mathfrak{T}_{min}(L_1 L_3), \mathfrak{T}_{min}(L_2)\}$$

Proof. Let $u \in L_1$, $v \in L_2^*$, $w \in L_3$. From the inequality (1) above, we obtain:

$$\begin{aligned}\mathfrak{T}(uvw) &= \frac{Ts(uvw)}{Tp(uvw)} \\ &= \frac{Ts(uw) + Ts(v)}{Tp(uw) + Tp(v)} \\ &\geq \min\left\{\frac{Ts(uw)}{Tp(uw)}, \frac{Ts(v)}{Tp(v)}\right\} \\ &\geq \min\left\{\mathfrak{T}_{min}(L_1 L_3), \mathfrak{T}_{min}(L_2^*)\right\}\end{aligned}$$

With Lemma 7, it holds:

$$\mathfrak{T}_{min}(L_1 L_2^* L_3) \geq \min\left\{\mathfrak{T}_{min}(L_1 L_3), \mathfrak{T}_{min}(L_2)\right\}$$

Conversely, $L_1 L_3 \subseteq L_1 L_2^* L_3$ hence $\mathfrak{T}_{min}(L_1 L_3) \geq \mathfrak{T}_{min}(L_1 L_2^* L_3)$. Finally, let $u \in L_1$, $v \in L_2$, $w \in L_3$. For any integer n, $v^n \in L_2^*$ and hence

$$\mathfrak{T}(uv^n w) \geq \mathfrak{T}_{min}(L_1 L_2^* L_3)$$

Moreover, we have $\mathfrak{T}(uv^n w) = \frac{Ts(uw)+n.Ts(v)}{Tp(uw)+n.Tp(v)}$. The limit, when $n \to \infty$, of $\mathfrak{T}(uv^n w)$ is thus equals to $\mathfrak{T}(v)$ and we deduce:

$$\mathfrak{T}(v) \geq \mathfrak{T}_{min}(L_1 L_2^* L_3)$$

Since this inequality is verified for any word $v \in L_2$ we obtain:

$$\mathfrak{T}_{min}(L_2) \geq \mathfrak{T}_{min}(L_1 L_2^* L_3)$$

This last inequality gathered with $\mathfrak{T}_{min}(L_1 L_3) \geq \mathfrak{T}_{min}(L_1 L_2^* L_3)$ implies that $\min\{\mathfrak{T}_{min}(L_2), \mathfrak{T}_{min}(L_1 L_3))\} \geq \mathfrak{T}_{min}(L_1 L_2^* L_3)$ and the lemma is proved. □

From Lemma 8, we deduce by an immediate induction:

Corollary 9. *Let $L_1, \cdots, L_k, R_0, \cdots, R_k \subseteq \mathcal{A}^*$ be non-empty languages. Then* $\mathfrak{T}_{min}(R_0 L_1^* R_1 L_2^* R_2 \cdots R_{k-1} L_k^* R_k) =$
$$\min\{\mathfrak{T}_{min}(R_0 R_1 \cdots R_k), \mathfrak{T}_{min}(L_1), \cdots, \mathfrak{T}_{min}(L_k)\}.$$

We have now all elements to establish our main result on the speedup of recognizable language.

Theorem 10. *Let $L \subseteq \mathcal{A}^*$ be a non-empty recognizable language. Then $\mathfrak{T}_{min}(L)$ can be computed in a modular way from any rational expression defining L.*

Proof. We prove this result by induction on the star height h of a rational expression defining L. If $h = 0$ then L is a finite language and $\mathfrak{T}_{min}(L)$ can be computed from the definition of $\mathfrak{T}$.

Assume that the result is true for rational expressions of star height at most equal to $h-1$ and let $L \subseteq \mathcal{A}^+$ be a recognizable language given by a rational expression of star height h. By definition of the star height, L is a union of languages defined by rational expressions of star height of at most $h-1$ and languages in the form $R_0 L_1^* R_1 L_2^* R_2 \cdots R_{k-1} L_k^* R_k$ where R_i, $0 \leq i \leq k$ and L_i, $1 \leq i \leq k$ are defined by rational expressions of star height of at most $h-1$. Using the induction hypothesis, Lemma 6 and Corollary 9, we deduce that $\mathfrak{T}_{min}(L)$ can be computed from the speedup of languages defined by rational expressions of star height at most equal to $h-1$ and the theorem is proved. □

From Proposition 5 we deduce immediately:

Corollary 11. *Let $T \subseteq \mathbb{M}^+(X, D)$ be a non-empty recognizable trace language. Then $\mathfrak{T}_{min}(T)$ can be computed in an effective way.*

Example 1 (continued): Let $L \subseteq X^*$ be the recognizable word language defined by the rational expression: $L = b[b(ac+ca)]^*(ac+ca)([b(aacc+acac+acca+caac+caca+ccaa)]^*b(ac+ca)b)^*b(ac+ca)a$.

Since L verifies $L = \varphi^{-1}(\varphi(L))$, $T = \varphi(L)$ is a recognizable trace language. It is not difficult to verify that

$$R = \mathcal{X}^{-1}(L) \cap FNF = C_b(C_b C_{a,c})^* C_{a,c}((C_b C_{a,c} C_{a,c})^* C_b C_{a,c} C_b)^* C_b C_{a,c} C_a$$

From Corollary 9 we get:

$$\mathfrak{T}_{min}(R) = \min\{\mathfrak{T}(C_b C_{a,c} C_b C_{a,c} C_a), \mathfrak{T}(C_b C_{a,c}), \mathfrak{T}_{min}((C_b C_{a,c} C_{a,c})^* C_b C_{a,c} C_b)\}$$

and hence, from Lemma 8 (or Corollary 9)

$$\begin{aligned}\mathfrak{T}_{min}(R) &= \min\{\mathfrak{T}(C_bC_{a,c}C_bC_{a,c}C_a), \mathfrak{T}(C_bC_{a,c}), \mathfrak{T}(C_bC_{a,c}C_{a,c}), \mathfrak{T}(C_bC_{a,c}C_b)\} \\ &= \min\{\tfrac{7}{5}, \tfrac{3}{2}, \tfrac{5}{3}, \tfrac{4}{3}\} \\ &= \tfrac{4}{3}\end{aligned}$$

Finally, from Proposition 1, we obtain $\mathfrak{T}_{min}(T) = \frac{4}{3}$.

5 Conclusion

We have defined naturally the speedup of a trace as the quotient of its sequential execution time, i.e. its length, by its parallel execution time. i.e. the number of steps in its Foata normal form. This definition has been generalized to trace languages.

Our main result consists in proving that this speedup can be computed in a modular way for any recognizable trace language. The algorithm of calculus has been successfully implemented [Cér92]. In this way, we can evaluate the gain between the sequential execution of a recognizable trace language and its fastest parallel execution.

In future work, we will investigate more complex definitions of the speedup of trace language. For instance, the probability to have a given trace in the langage could be an interesting parameter to take into account. In this case, the (modular) calculus of the speedup of a recognizable trace language remains an open problem.

References

[AR88] I.J. Aalbersberg and G. Rozenberg. Theory of traces. *Theoretical Computer Science*, 60:1–82, 1988.

[Ber79] J. Berstel. *Transductions and Context Free Languages*. Taubner, Stuttgart, 1979.

[Cér92] C. Cérin. Application de la théorie des traces à l'implantation et à la mesure d'algorithmes de distribution. Thèse, Université de Paris-Sud, 1992.

[CF69] P. Cartier and D. Foata. *Problèmes combinatoires de commutation et réarrangements*, volume 85 of *Lecture Notes in Mathematics*. Springer, Berlin-Heidelberg-New York, 1969.

[CP85] R. Cori and D. Perrin. Automates et commutations partielles. *R.A.I.R.O.-Informatique Théorique et Applications*, 19:21–32, 1985.

[Die90] V. Diekert. *Combinatorics on Traces*, volume 454 of *Lecture Notes in Computer Science*. Springer, Berlin-Heidelberg-New York, 1990.

[Maz77] A. Mazurkiewicz. Concurrent program schemes and their interpretations. DAIMI Rep. PB 78, Aarhus University, Aarhus, 1977.

[Maz87] A. Mazurkiewicz. Trace theory. In W. Brauer et al., editors, *Petri Nets, Applications and Relationship to other Models of Concurrency*, volume 255 of *Lecture Notes in Computer Science*, pages 279–324. Springer, Berlin-Heidelberg-New York, 1987.

[Och88] E. Ochmanski. On morphisms onf trace monoids. In *STACS'88*, volume 294 of *Lecture Notes in Computer Science*, pages 346–355. Springer, Berlin-Heidelberg-New York, 1988.

[Per89] D. Perrin. Partial commutations. In *Proceedings of the 16th International Colloquium on Automata, Languages and Programming (ICALP'89), Stresa (Italy) 1989*, volume 372 of *Lecture Notes in Computer Science*, pages 637–651. Springer, Berlin-Heidelberg-New York, 1989.

[Sah89] N. Saheb. Concurrency measure in commutation monoids. *Discrete Applied Mathematics*, 24:223–236, 1989.

May I Borrow Your Logic?*

Maura Cerioli[1] and José Meseguer[2]

[1] DISI–Dipartimento di Informatica e Scienze dell'Informazione, Università di Genova, Viale Benedetto XV, 16132 Genova, Italy, e-mail: cerioli@disi.unige.it
[2] SRI International, Menlo Park, CA 94025, and Center for the Study of Language and Information, Stanford University, Stanford, CA 94305
e-mail: meseguer@csl.sri.com

Abstract. It can be very advantageous to borrow key components of a logic for use in another logic. The advantages may be not only conceptual; due to the existence of software systems supporting mechanized reasoning in a given logic, it may be possible to reuse a system developed for one logic—for example, a theorem-prover—to obtain a new system for another. Translations between logics by appropriate mappings provide a first way of reusing tools of one logic in another. This paper generalizes this idea to the case where entire components—for example, the proof theory—of one of the logics involved may be completely missing, so that the appropriate mapping could not even be defined. The idea then is to borrow the missing components (as well as their associated tools if they exist) from a logic that has them in order to create the full-fledged logic and tools that we desire. The relevant structure is transported using maps that only involve a limited aspect of the two logics in question—for example, their model theory. The constructions accomplishing this kind of borrowing of logical structure are very general and simple. They only depend upon a few abstract properties that hold under very general conditions given a pair of categories linked by adjoint functors.

1 Introduction

The use of logic in computer science is undergoing vigorous growth. Since the applications are many, there are increasingly stronger interactions between the two fields that are having a profound impact on both of them. New logics are frequently been proposed, and new variants or adaptations of existing logics for new purposes are widespread.

This proliferation of logics—although certainly a sign of vitality and intellectual creativity—brings with it important conceptual challenges. In a sense, each logic is a different language and, as in the case of natural languages, there is often a

* Partially supported by Progetto Finalizzato Sistemi Informatici e Calcolo Parallelo of C.N.R. (Italy), MURST-40% Modelli e Specifiche di Sistemi Concorrenti, by the US Office of Naval Research under contracts N00014-90-C-0086 and N00014-92-C-0518, and by the Information Technology Promotion Agency, Japan, as a part of the R & D of Basic Technology for Future Industries "New Models for Software Architecture" sponsored by NEDO (New Energy and Industrial Technology Development Organization).

serious need to bridge the gap between different languages by means of appropriate translations, and the danger of serious confusion when translations are not correct. There is also a related need to understand the essential features shared by logics in general so that systematic methods can be developed to deal with these problems.

In computer science, the conceptual needs posed by the proliferation of logics were first addressed by Goguen and Burstall [8], who proposed their theory of ***institutions*** as a general framework for logics. The work on institutions has been further developed by their original proponents and by others [9, 10, 22, 23], and has influenced other notions proposed by different authors [15, 19, 7, 16, 13, 20, 5, 2]. Some of the notions proposed are closely related to institutions; however, in other cases the main intent is to substantially expand the primarily model-theoretic viewpoint provided by institutions to give an adequate treatment of proof-theoretic aspects such as entailment and proof structures.

Institutions arose out of work on the Clear specification language [3], in which the goal was to provide powerful modularity and parameterization mechanisms to structure and reuse formal specifications. Such reusability techniques have later been applied to a good number of specification and logical programming languages such as, for example, [12, 6, 21, 11, 17]. However, the need for reusability arises not only inside one logic—so that specifications or logical programs written in that logic can be reused—but also at the metalevel, in the sense that it can be greatly advantageous to reuse entire logics, or key components of such logics. The advantages may be not only conceptual, although of course this is important; due to the existence of software systems supporting mechanized reasoning in a given logic, it may be possible to reuse a system developed for one logic—for example, a theorem-prover—to obtain a new system for another.

Translations between logics by appropriate mappings—especially if they are *conservative* in the sense of [16]—may provide a first way of reusing tools of one logic in another, by translating the appropriate sentences or proofs and using the original tool on the translations. This paper generalizes this idea to the case where entire components—for example, the proof theory—of one of the logics involved may be completely missing, so that the appropriate mapping could not even be defined. The idea then is to borrow the missing components (as well as their associated tools if they exist) from a logic that has them in order to create, *ex nihilo* as it were, the full-fledged logic and tools that we desire. The relevant structure is transported using maps that only involve a limited aspect of the two logics in question—for example, their model theory.

The constructions accomplishing this kind of borrowing of logical structure are very general and simple. We show that they only depend upon a few abstract properties that hold under very general conditions given a pair of categories linked by adjoint functors. Therefore, the constructions capitalize on the fact that, as was shown in [16], the different components of a logic—entailment relation, model theory, and proof theory—are in a very precise technical sense *modular*, namely in that they can be added or deleted by means of constructions that are adjoint functors.

Specifically, we show that, given two institutions—i.e., the model-theoretic components of two logics—$\mathcal{I}$ and $\mathcal{I}'$ and a map of institutions $\mathcal{I} \longrightarrow \mathcal{I}'$, we can borrow the entailment relation or the proof theory (or both things) from $\mathcal{I}'$ if they exist to use them for $\mathcal{I}$, and that completeness, if it holds, is preserved by the borrowing.

Similarly, given two entailment systems $\mathcal{E}$ and $\mathcal{E}'$—i.e., the provability relations of two logics—and a map of entailment systems $\mathcal{E} \longrightarrow \mathcal{E}'$, we show that we can borrow the proof theory from $\mathcal{E}'$ if it exists to endow $\mathcal{E}$ with a proof theory. Finally, given two logics $\mathcal{L}$ and $\mathcal{L}'$—i.e., the provability relations and the model-theoretic components of two logics—and a map of logics $\mathcal{L} \longrightarrow \mathcal{L}'$, we show that we can borrow the proof theory from $\mathcal{L}'$ if it exists to endow $\mathcal{L}$ with a proof theory. Due to space limitations neither proofs nor examples are given; they can be found in [4].

2 General Logics

Since the significant examples of application of the general categorical construction are all largely based on the concepts of *institution* [8] and *general logic* [16], this section is devoted to recall some basic definitions and results from these theories. For a detailed discussion of these concepts see [10, 16].

Institutions cover the semantic aspects of a logical framework, providing formal counterparts for the notions of *signature*, *sentences*, *models* and *validity*, while entailment systems deal with the deductive part. Putting together an institution and a compatible entailment system, i.e. an entailment system that is sound w.r.t. the institution, a *logic* is obtained, where tools to deal with provability and with model theoretic aspects are both at hand. But, since entailment systems disregard the computational aspects, the concept of (structured) proofs is not formalized; thus *proof calculi* are introduced to cover also this feature.

Definition 1. [16] An *entailment system* $\mathcal{E} = (\mathbf{Sign}, Sen, \vdash)$ consists of a category **Sign** of *signatures*, a functor $Sen\colon \mathbf{Sign} \to \mathbf{Set}$ giving the set of *sentences* over a given signature, and for each Σ in **Sign** a binary relation $\vdash_\Sigma \subseteq \wp(Sen(\Sigma)) \times Sen(\Sigma)$ called *Σ-entailment* satisfying the following properties:

reflexivity: for any $\phi \in Sen(\Sigma)$, $\{\phi\} \vdash_\Sigma \phi$;
monotonicity: if $\Gamma \vdash_\Sigma \phi$ and $\Gamma \subseteq \Gamma'$, then $\Gamma' \vdash_\Sigma \phi$;
transitivity: if $\Gamma \vdash_\Sigma \phi_i$ for all $i \in I$ and $\Gamma \cup \{\phi_i \mid i \in I\} \vdash_\Sigma \psi$, then $\Gamma \vdash_\Sigma \psi$;
$\vdash$-translation: if $\Gamma \vdash_\Sigma \phi$, then for any $\sigma : \Sigma \to \Sigma'$ in **Sign**, $Sen(\sigma)(\Gamma) \vdash_{\Sigma'} Sen(\sigma)(\phi)$. □

Definition 2. [8] An *institution* $\mathcal{I} = (\mathbf{Sign}, Sen, Mod, \models)$ consists of a category **Sign** of *signatures*, a functor $Sen\colon \mathbf{Sign} \to \mathbf{Set}$ giving the set of *sentences* over a given signature, a functor $Mod\colon \mathbf{Sign}^{op} \to \mathbf{Cat}$ giving the category of *models* of a given signature, and for each Σ in **Sign** a *satisfaction* relation $\models_\Sigma \subseteq |Mod(\Sigma)| \times Sen(\Sigma)$, such that for each morphism $\sigma\colon \Sigma \to \Sigma'$ in **Sign**, the *Satisfaction Condition*

$$M' \models_{\Sigma'} Sen(\sigma)(\xi) \iff Mod(\sigma)(M') \models_\Sigma \xi$$

holds for each $M' \in |Mod(\Sigma')|$ and each $\xi \in Sen(\Sigma)$. □

Given an entailment system (respectively, an institution) its category $\mathbf{Th}_0$ of *theories* has as objects pairs $T = (\Sigma, \Gamma)$ with Σ a signature and Γ a set of sentences on Σ, and as morphisms $\sigma\colon (\Sigma, \Gamma) \to (\Sigma', \Gamma')$ the signature morphisms $\sigma\colon \Sigma \to \Sigma'$ s.t. $Sen(\sigma)(\Gamma) \subseteq \Gamma'$. Since a set Γ of sentences on Σ determines the full subcategory of models that satisfy all the sentences Γ, it is easy to show that the functor *Mod* extends to a functor $Mod\colon \mathbf{Th_0}^{op} \to \mathbf{Cat}$. Similarly, by assigning to each theory $T = (\Sigma, \Gamma)$ the sentences on its signature we can extend the functor $Sen\colon \mathbf{Sign} \to \mathbf{Set}$ to a functor $Sen\colon \mathbf{Th_0} \to \mathbf{Set}$.

Definition 3. [16] A *logic* is a 5-tuple $\mathcal{L} = (\mathbf{Sign}, Sen, Mod, \models, \vdash)$ such that $(\mathbf{Sign}, Sen, \vdash)$ is an entailment system, $(\mathbf{Sign}, Sen, Mod, \models)$ is an institution, and for each $\Sigma \in |\mathbf{Sign}|$, $\Gamma \subseteq Sen(\Sigma)$ and $\phi \in Sen(\Sigma)$, $\Gamma \vdash_\Sigma \phi \Rightarrow \Gamma \vdash_{\models_\Sigma} \phi$, where $\Gamma \vdash_{\models_\Sigma} \phi$ iff ($M \models_\Sigma \gamma$ for all $\gamma \in \Gamma$) implies $M \models_\Sigma \phi$. □

Definition 4. [16] A *proof calculus* is a 6-tuple $\mathcal{P} = (\mathbf{Sign}, Sen, \vdash, P, \mathrm{Pr}, \pi)$ such that: $(\mathbf{Sign}, Sen, \vdash)$ is an entailment system, $P\colon \mathbf{Th_0} \to \mathbf{Struct}_\mathcal{P}$ is a functor (for each theory T, the object $P(T) \in \mathbf{Struct}_\mathcal{P}$ is called its *proof-theoretic structure*), $\mathrm{Pr}\colon \mathbf{Struct}_\mathcal{P} \to \mathbf{Set}$ is a functor (for each theory T, the set $\mathrm{Pr}(P(T))$ is called its *set of proofs*, and *proofs* denotes the composite functor $\mathrm{Pr} \cdot P\colon \mathbf{Th_0} \to \mathbf{Set}$), $\pi : proofs \Rightarrow Sen$ is a natural transformation, such that for each theory $T = (\Sigma, \Gamma)$, the image of $\pi_T\colon proofs(T) \to Sen(T)$ is the set $\Gamma^\bullet$ of all sentences ϕ s.t. $\Gamma \vdash \phi$. □

Definition 5. [16] The 8-tuple $\mathcal{S} = (\mathbf{Sign}, Sen, Mod, \vdash, \models, P, \mathrm{Pr}, \pi)$ is a *logical system* if $(\mathbf{Sign}, Sen, Mod, \models, \vdash)$ is a logic, and $(\mathbf{Sign}, Sen, \vdash, P, \mathrm{Pr}, \pi)$ is a proof calculus. □

In [16] maps of entailment systems, institutions, logics, proof calculi, and logical systems are defined and are illustrated with examples; here the definitions are presented in summarized form.

Definition 6. [16] For $\mathcal{E} = (\mathbf{Sign}, Sen, \vdash)$ and $\mathcal{E}' = (\mathbf{Sign}', Sen', \vdash')$ entailment systems, a *map of entailment systems* $(\Phi, \alpha)\colon \mathcal{E} \to \mathcal{E}'$ consists of a natural transformation $\alpha\colon Sen \Rightarrow Sen' \cdot \Phi$ and an α-sensible functor[3] $\Phi\colon \mathbf{Th_0} \to \mathbf{Th'_0}$ satisfying the following property:

$$\Gamma \vdash_\Sigma \phi \;\Rightarrow\; \Gamma' \cup \alpha_\Sigma(\Gamma) \vdash'_{\Sigma'} \alpha_\Sigma(\phi),$$

where, by convention, $(\Sigma', \Gamma') = \Phi(\Sigma, \Gamma)$.

For $\mathcal{I} = (\mathbf{Sign}, Sen, Mod, \models)$ and $\mathcal{I}' = (\mathbf{Sign}', Sen', Mod', \models')$ institutions, a *map of institutions* $(\Phi, \alpha, \beta)\colon \mathcal{I} \to \mathcal{I}'$ consists of an α-sensible functor $\Phi\colon \mathbf{Th_0} \to \mathbf{Th'_0}$, and natural transformations $\alpha : Sen \Rightarrow Sen' \cdot \Phi$, and $\beta : Mod' \cdot \Phi \Rightarrow Mod$ such that for each $\Sigma \in |\mathbf{Sign}|$, each $\phi \in Sen(\Sigma)$, and each $M' \in |Mod'(\Phi(\Sigma, \emptyset))|$ the following property is satisfied:

$$M' \models'_{\Sigma'} \alpha_\Sigma(\phi) \;\Leftrightarrow\; \beta_{(\Sigma,\emptyset)}(M') \models_\Sigma \phi,$$

where Σ' is the signature of the theory $\Phi(\Sigma, \emptyset)$.

Given logics $\mathcal{L}$ and $\mathcal{L}'$, a *map of logics* $(\Phi, \alpha, \beta)\colon \mathcal{L} \to \mathcal{L}'$ is a map $(\Phi, \alpha, \beta) : inst(\mathcal{L}) \to inst(\mathcal{L}')$ of the underlying institutions s.t., in addition, $(\Phi, \alpha) : ent(\mathcal{L}) \to ent(\mathcal{L}')$ is a map of the underlying entailment systems.

Given proof calculi $\mathcal{P}$ and $\mathcal{P}'$, a *map of proof calculi* $(\Phi, \alpha, \gamma)\colon \mathcal{P} \to \mathcal{P}'$ consists of a map $(\Phi, \alpha)\colon ent(\mathcal{P}) \to ent(\mathcal{P}')$ of the underlying entailment systems together with a natural transformation $\gamma\colon proofs \Rightarrow proofs' \cdot \Phi$ such that $\pi'_\Phi \circ \gamma = \alpha \circ \pi$.

[3] We refer to [16] for the detailed definition of α-sensible functors.

Given logical systems $\mathcal{S}$ and $\mathcal{S}'$, a *map of logical systems* $(\Phi, \alpha, \beta, \gamma): \mathcal{S} \to \mathcal{S}'$ consists of a map of the underlying logics $(\Phi, \alpha, \beta): log(\mathcal{S}) \to log(\mathcal{S}')$ and a map of the underlying proof calculi $(\Phi, \alpha, \gamma): pcalc(\mathcal{S}) \to pcalc(\mathcal{S}')$. □

The relationships among the categories $\underline{Ent}$ of entailment systems, $\underline{Inst}$ of institutions, $\underline{Log}$ of logics, $\underline{PCalc}$ of proof calculi, and $\underline{LogSys}$ of logical systems can be illustrated by the following diagram, where all the arrows depicted are forgetful functors that "throw away" appropriate components of a logic. For example, the functor *inst* maps a logic $\mathcal{L} = (\mathbf{Sign}, Sen, Mod, \models, \vdash)$ to the institution $(\mathbf{Sign}, Sen, Mod, \models)$.

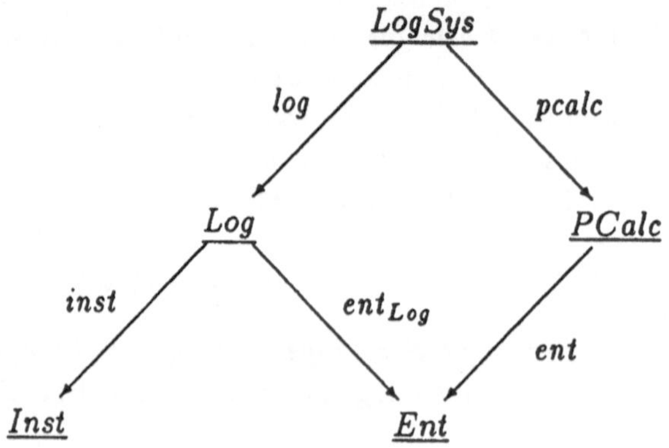

Proposition 7. *[16]*

1. *The functor* $(\_)^+: \underline{Inst} \to \underline{Log}$, *mapping* $\mathcal{I} = (\mathbf{Sign}, Sen, Mod, \models)$ *to* $\mathcal{I}^+ = (\mathbf{Sign}, Sen, Mod, \models, \vdash_{\models})$, *where* $\Gamma \vdash_{\models} \phi$ *iff* ($M \models_\Sigma \gamma$ *for all* $\gamma \in \Gamma$) *implies* $M \models_\Sigma \phi$, *and* ρ *to* $\rho^+ = \rho$ *for any map* ρ, *is the right adjoint of the forgetful functor inst:* $\underline{Log} \to \underline{Inst}$. *The unit of the adjunction for a logic* $\mathcal{L} = (\mathbf{Sign}, Sen, Mod, \models, \vdash)$ *is the map* $(Id_{\mathbf{Sign}}, Id_{Sen}, Id_{Mod})$.
2. *The functor* $(\_)^{\mathsf{l}}: \underline{Ent} \to \underline{PCalc}$, *defined by* $\mathcal{E}^{\mathsf{l}} = (\mathbf{Sign}, Sen, \vdash, thm, Id_{\mathbf{Set}}, j)$ *(where thm is the functor sending each theory to the set of its theorems, described in Section 2.2 of [16], and j is the natural subfunctor inclusion* $thm \subseteq Sen$*) and mapping* (Φ, α) *to* $(\Phi, \alpha, \alpha_{|thm})$, *where* $\alpha_{|thm}$ *is the restriction of* α *to the theorems of the domain, is the right adjoint of the forgetful functor ent:* $\underline{PCalc} \to \underline{Ent}$. *The unit of the adjunction for a proof calculus* $\mathcal{P} = (\mathbf{Sign}, Sen, \vdash, P, \mathrm{Pr}, \pi)$ *is the map* $(Id_{\mathbf{Sign}}, Id_{Sen}, \pi)$, *where* π *is now viewed as a natural transformation* $\pi: proofs \Rightarrow thm$.
3. *The functor* $(\_)^\circ: \underline{Log} \to \underline{LogSys}$, *mapping* $\mathcal{L} = (\mathbf{Sign}, Sen, Mod, \models, \vdash)$ *to* $\mathcal{L}^\circ = (\mathbf{Sign}, Sen, Mod, \vdash, \models, thm, Id_{\mathbf{Set}}, j)$, *and* (Φ, α, β) *to* $(\Phi, \alpha, \beta, \alpha_{|thm})$, *where* $\alpha_{|thm}$ *is the restriction of* α *to the theorems of the domain, is the right adjoint of the forgetful functor log:* $\underline{LogSys} \to \underline{Log}$. *The unit of the adjunction for a logical system* $\mathcal{S} = (\mathbf{Sign}, Sen, Mod, \vdash, \models, P, \mathrm{Pr}, \pi)$ *is the map* $(Id_{\mathbf{Sign}}, Id_{Sen}, Id_{Mod}, \pi)$, *where* π *is now viewed as a natural transformation* $\pi: proofs \Rightarrow thm$.
4. *The functor* $(\_)^\flat: \underline{Inst} \to \underline{LogSys}$ *mapping* $\mathcal{I} = (\mathbf{Sign}, Sen, Mod, \models)$ *to* $\mathcal{I}^\flat = (\mathbf{Sign}, Sen, Mod, \vdash_{\models}, thm, Id_{\mathbf{Set}}, j)$, *and* (Φ, α, β) *to* $(\Phi, \alpha, \beta, \alpha_{|thm})$, *is the right adjoint of the forgetful functor inst:* $\underline{LogSys} \to \underline{Inst}$. *The unit of the adjunction for a logical system* $\mathcal{S} = (\mathbf{Sign}, Sen, Mod, \vdash, \models, P, \mathrm{Pr}, \pi)$ *is the map* $(Id_{\mathbf{Sign}}, Id_{Sen}, Id_{Mod}, \pi)$. □

3 Transporting Structures Across Categories

The transportation of structure that we study will involve two categories of logical structures, **C** and **D**, and a couple of functors $U: \mathbf{D} \to \mathbf{C}$ and $R: \mathbf{C} \to \mathbf{D}$ with R right adjoint to U. In the applications that we will consider U, in spite of being a left adjoint, will have the flavor of a forgetful functor. The basic construction applies to arbitrary categories **C** and **D** with functors U and R as above, and consists in the process of transporting to an object $C \in |\mathbf{C}|$ the structure of an object $D \in |\mathbf{D}|$ via a map $c: C \to U(D)$. The transported structure is enjoyed by an object $\widetilde{C} \in |\mathbf{D}|$ which, roughly speaking, corresponds to C plus the features of D translated by c.

Theorem 8. *Let $R: \mathbf{C} \to \mathbf{D}$ be right adjoint to $U: \mathbf{D} \to \mathbf{C}$, with $\eta: 1_{\mathbf{D}} \to R \cdot U$ the unit of the adjunction, and let us denote by $\mathbf{D}'$ the comma category $\mathbf{D} \downarrow 1_{\mathbf{D}}$ (see e.g. [14]) and by $\mathbf{C}'$ the comma category $\mathbf{C} \downarrow U$. If for each object $C \in |\mathbf{C}|$, each object $D \in |\mathbf{D}|$, and each arrow $c \in \mathbf{C}(C, U(D))$ the pullback of $R(c)$ and η_D exists, then there is a functor $R': \mathbf{C}' \to \mathbf{D}'$, defined as follows:*

- *for each object $(c: C \to U(D), D) \in |\mathbf{C}'|$, the object $R'(c, D) = (\tilde{c}: \widetilde{C} \to D)$ is defined by the following pullback:*

$$\begin{array}{ccc} R(C) & \xrightarrow{R(c)} & R \cdot U(D) \\ \uparrow{\scriptstyle j_{c,D}} & & \uparrow{\scriptstyle \eta_D} \\ \widetilde{C} & \xrightarrow[\tilde{c}]{} & D \end{array}$$

- *for each arrow $(g, g') \in \mathbf{C}'((c_1: C_1 \to U(D_1), D_1), (c_2: C_2 \to U(D_2), D_2))$ the arrow $R'(g, g') = (g^!, g')$, where $g^!$ is the unique arrow $g^!: \widetilde{C_1} \to \widetilde{C_2}$ such that:*
  1. *$j_{c_2,D_2} \cdot g^! = R(g) \cdot j_{c_1,D_1}$;*
  2. *$\widetilde{c_2} \cdot g^! = g' \cdot \widetilde{c_1}$.*

If such a functor R' exists, then **C** *is said to* admit generalization under R and U. □

Theorem 9. *Under the same hypotheses as in Thm. 8, R' is the right adjoint of the functor $U': \mathbf{D}' \to \mathbf{C}'$, defined as follows:*

- *$U'(d: D \to D') = (U(d): U(D) \to U(D'), D')$ for each object $(d: D \to D') \in |\mathbf{D}'|$;*
- *$U'(f, f') = (U(f), f')$ for each arrow $(f, f') \in \mathbf{D}'(d_1, d_2)$.* □

Proposition 10. *Let $R_1: \mathbf{C} \to \mathbf{D}$ be right adjoint to $U_1: \mathbf{D} \to \mathbf{C}$, and let $R_2: \mathbf{D} \to \mathbf{E}$ be right adjoint to $U_2: \mathbf{E} \to \mathbf{D}$. If* **C** *admits generalization under R_1 and U_1, and* **D** *admits generalization under R_2 and U_2, then* **C** *admits generalization under $R_2 \cdot R_1$ and $U_1 \cdot U_2$.* □

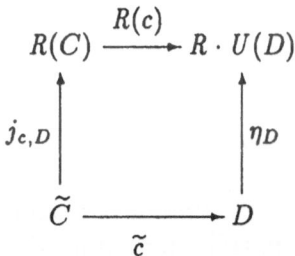

4 Endowing an Institution with an Entailment System

We apply here the general result to the adjunction between institutions and logics to borrow an entailment system from another logic using a map of institutions. This borrowing in fact preserves completeness. Although in general <u>*Log*</u> does not have pullbacks, they do exist in the particular case of pulling back the unit of the adjunction along the image under $(\_)^+$ of a map of institutions.

Lemma 11. *Let* $\mathcal{I} = (\mathbf{Sign}, Sen, Mod, \models)$ *be an institution,* $\mathcal{L}'$ *a logic, and* $\rho : \mathcal{I} \longrightarrow inst(\mathcal{L}')$ *a map of institutions. Denoting by* $\eta_{\mathcal{L}'}$ *the embedding of* $\mathcal{L}'$ *into* $(inst(\mathcal{L}'))^+$, *the following diagram, where* $\mathcal{L} = (\mathbf{Sign}, Sen, Mod, \models, \vdash)$ *and* $\vdash$ *is defined by:* $\Gamma \vdash_\Sigma \phi$ *iff both* $\alpha_\Sigma(\Gamma) \vdash'_{\Phi(\Sigma,\emptyset)} \alpha_\Sigma(\phi)$ *and* $\Gamma \models_\Sigma \phi$, *is a pullback.*

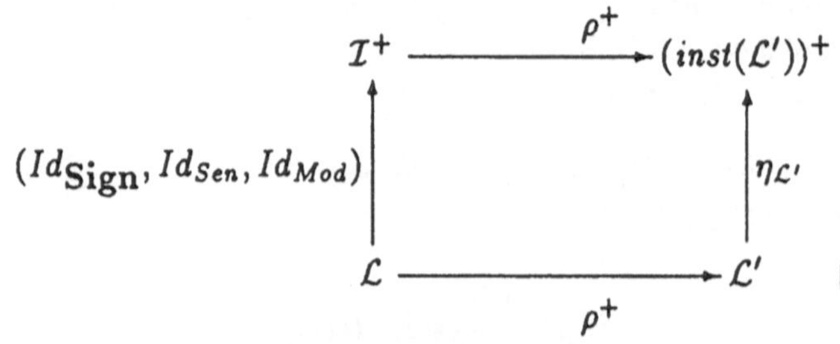

□

Proposition 12. *The category* <u>*Inst*</u> *admits generalization under* $(\_)^+$ *and the forgetful functor inst:* <u>*Log*</u> $\rightarrow$ <u>*Inst*</u>. □

As a direct application of this result one can obtain the construction of a many-sorted entailment system in [18], translating the Birkhoff system along a map of institutions implicitly defined; a more recent application, formulated in a style closer to ours, is the definition of an inference system for partial conditional higher-order types importing an entailment system for first-order conditional types in [1]. A key general feature of the construction is the preservation of completeness.

Proposition 13. *Using the notation of Lemma 11, if* $\mathcal{L}'$ *is complete, then* $\mathcal{L}$ *is complete, too.* □

5 Endowing an Entailment System with a Proof Calculus

Let us now apply the general result to the adjunction between entailment systems and proof calculi to show how an informative proof system can be added to an entailment system $\mathcal{E}$ provided we are given a proof calculus $\mathcal{P}$ and a translation of $\mathcal{E}$ into the entailment system underlying $\mathcal{P}$.

Lemma 14. *Given an entailment system* $\mathcal{E} = (\mathbf{Sign}, Sen, \vdash)$, *a proof calculus* $\mathcal{P}' = (\mathbf{Sign}', Sen', \vdash', P', Pr', \pi')$, *and a map of entailment systems* $(\Phi, \alpha) : \mathcal{E} \rightarrow ent(\mathcal{P}')$, *then the following diagram is a pullback*

$$
\begin{array}{ccc}
\mathcal{E}^{\mathsf{I}} & \xrightarrow{(\Phi,\alpha,\alpha_{|thm})} & (ent(\mathcal{P}'))^{\mathsf{I}} \\
\uparrow{\scriptstyle (Id_{\mathbf{Sign}}, Id_{Sen}, \pi)} & & \uparrow{\scriptstyle (Id_{\mathbf{Sign}'}, Id_{Sen'}, \pi')} \\
\mathcal{P} & \xrightarrow[(\Phi,\alpha,\gamma)]{} & \mathcal{P}'
\end{array}
$$

where $\mathcal{P} = (\mathbf{Sign}, Sen, \vdash, P, \mathrm{Pr}, \pi)$, *for* $P = proofs$ *and* $Pr = Id_{\mathbf{Set}}$, *and* π, γ, *proofs are defined by the following pullback square:*

$$
\begin{array}{ccc}
thm & \xrightarrow{\alpha_{|thm}} & thm' \cdot \Phi \\
\uparrow{\scriptstyle \pi} & & \uparrow{\scriptstyle \pi'_{\Phi}} \\
proofs & \xrightarrow[\gamma]{} & proofs' \cdot \Phi
\end{array}
\quad \square
$$

Proposition 15. *The category* Ent *admits generalization under the functors* $(\_)^{\mathsf{I}}$ *and ent.* □

6 Endowing a Logic or an Institution with a Logical System

Extending the constructions for proof calculi to logics, we can build a logical system on top of a logic by borrowing the missing proof calculus from another logical system. Using Prop. 10, we can compose this construction with the one enriching an institution with a logic, in order to get also a general construction enriching an institution with the models and the proof calculus of another logical system.

Lemma 16. *Given a logic* $\mathcal{L} = (\mathbf{Sign}, Sen, Mod, \models, \vdash)$, *a logical system* $\mathcal{S}' = (\mathbf{Sign}', Sen', Mod', \vdash', \models', P', Pr', \pi')$, *and a map of logics* $(\Phi, \alpha, \beta)\colon \mathcal{L} \to log(\mathcal{S}')$, *then the following diagram is a pullback*

$$
\begin{array}{ccc}
\mathcal{L}^{\circ} & \xrightarrow{(\Phi,\alpha,\beta,\alpha_{|thm})} & (log(\mathcal{S}'))^{\circ} \\
\uparrow{\scriptstyle (Id, Id, Id, \pi)} & & \uparrow{\scriptstyle (Id, Id, Id, \pi')} \\
\mathcal{S} & \xrightarrow[(\Phi,\alpha,\beta,\gamma)]{} & \mathcal{S}'
\end{array}
$$

where $\mathcal{S} = (\mathbf{Sign}, Sen, Mod, \vdash, \models, P, \mathrm{Pr}, \pi)$, *for* $P = proofs$ *and* $Pr = Id_{\mathbf{Set}}$, *and* π, γ, *proofs are defined as in Lemma 14.* □

Proposition 17. *The category* Log *admits generalization under* $(\_)^{\circ}$ *and log.* □

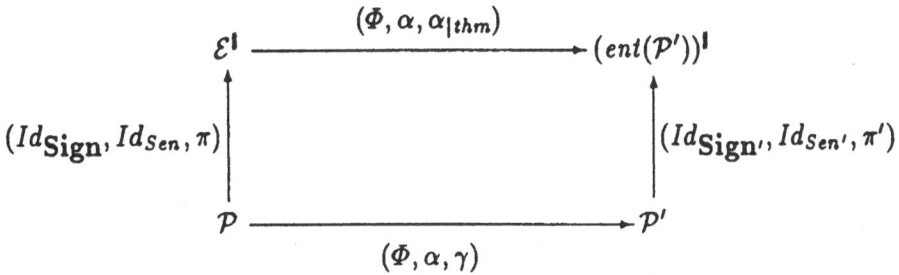

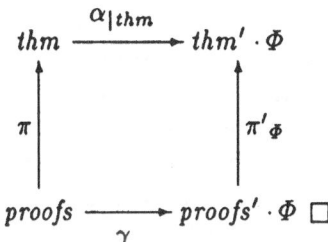

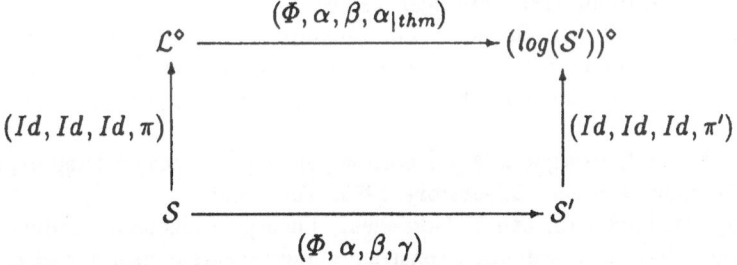

Proposition 18. *The category* $\underline{Inst}$ *admits generalization under the functors* $(\_)^\flat$ *and inst:* $\underline{LogSys} \rightarrow \underline{Inst}$. □

7 Concluding Remarks

We have presented a general result about transportation of structure across comma categories and have shown how it yields a general technique for borrowing logical components from one logic for use in another logic when a map between some basic component of both exists. In addition, we have shown that the constructions are particularly well behaved in that they preserve completeness when it exists.

Given the generality of the categorical techniques used, they could in principle be used not only for the notions of mapping between logics and logical components that we have considered, but also for other axiomatic notions of mapping and logical component proposed in the literature by other authors. Exploring the applicability or our techniques to those other notions seems a worthwhile research direction.

Another topic that deserves more research is studying what limits and colimits exist in categories of logics or in categories of logical components. We have made use of some positive results in this paper and are aware of some negative results as well. A systematic study would be very useful.

Acknowledgements

We are indebted to Narciso Martí-Oliet for his careful reading of the manuscript and for his many very helpful suggestions to improve the exposition. We cordially thank José Luiz Fiadeiro for his valuable suggestions that have also helped us improve the paper. We finally thank the anonymous referees for their positive suggestions and criticism. The diagrams have been realized using P. Taylor's macros.

References

1. E. Astesiano and M. Cerioli. Partial higher-order specifications. *Fundamenta Informaticae*, 16(2):101–126, 1992.
2. E. Astesiano and M. Cerioli. Relationships between logical frames. In M. Bidoit and C. Choppy, editors, *Recent Trends in Data Type Specification*, number 655 in LNCS, pages 126–143, Berlin, 1993. Springer Verlag.
3. R.M. Burstall and J.A. Goguen. The semantics of Clear, a specification language. In D. Bjørner, editor, *Proceedings of the 1979 Copenhagen Winter School on Abstract Software Specification*, number 86 in LNCS, pages 292–332, Berlin, 1980. Springer Verlag.
4. M. Cerioli and J. Meseguer. May I borrow your logic? Technical report, SRI International, Computer Science Laboratory, 1993. To appear.
5. H. Ehrig, M. Baldamus, and F. Cornelius. Theory of algebraic module specification including behavioural semantics, constraints and aspects of generalized morphisms. In *Proceedings of 2nd International Conference on Algebraic Methodology and Software Technology*, pages 101–125, Iowa City, Iowa, USA, 1991.
6. H. Ehrig and B. Mahr. *Fundamentals of Algebraic Specifications 1: Equations and Initial semantics*, volume 6 of *EATCS Monographs on Theoretical Computer Science*. Springer Verlag, Berlin, 1985.

7. J. Fiadeiro and A. Sernadas. Structuring theories on consequence. In D. Sannella and A. Tarlecki, editors, *Recent Trends in Data Type Specification*, number 332 in LNCS, pages 44–72, Berlin, 1987. Springer Verlag.
8. J.A. Goguen and R.M. Burstall. Introducing institutions. In E. Clarke and D. Kozen, editors, *Logics of Programming Workshop*, number 164 in LNCS, pages 221–255, Berlin, 1984. Springer Verlag.
9. J.A. Goguen and R.M. Burstall. A study in the foundations of programming methodology: Specifications, institutions, charters and parchments. In D. Pitt, S. Abramsky, A. Poigné, and D. Rydehard, editors, *Proceedings of Summer Workshop on Category Theory and Computer Programming*, number 240 in LNCS, pages 313–333, Berlin, 1986. Springer Verlag.
10. J.A. Goguen and R.M. Burstall. Institutions: Abstract model theory for specification and programming. *Journal of the ACM*, 39(1):95–146, 1992.
11. J.A. Goguen and J. Meseguer. Eqlog: Equality, types, and generic modules for logic programming. In D. DeGroot and G. Lindstrom, editors, *Logic Programming: Functions, Relations and Equations*, pages 295–363. Prentice-Hall, 1986.
12. J.A. Goguen, T. Winkler, J. Meseguer, K. Futatsugi, and J.-P. Jouannaud. Introducing OBJ. Technical Report SRI-CSL-92-03, SRI International, Computer Science Laboratory, 1992. To appear in J.A. Goguen, editor, *Applications of Algebraic Specification Using OBJ*, Cambridge University Press.
13. R. Harper, D. Sannella, and A. Tarlecki. Logic representation in LF. In D.H. Pitt et al., editor, *Category Theory and Computer Science*, number 389 in LNCS, pages 250–272. Springer Verlag, 1989.
14. S. MacLane. *Categories for the Working Mathematician*. Springer Verlag, 1971.
15. B. Mayoh. Galleries and institutions. Technical Report DAIMI PB - 191, Aarhus University, 1985.
16. J. Meseguer. General logics. In *Logic Colloquium '87*, pages 275–329, Amsterdam, 1989. North Holland.
17. J. Meseguer. A logical theory of concurrent objects and its realization in the Maude language. SRI Technical Report SRI-CSL-92-08, July 1992. To appear in G. Agha, P. Wegner, and A. Yonezawa, editors, *Research Directions in Object-Based Concurrency*, MIT Press, 1993.
18. P. Padawitz and M. Wirsing. Completeness of many-sorted equational logic revisited. *Bulletin EATCS*, (24), 1984.
19. A. Poigné. Foundations are rich institutions, but institutions are poor foundations. In H. Ehrig, H. Herrlich, Kreowski H.-J., and G. Preuß, editors, *Categorical Methods in Computer Science*, number 393 in LNCS, pages 82–101, Berlin, 1989. Springer Verlag.
20. A. Salibra and G. Scollo. A soft stairway to institutions. In *Recent Trends in Data Type Specification*, number 655 in LNCS, pages 310–329, Berlin, 1992. Springer Verlag.
21. D. Sannella and A. Tarlecki. On observational equivalence and specifications. *Journal of Comp. and Sys. Sciences*, 34:150–178, 1987.
22. A. Tarlecki. Free constructions in algebraic institutions. In M.P. Chytil and V. Koubek, editors, *Proceedings of Mathematical Foundation of Computer Science '84*, number 176 in LNCS, pages 526–534, Berlin, 1984. Springer Verlag.
23. A. Tarlecki. On the existence of free models in abstract algebraic institutions. *Theoretical Computer Science*, 37(3):269–304, 1985.

Approximate and Exact Deterministic Parallel Selection*

Shiva Chaudhuri,[1] Torben Hagerup,[1] and Rajeev Raman[2]

[1] Max-Planck-Institut für Informatik, Im Stadtwald, W-6600 Germany
[2] UMIACS, University of Maryland, College Park, MD 20742

Abstract. The selection problem of size n is, given a set of n elements drawn from an ordered universe and an integer r with $1 \leq r \leq n$, to identify the rth smallest element in the set. We study approximate and exact selection on deterministic concurrent-read concurrent-write parallel RAMs, where approximate selection with relative accuracy $\lambda > 0$ asks for any element whose true rank differs from r by at most λn. Our main results are: (1) For all $t \geq (\log\log n)^4$, approximate selection problems of size n can be solved in $O(t)$ time with optimal speedup with relative accuracy $2^{-t/(\log\log n)^4}$; no deterministic PRAM algorithm for approximate selection with a running time below $\Theta(\log n/\log\log n)$ was previously known. (2) Exact selection problems of size n can be solved in $O(\log n/\log\log n)$ time with $O(n\log\log n/\log n)$ processors. This running time is the best possible (using only a polynomial number of processors), and the number of processors is optimal for the given running time (optimal speedup); the best previous algorithm achieves optimal speedup with a running time of $O(\log n\log^* n/\log\log n)$.

1 Introduction

Selecting the element of prescribed rank from an ordered (but not sorted) set is an important and well-studied problem. Blum *et al.* [6] showed that selection from a set of size n can be performed in linear time $O(n)$ sequentially. Considerable research has gone into determining the parallel complexity of selection. Valiant [20] introduced the parallel comparison-tree (PCT) model and showed that any deterministic algorithm in the PCT model for finding the maximum of n elements using p processors requires $\Omega(n/p+\log(\log n/\log(1+p/n)))$ time. Since finding the maximum is a special case of selection, this lower bound applies to selection in general as well. Azar and Pippenger [4], building on the work of Ajtai *et al.* [1], gave a matching upper bound for the PCT model. Reischuk [19] showed that in the randomized PCT model, selection can be done in constant expected time using a linear number of processors, which is clearly optimal. The problem of selection in the PCT model has therefore been completely solved. The PCT model counts only comparisons, however, while processing of any other kind is considered free. For this reason, lower bounds obtained for the PCT model apply to all parallel comparison-based algorithms, but upper bounds do not carry over to other, more realistic, models of parallel computation.

In the PRAM model of computation, the upper bounds for the PCT model demonstrably cannot be matched. It follows as a corollary to the lower bound of

* Supported by the ESPRIT Basic Research Actions Program of the EC under contract No. 7141 (project ALCOM II). Authors' email addresses: shiva@mpi-sb.mpg.de, torben@mpi-sb.mpg.de, raman@umiacs.umd.edu.

Beame and Håstad [5] that any (randomized) algorithm that selects the rth smallest among n elements on a p-processor CRCW PRAM has an (expected) running time of $\Omega(\log r/\log\log p)$. In particular, finding the median requires $\Omega(\log n/\log\log n)$ time using any polynomial number of processors. It is not difficult to solve selection problems of size n in (the best possible) time $\Theta(\log n/\log\log n)$ on the CRCW PRAM, but straightforward algorithms for this task use a large number of processors. An important design goal is to get by with as few *operations* as possible, where, as usual, the number of operations executed by a parallel algorithm is defined as the product of the number of processors used and the number of time steps needed by the algorithm. The obvious sequential simulation of any parallel computation shows that the number of operations executed by a parallel algorithm for a given problem is always $\Omega(T)$, where T is the sequential complexity of the problem. A parallel algorithm that uses only $O(T)$ operations is said to have *optimal speedup* or to be *work-optimal*, because it employs the available processors in the most efficient manner possible (up to a constant factor). The development of work-optimal algorithms is one of the most important goals of current research in parallel computation.

In the special case of selection, the result of Blum *et al.* [6] implies that a work-optimal algorithm is one that executes $O(n)$ operations. Cole [9] gave a work-optimal deterministic CRCW PRAM algorithm for selection with a running time of $O(\log n \log^* n/\log\log n)$, and Dietz and Raman [11] recently described a work-optimal algorithm that selects the rth smallest among n elements in $O(\log\log n + \log r/\log\log n)$ time if $1 \le r \le n^{1/3}$, which is the fastest possible for this range of r. The problem of discovering a deterministic CRCW PRAM algorithm for general selection that combines work-optimality with the fastest running time of $O(\log n/\log\log n)$ has remained unsolved, and no progress was made on this front since the publication of Cole's paper.

Attempts have been made to circumvent the lower bounds mentioned above by replacing exact selection by approximate selection. Here, in addition to a target rank $r \in \{1,\dots,n\}$, an *accuracy parameter* $\lambda > 0$ is specified, and the task is to select an element whose rank is guaranteed to lie between $r - \lambda n$ and $r + \lambda n$, which we call *λ-selection for rank r*. Upper and lower bounds for the complexity of approximate selection in the PCT model were given by Alon and Azar [3]. In the PRAM setting, the lower bound of Beame and Håstad does not apply directly to approximate selection, although it can be used to place bounds on the accuracy obtainable with a given amount of resources (i.e., processors and time). Hagerup [13], extending a result of Goodrich [12], showed that approximate selection problems of size n can be solved in constant time with high probability on an n-processor CRCW PRAM for $\lambda = 1/(\log n)^{O(1)}$, which is the best possible accuracy for the stated time and processor bounds. On the other hand, no deterministic PRAM algorithms for approximate selection were previously known.

In this paper we describe deterministic CRCW PRAM algorithms for the problems of approximate and exact selection. Our main result (Corollary 12) is that for all $t \ge (\log\log n)^4$, approximate selection problems of size n can be solved with optimal speedup with relative accuracy $2^{-t/(\log\log n)^4}$. The minimum running time is hence $(\log\log n)^4$, but allowing more time yields a better accuracy, a tradeoff that has been observed before [15, 14, 16]. As an easy corollary of the main result, we derive a work-optimal algorithm for exact selection with a running time of $O(\log n/\log\log n)$ (Theorem 13), thereby solving the open problem left by Cole's paper.

2 Brute-Force Approximate Selection

Without loss of generality we can always assume that the input elements presented to a selection algorithm are distinct. Given an ordered set A, i.e., a subset of an ordered universe U, and an element $x \in U$, the *rank* of x in A, $rank_A(x)$, is defined as $|\{y \in A : y \leq x\}|$, and if A is nonempty, the *relative rank* of x in A, $\rho_A(x)$, is defined as $rank_A(x)/|A|$.

If a sufficient number of processors is available, a selection problem can be solved simply by independently computing the rank of each element and returning the unique element whose rank has the desired value. This reduces selection to counting (the number of smaller elements). Approximate selection similarly reduces to approximate counting. The following result, a special case of a theorem due to Hagerup [14, Theorem 6.1], deals with the latter problem.

Lemma 1. *Let $n, t \geq 4$ be given integers with $t \geq (\log\log n)^3$ and take $\lambda = 2^{-t\log\log\log n/(\log\log n)^3}$. Then, given n bits $b_1, \ldots, b_n$, an integer r with $\sum_{i=1}^n b_i \leq r \leq (1+\lambda)\sum_{i=1}^n b_i$ can be computed using $O(t)$ time and $O(n)$ operations.*

Lemma 2. *Let $n, t \geq 4$ be given integers with $t \geq (\log\log n)^3$ and take $\lambda = 2^{-t\log\log\log n/(\log\log n)^3}$. Then, given an ordered set A of size n and an integer r with $1 \leq r \leq n$, an element $x \in A$ with $(1-\lambda)r \leq rank_A(x) \leq (1+\lambda)r$ can be computed using $O(t)$ time and $O(n^2)$ operations.*

Proof. Let $A = \{x_1, \ldots, x_n\}$, compare x_i and x_j for all $i, j \in \{1, \ldots, n\}$ and express the result in the form of an $n \times n$ boolean matrix. Then apply the algorithm of Lemma 1 independently to each row of this matrix to compute integers $r_1, \ldots, r_n$ such that $rank_A(x_i) \leq r_i \leq (1+\lambda)rank_A(x_i)$, for $i = 1, \ldots, n$, and return any x_i with $r \leq r_i \leq (1+\lambda)r$.

The relation $r \leq r_i \leq (1+\lambda)r$ is satisfied for at least one element x_i, namely the one of rank r. On the other hand, it clearly is not satisfied for any element of rank larger than $(1+\lambda)r$, and since $(1+\lambda)(1-\lambda)r < r$, it cannot be satisfied for any element of rank smaller than $(1-\lambda)r$ either. The algorithm is therefore correct. □

3 Sampling

This section studies certain properties of sampling. Some of the arguments parallel ones made in [16].

We allow an element of a set A to be included more than once in a sample of A, i.e., the sample is a *multisubset* of A. Following [16], we define a *ranking function* on a nonempty multiset B as any function $rank_B$ that maps each element x outside of B to the number of elements $y \in B$ with $y \leq x$, and that maps B itself bijectively to $\{1, \ldots, |B|\}$ in an order-preserving fashion, i.e., for all $x, y \in B$, if $x < y$, then $rank_B(x) < rank_B(y)$ (in other words, a total order is imposed on B). We also define a *relative ranking function* on B as any function of the form $rank_B/|B|$, where $rank_B$ is a ranking function on B.

Definition 3. Let $\lambda \geq 0$. A nonempty multisubset B of an ordered set A is a λ-*sample* of A if for some relative ranking function ρ_B on B (and therefore for all such functions), $|\rho_A(x) - \rho_B(x)| \leq \lambda$, for all $x \in B$.

Lemma 4. *Let n and m be positive integers with m dividing n, let A be an ordered set of size n and let $\lambda \geq 0$. Suppose that x_l is obtained by λ-selecting from A for rank $l \cdot n/m$, for $l = 1, \ldots, m$. Then the multiset $\{x_1, \ldots, x_m\}$ is a λ-sample of A.*

Proof. By construction, $|\rho_A(x_l) - \frac{l}{m}| \leq \lambda$, for $l = 1, \ldots, m$. Let $rank_B$ be an arbitrary ranking function on $B = \{x_1, \ldots, x_m\}$ and fix $i \in \{1, \ldots, m\}$. Taking $j = rank_B(x_i)$, our goal is to show that $|\rho_A(x_i) - \frac{j}{m}| \leq \lambda$. We distinguish between three cases.

Case 1: $i = j$. The claim follows immediately.

Case 2: $i > j$. Since $rank_B$ is a bijection from B to $\{1, \ldots, m\}$, there must be some $l \in \{1, \ldots, j\}$ with $rank_B(x_l) > j$. But then $x_l \geq x_i$ and hence $\rho_A(x_i) - \frac{j}{m} \leq \rho_A(x_l) - \frac{l}{m} \leq \lambda$. On the other hand, $\rho_A(x_i) - \frac{j}{m} \geq \rho_A(x_i) - \frac{i}{m} \geq -\lambda$. The claim follows.

Case 3: $i < j$. Symmetrical to Case 2. □

Lemma 5. *Let $\lambda \geq 0$ and let m be a positive integer. Furthermore let $A_1, \ldots, A_m$ be disjoint sets of a common size, and let $B_1, \ldots, B_m$ be λ-samples of a common size of $A_1, \ldots, A_m$, respectively. Then $B = \bigcup_{i=1}^m B_i$ is a $(\lambda + 1/|B_1|)$-sample of $A = \bigcup_{i=1}^m A_i$.*

Proof. Let $rank_B$ be an arbitrary ranking function on B, and let $rank_{B_i}$ be a ranking function on B_i consistent with $rank_B$, for $i = 1, \ldots, m$, i.e., for all $x, y \in B_i$, $rank_{B_i}(x) < rank_{B_i}(y)$ if and only if $rank_B(x) < rank_B(y)$.

Let $x \in \bigcup_{i=1}^m B_i$, fix $i \in \{1, \ldots, m\}$ and let $r = rank_{B_i}(x)$. If $r < |B_i|$, the rank of x in A_i is no larger than the rank in A_i of the element in B_i of rank $r+1$ in B_i, i.e., no larger than

$$(r+1)\frac{|A_i|}{|B_i|} + \lambda|A_i| = (r+1)\frac{|A|}{|B|} + \lambda|A_i|.$$

This relation obviously also holds if $r = |B_i|$. Summing it for $i = 1, \ldots, m$, we obtain

$$rank_A(x) \leq rank_B(x)\frac{|A|}{|B|} + m\frac{|A|}{|B|} + \lambda|A| = rank_B(x)\frac{|A|}{|B|} + \left(\lambda + \frac{1}{|B_1|}\right)|A|.$$

Arguing similarly, one easily shows that $rank_A(x) \geq rank_B(x)\frac{|A|}{|B|} - \lambda|A|$. □

Lemma 6. *Let $n, k, t \geq 4$ be given integers with $t \geq (\log\log n)^3$ such that $(8k)^2$ divides n and let $\lambda = 2^{-t\log\log\log n/(\log\log n)^3}$. Then, given an ordered set A of size n, a $(\lambda/(12\log\log n) + 1/(8k))$-sample of A of size $n/(8k)$ can be computed using $O(t)$ time and $O(k^3 n)$ operations.*

Proof. Partition A into $n/(8k)^2$ groups of $(8k)^2$ elements each and use the algorithm of Lemma 2 in parallel for each group and for each $i \in \{1, \ldots, 8k\}$ to $(\lambda/(12\log\log n))$-select for rank $i \cdot 8k$. By Lemma 4, this produces a $(\lambda/(12\log\log n))$-sample of each group. Return as the final sample the union of the group samples, which, by Lemma 5, is a $(\lambda/(12\log\log n) + 1/(8k))$-sample of A. □

Lemma 7. *Let $\lambda, \lambda' \geq 0$ and let A be an ordered set. Suppose that B is a λ-sample of A and that C is a λ'-sample of B, where a total order is imposed on B by means of a relative ranking function ρ_B on B (otherwise it makes no sense to speak of a λ'-sample of B). Then C is a $(\lambda + \lambda')$-sample of A.*

Proof. Let ρ_C be a relative ranking function on C consistent with the order on B imposed by ρ_B (i.e., for all $x, y \in C$, if $\rho_B(x) < \rho_B(y)$, then $\rho_C(x) < \rho_C(y)$). Then for all $x \in C$, $|\rho_A(x) - \rho_C(x)| \leq |\rho_A(x) - \rho_B(x)| + |\rho_B(x) - \rho_C(x)| \leq \lambda + \lambda'$. □

4 Work-Optimal Approximate Selection

Our algorithm has the same top-level structure as that of Cole [9]; the details differ. The algorithm consists of two loops, one nested within the other. The outer loop makes progress by eliminating more and more elements from consideration. Since the remaining elements are always those between a lower and an upper limit, similarly as in the case of binary search, the outer loop can be thought of as narrowing down the "search interval". When the search interval has become so small that an arbitrary element in the search interval is an acceptable answer, the algorithm returns such an element and stops.

Each iteration of the outer loop is called a *stage*. A stage works by computing a sample B of the input set A of the current stage (i.e., of the elements in the current search interval) with the property that, on the one hand, B is sufficiently small to make brute-force selection from B according to Lemma 2 feasible and, on the other hand, B represents A faithfully in the sense that the relative rank of an element in B is a good approximation of its relative rank in A. The goal of the algorithm at this point is to select an element from A whose rank in A is within m of r, for some integers m and r. Approximate selection from B is then used to obtain two elements x_L and x_H of A with $x_L \leq x_H$ whose ranks are as close as possible, but guaranteed to be on either side of r. An approximation of the number of elements smaller than x_L is then computed and subtracted from r, after which the next stage operates on the elements between x_L and x_H with an absolute accuracy slightly smaller than m (to counteract the inaccuracy incurred in counting the number of elements smaller than x_L).

Computing the sample B is the task of the inner loop, whose single iterations are called *rounds*. The inner loop works by gradually thinning out a sample, which initially is the full input set A and at the end is the final sample B returned to the outer loop. The size of B always lies between $|A|^{1/4}$ and $|A|^{1/2}$, while the quality of B as a sample of A will depend on the processor advantage (number of processors per element) initially available for the computation of B. The processor advantage derives from the progress of the outer loop: When the search interval has narrowed down to the point where it contains only n/k elements, for some $k \geq 1$, the processor advantage available to the inner loop will be essentially k, which allows B to be computed as a $1/k$-sample of A. This in turn allows the search interval to be narrowed down by a factor of roughly k, so that the initial processor advantage available to the inner loop of the next stage will be about k^2. The processor advantage therefore increases doubly-exponentially. If λ is the relative accuracy of the top-level selection, the narrowing process stops when the number of elements in the search interval has dropped to about λn; by the above, this happens after $O(\log\log(1/\lambda))$ stages.

The gradual thinning-out in each stage mentioned above is done by means of repeated subsampling in $O(\log\log n)$ rounds and is characterized in the following lemma.

Lemma 8. *Let n and k be given powers of 2 with $2^6 \leq k \leq n^{1/8}$, let $t \geq (\log\log n)^4$ be a given integer and take $\lambda = 2^{-t\log\log\log n/(\log\log n)^4}$. Then, given an ordered set A of size n, a $\max\{\lambda, 1/k\}$-sample B of A with $n^{1/4} \leq |B| \leq n^{1/2}$ can be computed using $O(t)$ time and $O(k^3 n)$ operations.*

Proof. Take $k_1 = k$ and $B_0 = A$ and execute a number of *rounds*. In Round i, for $i = 1, 2, \ldots$, do the following: Spending $O(t/\log\log n)$ time, use the algorithm of Lemma 6 to compute a $(\lambda/(12\log\log n) + 1/(8k_i))$-sample of B_{i-1} of size $|B_{i-1}|/(8k_i)$, call this sample B_i and let k_{i+1} be the largest power of 2 no larger than $\min\{k_i^{4/3}, \sqrt{|B_i|}/8\}$. Take B as the first sample of size $\leq n^{1/2}$ encountered in this process.

If for some i we have $|B_i| > n^{1/2}$, but $k_i^{4/3} \geq \sqrt{|B_i|}/8$, we will have $k_{i+1} \geq \sqrt{|B_i|}/16$ and hence $|B_{i+1}| \leq 2\sqrt{|B_i|} \leq n^{1/2}$, i.e., the $(i+1)$st round will be the last. Let T be the number of rounds executed. By the observation just made, $k_{i+1} \geq k_i^{8/6}/2 \geq k_i^{7/6}$, for $i = 1, \ldots, T-2$ (recall that $k \geq 2^6$), so that $k_i \geq k^{(7/6)^{i-1}}$, for $i = 1, \ldots, T-1$. In particular, $k^{(7/6)^{T-2}} \leq n$, which can be shown to imply that $T \leq 6\log\log n$.

Since $k_{i+1}^3|B_i| \leq k_i^3|B_{i-1}|/8$, for $i = 1, \ldots, T-1$, it is easy to see that the total number of operations executed is indeed $O(k^3 n)$. We have $k_{i+1} \geq 2k_i$, for $i = 1, \ldots, T-2$, and either $k_T \geq 2k_{T-1}$ as well, or $k_T \geq \sqrt{|B_{T-1}|}/16 \geq n^{1/4}/16 \geq k$. In either case $\sum_{i=1}^{T}(1/k_i) \leq 3/k$. By Lemma 7, B is a λ'-sample of A, where $\lambda' = \sum_{i=1}^{T}(\lambda/(12\log\log n) + 1/(8k_i)) \leq \lambda/2 + 1/(2k) \leq \max\{\lambda, 1/k\}$. Finally observe that $|B| = |B_T| \geq \sqrt{|B_{T-1}|} \geq n^{1/4}$. □

The following two lemmas describe the reduction of the "search interval" performed in a single stage of the outer iteration, whereas the outer iteration as a whole is treated in the proof of Theorem 11. To select for rank r with *absolute* accuracy m is to select an element whose rank differs from r by at most m.

Lemma 9. *Let $n, m, t, q \geq 4$ be given integers with $t \geq (\log\log n)^3$, take $\lambda = 2^{-q}$ and $\mu = 2^{-t\log\log\log n/(\log\log n)^3}$ and suppose that $m \geq \mu n$. Then, given an ordered set A of size n and a λ-sample B of A with $n^{1/4} \leq |B| \leq n^{1/2}$, $O(t)$ time and $O(n)$ operations suffice to reduce the problem of selection from A with absolute accuracy m to that of selection from a set of size at most $10\lambda n$ with absolute accuracy $m - \mu n$.*

Proof. Let $\mu' = 2^{-\lceil t\lceil\log\log\log n\rceil/\lfloor\log\log n\rfloor^3\rceil}$. We can assume without loss of generality that $\mu' n \geq 9(2n^{3/4} + 1)$, since otherwise $t = \Omega(\log n)$, in which case the problem can be solved using Cole's selection algorithm [9].

Assume that the task is to select from A for rank r with absolute accuracy m. Let $\lambda' = \max\{\lambda, \mu'/9\}$ and $r_L = \lfloor(r/n - 2\lambda')|B|\rfloor$. If $r_L < 1$, compute x_L as $\min A$. Otherwise use Lemma 2 to λ'-select from B for rank r_L and let x_L be the resulting element.

We will show that the rank of x_L in A is at most r. Since this is obvious if $x_L = \min A$, assume that this is not the case. Then the rank of x_L in B is at most $r_L + \lambda'|B| \leq (r/n - \lambda')|B|$. Since B is a λ'-sample of A, the rank of x_L in A therefore is at most $(r/n - \lambda' + \lambda')|A| = r$.

Similarly, let $r_H = \lceil(r/n + 2\lambda')|B|\rceil$ and obtain x_H by λ'-selecting from B for rank r_H, unless $r_H > |B|$, in which case we take $x_H = \max A$. The rank of x_H in A is at least r.

Next partition A into the sets $A_L = \{x \in A : x < x_L\}$, $A_M = \{x \in A : x_L \leq x \leq x_H\}$ and $A_H = \{x \in A : x > x_H\}$, i.e., mark each element in A with the set to which it belongs. The rank of x_L in B is at least $\lfloor(r/n - 2\lambda')|B|\rfloor - \lambda'|B| \geq (r/n - 3\lambda' - 1/|B|)|B|$, and the rank of x_L in A therefore is at least $(r/n - 4\lambda' - 1/|B|)|A| \geq$

$r - 4\lambda' n - n^{3/4}$. Similarly, the rank of x_H in A is at most $r + 4\lambda' n + n^{3/4}$, and hence $|A_M| \le 8\lambda' n + 2n^{3/4} + 1 \le 9\lambda' n$.

If $\lambda' = \lambda$, $|A_M| \le 9\lambda n$, and the approximate compaction algorithm of [14] can be used to store the elements of A_M in an array of size at most $10\lambda n$, unused cells of which are filled with dummy elements with a value of ∞. We will show that selecting from A with absolute accuracy m reduces to selecting from A_M with absolute accuracy $m - \mu n$. First the algorithm of Lemma 1 can be used to compute an integer s with $|A_L| \le s \le (1+\mu)|A_L|$. It is now easy to see that if the rank of an element in A_M is within $m - \mu n$ of $r' = r - s$, then its rank in A is within m of r, as desired (in particular, the dummy elements do not change the rank of any real element).

If $\lambda' \ne \lambda$, we have $|A_M| \le 9\lambda' n = \mu' n \le m$, so that the rank in A of any element in A_M is within m of r; we can hence return an arbitrary element of A_M as the answer. □

Lemma 10. *Let n and k be given powers of 2 with $2^6 \le k \le n^{1/8}$ and let t and m be given integers with $t \ge \lceil \log k / \log\log\log n \rceil (\log\log n)^4$ and $m \ge \mu n$, where $\mu = 2^{-t\log\log\log n/(\log\log n)^3}$. Then $O(t)$ time and $O(k^3 n)$ operations suffice to reduce the problem of selection from A with absolute accuracy m to that of selection from a set of size at most $(10/k)n$ with absolute accuracy $m - \mu n$.*

Proof. Let $\lambda = 2^{-t\log\log\log n/(\log\log n)^4}$ and note that $\lambda \le 1/k$. The algorithm of Lemma 8 can be used to compute a $(1/k)$-sample B of A with $n^{1/4} \le |B| \le n^{1/2}$, and the claim can be seen to follow from Lemma 9, used with $q = \log k$. □

Theorem 11. *For all given integers $n \ge 16$ and $t \ge (\log\log n)^4$, approximate selection problems of size n can be solved with relative accuracy λ using $O(t)$ time, $O(n)$ operations and $O(n)$ space, where*

$$\lambda = \begin{cases} 2^{-2^{t/(\log\log n)^4}}, & \text{if } t < (\log\log n)^4 \log^{(4)} n; \\ 2^{-t\log\log\log n/(\log\log n)^4}, & \text{if } t \ge (\log\log n)^4 \log^{(4)} n. \end{cases}$$

Proof. Without loss of generality we can assume that $\lambda \ge 1/n$, since otherwise $t = \Omega(\log n)$, so that the problem can be solved using Cole's algorithm [9], and that n is a power of 2 larger than 160^{32}. Let $m = \lambda n$, so that the problem is to select from a set of size n with absolute accuracy m. As mentioned earlier, the main part of our algorithm consists of T *stages*, for some integer $T \ge 0$. For $i = 1, \ldots, T$, Stage i transforms a problem of selection from a set A_i of size n_i with absolute accuracy m_i to one of selection from a set A_{i+1} of size n_{i+1} with absolute accuracy m_{i+1}, where n_{i+1} is significantly smaller than n_i, while m_{i+1} is slightly smaller than m_i, and both n_i and n_{i+1} are powers of 2. In particular, $n_1 = n$ and $m_1 = m$. T is the smallest nonnegative integer such that $n_{T+1} \le m_{T+1}$ or $n_{T+1}^2 \le n$.

We now describe Stage i, for $i \in \{1, \ldots, T\}$. Let k_i be the smallest power of 2 larger than $40\lceil (n/n_i)^{1/16} \rceil$ and take $t_i = \lceil \log k_i / \lfloor \log\log\log n \rfloor \rceil \lceil \log\log n \rceil^4$. For $i = 1, \ldots, T$, $n_i > n^{1/2}$ and hence $k_i \le 160 n^{1/32} \le n^{1/16} \le n_i^{1/8}$. We will show below that $m_i \ge \mu n_i$, where $\mu = 2^{-t\log\log\log n/(\log\log n)^4 - 2}$. Spending $O(t_i + t/\log\log n)$ time, we can therefore use the algorithm of Lemma 10 to reduce the problem at hand to one of selection from a set A_{i+1} of size at most n_{i+1} with absolute accuracy $m_{i+1} = m_i - \mu n_i$, where n_{i+1} is the smallest power of 2 larger than $(10/k_i)n_i$. We increase the size of

A_{i+1} to exactly n_{i+1} by adding a suitable number of dummy elements with a value of ∞ and note that $n_{i+1} \le (20/k_i)n_i$.

For $i = 1, \ldots, T-1$,

$$\frac{n}{n_{i+1}} \ge \frac{n}{(20/k_i)n_i} \ge \frac{2n(n/n_i)^{1/16}}{n_i} = 2\left(\frac{n}{n_i}\right)^{17/16}.$$

In particular, $n_{i+1} \le n_i/2$, for $i = 1, \ldots, T-1$, so that $\sum_{i=1}^{T} n_i \le 2n$. Noting that $\mu \le \lambda/4$, we will now prove that $m_i \ge \mu n_i$, for $i = 1, \ldots, T$, as claimed above. Since $m = \lambda n \ge 4\mu n$, it suffices to show that $m_T \ge m/2$. But $m - m_T = \mu \sum_{i=1}^{T-1} n_i \le 2\mu n \le m/2$, as required.

After Stage T, the computation is finished in one of two ways. If $n_{T+1} \le m_{T+1}$, we simply return an arbitrary element of A_{T+1}, which is obviously correct. Otherwise, $n_{T+1}^2 \le n$, and we use the algorithm of Lemma 2 to select from A_{T+1} with absolute accuracy $m/2$, for which $O(t)$ time amply suffices. Since $m_T \ge m/2$, this is also correct.

It remains to estimate the time and the number of operations needed by the algorithm. We begin by bounding the number of stages. Since $n/n_{i+1} = \Theta((n/n_i)^{17/16})$, we have $k_{i+1} = \Theta((n/n_{i+1})^{1/16}) = \Theta((n/n_i)^{17/16^2}) = \Theta(k_i^{17/16})$, for $i = 1, \ldots, T-1$. Now if $T \ge 2$, $(20/k_{T-1})n_{T-1} \ge n_T \ge m/2$ and hence $k_{T-1} \le 40n_{T-1}/m \le 40/\lambda$, which implies that $T = O(\log\log(4/\lambda))$.

The processing after Stage T obviously uses $O(t)$ time and $O(n)$ operations. The total number of operations executed in Stages 1 through T is

$$O\Big(\sum_{i=1}^{T} k_i^3 n_i\Big) = O\Big(\sum_{i=1}^{T} (n/n_i)^{3/16} n_i\Big) = O\Big(n \sum_{i=1}^{T} (n_i/n)^{13/16}\Big) = O(n),$$

where the last relation follows from the rapid growth of the sequence $\{n/n_i\}_{i=1}^{T}$. The time needed by Stages 1 through T, finally, is

$$O\Big(\sum_{i=1}^{T}\Big(t_i + \frac{t}{\log\log n}\Big)\Big) = O\Big(\frac{tT}{\log\log n} + \sum_{i=1}^{T}\Big\lceil \frac{\log k_i}{\log\log\log n}\Big\rceil (\log\log n)^4\Big)$$

$$= O\Big(t + \Big(T + \frac{\sum_{i=1}^{T}\log k_i}{\log\log\log n}\Big)(\log\log n)^4\Big) = O\Big(t + \Big(T + \frac{\log k_T}{\log\log\log n}\Big)(\log\log n)^4\Big)$$

$$= O(t + (\log\log(4/\lambda) + \log(4/\lambda)/\log\log\log n)(\log\log n)^4).$$

A case analysis shows that the latter expression is $O(t)$. □

Corollary 12. *For all given integers $n \ge 4$ and $t \ge (\log\log n)^4$, approximate selection problems of size n can be solved with relative accuracy $2^{-t/(\log\log n)^4}$ using $O(t)$ time and $O(n)$ operations. In particular, for any constant $\lambda > 0$, λ-selection problems of size n can be solved in $O((\log\log n)^4)$ time with optimal speedup.*

5 Work-Optimal Exact Selection

Armed with a reasonably accurate work-optimal algorithm for approximate selection, it is an easy matter to derive a work-optimal algorithm for exact selection. The algorithm is similar to one stage of the algorithm of Theorem 11. We now provide the details.

Theorem 13. *For all integers $n \geq 4$, selection problems of size n can be solved using $O(\log n/\log\log n)$ time, $O(n)$ operations and $O(n)$ space.*

Proof. Suppose that the task is to select the element of rank r from a set A. Applied with $t = \Theta(\log n/\log\log n)$, the algorithm of Corollary 12 allows us to compute two elements $x_L, x_H \in A$ with $rank_A(x_L) \leq r \leq rank_A(x_H)$, but $rank_A(x_H) - rank_A(x_L) = O(n \cdot 2^{-\sqrt{\log n}})$. Next partition A into the sets $A_L = \{x \in A : x < x_L\}$, $A_M = \{x \in A : x_L \leq x \leq x_H\}$ and $A_H = \{x \in A : x > x_H\}$ as in the proof of Lemma 9. Then use exact prefix summation [10, Theorem 2.2.2] to determine $|A_L|$ and to store the elements of A_M in an array of size $|A_M| = O(n \cdot 2^{-\sqrt{\log n}})$. The remaining problem is to select the element of rank $r - |A_L|$ from A_M, which trivially reduces to sorting A_M. The latter task can be accomplished using the algorithm of Cole [8, Theorem 3], which, given the available processor advantage of $2^{\Theta(\sqrt{\log n})}$, runs in $O(\log n/\log\log n)$ time. □

6 Increasing the Accuracy

At the price of increasing the number of operations slightly, we can obtain a better accuracy than that provided by Theorem 11. Specifically, in $O(t)$ time, we achieve a relative accuracy of 2^{-t}, as compared to the approximately $2^{-t\log\log\log n/(\log\log n)^4}$ of Theorem 11. Our algorithm is based on the *AKS sorting network* [2] and further improvements and expositions of it [17, 18, 7]. In a manner first used in [1], we simulate the initial stages of the operation of the network in order to focus on an interesting subset of the elements.

For an input of size n, where n is a power of 2, the AKS network can be thought of as moving the elements between the nodes of a complete binary tree with n leaves numbered $1, \ldots, n$ from left to right. Say that an element is *addressed* to a node u in the tree if its rank is the number of a leaf descendant of u, and that it is a *stranger* at u otherwise.

The analysis of the AKS network in [7] can be shown to imply the following: For every integer i with $1 \leq i \leq \log n$, there is a positive integer T with $T = O(i)$, computable from i in constant time, such that after T stages of the operation of the network and for any node u at depth i in the tree, the number of keys at u is nonzero, but the fraction of strangers among these is at most 2^{-7}.

Let $n \geq 4$ be a power of 2 and let q be a positive integer. In order to use the above to λ-select from n elements, where $\lambda = 2^{-q}$, we proceed as follows: Choose the positive integer i minimal such that there is a node u at depth i with the property that every element addressed to u is a valid answer to the selection problem; it is not difficult to see that $i = O(q)$. Then compute T from i, run the AKS network for T stages and finally use the algorithm of Corollary 12 to select an element from the middle half of the elements stored at u. Since the strangers at u occupy the extreme ranks of relative size at most 2^{-7}, this produces an element addressed to u, and hence a correct answer. The time needed is $O(T + (\log\log n)^4) = O(q + (\log\log n)^4)$. We hence have

Theorem 14. *For all given integers $n \geq 4$ and $q \geq 1$, approximate selection problems of size n can be solved with relative accuracy 2^{-q} using $O(q+(\log\log n)^4)$ time, $O(qn)$ operations and $O(n)$ space.*

References

1. M. Ajtai, J. Komlós, W. L. Steiger, and E. Szemerédi. Optimal parallel selection has complexity $O(\log\log N)$. *Journal of Computer and System Sciences*, **38** (1989), pp. 125–133.
2. M. Ajtai, J. Komlós, and E. Szemerédi. An $O(n \log n)$ sorting network. In *Proc. 15th ACM STOC* (1983), pp. 1-9.
3. N. Alon and Y. Azar. Parallel Comparison Algorithms for Approximation Problems. *Combinatorica*, **11** (1991), pp. 97–122.
4. Y. Azar and N. Pippenger. Parallel selection. *Discrete Applied Mathematics*, **27** (1990), pp. 49–58.
5. P. Beame and J. Håstad. Optimal bounds for decision problems on the CRCW PRAM. *Journal of the ACM*, **36** (1989), pp. 643–670.
6. M. Blum, R. W. Floyd, V. Pratt, R. L. Rivest, and R. E. Tarjan. Time bounds for selection. *Journal of Computer and System Sciences*, **7** (1973), pp. 448–461.
7. V. Chvatal. Lecture notes on the new AKS sorting network. *DIMACS Technical Report 92-29*, 1992.
8. R. Cole. Parallel merge sort. *SIAM Journal on Computing*, **17** (1988), pp. 770–785.
9. R. Cole. An optimally efficient selection algorithm. *Information Processing Letters*, **26** (1988), pp. 295–299.
10. R. Cole and U. Vishkin. Faster optimal parallel prefix sums and list ranking. *Information and Computation*, **81** (1989), pp. 334–352.
11. P. F. Dietz and R. Raman. Heap construction on the CRCW PRAM. In preparation, 1993.
12. M. T. Goodrich. Using approximation algorithms to design parallel algorithms that may ignore processor allocation. In *Proc. 32nd IEEE FOCS* (1991), pp. 711-722.
13. T. Hagerup. The log-star revolution. In *Proc. 9th STACS* (1992), LNCS 577, pp. 259–278.
14. T. Hagerup. Fast deterministic processor allocation. In *Proc. 4th ACM-SIAM SODA* (1993), pp. 1–10.
15. T. Hagerup and R. Raman. Waste makes haste: Tight bounds for loose parallel sorting. In *Proc. 33rd IEEE FOCS* (1992), pp. 628–637.
16. T. Hagerup and R. Raman. Fast deterministic approximate and exact parallel sorting. In *Proc. 5th ACM SPAA* (1993), to appear.
17. M. S. Paterson. Improved sorting networks with $O(\log N)$ depth. *Algorithmica*, **5** (1990), pp. 75–92.
18. N. Pippenger. Communication Networks. In *Handbook of Theoretical Computer Science, Vol A, Algorithms and Complexity* (J. van Leeuwen, ed.). Elsevier/The MIT Press (1990), Chapter 15, pp. 805–833.
19. R. Reischuk. Probabilistic parallel algorithms for sorting and selection. *SIAM Journal on Computing*, **14** (1985), pp. 396–409.
20. L. G. Valiant. Parallelism in comparison problems. *SIAM Journal on Computing*, **4** (1975), pp. 348–355.

Defining Soft Sortedness by Abstract Interpretation

Jian Chen and John Staples

Software Verification Research Centre
The University of Queensland, Brisbane, Australia 4072

Abstract. Sorted languages can improve the expressiveness and efficiency of reasoning. A conventional sorted language typically includes well-sortedness rules amongst the rules for well-formedness. A major disadvantage of this approach is that many intuitively meaningful expressions are ill-sorted and hence not part of the language.
To overcome this limitation, soft sorting regards as well-formed, all first-order expressions of the corresponding unsorted language, and lets the semantics be the basis for defining the significance of the sort syntax. In this paper we show how soft sortedness can be defined by abstract interpretations which characterise semantic properties of softly sorted expressions.

1 Introduction

Symbols in a first-order language denote objects of an intended universe of discourse, and functions and relations on them. When a universe of discourse contains more than one sort of object, it is often useful to include in the language, notations for describing sort restrictions on those functions and predicates, so as to achieve more expressiveness and more efficient reasoning.

Approaches to sorting first-order languages can be classified into two categories. Each of these categories has both advantages and disadvantages.

The first category uses ***sorted languages***: syntactic sort restrictions ("signatures") on function and relation symbols are used to define narrower concepts of well-formed expressions than are used in unsorted languages. A conventional sorted language typically includes well-sortedness rules amongst the rules for well-formedness of expressions [7, 9, 12]. We call such approaches ***strict sorting***.

The second approach uses ordinary (unsorted) first-order languages and adds special predicates to characterise subsets of the universe of discourse [9, 10; see also 2].

Compared with the unsorted approach, strict sorting has many advantages: more explicit representation of sorting, decidable or more efficient checking of well-sortedness, and more efficient reasoning.

However the following example illustrates a limitation of strict sorting.

Example 1. Suppose there are two different ground sorts c and d whose values are disjoint domains from a universe of discourse U. Let p and q be two one-place predicate symbols such if an atomic formula p(x) or q(y) has the truth value *true* then the value of x or y is an object of the domain of c or an object of the domain of d respectively. Now consider the formula:

$$\mathtt{p(x)\ or\ q(x).}$$

In strict sorting, the formula p(x) is well-sorted when x is a sorted variable of sort c; the formula q(x) is well-sorted when x is a sorted variable of sort d. However (p(x) or q(x)) is ill-sorted for every sorted variable x as there is no object in the universe of discourse which is common to the domains of c and d. But intuitively, formally and in an unsorted language, the formula has a well-defined truth value. □

The main difference between the two approaches is whether sorting is supported at the syntactic or the semantic level. The unsorted approach is more general than strict sorting, as many researchers have observed, see for example [9]. On the other hand, from a computational point of view many properties of strict sorting are very useful.

This motivates us to investigate combining the best features of both the above approaches. The basic idea is to regard as well-formed, all first-order expressions of the corresponding unsorted language, and to let the semantics be the basis for defining the significance of the sort syntax. In [3], the approach is called *soft sorting* because it is in the same direction as Cartwright and Fagan's soft typing [2]. More recently Henglein [8] has taken a similar approach.

The motivation of Cartwright and Fagan's work and motivation of soft sorting are similar: soft typing aims to combine the best features of static and dynamic typing in the lambda calculus and functional programming; soft sorting aims to combine the best features of strictly sorted and unsorted first-order languages. Soft typing and soft sorting are also similar in that neither imposes additional well-formedness rules on expressions.

The two approaches are different in the sense that Cartwright and Fagan looked at higher-order expressions (functional programming languages are typically higher-order), while in soft sorting we focus on first-order languages. Our reasons for doing so are as follows.

1. Stronger results are available for first-order languages;
2. These stronger results are useful in logic programming languages, which are typically first-order;
3. They are potentially also relevant to set theory (a first-order theory) and its applications, for example to formal methods in computer science.

In soft sorting, there are no *a priori* sort restrictions in the definition of the language's expressions. Expressions whose arguments do not satisfy the specified sort restrictions nevertheless do have semantics. In particular, terms which do not satisfy the language's sort restrictions are interpreted as having an exceptional value *bad*. Atomic predicates which do not satisfy the language's sort restrictions are interpreted as having the truth value *false*, and on this basis all expressions, even those which fail to satisfy the language's sort restrictions, are interpreted. However, to maintain efficient reasoning as provided by strict sorting, it is important to have a suitable notion of sortedness of softly sorted expressions.

In this paper, we introduce soft sorting and discuss an approach in which the sortedness of softly sorted expressions is defined by abstract interpretations. For a systematic account of abstract interpretation, see [1]. In this paper we introduce only aspects needed for our purposes. We shall show that the sortedness characterises some semantic properties of those expressions.

The following example illustrates some features of soft sorting, as well as the motivation of our sortedness definitions in soft sorting.

Example 2. Consider the three formulas

```
F1:  p(x) & q(y),
F2:  p(x) & q(x),
F3:  p(x) or (not q(x)).
```

Suppose now that the arguments of p and q are required to be of different ground sorts c and d respectively, where the domains of c and d are disjoint.

First we consider these formulas in strict sorting. F1 is well-sorted, and hence is a well-formed formula, only when variables x and y are of sorts c and d respectively. F2 is ill-sorted, as there is no object common to the domains of c and d. F3 is ill-sorted, as in Example 1.

On the other hand, all the three formulas are well-formed in soft sorting.

1. Consider first F1. When the variables x and y are assigned objects in domains of c and d respectively, the sort restrictions of p and q are both satisfied. Hence the truth value of p(x) is determined by the interpretation of p over the domain of c; the truth value of q(y) is determined by the interpretation of q over the domain of d. Thus the truth value of F1 depends on the interpretations of p and q. On the other hand, when for example x is assigned an object not in the domain of c, the sort restriction of p is not satisfied. An atomic predicate in such a case is interpreted in soft sorting as having the truth value *false*. In particular the truth value of p(x), and therefore of F1, is *false*.
2. Consider now F2. As the places of p and q have distinct sorts, and as distinct sort domains are disjoint in many-sorting, then the sort restriction of at least one of p or q is not satisfied. Thus for every possible value of x, either p(x) or q(x) has the truth value *false*, and hence so does F2.
3. Likewise, for F3, when the value of x is an object in the domain of sort c, then q(x) has the truth value *false*. Thus F3 is *true* in this case. When the value of x is an object in the domain of sort d, then the value of p(x) is *false*. Thus the truth value of F3 is the same as that of (not q(x)). Finally, when the value of x is an object neither in the domain of c nor in that of d, both p(x) and q(x) are *false*, and so F3 is *true*. □

Similarly, in soft sorting the value of a term may be *bad* for all interpretations and variable valuations, or may depend on individual interpretations and variable valuations. This suggests the possibility of an abstract interpretation on expressions with respect to the sorting of their symbols.

In the case of terms it is convenient to use the two abstract values, "well-sorted" and "ill-sorted". Intuitively, ill-sortedness means that the term is known statically to be assigned *bad* in all interpretations. In the case of formulas it turns out to be more useful to recognise three abstract values, "well-sorted", "F-sorted" and "T-sorted". Intuitively, F-sortedness means that the formula is known statically to have the truth value *false* in all interpretations. Similarly T-sorted means that the formula is *true* in all interpretations.

We shall see that it is possible that the abstract value of an expression is not statically known. We call such an expression *well-sortable*, and we shall show later in this paper that every well-sortable expression does become well-sorted under suitable bindings of its sort variables.

The structure of this paper is as follows. In Section 2 we outline the syntax and semantics of softly sorted languages. We define the sortedness by abstract interpretations in Section 3. We give some basic results in Section 4. Finally we discuss some related issues and propose further research in Section 5. Because of the limitation of space, proofs of the results stated in this paper are not included here but can be found in [3].

2 Softly sorted first-order languages

2.1 A sort language for parametric many-sorting

By a sort language we mean an unsorted first-order language, intended to describe a sort structure. The intended interpretations of its terms are as non-empty disjoint subsets of a universe of discourse U. In our work we assume an distinguished object *bad* in U, which is used as an exceptional value for terms which may be regarded as ill-sorted. If no suitable object *bad* exists initially in U, U is augmented.

A *sort alphabet*, denoted $\mathcal{SA}$, is an non-empty finite set of *sort constructors*. Each sort constructor has an associated non-negative integer called its *arity*. Sort constructors of arity 0 are called *base sorts*. In soft sorting the distinguished base sort `wrong` is optional; where it occurs its intended interpretation is as the set $\{bad\}$. As this example illustrates, we use typewriter font for explicit formal notation. We assume that there is a denumerable set $\mathcal{SV}$ of *sort variables*.

The *sort language* $\mathcal{SL}$ based on a given sort alphabet $\mathcal{SA}$ is the set of *sorts*, recursively defined as follows.

1. Each base sort $\mathbf{c} \in \mathcal{SA}$ is a sort.
2. Each sort variable $\mathbf{u} \in \mathcal{SV}$ is a sort.
3. If $\mathbf{k} \in \mathcal{SA}$ has arity $n \geq 1$, and if $\mathbf{s}_i$, $i = 1, \ldots, n$ are sorts, then $\mathbf{k}(\mathbf{s}_1, \ldots, \mathbf{s}_n)$ is a sort.

A sort in which no sort variable occurs is called a *ground sort.*

A *sort substitution* is a finite set $\{\mathbf{u}_1/\mathbf{s}_1, \ldots, \mathbf{u}_n/\mathbf{s}_n\}$ of *bindings* $\mathbf{u}_i/\mathbf{s}_i$, for distinct variables $\mathbf{u}_i$, where $\mathbf{s}_i$ is a sort distinct from $\mathbf{u}_i$, $i = 1, \ldots, n$. A *ground sort substitution* is a sort substitution, all of whose values $\mathbf{s}_i$ are ground sorts. Sort substitutions are denoted Θ, Φ, Ψ, ... The *instance* $\mathbf{s}\Theta$ of $\mathbf{s}$ by Θ is the sort obtained from s by simultaneously replacing each occurrence of $\mathbf{u}$ by $\mathbf{s}'$ for each $\mathbf{u}/\mathbf{s}' \in \Theta$. *Sort variants*, the *solvability* of a set of sorts, *sort unifiers*, *sort solutions*, and the *solvability* of a set of sort substitutions are also defined as usual as for a first-order language [12].

A *domain universe* $\mathcal{S}$ with respect to an object universe U is a set of non-empty disjoint subsets of U such that the union of the elements of $\mathcal{S}$ is U. An interpretation *si* in $\mathcal{S}$ of $\mathcal{SL}$ is a one-one, onto function from ground sorts to $\mathcal{S}$, such that if $\mathbf{wrong} \in \mathcal{SA}$, then $si(\mathbf{wrong}) = \{bad\}$. The interpretation $si(\mathbf{k})$ of a sort constructor $\mathbf{k}$ which is implied by *si*, is as follows. If $\mathbf{k}$ is an n-ary sort constructor, $n \geq 1$, then $si(\mathbf{k})$ is a function $\mathcal{S}^n \to \mathcal{S}$. In particular, if $D_1, \ldots, D_n$ are in $\mathcal{S}$, then as *si* is one-one and onto there are unique ground sorts $\mathbf{s}_1, \ldots, \mathbf{s}_n$ such that $si(\mathbf{s}_i) = D_i$, and $si(\mathbf{k})(D_1, \ldots, D_n)$ is defined to be $si(\mathbf{k}(\mathbf{s}_1, \ldots, \mathbf{s}_n))$. A *sort variable valuation* is a function $\mathcal{SV} \to \mathcal{S}$. We write $\mathcal{SVV}$ for the set of sort variable valuations. We write *sv* for meta variables ranging over $\mathcal{SVV}$.

Intuitively, an interpretation of sorts is a function from sort variable valuations to the domain universe $\mathcal{S}$. More precisely, the *sort interpretation* si^* of sorts of $\mathcal{SL}$ defined by *si* is the function $\mathcal{SA} \to (\mathcal{SVV} \to \mathcal{S})$ defined recursively as follows.

1. For each base sort $\mathbf{c}$, $si^*(\mathbf{c})(sv) = si(\mathbf{c})$.
2. For each sort variable $\mathbf{u}$, $si^*(\mathbf{u})(sv) = sv(\mathbf{u})$.
3. For each sort $\mathbf{k}(\mathbf{s}_1, \ldots, \mathbf{s}_n)$,
   $si^*(\mathbf{k}(\mathbf{s}_1, \ldots, \mathbf{s}_n))(sv) = si(\mathbf{k})(si^*(\mathbf{s}_1)(sv), \ldots, si^*(\mathbf{s}_n)(sv))$.

Given the above syntax and semantics, the sort language has the following properties, for all domain universes over a given object universe (as proved in [3]).

- *Soundness and completeness of parametric many-sorting.* This means that two sorts are syntactically identical if and only if the domains assigned to them by every interpretation and every sort variable valuation, are the same.
- *Soundness and completeness of sort checking.* This means that two sorts are unifiable if and only if there exist an interpretation and a sort variable valuation under which the two sorts denote the same domain.

2.2 Softly sorted languages

Let $\mathcal{A}^0$ be an ordinary alphabet, called an *unsorted alphabet*, let $\mathcal{V}^0$ be a set of unsorted variables, and let $\mathcal{SL}$ be a sort language.

A sort declaration $\mathcal{R}$ for $\mathcal{A}^0$ is defined as follows.

1. $\mathcal{R}$ assigns to each constant symbol a tuple $< \mathbf{s} >$, sometimes abbreviated to $\mathbf{s}$, where $\mathbf{s}$ is a ground sort.

2. $\mathcal{R}$ assigns to each **f** with arity n, $n \geq 1$, a tuple $< \mathbf{s}_1, \ldots, \mathbf{s}_n, \mathbf{s} >$, sometimes abbreviated to $\mathbf{s}_1, \ldots, \mathbf{s}_n \rightarrow \mathbf{s}$.
3. $\mathcal{R}$ assigns to each predicate symbol **p** with arity $n \geq 1$, a tuple $< \mathbf{s}_1, \ldots, \mathbf{s}_n >$, sometimes abbreviated to $\mathbf{s}_1, \ldots, \mathbf{s}_n$.

All instances of the declared sort of a symbol are called *associated sorts* of the symbol. We call an unsorted symbol, together with one of its associated sorts, a *sorted symbol.*

Given an unsorted alphabet $\mathcal{A}^0$, a sort language $\mathcal{SL}$ and a sort declaration $\mathcal{R}$, a *sorted alphabet* $\mathcal{A}$ is the set of all sorted symbols derived from $\mathcal{A}^0$ and $\mathcal{SL}$ with respect to $\mathcal{R}$. It is optional to have in $\mathcal{A}^0$ a sorted constant **bad**, intuitively to denote *bad.* When **bad** occurs it is called the *abnormal constant,* in which case it must be declared with sort **wrong**.

Sometimes we write an associated sort of a sorted symbol as a subscript. For example the constant **bad** with associated sort **wrong** may be written $\mathbf{bad}_{\mathbf{wrong}}$.

Variables of $\mathcal{V}^0$ will also be assigned sorts. A variable has a unique most general sort assigned to it which we call its declared sort. Likewise, instances of the declared sort of a variable will be called associated sorts of the variable. A *sorted variable* is a variable symbol together with an associated sort. The set of all sorted variables is denoted $\mathcal{V}$. When we refer to *distinct* sorted variables we mean that the variable symbols are different.

Terms, *atoms*, and *formulas* are defined as usual for an unsorted first-order language [7]. We may call them *sorted* to emphasise that they are based on sorted symbols and variables, but strict sorting is not implied. The *logical connectives* are `not`, `or`, `&`, `->`, and `<->`. The *logical quantifiers* are written `forall` and `exists`. Terms and formulas may be referred to generically as *expressions.* A *softly sorted language* $\mathcal{L}$ is the set of expressions thus defined. Given a sort substitution Θ and an expression $\mathbf{E}$, $\mathbf{E}\Theta$ is the expression obtained by applying Θ to every associated sort of symbol occurring in $\mathbf{E}$.

The following definition of interpretation is fundamental to the concept of soft sorting.

An *interpretation* I of $\mathcal{L}$ in U, relative to an *si* whose domain is a domain universe with respect to U, is a function from the alphabet of $\mathcal{L}$, whose values are functions with domain $\mathcal{SVV}$, defined as follows.

1. For each sorted constant $\mathbf{a_s}$, $I(\mathbf{a_s})(sv)$ is an object in $si^*(\mathbf{s})(sv)$.
2. For each sorted function symbol $\mathbf{f}_{\mathbf{s}_1,\ldots,\mathbf{s}_n \rightarrow \mathbf{s}}$, $I(\mathbf{f}_{\mathbf{s}_1,\ldots,\mathbf{s}_n \rightarrow \mathbf{s}})(sv)$ is a mapping $U^n \rightarrow U$ such that for every n-tuple $< t_1, \ldots, t_n >$ of individual objects in $si^*(\mathbf{s}_1)(sv) \times \cdots \times si^*(\mathbf{s}_n)(sv)$, $I(\mathbf{f}_{\mathbf{s}_1,\ldots,\mathbf{s}_n \rightarrow \mathbf{s}})(sv)(t_1, \ldots, t_n) \in si^*(\mathbf{s})(sv)$, and for other tuples $< t_1, \ldots, t_n >$, $I(\mathbf{f}_{\mathbf{s}_1,\ldots,\mathbf{s}_n \rightarrow \mathbf{s}})(sv)(t_1, \ldots, t_n) = bad$.
3. For each sorted predicate symbol $\mathbf{p}_{\mathbf{s}_1,\ldots,\mathbf{s}_n}$, $I(\mathbf{p}(\mathbf{s}_1, \ldots, \mathbf{s}_n)(sv)$ is a mapping $U^n \rightarrow \{true, false\}$ such that for every n-tuple $< t_1, \ldots, t_n >$ of objects in $si^*(\mathbf{s}_1)(sv) \times \cdots \times si^*(\mathbf{s}_n)(sv)$, $I(\mathbf{p}_{\mathbf{s}_1,\ldots,\mathbf{s}_n})(sv)(t_1, \ldots, t_n) \in \{true, false\}$. At other $t_1, \ldots, t_n$, $I(\mathbf{p}_{\mathbf{s}_1,\ldots,\mathbf{s}_n})(sv)(t_1, \ldots, t_n) = false$.

A *sorted variable valuation* v relative to *si*, is a function $\mathcal{V} \rightarrow (\mathcal{SVV} \rightarrow U)$ such that for each sorted variable $\mathbf{x_s}$, $v(\mathbf{x_s})(sv)$ is an object in $si^*(\mathbf{s})(sv)$. We write

$\mathcal{VV}$ for the set of sorted variable valuations.

Accordingly, a *term interpretation* I^* based on I (relative to *si*) is a function on terms whose values are functions $\mathcal{SVV} \rightarrow (\mathcal{VV} \rightarrow U)$, and a *formula interpretation* I^* based on I (relative to *si*) is a function on formulas whose values are functions $\mathcal{SVV} \rightarrow (\mathcal{VV} \rightarrow \{true, false\})$.

3 Abstract interpretations of softly sorted expressions

Under a semantics as above, some softly sorted expressions have the same meaning in all sorted interpretations, as illustrated by the following example.

Example 3. Suppose there is a base sort `int` and a 1-place sort constructor `list`. Consider $\mathbf{t} = \mathtt{f}_{\mathtt{list(int)}\rightarrow\mathtt{int}}(\mathtt{x}_{\mathtt{int}})$. According to the semantics of softly sorted languages, $\mathbf{t}$ is assigned object *bad* in every interpretation. Consider $\mathbf{s} = \mathtt{f}_{\mathtt{list(int)}\rightarrow\mathtt{int}}(\mathtt{x}_{\mathtt{list(int)}})$. The meaning of $\mathbf{s}$ depends on individual interpretation of $\mathtt{f}$ and sorted variable valuation of $\mathtt{x}_{\mathtt{list(int)}}$. For the term $\mathbf{s}' = \mathtt{f}_{\mathtt{list(int)}\rightarrow\mathtt{int}}(\mathtt{x}_{\mathbf{u}})$, the meaning depends on the value of the sort variable $\mathbf{u}$. When the value of $\mathbf{u}$ is the domain of `list(int)`, the meaning of $\mathbf{s}'$ is similar to that of $\mathbf{s}$; when the value of $\mathbf{u}$ is a domain other than the domain of `list(int)`, the situation is similar to that of $\mathbf{t}$. □

From the above example, we see the possibility of static recognition of the meanings of some expressions. In this example, those expressions whose sort restrictions can never be satisfied, have fixed meaning. We characterise that by using appropriate abstract values. The values "well-sorted" and "ill-sorted" are convenient for terms. For formulas however it turns out to be more useful to recognise the three values "well-sorted", "F-sorted" and "T-sorted". Intuitively, F-sortedness means that the formula is known statically to have the truth value *false* in all interpretations. Similarly T-sorted means that the formula is known to be *true* in all interpretations. Both sorts of abstract value are defined for atoms, and in that case ill-sorted is equivalent to F-sorted.

Definition 1. The *well-sortedness* and *ill-sortedness* of terms and atoms are defined as follows.

1. A sorted constant $\mathbf{a_s}$ or variable $\mathbf{x_s}$ is well-sorted and is of sort $\mathbf{s}$.
2. If $\mathbf{t}_i$ is of sort $\mathbf{s}_i$, $i \in \{1, \ldots, n\}$, then $\mathbf{f}_{\mathbf{s}_1,\ldots,\mathbf{s}_n\rightarrow\mathbf{s}}(\mathbf{t}_1, \ldots, \mathbf{t}_n)$, is well-sorted and is of sort $\mathbf{s}$. If there does not exist any sort substitution Θ such that $\mathbf{t}_i\Theta$, is of sort $\mathbf{s}_i\Theta$, $i \in \{1, \ldots, n\}$, then $\mathbf{f}_{\mathbf{s}_1,\ldots,\mathbf{s}_n\rightarrow\mathbf{s}}(\mathbf{t}_1, \ldots, \mathbf{t}_n)$ is ill-sorted and is of sort `wrong`.
3. If $\mathbf{t}_i$ is of sort $\mathbf{s}_i$, $i \in \{1, \ldots, n\}$, then $\mathbf{p}_{\mathbf{s}_1,\ldots,\mathbf{s}_n}(\mathbf{t}_1, \ldots, \mathbf{t}_n)$ is well-sorted. If there does not exist any sort substitution Θ such that $\mathbf{t}_i\Theta$, $i \in \{1, \ldots, n\}$, is of sort $\mathbf{s}_i\Theta$, then $\mathbf{p}_{\mathbf{s}_1,\ldots,\mathbf{s}_n}(\mathbf{t}_1, \ldots, \mathbf{t}_n)$ is ill-sorted.

When a term is either well-sorted or ill-sorted, we say the sort of the term is defined.

Definition 2. The *well-sortedness*, *F-sortedness*, and *T-sortedness* of formulas are defined as follows.

1. Formula `true` is T-sorted and formula `false` is F-sorted.
2. If **F** is an atom **A** and **A** is well-sorted (ill-sorted), then **F** is well-sorted (F-sorted).
3. If **F** is (`not` **F**′), then: if **F**′ is well-sorted, then **F** is well-sorted; if **F**′ is F-sorted, then **F** is T-sorted; and if **F**′ is T-sorted, then **F** is F-sorted.
4. If **F** is ($\mathbf{F}_1$ `or` $\mathbf{F}_2$), then: if one of $\mathbf{F}_1$ and $\mathbf{F}_2$ is well-sorted and the other F-sorted, or if both of them are well-sorted, then **F** is well-sorted; if either of $\mathbf{F}_1$ or $\mathbf{F}_2$ is T-sorted, then **F** is T-sorted; if $\mathbf{F}_1$ and $\mathbf{F}_2$ are both F-sorted, then **F** is F-sorted; if $\mathbf{F}_1$ and $\mathbf{F}_2$ are not F-sorted and are not both well-sorted, then **F** is T-sorted.
5. If **F** is ($\mathbf{F}_1$ `&` $\mathbf{F}_2$), then: if $\mathbf{F}_1$ and $\mathbf{F}_2$ are both well-sorted, or if one of them is well-sorted and the other T-sorted, then **F** is well-sorted; if either $\mathbf{F}_1$ or $\mathbf{F}_2$ is F-sorted, then **F** is F-sorted; if $\mathbf{F}_1$ and $\mathbf{F}_2$ are both T-sorted, then **F** is T-sorted; if $\mathbf{F}_1$ and $\mathbf{F}_2$ are not T-sorted and are not both well-sorted, then **F** is F-sorted.
6. If **F** is ($\mathbf{F}_1$ `->` $\mathbf{F}_2$), then **F** is well-sorted, F-sorted, or T-sorted according as the formula ((`not` $\mathbf{F}_1$) `or` $\mathbf{F}_2$) is well-sorted, F-sorted, or T-sorted.
7. If **F** is ((`forall` $\mathbf{x_s}$) **F**′) or ((`exists` $\mathbf{x_s}$) **F**′), then **F** is well-sorted, F-sorted, or T-sorted according as **F**′ is well-sorted, F-sorted, or T-sorted.

Definition 3. A term or atom which is neither well-sorted nor ill-sorted is called *well-sortable*. A formula which is not well-sorted, F-sorted, or T-sorted is also called well-sortable.

This choice of name is justified by Proposition 11 in Section 4.

If a well-sortable E is well-sorted (ill-sorted if a term or atom; F-sorted, T-sorted if a formula) under a sort substitution Θ, then we call Θ a *well-sorter* (*ill-sorter* if a term or atom; *F-sorter, T-sorter* with if a formula) of E. We call Θ a *soft sorter*, or simply *sorter*, of E if $E\Theta$ is either well-sorted or (ill-sorted if a term or atom; F-sorted or T-sorted if a formula).

4 Some results

Proposition 4. *Let* **t** *be a term and let* **F** *be a formula.*

1. If there is no sort variable occurring in **t** *then* **t** *is either well-sorted or ill-sorted.*
2. If there is no sort variable occurring in **F** *then* **F** *is either well-sorted, F-sorted, or T-sorted.*

Proposition 5. *(Soundness of ground sorting). If* **t** *is a term without any sort variable, then, for every si and sv, and I and v relative to si,* $I^*(\mathbf{t})(sv)(v)$ *is in the domain* $si^*(\mathbf{s})(sv)$ *and* **s** *is the sort of* **t** *and is ground.*

Proposition 6. *(Completeness of ground sorting). Let* $\mathbf{t}$ *be a term without any sort variable. If there exists a ground sort* $\mathbf{s}$ *such that, for every* si *and* sv*, and* I *and* v *relative to* si*,*

$$I^*(\mathbf{t})(sv)(v) \in si^*(\mathbf{s})(sv),$$

then the sort of $\mathbf{t}$ *is defined and is* $\mathbf{s}$*.*

Proposition 7. *(Soundness of term sorting). If the sort of* $\mathbf{t}$ *is defined to be* $\mathbf{s}$ *then, for each sort interpretation* si*, each interpretation* I *relative to* si *and each each variable valuation* v*,* $I^*(\mathbf{t})(sv)(v)$ *is in the domain* $si^*(\mathbf{s})(sv)$*.*

Proposition 8. *A sort substitution instance of a well-sorted (ill-sorted) term or atom is also well-sorted (ill-sorted). A sort substitution instance of a well-sorted (F-sorted, T-sorted) formula is also well-sorted (F-sorted, T-sorted).*

Proposition 9. *An ill-sorted term is assigned the object bad in every sorted interpretation.*

Proposition 10. *An ill-sorted atom is assigned the truth value false in every sorted interpretation. An F-sorted (T-sorted) formula is assigned the truth value false (true) in every sorted interpretation.*

The following proposition justifies the name "well-sortable".

Proposition 11. *If a term* $\mathbf{t}$ *(a formula* $\mathbf{F}$*) is well-sortable, then there exists at least one sort substitution* Θ *such that* $\mathbf{t}\Theta$ *(*$\mathbf{F}\Theta$*) is well-sorted.*

5 Discussion

There have been many works in automated reasoning on the development of sorted logics, for example [5]. Sorted languages have been used in logic programming to support polymorphic sorting (typing), for example [9, 13]. These approaches are based on strict sorting. To be able to support more flexible systems, there are works in logic programming which use unsorted languages with sort predicates, such as in [10]. However the issue of determining sortedness is typically undecidable in such an approach [10]. Soft sorting is an alternative in combining the best features of both approaches [3, 4].

An issue closely related to soft sorting is to determine the sortedness of softly sorted expressions. It is shown in [3] that it is decidable whether a term (formula) is well-sorted, ill-sorted (F-sorted or T-sorted), or well-sortable. In case of a well-sortable term (formula), a finite complete representation of sorters for each of well-sortedness and ill-sortedness (F-sortedness and T-sortedness) is computable. However, the notion of most general sortings [9, 13], which is based on the notion of most general unifiers, is no longer sufficient for the representations of soft sorters. The computation of soft sorters involves solving sets of equations and inequations, and hence is based on the work in [6, 11]. A complete discussion of this issue is beyond the scope of this paper and can be found in [3].

In this paper, we have considered parametric polymorphism. It would be useful to investigate extensions to inclusion polymorphism in soft sorting in the future.

Acknowledgements

The authors are grateful to anonymous referees who pointed out two errors in an earlier version of this paper and whose comments were helpful in preparing the current version.

References

1. Abramsky, S. and Hankin, C. (eds.), *Abstract Interpretation of Declarative Languages*, Ellis Horwood, 1987.
2. Cartwright, R. and M. Fagan, "Soft Typing", in *Proceedings of the ACM SIGPLAN'91 Conference on Programming Language Design and Implementation*, pages 278-292, 1991.
3. Chen, J., "Soft Sorting: Integrating Static and Dynamic Sort Checking", PhD thesis, Department of Computer Science, The University of Queensland, Australia, 1992.
4. Chen, J. and J. Staples, "Soft Sorting in Logic Programming", in *Proceedings of ALPUK'92*, Springer-Verlag, pages 79-96, 1992.
5. Cohn, A. G., "A More Expressive Formulation of Many Sorted Logic", *JAR* **3**(1987) 113-200.
6. Comon, H., "Disunification: A Survey", in *Computational Logic: Essays in Honor of Alan Robinson*, Lassez, J.-L. and Plotkin, G. (eds.), The MIT Press, 1991, 322-359.
7. Enderton, H. B., *A Mathematical Introduction to Logic*, Academic Press, 1972.
8. Henglein, F., "Dynamic Typing", in *Proceedings of the 4th European Symposium on Programming (ESOP '92)*, pages 233-253, LNCS 582, Springer-Verlag, 1992.
9. Hill, P. M. and R. W. Topor, "A Semantics for Typed Logic Programs", in Pfenning, F., (ed.), *Types in Logic Programming*, The MIT Press, pages 1-62, 1992.
10. Kifer, M. and J. Wu, "A First-Order Theory of Typed and Polymorphism in Logic Programming", in *Proceedings of LICS '91*, pages 310-321, 1991.
11. Lassez, J.-L., M. J. Maher and K. Marriott, "Unification Revisited", in *Foundations of Deductive Databases and Logic Programming*, Minker, J. (ed.), Morgan Kaufmann Publishers, pages 587-625, 1988.
12. Lloyd, J. W., *Foundations of Logic Programming*, Springer-Verlag, 1987.
13. Mycroft, A. and R. A. O'Keefe, "A Polymorphic Type System for Prolog", *Artificial Intelligence* **23**, 295-307, 1984.

A Model for Real-Time Process Algebras (Extended Abstract)

Liang Chen*

Centre for Communications Research
University of Bristol
Bristol BS8 1TR, Great Britain

1 Introduction

In real-time systems, the interactions with the environment should satisfy some time constraints. It is not sufficient only to say events occur or eventually occur. There are lower and upper bounds on when events can occur relative to other events. Examples are fault-tolerant systems (time out) and safety critical systems (duration control). However, the traditional methods for reasoning about nondeterministic and concurrent systems, such as [Hen85, Mil80, Mil89, Win84], do not consider hard time aspects of systems. Instead they deal with the quantitative aspects of time in a qualitative way and ignore explicit time information. It is clear that purely qualitative specification and analysis of real-time systems are inadequate.

Tree semantics arise naturally as concurrency is simulated by nondeterministic interleaving. In [Mil80], Milner uses synchronization trees as an interleaving model for parallel computation in which processes communicate by mutual synchronization. In this paper, we extend synchronization trees with time. We make no assumption about the underlying nature of time, allowing it to be discrete or dense. The resulting trees, timed synchronization trees, are a general interleaving model of real-time systems. All constructions on timed synchronization trees are continuous with respect to a complete partial order. As a result, timed synchronization trees can be used as a model for a wide range of real-time process algebras [BB91, Che92a, MT90]. As an example, we define a denotational semantics for Timed CCS [Che92a].

2 A Description of the Model

Timed synchronization trees are certain kinds of rooted trees. The vertices of trees represent states and the edges represent event occurrences which cause changes of states. Edges and vertices are both labelled. The edges are labelled by elements of a nonempty set A of actions, which the machine under consideration may perform. The labels over edges show how the machine synchronizes with the environment. The vertices are labelled with time informations which represent time constraints over actions labelling the pathes from the root to the vertices and the maximal delay time of the machines when the machines arrive at the vertices. To define timed synchronization trees, we start with a definition of graphs.

* Most of the work was done when the author was in LFCS, University of Edinburgh. The author is supported by grant GR/G54399 of the Science and Engineering Research Council of the UK.

Definition 2.1 *A* graph *G is a pair* (V, E), *where V is a set of vertices and E, a subset of the set of unordered pairs of V, is a set of edges.*

A *path* from vertex v_1 to vertex v_{n+1} in G is a sequence of edges $e_1, \cdots, e_n$ of G, where $e_1 = \langle v_1, v_2 \rangle, \cdots, e_n = \langle v_n, v_{n+1} \rangle$ are all distinct. Its *length* is defined to be the length of the sequence. It is called a *cycle* when $n > 0$ and $v_1 = v_{n+1}$.

Definition 2.2 *A* tree *is a graph* (V, E) *with a special vertex* $r \in V$, the root of the tree, *such that*

1. *for any* $v \in V$ *there is a path from* r *to* v;
2. *there are no cycles in* (V, E).

Note that a tree contains at least one vertex, *the root of the tree.* We use (V, E, r) to represent a tree with the set V of vertices, the set E of edges and the root r. The *depth* of a vertex v in a tree (V, E, r), written $h(v)$, is defined to be the length of the path from v to r.

Definition 2.3 *An A*-labelled tree *is a tree* (V, E, r) *together with a labelling function* $l : E \longrightarrow A$.

We use (V, E, r, l) to represent an A-labelled tree with the tree (V, E, r) and the labelling function $l : E \longrightarrow A$. For convenience, we will say a labelled tree instead of an A-labelled tree when A is clear from the context.

To define timed synchronization trees, we presuppose a time domain $(\mathcal{T} \cup \{\infty\}, \leq)$, where $\mathcal{T}$ contains a least element 0 which represents the starting time and $\leq$ is a linear order over $\mathcal{T}$. We introduce an infinite time ∞ (where $\infty \notin \mathcal{T}$) which satisfies that $u \in \mathcal{T}$ implies $u \leq \infty$. We make no assumption about the underlying nature of time, allowing $\mathcal{T}$ to be $\aleph$, the set of natural numbers, or $\Re^{\geq 0}$, the set of non-negative reals. Let D^0 be $\mathcal{T} \cup \{\infty\}$ and D^i be $2^{\mathcal{T}^i \times D^0}$, where $i \in \aleph - \{0\}$ and $\mathcal{T}^i$ represents the set $\underbrace{\mathcal{T} \times \ldots \times \mathcal{T}}_{i}$. We write D^* in place of $\bigcup_{i \in \omega} D^i$ and $S_\perp$, for any set S, in place of $S \cup \{\perp\}$.

Definition 2.4 *A* timed A-labelled tree *is a labelled tree* (V, E, r, l) *together with a time function t which assigns to every vertex v of depth n an element of* $D^n_\perp$ *such that*

1. *if* $t(v) = \perp$ *and* $\langle v, v' \rangle \in E$, *where* $h(v') = h(v) + 1$, *then* $t(v') = \perp$. *Vertices labelled by* $\perp$ *are called open vertices.*
2. *if* $\langle v, v' \rangle \in E$, *where* $h(v') = h(v) + 1$, *and vertex* v' *is not an open vertex, then for any* $((u_1, \cdots, u_n), u) \in t(v')$, *there is a* $u' \in D^0$ *such that* $u_n \leq u'$ *and* $((u_1, \cdots, u_{n-1}), u') \in t(v)$.

Similarly, we will say a timed labelled tree instead of a timed A-labelled tree whenever A is understood from the context.

Elements of $D^*_\perp$ associated to vertices of a labelled tree record time constraints over actions labelling paths from the root to the vertices and the maximal delay time of the machine when it arrives at the vertices. For example, if a path from the root to a vertex v are labelled by actions $a_1, \cdots, a_n$ and an element labelling v contains $((u_1, \cdots, u_n), u)$, then to arrive at v, action a_i may occur at time u_i relative to the previous action, where $i = 1, \cdots, n$. When the machine arrives at v, its maximal delay time is u.

Open vertices describe parts of machines which are not fully defined. Time constraints associated to those open vertices cannot be elaborated. As a consequence, the successor vertices of open vertices must also be open vertices. Condition 2 of Definition 2.4 represents that time constraints associated to vertices on a path starting from the root should satisfy some consistency condition. The consistency condition may be understood as follows: if vertices v and v' are connected by an edge $\langle v, v' \rangle$, where $h(v) + 1 = h(v')$, and $((u_1, \cdots, u_n), u) \in t(v')$; let $a_1, \cdots, a_{n-1}$ label the path from the root r to v and a label the edge $\langle v, v' \rangle$, then $a_1, \cdots, a_{n-1}$ and a may happen at times $u_1, \cdots, u_{n-1}$ and u_n, respectively. When the machine arrives at the state v, it may delay at least time u_n before action a, as action a occurs at time u_n. So we have $((u_1, \cdots, u_{n-1}), u') \in t(v)$ for some $u' \geq u_n$.

We use $v \xrightarrow{a}_T v'$ to represent the edge between the vertices v and v' labelled by a of a tree T, where $h(v) + 1 = h(v')$. Let $\mathcal{P}(T)$ represent the set of all paths starting from the root of T. By convention, we use V, E, r, l and t to represent the sets of vertices and edges, the root, the labelling function and the time function of a timed labelled tree T, respectively.

Since an experimenter cannot see the states (only event occurrences can be observed from outside), we identify those timed labelled trees which are the same up to changes of the names of the vertices.

Definition 2.5 *Two timed labelled trees $T = (V, E, r, l, t)$ and $T' = (V', E', r', l', t')$ are isomorphic, written as $T = T'$, if there is a bijective map $f : V \longrightarrow V'$ such that*

1. $f(r) = r'$;
2. $v \xrightarrow{a}_T v'$ *if and only if* $f(v) \xrightarrow{a}_{T'} f(v')$; *and*
3. $t = f \circ t'$.

Our timed synchronization trees are just isomorphic classes of timed labelled trees.

Definition 2.6 *A* timed synchronization tree $[V, E, r, l, t]$ *is an isomorphic class of a timed labelled tree* (V, E, r, l, t).

Let Ts represent the set of all timed synchronization trees, ranged over by T. For any timed synchronization tree $[V, E, r, l, t]$, we regard the labelled tree (V, E, r, l) as its underlying tree structure.

Timed synchronization trees can be considered as a CPO. Let $\sqsubseteq$ be a partial order over $D^*_\perp$ satisfying that for any S, $S' \in D^*_\perp$, $S \sqsubseteq S'$ implies $S = \perp$ or $S = S'$. A partial order $\sqsubseteq$ over the set $[V \longrightarrow D^*_\perp]$ of continuous time functions is the induced pointwise ordering, where the induced pointwise ordering of the set $[D \longrightarrow E]$ of continuous functions is defined as follows: for any f, $g \in [D \longrightarrow E]$, $f \sqsubseteq g$ if and only if for any $d \in D$. $f(d) \sqsubseteq g(d)$.

Definition 2.7 (A Partial Order over *Ts*) *For any $T, T' \in Ts$, we say $T \sqsubseteq T'$ if for any $(V, E, r, l, t) \in T$, there is a $(V', E', r', l', t') \in T'$ such that*

$$\begin{array}{llll} (1) & V \subseteq V' & (4) & l = l'|_E \\ (2) & E \subseteq E' & (5) & t \sqsubseteq t'|_V \\ (3) & r = r' & & \end{array}$$

where for any function f, we use $f|_S$ to represent the restriction of f to the set S.

Theorem 2.8 $(Ts, \sqsubseteq)$ *is a* CPO.

3 Constructions

In this section, we consider some constructions over *Ts*.

3.1 Constants

Let Ω be the timed synchronization tree $[\{v\}, \emptyset, v, \emptyset, \{(v, \perp)\}]$, which is the least element of *Ts*, and *NIL* be the timed synchronization tree $[\{v\}, \emptyset, v, \emptyset, \{(v, \infty)\}]$.

3.2 Prefix

In real-time systems, an important notion is time dependency. It says that time for actions may depend on happening time of their previous actions. To capture the notion, we first define a time synchronization tree with a parameter as an isomorphism class of a labelled tree T with a time function which assigns to every vertex v of depth i a function $f_v : T \longrightarrow D^i_\perp$ such that if $v \rightarrow_T v'$ then for any $u \in T$, $f_v(u)$ and $f_{v'}(u)$ satisfy conditions 1 and 2 of Definition 2.4. A timed synchronization tree with a parameter x can be regarded as a family of timed synchronization trees which are indexed by the parameter x. All timed synchronization trees in the family have the same underlying tree structure, i.e. they represent the same labelled tree if we ignore their time function. and only time constraints associated to every vertices may contain the parameter x. We then generalise the notion of timed synchronization trees with a parameter to a notion of timed synchronization trees with n parameters.

Definition 3.1 *For any $a \in A$, $S \subseteq T$ which has a least upper bound in D^0, and a timed synchronization tree T, possibly with parameters, the prefix $a(S).(\lambda x.T)$ represents a new timed synchronization tree T', possibly with parameters, such that*

- $V' = \{a\} \uplus V$
- $E' = \{\langle (v, 1), (v', 1) \rangle : \langle v, v' \rangle \in E\} \uplus \{\langle (a, 0), (r, 1) \rangle\}$
- $r' = (a, 0)$
- $l' = \lambda(e, i).\begin{cases} a & i = 1 \& e = \langle (a,0), (r,1) \rangle \\ l(\langle v, v' \rangle) & i = 0 \& e = \langle (v,1), (v',1) \rangle \end{cases}$
- $t' = \lambda(v, i).\begin{cases} Sup(S) & (v, i) = r' \\ \perp & t_u(v) = \perp \\ & \textit{where } t_u(v) = (t\{u/x\})(v) \;\&\; u \in S \\ \{\langle u \rangle s : u \in S \wedge s \in t_u(v)\} & \textit{otherwise} \end{cases}$

where $Sup(S)$ represents the least upper bound of S. By convention, $Sup(\emptyset) = 0$.

Note that we use $\lambda x.T$ in the prefix $a(S)(\lambda x.T)$ to emphasize the parameter x of T which will be instantiated next. We still use Ts to represent the set of all timed synchronization trees, possibly with parameters. Let $\sqsubseteq$ over timed synchronization trees with parameters be the induced pointwise ordering, i.e. for any timed synchronization trees T and T', where T and T' contain at most parameters $x_1, \cdots, x_n$, we say $T \sqsubseteq T'$ when $T\{u_1/x_1, \cdots, u_n/x_n\} \sqsubseteq T'\{u_1/x_1, \cdots, u_n/x_n\}$ for any $u_1, \cdots, u_n \in T$.

Theorem 3.2 *$a(S) : Ts \longrightarrow Ts$ is continuous with respect to $\sqsubseteq$.*

3.3 Summation

The sum of two timed synchronization trees T and T' represents a new timed synchronization tree resulted by putting T and T' together at their roots.

Definition 3.3 *For any timed synchronization trees T_1 and T_2, possibly with parameters, $T_1 + T_2$ represents a new timed synchronization tree T, where*

- $V = (V_1 - \{r_1\}) \uplus (V_2 - \{r_2\}) \uplus \{(r_1, r_2)\}$
- $E = \{\langle (v,0), (v',0)\rangle : v \neq r_1 \& \langle v, v'\rangle \in E_1\}$
  $\uplus \{\langle (v,1), (v',1)\rangle : v \neq r_2 \& \langle v, v'\rangle \in E_2\}$
  $\uplus \{\langle ((r_1, r_2), 2), (v, i)\rangle : \langle r_{i+1}, v\rangle \in E_1 \vee \langle r_{i+1}, v\rangle \in E_2\}$
- $r = ((r_1, r_2), 2)$
- $l = \lambda(e,i). \begin{cases} l_1(\langle v, v'\rangle) & i = 0 \& e = \langle (v,0), (v',0)\rangle \\ l_2(\langle v, v'\rangle) & i = 1 \& e = \langle (v,1), (v',1)\rangle \\ l_{j+1}(\langle r_{j+1}, v\rangle) & i = 2 \& e = (r, (v,j)) \& v \in V_{j+1} \& j = 0, 1 \end{cases}$
- $t = \lambda(v,i). \begin{cases} max(t_1(r_1), t_2(r_2)) & (v,i) = r \\ t_i(v) & v \in V_i \,\&\, i = 0, 1 \end{cases}$

where, by convention, $max(u, \perp) = u$.

Remark If we use *min* to replace *max* in the definition of summation, we get Moller and Tofts' strong summation [MT90]. □

Clearly the summation operation is well defined. Moreover we have the following theorem.

Theorem 3.4 *$+ : Ts \times Ts \longrightarrow Ts$ is continuous with respect to $\sqsubseteq$.*

3.4 Restriction

The restriction of a specific action $\lambda \in A$, where $\lambda \neq \tau$, on a timed synchronization tree T represents a new timed synchronization tree resulted by pruning away all edges labelled by λ and $\overline{\lambda}$ together with their successors.

Definition 3.5 *Let T be a timed synchronization tree, possibly with parameters, and $\lambda \in A$, where $\lambda \neq \tau$, $T \backslash \lambda$ represents a new timed synchronization tree T', where*

- $V' = \{v$: *there is a path* $e_1, \cdots, e_n$ *from the root* r *to* v *such that for any* $i = 1, \cdots, n,\ l(e_i) \neq \lambda$ *and* $l(e_i) \neq \bar{\lambda}\}$
- $E' = E \cap V' \times V'$
- $r' = r$
- $l' = l|_{E'}$
- $t' = t|_{V'}$

Clearly the restriction operator is well defined. Moreover we have the following theorem.

Theorem 3.6 $\backslash\lambda : Ts \longrightarrow Ts$ *is continuous with respect to* $\sqsubseteq$.

3.5 Parallel

We can also define a parallel operation on timed synchronization trees.

Definition 3.7 *For any* T_0, $T_1 \in Ts$, *the interleaving* $\mathcal{I}(T_0, T_1)$ *of their vertices is defined as*

$$\mathcal{I}(T_0, T_1) = \{v : \pi_i(v) \in \mathcal{P}(T_i)\ \&\ v \in C^*\ \&\ i = 0, 1\}$$

where $C = E_0 \uplus E_1 \uplus \{(e, e') : e \in E_0\ \&\ e' \in E_1\ \&\ \overline{l_0(e)} = l_1(e') \neq \tau\}$, $\mathcal{P}(T)$ *is the set of all path starting from the root in* T *and we first define* $\pi_i : C \longrightarrow E_i$, $i = 0, 1$, *as*

$$\pi_i = \lambda x. \begin{cases} \langle\rangle & x \in E_{1-i} \\ \langle x \rangle & x \in E_i \\ \langle x_i \rangle & x = (x_0, x_1) \in E_0 \times E_1 \end{cases}$$

then we generalize $\pi_i : C^* \longrightarrow E_i^*$ *as follows*

$$\pi_i(\langle e_1 \cdots e_n \rangle) = \pi_i(e_1) \cdots \pi_i(e_n)$$

Clearly, for any timed synchronization trees T_1 and T_2, the interleaving $\mathcal{I}(T_1, T_2)$ of their vertices contains the null sequence $\langle\rangle$ and is closed under the initial subsequence relation.

Definition 3.8 *For any* T_0, $T_1 \in Ts$, *their parallel composition* $T_0 \,|\, T_1$ *represents a new timed synchronization tree* T, *where*

- $V = \mathcal{I}(T_0, T_1)$
- $E = \{\langle v, v' \rangle : v, v' \in V\ \&\ \exists \alpha \in C.\ v\langle \alpha \rangle = v'\}$
- $r = \langle\rangle$
- $l = \lambda e. \begin{cases} l_i(\alpha) & e = \langle v, v' \rangle\ \&\ v\langle(\alpha, i)\rangle = v'\ \&\ i = 0, 1 \\ \tau & e = \langle v, v' \rangle\ \&\ v\langle(\alpha, 2)\rangle = v' \end{cases}$
- $t = \lambda v.S$

  where if $t_i(\pi_i(v)) = \bot$ $(i = 0, 1)$ *then* $S = \bot$;
  otherwise, let $v = \langle \alpha_1, \cdots, \alpha_n \rangle$, *then* S *is the least set which satisfies*
  (1) $(u_1, \cdots, u_n, u) \in S$ *implies that*

$$\Big(\sum_{1 \leq j \leq k_1} u_j, \cdots, \sum_{k_{l-1} < j \leq k_l} u_j, u'\Big) \in t_0(\pi_0(v))$$

$$(\sum_{1\le j\le k'_1} u_j,\cdots,\sum_{k'_{m-1}<j\le k'_m} u_j,u'')\in t_1(\pi_1(v))$$

and

$$u=min(u'-\sum_{k_l<j\le n} u_j,u''-\sum_{k'_m<j\le n} u_j)$$

for some u', $u''\in D^0$, *where* $\{k_1,\cdots,k_l\}$ *and* $\{k'_1,\cdots,k'_m\}$ $(k_1\le\cdots\le k_l$ *and* $k'_1\le\cdots\le k'_m)$ *satisfy that* $\forall j\in\{k_1,\cdots,k_l\}.\ \pi_0(\alpha_j)\ne\langle\rangle\ \&\ \forall i\in\{1,\cdots,n\}-\{k_1,\cdots,k_l\}.\ \pi_0(\alpha_i)=\langle\rangle\ \&\ \forall j\in\{k'_1,\cdots,k'_m\}.\ \pi_1(\alpha_j)\ne\langle\rangle\ \&\ \forall i\in\{1,\cdots,n\}-\{k'_1,\cdots,k'_m\}.\ \pi_1(\alpha_i)=\langle\rangle$; *and*
(2) $(u'_1,\cdots,u'_l,u')\in t_0(\pi_0(v))$ *and* $(u''_1,\cdots,u''_m.u'')\in t_1(\pi_1(v))$ *implies that there are* $(u_1,\cdots,u_n,u)\in t(v)$, *such that*

$$u'_1=\sum_{1\le j\le k_1} u_j,\cdots,u'_l=\sum_{k_{l-1}<j\le k_l} u_j$$

$$u''_1=\sum_{1\le j\le k'_1} u_j,\cdots,u''_l=\sum_{k'_{m-1}<j\le k'_m} u_j$$

and

$$u=min(u'-\sum_{k_l<j\le n} u_j,u''-\sum_{k'_m<j\le n} u_j)$$

where $k_1,\cdots.k_l,k'_1,\cdots.k'_m$ *are defined as in* (1) *and* n *is the size of the set* $\{k_1,\cdots,k_l\}\cup\{k'_1,\cdots,k'_m\}$.

It is easy to check that the parallel operation is well defined. Moreover we have the following theorem.

Theorem 3.9 $|: Ts\times Ts\longrightarrow Ts$ *is continuous with respect to* $\sqsubseteq$.

3.6 Relabelling

By relabelling a tree T with a relabelling function $R: A\longrightarrow A$ we get a new tree which is resulted from T by changing the labels of the edges of T using the relabelling function R.

Definition 3.10 *Let* $[V,E,r,l,t]\in Ts$ *and* $R: A\longrightarrow A$ *be a relabelling function,* $[V,E,r,l,t][R]$ *represents a new tree* $[V,E,r,l\circ R,t]$.

It is easy to check that relabelling operation is well defined. Moreover we have the following theorem.

Theorem 3.11 $[R]: Ts\longrightarrow Ts$ *is continuous with respect to* $\sqsubseteq$.

4 An Equivalence over Ts

Trees are concrete structures and contain detailed information. They distinguish too many things. For example, a tree T in general is not equal to a tree $T+T$, although it is hard to imagine any programming contexts in which their behaviour could be distinguished. In this section, we define an equivalence on Ts based on a notion of behaviour of timed synchronization trees. The equivalence identifies those trees which have the same behaviour.

Definition 4.1 *Given T, $T' \in Ts$, we write $T \xrightarrow{a}_u T'$, where $a \in A$ and $u \in T$, when $r \xrightarrow{a}_T r'$ and $u \in \{u' : (u', u'') \in t(r')\}$ together with*

1. $V' = \{v : \text{there is a path between } r' \text{ and } v \text{ in } (V, E - \{(r, r')\})\}$
2. $E' = E \cap V' \times V'$
3. $l' = l|_{E'}$
4. $t' = \lambda v. \begin{cases} \bot & t(v) = \bot \\ \{(u_2, \cdots, u_{n+1}, u') : (u, u_2, \cdots, u_{n+1}, u') \in t(v)\} & \text{otherwise} \end{cases}$

We can understand $T \xrightarrow{a}_u T'$ as that the machine represented by T performs an action a at time u and then evolves to a state represented by T'. A timed synchronization tree is closed if it contains no open vertices. The presence of an open vertex indicates that the tree is not fully defined.

Definition 4.2 *A binary relation $\mathcal{R}$ over closed timed synchronization trees is a tree bisimulation if $(T_1, T_2) \in \mathcal{R}$ implies that for any $a \in A$ and $u \in T$*

1. *if $T_1 \xrightarrow{a}_u T_1'$, then $T_2 \xrightarrow{a}_u T_2'$ and $(T_1', T_2') \in \mathcal{R}$ for some T_2';*
2. *if $T_2 \xrightarrow{a}_u T_2'$, then $T_1 \xrightarrow{a}_u T_1'$ and $(T_1', T_2') \in \mathcal{R}$ for some T_1'; and*
3. *$t_1(r_1) = t_2(r_2)$.*

Two closed trees T and T' are bisimilar, written as $T \simeq T'$, if there is a tree bisimulation $\mathcal{R}$ such that $(T, T') \in \mathcal{R}$. It is easy to see that $\simeq$ itself is a tree bisimulation, the largest tree bisimulation. Moreover it is an equivalence relation.

Since $(Ts, \sqsubseteq)$ is a CPO and all operations are continuous with respect to $\sqsubseteq$, we may define infinite timed synchronization trees along the standard line. As an example, we use $\mu X.T$ to represent the limit of an ω-chain $T^{(0)}, \cdots, T^{(n+1)}, \cdots$, where

$$T^{(0)} = \Omega \quad \text{and} \quad T^{(n+1)} = T\{T^{(n)}/X\}$$

Lemma 4.3 *Given timed synchronization trees T_1 and T_2, if $T_1 \sqsubseteq T_2$ and $T_1 \xrightarrow{a}_u T_1'$, then $T_2 \xrightarrow{a}_u T_2'$ and $T_1' \sqsubseteq T_2'$ for some tree T_2'.*

With Lemma 4.3, we can now define behaviours of an infinite timed synchronization tree.

Definition 4.4 *For an infinite tree $\mu X.T$, we say $\mu X.T \xrightarrow{a}_u T'$ if there is a $n \in \omega$ such that for any $i \geq n$*

$$T^{(i)} \xrightarrow{a}_u T_i' \quad \text{and} \quad T' = \bigsqcup_{\substack{i \in \omega \\ i \geq n}} T_i'$$

where $T^{(0)} = \Omega$ and $T^{(i+1)} = T\{T^{(i)}/X\}$ for every $i \in \omega$.

Proposition 4.5 *$\mu X.T \xrightarrow{a}_u T'$ if and only if $T\{\mu X.T/X\} \xrightarrow{a}_u T'$.*

5 Using Model as a Semantics

We have shown how Ts can be considered as a CPO. Now we can interpret real-time process algebras in Ts in a straightforward way. As an example, we define a denotational semantics for Timed CCS [Che92a]. For completeness sake, we brief Timed CCS before giving its denotational semantics.

Let $Act = \Lambda \cup \{\tau\}$ be a set of action names, ranged over by a, b. We presuppose a structure on Λ that it can be partitioned into Γ, a set of names, and $\bar{\Gamma} = \{\bar{a} \mid a \in \Gamma\}$, a set of co-names, with the provision that $\bar{\bar{a}} = a$. We call actions a and $\bar{a}$ a pair of complementary actions which forms the basis of communications in Timed CCS, analogous to CCS. A relabelling function $S : Act \longrightarrow Act$ satisfies $\overline{S(a)} = S(\bar{a})$ and $S(\tau) = \tau$. We presuppose a set V_t of time variables, ranged over by x, and a set V_p of process variables, ranged over by X. Time expressions are linear arithmetic expressions over T which may contain max and min operators. The process expressions of Timed CCS, ranged over by P, Q, are defined by the following BNF expression

$$P ::= X \mid nil \mid a(x)_e^{e'}.P \mid P + Q \mid P \mid Q \mid P \backslash b \mid P[S] \mid \mu X.P$$

where $b \neq \tau$, e and e' are time expressions or e' is the infinite time ∞.

Process nil cannot do any action, but idles indefinitely. Process $a(x)_e^{e'}.P$ can perform action a at any time of the interval $e \leq t \leq e'$. After performs action a at time t, it evolves to a process $P\{t/x\}$, where $P\{t/x\}$ is the result of substituting all free occurrences of x in P. The time variable x of P in $a(x)_e^{e'}.P$ allows us to express the notion of time dependency. Summation $P+Q$ represents choice between processes P and Q. The choice is made at the time of the first action of P or Q, or at the time when only one process can idle. In the late case, the process which cannot delay is dropped from the further computation. Process $P \mid Q$ represents the parallel composition of processes P and Q. Each of them may perform actions independently, or they may synchronize on complementary actions which represent communications between them. Parallel composition is synchronous with respect to time progress, i.e. the parallel composition $P \mid Q$ of processes P and Q can delay time t only when both P and Q can.

We say a process P is weakly guarded if every process variable of P is weakly guarded in P, where a process variable X is weakly guarded in P if every occurrence of X is in some subterm of form $a(x)_e^{e'}.Q$ of P. For example, process $a(x)_0^{\infty}.X + b(x)_0^{20}.nil$ is weakly guarded, but process $a(x)_0^{\infty}.X + X$ is not.

To deal with process variables, we introduce a notion of environments which are mappings from V_p to Ts. Let Env be the set of environments, ranged over by ρ. For any environment ρ, $\rho\{T/X\}$ represents a new environment which differs from ρ only at X, where it is defined to be T. The denotational semantics is given by a mapping

$$\mathcal{M} : \mathcal{P} \longrightarrow (Env \longrightarrow Ts)$$

which is defined by induction on processes.

Definition 5.1 *For any weakly guarded process $P \in \mathcal{P}$ and environment ρ, the denotation of P with respect to ρ, written as $\mathcal{M}(P)\rho$, is defined by induction on P as follows:*

(1) $\mathcal{M}(X)\rho = \rho(X)$
(2) $\mathcal{M}(nil)\rho = NIL$
(3) $\mathcal{M}(a(x)_e^{e'}.P)\rho = a(\{t \,:\, e \leq t \leq e'\})(\lambda x.\mathcal{M}(P)\rho)$
(4) $\mathcal{M}(P+Q)\rho = \mathcal{M}(P)\rho + \mathcal{M}(Q)\rho$
(5) $\mathcal{M}(P\,|\,Q)\rho = \mathcal{M}(P)\rho\,|\,\mathcal{M}(Q)\rho$
(6) $\mathcal{M}(P\backslash b)\rho = (\mathcal{M}(P)\rho)\backslash b$
(7) $\mathcal{M}(P[S])\rho = (\mathcal{M}(P)\rho)[S]$
(8) $\mathcal{M}(\mu X.P)\rho = \mu T.(\mathcal{M}(P)\rho\{T/X\})$

For any weakly guarded agent P, $\mathcal{M}(P)\rho$ is a closed timed synchronization tree. If P is closed with respect to process variables, then $\mathcal{M}(P)$ is a constant function from *Env* to *Ts*, which may be regarded as an element of *Ts*.

Acknowledgements: I would like to thank Stuart Anderson, Robin Milner, Faron Moller, Alistair Munro and John Power for many invaluable discussions and suggestions, especially to Robin Milner for his suggestion on studying models of timed calculi at my thesis proposal committee. Also, I would like to thank anonymous referees for their useful comments.

References

[ACM92] S. Anderson, L. Chen & F. Moller, *Observing Causality in Real-Timed Calculi*, Preliminary Report, LFCS, University of Edinburgh, 1992

[BB91] J.C.M. Baeten & J.A. Bergstra, *Real Time Process Algebra*, Formal Aspects of Computing, Vol 3. No 2, pp142-188, 1991

[Che91] L. Chen, *Decidability and Completeness in Real-Time Processes*, Technical Report ECS-LFCS-91-185, Edinburgh University, 1991

[Che92a] L. Chen, *An Interleaving Model for Real-Time Systems*, in Proc. of Symp. of Logical Foundations of Computer Science, Lecture Notes in Computer Science 620, pp 81-92, Springer-Verlag, 1992

[Che92b] L. Chen *Timed Processes: Models, Axioms and Decidability*, Ph.D Thesis, University of Edinburgh, 1992

[Che93] L. Chen, *Axiomatising Real-Timed Processes*, Proc. of MFPS'93, Lecture Notes in Computer Science, Springer-Verlag, 1993

[CM93] L. Chen & A. Munro. *Applications of Modal Logic for the Specifications of Real-Time Systems*, Proc. of FME'93, Lecture Notes in Computer Science 670, pp 235-249, Springer-Verlag, 1993

[Hen85] M. Hennessy, *Acceptance Trees*, Journal of the Association for Computing Machinery, Vol 32, No 4, pp 896-928, 1985

[Hen88] M. Hennessy, **Algebraic Theory of Processes**, The MIT Press, 1988

[Mil80] R. Milner, *A Calculus of Communicating systems*, Lecture Notes in Computer Science 92, Springer-Verlag, 1980

[Mil89] R. Milner, **Communication and Concurrency**, Prentice-Hall International, 1989

[MT90] F. Moller & C. Tofts, *A Temporal Calculus of Communicating System*, Proc. of CONCUR'90, Lecture Notes in Computer Science 458, pp 401-415, 1990

[RR88] R. Reed & A. W. Roscoe, *A Timed Model for Communicating Sequential Processes*, Theoretical Computer Science 58, pp 249-261. 1988

[Win84] G. Winskel, *Synchronization Trees*, Theoretical Computer Science 34, pp32-82, 1984

Data Encapsulation and Modularity: Three Views of Inheritance

J.F.Costa, A.Sernadas & C.Sernadas

INESC & Dept. Matemática - IST
Apartado 10105, 1017 Lisboa Codex, PORTUGAL
fgc@inesc.pt

Abstract. A semantic domain based on state-machines is proposed for object-orientation in order to clarify the most important constructions: aggregation, interconnection and specialization. Three kinds of specialization are discussed: subtyping (specialization without side-effects and no non-monotonic overriding); monotonic specialization (possibly with side-effects but still with monotonic overriding only); and non-monotonic specialization (possibly with side-effects and non-monotonic overriding).

1.Introduction

The advantages of the object-oriented approach to software engineering are by now well understood. Many object-oriented languages, systems and methods have been proposed and some of them are now extensively used. To name just a few, there are programming languages like Smalltalk, C++ and Eiffel, database systems like GemStone, O_2, IRIS and ORION, development methods like GOOD, MOOD and HOOD, as well as high-level specification languages [Goguen 91, Jungclaus *et al* 91, SernadasA *et al* 89]. Moreover, some effort has been dedicated to the theoretical foundations of this promising paradigm. The overview [Ehrich *et al* 92] points to several schools of thought in this area: *e.g.* ADT-based like [Goguen 91] and process-based like [Ehrich *et al* 91]. Higher programmer productivity through software reuse is paramount among the claimed advantages of object orientation. A rich mechanism of inheritance, including both specialization and abstraction, is essential for supporting the envisaged levels of software reuse. Another essential ingredient is a general mechanism for putting objects together, that is, for setting-up communities of interacting objects. Therefore, a semantic foundation of object-orientation should address all these issues: specialization, abstraction and interconnection. Herein, we adopt the view that an object is a state-machine, adapting the notion of reactive system proposed in [Manna and Pnueli 81]. Then, we develop the notion of object morphism in order to explain the basic relationhips between objects: *is-a* and *is-part-of*. To this end we apply the program outlined in [Costa *et al* 92] to objects as state-machines. The advantages of the new semantic domain are all related to the abstract view of the state: both the behaviour (process) and the memory structure (slot valuation) of the object appear as secondary concepts on top of the notion of state. By avoiding concrete states we arrive at a cleaner notion of object morphism that makes possible a unified semantic domain explaining all basic relationships between objects. Namely, we explain three kinds of specialization (strict, with side-effects, and with overriding), the part-of relationship between an object and each of its components and two basic kinds of mechanisms for interconnecting objects (component sharing, of which event sharing is a special case, as well as event calling). To our knowledge this is the first semantic domain for objects that incorporates these features.

In Section 2 the basic category of objects as state-machines is introduced. Section 3 addresses subtyping (strict specialization). Section 4 deals with object interconnection. Finally, in Section 5 we deal with liberal forms of specialization. We make moderate use of the theory of categories. The reader may find all the necessary category-theoretic notions in the book by Jiří Adámek *et al* [Adámek *et al* 90].

2.Objects

We start with an extension of the model found in [Winskel 84]. By Pfn we denote the category of sets and partial functions. Let V be a fixed class of sets.

Definition 2.1. A *slot alphabet* is a pair Att=$<A,\partial>$ where A is a set and ∂ is a total map from A into V. ◇

The idea is that, for each slot *a* in A, ∂ returns its codomain, i.e., an element of V. In this way $\partial(a)$ is the set of the possible values that slot *a* may have.

Definition 2.2. An *object* ob=$<E,Att,S,Q,B,\mu,\delta>$ is a septuple consisting of a set E, a slot alphabet Att=$<A,\partial>$, a set S, two subsets Q and B of S, a family $\mu=\{\mu_a\}_{a\in A}$ of total functions $\mu_a:S\to 2^{\partial(a)}$ and a family $\delta=\{\delta_e\}_{e\in E}$ of total functions $\delta_e:S\to 2^S$. ◇

The elements of E are called *events*, the elements of S *states*, the elements of Q *quiescent states* and the elements of B birth (*initial*) *states*. Each *valuation function* $\mu_a:S\to 2^{\partial(a)}$ gives the set of values that *a* may have in a given state. If $\mu_a(s)$ is empty we say that *the slot a is undefined in state s*. It may also happen that a slot is given more than one value. Finally, each function $\delta_e:S\to 2^S$ is called the *transition function for e*: gives the set of states reachable from a given state by the occurrence of event *e*.

Consider a *bank* account as object *acc*, with $A_{acc}=\{\text{balance}\}$, $\partial_{acc}(\text{balance})=\mathbf{Z}_\perp$†, $E_{acc}=\{\text{open}\}\cup\{\text{withdrawal(i)},\text{deposit(i)}:i\in\text{integer}\}$. The set of states is $\mathbf{Z}_\perp$. The set of quiescent states is also $\mathbf{Z}_\perp$. The set of initial states is $\{\perp\}$. We have a unique valuation function:

$$\mu_{acc,balance} = \lambda s.\{s\}$$

and transition functions are given as follows:

$$\delta_{acc,open} = \lambda s.\text{if } s=\perp \text{ then } \{1000\} \text{ else } \emptyset$$

$$\delta_{acc,deposit(i)} = \lambda s.\text{if } s=\perp \text{ then } \emptyset \text{ else } \{s+i\}$$

$$\delta_{acc,withdrawal(i)} = \lambda s.\text{if } s=\perp \text{ then } \emptyset \text{ else } \{s-i\}$$

It is now possible to establish a suitable *category of objects* by choosing an appropriate notion of *object morphism*.

---

† By $\mathbf{Z}_\perp$ we mean the set of integers plus the distinguished element $\perp$.

Definition 2.3. A *slot morphism* $f:<A_1,\partial_1>\rightarrow<A_2,\partial_2>$ is a total map $f_A:A_1\rightarrow A_2$ such that $\partial_1(a)=\partial_2(f_A(a))$, for every $a\in A_1$. ◇

Proposition 2.4. The slot alphabets and slot morphisms constitute a complete and cocomplete category Slt (called the *category of slots*). ◇

Now, we want to ensure that if we extend a given object with more slots and events then the additional events should have no effects on the old slots (*no side-effects*). Clearly, that should be the case when an object is part of a composite object: the events of the other parts should not affect within the whole the slots of the part. The same is to be expected in the case of a strict form of specialization without side-effects: the additional events of the specialization should not affect the inherited slots. Finally, in both cases, the effects of original events on the original slots should be the same in the extended object (*no overriding*). Assuming a *frame constraint* (no side-effects and no overriding) we arrive at the following definition:

Definition 2.5. A *frame morphism*

$$h:<E_1,<A_1,\partial_1>,S_1,Q_1,B_1,\mu_1,\delta_1>\rightarrow<E_2,<A_2,\partial_2>,S_2,Q_2,B_2,\mu_2,\delta_2>$$

is a triple $<h_E,h_A,h_S>$ where $h_E:E_2\rightarrow E_1$ is a partial function, $h_A:A_1\rightarrow A_2$ is a slot morphism and $h_S:S_2\rightarrow S_1$ is a total function such that the following conditions hold: (a) $Q_2\subseteq h_S^{-1}(Q_1)$, (b) $B_2\subseteq h_S^{-1}(B_1)$, (c) $\mu_{1,a}(h_S(s))\subseteq\mu_{2,h_A(a)}(s)$, for every $a\in A_1$, $s\in S_2$, and (d) for every $e\in E_2$, $s\in S_2$, if $h_E(e)\downarrow$ then $\delta_{2,e}(s)\subseteq h_S^{-1}(\delta_{1,h_E(e)}(h_S(s)))$ else $\delta_{2,e}(s)\subseteq h_S^{-1}(h_S(\{s\}))$. ◇

From now on, when $h_E(e)\uparrow$ we write $\delta_{h_E(e)}$ for the function $\delta_\perp:s\mapsto\{s\}$. We also write $\delta_{2,e}\subseteq h_S^{-1}\delta_{1,h_E(e)}h_S$ for $\delta_{2,e}(s)\subseteq h_S^{-1}(\delta_{1,h_E(e)}(h_S(s)))$ for every $e\in E_2$ and for every $s\in S_2$.

Since the morphism condition should reflect the envisaged frame constraint, the reader might expect equality instead of inclusion. However, that would be too strong. The resulting category would not be cocomplete. It is easier to work with the more relaxed notion of morphism and recapture the frame constraint by the interplay between the morphism constraint (inclusion) and the universal property of the colimits as explained below in the case of object aggregation.

Proposition 2.6. Objects and frame morphisms constitute a cocomplete, concrete category FrOb over $Pfn^{op}\times Slt\times Set^{op}$. ◇

Definition 2.7. Let $ob=<E,<A,\partial>,S,Q,B,\mu,\delta>$ be an object. *Menu* of ob is the total function $M_{ob}:S\rightarrow 2^E$ defined by $M_{ob}=\lambda s.\{e\in E:\delta_e(s)\neq\emptyset\}$. ◇

Clearly, $M_{ob}(s)$ is the set of enabled events at s. Starting with a given intial state we may establish a life-cycle λ of the object by successively picking up an enabled event and determining the subsequent state using δ until a quiescent state is reached. We might call the set of all such life-cycles the process induced by the object, recovering the model used e.g. in [Costa *et al* 92]. As we shall see in Section 4, as expected, coproducts in FrOb do correspond to parallel composition of the induced processes.

3.Subtyping

At this point we are ready to *start* the discussion of what is the meaning of ob' *is-a* ob, that is, ob' *is a specialization of* ob. Intuitively, we would like it to be: ob' can be taken as ob if we forget a few details. For instance, a savings account is an account in this sense. The former may have a few more slots and events, more specific behaviour and valuation, but it should be possible to forget all that and recognize a simple account. Therefore, a *strict is-a* relationship should correspond to a frame object morphism in the opposite direction. In this strict sense, ob' *is-a* ob iff there is a *frame inclusion morphism*† from ob into ob'. We classify this relationship as strict because as we shall see it disallows both side-effects and overriding. For instance, we would say that a savings account sav_acc *is (strictly) an* acc iff there is a frame inclusion from acc into sav_acc: $E_{sav\_acc} = \{open, new\_day, start\_saving, end\_saving\} \cup \{withdrawal(i), deposit(i) : i \in integer\}$, $A_{sav\_acc} = \{balance, saving?, days\}$, $S_{sav\_acc} = \mathbf{Z}_\perp \times \mathbf{Z}_\perp \times \mathbf{Z}_\perp$, $Q_{sav\_acc} = \mathbf{Z}_\perp \times \mathbf{Z}_\perp \times \mathbf{Z}_\perp$, $B_{sav\_acc} = \{<\perp,\perp,\perp>\}$. With respect to behaviour, the morphism requirement is very simple and natural: any possible menu at a given state of sav_acc should degenerate into a submenu of acc at the projected state when we forget the specific events. Note that we are free to impose further constraints on the behaviour of sav_acc even with respect to the events inherited from acc. For instance, we might impose that if the sav_acc is in the saving mode (i.e., if saving?=1) then no withdrawals are allowed (only deposits). But, in order to illustrate another aspect discussed below, we prefer instead to allow a withdrawal to happen even when acc is in the saving mode and consider that in that event the mode is changed back to non-saving. The slot days counts the number of days since the beginning of the saving period and is not changed when sav_acc is not in the saving mode. The slot saving? indicates if acc is in the saving mode (saving?=1) or not (saving?=0). Thus††:

$$\mu_{sav\_acc,balance} = \lambda s.\{\pi_1(s)\}$$

$$\mu_{sav\_acc,saving?} = \lambda s.\{\pi_2(s)\}$$

$$\mu_{sav\_acc,days} = \lambda s.\{\pi_3(s)\}$$

$$\delta_{sav\_acc,open} = \lambda s.\text{if } s=<\perp,\perp,\perp> \text{ then } \{<1000,0,\perp>\} \text{ else } \emptyset$$

$$\delta_{sav\_acc,deposit(i)} = \lambda s.\text{if } s=<\perp,\perp,\perp> \text{ then } \emptyset \text{ else } \{<\pi_1(s)+i,\pi_2(s),\pi_3(s)>\}$$

$$\delta_{sav\_acc,withdrawal(i)} = \lambda s.\text{if } s=<\perp,\perp,\perp> \text{ then } \emptyset \text{ else } \{<\pi_1(s)-i,0,\pi_3(s)>\}$$

$$\delta_{sav\_acc,start\_saving} = \lambda s.\text{if } s=<\perp,\perp,\perp> \text{ then } \emptyset \text{ else } \{<\pi_1(s),1,0>\}$$

$$\delta_{sav\_acc,end\_saving} = \lambda s.\text{if } s=<\perp,\perp,\perp> \text{ then } \emptyset \text{ else } \{<\pi_1(s),0,\pi_3(s)>\}$$

$$\delta_{sav\_acc,new\_day} = \lambda s.\text{if } s=<\perp,\perp,\perp> \text{ then } \emptyset \text{ else } \{<\pi_1(s),\pi_2(s),\pi_3(s)+1>\}$$

† By frame inclusion morphism f from ob into ob' we mean a morphism induced by an inverse of an inclusion on events, an inclusion on slots and a surjective map on states.

†† We use π_i for the i-th projection of a product.

Note that the original events of account are allowed to change the specific slots of sav_acc — namely withdrawals may change the extra slots. On the contrary, the new events of sav_acc do not affect the values of the old slots inherited from acc (no side-effects!) because of the frame morphism constraint on functional objects. *E.g.*,

$$\pi_1(\delta_{sav\_acc \cdot start\_saving}(s))=\delta_{acc,\pi_1(start\_saving)}(\pi_1(s))=\{\pi_1(s)\}$$

and this equalily reads as follows: the set of sucessor states of s in sav_acc are all mapped onto the state $\pi_1(s)$; thus the balance at state s ($=\pi_1(s)$) coincide with the balance at any state belonging to $\delta_{sav\_acc \cdot start\_saving}(s)$.

Moreover, the effects of the old events on the old slots inherited from acc are the same in sav_acc (no overriding!) again because of the frame morphism constraint on functional objects. Eg, if $s \in S_{sav\_acc}$ is different from $<\perp,\perp,\perp>$, then, we have $\{<\pi_1(s)+i>\}= \delta_{acc \cdot deposit(i)}(\pi_1(s))=\pi_1(\delta_{sav\_acc \cdot deposit(i)}(s))=\pi_1(<\pi_1(s)+i,\pi_2(s),\pi_3(s)>)=\{<\pi_1(s)+i>\}$.

This is a strict version of the *is-a* relationship. We shall provide more liberal versions in a subsequent section of this paper.

4.Interaction

We just saw that frame morphisms explain only a rather strict form of specialization: subtyping. Therefore, for inheritance purposes we need to consider other more general morphisms. However, the frame morphisms are precisely what is needed to explain object aggregation! Indeed, the slots of an aggregation component should not be affected by the events of the other parts (no side-effects!). And the effects of the events of an aggregation component should be the same in the whole (no overriding!).

Therefore, we are now ready to characterize object *aggregation* (without and with interaction) using some colimits in the category FrOb, keeping in mind that wrt behaviour it should correspond to parallel composition. For instance, binary coproducts provide the disjoint aggregation of the argument objects: disjoint parallel composition of behaviours encoded within the product of transition functions, plus addition of valuation maps.

As an illustration consider the coproduct of two bank accounts (two isormorphic copies of acc): $acc_1 || acc_2$. Note that we are extending the traditional notation for parallel composition of processes to object aggregation, since they are so closely related. As expected these two independent objects can be put together without affecting each other. Clearly, there must be a frame injection morphism from acc_i into $acc_1 || acc_2$, for i=1,2.

Now consider a client cl owning an account acc: $E_{cl}=\{open\_acc, close\_acc\} \cup \{become\_client(i), withdrawal(i), deposit(i): i \in integer\}$, $A_{cl}=\{number\}$, $S_{cl}=\mathbf{Z}_\perp$, $Q_{cl}=\mathbf{Z}_\perp$, and $B_{cl}=\{\perp\}$. We also have:

$$\mu_{cl,number} = \lambda s.\{s\}$$

$$\delta_{cl,become\_client(i)} = \lambda s.\text{if } s=\perp \text{ then } \{i\} \text{ else } \emptyset$$

$$\delta_{cl,open\_acc} = \lambda s.\text{if } s=\perp \text{ then } \emptyset \text{ else } \{s\}$$

$\delta_{cl,close\_acc} = \lambda s.\text{if } s{=}\bot \text{ then } \varnothing \text{ else } \{s\}$

$\delta_{cl,withdrawal(i)} = \lambda s.\text{if } s{=}\bot \text{ then } \varnothing \text{ else } \{s\}$

$\delta_{cl,deposit(i)} = \lambda s.\text{if } s{=}\bot \text{ then } \varnothing \text{ else } \{s\}$

We want to establish their joint behaviour in the presence of some interaction. For instance, the account is opened by the client. Assuming that the objects acc and cl are as specified before, we would like to impose the following set of interaction (in)equations:

ρ = {	cl.open_acc	*calls*	acc.open,	
	cl.deposit(i)	*calls*	acc.deposit(i),	
	cl.withdrawal(i)	*calls*	acc.withdrawal(i),	
	cl.close_acc	*calls*	acc.close	}

That is, open, deposit, close and withdrawal actions are to be related in some way by the two objects. The aggregation $cl\|_\rho acc$ displays the joint behaviour of these two interacting objects. Even in such a case of interacting components, each component *is-part-of* the whole. For instance, acc *is-part-of* $cl\|_\rho acc$.

Another form of interaction can be achieved by *slot sharing*. However it should be stressed that, in order to obtain the envisaged joint behaviour, all the events affecting the slot must also be shared. That is, it is only possible to share whole objects. This seems to be essential to object orientation.

Proposition 4.1. The coproduct of objects $ob_1{=}\langle E_1,\langle A_1,\partial_1\rangle,S_1,Q_1,B_1,\mu_1,\delta_1\rangle$ and $ob_2{=}\langle E_2,\langle A_2,\partial_2\rangle,S_2,Q_2,B_2,\mu_2,\delta_2\rangle$ is, up to an isomorphism, a 2-sink with vertex $ob_1\|ob_2{=}\langle E_1{\times_{Pfn}}E_2,\langle A_1{+_{Set}}A_2,[\partial_1,\partial_2]\rangle,S_1{\times_{Set}}S_2,Q_1{\times_{Set}}Q_2,B_1{\times_{Set}}B_2,\mu_1{+}\mu_2,\delta_1{\times}\delta_2\rangle$, where $[\partial_1,\partial_2]$ is the universal arrow from the coproduct $A_1{+_{Set}}A_2$ into the set of carriers V, $(\mu_1+\mu_2)_a = \lambda s{:}S_1{\times_{Set}}S_2.\cup_{x\in inj1_A^{-1}(a)}\mu_{1,x}(\pi_1(s))\cup\cup_{x\in inj2_A^{-1}(a)}\mu_{2,x}(\pi_2(s))$ and $(\delta_1\times\delta_2)_e = \lambda s{:}S_1{\times_{Set}}S_2.\delta_{1,e}(\pi_1(s)){\times_{Set}}\delta_{2,e}(\pi_2(s))$. ◇

The aggregation (coproduct) of two objects induces the parallel composition of their induced processes. Indeed, the product of menus leads to the interleaving of the life-cycles as expected, *i.e.*, $M_{ob_1+_{FrOb}ob_2}(s){=}M_{ob_1}(\pi_1(s)){\times_{Pfn}}M_{ob_2}(\pi_2(s))$. Therefore, the event set of $ob_1\|ob_2$ should correspond to the categorial product in Pfn of the event sets of ob_1 and ob_2. An element $a|b$ of $M_{ob_1}(\pi_1(s)){\times_{Set}}M_{ob_2}(\pi_2(s))$ stands for concurrent execution of *a* by ob_1 and *b* by ob_2. An element *a* of $M_{ob_1}(\pi_1(s))$ stands for the execution of *a* by ob_1 while ob_2 remains idle. And an element *b* of $M_{ob_2}(\pi_2(s)))$ stands for the execution of *b* by ob_2 while ob_1 remains idle. That is, the object $ob_1\|ob_2$ would be able, at each instant, to be involved either in an event of ob_1 (while ob_2 remains idle), or in an event of ob_2 (ob_1 remaining idle), or in both an event of ob_1 and one of ob_2.

Some concepts of category theory are useful in order to provide the required techniques for object synchronization. Here we follow the technique of fibrations used in [Winskel 89].

Proposition 4.2. The forgetful functor υ:FrOb→Pfn^{op} that maps each object onto its set of events is a cofibration. ◇

For each morphism $f:Z_2 \rightarrow Z_1$ in Pfn, the cofibration υ:FrOb→Pfn^{op} induces a canonical functor $f^{>}:\upsilon^{-1}(Z_1)\rightarrow\upsilon^{-1}(Z_2)$ from the fiber over Z_1 into the fiber over Z_2.

Definition 4.3. An *interaction structure* is a triple $<E_1,E_2,\ll>$ where E_1 and E_2 are sets, and $\ll\subseteq E_1\times E_2\cup E_2\times E_1$ is a one to one binary relation, called the *interaction relation.* ◇

Each interaction structure $<E_1,E_2,\ll>$ determines a subset E of the categorial product $E_1\times_{Pfn}E_2$ ($=E_1+_{Set}E_1\times_{Set}E_2+_{Set}E_2$) in Pfn given by

$$E=\{e\in E_1\times_{Pfn}E_2\text{: } (\pi_1(e)\notin \mathrm{dom}(\ll) \text{ and } \pi_2(e)\notin \mathrm{dom}(\ll)) \text{ or } \pi_1(e)\ll\gg\pi_2(e)\}^{\dagger},$$

where dom denotes the domain of «, and an inclusion map $\subseteq_{\ll}:E\rightarrow E_1\times_{Pfn}E_2$. This means that if *a«b* then the happening of *a* leads to the happening of *b* — *a* is not allowed to happen alone or even to happen with any other event different from *b*.

An interaction structure endowed with a symmetric interaction relation is said to be a *sharing structure* and the underlying relation is called a *sharing relation.* Otherwise they are said to be a *calling structure* and a *calling relation*, respectively.

Definition 4.4. Given two objects ob_1 with events in E_1 and ob_2 with events in E_2, and an interaction relation «, we define $ob_1\|_{\ll}ob_2=(\subseteq_{\ll})^{>}(ob_1+ob_2)$. ◇

5.Liberal Specializations

In order to allow side-effects we propose the following definition of object morphism (relaxing the frame constraint):

Definition 5.1. An *object morphism*

$$h:<E_1,<A_1,\partial_1>,S_1,Q_1,B_1,\mu_1,\delta_1>\rightarrow<E_2,<A_2,\partial_2>,S_2,Q_2,B_2,\mu_2,\delta_2>$$

is a triple $<h_E,h_A,h_S>$ where $h_E:E_2\rightarrow E_1$ is a partial function, $h_A:A_1\rightarrow A_2$ is a slot morphism and $h_S:S_2\rightarrow S_1$ is a total function such that the following conditions hold: (a) $Q_2\subseteq h_S^{-1}(Q_1)$, (b) $B_2\subseteq h_S^{-1}(B_1)$, (c) $\mu_{1,a}(h_S(s))\subseteq\mu_{2,h_A(a)}(s)$, for every $a\in A_1$, $s\in S_2$, and (d) for every $e\in E_2$, $s\in S_2$, if $h_E(e)\downarrow$ then $\delta_{2,e}(s)\subseteq h_S^{-1}(\delta_{1,h_E(e)}(h_S(s)))$. ◇

Proposition 5.2. Objects and object morphisms constitute a complete and cocomplete, concrete category Ob over $Pfn^{op}\times Slt\times Set^{op}$. Moreover, FrOb is a wide subcategory of Ob. ◇

Fortunately, Ob is also finitely complete (limits within Ob are also important for overriding as we shall see).

---

† By $\pi_1(e)\ll\gg\pi_2(e)$ we mean $\pi_1(e)\ll\pi_2(e)$ or $\pi_2(e)\ll\pi_1(e)$.

For instance, assume that we would like to have a fee (say of 50) being automatically taken from a savings account whenever a saving period starts (side-effect of start_saving on balance). We would like to impose

$$\delta_{sav\_acc,start\_saving} = \lambda s.\text{if } s=<\perp,\perp,\perp> \text{ then } \varnothing \text{ else } \{<\pi_1(s)\text{-}50,\{1\},\{0\}>\}$$

The existence of an *inclusion* in Ob from acc into sav_acc tells us that if we choose a slot a in A_{acc} and an inherited event e in E_{sav_acc} then $\delta_{acc,e}\pi_1=\pi_1\delta_{sav\_acc,e}$. However, nothing is said about what happens if we choose a new event of E_{sav_acc}. Thus morphisms in Ob are particularly suitable for providing the semantics of inheritance (side-effects being possible).

With respect to behaviour, the morphism condition remains as discussed in FrOb: any possible *menu* of sav_acc should degenerate into a possible *menu* of acc when we forget the new events of sav_acc.

Morphisms of Ob are liberal enough to explain specialization with side-effects. But they completely disallow non-monotonic overriding in the sense that the effects of inherited events on inherited slots are not changed at all. In order to cope with non-monotonic overriding we need to consider even more relaxed morphisms. Since the effects in the specialization may be rather different from those in the original object, a decoupling is needed between the two objects, suggesting the use of partial morphisms. Indeed, a partial morphism from ob into ob' is a morphism from a subobject of ob into ob'. That subobject characterizes what is kept from ob in ob'. The rest may be changed (overriden).

Let M be a class of morphisms in category A. A 2-source ($X\overset{m}{\leftarrow}\bullet\overset{f}{\rightarrow}Y$) with $m\in M$ is called an M-partial morphism from X into Y.

Definition 5.3. Let M be the class of $Pfn^{op}\times Slt\times Set^{op}$ identity morphisms. A 2-source in Ob, $ob_1\overset{\iota}{\leftarrow}x\overset{f}{\rightarrow}ob_2$, with $\iota\in M$ is called a *partial morphism* with source ob_1 and target ob_2. ◇

For a partial morphism $ob_1\overset{\iota}{\leftarrow}x\overset{f}{\rightarrow}ob_2$ within Ob we just write $ob_1\leftarrow x\overset{f}{\rightarrow}ob_2$ or even $f{:}ob_1\nrightarrow ob_2$.

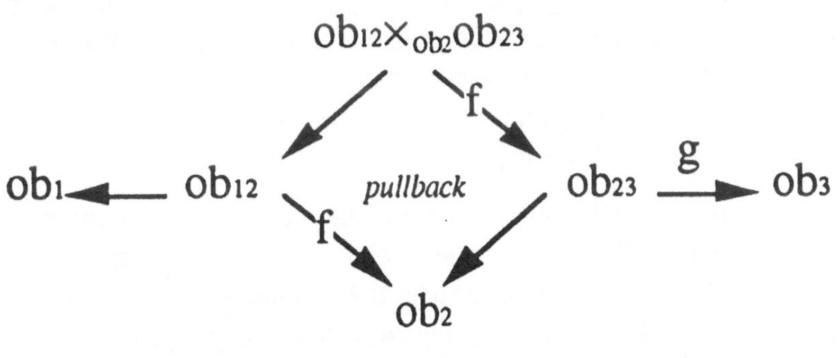

Fig. 5.1 Composition of partial morphisms.

Let us now define *composition of partial morphisms*. But first recall that in every category $x_1\leftarrow x_1\overset{f}{\rightarrow}x_2$ is a pullback of $x_1\overset{f}{\rightarrow}x_2\leftarrow x_2$ and $x_1\rightarrow x_1\overset{f}{\leftarrow}x_2$ is a pushout of

$x_1 \overset{f}{\leftarrow} x_2 \rightarrow x_2$. Thus, if $ob_1 \leftarrow ob_{12} \overset{f}{\rightarrow} ob_2$ and $ob_2 \leftarrow ob_{23} \overset{g}{\rightarrow} ob_3$ are partial morphisms within Ob, then a pullback of $ob_{12} \overset{f}{\rightarrow} ob_2 \leftarrow ob_{23}$ can be obtained by providing the initial lifting of the source [see Adámek *el al* 90]

$$<E_{12},A_{12},S_{12}> \leftarrow <E_{12},A_{12},S_{12}> \overset{f}{\rightarrow} <E_{23},A_{23},S_{23}>$$

in $Pfn^{op} \times Slt \times Set^{op}$. Let $ob_{12} \times_{ob_2} ob_{23}$ be the vertex of the corresponding initial source in Ob.

Definition 5.4. Composition of two partial morphisms $ob_1 \leftarrow ob_{12} \overset{f}{\rightarrow} ob_2$ and $ob_2 \leftarrow ob_{23} \overset{g}{\rightarrow} ob_3$ is the partial morphism $ob_1 \leftarrow ob_{12} \times_{ob_2} ob_{23} \overset{g \circ f}{\rightarrow} ob_3$. ◇

Proposition 5.5. Objects and partial morphisms constitute a category POb over $Pfn^{op} \times Slt \times Set^{op}$.

Proof: Partial morphisms compose and this composition is associative. Identity morphisms are 2-sources of the kind $ob \overset{id}{\leftarrow} ob \overset{id}{\rightarrow} ob$. ◇

We already saw an example of a rather restricted form of overriding: the withdrawal event of the acc object is redefined to some extent in the sav_acc object. Indeed, in the latter an withdrawal has more effects than in the former — namely, it also affects the values of the additional slot saving?.

The semantics of specialization given above within Ob does allow for this kind of event overriding: an event when inherited may have further effects, but the effects on the inherited slots must remain the same. We could say that we have so far only *monotonic* overriding. Any stronger form of specialization will also require a more liberal overriding mechanism, allowing, for instance, a different treatment of the inherited slots by the inherited events in the specialized object. In order to provide the semantics of such a *non-monotonic overriding* mechanism we have to consider morphisms in POb.

A suitable partial morphism $h:acc \nrightarrow sav\_acc$ introduces a *subobject* acc' of acc that coincides with acc with the exception of the map $\delta_{acc',e}$ corresponding to the method we may want to override. In acc' we must impose $\delta_{acc',e} = \lambda s.S_{acc'}$ for those maps: then for all slots *a* of acc, the maps $\delta_{acc,a}$ and $\delta_{sav\_acc,a}$ are no longer necessarily the same. Therefore, if sav_acc *is-an* acc, that is, if there is a partial morphism $h:acc \nrightarrow sav\_acc$, then we may have method overriding: whenever some rules of acc are to be rewritten in sav_acc, the partial morphism $acc \leftarrow acc' \overset{h}{\rightarrow} sav\_acc$ should be chosen so that $\delta_{acc',e} = \lambda s.S_{acc'}$ in acc' for each method *e* to be overridden.

It should be stressed that both source and target of a partial morphism can be deterministic. However, the auxiliary middle object fails to be deterministic! In practice, we can go on defining only deterministic objects (and imposing configurations based upon inclusions and injections of Ob) if we disregard such auxiliary objects.

6.Further Work

In this paper we do not distinguish between objects and aspects, contrarily to what is proposed in [Ehrich *et al* 92]. Therefore, we ignore the problem of having the same object with different aspects (*i.e.*, the same savings account x with several aspects, such as x.ACCOUNT and x.SAVINGS_ACCOUNT). According to the terminology in [Ehrich *et al* 92], we may look at this paper as a study of object template inheritance and aggregation. An immediate research concern is to bring the results presented herein to bear on a suitable framework of object identifiers, aspects and classes.

Acknowledgements

This work was partially supported by the JNICT Project PMCT/C/TIT/178/90 (FAC3) and by the ESDI Project OBLOG. The authors are grateful to their colleagues in these two projects for many rewarding interactions.

References

[Adámek *et al* 90]
J.Adámek, H.Herrlich and G.Strecker, *Abstract and Concrete Categories*, Wiley, 1990

[Costa *et al* 92a]
J.F.Costa, A.Sernadas and C.Sernadas, "Object Inheritance Beyond Subtyping", INESC Report, 1992, to appear in *Acta Informatica*

[Ehrich *et al* 91]
H.-D.Ehrich, J.Goguen and A.Sernadas, "A Categorial Theory of Objects as Observed Processes", in J.W. deBakker, W.P.deRoever and G.Rozenberg (eds), *Proc. of the REX90/Workshop on Foundations of Object-Oriented Languages*, LNCS 489, Springer-Verlag, 1991, 203-228

[Ehrich *et al* 92]
H.-D.Ehrich, G.Saake and A.Sernadas, "Concepts of Object-Orientation", to appear in *Proc. of the 2nd IS/KI Workshop*, Ulm, 1992

[Goguen 91]
J.Goguen, "Types as Theories", *Proc. Conf. on Topology and Category Theory in Computer Science*, Oxford University Press, 1991, 357-390

[Jungclaus *et al* 91]
R.Jungclaus, G.Saake, T.Hartmann and C.Sernadas, *Object-Oriented Specification of Information Systems: The TROLL Language*, Informatik-Berichte, Tech. Univ. Braunschweig, 1991, *to appear*

[Manna and Pnueli 81]
Z.Manna and A.Pnueli, "The Temporal Framework for Concurrent Programs", in R.Boyer and J.Moore (eds), *The Correcteness Problem in Computer Science*, Academic Press, 1981, 215-274

[SernadasA *et al* 89]
A.Sernadas, J.Fiadeiro, C.Sernadas and H.-D.-Ehrich, "Basic Building Blocks of Information Systems", in E.Falkenberg and P.Lindgreen (eds), *Information System Concepts: An In-depth Analysis*, North-Holland, 1989, 225-246

[Winskel 84]
G.Winskel, "Synchronization trees", in *Theoretical Computer Science* 34, 1984

[Winskel 89]
G.Winskel, "An Introduction to Event Structures", in J.W.deBakker, W.P.de Roever and G.Rozenberg (eds), *Linear Time, Branching Time and Partial Order in Logics and Models for Concurrency*, LNCS 354, Springer-Verlag, 1989, 364-397

Image Compression Using Weighted Finite Automata

Karel Culik II[1] and Jarkko Kari[2]

[1] Department of Computer Science, University of South Carolina, Columbia, S.C. 29208, USA
[2] Department of Mathematics, University of Turku, Turku, 20500 Finland

Abstract. We introduce Weighted Finite Automata (WFA) as a tool to define real functions, in particular, the greyness functions of grey-tone images. Mathematical properties and the definition power of WFA have been studied by Culik and Karhumäki. Their generative power is incomparable with Barnsley's Iterative Function Systems. Here, we give an automatic encoding algorithm that converts an arbitrary grey-tone-image (a digitized photograph) into a WFA that can regenerate it (with or without information loss). The WFA seems to be the first image definition tool with such a relatively simple encoding algorithm.

1 Introduction

We introduce weighted finite automata (WFA) that define texture (grey-tone or color) images. We describe an efficient algorithm for generating an image described by WFA and most importantly an automatic encoding algorithm for WFA that is an algorithm which given as input a finite resolution image (in pixel forms) produces as output a relatively small WFA from which a good approximation of the original image can be efficiently regenerated.

The WFA's are a modification of probabilistic finite generators from [7]. A nonprobabilistic version of equivalent devices, describing black and white images, were independently studied in [4, 5, 16, 17].

Another well known tool for definition of texture images are the Iterative Function Systems (IFS) [1, 2, 3]. Decoding and automatic encoding algorithms and dedicated hardware for IFS have been developed by the Iterative Systems Inc. Algorithms based on extensions of IFS were developed in [14, 15]. The WFA seems to be the first "fractal type" image definition tool for which there is a simple automatic encoding method.

Mutually recursive function systems (MRFS) have been studied in [8, 9] where also an efficient decoding algorithm for them has been described. Both IFS and probabilistic finite generators are special cases of MRFS, which constitute a very powerful tool for designing images, however, no automatic encoding procedure for them is known.

In Section 2 we review how functions specify various types of images. Then we introduce the WFA and mention some elementary properties. In Section 3 we give the important encoding algorithm for unrestricted WFA. First we give the "theoretical version" of our algorithm. Here we consider a multi-resolution image (average preserving function) with unlimited detail. In this case we actually have only a procedure which terminates if the given functions can be defined by a WFA and construct such a WFA. Finally in Section 4 we give the "practical algorithm" that given a finite resolution image I always produces a WFA which defines an approximation of I. We discuss various practical issues of implementation and optimization of this algorithm and show some examples obtained by our implementation for both one-dimensional functions and two-dimensional functions (images).

2 Images and Weighted Finite Automata.

By a finite-resolution image we mean a digitized grey scale picture that consists of 2^m by 2^m pixels (typically $7 \leq m \leq 11$) each of which takes a real value (practically digitized to a value between 0 and $2^n - 1$, typically $n = 8$). By a multi-resolution image, we mean a collection of compatible 2^n by 2^n resolution images for $n = 0, 1, \ldots$. We will assign to each pixel at 2^n by 2^n resolution a word of the length n over the alphabet $\Sigma = \{(0,0), (0,1), (1,0), (1,1)\}$, hence in our formalism a multi-resolution image is a real function on $\Sigma^\star$. The compatibility of the different resolutions is formalized by requiring that $F : \Sigma^\star \rightarrow \mathbb{R}$ is an average preserving function. A function $F : \Sigma^\star \rightarrow \mathbb{R}$ is average preserving (ap) if

$$f(w) = \frac{1}{4}[f(w(0,0)) + f(w(0,1)) + f(w(1,0)) + f(w(1,1))] \tag{1}$$

for each $w \in \Sigma^\star$.

An ap-function f can be represented by an infinite labeled quadtree. The root is labeled by $f(\varepsilon)$, its sons by $f((0,0)), f((0,1)), f((1,0)), f((1,1))$ from left-to-right, etc. Intuitively, $f(w)$ is the average greyness of the subsquare w for a given grey-tone image.

By an infinite-resolution image we mean a local-greyness function $f : [0,1]^2 \rightarrow \mathbb{R}$. For every local-greyness function f we can find the corresponding multi-resolution function $F : \Sigma^\star \rightarrow \mathbb{R}$ by computing $F(w)$ as the integral over the square with the address w, for each $w \in \Sigma^\star$. Conversely, for a point $p \in [0,1]^2$, $f(p)$ is the limit of the pixel values containing p, if such a limit exists. Thus not every multi-resolution image can be converted into an infinite-resolution image.

For a color image in rgb (red-green-blue) representation we need 3 greytone images one for each basic color.

The distance between two functions f and g (the error of approximating f by g) is usually measured by

$$||f - g||_p = \int_0^1 \int_0^1 |f(x,y) - g(x,y)|^p dx dy \tag{2}$$

In practice, one desires a metric that parallels the human perception, i.e. that the image differences that seem larger to the human eye are mathematically large and those perceived as insignificant are mathematically small. This seem to be achieved best with the choice $p = 1$ in (2). Therefore we will consider an average absolute error also in the case of finite resolution images.

Now, we explain how we address subsquares of $[0,1]^2$ by words over Σ. Let $\Sigma = \{(0,0),(0,1),(1,0),(1,1)\}, w \in \Sigma^*, w = (a_1,b_1)(a_2,b_2)\dots(a_k,b_k)$. We say that w is the address of the subsquare of X with bottom-left corner $(\sum_{i=1}^{k} \frac{a_i}{2^i}, \sum_{i=1}^{k} \frac{b_i}{2^i})$ and the side of the length 2^{-k}. For example, the whole $[0,1]^2$ has the address ε and the black square in the Figure 1 has the address $(1,1)(1,0)(0,0)$.

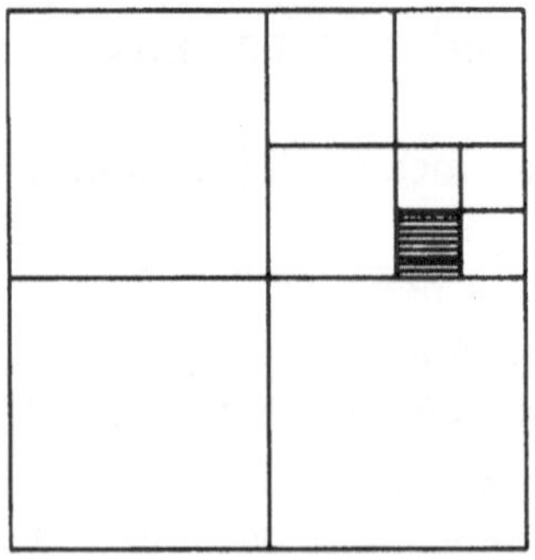

Fig. 1. The subsquare specified by the string $(1,1)(1,0)(0,0)$

A weighted finite automaton (WFA) is a 5-tuple $A = (Q, \Sigma, f, \alpha, \beta)$ where

1. Q is a finite set of *states*,
2. Σ is a finite alphabet (here the alphabet $\Sigma = \{(0,0),(0,1),(1,0),(1,1)\}$),
3. $f : Q \times \Sigma \times Q \to [-\infty, \infty]$ is the *weight function*,
4. $\alpha : Q \to [-\infty, \infty]$ is the *initial distribution*,
5. $\beta : Q \to [-\infty, \infty]$ is the *final distribution*.

We will say that $(p, a, q) \in Q \times \Sigma \times Q$ is a transition of A if $f(p, a, q) \neq 0$. a is called the label of the transition.

A WFA A defines a distribution function $\delta : Q \times \Sigma^* \to [-\infty, \infty]$ as follows.

(i) $\delta(q, \varepsilon) = \alpha(q)$ for each $q \in Q$.
(ii) $\delta(q, wa) = \sum_{p \in Q} \delta(p, w) \cdot f(p, a, q)$ for each $q \in Q$, $w \in \Sigma^*$ and $a \in \Sigma$.

WFA A defines the function $\phi_A : \Sigma^* \to [-\infty, \infty]$ specified by

$$\phi_A(w) = \sum_{q \in Q} \beta(q)\delta(q, w).$$

$\phi_A(w)$ can be written explicitly as follows :

$$\phi_A(w) = \sum_{q_0,\dots,q_n \in Q} \alpha(q_0) \cdot f(q_0, a_1, q_1) \cdot f(q_1, a_2, q_2) \cdot \dots \cdot f(q_{n-1}, a_n, q_n) \cdot \beta(q_n),$$

if $w = a_1 a_2 \ldots a_n$. Intuitively, $\phi_A(w)$ is obtained by taking all paths in the automaton, whose labels form word w. The weight of such a path is obtained by multiplying the weights of the transitions on the path, the initial distribution on the first state and the final distribution on the last state of the path. $\phi_A(w)$ is the sum of the weights of all such paths.

A WFA A is *average preserving* (ap-WFA) iff for all $p \in Q$

$$\sum_{a \in \Sigma, q \in Q} f(p, a, q) \cdot \beta(q) = 4 \cdot \beta(p). \tag{3}$$

Theorem 1 *Let A be a WFA. If A is average preserving then ϕ_A is average preserving.*

Proof. Let $w \in \Sigma$ be arbitrary. Because

$$\phi_A(wa) = \sum_{q \in Q} \beta(q)\delta(q, wa) = \sum_{p, q \in Q} \beta(q) \cdot f(p, a, q) \cdot \delta(p, w)$$

we have

$$\begin{aligned} \sum_{a \in \Sigma} \phi_A(wa) &= \sum_{p \in Q} \delta(p, w) \cdot \left(\sum_{a \in \Sigma, q \in Q} f(p, a, q) \cdot \beta(q) \right) \\ &= \sum_{p \in Q} \delta(p, w) \cdot 4\beta(p) \\ &= 4 \cdot \phi_A(w). \end{aligned}$$

In the second equality we have used (3). This proves that ϕ_A satisfies (1) for every word $w \in \Sigma$, which means that ϕ_A is average preserving. □

3 The encoding algorithm for WFA.

In a practical situation we are given an image, e.g. a grey-tone or color photograph, with certain finite resolution. In the terms of a quadtree we are given all the values at one level, say level k. By computing for each parent the average value of the labels of all its children, for all the nodes at the higher levels, we get the labels everywhere above the given resolution, and leave "don't cares" below it. By assigning a different state at level k and higher we trivially get (too large) ap-WFA that (perfectly) defines the given image. The practical problem therefore is not whether it is possible to encode a given image but rather whether we can get a good trade-off between the size of the automaton and the quality of the regenerated approximation of the given image. This trade-off, of course, depends on the "nonrandomness" or "regularity" of the image. It is well known fact in the descriptive (Kolmogorov) complexity of strings [6] that most strings are algorithmically random and cannot be encoded (compressed) by a shorter program. However, the "interesting strings" are not random and possibly can be compressed. The same holds for images.

First we consider briefly the deterministic WFA. We say that a weighted automaton $A = (Q, \Sigma, f, \alpha, \beta)$ is deterministic if the underlying finite automaton obtained by omiting the weights is deterministic, formally if for each pair $(q, a) \in Q \times \Sigma$ there is at most one $p \in Q$ such that $f(q, a, p) \neq 0$.

Besides the existence of a short description of an image, it is also important that this description can be easily found from the given image (existence of an efficient encoding program) and that it is easy to regenerate the image from the description (existence of an efficient decoding program). Both of these requirements are excellently satisfied by the deterministic ap-WFA. An encoding algorithm restricted to the deterministic WFA is not difficult to construct, see the full version of this paper [12]. The decoding (for resolution corresponding to the k-th level of the quadtree) can be done by simply multiplying the weights on every path from the root to a node at k-level (a pixel).

However, experiments have shown that deterministic ap-WFA are not powerful enough for encoding practical images (e.g. photographs) and there is a good theoretical reason for it. Indeed, in contrast to (nonprobabilistic) generators for black and white images discussed in [7], the nondeterministic ap-WFA are much more powerful than the deterministic one. We show a simple example of linear greyness function that can be generated by a nondeterministic but not by a deterministic ap-WFA. Actually, the deterministic ap-WFA can generate only countable unions of fractals or constant level greyness functions but not smoothly growing greyness functions. For a detailed discussion see [11].

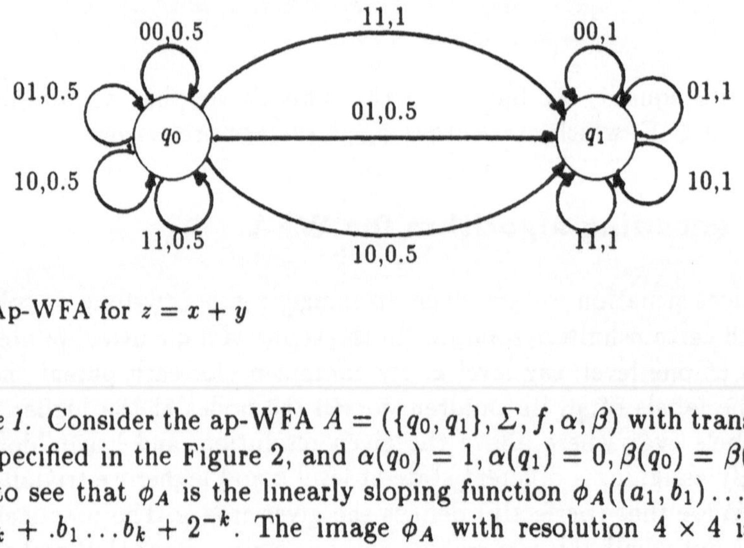

Fig. 2. Ap-WFA for $z = x + y$

Example 1. Consider the ap-WFA $A = (\{q_0, q_1\}, \Sigma, f, \alpha, \beta)$ with transition function f specified in the Figure 2, and $\alpha(q_0) = 1, \alpha(q_1) = 0, \beta(q_0) = \beta(q_1) = 1$. It is easy to see that ϕ_A is the linearly sloping function $\phi_A((a_1, b_1)\dots(a_k, b_k)) = .a_1 \dots a_k + .b_1 \dots b_k + 2^{-k}$. The image ϕ_A with resolution 4×4 is shown in Figure 3.

Clearly, the average-preserving function ϕ_A specified on the dense discrete subset of $[0, 1] \times [0, 1]$ converges to the local greyness function $\phi(x, y) = x + y$ defined everywhere on $[0, 1] \times [0, 1]$.

A fast decoding algorithm for unrestricted WFA is shown in [12], here we will address only the most difficult problem, namely, the construction of an encoding algorithm for unrestricted (nondeterministic) ap-WFA.

1	$\frac{5}{4}$	$\frac{3}{2}$	$\frac{7}{4}$
$\frac{3}{4}$	1	$\frac{5}{4}$	$\frac{3}{2}$
$\frac{1}{2}$	$\frac{3}{4}$	1	$\frac{5}{4}$
$\frac{1}{4}$	$\frac{1}{2}$	$\frac{3}{4}$	1

Fig. 3. The image ϕ_A with resolution 4×4

Now we prepare the tools used in the encoding algorithm. We consider the set of functions $\phi : \Sigma^\star \to \mathbb{R}$ as a vector space. The operations of sum and multiplication with a real number are defined in a natural way :

$$(\phi_1 + \phi_2)(w) = \phi_1(w) + \phi_2(w), \text{ for any } \phi_1, \phi_2 : \Sigma^\star \to \mathbb{R} \text{ and } w \in \Sigma^\star,$$
$$(c\phi)(w) = c\phi(w), \text{ for any function } \phi : \Sigma^\star \to \mathbb{R} \text{ and real number } c.$$

The set of ap-functions forms a linear sub-space, because any linear combination of ap-functions is average preserving. The sum of two ap-functions represents the image obtained by summing up the greynesses of the two images, and the multiplication with a real number corresponds to the change of the contrast.

Now we are ready to present the encoding algorithm. First we give the "theoretical version" of our algorithm, that is we consider a multi-resolution image (average preserving function) with unlimited detail. We assume that the executor of the algorithm has the ability to recognize that there are two (scaled) copies of the same image, of possibly different contrast, in two subsquares and determine the relative contrast. We assume that we can find out if a given ap-function can be expressed as a linear combination of elements from a given finite collection of ap-functions. We also assume that we can compute the coefficients of such a linear combination.

In the algorithm we use the following notation. For an ap-function ϕ and word $w \in \Sigma^\star$, ϕ_w denotes the ap-function defined by

$$\phi_w(u) = \phi(wu) \text{ for all } u \in \Sigma^\star.$$

The function ϕ_w represents thus the image that the ap-function ϕ defines on the subsquare w. It is obtained from ϕ by zooming in the subsquare.

Algorithm

Let $\Sigma = \{(0,0),(0,1),(1,0),(1,1)\}$. For an image given by ap-function $\phi : \Sigma^\star \to \mathbb{R}$, an ap-WFA $A = (\{q_0, \ldots, q_N\}, \Sigma, f, \alpha, \beta)$ such that $\phi_A = \phi$ is constructed, provided such an ap-WFA exists. During the construction we use

N, the index of the last state created,

i, the index of the first unprocessed state,

$\gamma : Q \to \Sigma^\star$, a mapping of states to subsquares.

1. Set $N = 0, i = 0, \beta(q_0) = \phi(\varepsilon), \gamma(q_0) = \varepsilon$.
2. Process q_i, that is for $w = \gamma(q_i)$ and each $a = (0,0), (0,1), (1,0), (1,1)$ do
   (a) if there are $c_0, \ldots, c_N$ such that $\phi_{wa} = c_0\phi_0 + \ldots + c_N\phi_N$ where $\phi_j = \phi_{\gamma(q_j)}$ for $j = 0, \ldots, N$ then set $f(q_i, a, q_j) = c_j$ for $j = 0, \ldots, N$,
   (b) otherwise set $\gamma(q_{N+1}) = wa$, $\beta(q_{N+1}) = \phi(wa)$, $f(q_i, a, q_{N+1}) = 1$, $N = N + 1$.
3. Set $i = i + 1$, if $i \leq N$, then goto 2.
4. Set $\alpha(q_0) = 1, \alpha(q_j) = 0$ for $j = 1, \ldots, N$.

It is easy to see that the algorithm terminates if and only if the set $\{\phi_w | w \in \Sigma^*\}$ generates a linear space of finite dimension. The number of states in the ap-WFA produced by Algorithm 3 is the same as the dimension of the linear space. According to the next theorem the algorithm gives an ap-WFA with the minimal number of states defining the given image exactly, provided such an automaton exists. The theorem contains also a converse statement of Theorem 1: Any ap-function that can be defined using an arbitrary WFA can be defined using an ap-WFA. No ap-functions definable by WFA are lost when one considers only ap-WFA.

Theorem 2 *Let $\phi : \Sigma^* \to \mathbb{R}$ be a function.*

(i) The function ϕ can be defined using a WFA if and only if the set of functions defined in all the subsquares of Σ^* by ϕ (that is, the set $\{\phi_w | w \in \Sigma^*\}$) generates a linear space of finite dimension d.

(ii) If ϕ is defined by a WFA and is average preserving, then it can be defined by an ap-WFA, and the ap-WFA produced by Algorithm 3 has the smallest number of states (namely d states) among all WFA defining ϕ.

Proof. see [12].

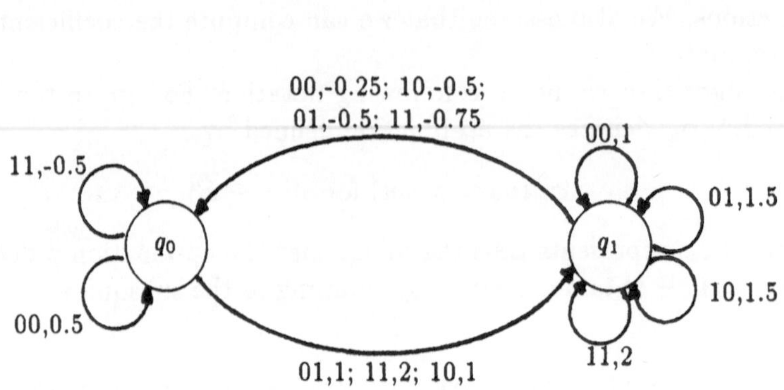

Fig. 4. The ap-WFA for $z = x + y$ produced by the encoding algorithm

Example 2. Consider the linearly sloping ap-function ϕ_A introduced in Example 2. Let us apply Algorithm 3 to find a minimal ap-WFA generating ϕ_A.

First state q_0 is assigned to the square ε and we define $\beta(q_0) = 1$. Consider then the four subsquares $(0,0)$, $(0,1)$, $(1,0)$ and $(1,1)$. The image in the first subsquare $(0,0)$ can be expressed as $\frac{1}{2} \cdot \phi_A$ (it is obtained from the original image ϕ_A by decreasing the contrast by one half) so that we define $f(q_0, (0,0), q_0) = 0.5$.

The image in the second subsquare $(0,1)$ cannot be expressed as a linear combination of ϕ_A so that we have to use a second state q_1. Define $f(q_0, (0,1), q_1) = 1$ and $\beta(q_1) = 1$ (the average greyness of the subsquare $(0,1)$ is 1). Let ϕ_1 denote the image given by ϕ_A into the square $(0,1)$.

The image in the third subsquare $(1,0)$ is the same as in the second subsquare, so that $f(q_0, (1,0), q_1) = 1$. In the fourth quadrant $(1,1)$ we have image which can be expressed as $2 \cdot \phi_1 - \frac{1}{2} \cdot \phi_A$. We define $f(q_0, (1,1), q_0) = -\frac{1}{2}$ and $f(q_0, (1,1), q_1) = 2$. The outgoing transitions from state q_0 are now ready.

Consider then the images in the squares $(0,1)(0,0)$, $(0,1)(0,1)$, $(0,1)(1,0)$ and $(0,1)(1,1)$. They can be expressed as $\phi_1 - \frac{1}{4} \cdot \phi_A$, $\frac{3}{2} \cdot \phi_1 - \frac{1}{2} \cdot \phi_A$, $\frac{3}{2} \cdot \phi_1 - \frac{1}{2} \cdot \phi_A$ and $2 \cdot \phi_1 - \frac{3}{4} \cdot \phi_A$, respectively. This gives us the ap-WFA of figure 4.

The initial distribution α is defined by $\alpha(q_0) = 1$ and $\alpha(q_1) = 0$.

4 About practical implementation of the encoding algorithm

There are several aspects that have to be taken into account when implementing Algorithm 3, the most important of which are the following ones:

1. Although the ap-WFA produced by Algorithm 3 is minimal in respect of the number of states, it is not necessarily minimal in respect of the number of edges. In general, the same image can be generated with several automata, all having the minimal number of states but varying greatly in the number of edges. Modifying Algorithm 3 in order to make it produce ap-WFA with small number of edges, without sacrificing the efficiency of the algorithm is an important problem.
2. It is not usually a good strategy to require the algorithm to produce an ap-WFA that generates exactly the original image. Any image of finite resolution can be generated exactly using an ap-WFA, but in most cases the automaton will be large. However, if one allows a small error in the reproduced image then one can usually find a much smaller automaton. How to find a small automaton that generates an image which is close to the original one, is another problem.

Let us briefly describe the modifications we have made to Algorithm 3 in order to solve the two problems above.

We change Step 2(a) in Algorithm 3 by removing edges with small weights. If the absolute value of weight c_j in Step 2(a) is small then we omit the edge from state q_i to state q_j by defining

$$f(q_i, a, q_j) = 0.$$

Note that this change is related to both of the two problems above because the number of edges in the automaton is reduced (problem 1) but the automaton obtained does not generate any longer the original image exactly, but some image close to the original one (problem 2). The algorithm gets as input a positive real number e indicating how big error is allowed in the regenerated image. An edge is removed only if the resulting error is smaller than e. Note that the weight of the edge alone is not sufficient in determining its effect on the final image — the weights of all paths leading to the state q_i where the edge comes from must be taken into account. If these weights are bigger then the state q_i affects the image more, and only smaller edges from q_i may be removed.

This modification is not guaranteed to solve problems 1 and 2 in an optimal way. Especially problem 1 seems difficult if we want the implementation to be efficient.

A natural question concerning Algorithm 3 is how does one efficiently find out in step 2 whether the ap-function ϕ' can be expressed as a linear combination of the ap-functions $\phi_0, \phi_1, \ldots, \phi_N$, and what are the coefficients of such a linear combination. In practice the ap-functions are represented as quadtrees of some finite depth. If the original image is given with resolution $2^k \times 2^k$ pixels, then the labeled quadtree representing the image is given up to depth k. The nodes that are deeper than k are "don't care" nodes — the values on these nodes is unknown. Algorithm 3 should produce an ap-WFA that defines an ap-function, which (almost) agrees with the given quadtree up to depth k.

In this case we can use an inner product defined in the space of quadtrees (ap-functions) of depth k. The inner product of ap-functions f and g is

$$(f,g) = \sum_{w \in \Sigma^k} f(w)g(w).$$

(This corresponds to the normal inner product in the isomorfic linear space $\boldsymbol{R}^{4^k}$.)

Using this inner product we can always find an ortogonal basis for the linear subspace generated by the ap-functions $\phi_0, \phi_1, \ldots, \phi_N$ in Step 2 of the algorithm. To find out if the ap-function ϕ' can be expressed as a linear combination of the ap-functions $\phi_0, \phi_1, \ldots, \phi_N$ we can apply the standard methods of linear algebra using the ortogonal basis instead of the original ap-functions. Note that the depth k of the known nodes of the quadtrees decreases as the algorithm proceeds. The ap-function ϕ_w is known only up to the depth $k - |w|$. This means that in step 2 the computations (calculating the inner products and finding an ortogonal basis) are done using all ap-functions only up to the depth $k - |wa|$, forgetting their values on smaller squares.

5 Conclusions

Examples of test runs of our current encoding algorithm are in Fig. 5. The original photographs (resolution 256 x 256) are shown on the left and the regenerated images (same resolution) on the right-hand side. The WFA produced by the encoding program have about 50 states and half full matrices. Both the encoding

Fig. 5. Photographs and the regenerated images

and the decoding take only tens of seconds on a DEC-workstation therefore we can afford to make the encoding program more sophisticated. We expect to achieve a higher compression ratio and simultaneously higher quality of the regenerated images.

Even more promising is the current work [10] combining the WFA with the wavelet transform technique [18, 13]. It turns out that we can construct a WFA with almost tree-structure such that each state represents one wavelet function from the family of Haar's wavelets forming a basis for the wavelet transform. The coefficients of a wavelet transform are then used as an initial distribution of this fixed WFA to specify a given function (image). Such WFA is then simplified to achieve further compression. This method combines the advantages of the "classical" wavelet compression method with the "fractal" approach and works especially well for functions (images) that are repetitive, fractal or (piecewise) smooth functions.

References

1. M. F. Barnsley, *Fractal Everywhere*, Academic Press, (1988).
2. M. F. Barnsley, J. H. Elton and D. P. Hardin, Recurrent Iterated Function Systems, *Constructive Approximation* 5, 3-31 (1989).
3. M. F. Barnsley, A. Jacquin, L. Reuter and A. D. Sloan, Harnessing Chaos for Image Synthesis, *Computer Graphics*, SIGGARPH 1988 Conference proceedings (1988).
4. J. Berstel and A. Nait Abdullah, Quadtrees Generated by Finite Automata, AFCET 61-62, 167-175 (1989).
5. J. Berstel and M. Morcrette, Compact Representation of Pattern by Finite Automata, Res.Rep. 89-66, Institut de Programmation, Université Paris 7 (1989).
6. G. J. Chaitin, Algorithmic Information Theory, *IBM Journal of Research and Development* 21, 350-359 (1977).
7. K. Culik II and S. Dube, Rational and Affine Expressions for Image Description, *Discrete Applied Mathematics*, to appear.
8. K. Culik II and S. Dube, Affine Automata and Related Techniques for Generation of Complex Images, *Theoretical Computer Science*, to appear. Preliminary Version in Proceedings of MFCS'1990, *Lecture Notes in Computer Science* 452, Springer-Verlag, 224-231 (1990).
9. K. Culik II and S. Dube, Balancing Order and Chaos in Image Generation, Proceedings of the 18th International Colloquium on Automata, Languages and Programming, Madrid, Spain, July 1991, in *Lecture Notes in Computer Science* 510, 600-614, Springer-Verlag (1991).
10. K. Culik II and S. Dube, On Combining Weighted Finite Automata and Wavelet Transforms in Data Compression, Proceedings of STACS 1993. Lecture Notes in Computer Science, to appear.
11. K. Culik II and J. Karhumäki, Automata Computing Real Functions, Tech. Report TR 9105, University of South Carolina, Columbia (1991).
12. K. Culik II and J. Kari, Image Compression using Weighted Finite Automata, Technical Report TR 9202, Univ. of South Carolina (1992).
13. R.A.DeVore, B.Jawerth and B.J.Lucier, Image Compression through Wavelet Transform Coding, *IEEE Transactions on Information Theory* 38, 719-746 (1992).
14. Y. Fisher, E. W. Jacobs and R. D. Boss, Fractal Image Compression Using Iterated Transforms: in: *Data Compression*, ed. J. Storel, Kluwer Academic Publ., Norwall, MA. (1992).
15. E. W. Jacobs, Y. Fisher and R. D. Boss, Image Compression: A Study of the Iterated Transform Method, *Signal Processing*, to appear.
16. J. Shallit and J. Stolfi, Two Methods for Generating Fractals, *Comput. and Graphics* 13, 185-191 (1989).
17. L. Staiger, Quadtrees and the Hausdorff Dimension of Pictures, Workshop on Geometrical Problems of Image Processing, Georgental GDR, 173-178 (1989).
18. G. Strang, Wavelets and Dilation Equations: A Brief Introduction, *SIAM Review* 31, 614-627 (1989).

Filter Models for a Parallel and Non Deterministic λ-calculus*

(Extended Abstract)

Mariangiola Dezani-Ciancaglini[1] Ugo de'Liguoro[1] Adolfo Piperno[2]

[1] Dipartimento di Informatica, Università di Torino
Corso Svizzera 185, 10149 Torino, Italy
{dezani,deligu}@pianeta.di.unito.it

[2] Dipartimento di Scienze dell'Informazione
Università di Roma "La Sapienza"
Via Salaria 113, 00198 Roma, Italy
piperno@dsi-next1.ing.uniroma1.it

Abstract. The distinction between the conjunctive nature of non-determinism as opposed to the disjunctive character of parallelism constitutes the motivation and the starting point of the present work. λ-calculus is extended with both a non-deterministic choice and a parallel operator; a notion of reduction is introduced, extending β-reduction of the classical calculus.

We study type assignment systems for this calculus, together with a denotational semantics which is initially defined constructing a set semimodel via simple types. We enrich the type system with intersection and union types, dually reflecting the disjunctive and conjunctive behaviour of the operators; we build a filter model whose local structure is compared with a Morris-style operational semantics.

1 Introduction

A variety of non-deterministic and parallel operators have been added to the λ-calculus to study some features of non determinism and/or concurrency in the setting of functional languages (see e.g. [3, 12, 13, 15, 16]). Also they have been considered (possibly with other operators) to gain definability of combinators like Plotkin's parallel-or [17] that is to increase the power of the λ-calculus to detect convergency internally (see [7] and also [20]: these studies are however in a lazy perspective).

Our interest is to study these concepts on their own, to investigate their nature and to understand their logical and operational properties. To this aim we introduce a pure (that is: type-free) λ-calculus whose syntax includes an internal choice operator '+' and a synchronous parallel operator '||'. The idea is to have operators which, with respect to the fundamental properties considered in the λ-calculus theory such as being solvable or strongly normalizable have a

* This work has been partially supported by grants from ESPRIT-BRA 7232 GENTZEN.

conjunctive and a disjunctive semantics, respectively. This means that $M + N$ will be, for example solvable if and only if both M and N are, while $M \| N$ enjoys this property when M or N does.

In polymorphic type disciplines, types are seen as syntactic counterparts of properties of terms, so that it is expected that in a suitable type assignment system the conjunctive and disjunctive nature of the operators $+$ and $\|$ is (dually) mirrored. This can be done either in a weaker form, using simple types, or, more explicitly, using union and intersection types. In both cases we build a "logical domain" in the sense of both filter models (see [6]) and Abramsky's theory [2]. This gives us two abstract interpretations of terms, the first one extending the semimodel interpretation of [19], the second one giving an instance of what we call a λ-lattice, that is a structure which is at the same time a λ-model and a lattice, interpreting $+$ as the meet, and $\|$ as the join.

In a final section we discuss the solvability theory in our setting. We know that both denotational semantics proposed above induce (pre)-congruences which are included in this theory, but neither of them coincides with it. At present it is not clear to the authors how to modify the type theory to get a filter model whose local structure is the solvability theory; however we conjecture that, considering a lazy theory (and hence a lazy reduction relation), a fully abstract model could be constructed with the present tools.

Because of the limited space most technical proofs are omitted or simply sketched. For an account of them the reader is referred to the full version of this paper [8].

2 Parallel and Non Deterministic λ-calculus

We extend the syntax of λ-calculus with a non deterministic choice operator $+$ and a parallel combinator $\|$:

$$M ::= x \mid \lambda x.M \mid (MN) \mid (M + N) \mid (M\|N).$$

We call $\Lambda_{+\|}$ the set of terms. For any $M \in \Lambda_{+\|}$, $FV(M)$ denotes the set of free variables of M. To turn this syntax into a calculus we introduce a reduction relation which extends β-reduction so that $+$ is an (internal) choice operator and $\|$ a synchronous evaluator of its operands; moreover $\|$ has the feature of passing to the operands M and N any argument to which $(M\|N)$ applies.

Definition 1. The reduction relation $\longrightarrow \subseteq \Lambda^2_{+\|}$ is inductively defined by:

(β) $(\lambda x.M)N \longrightarrow M[N/x]$; (μ) $M \longrightarrow N \Rightarrow LM \longrightarrow LN$;
$(+)$ $M + N \longrightarrow M$, $M + N \longrightarrow N$; (ν) $M \longrightarrow N \Rightarrow ML \longrightarrow NL$;
$(\|_1)$ $(M\|N)L \longrightarrow (ML)\|(NL)$; (ξ) $M \longrightarrow N \Rightarrow \lambda x.M \longrightarrow \lambda x.N$;
$(\|_2)$ $M \longrightarrow M', N \longrightarrow N' \Rightarrow M\|N \longrightarrow M'\|N'$.

With $\stackrel{*}{\longrightarrow}$ we denote the reflexive and transitive closure of $\longrightarrow$. We finally denote by $\longrightarrow_h$ the relation obtained by dropping rule (μ).

The idea underlying this definition is to have, in the λ-calculus syntax, a kind of conjunction and disjunction operators. Suppose any term which is strongly normalizing is considered convergent: hence $M + N$ is convergent iff both M and N are; on the other hand, because of the synchronous character of the rule ($\|_2$), $M\|N$ will be convergent iff either M or N is strongly normalizing. We insist on this notion, a sort of ***must convergency*** concept, since with ***may convergency*** the distinction between the two operators disappears (compare, for the non-deterministic choice, [12, 13]).

Our concern in the next sections will be to substantiate these intuitions, introducing types as formal counterparts of properties on one hand, and, on the other, defining equivalence relations among terms allowing formal reasoning about their meaning and behaviour.

3 Simple Types and Semimodels

Types are thought of as properties of terms. A basic requirement is that such properties should be preserved under reduction, so that we can safely assume that they concern the result (if any) of the computation of the term to which they are assigned. No question arises for classical rules concerning application and functional abstractions, but a few words are in order for $+$ and $\|$.

One should realize that type assignment systems are mainly used to express normalizability properties of terms (see e.g. [11]). In the case of $+$ we know that the term $M+N$ can be reduced to both M and N, so that to ensure the subject reduction property we have to prove that both M and N have the same type σ to conclude that $M + N$ has type σ. This is also the choice of [1]. Instead for $M\|N$ to be normalizable it suffices that either M or N normalizes. Extending this notion to arbitrary properties, it follows that one is entitled to type $M\|N$ with σ as soon as M or N (or both) can be typed with σ. See [7] for further motivations.

Definition 2 (The System $\mathcal{B}$). Let the set *Type* of types be defined by the $\sigma ::= t \mid \sigma \rightarrow \tau$, where t stands for a denumerable collection of type variables. A ***statement*** is an expression of the form $M : \sigma$, where M is a λ-term and σ a type. A ***basis*** Γ is a set of statements such that subjects are pairwise distinct variables. Then the axioms and rules of the basic assignment system $\mathcal{B}$ are the following:

$$(\mathrm{Ax})\quad \Gamma, x : \sigma \vdash x : \sigma$$

$$(\rightarrow \mathrm{I})\,\frac{\Gamma, x : \sigma \vdash M : \tau}{\Gamma \vdash \lambda x.M : \sigma \rightarrow \tau} \qquad (\rightarrow \mathrm{E})\,\frac{\Gamma \vdash M : \sigma \rightarrow \tau \quad \Gamma \vdash N : \sigma}{\Gamma \vdash MN : \tau}$$

$$(\| \,\mathrm{I})\,\frac{\Gamma \vdash M : \sigma}{\Gamma \vdash M\|N : \sigma} \qquad \frac{\Gamma \vdash N : \sigma}{\Gamma \vdash M\|N : \sigma} \qquad (+\,\mathrm{I})\,\frac{\Gamma \vdash M : \sigma \quad \Gamma \vdash N : \sigma}{\Gamma \vdash M + N : \sigma}$$

If $\Gamma \vdash M : \sigma$ is derivable in $\mathcal{B}$, we write $\Gamma \vdash_{\mathcal{B}} M : \sigma$.

In this system, as in Curry's original one, there is a correspondence between the main constructor of the subject of the conclusion in each rule and the rule itself; this does not hold for the type. However, classical terms (i.e. those without occurences of $+$ and $||$) have just their simple types. This property results in a simple theory of the type assignment system.

A routine induction on derivations in $\mathcal{B}$ shows:

Lemma 3 (Structural Properties of Deductions).

(i) $\Gamma \vdash_{\mathcal{B}} x : \tau \Leftrightarrow x : \tau \in \Gamma$;

(ii) $\Gamma \vdash_{\mathcal{B}} \lambda x.M : \sigma \to \tau \Leftrightarrow \Gamma, x : \sigma \vdash_{\mathcal{B}} M : \tau$;

(iii) $\Gamma \vdash_{\mathcal{B}} MN : \tau \Leftrightarrow \Gamma \vdash_{\mathcal{B}} M : \sigma \to \tau$ and $\Gamma \vdash_{\mathcal{B}} N : \sigma$ for some σ;

(iv) $\Gamma \vdash_{\mathcal{B}} M + N : \sigma \Leftrightarrow \Gamma \vdash_{\mathcal{B}} M : \sigma$ and $\Gamma \vdash_{\mathcal{B}} N : \sigma$;

(v) $\Gamma \vdash_{\mathcal{B}} M||N : \sigma \Leftrightarrow \Gamma \vdash_{\mathcal{B}} M : \sigma$ or $\Gamma \vdash_{\mathcal{B}} N : \sigma$.

Using this lemma it is easy to prove the following corollary by induction on the definition of $\longrightarrow$.

Corollary 4 (Subject Reduction of $\mathcal{B}$). $\Gamma \vdash_{\mathcal{B}} M{:}\sigma, M \overset{*}{\longrightarrow} N \Rightarrow \Gamma \vdash_{\mathcal{B}} N{:}\sigma$.

For the classical λ-calculus, a filter model construction with simple types, even considering as a "filter" any set of types, does not yield a λ-model (see e.g. [10]). Indeed the best one can obtain is a *semimodel* in the sense of [19], i.e. a model in which interreducible terms are equal, but in general convertible terms are not. This is attractive in the present case since the symmetric closure of $\overset{*}{\longrightarrow}$ is clearly the trivial relation. Adapting Plotkin's definition to the present context (see also [1]) we introduce the following notion:

Definition 5. A *semimodel* is a structure $\mathcal{P} = \langle P, \sqsubseteq, \cdot, \sqcap, \sqcup, [\![\cdot]\!]^{\mathcal{P}} \rangle$ where $\langle P, \sqsubseteq \rangle$ is a poset, and $\cdot, \sqcap, \sqcup$ are binary monotonic operations that satisfy the following requirements: $d \sqcap e \sqsubseteq d$, $d \sqcap e \sqsubseteq e$, $d \sqsubseteq d \sqcup e$, $e \sqsubseteq d \sqcup e$ and $(d \sqcup d') \cdot e \sqsubseteq (d \cdot e) \sqcup (d' \cdot e)$. Finally $[\![\cdot]\!]^{\mathcal{P}} : \Lambda_{+||} \times Env \to P$, where $Env = \{\rho \mid \rho : TermVar \to P\}$, is such that:

(a) $[\![x]\!]^{\mathcal{P}}_{\rho} = \rho(x)$;

(b) $[\![MN]\!]^{\mathcal{P}}_{\rho} = [\![M]\!]^{\mathcal{P}}_{\rho} \cdot [\![N]\!]^{\mathcal{P}}_{\rho}$;

(c) $\rho \lceil FV(M) = \rho' \lceil FV(M) \Rightarrow [\![M]\!]^{\mathcal{P}}_{\rho} = [\![M]\!]^{\mathcal{P}}_{\rho'}$;

(d) $\forall d \in P.\ [\![M]\!]^{\mathcal{P}}_{\rho[d/x]} \sqsubseteq [\![N]\!]^{\mathcal{P}}_{\rho[d/x]} \Rightarrow [\![\lambda x.M]\!]^{\mathcal{P}}_{\rho} \sqsubseteq [\![\lambda x.N]\!]^{\mathcal{P}}_{\rho}$;

(e) $[\![\lambda x.M]\!]^{\mathcal{P}}_{\rho} \cdot d \sqsubseteq [\![M]\!]^{\mathcal{P}}_{\rho[d/x]}$;

(f) $[\![M + N]\!]^{\mathcal{P}}_{\rho} = [\![M]\!]^{\mathcal{P}}_{\rho} \sqcap [\![N]\!]^{\mathcal{P}}_{\rho}$;

(g) $[\![M||N]\!]^{\mathcal{P}}_{\rho} = [\![M]\!]^{\mathcal{P}}_{\rho} \sqcup [\![N]\!]^{\mathcal{P}}_{\rho}$.

Semimodels interpret the reduction relation, as it is stated in the following proposition, which can be proved by induction on the definition of $\longrightarrow$.

Proposition 6. $M \overset{*}{\longrightarrow} N \Rightarrow \forall \rho.\ [\![M]\!]^{\mathcal{P}}_{\rho} \sqsubseteq [\![N]\!]^{\mathcal{P}}_{\rho}$ *for all semimodels* $\mathcal{P}$.

In the case of the classical calculus one has $\Leftrightarrow$ (see [19]). Here instead completeness with respect to reduction does not hold: e.g. we have, by definition, that $\forall \rho \in Env.\ [\![M]\!]_{\rho} \sqsubseteq [\![M||N]\!]_{\rho}$ but we do not have $M \overset{*}{\longrightarrow} M||N$. This does not seem to be unfortunate; indeed we are looking for a partial order (and its relative equivalence) which is, in a sense, more abstract than simple reducibility.

As expected, the type assignment $\mathcal{B}$ induces a semimodel.

Proposition 7. *Let* $a \cdot b = \{\tau \in Type \mid \exists \sigma \in b.\ \sigma \to \tau \in a\}$ *and* $[\![M]\!]^{\mathcal{B}}_{\rho} = \{\sigma \mid \Gamma \vdash_{\mathcal{B}} M : \sigma$ *for some* $\Gamma \subseteq \{x : \tau \mid \tau \in \rho(x)\}\}$. *The structure* $\langle \wp(Type), \subseteq, \cdot, \cap, \cup, [\![\cdot]\!]^{\mathcal{B}} \rangle$ *is a semimodel (the* set semimodel*).*

The definition of interpretation of a term in the set semimodel can be given in a compositional way. For example for all ρ: $[\![M+N]\!]_\rho = [\![M]\!]_\rho \cap [\![N]\!]_\rho$ and $[\![M\|N]\!]_\rho = [\![M]\!]_\rho \cup [\![N]\!]_\rho$.

To interpret types over a given semimodel we use the ***simple semantics*** of types ([9, 19]).

Definition 8. A *type structure* over $\mathcal{P}$ is a pair $\langle \mathcal{T}, \Rightarrow \rangle$ where:
(i) $\mathcal{T} \subseteq \{X \in \wp(P) \mid X \text{ is upwards closed}\}$;
(ii) $X, Y \in \mathcal{T}$ implies $(X \Rightarrow Y) \in \mathcal{T}$, where
$$X \Rightarrow Y = \{d \in P \mid \forall e \in X.\, d \cdot e \in Y\},$$
which is well defined because of the monotonicity of the application. A *type environment* is a map η from type variables to $\mathcal{T}$; define $[\![\sigma]\!]^{\mathcal{T}}_\eta \in \mathcal{T}$ by
$$[\![t]\!]^{\mathcal{T}}_\eta = \eta(t) \quad \text{and} \quad [\![\sigma \to \tau]\!]^{\mathcal{T}}_\eta = [\![\sigma]\!]^{\mathcal{T}}_\eta \Rightarrow [\![\tau]\!]^{\mathcal{T}}_\eta .$$
Now a basis Γ *satisfies* ρ and η iff, for all $x : \tau \in \Gamma$, $\rho(x) \in [\![\tau]\!]^{\mathcal{T}}_\eta$; then:
$$\Gamma \models M : \sigma \;\Leftrightarrow\; \forall \mathcal{P}, \mathcal{T} \text{ over } \mathcal{P},\ \forall \rho, \eta.\ \Gamma \text{ satisfies } \rho, \eta \Rightarrow [\![M]\!]^{\mathcal{P}}_\rho \in [\![\sigma]\!]^{\mathcal{T}}_\eta .$$

Theorem 9 (Completeness of $\mathcal{B}$). $\Gamma \vdash_{\mathcal{B}} M : \sigma \;\Leftrightarrow\; \Gamma \models M : \sigma$.

Proof. This proof essentially adapts Plotkin's completeness proof in [19].
($\Rightarrow$) Simple induction on the derivation of $\Gamma \vdash M : \sigma$.
($\Leftarrow$) Using the set semimodel. Indeed the pair $\langle \mathcal{T}, \Rightarrow \rangle$ is a type structure for it, where $\mathcal{T} = \{\chi_\sigma\}_{\sigma \in Type}$, $\chi_\sigma = \{a \subseteq Type \mid \sigma \in a\}$ and $\chi_\sigma \Rightarrow \chi_\tau = \chi_{\sigma\to\tau}$. Taking η and ρ such that $\rho(x) = \{\sigma \mid x : \sigma \in \Gamma\}$ for every term variable x and $\eta(t) = \chi_t$ for every type variable t, one has that $[\![\sigma]\!]_\eta = \chi_\sigma$ for all $\sigma \in Type$ and $[\![M]\!]^{\mathcal{B}}_\rho \in [\![\sigma]\!]^{\mathcal{T}}_\eta$ which implies $\Gamma \vdash_{\mathcal{B}} M : \sigma$.

The set semimodel allows to define a preorder over terms which is a precongruence:
$$M \sqsubseteq^{\mathcal{B}} N \Leftrightarrow_{def} \forall \rho.\ [\![M]\!]^{\mathcal{B}}_\rho \subseteq [\![N]\!]^{\mathcal{B}}_\rho .$$
We list in the following proposition the main (in)-equations holding in the set semimodel semantics.

Proposition 10. *Let* $\simeq^{\mathcal{B}} = \sqsubseteq^{\mathcal{B}} \cap \sqsupseteq^{\mathcal{B}}$*, then:*
(i) $(\lambda x.M)N \sqsubseteq^{\mathcal{B}} M[N/x]$;
(ii) $(M+N)L \sqsubseteq^{\mathcal{B}} ML + NL$;
(iii) $L(M+N) \sqsubseteq^{\mathcal{B}} LM + LN$;
(iv) $(M\|N)L \simeq^{\mathcal{B}} ML\|NL$;
(v) $L(M\|N) \simeq^{\mathcal{B}} LM\|LN$;
(vi) $\lambda x.(M+N) \simeq^{\mathcal{B}} \lambda x.M + \lambda x.N$;
(vii) $\lambda x.(M\|N) \simeq^{\mathcal{B}} \lambda x.M\|\lambda x.N$,
where the inequalities (i), (ii) and (iii) are in general proper.

Proof. The positive statements are straightforward consequences of the completeness theorem. For the negative parts observe that (i) essentially claims that the set semimodel is not a λ-model. To see (ii) let $\Gamma = \{x : \sigma_1 \to \tau, y : \sigma_2 \to \tau, z : \sigma_1, v : \sigma_2\}$ where $\sigma_1 \not\equiv \sigma_2$. Then $\Gamma \vdash_{\mathcal{B}} x(z\|v) + y(z\|v) : \tau$, but $\Gamma \not\vdash_{\mathcal{B}} (x+y)(z\|v) : \tau$ since $x+y$ has no type. Similarly for (iii) we have that $\Gamma \vdash_{\mathcal{B}} (x\|y)z + (x\|y)v : \tau$, but $\Gamma \not\vdash_{\mathcal{B}} (x\|y)(z+v) : \tau$ since $z+v$ has no type.

4 Intersection, Union Types and λ-lattices

In this section we extend the notion of filter-model introduced in [6] to our calculus, the aim being this time to build a model in which usual λ-calculus equations hold and which fits better the operational behaviour of $+$ and $||$. Let us redefine the syntax of types according to the following:

$$\sigma ::= t \mid \omega \mid \sigma \to \tau \mid \sigma \wedge \tau \mid \sigma \vee \tau$$

and call again *Type* the resulting set.

Definition 11. Let $\leq$ be the smallest partial order over types s.t. $\langle Type, \leq \rangle$ is a distributive lattice, in which $\wedge$ is the meet, $\vee$ is the join and ω is the top, and moreover the arrow satisfies:

(i) $\omega \leq \omega \to \omega$;
(iii) $\sigma \geq \sigma', \tau \leq \tau' \Rightarrow \sigma \to \tau \leq \sigma' \to \tau'$.
(ii) $(\sigma \to \rho) \wedge (\sigma \to \tau) \leq \sigma \to (\rho \wedge \tau)$;

Definition 12. The system $\mathcal{L}$ is obtained by adding to the rules in the basic system $\mathcal{B}$ the following axiom and rules:

$$(\omega)\quad \Gamma \vdash M : \omega \qquad (\wedge\text{I})\ \frac{\Gamma \vdash M : \sigma \quad \Gamma \vdash M : \tau}{\Gamma \vdash M : \sigma \wedge \tau}$$

$$(\vee\text{I})\ \frac{\Gamma \vdash M : \sigma}{\Gamma \vdash M : \sigma \vee \tau} \qquad \frac{\Gamma \vdash M : \tau}{\Gamma \vdash M : \sigma \vee \tau} \qquad (\leq)\ \frac{\Gamma \vdash M : \sigma \quad \sigma \leq \tau}{\Gamma \vdash M : \tau}$$

If $\Gamma \vdash M : \sigma$ is derivable in the system $\mathcal{L}$ we write $\Gamma \vdash_{\mathcal{L}} M : \sigma$.

Of course rule $(\vee\text{I})$ is redundant. The following rules are admissible:

$$\frac{\Gamma \vdash M : \sigma \quad \Gamma \vdash N : \tau}{\Gamma \vdash M + N : \sigma \vee \tau} \qquad \frac{\Gamma \vdash M : \sigma \quad \Gamma \vdash N : \tau}{\Gamma \vdash M || N : \sigma \wedge \tau} \qquad \frac{\Gamma, x : \sigma \vdash M : \tau \quad \sigma' \leq \sigma}{\Gamma, x : \sigma' \vdash M : \tau}.$$

Remark. Notice that $\sigma_1 \vee \sigma_2 \to \tau \leq (\sigma_1 \to \tau) \wedge (\sigma_2 \to \tau)$, but the converse does not hold. The equality is derivable, for example, in the system proposed in [4], where only pure λ-terms are considered. In presence of $+$ and of the corresponding typing rule, by postulating $(\sigma_1 \to \tau) \wedge (\sigma_2 \to \tau) \leq \sigma_1 \vee \sigma_2 \to \tau$ we would lose the subject reduction property. In fact we have: $\vdash_{\mathcal{L}} \lambda x.xx\mathbf{KI}(\Delta\Delta) : (((\nu \to \omega \to \nu) \wedge \nu) \to \mu) \wedge ((\omega \to \nu \to \nu) \to \mu)$ and $\vdash_{\mathcal{L}} \mathbf{K} + \mathbf{O} : ((\nu \to \omega \to \nu) \wedge \nu) \vee (\omega \to \nu \to \nu)$ where $\mathbf{K} \equiv \lambda xy.x$, $\mathbf{I} \equiv \lambda x.x$, $\Delta \equiv \lambda x.xx$, $\mathbf{O} \equiv \lambda xy.y$, $\mu \equiv t \to t$ and $\nu \equiv \mu \to \omega \to \mu$. Therefore, since $((\nu \to \omega \to \nu) \wedge \nu \to \mu) \wedge ((\omega \to \nu \to \nu) \to \mu) \leq ((\nu \to \omega \to \nu) \wedge \nu) \vee (\omega \to \nu \to \nu) \to \mu$ becomes valid, we could derive: $\vdash_{\mathcal{L}} (\lambda x.xx\mathbf{KI}(\Delta\Delta))(\mathbf{K}+\mathbf{O}) : \mu$; but $(\lambda x.xx\mathbf{KI}(\Delta\Delta))(\mathbf{K}+\mathbf{O})$ reduces to $\Delta\Delta$, which is unsolvable and has only types equivalent to ω (see [8]).

Lemma 13 (Structural Properties of Deductions).

(i) $\Gamma \vdash_{\mathcal{L}} x : \tau \Leftrightarrow x : \sigma \in \Gamma$ for some $\sigma \leq \tau$;
(ii) $\Gamma \vdash_{\mathcal{L}} \lambda x.M : \sigma \to \tau \Leftrightarrow \Gamma, x : \sigma \vdash_{\mathcal{L}} M : \tau$;
(iii) $\Gamma \vdash MN : \tau \Leftrightarrow \Gamma \vdash_{\mathcal{L}} M : \sigma \to \tau$ and $\Gamma \vdash_{\mathcal{L}} N : \sigma$ for some σ;
(iv) $\Gamma \vdash_{\mathcal{L}} M + N : \sigma \Leftrightarrow \Gamma \vdash_{\mathcal{L}} M : \sigma$ and $\Gamma \vdash_{\mathcal{L}} N : \sigma$;
(v) $\Gamma \vdash_{\mathcal{L}} M||N : \tau \Leftrightarrow \Gamma \vdash_{\mathcal{L}} M : \sigma$ and $\Gamma \vdash_{\mathcal{L}} N : \sigma'$ for some σ, σ' such that $\sigma \wedge \sigma' \leq \tau$.

The invariance of types under subject reduction is now an easy consequence of previous Lemma.

Corollary 14 (Subject Reduction of $\mathcal{L}$). $\Gamma \vdash_{\mathcal{L}} M{:}\sigma, M \xrightarrow{*} N \Rightarrow \Gamma \vdash_{\mathcal{L}} N{:}\sigma$.

Given the usual notion of *filter*, rules $(\omega), (\leq)$ and $(\wedge I)$ imply that, for any Γ and M, $\{\sigma \mid \Gamma \vdash_{\mathcal{L}} M : \sigma\}$ is a filter. A filter model construction as in [6] can be carried out. If X is a subset of any poset, then $\uparrow X$ is its upward closure.

As the set semimodel suggests, when interpreting our calculus we naturally get lattices. For the sake of our construction, we have to make precise what is a model of this calculus, which is done by incorporating the notion of lattice into that of λ-model of [10].

Definition 15. A *λ-lattice* is a structure $\mathcal{D} = \langle D, \sqsubseteq, \cdot, \sqcap, \sqcup, [\![\cdot]\!]^{\mathcal{D}}\rangle$ where:

(i) $\langle D, \sqsubseteq, \sqcap, \sqcup\rangle$ is a lattice;
(ii) $\cdot : D \times D \to D$ is monotonic;
(iii) $\langle D, \cdot, [\![\cdot]\!]^{\mathcal{D}}\rangle$ is a λ-model;
(iv) $[\![M+N]\!]^{\mathcal{D}}_{\rho} = [\![M]\!]^{\mathcal{D}}_{\rho} \sqcap [\![N]\!]^{\mathcal{D}}_{\rho}$;
(v) $[\![M \| N]\!]^{\mathcal{D}}_{\rho} = [\![M]\!]^{\mathcal{D}}_{\rho} \sqcup [\![N]\!]^{\mathcal{D}}_{\rho}$.

As usual we denote by ρ the generic term environment, being a map from term variables to D.

Theorem 16. *Let $\mathcal{F}(Type)$ be the set of filters over Type and define, for $f, f' \in \mathcal{F}(Type)$: $f \bar{\cup} f' = \uparrow\{\sigma \wedge \tau \mid \sigma \in f, \tau \in f'\}$, $f \cdot f' = \{\tau \mid \exists \sigma \in f'. \sigma \to \tau \in f\}$. Then $f \bar{\cup} f', f \cdot f' \in \mathcal{F}(Type)$. Moreover the structure $\langle \mathcal{F}(Type), \subseteq, \cdot, \cap, \bar{\cup}, [\![\cdot]\!]^{\mathcal{L}}\rangle$, where $[\![M]\!]^{\mathcal{L}}_{\rho} = \{\sigma \mid \Gamma \vdash_{\mathcal{L}} M : \sigma$ for some $\Gamma \subseteq \{x : \sigma \mid \sigma \in \rho(x)\}\}$, is a λ-lattice (the* filter λ-lattice*).*

Proof. $f \bar{\cup} f'$ is the least filter including $f \cup f'$. It is easy to see that $f \cdot f'$ is a filter too: hence $\langle \mathcal{F}(Type), \cap, \bar{\cup}\rangle$ is a lattice. The application $\cdot$ is clearly monotonic. Now lemma 13 implies that $\mathcal{F}(Type)$ is a λ-lattice.

In the filter λ-lattice we have again a clear characterization of the meaning of $M+N$ and $M\|N$, in terms of the meaning of M and N: $[\![M+N]\!]_{\rho} = [\![M]\!]_{\rho} \cap [\![N]\!]_{\rho}$ and $[\![M\|N]\!]_{\rho} = [\![M]\!]_{\rho} \bar{\cup} [\![N]\!]_{\rho}$, for all ρ.

Definition 17. Let $\mathcal{D} = \langle D, \sqsubseteq, \cdot, \sqcap, \sqcup, [\![\cdot]\!]^{\mathcal{D}}\rangle$ be a λ-lattice. Then a type structure over $\mathcal{D}$ is a pair $\langle \mathcal{T}, \Rightarrow\rangle$ such that $\mathcal{T}$ is a sublattice of the lattice of filters over D, $D \in \mathcal{T}$ and $\mathcal{T}$ is closed under $\Rightarrow$ (see Definition 8). The map $[\![\cdot]\!]^{\mathcal{T}}$, interpreting types over $\mathcal{T}$, is defined as in Definition 8, adding three clauses:

(iii) $[\![\omega]\!]^{\mathcal{T}}_{\eta} = D$;
(iv) $[\![\sigma \wedge \tau]\!]^{\mathcal{T}}_{\eta} = [\![\sigma]\!]^{\mathcal{T}}_{\eta} \cap [\![\tau]\!]^{\mathcal{T}}_{\eta}$;
(v) $[\![\sigma \vee \tau]\!]^{\mathcal{T}}_{\eta} = [\![\sigma]\!]^{\mathcal{T}}_{\eta} \bar{\cup} [\![\tau]\!]^{\mathcal{T}}_{\eta} =_{def} \uparrow \{d \sqcap d' \mid d \in [\![\sigma]\!]^{\mathcal{T}}_{\eta}, d' \in [\![\tau]\!]^{\mathcal{T}}_{\eta}\}$,

with some overloading of $\bar{\cup}$.

In case of the filter λ-lattice of Theorem 16, the interpretation of a type turns out to be a filter of filters of types. Since the lattice of types is distributive, the lattice of filters forming the filter λ-lattice is distributive too, hence the upward closure in clause (v) above is in this case redundant. The following proposition is proved by routine calculations.

Proposition 18. *Let $\chi_{\sigma} = \{f \in \mathcal{F}(Type) \mid \sigma \in f\}$. Then $\langle\{\chi_{\sigma} \mid \sigma \in Type\}, \Rightarrow\rangle$ (where $\Rightarrow$ is defined as in Definition 8) is a type structure over the filter λ-lattice.*

Moreover it satisfies the following equations:
(i) $\chi_\omega = \mathcal{F}(Type)$; (iii) $\chi_{\sigma\wedge\tau} = \chi_\sigma \cap \chi_\tau$;
(ii) $\chi_{\sigma\to\tau} = (\chi_\sigma \Rightarrow \chi_\tau)$; (iv) $\chi_{\sigma\vee\tau} = \chi_\sigma \bar{\cup} \chi_\tau = \{f \cap f' \mid f \in \chi_\sigma, f' \in \chi_\tau\}$.

As for system $\mathcal{B}$, the immediate consequence of Theorem 16 and of Proposition 18 is completeness. Redefining $\models$ in the obvious way, this is stated as follows.

Corollary 19 (Completeness of $\mathcal{L}$). $\Gamma \vdash_{\mathcal{L}} M : \sigma \;\Leftrightarrow\; \Gamma \models M : \sigma$.

The precongruence on terms induced by the filter λ-lattice is defined as follows:

$$M \sqsubseteq^{\mathcal{L}} N \Leftrightarrow_{def} \forall \rho.\ [\![M]\!]^{\mathcal{L}}_\rho \subseteq [\![N]\!]^{\mathcal{L}}_\rho.$$

It enjoys the following properties.

Proposition 20. *Let* $\simeq^{\mathcal{L}} = \sqsubseteq^{\mathcal{L}} \cap \sqsupseteq^{\mathcal{L}}$, *then it is a congruence such that:*
(i) $(\lambda x.M)N \simeq^{\mathcal{L}} M[N/x]$;
(ii) $(M+N)L \simeq^{\mathcal{L}} ML+NL$;
(iii) $L(M+N) \sqsubseteq^{\mathcal{L}} LM+LN$;
(iv) $(M\|N)L \simeq^{\mathcal{L}} ML\|NL$;
(v) $LM\|LN \sqsubseteq^{\mathcal{L}} L(M\|N)$;
(vi) $\lambda x.(M+N) \sqsubseteq^{\mathcal{L}} \lambda x.M + \lambda x.N$;
(vii) $\lambda x.(M\|N) \simeq^{\mathcal{L}} \lambda x.M\|\lambda x.N$.
where inequations (iii), (v) and (vi) are in general proper.

Proof. The equalities follow from Theorem 16 and the properties of $\leq$. For the negative part of (iii) let $\Delta \equiv \lambda x.xx$, $P \equiv \lambda x.x(\lambda yzv.v)\Delta$ and $Q \equiv \lambda x.\Delta$. Then $\Delta P + \Delta Q$ has type $(\sigma \wedge (\sigma \to \tau)) \to \tau$ while ω is the only type which can be deduced for $\Delta(P+Q)$.
For the negative part of (vi) we have for example $\vdash_{\mathcal{L}} (\lambda x.x) + (\lambda x.xx) : (\sigma \to \sigma) \vee (\sigma \wedge (\sigma \to \tau) \to \tau)$, but this type cannot be deduced for $\lambda x.(x + xx)$.

We observe that parts (ii) and (iii) of this Proposition imply that we have a (properly) semilinear applicative structure as defined in [12, 13]. This is true for $\simeq^{\mathcal{O}}$, but was not true for $\simeq^{\mathcal{B}}$. It is worth to stress that, without the union type constructor, this cannot be achieved (see [1]). From this Proposition and from Proposition 10 it is also clear that the theories induced by $\simeq^{\mathcal{B}}$ and $\simeq^{\mathcal{L}}$ are incomparable.

5 Solvability and Adequacy

Even if the set semimodel semantics deserves some interest, it is not satisfactory to deal with models in which the β equation does not hold; on the contrary, so far, λ-lattices do not have an "operational" instance.

If one is interested in a (possibly conservative) extension of a λ-theory, the suitable "contextual theory"(see [14, 5]) seems to be a natural candidate. Let us define as ***solvable*** those terms which do not admit an infinite head reduction, that is:

$$\mathsf{SOL} = \{M \in \Lambda_{+\|} \mid \exists n \in \mathsf{N}\ \forall N \in \Lambda_{+\|},\ m \in \mathsf{N}.\ M \xrightarrow{m}_h N \Rightarrow m \leq n\},$$

where $\xrightarrow{m}_h$ is the m-times composition of $\longrightarrow_h$ with itself. Note that this is different from having all reducts with a head normal form: as a counterexample take the unsolvable $F0$, where $F \equiv \mathbf{Y}(\lambda fx.x + f(\mathsf{Succ}\ x))$ and 0 and Succ are the zero and successor of Church numerals respectively.

Now for any $M, N \in \Lambda_{+\|}$ we define:

$$M \sqsubseteq^{\mathcal{O}} N \Leftrightarrow_{def} \forall C[\,] \in \Lambda_{+\|}[\,].\ C[M] \in \mathsf{SOL} \Rightarrow C[N] \in \mathsf{SOL}.$$

Accordingly,

$$M \simeq^{\mathcal{O}} N \Leftrightarrow_{def} M \sqsubseteq^{\mathcal{O}} N \sqsubseteq^{\mathcal{O}} M.$$

Clearly the relation $\sqsubseteq^{\mathcal{O}}$ is a precongruence. The set SOL, when restricted to pure λ-terms, is the set of terms having a head normal form (that is solvable in the classical sense), hence the restriction of $\simeq^{\mathcal{O}}$ is the λ-theory of D_∞.

Proposition 21. *We list some (in)-equations according to* $\sqsubseteq^{\mathcal{O}}$*:*

(i) $(\lambda x.M)N \simeq^{\mathcal{O}} M[N/x]$;
(ii) $\lambda x.(M + N) \simeq^{\mathcal{O}} \lambda x.M + \lambda x.N$;
(iii) $\lambda x.(M\|N) \simeq^{\mathcal{O}} \lambda x.M\|\lambda x.N$;
(iv) $M + N \sqsubseteq^{\mathcal{O}} M, N$;
(v) $L \sqsubseteq^{\mathcal{O}} M, N \Rightarrow L \sqsubseteq^{\mathcal{O}} M + N$;
(vi) $M, N \sqsubseteq^{\mathcal{O}} M\|N$;
(vii) $M, N \sqsubseteq^{\mathcal{O}} L \Rightarrow M\|N \sqsubseteq^{\mathcal{O}} L$.

Proof. (v): Idempotence of + is easily proved. Now, given an arbitrary context $C[\ ]$, let $C'[\ ] \equiv C[[\ ]+L]$ and $C''[\ ] \equiv C[M+[\ ]]$. If $L \sqsubseteq^{\mathcal{O}} M, N$, then $C[L] \in \mathsf{SOL} \Rightarrow C[L+L] \equiv C'[L] \in \mathsf{SOL} \Rightarrow C'[M] \equiv C''[L] \in \mathsf{SOL} \Rightarrow C''[N] \equiv C[M+N] \in \mathsf{SOL}$.
(vii): Similarly, $\|$ is idempotent, hence given an arbitrary context $C[\,]$ let $C'[\,] \equiv C[[\,]\|N]$ and $C''[\,] \equiv C[L\|[\,]]$. If $M, N \sqsubseteq^{\mathcal{O}} L$, then $C[M\|N] \equiv C'[M] \in \mathsf{SOL} \Rightarrow C'[L] \equiv C''[N] \in \mathsf{SOL} \Rightarrow C''[L] \equiv C[L\|L] \in \mathsf{SOL} \Rightarrow C[L] \in \mathsf{SOL}$.

As an immediate consequence of Proposition 21, we obtain a term model which is a λ-lattice.

Proposition 22. *For* $M \in \Lambda_{+\|}$, *define* $[M] = \{M' \in \Lambda_{+\|} \mid M \simeq^{\mathcal{O}} M'\}$, $[M] \cdot [N] = [MN]$ *and* $[M] \sqsubseteq [N]$ *iff* $M \sqsubseteq^{\mathcal{O}} N$. *Then these are well defined and induce a λ-lattice, where* $[\![M]\!]_\rho = [M[\vec{N}/\vec{x}]]$ *when* $FV(M) = \vec{x}$ *and* $\rho(\vec{x}) = [\vec{N}]$.

The existence of the term model implies an adequacy result.

Corollary 23. $\forall M, N \in \Lambda_{+\|}.\ (\forall \lambda\text{-}lattice\ \mathcal{D},\ \rho.\ [\![M]\!]^{\mathcal{D}}_\rho \sqsubseteq [\![N]\!]^{\mathcal{D}}_\rho) \Rightarrow M \sqsubseteq^{\mathcal{O}} N$.

The adequacy is also true for the filter λ-lattice.

Theorem 24 Adequacy Theorem. *The filter λ-lattice is adequate, i.e.:*

$$M \sqsubseteq^{\mathcal{L}} N \Rightarrow M \sqsubseteq^{\mathcal{O}} N.$$

Proof. It can be proved by standard techniques that a term is typable with a type $\neq \omega$ iff it is solvable. Now, since $\sqsubseteq^{\mathcal{L}}$ is a precongruence, we immediately have that $M \sqsubseteq^{\mathcal{L}} N \Rightarrow \forall \Gamma, \sigma, C[\,].\ [\Gamma \vdash_{\mathcal{L}} C[M] : \sigma \Rightarrow \Gamma \vdash_{\mathcal{L}} C[N] : \sigma]$. It follows that $C[M] \in \mathsf{SOL} \Rightarrow \exists \Gamma, \sigma \neq \omega.\ \Gamma \vdash_{\mathcal{L}} C[M] : \sigma \Rightarrow \exists \Gamma, \sigma \neq \omega.\ \Gamma \vdash_{\mathcal{L}} C[N] : \sigma \Rightarrow C[N] \in \mathsf{SOL}$.

Comparing Proposition 20 with Proposition 21 it is clear that the filter λ-lattice is not fully abstract.

References

1. M. Abadi, "A Semantics for Static Type Inference in a Nondeterministic Language", to appear in *Information and Computation.*
2. S. Abramsky, "Domain Theory in Logical Form", *Ann. of Pure and Appl. Logics*, 51, 1991, 1-77.
3. E.A. Ashcroft, M.C.B. Hennessy, "A Mathematical Semantics for a Non Deterministic Typed Lambda Calculus", *TCS* 11, 1980, 227-245.
4. F. Barbanera, M. Dezani-Ciancaglini, "Intersection and Union Types", TACS'91, *LNCS* 526, 1991, 651-674.
5. H.P. Barendregt, *The Lambda-Calculus: Its Syntax and Semantics*, North-Holland, Amsterdam 1984.
6. H. Barendregt, M. Coppo, M. Dezani-Ciancaglini, "A Filter Lambda Model and the Completeness of Type Assignment", *J.Symbolic Logic* 48, 1983, 931-940.
7. G. Boudol, "A Lambda Calculus for (Strict) Parallel Functions", INRIA -Sophia Antipolis Tech. Rep. 1387, 1991, to appear in *Information and Computation.*
8. M. Dezani-Ciancaglini, U. de'Liguoro, A. Piperno, "Filter Models for a Parallel and Non Deterministic λ-calculus", Internal report, Un. di Roma I, 1993.
9. R. Hindley, "The Completeness Theorem for Typing λ-terms", *TCS* 22, 1983, 1-17.
10. R. Hindley, G. Longo, "Lambda Calculus Models and Extensionality", *Z. Math. Logik* 26, 1980, 289-310.
11. J.L. Krivine, *Lambda-calcul, types et modèles*, Masson, Paris 1990.
12. U. de' Liguoro, "Non-deterministic Untyped λ-calculus", PhD Thesis, Un. di Roma I, 1991.
13. U. de' Liguoro, A. Piperno, "Must Preorder in Non-deterministic Untyped λ-calculus", CAAP '92, *LNCS* 582, 1992, 203-220.
14. J.H. Morris, *Lambda Calculus Models of Programming Languages*, Dissertation, M.I.T. 1968.
15. C.-H.L. Ong, "Concurrent Lambda Calculus and a General Precongruence Theorem for Applicative Bisimulations", Draft, Cambridge Un., 1992.
16. C.-H.L. Ong, "Non-Determinism in a Functional Setting", to appear in *LICS '93.*
17. G.D. Plotkin, "LCF considered as a programming language", *TCS*, 5, 1977, 223-256.
18. G.D. Plotkin, "A Powerdomain Construction", *SIAM J. of Comp.* 3, 1976, 452-487.
19. G. D. Plotkin, "A Semantics for Static Type Inference", to appear in *Information and Computation.* A preliminary version in TACS'91, *LNCS* 526, 1991, 1-17.
20. D. Sangiorgi, "The Lazy λ-calculus in a Concurrency Scenario", *LICS'92*, 1992, 102-109.
21. M.B. Smith, "A Power Domains", *J. Comp. Sys. Sci.* 16, 1978, 23-36.

Real Number Computability and Domain Theory

Pietro Di Gianantonio

dip. di Matematica e Informatica, Università di Udine
via Zanon 6 I-33100 Udine Italy

Abstract. We propose a possible implementation, using lazy functional programming, of the exact computation on real numbers. Using domain theory we can analyze this kind of computation and give a definition of computability for the functions on the real number. This definition turns out to be equivalent to other definitions given in the literature using different methods.
Domain theory is a useful tool to study higher order computability on the real numbers. An interesting connection between Scott Topology and the topologies on the real line and on the space of the real functions is stated. The main original result in this work is the proof that every computable functional on real numbers is continuous w.r.t. the compact open topology.

1 The Representation of the Real Numbers

It is well known that one can compute over potentially infinite data structures (e.g. infinite strings, infinite trees) using functional programming languages with lazy evaluation. It is therefore possible to represent the real numbers by infinite strings of digits and to write programs that implement the usual analytic functions with an arbitrary precision.

For example in this context a program for addition is a program that:

1. receives as input two strings of digits representing the two summands,
2. reads sequentially the strings starting from the most significant digits and
3. at the same time, generates sequentially the digits of the result starting from the most significant digits.

It is a well known fact that the standard decimal and binary representations are not suitable for this kind of computation [1, 3, 16, 18, 20]. Using these representations even the most fundamental functions such as addition or multiplication are not computable. Here is a simple example that illustrates the inadequacy of this representation. We show that no algorithm can compute the multiplication by 3. An hypothetical algorithm for this function is not able to generate the first digit of the result when receives as input the value 0.333... . In this case the possible results are two 1.000... and 0.999... . If the algorithm generates 1 as first digit, this happen after the algorithm has read a finite number of digits of the result. Let us suppose that the first n are examined. In this case the algorithm however cannot generate the exact result when receives as input the string 0.33....32 (the exact result is 0.99....96 but the algorithm has already generated as first digit 1). An analogous consideration can be done if the we suppose that the first generated digit is 0. Similar examples show that also the other arithmetic operations are not computable. It is clear that the problem presented above is not caused by the choice of the basis 10 for the representation of the real numbers. Any other basis will have the same problems.

A simple way to overcome these difficulties is to change the representation by the introduction of negative digits [3, 20]. The standard interpretation can be extended to strings of positive and negative digits. For example the string $0.\langle +4\rangle\langle -5\rangle\langle -3\rangle\langle +2\rangle$ represents the rational number

$$(+4 \times 10^{-1}) + (-5 \times 10^{-2}) + (-3 \times 10^{-3}) + (+2 \times 10^{-4}).$$

We can now reconsider the previous example and show how the introduction of negative digits solve the difficulty. The algorithm for the multiplication by 3 can safely generate 1, as first digit, after having read the first two digits of the string 0.333... . We can easily observe that if the input becomes 0.3(-9)(-9)... = 0.2 the output can become 1.(-4)000... = 0.6. If the input becomes 0.3999... = 0.4 the output can become 1.2000... .

For the sake of simplicity in the following of the article we will consider the simplest negative digit representation. Namely the binary representation with a negative digit added to it. We thus represent real numbers by infinite strings of the three digits $+1, -1, 0$. We will abbreviate the digit $+1$ with $+$ and -1 with $-$. At the beginning we will be concerned only with the strings in the form 0."first digit" "second digit"... . With these sequences it is possible to represent the reals contained in the interval $[-1, 1]$.

2 A domain for the real numbers

As remarked in the previous section the correspondence between infinite strings and real numbers is the obvious one. More formally we associate to the infinite string $\langle 0.a_1 a_2 \ldots\rangle$ the real number:

$$[\![\langle 0.a_1 a_2 \ldots\rangle]\!]_R = \sum_{i=1}^{\omega} a_i \times 2^{-i}$$

In order to analyze the computation on real numbers we need a structure in which to interpret the syntactic objects: the strings of digits. The idea is to embed the real line in a larger structure in which finite and infinite strings have a semantic interpretation. We look for this bigger structure in the category of the ω-algebraic cpo's.

We first consider the finite strings. We introduce the interpretation function $[\![\,]\!]_I$ for the finite strings in the following way:

Definition 1. The function $[\![\ ]\!]_I$ from finite strings to closed rational intervals is defined by:

$$[\![\langle 0.a_1 \dots a_n\rangle]\!]_I = [((\sum_{i=1}^{n} a_i \times 2^{-i}) - 2^{-n}), ((\sum_{i=1}^{n} a_i \times 2^{-i}) + 2^{-n}] =$$

$$= [[\![\langle 0.a_1 \dots a_n -^{\omega}\rangle]\!]_R, [\![\langle 0.a_1 \dots a_n +^{\omega}\rangle]\!]_R]$$

(where with a^{ω} we indicate the infinite string $aaa\dots$)

The left and the right limit of the interval $[\![\langle 0.a_1 \dots a_n\rangle]\!]_I$ represent respectively the smallest and the biggest number that can be represented by a string starting with $\langle a_1 \dots a_n\rangle$ (respectively the numbers $[\![\langle a_1 \dots a_n -^{\omega}\rangle]\!]_R$ and $[\![\langle a_1 \dots a_n +^{\omega}\rangle]\!]_R$). For example: $[\![\langle 0\rangle]\!]_I$ denotes $[-1/2, 1/2]$, $[\![\langle ++\rangle]\!]_I$ denotes $[1/2, 1]$ and $[\![\langle 0+0\rangle]\!]_I$ denotes $[1/8, 3/8]$.

The function $[\![\ ]\!]_I$ induces the obvious equivalence relation on the finite strings:
$a \equiv b$ iff $[\![a]\!]_I = [\![b]\!]_I$
Two string are equivalent if they give the same information. We indicate with $[a]$ the equivalence class containing a.

Definition 2. Let I be the information order between equivalence classes of strings defined as follows: $[a] \sqsubseteq [b]$ iff $[\![a]\!]_I \supseteq [\![b]\!]_I$ (as subsets of real numbers) i.e. $[a] \sqsubseteq [b]$ if and only if the information given by a is smaller than the information given by b.

We can now make a few simple observations.

1. if a is a substring of b then $[\![b]\!]_I \subseteq [\![a]\!]_I$ (so $[a] \sqsubseteq [b]$)
2. $\langle a_1 \dots a_n\rangle \equiv \langle b_1 \dots b_m\rangle$ iff $n = m$ and $\sum_{i=1}^{n} a_i \times 2^{-i} = \sum_{i=1}^{m} b_i \times 2^{-i}$
3. for every string $\langle a_1 \dots a_n\rangle$, $[\![\langle a_1 \dots a_n\rangle]\!]_I$ can be written in the form $[(m-1)/2^n, (m+1)/2^n]$ where n is the number of digits in the string, m is an integer and $-2^n < m < 2^n$ (more precisely $m = \sum_{i=1}^{n} a_{n-i} \times 2^{-i}$). Moreover every interval on this form can be obtain as interpretation of a suitable string.

The rational numbers in the form $m/2^n$ with $m, n \in Z$ are call rational dyadic numbers. We call dyadic intervals the rational interval in the form:
$[(m-1)/2^n, (m+1)/2^n]$ with $m, n \in Z$.
Figure 1 shows the diagram representing I: Equivalently I can be thought as representing the superset relation between the dyadic intervals in the form:
$[m - 1/2^{-n}, m + 1/2^{-n}]$ with $n \in I\!N$, $m \in Z$ and $-2^n < m < 2^n$
In order to give a semantic to finite and infinite strings we need to consider the completion of I [13].

Definition 3. Small Reals Domain (SRD) is the partial order obtained by the ideal completion of I.

Proposition 4. *SRD is a bounded complete algebraic cpo.*

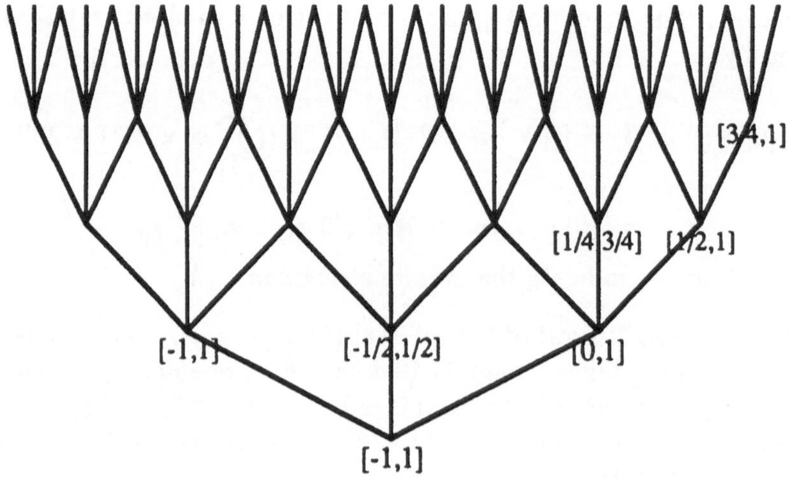

Fig. 1.

Definition 5. The interpretation functions $[\![\,]\!]_{SRD}$ from strings of digits to SRD is defined by:
for finite strings $[\![\langle 0.a_1 \ldots a_i\rangle]\!]_{SRD}$ is the principal ideal generated by $[\![\langle 0.a_1 \ldots a_i\rangle]\!]_I$, for the infinite strings

$$[\![\langle 0.a_1, a_2 \ldots\rangle]\!]_{SRD} := \bigsqcup_{i \in N} [\![\langle 0.a_1 \ldots a_i\rangle]\!]_{SRD}$$

What we hope now is to have the following correspondence: for every pair of infinite strings α, β

$$(([\![\alpha]\!]_R = [\![\beta]\!]_R) \Leftrightarrow ([\![\alpha]\!]_{SRD} = [\![\beta]\!]_{SRD}))$$

that is, two strings denote the same real iff they denote the same element in SRD. Unfortunately this is not the case. If two strings denote the same element in SRD then they denote the same number. But the opposite is not always true. For every dyadic number d we can divide the strings representing it in three non-empty classes: the strings eventually ending with a series of 0, the ones ending with a series of + and the ones ending with a series of −. The elements contained in each class receive an identical representation in SRD. We call respectively $\overline{d}$, $\overline{d^-}$ and $\overline{d^+}$ these representations in SRD. $\overline{d}$, $\overline{d^-}$ and $\overline{d^+}$ are distinct elements of SRD In the SRD order we have: $\overline{d} \sqsubseteq \overline{d^-}$, $\overline{d} \sqsubseteq \overline{d^+}$ and $\overline{d^-}$ incomparable with $\overline{d^+}$

For the not dyadic numbers, SRD behaves properly. For each pair of infinite strings α, β

$$([\![\alpha]\!]_R = [\![\beta]\!]_R = r \ \wedge \ r \text{ not dyadic }) \ \Rightarrow \ [\![\alpha]\!]_{SRD} = [\![\beta]\!]_{SRD}.$$

Moreover any infinite point in SRD is the representation of a real number. So the infinite elements in SRD look like the interval $[-1, 1]$ but for each dyadic numbers which are triplicated.

Figure 2 shows the diagram representing SRD.

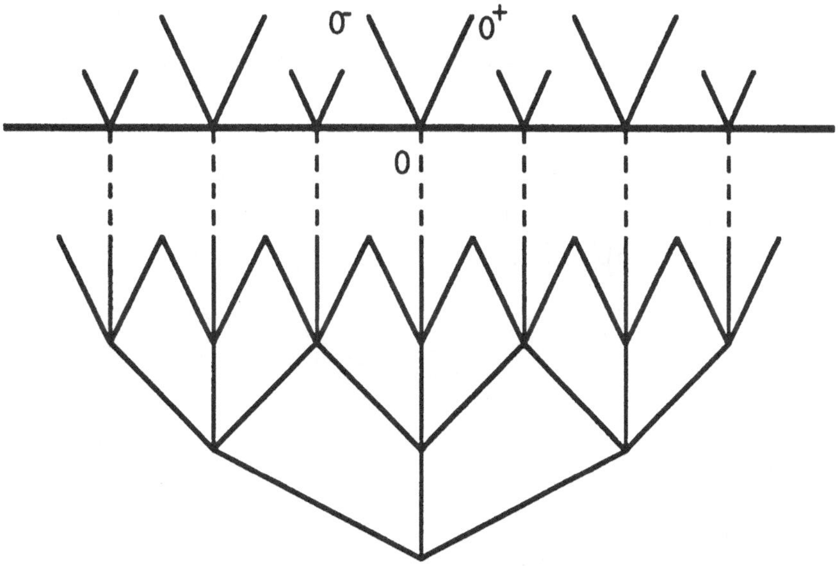

Fig. 2.

In the next section we will show how to solve the problem of multiple representations by means of a retract construction. Since *SRD* is an algebraic cpo we can apply to it the standard machinery for defining computability above it. We obtain a definition of the notion of computability for real numbers.

2.1 The whole real line

One can extend the string representation to the whole real line in several ways. We represent an arbitrary real by a pair consisting of an integer and a small real.

If we repeat the previous construction we obtain a cpo represented by the diagram in figure 3:

This cpo will be called Real Domain (*RD*). *RD* will be the main subject in the rest of the article. The main properties of *RD* are:

Proposition 6. *RD is a bounded complete algebraic cpo.*

Proposition 7. *Let* $\langle\ \rangle$ *be any coding function of n tuples of integer. We can define an effective enumeration of the finite elements of RD in the following way:*

$$e(0) = \perp$$

$$e(\langle m, m', n\rangle + 1) = [(m - m' - 1)/2^n, (m - m' + 1)/2^n]$$

(RD, e) *is an effective Scott domain.*

The proof is easy.

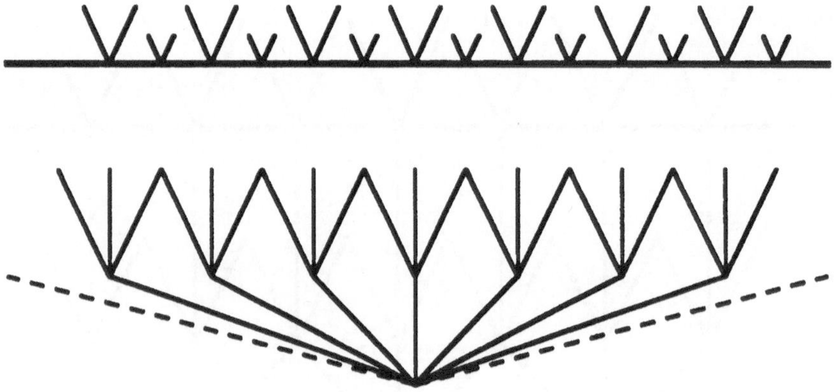

Fig. 3.

3 Topology

In this section we want to state more clearly the relations between RD and the real line. These relations will be expressed in terms of topology. Let us denote by $RD^{\dagger}$ the subspace of RD consisting of the infinite elements. It is possible to prove that the real line is a retract of $RD^{\dagger}$ via a couple of functions $q_r : RD^{\dagger} \to \mathbb{R}$ and $e_r : \mathbb{R} \to RD^{\dagger}$. The function q_r associates to each element of $RD^{\dagger}$ the corresponding real number. e_r can be thought as the function choosing a canonical representation in $RD^{\dagger}$ for each real number. Using the previous result we can associate a partial real function $\overline{f} : \mathbb{R} \to \mathbb{R}$ to each Scott continuous function $f : RD \to RD$. $\overline{f}$ is defined by $\overline{f} := q_r \circ f \circ e_r$, $\overline{f}$ is partial because q_r is not always defined. $\overline{f}$ is the function on reals representing f. In this way we can use the computability theory on domains to characterize the computable functions between real numbers. We define a partial function $g : \mathbb{R} \to \mathbb{R}$ is computable if there exits a computable function $f : RD \to RD$ such that $g = \overline{f}$. This definition of computable real function coincides with the other definitions given, with different methods in the literature [6, 7, 8, 9, 10, 11, 18].

It is easy to show that for every Scott continuous function f, the function $\overline{f}$ is continuous. In this way we obtain a new proof of the following classical result in computable analysis.

Theorem 8. *Computable functions on real numbers are continuous w. r. to the Euclidean topology.*

We now generalize the results presented above. In particular we want to consider not only functions in a single argument but a larger class of functions and functionals. Functions and functionals are characterized by a type. The types we will consider are the ones generated by the grammar:
$\sigma := r \mid \sigma \to \sigma$
where r is the only type constant, it is the type of the reals. The type structure defined here is a minimal one. Notice that there is no product type. Functions of

several arguments can be retyped as functions of a single argument by currying: $(\sigma_1 \times \sigma_2) \to \sigma := \sigma_1 \to (\sigma_2 \to \sigma)$.
Each type is characterized by a rank. The rank of a type σ is defined by:
$\partial(r) = 0$
$\partial(\sigma \to \sigma') = max\{\partial(\sigma) + 1, \partial(\sigma')\}$
The rank of a type measures how "higher order" the type really is. Note that all types σ with $\partial(\sigma) = 1$ have form $\sigma = r \to (r \to \ldots r)\ldots)$.

In the following of the article a good number of definitions will be given by structural induction on types. We will omit to mention it explicitly. Starting from $\mathbb{R}$ we can associate a topological space to each type σ with $\partial(\sigma) \leq 1$. We indicate with $\mathbb{R}_\sigma$ this space,
$\mathbb{R}_r = \mathbb{R}$
$\mathbb{R}_{r\to\sigma} = \{f : \mathbb{R} \to \mathbb{R}_\sigma \mid \text{f total continuous function}\}$
The topology on $\mathbb{R}_{r\to\sigma}$ is the compact-open topology. In a space of continuous functions $S \to T$, the compact-open topology is the topology having a prebase consisting of the sets of form $\{f \mid f(c) \subseteq o\}$, where c is a compact set of S and o is an open set of T. In analysis the compact open topology is the topology normally associated to the space of the real continuous functions.

We do not associate a topological space to the types of rank 2, in correspondence of these types we define sets of functionals:
$\mathbb{R}_{\sigma_1\to(\ldots(\sigma_n\to r)\ldots)} = \{F : \mathbb{R}_{\sigma_1} \to (\ldots(\mathbb{R}_{\sigma_n} \to \mathbb{R})\ldots) \mid \lambda\langle x_1,\ldots,x_n\rangle.F(x_1)\ldots(x_n) : (\mathbb{R}_{\sigma_1} \times \ldots \times \mathbb{R}_{\sigma_n}) \to \mathbb{R} \text{ is a continuous function }\}$.

We do not define any set of real functionals associated to the type of order bigger than two. This because the results we are going to present cannot be extended to these types.

Starting from RD we can associate an effective Scott domain to each type. RD_σ is defined in the obvious way:
$RD_r := RD$
$RD_{\sigma\to\sigma'}$ is the Scott domain of the Scott continuous functions from RD_σ to $RD_{\sigma'}$.

In RD not every element represents a real number, some elements in RD are just a finite approximation of real numbers, an example is the interval $[-1, 1]$. Similarly not every element in RD_σ will represent an element in $\mathbb{R}_\sigma$ (a totally defined real continuous function). We indicate with $RD^\dagger_\sigma$ the subspace of RD_σ whose elements are going to represent the elements of $\mathbb{R}_\sigma$.

Formally:
$RD^\dagger_r = \{s \in RD \mid x \text{ is an infinite element of } RD\}$
$RD^\dagger_{\sigma\to\sigma'} := \{g \in RD_{\sigma\to\sigma'} \mid g(RD^\dagger_\sigma) \subseteq RD^\dagger_{\sigma'}\}$
The topology on $RD^\dagger_\sigma$ is the subset induced topology.

We have the following result:

Theorem 9. *For each type σ with $\partial(\sigma) \leq 1$, $\mathbb{R}_\sigma$ is a retract of $RD^\dagger_\sigma$. The retract functions $q_\sigma : RD^\dagger_\sigma \to \mathbb{R}_\sigma$ and $e_\sigma : \mathbb{R}_\sigma \to RD^\dagger_\sigma$ are defined by:*
$q_r(\overline{x}) := x$ if x is not a dyadic number
$q_r(\overline{x}) := q_r(\overline{x^+}) := q_r(\overline{x^-}) := x$ if x is a dyadic number

$e_r(x) := \overline{x}$
$q_\sigma(g) := q_{\sigma_1} \circ g \circ e_r$
$e_\sigma(f)$ *is the continuous extension of the function* $f' : RD^\circ \to RD_\sigma$ *defined by:*

$$f'(s^\circ) = \begin{cases} \bot & \text{if } s^\circ = \bot \\ \bigsqcup\{t^\circ \mid e_{\sigma_1} \circ f \circ q_r(\overline{\overset{\vee}{s^\circ}}) \subseteq \overset{\vee}{t^\circ}\} & \text{otherwise} \end{cases}$$

where, if $s_\sigma \in RD_\sigma$ *then* $\overset{\vee}{s}_\sigma = \{t_\sigma | t_\sigma \in RD_\sigma, s_\sigma \leq t_\sigma\}$ *and* $\overline{\overset{\vee}{s}}_\sigma = \overset{\vee}{s}_\sigma \cap RD^\dagger\sigma$

The proof is omitted.

e_σ and q_σ are the natural generalization of the functions e_r and q_r. That is: q_σ associates to each element of $RD^\dagger_\sigma$ the element of $\mathbb{R}_\sigma$ represented by it. e_σ chooses for each element in $\mathbb{R}_\sigma$ a canonical representation of it in $RD^\dagger_\sigma$.

We discuss now the problem of defining an effective method that given an element f in $RD^\dagger$, returns the canonical representation of the equivalence class of f. Such a method exists if the function $e_\sigma \circ q_\sigma : RD^\dagger_\sigma \to RD^\dagger_\sigma$ can be extended to a continuous and effective function $c_\sigma : RD_\sigma \to RD_\sigma$.

Notation Given $s^\circ \in RD^\circ_\sigma$ we say that the finite set $I = \{s^\circ_1, \ldots, s^\circ_n | s^\circ_i \in RD^\circ_\sigma\}$ is a *partition covering* of s° if $\forall y \in \overline{\overset{\vee}{s}}_\sigma\ \exists s^\circ_i \in I$ s.t. $y \in \overline{\overset{\vee}{s^\circ}}_\sigma$

Theorem 10. *For every type* $\sigma = r \to \sigma_1$ *with* $\partial(\sigma) \leq 1$ *there exists a continuous and effective function* $c_\sigma : RD_\sigma \to RD_\sigma$ *such that* $c_\sigma|_{RD^\dagger_\sigma} = e_\sigma \circ q_\sigma$.
c_σ *is the continuous extension of the function* c'_σ *defined on the finite elements by:*
$c'_r([a,b]) = \bigsqcup\{[\![c,d]\!] | c < a, b < d\}$

$$c'_{r\to\sigma_1}(g^\circ)(s^\circ) = \begin{cases} \bot & \text{if } s^\circ = \bot \\ \bigsqcup_{\substack{I \text{ partition} \\ \text{covering of } s^\circ}} \{\sqcap_{s^\circ_i \in I}\{c'_\sigma \circ g^\circ \circ c^\circ_r(s^\circ_i)\}\} & \text{otherwise} \end{cases}$$

The proof is omitted.

The retract construction cannot be extended to spaces of functions having rank strictly larger than 1. In fact, giving a type σ having rank 2, although it is possible to define a function $q_\sigma : RD^\dagger_\sigma \to \mathbb{R}_\sigma$ that associates, in a natural way, to each element in $RD^\dagger_\sigma$ the function on real numbers represented by it, there is no topology on $\mathbb{R}_\sigma$ and no function $e_\sigma : \mathbb{R}_\sigma \to RD^\dagger_\sigma$ which makes the pair of functions q_σ, e_σ a retraction. As far as the functions having rank 2 are concerned, we give the following result.

Theorem 11. *For every type* σ *of rank 2 there exist two function* $q_\sigma : RD^\dagger_\sigma \to \mathbb{R}_\sigma$ *and* $e_\sigma : \mathbb{R}_\sigma \to RD^\dagger_\sigma$ *such that* $q_\sigma \circ e_\sigma = Id_{\mathbb{R}_\sigma}$.

The functions are defined by structural induction as follows:
$q_{\sigma\to\tau} := \lambda G.q_\tau \circ G \circ e_\sigma$
$e_{\sigma\to\tau}(F)$ *is the continuous extension of the function* $\overline{F} : RD^\circ_\sigma \to RD_\tau$

$$\overline{F}(s^\circ) = \begin{cases} \bigsqcup\{t^\circ \mid e_\sigma \circ f \circ q_r(\overline{\overset{\vee}{s^\circ}}) \subseteq \overset{\vee}{t^\circ}\} & \text{if } \sigma = r \\ \bigsqcup\{t^\circ \mid e_\sigma \circ f \circ q_r(\overline{\overset{\vee}{c_\sigma(s^\circ)}}) \subseteq \overset{\vee}{t^\circ}\} & \text{if } \partial(\sigma) = 1 \end{cases}$$

Also in this case q_σ associates to each functional on RD the functional on $\mathbb{R}$ represented by it, and e_σ chooses a canonical representation for each continuous functional on $\mathbb{R}$. The fact that $q_\sigma \circ e_\sigma = Id$ implies that all continuous functionals on $\mathbb{R}$ can be represented by a proper functionals on RD. A new definition of computability for functionals can now be given:

Definition 12. An element $f \in \mathbb{R}_\sigma$ is computable if there exists a computable element $g \in RD_\sigma$ such that $f = q_\sigma(g)$.

From the above results a very interesting property of computable functionals it follows

Theorem 13. *Every computable function or functional on real numbers is continuous, when we consider the space of the functions endowed with the compact open topology.*

This is a useful criteria for determining the non computability of functionals. Starting from this result it is easy to demonstrate that the following functionals are not computable.

1. Derivative.
$$F(f) = \frac{df}{dx}$$
2. The functional that given a function f and an interval [a,b] yields the point x in $[a,b]$ where the value f(x) is minimum.
$$F(f,[a,b]) = x \text{ if } f(x) = \min\{f(y) \mid y \in [a,b]\}$$
3. The functional that given a function f and an interval $[a,b]$ yields a point x in $[a,b]$ where the value $f(x)$ is zero or is equal to a if such a value does not exist.
$$F(f,[a,b]) = \begin{cases} x & \text{if } f(x) = 0 \text{ and } x \in [a,b] \\ a & \text{if } \forall x \in [a,b].\ f(x) \neq 0 \end{cases}$$

So we have thus an important topological characterization of the computable functionals on the real numbers.

Acknowledgements

Thanks are due to Michael Smyth and Furio Honsell for helpful discussions.

References

1. L.E.J. Brouwer, "Beweis, dass jede volle Funktion gleichmässig stetig ist" Proc. Amsterdam 27 (1924) 189-194.
2. M.J. Beeson, "Foundation of Constructive Mathematics" Spriger-Verlag, Berlin, 1985.

3. H.-J. Boehm, R. Cartwright, M. Riggle, and M.J. O'Donell, "Exact Real Arithmetic: A Case Study in Higher Order Programming." 1986 ACM Symposium on Lisp and Functional Programming.
4. H.-J. Boehm, R. Cartwright, "Exact Real Arithmetic: Formulating Real Numbers as Functions" in "Research Topics in Functional Programming" David Turner editor, Addison-Wesley, 1990, pp. 43-64
5. D. Bridges and E. Bishop, "Constructive Analysis." Springer-Verlag, Berlin, 1985.
6. A. Grzegorczyk, "On the Definition of Computable Real Continuous Functions." Fund. Math. 44 (1957) 61-77.
7. K. Ko and Friedmann, "Computational Complexity of Real Functions." Theoret. Comput. Sci. 20 (1982) 323-352.
8. D. Lacombe, "Quelques procédés de définitions en topologie recursif." in: Constructivity in Mathematics, North-Holland (1959) 129-158.
9. P. Martin-Löf, "Note on Constructive Mathematics." Almqvist and Wiksell, Stockholm (1970).
10. J. Myhill, "What is a Real Number?" America Mathematical Monthly (1979) 748-754.
11. H.G. Rice, "Recursive Real Numbers." Proc. Amer. Math. Soc 5 (1954) 784-791.
12. D. Scott, "Outline of the Mathematical Theory of Computation." Proc. 4th Princeton Conference on Information Science (1970).
13. D. Scott, "Data Types as Lattices." SIAM J. Comput. 5 (1976) 522-587.
14. A.S. Troelstra and D. van Dalen, "Constructivism in Mathematics." North-Holland, Amsterdam (1988).
15. A.M. Turing, "On Computable Numbers, with an Application to the Entscheidungs Problem." Proc. London Math. Soc. 42 (1937) 230-265.
16. J. Vuillemin, "Exact Real Computer Arithmetic with Continued Fraction." Proc. A.C.M. conference on Lisp and functional Programming (1988) 14-27.
17. K. Weihrauch, U. Schreiber, "Embeding Metric Spaces into cpo's" Theoret. Comp. Sci. 16 (1981) 5-34.
18. K. Weihrauch and C. Kreitz, "Representation of the Real Numbers and of the Open Subsets of the Set of Real Numbers." Annals of Pure and Applied Logic 35 (1987) 247-260.
19. K. Weihrauch, "Computability." Springer-Verlag, Berlin, 1987.
20. E. Wiedmer, "Computing with Infinite Objects." Theoret. Comp. Sci. 10 (1980) 133-155.

Lambda Substitution Algebras

Zinovy Diskin Ilya Beylin

Frame Inform Systems University of Latvia
bac@frame.riga.lv *ilya@ mii.lu.lv*
Riga, Latvia

Abstract. In the paper an algebraic metatheory of type-free λ-calculus is developed. Our version is based on lambda substitution algebras (λSAs), which are just SAs introduced by Feldman (for algebraizing equational logic) enriched with a countable family of unary operations of λ-abstraction and a binary operation of application. Two representation theorems, syntactical and semantic, are proved, what directly provides completeness theorems.

The basic paradigm of algebraic logic consists in relating a class **C** of algebras to the logic L in question in such a way that L-models could be considered as homomorphisms from syntactically generated **C**-algebras into **C**-algebras arising from L-semantics. Moreover, L-theories are connected with kernels of these homomorphisms in a definite way, in particular, for equational-like logics theories coincide with kernels. Such a machinery gives rise to an algebraic metatheory of the logic L and we shall say that this metatheory is *constituted* by the class **C**.

The present paper gives an instance of applying the general methodology above described to type-free λ-calculus and its title refers just to the constituting class of the metatheory we have adopted. Certainly, there are known several classes of algebras intended to be an algebraic counterpart of type-free λ-calculus: λ-algebras, combinatory models, λ-models (here and further we follow, on the whole, the terminology adopted by Meyer [M82]) constituting corresponding metatheories (see, for example, [B81]). However, in contrast to these approaches, we prefer to regard type-free λ-calculus as an immediate (as much as possible) generalization of one-sorted equational logic, L_{eq}, and to develop its metatheory on the ground of some algebraic metatheory of L_{eq}. For such a prototype theory we take algebraic treatment of L_{eq} by means of *substitution algebras* (abbreviated to SA in the singular and SAs in the plural) described by Feldman in [F82] and further developed by Cīrulis in [C88] (in particular, in contrast to Feldman, Cīrulis gives a purely equational axiomatic description of SAs).

According to this treatment, given a non-empty set D, the set of ω-ary operations on D, $\mathrm{Op}_\omega D$, is equipped with a countable family of binary operations of composition, s_i, $i<\omega$, while projections, π_i, $i<\omega$, are regarded as constants (see construction 3.0). In this way one comes to the algebra $\mathcal{Op}_\omega D = (\mathrm{Op}_\omega D, \pi_i, s_i)_{i<\omega}$, and any its subalgebra is called *a concrete* SA_ω, that is, *a concrete substitution algebra of dimension* ω. (Further we shall usually omit the subscript ω and use the abbreviation SA as a denotation for the class of all substitution algebras as well).

In syntax, given a one-sorted signature Ω of operation symbols and a countable set $V=\{v_i, i<\omega\}$ of variables, the set of all Ω-terms over V, $\mathrm{Tm}\Omega$, can be also converted into an SA by setting $s_i(u,w)=w[u/v_i]$. Finally, every Ω-algebra is nothing but an SA-homomorphism μ: $\mathcal{T}m\Omega \to \mathcal{Op}D$ for some non-empty set D. Moreover, equational theories are nothing but kernels of such homomorphisms.

The goal of the present paper is to adapt the above outlined machinery for the situation of type-free λ-calculus. To this end we enrich the SA-signature with a countable family of unary operations of λ-abstraction and a binary operation of application and then add to the Feldman-Cīrulis' list of equations axiomatizing SA new equations reflecting interaction of λ-quantifiers with substitutions and α-conversion as well. We call these new SAs *lambda substitution algebras (of dimension* ω*)*, λSAs for short. By adding one (two) more family(ies) of equations reflecting β (and η)-conversion, one comes to $\lambda\beta(\eta)$SAs.

Similarly to above described algebraization of equational logic, λSAs appear quite naturally from λ-calculus syntax (as term models) and semantics (as closed sets of operations over functional domains in the spirit of Meyer [M82]) so that set-theoretical interpretations of λ-terms can be regarded as homomorphisms into semantic λSAs while λ-theories prove to be kernels of such homomorphisms. (It is worth noting that while Meyer's notion of environment model is (we quote Meyer himself) "so entangled with the syntax of λ-terms that it is hard to visualize what models look like", we state purely semantic formulation of required closure conditions for sets of operations over domains and just in this way we come to semantic $\lambda\beta$SAs of "real" operations).

Among all λSAs we single out the so called *locally finite* (l.f.) λSAs arising from the point that ordinary λ-terms depend on only finite number of variables — this construction is well known in algebraic logic (l.f. substitution, polyadic, cylindric etc. algebras). The main result of the paper states that any l.f. λβ(η)SA can be presented either semantically, as a λSA of set-theoretical operations over a suitable functional (extensional) domain, or syntactically, as the λSA of a certain λβ(η)-theory T over a suitable set of constants. Moreover, by taking into consideration morphisms of λ-theories, of λSAs and of functional domains, one can show that the categories of all λβ(η)-theories, of all l.f. λβ(η)SAs and of all (extensional) environment models are equivalent (see [D91] for the first equivalence), in other words, λβ(η)-theories, l.f. λβ(η)SAs and (extensional) environment models can be regarded as alternative formulations of the same concept.

Thus, we start from the SA-metatheory of L_{eq} (one-sorted equational logic) and come to the λSA-metatheory of λ-calculus. Such a way of constructing λ-calculus metatheory has already been used by Obtułowicz and Wiweger ([OW82]) who began at the categorical L_{eq}-metatheory constituted by algebraic theories in the sense of Lawvere and came to the notion of Church algebraic theory. Both ways are closely connected: it can be shown that categories of l.f. SAs and Lawvere algebraic theories are equivalent as well as categories of l.f. λβSAs and Church algebraic theories are. Moreover, for any given λβSA, $\mathcal{A}$, the Lawvere theory corresponding to the SA-reduct of $\mathcal{A}$ is just the Lawvere reduct of the Church theory corresponding to $\mathcal{A}$. (Further in the paper we shall not concern the question how λSAs are related to other algebraic versions of λ-calculus; if one wishes, this can be done through the very λ-calculus).

All these results are rather natural as well as somewhat tedious. What is much more interesting, and, maybe, the only interesting thing in the λSA-approach is the inverse problem of constructing a λ-calculus source of the whole variety of λSAs without l.f.-restriction, i.e. constructing a λ-calculus with terms depending on infinitely many variables. Another interesting generalization consists in constructing an algebraic SA-version of λα-calculus, i.e., λ-calculus without β-conversion. In fact, λα-calculus is only a formal machinery for bounding variables (with λ-quantifiers) and substituting (for free variables) but is not connected with any pithy logic as such. The latter is just given by β-conversion or η-conversions (or something else). Some suggestions on these two generalizations are contained in the concluding section.

1 Lambda Substitution Algebras

1.1 Definition. Let $\Sigma_\omega=(\upsilon_i,s_i,',\lambda_i)_{i<\omega}$ be a signature of function symbols such that all υ_i are nullary, $'$ and all s_i are binary and all λ_i are unary. *A lambda substitution algebra (of dimension ω)*, λSA, is a Σ_ω-algebra $\mathcal{A}=(A,\upsilon_i,s_i,',\lambda_i)_{i<\omega}$ such that the following identities are fulfilled for all $i,j,k<\omega$, $k\neq i\neq j$, and any $a,b,c\in A$ (instead of $s_i(b,a)$ we write a more suggestive $[b/i]a$ and for better readability of the identities below we use the abbreviation $\mathbf{a}_{ki}$ for the expression $[\upsilon_k/i]a$):

(S1)$_i$ $\quad [\upsilon_i/i]a = a;$

(S2)$_i$ $\quad [a/i]\upsilon_i = a;$

(S3)$_{ij}$ $\quad [a/i]\upsilon_j = \upsilon_j;$

(S4)$_i$ $\quad [a/i][b/i]c = [[a/i]b/i]c;$

(S5)$_{ijk}$ $\quad [\mathbf{a}_{ki}/j][b/i]c = [[\mathbf{a}_{ki}/j]b/i][\mathbf{a}_{ki}/j]c;$

(SA)$_i$ $\quad [b/i](a1'a2) = [b/i]a1'[b/i]a2;$

(SL1)$_i$ $\quad [b/i]\,\lambda_i a = \lambda_i a;$

(SL2)$_{ijk}$ $\quad [\mathbf{a}_{ki}/j]\,\lambda_i b = \lambda_i[\mathbf{a}_{ki}/j]b;$

$(\alpha)_{ijk}$ $\quad \lambda_j \mathbf{a}_{ki} = \lambda_i[\mathrm{v}_i/j]\mathbf{a}_{ki}.$

A $\lambda\beta$SA is a λSA satisfying the following identities:

$(\beta)_i$ $\quad (\lambda_i a)'b = [b/i]a.$

A $\lambda\beta\eta$SA is a $\lambda\beta$SA satisfying, in addition, identities:,

$(\eta)_{ik}$ $\quad \lambda_i(\mathbf{a}_{ki}'i) = \mathbf{a}_{ki}.$

It is worth noting that (S4) and (S5) are derivable in the presence of (β)-identities (see [D91]).

Note also that, owing to (S2) and (S3), $|A|>1$ iff $i\neq j$ implies $\upsilon_i\neq\upsilon_j$ and we shall always assume this condition is fulfilled. So the mapping $i \longmapsto \upsilon_i$ states an isomorphism of ω onto the set $V\mathcal{A} := \{\upsilon_i\colon i<\omega\}$ of *$\mathcal{A}$-variables* and we shall often identify i and υ_i.

The $(\upsilon_i,s_i)_{i<\omega}$-reduct of a λSA is nothing but an SA introduced by Feldman [F82] (or a term system by Cīrulis [C88]).

1.2 Definition. Following Feldman [F82], given a λSA $\mathcal{A}$, for each element $a\in A$ we introduce *the dimension set Δa of a*,

$$\Delta a := \{i<\omega\colon [b/i]a \neq a \text{ for some } b\in A\},$$

and say that *a is independent on i (the i-th variable)* if $i\notin\Delta a$.

From (S4)$_i$ and (S3)$_{ij}$, $i\neq j$, it follows that, given $c\in A$, $[a/i][\upsilon_j/i]c = [\upsilon_j/i]c$ for all $a\in A$, i.e., for any c, $i\notin\Delta[\upsilon_j/i]c$ if $i\neq j$. Therefore, in the above list of identities, $\mathbf{a}_{ki}$ denotes an arbitrary element independent on i. Also, $i\notin\Delta\lambda_i a$ by (SL1)$_i$.

1.3 Definition. Given a λSA $\mathcal{A}$, we call an element $a\in A$ *finitary* if Δa is finite; a λSA $\mathcal{A}$ itself is said to be *locally* finite (l.f.) if Δa is finite for all $a \in A$.

Given a λSA $\mathcal{A}$, for any $Y \subseteq \omega$ we introduce the set $A[Y] := \{a \in A\colon \Delta a \subseteq Y\}$ and call elements of $A[\emptyset]$ *closed.* We shall also write $A[i_1 \ldots i_n]$ for $A[\{i_1, \ldots, i_n\}]$.

For a given λSA $\mathcal{A}$, we shall call the set $A_{fin} := \{a \in A\colon \Delta a$ is finite$\}$ *the locally finite part of* $\mathcal{A}$.

A more detailed examination of the structure of λSAs gives the following results (see [F82],[C88],[D91] for proofs).

1.4 Proposition. If $\mathcal{A}=(A, v_i, s_i, ', \lambda_i)_{i<\omega}$ is a λSA, then for all $i<\omega$, $a \in A$:

(o) $i \notin \Delta a$ iff $[v_j/i]a = a$ for some $j \neq i$,

(i) $\Delta v_i = \{i\}$,

(ii) $\Delta([b/i]a) \subseteq (\Delta a - \{i\}) \cup \Delta b$,

(iii) $\Delta(a1'a2) \subseteq \Delta a1 \cup \Delta a2$,

(iv) $\Delta(\lambda_i a) \subseteq \Delta a - \{i\}$,

(v) A_{fin} is the subalgebra generated by the set $A[\emptyset]$.

1.5 Proposition. For any homomorphism $h\colon \mathcal{A} \to \mathcal{B}$ of λSAs and any $a \in A$ one has $\Delta ha \subseteq \Delta a$.

1.6 Construction. Let $\mathcal{A}$ be a λSA, $I=\{i_1, \ldots, i_m\} \subset \omega$ and $a, a_1, \ldots, a_m \in A_{fin}$. Let us choose a set $J=\{j_1, \ldots, j_m\} \subset \omega$ such that $J \cap (\Delta a \cup \Delta a_1 \cup \ldots \cup \Delta a_m \cup I) = \emptyset$ and consider the term

$$[a_m/j_m] \ldots [a_1/j_1][j_m/i_m] \ldots [j_1/i_1]a.$$

The identities (S1)...(S5) provide that this term does not depend on the choice and enumeration of J. Hence, the term can be denoted by $[a_1/i_1, \ldots, a_m/i_m]a$ where the order of terms a_k/i_k, $k \leq m$, inside of square brackets does not matter. Thus, for any finitary a, i.e. $a \in A[I]$ for some finite set $I=\{i_1, \ldots, i_m\} \subset \omega$, and any substitution $\rho\colon \omega \longrightarrow A_{fin}$ there is defined unambiguously the element $\rho^*a := [\rho i_1/i_1, \ldots, \rho i_m/i_m]a$. In addition, $\Delta \rho^* a \subseteq \bigcup\{\Delta \rho i_k\colon k=1, \ldots, m\}$.

2 Lambda Substitution Algebras and Lambda Theories

In this section we regard the class λSA as an algebraic semantics for λ-calculus and our main purpose is to prove the corresponding completeness theorem.

2.1 Definition. Let C be a set of constants and $\Lambda(C)$ denote the set of all λ-terms over C. *An algebraic (through* λSA*) C-model* is defined to be a pair $(\mathcal{A}, \mu)$ with $\mathcal{A}$ a λSA and μ a mapping $C \to A[\emptyset]$. We can extend μ to the mapping $\bar{\mu}\colon \Lambda(C) \to A$ by the evident induction on the structure of λ-terms and, for $u, w \in \Lambda(C)$, write $(\mathcal{A}, \mu) \models u=w$ if $\bar{\mu}u = \bar{\mu}w$. The consequence relation on the set $\Lambda(C) \times \Lambda(C)$ generated by such algebraic models in the ordinary way will be denoted by $\models_{\lambda(\beta,\beta\eta)SA}$ in accordance with the name of the class of λSAs from which we take algebras $\mathcal{A}$'s.

2.2 Algebraic completeness theorems. For any $\Gamma\cup\{u=w\}\subset \Lambda(C)\times\Lambda(C)$ one has $\Gamma\models_{\lambda(\beta,\beta\eta)SA}u=w$ iff $\Gamma\vdash_{\lambda(\beta,\beta\eta)}u=w$.

Proof (sketch). Soundness is stated by routine induction on λ-terms which shows that $\bar{\mu}$ is compatible also with substitutions and, hence, all axioms are λSA-valid. To prove completeness, let $T\subseteq\Lambda(C)\times\Lambda(C)$ be a λα-theory, i.e., a set closed under λα-deducibility. Evidently, T is an equivalence and the quotient set $\Lambda(C)/T$ can be equipped with the structure of a Σ-algebra and, moreover, routine verification shows that it is an l.f. λSA; we designate this algebra as **A**(T). (Note that $\Delta(u/T)\subseteq FV(u)$ but, in general, the equality does not hold). In addition, **A**(T) is a λβ(η)SA iff T is actually a λβ(η)-theory. Now, it remains to consider the model $(\Lambda(C)/T, \mu)$ where T is the theory generated by Γ, and μ is the natural surjection $C\to C/T$.

2.3 Syntactical representation theorems. Any l.f. λ(β,βη)SA can be presented as a syntactical λSA, **A**(T), for some λβ(η)-theory T over a suitable set of constants.

Proof (sketch). Given a λSA $\mathcal{A}$, let us consider the λ-calculus over the set of closed elements, $\Lambda(A[\varnothing])$. Let μ be the identity inclusion $A[\varnothing]\hookrightarrow A$ and let $T(\mathcal{A})$ denote the kernel of $\bar{\mu}$, $T(\mathcal{A})=\{u=w\colon \bar{\mu}u=\bar{\mu}w\}$. A direct check shows that $T(\mathcal{A})$ is λα-closed and, moreover, **A**$(T(\mathcal{A}))\cong \mathcal{A}_{fin}$. In addition, $T(\mathcal{A})$ is a λβ(η)-theory iff $\mathcal{A}$ is actually a λβ(η)SA.

3 Lambda Substitution Algebras and Environment Models

While in the previous section we were dealing with algebraic semantics, now we turn to the "real" set-theoretical semantics of λ-calculus. Following Meyer, we treat the latter by means of functional domains (λβ-sets in our terminology). So, our first purpose in this section is to present the notion of environment model in an algebraic fashion as a homomorphism from the Σ-algebra of Λ(*C*)-terms over a given set of constants, *C*, into a Σ-algebra of operations over the functional domain in question. We call such Σ-algebras *concrete*, and it is easy to verify that they turn out to be λβSAs (λβηSAs, if the domain is extensional). Then, we prove that any l.f. λβSA can be presented as a concrete one, hence, the algebraic completeness theorem 2.2 provides the set-theoretical (via environment models) completeness theorem.

We begin with a more simple situation of ordinary SAs of operations without λ-abstraction.

3.0 Construction. Let D be a non-empty set. By OpD we designate the set of all ω-ary operations on D, i.e. maps from D^{ω} into D. In fact, OpD contains also all finitary operations which (being considered as elements of OpD) depend actually on finite number of arguments only, while other arguments are dummy. Tuples from D^{ω} in our context will be referred to as *environments* and denoted by ρ,σ,τ etc., the very operations — by u,w etc. Given $i<\omega$ and an element $d\in D$, for every environment ρ there is defined another environment, denoted by $\rho\{d/i\}$, coinciding with ρ for all $j<\omega$ different from i while $(\rho\{d/i\})i=d$.
The following standard operations are defined on the set OpD:
(O1)′ projections, $\pi_i\in \mathrm{Op}D$, $\pi_i\rho := \rho i$, $i<\omega$;
(O1)″ binary compositions, $s_i(u,w)\rho := w(\rho\{u\rho/i\}), i<\omega$.

A set $O\subseteq \mathrm{Op}D$ is said to be *closed* if it contains all projections and closed under all compositions. In such a case, O can be converted into the algebra $\mathcal{O}=(O,\pi_i,s_i)_{i<\omega}$, in fact, a finitary (as compositions are binary) SA of infinitary operations. Its locally finite part consists of finitary operations.

Now we want to generalize the above construction for the case of application and λ-abstraction.

3.1 Construction. A $\lambda\beta$*-set* is defined to be a triple $\mathcal{D}=(D,\Psi,\Phi)$ with D a non-empty set called *a domain*, Ψ a map from a certain set of functions over D, $[D\to D]$, into the very D and Φ a map from D onto $[D\to D]$ such that
(FDβ) $\Phi\Psi f = f$ for all $f\in[D\to D]$;

Given a $\lambda\beta$-set $\mathcal{D}$, a set $O\subseteq \mathrm{Op}D$ is said to be $\lambda\beta$*-closed* iff it satisfies the following closure conditions (we shall use the symbol Λ for the ordinary set-theoretical (meta) abstraction):
(C1) O is closed under (O1)′,(O2)″;
(C2) if $u,w\in O$ then $(\Lambda\rho\in D^{\omega}.(\Phi u\rho)w\rho)\in O$;
(C3) if $u\in O$ then, for any fixed $i<\omega$, $\rho\in D^{\omega}$,
$\Lambda d\in D.u(\rho\{d/i\}) \in [D\to D]$.

Thus, if $\mathcal{D}$ is a $\lambda\beta$-set and $O\subseteq \mathrm{Op}D$ is $\lambda\beta$-closed, then, besides projections and compositions, the following operations can be defined on O:
(O2) application, $(u'w)\rho := (\Phi u\rho)w\rho$;
(O3) λ-abstractions, $(\lambda_i u)\rho := \Psi(\Lambda d\in D.u(\rho\{d/i\}))$, for all $i<\omega$.

Hence, O can be converted into a Σ-algebra $\mathcal{O}=(O,\pi_i,s_i,{}',\lambda_i)_{i<\omega}$. Moreover, direct checking states that any such an algebra is actually a $\lambda\beta$SA in the sense of definition 1.1 and, in addition, it is a $\lambda\beta\eta$SA iff Φ is one-one. The latter point motivates the name *λβη-set* for a λβ-set $\mathcal{D}=(D,\Psi,\Phi)$ with injective Φ.

3.2 Definition. *A* λβ(η)SA *of operations* or *a concrete* λβ(η)SA is a pair $(\mathcal{D},\mathcal{O})$ with $\mathcal{D}$ a λβ(η)-set and $\mathcal{O}$ the λSA over some λβ-closed set $O\subset \mathrm{Op}D$.

Given a set of constants, C, *an environment C-model* is a triple $\mathcal{E}=(\mathcal{D},\mathcal{O},\mu)$ with $(\mathcal{D},\mathcal{O})$ as above and μ a mapping $C\longrightarrow D$ such that for every $d\in\mu(C)$ the constant operation $\#d := \Lambda\rho\in D^{\omega}.d$ belongs to O. By composing the mapping $\#$: $\mu(C) \longrightarrow O[\emptyset]$ with μ, each environment C-model gives rise to the algebraic C-model $(\mathcal{O},\#\mu)$ (definition 2.1), and we define $\mathcal{E}\models u=w$ iff $(\mathcal{O},\#\mu)\models u=w$. It is easy to see that Meyer's and our notions of environment model coincide; we shall designate the class of all (extensional) environment models as (η)EM.

Our next goal is to prove that, conversely, every algebraic model can be presented as an environment model. To this end we need a procedure of extracting a concrete λβSA from any abstract λβSA.

3.3 Construction. Let $\mathcal{A}$ be a λβSA. First of all, we connect with $\mathcal{A}$ a certain λβ-set $\mathcal{D}(\mathcal{A})=(D,\Psi,\Phi)$ as follows.

Let $D=A_{fin}$, $\Phi a=\Lambda b\in D.a'b$ and $[D\to D] = \{\Phi a\colon a\in A_{fin}\}$. Owing to the (α)-identities, $\Phi a1=\Phi a2$ implies $\lambda_i(a1'v_i)=\lambda_j(a2'v_j)$ for any $a1,a2\in D$ and $i,j\notin\Delta a1 \cup \Delta a2$. This justifies defining $\Psi(\Phi a)$ by $\lambda_i(a'v_i)$ for some $i\notin\Delta a$ and it can be easily verified that $\Phi\Psi\Phi a = \Phi a$ for all $a\in A$.

Now, if $a\in D$, then let $|a|$ be the mapping $D^{\omega}\longrightarrow D$ defined as follows: $|a|\rho =\rho^{*}a$ for any $\rho\in D^{\omega}$ (see construction 1.6). Evidently, $|a|=|b|$ implies $a=|a|\iota=|b|\iota=b$ where ι is the identical environment $\iota j=v_j$, $j<\omega$. Let O be the set $\{|a|\colon a\in D\}$; evidently, it satisfies condition (C1). It can be also shown that, for arbitrary $|a|,|b| \in O$ and $\rho \in D^{\omega}$, we have $(\Phi|a|\rho)(|b|\rho) = |a'b|\rho$ and, for any fixed $i<\omega$ and $\rho\in D^{\omega}$, there is $j<\omega$ such that $|a|(\rho\{b/i\}) = \Phi(\lambda_j(\rho\{v_j/i\})^{*}a)b$ for all $b\in D$. These equations show that O satisfies closure conditions (C2),(C3) and, hence, can be equipped with the structure of a λSA. Moreover, the mapping $|\_|$ turns out to be a homomorphism. Thus, we obtain

3.4 Semantic representation theorems. With any abstract λβSA, $\mathcal{A}$, there is correlated a certain concrete λβSA, $(\mathcal{D},\mathcal{O})$, such that $D=A_{fin}$ and $\mathcal{O} \cong \mathcal{A}_{fin}$; in addition, $\mathcal{D}$ is a λβη-set iff $\mathcal{A}$ is a λβηSA. Moreover, if μ: $C\longrightarrow A[\emptyset]$ is an algebraic C-model, then $\mathcal{E}=(\mathcal{D},\mathcal{O},\mu)$ is an environment C-model and $\mathcal{E}\models u=w$ iff $(\mathcal{A},\mu)\models u=w$ for any $u,w\in\Lambda(C)$.

Now, from algebraic completeness theorems 2.2 it follows

3.5 Completeness theorems. For any set of λ-formulas $\Gamma\cup\{u=w\}$ one has:

$$\Gamma\models_{(\eta)\mathrm{EM}} u=w \quad \text{iff} \quad \Gamma\models_{\lambda\beta(\eta)\mathrm{SA}} u=w \quad \text{iff} \quad \Gamma\vdash_{\lambda\beta(\eta)} u=w.$$

3.6 Remark. By taking into consideration morphisms, one can construct categories of λ-theories, of algebraic and of environment models. Then completeness theorems mean nothing but embedability of categories built from theories into corresponding semantic categories. Note, the results on categorical equivalence mentioned in introduction are strictly stronger since they state also the inverse embedability of semantic categories into syntactical ones.

4 Towards Generalizations

4.1 Infinitary λ-calculus. As we have seen, different kinds of l.f. λSAs are precise algebraic counterparts of the corresponding λ-calculi and the local finiteness restriction is essential. Now the following question arises naturally: what is a λ-calculus source of the whole class λSA without this restriction, i.e. what is λ-calculus with terms depending on infinitely many variables? (Note that such terms are converted into infinitary operations over the carrying domain of a given environment model). In this context, the following question is interesting and quite natural from the view point of universal algebra: whether there is an identity which holds for all l.f. λβSA but does not hold for some non l.f. λβSA. If such an identity really exists (and we conjecture that this is the case), then the difference between ordinary λβ(η)-calculus and its infinitary modification can be captured by means of equational metatheory.

At another point, as soon as we turn to infinitary λ-calculus, the finitary λSAs we have defined become somewhat unnatural tool since finitary operations acting on closed elements cannot reach elements with infinite dimension sets (Proposition 1.4(v)). A more natural structure is a modification of λSA by infinitary, $(\gamma+1)$-ary for all $\gamma\leq\omega$, applications and substitutions, and unary (as before) λ-quantifiers indexed, however, by tuples $\mathbf{i}\in\omega^{\gamma}$. In other words, if $\mathcal{A}$ is such an algebra then, for all $\gamma\leq\omega$, $\mathbf{i}\in\omega^{\gamma}$, $\rho\in A^{\gamma}$ and $a\in A$, there are defined $s_{\mathbf{i}}(\rho,a)$, $\lambda_{\mathbf{i}}a$ and $a'\rho$ (we omit the index i at $'$). In this way we obtain an algebraic version of infinitary λ-calculus with infinitary λ-quantification, application and substituting. A categorical counterpart of this version is an infinitarily reflexive object in something like an infinitary Cartesian closed category with finite and countable products and exponentials.

4.2 General α-calculi. Another interesting possible development of the λSA framework is a general algebraic theory of *α-calculi*, that is, a theory of interaction between quantifiers similar to λ-quantifier and terms-in-terms substitutions (in contrast to the general theory of interaction between ∃,∀-like quantifiers and terms-in-formulas substitutions considered by Pigozzi and Salibra [PS92]). The syntactical aspect of the problem is easy and has already decided in section 2. In semantics, we must modify the notion of λβ-set (functional domain) and introduce the notion of α-set. In more detail, *an α-set* is defined to be a quadruple $\mathcal{D}=(D,\Psi,\Phi,Q)$ with (D,Ψ,Φ) as in construction 3.1 but without imposing condition (FDβ), and Q is a mapping $D^{\omega}\times D\times\omega \longrightarrow D^{\omega}$ intended to replace the mapping $(\rho,a,i) \longmapsto \rho\{a/i\}$ used above in points (C3) and (O3) of the construction. Certainly, the mapping Q must be connected with substitutions in a definite way (α-conversion). Thus, a λβ-set is an α-set with $Q(\rho,a,i) = \rho\{a/i\}$. We conjecture that any l.f. λSA can be semantically presented via some α-set.

References

[B81] Barendregt,H.P. *The Lambda Calculus: its Syntax and Semantics.* Studies in Logic,103,N-Holland,1982.

[C88] Cīrulis,J. *An Algebraization of First-order Logic with Terms.* In "Algebraic logic (Proc. Conf. in Budapest, 1988)", ed. H.Andréka,J.D.Monk,I.Németi, N-Holland,1991, pp.125-146

[D91] Diskin,Z.B. *Lambda Term Systems.* Submitted to Z.Math.Logik und Grundl.Math.

[F82] Feldman,N. *Axiomatization of Polynomial Substitution Algebras.* J.Symbolic Logic,**47**,3(1982), 481-492.

[M82] Meyer,A.R. *What is a Model of the Lambda Calculus?* Information and Control,**52**,1(1982), 87-122.

[OW82] Obtułowicz,A. and Wiweger,A., *Categorical, Functional and Algebraic Aspects of the Type-free Lambda Calculus.* in: "Universal algebra and Applications", Banach Center Publications, **9**(1982), 399-422.

[PS92] Pigozzi,D. and Salibra,A. *Polyadic Algebras over Nonclassical Logics.* Manuscript.

Global properties of 2D cellular automata: some complexity results*

Bruno Durand

Laboratoire de l'Informatique du Parallélisme
Unité de Recherche Associée 1398 du CNRS
Ecole Normale Supérieure de Lyon
46, Allée d'Italie
69364 Lyon Cedex 07
France

Abstract. In this paper, we prove the co-NP-completeness of the following decision problem: "given a 2-dimensional cellular automaton $\mathcal{A}$ (even with Von Neumann neighborhood), is $\mathcal{A}$ injective when restricted to finite configurations not greater than its length?" In order to prove this result, we introduce two decision problems concerning respectively Turing Machines and tilings that we prove NP-complete. Then, we transform problems concerning tilings into problems concerning cellular automata.

1 Introduction

Cellular automata (CA) are often used for modeling complex natural systems with many very simple cells interacting locally with each other. Possible evolutions of cellular automata have been extensively studied in order to analyze evolutions of such natural systems. Problems like bijectivity or surjectivity of CA are very basic because they correspond to physical notions: conservation of information (which corresponds to physical reversibility) or reachability of all states.

In 1962-63, Moore and Myhill proved the so-called "garden of Eden" theorem which proves that surjectivity is equivalent to injectivity on finite configurations [6, 5]. Richardson proved in 1972 [7] that if a CA realizes a bijective function, then there exists another CA called its inverse that realizes the inverse function. The same year, Amoroso and Patt proved that the reversibility (or the surjectivity) of one-dimensional CA is decidable [1]. One-dimensional CA work on a bi-infinite line of cells, two-dimensional CA work on a plane tiled with square cells, etc...

Recently, Jarrko Kari proved that the reversibility of two-dimensional CA fails to be decidable [4, 3]. An easy consequence of this result is that the reverse CA of a reversible CA cannot be found by algorithm: its size can be greater

* This work was partially supported by the Esprit Basic Research Action "Algebraic and Semantical Methods In Computer Science" and by the PRC "Mathématique et Informatique".

than any computable function of the size of the reversible CA. The proof of this result consists in transforming the tiling problem of the plane which has been proved undecidable in 1966 by Berger [2, 8] into the reversibility problem on an adequate family of CA.

The main goal of this paper is to prove that if we restrain the field of action of the two-dimensional CA to finite configuration bounded in size, it is still "difficult" to prove that the CA is (or is not) reversible. As far as we know, it would be the first complexity result concerning a global property of 2D CA. We prove that the decision problem presented above belongs to the class of co-NP-hard problems or to the class of co-NP-complete problems if the size of the considered finite configurations are supposed bounded by the size of the representation of the considered CA. Both of these complexity classes are supposed to contain only intractable decision problems...

Our result also holds for k-dimensional CA where $k \geq 2$. But in the case of one-dimensional CA, the reversibility problem can be solved in polynomial time in the size of the transition table (see [9] for another point of view on this problem).

We prove our result by introducing a very adequate set of tiles having an ad-hoc property. J. Kari has proved the undecidability of the surjectivity problem for two-dimensional CA [3] by introducing a very complicated set of tiles. We keep the ideas of the construction of J. Kari but the tile set we use is much simpler. With this construction, we reduce decision problems concerning tilings into decision problem concerning CA.

In the following section, we give the usual definitions of cellular automata, tilings, and present well-known theorems which are related to our topics. In the next section we prove our main complexity result: the problem that we call CA-FINITE-INJECTIVE is co-NP-complete. Our proof consists in a reduction of a problem concerning finite tilings that we call FINITE-TILING. We prove this problem NP-complete by a reduction of another problem concerning the minimal computing time of non-deterministic Turing machines: NDTM-TIME. The reduction of FINITE-TILING into CA-FINITE-INJECTIVE is not so simple as the previous one. Anyway, all technical aspects of the proof are contained in the construction of a special set of tiles which verifies specific properties. The construction is not difficult by itself nor are the properties, but in order to prove the properties, we have to check it for many kinds of tiles which is very long and tedious.

2 Definitions and basic properties

2.1 Cellular automata

Cellular automata are formally defined as quadruplets (n, S, N, f):

- The set of *states* is a finite set $S = \{s_1, s_2, \ldots, s_k\}$.
- A *configuration* is an application from $\mathbb{Z}^n$ to S. The set of all the configuration is $S^{\mathbb{Z}^n}$ and the *dimension* of the space is n. If we imagine $\mathbb{Z}^n$ as being

a tiling of $\mathbb{R}^n$ by identical hypercubes, each tile forms a cell which has an associated state.

- The *neighborhood* N of a cellular automaton is an l-tuple of distinct vectors of $\mathbb{Z}^n$. For us, $N = \{v_1, \ldots, v_l\}$ and $\forall (i,j) \in \{1, \ldots, l\}^2$,

$$i \neq j \Rightarrow v_i \neq v_j.$$

  The v_i's are the relative positions of the cells in the neighborhood with respect to a given center cell. The states of the neighbor cells are the states used to compute the new state of the center cell.
- The *local function* of the cellular automaton $f : S^l \mapsto S$ gives the local transition rule.
- The *global function* G of the cellular automaton is defined via f on the set of configurations $S^{\mathbb{Z}^n}$: $G(c) = c' \Leftrightarrow \forall i \in \mathbb{Z}^n$,

$$c'(i) = f(c(i + v_1), \ldots, c(i + v_l)).$$

In the following, we consider 2-dimensional CA i.e. CA for which $n = 2$. Remark that two distinct cellular automata do not differ by the definition of their global function G: they are only characterized by N and f.

Sometimes, a *quiescent* state q is distinguished in S. An everywhere quiescent configuration has to remain unchanged when the CA is applied i.e. $f(q, q, \ldots, q) = q$. A *finite configuration* is an almost everywhere quiescent configuration. If i and j are integers such that all the non-quiescent cells of the configuration are located inside a square of size $i \times j$, then, in the following, we say that the size of the finite configuration is smaller than (or equal to) $i \times j$.

Size: *The size necessary to code a cellular automaton is $\mathcal{O}(s^l . \log s)$ where s is its number of states and l the number of elements of its neighborhood.*

The size of the representation of the CA is exactly the sum of the size of its local transition function and of the size of its neighborhood. The local transition function is only a l-dimensional table. Hence its size is $\mathcal{O}(s^l . \log s)$. The size of the neighborhood is the size of the coding of the coordinates of each neighbor cell. We assume in the following that this last size is lower than the size of the transition table. If it is not the case, it means that the neighbors of a cell are very far from it (about 2^{s^l} cells far away) hence a single iteration of the CA is intractable! The reductions we present are still valid, but the decision problems do not belong, a priori, to co-NP.

2.2 Tilings

A tile is a square the sides of which are colored. The colors belong to a finite set C called the *color set*. A set of tiles T is a subset of C^4. All the tiles have the same (unit) size. A tiling of the plane is *valid* if and only if all pairs of adjacent sides have the same color.

Notice that it is not allowed to turn tiles and that coloring the sides and the corners would lead to the same notion. Furthermore, we could have affected to each tile a single color and say that a tiling of the plane is valid if and only if a given relation holds between each cell and its neighbors.

Theorem 1. *Given a tile set, it is undecidable to know whether this tile set can be used to tile the plane.*

This well-known theorem is due to Berger [2] in 1966 and a simplified proof was given in 1971 by Robinson [8].

We can also define *finite tilings*. We assume that the set of colors contains a special "blank" color and that the set of tiles contains a "blank" tile i.e. a tile whose all sides are blank. A finite tiling is an almost everywhere blank tiling of the plane. If there exist two integers i and j such that all the non-blank tiles of the tiling are located inside a square of size $i \times j$, then, we say that the size of the finite tiling is smaller than (or equal to) $i \times j$. Notice that inside the $i \times j$ square, there can be blank and non-blank tiles. If there is at least one non-blank tile, then the tiling is called non-trivial.

Another undecidability result is known and can be proved simply by using a construction presented in [3] which reduces the undecidability of the halting problem for Turing machines to the following problem:

Theorem 2. *Given a tile set with a blank tile, it is undecidable whether this tile set can be used to form a valid finite non-trivial tiling of the plane.*

3 Complexity results

In this section, we prove that the problem of deciding if a given CA is reversible restricted to finite configurations with size lower than $n \times n$ (where n is the size of the CA) is co-NP-complete. We call this problem CA-FINITE-INJECTIVE. In order to prove this result, we introduce two decision problems we prove NP-complete: NDTM-TIME and FINITE-TILING.

3.1 Intermediate problems

NDTM-TIME

INSTANCE: A *Non-Deterministic* Turing Machine. An integer n lower than the number of states of the machine.

QUESTION: Is there a computation of the machine beginning on an empty tape and halting after less than n steps.

FINITE-TILING

INSTANCE: Finite set C of colors with a blank color, collection $T \in C^4$ of tiles including a blank tile. A positive integer $n \leq |C|$.

QUESTION: Is there a finite non-trivial tiling of the plane and a $n \times n$ square such that all non-blank tiles are located inside the square.

Theorem 3. NDTM-TIME *and* FINITE-TILING *are* NP-*complete.*

Proof. The first problem (NDTM-TIME) is inherently NP-complete and it is not difficult to prove it.

We prove FINITE-TILING NP-complete by a reduction of NDTM-time to this problem. First remark that this problem trivialy belongs to NP. Consider an instance of NDTM-TIME i.e. a NDTM τ and an integer n lower than the number of states $\sigma(\tau)$ of the machine. We construct a set of tiles associated to τ as it was done in the case of deterministic Turing Machine in [3]. With these tiles, we simulate the space $\times$ time diagram of a computation of the NDTM τ. There exists a non-trivial tiling of the plane of length lower than $n \times n$ if and only if there exists a computation of τ that ends after less than $n-2$ steps. The transformation is polynomial, hence FINITE-TILING is NP-complete. □

3.2 From FINITE-TILING to CA-FINITE-INJECTIVE

A transformation between tilings and two-dimensional cellular automata has been first presented by Jarrko Kari in [4] and a more complete proof can be found in [3]. The main idea of the transformation is to introduce a special set of tiles which have a very special property called by J. Kari finite plane filling property. We introduce another set of tiles, simpler than J. Kari's, which satisfies a slightly more restrictive property. We shall refer to this tile set as $\mathcal{D}$. With the help of $\mathcal{D}$, for each tile set $\mathcal{T}$, we construct a cellular automaton $\mathcal{A}_{\mathcal{T}}$ in order to reduce FINITE-TILING to CA-FINITE-INJECTIVE.

CA-FINITE-INJECTIVE

INSTANCE: A two-dimensional cellular automaton $\mathcal{A}$ with Von Neumann neighborhood. Two integer p and q smaller than the size of $\mathcal{A}$.

QUESTION: Is $\mathcal{A}$ injective restricted to all finite configurations smaller than $p \times q$.

Theorem 4. CA-FINITE-INJECTIVE *is co-NP-complete.*

Before proving this theorem, we introduce our tile set $\mathcal{D}$ and its properties. The sides contain a color ("blank", "border", "odd", "even", or "the-end"), a label (N, S, E, W, $N+$, $S+$, $E+$, $W+$, or ω), and possibly an arrow. With this set of tiles, a tiling is considered as valid if and only if all pairs of adjacent sides have the same color, the same label, and for each arrow, the head points out on the tail of the arrow of the adjacent cell.

This tile set $\mathcal{D}$ is described in Fig. 1. In this figure, the tiles are "generic" because the labels are not represented. Each tile can have different labels defined below.

The labels are there to force that, in a valid tiling, inside a rectangle bordered with "border" tiles, there exists a unique cell labeled with ($N+$, $E+$, $S+$, $W+$) (see Fig. 2 and lemma 7). The tiles of Fig. 1 with no arrow have their sides labeled by ω. The four tiles in the center of Fig. 1 have their North, East, South, and West sides labeled by ($N+$, $E+$, $S+$, $W+$), (X, Y, X, Y), (N, $Y+$, S, $Y+$) or ($X+$, E, $X+$, W) where X is N or S, and Y is E or W. The four upper tiles

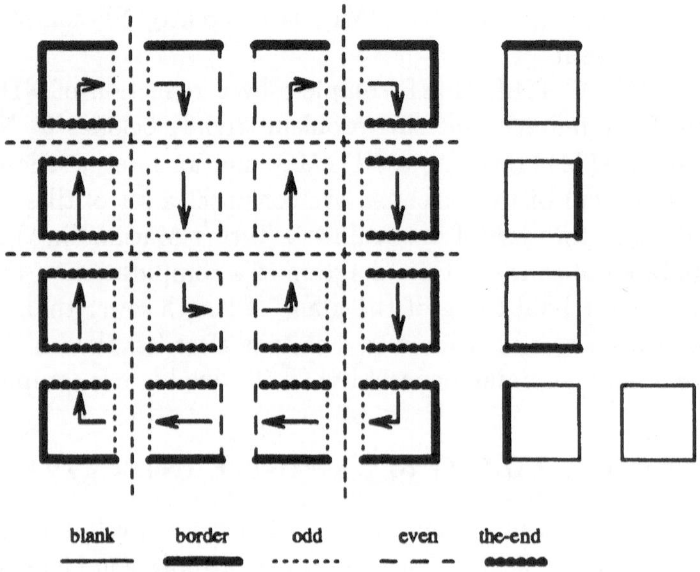

Fig. 1. The generic tiles of $\mathcal{D}$

have their south side labeled by N or $N+$, and the other sides labeled by ω. The four left tiles have their west side labeled by E or $E+$, and the other sides labeled by ω, etc...

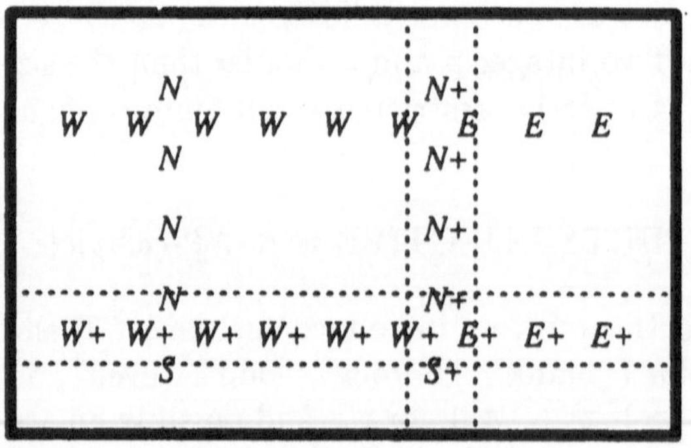

Fig. 2. The labels in a basic rectangle

Definition 5. A *basic rectangle* of size $p \times q$ is a finite valid tiling of the plane of size $p \times q$ with no side labeled "blank" or "border" inside the rectangle.

See Fig. 3 for a description of the path given by the arrows in a basic rectangle.

Lemma 6. *For every integers p, and q, both greater than 3, there exists a finite valid tiling of the plane by $\mathcal{D}$ of size $p \times 2q$. Each valid finite tiling of the plane consists in a finite number of juxtaposed basic rectangles.*

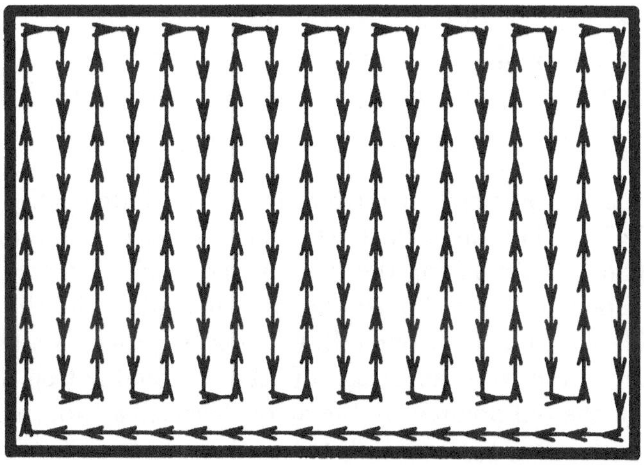

Fig. 3. The path in a basic rectangle

Lemma 7. *Consider a basic rectangle. Then the path defined by the arrows of the cells forms a loop which visits one time each tile of the inside of the rectangle. Inside the rectangle, there exists a unique cell labeled ($N+$, $E+$, $S+$, $W+$).*

Lemma 8. *Consider a* finite *tiling (valid or not). If the tiling is correct on each cell of a path, then this path forms a loop and visits every tiles of a basic rectangle.*

Consider an instance of FINITE-TILING i.e. a finite set C of colors with a blank color, a collection $\mathcal{T} \in C^4$ of tiles including a blank tile. A positive integer $n \leq |C|$. We construct a cellular automaton $\mathcal{A}_\mathcal{T} = (S, N, f_\mathcal{T})$ defined as follows:

- The *state* set $S \subset \mathcal{D} \times \mathcal{T} \times \{0,1\}$. S contains all triplets (d, t, α) of $\mathcal{D} \times \mathcal{T} \times \{0,1\}$ under the following restrictions:
  - if one of the sides of d is "blank" or "border", then t is the blank tile of $\mathcal{T}$.
  - if d is labeled by ($N+$, $E+$, $S+$, $W+$), then t *is not* the blank tile of $\mathcal{T}$.
- The *neighborhood* N is the Von Neumann neighborhood i.e.

$$N = \{(0,0), (0,1), (0,-1), (1,0), (-1,0)\}$$

- The *local rule* $f_\mathcal{T}$ applied on a cell the state of which is (d, t, α) may change only the bit component α. At each cell both the tilings $\mathcal{D}$ and $\mathcal{T}$ are checked. If there is a tiling error, or if the cell contains no arrow, then the state of the cell is not altered. Else, there is no tiling error in the concerned cell, and the bit component is changed by performing an "exclusive or" operation with the bit attached to the cell pointed by the direction of the $\mathcal{D}$-component.

The quiescent state of $\mathcal{A}_\mathcal{T}$ is (blank, blank,0).

We present now the basic theorem which provides a link between tilings and cellular automata.

Theorem 9. *Let n be an integer greater or equal to 3 and $\mathcal{T}$ be a set of tiles. The cellular automaton $\mathcal{A}_\mathcal{T}$ is* not *injective restricted to finite configurations of size lower than $2n \times 2n$ if and only if the tile set $\mathcal{T}$ can be used to form a finite non-trivial tiling of the plane of size lower than $(2n-4) \times (2n-4)$.*

Proof. Assume that $\mathcal{A}_\mathcal{T}$ is not injective restricted to finite configurations of size lower than $2n \times 2n$. Then there exist two different finite configurations c and c' of size lower than $2n \times 2n$ having the same image by $\mathcal{A}_\mathcal{T}$. Remark that only the bits can be different in c and c' since $\mathcal{A}_\mathcal{T}$ does not affect the tiles components: c and c' are different in at least one cell. On this cell, there is an arrow and the tilings are correct, otherwise the images of c and c' could not be the same. Thus c and c' differ in the cell pointed by the arrow because a "xor" is performed by $\mathcal{A}_\mathcal{T}$. By finite induction, by lemma 8, the path such constructed forms a loop and there exists a basic rectangle of $\mathcal{D}$ on which the tiling of $\mathcal{T}$ is correct. By lemma 7, the borders of the rectangle are blank in the state component of $\mathcal{T}$, hence we can construct a finite tiling with $\mathcal{T}$. The tiling is not trivial because $\mathcal{T}$ is not blank on the cell labeled by $(N+,\ E+,\ S+,\ W+)$ in $\mathcal{D}$ (lemma 6). The size of the tiling is at most $(2n-4) \times (2n-4)$.

Conversely assume that there exists a finite non-trivial tiling of the plane by $\mathcal{T}$ of size lower than $(2n-4) \times (2n-4)$. We put this tiling inside a $2n \times 2n$ rectangle tiled by $\mathcal{D}$. The tiling is not trivial thus there exists a non-blank tile on which we can put the tile of $\mathcal{D}$ labeled by $(N+,\ E+,\ S+,\ W+)$. We define two configurations c and c' of size $2n \times 2n$: c is obtained by adding the bit component 0 everywhere. For c', we add to the two tilings the bit component 1 on the cells whose $\mathcal{D}$-component has an arrow, and 0 elsewhere. As both the tilings are correct, $\mathcal{A}_\mathcal{T}$ performs an "exclusive or" on the loop of the rectangle and both c and c' have the same image c. Hence $\mathcal{A}_\mathcal{T}$ restricted to finite configurations of size lower than $2n \times 2n$ is not injective. □

Corollary 10. *Given a set of tiles $\mathcal{T}$ and an even integer n, $\mathcal{T}$ can be used to form a finite non-trivial tiling of the plane of size lower than $n \times n$ if and only if the cellular automaton $\mathcal{A}_\mathcal{T}$ is not injective when restricted to finite configurations of size lower than $(n+4) \times (n+4)$.*

Proof of theorem 4. With the previous result, it is very easy to prove that CA-FINITE-INJECTIVE is co-NP-complete. We prove a stronger result: CA-FINITE-INJECTIVE is co-NP-complete with a restriction on the instance: $p = q\ (= n)$ and $n \geq 2$ is an even integer.

CA-FINITE-INJECTIVE with or without the restriction mentionned above is in co-NP because one can check in time polynomial in the size of the cellular automaton $\mathcal{A}$ if two finite configurations smaller than $n \times n$ have the same image by $\mathcal{A}$. So we only have to prove that the reduction presented from FINITE-TILING to CA-FINITE-INJECTIVE with the help of the automaton $\mathcal{A}_\mathcal{T}$ (corollary 10) has a polynomial cost. If c denotes the number of colors used by the tile set $\mathcal{T}$ and t its number of tiles, then $t \leq c^4$. If we call d the size of the tile set $\mathcal{D}$ then the size of the state set S of $\mathcal{A}_\mathcal{T}$ is at most $2dt$. In the case of $\mathcal{D}$, the number d of its tiles is exactly 61. The size of the neighborhood is 5 hence the size of $\mathcal{A}_{\mathcal{T}}$ is $\mathcal{O}((2dt)^5.\log(2dt))$ which is bounded by a polynomial function of t. As each element of the transition table can be computed in constant time, the reduction between the two problems is polynomial. Notice that we have proved that under a polynomial reduction, the answer to the problem FINITE-TILING applied to the instance $\mathcal{T}$ is "yes" if and only if the answer to the problem CA-FINITE-INJECTIVE applied to the instance $\mathcal{A}_{\mathcal{T}}$ is "no". It explains why CA-FINITE-INJECTIVE is co-NP-complete and not NP-complete. □

4 Conclusion

In this paper, we have assumed that the size of the representation of a cellular automaton is the size of its transition table. But these tables are not always given extensively: a program which computes the value of the transition table on a given entry is often furnished. Thus it is natural to wonder if our complexity results remain true if we define the size of a CA as the length of the smallest program (or of the smallest Turing Machine) which computes its transition table. Maybe reversible CA given by "simple" algorithms, when restricted to finite bounded configurations, have their inverses given by "simple" algorithms too.

References

1. S. Amoroso and Y.N. Patt. Decision procedures for surjectivity and injectivity of parallel maps for tesselation structures. *J. Comp. Syst. Sci.*, 6:448–464, 1972.
2. R. Berger. The undecidability of the domino problem. *Memoirs of the American Mathematical Society*, 66, 1966.
3. J. Kari. Reversibility and surjectivity problems of cellular automata. *to appear in Journal of Computer and System Sciences.*
4. J. Kari. Reversability of 2d cellular automata is undecidable. *Physica*, D 45:379–385, 1990.
5. E.F. Moore. Machine models of self-reproduction. *Proc. Symp. Apl. Math.*, 14:13–33, 1962.
6. J. Myhill. The converse to Moore's garden-of-eden theorem. *Proc. Am. Math. Soc.*, 14:685–686, 1963.
7. D. Richardson. Tesselations with local transformations. *Journal of Computer and System Sciences*, 6:373–388, 1972.
8. R.M. Robinson. Undecidability and nonperiodicity for tilings of the plane. *Inventiones Mathematicae*, 12:177–209, 1971.
9. K. Sutner. De Bruijn graphs and linear cellular automata. *Complex Systems*, 5:19–30, 1991.

Completeness Results for Linear Logic on Petri Nets

(Extended Abstract)

Uffe Engberg Glynn Winskel

Computer Science Department*, Aarhus University, Ny Munkegade, DK-8000 Aarhus C, Denmark

Abstract

Completeness is shown for several versions of Girard's linear logic with respect to Petri nets as the class of models. The strongest logic considered is intuitionistic linear logic, with $\otimes$, $\multimap$, $\&$, $\oplus$ and the exponential ! ("of course"), and forms of quantification. This logic is shown sound and complete with respect to *atomic nets* (these include nets in which every transition leads to a nonempty multiset of places). The logic is remarkably expressive, enabling descriptions of the kinds of properties one might wish to show of nets; in particular, negative properties, asserting the impossibility of an assertion, can also be expressed.

1 Introduction

In [EW90] it was shown how Petri nets can naturally be made into models of Girard's linear logic in such a way that many properties one might wish to state of nets become expressible in linear logic. We refer the reader to [EW90] for more background and a discussion of other work. That paper left open the important question of completeness for the logic with respect to nets as a model. The question is settled here.

We restrict attention to Girard's intuitionistic linear logic. Our strongest result is for the full logic described in [GL87, Laf88], viz. it includes

$$\otimes,\ \multimap,\ \oplus,\ \&,\ \text{and}\ !$$

though at a cost, to the purity of the linear logic, of adding quantification over markings and axioms special to the net semantics. For this strongest completeness result, a slight restriction is also made to the Petri nets considered as models; they should be *atomic* (see definition 17), but fortunately this restriction is one generally met, and even often enforced, in working with Petri nets. The step of considering only atomic nets as models has two important pay-offs: one is that the exponential $!A$ becomes definable as $A \& \mathbf{1}$, where $\mathbf{1}$ is the unit of $\otimes$; the other is that we can say internally, within the logic, that an assertion is *not* satisfied—the possibility of asserting such negative properties boosts the logic's expressive power considerably. We can achieve completeness for more modest fragments of the logic without extra axioms and with respect to the entire class of nets as models (see section 5). Detailed proofs appear in [EW93].

The work here contrasts with other approaches to linear logic on Petri nets in that they either apply only to much smaller fragments of the logic such as the $\otimes$-fragment (cf. [GG89]), or use the transitions of a Petri net to freely generate a linear-logic theory (cf. [MOM91]), in which case the logic becomes rather inexpressive, and in particular cannot capture negative properties, or they don't address completeness at all.

While it is claimed that this paper together with [EW90] help in the understanding of linear logic, it is also hoped that, through these results, linear logic will come to be of use in reasoning about Petri nets and, through them, in concurrent computation.

2 Linear Intuitionistic Logic

The connectives of linear intuitionistic logic are:

*e-mail address: {engberg,gwinskel}@daimi.aau.dk, fax: ++45 86 13 57 25

- $\otimes$ tensor, with unit $\mathbf{1}$, called one,
- $\&$ conjunction, with unit T, called true,
- $\oplus$ disjunction, with unit F, called false.

We take as the definition of linear intuitionistic logic the proof rules presented in [GL87, Laf88]:

Structural rules

$$\frac{}{A \vdash A}(\text{identity}) \qquad \frac{\Gamma \vdash A \quad \Delta, A \vdash B}{\Gamma, \Delta \vdash B}(\text{cut}) \qquad \frac{\Gamma, A, B, \Delta \vdash C}{\Gamma, B, A, \Delta \vdash C}(\text{exchange})$$

Logical rules

$$\frac{\Gamma \vdash A \quad \Delta \vdash B}{\Gamma, \Delta \vdash A \otimes B}(\vdash\otimes) \qquad \frac{\Gamma, A, B \vdash C}{\Gamma, A \otimes B \vdash C}(\otimes\vdash) \qquad \frac{\Gamma \vdash A}{\Gamma, 1 \vdash A} \qquad \frac{}{\vdash 1}$$

$$\frac{\Gamma \vdash A \quad \Gamma \vdash B}{\Gamma \vdash A \,\&\, B}(\vdash\&) \qquad \frac{\Gamma, A \vdash C}{\Gamma, A \,\&\, B \vdash C}(l\&\vdash) \qquad \frac{\Gamma, B \vdash C}{\Gamma, A \,\&\, B \vdash C}(r\&\vdash) \qquad \frac{}{\Gamma \vdash \mathsf{T}}$$

$$\frac{\Gamma \vdash A}{\Gamma \vdash A \oplus B}(\vdash\oplus l) \qquad \frac{\Gamma \vdash B}{\Gamma \vdash A \oplus B}(\vdash\oplus r) \qquad \frac{\Gamma, A \vdash C \quad \Gamma, B \vdash C}{\Gamma, A \oplus B \vdash C}(\oplus\vdash) \qquad \frac{}{\Gamma, \mathsf{F} \vdash A}$$

$$\frac{\Gamma, A \vdash B}{\Gamma \vdash A \multimap B}(\vdash\multimap) \qquad \frac{\Gamma \vdash A \quad \Delta, B \vdash C}{\Gamma, \Delta, A \multimap B \vdash C}(\multimap\vdash)$$

where we use Γ as an abbreviation for a (possibly empty) sequence $A_1, \ldots, A_n$ of assumption formulae.

The absence of the rules for thinning and contraction is compensated, to some extent, by the addition of the logical operator "of course". In [GL87, Laf88] this operator is presented with the following proof rules (stronger than those in [Gir87]):

"Of course" rules

$$\frac{}{!A \vdash A} \qquad \frac{}{!A \vdash 1} \qquad \frac{}{!A \vdash !A \otimes !A} \qquad \frac{B \vdash A \quad B \vdash 1 \quad B \vdash B \otimes B}{B \vdash !A}$$

Given a proposition A, the assertion of $!A$ has the possibility of being instantiated by the proposition A, the unit 1 or $!A \otimes !A$, and thus of arbitrarily many assertions of $!A$.

The proof rules are sound when interpreted according to the algebraic semantics for linear intuitionistic logic provided by quantales (commutative monoids on complete join semilattices), see [EW90].

3 Petri Nets

Petri nets are a model of processes (or systems) in terms of types of resources, represented by *places* which can hold to arbitrary nonnegative multiplicity, and how those resources are consumed or produced by actions, represented by *transitions*. They are described using the notation of multisets.

A *multiset* over a set P is a function, $M : P \longrightarrow \mathbb{N}$. We shall henceforth only be concerned with *finite* multisets, i.e. $\{a \in P \mid M(a) \neq 0\}$ finite. With addition, $+$, of multisets defined by $(M + M')(a) = M(a) + M'(a)$ for all $a \in P$, multisets over P form a (free) commutative monoid with $\underline{0}$ ($\forall a \in P. \underline{0}(a) = 0$), the empty multiset, as unit.

We take a Petri net N to consist of $(P, T, {}^\bullet(-), (-)^\bullet)$, where P, a set of *places*, and T, a set of *transitions*, are accompanied by maps ${}^\bullet(-), (-)^\bullet$ on transitions T which for each $t \in T$ give a multiset of P, called the pre- and post (multi)set of t respectively. For the moment there are none of the usual restrictions on the net, such as absence of isolated elements, and in particular transitions with empty pre sets and/or post sets will be allowed. And we are actually considering nets with unconstrained capacity.

A Petri net possesses a notion of state, intuitively corresponding to a finite distribution of resources, formalized in the definition of a marking. A *marking* of N will simply be a finite multiset over P. We use $\mathcal{M}$ to denote the *set of markings* of the net, understood from the context. Sometimes nets are associated with an initial marking M_0. The behaviour of a net is expressed by saying how markings change as transitions occur (or fire). For markings, M, M', and a transition $t \in T$, $M\,[t\rangle\,M'$ stands for t fires from M to M'; that is the firing relation $[t\rangle$ is given by

$$M\,[t\rangle\,M' \quad \text{iff} \quad \exists M'' \in \mathcal{M}. M = M'' + {}^\bullet t \text{ and } t^\bullet + M'' = M' \ .$$

So t is enabled at M if there is an $M' \in \mathcal{M}$ such that $M\,[t\rangle\,M'$. We shall write $M \rightarrow M'$ for the reachability relation, the reflexive and transitive closure of the firing relations. We shall use $\downarrow(M)$ to denote the set of markings which can reach M. We will generally call this set the downwards closure of M. It is defined by $\downarrow(M) = \{M' \in \mathcal{M} \mid M' \to M\}$.

Petri nets can be presented by using the well-known graphical notation, which we will use in an example. Places are represented by circles, transitions as squares, and arcs of appropriate multiplicities used to indicate the pre and post sets. The formal definitions can then be brought to life in the so called "token game" where markings are visualized as consisting of a distribution of tokens over places; the number of tokens residing on a place expresses the multiplicity to which it holds according to the marking. The tokens are consumed and produced as transitions occur. A basic reference for Petri nets is [Rei85].

4 Interpretation

For simplicity we consider a linear logic language where the atomic propositions are places of nets. I.e. formulae are given by:

$$
\begin{array}{lll}
A ::= \mathsf{T} \mid \mathsf{F} \mid 1 & & \text{constants} \\
\quad \mid a & & \text{atoms} \\
\quad \mid A \otimes A \mid A \multimap A & & \text{multiplicative connectives} \\
\quad \mid A \,\&\, A \mid A \oplus A & & \text{additive connectives} \\
\quad \mid\, !A & & \text{exponential connective}
\end{array}
$$

We make the choice of interpreting an atomic proposition as the downwards closure of the associated place, but we could just as well have used the downwards closure of some marking without altering our results. This choice is consistent with the following intuitive understanding: the denotation of an assertion is to be thought of as the set of requirements sufficient to establish it. This reading will be discussed further shortly, after presenting the semantics. More abstractly, we are giving a semantics in a quantale $\mathcal{Q}$ consisting of downwards-closed subsets of markings with respect to reachability, ordered by inclusion, with a binary operation given by

$$q_1 \otimes q_2 =_{def} \{M \mid \exists M_1 \in q_1, M_2 \in q_2.\ M \to M_1 + M_2\} \ .$$

With respect to a net N, linear logic formulae are interpreted as follows. The denotation of an assertion can be thought of as consisting of the set of markings which satisfy it.

$$
\begin{array}{ll}
[\![\mathsf{T}]\!]_N & = \mathcal{M} \\
[\![\mathsf{F}]\!]_N & = \emptyset \\
[\![1]\!]_N & = \{M \mid M \to \underline{0}\} \\
[\![a]\!]_N & = \{M \mid M \to a\} \\
[\![A \otimes B]\!]_N & = \{M \mid \exists M_A \in [\![A]\!]_N, M_B \in [\![B]\!]_N.\ M \to M_A + M_B\} \\
[\![A \multimap B]\!]_N & = \{M \mid \forall M_A \in [\![A]\!]_N.\ M + M_A \in [\![B]\!]_N\} \\
[\![A \,\&\, B]\!]_N & = [\![A]\!]_N \cap [\![B]\!]_N \\
[\![A \oplus B]\!]_N & = [\![A]\!]_N \cup [\![B]\!]_N \\
[\![!A]\!]_N & = \bigcup\{q \in \mathcal{Q} \mid q \text{ a postfixed point of } x \mapsto 1 \cap [\![A]\!]_N \cap (x \otimes x)\}
\end{array}
$$

The final clause gives the denotation of $!A$ as a maximum fixed point. The above definitions correspond to the quantale semantics which is determined once we fix the interpretation of atoms (see [EW90]). The semantics of [Bro89] is similar, but somehow dual to that here.

Because of the interpretation of 1, validity of an assertion A for the given net, N, is defined by

$$\models_N A \quad \text{iff} \quad \underline{0} \in [\![A]\!]_N \ .$$

Semantic entailment between assertions A and B is given by

$$A \models_N B \quad \text{iff} \quad [\![A]\!]_N \subseteq [\![B]\!]_N \ .$$

Because of the interpretation of linear implication, this is equivalent to $\models_N A \multimap B$.

For $\Gamma = A_1, \ldots, A_n$ denote $A_1 \otimes \cdots \otimes A_n$ by $\bigotimes\Gamma$. We write $\Gamma \models_N B$ for $\bigotimes\Gamma \models_N B$.
General validity, $\models A$, of an assertion A is defined by

$$\models A \quad \text{iff} \quad \models_N A, \text{ for every net } N$$

and with respect to entailment: $\Gamma \models B$ iff $\Gamma \models_N B$, for every net N.

As a special case that quantale semantics is sound, we have the soundness result:

Theorem 1 *If* $\Gamma \vdash A$ *then* $\Gamma \models A$.

So we see that with respect to a Petri net, an assertion A is denoted by a set of markings $[A]_N$. As we have discussed, a marking of net can be viewed as a distribution of resources. When $M \in [A]_N$ we can think of the marking M as a distribution of resources sufficient to establish A according to the net; in this sense the marking M is one of the (in general many) requirements sufficient to establish A. The meaning of an assertion A is specified by saying what requirements are sufficient to establish it—this is the content of the denotation $[A]_N$. Accordingly, a net satisfies an assertion A when $\underline{0} \in [A]_N$, expressing that A can be established with no resources.

This reading squares with the fact that assertions denote subsets of markings which are downwards closed with respect to the reachability relation of the net; if $M \in [A]_N$, so M is a requirement sufficient to establish A, and $M' \rightarrow M$ so we can obtain M for M', then so also is M' a sufficient requirement of A. Casting an eye over the definition of the semantics of assertions we can read, for example, the definition of $[a]_N$, for an atom a, as expressing that a sufficient requirement of a is any marking from which the (singleton) marking a can be reached according to the net. Similarly, the sufficient requirements of $A \,\&\, B$ are precisely those which are sufficient requirements of both A and of B. An element of $[A \multimap B]_N$ can be seen as what is required, in addition to any requirement of A, in order to establish B. There are similar restatements of the semantics for the other connectives as well.

This understanding should be born in mind when considering the example that follows, where we make use of the fact that $\otimes$, $\&$ and $\oplus$ are associative and assume the precedence: $\multimap\, < \,\&, \oplus < \otimes$.

Notation: For a multiset, M, of assertions of our logic, we associate the formula $\widehat{M}$ which when M is nonempty is given by

$$\bigotimes_{M(A)\neq 0} A^{M(A)} \quad \text{where } A^0 = 1 \text{ and } A^n = \overbrace{A \otimes \cdots \otimes A}^{n}, \text{ for } n > 0$$

and otherwise, when $M = \underline{0}$, is given by the formula 1. We only distinguish M and $\widehat{M}$ in a few crucial statements.

We can then express that one marking, M', is reachable from another M:

Proposition 2 *For any multisets of atoms* M *and* M', $M \rightarrow M'$ *in the net* N *iff* $\models_N M \multimap M'$.

Example 3 (Mutual exclusion) Consider the net N:

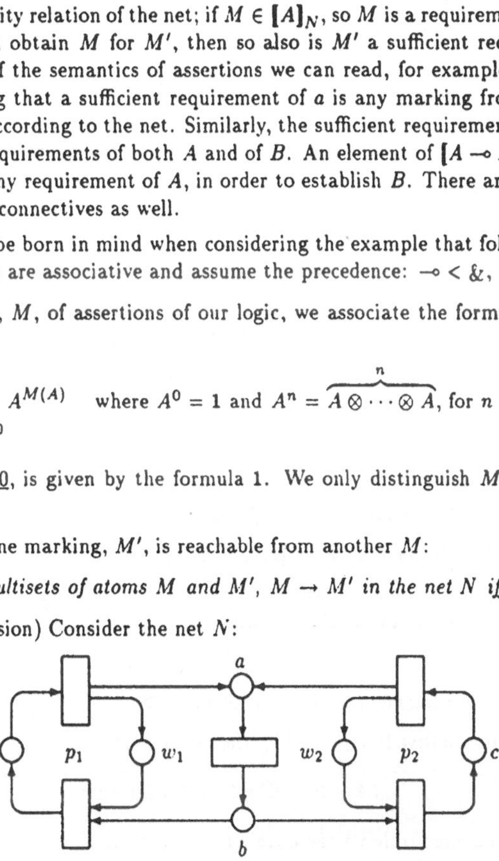

where the marking of the place w_1 indicates that the first process, p_1, is working outside its critical region, c_1, and similarly for the other process, p_2. The resource corresponding to b is used to ensure mutual exclusion of the critical regions and after a process has been in its critical region it returns a resource, a, which then is prepared (transformed into b) for the next turn. The initial marking, M_0, will be $M_0 = b \otimes w_1 \otimes w_2$. We can now express that e.g. p_1 can enter its critical region (from the initial marking) by: $\models_N M_0 \multimap c_1 \otimes \top$. However this does not ensure that no undesired tokens are present, so it is better to express it: $\models_N M_0 \multimap c_1 \otimes w_2$. If the system is in a "working state" then both processes have the possibility of entering their critical section: $\models_N w_1 \otimes (a \oplus b) \otimes w_2 \multimap c_1 \otimes w_2 \,\&\, w_1 \otimes c_2$. The property, that when p_1 is in its critical section and p_2 is working it is possible that p_2 can later come into its critical section with p_1 working, is expressed by: $\models_N c_1 \otimes w_2 \multimap w_1 \otimes c_2$. Similar other "positive" properties can be expressed. Shortly we shall see how to express the "negative" property that both processes cannot be in their critical regions at the same time.

5 Elementary Completeness Results

In this section we are concerned with completeness of different fragments of linear logic without exponentials.

5.1 Completeness for the $\oplus$-free Fragment

Restrict the syntax to the fragment:

$$A ::= \top \mid 1 \mid a \mid A_1 \,\&\, A_2 \mid A_1 \otimes A_2 \mid A_1 \multimap A_2 \qquad (\oplus\text{-free})$$

where a ranges over atoms. For the $\oplus$-free fragment we construct a net N where the places are formulae and the transitions essentially correspond to the provable sequents. I.e.

- *Places* are assertions of ($\oplus$-free) above.
- *Transitions* are pairs (M, M') of multisets of places for which $\widehat{M} \vdash \widehat{M'}$ with pre- and postset maps ${}^\bullet(M, M') = M$ and $(M, M')^\bullet = M'$.

Lemma 4 *For markings M, M' of the net N, $M \rightarrow M'$ in the net iff $\widehat{M} \vdash \widehat{M'}$.*

Lemma 5 *For the $\oplus$-free fragment we have: $[\![A]\!]_N = \{M \mid \widehat{M} \vdash A\}$.*

Because $\models_N A$ follows from $\models A$, and the fragment contains implication we deduce:

Theorem 6 *For the $\oplus$-free fragment we have: $\Gamma \models A$ iff $\Gamma \vdash A$.*

5.2 Completeness for the $\multimap$-free Fragment

We can obtain completeness for the $\multimap$-free fragment of propositional intuitionistic logic. Its syntax:

$$A ::= \top \mid \mathsf{F} \mid 1 \mid a \mid A_1 \oplus A_2 \mid A_1 \,\&\, A_2 \mid A_1 \otimes A_2 \qquad (\multimap\text{-free})$$

where a ranges over atoms. With a similar construction to that in the previous section we can obtain a rather weak form of completeness for the $\multimap$-free fragment.

Lemma 7 *For the $\multimap$-free fragment we have $[\![A]\!]_N \subseteq \{M \mid M \vdash A\}$.*

Proof Induction on the structure of A. All the cases except $A \equiv \mathsf{F}$ and $A \equiv A_1 \oplus A_2$ are handled exactly as the $\subseteq$-part of lemma 5 (notice the weaker hypothesis). □

As a corollary we have:

Theorem 8 *For the $\multimap$-free fragment we have: $\models A$ iff $\vdash A$.*

We have not used the distributive law yielded by the net semantics:

$$(A \oplus B) \,\&\, C \vdash (A \,\&\, C) \oplus (B \,\&\, C) \qquad (\&\text{-}\oplus\text{-dist.})$$

With this as an additional proof rule we can obtain a stronger completeness result for the $\multimap$-free fragment of propositional intuitionistic logic.

To show completeness we construct a net with places (and markings) identified with assertions in the $\oplus$-free subfragment:

$$A ::= \top \mid 1 \mid a \mid A_1 \,\&\, A_2 \mid A_1 \otimes A_2 \qquad (\multimap\text{-}\oplus\text{-free})$$

We will just call it the $\oplus$-free fragment in the rest of this section. Construct a net N where:

- *Places* are assertions in the $\oplus$-free fragment.
- *Transitions* are pairs (M, M') of multisets of places for which $\widehat{M} \vdash \widehat{M'}$.

Lemma 9 *For markings M, M' of the net, $M \rightarrow M'$ in the net iff M, M' $\oplus$-free and $M \vdash M'$ in the logic.*

Lemma 10 (Decomposition lemma). *For any $\multimap$-free assertion A there is a finite set I indexing $\multimap$-$\oplus$-free assertions M_i, such that*

$$A \dashv\vdash \bigoplus_{i \in I} M_i \ .$$

Proof The proof proceeds by structural induction on the assertion A. □

Lemma 11 *Let $\Gamma = B_1, \ldots, B_n$, possibly empty, be list of assumptions in the $\oplus$-free fragment above. Then,*

$$\Gamma \nvdash \mathsf{F} \quad \textit{and} \quad \textit{if } \Gamma \vdash C \oplus D \textit{ then } \Gamma \vdash C \textit{ or } \Gamma \vdash D.$$

Proof By cut-elimination any proof of a sequent can be replaced by a cut-free proof. The above lemma follows by induction or the size of cut-free proofs. □

Lemma 12 *For any $\multimap$-free assertion A, $[A]_N = \{M \mid M \text{ is } \oplus\text{-free}, \ M \vdash A\}$.*

Proof The proof proceeds by structural induction on the assertion A. □

Corollary 13 $\models_N A$ *iff* $\vdash A$, *for any $\multimap$-free assertion A.*

Thus we have completeness.

Because we only use the decomposition lemma (lemma 10) for the $\otimes$ case of the structural induction in lemma 12, we also get completeness for the larger fragment of assertions B given by:

$$B ::= A \mid A \multimap B \mid \top \mid \mathsf{F} \mid \mathbf{1} \mid a \mid B_1 \mathbin{\&} B_2 \mid B_1 \oplus B_2$$

where A lie in the $\multimap$-free fragment and a, as usual, ranges over atoms.

Lemma 14 *For the larger fragment, $[B]_N = \{M \mid M \ \oplus\text{-free}, \ M \vdash B\}$.*

Proof The proof proceeds by structural induction, as above in lemma 12, but for a new case where the assertion has the form $A \multimap B$. □

Corollary 15 *For the larger fragment, $\models B$ iff $\vdash B$ with the additional $\&$-$\oplus$-distributivity law.*

Theorem 16 *For the $\multimap$-free fragment, $\Gamma \models A$ iff $\Gamma \vdash A$ with the additional $\&$-$\oplus$-distributive law.*

Proof Corollary 15 gives $\models \bigotimes \Gamma \multimap A$ iff $\vdash \bigotimes \Gamma \multimap A$. Hence $\Gamma \models A$ iff $\Gamma \vdash A$. □

6 Quantification and Atomic Nets

Definition 17 *A net is* atomic *iff whenever $M \rightarrow \underline{0}$ then $\underline{0} \rightarrow M$, for any marking M.* □

This corresponds to $\mathbf{1}$ being atomic in the associated quantale—see the remark following the semantics of linear logic formulae in a net.

An interesting consequence of dealing with an atomic net N is, that whatever property we could state before in terms of validity of a closed formula A, can now be stated negatively as $\models_N A \mathbin{\&} \mathbf{1} \multimap \mathsf{F}$. Precisely:

Proposition 18 *For an atomic net N and a closed formula A, $\models_N A \mathbin{\&} \mathbf{1} \multimap \mathsf{F}$ iff $\not\models_N A$.*

Abbreviating $A \mathbin{\&} \mathbf{1} \multimap \mathsf{F}$ by $\sim\!A$ and combining this proposition with proposition 2 we can express that a marking M' cannot be reached from another M:

Corollary 19 *For multiset of atoms M and M', $M \not\rightarrow M'$ in an atomic net N iff $\models_N \sim\!(M \multimap M')$.*

Example 3 (continued)
We can now express that the processes, p_1 and p_2 cannot get into their critical regions at the same time. We might try $\models_N \sim\!(M_0 \multimap c_1 \otimes c_2)$. This is not quite right however, since $\models_N \sim\!(M_0 \multimap c_1 \otimes c_2)$ merely states that the two processes cannot be in their critical regions at the same time when no other tokens are present; the correct statement is $\models_N \sim\!(M_0 \multimap c_1 \otimes c_2 \otimes \top)$.

We are also able to express that a finite net N is 1-safe by $\models_N \sim\!(M_0 \multimap (\bigoplus_{a \in P} a \otimes a) \otimes \top)$. That a transition t is M-dead in a net N, i.e. $\forall M' \in [M\rangle.\ M' [t\!\!\!/\rangle$, is expressed by $\models_N \sim\!(M \multimap {}^\bullet t \otimes \top)$.

Notice that $A \mathbin{\&} \mathbf{1}$ plays the role of the exponential $!A$, and indeed according the net semantics, when the net N is atomic $[!A]_N = [A \mathbin{\&} \mathbf{1}]_N$.

Syntax

Assume a countable set of atoms. Define the assertions over the atoms to be:

$$A ::= \mathsf{T} \mid \mathsf{F} \mid 1 \mid a \mid x \mid A_1 \otimes A_2 \mid A_1 \multimap A_2 \mid A_1 \,\&\, A_2 \mid A_1 \oplus A_2 \mid \bigoplus_x A \mid \mathop{\&}_x A$$

where a ranges over the atoms and x ranges over countably many variables. The new constructions $\bigoplus_x A$ and $\&_x A$ are forms of existential and universal quantification and bind accordingly. We adopt the traditional notions of free and bound variable and in particular use $\mathrm{FV}(A)$ for the set of free variables in A, and more generally $\mathrm{FV}(A_1, \ldots, A_n)$ for $\mathrm{FV}(A_1) \cup \cdots \cup \mathrm{FV}(A_n)$. The variables x are to be thought of as standing for markings of a net.

Semantics

Given a net N, with markings (i.e. finite multisets of places) $\mathcal{M}$, a (marking) environment is a function ρ from variables to markings $\mathcal{M}$. Because of the presence of free variables we define the meaning of an assertion with respect to a marking environment. In particular,

$$\begin{aligned} [\![\bigoplus_x A]\!]_N\rho &= \textstyle\bigcup_{M\in\mathcal{M}} [\![A]\!]_N\rho[M/x], \\ [\![\&_x A]\!]_N\rho &= \textstyle\bigcap_{M\in\mathcal{M}} [\![A]\!]_N\rho[M/x], \\ [\![x]\!]_N\rho &= \{M \in \mathcal{M} \mid M \to \rho(x)\}. \end{aligned}$$

Atoms are interpreted as places of the net as in section 4 and similarly validity of a closed assertion A for the given net, N, can be expressed by:

$$\models_N A \quad \text{iff} \quad \underline{0} \in [\![A]\!]_N\rho\ .$$

This is generalized to open terms by taking the universal closure:

$$\models_N A \quad \text{iff} \quad \underline{0} \in [\![\mathop{\&}_{x_1} \cdots \mathop{\&}_{x_k} A]\!]_N\rho$$

where A has free variables $x_1, \ldots, x_k$ (here ρ can be arbitrary as $\&_{x_1} \cdots \&_{x_k} A$ is closed).

Let T be a subset of closed assertions in the original syntax. Define

$B_1, \ldots, B_n \models_T A$ iff for all atomic nets N such that $(\forall B \in T.\ \models_N B)$, $\models_N (B_1 \otimes \cdots \otimes B_n \multimap A)$.

Before proceeding with the proof rules we show how the new constructions can be used to express liveness. A transition t is life iff $\forall M \in [M_0\rangle \exists M' \in [M\rangle.\ M'\ [t\rangle$. This is expressed: $\models_N \&_x((M_0 \multimap x) \& 1 \multimap (x \multimap {}^\bullet t \otimes \mathsf{T}))$.

Proof rules

The proof rules are those of section 2 (without exponentials—they will become definable in the purely propositional logic), together with:

$$\frac{\Gamma \vdash A}{\Gamma[\theta] \vdash A[\theta]} \qquad \text{(Subst.)}$$

where θ is a substitution of *marking terms* (i.e. assertions built up from variables, atoms and 1 purely by $\otimes$)—the usual care to avoid capture of free variables applies here.

$$\frac{\Gamma \vdash A[M/x]}{\Gamma \vdash \bigoplus_x A} \ \text{where } M \text{ is a marking term} \quad (\vdash\oplus) \qquad\qquad \frac{\Gamma, B \vdash A}{\Gamma, \bigoplus_x B \vdash A}\ x \notin \mathrm{FV}(\Gamma, A) \quad (\oplus\vdash)$$

$$\frac{\Gamma \vdash A}{\Gamma \vdash \&_x A}\ x \notin \mathrm{FV}(\Gamma) \quad (\vdash\&) \qquad\qquad \frac{\Gamma, B \vdash A}{\Gamma, \&_x B \vdash A} \quad (\&\vdash)$$

In addition we have the following axioms valid of nets:

$$(A_1 \oplus A_2) \,\&\, B \vdash (A_1 \,\&\, B) \oplus (A_2 \,\&\, B) \qquad (\&\text{-}\oplus\text{-dist.})$$

$$(\bigoplus_x A) \,\&\, B \vdash \bigoplus_x (A \,\&\, B) \text{ where } x \notin FV(B) \qquad (\&\text{-}\bigoplus\text{-dist.})$$

In fact, in the presence of the atomicity, basis and primeness axioms, these distributivity laws are derivable from those in the special case where B is $\mathbf{1}$. The other distributive law, $\&_x(A \oplus B) \vdash (\&_x A) \oplus B$ where $x \notin FV(B)$, is also derivable (for general B).

$$\vdash (\&_x B) \oplus \bigoplus_x ((B \,\&\, \mathbf{1}) \multimap \mathsf{F}) \qquad \text{(Atomicity)}$$

These entail sequents of the following form (by taking the variable x to not appear in B): $\vdash B \oplus ((B \,\&\, \mathbf{1}) \multimap \mathsf{F})$. These hold because in an atomic net the denotation of a formula $B \,\&\, \mathbf{1}$, in an environment for its free variables, only has two possibilities, to be the denotation of F or the denotation of $\mathbf{1}$.

$$A \vdash \bigoplus_x x \otimes ((x \multimap A) \,\&\, \mathbf{1}) \quad \text{where } x \notin FV(A) \qquad \text{(Basis)}$$

These hold in an atomic net because there an assertion is denoted by a set of markings; notice how the expression $(x \multimap A) \,\&\, \mathbf{1}$ is equivalent to $\mathbf{1}$ in the case the marking x satisfies A and F otherwise, so the effect in the whole expression is only to make a contribution of x when this satisfies A.

$$(x \multimap \mathsf{F}) \vdash \mathsf{F}$$
$$(x \multimap B \oplus C) \vdash (x \multimap B) \oplus (x \multimap C) \qquad \text{(Primeness)}$$
$$(x \multimap \bigoplus_y A) \vdash \bigoplus_y (x \multimap A) \text{ where } y \text{ and } x \text{ are distinct.}$$

These axioms hold because if a marking is contained in union, denoting a disjunction, then it is clearly in a component of the union.

The soundness of the basis and atomicity axioms follows from the fact that, in an atomic net,

$$[\![A \,\&\, \mathbf{1}]\!]_N \rho = \begin{cases} \mathbf{1} & \text{if } \mathbf{1} \subseteq [\![A]\!]_N \rho \\ \mathsf{F} & \text{otherwise} \end{cases} \quad \text{and} \quad [\![A \,\&\, \mathbf{1} \multimap \mathsf{F}]\!]_N \rho = \begin{cases} \mathsf{F} & \text{if } \mathbf{1} \subseteq [\![A]\!]_N \rho \\ \mathsf{T} & \text{otherwise.} \end{cases}$$

We have already remarked that in an atomic net, an exponential $!A$ is represented by $A \,\&\, \mathbf{1}$. In fact from the atomicity axioms and the rules for exponentials, there is a fairly direct proof of their equivalence, yielding

$$!A \dashv\vdash A \,\&\, \mathbf{1}$$

—the syntax of exponentials can be eliminated in favour of the purely propositional connectives.

In constructing prime theories we follow Henkin and extend the original syntax to include new atoms drawn from

$$c_0, c_1, \ldots, c_n, \ldots$$

a countably infinite enumeration of atoms not already present in the syntax. Suppose C is a subset of $\{c_i \mid i \in \omega\}$. Suppose Γ, A are assertions from the syntax extended by C, and that $\mathcal{F}$ is a theory (i.e. a subset of assertions) of the extended syntax. We use

$$\Gamma \vdash^C_{\mathcal{F}} A$$

to mean the sequent is provable in the proof system for the extended syntax, using the assertions in $\mathcal{F}$ as axioms. A judgment $\Gamma \vdash A$ means a sequent is provable in the proof system of the original assertion language, without extra atoms.

Lemma 20 *Let B be a closed assertion and $\mathcal{F}$ a theory in a syntax extended by atoms $C \subseteq \{c_i \mid i \in \omega\}$. Then*

$$\Gamma \vdash^C_{\mathcal{F} \cup \{B\}} A \quad \textit{iff} \quad \Gamma, B \,\&\, \mathbf{1} \vdash^C_{\mathcal{F}} A \ .$$

Definition 21 *Let $C \subseteq \{c_i \mid i \in \omega\}$. A subset $\mathcal{F}$ of closed assertions, in the syntax extended by atoms C, is called a* prime theory *iff*

(i) $\mathsf{F} \notin \mathcal{F}$,
$A_1 \oplus A_2 \in \mathcal{F} \Rightarrow A_1 \in \mathcal{F}$ *or* $A_2 \in \mathcal{F}$,
$\bigoplus_x A \in \mathcal{F} \Rightarrow A[M/x] \in \mathcal{F}$, *for some (necessarily closed) marking term M.*

(ii) $\mathcal{F}$ *is deductively closed, i.e. A closed and* $\vdash^C_{\mathcal{F}} A \Rightarrow A \in \mathcal{F}$.

□

Lemma 22 (Existence of prime theories). *Let A be an assertion and T a subset of closed assertions in the original syntax, for which*

$$\not\vdash_T A \ .$$

Then, there is a prime theory $\mathcal{F}$, consisting of assertions over the syntax extended by some $C \subseteq \{c_i \mid i \in \omega\}$, such that

$$T \subseteq \mathcal{F} \text{ and } \not\vdash^C_{\mathcal{F}} A \ .$$

Proof As the atoms and variables form countable sets we can enumerate all the assertions

$$A_0, A_1, \ldots, A_n, \ldots$$

of the syntax extended by atoms $\{c_i \mid i \in \omega\}$.

By induction, for $n \in \omega$, we define a chain of deductively - closed theories $\mathcal{F}_n$ with new atoms C_n, such that

$$T \subseteq \mathcal{F}_n \text{ and } \not\vdash^{C_n}_{\mathcal{F}_n} A \ .$$

Take $\mathcal{F}_0 = \{B \mid \vdash_T B\}$. Clearly $T \subseteq \mathcal{F}_0$ and, by assumption, $\not\vdash_{\mathcal{F}_0} A$.

Assuming $\mathcal{F}_n$ is deductively closed, includes T, and $\not\vdash^{C_n}_{\mathcal{F}_n} A$, define $\mathcal{F}_{n+1}$ according to the following cases:

(i) n is even, and there is an assertion $B_1 \oplus B_2 \in \mathcal{F}_n$ with $B_1 \notin \mathcal{F}_n, B_2 \notin \mathcal{F}_n$.

(ii) n is odd, and there is an assertion $\bigoplus_x B \in \mathcal{F}_n$ with $B[M/x] \notin \mathcal{F}_n$ for any closed marking terms M in the syntax over atoms extended by C_n.

(iii) neither (i) nor (ii) applies.

In case (iii), define $C_{n+1} = C_n$ and $\mathcal{F}_{n+1} = \mathcal{F}_n$.

In case (i), take the earliest assertion in the enumeration $B_1 \oplus B_2 \in \mathcal{F}_n$ and $B_1 \notin \mathcal{F}_n$ and $B_2 \notin \mathcal{F}_n$. As $\mathcal{F}_n$ is deductively-closed, $(B_1 \oplus B_2) \,\&\, 1 \in \mathcal{F}_n$, so

$$(B_1 \,\&\, 1) \oplus (B_2 \,\&\, 1) \in \mathcal{F}_n$$

by the &-⊕-distributivity law. Suppose

$$\vdash^{C_n}_{\mathcal{F}_n, B_1} A \text{ and } \vdash^{C_n}_{\mathcal{F}_n, B_2} A \ .$$

Then, by lemma 20,

$$B_1 \,\&\, 1 \vdash^{C_n}_{\mathcal{F}_n} A \text{ and } B_2 \,\&\, 1 \vdash^{C_n}_{\mathcal{F}_n} A \ .$$

Hence $(B_1 \,\&\, 1) \oplus (B_2 \,\&\, 1) \vdash^{C_n}_{\mathcal{F}_n} A$. But this implies $\vdash^{C_n}_{\mathcal{F}_n} A$, a contradiction. Thus

$$\not\vdash^{C_n}_{\mathcal{F}_n, B_1} A \text{ or } \not\vdash^{C_n}_{\mathcal{F}_n, B_2} A \ .$$

Supposing, for instance, $\not\vdash^{C_n}_{\mathcal{F}_n, B_1}, A$, take $\mathcal{F}_{n+1}$ to be $\{D \text{ closed} \mid \vdash^{C_n}_{\mathcal{F}_n, B_1}, D\}$ and $C_{n+1} = C_n$.

In case (ii), take the earliest, according to the enumeration, $\bigoplus_x B \in \mathcal{F}_n$ for which $B[M/x] \notin \mathcal{F}_n$ for all marking terms M, and where x is not a free variable of A. As $\mathcal{F}_n$ is deductively-closed, $(\bigoplus_x B) \,\&\, 1 \in \mathcal{F}_n$, so

$$\bigoplus_x (B \,\&\, 1) \in \mathcal{F}_n \ .$$

Let c be the first new atom in the list $c_0, c_1, \ldots$ which is not in C_n. Define $C_{n+1} = C_n \cup \{c\}$ and $\mathcal{F}_{n+1}$ to consist of all closed assertions in the deductive closure of $\mathcal{F}_n \cup \{B[c/x]\}$, i.e.

$$\mathcal{F}_{n+1} = \{D \text{ closed} \mid \vdash^{C_{n+1}}_{\mathcal{F}_n, B[c/x]} D\} \ .$$

We must check that $\not\vdash^{C_{n+1}}_{\mathcal{F}_{n+1}} A$. To this end, assume otherwise, that $\vdash^{C_{n+1}}_{\mathcal{F}_n, B[c/x]} A$. Then, by lemma 20,

$$B[c/x] \,\&\, 1 \vdash^{C_{n+1}}_{\mathcal{F}_n} A \ .$$

As c does not appear in C_n or $\mathcal{F}_n$,

$$B \,\&\, 1 \vdash^{C_n}_{\mathcal{F}_n} A \ .$$

To obtain the proof of this sequent, replace all occurrences of the new atom c in the proof of $B[c/x] \& 1 \vdash^{C_{n+1}}_{\mathcal{F}_n} A$ by a new variable—one which does not appear anywhere in the proof—and finally use (Subst.) to replace this variable by x using the fact that renaming bound variables preserves logical equivalence. But now we can deduce

$$\bigoplus_x (B \,\&\, 1) \vdash^{C_n}_{\mathcal{F}_n} A \ .$$

But $\bigoplus_x(B \,\&\, 1) \in \mathcal{F}_n$ and $\mathcal{F}_n$ is deductively-closed making $\vdash^{C_n}_{\mathcal{F}_n} A$, a contradiction. Thus $\nvdash^{C_{n+1}}_{\mathcal{F}_{n+1}} A$, as required.

In this way, we inductively define a chain of theories $\mathcal{F}_n$ over the syntax extended by C_n, such that

$$C_n \subseteq C_{n+1} \text{ and } \mathcal{F}_n \subseteq \mathcal{F}_{n+1},$$

with $T \subseteq \mathcal{F}_0$. Finally take $C = \bigcup_{n \in \omega} C_n$ and $\mathcal{F} = \bigcup_{n\in\omega} \mathcal{F}_n$ to form the required prime theory. □

Assume a prime theory $\mathcal{F}$ with additional atoms C. Construct a net N from $\mathcal{F}$ by taking:

- *Places* to be the original those atoms, including those of C.
- *Transitions* as those pairs (M, M') of multisets of places for which $\widehat{M} \vdash^{C}_{\mathcal{F}} \widehat{M'}$.

We use $\mathcal{M}$ to represent the set of all markings of the net N. (Note the markings coincide with the closed marking terms of $\mathcal{F}$.)

Lemma 23 *N is an atomic net for which $M \to M'$ in N iff $M \vdash^{C}_{\mathcal{F}} M'$, for $M, M' \in \mathcal{M}$.*

We need the following facts:

Lemma 24

(i) Let B be an assertion with $\mathrm{FV}(B) \subseteq \{x\}$. Let the assertions Γ not include x as a free variable. Then

$$(\forall M \in \mathcal{M}.\ \Gamma \vdash^{C}_{\mathcal{F}} B[M/x]) \Rightarrow \Gamma \vdash^{C}_{\mathcal{F}} \mathop{\&}_x B \ .$$

(ii) Let B be an assertion with $\mathrm{FV}(B) \subseteq \{x_1, \ldots, x_k\}$, and Γ be assertions in which $x_1, \ldots, x_k$ are not free. Then

$$(\forall M_1, \ldots, M_k \in \mathcal{M}.\ \Gamma \vdash^{C}_{\mathcal{F}} B[M_1/x_1, \ldots, M_k/x_k]) \Rightarrow \Gamma \vdash^{C}_{\mathcal{F}} B \ .$$

Lemma 25 *For assertions Γ, B and A suppose $\Gamma, B[M/x] \vdash^{C}_{\mathcal{F}} A$ for all $M \in \mathcal{M}$. Assume $\mathrm{FV}(B) \subseteq \{x\}$. Then*

$$\Gamma, \bigoplus_x B \vdash^{C}_{\mathcal{F}} A \ .$$

Now we can relate semantics in the net N to provability in the prime theory $\mathcal{F}$:

Lemma 26 *For any assertion A, for any marking environment ρ, $[\![A]\!]_N \rho = \{M \in \mathcal{M} \mid M \vdash^{C}_{\mathcal{F}} A[\rho]\}$.*

Proof By structural induction on A. We show one case:

$A \equiv A_1 \multimap A_2$: $M \in [\![A_1 \multimap A_2]\!]_N \rho \Leftrightarrow \forall M_1 \in [\![A_1]\!]_N \rho.\ M + M_1 \in [\![A_2]\!]_N \rho.$
$\Leftrightarrow \forall M_1 \vdash_{\mathcal{F}} A_1[\rho].\ M \otimes M_1 \vdash^{C}_{\mathcal{F}} A_2[\rho]$ by induction,
$\Leftrightarrow \forall M_1 \in \mathcal{M}.\ M \otimes M_1 \otimes ((M_1 \multimap A_1[\rho]) \,\&\, 1) \vdash^{C}_{\mathcal{F}} A_2[\rho]$

where the last equivalence relies on atomicity and the fact that $\mathcal{F}$ is a prime theory. In more detail, writing $A_1' \equiv A_1[\rho], A_2' \equiv A_2[\rho]$, we have

$$\text{(i)} \vdash^{C}_{\mathcal{F}} M_1 \multimap A_1' \quad \text{or} \quad \text{(ii)} \vdash^{C}_{\mathcal{F}} (M_1 \multimap A_1') \,\&\, 1 \multimap \mathsf{F}$$

for any $M_1 \in \mathcal{M}$. For case (i), $(M_1 \multimap A_1') \,\&\, 1 \dashv\vdash^{C}_{\mathcal{F}} 1$. In case (ii), $(M_1 \multimap A_1') \,\&\, 1 \dashv\vdash^{C}_{\mathcal{F}} \mathsf{F}$.

It follows, by considering the two cases, that for any $M \in \mathcal{M}'$

$$(M_1 \vdash^{C}_{\mathcal{F}} A'_1 \Rightarrow M \otimes M_1 \vdash^{C}_{\mathcal{F}} A'_2) \quad \text{iff} \quad M \otimes M_1 \otimes ((M_1 \multimap A'_1) \,\&\, \mathbf{1}) \vdash^{C}_{\mathcal{F}} A'_2 \ .$$

Now note that by lemma 25, $\forall M_1 \in \mathcal{M}.\ M \otimes M_1 \otimes ((M_1 \multimap A'_1) \,\&\, \mathbf{1}) \vdash^{C}_{\mathcal{F}} A'_2$

$\Leftrightarrow M \otimes \bigoplus_{x_1} x_1 \otimes ((x_1 \multimap A'_1) \,\&\, \mathbf{1})) \vdash^{C}_{\mathcal{F}} A'_2,$

$\Leftrightarrow M \otimes A'_1 \vdash^{C}_{\mathcal{F}} A'_1$ as $A'_1 \dashv\vdash^{C}_{\mathcal{F}} \bigoplus_x x_1 \otimes ((x_1 \multimap A'_1) \,\&\, \mathbf{1}),$

$\Leftrightarrow M \vdash^{C}_{\mathcal{F}} A'_1 \multimap A'_2$ i.e. $M \vdash^{C}_{\mathcal{F}} A_1[\rho] \multimap A_2[\rho]$, as required. □

Theorem 27 (Completeness).
Let A, B and $\Gamma \equiv B_1, \ldots, B_n$ be assertions and T consist of closed assertions in the original syntax. Then,

$$\Gamma \models_T A \quad \textit{iff} \quad \Gamma \vdash_T A \ .$$

Proof As $\multimap$ is present as a constructor on assertions it is clearly sufficient to show

$$\models_T A \text{ iff } \vdash_T A \ .$$

The "if" direction is shown by induction on the proof of $\vdash_T A$. To show the "only if" direction we prove its contraposition:

$$\nvdash_T A \Rightarrow \not\models_T A \ .$$

Suppose A is an assertion with free variables $x_1, \ldots, x_k$. Suppose $\nvdash_T A$. Then there is a prime theory $\mathcal{F} \supseteq T$ over additional constants C such that $\nvdash^{C}_{\mathcal{F}} A$. Let N be the net constructed from $\mathcal{F}$; let $\mathcal{M}$ be the set consisting of its markings. As we now show, $\not\models_N A$. Suppose otherwise, that $\models_N A$. Recall

$$[\![A]\!]_N \rho = \{M \in \mathcal{M} \mid M \vdash^{C}_{\mathcal{F}} A[\rho]\}$$

by lemma 26. Hence as $\models_N A$ means $\mathbf{0} \in [\![A]\!]_N \rho$ for all environments ρ, we see $\vdash^{C}_{\mathcal{F}} A[\rho]$ for all environments. Therefore $\vdash^{C}_{\mathcal{F}} A[M_1/x_1, \ldots, M_k/x_k]$ for all $M_1, \ldots, M_k \in \mathcal{M}$. Hence by (ii) of lemma 24, $\vdash^{C}_{\mathcal{F}} A$, a contradiction. Thus N is an atomic net satisfying all axioms of the theory T and yet $\not\models_N A$. Hence $\not\models_T A$, as required. □

References

[Bro89] Carolyn Brown. Relating Petri Nets to Formulae of Linear Logic. Technical Report ECS LFCS 89-87, University of Edinburgh, 1989.

[EW90] Uffe Henrik Engberg and Glynn Winskel. Petri Nets as Models of Linear Logic. In *CAAP '90, Coll. on Trees in Algebra and Programming Copenhagen, Denmark, May 15–18*, pages 147–161. Springer-Verlag (*LNCS* 431), 1990.

[EW93] Uffe Henrik Engberg and Glynn Winskel. Completeness Results for Linear Logic on Petri Nets. Technical Report DAIMI-PB 435, Department of Computer Science, Aarhus University, April 1993.

[GG89] Carl Gunter and Vijay Gehlot. Nets as Tensor Theories. Technical Report MS-CIS-89-68, University of Pennsylvania, October 1989.

[Gir87] Jean-Yves Girard. Linear Logic. *Theoretical Computer Science*, 50(1):1–102, 1987.

[GL87] Jean-Yves Girard and Yves Lafont. Linear Logic and Lazy Computation. In *Proc. TAPSOFT 87 (Pisa), vol. 2*, pages 52–66. Springer-Verlag (*LNCS* 250), 1987.

[Laf88] Yves Lafont. The Linear Abstract Machine. *Theoretical Computer Science*, 59:157–180, 1988.

[MOM91] Narciso Martí-Oliet and José Meseguer. From Petri Nets to Linear Logic: a Survey. *International Journal of Foundations of Computer Science*, 2(4):297–399, 1991.

[Rei85] Wolfgang Reisig. *Petri Nets, An Introduction*, volume 4 of *EATCS Monographs on Theoretical Computer Science*. Springer-Verlag, 1985.

An Expressive Logic for Basic Process Algebra

A. Fantechi[1], S. Gnesi[2] and V. Perticaroli[3]

[1] Dipatrtimento di Ingegneria dell'Informazione, Università di Pisa, Italy
[2] IEI-CNR, Pisa, Italy
[3] SIP, Bologna, Italy

Abstract. In this paper we present a branching time temporal logic for Basic Process Algebra. This logic is an enrichment of the branching temporal logic CTL with a branching sequential composition operator, named *chop branching*. The logic so obtained is proved to be expressive with respect to the bisimulation semantics defined on $BPA_{\delta,rec}$ terms, and is able to describe context-free properties of systems.

1 Introduction

In this paper we present a temporal logic for Basic Process Algebra $_{\delta,\,rec}$, hereafter $BPA_{\delta,rec}$, a simplified version of the Algebra of Communicating Processes, which has the process constant δ and the operators of sequential composition, alternative composition and recursion as primitives [4,5]. A bisimulation equivalence relation is defined on $BPA_{\delta,rec}$ terms; in accordance with this, our aim is to give a logic for $BPA_{\delta,rec}$ which is expressive with respect to bisimulation equivalence, i.e. a logic which totally agrees with the overall behaviour of a process, modulo bisimulation.

The logic for $BPA_{\delta,rec}$ hereafter called $\nu CTL + C_Q$, will be defined starting from the branching time temporal logic CTL [8] enriching it with a maximal fixed point operator and with a new logic operator, *chop branching*, an extension of the linear chop operator defined in [3].

This extension is needed because in [10] we had proved that CTL, even enriched with a fixed point operator (thus becoming equipotent to μ-calculus [16]) is not expressive enough to give the meaning to a non-idempotent composition operator, such the sequential composition operator of $BPA_{\delta,rec}$.

Work has already been done to define expressive logics for processes algebras and our work extends the results of [11], to cover more than regular processes: In fact the sequential composition operator permits context-free processes to be expressed and accordingly, $\nu CTL+C_Q$ enables context-free properties of systems to be described.

The paper is organized as follows: Section 2 describes $BPA_{\delta,rec}$; Section 3 presents its operational semantics; then, in Section 4, we define the logic $\nu CTL+$

The third author contributed to the paper when she was a student at University of Pisa

C_Q which is shown to be adequate and expressive for $BPA_{\delta,rec}$ with respect to bisimulation equivalence in Section 5.

2 Basic Process Algebra$_{\delta,\ rec}$

Signature:

sorts $\mathcal{A}$—set of atomic actions
$\mathcal{P}$—set of processes; $\mathcal{A} \subseteq \mathcal{P}$
Var—infinite numerable set of process variables; $Var \subseteq \mathcal{P}$
functions $+ : \mathcal{P} \times \mathcal{P} \to \mathcal{P}$—alternative composition
$\cdot : \mathcal{P} \times \mathcal{P} \to \mathcal{P}$—sequential composition
constants $\delta \in \mathcal{A}$—deadlock

Basic Process Algebra$_{\delta,\ rec}$—$BPA_{\delta,rec}$— starts from a collection $\mathcal{A} = \{a, b, c, \ldots\}$ of given objects, called atomic actions [4,5]. These actions usually have no duration and form the basic building blocks of our systems.

The two compositional operators we consider are "$\cdot$", denoting sequential composition, and "+" for alternative composition. If x and y are two processes, then $x \cdot y$ is the process that starts the execution of y after the completion of x, and $x + y$ is the process that chooses to execute either x or y.

A vital element in the present set-up of process algebra is the process δ, signifying "deadlock". The process $a \cdot b$ performs its two steps and then stops, silently and happily; but the process $a \cdot b \cdot \delta$ deadlocks after the actions a and b: it wants to do a proper action but it cannot.

Let us now give an operational characterization of the semantics of $BPA_{\delta,rec}$ For the sake of simplicity, we derive from the signature of $BPA_{\delta,rec}$ a language of behaviour expressions which define processes of the process algebra, with the syntax given in Table 1, in which we explicit recursion through the use of a *rec* operator; B, B_1, B_2 are generic behaviour expressions, X is a recursion variable and B in $recX.B$ is a guarded expression containing X.

$BPA_{\delta,rec}$ can express context-free processes by means of the sequential composition operator and the recursion; for example, the term:

$$recX.(a \cdot X \cdot b + a \cdot b);$$

defines the context free process which performs equal numbers of a and b.

The operational semantics for this language is given in two steps: first a Labelled Transition System (hereafter LTS) is associated with a process, then an equivalence relation is defined on the transition systems [12].

A LTS is a triple $(\mathcal{P}, \sum, \{R_x, x \in \sum\})$ such that:

- $\mathcal{P}$ is a set of processes,
- $\sum$ is a set of actions,
- $R_x \subseteq \mathcal{P} \times \mathcal{P}$.

We will use the notation $p - g \longrightarrow q$ to mean that $(p, q) \in R_g$ and we will say that p is able to perform action g and transform in q. We will also use p, q (with indexes) as ranging over processes.

In Table 1 the operational semantics of $\text{BPA}_{\delta,rec}$ is shown, where the transitions of processes are labelled by the atomic actions in a set $\mathcal{A}$ (marked as a) or by the successful termination action $\surd$, not user definable; therefore, we have $\sum = \mathcal{A} \cup \{\surd\}$. The notation $[x_1/y_1, \ldots, x_n/y_n]$ means that y_i is replaced by x_i for $i = 1, \ldots, n$ and $a\surd$ means that an action a or $\surd$ is performed by the process.

A simple way to represent an LTS is by means of a tree, called *derivation tree* [17], which provides all its possible execution sequences. Because recursive processes can be defined in $\text{BPA}_{\delta,rec}$, the derivation trees can be of infinite depth; however, because the processes have only guarded recursion, these trees will be finitely branching. Several equivalence relations are defined on LTSs ; among them we consider the *bisimulation equivalence* [18] and so we will refer to the operational semantics of $\text{BPA}_{\delta,rec}$ as bisimulation semantics.

3 A Branching Temporal Logic for $\text{BPA}_{\delta,\, rec}$

As we are interested in reasoning in a temporal logic $\mathcal{L}$ on terms of a process language $\mathcal{P}$, with an associated operational semantics, we need to associate the sets of properties (formulae in $\mathcal{L}$) with processes in $\mathcal{P}$ which they satisfy. This entails defining the satisfaction relation $\models \subseteq \mathcal{P} \times \mathcal{L}$, written $p \models \Phi$ and read "p satisfies the property Φ". The definition of this relation induces an equivalence $\equiv_{\mathcal{L}}$ between processes which enjoys the same properties. Formally, we define:

$$p \equiv_{\mathcal{L}} q \text{ if and only if } \mathcal{F}(p) = \mathcal{F}(q)$$

where:

$$\mathcal{F}(p) = \{\Phi : \Phi \in \mathcal{L} \wedge p \models \Phi\}$$

Now, if $\approx \subseteq \mathcal{P} \times \mathcal{P}$ is the equivalence defined on the operational semantics of the process language $\mathcal{P}$, it is possible to define a relation between $\approx$ and $\equiv_{\mathcal{L}}$:

A logic $\mathcal{L}$ is ***adequate*** [20] w.r.t. an equivalence ($\approx$) defined on a given process language $\mathcal{P}$, if for every pair of processes p and q:

$$p \equiv_{\mathcal{L}} q \text{ if and only if } p \approx q.$$

In [6] it is shown that logic CTL [8] is adequate for strong bisimulation on Kripke structures. We can extend this result to strong bisimulation on $\text{BPA}_{\delta,rec}$ by relating the models of the logic and Labelled Transitions Systems: they differ only because in the former labels are associated with states while in the latter they are associated with arcs. We can transform each other by shifting the information from states to arcs or viceversa [15]. Therefore, CTL is adequate w.r.t. strong bisimulation on $\text{BPA}_{\delta,rec}$ terms. This means that it is always possible to write a CTL formula which distiguishes two non-bisimilar $\text{BPA}_{\delta,rec}$processes.

Adequacy, however, relates process languages and logics weakly: in fact CTL is not able totally to agree with the behaviour of a process, whose expressive

Table 1. Operational Semantics of $BPA_{\delta,rec}$

Syntax	*Operator*	*Operational semantics*	*Informal meaning*
δ	deadlock		denotes a process which cannot perform any action
ϵ	successful termination	$\epsilon - \surd \rightarrow \delta$	models the process able only to emit the successful termination signal $\surd$
a	action	$a - a \rightarrow \epsilon$	models a process which can perform the action a
$B_1 + B_2$	alternative composition	$\frac{B_1 - a\surd \rightarrow B_1'}{B_1 + B_2 - a\surd \rightarrow B_1'}$ $\frac{B_2 - a\surd \rightarrow B_2'}{B_1 + B_2 - a\surd \rightarrow B_2'}$	the actions of the process are the set of possible actions of B_1 and B_2
$B_1 \cdot B_2$	sequential composition	$\frac{B_1 - a \rightarrow B_1'}{B_1 \cdot B_2 - a \rightarrow B_1' \cdot B_2}$ $\frac{B_1 - \surd \rightarrow \delta \quad B_2 - a\surd \rightarrow B_2'}{B_1 \cdot B_2 - a\surd \rightarrow B_2'}$ $\frac{B_1 - a \rightarrow \delta}{B_1 \cdot B_2 - a \rightarrow \delta}$	this operator allows sequential process composition
$recX.B$	recursion	$\frac{B[recX.B/X] - a\surd \rightarrow B'}{recX.B - a\surd \rightarrow B'}$	the actions of the process are those of the behaviour expression B which replaces the recursion variable X with $recX.B$

power is higher than regular processes [9]. Indeed CTL is strictly contained into μ-calculus which can at most state regular properties of systems. This implies that the notion of adequacy of a logic with respect to a process language does not relate their expressive power.

With respect to this, we can define a stronger relation between a process language and a logic: informally a logic $\mathcal{L}$ is fully expressive (a refinement of the *expressiveness* relation given in [20]) w.r.t. an equivalence relation $\approx$, defined on a process language $\mathcal{P}$, if and only if it is possible to define a semantics $\mathcal{M}$ for $\mathcal{P}$ such that the denotation domain of $\mathcal{M}$ is $\mathcal{L}$, and $\mathcal{M}$ is fully abstract with respect to the operational semantics which define $\approx$.

A logic $\mathcal{L}$ is *fully expressive* w.r.t. an equivalence relation ($\approx$) defined on a process language $\mathcal{P}$, if it is adequate and if for every process $p \in \mathcal{P}$ there exists a *denotational logic semantics* $\mathcal{M}(p) \in \mathcal{L}$ such that:

$$\mathcal{M}(p) \equiv \mathcal{M}(q) \text{ if and only if } p \approx q$$

The full expressiveness relation permits us to associate a temporal logic formula (*temporal semantics* or *characteristic formula* [19,21]) with every process, and therefore the expressive power of the target logic should be at least equal to that of the process language.

CTL is thus extended with the addition of a maximal fixed point constructor and a new logic operator, *chop branching* C_Q (more precisely, C_Q is an indexed family of operators). The first operator, the fixed point, is needed to express recursion; note that $CTL + \nu$ has the same expressive power as μ-calculus, i.e. regular processes; to reach the expressive power of context-free processes we define the *chop branching* operator. This operator is an extension of the linear chop operator defined in [3] to express sequential composition in a linear temporal logic. We will call the logic so obtained $\nu CTL + C_Q$.

The atomic formulae of the logic are borrowed from the action set $\mathcal{A} \cup \{\surd\} \cup \{false, true\}$ (we will call them action predicates). The operators that we will use are the usual first order logic connectives: $\neg, \vee, \wedge$; together with the usual temporal operators: $AX, EX, AX!, EX!$, a maximal fixed point constructor: $\nu x.\Phi(x)$, and the new logic operator chop branching: C_Q.

In order to define the semantics of temporal logic formulae, we need to introduce the concept of a model. A model is a 4-uple (S, s_0, R, L), which define a Kripke structure M, where:

- S is a non-empty set of states;
- $s_0 \in S$ is the initial state or "root" of S;
- R is a relation on S which define the structure of the model; the properties of R are:
  (i) $\forall s \in S$, $(s,s) \notin R$ and $(s, s_0) \notin R$;
  (ii) $\forall s, s' \in S$, $(s,s') \in R \Rightarrow (\forall s'' \in S - \{s\}$, $(s'', s') \notin R)$;
  (iii) $\forall s \in S - \{s_0\}, \exists s_1, \ldots, s_n \in S$ such that $(s_0, s_1), (s_1, s_2), \ldots, (s_{n-1}, s_n), (s_n, s) \in R$;
- $L : S \longrightarrow 2^{\text{Prop}}$ is a mapping from S to the powerset of Prop (the set of atomic propositions, Prop $= \mathcal{A} \cup \{\surd\} \cup \{true, false\}$).

The formulae of temporal logic are interpreted on the states of a Kripke structure M; by $s \models_M \Phi$ we indicate that a state s in a model M satisfies a formula Φ.

Informally, $EX\,\Phi$ means that *if next states exist* then a *next state exists* which satisfies Φ. The formula $EX!\Phi$ means that a *next state exists* which satisfies Φ. Moreover, $AX\,\Phi$ means that *if next states exist* then *all next states* satisfy Φ; the formula $AX!\Phi$ means that a *next state exists* and *all next states* satisfy Φ. In addition every state satisfies *true* and no state satisfies *false*.

The maximal fixed point constructor $\nu x.\Phi(x)$ denotes the maximal solution to $x \Rightarrow \Phi(x)$, which exists if the function $\Phi(x)$ is monotonic. The monotonicity requirement can be ensured by the appearance of x in Φ under an even number of negations [2].

The chop branching operator is an extension, in branching temporal logic, of the chop operator introduced in [3] for linear temporal logic. The formula $\Phi C_Q \Psi$, where Q is an atomic proposition, holds, for a state s of a model M, if the submodel of M whose root is s is such that:

- there exists a family of states $\{s_i\}_{i\in I}$ of M reachable from s, all satisfying Ψ;
- if all the submodels $\{M_i\}_{i\in I}$ having as roots the states $\{s_i\}_{i\in I}$ are removed from M, and substituted by the final states $\{f_i\}_{i\in I}$ all satisfying Q, then we obtain a new model M_0 with root s, in which s satisfies Φ.

Now we show formally, by means of the satisfaction relation, the meaning of the $\nu CTL + C_Q$ operators.

• propositional connectives

$s \models_M Q$ iff $Q \in L(s)$
$s \models_M \neg\Phi$ iff not $(s \models_M \Phi)$
$s \models_M \Phi \vee \Psi$ iff $(s \models_M \Phi)$ or $(s \models_M \Psi)$
$s \models_M \Phi \wedge \Psi$ iff $(s \models_M \Phi)$ and $(s \models_M \Psi)$

• branching temporal operators

$s \models_M AX\Phi$ iff $R(s) = \{\}$ or for all $s' \in R(s)$, $s' \models_M \Phi$
$s \models_M EX\Phi$ iff $R(s) = \{\}$ or exists $s' \in R(s) : s' \models_M \Phi$
$s \models_M AX!\Phi$ iff $R(s) \neq \{\}$ and for all $s' \in R(s)$, $s' \models_M \Phi$
$s \models_M EX!\Phi$ iff $R(s) \neq \{\}$ and exists $s' \in R(s) : s' \models_M \Phi$

• maximal fixed point constructor

$s \models_M \nu x.\Phi(x)$ iff exists $\phi \in \{\psi \mid \psi \Rightarrow \Phi(\psi)\}$ such that $s \models_M \phi$

• chop branching

$s \models_M \Phi C_Q \Psi$ iff a possibly empty set of indexes I exists, a model

$$\mathrm{M}_0 = (S_0 \cup \{s_i\}_{i\in I} \cup \{f_i\}_{i\in I}, s, R_0, L_0)$$

and a family of submodels, of cardinality equal to the number of elements of I,

$$\{\mathrm{M}_i\}_{i\in I} = \{(S_i, s_i, R_i, L_i)\}_{i\in I}$$

such that:

- $s \models_{M_0} \Phi \ \wedge_{i\in I} \ s_i \models_{M_i} \Psi$
- $S \supseteq S_0 \cup \{s_i\}_{i\in I}$;
- $S \cap \{f_i\}_{i\in I} = \{\}$;
- $\neg\exists s' \in S_0 \cup \{s_i\}_{i\in I} : (s_i, s') \in R_0$ for all $i \in I$;

- $\forall s' \in S$ such that s' is reachable from s by means of R then either $s' \in S_0 \cup \{s_i\}_{i\in I}$ or $s' \in S_i$ for some $i \in I$. The f_i states, labelled by Q, are not in M, since their place is taken by the whole M_i models.
- $R_0 = \{(s', s'') \in R : s', s'' \in S_0 \cup \{s_i\}_{i\in I}\} \cup \{(s_i, f_i), \text{ for } i \in I\}$;
(the last two items imply that from the states s_i of M_0 we can move only to the states f_i);
- $L(s') \neq Q$ if $s' \in S_0 \cup \{s_i\}_{i\in I}$;
- $L_0(s') = \begin{cases} L(s') \text{ if } s' \in S_0 \cup \{s_i\}_{i\in I} \\ \{Q\} \text{ if } s' = f_i \text{ for } i \in I \end{cases}$
- $S_i \cap (S_0 \cup \{s_i\}_{i\in I} \cup \{f_i\}_{i\in I}) = \{s_i\}$, for all $i \in I$;
- $S_i \subseteq S$ for all $i \in I$;
- $\cap_{i\in I} S_i = \{\}$;
- $R_i = \{(s', s'') \in R : s', s'' \in S_i\}$ for all $i \in I$;
- $L_i(s') = L(s')$ for $s' \in S_i$ and for $i \in I$;

4 Expressiveness of $\nu CTL + C_Q$ for $\text{BPA}_{\delta, rec}$

Theorem 4.1 *$\nu CTL + C_Q$ is adequate for $\text{BPA}_{\delta,\text{rec}}$ with respect to the bisimulation equivalence.*

Proof. The proof shows that adequacy is not lost when adding the fixed point and the chop branching operator to CTL. As far the fixed point operator is concerned we can assert that adequacy of νCTL with respect to bisimulation can be directly derived from the adequacy of μ-calculus. In fact these two logics have the same expressive power. The chop branching operator when added to νCTL does not refine the equivalence because it cannot modify the structure of a tree by commmuting actions on its branches or by introducing or deleting branches, but it only joins trees sequentially.

Theorem 4.2 *$\nu CTL + C_Q$ is fully expressive for $\text{BPA}_{\delta,\text{rec}}$ with respect to the bisimulation equivalence.*

Proof (by construction): The theorem is proved by giving a temporal semantics to $\text{BPA}_{\delta,rec}$ where the domain of denotations is $\nu CTL + C_Q$ in order to produce a semantics fully abstract [13] with respect to bisimulation equivalence. The branching temporal semantics of $\text{BPA}_{\delta,rec}$ is shown below. It is defined by means of the semantics function $\mathcal{M}$ and the auxiliary functions $\mathcal{S}$, $\mathcal{T}$, needed respectively to define the set of outgoing transitions and the set of allowable transitions at a given time.

Hence the type of the semantic functions is:

$$\mathcal{M} : \text{BPA}_{\delta,rec} \longrightarrow \text{logic formulae}$$
$$\mathcal{S} : \text{BPA}_{\delta,rec} \longrightarrow \text{logic formulae}$$
$$\mathcal{T} : \text{BPA}_{\delta,rec} \longrightarrow \text{logic formulae}$$

For all $B \neq recX.B,\ B \neq X,\ B \neq B_1 \cdot B_2$ in $\mathrm{BPA}_{\delta,rec}$ we have:

$$\mathcal{M}(B) = \mathcal{S}(B) \wedge AX\, \mathcal{T}(B)$$

moreover we define:

$$\mathcal{M}(recX.B) = \nu x.\mathcal{M}(B) \qquad \mathcal{M}(X) = x$$
$$\mathcal{M}(B_1 \cdot B_2) = \mathcal{M}(B_1) C_{\surd} \mathcal{M}(B_2)$$

- deadlock, inaction

$\mathcal{S}(\delta) = \mathit{true} \qquad \mathcal{T}(\delta) = \mathit{false}$

The meaning of an inaction behaviour is therefore:

$$\mathcal{M}(\delta) = \mathit{true} \wedge AX\, \mathit{false} = AX\, \mathit{false};$$

the states satisfying this logic formula are states without successors.

- successful termination

$\mathcal{S}(\epsilon) = EX!(\surd \wedge \mathcal{M}(\delta)) \qquad \mathcal{T}(\epsilon) = \surd \wedge \mathcal{M}(\delta)$

The meaning of successful termination is:

$$\mathcal{M}(\epsilon) = EX!(\surd \wedge \mathcal{M}(\delta)) \wedge AX\,(\surd \wedge \mathcal{M}(\delta)).$$

The meaning $\mathcal{M}(\epsilon)$ is true for states which admit successors verifying $\mathcal{M}(\delta)$ and to which we arrive through the execution of the special action $\surd$. The logic formula $AX\,(\surd \wedge \mathcal{M}(\delta))$ guarantees that different successors cannot exist.

- action

$\mathcal{S}(a) = EX!(a \wedge \mathcal{M}(\epsilon)) \qquad \mathcal{T}(a) = a \wedge \mathcal{M}(\epsilon)$

The meaning of the action operator is the same of the previous operator, but considering the action a instead of $\surd$:

$$\mathcal{M}(a) = EX!(a \wedge \mathcal{M}(\epsilon)) \wedge AX\,(a \wedge \mathcal{M}(\epsilon)).$$

- alternative composition

$\mathcal{S}(B_1 + B_2) = \mathcal{S}(B_1) \wedge \mathcal{S}(B_2) \qquad \mathcal{T}(B_1 + B_2) = \mathcal{T}(B_1) \vee \mathcal{T}(B_2)$

The meaning of the alternative composition behaviour is:

$$\mathcal{M}(B_1 + B_2) = \mathcal{S}(B_1) \wedge \mathcal{S}(B_2) \wedge AX\,(\mathcal{T}(B_1) \vee \mathcal{T}(B_2)).$$

The meaning of the choice operator is true for states admitting both behaviours described by B_1 and B_2.

- sequential composition

$$\mathcal{S}(B_1 \cdot B_2) = \mathcal{S}(B_1) C_{\surd} \mathcal{M}(B_2) \qquad \mathcal{T}(B_1 \cdot B_2) = \mathcal{T}(B_1) C_{\surd} \mathcal{M}(B_2)$$

The meaning of the behaviour expression $B_1 \cdot B_2$ is true for states verifying the logic formula $\mathcal{M}(B_1) C_{\surd} \mathcal{M}(B_2)$. The $C_{\surd}$ operator simulates the behaviour of the sequential composition construct.

• recursion

$\mathcal{S}(recX.B) = \mathcal{S}(B)\,[\nu x.\mathcal{M}(B)/x] \qquad \mathcal{T}(recX.B) = \mathcal{T}(B)\,[\nu x.\mathcal{M}(B)/x]$
The meaning of $recX.B$ is the maximal solution of the equation $x = \mathcal{M}(B)$.

The semantics $\mathcal{M}$ is fully abstract with respect to bisimulation semantics of $\text{BPA}_{\delta,rec}$. Since $\mathcal{M}$ has been defined compositionally, the fully abstractness relation reduces to prove: $\mathcal{M}(p) \equiv_{\mathcal{L}} \mathcal{M}(q)$ *if and only if* $p \approx q$.

The if-direction can be proved by showing that the $\mathcal{M}$ function respects the laws of bisimulation on $\text{BPA}_{\delta,rec}$; the other implication by showing the correspondence between the transitions of a generic process and its $\mathcal{M}$-formula using inductive arguments.

5 Conclusions

We have given a temporal logic for $\text{BPA}_{\delta,rec}$ fully expressive w.r.t. bisimulation equivalence. By means of this logic it is possible to associate to each $\text{BPA}_{\delta,rec}$ process a $\nu CTL + C_Q$ formula, the characterisic formula, that exhibits all the properties satisfied by the process. The advantage of this is to reduce the verification that a process satisfies a property (expressed by a temporal logic formula) to check that the characterisic formula implies the given property formula.

The logic we have presented is more expressive than temporal or modal logics such as νCTL and μ-calculus; in fact, the addition of the chop operator permits to express context-free properties of processes. The interest in this logic is not limited to the temporal semantics presented in this paper, but it provides a general specification language which can express the behaviour of processes defined by means of recursion and sequential composition. So far, the verification of properties for context-free processes has been limited to regular properties: In [7] an algorithm is presented which performs model checking for context-free processes with μ-calculus as the target logic. With our approach the verification area can be extended to cover more powerful properties, using the logical implication. Obviously, the complexity of the calculus won't always be satisfactory.

We therefore intend to study the logic $\nu CTL + C_Q$ in more detail in order improve the verification phase. In particular we will focus ourselves on the applicability of model checking algorithms and other verification techniques based on the term structure rather than on the state space, such as those used for equivalence checking in [1,7,14].

Acknowledgement

Thanks to Gioia Ristori for suggestions and interesting discussions.

References

1. J. C. M. Baeten, J. A. Bergstra, J. W. Klop: *Decidibility of bisimulation equivalence for processes generating context-free languages.* Tech. Rep. CS-R8632, CWI, 1987.
2. B. Banieqbal, H. Barringer: *Temporal Logic Fixed Point Calculus.* Proc. Colloquium on Temporal Logic and Specification, LNCS 398, pp. 62-74, 1989.
3. H. Barringer, R. Kuiper, A. Pnueli: *Now You May Compose Temporal Logic Specifications.* Proceedings 16th ACM Symposium on the Theory of Computing, 1984.
4. J. A. Bergstra, J. W. Klop: *Process algebra: specification and verification in bisimulation semantics.* In: CWI Monograph 4, Proc. of the CWI Symposium Mathematics and Computer Science II, North-Holland, pp. 61-94, Amsterdam 1986.
5. J. A. Bergstra & J. W. Klop: *Process theory based on bisimulation semantics.* LNCS 354, pp. 50-122, 1988.
6. M. C. Browne, E. M. Clarke, O. Grümberg: *Characterizing Finite Kripke Structures in Propositional Temporal Logic.* Theor. Comp. Sci. vol. 59, pp. 115-131, 1988.
7. O. Burkart, B. Steffen: *Model Checking for Context-Free Processes.* Proceedings CONCUR 92, LNCS 630, pp.123-137, 1992.
8. E. A. Emerson, J. Y. Halpern: *"Sometimes" and "Not Never" Rivisited: On Branching versus Linear Time Temporal Logic.* Journal of ACM vol. 33, pp.151-178, 1986.
9. E. A. Emerson, C. Lei: *Efficient Model Checking in Fragments of the Propositional Mu-Calculus,* Proc. Symposium on Logics in Computer Science, 1986, pp. 267-278.
10. A. Fantechi, S. Gnesi, G. Ristori: *Compositionality and Bisimulation: a Negative Result.* Information Processing Letters, vol. 39, pp. 109-114, 1991.
11. S. Graf, J. Sifakis: *A Logic for Specification and Proof of Regular Controllable Processes of CCS.* Acta Informatica, vol 23, pp. 507-527, 1986.
12. J. F. Groote, F. W. Vaandrager: *Structured Operational Semantics and Bisimulation as a Congruence.* Proc. 16th ICALP, LNCS 372, pp.423-438, 1989.
13. M. Hennessy: *Algebraic Theory of Processes.*MIT Press, Cambridge, 1988.
14. H. Hüttel, C. Stirling: *Actions Speak Louder than Words: Proving Bisimilarity of Context-Free Processes.* Proc. LICS 91, Computer Society Press, 1991.
15. B. Jonsson, H. Khan, J. Parrow: *Implementing a Model Checking Algorithm by Adapting Existing Automated Tools.* In: Automatic Verification Methods for Finite State Systems, LNCS 407, pp. 179-188, 1990.
16. D. Kozen: *Results on the Propositional μ-calculus* . Theor. Comp. Sci. vol. 27, pp. 333-354, 1983.
17. R. Milner: *A calculus of communicating system.* LNCS 92, 1980.
18. D. M. R. Park: *Concurrency and automata on infinite sequences.* Proc. 5th GI Conference, LNCS 104, pp. 167-183 1981.
19. A. Pnueli: *The Temporal Semantics of Concurrent Programs* Theor. Comp. Sci., vol.13, 1981, pp.45-60.
20. A. Pnueli: *Linear and Branching Structures in the Semantics and Logic of Reactive Systems.* Proc. 12th ICALP, LNCS 194, 1985.
21. B. Steffen: *Characteristic Formulae.* Proc. 16th ICALP, LNCS 372, pp.723-732, 1989.

The Complexity of Finding Replicas Using Equality Tests*

Gudmund Skovbjerg Frandsen, Peter Bro Miltersen and Sven Skyum

Computer Science Department, Aarhus University, 8000 Aarhus C, Denmark

Abstract. We prove (for fixed k) that at least $\frac{1}{k-1}\binom{n}{2} - O(n)$ equality tests and no more than $\frac{2}{k}\binom{n}{2} + O(n)$ equality tests are needed in the worst case to determine whether a given set of n elements contains a subset of k identical elements. The upper bound is an improvement by a factor 2 compared to known results. We give tighter bounds for $k = 3$.

1 Introduction

When given n elements it is often possible to sort them according to bit string representation and hence decide which elements are identical in time $O(n \log n)$.

However, Campbell and McNeill [2] mention applications, where an equivalence relation is available, but it can not be extended to a total order on the bit representation. We mention one of their many examples. Consider a database of faces obtained by electronic scanning from different angles, where two images are equivalent if they represent the same person. If we want to verify that a set of images contains no duplicates, we may let a human being look at all possible image pairs, but it would seem to be of little help to sort the images according to bit representation.

This type of problem motivates the study of a computational model, where the only allowed operation is an equality test on pairs. We define two complexity measures:

Definition 1 Existence. For $2 \leq k \leq n$, let $E(k, n)$ denote the minimum number of equality tests that an algorithm must use in the worst case, to determine whether there exists k identical elements in a list of n elements.

Definition 2 Classification. Similarly, define $C(k, n)$ to be the minimum number of equality tests that an algorithm must use in the worst case to find a representative for and the cardinality of each equivalence class with at least k members in a list of n elements.

Clearly, $E(2, n) = C(2, n) = \binom{n}{2}$. This corresponds to the situation in the example above. If there are no duplicate images, that fact can be known to the algorithm only after comparison of all pairs.

However, the problem becomes nontrivial for $k > 2$. When checking the existence of triplicates, it is not necessary to compare all pairs of elements. We have managed

* This research was supported by the ESPRIT II BRA Programme of the EC under contract # 7141 (ALCOM II) and by CCI-Europe.

to prove that $\frac{7}{12}\binom{n}{2} - O(n) \leq E(3,n) = C(3,n) \leq \frac{3}{5}\binom{n}{2} + 1$. This result determines the fraction of pairs that must be compared in the worst case up to $\frac{3}{5} - \frac{7}{12} = \frac{1}{60}$.

For general k the literature contains a single result, $C(k,n) \leq 2\lfloor\frac{n}{k}\rfloor n$ [2]. We have improved this result by almost a factor 2, and obtain $C(k,n) \leq (1 + \lfloor\frac{n}{k}\rfloor)n$. We have also found a general lower bound. Our results amount to $\frac{1}{k-1}\binom{n}{2} - \frac{n}{2} \leq E(k,n) \leq C(k,n) \leq \frac{2}{k}\binom{n}{2} + \frac{3n}{2}$, i.e. we determine the fraction of pairs that must be compared to within a factor less than 2. It is obvious that $E(k,n) \leq C(k,n)$, but we have not been able to prove that $E(k,n) = C(k,n)$ in general. Similarly, it is a trivial observation that $C(k,n)$ is nonincreasing in k, but we have no such monotonicity knowledge of $E(k,n)$.

Earlier special results have focused on determining a majority element, and the exact complexity $E(\lceil\frac{n+1}{2}\rceil, n) = C(\lceil\frac{n+1}{2}\rceil, n) = \lfloor\frac{3}{2}(n-1)\rfloor$ was found independently by several people [3, 4]. The interest in deciding the existence of a majority element was motivated by the design of fault tolerant computer systems where a majority of single processors must agree on the output [6]. The linear result is also interesting in a model where the elements are ordered [5].

In a restricted version of the majority problem, it is known in advance that there are precisely two distinct equivalence classes. It requires and suffices to make $n - B(n)$ equality tests to find a representative of the larger class, when n is odd ($B(n)$ denotes the number of ones in the binary representation of n) [7].

We have divided our results into two sections. In the first of these, we present our results for the special case $k = 3$, and in the last section we present our results for the general case.

2 The Complexity of Finding Triplicates

We start by proving that finding the cardinality of all equivalence classes with at least three elements is no harder that determining whether any such class exists. We continue by presenting and analysing a simple algorithm that determines the existence of a triplicate element.

The main result of this section is a lower bound for the same problem, where the proof uses an adversary argument combined with a graph theoretic view on the problem.

Theorem 3.

$$E(3,n) = C(3,n)$$

Proof. Given n, let A be an algorithm that decides whether there exists 3 identical elements in an input consisting of n elements, using $E(3,n)$ equality tests in the worst case.

A can be regarded as a black box that once in a while poses an equality query. When fed the proper answers, it will eventually tell whether the input contains 3 identical elements or not. Since A never poses more than $E(3,n)$ queries it probably does some very efficient accounting inside the black box. Without knowledge of the nature of this internal accounting, we can nevertheless use A to construct an algorithm B that uses at most $E(3,n)$ equality queries to decide the cardinality of each equivalence class containing three or more elements and B will find a representative for each such class. Hence, the existence of B implies the theorem.

B starts the black box A. Each time A wants to know whether two elements x, y are identical then B makes the corresponding equality test, but B sometimes supplies a wrong answer to the black box A. B's strategy for deciding whether to supply A with the correct or the wrong answer is the following:

1. If the correct answer is "$\neq$" then this answer is passed on to A.
2. If the correct answer is "$=$" then B's action depends on the information A has received through earlier queries. If A when combining the correct answer "$=$" with old information may deduce the existence of 3 or more identical elements then B supplies the wrong answer "$\neq$" to A. Otherwise B gives A the correct answer "$=$".

When A stops, it must possess enough (possibly false) information to deduce that there are not three identical elements in the input. We claim that B at this point in time possesses enough (correct) information to determine all equivalence classes with three or more elements.

Let $X = \{x_1, \ldots, x_l\}$ be an equivalence class, $l \geq 3$. If A poses a query of the form $x_i?z$, where $z \notin X$, then A receives the correct answer "$\neq$" according to B's strategy, i.e. every equivalence class known to A is a subclass of a true equivalence class. Since $l \geq 3$, A must make at least one query of the form $x_i?x_j$. Without loss of generality, we may assume that the first such query is $x_1?x_2$. According to B's strategy, A receives the correct answer "$=$". Since A eventually concludes that $X' = \{x_1, x_2\}$ is a maximal equivalence class, then A must pose queries to reveal the relation between X' and any other equivalence class that it knows of. Since the class division known to A is a refinement of the true class division, B will through these queries obtain full knowledge of X. □

Theorem 4.

$$E(3,n) \leq \lceil \frac{3}{10} n(n-1) \rceil \leq \frac{3}{5}\binom{n}{2} + 1$$

Proof. We describe an algorithm that takes as input a list of elements and determines whether three of them are identical. The algorithm has two phases.

In the first phase, we take the input elements one by one and insert them in the smaller one of two sets $U_i, (i = 1, 2)$ such that we keep the sets equal sized, $|U_1| = |U_2| + \delta$, where $\delta \in \{0, 1\}$. We do not make any cross comparisons of U_1 and U_2, and hence we have no knowledge of $U_1 \cap U_2$. Each U_i is maintained as the disjoint union of two subsets: $U_i = S_i \cup D_i$, $S_i \cap D_i = \emptyset$, where S_i are the elements that have been seen only once, and D_i are the elements that we have seen twice. When inserting an element e into U_i, we first compare e to S_i. If $e \in S_i$ then we move e from S_i to D_i. Otherwise we compare e to D_i. If $e \in D_i$ then we have seen e thrice and the algorithm halts, otherwise e was not in U_i and is inserted into S_i.

If the algorithm finishes the first phase without halting, we must compare U_1 to U_2 in the second phase. We need not make a complete pairwise comparison. If there is three identical elements then there is a representative in $D_1 \cup D_2$. Hence it suffices to compare S_1 to D_2, D_1 to D_2 and D_1 to S_2.

To prove the upper bound, we need to make a count of the number of equality tests that the algorithm uses in the worst case. Let $s_i = |S_i|$ and $u_i = |U_i|$. In phase one we may count the comparisons made when inserting into the two U_i's separately for each i. The worst sequence of events consists in first inserting u_i distinct elements into U_i followed by the insertion of $u_i - s_i$ duplicates. This uses a maximum number of comparisons

$$\binom{u_i}{2} + \binom{u_i+1}{2} - \binom{s_i+1}{2} = u_i^2 - \binom{s_i+1}{2}.$$

In phase two we need to make at most $u_1u_2 - s_1s_2$ comparisons.

We may express the total number of comparisons made in terms of $s = s_1 + s_2$ only, when using that $u_1 - u_2 = \delta$ ($\delta \in \{0,1\}$) and $u_1 + u_2 = \frac{1}{2}(n+s)$:

$$\begin{aligned} & u_1u_2 - s_1s_2 + \textstyle\sum_{i=1}^{2}[u_i^2 - \binom{s_i+1}{2}] \\ &= (u_1+u_2)^2 - u_1u_2 - \tfrac{1}{2}s(s+1) \\ &= -\tfrac{5}{16}s^2 + (\tfrac{3}{8}n - \tfrac{1}{2})s + \tfrac{3}{16}n^2 + \tfrac{1}{4}\delta^2 \end{aligned}$$

The latter expression takes its maximum for $s = \frac{1}{5}(3n-4)$ and the integral part of the maximum expression value is at most $\lceil \frac{3}{10}n(n-1) \rceil$, which is an upper bound for $E(3,n)$. □

Theorem 5.

$$E(3,n) \geq \frac{7}{24}n^2 - O(n) = \frac{7}{12}\binom{n}{2} - O(n)$$

Proof. We prove this lower bound by means of an adversary argument. The adversary maintains a complete graph on n vertices, corresponding to the n elements that the algorithm may compare pairwise in queries.

Initially, all edges in the graph are *black*. When answering a query $(x?y)$ the adversary paints the corresponding edge (x,y) *green* if the answer given is $x = y$ and otherwise the edge is painted *red*.

The adversary will always give answers that are consistent with no triple of elements being identical, i.e. the green edges will form a matching. On the other hand the information will for quite some time be consistent with the existence of three identical elements, i.e. the graph contains a 3-clique with no red edges for a long time.

The adversary will maintain a partition of the vertices in two classes, the green vertices G and the red vertices R. A vertex with an adjacent green edge is green and all other vertices are red.

In describing the adversary strategy, we need only consider the case, where the algorithm queries the relation $(x?y)$ for a black edge (x,y). If there would arise a clique of size $\frac{n}{3}$ with all vertices and edges red in case the edge (x,y) were painted red, then the adversary answers *equal* and otherwise it answers *distinct*.

The strategy preserves the following invariants:

1. Let C be a maximumsize clique with all vertices and edges being red. Let $c = |C|$. It is always the case that $c < n/3$ and if $G \neq \emptyset$ then $c \geq n/3 - 2$. (To see the latter inequality, observe that immediately following the latest green-painting of an edge, $c \geq n/3 - 2$).

2. Let γ and ρ be the number of green and red *edges* respectively and let r be the number of red *vertices*. Then the number of distinct queries answered by the adversary is $\gamma + \rho$ and $\gamma = \frac{1}{2}(n - r)$.

We want to compute a lower bound for the number of queries made by an algorithm that verifies the nonexistence of three identical elements. Hence, it suffices to compute a lower bound for $\gamma + \rho$ when every 3-clique in the graph contains a red edge and the graph satisfies the invariants above. Clearly, $\gamma \leq n/2$ and since we want to prove a lower bound that hides linear terms under O-notation, we ignore γ and concentrate on ρ.

We shall find the number of red edges as a sum of four numbers, and for counting purposes we will regard a matched pair of green vertices as a supervertex.

1. Each pair of green supervertices are connected with a red edge, i.e. $\rho_1 = \frac{1}{2}\gamma(\gamma - 1)$.
2. Each red vertex is connected to each green supervertex with a red edge, i.e. $\rho_2 = r\gamma$.
3. We have not counted all red edges incident to green vertices. Two green super vertices may be connected by up to 4 red edges, and a red vertex and a green supervertex may be connected by 1 or 2 red edges. Our adversary strategy guarantees the existence of at least $n/3 - 2$ "extra" red edges for each green edge, i.e. $\rho_3 \geq (\frac{n}{3} - 2)\gamma$.
4. Finally, we count the red edges connecting red vertices. Note that the clique of red vertices contains only red and black edges, and each 3-clique in it must have at least one red edge. If x is a red vertex then let $N_b(x)$ ($N_r(x)$) denote all the red vertices that are connected to x by a black (red) edge. Observe that $N_b(x)$ must form a clique of red edges only, since we could otherwise find a 3-clique with black edges only. From the invariant, we therefore know that $|N_b(x)| \leq c$, and consequently $|N_r(x)| \geq r - 1 - c$. If x lies in the clique C then we know in addition that $|N_r(x)| \geq c - 1$. We may combine this information and obtain for $x \in C$: $|N_r(x)| \geq \max(r - 1 - c, c - 1) \geq \frac{1}{2}[(r - 1 - c) + (c - 1)] = \frac{r}{2} - 1$. Adding up, we get $\rho_4 \geq \frac{1}{2}[c(\frac{r}{2} - 1) + (r - c)(r - c - 1)]$.

We may eliminate low order terms from the bound of ρ to obtain:

$$\begin{aligned}\rho &= \textstyle\sum_{i=1}^{4} \rho_i \\ &\geq \tfrac{1}{2}\gamma(\gamma - 1) + r\gamma + (\tfrac{n}{3} - 2)\gamma + \tfrac{1}{2}[c(\tfrac{r}{2} - 1) + (r - c)(r - c - 1)] \\ &= \tfrac{1}{2}\gamma^2 + r\gamma + \tfrac{n}{3}\gamma + \tfrac{1}{2}r^2 + \tfrac{1}{2}c^2 - \tfrac{3}{4}rc - O(n)\end{aligned}$$

We would like to eliminate c from this bound. From the invariant, we know that $\frac{n}{3} - 2 \leq c < \frac{n}{3}$, provided that $\gamma > 0$. If $\gamma = 0$, then $r = n$ and we find that $\rho \geq \frac{1}{2}c^2 - \frac{3n}{4}c + \frac{n^2}{2} - O(n)$. For $c < \frac{n}{3}$ the value of this expression is bounded from below by the value at $c = \frac{n}{3}$, which is $\frac{11}{36}n^2 - O(n)$. Hence, if $\gamma = 0$ then the algorithm has used at least as many equality tests as claimed in the statement of the theorem ($\frac{11}{36} > \frac{7}{24}$). In the following, we assume that $\gamma > 0$, which enable us to express the bound on ρ in terms of r only when using $c = \frac{n}{3} - O(1)$ and $\gamma = \frac{1}{2}(n - r)$:

$$\begin{aligned}\rho &\geq \tfrac{1}{8}(n - r)^2 + \tfrac{1}{2}r(n - r) + \tfrac{n}{6}(n - r) + \tfrac{1}{2}r^2 + \tfrac{1}{18}n^2 - \tfrac{1}{4}rn - O(n) \\ &= \tfrac{1}{8}r^2 - \tfrac{n}{6}r + \tfrac{25n^2}{72} - O(n)\end{aligned}$$

This expression takes its minimum value at $r = \frac{2n}{3}$, where the value is $\frac{7}{24}n^2 - O(n)$, which is a lower bound for $E(3,n)$.

The analysis of our adversary strategy is optimal, in the sense that the algorithm in the proof of theorem 4 uses at most $\frac{7}{24}n^2 + O(n)$ equality tests, when run against the adversary.

The essence of the adversary strategy is to answer "$\neq$", when there would otherwise arise a red clique of size $\frac{n}{3}$. The choice of $\frac{n}{3}$ gives the best lower bound among all possible fixed fractions of n. Hence, a significantly different adversary strategy is required for a possible improvement of the lower bound. □

3 The Complexity of Finding Replicas in General

We present a general lower bound with a simple proof that is based on Turan's theorem. The main result of this section is an efficient algorithm for the general case. The analysis of the algorithm is based on a rather intricate amortization argument.

Theorem 6.

$$E(k,n) \geq \frac{n}{2}\left(\frac{n}{k-1} - 1\right) + k - 2 \geq \frac{1}{k-1}\binom{n}{2} - \frac{n}{2}$$

Proof. We rephrase our problem as a graph problem, and use an adversary argument combined with a result from extremal graph theory to prove the lower bound.

Observe that $E(k,n)$ is the minimum number of probes into the incidence matrix for a transitive graph G that will decide whether G contains a k-clique.

Our adversary strategy is the following: When an algorithm asks for an entry in the incidence matrix, we let the adversary answer that the edge is *not* present. So the algorithm will obtain information that is always consistent with the graph having no k-clique, but for quite some time the information will also be consistent with the graph having a k-clique. We need a lower bound on the number of probes for which the information remains nonconclusive. It is implied by the following result:

Fact 7 Consequence of Turan's theorem [1, p. 293]. *Let* $t_q(n) = \binom{n}{2} - \sum_{i=0}^{q-1}\binom{n_i}{2}$, *where* $n_i = \lfloor\frac{n+i}{q}\rfloor$. *Every graph with n vertices and more than* $t_{k-1}(n)$ *edges contains a k-clique.*

The adversary strategy combined with the fact gives us the following lower bound:

$$E(k,n) \geq \binom{n}{2} - t_{k-1}(n).$$

When letting $r = n - (k-1)\lfloor\frac{n}{k-1}\rfloor$, we may compute

$$\begin{aligned} E(k,n) &\geq \textstyle\sum_{i=0}^{k-2}\binom{n_i}{2} \\ &= (k-1-r)\binom{\frac{n-r}{k-1}}{2} + r\binom{\frac{n-r}{k-1}+1}{2} \\ &= \tfrac{(n-r)(n-(k-1-r))}{2(k-1)} \\ &\geq \tfrac{n}{2}\left(\tfrac{n}{k-1} - 1\right) \end{aligned}$$

The lower bound can be increased with $k-2$, if the adversary changes its strategy just before giving conclusive information. The adversary may then answer *equal* to at least $k-2$ queries without letting the algorithm resolve the situation. □

Theorem 8.

$$C(k,n) \le (\lfloor \frac{n}{k} \rfloor + 1)n \le \frac{2}{k}\binom{n}{2} + \frac{3}{2}n$$

Proof. Let $l = \lfloor \frac{n}{k} \rfloor$, i.e. $\frac{n}{l+1} < k \le \frac{n}{l}$. We shall describe an algorithm that receives at most ln "$\neq$"-answers to equality queries. This implies the theorem, since by proper accounting the algorithm will have complete information if it receives $n-1$ "="-answers to equality queries.

The algorithm works in two phases. In the first phase, the algorithm finds up to l elements satisfying that any equivalence class with at least k elements must have a representative among these l elements. In the second phase the abundance of the l candidates are checked.

In the first phase the algorithm inserts the elements one by one into a sequence of buckets $B_1, B_2, \ldots, B_m$, where m is initially 0 and increases according to need. All elements in a single bucket are identical and if the elements of two distinct buckets B_i, B_j are identical then their indices are at least $l+1$ apart, $|i-j| \ge l+1$. All buckets are nonempty, but only the first l buckets may contain two or more elements. Formally, if s_i is the number of elements in B_i, one of which is x_i, then we maintain the following invariants:

$$s_i \ge 1 \text{ for } 1 \le i \le \min(l,m)$$

$$s_i = 1 \text{ for } l < i \le m$$

$$1 \le |i-j| \le l \text{ implies that } x_i \ne x_j$$

When considering an unused element z we test whether it is identical to any of $x_1, x_2, \ldots, x_f$, where $f = \min(l,m)$. There are three cases:

1. If $z = x_i$, $i \le f$ then we add z to B_i (the invariants still hold).
2. If $m < l$ and z is distinct from all of $x_1, x_2, \ldots, x_m$ then we insert a new bucket B_{m+1} containing the single element z (invariants are preserved).
3. If $m \ge l$ and z is distinct from all of $x_1, x_2, \ldots, x_l$, the action is more complicated. This time we shift all buckets one up, and insert a new first bucket containing the single element z. At this point there may be more than one element in the updated B_{l+1}-bucket, which is not allowed according to the invariant. Exceeding elements from B_{l+1} are removed and inserted in a new first bucket following yet another shift. This continues until the invariant is satisfied. Formally, if $s_i \ge 2$ for all $1 \le i \le l$ then let $r = 0$ otherwise let r be maximum such that $1 \le r \le l$ and $s_r = 1$. If x_i', s_i', m' refers to the updated sequence of buckets then we have
   (a) $m' = m + l - r + 1$.
   (b) $x'_{i+l-r+1} = x_i$ and $s'_{i+l-r+1} = 1$ for $r \le i \le m$.
   (c) $x'_{i+l-r+1} = x_i$ and $s'_{i+l-r+1} = s_i$ for $1 \le i \le r-1$.
   (d) $x'_{i+l-r} = z$ and $s'_{i+l-r} = 1$.
   (e) $x'_{i-r} = x_i$ and $s'_{i-r} = s_i - 1$, for $r+1 \le i \le l$.
   One may check that this action preserves the invariants.

We have not quite finished the description of the first phase. When comparing z to $x_1, x_2, \ldots, x_f$, the comparison order is irrelevant, if no equality is found. However, if equality is found, it will be essential to the later complexity analysis that the following comparison order has been used: We compare z to an element of a larger bucket before comparing to an element of a smaller bucket, and among equal sized buckets, we compare z to an element from a bucket with small index before comparing z to an element from a bucket with large index. We stop comparing once equality is found.

Before describing the exact actions taken in the second phase, we make the following observation:

> Let $l \leq e \leq m$. Then the buckets $B_{e+1}, B_{e+2}, \ldots, B_m$ each contain exactly one element, and there are at least l buckets between two buckets with identical elements according to the invariant. This implies that if the listed buckets contain $q \geq 1$ copies of a specific element z, then $m-e \geq (q-1)(l+1)+1$, or equivalently $\phi(e,q) \geq 0$ where ϕ is defined as $\phi(e,q) = m - e - (q-1)(l+1) - 1$.
> Observe that $\phi(e+(l+1), q-1) = \phi(e,q)$ and $\phi(e+1,q) = \phi(e,q) - 1$. We may now determine the number of copies of z (if it is at least q) in $B_{e+1}, \ldots, B_m$ as follows: While $\phi(e,q) \geq 0$ and $e < m$ compare z to x_{e+1}. If $z = x_{e+1}$ then $x_{e+2}, \ldots, x_{e+(l+1)}$ are all distinct from z and we increment e by $l+1$ and decrement q by one, which leaves $\phi(e,q)$ unchanged. If $z \neq x_{e+1}$ then we increment e by one and leave q unchanged, which decrements $\phi(e,q)$ by one.
> The sketched algorithm determines whether there are q copies of z in $B_{e+1}, \ldots, B_m$, and if so the algorithm finds the exact number of copies. It receives at most $\max(0, \phi(e,q)+1) = \max(0, m-e-(q-1)(l+1))$ "$\neq$"-answers.

This observation implies that we need only check the abundance of the elements in the first f buckets for $f = \min(l,m)$, since an element z distinct from all of $x_1, x_2, \ldots, x_l$ can only occur in k copies or more if $\phi(l,k) \geq 0$, i.e. if $n \geq m \geq k(l+1)$, which is contradictory to the definition of l.

We check the abundance of each of the f candidates $x_1, x_2, \ldots, x_f$ in turn. Let us consider the checking of an arbitrary one of these, $z = x_i$. By the invariant, we know that there are precisely s_i copies of z in the buckets $B_1, \ldots, B_{l+i}$. Hence, if $m \leq l+i$ then we are done and otherwise we check whether there are at least $k - s_i$ copies of z in the buckets $B_{l+i+1}, \ldots, B_m$ using the algorithm described in the observation.

We have now presented the algorithm and argued its correctness. We need still analyse its complexity. We argued earlier that it suffices to show that the algorithm never makes more than $l \cdot n$ comparisons that result in the answer "$\neq$". We compute the cost of the algorithm separately for the two phases:

Let C_1 be the number of "$\neq$"-answers that the algorithm receives when inserting the first v elements during the first phase, and let C_2 be the number of "$\neq$"-answers that the algorithm receives during the second phase, if it was executed directly following insertion of the first v elements in the first phase.

Let m, s_i denote the values taken by these variables following the insertion of the first v elements. Let $f = \min(l,m)$, let $t = \sum_{i=1}^{f} s_i$, let

$$F = \sum_{1 \leq i < j \leq f} \min(s_i, s_j)$$

and let

$$G = \#\{(i,j) | 1 \leq i < j \leq f \text{ and } s_i < s_j\}.$$

Then we have

Lemma 9.

$$C_1 \leq vl - tl - \binom{f}{2} + 2F + G$$

Lemma 10. *If $f < l$ then $C_2 = 0$ and if $f = l$ then*

$$C_2 \leq tl + \binom{f}{2} - 2F - G$$

We note that the total cost of the algorithm is $C_1 + C_2 \leq ln$ by the lemmas. Hence the theorem follows, when we have proved the lemmas:

Proof of lemma 9. We prove the inequality by induction on v. If $v = 1$ then $t = v = f = 1$ and the statement reduces to $C_1 \leq 0$, which is true.

Assume that the inequality describes a correct upper bound for C_1. When inserting the $(v+1)$'th element z, the algorithm distinguishes three cases. We will argue that the invariant is preserved in each case. In the following we let unprimed variables refer to the situation before the insertion of z and we let primed variables refer to the situation after the insertion:

1. z is added to B_i. Let $J_1 = \{j | 1 \leq j \leq f \text{ and } s_i < s_j\}$ and let $J_2 = \{j | 1 \leq j < i \text{ and } s_i = s_j\}$. By the comparison order, used by the algorithm, z is found distinct to other elements in precisely $\#J_1 + \#J_2$ comparisons, before the insertion into B_i. We note that $v' - t' = v - t$, $f' = f$, $F' = F + \#J_1$ and $G' \geq G + \#J_2 - \#J_1$. Hence, C_1' satisfies the stated inequality.
2. $m < l$ and z is found distinct from all of $x_1, \ldots, x_f$. We get f "$\neq$"-answers. $v' - t' = v - t = 0$, $f' = f + 1$, $F' = F + f$ and $G' = G$. Hence, C_1' satisfies the inequality in this case too.
3. $m \geq l$ and z is distinct from all of $x_1, \ldots, x_l$. We get l "$\neq$"-answers. Let r be defined as in the description of the algorithm. $v' = v + 1$, $t' = t - r$, $f' = f = l$ and $2F' + G' \geq 2F + G - (l - r)r$. Hence, in all three cases C_1' satisfies the inequality of the lemma.

Proof of lemma 10. The case of $f < l$ is trivial, hence we assume that $f = l$. By the observation in the description of the second phase, we know that

$$\begin{aligned}
C_2 &\le \textstyle\sum_{i=1}^{f} \max(0, \phi(l+i, k-s_i)+1) \\
&= \textstyle\sum_{i=1}^{f} \max(0, m-l-i-(k-s_i-1)(l+1)) \\
&= \textstyle\sum_{i=1}^{f} \max(0, v-t-i-(k-s_i-1)(l+1)) \\
&= \textstyle\sum_{i=1}^{f} \max(0, v-k(l+1)-t-i+(s_i+1)(l+1)) \\
&\le \textstyle\sum_{i=1}^{f} \max(0, -1-t-i+(s_i+1)(l+1)) \\
&= \textstyle\sum_{i=1}^{f} \max(0, s_i l+(l-i)-(t-s_i)) \\
&= \textstyle\sum_{i=1}^{f} \max(0, s_i l+(l-i)-\sum_{1\le j<i} s_j-\sum_{i<j\le f} s_j) \\
&\le \textstyle\sum_{i=1}^{f} \max(0, s_i l+(l-i) \\
&\quad -\textstyle\sum_{1\le j<i}\min(s_i,s_j)-\sum_{i<j\le f}\min(s_i+1,s_j)) \\
&= \textstyle\sum_{i=1}^{f} [s_i l+(l-i) \\
&\quad -\textstyle\sum_{1\le j<i}\min(s_i,s_j)-\sum_{i<j\le f}\min(s_i+1,s_j)] \\
&= tl+\binom{f}{2}+f(l-f)-2\textstyle\sum_{1\le i<j\le f}\min(s_i,s_j) \\
&\quad -\#\{(i,j) \mid 1\le i<j\le f \text{ and } s_i<s_j\}
\end{aligned}$$

The term $f(l-f)$ disappears, since we have assumed $l = f$. We have thus proved the lemma. □

References

1. Bollobás, B., *Extremal Graph Theory.* Academic Press, 1978.
2. Campbell, D. and McNeill, T., Finding a majority when sorting is not available. *The Computer Journal* 34 (1991) 186.
3. Fischer, M. J. and Salzburg S. L., Solution to problem 81-5. *Journal of Algorithms* 3 (1982) 376–379.
4. Matula, D. W., An Optimal Algorithm for the Majority Problem. Manuscript, Southern Methodist University, Texas, 1990.
5. Misra, J. and Gries, D., Finding Repeated Elements. *Science of Computer Programming* 2 (1982) 143–152.
6. Moore, J., Problem 81-5. *Journal of Algorithms* 2 (1981) 208–209.
7. Saks, M. E. and Werman, M., On Computing Majority by Comparisons. *Combinatorica* 11 (1991) 383–387.

A Complete Axiomatization for Branching Bisimulation Congruence of Finite-State Behaviours

R.J. van Glabbeek*
Computer Science Department, Stanford University
Stanford, CA 94305, USA.
rvg@cs.stanford.edu

This paper offers a complete inference system for branching bisimulation congruence on a basic sublanguage of CCS for representing regular processes with silent moves. Moreover, complete axiomatizations are provided for the guarded expressions in this language, representing the divergence-free processes, and for the recursion-free expressions, representing the finite processes. Furthermore it is argued that in abstract interleaving semantics (at least for finite processes) branching bisimulation congruence is the finest reasonable congruence possible. The argument is that for closed recursion-free process expressions, in the presence of some standard process algebra operations like partially synchronous parallel composition and relabelling, branching bisimulation congruence is completely axiomatized by the usual axioms for strong congruence together with Milner's first τ-law $a\tau X = aX$.

1 Introduction

An important class of mathematical models for concurrent systems are the *term models*, in which a *process* or behaviour (of a system) is represented as a congruence class of expressions in a *system description language*. The best known system description language is MILNER's *Calculus of Communicating Sytems* (CCS), and the best known congruence on CCS expressions[1] is *bisimulation congruence* [7]. The choice of bisimulation congruence was originally motivated by a notion of *observability*: "processes are equal iff they are indistinguishable by any experiment based on observation" [7]. However, since the appearance of bisimulation congruence, many alternative notions of observability, or *testing scenarios*, have been proposed, all leading to different—and invariably coarser—congruences. See VAN GLABBEEK [3] for an overview. What makes bisimulation congruence special among all these alternatives is not so much the underlying notion of observation, but the fact that it is the *finest* reasonable congruence. To be precise, this is the case in *interleaving semantics*, where the "concurrent occurrence of two observable actions is not distinguished from their occurrence in arbitrary sequence" [7]. In non-interleaving semantics one finds finer congruences, but the finest ones are just variations of bisimulation congruence that take *causal dependence* between action occurrences explicitly into account. What makes bisimulation congruence the finest reasonable congruence are two properties:

- Any two bisimilar process expressions have the same internal structure, to be precise the same *branching structure*. As the observable behaviour of processes according to any alternative (interleaving based) testing scenario is completely determined by their branching structure, it follows that other observable congruences must be coarser.
- Finer equivalences than bisimulation congruence (such as *tree equivalence* or *graph isomorphism*) suffer from serious drawbacks such as a higher complexity (to decide the equivalence of finite-state behaviours) and the inequivalence of standard operational and denotational interpretations of CCS-like system description languages.

*This work was supported by ONR under grant number N00014-92-J-1974.
[1]In this first paragraph I restrict myself to expressions that model behaviours without hidden moves (τ-actions).

A crucial tool in practical applications of system description languages like CCS, especially for verification purposes, is an *abstraction* mechanism. Abstraction is usually performed by turning actions that are considered unimportant into the *invisible* action τ. Then, a system that after some activity reaches a state from which only an invisible action is possible, leading to another state, is considered equivalent to an otherwise identical system, that after said activity immediately reaches the other state. Thus the mechanism of abstraction by hiding of irrelevant or unobservable actions needs support from the congruence notion employed.

There are many ways to extend bisimulation congruence to processes with hidden moves. The simplest generalization is *strong (bisimulation) equivalence*, in which τ-actions are treated no different than visible actions. For this reason strong congruence is not *abstract* in the sense stipulated above. Another option is to take the testing scenario underlying bisimulation equivalence as primary, incorporating the unobservable nature of hidden moves. This yields MILNER's notion of *weak (bisimulation) congruence* [7], also called *observation congruence*, in spite of the rather far-going assumptions about the capabilities of observers that need to be made for weak congruence to be truly observable. In VAN GLABBEEK & WEIJLAND [4] another generalization of bisimulation congruence was proposed. *Branching (bisimulation) congruence* is not so much motivated in terms of its testing scenario (although it has one that is arguably only twice as contrived as that of weak bisimulation congruence), but generalizes the property of bisimulation congruence of being the finest reasonable interleaving congruence to an abstract setting. To be precise: it preserves the branching structure of processes (unlike weak congruence) [4], and (at least for finite processes) any finer or incomparable abstract version of bisimulation congruence violates the *expansion theorem* [7], that is characteristic for interleaving semantics.

Besides a substantiation of the last claim, this paper offers a complete axiomatization of branching bisimulation congruence for a sublanguage BCCS$^\omega$ of CCS, only containing operators for action, inaction, choice and recursion. The B stands for *Basic* and ω is a strict upper bound for the number of arguments of the choice and recursion operators. This language represents all and only the *regular processes* or *finite-state behaviours.* Moreover, complete axiomatizations for two sublanguages are given: the language of *recursion-free* BCCS$^\omega$ expressions, representing the *finite* processes, and the language of *guarded* BCCS$^\omega$ expressions, representing the *divergence-free* processes, where a processes is *divergent* if it has a state from which an infinite sequence of hidden moves is possible.

A complete axiomatization for strong congruence on BCCS$^\omega$ was provided in MILNER [5]. It consisted of the axioms E1-4, A0-3 and R1-3 of Section 4 (as well as α-conversion, which is derivable). A complete axiomatization for weak congruence on a slightly different language was first provided in BERGSTRA & KLOP [2]. A more aesthetic axiomatization (on BCCS$^\omega$), partly inspired by the one in [2], was given in MILNER [6]. It consisted of the axioms for strong congruence, 3 so-called τ-laws, and 2 extra axioms for unguarded recursion (besides R3). The present axiomatization counts, besides the axioms for strong congruence, only one τ-law, but 3 extra axioms for unguarded recursion (all weaker than the axioms for weak congruence). In all three cases the axioms for unguarded recursion can be dropped to obtain complete axiomatizations for guarded expressions, and on top of that R1 and R2 can be dropped to obtain complete axiomatizations for finite processes.

Milner's completeness proof was delivered in five steps:

(a) Any expression can be converted into a guarded one.

(b) Any guarded expression provably satisfies a standard guarded set of equations.

(c) Any standard guarded set of equations can be converted into a saturated one (preserving the property of being provably satisfied by an expression).

(d) Two congruent processes that each provably satisfy a saturated standard guarded set of equations, provably satisfy a common guarded set of equations.

(e) If two guarded expressions satisfy the same guarded set of equations, they are provably equal.

Steps (b) and (e) only use the axioms for strong congruence, and can thus be applied in the setting of branching bisimulation as well. Step (a) can be made completely analogous, even though the present axioms for unguarded recursion are much more complicated (in particular, the side-condition of R4 can not be eliminated, as could be done for the corresponding axiom R5 in [6]). Step (c) must be skipped as saturation is unsound in branching bisimulation semantics, and therefore step (d) needs to be made more subtle. But the absence of step (c) makes it possible to incorporate step (b) into step (d) at no extra cost.

The completeness theorem for branching congruence on recursion-free process expressions, at least the closed ones, was already proven in [4] by the method of graph transformations, due to BERGSTRA & KLOP [1]. The present proof is distinctly shorter. On the other hand, the method of graph transformations, once mastered, tends to deliver completeness proofs on finite closed terms for arbitrary interleaving equivalences almost instantaneously, whereas the method used here seems rather bisimulation oriented and requires more thought.

2 A language for finite-state behaviours

Let the nonempty set A of *visible actions* and the disjoint infinite set V of variables be given. Let $\tau \notin A$ be the *invisible action* or *hidden move* and write $A_\tau = A \cup \{\tau\}$.

Definition 1 The set $\mathcal{E}$ of *process expressions* over BCCS$^\omega$ is given by

X	$\in \mathcal{E}$	for $X \in V$	(variable)
0	$\in \mathcal{E}$		(inaction)
aE	$\in \mathcal{E}$	for $a \in A_\tau$ and $E \in \mathcal{E}$	(action)
$E+F$	$\in \mathcal{E}$	for $E, F \in \mathcal{E}$	(choice)
μXE	$\in \mathcal{E}$	for $X \in V_S$ and $E \in \mathcal{E}$	(recursion)

The expression 0 represents a process that is unable to perform any action. aE represents a process that first performs the action a and then proceeds as E. $E+F$ represents a process that will behave as either E or F, and μXE represents a solution of the equation $X = E$.

Definition 2 An occurrence of a variable X in an expression $E \in \mathcal{E}$ is *bound* if it occurs in a subexpression of the form μXF. Otherwise it is *free*. E is *open* if it contains a free occurrence of a variable, and *closed* otherwise. $E\{F/X\}$ denotes the result of substituting F for all free occurrences of X in E, if necessary[2] renaming bound variables in E in order to ensure that no free occurrence of a variable in F becomes bound in $E\{F/X\}$. Likewise $E\{E_X/X\}_{X \in V'}$, for $V' \subseteq V$, denotes the result of simultaneously substituting E_X for X in the same fashion.

Definition 3 The transition relation $\longrightarrow \subseteq \mathcal{E} \times (A_\tau \cup V) \times \mathcal{E}$ is the smallest relation satisfying

- $X \xrightarrow{X} 0$ for $X \in V$
- $aE \xrightarrow{a} E$ for $a \in A_\tau$
- if $E \xrightarrow{x} G$ or $F \xrightarrow{x} G$ then $E + F \xrightarrow{x} G$
- if $E\{\mu XE/X\} \xrightarrow{x} F$ then $\mu XE \xrightarrow{x} F$

Here $E \xrightarrow{a} F$ for $a \in A_\tau$ means that the system represented by E can perform the action a, thereby evolving into F, and $E \xrightarrow{X} 0$ means that the system represented by E has the possibility to continue as whatever system is substituted for the variable X.

Definition 4 Let $E \in \mathcal{E}$. The set $\mathcal{E}_E$ of process expressions *reachable* from E is defined as the smallest subset of $\mathcal{E}$ satisfying $E \in \mathcal{E}_E$ and if $F \xrightarrow{a} G$ with $a \in A_\tau$ and $F \in \mathcal{E}_E$ then $G \in \mathcal{E}_E$.

Proposition 1 $\mathcal{E}_E$ is finite for $E \in \mathcal{E}$.

[2]Renaming is necessary if a free occurrence of X appears in a subterm μYG of E with Y occurring free in F.

Proof: Consider the transition relation $\rightarrow \subseteq \mathcal{E}' \times (A_\tau \dot{\cup} \{+,\mu\}) \times \mathcal{E}'$, given by

- $aE \stackrel{a}{\rightarrow} E$
- $E+F \stackrel{+}{\rightarrow} E$ and $E+F \stackrel{+}{\rightarrow} F$
- $\mu XE \stackrel{\mu}{\rightarrow} E\{\mu XE/X\}$

Here $\mathcal{E}'$ is defined as $\mathcal{E}$, except that every operator symbol $(X, 0, a, +, \mu X)$ in an expression $E \in \mathcal{E}'$ is coloured either red or black. Furthermore, if in a subexpression aF, $F+G$ or μXF of E the leading operator a, $+$ or μX is coloured black, the entire subexpression must be black. Whether an occurrence of a variable is free or bound does not depend on its colour. Substitution on $\mathcal{E}'$ is defined such that $E\{F/X\}$ means $E\{black(F)/X\}$ (i.e. a black version of F is substituted for any free red or black occurrence of X), and renaming of bound variables doesn't change their colour. Furthermore colours are preserved under transitions.

Choose $E \in \mathcal{E}$ and let $\mathcal{E}'_E$ be the set of coloured expressions in $\mathcal{E}'$ that are reachable by $\rightarrow$ from $red(E)$. If $F \stackrel{a}{\longrightarrow} F'$ for $F, F' \in \mathcal{E}$, and $F_0 \in \mathcal{E}'$ is a coloured version of F, then there must be $F_1, \ldots, F_{n+1} \in \mathcal{E}'$ with $n \in \mathbb{N}$ such that $F_{i-1} \stackrel{+}{\rightarrow} F_i$ or $F_{i-1} \stackrel{\mu}{\rightarrow} F_i$ for $i = 1, \ldots, n$, $F_n \stackrel{a}{\rightarrow} F_{n+1}$ and F_{n+1} is a coloured version of F'. Thus for any $F \in \mathcal{E}_E$ a coloured version appears in $\mathcal{E}'_E$, and it suffices to proof that $\mathcal{E}'_E$ is finite, or becomes finite after forgetting the colours.

Observe that if an expression F is partly red and $F \rightarrow F'$ then the red part of F' is smaller than the red part of F. Thus there are only finitely many expressions in $\mathcal{E}'_E$ that are partly red.

Furthermore observe that for any $F \in \mathcal{E}'_E$, if F contains a subexpression μYG with μY red, then no black subexpression of G contains a free occurrence of Y. This property is trivially true for $red(E)$, trivially preserved under $\stackrel{a}{\rightarrow}$ and $\stackrel{+}{\rightarrow}$, and preserved under $\stackrel{\mu}{\rightarrow}$ by the renaming-of-bound-variables convention of Definition 2. It follows that if $F \in \mathcal{E}'_E$ is partly red and $F \rightarrow F'$, then the black subexpressions of F that are inherited by F'—unlike the red ones—are unchanged in F'. Thus if $H \in \mathcal{E}'_E$ is partly red, $H \rightarrow H'$ and H' is completely black, then H' has the form μXG and has been generated by a derivation $\mu XG \stackrel{\mu}{\rightarrow} G\{\mu XG/X\}$. Hence the black term $H = \mu XG \in \mathcal{E}'_E$ also occurs as a partly red term $\mu XG \in \mathcal{E}'_E$.

It follows that $\mathcal{E}_E$ is finite. In fact $\mathcal{E}_E$ contains at most one element more than it has subexpressions of the form aF. □

Definition 5 A free occurrence of a variable X in an expression $E \in \mathcal{E}$ is *guarded* if it occurs in a subexpression of the form aF with $a \in A$ (i.e. $a \neq \tau$). X is *(un)guarded* in E if (not) every free occurrence of X in E is guarded. A process expression $E \in \mathcal{E}$ is *guarded* if for every subexpression μXF, X is guarded in F. Let $\mathcal{E}^g \subseteq \mathcal{E}$ be the set of guarded process expressions over BCCS$^\omega$.

Definition 6 A process expression $E \in \mathcal{E}$ is called *finite* or, more accurately, *recursion-free* if it has no subexpression of the form μXF. Let $\mathcal{E}^f \subseteq \mathcal{E}^g \subseteq \mathcal{E}$ be the set of finite process expressions over BCCS$^\omega$.

Lemma 1 If $E \in \mathcal{E}^f$, then the relation $\longrightarrow$ is well-founded in $\mathcal{E}_E$. This means that there are no $F_i \in \mathcal{E}_E$ and $a_i \in A_\tau$ for $i \in \mathbb{N}$ with $F_i \stackrel{a_i}{\longrightarrow} F_{i+1}$ for $i \in \mathbb{N}$.

Proof: If $F \in \mathcal{E}^f$ and $F \stackrel{a}{\longrightarrow} F'$ then $F' \in \mathcal{E}^f$ and F' is smaller than F. □

Lemma 2 If $E \in \mathcal{E}^g$, then the relation $\stackrel{\tau}{\longrightarrow}$ is well-founded in $\mathcal{E}_E$.

Proof: First note that if E is guarded and $F \in \mathcal{E}_E$ then F is guarded. This follows with a straightforward induction on derivations. For $F \in \mathcal{E}$, let $F^\bullet$ be F, in which every occurrence of a subterm aG with $a \in A$ is replaced by 0. Note that if F is guarded then $F^\bullet$ is guarded, and if $F \stackrel{\tau}{\longrightarrow} G$ then $F^\bullet \stackrel{\tau}{\longrightarrow} G^\bullet$. Now suppose there is an infinite path $F_0 \stackrel{\tau}{\longrightarrow} F_1 \stackrel{\tau}{\longrightarrow} F_2 \stackrel{\tau}{\longrightarrow} \cdots$ as denied in the lemma. Then there must be an infinite path $F_0^\bullet \stackrel{\tau}{\longrightarrow} F_1^\bullet \stackrel{\tau}{\longrightarrow} F_2^\bullet \stackrel{\tau}{\longrightarrow} \cdots$, only passing through guarded process expressions without subexpressions of the form aG for $a \in A$. But if H is such an expression and $H \stackrel{\tau}{\longrightarrow} H'$, then H' is smaller than H, yielding a contradiction. □

Write $E \Longrightarrow E'$ if there are $E_0, \ldots, E_n \in \mathcal{E}$ with $E = E_0 \xrightarrow{\tau} E_1 \xrightarrow{\tau} \cdots \xrightarrow{\tau} E_n = E'$.

Lemma 3 $X \in V$ is unguarded in $E \in \mathcal{E}$ iff $E \Longrightarrow E' \xrightarrow{X} 0$.
Proof: Straightforward. □

Definition 7 Renaming of bound variables is called *α-conversion*. Write $E =_\alpha F$ if $E, F \in \mathcal{E}$ only differ by α-conversion.

Lemma 4 Let $x \in A_\tau \cup V$.

1. $H \xrightarrow{X} 0 \wedge E \xrightarrow{x} F \;\Rightarrow\; H\{E/X\} \xrightarrow{x} F$
2. $H \xrightarrow{x} H' \wedge x \neq X \;\Rightarrow\; H\{E/X\} \xrightarrow{x} H''\{E/X\}$ with $H' =_\alpha H''$
3. $H\{E/X\} \xrightarrow{x} F \;\Rightarrow\; (H \xrightarrow{X} 0 \wedge E \xrightarrow{x} F) \vee (x \neq X \wedge H \xrightarrow{x} H' =_\alpha H'' \wedge F = H''\{E/X\})$

Proof: 1. and 2. are straightforward by induction on inference. I will prove 3. by induction on the inference of $H\{E/X\} \xrightarrow{x} F$. In case $H = X$ the first alternative applies: $H \xrightarrow{X} 0 \wedge E \xrightarrow{x} F$. The cases $F = Y \neq X$, $H = aG$, $H = H_1 + H_2$ and $H = \mu XG$ are straightforward, so assume $H = \mu YG$ with $Y \neq X$. Let $\tilde{H} = \mu\tilde{Y}\tilde{G}$ be the result of renaming bound variables in H, as described in Definition 2. Now, by a shorter inference $\tilde{G}\{E/X\}\{\tilde{H}\{E/X\}/\tilde{Y}\} = \tilde{G}\{\tilde{H}/\tilde{Y}\}\{E/X\} \xrightarrow{x} F$, so by induction $(\tilde{H} \xrightarrow{X} 0 \wedge E \xrightarrow{x} F) \vee (x \neq X \wedge \tilde{H} \xrightarrow{x} \tilde{H}' =_\alpha H'' \wedge F = H''\{E/X\})$, from which the desired conclusion follows. □

3 Branching bisimulation congruence

Definition 8 A *branching bisimulation* is a symmetric relation $\mathcal{R} \subseteq \mathcal{E} \times \mathcal{E}$ such that $\forall x \in A_\tau \cup V$:

$$\text{if } (E,F) \in \mathcal{R} \wedge E \xrightarrow{x} E' \text{ then } \begin{array}{l} x = \tau \text{ and } (E',F) \in \mathcal{R} \\ \text{or } \exists F'', F' : F \Longrightarrow F'' \xrightarrow{x} F' \wedge (E, F'') \in \mathcal{R} \wedge (E', F') \in \mathcal{R}. \end{array}$$

Two expressions E and F are *branching (bisimulation) equivalent*—notation $E \leftrightarrow_b F$—if there exists a branching bisimulation $\mathcal{R}$ with $(E,F) \in \mathcal{R}$.

For further motivation of branching bisimulation equivalence see VAN GLABBEEK & WEIJLAND [4]. The consise definition above is possible thanks to the following lemma.

Lemma 5 If $E \leftrightarrow_b F$, $E \leftrightarrow_b F''$ and $F \Longrightarrow F''$, then $E \leftrightarrow_b F'$ for any F' with $F \Longrightarrow F' \Longrightarrow F''$.
Proof: In [4]. □

It is more common to use Definition 8 for closed process expressions only, thereby avoiding the use of the transitions $\xrightarrow{X}$, and to extend the definition to open process expressions by

$$E \leftrightarrow_b F \text{ iff for all closed process expressions } G,\ E\{G/X\} \leftrightarrow_b F\{G/X\}$$

By Propositions 2 and 3 below both approaches yield the same equivalence relation. The way of defining $\leftrightarrow_b$ on open process expressions employed here is a mild variation of the way weak equivalence was defined in MILNER [6]. It does not carry over to full CCS.

Proposition 2 $\leftrightarrow_b \subseteq \mathcal{E} \times \mathcal{E}$ is a bisimulation and an equivalence, satisfying, for $E, F, G \in \mathcal{E}$

$$E \leftrightarrow_b F \Rightarrow E\{G/X\} \leftrightarrow_b F\{G/X\}.$$

Proof: The identity relation $\mathrm{Id}_\mathcal{E}$ is a branching bisimulation and if $\mathcal{R}$ and $\mathcal{S}$ are branching bisimulations, then so are $\mathcal{R}^{-1}$ and $\mathcal{R} \circ \mathcal{S} = \{(E,F) \mid \exists G \in \mathcal{E} \text{ with } (E,G) \in \mathcal{R} \text{ and } (G,F) \in \mathcal{S}\}$. Hence $\leftrightarrow_b$ is an equivalence.

If $\mathcal{R}_i$ $(i \in I)$ are branching bisimulations, so is $\bigcup_{i \in I} \mathcal{R}_i$. Thus $\leftrightarrow_b = \bigcup\{\mathcal{R} \mid \mathcal{R} \text{ is a bisimulation}\}$ is a branching bisimulation.

$\{(E\{G/X\}, F\{G/X\}) \mid E \leftrightarrow_b F,\ G \in \mathcal{E}\} \cup \mathrm{Id}_\mathcal{E}$ is a bisimulation by Lemma 4 (using $=_\alpha \subseteq \leftrightarrow_b$).□

Proposition 3 If $E\{G/X\} \underline{\leftrightarrow}_b F\{G/X\}$ for all closed process expressions G, then $E \underline{\leftrightarrow}_b F$.

Proof: As A is nonempty, there is an $a \in A$. It is easy to see that $a^m \not\underline{\leftrightarrow}_b a^n$ for $m \neq n$, where $a^n = aa \cdots a0$ with n a's. Thus, by Proposition 1, for given E and F it is possible to choose $n \in \mathbb{N}$ such that $a^{n-1} \not\underline{\leftrightarrow}_b H$ and thus $a^{n-1} \not\underline{\leftrightarrow}_b H\{a^n/X\}$ for $H \in \mathcal{E}_E \cup \mathcal{E}_F$. By assumption $E\{a^n/X\} \underline{\leftrightarrow}_b F\{a^n/X\}$. It suffices to prove that $\{(E', F') \subseteq \mathcal{E}_E \times \mathcal{E}_F \mid E'\{a^n/X\} \underline{\leftrightarrow}_b F'\{a^n/X\}\}$ is a branching bisimulation, which is a straightforward application of Lemma 4. □

The following is a powerful tool for establishing statements $E \underline{\leftrightarrow}_b F$. It is analogous to MILNER's notions of *strong bisimulation up to* $\underline{\leftrightarrow}_s$ and *weak bisimulation up to* $\underline{\leftrightarrow}_w$. As for weak bisimulation up to $\underline{\leftrightarrow}_w$, versions of the notion below without the double arrow in the premises are easily seen to be unsound [7].

Definition 9 A *branching bisimulation up to* $\underline{\leftrightarrow}_b$ is a symmetric relation $\mathcal{R} \subseteq \mathcal{E} \times \mathcal{E}$ such that if $E\mathcal{R}F$ and $E \Longrightarrow E' \xrightarrow{x} E''$ with $E \underline{\leftrightarrow}_b E'$ and $x \neq \tau \vee E' \not\underline{\leftrightarrow}_b E''$ then $\exists E_1', E_1'', F_1', F_1'', F', F''$ such that

$$\begin{array}{ccccccc} E & & & \mathcal{R} & & & F \\ \Downarrow & & & & & & \Downarrow \\ E' & \underline{\leftrightarrow}_b & E_1' & \mathcal{R} & F_1' & \underline{\leftrightarrow}_b & F' \\ \downarrow x & & & & & & \downarrow x \\ E'' & \underline{\leftrightarrow}_b & E_1'' & \mathcal{R} & F_1'' & \underline{\leftrightarrow}_b & F'' \end{array}$$

Proposition 4 If R is a branching bisimulation up to $\underline{\leftrightarrow}_b$ and $E\mathcal{R}F$, then $E \underline{\leftrightarrow}_b F$.

Proof: It suffices to prove that the relation $\underline{\leftrightarrow}_b \mathcal{R} \underline{\leftrightarrow}_b = \{(E_0, F_0) \mid \exists E, F : E_0 \underline{\leftrightarrow}_b E\mathcal{R}F \underline{\leftrightarrow}_b F_0\}$ is a branching bisimulation. So suppose E_0, E, F and F_0 are as indicated, and $E_0 \xrightarrow{x} E_0''$. Then either $x = \tau$ and $E_0'' \underline{\leftrightarrow}_b E$, which completes the proof, or there are E' and E'' with $E \Longrightarrow E' \xrightarrow{x} E''$, $E' \underline{\leftrightarrow}_b E_0 \underline{\leftrightarrow}_b E$ and $E'' \underline{\leftrightarrow}_b E_0'' (\not\underline{\leftrightarrow}_b E$ if $x = \tau)$. In the latter case apply Definition 9, and use that $F_0 \underline{\leftrightarrow}_b F \Longrightarrow F' \xrightarrow{x} F''$ implies $x = \tau \wedge F'' \underline{\leftrightarrow}_b F_0$ or $F_0 \Longrightarrow F_0' \xrightarrow{x} F_0''$ with $F' \underline{\leftrightarrow}_b F_0'$ and $F'' \underline{\leftrightarrow}_b F_0''$ by Definition 8 (and in one case Lemma 5 to find F_0'). □

Just like weak bisimulation equivalence, branching equivalence is not a congruence on BCCS$^\omega$. Also the simplest counterexample is the same: $a \underline{\leftrightarrow}_b \tau a$ but, for $b \neq a$, $a + b \not\underline{\leftrightarrow}_b \tau a + b$. Here, as usual, $a0$ is abbreviated by a and action prefixing binds stronger than choice. Milner selected weak bisimulation congruence to be the largest (= coarsest) congruence contained in weak equivalence, and the same solution is applied here. Just like weak congruence, branching congruence has a nice characterization, showing that it is close to the original equivalence.

Definition 10 Two expressions E and F are *rooted branching bisimulation equivalent* or *branching (bisimulation) congruent*—notation $E \underline{\leftrightarrow}_{rb} F$—if $\forall x \in A_\tau \cup V$:

$$E \xrightarrow{x} E' \text{ implies } \exists F' : F \xrightarrow{x} F' \wedge E' \underline{\leftrightarrow}_b F'$$
$$F \xrightarrow{x} F' \text{ implies } \exists E' : E \xrightarrow{x} E' \wedge E' \underline{\leftrightarrow}_b F'.$$

Proposition 5 (Congruence) $\underline{\leftrightarrow}_{rb}$ is an equivalence relation such that

$$\text{if } E = F \text{ then } aE = aF,\ \ E + G = F + G,\ \ G + E = G + F \text{ and } \mu XE = \mu XF.$$

Moreover it is the coarsest relation with these properties contained in $\underline{\leftrightarrow}_b$.

Proof: Similar to the congruence proofs for strong and weak bisimulation congruence in [7]. □

The following shows that the definition of $\underline{\leftrightarrow}_{rb}$ for open expressions yields the same notion as the standard approach based on substitution of closed terms.

Proposition 6 Let $E, F \in \mathcal{E}$. Then $E \mathbin{\underline{\leftrightarrow}}_b F$ implies $E\{G/X\} \mathbin{\underline{\leftrightarrow}}_b F\{G/X\}$ for $G \in \mathcal{E}$, and if $E\{G/X\} \mathbin{\underline{\leftrightarrow}}_b F\{G/X\}$ for closed $G \in \mathcal{E}$, then $E \mathbin{\underline{\leftrightarrow}}_b F$.

Proof: Straightforward with Lemma 4, using Propositions 2 and 3 and the same G as before. $\square$

MILNER [7] listed two results that show how close weak equivalence and congruence are to each other. The first was that for *stable* processes (processes without outgoing τ-transitions) the equivalence and congruence coincide. This result carries over to branching bisimulation, as follows immediately from the definitions. The second result says that in each weak bisimulation equivalence class there are at most two congruence classes, with representatives E and τE for some $E \in \mathcal{E}$. This is not true for branching bisimulation, indicating that branching equivalence and congruence are less close than weak equivalence and congruence. However, a corollary of this property does hold, showing that the distance is still reasonable.

Proposition 7 $E \mathbin{\underline{\leftrightarrow}}_b F \Leftrightarrow \tau E \mathbin{\underline{\leftrightarrow}}_{rb} \tau F$.
Proof: Immediate from Definition 10. $\square$

This proposition effectively turns any complete axiomatization for $\mathbin{\underline{\leftrightarrow}}_{rb}$ into one for $\mathbin{\underline{\leftrightarrow}}_b$.

4 The axioms

The following set of axioms will be proven to be sound and complete for $\mathbin{\underline{\leftrightarrow}}_{rb}$. The entries below are actually axiom schemes, in metavariables $E, F, G \in \mathcal{E}$, $X \in V$ and (in the axiom B) $a \in A_\tau$. This means that there is an axiom for every choice of E, F, G, X and a. The axiom schemes E1-3 and A1-4 could be replaced by single axioms, by using real variables X, Y and Z instead of the metavariables E, F and G, and adding the law of substitution: if $E = F$ then $E\{G/X\} = F\{G/X\}$, which is sound by Proposition 6. However, this would not work for R1-6, since the bound variable X is allowed to occur in E, F and G. The axioms $\mu X E = \mu Y(E\{Y/X\})$ (α-conversion) are derivable from R1-6, using Theorem 3 and R2.

E1 $E = E$
E2   if $E = F$ then $F = E$
E3   if $E = F$ and $F = G$ then $E = G$
E4   if $E = F$ then $aE = aF$, $E + G = F + G$, $G + E = G + F$, and $\mu X E = \mu X F$

A0 $E + 0 = E$
A1 $E + F = F + E$
A2 $E + (F + G) = (E + F) + G$
A3 $E + E = E$

B $a(\tau(E + F) + E) = a(E + F)$ for $a \in A_\tau$

R1 $\mu X E = E\{\mu X E/X\}$
R2   if $F = E\{F/X\}$ then $F = \mu X E$, provided X is guarded in E
R3 $\mu X(X + E) = \mu X E$
R4 $\mu X(\tau(\tau E + F) + G) = \mu X(\tau(E + F) + G)$, provided X is unguarded in E
R5 $\mu X(\tau(X + E) + \tau(X + F) + G) = \mu X(\tau(X + E + F) + G)$
R6 $\mu X(\tau(X + E) + F) = \mu X(\tau(E + F) + F)$

One writes $T \vdash E = F$, with T a list of axiom names, if the equation $E = F$ is derivable from the axioms in T. Moreover, in this paper the convention is adopted that the axioms E1-4 and A0-3 are always in T, even if not explicitly listed. In the next 3 sections I will establish the following completeness theorems.

- For $E, F \in \mathcal{E}^g$: $\quad E \mathbin{\underline{\leftrightarrow}}_{rb} F \;\Leftrightarrow\; \mathrm{BR1\text{-}2} \vdash E = F$

- For $E, F \in \mathcal{E}^f$: $\quad E \mathbin{\underline{\leftrightarrow}}_{rb} F \;\Leftrightarrow\; \mathrm{B} \vdash E = F$

- For $E, F \in \mathcal{E}$: $\quad E \mathbin{\underline{\leftrightarrow}}_{rb} F \;\Leftrightarrow\; \mathrm{BR} \vdash E = F$

The rest of this section will be devoted to the soundness of the axioms.

Soundness: The soundness of E1-4 is established in Proposition 5. As far as R1 concerns, one has $\mu X E \xrightarrow{x} F \;\Leftrightarrow\; E\{\mu X E/X\} \xrightarrow{x} F$ from which it follows that $\mu X E \mathbin{\underline{\leftrightarrow}}_{rb} E\{\mu X E/X\}$ (the terms are even *strongly* bisimilar). In the same way the soundness of A0-4 is established. By inspection of their outgoing transitions, it follows that $\{(\tau(E+F)+E, E+F)\} \cup \mathrm{Id}_{\mathcal{E}}$ is a branching bisimulation and hence $a(\tau(E+F)+E) \mathbin{\underline{\leftrightarrow}}_{rb} a(E+F)$.

Proposition 8 If $F \mathbin{\underline{\leftrightarrow}}_{rb} E\{F/X\}$ then $F \mathbin{\underline{\leftrightarrow}}_{rb} \mu X E$, provided X is guarded in E.

Proof: For $G, H \in \mathcal{E}$ write $H(G)$ for $H\{G/X\}$. Let $E, F, G \in \mathcal{E}$, such that X is guarded in E, $F \mathbin{\underline{\leftrightarrow}}_{rb} E(F)$ and $G \mathbin{\underline{\leftrightarrow}}_{rb} E(G)$. I will show that the symmetric closure of $\{(H(E(F)), H(E(G))) \mid H \in \mathcal{E}\}$ is a bisimulation up to $\mathbin{\underline{\leftrightarrow}}_b$. So suppose that $H(E(F)) \Longrightarrow K' \xrightarrow{x} K''$ (in this proof one doesn't even need to assume that $H(E(F)) \mathbin{\underline{\leftrightarrow}}_b K'$ and $x \neq \tau \vee K' \not\mathbin{\underline{\leftrightarrow}}_b K''$). As X is guarded in E and hence in $H(E)$, it cannot be that $H(E) \Longrightarrow \xrightarrow{X} 0$, by Lemma 3. Thus K' and K'' are of the form $H'(F)$ and $H''(F)$ by Lemma 4.3, and by Lemma 4.2 $H(E(G)) \Longrightarrow H'''(G) \xrightarrow{x} H''''(G)$ with $H''' =_\alpha H'$ and $H'''' =_\alpha H''$. Furthermore, by Proposition 5, $H'(E(F)) \mathbin{\underline{\leftrightarrow}}_b H'(F)$, $H'''(G) \mathbin{\underline{\leftrightarrow}}_b H'(E(G))$, $H''(E(F)) \mathbin{\underline{\leftrightarrow}}_b H''(F)$ and $H''''(G) \mathbin{\underline{\leftrightarrow}}_b H''(E(G))$. The requirement starting with $H(E(G))$ follows by symmetry, so the relation is a branching bisimulation up to $\mathbin{\underline{\leftrightarrow}}_b$ and by Proposition 4 $H(E(F)) \mathbin{\underline{\leftrightarrow}}_b H(E(G))$ for $H \in \mathcal{E}$. Using this, a repeat of the argument above with $K' = H(E(F))$ gives $H(E(F)) \mathbin{\underline{\leftrightarrow}}_{rb} H(E(G))$, so in particular $E(F) \mathbin{\underline{\leftrightarrow}}_{rb} E(G)$, and hence $F \mathbin{\underline{\leftrightarrow}}_{rb} G$. Finally take $G = \mu X E$. $\qquad\square$

Proposition 9 $\mu X(\tau(\tau E + F) + G) \mathbin{\underline{\leftrightarrow}}_{rb} \mu X(\tau(E+F)+G)$, provided X is unguarded in E.

Proof: By Lemma 3 there are $E_0, \ldots, E_n$ such that $\tau E + F \xrightarrow{\tau} E_0 \xrightarrow{\tau} E_1 \xrightarrow{\tau} \cdots E_n \xrightarrow{X}$ with $E_0 = E$ and $n \in \mathbb{N}$. Write E'_{-1} for $\tau E + F$ and E''_0 for $E + F$. Then by Lemma 4

$$L \xrightarrow{\tau} E'_{-1}\{L/X\} \xrightarrow{\tau} E'_0\{L/X\} \xrightarrow{\tau} E'_1\{L/X\} \xrightarrow{\tau} \cdots E'_n\{L/X\} \xrightarrow{\tau} E'_{-1}\{L/X\}$$

for certain $E'_i =_\alpha E_i$ $(i = 0, \ldots, n)$ and

$$R \xrightarrow{\tau} E''_0\{R/X\} \xrightarrow{\tau} E''_1\{R/X\} \xrightarrow{\tau} \cdots E''_n\{R/X\} \xrightarrow{\tau} E''_0\{R/X\}$$

for certain $E''_j =_\alpha E_j$ $(j = 1, \ldots, n)$. Let $\mathcal{R} \subseteq \mathcal{E} \times \mathcal{E}$ be the symmetric closure of

$$\{(H\{L/X\}, H'\{R/X\}) \mid H =_\alpha H'\} \cup \{(E'_i\{L/X\}, E''_j\{R/X\}) \mid -1 \le i \le n, \; 0 \le j \le n\}$$

Then $\mathcal{R}$ is a branching bisimulation and $L \mathbin{\underline{\leftrightarrow}}_{rb} R$ by Lemma 4. $\qquad\square$

Proposition 10 $\mu X(\tau(X+E) + \tau(X+F) + G) \mathbin{\underline{\leftrightarrow}}_{rb} \mu X(\tau(X+E+F)+G)$.

Proof: The closure under symmetry and α-recursion of $\{(H\{L/X\}, H\{R/X\}) \mid H \in \mathcal{E}\} \cup$

$$\{(\tau(X+E)\{L/X\}, \tau(X+E+F)\{R/X\}) \cup \{(\tau(X+F)\{L/X\}, \tau(X+E+F)\{R/X\})\}$$

is a branching bisimulation. $\qquad\square$

In the same way one proves the soundness of R3 and R6.

Proposition 11 $\mu X(X + E) \mathbin{\underline{\leftrightarrow}}_{rb} \mu X E$. $\qquad\square$

Proposition 12 $\mu X(\tau(X+E)+F) \mathbin{\underline{\leftrightarrow}}_{rb} \mu X(\tau(E+F)+F)$. $\qquad\square$

Corollary 1 (Soundness) For $E, F \in \mathcal{E}$: $\mathrm{BR} \vdash E = F \;\Rightarrow\; E \mathbin{\underline{\leftrightarrow}}_{rb} F$.

5 Completeness for guarded process expressions

Let, for $S = \{E_1, \ldots, E_n\}$, $\sum S$ be an abbreviation for $E_1 + \cdots + E_n$. This notation is justified by the axioms A0-3.

Lemma 6 For $E \in \mathcal{E}^g$, $\text{R1} \vdash E = \sum\{aE' \mid E \xrightarrow{a} E'\} + \sum\{W \mid E \xrightarrow{W} 0\}$.

Proof: By induction on the number of recursion operators in E, not counting the ones that occur in a subterm aG. If this number is 0, then E has the form $\sum_{i \in I} a_i E_i + \sum_{j \in J} W_j$ with $a_i \in A_\tau$ and $W_j \in V$ (the so-called *head normal form*) and the statement holds trivially. Otherwise E has a summand μXF, which can be replaced by $F\{\mu XF/X\}$ using R1, yielding E''. As E is guarded, E' has less recursion operators that don't occur in a subterm aG, so by induction $\text{R1} \vdash E'' = \sum\{aE' \mid E'' \xrightarrow{a} E'\} + \sum\{W \mid E'' \xrightarrow{W} 0\}$. As $E'' \xrightarrow{x} E' \Leftrightarrow E \xrightarrow{x} E'$ for $x \in A_\tau \cup V$, the statement follows. □

Definition 11 A *recursive specification* S is a set of equations $\{X = S_X \mid X \in V_S\}$ with $V_S \subseteq V$ and $S_X \in \mathcal{E}$ for $X \in V_S$. $E \in \mathcal{E}$ *T-provably satisfies* the recursive specification S *in* the variable $X_0 \in V_S$ if there are expressions E_X for $X \in V_S$ with $E = E_{X_0}$, such that for $X \in V_S$

$$T \vdash E_X = S_X\{E_Y/Y\}_{Y \in V_S}.$$

Definition 12 Let S be a recursive specification. The relations $\xrightarrow{o} \subseteq V_S \times V_S$ and $\xrightarrow{u} \subseteq V_S \times V_S$ are defined by

- $X \xrightarrow{o} Y$ if Y occurs free in S_X
- $X \xrightarrow{u} Y$ if Y occurs free and unguarded in S_X

Now S is called *well-founded* if $\xrightarrow{o}$ is well-founded on V_S, and *guarded* if $\xrightarrow{u}$ is well-founded on V_S.

Proposition 13 (Unique solutions) If S is a finite guarded recursive specification and $X_0 \in V_S$, then there is an expression E which R1-provably satisfies S in X_0. Moreover if there are two such expressions E and F, then $\text{R2} \vdash E = F$.

Proof: In MILNER [6]. □

Theorem 1 Let $E_0, F_0 \in \mathcal{E}^g$ with $E_0 \leftrightarrow_{rb} F_0$. Then there is a finite guarded recursive specification S BR1-provably satisfied in the same variable $X_0 \in V_S$ by both E_0 and F_0.

Proof: Take a fresh set of variables $V_S = \{X_{EF} \mid E \in \mathcal{E}_{E_0}, F \in \mathcal{E}_{F_0}, E \leftrightarrow_b F\}$. $X_0 = X_{E_0F_0}$. Now for $X_{EF} \in V_S$, S contains the equation

$$X_{EF} = \sum\{W \mid E \xrightarrow{W} 0 \text{ and } F \xrightarrow{W} 0\} + \sum\{aX_{E'F'} \mid E \xrightarrow{a} E',\ F \xrightarrow{a} F' \text{ and } E' \leftrightarrow_b F'\} +$$

$$\sum\{\tau X_{E'F} \mid X_{EF} \neq X_0,\ E \xrightarrow{\tau} E' \text{ and } E' \leftrightarrow_b F\} + \sum\{\tau X_{EF'} \mid X_{EF} \neq X_0,\ F \xrightarrow{\tau} F' \text{ and } E \leftrightarrow_b F'\}.$$

Using that $X_{EF} \xrightarrow{u} X_{E'F'}$ iff $S_{X_{EF}}$ has a summand $\tau X_{E'F'}$, it is easy to show that any infinite u-path $X_{EF} \xrightarrow{u} X_{E'F'} \xrightarrow{u} \cdots$ implies an infinite τ-path $E \xrightarrow{\tau} E' \xrightarrow{\tau} \cdots$ or $F \xrightarrow{\tau} F' \xrightarrow{\tau} \cdots$, which cannot exist by Lemma 2 since E_0 and F_0 are guarded. Hence S is a guarded recursive specification. Moreover S is finite by Proposition 1. It remains to be established that E_0 BR1-provably satisfies S in X_0. The same statement for F_0 then follows by symmetry.

For $X_{EF} \in V_S$, let H_{EF} be the expression $\sum\{W \mid E \xrightarrow{W} 0 \text{ and } F \xrightarrow{W} 0\}+$

$$+\sum\{aE' \mid E \xrightarrow{a} E',\ F \xrightarrow{a} F' \text{ and } E' \leftrightarrow_b F'\} + \sum\{\tau E' \mid X_{EF} \neq X_0,\ E \xrightarrow{\tau} E' \text{ and } E' \leftrightarrow_b F\}$$

and define the expression G_{EF} by

$$G_{EF} = \begin{cases} H_{EF} + \tau E & \text{if } X_{EF} \neq X_0 \text{ and } \exists F' \text{ with } F \xrightarrow{\tau} F' \text{ and } E \mathbin{\underline{\leftrightarrow}}_b F' \\ E & \text{otherwise.} \end{cases}$$

It follows from Lemma 6 that $\text{R1} \vdash E = E + H_{EF}$ and hence $\text{BR1} \vdash a(H_{EF} + \tau E) = aE$. Thus

$$\text{BR1} \vdash aG_{EF} = aE \quad \text{for } a \in A_\tau. \tag{1}$$

It suffices to prove that for $X_{EF} \in V_S$

$$\text{BR1} \vdash G_{EF} = \sum \{W \mid E \xrightarrow{W} 0 \text{ and } F \xrightarrow{W} 0\} + \sum \{aG_{E'F'} \mid E \xrightarrow{a} E', \ F \xrightarrow{a} F' \text{ and } E' \mathbin{\underline{\leftrightarrow}}_b F'\} +$$

$$\sum \{\tau G_{E'F} \mid X_{EF} \neq X_0, \ E \xrightarrow{\tau} E' \text{ and } E' \mathbin{\underline{\leftrightarrow}}_b F\} + \sum \{\tau G_{EF'} \mid X_{EF} \neq X_0, \ F \xrightarrow{\tau} F' \text{ and } E \mathbin{\underline{\leftrightarrow}}_b F'\}.$$

By (1) this is equivalent to

$$\text{BR1} \vdash G_{EF} = H_{EF} + \sum \{\tau E \mid X_{EF} \neq X_0, \ F \xrightarrow{\tau} F' \text{ and } E \mathbin{\underline{\leftrightarrow}}_b F'\}. \tag{2}$$

In case $X_{EF} \neq X_0$ and $\exists F'$ with $F \xrightarrow{\tau} F'$ and $E \mathbin{\underline{\leftrightarrow}}_b F'$, this follows from the definition of G_{EF}. In case $X_{EF} \neq X_0$ and $\nexists F'$ with $F \xrightarrow{\tau} F'$ and $E \mathbin{\underline{\leftrightarrow}}_b F'$, (2) reduces to $\text{BR1} \vdash E = H_{EF}$, and by Lemma 6 if suffices to establish, for $x \in A_\tau \cup V$, that

$$\text{if } E \xrightarrow{x} E' \text{ then } \begin{array}{l} x = \tau \text{ and } E' \mathbin{\underline{\leftrightarrow}}_b F \\ \text{or } \exists F' : F \xrightarrow{x} F' \wedge E' \mathbin{\underline{\leftrightarrow}}_b F'. \end{array}$$

But this follows from $E \mathbin{\underline{\leftrightarrow}}_b F$, using that if $F \xrightarrow{\tau} F_1 \implies F''$ with $F'' \mathbin{\underline{\leftrightarrow}}_b E \mathbin{\underline{\leftrightarrow}}_b F$, then $E \mathbin{\underline{\leftrightarrow}}_b F_1$ by Lemma 5, violating the assumptions. Finally, in case $X_{EF} = X_0$, (2) also reduces to $\text{BR1} \vdash E = H_{EF}$, and this time I have to establish, for $x \in A_\tau \cup V$, that

$$\text{if } E \xrightarrow{x} E' \text{ then } \exists F' : F \xrightarrow{x} F' \wedge E' \mathbin{\underline{\leftrightarrow}}_b F',$$

which follows immediately from $E \mathbin{\underline{\leftrightarrow}}_{rb} F$. □

Corollary 2 (Completeness) For $E, F \in \mathcal{E}^g$: $E \mathbin{\underline{\leftrightarrow}}_{rb} F \Leftrightarrow \text{BR1-2} \vdash E = F$.

6 Completeness for finite process expressions

Theorem 2 Let $E_0, F_0 \in \mathcal{E}^f$ with $E_0 \mathbin{\underline{\leftrightarrow}}_{rb} F_0$. Then there is a finite well-founded recursive specification S B-provably satisfied in the same variable $X_0 \in V_S$ by both E_0 and F_0.

Proof: The construction of S is exactly as in the proof of Theorem 1. Using that $X_{EF} \xrightarrow{o} X_{E'F'}$ iff $S_{X_{EF}}$ has a summand $aX_{E'F'}$ with $a \in A_\tau$, it is easy to show that any infinite o-path $X_{EF} \xrightarrow{o} X_{E'F'} \xrightarrow{o} \cdots$ implies an infinite path $E \xrightarrow{a_1} E' \xrightarrow{a_2} \cdots$ or $F \xrightarrow{b_1} F' \xrightarrow{b_2} \cdots$, which cannot exist by Lemma 1 since E_0 and F_0 are finite. Hence S is a well-founded recursive specification. The proof that S is finite and provably satisfies both E_0 and F_0 in X_0 is exactly as before, except that Lemma 6 is not needed, as recursion-free process expression are already in head normal form and therefore satisfy

$$\vdash E = \sum \{aE' \mid E \xrightarrow{a} E'\} + \sum \{W \mid E \xrightarrow{W} 0\}$$

without using axiom R1. As this was the only call for this axiom in the proof of Theorem 1 it follows that S is B-provably satisfied in $X_0 \in V_S$ by both E_0 and F_0. □

Proposition 14 (Unique solutions) If S is a finite well-founded recursive specification and $X_0 \in V_S$, then there is an expression E which provably satisfies S in X_0. Moreover if there are two such expressions E and F, then $\vdash E = F$.

Proof: This is a matter of repeatedly substituting S_Y for Y ($Y \in V_S$) in the equations of S. A detailed proof is omitted due to lack of space. It can be found in the technical report version of this paper. Note that this proposition does not follow from Proposition 13, as one may not use axioms R1 and R2. □

Corollary 3 (Completeness) For $E, F \in \mathcal{E}^f$: $E \mathbin{\underline{\leftrightarrow}}_{rb} F \Leftrightarrow \text{B} \vdash E = F$.

7 Completeness for all process expressions

Theorem 3 For every $E \in \mathcal{E}$ there exists a guarded expression E' with R1,3-6 $\vdash E = E'$.

Proof: It suffices to prove this for expressions of the form $E = \mu XF$. Following Milner, I prove a stronger result by induction on the depth of nesting of recursions in F, namely

For every $F \in \mathcal{E}$, there exists a guarded expression F' for which

- X is guarded in F'
- No free unguarded occurrence of any variable in F' lies within a recursion in F'
- R1,3-6 $\vdash \mu XF = \mu XF'$.

Assume that this property holds for every G whose recursion depth is less than that of F. Then for each recursion μYG in F that lies within no other recursion in F, there must be a guarded expression G' such that Y is guarded in G', no free unguarded occurrence of any variable in G' lies within a recursion in G', and R1,3-6 $\vdash \mu YG = \mu YG'$. These conditions ensure that no free unguarded occurrence of any variable in $G'\{\mu YG'/Y\}$ lies within a recursion in this expression.

Let F_1 be the result of simultaneously replacing every such top-level recursion μYG in F by $G'\{\mu YG'/Y\}$. Clearly F_1 is guarded, R1,3-6 $\vdash F = F_1$, and no free unguarded occurrence of any variable in F_1 lies within a recursion in F_1. In converting F_1 to F' such that R1,3-6 $\vdash \mu XF_1 = \mu XF'$, it remains only to remove all free unguarded occurrences of X from F_1, knowing that they do not lie within recursions. Here the axioms R3-6 are applied.

First any free unguarded occurrence of X that is not in the scope of a τ prefixing operator can be removed by R3. Next for any free unguarded occurrence of X that is in the scope of 2 or more τ's, this number can be lowered by application of R4. Applying R4 from left to right does not change the number of free unguarded occurrences of X, and does not raise the number of τ's scoping any particular such occurrence. So after finitely many applications all free unguarded occurrences of X are in the scope of exactly one τ operator, and applying R5 makes that they are all in the scope of the same τ. Finally by A3 at most one such occurrence remains, and this one is eliminated by R6. □

Corollary 4 For $E, F \in \mathcal{E}$: $E \leftrightarrow_{rb} F \Leftrightarrow \text{BR} \vdash E = F$.

8 Concluding remarks

The notion of branching bisimulation congruence employed here

- equates livelock and deadlock: $\mu X(\tau X) = \tau 0$
- does not equate divergence and livelock: $\mu X(\tau X + E) \neq \mu X(\tau X) + E$
- abstracts from divergence: $\mu X(\tau X + \tau E) = \tau E$
- and chooses minimal solutions in case of underspecification: $\mu XX = 0$,

just as Milner's standard version of weak bisimulation congruence. As in the case of weak congruence, there are alternative versions of branching congruence where these choices are made differently [4]. Complete axiomatizations for these notions remain to be provided. For weak bisimulation, such work has been done in D.J. Walker (1990): *Bisimulation and divergence*, I&C 85(2), pp. 202-241.

For arbitrary cardinals κ, one could define the language BCCS$^\kappa$ by allowing sets of expressions as argument of a choice operator $\sum$, and functions $V_S \to \mathcal{E}$ for $V_S \subseteq V$ as argument of a recursion operator μ, as long as the size of these sets and functions is less than κ. Such a language would represent all and only the behavious with less than κ states. In generalizing the completeness theorem for guarded BCCS$^\kappa$ expressions, one has to reformulate most axioms in an obvious way to deal with the new operators, slightly adapt the proof of Lemma 6 and make sure that there are at least κ variables in order for the first act in the proof of Theorem 1 to be possible. But nothing in my proof essentially depends on finiteness, and the result generalizes smoothly to guarded infinite-state behaviours. One could even take V to be a proper class and do away with all cardinality restrictions. Of course these axiomatizations are not effective, as some axioms have infinitely many premisis. The case for unguarded expressions does not generalize in this way, as not every unguarded BCCS$^\kappa$ expression is branching congruent with a guarded one.

By combining the axioms presented here with the complete axiomatizations for strong bisimulation that allow closed CCS, CSP and ACP expressions to be converted into head normal form, one obtains complete axiomatizations for closed terms in the language BCCS$^\omega$ to which the ACCSP operators have been added, provided that they do not occur in the scope of recursion operators (cf. MILNER [6]). Remarkably, in this setting the axiom B can be simplified to $a\tau X = aX$.

Theorem 4 Every closed instance of B is derivable from $a\tau X = aX$.

Proof: (sketch) $a(bc+cb)+cab = ab\|c = a\tau b\|c = a(\tau(bc+cb)+c\tau b)+ca\tau b$. Placing both sides in CSP's synchronous composition with $a(b+c)$ yields $a(b+c) = a(\tau(b+c)+c)$. In this proof b can be replaced by $\sum_{i\in I} b_i$ and similarly for c. Now a parallel composition with $a(\sum_{i\in I} b_iE_i + \sum_{j\in J} c_jF_j)$, in which synchronization is required (only) for a, b_i and c_j yields

$$a(\sum_{i\in I} b_iE_i + \sum_{j\in J} c_jF_j) = a(\tau(\sum_{i\in I} b_iE_i + \sum_{j\in J} c_jF_j) + \sum_{j\in J} c_jF_j).$$

(In fact, one needs to assume here that the b_i and c_j are pairwise distinct, and do not occur in E_h and F_k, but this restriction can be removed with a relabelling.) □

If one would now require an abstract congruence to satisfy $a\tau X = aX$, and an interleaving congruence to be a congruence for all the operators needed above and to satisfy the equations needed above (which are standard and already satisfied by strong congruence), and if one agrees that any finite process is representable by an expression $\sum a_iE_i$, then it follows that for finite processes branching bisimulation is the finest abstract interleaving congruence that generalizes τ-less bisimulation.

References

[1] J.A. BERGSTRA & J.W. KLOP (1985): *Algebra of communicating processes with abstraction. Theoretical Computer Science* 37(1), pp. 77–121.

[2] J.A. BERGSTRA & J.W. KLOP (1988): *A complete inference system for regular processes with silent moves.* In F.R. Drake & J.K. Truss, editors: *Proceedings Logic Colloquium 1986,* Hull, North-Holland, pp. 21–81. First appeared as: Report CS-R8420, CWI, Amsterdam, 1984.

[3] R.J. VAN GLABBEEK (1990): *The linear time - branching time spectrum.* In J.C.M. Baeten & J.W. Klop, editors: *Proceedings CONCUR 90,* Amsterdam, LNCS 458, pp. 278–297.

[4] R.J. VAN GLABBEEK & W.P. WEIJLAND (1990): *Branching time and abstraction in bisimulation semantics.* Technical Report TUM-I9052, SFB-Bericht Nr. 342/29/90 A, Institut für Informatik, Technische Universität München, Munich, Germany. Extended abstract in G.X. Ritter, editor: *Information Processing 89,* Proceedings of the IFIP 11th World Computer Congress, San Fransisco, USA 1989, Elsevier Science Publishers B.V. (North-Holland), 1989, pp. 613-618.

[5] R. MILNER (1984): *A complete inference system for a class of regular behaviours. Journal of Computer and System Sciences* 28, pp. 439–466.

[6] R. MILNER (1989): *A complete axiomatisation for observational congruence of finite-state behaviours. Information and Computation* 81, pp. 227–247.

[7] R. MILNER (1990): *Operational and algebraic semantics of concurrent processes.* In J. van Leeuwen, editor: *Handbook of Theoretical Computer Science,* chapter 19, Elsevier Science Publishers B.V. (North-Holland), pp. 1201–1242. Alternatively see *Communication and Concurrency,* Prentice-Hall International, Englewood Cliffs, 1989, of which an earlier version appeared as *A Calculus of Communicating Systems,* LNCS 92, Springer-Verlag, 1980.

Object Oriented Application Flow Graphs and their Semantics

Erik de Haas and Peter van Emde Boas

ILLC; University of Amsterdam, Plantage Muidergracht 24,
NL-1018TV Amsterdam, The Netherlands

Abstract. In this paper we present a language called OOAFG and its semantics. OOAFG intends to express parallelism in the context of object identity. We believe that the approach to parallelism we take here is a kind of parallelism that is relatively unexplored which connects both to the world of Object Oriented programming, and to the dataflow paradigm.

1 Introduction

The purpose of this paper is the presentation of an operational semantic model for a programming system which may be considered to be a hybrid between the dataflow and the Object Oriented paradigms. As such it represents an attempt to introduce a notion of objects and object identity within the dataflow model and at the same time a probably new approach for introducing parallelism in the world of OO programming.

By way of motivating example consider the simple model of an order handling department. When an order from client A for the item B has been entered at the input desk of the department two subdivisions become active: the storage room will check whether item B is deliverable, and the financial department will check the credit status for client A ; subsequently when both actions are completed with positive result the shipping department will deliver the ordered goods and perform the billing activities. If one of the procedures has a negative outcome some other appropriate action will be taken.

In this application there is room for parallel processing: the storage room and the financial division may work in parallel, processing the same order. Moreover the order department may be involved in processing several orders at the same time, with quite different processing times for the various subtasks. It is an attractive idea therefore to model the department as described above by a dataflow graph as indicated in the picture below. In this model the orders are modelled by tokens which are accepted at the input, split in two tokens for processing by the two parallel subtasks and finally merged at the shipping department. However, as soon as it is realized that the subprocesses may deliver their results out of order, it becomes evident that a pure dataflow approach will also model the course of events where the item B1 of the first order will be combined with the credit status of the client A2 from the second order, resulting into an unintended shipment of item B1 to A2. The pure dataflow model lacks the notion of object identity required for keeping the two subtasks connected. An alternative would be to represent the original order by an object in the sense of the Object Oriented paradigm. Being entered, both checks and the final shipment will become methods for the object class representing the orders. But in this model the potential parallelism is disregarded; an object can be involved in performing a single method only; parallelism results from having several objects being active at the same time, behaving in an independent way and sending messages

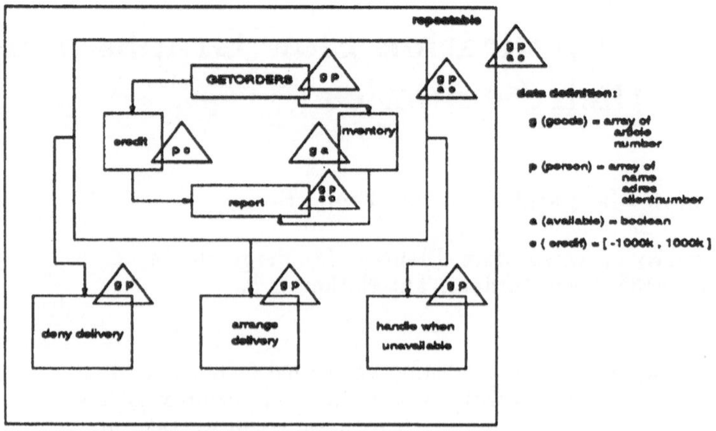

Fig. 1. Example of an OOAFG

to each other, but within the the thread of the object itself the flow of control is sequential. The idea of an object branching into parts which become active independently and which are subsequently merged into the original object is not available in a traditional language like Smalltalk ([GB83]). Also in the Parallel OO languages POOL ([Ame87]) and PROCOL ([BL91]) this form of parallelism is absent.

The model of Object Oriented Application Flow Graphs represents a first step towards merging the dataflow and the OO paradigms by solving the two problems mentioned above. From the OO paradigm we incorporate the traditional notions of data structures, encapsulation, object identity and inheritance whereas the dynamic behaviour of an object consisting of a sequence of method invocations and/or acceptances is substituted by the data-flow originated traversal of an application flow graph. The nodes in this graph, called *components* in this paper will replace the methods of the OO world. Moreover we will allow these application flow graphs to be hierachically composed: the nodes in the graph either represent atomic actions, or they can be expanded into a graph consisting of subcomponents. By introducing this hierarchical structure we are in fact linking with a third less known paradigm; the paradigm of *Application Flow Graphs*. A further extension is that components may behave in a nondeterministic way. Finally we allow for nodes to be marked as repeatable, thus allowing for a form of iteration.

Our model is described by means of an operational semantics using a Plotkin style transition system ([HP79], [Plo81]). Notwithstanding the potential complexity caused by the merging of the two paradigms it turns out that the actual semantic model is quite simple: four rules suffice for describing the intended transition system modelling the overall flow of control. Note however that in our model the transitions for atomic actions are left unspecified.

Certainly the theory presented in this paper is not the first attempt for providing a formal semantics for a (parallel) OO language. For the language POOL mentioned before extensive work on its mathematical foundation has been performed, and this work has served as a source for inspiration for our paper ([Ame86b], [ABKR89], [Rut88]); however the computational models of POOL and OOAFG are quite different. A large amount of foundational work on the semantics for objects, classes and inheritance has been performed in the wake of the already much older theory of abstract data types (see for example [CW85]), but these problems are orthogonal to the problems of adding parallelism to object bahavior studied in this paper. Also we believe that the operational aspects of our model should be described using a simple and appropriate formalism like the nowadays standard tool of transition systems. It will been seen that our specification doesn't require the use of powerful mathematical formalisms like full first order logic or category theory.

The remainder of this paper is structured as follows. In section 2 we present a sketch of (our perspective on) the three paradigms which are combined within OOAFG. Section 3 presents the description of our language OOAFG and its features. In section 4 we present the transition system for the operational semantics of OOAFG. In the final section we mention some possible extensions.

2 Three Paradigms (and one perspective)

There exist many points of view on what is meant by the Object Oriented style of simulating the real world as in programming and modelling. One concept most of these points of view have in common is that if one knows what an 'object' is, we can define Object Oriented programming or designing as a style of programming in which a whole system is described as consisting of a collection of objects that communicate with each other. In a first approximation an object can be described as an integrated unit of data and procedures acting on these data.

A methodology could safely be called Object Oriented if it is possible to describe its entities in terms of objects, classes and methods and preferably supports a feature like inheritance for its objects and classes. Furthermore the objects must interact with each other via messages. This definition largely coincides with the definitions in [Mey89],[Ame86a] and [Weg89].

The central structure in a programming language following the Data Flow paradigm is a *directed graph* (often called *net*) which ties up the data flow. The graph consists of *nodes* and *edges*. The following phrases characterize the data flow paradigm: *data items (tokens) are sent over the edges by the nodes; A node has incomming (input) and outgoing (output) edges: it receives data items from its incomming edges and puts data items on its outgoing edges; A node can perform a firing: it removes data items from its incomming edges, performs some computation and puts data items on its outgoing edges; A node can perform a firing depending on the availability of tokens on its incomming edges.* Consult for a nice and formal approach to Data Flow structures for example [Kah74] or [Kok88] and for an overview consult [TBH82] or [MR90].

We consider a graphical structure called *Application Flow Graph* (AFG) originated from the data flow paradigm for programming languages

AFG relates executable blocks, called *components*. Globally seen an AFG consists of a collection of components, that are related to each other as nodes in a directed graph. A component itself can be a graph of (sub)components. This way one can construct a layered AFG. At the lowest level in the layering the components should be executable. The directed edges in the graph represent the control and data flow. Along the edges *data packets* are send from one component to the other.

For defining the language OOAFG we will use the following perspective on objects, classes and inheritence:

An Object is a data packet (state data) together with a collection of components (methods). We model the data packets by saying that data packets consist of a number of *fields* where each field has a *type*. The content of a data packet are the values stored in those fields. We associate with each data packet a *data definition* that defines the structure of the data packet by enumerating its fields.

A component processes some data. We say that each component C *covers* a collection of fields, denoted by $f(C) = \{a, b, c, \ldots\}$. We call C a method of a data object α if the data packet (state data) of α contains all the fields that C processes. In other words: C is a method of a data object α if the data definition D of α contains the cover of C (i.e. $f(C) \subseteq D$). This way we associate methods with state data, together forming the notion of object in the OO paradigm. It is easy to see that it is natural to associate a component C with a data object α if the collection of fields covered by C is contained in α, because then the object α contains the necessary ingredients to be processed by C and C meets the requirements to be a method for α because it can process the data of the right sort.

Now we have defined the objects the classes are easily defined being the data definitions for the data objects. The data definitions define the structure of the (state) data of an object and also determine which methods (components) are associated with that object. We can construct arbitrary classes with their own set of methods, because we are free to use different fields for the same types of data.

We also have inheritance. A class D (i.e. a data definition) inherits the components (i.e. methods) from all classes that consist of a subset of fields of D. So for example the class $D = \{a, b, c\}$ inherits the properties of class $D' = \{a, b\}$ and from $D" = \{b, c\}$ and from $D"' = \{a\}$ etc.. In this objective class A is a superclass of B if and only if A is a subset of B. This is a simple kind of inheritance.

3 Definition of Object Oriented Flow Graphs (OOAFG)

A specification written in the language of OOAFG consists of *definitions of data objects* and *graphs* that are built up from *components* and *directed edges*. In our language we have a number of features: layering, parallelism, choice, iteration and synchronization.

Definition 3.1 A definition of a data object (in short *data definition*) D defines a class of data objects consisting of a collection of *fields* with their *type*, i.e. $D \subseteq$ **FIELDS** $\times$ **TYPES** where **FIELDS** is the set of all fields and **TYPES** is the set of all types. ♣

Data objects travel through a structure graph or in other words a graph gives the operation sequence of a data object.

Definition 3.2 An (OO) application flow graph (OOAFG) consists of specification of a component. It is sufficient to see a specification of an application as the specification of a component, because an application itself is seen as a component consisting of subcomponents.

An OOAFG for a structural component C looks as follows:

- it contains a **data definition** $D_C \subseteq$ **FIELDS** $\times$ **TYPES**
- it contains a **structure graph** $G_C = (V_C, E_C)$
- it contains an OOAFG for all the nodes of G_C (i.e. subcomponents of C) and a mapping g_C that maps the nodes of G_C to their OOAFGs.
- it contains a function f_C representing a cover of the data definition of C by mapping nodes of G_C to fields of D_C.

- it optionally contains of the marker *repeatable*
- an OOAFG is a *set*

An OOAFG for an executable component C looks as follows:

- it contains a **data definition** $D \subseteq$ **FIELDS** $\times$ **TYPES**
- it contains a structure graph $G_C = (\emptyset, \emptyset)$
- it contains an executable **program** π_C
- it optionally contains the marker *repeatable*
- it is a set ♣

The structure graphs consist of *nodes* denoting the *sub components* of the component and *directed edges* denoting the control and data flow. The sub components themselves have an OOAFG specification. A node can be marked repeatable. A *cover* denotes which component (node) processes which part of a data object (fields). We represent a sub cover by a function $f : V \rightarrow P(\textbf{FIELDS})$ where V is the set of nodes of an OOAFG-graph and $P(\textbf{FIELDS})$ the set of all sets of fields of data objects. The function f maps a node to the set of fields this node (component) processes. We will demand that for each component the cover agrees with its data definition , i.e. $f_C(C') = D_{C'}$ where $D_{C'}$ is the data definition in the OOAFG specification of component C'.

If C' is a sub component of C we will call C the *outer component* of C'. We will often write $f(C')$ to denote the collection of fields associated to (sub) component C' instead of $f_C(C')$.

We will assume that we can distinguish between structural and executable components. The requirement that an OOAFG specification is a set is necessary and sufficient to guarantee wellfoundedness of the definition of the OOAFG specifications. We will denote the set of Object Oriented Application Structuring specifications (i.e. the language) by **OOAFG**

4 Semantics of OOAFG

Below we will present the semantics of OOAFG. First we need some definitions.

Definition 4.1 In the course of this text we will make use of the *variant notation* to indicate a change in some function.

Let ρ be a (possibly partial) function then $\rho\{y/x\}$ is defined by:

$$\rho\{y/x\}(z) = \begin{cases} \rho(z) & \text{if } z \neq x \text{ (possibly undefined)} \\ y & \text{if } z = x \end{cases}$$

If $A = \{(x_1, y_1), ..., (x_n, y_n)\}$ and $x_i \neq x_j$ (for $1 \leq i, j \leq n$ and $i \neq j$) is some finite collection of pairs, we mean by $\rho\{A\}$ the following variant of ρ: $\rho\{y_1/x_1\}...\{y_n/x_n\}$ Observe that this definition is independent of the ordering of the tuples (x_i, y_j) ♣

Definition 4.2 The labels identify a data object. α, β, ... denote *labels*. We denote the set of all labels by **Labels**. A *name of a component* is a statement. We define labeled statements as being tuples $< \alpha, s >$ where α is some label. We denote by **LSTAT** the set of labeled statements. ♣

Definition 4.3 To keep track of the values in the fields of the data objects, we will use a function of the following signature:

$$\mathbf{Labels} \mapsto (\mathbf{FIELDS} \mapsto \mathbf{VALUES})$$

This function assigns to each data object a function that maps the fields of a data object to a value. We will call a function of the above signature a *data state*. We denote the set of all data states by Δ with typical element δ. We will call $\delta(\alpha)$: **FIELDS** $\mapsto$ **VALUES** the data state for object α.

A change in the data state is given by the (partial) function $C(\alpha, \delta)$: **FIELDS** $\mapsto$ **VALUES.** ♣

To denote a change in the data state we use the variant notation. Given a component name C an object α and a data state δ, $C(\alpha, \delta)$ determines a set of field/value pairs[1] that give the changes in the values of some fields of object α. The domain of $C(\alpha, \delta)$ is a subset of its cover $f(C)$.

Because $C(\alpha, \delta)$ is a function it holds that $(x, y) \in C(\alpha, \delta)$ and $(x, z) \in C(\alpha, \delta)$ only if $y = z$, so we may assume for $C(\alpha, \delta) = \{(x_1, y_1), (x_2, y_2), \ldots\}$ that all x_i are different. Thus we can denote with the variant notation a change in the data state for an object α by $\delta(\alpha)\{C(\alpha, \delta)\}$. A change in a data state δ is then denoted by $\delta\{\delta(\alpha)\{C(\alpha, \delta)\}/\alpha\}$, meaning that the data state is changed in such a way that the function that δ assigns to an object α is updated with $C(\alpha, \delta)$. We will abbreviate this by $\delta\{\uparrow C(\alpha, \delta)\}$.

We will also use a function Initial(α), that assigns to fields of α an initial value. We will handle this function in the same manner as we did with $C(\alpha, \delta)$.

Furthermore we define a predicate called *changes*. We say *changes*(C, field$_a$) only if C covers field$_a$ (i.e. field$_a \in f(C)$). If field$_a \in f(C)$ then field$_a$ can be in the domain of $C(\alpha, \delta)$.

Definition 4.4 A *configuration* in our transition system will consists of three parts: A set of *labeled statements*, a *data state* that assigns values to the fields of data objects, and an *OOAFG specification*. In other words[2]

$$\mathbf{CONFIG} = P_{fin}(\mathbf{LSTAT}) \times \Delta \times \mathbf{OOAFG} \quad ♣$$

We construct a partial ordering on the components of an OOAFG- specification. That is on the components of the structure graph and on the components of the structure graphs of the components of this structure graph etc. etc..

Because we allow iteration it is possible that more than once the same component can be executed. In order to make things go well, we have to define some ordering between the subcomponents of a component and the component itself. We will put subcomponents *before* their outer component.

Definition 4.5 Let AV_{CP} denote the set of all the sub components of CP and of all the sub components of these sub components etc. etc.. Let AE_{CP} denote the set of all the edges of G_{CP} and of all the edges of the the graphs of the sub components of CP etc. etc. . We construct a partial ordering $(AV_{CP}, \preceq)$ directly from its OOAFG-specification. We define the relation $\preceq$ on AV_{CP} as follows:

for all $C, C' \in AV_{CP}$

---

[1] A function f can be seen as a set of pairs where $(x, y) \in f$ iff $f(x) = y$

[2] P_{fin}(**LSTAT**) denotes the set of all finite subsets of **LSTAT**. We will often write a configuration down like this: $< X \cup \{< \alpha, C >\}, \delta, \text{ooafg} >$. We assume then that $< \alpha, C > \notin X$. We can call $< \alpha, C > \in X$ a *component instance of component C for the object α.*

1. $C \preceq C'$ if $(C, C') \in AE_{CP}$
2. $C \preceq C$
3. $C \preceq C'$ if there exists a $C'' \in AV_{CP}$ such that $C \preceq C''$ and $C'' \preceq C'$
4. if $C \preceq C'$ and $C \neq C'$ then
   (a) for all sub components $C_i \in V_C$ holds $C_i \preceq C'$
   (b) for all sub components $C'_i \in V_{C'}$ holds $C \preceq C'_i$
5. for all $C' \in V_C$ holds $C' \preceq C$ ♣

It is easily proven to that $(AV_{CP}, \preceq)$ is reflexive, transitive and antisymetric.

We will use the partial order to determine which component is to be executed before or after or in parallel to which other component. We say $C \preceq C'$ if C is to be executed before C', $C' \preceq C$ if C is to be executed after C' and $C \bot C'$ if C is to be executed in parallel with C'.

The following assertion holds for a component to be allowed to be executed: A data object α can execute a component C considering a set of labeled statements X *iff* there does not exist any component C' to be executed in the set of labeled statements X by the same data object α that precedes α in X. In other words $< \alpha, C >$ can be executed considering the configuration $< X \cup \{< \alpha, C >\}, \delta, \text{ooafg} >$ if $\neg\exists < \alpha, C' > \in X(C' \preceq C \wedge C' \neq C)$

We only allow changes in the values of a data object when it is *safe* to make them. It is safe to change some part of a data object only if it is not possible that the same part of that data object can be executed at the same time, for then we have inconsistent data. Considering a configuration $< X \cup \{< \alpha, C >\}, \delta, \text{ooafg} >$, it is safe to execute $< \alpha, C >$ (i.e. it is safe for a component C to change some part of the data object α) if in X there does not exist any collateral component C' $(C \bot C')$, that is labeled with the same label (i.e. same data object α) and covers a field of the data object that is also covered by C. So

Definition 4.6 $\text{safe}(C(\alpha, \delta), X) \Leftrightarrow$
$\forall < \alpha, C' > \in X(C \bot C' \rightarrow \neg\exists \text{field}_a \in f(C)(\text{changes}(C', \text{field}_a))$ ♣

As stated before in OOAFG we allow nondeterminism. In other words if a node has several outgoing edges, only a subset of these edges has to be traveled along by a data object that has visited that node. We will describe this phenomenon by nondeterministically choosing a connected sub graph of a structure graph of a structural component when we want to execute it. Traveling through the connected sub-structure-graph is equivalent with not traveling along all edges in the whole graph. We will call this connected sub graph a ***sub-structure-graph.***

Definition 4.7 Let G be the OOAFG specification for some component C consisting of a structure graph G_C. We distinguish a subset of the nodes in G_C, the set of *begin nodes of* C or *begin components of* C, being those nodes that have no incomming edges in structure graph G_C. H_C is a *sub-structure-graph of* G_C if it satisfies the following condition:

- Let $G_C = (V_C, E_C)$ and let I_C denote the set of *begin nodes* of G_C. Let $H_C = (V_H, E_H)$ and I_H the set of begin nodes for H_C. Then
  1. $I_H \subseteq I_C$
  2. H_C is a connected sub graph of G_C such that all the nodes of H_C are on a directed path starting from a begin node. (i.e.$\forall v \in V \exists b \in I_H \exists v_1, ..., v_n \in V[(b, v_1), (v_1, v_2), ..., (v_{n-1}, v_n), (v_n, v) \in V]$) ♣

Definition 4.8 Now we give the *transition system* that describes OOAFG. Let 'ooafg' be the OOAFG-specification of component CP.

1. Executing a structural component amounts to replace the instance of the structural component by the instances of the components of a sub-structure-graph of the structural component. A component C can be executed if there do not exist any other component instances of the same data object that precedes C.

$$< X \cup \{< \alpha, C >\}, \delta, \text{ooafg} > \rightarrow < X \cup Y, \delta, \text{ooafg} >$$

where C is structural $\wedge \neg\exists C'(< \alpha, C' >\in X \wedge C' \preceq C \wedge C' \neq C) \wedge Y = \bigcup_{z \in Z} < \alpha, z >$ where Z is the set of nodes of a sub-structure-graph of component C

2. Executing an executable component amounts to process the data fields of the executing data object. The executable component C can be executed if it is safe to do so and if there does not exist any component instance for the same data object that should be executed before C.

$$< X \cup \{< \alpha, C >\}, \delta, \text{ooafg} > \rightarrow < X, \delta', \text{ooafg} >$$

where C is executable $\wedge \neg\exists C'(< \alpha, C' >\in X \wedge C' \preceq C \wedge C' \neq C) \wedge \delta' = \delta\{\uparrow C(\alpha, \delta)\} \wedge \text{safe}(C(\alpha, \delta), X)$

3. Iterating a structural component amounts to executing a component and create an other instance of this component. Note that a structural component succeeds its sub components in the ordering.

$$< X \cup \{< \alpha, C >\}, \delta, \text{ooafg} > \rightarrow < X \cup Y \cup \{< \alpha, C >\}, \delta, \text{ooafg} >$$

where C is structural and repeatable $\wedge \neg\exists < \alpha, C' >\in X(C' \preceq C \wedge C' \neq C)$ $\wedge\, Y = \bigcup_{z \in Z}\{< \alpha, z >\}$ where Z is the set of nodes of a sub-structure-graph of component C

4. For iteration of an executable component the same holds as in axiom 3

$$< X \cup \{< \alpha, C >\}, \delta, \text{ooafg} > \rightarrow < X \cup \{< \alpha, C >\}, \delta', \text{ooafg} >$$

where C is executable and repeatable $\wedge \neg\exists < \alpha, C' >\in X(C' \preceq C \wedge C' \neq C) \wedge$ $\text{safe}(C(\alpha, \delta), X) \wedge \delta' = \delta\{\uparrow C(\alpha, \delta)\}$ ♣

Before we can give the semantics of an OOAFG specification we need some definitions.

- We call $< \emptyset, \delta, \text{ooafg} >$ a *final configuration.* This is justified by the observation that assuming some fixed domain of data objects , in a configuration of the form $< \emptyset, \delta, \text{ooafg} >$ all the data objects that satisfy the OOAFG specification *ooafg* are either already totally processed or not processed at all.
- With $config_1 \xrightarrow{*} config_2$ we mean that configuration $config_1$ can be transferred to $config_2$ in *zero or more* transition steps ($\rightarrow$). Naturally all the transition steps are derived from (given by) the transition system.
- With $\mathbf{CONFIG}_\perp$ we denote the set $\mathbf{CONFIG} \cup \{\perp.\}$.

Now we can give the semantics of an OOAFG specification in terms of a non-deterministic configuration transformation function:

$$\mathcal{O} : \mathbf{OOAFG} \mapsto (\mathbf{CONFIG} \mapsto \mathcal{P}(\mathbf{CONFIG}_{\perp}))$$

where:

$$\mathcal{O}[\text{ooafg}](< X, \delta, \text{ooafg} >) =$$

$$\{< \emptyset, \delta', \text{ooafg} > \in \mathbf{CONFIG} \mid < X, \delta, \text{ooafg} > \stackrel{*}{\rightarrow} < \emptyset, \delta', \text{ooafg} >\} \cup$$

$$\{\perp \mid \text{there exists an infinite sequence } < X, \delta, \text{ooafg} > \rightarrow \ldots \rightarrow < X_n, \delta_n, \text{ooafg} > \rightarrow \ldots\}.$$

Because we did not demand the number of data objects to be finite and because we allow unguarded loops in the OOAFG specification, a configuration *config* will in general be mapped to an infinite subset of $\mathbf{CONFIG}_{\perp}$ by the function $\mathcal{O}[\text{ooafg}]$ that describes an OOAFG specification. If we fix the number of data objects, it is easily shown that the collection of configurations that can be reached in *one* transition step starting at some specific configuration is finite. Königs lemma then shows that $\mathcal{O}[\text{ooafg}](\mathit{config})$ contains $\perp$ or is finite.

The configuration transformation function $\mathcal{O}[\text{ooafg}]$ associates with every configuration a set of end configurations. This way a meaning is given to a configuration. If we want to obtain a proper meaning for an OOAFG specification, we will have to consider a proper configuration that denotes an initial state of an environment that runs an OOAFG specification.

5 Extensions to the semantics of OOAFG

There are several extensions on the semantics of OOAFG imaginable in order to enlarge the understanding of OOAFG and give directions towards a broader theoretical fundamentals for the concepts of OOAFG.

In the transition system for describing OOAFG we treated parallelism as interleaving of actions (an action is in this context a change of the data state). We also constructed a semantics for OOAFG in which we can express parallelism as *true* parallelism (i.e. more that one action can take place at one moment in time) or even maximal parallelism (all the actions that could possibly take place in parallel do take place at exactly the same time).

An other variant for the semantics of OOAFG we constructed, is to describe the semantics is term of *histories* in the spirit of [BKMOZ85]. This approach gives the possibility to model side effects of the execution of a component (i.e. a component does not only change the data state but also produces some side effect like printing or bleeping).

We also studied the semantics of a subset of the flow graphs, the series parallel graphs. These graphs are interesting because these graphs have a linear syntax where the description of such a graph can be mapped to a mathematical description of a graph in a compositional manner.

We will not present these three extended semantics in this paper, due to lack of space.

6 Conclusion & Acknowledgements

It is proved to be possible to obtain with simple means an operational semantics for the language OOAFG, which was designed to express parallelism in the context of object identity. Standard Object Oriented languages do not allow objects to execute more then one of their methods in parallel. OOAFG intends to express this kind of parallelism.

A nice observation is that the principle that motivated OOAFG can be described in only four axioms. One could say that this principle is naturally formalizable, and therefore not difficult to comprehend.

We would like to thank the members of ESAT IBM Uithoorn and especially Ghica van Emde Boas and Gilles Schreuder, who largely contributed to the ideas presented in this paper.

References

[Ame86a] P. America, *Object Oriented programming: a theoreticians introduction*, Bulletin of the European Association for Theoretical Computer Science, 29, 1986, pp.69-84.

[Ame87] P. America, *POOL-T: A parallel object oriented Language*, in A. Yonezawa, M. Tokoro (Eds.), Object Oriented Concurrent Programming, MIT Press, 1987, pp.199-220.

[ABKR86] P. America, J.W. de Bakker, J.N. Kok, J.Rutten, *Operational semantics for a parallel object oriented language*, in Conference Record of the 13th Symposium on Principles of Programming Languages (POPL), St. Petersburg Florida, 1986, pp.194-208.

[ABKR89] P.America, J.W. de Bakker, J.Rutten, J.N. Kok, *Denotational semantics of a parallel object oriented language*, Information and Computation vol. 83, pp.152-205, 1989

[BL91] J. van den Bos, C. Laffra, *PROCOL, a Concurrent Object Oriented Language with Protocols, deligation and constraints*, Acta Informatica, Vol.28, fasc.6, pp.511-538, 1991.

[BKMOZ85] J.W. de Bakker, J.N. Kok, J.-J.Ch. Meyer, E.-R. Olderog, J.I. Zucker, *Contrasting themes in the semantics of imperative concurrency*, Current Trends in Concurrency (J.W. de Bakker e.a. eds.), LNCS 244, Springer 1985.

[BJ66] C.Böhm, G. Jacopini, *Flow-diagrams, Turing Machines, and Languages with Only Two Formation Rules*, Comm. ACM 9 5 ,May 1966, pp. 366-371.

[CW85] L. Cardelli, P. Wegner, *On understanding Types, Data Abstractions and Polymorphism*, Computing Surveys, vol. 17, nr. 4, December 1985, pp.471-522.

[GR83] A. Goldberg, D, Robson, *Smalltalk-80: The language and its implementation*, Addison-Wesley, Reading, MA, 1983.

[HP79] M.C.B. Hennessy, G.D. Plotkin, *Full abstraction for a simple parallel programming language*, Proceedings of the 8th MFCS (J. Becvar ed.), LNCS 74 Springer 1979, pp.108-120.

[Kah74] G. Kahn, *The semantics of a simple language for parallel programming*, Proceedings Information Processing (Rosenfeld ed.), pp.471-475, North Holland, 1977.

[Kok88] J.N. Kok, *Data Flow semantics*, Technical report CS-R8835, Centre for Mathematics and Computer Science, Amsterdam, 1988.

[Mey88] Bertrand Meyer, *Object Oriented Software Construction*, Prentice Hall 1988, ISBN 0-13-629049-3.

[MR90] T.J. Marlow, B.G. Ryder, *Properties of dataflow frameworks*, Acta Informatica, vol.28, fasc.2, pp.121-163, 1990.

[Plo81] G.D. Plotkin, *A structural approach to operational semantics*, Technical Report DAIMI FN-19, Aarhuis University, Computer Science department, 1981.

[Rut88] J. Rutten, *Semantic correctness for a parallel object oriented language*, Report CS-R8843, Centre for Mathematics and Computer Science, Amsterdam, November 1987.

[TBH82] P.C. Treleaven, D.R. Brownbridge, R.P. Hopkins, *Data driven and demand driven computer architecture*, Computing Surveys 14(1), March 1982.

[Weg89] Peter Wegner, *Learning the Language*, BYTE, march 1989.

Some Hierarchies for the Communication Complexity Measures of Cooperating Grammar Systems

Juraj Hromkovic[1], Jarkko Kari[2], Lila Kari[2]

Abstract. We investigate here the descriptional and the computational complexity of parallel communicating grammar systems (PCGS). A new descriptional complexity measure - the communication structure of the PCGS - is introduced and related to the communication complexity (the number of communications). Several hierarchies resulting from these complexity measures and some relations between the measures are established. The results are obtained due to the development of two lower-bound proof techniques for PCGS. The first one is a generalization of pumping lemmas from formal language theory and the second one reduces the lower bound problem for some PCGS to the proof of lower bounds on the number of reversals of certain sequential computing models.

1 Introduction

Parallel Communicating Grammar Systems (PCGS) represent one of the several attempts that have been made for finding a suitable model for parallel computing (see [4] for an algebraic and [6], [1] for an automata theoretical approach). PCGS have been introduced in [12] as a grammatical model in this aim, trying to involve as few as possible non-syntactic components.

A PCGS of degree n consists of n separate usual Chomsky grammars, working simultaneously, each of them starting from its own axiom; furthermore, each grammar i can ask from the grammar j the string generated so far. The result of this communication is that grammar i includes in its own string the string generated by grammar j, and that grammar j returns to its axiom and resumes working. One of the grammars is distinguished as a master grammar and the terminal strings generated by it constitute the language generated by the PCGS.

Many variants of PCGS can be defined, depending on the communication protocol (see [8]), on the type of the grammars involved (see [12], [9]), and so on. In [12], [9], [11], [10] and [8], [14] various properties of PCGS have been investigated, including the generative power, closure under basic operations, complexity, and efficiency. In this paper we restrict ourselves to the study of PCGS composed of *regular grammars*. As no confusion will arise, in the sequel we will use the more general term PCGS when referring to these particular PCGS consisting of regular grammars.

[1] Department of Mathematics and Computer Science, University of Paderborn, 4790 Paderborn, Germany

[2] Academy of Finland and Department of Mathematics, University of Turku, 20500 Turku, Finland, email- santean@sara.cc.utu.fi

The most investigated complexity measure for PCGS has been the number of grammars the PCGS consists of, which is clearly a descriptional complexity measure. Here we propose for investigation two further complexity measures. One is the communication structure of the PCGS (the shape of the graph consisting of the communication links between the grammars) which can be considered as an alternative descriptional complexity measure to the number of grammars. This measure may be essential for the computational power of the PCGS, as showed also by results established in this paper. Here we consider mostly the following graphs as communication structures: linear arrays, rings, trees and directed acyclic graphs. The second complexity measure proposed here is the number of communications between the grammars during the generation procedure. This measure is obviously a computational complexity measure which is considered as a function of the length of the generated word. Here we investigate these complexity measures and the relations between them.

First, in Section 3, we relate these complexity measures to some sequential complexity measures. It is shown that PCGS with tree communications structure and $f(n)$ communication complexity can be simulated in real-time by one-way nondeterministic multicounter machines with at most $2\,f(n)$ reversals. PCGS with acyclic communication structure can be simulated in linear time by nondeterministic off-line multitape Turing machines.

The first simulation result is used in Section 4 to prove some lower bounds on the communication complexity of tree-PCGS. The lower bounds are achieved due to the modification of the lower bound proof technique on the number of reversals of multicounter machines developed in [5], [2].

The consequences are not only some hierarchies of communication complexity but also the fact that for tree-PCGS the increase of descriptional complexity cannot compensate for some small decreases of communication complexity.

Section 5, devoted to descriptional complexity measures, involves pumping lemmas for PCGS with tree structure, ring structure and with acyclic structures. This enables to obtain several strong hierarchies on the number of grammars of such PCGS.

2 Definitions and Notations

We assume the reader familiar with basic definitions and notations in formal language theory (see [13]) and we specify only some notions related to PCGS.

For a vocabulary V, we denote by V^* the free monoid generated by V under the operation of concatenation, and by λ the null element. For $x \in V^*$, $|x|$ is the length of x and if K is a set, $|x|_K$ denotes the number of occurrences of letters of K in x.

All the grammars appearing in this paper are assumed to be regular, that is, with productions of the form $A \longrightarrow wB$, and $A \longrightarrow w$, where A, B are nonterminals and w is a terminal word or the empty word.

Definition 1 *A PCGS of degree n, $n \geq 1$, is an n-tuple*

$$\pi = (G_1, G_2, \ldots, G_n),$$

where

- *$G_i = (V_{N,i}, \Sigma, S_i, P_i)$, $1 \leq i \leq n$, are regular Chomsky grammars satisfying $V_{N,i} \cap \Sigma = \emptyset$ for all $i \in \{1, 2, \ldots, n\}$;*
- *there exists a set $K \subseteq \{Q_1, Q_2, \ldots, Q_n\}$ of special symbols, called communication symbols, $K \subseteq \bigcup_{i=1}^{n} V_{N,i}$, used in communications as will be shown below.*

The communication protocol in a PCGS π is determined by its *communication graph*. The vertices of this directed graph are labeled by $G_1, \ldots, G_n$. Moreover, for $i \neq j$ there exists an arc starting with G_i and ending with G_j in the communication graph iff the communication symbol Q_j belongs to the nonterminal vocabulary of G_i.

An n-tuple $(x_1, \ldots, x_n)$ where $x_i \in \Sigma^*(V_{N,i} \cup \lambda)$, $1 \leq i \leq n$, is called a *configuration*. The elements x_i, $1 \leq i \leq n$, will be called *components* of the configuration.

We say that the configuration $(x_1, \ldots, x_n)$ directly derives $(y_1, \ldots, y_n)$ and write $(x_1, \ldots, x_n) \Longrightarrow (y_1, \ldots, y_n)$, if one of the next two cases holds:

(i) $|x_i|_K = 0$ for all i, $1 \leq i \leq n$, and, for each i, either x_i contains a nonterminal and $x_i \Longrightarrow y_i$ in G_i or x_i is a terminal word and $x_i = y_i$.

(ii) $|x_i|_K > 0$ for some i, $1 \leq i \leq n$.
For each such i we write $x_i = z_i Q_j$, where $z_i \in \Sigma^*$.
(a) If $|x_j|_K = 0$ then $y_i = z_i x_j$ and $y_j = S_j$.
(b) If $|x_j|_K > 0$ then $y_i = x_i$.
For all the remaining indexes l, that is, for those indexes l, $1 \leq l \leq n$, for which x_l does not contain communication symbols and Q_l has not occurred in any of the x_i, $1 \leq i \leq n$, we put $y_l = x_l$.

Informally, an n-tuple $(x_1, x_2, \ldots, x_n)$ directly yields $(y_1, y_2, \ldots, y_n)$ if either no communication symbol appears in $x_1, \ldots, x_n$ and we have a componentwise derivation, $x_i \Longrightarrow y_i$ in G_i, for each i, $1 \leq i \leq n$, or communication symbols appear and we perform a *communication step*, as these symbols impose: each occurrence of Q_{i_j} in x_i is replaced by x_{i_j}, provided x_{i_j} does not contain further communication symbols.

A derivation consists of *rewriting steps* and *communication steps*.

The derivation relation, denoted $\Longrightarrow^*$, is the reflexive transitive closure of the relation $\Longrightarrow$. The language generated by the system consists of the terminal strings generated by the *master grammar*, G_1, regardless the other components (terminal or not).

Definition 2 $L(\pi) = \{\alpha \in \Sigma^* \mid (S_1, \ldots, S_n) \Longrightarrow^* (\alpha, \beta_2, \ldots, \beta_n)\}$.

Of special interest are the *centralized PCGS*, denoted by c-PCGS. In this case, only the master grammar can ask for the strings generated by the others. The communication graph is therefore a tree (star) consisting of a father and its sons.

Definition 3 *A dag-PCGS (tree-, two-way array-, one-way array-, two-way ring-, one-way ring- PCGS) is a PCGS whose communication graph is a directed acyclic graph (respectively tree, two-way linear array, one-way linear array, two-way ring, one-way ring).*

Denote by $x - PCGS_n$ the class of PCGS's of degree n whose communication graph is of type x, where $x \in$ {c, dag, tree, two-way array, one-way array, two-way ring, one-way ring}. Moreover, denote by $\mathcal{L}(x - PCGS_n)$ the family of languages generated by $x - PCGS$'s of degree n whose communication graph is of type x, where x is as before.

If x denotes one of the above communication graphs, $x - PCGS_n(f(m))$ will denote the class of PCGS's with communication graph of shape x and using at most $f(m)$ communication steps to generate any word of length m. (Note that $0 \leq f(m) \leq m$.) As above, $\mathcal{L}(x - PCGS_n(f(m)))$ will denote the family of languages generated by PCGS of this type.

Let us give now a simple example that shows the generative power of PCGS.

Example 1. Let π be the PCGS $\pi = (G_1, G_2, G_3)$ where

$$\begin{aligned} G_1 = (\{S_1, S_1', S_2, S_3, Q_2, Q_3\}, \{a, b, c\}, S_1, \{S_1 \longrightarrow abc,\\ S_1 \longrightarrow a^2b^2c^2, S_1 \longrightarrow a^3b^3c^3, S_1 \longrightarrow aS_1', S_1' \longrightarrow aS_1',\\ S_1' \longrightarrow a^3Q_2, S_2 \longrightarrow b^2Q_3, S_3 \longrightarrow c\}),\\ G_2 = (\{S_2\}, \{b\}, S_2, \{S_2 \longrightarrow bS_2\}),\\ G_3 = (\{S_3\}, \{c\}, S_3, \{S_3 \longrightarrow cS_3\}). \end{aligned}$$

This is a regular centralized PCGS of degree 3 and it is easy to see that we have

$$L(\pi) = \{a^nb^nc^n | n \geq 1\},$$

which is a non-context-free language. □

Let us now informally define one-way nondeterministic multicounter machines. The formal definition can be found in [3]. A multicounter machine consists of a finite state control, a one-way reading head which reads the input from the input tape, and a finite number of counters. We regard a counter as an arithmetic register containing an integer which may be positive or zero. In one step, a multicounter machine may increment or decrement a counter by 1. The action or the choice of actions of the machine is determined by the input symbol currently scanned, the state of the machine and the sign of each counter: positive or zero. A reversal is a change from increasing to decreasing contents of a counter or viceversa. The machine starts with all counters empty and accepts if it reaches a final state.

3 Characterization of PCGS by Sequential Complexity Measures

In this section we shall characterize the families of languages generated by PCGS by some sequential complexity classes. These characterizations will depend on the communication structure of PCGS and on the communication complexity of PCGS. This enables us to obtain some hierarchies for the communication complexity measures of PCGS as consequences of some hierarchies for sequential complexity measures.

Let us start first with the characterization of tree-PCGS by linear-time nondeterministic multicounter machines.

Lemma 1. *Let π be a tree-PCGS$_m(f(n))$ for some positive integer m and for some function f : $\mathbf{N} \longrightarrow \mathbf{N}$. Then there exists a linear-time nondeterministic $(m-1)$-counter automaton M recognizing $L(\pi)$, with $2\,f(n)$ reversals and $f(n)$ zerotests.*

Proof. Let $\pi = (G_1, \ldots, G_m)$ be a tree-PCGS$_m(f(n))$. The simulation of π by a real-time 1MC$(m-1)$ machine M is based on the following idea. The finite control of M is used to store the description of all regular grammars $G_1, \ldots, G_m$ and to simulate always the rewriting of one of the grammars which is responsible for the input part exactly scanned.

M uses its counters $C_2, C_3, \ldots, C_m$ in the following way which secures that none of the grammars $G_1, \ldots, G_m$ is used longer than possible in actual situations (configurations). In each configuration of M and for each $i \in \{2, \ldots, m\}$ the number $c(C_i)$ stored in C_i is the difference between the number of the rewriting steps of G_i already simulated by M and the number of simulated rewriting steps of the father of G_i in the communication tree (this means that if G_i is asked by its father to give its generated word then this word is generated by G_i in at most $c(C_i)$ steps).

Now let us describe the simulation. M nondeterministically simulates the work of π by using its finite control to alternatively simulate the work of $G_1, \ldots, G_m$ and checking in real-time whether the generated word is exactly the word laying on the input tape. The simulation starts by simulating the work of G_1 and with the simultaneous comparison of the generated terminals with the corresponding terminals on the input tape. During this procedure M increases after each simulated rewriting step of G_1 the content of all counters assigned to the sons of G_1 and does not change the content of any other counter. This simulation procedure ends when a communication nonterminal Q_i (for some i) is generated. Then M starts to simulate the generation procedure of G_i from the initial nonterminal of G_i. Now, in each simulation step of M the content of the counter C_i is decreased by 1 and the contents of all counters of the sons of G_i are increased by 1. If G_i rewrites its nonterminal in a terminal word, then M halts and it accepts the input word iff the whole input word has been read. If C_i is empty and G_i has produced a nonterminal A in the last step then the control is given to the father of G_i (G_1) which continues to rewrite from the nonterminal A (if A is not a nonterminal of G_1, then M rejects the input). If G_i has produced a communication symbol Q_j for some j, then the son G_j of G_i is required to continue to generate the input word. Now the simulation continues recursively as described above.

Obviously, the number of reversals is bounded by $2\,f(n)$ and the number of zerotests is bounded by $f(n)$ because the content of a counter C_i starts to be decreased iff the communication symbol Q_i was produced.

Clearly, if there are no rules $A \longrightarrow B$, where both A and B are nonterminals, then M works in real-time. If such rules may be used, then the simulation works in linear time because there exists a constant d such that for each word $w \in L(\pi)$ there exists a derivation of w which generates in each d steps at least one terminal symbol. □

Realizing the facts that each 1-multicounter-machine can be simulated in the same time by an off-line multitape Turing machine, and that the contents of counters of M from Lemma 1 is in $O(|w|)$ for any input w, we get the following result.

Theorem 2. $\mathcal{L}$*(tree-PCGS)* $\subseteq$ *NTIME*$(n) \cap$ *NSPACE*$(\log_2 n)$.

The following theorem shows that there is a simulation of dag-PCGS by an off-line nondeterministic multitape Turing machine, working in linear time.

Theorem 3. $\mathcal{L}(\textit{dag-PCGS}) \subseteq \textit{NTIME}(n)$.

Finally, we let open the problem whether the general PCGS can be simulated nondeterministically in linear time. Some effort in this direction has been made in [15], [7], where some PCGS with cycles in communication structures and with some additional restrictions are simulated nondeterministically in linear time.

Another interesting question is whether $\mathcal{L}(PCGS) \subseteq NLOG$. If YES, then each PCGS can be simulated deterministically in polynomial time because $NLOG \subseteq P$. We only know as a consequence of Theorem 2 that $\mathcal{L}(\text{tree-}PCGS)$ is included in P.

4 Communication Complexity Hierarchies

In this section we shall use the simulation result from Lemma 1 to get some strong hierarchies on the number of communication steps for tree-PCGS and its subclasses. Following Lemma 1 we have that $L \in \mathcal{L}(\text{tree} - PCGS_m(f(n)))$ implies $L = L(M)$ for a real-time nondeterministic $(m-1)$-counter automaton M with $2\, f(n)$ reversals. Following the proof of Lemma 1 we see that M has the following property.

(i) For any computation part D of M containing no reversal, the counters can be divided into three sets, $S_1 = \{$the counters whose contents is never changed in $D\}$, $S_2 = \{$the counters whose content is increased in $D\}$, and $S_3 = \{$the counters whose content is decreased in $D\}$, such that for each step of D one of the following conditions holds:

1. either no counter changes its content in the given step, or
2. the counters from S_1 do not change their contents, each counter in S_2 increases its content by 1, and each counter in S_3 decreases its content by 1.

So, the property (i) of D means that, for any subpart D' of D, there exists a constant d' such that the volume of the change of the content of any counter in D' is either $+d'$, or $-d'$, or 0.

Now we will use (i) to get the following result.

Let $L = \{a^{i_1}b^{i_1}a^{i_2}b^{i_2}\ldots a^{i_k}b^{i_k}c \mid k \geq 1, i_j \in \mathrm{N} \text{ for } j \in \{1,\ldots,k\}\}$.

Lemma 4. *$L \in \mathcal{L}(c\text{-}PCGS_2(n)) - \bigcup_{m\in\mathrm{N}}\mathcal{L}(\text{ tree-}PCGS_m(f(n)))$ for any $f(n) \notin \Omega(n)$.*

Following Lemma 4 we get the following hierarchies on the communication complexity.

Theorem 5. *For any function $f : \mathrm{N} \longrightarrow \mathrm{N}$, $f(n) \notin \Omega(n)$, and any $m \in \mathrm{N}$, $m \geq 2$:*

$$\mathcal{L}(\textit{one-way-array-}PCGS_m(f(n))) \subset \mathcal{L}(\textit{one-way-array-}PCGS_m(n)),$$

$$\mathcal{L}(c\text{-}PCGS_m(f(n))) \subset \mathcal{L}(c\text{-}PCGS_m(n)),$$

$$\mathcal{L}(\textit{tree-}PCGS_m(f(n))) \subset \mathcal{L}(\textit{tree-}PCGS_m(n)).$$

Besides Theorem 5, Lemma 5 claims a more important result namely that no increase of the number of grammars and no increase of communication links in tree communication structure (i.e. no increase of the descriptional complexity under the tree communication structure) can compensate for the decrease of the number of communication steps (i.e. computational complexity).

Now we shall deal with PCGS whose communication complexity is bounded by a constant. Let

$$L_k = \{a^{i_1}b^{i_1}a^{i_2}b^{i_1+i_2}\dots a^{i_k}b^{i_1+i_2+\dots+i_k}c \mid i_j \in \mathrm{N} \text{ for } j = 1,\dots,k\},$$

for any $k \in \mathrm{N}$.

Lemma 6. $L_k \in \mathcal{L}(c\text{-}PCGS_{k+1}(k)) - \cup_{m\in\mathrm{N}}\mathcal{L}(\text{ } tree\text{-}PCGS_m(k-1))$.

Theorem 7. *For any positive integer k and any $X \in \{c,\ tree,\ one\text{-}way\ array\}$ we have*

$$\mathcal{L}(X\text{-}PCGS_{k+1}(k-1)) \subset \mathcal{L}(X\text{-}PCGS_{k+1}(k)) \text{ and}$$

$$\cup_{m\in\mathrm{N}}\mathcal{L}(X\text{-}PCGS_m(k-1)) \subset \cup_{m\in\mathrm{N}}\mathcal{L}(X\text{-}PCGS_m(k)).$$

An open problem is to prove hierarchies for more complicated communication structures. Some results in this direction have been recently established in [7].

5 Pumping Lemmas and Infinite Hierarchies

In this section descriptional complexity measures of PCGS are investigated. For PCGS with communication structures tree and dag, strong hierarchies on the number of grammars are proved. To obtain them, some pumping lemmas as lower bound proof techniques are established. In the case of PCGS with communication structures arrays and rings, no such pumping lemmas are known. However, the infinity of the hierarchies of such PCGS on the number of grammars is obtained as a consequence of the following stronger result. There exist languages that can be generated by two-way array-PCGS, two-way ring-PCGS and one-way ring-PCGS but cannot be generated by *any* PCGS of smaller degree, regardless of the complexity of its communication graph. This also shows that in some cases the increase in the descriptional complexity (the number of grammars the PCGS consists of) cannot be compensated by any increase in the complexity of the communication graph.

Before entering the proof of the pumping lemmas, an ordering of the vertices in a directed acyclic graph is needed.

Proposition 8. *Let $G = (X, \Gamma)$ be a dag, where X is the set of vertices and Γ the set of arcs. We can construct a function $f : X \longrightarrow \mathrm{N}$ such that for all $x, y \in X$ we have:*

$$f(x) \geq f(y) \text{ implies that there is no path from } y \text{ to } x \text{ in the graph } G.$$

The classical proof of the pumping lemma for regular languages is based on finding, along a sufficiently long derivation, two "similar" sentential forms. "Similar" means that the two sentential forms contain the same nonterminal, a fact that allows us to iterate the subderivation between them arbitrarily many times.

We will use an analogous procedure for dag-PCGS. The difference will be that, due to the communications we need a stronger notion of "similarity". The first request will obviously be that the correspondent components of the two "similar" configurations contain the same nonterminal. Moreover, we will require that, in case communications are involved, also the terminal strings are identical.

Definition 4 *Let* $c_1 = (x_1A_1, \ldots, x_nA_n)$ *and* $c_2 = (y_1B_1, \ldots, y_nB_n)$ *be two configurations where* x_i, y_i *are terminal strings and* A_i, B_i *are nonterminals or* λ, *for* $1 \leq i \leq n$.

The configurations are called equivalent, and we write $c_1 \equiv c_2$ *if* $A_i = B_i$ *for each* i, $1 \leq i \leq n$.

Clearly, $\equiv$ is an equivalence relation.

Let us consider a derivation according to π, $D : c \Longrightarrow^* c_1 \Longrightarrow^* c_2 \Longrightarrow^* c'$, where c_1 and c_2 are defined as in the previous definition.

Definition 5 *The configurations* c_1 *and* c_2 *are called D-similar iff*

(i) c_1 *and* c_2 *are equivalent,*

(ii) if a communication symbol Q_i, $1 \leq i \leq n$, *is used in the derivation D between* c_1 *and* c_2, *then* $x_i = y_i$.

We are now in position to prove the pumping lemma for dag-PCGS. For the sake of clarity, the proof is split in two parts. The first result claims that in any sufficiently long derivation according to a dag-PCGS we can find two "similar" configurations.

Lemma 9. *Let* π *be a dag-PCGS. There exists a constant* $q \in \mathbf{N}$ *such that in any derivation D according to* π *whose length is at least* q, *there are two D-similar configurations.*

The following pumping lemma shows that any sufficiently long word generated by a dag-PCGS can be decomposed such that, by simultaneously pumping a number of its subwords, we obtain words that still belong to the language. Due to the dag structure of the communication graph which allows a string to be read by more than one grammar (a vertex can have more fathers), the number of the pumped subwords can be arbitrarily large. However, the number of *distinct* pumped subwords is bounded by the degree of the dag-PCGS.

Lemma 10. (Pumping lemma for dag-PCGS) *Let L be a language generated by a dag-PCGS of degree* $n > 1$. *There exists a natural number N such that every word* $\alpha \in L$ *whose length is greater than N can be decomposed as*

$$\alpha = \alpha_1\beta_1 \ldots \alpha_m\beta_m\alpha_{m+1},$$

where $\beta_i \neq \lambda$ *for every* i, $1 \leq i \leq m$, *and* $1 \leq card\{\beta_1, \ldots, \beta_m\} \leq n$. *Moreover, for all* $s \geq 0$ *the word*

$$\alpha_1\beta_1^s \ldots \alpha_m\beta_m^s\alpha_{m+1}$$

belongs to L.

Proof. Let $\pi = (G_1, \ldots, G_n)$ be a dag-PCGS, where $G_i = (V_{N,i}, \Sigma, S_i, P_i)$. Denote by z the maximum length of the right sides of all productions.

Claim. The length of any component of a configuration produced by π starting from the axiom in k derivation steps is at most $z \cdot 2^{k-1}$.

The claim will be proved by induction on k.

If $k = 1$ then the claim obviously holds as π can produce in one step only words of length at most z.

$k \mapsto k+1$. Let us consider a derivation according to π which starts from the axiom and has $k+1$ steps. In the $(k+1)$th step, the length of any component α is:

$$|\alpha| \leq |\alpha'| + \max\{z, |\alpha'|\} \leq 2 \cdot |\alpha'| = z \cdot 2^k.$$

where $|\alpha'|$ denotes the maximum length of any component of a configuration that can be obtained after k derivation steps, starting from the axiom. The proof of the claim is complete.

If we choose now $N = z \cdot 2^{q-1}$, where q is the number defined in Lemma 9 and a word α whose length is greater than N, then a minimal derivation D of α contains at least q steps.

According to the Lemma 9, during this derivation occur at least two D-similar configurations c_1 and c_2 as shown below:

$$\begin{aligned}(S_1, S_2, \ldots, S_n) &\Longrightarrow^* c_1 = (x_1A_1, x_2A_2, \ldots, x_nA_n)\\ &\Longrightarrow^* c_2 = (x_1z_1A_1, x_2z_2A_2, \ldots, x_nz_nA_n)\\ &\Longrightarrow^* (\alpha, \ldots\ldots).\end{aligned}$$

If all the strings x_iz_i which occur in c_2 and become later subwords of α have the property $z_i = \lambda$ then D is not minimal. Indeed, if this would be the case, the subderivation between c_1 and c_2 could be eliminated – a contradiction with the minimality of D.

Consequently, there exist $i_1, \ldots, i_k \in \{1, \ldots, n\}$, such that

$$\alpha = \alpha_1 x_{i_1} z_{i_1} \alpha_2 x_{i_2} z_{i_2} \ldots \alpha_k x_{i_k} z_{i_k} \alpha_{k+1}$$

$z_{i_j} \neq \lambda$, $1 \leq j \leq k$, and $x_{i_j} z_{i_j}$, $1 \leq j \leq k$, are exactly the terminal strings that have appeared in the components with the corresponding index of c_2. Observe that we do not necessarily have $i_j \neq i_p$ for $j \neq p$, $1 \leq j, p \leq k$. Indeed, because of possible communications, the same string $x_{i_j} z_{i_j}$ originating from the i_j-component of c_2 can appear several times in α.

By iterating the subderivation between the two D-similar configurations c_1 and c_2 s times, for an arbitrary s, we obtain a valid derivation for

$$\alpha^{(s)} = \alpha_1 x_{i_1} z_{i_1}^s \alpha_2 x_{i_2} z_{i_2}^s \ldots \alpha_k x_{i_k} z_{i_k}^s \alpha_{k+1}.$$

The word $\alpha^{(s)}$ therefore belongs to L for all natural numbers $s > 0$. The derivation between c_1 and c_2 can also be omitted and therefore also $\alpha^{(0)}$ belongs to L.

Note that we do not give an upper bound for k. This follows from the fact that in a dag a vertex can have more fathers. Consequently, a component x_iz_i can be read by more than one grammar and thus appear more than once in α. However, the number of *different* words z_{i_j} is at most n. Indeed when iterating the subderivation $c_1 \Longrightarrow^* c_2$, we can only pump the z_i's already existing in some components of c_2, that is, at most n different ones. As explained before, because of the communications steps that occur after c_2, some of the words z_i^s can appear several times in $\alpha^{(s)}$. □

An analogous pumping lemma can be obtained for tree-PCGS, but in this case the number of pumped positions is bounded by the number of grammars of the tree-PCGS.

Lemma 11. (Pumping lemma for tree-PCGS) *Let L be a language generated by a tree-PCGS. There exists a natural number N such that every word $\alpha \in L$ whose length is greater than N can be decomposed as*

$$\alpha = \alpha_1 \beta_1 \ldots \alpha_m \beta_m \alpha_{m+1},$$

where $1 \leq m \leq n$, $\beta_i \neq \lambda$ for every i, $1 \leq i \leq m$, and the word

$$\alpha_1 \beta_1^s \ldots \alpha_m \beta_m^s \alpha_{m+1}$$

belongs to L for all $s \geq 0$.

As a consequence of Lemma 10, we can obtain a language that can be generated by a tree-PCGS but cannot be generated by any dag-PCGS of smaller degree.

Theorem 12. *For all $n > 1$, $\mathcal{L}(tree\text{-}PCGS_n) - \mathcal{L}(dag\text{-}PCGS_{n-1}) \neq \emptyset$.*

Proof. Consider the language $L_n = \{a_1^{k+1} a_2^{k+2} \ldots a_n^{k+n} \mid k \geq 0\}$. □

The following infinite hierarchies are obtained as consequences of the preceding result.

Corollary 13. *The hierarchy $\{\mathcal{L}(dag\text{-}PCGS_n)\}_{n \geq 1}$ is infinite.*

Corollary 14. *The hierarchy $\{\mathcal{L}(tree\text{-}PCGS_n)\}_{n \geq 1}$ is infinite.*

In the remaining part of this section we will consider some PCGS with communication structures for which no pumping lemmas are known, namely two-way array, two-way ring and one-way ring-PCGS. The following theorem provides a language that can be generated by a two-way array-PCGS but cannot be generated by *any* PCGS of smaller degree. This shows that in some cases the increase in descriptional complexity cannot be compensated by an increase in the complexity of the communication structure.

Theorem 15. *For all $m \geq 1$,*

$$\mathcal{L}(two\text{-}way\ array\text{-}PCGS_{m+1}) - \mathcal{L}(two\text{-}way\ array\text{-}PCGS_m) \neq \emptyset.$$

Proof. Consider the language

$$L_m = \{a_1^n a_2^n \ldots a_{2m}^n \mid n \geq 1\}.$$

We can show a stronger result than the one stated in the theorem. For all $m > 1$ there exists the language L_m that can be generated by a two-way array PCGS of degree $m + 1$ but cannot be generated by *any* PCGS of smaller degree. □

Corollary 16. *The hierarchy* $\{\mathcal{L}(\textit{two-way array-PCGS}_n)\}_{n\geq 1}$ *is infinite.*

Corollary 17. *The hierarchy* $\{\mathcal{L}(\textit{two-way ring-PCGS}_n)\}_{n\geq 1}$ *is infinite.*

The language used in the proof of Theorem 15 can be used to show that the hierarchy of one-way ring-PCGS, relative to the number of the grammars in the PCGS, is infinite. When constructing the one-way ring-PCGS which generates the language, special care has to be payed to synchronization problems.

Theorem 18. *For all* $m \geq 1$,

$$\mathcal{L}(\textit{one-way ring-PCGS}_{m+1}) - \mathcal{L}(\textit{one-way ring-PCGS}_m) \neq \emptyset.$$

Corollary 19. *The hierarchy* $\{\mathcal{L}(\textit{one-way ring-PCGS}_n)\}_{n\geq 1}$ *is infinite.*

The study of hierarchies on the number of grammars for PCGS with other communication structures (planar graphs, hypercubes, etc) remains open.

References

1. K.Culik, J.Gruska, A.Salomaa. Systolic trellis automata. *International Journal of Computer Mathematics* **15** and **16**(1984).
2. P.Duris, J.Hromkovic: Zerotesting bounded multicounter machines. *Kybernetika* 23(1987), No.1, 13-18.
3. S.Ginsburg: *Algebraic and Automata-Theoretic Properties of Formal Languages.* North-Holland Publ.Comp., Amsterdam 1975.
4. C.A.R.Hoare. Communicating sequential processes. *Comm. ACM* **21** vol. 8 (1978).
5. J.Hromkovic: Hierarchy of reversal bounded one-way multicounter machines. *Kybernetica* 22(1986), No.2, 200-206.
6. J.Kari. Decision problems concerning cellular automata. *University of Turku, PhD Thesis* (1990).
7. D.Pardubska. The communication hierarchies of parallel communicating systems. *Proceedings of IMYCS'92*, to appear.
8. Gh.Paun. On the power of synchronization in parallel communicating grammar systems. *Stud. Cerc. Matem.* **41** vol.3 (1989).
9. Gh.Paun. Parallel communicating grammar systems: the context-free case. *Found. Control Engineering* **14** vol.1 (1989).
10. Gh.Paun. On the syntactic complexity of parallel communicating grammar systems. *Kybernetika*, **28**(1992), 155-166.
11. Gh.Paun, L.Santean. Further remarks on parallel communicating grammar systems. *International Journal of Computer Mathematics* **35** (1990).
12. Gh.Paun, L.Santean. Parallel communicating grammar systems: the regular case. *Ann. Univ. Buc. Ser. Mat.-Inform.* **37** vol.2 (1989).
13. A.Salomaa. *Formal Languages.* Academic Press New York London (1973).
14. L.Santean, J.Kari:The impact of the number of cooperating grammars on the generative power, *Theoretical Computer Science*, **98**, 2(1992), 249-263.
15. D.Wierzchula: Systeme von parallellen Grammatiken (in German). Diploma thesis, Dept. of Mathematics and Computer Science, University of Paderborn, 1991.

Efficient parallel graph algorithms based on open ear decomposition

Louis Ibarra*
Dept. of Computer Science
Rutgers University
New Brunswick, NJ 08903

Dana Richards
Division of Computer and
Computation Research
National Science Foundation
Washington, D.C. 20550

Abstract. We present a new technique called "disjoint decreasing ear paths", which is based on a graph's open ear decomposition. We apply this technique in CRCW PRAM parallel algorithms for the two vertex disjoint $s-t$ paths problem and the maximal path problem in planar graphs. These run in O(log n) time with $n+m$ processors and O($\log^2 n$) time with O(n) processors, respectively, where the graph has n vertices and m edges.

1 Introduction

The development of parallel algorithms for the PRAM has shown that many techniques which are very useful in sequential algorithms are often not amenable to efficient implementation in parallel algorithms. Consequently, many researchers believe that new techniques and paradigms must be developed in order to obtain efficient parallel algorithms [11, 13]. In this paper, we present parallel algorithms based on the technique of *open ear decomposition*. This technique has been used in efficient parallel algorithms for a variety of graph problems, including st-numbering [13], triconnected components [6, 14, 16], 4-connected components [9] and planarity testing [15].

Let $G=(V,E)$ be an undirected graph with no loops or multiple edges and let $n=|V|$ and $m=|E|$. Throughout this paper, every path and cycle is simple. An *ear decomposition* of G is a partition of the graph's edge set into an ordered collection of paths $E_0, E_1, \ldots, E_r$ called *ears*, such that following properties hold:

- E_0 is a cycle called the *root ear*.
- Each endpoint of $E_i, i>0$, is contained in some $E_j, j<i$.
- No internal vertex of $E_i, i>0$, is contained in any $E_j, j<i$.

If the endpoints of an ear do not coincide, then the ear is *open*; otherwise, the ear is *closed*. An *open ear decomposition* of a graph G is an ear decomposition of G in which every ear $E_i, i>0$, is open. Some observations, which are true for both open and closed ear decompositions, follow from the definitions:

- An edge is contained in exactly one ear.

* Supported by NASA Graduate Student Researchers Program.

- A vertex is contained in one or more ears.
- A vertex is an internal vertex of exactly one ear, where we consider the root ear's internal vertices to be all its vertices.

Whitney first studied open ear decompositions and showed that a graph G has an open ear decomposition if and only if G is biconnected [20]. Lovász showed that the problem of computing an open ear decomposition in parallel is in NC [12].

We use the parallel random access machine (PRAM) model of parallel computation, which may be exclusive read exclusive write (EREW), concurrent read exclusive write (CREW), or concurrent read concurrent write (CRCW) [5]. We will use the following algorithms extensively (and often implicitly). First, the "pointer doubling" technique solves the list ranking and list prefix problems in O(log n) time with n processors on an EREW PRAM [5]. Second, computing the connected components of a graph requires O(log n) time with (asymptotically fewer than) O($n+m$) processors on the CRCW PRAM [4]. Third, computing the biconnected components of a graph requires O(log n) time with O($n+m$) processors on the CRCW PRAM [18, 5]. Fourth, computing an open ear decomposition requires O(log n) time with $n+m$ processors on the CRCW PRAM [13].

This paper is organized as follows. In section 2, we present a novel problem in a graph's open ear decomposition, known as *disjoint decreasing ear paths*. (Throughout this paper, *disjoint* means *vertex disjoint*.) We give a parallel algorithm for this problem which runs in O(log n) time with n processors on the CREW PRAM. We demonstrate its utility in sections 3 and 4, where we apply the algorithm in two efficient parallel graph algorithms for the CRCW PRAM. In section 3, we solve the two disjoint $s-t$ paths problem in O(log n) time with $n+m$ processors. This problem can also be solved with st-numbering, which can be computed with the same time and processor bounds [13]. In section 4, we solve the maximal path in planar graphs problem in O($\log^2 n$) time with O(n) processors. This improves the time and processor bounds on the previous algorithm for the problem [2], but algorithms for depth-first search in planar graphs achieve the same bounds [7, 17]. However, our approach to both problems is considerably different. In the first algorithm, we avoid computing an st-numbering and instead compute disjoint decreasing ear paths in an open ear decomposition. We believe that this is a intuitively simpler construction than st-numbering. (The algorithm also solves the problem of computing an open ear decomposition with any two vertices in the root ear. We do not know of any previous algorithm for this problem.) In the second algorithm, we do not use the fact that every planar graph has a vertex of degree at most five, as in [2], nor do we compute cyclic separators, as in [7, 17].

2 An O(log n) time algorithm to compute disjoint decreasing ear paths

In this section, we define a novel problem in a graph's open ear decomposition and give a parallel algorithm which solves it in O(log n) time with n processors on the CREW PRAM.

We give several definitions before presenting the problem. Let G be a graph with open ear decomposition D. If ear E has its endpoints internally contained in different ears, then we say the endpoint of E internally contained in the ear with smaller (resp. greater) index is the *smaller* (resp. *greater*) endpoint. Let u be any vertex of G and let E_u be the ear which internally contains u. Suppose each ear selects one of its endpoints by some, as yet unspecified, criterion and designates that endpoint to be its *exit vertex*. Let u be any vertex and let E_u be the ear which internally contains u. We can compute a path from u to the root ear as follows. We extend the path from u to E_u's exit vertex v, which is internally contained in ear E_v. We then extend the path from v to E_v's exit vertex w, which is internally contained in ear E_w, and so on. Since the ear index decreases after each extension, this procedure halts with a path whose terminal vertex is contained in the root ear. We call this path a *decreasing ear path from u*. Notice that given a set of exit vertices, there is a unique decreasing ear path from each vertex. However, different sets of exit vertices may yield different decreasing ear paths from each vertex. We will specify a particular decreasing ear path by its sequence of exit vertices. More precisely, we specify decreasing ear path P from vertex u by a successor function $\mathcal{P}$ defined on P's exit vertices. (For the purpose of this definition, we consider the initial vertex u to be an exit vertex of P.) Thus, $\mathcal{P}(e_i) = e_{i+1}$, where e_i is the i-th exit vertex of P, $i \geq 1$.

The *disjoint decreasing ear paths problem* is: Given a graph G with open ear decomposition D and specified (distinct) vertices x, y, compute a decreasing ear path P from x and a decreasing ear path Q from y such that P and Q are disjoint. The following theorem shows that a solution to this problem always exists; the proof is straightforward by induction on the number of ears.

Theorem 1 *Let G be a graph with open ear decomposition D and specified (distinct) vertices x, y of G. There exists a decreasing ear path P from x and a decreasing ear path Q from y, such that P and Q are disjoint.*

We will use the following terminology. Let P and Q be decreasing ear paths. Consider the exit vertex sequences of P and Q:

$P : p_1 \ldots p_{n_P}$

$Q : q_1 \ldots q_{n_Q}$

Let p_i and q_j be exit vertices of P and Q, respectively. If $p_i = q_j$, then we say p_i, q_j are *common* exit vertices. If $p_i \neq q_j$ and p_i, q_j are internally contained in the same ear, then consider the subpath of P from p_i to p_{i+1} and the subpath of Q from q_j to q_{j+1}. If these subpaths have vertices in common, then we say p_i, q_j are *crossing* exit vertices. Notice that a common (resp. crossing) exit vertex of P is associated with exactly one common (resp. crossing) exit vertex of Q and vice-versa. Furthermore, if p_i, q_j and p_k, q_l are common or crossing vertex pairs so that p_i, q_j are distinct from p_k, q_l, then either $i < k$, $j < l$ or $i > k$, $j > l$.

The algorithm for the disjoint decreasing ear paths problem has five steps. Step 1 computes a decreasing ear path P from x. Steps 2 and 3 compute a decreasing ear path Q from y, making a preliminary attempt to avoid using the exit vertices of P as the exit vertices of Q. If the resulting paths are not disjoint, then steps 4 and 5 "reroute" the paths in order to obtain disjoint decreasing ear paths. One processor is assigned to each vertex of the graph.

Algorithm (Compute disjoint decreasing ear paths)
Input: A graph G, an open ear decomposition of G, specified (distinct) vertices x, y.
Output: A decreasing ear path P from x and a decreasing ear path Q from y, such that P and Q are disjoint.

Step 1 (Compute a decreasing ear path P from x)

Each non-root ear E_i arbitrarily designates one of its endpoints as its exit vertex, with the exception that if one of its endpoints is y, then the other endpoint is designated. The decreasing ear path P from x is computed, so that each exit vertex of P has a $\mathcal{P}$ successor. Each exit vertex of P is marked.

Step 2 (Choose new exit vertices)

Each non-root ear E_i notes whether its endpoints are marked. If neither endpoint is marked, then E_i arbitrarily designates one of its endpoints as its exit vertex. If exactly one of its endpoints is marked, then E_i designates the unmarked endpoint as its exit vertex. If both of E_i's endpoints are marked, then E_i designates its greater endpoint as its exit vertex. We call an ear with both endpoints marked a *double marked ear*. Since P is a decreasing ear path, then every double marked ear has its endpoints internally contained in distinct ears.

Step 3 (Compute a decreasing ear path Q from y)

The decreasing ear path Q from y is computed, so that each exit vertex in Q has a $\mathcal{Q}$ successor. If P and Q are disjoint, then the algorithm halts with the desired output. If P and Q have no common exit vertices, then let $P' = P$ and $Q' = Q$ and proceed to step 5.

Comments

Step 4 eliminates the common exit vertex pairs by "rerouting" P and Q to obtain decreasing ear paths P' and Q' from x and y with no common exit vertices. This "rerouting" is done by defining an $\mathcal{R}$ successor for each exit vertex of P and Q; these successors will specify both P' and Q'. Step 5 eliminates the crossing exit vertex pairs by redefining the $\mathcal{R}$ successors of each crossing exit vertex pair. The resulting $\mathcal{R}$ successors will specify disjoint decreasing ear paths P'' and Q'' from x and y.

We illustrate the main idea used in the rerouting in step 4 by considering the case when P, Q have exactly one pair p_i, q_j of common exit vertices. Since q_j is an exit vertex of P, then we know that the ear which internally contains q_j's $\mathcal{Q}$ predecessor is a double marked ear. (Otherwise, the ear would have chosen an unmarked endpoint as its exit vertex and q_j would not be an exit vertex of Q.) Let E be this ear, which has $p_i = q_j$ as its greater endpoint and exit vertex p_k as its smaller endpoint, where p_i, p_k are exit vertices of P (see figure 1). Since p_i is the greater endpoint of E, p_i appears before p_k in the exit vertex sequence of P and so $i < k$. Thus, the exit vertex sequences of P and Q are:

$P : p_1 \dots p_i \dots p_k \dots p_{n_P}$

$Q : q_1 \dots q_{j-1} q_j q_{j+1} \dots q_{n_Q}$

We define $\mathcal{R}$ successors for the exit vertices of P and Q as follows. If exit vertex v is contained only in P (resp. Q), then $\mathcal{R}(v)$ is $\mathcal{P}(v)$ (resp. $\mathcal{Q}(v)$). We define $\mathcal{R}(p_i = q_j)$

to be q_{j+1} and we redefine $\mathcal{R}(q_{j-1})$ to be p_k. Since each exit vertex has a unique $\mathcal{R}$ successor, then the $\mathcal{R}$ successors define decreasing ear paths P' and Q' (see figure 2):

$P' : p_1 \ldots p_i q_{j+1} \ldots q_{n_Q}$

$Q' : q_1 \ldots q_{j-1} p_k \ldots p_{n_P}$

Observe that only P' contains $p_i = q_j$ and only Q' contains p_k. Since the $\mathcal{R}$ successors of $p_i = q_j$ are contained only in Q and the $\mathcal{R}$ successors of p_k are contained only in P, then P' and Q' have no common exit vertices. Since vertices $p_{i+1}, \ldots, p_{k-1}$ are not contained in either P' or Q', we may consider them discarded.

The rerouting for the case when P and Q may have many common exit vertices is more complicated and is described in step 4.

Step 4 (Eliminate common exit vertex pairs)

We say a double marked ear whose greater endpoint is a common exit vertex of P and Q is a *detour* ear. Since Q is a decreasing ear path, no exit vertex of P is the greater endpoint of more than one detour ear. Hence, we may totally order the detour ears with respect to the order in which their greater endpoints appear on P. If E, F are detour ears, then we say E *precedes* F if E's greater endpoint precedes F's greater endpoint on P. We will call this ordering of detour ears the *ear ordering* (see figure 3).

Step 4a. Each detour ear indicates that its greater endpoint is an *escape* vertex. Using pointer doubling, every vertex v on P defines Escape(v) to be the first escape vertex on P not preceding v.

Step 4b. Each detour ear E designates its *tail* ear to be the detour ear which contains Escape(s) as its greater endpoint, where s is the smaller endpoint of E. Since a detour ear precedes its tail ear in the ear ordering, the tail function defines a forest of detour ears (see figure 4).

Step 4c. We define the $\mathcal{R}$ successors as follows. If exit vertex v is contained only in P (resp. Q), then $\mathcal{R}(v)$ is $\mathcal{P}(v)$ (resp. $\mathcal{Q}(v)$). If exit vertex v is common to P and Q, then $\mathcal{R}(v)$ is $\mathcal{Q}(v)$. Let E be a detour ear and let $p_i = q_j$ be its greater endpoint and p_k its smaller endpoint, so that q_{j-1} is an internal vertex of E and p_i, p_k are exit vertices of E with $i < k$. (So $\mathcal{R}(p_i) = \mathcal{R}(q_j) = q_{j+1}$.) Now some tree T in the forest of detour ears contains the first detour ear in the ear ordering and it contains the ear as a leaf. The path in T from this leaf to the root of T is computed. If E is on this path, then E redefines $\mathcal{R}(q_{j-1})$ to be p_k and we say E is *routed down*; otherwise, E does nothing and we say E is *routed up* (in this case, $\mathcal{R}(q_{j-1})$ remains $p_i = q_j$).

Every exit vertex of P or Q has exactly one $\mathcal{R}$ successor, which is internally contained in an ear with smaller index. Hence, $\mathcal{R}$ defines a decreasing ear path P' from x and a decreasing ear path Q' from y (see figure 5).

Lemma 1 *The decreasing ear paths P' and Q' have no common exit vertices.*

Proof In the full paper [8].

Step 5 (Eliminate crossing exit vertex pairs)

We now have decreasing ear paths P' and Q' with no common exit vertices. If P' and Q' have no crossing exit vertex pairs, then P' and Q' are disjoint and the algorithm halts with the desired output. Otherwise, let p_i, q_j be a crossing exit vertex pair of P', Q', respectively. Let p_i, q_j be internally contained in ear E, which has endpoints e_1, e_2. Since $\mathcal{R}(p_i)$ is e_2 and $\mathcal{R}(q_j)$ is e_1, E exchanges the $\mathcal{R}$ successors so that $\mathcal{R}(p_i)$ is e_1 and $\mathcal{R}(q_j)$ is e_2. This "unravels" the subpaths from p_i to $\mathcal{R}(p_i)$ and from q_j to $\mathcal{R}(q_j)$, so that they are disjoint. Each crossing exit vertex pair is internally contained in an ear and each such ear performs this exchange. Hence, the resulting paths P'' and Q'' have no common or crossing exit vertices and so P'' and Q'' are disjoint.

Theorem 2 *There is an algorithm that, given a graph G and an open ear decomposition of G, computes disjoint decreasing ear paths from any two vertices of G in $O(\log n)$ time with n processors on a CREW PRAM.*

3 An O(log n) time algorithm for the two disjoint $s-t$ paths problem

In this section, we use the disjoint decreasing ear paths algorithm to solve the two disjoint $s-t$ paths problem. Since it suffices to consider the problem for biconnected graphs, we assume that the input graph is biconnected.

The *two disjoint $s-t$ paths problem* is: Given two specified vertices s, t in a graph G, compute two disjoint paths from s to t. The algorithm computes an open ear decomposition with s in the root ear [13] and then computes disjoint decreasing ear paths from the ear which internally contains t. These paths are used to form a cycle containing both s and t; this cycle becomes the root ear of a new open ear decomposition. The running time is O(log n) with $n+m$ processors on the CRCW PRAM. Details are in the full paper [8]. This algorithm also solves the problem of computing an open ear decomposition with any two vertices in the root ear. We are not aware of any previous algorithm for this problem.

The two disjoint $s-t$ paths problem can also be solved with st-numbering, which can be computed with the same time and processor bounds [13]. However, our algorithm takes a considerably different approach to the problem: we avoid computing an st-numbering and instead compute disjoint decreasing ear paths in an open ear decomposition.

4 An O(log n) time algorithm to compute a maximal path in planar graphs

In this section, we use the disjoint decreasing ear path algorithm to compute a maximal path in a planar graph in $O(\log^2 n)$ time with $O(n)$ processors on a CRCW PRAM.

We will use the following terminology. Let G be a graph. A ***disconnecting path*** of G is a path P such that $G-P$ contains two or more components. A ***splitting path*** of G is a disconnecting path P such that the terminal vertex of P is adjacent to at least two components of $G-P$. A ***disconnecting cycle*** of G is a cycle C such that $G-C$

contains two or more components. A *maximal path* of G is a path P such that every neighbor of the terminal vertex of P is contained in P. In other words, a maximal path is a path that cannot be extended without encountering a vertex already on the path. The *maximal path problem* is: Given a graph G and a specified vertex r, find a maximal path from r. This problem was apparently first studied by Anderson, who gave the following parallel algorithm for it [1]. Suppose we have an algorithm to find a splitting path from r in time $T(n)$. By finding a splitting path P from r and recursing on the smallest component in $G - P$, we compute a maximal path from r in time $(\log n)T(n)$.

In the algorithm by Anderson for general graphs [1], the splitting path is found using a randomized algorithm for maximum matching. In the algorithm by Anderson and Mayr for planar graphs [2], the splitting path is found by using the fact that every planar graph has a vertex of degree at most five. In our algorithm for planar graphs, we find a splitting path as follows. The open ear decomposition algorithm and disjoint decreasing ear paths algorithm are used to compute a disconnecting cycle X in G. A disconnecting path Y from r is then computed by finding a path from r to X and then extending this path through every vertex of X. Now either Y is a splitting path or some subpath of Y is a splitting path, so a splitting path Z from r is readily found. The algorithm recurses on the smallest component in $G - Z$ to obtain a maximal path from r. Computing the splitting path requires $O(\log n)$ time and thus the algorithm runs in $O(\log^2 n)$ time.

The algorithm uses the following procedure of Anderson [1] as a subroutine, which we call the *path-building procedure.* Suppose U is a (possibly null) path from r and suppose the vertices not in U have been partitioned into a constant number of disjoint paths. We call these paths the *untraversed* paths. *Step one.* Extend U to any vertex in any untraversed path V. This can be done by computing connected components. *Step two.* Extend U through the maximum possible number of vertices of V. In other words, if x and y are the initial and terminal vertices of V, then extend U to x or y, whichever produces the "longer" extension. Replace V with its untraversed subpath. *Step three.* If U is a maximal path, then halt; otherwise, repeat steps one and two. In each iteration, the length of some untraversed path is reduced by at least half. Hence, the procedure terminates with a maximal path in $O(\log^2 n)$ time and uses $O(n)$ processors.

The algorithm is complicated, although most of the complexity is in steps 1 to 5, which compute a disconnecting cycle. In step 6, we find a splitting path and recurse. Due to space limitations, we only sketch the algorithm; details are in the full paper [8]. Notice that if G is not biconnected, then a path from r to an articulation point in r's biconnected component is a splitting path. Hence, it suffices to consider the problem for biconnected graphs and so we assume that the input graph is biconnected.

Algorithm (Compute a maximal path in a planar graph)
Input: A biconnected planar graph G and a specified start vertex r.
Output: A maximal path in G from r.

Step 1 (Compute an open ear decomposition)

Choose any edge incident on r and compute an open ear decomposition from this edge using the algorithm of Maon, Schieber, Vishkin [13]. This yields an open ear decomposition with r in the root ear. Let $E_i, i > 0$, be an ear with endpoints e_1, e_2 internally contained in ears F_1, F_2, respectively. We say that E_i is *good* if (1) F_1 and F_2 are distinct ears, (2) neither e_1 nor e_2 is an endpoint of F_1 or F_2. (Thus, E_i is not good if and only if F_1, F_2 are identical, or e_1 is the endpoint of F_2, or e_2 is an endpoint of F_1.) If a good ear exists, then let E be an arbitrary good ear and proceed to step 2. If no good ear exists, then every ear $E_i, i > 0$, has both endpoints contained in some ear $E_j, j < i$. (In other words, D is a *tree* open ear decomposition [10].) In this case, we can readily compute a maximal path from r directly.

Step 2 (Compute disjoint decreasing ear paths)

Compute disjoint decreasing ear paths P and Q from e_1 and e_2, respectively, using the algorithm from section 2. Now there are two possible cycles formed by traversing P, E, Q, and the root ear E_0 in this order. If either of these cycles or E_0 is a disconnecting cycle, then proceed to step 6.

Comments

Consider a planar embedding of G with (at least one edge of) E incident on the outer face, which exists by stereographic projection [3]. We will appeal to this embedding to show correctness, but the algorithm itself does not require an embedding as input. Notice E, P, Q and E_0 partition the plane into two bounded regions and one unbounded region. Let R_0 be the bounded region bounded by E_0. Let R_1 be the other bounded region. Let R_2 be the unbounded region (see figure 6). Let C_1, C_2 be the cycles which form the boundaries of R_1, R_2, respectively. Since the algorithm checked whether E_0, C_1 or C_2 are disconnecting cycles in step 2, then we assume in steps 3, 4, 5 that none of these cycles is a disconnecting cycle. By property (1) of good ears, F_1 and F_2 cannot both be the root ear. Let F_2 be distinct from the root ear, so that e_2 is not contained in the root ear. Let the endpoints of F_2 be f_1, f_2. By property (2) of good ears, the vertices e_1, e_2, f_1, f_2 are all distinct.

Step 3 (Find an endpoint of F_2 that is contained in neither P nor Q)

Since Q is a decreasing ear path from an internal vertex of F_2, then Q must contain either f_1 or f_2. Since P and Q are disjoint, then we have the following three cases.

case 1. P contains neither of f_1, f_2 and Q contains one of f_1, f_2. Since the desired endpoint exists, we proceed to step 4.

case 2. P contains neither of f_1, f_2 and Q contains both f_1, f_2. Let f_1 precede f_2 in Q. Then we can redefine Q so that f_1 is contained in neither P nor Q. Proceed to step 4.

case 3. P contains f_1 and Q contains f_2. Let the endpoints of F_1 be g_1, g_2. Suppose P contains g_1 and Q contains g_2. (The other cases are handled as in case 1 or 2.) Then the embedding of G must have the shape shown in figure 7 (or a analogous shape). If every vertex of G is contained in E, P, Q, X, Y or E_0, then we use the path-building procedure given before the algorithm to compute a maximal path from r. Suppose v is a vertex not contained in E, P, Q, X, Y or E_0. Then v is embedded in the interior of R_0, R_1 or R_2. In every case, the boundary of one of the regions shown in figure 7 is a disconnecting cycle. Proceed to step 6.

Step 4 (Compute another decreasing ear path)

We have the following. There are disjoint decreasing ear paths P and Q from e_1 and e_2, respectively, such that there is an endpoint of F_2 contained in neither P nor Q. Let f_2 be contained in neither P nor Q. Thus, f_1 is contained in Q.

Compute the decreasing ear path from e_2 through f_2. Let Q' be the maximal subpath of this path such that Q' starts at e_2 and contains at most one vertex of P and Q, not counting e_2 itself. (If the decreasing ear path from e_2 contains no vertices of P or Q besides e_2, then Q' is the entire path.)

Since E and F_2 are embedded outside of R_0, then f_2 is embedded in R_1 or R_2. In the first case, R_1 is partitioned into two bounded regions by Q' (see figure 8). In the second case, R_2 is partitioned into a bounded and an unbounded region by Q' (see figure 9).

Step 5 (Find a disconnecting cycle)

If every vertex of G is contained in E, P, Q, Q' or E_0, then we use the path-building procedure to compute a maximal path from r. Suppose v is a vertex that is not contained in E, P, Q, Q' or E_0. Then v is embedded in the interior of R_0, R_1 or R_2. In every case, the boundary of one of the regions is a disconnecting cycle.

Step 6 (Find a splitting path and recurse)

We now have a disconnecting cycle X in G. We can readily find a disconnecting path Y from r, and then compute a splitting path Z from r. Recurse on the smallest component H in $G - Z$.

Theorem 3 *There is an algorithm that, given a planar graph G, computes a maximal path in G in $O(\log^2 n)$ time with $O(n)$ processors on a CRCW PRAM.*

The Anderson and Mayr algorithm [2] for the maximal path problem in planar graphs runs in $O(\log^3 n)$ time with $O(n^2)$ processors on a CREW PRAM. Since any CRCW algorithm can be simulated on a CREW with only logarithmic slowdown [5, 19], our algorithm achieves the same running time on the CREW using far fewer processors. However, a $O(\log^2 n)$ time algorithm with $O(n)$ processors on a CRCW PRAM for depth-first search in planar graphs has been given by He and Yesha [7] and Shannon [17]. Since any path from the root to a leaf of a depth-first search tree is a maximal path, then these results imply an algorithm for the maximal path problem with the same time and processor bounds as the algorithm presented in this paper.

Conclusion

This paper has introduced the technique of disjoint decreasing ear paths and applied it in two parallel graph algorithms. The disjoint decreasing ear paths problem has a natural relationship with open ear decomposition. Consequently, we believe the technique may be useful in future parallel graph algorithms, especially in light of the numerous existing algorithms based on open ear decomposition.

References

1. R. Anderson. A parallel algorithm for the maximal path problem. *Combinatorica*, 7:315–326, 1987.
2. R. Anderson and E. Mayr. Parallelism and the maximal path problem. *Information Processing Letters*, 24:121–126, 1987.
3. J. Bondy and U. Murty. *Graph Theory with Applications*. North-Holland, Amsterdam, 1976.
4. R. Cole and U. Vishkin. Approximate parallel scheduling ii: Applications to logarithmic-time optimal parallel graph algorithms. *Information and Computation*, 92:1–47, 1991.
5. T. H. Corman, C. E. Leiserson, and R. L. Rivest. *Introduction to Algorithms*. MIT Press and McGraw-Hill Book Co., Cambridge, MA and New York, NY, 1990.
6. D. Fussell, V. Ramachandran, and R. Thurimella. Finding triconnected components by local replacements. In *Proc. 16th ICALP, Lecture Notes in Computer Science 372*, pages 379–393, New York, 1989. Springer-Verlag.
7. X. He and Y. Yesha. A nearly optimal parallel algorithm for constructing depth first search trees in planar graphs. *SIAM Journal on Computing*, 17(3):486–491, 1988.
8. L. Ibarra and D. Richards. Efficient parallel algorithms based on open ear decompositions. Submitted to Parallel Computing.
9. A. Kanevsky and V. Ramachandran. Improved algorithms for four-connectivity. *Journal of Computer and System Sciences*, 42:288–306, 1991.
10. S. Khuller. Ear decompositions. *SigAct News*, 20(1):128, 1989.
11. S. Khuller and B. Schieber. Efficient parallel algorithms for testing connectivity and finding disjoint s-t paths in graphs. In *Proc. 30th Annual IEEE Symp. on Foundations of Computer Science*, pages 288–293, 1989.
12. L. Lovász. Computing ears and branchings in parallel. In *Proc. 26th Annual IEEE Symp. on Foundations of Computer Science*, pages 464–467, 1985.
13. Y. Maon, B. Schieber, and U. Vishkin. Parallel ear decomposition search (eds) and st-numbering in graphs. *Theoretical Computer Science*, 47:277–298, 1986.
14. G. L. Miller and V. Ramachandran. A new graph triconnectivity algorithm and its parallelization. In *Proc. 19th Annual ACM Symp. on Theory of Computing*, pages 335–344, 1987.
15. V. Ramachandran and J. Reif. An optimal parallel algorithm for graph planarity. In *Proc. 30th Annual IEEE Symp. on Foundations of Computer Science*, pages 282–287, 1989.
16. V. Ramachandran and U. Vishkin. Efficient parallel triconnectivity in logarithmic time. In *3rd Aegean Workshop on Computing, Lecture Notes in Computer Science 319*, pages 33–42, New York, 1988. Springer-Verlag.
17. G. E. Shannon. A linear-processor algorithm for depth-first search in planar graphs. *Information Processing Letters*, 29:119–123, 1988.
18. R. E. Tarjan and U. Vishkin. An efficient parallel biconnectivity algorithm. *SIAM Journal on Computing*, 14(4):862–874, 1985.
19. U. Vishkin. Implementation of simultaneous memory address access in models that forbid it. *Journal of Algorithms*, 4(1):45–50, 1983.
20. H. Whitney. Non-separable and planar graphs. *Transactions of the American Mathematical Society*, 34:339–362, 1932.

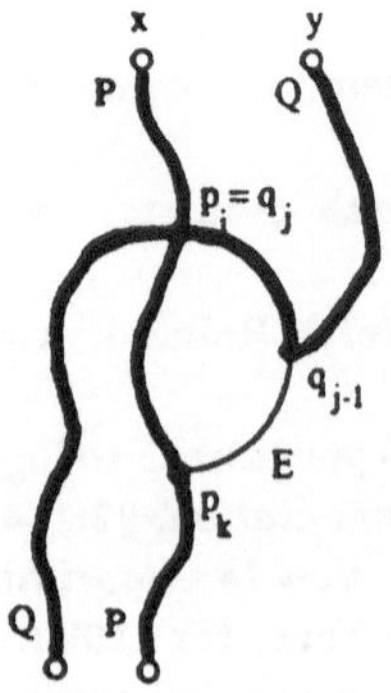

Figure 1

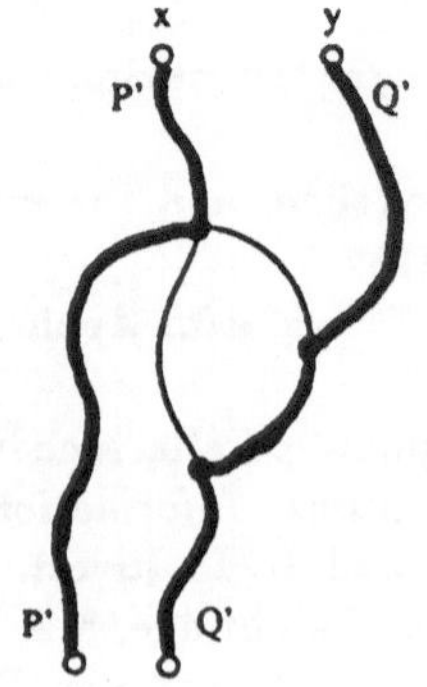

Figure 2

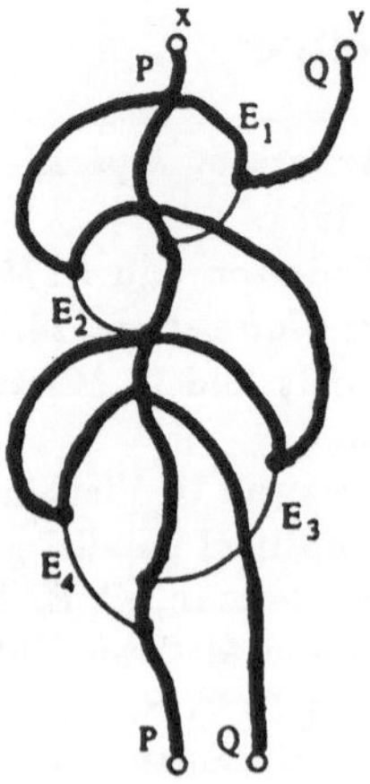

$E_1 < E_2 < E_3 < E_4$ in the ear ordering

Figure 3

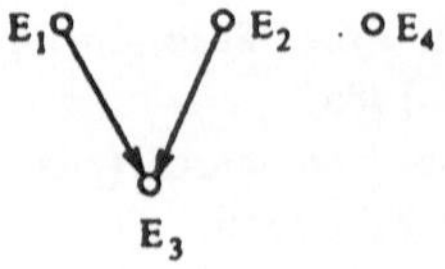

Figure 4

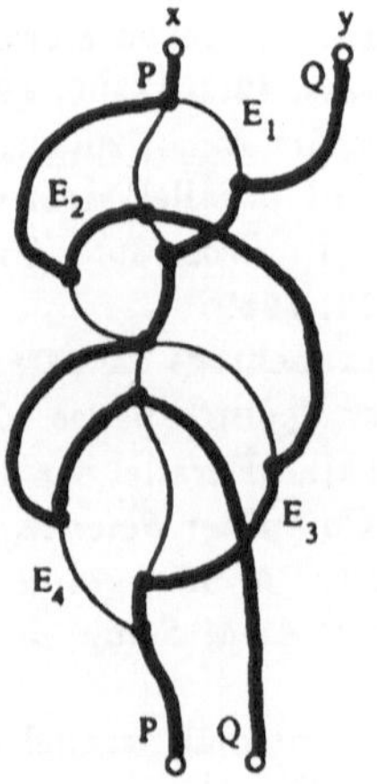

Figure 5

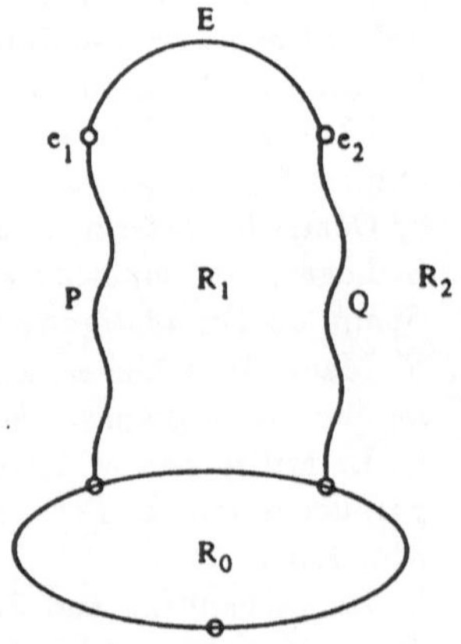

Figure 6

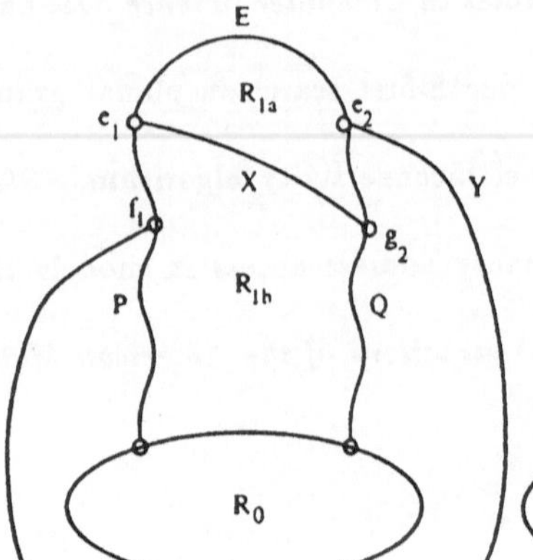

Figure 7

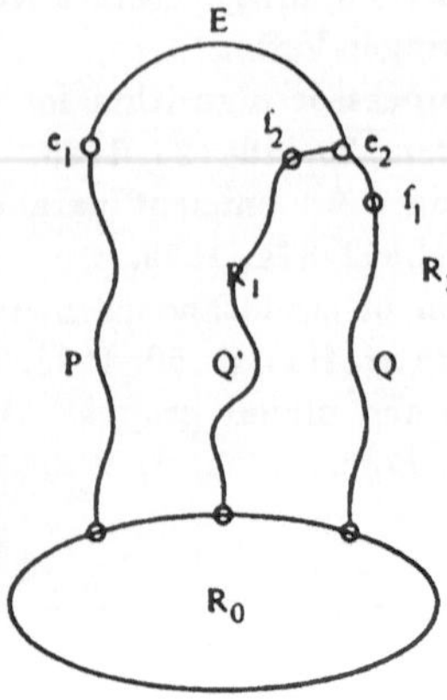

Figure 8

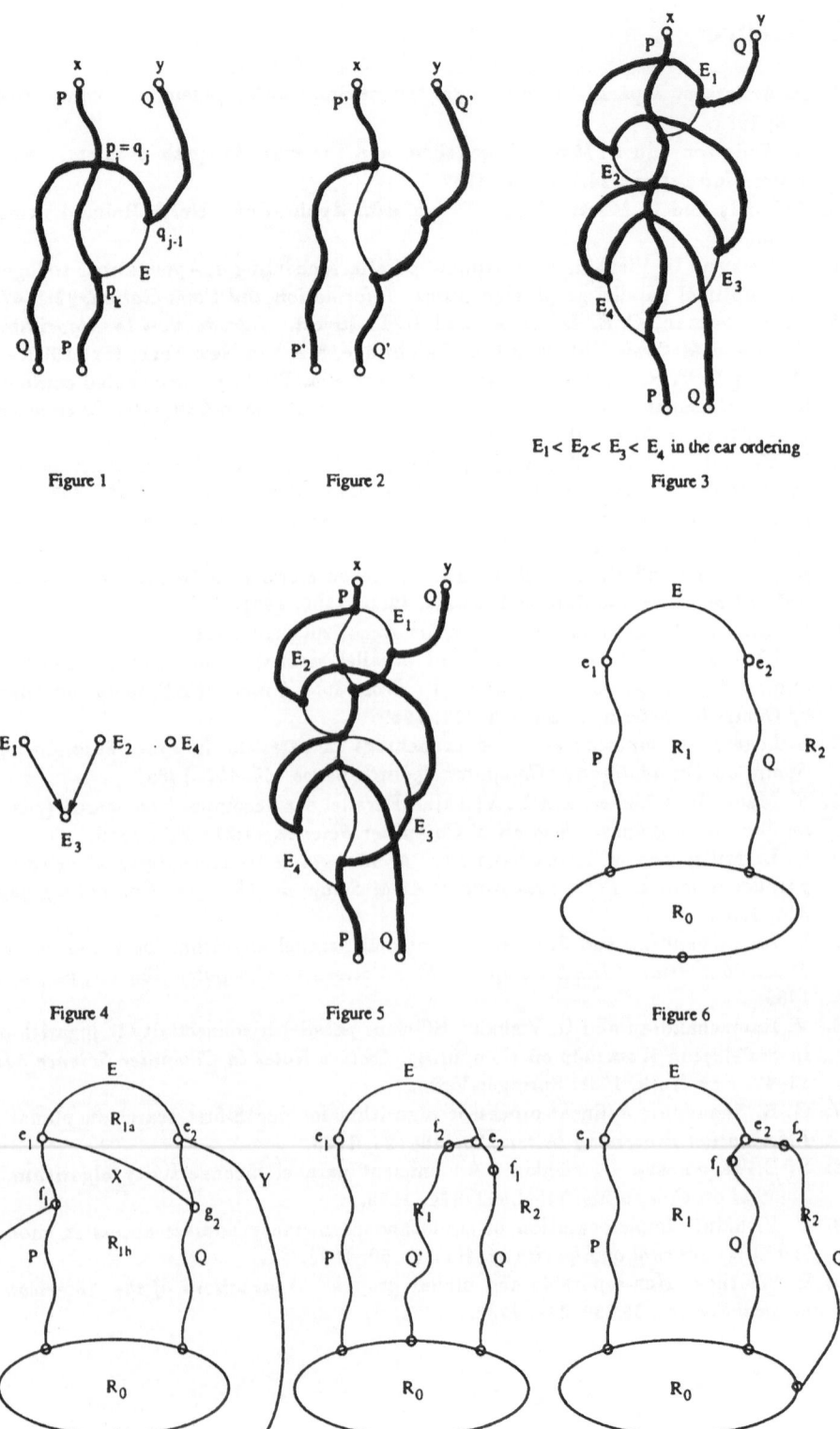

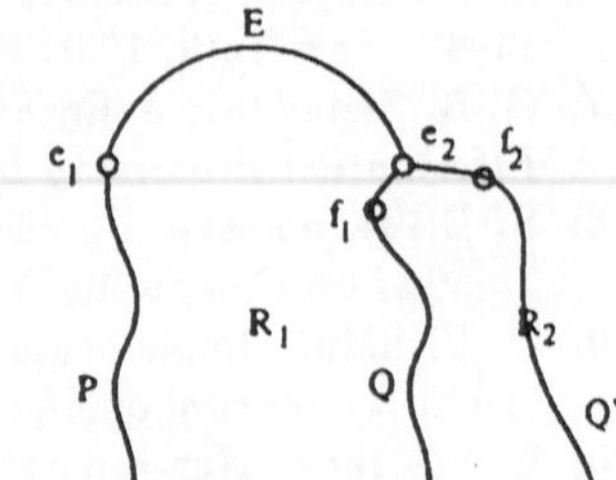

Figure 9

On the Communication Complexity of Parallel Computation*

Oscar H. Ibarra and Nicholas Trân

Department of Computer Science
University of California
Santa Barbara, CA 93106 USA

Abstract. We argue that the synchronized alternating finite automaton (introduced by Hromkovič et al.) is a viable model for studying the communication complexity of parallel computation. This motivates our study of the classes of languages accepted by synchronized alternating finite automata (SAFA) whose messages are bounded in length by a function $m(n)$. We establish the lower bounds on $m(n)$ for some types of SAFA to accept nonregular languages; we also show that these bounds are tight. Next, we establish dense hierarchies of these machines on $m(n)$, and finally we give a characterization of NP in terms of bounded-message multihead SAFA.

1 Introduction

Communication complexity theory studies the communication requirement by a system of cooperating computers to compute a function. Originally motivated by problems concerning VLSI lower bounds, communication complexity has become an extensively studied topic that yields many interesting results [Yao79, AUY83, BFS86]. The model of computation assumed by most work on this subject is a system of two processors of unlimited computational power. Initially, each processor is given one half of the input to a Boolean function $f(x, y)$ in $2n$ variables, and the goal is for the two processors to compute the value of f collaboratively by exchanging information according to some protocol. The communication complexity of $f(x, y)$ is the minimal number of exchanged bits that is required by some protocol. For example, [BFS86] defines P^{cc} to be the class of Boolean functions computable with polylogarithmic communication, and NP^{cc} (BPP^{cc}) to be the corresponding class when nondeterministic (bounded-error probabilistic) protocols are allowed. (Since no more than n bits need to be exchanged in order for the two processors to compute any function f, only sublinear communication complexity classes are of interest.) Surprisingly, it can be shown that $P^{cc} = NP^{cc} \cap$ co-$NP^{cc} \neq NP^{cc}$ [Yao79, AUY83] and that BPP^{cc} and NP^{cc} are incomparable [Yao79, BFS86].

We feel that the underlying model considered in these studies is more characteristic of distributed computation than of truly parallel computation, however. For one thing, an ideal model of parallel computation should support a possibly unbounded set of relatively weak processors instead of a few powerful ones to remove the bottleneck on access to a processor. The ideal model also should not assume a fixed

* Research supported in part by NSF Grant CCR89-18409.

communication network that connects the processors to avoid artificial constraints imposed by the network topology. Although almost diametric to the characteristics of traditional models of parallel computation, these requirements are not unrealistic; in fact they underlie the designing philosophy of the Connection Machine, a non-von Neumann massively parallel computer proposed and constructed by W. D. Hillis [Hil85].

In this paper, we study the communication complexity of parallel computation using a model that satisfies the above requirements. The model we use is called the *synchronized alternating finite automaton*, first introduced in [Hro86] to study the effect of allowing interprocess communication via synchronization. Synchronized alternating finite automata are simply alternating finite automata [CKS81] whose processes, as finite automata, also generate outputs or *messages*. A synchronized alternating finite automaton accepts an input if there is an accepting computation tree on that input in the usual sense for alternating finite automata, and furthermore, the message generated by every branch in this computation tree is a prefix of a common message. Our model departs from the traditional communication model in two ways. First, every process of a synchronized finite automaton has access to the whole input, but each is so weak that it cannot perform any meaningful computation alone. Secondly, we define the communication complexity of a string x in some language L (instead of an input to a function) to be the minimum number of bits in the longest message of some accepting tree of M, where M is a synchronized finite automaton accepting L.

The ability to use messages for interprocess communication makes synchronized alternating finite automata a powerful computing model. We digress for a moment to mention some interesting results known about them. For example, [HKRS89] showed that $\mathcal{L}(\text{2SAFA}) = \text{NSPACE}(n)$, i.e. two-way synchronized alternating finite automata recognize exactly the class of context-sensitive languages. This result was extended in [IT92b] to $\mathcal{L}(\text{1SAFA}) = \text{NSPACE}(n)$, and later in [HRS89] to $\mathcal{L}(\text{1SAFA}(k\text{-heads})) = \mathcal{L}(\text{2SAFA}(k\text{-heads})) = \text{NSPACE}(n^k)$ for every $k \geq 1$, so that $\bigcup_{k=1}^{\infty} \mathcal{L}(\text{SAFA}(k\text{-heads})) = \text{PSPACE}$. [HKRS89] also nicely characterized the restricted versions of one-way and two-way synchronized finite automata that have a constant number of processes to be one-way and two-way multihead nondeterministic finite automata, i.e. $\mathcal{L}(\text{1SAFA}(k\text{-procs})) = \mathcal{L}(\text{1NFA}(k\text{-heads}))$ and $\mathcal{L}(\text{2SAFA}(k\text{-procs})) = \mathcal{L}(\text{2NFA}(k\text{-heads}))$ for every $k \geq 1$. [IT92a] studied synchronized multihead finite automata with only universal states and showed that synchronization did not increase the computational power of these machines, i.e. $\bigcup_{k=1}^{\infty} \mathcal{L}(\text{1SUFA}(k\text{-heads})) = \bigcup_{k=1}^{\infty} \mathcal{L}(\text{1UFA}(k\text{-heads}))$, and $\bigcup_{k=1}^{\infty} \mathcal{L}(\text{2SUFA}(k\text{-heads})) = \bigcup_{k=1}^{\infty} \mathcal{L}(\text{2UFA}(k\text{-heads})) = \text{NLOG}$. Indeed, many fascinating results are also known about the general case of synchronized alternating Turing machines; see [IT92a] for a summarizing account.

Specifically, in this paper we investigate the complexity classes of languages accepted by a synchronized alternating finite automaton whose messages are bounded in length by some function $m(n)$. We establish lower bounds on the message length necessary for a synchronized finite automaton to accept a nonregular language. In particular, we show that a 1SUFA or a 2SUFA needs messages of length at least $\sqrt[k]{|x|}$ for some $k \geq 1$ infinitely often, and a 1SAFA or a 2ASA needs messages of length at least $\log\log|x|$ infinitely often, to accept a nonregular language. We also produce nonregular languages that witness the tightness of these bounds in all cases.

We also establish hierarchies on the function bounding the message length for 1SUFA, 2SUFA, 1SAFA(k-procs), and 2SAFA(k-procs). We show that if $m(n) \in o(n^{a/b})$ for some rational number $0 < a/b \leq 1$, then $\mathcal{L}$(1SUFA($m(n)$-msg)) $\subsetneq$ $\mathcal{L}$(1SUFA($n^{a/b}$-msg)) and $\mathcal{L}$(2SUFA($m(n)$-msg)) $\subsetneq$ $\mathcal{L}$(2SUFA($n^{a/b}$-msg)). For the nondeterministic cases, we show that if $f(n) \in o(n)$ is message-constructible by a 1SAFA(k-procs, $f(n)$-msg) for some $k \geq 2$, then $\mathcal{L}$(1SAFA(k+1-procs, $f(n)$-msg)) $\not\subseteq$ $\mathcal{L}$(2SAFA(l-procs, $g(n)$-msg)) for any $g(n) \in o(f(n))$ and $l \geq 2$. Note that these results are complete in the sense that $\mathcal{L}$(1SUFA) = $\mathcal{L}$(1SUFA($O(n)$-msg)), $\mathcal{L}$(2SUFA) = $\mathcal{L}$(2SUFA($O(n)$-msg)), and $\mathcal{L}$(1SAFA(k-procs)) = $\mathcal{L}$(1SAFA(k-procs, $O(n)$)) for every $k \geq 1$.

Finally, we give a characterization of NP in terms of bounded-message multihead synchronized alternating finite automata. We show that NP = $\bigcup_{k\geq 1}$ $\mathcal{L}$(1SAFA(k-heads, n^k-msg)) = $\bigcup_{k\geq 1}$ $\mathcal{L}$(2SAFA(k-heads, n^k-msg)). It is interesting to note that this result recasts the question P $\stackrel{?}{=}$ NP $\stackrel{?}{=}$ PSPACE in terms of multihead synchronized alternating finite automata, namely, $\bigcup_{k\geq 1}$ $\mathcal{L}$(2SAFA(0-msg, k-heads)) $\stackrel{?}{=}$ $\bigcup_{k\geq 1}$ $\mathcal{L}$(2SAFA(poly-msg, k-heads)) $\stackrel{?}{=}$ $\bigcup_{k\geq 1}\mathcal{L}$(2SAFA($k$-heads)). Using the same technique we can show that $\bigcup_{k\geq 1}$ NTM(n-space, n^k-time) $=\bigcup_{k\geq 1}$ $\mathcal{L}$(1SAFA(n^k-msg)) = $\bigcup_{k\geq 1}$ $\mathcal{L}$(2SAFA(n^k-msg)).

The rest of this paper is organized as follows. Section 2 gives definitions concerning synchronized alternating finite automata. Section 3 establishes the lower bounds on the message length for SUFA and SAFA to accept nonregular languages, and shows that these bounds are tight. Section 4 gives proper hierarchies on the function bounding the message length for 1SUFA, 2SUFA, 1SAFA(k-procs) and 2SAFA(k-procs). Finally, Section 5 provides a characterization of NP in terms of bounded-message multihead synchronized finite automata. Due to space constraint some proofs are omitted, but they will appear in the full version of this paper.

2 Preliminaries

Synchronized alternating finite automata are formally defined in [IT92b]; we only describe their salient features here. An SAFA M is an alternating finite automaton augmented with a one-way write-only output tape and an output alphabet Σ. M accepts x if there is an accepting computation tree T of M on x (in the usual sense for AFA as defined in [CKS81]), and furthermore, each sequence of output symbols (called message) associated with a path in T is a prefix of a common string in Σ^*. Intuitively, the messages serve to synchronize information among the processes at various stages of the computation.

We will be concerned with different versions of synchronized alternating finite automata in this paper. 1SUFA, 2SUFA, 1SAFA, 2SAFA denote the one-way and two-way input head versions of universal states only and general synchronized alternating finite automata. 1SUFA(k-procs), 2SUFA(k-procs), 1SAFA(k-procs), 2SAFA(k-procs) are restrictions of synchronized finite automata whose computation trees on any input have at most k leaves. In the same vein, 1SUFA(m(n)-msg), 2SUFA(m(n)-msg), 1SAFA(m(n)-msg), 2SAFA(m(n)-msg) are restrictions such that on accepted inputs of length n, there is an accepting computation tree whose longest message is bounded in length by $m(n)$.

Let M be a synchronized alternating finite automaton of any kind. In the discussions that follow, $T_M(x)$ denotes a computation tree of M on input x. Note that if M has only universal state, then there is only one $T_M(x)$. For each computation tree $T_M(x)$, let $Msg(T_M(x))$ denote the length of the longest message generated by some process of $T_M(x)$. If $T_M(x)$ is unique, then we simply say $Msg_M(x)$. For example, L is a language accepted by a 1SAFA(m(n)-msg) M iff for every $x \in L$, there is an accepting computation tree $T_M(x)$ such that $Msg(T_M(x)) \leq m(|x|)$. If in addition, there exist infinitely many $x \in L$ such that $Msg(T_M(x)) = m(|x|)$ for every accepting computation tree $T_M(x)$, then we say $m(n)$ is *message-constructible* by M.

For two binary strings x and y, we say $x < y$ if x precedes y in lexicographical order, and finally, REG denotes the class of regular languages.

3 Lower Bounds on the Message Length

Proposition 1. *$\mathcal{L}$(1SAFA(c-msg)) = $\mathcal{L}$(2SAFA(c-msg)) = REG for every $c \geq 1$.*

Proof. Follows from the fact that a 2AFA M can simulate a 2SAFA(c-msg) M' by guessing and storing a sequence of c synchronizing symbols in its finite control for later verification, and that 2AFA's accept only regular sets.

Our first theorem shows that messages of linear length are necessary for SUFA(k-procs) to accept nonregular languages.

Theorem 2. *$\mathcal{L}$(1SUFA(k-procs, m(n)-msg)) = $\mathcal{L}$(2SUFA(k-procs, m(n)-msg)) = REG for every $k \geq 2$ and $m(n) \in o(n)$.*

The bound given in Theorem 2 is tight, since there is a 1SUFA(2-procs, n-msg) that accepts $L = \{0^i1^i : i \geq 0\}$. This bound is also complete in the sense that SUFA(k-procs) = SUFA(k-procs, O(n)-msg) for every $k \geq 1$. Next, we derive the corresponding bound for SUFA, using a cut-and-paste argument.

Theorem 3. *$\mathcal{L}$(1SUFA(m(n)-msg)) = $\mathcal{L}$(2SUFA(m(n)-msg)) = REG for $m(n) \in \bigcap_{k\geq 1} o(\sqrt[k]{n})$.*

Proof. Let $m(n) \in \bigcap_{k\geq 1} o(\sqrt[k]{n})$, and suppose there is some 2SUFA(m(n)-msg) M with q states that accepts a nonregular language L. We may assume without loss of generality that every halting process of M halts on the right endmarker in an accepting state (hence M rejects only if one of its processes loops forever, or if there is a deadlock). By Fact 1, for every $l \geq 1$, there is some $x \in L$ such that $|Msg_M(x)| \geq l$.

Let x be a string in L such that $m(|x|) < \sqrt[4q]{|x|/2^{4q^2}} - 1$, and $|Msg_M(w)| < |Msg_M(x)|$ for $w < x$ in L. Associate with each partition $y;z$ of x (; is the boundary between y and z) a sequence of $4q$ integers $|s_{left,1}(y;z)|, \ldots, |s_{left,q}(y;z)|, |s_{right,1}(y;z)|, \ldots, |s_{right,q}(y;z)|, |t_{left,1}(y;z)|, \ldots, |t_{left,q}(y;z)|, |t_{right,1}(y;z)|, \ldots, |t_{right,q}(y;z)|$, and a set $S(y;z)$.

Each $s_{d,j}(y;z)$ denotes the longest message generated by some process of $T_M(x)$ since the beginning of the computation until it crosses $y;z$ at some point in state j from direction d. If there is no such process then $|s_{d,j}(y;z)| = -\infty$. Similarly, each $t_{d,j}(y;z)$ denotes the longest message generated by some process of $T_M(x)$ after it crosses $y;z$ in state j from direction d until the end of the computation. Note that if $|s_{d,j}(y;z)| = -\infty$ iff $|t_{d,j}(y;z)| = -\infty$. Finally, $S(y;z)$ consists of elements of the form (s_1, d_1, s_2, d_2) such that there is a process of $T_M(x)$ which at some point enters z in state s_1 from direction d_1 and eventually emerges from z in state s_2 from direction d_2.

Since there are at most $2^{4q^2}(m(|x|)+1)^{4q} < |x|$ different such combinations of $4q$ integers and a set, x can be written as uvw, $|v| \geq 1$, so that the information associated with $u;vw$ and $uv;w$ are identical, i.e. $|s_{d,j}(u;vw)| = |s_{d,j}(uv;w)|$, $|t_{d,j}(u;vw)| = |t_{d,j}(uv;w)|$ for each direction d and state j, and $S(u;vw) = S(uv;w)$. Furthermore, if $|s_{d,j}(u;vw)| \neq -\infty$ then there are two processes of the accepting computation tree $T_M(x)$ that actually generate the messages $s_{d,j}(u;vw)t_{d,j}(u;vw)$ and $s_{d,j}(uv;w)t_{d,j}(uv;w)$ respectively. Hence it follows that $s_{d,j}(u;vw) = s_{d,j}(uv;w)$ and $t_{d,j}(u;vw) = t_{d,j}(uv;w)$ for each direction d and state j.

Also, note that if some process of $T_M(x)$ crosses over $uv;w$ from the left in state r_1, generates m, and then crosses over $uv;w$ again from the right in state r_2, then mt_{right,r_2} is a prefix of t_{left,r_1}. Similarly, if some process of $T_M(x)$ crosses over $u;vw$ from the right in state r_1, generates m, and then crosses over $u;vw$ again from the left in state r_2, then mt_{left,r_2} is a prefix of t_{right,t_1}.

Let $x' = uw$. We will show that M also accepts x', and $Msg_M(x') = Msg_M(x)$. Consider a process p' of $T_M(x')$. Since x is in L and $S(u;vw) = S(uv;w)$, p' must halt in an accepting state. Let r_i be the state that p' is in when it crosses over $u;w$ the ith time, and write the message p' generates over the whole computation as $m = m_1 m_2 \ldots m_{2k-1}$ for some $k \leq q$, where m_i is the message p' generates between the $(i-1)th$ and ith crossings.

Clearly m_1 is a prefix of the message $m_1 t_{left,r_1}(u;vw)$ generated by a process p_1 of $T_M(x)$. Since p_1 crosses $u;vw$ from the left in state r_1, there is another process p_2 of $T_M(x)$ that crosses $uv;w$ from the left in state r_1, generates m_2, crosses $uv;w$ again from the right in state r_2, and generates $t_{right,r_2}(uv;w)$ for the rest of the computation. Since $m_2 t_{right,r_2}(uv;w)$ is a prefix of $t_{left,r_1}(uv;w)$, $m_2 t_{right,r_2}(uv;w)$ is a prefix of $t_{left,r_1}(u;vw)$. Similarly, since p_2 crosses $uv;w$ from the right in state r_2, there is another process p_3 of $T_M(x)$ that crosses $u;vw$ from the right in state r_2, generates m_3, crosses $u;vw$ again from the left in state r_3, and generates $t_{left,r_3}(u;vw)$ for the rest of the computation. Since $t_{right,r_2}(uv;w) = t_{right,r_2}(u;vw)$, $m_3 t_{left,r_3}(u;vw)$ is a prefix of $t_{right,r_2}(uv;w)$. Continue this argument until we have m_{2k-1} is a prefix of $t_{left,r_{2k-2}}$. Then m is a prefix of $m_1 t_{left,r_1}(u;vw)$, which is a prefix of $Msg(x)$. Since p' is an arbitrary process of $T_M(x')$, this shows that $T_M(x')$ is an accepting computation tree.

Now, suppose p_1 is a process of $T_M(x)$ that actually generates $Msg_M(x)$, and suppose after generating the message m_1, p_1 crosses over $u;vw$ in state r_1 the first time (from the left). Then for the rest of the computation, p_1 must generate $t_{left,r_1}(u;vw)$, or else there is another process of $T_M(x)$ that generates a longer message than p_1 does. But since $t_{left,r_1}(u;vw) = t_{left,r_1}(uv;w)$, there is a process p_2 of $T_M(x)$ that at some point crosses over $uv;w$ in state r_1 from the left and generates $t_{left,r_1}(uv;w)$

for the rest of the computation. Again, suppose after generating the message m_2, p_2 crosses $uv;w$ the next time in state r_2. Then for the rest of the computation, p_2 must generate $t_{right,r_2}(uv;w)$, and therefore there is another process p_3 of $T_M(x)$ that at some point crosses over $u;vw$ in state r_2 from the right and generates $t_{right,r_1}(u;vw)$ for the rest of the computation. Repeat the argument until for some $k \leq 2q-1$, $m_{k+1} = t_{left,r_k}(u;vw)$ or $m_{k+1} = t_{right,r_k}(uv;w)$. Then from p_1, p_2, ..., p_k, we can construct a process of $T_M(x')$ that crosses $u;w$ the first k times in states r_1, r_2, ..., r_k and generates exactly $Msg_M(x)$. Hence $Msg_M(x) = Msg_M(x')$.

But this contradicts the assumption on x, since $x' < x$ is in L, and $Msg_M(x) = Msg(x')$. Hence L cannot be nonregular, and this proves the theorem.

Theorem 4. *For every rational number $0 < a/b \leq 1$, there is a 1SUFA M that accepts a nonregular language L and $|Msg_M(x)| = |x|^{a/b}$ for infinitely many $x \in L$. Hence, the bound given in Theorem 3 is tight.*

Proof. We only prove the theorem for the case $a/b = 2/3$. The technique can be easily generalized to prove the general case.

Let $L_3 = \{(\$(\#0^m)^m)^m : m \geq 1\}$. Then there is a 1SUFA M_3 that recognizes L_3 as follows. M_3 first splits into three processes. One is in state q_1, one is in state q_4, and one makes a pass over the input in state q_0. During this scan, if the process in state q_0 reads a \$ (#), it splits into two processes. One is in state q_2 (q_3 respectively); the other continues moving to the right in state q_0. When the process in state q_0 reaches the right endmarker, it halts and accepts.

The process in state q_1 moves to the right and generates an s-symbol 0 every time it scans a \$, until it reaches the right endmarker. It then generates an s-symbol E and halts in acceptance. Similarly, each process in state q_2 moves to the right and generates a 0 every time it scans a #, until it scans another \$; at that point, it generates an E and halts in acceptance. Each process in state q_3 moves to the right and generates a 0 every time it scans a 0, until it scans another #; at that point, it generates an E and halts in acceptance.

Finally, the process in state q_4 moves to the right and generates a 0 every time it scans a 0 until it reaches a #. Then it generates an E; next it continues to move to the right, generating a 0 for every 0 it scans until it reads the first \$. Then it halts and accepts.

Clearly, an input w is in L_3 iff it is accepted by M_3. Furthermore, the process in state q_0 generates no s-symbols, each process in state q_1, q_2, and q_3 generates exactly $|w|^{1/3}$ s-symbols, and the process in state q_4 generates exactly $|w|^{2/3}$ s-symbols if consecutive symbols # and \$ in the input are compressed into one composite input symbol. Furthermore, since L_3 is nonregular, this shows that the bound given in Theorem 3 is tight for $a/b = 2/3$.

Theorem 5. *There is a 1SAFA(2-procs,loglog(n)-msg) M that accepts a nonregular language.*

Proof. For every $i \geq 1$, define the string $bin(i)$ to be the binary representation of i such that the rightmost bit is the most significant bit and is always 1. Note that $|bin(i)| \leq \log(i) + 1$. For every binary string of positive length $s = b_1 b_2 \ldots b_n$, define

$ind(s) = bin(1)b_1\$bin(2)b_2\$\ldots\$bin(n)b_n\$$. Now define a language

$$L = \{ind(bin(1))\#ind(bin(2))\#\ldots\#ind(bin(k))\# : k \geq 1\}.$$

Then we claim that there is a 1SAFA(2-procs, loglog(n)-msg) M that accepts $\overline{L}$.

Every $x \in \overline{L}$ can be written as $ind(bin(1))\#ind(bin(2))\#\ldots\#ind(bin(i-1))\#y$, for some $i \geq 1$ and $y \neq ind(bin(i))$, such that one of the following must hold:

1. $i = 1$ and $11\$\#$ is not a prefix of y,
2. neither $11\$$ nor $10\$$ is a prefix of y,
3. $y = bin(1)b_1\$bin(2)b_2\$\ldots\$bin(j-1)b_{j-1}\$z$ for some $j \leq |bin(i)|$, and neither $bin(j)0\$$ nor $bin(j)1\$$ is a prefix of z, or
4. $y = bin(1)b_1\$bin(2)b_2\$\ldots\$bin(j)b_j\$z$ for some $j \leq |bin(i)|$, and b_j is not the jth bit of $bin(i)$.
5. $y = ind(bin(i))z$ and $\#$ is not a prefix of z

M can easily verify (1) or (2) with a DFA. To verify (3) M splits into two processes p_1 and p_2. Process p_1 nondeterministically locates $bin(j-1)$ and generates the message $m = bin(j)E$. Process p_2 guesses and generates m while scanning z to make sure either that $bin(j)$ is not a prefix of z or that $bin(j)$ *is* a prefix of z but the next two symbols of z are not $0\$$ or $1\$$. If p_2 cannot verify either condition, then it halts in a rejecting state.

To verify (4), one process p_1 of M nondeterministically locates the subword $ind(bin(i-1))$, determines the jth bit b of $bin(i)$ for some $j \leq |bin(i)|$, and then generates the message $m = bin(j)bE$. Process p_2 nondeterministically locates a subword $bin(k)b_k$ of y, guesses and generates m while scanning $bin(k)b_k$ to make sure that $bin(k) = bin(j)$ and $b_k \neq b$. Again, if p_2 fails to verify this condition, then it halts in a rejecting state.

Finally, to verify (5), one process p_1 of M nondeterministically locates the subword $ind(bin(i-1))$, determines $|bin(i)|$, and generates the message $m = bin(|bin(i)|)E$. Process p_2 nondeterministically locates a subword $bin(k)b_k$ of y, guesses and generates m, while scanning $bin(k)$ to make sure that $bin(k) = bin(|bin(i)|)$ and that the symbol following $bin(k)b_k$ in y is not $\#$.

In all cases, M uses messages of length at most $\log\log(|x|)$.

The $\log\log(n)$ bound on message length is tight for SAFA, since it can be shown that

Theorem 6. *$\mathcal{L}$(1SAFA(k-procs, m(n)-msg)) = $\mathcal{L}$(2SAFA(k-procs, m(n)-msg)) = $\mathcal{L}$(1SAFA(m(n)-msg)) = $\mathcal{L}$(2SAFA(m(n)-msg)) = REG for* $m(n) \in o(\log\log(n))$.

4 Hierarchies on the Message Length

Theorem 7. *Suppose $m(n) \in o(n^{a/b})$ for some rational number $0 < a/b \leq 1$. Then $\mathcal{L}$(1SUFA($n^{a/b}$-msg)) − $\mathcal{L}$(2SUFA(m(n)-msg)) $\neq \emptyset$.*

Proof. Let $m(n) \in o(n^{a/b})$ for some rational number $0 < a/b \leq 1$. We show that there is a language $L \in \mathcal{L}(1SUFA(n^{a/b}\text{-}msg)) - \mathcal{L}(2SUFA(m(n)\text{-}msg))$.

By Theorem 4 there is a nonregular language K accepted by a 1SUFA($n^{a/b}$-msg) M so that $|Msg_M(x)| = |x|^{a/b}$ for infinitely many $x \in K$. Let c_M be the maximum number of children that a node in any computation tree of M can have, and let $d = \max(2^{b/a}, c_M)$. Clearly c_M is a constant depending only on M. For each $x = b_1 b_2 \ldots b_n \in K$, define $pad(x) = b_1\$0^{r_1}\$0^{d-r_1}\$b_2\$0^{r_2}\$0^{d-r_2}\$\ldots\$b_n\$0^{r_n}\$0^{d-r_n}$, where $b_i \in \{0,1\}$ and $r_i \leq c_M$ for each i, so that the computation path of $T_M(x)$ specified by $r_1, r_2, \ldots, r_n$ generates $Msg_M(x)$. Define $L = \{pad(x)\#y\#0^{|y|} : x \in L \;\&\; y = Msg_M(x)\}$. Then there is a 1SUFA($n^{a/b}$-msg) N that accepts L as follows.

On input $w = x\#y\#0^z$, N first splits into three processes p_1, p_2, and p_3. Process p_1 moves past the first symbol # and then generates the message $yE0^zE$, where E is a new s-symbol not used by M. Process p_2 verifies that x has the correct form to be $pad(v)$ for some v; if not p_2 rejects. Process p_3 simulates M on v faithfully; furthermore, at the end of the simulation, the unique subprocess of p_3 specified by the path embedded in $pad(v)$ also generates $E0^{|y|}E$.

If $w \in L$, then $w = pad(x)\#Msg_M(x)\#0^{|Msg_M(x)|}$ for some $x \in K$, and hence all processes of $T_N(w)$ accept, and the messages generated by the children of process p_3 are prefixes of the message generated by the subprocess determined by $pad(x)$, which is the same as the mespsage generated by process p_1. Thus N accepts w. On the other hand, if $w = x\#y\#z \notin L$, then either x is not of the correct form, or x without the padding \$'s and 0's is not a string in L, or the path embedded in x does not generate $Msg_M(x)$, or $y \neq Msg_M(x)$, or $z \neq 0^{|Msg_M(x)|}$. In all cases, either some process of $T_N(w)$ rejects, or two processes of $T_N(w)$ deadlock.

Thus N accepts w iff $w \in L$. Furthermore, N on w generates message of length at most $|w|^{a/b}$, and so $L \in \mathcal{L}(1SUFA(n^{a/b}\text{-}msg))$.

Now suppose $m(n) \in o(n^{a/b})$ and there is some 2SUFA($m(n)$-msg) N' with q states that recognizes L. Let $w = pad(x)\#Msg_M(x)\#0^l \in L$, where $l = |Msg_M(x)| = |x|^{a/b}$, such that $|x|$ is random, and $m(|w|) = m(2^{b/a}|x|) < |x|^{a/b}/(2q^2+1) - 3$, and consider a process p in the accepting computation tree $T_{N'}(w)$. Associate with each partition $0^a; 0^{l-a}$ of the input suffix 0^l of w a set $S(a)$, so that the triple (s, i) is in $S(a)$ iff p crosses over the border in state s the *ith* time. Since p can make at most $2q$ crossings, there are at most $2q^2$ different such sets.

Let $e = 2q^2 + 1$ and write 0^l as $0^e 0^{l-2e} 0^e$. Then p generates no more than $2q$ s-symbols while on the segment 0^{l-2e}, or else an easy argument using crossing sequences shows that there is another process of $T_{N'}(w)$ that makes at least $(l - 3e)/e = |x|^{a/b}/(2q^2+1) - 3$, contradicting the choice of w. Hence p generates at most $6q$ s-symbols while on the input suffix 0^l. But then $|x|$ can be recovered by a program of constant size that simulates M on the input portion 0^l. This contradicts the randomness of $|x|$. Hence N does not recognize L.

Corollary 8. *Suppose $m(n) \in o(n^{a/b})$ for some rational number $0 < a/b \leq 1$. Then $\mathcal{L}(1SUFA(m(n)\text{-}msg)) \subsetneq \mathcal{L}(1SUFA(n^{a/b}\text{-}msg))$, and $\mathcal{L}(2SUFA(m(n)\text{-}msg)) \subsetneq \mathcal{L}(2SUFA(n^{a/b}\text{-}msg))$.*

Similarly, we can show the following

Theorem 9. *Suppose $f(n) \in o(n)$ is message-constructible by a 1SAFA(k-procs, $f(n)$-msg) for some $k \geq 2$. Then $\mathcal{L}(1SAFA(k+1\text{-}procs, f(n)\text{-}msg)) \not\subseteq \mathcal{L}(2SAFA(l\text{-}procs, g(n)\text{-}msg))$ for any $g(n) \in o(f(n))$ and $l \geq 2$.*

5 A Characterization of NP

We first show that NP machines can be simulated by multihead bounded-message SUFA using a simulation of nondeterministic TMs by SATMs similar to the one described in [HKRS89].

Lemma 10. *$NP \subseteq \bigcup_{k\geq 1} \mathcal{L}$(1SAFA(k-heads, n^k-msg)).*

Proof. Suppose L is in NP, and let M be a one-tape nondeterministic Turing machine M that accepts L within n^k steps. We construct a 1SAFA(k-heads, n^{k+1}-msg) N that accepts L as follows. On input x of length n, N splits into n^k processes $p_0, p_1, \ldots, p_{n^k-1}$ so that for every $0 \leq i < n^k$, the positions of the k heads of process i together represent the integer i in n-ary notation. Process p_i will be used to store the content of the ith tape square of M for $0 \leq i < n^k$. Initially, $p_0, p_1, \ldots, p_{n-1}$ store the first, second, ..., nth symbols respectively in their finite controls, and the other processes store the blank symbol. Furthermore, p_0 starts in state q_0, the starting state of M, and other processes start in a special state S that is not a state of M.

At any time during the simulation, only one process of M is *not* in state S. This distinguished process represents the current input head position of M and decides the next move of M to be simulated. Suppose at some point p_d is the distinguished process, and the next action of M is to replace the dth tape symbol with s and then move the input head to the right in state q. To do this, p_d spawns an exact copy p'_d of itself, which then generates deterministically the message $Bl_1^{d+1}\#l_2^{d+1}\#\ldots\#l_k^{d+1}E$ where l_i^{d+1} is the distance between the ith input head of p_{d+1} and the right marker, or a special symbol if the input head is on the leftmost square.

In the meantime, each of the other processes representing tape squares of M decides whether it is the right neighbor of p_d. If some process p_e decides that it is the right neighbor, then it spawns an exact copy p'_e, which generates deterministically the message $Bl_1^e\#l_2^e\#\ldots\#l_k^eE$ and accepts. On the other hand, if p_e decides that it is not the right neighbor, then p_e spawns an exact copy p'_e, which nondeterministically generates $Bl_1^{d+1}\#l_2^{d+1}\#\ldots\#l_k^{d+1}E$, and during the process, verifies that for some i, $l_i^e \neq l_i^{d+1}$; p'_e accepts iff there is such an i.

Also, the original processes $p_1, p_2, \ldots, p_{n^k-1}$ nondeterministically generate the message $Bl_1^{d+1}\#l_2^{d+1}\#\ldots\#l_k^{d+1}E$ without moving their input heads. Furthermore, after generating the symbol E, p_d stores the symbol s in its finite control and then enters state S, and the processes that decide they are the right neighbor of p_d enter state q. This finishes N's simulation of a single step of M moving to the right. The simulation is similar when the input head of M moves to the left.

It is clear that if M accepts an input x, then there is an accepting computation of N that correctly simulates an accepting computation of M on x. On the other hand, if there is an accepting computation of N on x, then since in every step, only either the right or the left neighbor of the distinguished process can become distinguished in the next step, it follows that M accepts x. Hence $L(M) = L(N)$.

Finally, note that N generates messages of length at most $O(n)$ to simulate a step of M, and M needs at most n^k steps, so N generates messages of length at most $O(n^{k+1})$, which can be compressed to have length exactly n^{k+1}, on inputs of length n. Hence $L(M) \in \mathcal{L}$(1SAFA($k+1$-heads, n^{k+1}-msg)).

Lemma 11. $\bigcup_{k\geq 1}$ *$\mathcal{L}$(2SAFA(k-heads, n^k-msg))* $\subseteq$ *NP.*

Proof. Suppose L is accepted by a 2SAFA(k-heads, n^k-msg) M for some $k \geq 1$. Define $L' = \{x\#y : x \in L$ & for some accepting computation tree $T_M(x)$, $Msg(T_M(x))$ is a prefix of $y\}$. L' can be accepted by a 2AFA($k+1$-heads) M' as follows: on input $x\#y$, M' uses k heads to simulate M on x faithfully and the remaining head to verify that every process shares the same common synchronizing sequence. Clearly $x \in L \Leftrightarrow \exists |y| \leq |x|^k [x\#y \in L']$. But since L' is accepted by a multihead alternating finite automata, $L' \in P$ [Kin88], and hence $L \in NP$.

Theorem 12. *$NP = \bigcup_{k\geq 1} \mathcal{L}$(1SAFA(k-heads, n^k-msg)) $= \bigcup_{k\geq 1} \mathcal{L}$(2SAFA(k-heads, n^k-msg)).*

Using the same technique, we can show that

Theorem 13. $\bigcup_{k\geq 1}$ *NTM(n-space, n^k-time)* $= \bigcup_{k\geq 1}$ *$\mathcal{L}$(1SAFA(n^k-msg))* $= \bigcup_{k\geq 1}$ *$\mathcal{L}$(2SAFA(n^k-msg))*

References

[AUY83] A. V. Aho, J. D. Ullman, and M. Yannakakis, *On notions of information transfer in VLSI circuits*, in Proc. 15th Symp. on Theory of Computing, ACM, 1983, pp. 133–138.

[BFS86] L. Babai, P. Frankl, and J. Simon, *Complexity classes in communication complexity theory (preliminary version)*, in Proc. 27th Ann. Symp. on Foundations of Computer Science, IEEE, 1986, pp. 337–347.

[CKS81] A. K. Chandra, D. K. Kozen, and J. Stockmeyer, *Alternation*, J. ACM, 28 (1981), pp. 114–133.

[Hil85] W. D. Hillis, *The Connection Machine*, PhD thesis, Massachusetts Institute of Technology, 1985.

[HKRS89] J. Hromkovič, J. Karhumäki, B. Rovan, and A. Slobodová, *On the power of synchronization in parallel computations*, tech. report, Comenius University, Bratislava, Czechoslovakia, 1989.

[Hro86] J. Hromkovič, *How to organize the communication among parallel processes in alternating computations.* Manuscript, January 1986.

[HRS89] J. Hromkovič, B. Rovan, and A. Slobodová, *Deterministic versus nondeterministic space in terms of synchronized alternating machines*, tech. report, Comenius University, Bratislava, Czechoslovakia, 1989.

[IT92a] O. H. Ibarra and N. Q. Trân, *New results concerning synchronized finite automata*, in Proc. 19th ICALP, Vienna 1992, Lecture Notes in Computer Science 623, Springer-Verlag, 1992, pp. 126–137.

[IT92b] ——, *On space-bounded synchronized alternating Turing machines*, Theoretical Computer Science, 99 (1992), pp. 243–264.

[Kin88] K. N. King, *Alternating multihead finite automata*, Theoretical Computer Science, 61 (1988), pp. 149–174.

[Yao79] A. C.-C. Yao, *Some complexity questions related to distributed computing*, in Proc. 11th Symp. on Theory of Computing, ACM, 1979, pp. 209–213.

A Taxonomy of Forgetting Automata*

Petr Jančar+, František Mráz*, Martin Plátek*

+) Department of Computer Science, University of Ostrava,
Dvořákova 7, 701 03 Ostrava, Czechland

*) Department of Computer Science, Charles University,
Malostranské nám. 25, 118 00 Praha 1, Czechland,
e-mail: mrazf@cspguk11.bitnet, platek@cspguk11.bitnet

Abstract. Forgetting automata are nondeterministic linear bounded automata whose rewriting capability is restricted as follows: each cell of the tape can only be "erased" (rewritten by a special symbol) or completely "deleted".

We consider all classes of languages corresponding to various combinations of operations (erasing and deleting combined with moving the head), classify them according to the Chomsky hierarchy and show (some) other relations among them.

1 Introduction

Our motivation mainly comes from linguistics. We illustrate it by the following simple example: parsing a sentence can consist in the stepwise leaving out the words whose absence does not affect correctness of the sentence, e.g.

"The little boy ran quickly away",
"The ____ boy ran quickly away",
"The ____ boy ran _____ away",

It leads us to considering linear bounded automata with limited rewriting ability: any symbol can only be rewritten by a special symbol (meaning the emptiness of the relevant cell); we refer to the operation as *ERASING*.

In fact, erasing automata were also considered by von Braunmühl and Verbeek in [BV79]. The motivation was different: they introduced a new storage medium - the finite-change tape - which is, roughly speaking, "something between time and space"; the erasing automaton is a special case then.

In our example, the segments of erased symbols are not important and can be deleted completely

("The boy ran quickly away", "The boy ran away", ...).

To model it directly, we also consider an operation called *DELETING*, which "cuts out" the cell (and pastes the two remaining parts of the tape together).

* The research presented in this paper was carried out in connection with the COST Project No. 2824 "Language Processing Technologies for Slavic Languages".

To make *DELETING* more natural, we use the model of list automata; a list automaton, uses a (doubly linked) list of items (cells) rather than the usual tape.

Notice that *ERASING* is more general than *DELETING*; unlike deleted items, erased items can still carry information.

We consider all combinations of operations "move (the head) without changing the item to the left (right)" , "erase (delete) the item and move left (right)". First we classify the corresponding classes of languages according to the Chomsky hierarchy. Doing this, we use some of the previous results ([PV86], [JMP92a]).

In addition, we show some other relations among the considered classes.

Among others, we get an interesting characterization of context-free languages and show that (at least in some contexts) *ERASING* is really more powerful than *DELETING*.

The results are given in Section 2. Easy ideas (combined with some references) are shown together with the results, two longer proofs are given in Section 3. Section 4 contains some of the open questions and our conjectures.

Remark . A preliminary version of a part of this work was published in [JMP92b]. Some longer proofs could be found in [JMP93] and [J93].

2 Definitions and results

An *F-automaton* (forgetting automaton) F has a finite state control unit with one head moving on a linear (doubly linked) list of items (cells); each item contains a symbol from a finite alphabet. In the initial configuration, the control unit is in a fixed (initial) state, the list contains an input word bounded by special sentinels #, \$ and the head scans the item with the left sentinel #.

The computation of F is controlled by a finite set of instructions of the form $[q, a] \rightarrow [q_1, op]$, with the following meaning: according to the actual state q and the scanned symbol a, F may change the state to q_1 and perform op, one of the following six operations:

- MV_R, MV_L — moving the head one item to the right (left),
- ER_R, ER_L — erasing, i.e. rewriting the contents of the scanned item with a special symbol, say $*$ and moving the head one item to the right (left)
- DL_R, DL_L — deleting the item from the list moving the head one item to the right (left)

Generally, F is *nondeterministic* (more than one instruction can be applicable at the same time).

A configuration (situation) of F is represented by triple (ua, q, w), where q is the current state of F, a the scanned symbol, u the string contained in the list on the left from a, w the string contained in the list on the right from a.

An input *word is accepted by F* if there is a computation starting in the initial configuration which achieves a configuration with the control unit being in one of accepting states.

$L(F)$ denotes the language consisting of all words accepted by F; we say that F *recognizes* $L(F)$.

By $[O]$, where O is a subset of $\{MV_R, MV_L, ER_R, ER_L, DL_R, DL_L\}$, we denote the class of languages recognisable by F-automata using operations from O only. (We write $[Op_1, Op_2, \ldots, Op_n]$ instead of $[\{Op_1, Op_2, \ldots, Op_n\}]$).

The couple MV_R, MV_L we abbreviate by MV; similarly for ER, DL.

For the situations with the head scanning # ($) we use the following technical assumption:

- only MV_R (MV_L)-instructions can be applicable, and
- MV_R (MV_L)-instructions are in these situations always allowed also if MV_R (MV_L) is not contained in considered O.

There are 6 operations; hence there are $6+15+20+15+6+1 = 63$ combinations of operations to be considered.

Remark 1. It is easy to see that

a) If $MV_R, ER_R, ER_L \in O$ ($MV_R, DL_R, DL_L \in O$) then we can leave out ER_R (DL_R) without changing $[O]$.
Similarly for MV_L, ER_R, ER_L (MV_L, DL_R, DL_L) and ER_L (DL_L). (Hence e.g. $[MV, ER] = [MV, ER_R] = [MV, ER_L]$.) (For similar reasons, we do not need to consider "erasing without moving the head").

b) If $ER_R, DL_R, DL_L \in O$ and $MV_R \notin O$ then we can leave out DL_R without changing $[O]$.

In what follows we use it implicitly.

Now we give the classification of classes $[O]$, for all possible O, according to the Chomsky hierarchy. First notice that $[MV, ER, DL]$ (and hence each $[O]$) is contained in CSL (the class of context-sensitive languages) because an F-automaton is a special case of a linear bounded automaton.

Moreover the next theorem holds.

Theorem 2. *$[MV, ER, DL]$ is strictly contained in $DCSL$ (D stands for "deterministic").*

Proof. It is based on common method of diagonalization. The construction of "diagonalizing" deterministic linear bounded automaton ($DLBA$) is based on Savitch's idea from [S70] of simulating a nondeterministic LBA by a deterministic quadratic-space bouded $DLBA$ and was outlined in [BV79].

This proof could be found in [J93] or in [JMP93].

(We use $A \longrightarrow B$ to denote trivial or "easily-seen" inclusion A is a subclass of B.

At the right side, we show the numbers of combinations out of the mentioned 63, which correspond to the relevant classes.)

1. Trivial classes:
   - All $[O]$, where no X_R-operation ($X \in \{MV, ER, DL\}$) is contained in O (7 comb.)

2. **Classes equal to the class of regular languages (REG):**
   - **All $[O]$, where only X_R-operations are in O** **(7 comb.)**
   - **All $[O]$, where DL_R is the only X_R-operation in O and some X_L-operation is in O (proofs trivial)** **(7 comb.)**
   - **$[MV]$, $[MV_L, ER]$, $[ER]$** **(4 comb.)**

   (proofs as in the case of two-way finite automata; cf.[S59])
3. **Classes strictly between REG and CFL (CFL denoting the class of context-free languages): $[ER_R, DL] \longrightarrow [MV_R, DL]$** **(4 comb.)**

   Proof. **An F-automaton F_1 with ER_R, DL can be simulated by an F-automaton F_2 with MV_R, DL: after each performing DL_L, F_2 "pretends" that it reads $*$ (the special symbol for erasing) and thus is able to behave exactly as F_1.**
   It is easy to see that $REG \longrightarrow [ER_R, DL]$, $\{a^nb^n\} \in [ER_R, DL]$ and any language from $[MV_R, DL]$ can be recognized by some push-down automaton.
   Using certain pumping lemmas, similar as in [PV86], it can be shown that the language $\{c^{n_1}a_{i_1}c^{n_1}c^{n_2}a_{i_2}c^{n_2}...c^{n_k}a_{i_k}c^{n_k}sd_{i_k}d_{i_{k-1}}...d_{i_1}; k \geq 1, i_j \in \{1,2\}, n_j > 0 \text{ for } 1 \leq j \leq k\}$ is not in $[MV_R, DL]$.)
   For the proof see [JMP93].
4. **Classes equal to CFL:**
   - **$[MV_R, ER]$, $[MV_R, ER_R, DL]$** **(4 comb.)**

     The fact that CFL is contained in both the classes can be derived from [JMP92a]. There we show how to recognize any context-free language by an F-automaton with operations MV, ER (despite of the opposite hypothesis in [BV79]); it holds even if each item has to be erased at the second visit of the head at latest.
     It can be checked easily that the F-automaton described in [JMP92a] does not need MV_L and, in addition, ER_L can be replaced by DL_L.
     An F-automaton with operations MV_R, ER_R, DL only is easy to simulate by a pushdown automaton, which keeps the part of the list of F left to the the head of F on the pushdown. So $[MV_R, ER_R, DL]$ equals to CFL.
     According to Remark 1a) $[MV_R, ER] = [MV_R, ER_L]$. For a F-automaton with operations MV_R, ER_L there exists a pushdown automaton simulating it. The construction is left out and could be found in [JMP93].
5. **Classes strictly between REG and CSL incomparable with CFL (CSL denoting the class of context-sensitive languages):**
   - **$[ER, DL_R] \longrightarrow [ER, DL] \longrightarrow [MV_L, ER, DL]$** **(9 comb.)**

     Proof. **It is clear that $REG \longrightarrow [ER, DL_R]$ and $\{a^nb^n\} \in [ER, DL_R]$.**
     In addition, $\{a^k; k = 2^n \text{ for some } n\} \in [ER, DL_R]$ (the automaton is able to shorten the list between sentinels to one half by deleting each "even" item, repeat the process etc.), on the other hand, $\{wcw^R; w \in \{a,b\}^*\}$ is not in $[MV_L, ER, DL]$ - cf. Theorem 3 (w^R is w written in the reversed order).
6. **Classes strictly between CFL and CSL:**
   $[MV_R, ER, DL_R] \longrightarrow [MV_R, ER, DL]$ **(6 comb.)**

   $$\searrow$$

   $[MV, ER] \longrightarrow [MV, ER, DL]$ **(12 comb.)**

Proof. CFL is strictly contained in $[MV, ER]$ and $[MV_R, ER, DL_R]$: cf. Point 4 above and notice that $\{a^n b^n c^n; n \geq 0\}$ is in both the classes (for $[MV, ER]$ it is immediate; an F-automaton with MV_R, ER, DL_R can by stepwise erasing b's and deleting a's compare the numbers of a's and b's and then by stepwise deleting compare the numbers of b's and c's).
Again cf.Remark 1.

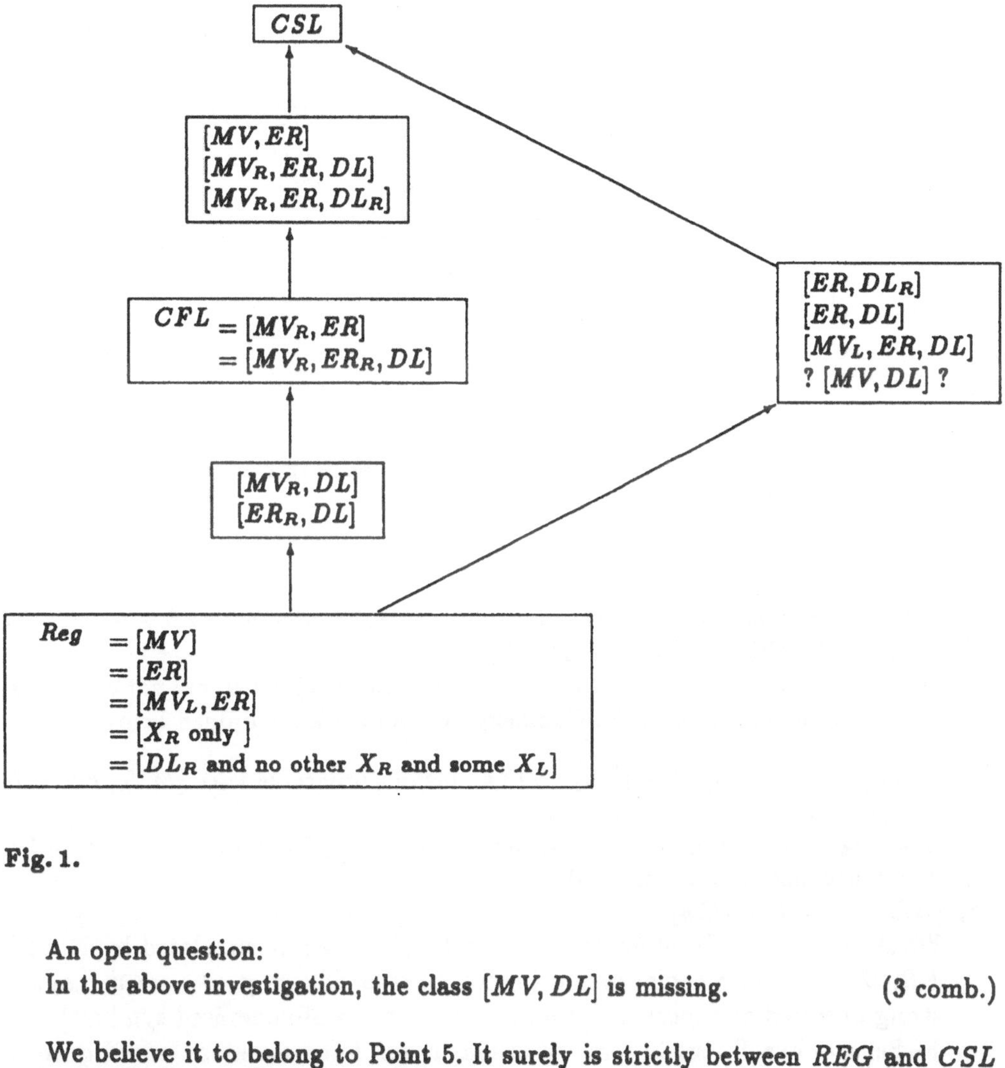

Fig. 1.

An open question:
In the above investigation, the class $[MV, DL]$ is missing. (3 comb.)

We believe it to belong to Point 5. It surely is strictly between REG and CSL and, in addition, $\{a^{2^k}; k \geq 0\} \in [MV, DL]$.

We suppose that there is a context-free language which is not in $[MV, DL]$ but we have no exact proof for it.

The shown classification relative to the Chomsky hierarchy is illustrated by Figure 1.

10 classes are different from REG, CFL and CSL.

In what follows we show some relations among them. The results are summarized in Figure 2.

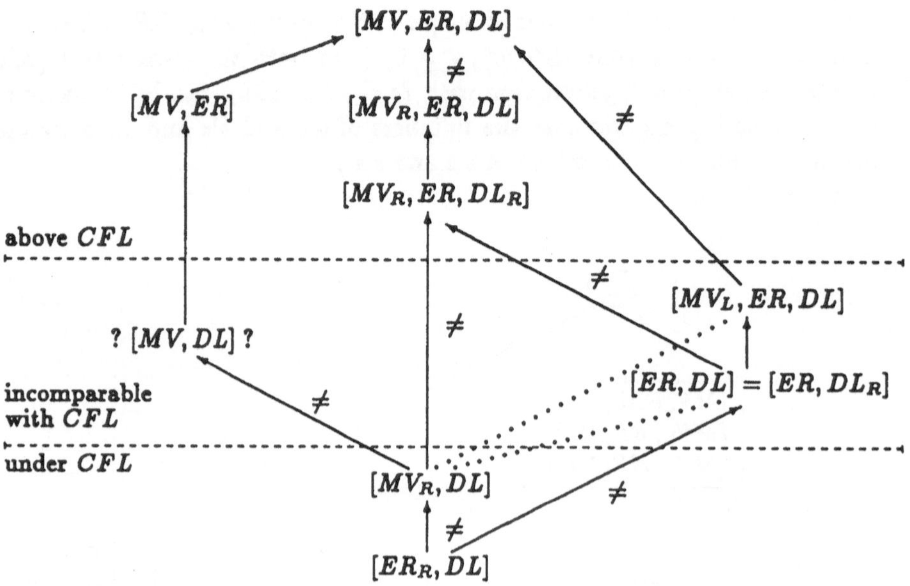

Fig. 2.

(If A is connected with the full line with B and A is below B then A is a subclass of B; if the line is marked by $\neq$, the inclusion is strict.
The classes connected with a dotted line are incomparable.)

The depicted inclusions are either trivial or were shown above or follow from the ability of *ERASING* to simulate *DELETING*.

The other results (strict inclusions and incomparability) which are not clear from the classification according to the Chomsky hierarchy are mentioned below.

a) $[MV_R, DL]$ is not equal to $[ER_R, DL]$ and is incomparable with $[MV_L, ER, DL]$ and $[ER, DL]$:
Language $\{wcw^R; w \in \{a,b\}^*\}$ obviously is in $[MV_R, DL]$ but is not in any other mentioned class (cf. Theorem 3).

b) $[ER, DL] = [ER, DL_R]$
We sketch how an F-automaton F_1 with ER, DL can be simulated by F_2 with ER, DL_R: during a computation, the list looks as follows $\# * * ... * a_1 a_2 ... a_k \$$ (a string of erased symbols concatenated with a string of nonerased symbols).
F_2 behaves like F_1 but in some situations it can have one more erased symbol; in such a case, it keeps in a one-symbol memory what symbol should stand instead of the last $*$. If F_2 reads $*$, it guesses whether it is the last $*$ and behaves according to the guess and (possibly) to the content of the memory:

- if F_1 performs ER_R or DL_R, F_2 performs it too and verifies that the guess was correct (now the last $*$ really means $*$ and the memory is cleared),
- if F_1 performs ER_L (DL_L), F_2 performs ER_R (DL_R), verifies correctness of the guess, does ER_L, ER_L (ER_L) and stores the meaning of the last $*$ if

it is needed.

c) $[MV_R, ER, DL]$ is not equal to $[MV, ER, DL]$:
it is shown using the language $\{wcw; w \in \{a,b\}^*\}$. It is clear that it is in $[MV, ER]$ and in Section 3.2 we show that it is not in $[MV_R, ER, DL]$.

3 Proofs

3.1

The next theorem is proved by the common method of "crossing" sequences.

Theorem 3. $L_{sym} = \{wcw^R; w \in \{a,b\}^*\}$ is not in $[MV_L, ER, DL]$.

Proof. Suppose there is an F-automaton F "without" MV_R which recognizes L_{sym}.

Notice that, for any n, there are 2^n words wcw^R , where $length(w) = n$. All of them have to be accepted by F.

It is clear that, for any such wcw^R, F has to read the whole input word before accepting; hence there is always the first situation when F reads c. At this moment, F is in some of its r states, the content of the list from the left sentinel to the head is a string of i erased symbols, where $0 \leq i \leq n$, and the list from the head to the right sentinel is unchanged.

From this, it is straightforward to conclude that, for n such that $2^n > r.(n+1)$, there are two different words w_1, w_2 such that $w_1cw_2^R$ is accepted as well; thus we have a contradiction.

3.2

Theorem 4. The language $L = \{wcw; w \in \{a,b\}^*\}$ is not in $[MV_R, ER, DL]$.

Proof. By contradiction. Let F be an F-automaton with operations MV_R, ER ,DL recognizing the language L and s denote the number of states of M.

Let n be a positive integer and A_n denote a set of accepting computations containing exactly one arbitrary accepting computation for every word of length of $2n+1$ in L (note that M is nondeterministic).

Claim 5. a) The size of A_n is $|A_n| = 2^n$.
Proof obvious.

b) In every computation C from A_n all (items holding) symbols of the accepted word are visited by M during C.
Proof: Otherwise the visited part of the corresponding input word is not in L and C is an accepting computation for it.

c) Let w be a word from L of length $2n+1$. Replacing any part u of length at most n of w by a different word of the same length as u we get a word outside L.
Proof obvious.

Let C be a computation of F on some input word w. A contiguous sequence of symbols from w which were deleted or erased during the computation C we will call a gap after C. Let E(C) denote the maximal length of a gap in C.

Let k be a positive integer $0 < k < \frac{n}{2}$.

Let $A_n^k = \{C; C \in A_n \text{ and} E(C) \geq k\}$ i.e. subset of A_n of computations which have a gap of length at least k.

Claim 6. $|A_n - A_n^k| \leq sk^2 2^k$.

Proof. Let $S(C)$ denote the configuration (situation) in a computation C from $A_n - A_n^k$ on input word w immediately after the last move to the right (by MV_R, ER_R or DL_R) from some item containing a symbol originally left to the symbol c in w. I.e. after this move F will never enter a symbol originally left to the c in w. According the Claim 5b) F must visit the last symbol of w and because of $E(C) < k < \frac{n}{2}$, F cannot move left so far again.

Let $G(C)$ denote the gap after the initial part C_1 of computation C with the last configuration $S(C)$, such that the left margin of $G(C)$ is positioned on the place of symbol c in the corresponding input word w and the symbol right to the $G(C)$ was not visited yet. $G(C)$ is empty when the item containing c is visited for the first time in $S(C)$.

After $S(C)$ the automaton F will not visit any item left to the c. So the influence of the first half of w is limited to the gap $G(C)$ until $S(C)$. We need a characterization of transport of information from the lefthand side of c to the righthand side of c during the whole computation C.

We denote it by $CH(C)$ and it is composed of:

a) the state of F in $S(C)$;
b) the length of the gap $G(C)$;
c) the number of erased items in the gap $G(C)$ in the configuration $S(C)$.
d) the contents of the gap $G(C)$ (in the original input word).

There are at most $sk^2 2^k$ different characteristics $CH(C)$ for C from $A_n - A_n^k$.

Supposing $A_n - A_n^k > sk^2 2^k$ we get that there exist two accepting computations C_1, C_2 on different input words w_1,w_2 with the same characteristic. By exchange of the parts of w_1, w_2 not visited till $S(C_1)$ resp. $S(C_2)$ we get two accepting computations for two words which are not in L - a contradiction (note that w_1, w_2 differ in these non-visited parts and cf. Claim 5c)).

Let C be a computation from A_n^k on input word w and $S'(C)$ denote the configuration of F in C in which a gap of length at least k for the first time arose. Let $G'(C)$ denote this gap. The length of $G'(C)$ is at most $2k - 1$ (when two gaps of length $k - 1$ are joined through erasing one item separating them).

We characterize now the "events" during C until the configuration $S'(C)$, which occur in the area of $G'(C)$.

We denote this characterization by $CH'(C)$. The $CH'(C)$ is composed from:

a) the state of F in $S'(C)$;
b) the state of F in which the leftmost symbol of the gap $G'(C)$ (in w) was visited for the first time;

c) the symbol left to the leftmost symbol of $G'(C)$ in w;
d) the position of $G'(C)$ in w;
e) the length of $G'(C)$;
f) the number of erased symbols in $G'(C)$ in the situation $S'(C)$;
g) the position of the head of F in $S'(C)$ relative to the leftmost symbol of $G'(C)$ in $S'(C)$.

There are at most

$$s.s.3.(2n-1).(2k-1).(2k-1).(2k-1) < 48s^2k^3n$$

different characteristics $CH'(C)$ for C in A_n^k. According to the Claim 6 there exists a characteristic such that there are at least

$$\frac{A_n^k}{48s^2k^3n} > \frac{2^n - sk^2 2^k}{48s^2k^3n}$$

accepting computations from A_n^k with this characteristic. The number of words of length $2n+1$ from L, which differ outside a gap G of length at least k, is at most 2^{n-k+1} (the $+1$ term is because of gaps containing the symbol c). There exist positive integers n,k such that

$$\frac{2^n - sk^2 2^k}{48s^2k^3n} > 2^{n-k+1}$$

holds. So there are two accepting computations C_1, C_2 for two words w_1,w_2 from L such that C_1, C_2 have the same characteristic and original "contents" u_1, resp. u_2, of the gap $G'(C_1)$, resp. $G'(C_2)$, in w_1, resp. w_2, are different.

Let $w_1 = xu_1y$, where x, y are from $\{a,b,c\}^*$ and u_1 is on the place of the gap $G'(C_1)$ in $S'(C_1)$. Then we can compose an accepting computation C on the word xu_2y which shows a contradiction.

The initial part of C before F enters the leftmost symbol of u_2 is the same as in C_1. The middle part till the gap u_2 is created is the same as in C_2 because in this part of computation F will operate on u_2 and the last symbol of x only (this symbol could not be erased or deleted till now, otherwise it would be in the gap u_1, resp. u_2). But the last symbol of x is the same as the symbol left to u_2 in w_2 (item c) in the characteristic). So the corresponding sequence of steps is the same as in C_2. The situation $S'(C)$ (contents of the working list, position of the head and the state of F) is equal to $S'(C_1)$, so the final part of the computation C can be the same as in C_1. So we have shown that there is an accepting computation for the word xu_2y which is not in L - a contradiction to the existence of F.

4 Open questions and conjectures

We have already mentioned the question whether there is a context-free language which is not in $[MV, DL]$. We conjecture that the language generated by the following grammar can be an example:

$S_{pr} \rightarrow ST$
$T \rightarrow ATa \mid BTb \mid a \mid b$

$S \to aSA \mid bSB \mid a \mid b$
$A \to aAa \mid bAb \mid a$
$B \to aBa \mid bBb \mid b$

It would also show that $[MV, ER]$ is strictly larger than $[MV, DL]$, which is another interesting open question.

Many other open questions (maybe not so interesting) can be derived from Figure 2. For example, our another conjecture, that the language $\{a^k; k = n + 2^n$ for some $n \geq 0\}$ is not in $[MV, DL]$, would not only show the strict inclusion $[MV, DL] \longrightarrow [MV, ER]$ but also the incomparability of $[MV, DL]$ and $[ER, DL]$.

5 References

[BV79] von Braunmühl B., Verbeek R.: *Finite change automata*. Proceedings of the Fourth GI Conference on Theoretical Computer Science, Lecture Notes in Computer Science, Vol.67, Springer-Verlag, Berlin, 1979, pp. 91–100

[J93] Jančar P.: *Nondeterministic Forgetting Automata are Less Powerful than Deterministic Linear Bounded Automata*, Acta Math. et Inf. Univ. Ostraviensis, 1, Ostrava, 1993 (to appear)

[JMP92a] Jančar P., Mráz F., Plátek M.: *Characterization of Context-Free Languages by Erasing Automata,* in Proceedings of the 17th International Symposium on Mathematical Foundations of Computer Science 1992, Lecture Notes in Computer Science, Vol. 629, Springer-Verlag, Berlin 1992, pp.305–314

[JMP92b] Jančar P., Mráz F., Plátek M.: *Forgetting automata and the Chomsky hierarchy*, in Proc. SOFSEM '92, Ždiar, Slovakia, November 1992, pp. 41-44

[JMP93] Jančar P., Mráz F., Plátek M.: *A Taxonomy of Forgetting automata*, Technical Rep. No. 101, Department of Computer Science, Charles University, Prague, May 1993

[PV86] Plátek M., Vogel J.: *Deterministic List Automata and Erasing Graphs*, The Prague bulletin of mathematical linguistics 45, 1986

[P92] Plátek M.: *Syntactic Error Recovery with Formal Guarantees I.*, Technical Rep. No. 100, Department of Computer Science, Charles University, Prague, April 1992

[S59] Sheperdson, J.C.: *The Reduction of two-way automata to one way automata*, IBM J. Res. Develop. 3, 1959, pp. 198–200

[S70] Savitch, W.J.: *Relationships Between Nondeterministic and Deterministic tape Complexities*, Jurnal of Computer and System Sciences 4, 1970, pp. 177–192

Hybrid Parallel Programming and Implementation of Synchronised Communication

He Jifeng*
Oxford University Computing Laboratory
Programming Research Group
11 Keble Road, Oxford OX1 3QD, England

May 3, 1993

Abstract

An occam program is usually translated into a machine program executed in parallel with a set of system processes such as communication protocol and scheduler, where the target program appears in a form which cannot be adequately modelled in a purely communication-based parallel language since concurrent components share variables. This paper presents a mathematical theory for a hybrid language equipped with a parallel construct, whose sub-processes can communicate with each other via both channels and shared variables. We examine the algebraic laws of the language, and show how they can help in the implementation of concurrency and synchronised communication of occam.

1 Introduction

An **occam** program can be implemented by a machine program executed in a transputer in parallel with a set of system processes such as communication protocol and scheduler [8], where the target program appears in forms which cannot be adequately modelled in a purely communication-based parallel language since parallel components share variables [14]. In order to capture the behaviour of such a system, it is worthwhile to investigate a mathematical theory for hybrid parallel systems. A hybrid parallel language contains a parallel construct whose sub-processes can communicate with each other via both channels and shared variables. It becomes a more active and tractable research area due largely to the fact that the increasing complexity of computer hardware requires a sound basis for the design and specification of micro-processor architectures, and the further maturity of formal verification techniques [1, 15, 16].

*supported by the ESPRIT Basic Research Action ProCoS II

In this paper we examine a hybrid parallel language, which is a mild extention of **occam** with await-guarded command and a hybrid parallel operator as new constructs. Among many orthogonal issues in the implementation of concurrency and synchronisation of **occam**, we demonstrate how to realise internal channels and synchronisation in a uni-processor implementation, and how a process can interact with its environment via a simple communication protocol in the case of multiple-processor.

The results reported in this paper are part of a wide programme of research devoted to establish links between different notations, methods, and computational paradigms. The objectives are to aid in the transformation of specifications, designs and programs from simple mathematical abstractions to efficient computations in either hardware or software or both [4, 11, 9, 12]. Other possible applications include the design and construction of systems involving a mixture of language and computational paradigms.

2 Hybrid Parallel Language

The language defined in this paper extends **occam** by including await-guarded process and hybrid parallel as new constructs. In the following BNF-style syntax, P and SP are drawn from some set of process variables, ch from some set of channel variables. We use b and c to stand for Boolean expressions, and v and e for a list of program variables and expressions respectively.

$$
\begin{array}{rcl}
P & ::= & \texttt{STOP} \mid \texttt{SKIP} \mid \texttt{CHAOS} \mid IO \mid v := e \mid \\
 & & P \sqcap P \mid P;P \mid P \lhd b \rhd P \mid [\![SP]\!] \mid \mathbf{local}(v, P) \mid \\
 & & G \mid \,\Box\, (\underline{G}) \mid P \| P \mid \mathbf{var}\, v\, ; P; \mathbf{end}\, v \mid b * P \\
G & ::= & g \rightarrow P \\
g & ::= & c\&SP \mid b\&IO \\
IO & ::= & ch?v \mid ch!e \\
SP & ::= & \texttt{SKIP} \mid \texttt{CHAOS} \mid v := e \mid [\![SP]\!] \mid (SP \sqcap SP) \mid \\
 & & SP;SP \mid SP \lhd c \rhd SP \mid c * SP
\end{array}
$$

where $\underline{G}$ represents a finite list of guarded processes G.

Informally, the process terms stand for the following processes:

- `SKIP` does nothing and terminates successfully.
- `STOP` is the canonical deadlock process which can make no further progress.
- `CHAOS` is the worst program, and no interference from the environment can rescue it.

- $ch?v$ is a process which is willing to receive a message from the channel ch, and assigns it to the variable v. After that input event takes places, it terminates successfully.

- $ch!e$ is a process which is ready to send the value of the expression e to the channel ch at the very beginning. It terminates after engaging the output event.

- $v := e$ is executed by evaluating all the expressions in the list e and then assigning the values of each expression to the variable at the same position in the list v. The execution of an assignment is treated as an atomic action, i.e., the interfernce from its environment is forbidden. $v := e$ is a *local assignment* if neither v nor e mentions shared variables.

- If P and Q are processes, then $P \sqcap Q$ represents a non-deterministic choice between P and Q. This choice is made invisible to the environment.

- $(P; Q)$ is the sequential composition of P and Q.

- Let P and Q be processes, and b a Boolean expression. $P \lhd b \rhd Q$ represents a conditional which chooses to execute P if b is true initially, but execute Q instead if b is false. To facilitate the algebraic reasoning it is postulated that b does not contain shared variables. Consequently, the interference on shared variables, which happens after the choice is made, has no effect on the value of b. As mentioned in [18], this constraints has no effect on the expression power of the language. In the rest of this paper we will use b (possibly with subscript) to stand for Boolean expressions without shared variables. To simplify mathematical reasoning, the notation $P \unlhd c \unrhd Q$ is introduced to stand for uninterrupted conditional which behaves like conditional $P \lhd c \rhd Q$ except that the condition c may contain shared variables, and the interference is forbidden before the corresponding alternative starts its execution. We will use $<< c >>$ to stand for both $< c >$ and $\unlhd c \unrhd$ later.

- **local**(v, P) localizes the variable v in P. As a result, the environment can no longer access the variable v.

- $b\&IO \rightarrow Q$ is a process which behaves like the sequential composition of a communication command IO and the process Q if b is true initially. If b is false, it behaves like STOP.

- $c\&SP \rightarrow Q$ is a await-guared process which starts its execution only when c becomes true, and then behaves the same as the sequential composition of SP and Q, except that the execution of SP is treated as an atomic action, and no interference is allowed. If c is false, the process is suspended.

- $\Box(\underline{G})$ is an arbitration with symmetrically combined branches $\underline{G}$. As soon as branch G_i becomes ready, the arbitration may choose to execute that branch.

- $P \| Q$ is a concurrent composition of P and Q where parallel components can talk to each other by either internal links between processes or shared variables. The whole system terminates only when both P and Q terminate.
- $b * P$ represents an iteration. It is executed by first evaluating b, if b is false, execution terminates successfully. But if b is true, it proceeds to execute $(P\,;\, b * P)$.
- $\mathbf{var}\, v\,;\, P\,;\, \mathbf{end}\, v$ behaves like P except that the variable v becomes local to P, and its value is hidden from the environment.
- $[\![SP]\!]$ is a program which behaves like SP but its execution is protected from the interference of its environment.

Denotational semantics of [6] map each process into a domain with a partial order according to which one process is greater than another if it is better defined, or more predictable. Unless indicated explicitly, we make no formal distinction between the text of a program and its value (semantics). If P and Q are processes, we will write $P \sqsubseteq Q$ when the semantic value of P is less than that of Q. Processes P and Q are said to be *equivalent*, denoted $P = Q$, if both $P \sqsubseteq Q$ and $Q \sqsubseteq P$ hold.

The algebraic properties of the language are widely used in the proof of the main theorems of this paper. The great advantage of algebra is that it uses equational (and inequational) reasoning, with a single conceptual framework and notation throughout. The set of laws is useful for the derivation of designs and for the optimization of programs [2]. It also helps in comparing and classifying the variety of possible languages [7]. The sufficiency of such a set of algebraic laws can be established by an appropriate kind of normal form. Because of the limit of space, we omit the section of algebraic laws, the reader is refereed to [6].

The proof of the main theorems of this paper makes use of the well-known concept of systactic approximation, which gives a pre-order on the systax of a language. The order is a very simple one, based on the idea that replacing part of a program by the least defined program produces an approximation, and that unfolding a loop produces an approximation. It can be shown that the semantics of a program is the least upper bound of its finite approximations. Finite programs are easy to reason about algebraically, but they are not very useful in practice. The notation of finite syntactic approximation allows us to apply the results on finite programs to general programs [5].

3 Uni-processor Implementation of Channel

The programming language occam, which deals with the state of program variables, has concurrency and synchronisation. Synchronisation takes place between pair of processes along channels, each of which is used exclusively for communication between a particular pair. Values may only be passed along channels in one direction and so the user processes may be assigned the status of inputting or outputing along that channel. The concept of communication is abstract and allows a channel to be implemented in various ways [10, 11],

- As store location(s) and a program.
- As a microprogram instead of a program.
- As a parallel path with handshanking signals.
- As a series path, the communicating process breaking the data into pieces.

In CSP and the hybrid parallel language, processes can be prepared to perform a number of alternative input or output actions. In order to admit an efficient distributed implementation it is necessary to put some restrictions on the language . The crucial step is to break the symmetry between communicating processes. This can be done by placing restriction on the language, namely by ensuring that if a process is prepared to output, then it is not prepared to engage in any different alternatives. In other words, synchronisation must take place between an output process and a process which is prepared to engage in a number of alternative input actions or internal actions.

This section investigates the uni-processor implementation of channel communication. In this case, an internal channels is usually implemented by locations in store (variables) and a program. Consequently, a communication-based parallel program is translated into a hybrid parallel program since parallel components will share variables which are introduced to represent the internal channels.

First we treat an internal channel ch as a pair ($ch?$, $ch!$) of ports. Since both ports of the channel ch reside in the same processor, it is convinent to represent them by the same set of locations in store. Let Ψ be a symbol table which allocate a triple $(in_{ch},\ out_{ch},\ buf_{ch})$ of locations for channel ports, where

- out_{ch} is a channel flag, which is true if the process is ready to send a message to the channel ch.
- in_{ch} is a channel flag, which becomes true only when the process has agreed to input from the channel ch.
- buf_{ch} is a variable being used to store the message.

For simplicity, the subscripts of those channel variables will be dropped later.

The programs for implementing input $ch?x$ and output $ch!e$ are defined by

$$\mathbf{input}_{\Psi}(ch,\ x) \ \stackrel{def}{=}\ out\&\ \mathtt{SKIP} \rightarrow [ch?x]$$

$$[ch?x] \stackrel{def}{=} in := true\,;\, \neg out\&(x,\ in := buf,\ false) \rightarrow \mathtt{SKIP}$$

$$\mathbf{output}_{\Psi}(ch,\ e) \stackrel{def}{=} out := true\,;\,[ch!e]$$

$$[ch!e] \stackrel{def}{=} in\&(buf,\ out := e,\ false) \rightarrow (\neg in\&\ \mathtt{SKIP} \rightarrow \mathtt{SKIP})$$

The sender $\mathbf{output}_{\Psi}$ starts the link connection phase by turning the flag *out* on. The receiver $\mathbf{input}_{\Psi}$ agrees to use the channel *ch* for an input by setting $in = true$ after it detects that the channel flag *out* is already on. On receipt of the handshanking signal (*true*), $\mathbf{output}_{\Psi}$ is then allowed to store the value of *e* into *buf* and deliver a completion signal to the receiver by setting $out = false$. Finally, the process $\mathbf{input}_{\Psi}$ receives the message from the variable *buf* and send its acknowledgement (*false*) back to the sender.

Assume that $\{in,\ out,\ buf\} \cap Vars(\alpha.P) = \emptyset$ and $ch \in Chans(\alpha.P)$, where αP is the interface of P consisting the sets of variable names and channel mames used by P. P can be transformed into a process with *in*, *out*, *buf* as new shared variables by applying the mapping Ψ_{ch} to the text of P

$$\Psi_{ch}(\lfloor P \rfloor) \stackrel{def}{=} P$$

P is a primitive or atomic command but not a communication command

$$\Psi_{ch}(\lfloor IO \rfloor) \stackrel{def}{=} IO$$

if IO does not use channel ch

$$\Psi_{ch}(\lfloor ch?x \rfloor) \stackrel{def}{=} \mathbf{input}_{\Psi}(ch,\ x)$$

$$\Psi_{ch}(\lfloor ch!e \rfloor) \stackrel{def}{=} \mathbf{output}_{\Psi}(ch,\ e)$$

$$\Psi_{ch}(\lfloor c\&SP \rightarrow Q \rfloor) \stackrel{def}{=} c\&\Psi_{ch}(\lfloor SP \rfloor) \rightarrow \Psi_{ch}(\lfloor Q \rfloor)$$

$$\Psi_{ch}(\lfloor b\&IO \rightarrow Q \rfloor) \stackrel{def}{=} c\&IO \rightarrow \Psi_{c}(\lfloor Q \rfloor)$$

if IO does not use channel ch

$$\Psi_{ch}(\lfloor b\&ch?x \rightarrow Q \rfloor) \stackrel{def}{=} (b \wedge out)\&\ \mathtt{SKIP} \rightarrow (\mathbf{input}_{\Psi}(ch,\ x)\,;\, \Psi_{c}(\lfloor Q \rfloor))$$

$$\Psi_{ch}(\lfloor Q_1\, bop\, Q_2 \rfloor) \stackrel{def}{=} \Psi_{ch}(\lfloor Q_1 \rfloor)\, bop\, \Psi_{ch}(\lfloor Q_2 \rfloor)$$

where $bop \in \{;,\ \sqcap,\ \|\}$

$$\Psi_{ch}(\lfloor P \lhd b \rhd Q \rfloor) \stackrel{def}{=} \Psi_{ch}(\lfloor P \rfloor) \lhd b \rhd \Psi_{ch}(\lfloor Q \rfloor)$$

$$\Psi_{ch}(\lfloor b * Q \rfloor) \stackrel{def}{=} b * \Psi_{ch}(\lfloor Q \rfloor)$$

$$\Psi_{ch}(\lfloor \square(\underline{G}) \rfloor) \stackrel{def}{=} \square(\underline{\Psi_{ch}(G)})$$

$$\Psi_{ch}(\lfloor \mathbf{local}(v, Q) \rfloor) \stackrel{def}{=} \mathbf{local}(v, \Psi_{ch}(\lfloor Q \rfloor))$$

$$\Psi_{ch}(\lfloor \mathbf{var}\, v \star Q \star \mathbf{end}\, v \rfloor) \stackrel{def}{=} \mathbf{var}\, v\, ; \Psi_{ch}(\lfloor Q \rfloor)\, ; \mathbf{end}\, v$$

Assume that *ch* is a channel connecting processes P and Q, and that neither P nor Q refers to free variables *in*, *out* and *buf*, we define

$$\begin{array}{lcl} \mathbf{I}\,(\lfloor P \,||\, Q \rfloor) & \stackrel{def}{=} & \mathbf{var}\, in,\, out,\, buf\, ; \\ & & in,\, out := false,\, false\, ; \\ & & \Psi_{ch}(\lfloor P \rfloor) \,||\, \Psi_{ch}(\lfloor Q \rfloor)\, ; \\ & & (in,\, out \,=\, false,\, false)_{\perp}\, ; \\ & & \mathbf{end}\, in,\, out,\, buf \end{array}$$

Theorem (Implementation of Channel by Locations)
If $ch \in Chans(P) \cap Chans(Q)$, and no communication in either P or Q involves shared variables, and variables *in*, *out*, *buf* do not appear neither in P nor in Q, then

$$P \,||\, Q \;=\; \mathbf{I}\,(\lfloor P \,||\, Q \rfloor)$$

4 Communication Protocol

This section discusses a multi-processor implementation of occam concurrent programs. A channel connecting parallel components will be implemented as a parallel path with handshanking signals.

Let P and Q be processes with $Vars(P) \cap Vars(Q) = \emptyset$. One can implement $P||Q$ by placing P and Q in two processor and implementing the internal channel *ch* by physical links between these two processors. Since ports *ch*? and *ch*! reside in two processors, they will be assigned two disjoint set of locations. The symbol table Ψ will allocate a triple of locations in one processor to the input port *ch*?, and another triple of locations in the second processor to the output port *ch*!, where we assume that

$$\begin{array}{lcl} \Psi(ch?) & = & < in_1,\, out_1,\, buf_1 > \\ \Psi(ch!) & = & < in_2,\, out_2,\, buf_2 > \end{array}$$

We connect ports *ch*? and *ch*! by two physical links (*say* $<$ *data*, *ack* $>$), where *data* is one-way link used to deliver the message, and *ack* for the acknowledgement. A simple communication protocol is introduced below for handshanking signal and data transmission. On the transmitting end of the channel ch, the output port $ch!$ is associated with a process $SEND$, while on the receiving end the port $ch?$ is attached by a process REC. When a process is ready to deliver a message to the channel ch, it first sets the outputting flag out_2 true. $SEND$ then initiates the handshanking phase by transmitting the link connection signal (the value of the flag out_2) to REC, and waits for the response. If the inputting process is willing to communicate via ch, it instructs REC to send a *yes* response (the value of the flag in_1) to $SEND$. Upon receipt of a *yes* response, $SEND$ considers that the link is set up and enters the data-transfer phase. It delivers the data stored in buf_2 and the communication completion signal (the value of out_2) to REC. After receiving the message and the completion signal the process REC closes the link by sending the disconnection signal (the value of in_1) to $SEND$.

Formally REC and $SEND$ are defined by

$$\begin{aligned}
REC(ch) &\stackrel{def}{=} \Box(data?out_1 \rightarrow REC_1(ch),\ finish_1\&\ \mathtt{SKIP} \rightarrow \mathtt{SKIP}) \\
REC_1(ch) &\stackrel{def}{=} in_1\&\ \mathtt{SKIP} \rightarrow REC_2(ch) \\
REC_2(ch) &\stackrel{def}{=} ack!in_1 \rightarrow REC_3(ch) \\
REC_3(ch) &\stackrel{def}{=} data?buf_1 \rightarrow (data?out_1\ ;\ REC_4(ch)) \\
REC_4(ch) &\stackrel{def}{=} \neg in_1\&\ \mathtt{SKIP} \rightarrow (ack!in_1\ ;\ REC(ch))
\end{aligned}$$

$$\begin{aligned}
SEND(ch) &\stackrel{def}{=} \Box(out_2\&\ \mathtt{SKIP} \rightarrow SEND_1(ch),\ finish_2\&\ \mathtt{SKIP} \rightarrow \mathtt{SKIP}) \\
SEND_1(ch) &\stackrel{def}{=} data!out_2 \rightarrow SEND_2(ch) \\
SEND_2(ch) &\stackrel{def}{=} ack?in_2 \rightarrow SEND_3(ch) \\
SEND_3(ch) &\stackrel{def}{=} \neg out_2\&\ \mathtt{SKIP} \rightarrow (data!buf_2\ ;\ data!out_2\ ;\ SEND_4(ch)) \\
SEND_4(ch) &\stackrel{def}{=} ack?in_2 \rightarrow SEND(ch)
\end{aligned}$$

where the variables $finish_1$ and $finish_2$ represent the status of the inputting process and outputting process respectively, it becomes true only when the corresponding process terminates.

Assume that $ch \in InputChans(P) \cap OutChans(Q)$ define

$$\mathbf{PI}\ (\lfloor P \,||\, Q \rfloor) \stackrel{def}{=} \Phi_{ch}(\lfloor P \rfloor) \,||\, \Phi_{ch}(\lfloor Q \rfloor)$$

where

$$\begin{aligned}
\Phi_{ch}(\lfloor P \rfloor) \stackrel{def}{=}\ & \mathbf{var}\ finish_1,\ in_1,\ out_1,\ buf_1\ ; \\
& finish_1,\ in_1,\ out_1 := false,\ false,\ false\ ;
\end{aligned}$$

$$((\Psi_{ch}(\lfloor P \rfloor)\,;\, finish_1 := true) \;||\; REC(ch))\,;$$
$$(finish_1,\, in_1,\, out_1 = true,\, false,\, false)_{\perp}\,;$$
$$\textbf{end}\, finish_1,\, in_1,\, out_1,\, buf_1$$

$$\Phi_{ch}(\lfloor Q \rfloor) \stackrel{def}{=} (\textbf{var}\, finish_2,\, in_2,\, out_2,\, buf_2\,;$$
$$finish_2,\, in_2,\, out_2 := false,\, false,\, false\,;$$
$$((\Psi_{ch}(\lfloor Q \rfloor)\, finish_2 := true) \;||\; SEND(ch))\,;$$
$$(finish_2,\, in_2,\, out_2 = true,\, false,\, false)_{\perp}\,;$$
$$\textbf{end}\, finish_2,\, in_2,\, out_2,\, buf_2$$

Here we postulate that

- $Vars(P) \cap Vars(Q) = \emptyset$.
- $ch \in InputChans(P) \cap OutputChans(Q)$.
- $\{finish_1,\, in_1,\, out_1,\, buf_1\} \cap Vars(P) = \emptyset$.
- $\{finish_2,\, in_2,\, out_2,\, buf_2\} \cap Vars(Q) = \emptyset$.
- $\{finish_1,\, in_1,\, out_1,\, buf_1\} \cap \{finish_2,\, in_2,\, out_2,\, buf_2\} = \emptyset$.

Theorem (Implementation of Channel by a Communication Protocol)

$$P \,||\, Q \;=\; \mathbf{PI}\,(\lfloor P \,||\, Q \rfloor)$$

References

[1] M. Abadi and L. Lamport. *Composing Specifications* LNCS 430, 1–42, (1990).

[2] R.S. Bird. *Lectures on Correctness Functional Programming.* Technical Monograph PRG-69, Oxford University Computing Laboratory, (1989).

[3] S.D. Brookes, A.W. Roscoe. *An Improved Failure Model For Communicating Processes.* LNCS 197, 281–305, (1985).

[4] M. Gordan. *Proving a computer correct.* Technical Report 42, University of Cambridge Computing Laboratory, (1983).

[5] I. Gaessarian. *Algebraic Semantics.* LNCS 99, (1981).

[6] He Jifeng. *Introduction of Hybrid Parallel Programming.* ProCoS Technical Report, Oxford University Computing Laboratory, (1992).

[7] C.A.R. Hoare. *The Varieties of Programming Languages.* TAPSOFT Proceedings, Springer-Verlag, LNCS 351, 1–18, (1989).

[8] INMOS Ltd. *The occam programming manual.* Prentice-Hall, (1984).

[9] A.J. Martin. *Compiling communicating process into delay-insensitive VLSI circuits.* Distributed Computing 1 (4), (1986).

[10] D. May. *OCCAM and Transputer.* In Developments in Concurrency and Communication, 65–87, (1991).

[11] D. May. *Compiling OCCAM into Silicon.* In Developments in Concurrency and Communication, 88–107, (1991).

[12] D. May and R. Shepherd. *Communicating Process Computers.* Communicating Process Architecture, 31–44, Prentice-Hall, (1988).

[13] G. Nelson and M. Manasse. *The Proof of a Second Step of a Factored Compiler.* Lecture Notes for International Summer School on Programming and Mathematical Method, Marktoberdorf, Germany, (1990).

[14] S. Owicki. *Axiomatic Proof Techniques for Parallel Programs.* PhD thesis, Department of Computer Science, Cornell University, (1975).

[15] A. Pnueli. *The temporal semantics of concurrent programs.* LNCS 70, 1–20, (1979).

[16] L. Pomello. *Refinement of Concurrent System Based on Local State Transformation.* LNCS 430, 641–669, (1990).

[17] A.W. Roscoe and C.A.R. Hoare. *The Laws of occam Programming.* Theoretical Computer Science, (1988).

[18] K. Stolen. *Development of Parallel Programs on Shared-Data Structure.* Thesis DPhil. Manchester University, (1990).

Proof Systems for Cause Based Equivalences

Astrid Kiehn

Technische Universität München, Arcisstr.21, D-8000 München 2,
kiehn@informatik.tu-muenchen.de

Abstract. Two proof systems for the congruences of causal bisimulation and location equivalence of finite CCS are proposed. They generalize the known axiomatizations for restriction free processes which are based on the two merge operators of ACP. Since the proof systems only differ in three equations they provide a simple means of comparison.

1 Introduction

The study of non-interleaving equivalences for process algebras has been a major branch of research in semantics over the last years. Roughly speaking non-interleaving equivalences aim at distinguishing between processes if at some point their behaviour can only be compared if independent actions of one process can at most be matched with mutually dependent actions of the other. *Independence* is defined in terms of *causality*, *locality*, *robustness with respect to action refinement* or *with respect to splitting of actions*. Accordingly there is a wide variety of non-interleaving equivalences. One way of studying equivalences is to axiomatize them. This means to exhibit a set of equations and to prove for them that two processes can equationally be transformed into each other if and only if the processes are equivalent (congruent to be precise). Axiomatizations provide insight into what kind of transformations preserve equivalence. Additionally they can serve as an elegant means for comparison. For example in [3] it is shown that causal bisimulations ([7]), location equivalence ([6]) and ST equivalence ([19], [1]) can be axiomatized by the same set of equations if one restricts CCS ([16]) to finite, restriction and renaming free processes disallowing communication. For the same language enriched by the ability to communicate there is an axiomatization of ST equivalence in [1], of location equivalence in [5] and of causal bisimulations in [15]. The respective sets of equations only differ in two or three equations. Thus the differences and common features are much easier understood by inspecting the underlying axiomatizations than the fairly different operational definitions.

The aim of this paper is to provide a similar framework for comparing equivalences for whole of finite CCS. There are axiomatizations for this language, for example in [7] for causal bisimulations, in [6] for location equivalence and two unifying theories are suggested in [8] and [9] with *observation trees* and *proved trees*. These axiomatizations rely on an expansion law similarly as for observation equivalence. Due to this expansion law a concurrent process is reducible to a sequential process where the independence information is retrievable from its actions. This is possible as all these paper introduce what we call *compound actions* here. Compound actions are either actions enriched with some information on their history (used in [7], [6] and [9]) or actions which are composed of other actions, so-called computations (used in [8] and [18]). Noteworthy the axiomatizations for restriction free processes do not require compound actions. They are based on a different proof technique with a different expansion law using the two auxiliary operators left merge and communication merge known from [4] and [11]. There have been two approaches to generalize this technique for processes containing restriction. [10] contains an elegant axiomatization of location equivalence at the expense of losing commutativity and associativity of parallel composition. In [12] a proof system for ST equivalence has been suggested. It makes only limited use of the two merge operators and has rather complex rules. In this paper we show that the technique commonly used to axiomatize equivalences for restriction free processes can be generalized in a rather straightforward way if left merge and communication merge are taken as first class operators. As a result we obtain two proof systems which do not need compound actions and whose equations are natural extensions of the restriction free case.

We consider location equivalence ([6]) and causal bisimulations ([7]). To start out from transition systems over the same language we exploit the characterizations given in [14] which show that both equivalences can be defined by means of the same set of operators. The respective equivalences are called local cause (*lc*) and global cause (*gc*) equivalence. The main difference between them is that in case of global causality a communication can introduce new causal dependencies in a term. As a consequence the proof systems only differ in the equations concerning communication.

The paper is structured as followed. The first section gives a brief introduction to the local and global cause semantics. Section 3 presents the proof systems and the last section outlines the main steps of the completeness proofs.

2 Basic Definitions

We consider finite CCS ([16]), extended by a cause prefixing operator $\Gamma \triangleright P$ and the two auxiliary operators left merge and communication merge. The first operator is needed to reason about causal dependencies, the latter to obtain a finite proof system at least for restriction free processes (cf. [17]). Let CCS denote the set of closed terms and let CCS(x) denote the set of processes which additionally may contain variables $x, y, z, \ldots$. Actions are taken from the set Act ranged over by $a, b \ldots$. Act_τ additionally contains τ and is ranged over by $\mu, \nu, \ldots$. C is the countable set of causes where $C \cap Act_\tau = \emptyset$. The set of all processes $\mathbb{P}_{\mathcal{CG}}(X)$ is defined by the following abstract syntax.

$$T ::= \Gamma \triangleright p \mid X \mid T+T \mid T|T \mid T[f] \mid T\backslash b \mid T \lfloor T \mid T|_c T$$

where $p \in$ CCS(x), $\Gamma \subseteq C$, Γ is finite and X is a variable. As in [7] we assume that $\Gamma \triangleright$ distributes over all operators except action prefixing. For example $\Gamma \triangleright (a.nil \mid (b.nil + \tau.nil))$ reduces to $\Gamma \triangleright a.nil \mid (\Gamma \triangleright b.nil + \Gamma \triangleright \tau.nil)$. The subset of $\mathbb{P}_{\mathcal{CG}}(X)$ of processes without variables is $\mathbb{P}_{\mathcal{CG}}$. Processes in $\mathbb{P}_{\mathcal{CG}}$ are ranged over by capital letters $P, Q, \ldots$ and those in CCS by small letters $p, q, \ldots$.

The set of causes in a term $P \in \mathbb{P}_{\mathcal{CG}}(X)$ is given by $cau(P)$ and $cau(P_1, P_2)$ denotes the union of $cau(P_1)$ and $cau(P_2)$. Often a cause l is replaced by a set of causes Γ and this is written as $P[l \rightarrow \Gamma]$. In general we will use Greeks letters $\alpha, \beta, \ldots$ for actions which are restricted. Processes $P \setminus \alpha_1 \setminus \alpha_2 \cdots \setminus \alpha_n$ are rendered to $P \setminus \{\alpha_1, \ldots, \alpha_n\}$. In cases of large sums and parallel compositions we write

$$\sum_{i \in \{1,\ldots,n\}} p_i \quad \text{for} \quad p_1 + \cdots + p_n \qquad \text{and} \qquad \prod_{i \in \{1,\ldots,n\}} p_i \quad \text{for} \quad p_1 \mid \cdots \mid p_n.$$

All these notations are justified by the operational semantics. Finally *nil* is omitted if it is prefixed by an action.

We introduce two semantic equivalences for $\mathbb{P}_{\mathcal{LG}}$ processes. The first one is called *local cause* (*lc*) *equivalence*. Local causality is observed according to the structure of a process term. An executed action causes a later action or process if and only if the latter is syntactically in the scope of the former. For example in $a.(b \mid c) \mid d$ action a will cause actions b and c but not d. This remains valid if the process additionally enforces a communication as in $(a.\alpha.(b \mid c) \mid \bar{\alpha}.d) \setminus \alpha$. Thus these two processes are local cause equivalent. The second semantic equivalence is *global cause* (*gc*) *equivalence*. Global causality is induced by the dynamic flow of control. As in the case of local causality an action or a process is caused by an occurring action if the former is in the latter's scope. Additionally actions of a parallel component may act as causes if their causes are transmitted with a communication. For example the two processes above are not global cause equivalent as d in the second process is also caused by a. The two cause equivalences are incomparable. The following two processes give a counterexample for the other implication. $(a.\alpha.b \mid \bar{\alpha}) \setminus \alpha \overset{\not\approx_{lc}}{\approx_{gc}} (a.\alpha. \mid \bar{\alpha}.b) \setminus \alpha$. In the latter process b is locally independent of a but not globally. To be able to argue about actions which have been executed in the past *causes* are introduced. They are inserted into a term with the execution of an action uniquely for each event. Actions in the scope of a cause will be caused by it. The difference in the definitions of *lc* and *gc* equivalence lies in the underlying transition systems. In both cases there is a transition system for τ actions and one for visible actions $a, b, c, \ldots$. This separate treatment is due to the underlying assumption that only visible actions have visible causes which may be observed. Thus τ transitions as usually are of the form $P \overset{\tau}{\Longrightarrow} P'$ while transitions of visible actions $P \underset{\Gamma,l}{\overset{a}{\Longrightarrow}} P'$ additionally yield a set of causes Γ and a new cause l representing the a action. The rules for visible transitions are given in Figure 1. They are valid for both semantics. Axiom (LG1) extends the standard CCS axiom in that a new cause l representing the executed a action is added to the causes of P'. Γ and l are both observed with the transition as they are reported in the arc underscription. Rules (LG2) – (LG8) are as in standard CCS. The precondition $l \notin cau(Q)$ ensures that the new cause l indeed is unique. The transition systems for τ moves only differ in the rules for communication. The axiom is $\Gamma \rhd \tau.p \overset{\tau}{\Longrightarrow} \Gamma \rhd p$, the remaining rules not concerning communication are as for visible transitions. The rules for communication are defined in terms of the cause transition system for visible moves, so the preconditions are $P \underset{\Gamma,l}{\overset{a}{\Longrightarrow}} P'$ and $Q \underset{\Delta,k}{\overset{\bar{a}}{\Longrightarrow}} Q'$. One point to be implemented with these rules is that the new causes l and k introduced for the communicating actions a and $\bar{a}$ have to vanish, as due to the communication a and $\bar{a}$ become an invisible τ action. In case of local causality they are simply erased.

$$\text{(LC1) } P \xRightarrow[\Gamma,l]{a}{}^{(lc)} P',\quad Q \xRightarrow[\Delta,k]{\bar{a}}{}^{(lc)} Q' \text{ implies } P \mid Q \xRightarrow{\tau}{}^{(lc)} P'[l \to \emptyset] \mid Q'[k \to \emptyset],$$

$$\text{(LC2) } P \xRightarrow[\Gamma,k]{a}{}^{(lc)} P',\quad Q \xRightarrow[\Delta,k]{\bar{a}}{}^{(lc)} Q' \text{ implies } P \mid_c Q \xRightarrow{\tau}{}^{(lc)} P'[l \to \emptyset] \mid Q'[k \to \emptyset].$$

If global causality is to be observed one has additionally to ensure that actions in the scope of a (of $\bar{a}$ respectively) will due to the communication also depend on the causes of $\bar{a}$ (a). Both points are guaranteed with the replacement of l by a copy of the causes of $\bar{a}$ and the replacement of k by a copy of the causes of a.

$$\text{(GC1) } P \xRightarrow[\Gamma,l]{a}{}^{(gc)} P', Q \xRightarrow[\Delta,k]{\bar{a}}{}^{(gc)} Q' \text{ implies } P \mid Q \xRightarrow{\tau}{}^{(gc)} P'[l \to \Delta] \mid Q'[k \to \Gamma]$$

$$\text{(GC2) } P \xRightarrow[\Gamma,l]{a}{}^{(gc)} P', Q \xRightarrow[\Delta,k]{\bar{a}}{}^{(gc)} Q' \text{ implies } P \mid_c Q \xRightarrow{\tau}{}^{(gc)} P'[l \to \Delta] \mid Q'[k \to \Gamma]$$

Based on the these transition systems we define local and global cause equivalence as a refinements of weak bisimulation equivalence.

Definition 1. [Local/Global Cause Equivalence, Congruence]
A symmetric relation $R \subseteq \mathbb{P}_{\mathcal{CG}} \times \mathbb{P}_{\mathcal{CG}}$ is called a *local cause bisimulation* iff $R \subseteq G(R)$ where $(P,Q) \in G(R)$ iff

(*i*) $P \xRightarrow{\tau}{}^{(lc)} P'$ implies $Q \xRightarrow{\tau}{}^{(lc)} Q'$ for some $Q' \in \mathbb{P}_{\mathcal{CG}}$ s.t. $(P',Q') \in R$,

(*ii*) $P \xRightarrow[\Gamma,l]{a}{}^{(lc)} P', l \notin cau(P,Q)$, implies

$$Q \xRightarrow[\Gamma,l]{a}{}^{(lc)} Q' \text{ for some } Q' \in \mathbb{P}_{\mathcal{CG}} \text{ s.t. } (P',Q') \in R.$$

Two processes P and Q are said to be *local cause (lc) equivalent*, $P \approx_{lc} Q$, iff there is a local cause bisimulation R such that $(P,Q) \in R$. P and Q are *local cause* (*lc*) *congruent*, $P \approx_{lc}^c Q$, if P and Q are local cause equivalent in any allowed context, i.e. for all contexts $C(\ )$ with $C(P), C(Q) \in \mathbb{P}_{\mathcal{CG}}$: $C(P) \approx_{lc} C(Q)$.

Global cause (gc) equivalence, $\approx_{gc}$, and *global cause (gc) congruence*, $\approx_{gc}^c$, are obtained from these definitions by replacing local transitions by global transitions.

For technical reasons it will be convenient to have alternative definitions for $\approx_{lc}$, $\approx_{gc}$, $\approx_{lc}^c$ and $\approx_{gc}^c$. The proofs of these characterizations are similar to those for observation equivalence, see [13] and [16].
Note: As in later statements in the following lemma we omit the indices *lc* and *gc* meaning that it is valid for both semantics.

Lemma 2. [Characterization of $\approx_{lc}^c$, $\approx_{gc}^c$]
Let $P, Q \in \mathbb{P}_{\mathcal{CG}}$. Then $P \approx^c Q$ if and only if $P \approx Q$ and

(i) $P \xRightarrow{\tau} P'$ *implies* $Q \xRightarrow{\tau} Q'$ *for some* $Q' \in \mathbb{P}_{\mathcal{CG}}$ *such that* $P' \approx Q'$,

(ii) $Q \xRightarrow{\tau} Q'$ *implies* $P \xRightarrow{\tau} P'$ *for some* $P' \in \mathbb{P}_{\mathcal{CG}}$ *such that* $P' \approx Q'$.

The second characterization looks slightly different than the standard one in [16]. Instead of prefixing processes with τ, we add $\emptyset \triangleright \tau.nil$ as a new parallel component which has priority over the other process as $\not|$ is used. Thus operationally it has the same affect; the first move must be a τ transition. However, due to the 'prefixing by

For each $a \in Act$ let $\xrightarrow[\Gamma,l]{a} \subseteq (\mathbb{P}_{\mathcal{LG}} \times \mathbb{P}_{\mathcal{LG}})$ be the least binary relation such that $\bigcup \xrightarrow[\Gamma,l]{a}$ satisfies the following axiom and rules.

(LG1) $\quad \Gamma \rhd a.p,\ l \notin \Gamma \xrightarrow[\Gamma,l]{a} \Gamma \cup \{l\} \rhd p \qquad l \in C$

(LG2) $\quad P \xrightarrow[\Gamma,l]{a} P'$ implies $P + Q \xrightarrow[\Gamma,l]{a} P'$, $Q + P \xrightarrow[\Gamma,l]{a} P'$

(LG3) $\quad P \xrightarrow[\Gamma,l]{a} P',\ l \notin cau(Q)$ implies $P \mid Q \xrightarrow[\Gamma,l]{a} P' \mid Q$, $Q \mid P \xrightarrow[\Gamma,l]{a} Q \mid P'$

(LG4) $\quad P \xrightarrow[\Gamma,l]{a} P',\ l \notin cau(Q)$ implies $P \not\mid Q \xrightarrow[\Gamma,l]{a} P' \mid Q$

(LG5) $\quad P \xrightarrow[\Gamma,l]{a} P'$ implies $P[f] \xrightarrow[\Gamma,l]{f(a)} P'[f]$

(LG6) $\quad P \xrightarrow[\Gamma,l]{a} P'$ implies $P \backslash b \xrightarrow[\Gamma,l]{a} P' \backslash b,\ a \notin \{b, \bar{b}\}$

(LG7) $\quad P \xrightarrow[\Gamma,l]{a} P',\quad P' \xrightarrow{\tau} P''$ implies $P \xrightarrow[\Gamma,l]{a} P''$

(LG8) $\quad P \xrightarrow{\tau} P',\quad P' \xrightarrow[\Gamma,l]{a} P''$ implies $P \xrightarrow[\Gamma,l]{a} P''$

Fig. 1. Visible Transitions

left merge' the clear separation between past and future that is causes may prefix actions but not vice versa is preserved.

Lemma 3. **[Characterization of $\approx_{lc}$, $\approx_{gc}$]** *Let $P, Q \in \mathbb{P}_{\mathcal{LG}}$. Then $P \approx Q$ if and only if $P \approx^c Q$ or $P \approx^c \emptyset \rhd \tau \not\mid Q$ or $\emptyset \rhd \tau \not\mid P \approx^c Q$.*

3 The Proof Systems

We here present two proof systems which allows one to derive $P =_{LC} Q$ if and only if $P \approx_{lc}^c Q$ and $P =_{GC} Q$ if and only if $P \approx_{gc}^c Q$. We call them LC and GC Proof System respectively. Like the underlying operational semantics they are closely related. They consist of a set of equations and a set of rules which describe how the equations may be used to derive equality of processes. The rules are the same for both proof systems. Apart from the standard ones for reflexivity, symmetry, transitivity, instantiations of equations and rules to obtain the congruence for all operators there is the new rule (LM):

$$\frac{(\Gamma \cup \{l\} \rhd x \mid X) \setminus A = (\Gamma \cup \{l\} \rhd y \mid Y) \setminus A}{(\Gamma \rhd a.x \lfloor X) \setminus A = (\Gamma \rhd a.y \lfloor Y) \setminus A}, l \notin cau(X,Y)$$

The equations of the proof systems can be divided into three groups. First there is a set of restriction and relabelling laws which in the completeness proof are needed to obtain normal forms. These laws are standard (cf. [16], [12]) and for lack of space they are omitted here. The second group CON, Figure 2, contains basic equations for non-interleaving equivalences. They are typical for the axiomatizations based on the two merge operators, see [2], [12] and [5]. In this group the fifth communication law corresponding to

$$(a.x \lfloor x') \mid_c (b.y \lfloor y') = \begin{cases} \tau \lfloor (x \mid y \mid x' \mid y') & \text{if } a = \bar{b} \\ nil & \text{otherwise} \end{cases}$$

in the restriction free case is missing. Depending on whether local or global causality is considered this law has a different shape. It is therefore in the respective group $LCau$ and $GCau$ which is characteristic for the particular underlying congruence.

Monoid Laws:

$$\begin{aligned} X + (Y + Z) &= (X + Y) + Z \\ X + Y &= Y + X \\ X + \Gamma \rhd nil &= X \\ X + X &= X \end{aligned}$$

Communication Laws :

$$\begin{aligned} (\Gamma \rhd \tau \lfloor X) \mid_c Y &= X \mid_c Y \\ (X + Y) \mid_c Z &= (X \mid_c Z) + (Y \mid_c Z) \\ X \mid_c Y &= Y \mid_c X \\ X \mid_c \Gamma \rhd nil &= \Gamma \rhd nil \end{aligned}$$

Parallel Laws:

$$\begin{aligned} X \mid Y &= X \lfloor Y + Y \lfloor X + X \mid_c Y \\ (X + Y) \lfloor Z &= X \lfloor Z + Y \lfloor Z \\ (X \lfloor Y) \lfloor Z &= X \lfloor (Y \mid Z) \\ X \lfloor \Gamma \rhd nil &= X \\ \Gamma \rhd nil \lfloor X &= \Gamma \rhd nil \end{aligned}$$

τ Laws :

$$\begin{aligned} \Gamma \rhd \tau.x &= \emptyset \rhd \tau \lfloor \Gamma \rhd x \\ \Gamma \rhd \mu.x &= \Gamma \rhd \mu.\tau.x \\ X \lfloor Y &= X \lfloor (\emptyset \rhd \tau \lfloor Y) \\ \Gamma \rhd \tau \lfloor X &= \Gamma \rhd \tau \lfloor X + X \\ X \lfloor (Y + \Gamma \rhd \tau \lfloor Z) &= X \lfloor (Y + \Gamma \rhd \tau \lfloor Z) + X \lfloor Z \end{aligned}$$

Fig. 2. Equations CON.

We now turn to the equations distinctive for local and global causality. For local causality this is the set $LCau$ consisting of the fifth communication law and one absorption where $X \sqsupseteq Y$ stands for $X = X + Y$. Both laws do not hold for global causality. For example with (LC5) one can prove $\{k\} \rhd b \mid_c \emptyset \rhd \bar{b}.c = \emptyset \rhd \tau \lfloor (\{k\} \rhd nil \mid \emptyset \rhd c)$ but

$$\{k\} \rhd b \mid_c \emptyset \rhd \bar{b}.c \xrightarrow[\{k\},l]{c}{}^{(gc)} \{k\} \rhd nil \mid \{k,l\} \rhd nil$$

and

$$\emptyset \rhd \tau \lfloor (\{k\} \rhd nil \mid \emptyset \rhd c) \xrightarrow[\emptyset,l]{c}{}^{(gc)} \emptyset \rhd nil \mid (\{k\} \rhd nil \mid \{l\} \rhd nil).$$

$$\text{(LC5)} \quad (\Gamma \rhd a.x \lfloor X) \mid_c (\Gamma' \rhd b.y \lfloor Y) = \begin{cases} \emptyset \rhd \tau \lfloor (\;\Gamma \rhd x \mid X \mid \;\; \Gamma' \rhd y \mid Y) & \text{if } a = \bar{b} \\ nil & \text{otherwise} \end{cases}$$

$$(LAbs) \qquad \begin{array}{c} \Gamma \rhd a.(x + b.x_0 \lfloor x_0') \lfloor \\ (\Gamma' \rhd (y + \bar{b}.y_0 \lfloor y_0')) \end{array} \sqsupseteq \begin{array}{c} \Gamma \rhd a.(x_0 \mid x_0') \lfloor \\ (\Gamma' \rhd (y_0 \mid y_0')) \end{array}$$

Fig. 3. LCAU – Characteristic Equations for Local Causality.

The absorptions treat the case of that the performance of a visible action a is immediately followed by a communication. In case of local causality this communication does not affect the causes in a term. If global causality is considered instead the causes of the subprocesses taking part in the communication have to be distributed to the communicating partners. Additionally it has to be ensured that the new cause representing a is also forwarded to them. This cause transfer is the only role of the α and β actions in the absorptions. On the whole one has to distinguish between four cases of causal dependencies. Two of them are derivable. All laws in GCAU are characteristic for global causality as their validity depends on the exchange of causes with a communication.

In the rest of the paper we are concerned with showing completeness of the proof systems that is for all closed (finite) terms P, Q with $P \approx_{lc}^{c} Q$ we can derive with LC Proof System $P =_{LC} Q$ (and respectively for $\approx_{gc}^{c}$).

4 Completeness of the Proof Systems

To be able to argue about processes of a regular structure we first show that every process can equationally be transformed into a normal form. This normal form can be seen as a generalization of the normal form (nf) used for restriction free processes $\sum\limits_{i \in I} a_i.p_i \lfloor p_i' + \sum\limits_{j \in J} \tau.p_j$ where p_i, p_i' and p_j are normal forms again. Note that we use 'prefixing by left merge' for the τ summands as causes may not be prefixed by actions.

Definition 4. **[(Restricted) Cause Normal Form]**
Let p_i be nf's, let I and J be disjoint index sets. Then

$$\hat{P} = \sum_{i \in I} \Gamma_i \rhd a_i.p_i \lfloor P_i + \sum_{j \in J} \emptyset \rhd \tau.nil \lfloor P_j$$

is said to be a *cause normal form* (cnf) if P_i and P_j are cause normal forms.
$P \setminus A$ is called a *restricted cause normal form* ($rcnf$) if P is a cnf.
The conventions for empty index sets are as usual.

(GC5)

$$(\Gamma \rhd a.x \,\slashed{Y}\, X) \,|_c\, (\Gamma' \rhd b.y \,\slashed{Y}\, Y) = \begin{cases} \emptyset \rhd \tau \,\slashed{Y}\, (\begin{matrix} \Gamma \cup \Gamma' \rhd x \mid X \mid \\ \Gamma \cup \Gamma' \rhd y \mid Y) \end{matrix} & \text{if } a = \bar{b} \\ nil & \text{otherwise} \end{cases}$$

(GAbs1)

$$\begin{array}{c}
(\Gamma \rhd a.(\alpha \mid x) \,\slashed{Y} \\
\quad (\Gamma_0 \rhd \bar{a}.(b.y_0 \,\slashed{Y}\, y_0' + z_0) \mid \\
\quad\quad \Gamma_1 \rhd \quad (\bar{b}.y_1 \,\slashed{Y}\, y_1' + z_1))) \\
\quad \backslash \alpha
\end{array} \quad \sqsupseteq \quad
\begin{array}{c}
(\Gamma \rhd a.(\alpha \mid \beta_0 \mid \beta_1 \mid x) \,\slashed{Y} \\
\quad (\Gamma_0 \cup \Gamma_1 \rhd \bar{\beta}_0.y_0 \mid \Gamma_0 \rhd \bar{a}.y_0' \mid \\
\quad\quad \Gamma_0 \cup \Gamma_1 \rhd \bar{\beta}_1.y_1' \mid \Gamma_1 \rhd y_1' \quad)) \\
\quad \backslash \alpha, \beta_0, \beta_1
\end{array}$$

where α, β_0 and β_1 do not occur anywhere else in the terms

(GAbs2)

$$\begin{array}{c}
(\Gamma \rhd a.(\alpha_0 \mid \alpha_1 \mid x) \,\slashed{Y} \\
\quad (\Gamma_0 \rhd \bar{\alpha}_0.(b.y_0 \,\slashed{Y}\, y_0' + z_0) \mid \\
\quad\quad \Gamma_1 \rhd \bar{\alpha}_1.(\bar{b}.y_1 \,\slashed{Y}\, y_1' + z_1))) \\
\quad \backslash \alpha_0, \alpha_1
\end{array} \quad \sqsupseteq \quad
\begin{array}{c}
(\Gamma \rhd a.(\alpha_0 \mid \alpha_1 \mid \beta_0 \mid \beta_1 \mid x) \,\slashed{Y} \\
\quad (\Gamma_0 \cup \Gamma_1 \rhd \bar{\beta}_0.y_0 \mid \Gamma_0 \rhd \bar{\alpha}_0.y_0' \mid \\
\quad\quad \Gamma_0 \cup \Gamma_1 \rhd \bar{\beta}_1.y_1' \mid \Gamma_1 \rhd \bar{\alpha}_1.y_1')) \\
\quad \backslash \alpha_0, \alpha_1 \beta_0, \beta_1
\end{array}$$

where α, α_0, α_1, β_0 and β_1 do not occur anywhere else in the terms

---

Fig. 4. GCAU – Characteristic Equations for Global Causality.

For a particular process the outcome of the normalization can be different for local and global causality. This is due to the fact that with the normalization the fifth communication law may be used which is different for local and global causality. This is the only point where the normalizations vary.

Lemma 5. [Normalization] *For each $P \in \mathbb{P}_{\mathcal{LG}}$ there is an rcnf $\hat{P}$ such that $P = \hat{P}$.*

From the derivate of a *rcnf* a process may be constructed that equationally can be absorbed into the *rcnf*. This is the assertion of the two absorption lemmas. For τ derivatives the absorption lemma has a familiar form.

Lemma 6. [τ–Absorption] *Let P be a rcnf.*
If $P \overset{\tau}{\Longrightarrow} P'$ then $P = P + \emptyset \rhd \tau \,\slashed{Y}\, P'$.

For the absorption of derivatives from a visible transition the causes of the executed action have to be taken into account. In case of local causality the term to be absorbed is —loosely spoken— obtained from re-replacing the new cause by the executed action.

Lemma 7. [General Absorption for Local Causality]
Let P be a rcnf.
If $P \overset{\bullet}{\underset{\Gamma,l}{\Longrightarrow}}^{(lc)} P'$ *then* $P' =_{LC} (\prod_{k\in K} \Gamma \cup \{l\} \rhd q_k \mid \prod_{m\in M} \Delta_m \rhd q_m \mid S) \setminus A$
where $l \notin \bigcup \Delta_m \cup cau(S)$ *and* q_k, q_m *are nf's,* S *is a cnf, and*

$$P \quad =_{LC} \quad P + (\Gamma \rhd a. \prod_{k\in K} q_k \nmid (\prod_{m\in M} \Delta_m \rhd q_m \mid S)) \setminus A.$$

In case of global causality the term to be absorbed is more involved. Processes q_{k_1} and q_{k_2}, $k_1, k_2 \in K$, may have different causes which also are not necessarily causes of the executed a action. Such situations arise from communications. The solution is to add $\Gamma \rhd a$ as a new parallel component (with priority) which by means of new restricted communication channels $\alpha_1, \ldots, \alpha_{|K|}$ transfers Γ and the 'future' cause of a to the q_k processes.

Lemma 8. [General Absorption for Global Causality] *Let P be a rcnf.*
If $P \overset{\bullet}{\underset{\Gamma,l}{\Longrightarrow}}^{(gc)} P'$ *and* $P' =_{GC} (\prod_{k\in K} \Delta_k \cup \{l\} \rhd q_k \mid \prod_{m\in M} \Delta_m \rhd q_m \mid S) \setminus A$, *where* $l \notin \bigcup \Delta_k \cup \bigcup \Delta_m \cup cau(S)$ *and* q_k, q_m *are nf's,* S *is a cnf, then*

$$\begin{aligned} P \quad =_{GC} \quad & P + (\Gamma \rhd a.(\alpha_1 \mid \alpha_2 \mid \cdots \mid \alpha_{|K|}) \nmid \\ & \qquad (\prod_{k\in K} \Delta_k \rhd \bar{\alpha}_k.q_k \mid \prod_{m\in M} \Delta_m \rhd q_m \mid S)) \setminus A \cup \{\alpha_1, \ldots, \alpha_{|K|}\} \end{aligned}$$

where $\alpha_1, \ldots, \alpha_{|K|}$ *are* $|K|$ *distinct actions not occurring in* P.

The completeness proof is similar to the restriction free case. However one here comes across the problem that from (in a simplified form) $(\{l\} \rhd p \mid p')\setminus A \approx (\{l\} \rhd q \mid q')\setminus B$ one has to infer $(a.p \nmid p')\setminus A = (a.q \nmid q')\setminus B$. In absence of restriction it can be solved by applying a decomposition lemma. It yields $p \approx p'$ and $q \approx q'$ from which by means of Lemma 3, induction and congruence properties $a.p \nmid p' = a.q \nmid q'$ can be derived. As we do not have such a decomposition lemma in the general setting it is here where Rule (LM) comes into play. So all together one needs the normalization (which yields a *rcnf* of equal or smaller depth), the absorption lemmas, Rule (LM) and the characterizations of $\approx$ and $\approx^c$.

Theorem 9. [Completeness] *Let* $P, Q \in \mathbb{P}_{\mathcal{LG}}$. *Then* $P \approx^c Q$ *if and only if* $P = Q$.

Acknowledgement I particularly like to thank the referee who suggested to distribute causes immediately. This change led to some simplifications.

References

1. L. Aceto. *Action Refinement in Process Algebra.* Ph.d. thesis, University of Sussex, 1991.

2. L. Aceto. A static view of localities. Report 1483, INRIA, 1991.
3. L. Aceto. Relating distributed, temporal and causal observations of simple processes. Technical Memo HPL-PSC-92-32, Hewlett-Packard Laboratories, Pisa Science Center, 1992.
4. J. Bergstra and J.W. Klop. Algebra of communicating processes with abstraction. *Theoretical Computer Science*, (37):77-121, 1985.
5. G. Boudol, I. Castellani, M. Hennessy, and A. Kiehn. Observing localities. Report 4/91, University of Sussex, 1991. Extended abstract in the proceedings of MFCS 91, LNCS 520.
6. G. Boudol, I. Castellani, M. Hennessy, and A. Kiehn. A theory of processes with localities. Report 13/91, University of Sussex, 1991. Extended abstract in the proceedings of CONCUR 92.
7. P. Darondeau and P. Degano. Causal trees. In *Proceedings of ICALP 88*, number 372 in Lecture Notes in Computer Science, pages 234-248, 1989.
8. P. Degano, R. De Nicola, and U. Montanari. Universal axioms for bisimulations. Report TR-9/92, University of Pisa, 1992.
9. P. Degano and C. Priami. Proved trees. In W. Kuich, editor, *Proceedings of ICALP 92*, number 623 in Lecture Notes in Computer Science, pages 629-640. Springer-Verlag, 1992.
10. G. Ferrari, R. Gorrieri, and U. Montanari. Parametric laws for concurrency. In *Proceedings of Tapsoft 91*, number 494 in Lecture Notes in Computer Science, 1991.
11. M. Hennessy. Axiomatising finite concurrent processes. *SIAM Journal of Computing*, 17(5):997-1017, 1988.
12. M. Hennessy. A proof system for weak ST-bisimulation over a finite process algebra. Report 6/91, University of Sussex, 1991.
13. M. Hennessy and R. Milner. Algebraic laws for nondeterminism and concurrency. *Journal of the Association for Computing Machinery*, 32(1):137-161, 1985.
14. A. Kiehn. Comparing locality and causality based equivalences (Revision of: Local and global causes. Report TUM-I9132, Technische Universität München, 1992).
15. A. Kiehn. Proof systems for cause based equivalences. full paper, 1993.
16. R. Milner. *Communication and Concurrency*. Prentice-Hall, 1989.
17. F. Moller. *Axioms for Concurrency*. Ph.d. thesis, University of Edinburgh, 1989.
18. U. Montanari and D. Yankelevich. A parametric approach to localities. In W. Kuich, editor, *Proceedings of ICALP 92*, number 623 in Lecture Notes in Computer Science, pages 617-628. Springer-Verlag, 1992.
19. R.J. van Glabbeek and F.W. Vaandrager. Petri net models for algebraic theories of concurrency. In J.W. de Bakker, A.J. Nijman, and P.C. Treleaven, editors, *Prooceedings PARLE conference*, number 259 in Lecture Notes in Computer Science, pages 224-242. Springer-Verlag, 1987.

A Uniform Universal CREW PRAM

Bruno Martin *

LIP-IMAG, URA CNRS n° 1398
Ecole Normale Supérieure de Lyon,
46 Allée d'Italie, 69364 Lyon Cedex 07, France.

Abstract. The universality of the Parallel Random-Access Machines is usually defined by simulating universal Turing machines or boolean networks. These definitions are well-suited if we are interested in evaluating the complexity of algorithms but it is not as good if we want to deal with computability. We propose in this paper another definition for the universality of the Parallel Random-Access Machines based on cellular automata and we discuss the advantages and the drawbacks of this simulation. We prove that there exists a Concurrent-Read Exclusive-Write Parallel Random-Access Machine which is capable of simulating any given cellular automaton in constant time. We then derive to the definition of complexity classes for the Parallel Random-Access Machines and for cellular automata.

Introduction

The field of parallel computation is going through a period of unrest. While most theoretical computer science is busy designing and evaluating algorithms on Parallel Random-Access Machines, one of the first problem to be solved is maybe the universality for the models of parallel computation. This kind of problem seems to be quite new in this area of the research in theoretical computer science devoted to parallelism. Very few papers speak about the computational power of the parallel machines. They are often known to be universal by simulating the most classical models of computation such as Turing machines or boolean networks. These two models have both drawbacks when applied to parallel machines. The first one implies a sequentialization of the parallel machine and the second one introduces the difficult notion of uniformity. Let us briefly recall the notion of *uniformity*; a circuit is defined for bounded inputs. So, to speak of problem resolution (that is, computations are defined for any size of the inputs) we need the definition of the circuits capable of solving a problem for any input size. We get then a family of circuits $\{C_n\}$ for solving a given problem. A common definition of the uniformity is the *logspace uniformity* which means that the description of the nth circuit C_n can be generated by a Turing machine using $O(\log n)$ workspace. Then, the complexity of a problem solvable by circuits must use the definition of the uniformity.

In the present paper, we introduce a proper uniform intrinsically parallel model of computation, the *cellular automata* model known to be near the single program

* This work was supported by Esprit Basic Research Action "*Algebraic and Syntactic Methods In Computer Science*" and by the *Programme de Recherches Coordonnées Mathématiques-Informatique*

many data models of parallel machines. We will show that there exists a concurrent-read exclusive-write parallel random-access machine which simulates in constant-time and with little space any given unidimensional cellular automaton. Thus, we prove directly the equivalence between the PRAM model and the cellular automata model and precise in which sense the two models can be compared.

From this point, it is possible to apply well-known results of the theory of computability to find out complexity measures and complexity classes relative to these models. We will discuss the PLINEAR class, which seems well-suited for the cellular automata model.

1 Definitions

1.1 Parallel Random-Access Machines

The concept of *Parallel Random-Access Machine* (PRAM for short) has been introduced to include the processing of huge data. In this model, the time is proved equivalent, within a polynomial, to the space of the Turing Machine [2]. This machine is capable to wake up a large (possibly infinite) number of processors, all operating both on a private memory and on a global shared memory. However, this machine is not restricted to execute the same instruction in all active processors at each unit of time. For this model, a memory word can have a constant size, be logarithmic in the size of the used memory, or be unlimited. A ***uniform cost*** Parallel Random-Access Machine is a PRAM for which every instruction takes a unit of time and in a ***logarithmic-cost*** one, instruction time is the sum of the sizes of the values involved and the size of the address of the operand, all in bits. More formally, a Parallel Random-Access Machine consists in an unbounded set of processors $p_0, p_1, p_2 \dots$ and an unbounded global memory $x_0, x_1, x_2, \dots$ and a finite program. Each processor p_i has an unbounded local memory $y_0, y_1, y_2, \dots$. Register y_0 is called the accumulator of the processor. Each processor has a program counter and a flag indicating whether the processor is running or not. A program consists of possibly labeled instructions chosen from the following list:

$y_i :=$ constant;
$y_i := y_i + y_k$;
$y_i := \lceil y_i/2 \rceil$;
$y_i := y_{y_i}$;
accept;
goto m if $y_i > 0$;
$y_i := x_{y_i}$;
$x_{y_i} := y_i$;
fork m;

The parallelism is achieved by the "fork m" instruction. When a processor p_i executes a fork instruction, it selects the first inactive processor p_j, clears the local memory of p_j and fetches the accumulator of p_i onto the accumulator of p_j. Then, p_j starts its computation in the label m of the program. The communications between processors take place through the global shared memory. In the Parallel Random-Access Machine model we allow simultaneous reads in the local memory by different processors. If the two processors try to write into the same position of the global memory, the machine immediately halts and rejects. Several processors may read a position of the global memory while one writes into it; in this case, all the reads are performed before the value of the position is modified.

To deal with different degrees of concurrence when reading from or writing to the global memory, there are some variants of the Parallel Random-Access Machines:

- EREW PRAM; a model with exclusive read and exclusive write. Neither concurrent reading nor concurrent writing is allowed.
- CREW PRAM; a model with concurrent read and exclusive write. Concurrent reading is allowed but not concurrent writing.
- CRCW PRAM; a model with concurrent read and concurrent write. Both kinds of concurrency are permitted.

Many varieties of CRCW PRAM have been defined, differing in the way they resolve the write conflicts. The subject is still under study [4] and leads to difficult simulations of the distinct CRCW machines.

1.2 Cellular Automata

About 1953, John von Neumann suggests the use of cellular automata as a device for parallel computation. Then, cellular automata have been more or less neglected because of the lack of technology. Interest in cellular automata has been renewed with the resurgence of massively parallel computing with the hope of being a realistic model. It is in fact the case modulo the great amount of processors required (Feldman and Shapiro [3]) propose the definition of spatial machines which look a little bit like cellular automata in several aspects but with a smallest number of processors). Here we will deal with one dimensional cellular automata as computational arrays. Their formal definition is as follows:

Definition 1. A cellular automaton is a right infinite array of identical cells, indexed by $\mathbb{N}$, the set of integers. Each cell is a finite state machine $C = (Q, \delta)$ where:

- Q is a finite set, the set of the states;
- δ is a mapping, $\delta : Q^3 \rightarrow Q$

Then, all the cells evolve synchronously according to the local transition function δ, which has the following meaning: the state of the ith cell at time t is a function of its own state at the previous time and of its left and right neighbors at the previous time. Thus, if we denote by $c(i,t)$ the content of cell i at time t, we get the following equality:

$$c(i,t) = \delta(c(i-1,t-1), c(i,t-1), c(i+1,t-1))$$

We have, in addition, a special state q called *quiescent* with the property $\delta(q,q,q) = q$. We observe that the set of the states has no structure. Below we introduce a variant of cellular automata allowing a structured set of states. In order to do that, we indentify the states to non-negative integers. This numbering allows to define the local transition function simply as a mapping $f : \mathbb{N} \rightarrow \mathbb{N}$. This leads to the definition of *totalistic* cellular automata introduced by J. Albert and K. Culik in [1].

Definition 2. A cellular automaton with set of states Q and transition function $\delta : Q^3 \rightarrow Q$ is called *totalistic* if $Q \subset \mathbb{N}$ and there exists $f : \mathbb{N} \rightarrow \mathbb{N}$ such that:

$$\forall a, b, c \in Q, \delta(a, b, c) = f(a + b + c)$$

We observe that such a totalistic cellular automaton is capable of differentiating its own state from its left or right neighbor state. This comes from the following lemma proved in [1] which allows the translation of any cellular automaton into a totalistic.

Lemma 3. *For any cellular automaton A, there exists a totalistic one T which simulates A without loss of time and has at most four times as many states.*

1.3 Universality

There are basically two ways of defining universal machines. The first one, called *computation universality*, states that a machine is universal if any computation with respect to the Church-Turing thesis can be achieved. This is, for instance, the usual way to prove the universality of the Parallel Random-Access Machines. To do that one has just to simulate an universal machine (say an universal boolean network). This kind of simulation can be found in the paper of R. M. Karp and V. Ramachandran [5]. The difficult point is then to design the universal boolean network.

The second one, called *intrinsic universality*, asserts that a machine is intrinsically universal if it is capable of simulating any other machine within the same model. In this way the simulation is "fair" as referring to another model is not required and all the notions are self-contained in the model. From our point of view, it is the best definition for the universality of a model.

We will prove in this paper that there exists a computation universal CREW PRAM which is capable of simulating any given cellular automaton in totalistic form and, in particular, an intrinsic universal cellular automaton. Before entering the details of the simulation of totalistic cellular automata by a CREW PRAM we have to recall the functioning of an intrinsic universal cellular automaton.

2 Intrinsic universality of cellular automata

Self referring (*intrinsic*) universal computation by a one dimensional cellular automaton has been solved by J. Albert and K. Culik in [1] by the description of normal forms construction for cellular automata, their totalistic definition. They use an encoding of this type of cellular automata together with an initial configuration as an input for the universal machine. The description of this intrinsic universal cellular automaton with fourteen internal states can be found in [1]. We have decreased the time of the simulation from a quadratic time to a quasi-linear time in [8, 10].

Intuitively, the universal computation simply consists in the simulation of a given totalistic cellular automaton A together with an input x (the content of the non-quiescent cells of A at time 0 also called *initial configuration*). Recalling that A is under totalistic form, implying that the new state is obtained by sums, one *beat* of each simulated cell of A by the universal machine U consists in:

- sending to the left and right simulated cells the content of the current "cell";

- summing the content of the current "cell" with the contents of the two cells received;
- looking for the image of this sum by means of the transition function f_A;
- replacing the content of the current "cell" by the new value.

Then, we can state:

Theorem 4. *There exists a universal cellular automaton U with thousands of states which can simulate any given totalistic cellular automaton A with any initial configuration in quasi-linear time.*

The great number of states comes from the fact that a cartesian product of some cellular automata have been used to design the universal cellular automaton U. Each state of U is then a 4-tuple; three elements are taking their values among 18 symbols and one among 6 symbols.

3 A uniform universal PRAM

From the viewpoint of the previous sections, the design of a uniform universal Parallel Random-Access Machine may simply consists in the simulation of a universal cellular automaton. We describe such a simulation in the next section.

3.1 Simulation of an arbitrary cellular automaton

Basically, the universality of the Parallel Random-Access Machines is considered with circuits or alternating Turing machines as reference models [2, 5]. As both of them support universal computation, it shows clearly that the Parallel Random-Access Machines are capable of universal computation. But the two reference models mentioned above are defined with bounded input for the circuits and with sequential reference for the alternating Turing machines. We will propose here another simulation of a model based on the intrinsically universal cellular automaton defined above, summarized in the next theorem:

Theorem 5. *There exists a CREW PRAM with uniform cost which can simulate any given totalistic cellular automaton A with any initial configuration in constant time.*

Proof The idea of the simulation is suggested by figure 1 which describes the way the CREW PRAM simulates the evolution of a (universal) cellular automaton A. The content of the cells of A are contained in the d_i's to which a processor p_i is associated. The local transition function of A is contained in the shared memory in its totalistic form. The behavior of the Parallel Random-Access Machine P is the following:

- each processor p_i writes d_i on the registers according to the arrows of figure 1;
- each processor p_i sums the content of the registers and puts the result r instead of the content of d_i; clearly, $1 \leq r \leq 3 \times \#\ \text{states}(A)$;

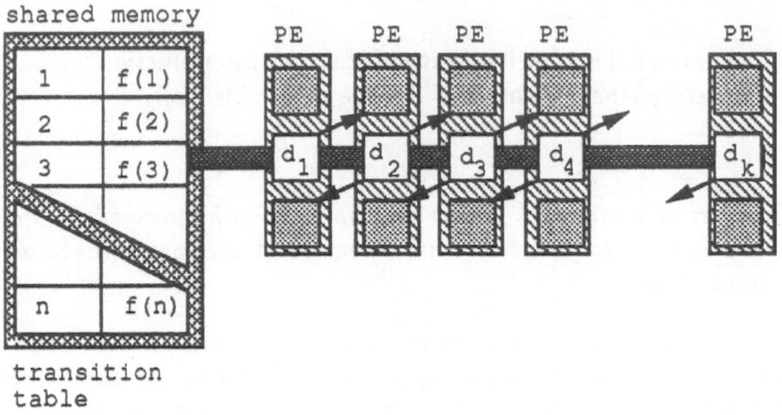

Fig. 1. Simulation of a (universal) CA by a CREW PRAM

- each processor p_i looks in the shared transition table at index r to find the image of r by by the transition function f_A of A and writes the result $f_A(r)$ instead of r;
- iterate the points above until the result is written on the d_i's if this event ever occurs.

We give below the details of the simulation of an arbitrary cellular automaton A by the PRAM P. Let us assume that the PRAM has infinitely many processors. The first thing to do is to associate the registers to the processors. Each of them contains the following informations. Each data item of the cells of A is contained in one of the registers of the processor associated to cell i, the next registers all contain number zero representing the quiescent state;

The local transition function of A is contained in the global shared memory as a table with index entries corresponding to the possible values of the sum of the cells. in such a way, accessing an image state by the transition function when the sums have been computed can be achieved in constant time. It only requires to access one element of the table. Then, the program of processor p_i for any i is the following:

1 $y_3 := i + (3.k + 2)$;
2 $y_0 := x_{y_0}$;
3 $x_{y_3} := y_0$;
4 $y_1 := x_{y_3} - 1$;
5 $y_0 := y_0 + y_1$;
6 $y_1 := x_{y_3} + 1$;
7 $y_0 := y_0 + y_1$;
8 goto 2 if $y_0 > 0$;

Observe that a "new" cell is created when its content changes from zero to another value. Furthermore, only one cell can be created at each iteration. That is, the number of the cells of the cellular automaton can be bounded. This bound can be considered as the space of a Turing machine. □

This new proof of the universality of a PRAM shows clearly that we just require a CREW PRAM because there are no concurrent writes, but only concurrent reads of the table of the transition function. Moreover, the process defined above simulates one step of A in constant time. Then, the Parallel Random-Access Machines can be proved capable of universal computation with this simulation and theorem 4 thus defining a sort of CA-PRAM which is a uniform single program stream and multiple data stream kind of parallel model.

As we have inherited the definition of universality from cellular automata we can also define the notion of computation for the CA-PRAMs.

3.2 Computation with CA-PRAM

Notions of parallel computations are not defined very often. For the purpose of this paper we will take a classical definition of the computation on cellular automata which says that the result of a computation for a cellular automaton remains as a unit-time periodical configuration. That is, let φ be the partial recursive function computed by cellular automaton A. We say that y is the result of $\varphi(x)$ if y appears as unchanging configuration during the evolution of A on the initial configuration x. That is, there exists a non-negative integer τ such that

$$\forall t > \tau, \Delta^{(t)}(x) = x^{\tau} = y = \varphi(x)$$

where the global function Δ is a mapping from $Q^{\mathbb{N}} \to Q^{\mathbb{N}}$ obtained by parallel application of the local transition function δ of cellular automaton A and with $\Delta^{(t)}$ denoting the tth iterate of mapping Δ. Moreover, if such a time does not exist, we say that the function is *divergent*.

Half of the work is then done to define a notion of computation on the Parallel Random-Access Machines. It remains to give a natural way to check the halting of the CA-PRAM. Thus, in order to decide if the if the result has been computed, we have to store the entire configuration between each beat and check if the new one is the same as the the old one stored. Thus, we need to add one more register to each processor and store the configuration between each beat of the simulation of the cellular automaton. One can easily derive from the parallel prefix algorithm [5] an effective procedure to get the comparison between the values newly computed and the values previously computed and broadcast a special state to the processors to make them continue their simulation. The two operations required are both computed in logarithmic-time on an EREW-PRAM and hence also in logarithmic-time on a CREW-PRAM.

Then, it is possible to define properly a the computation on a PRAM by means of the process described above. If we assume that we have a log-cost PRAM instead of the uniform-cost PRAM, each step of the simulation can be achieved in constant time. Thus, we have defined properly the CA-PRAM model of parallel computation.

3.3 Consequences

With this definition of the CA-PRAMs, it is possible to get some more results in the field of the theory of computation. To that end, we introduce the definition of *acceptable programming systems* as given in the book of M. Machtey and P. Young [6]:

Definition 6. A *programming system* is a listing $\varphi_0, \varphi_1, \ldots$ which includes all of the partial recursive functions (of one argument over $\mathbb{N}$). A programming system is *universal* if the partial function φ_{univ} such that $\forall i \forall x$, $\varphi_{univ}(i, x) = \varphi_i(x)$ is itself a partial recursive function. A universal programming system $\varphi_0, \varphi_1, \ldots$ is *acceptable* if there exists a total recursive function c for the composition such that $\forall i \forall j$, $\varphi_{c(i,j)} = \varphi_i \circ \varphi_j$.

Programming systems are often referred to as indexings of the partial recursive functions. An important and useful property of reasonable (i.e. satisfying the definition above) programming systems is the ability to modify programs so that the input parameters are held constant. We have proved in [9] that cellular automata form an acceptable programming system and thus, with our definition we can also state:

Theorem 7. *The CA-PRAM model forms an acceptable programming system.*

As a consequence [6, 12], we obtain a S-m-n theorem for the CA-PRAMs, a Rice theorem, the Rogers isomorphism [11].

Conclusion

Another interesting aspect of the definition of the CA-PRAMs is the way we can obtain complexity classes for the model. To do that, let us introduce briefly the definition of a complexity class, that we will call PLINEAR similarly to the LINEAR class usually defined for the Random-Access Machines (sequential). Then, PLINEAR is the class composed by the CA-PRAM programs which stop within a time linearly depending upon the input. More formally, let CASIZE denote the maximum number of non-quiescent cells of a cellular automaton A and CATIME the minimal time for which A halts with the result of the computation. We can then define formally PLINEAR=CATIME(A)=O(CASIZE(A)). This complexity class has been more or less studied for cellular automata by J. Mahajan and K. Krithivasan in [7] where hierarchies are constructed and could be applied to the CA-PRAMs.

With this complexity class, we observe that most of the programs are inside this class. Unfortunately, we do not have usual consistent examples which can be solved by CA-PRAMs in linear time for the moment. But this class seems to be a good class for cellular automata and thus for the CA-PRAMs.

If this assumption is founded, we can rely on the PLINEAR class to compare RAM programs and CA-PRAM programs with the same complexity class which would be much more comfortable than the usual translations from a class to another.

References

1. J. Albert and K. Culik II. A simple universal cellular automaton and its one-way and totalistic version. *Complex Systems*, 1:1–16, 1987.
2. J. L. Balcázar, J. Díaz, and J. Gabarró. *Structural Complexity II.* Springer Verlag, 1990.

3. Y. Feldman and E. Shapiro. Spatial machines, a more realistic approach to parallel computation. *Communications of the ACM*, 35(10):61–73, 1992.
4. T. Hagerhup. Fast and optimal simulations between CRCW PRAMs. In *STACS '92*, Lecture Notes in Computer Science, pages 45–56. Springer Verlag, 1992.
5. R. M. Karp and V. Ramachandran. Parallel algorithms for shared-memory machines. In *Handbook of Theoretical Computer Science*, volume A, chapter 17. Elsevier, 1990.
6. M. Machtey and P. Young. *An introduction to the general theory of algorithms*. Theory of computation series, North Holland, 1978.
7. J. Mahajan and K. Krithivasan. Relativised cellular automata and complexity classes. In S. Biswas and K. V. Nori, editors, *FSTCS*, Lecture Notes in Computer Science. Springer Verlag, 1991.
8. B. Martin. Efficient unidimensional universal cellular automaton. In *Proceedings of the Mathematical Foundations of Computer Science*. Springer Verlag, August 1992.
9. B. Martin. *Construction modulaire d'automates cellulaires*. PhD thesis, Ecole Normale Supérieure de Lyon, 1993.
10. B. Martin. A universal cellular automaton in quasi-linear time and its s-m-n form. *Theoretical Computer Science*, 123, 1994. To be published.
11. H. Rogers. *Theory of recursive functions and effective computability*. Mc Graw-Hill, 1967.
12. R. Sommerhadler and S.C. van Westrhenen. *The Theory of Computability, Programs, Machines, Effectiveness and Feasibility*. Addison Wesley, 1988.

Observing Located Concurrency

David Murphy

University of Birmingham

Abstract. We present a process algebra with an explicit notion of location, and give an operational semantics for it that distinguishes between processes with different distributions. We then introduce a denotational semantics parameterised by a topology over the set of locations; this topology allows observers to regard some locations as indistinguishable. We show that the denotational semantics is fully abstract if the topology satisfies the separation axiom T_1, and that it coincides with the usual interleaving operational semantics if it is indiscrete, thus giving a criteria for when a given notion of 'indistinguishable location' corresponds to completely distributed or interleaved settings.

The algebra we consider is then extended to allow communication between different locations. A natural communication operator gives rise to a form of expansion theorem which allows us to extend full abstraction to this setting.

Introduction

The very term 'distributed system' implies an interest in location; for a system to be distributed, it must have components in more than one place. This geographical distribution may be unimportant, as in a tightly-coupled two-processor machine where we do not care which slot a given processor occupies, or it may be of considerable interest, for instance if the efficiency of communication between processors depends on the distance that separates them.

We contend that location information is *often* of interest, and thus we shall distinguish computations that take place *locally* from those that are distributed around a network of machines. Furthermore, we shall distinguish between various distributed computations according to *where* as well as *what* subcomputations happen.

Let us consider a concrete situation. I am sitting at my workstation in Sankt Augustin, and I wish to send a message to my girlfriend in London. To communicate with her, my message passes to the main GMD machine *gmdzi*, then to the main German e-mail gateway in Dortmund (otherwise *Dort*), thence to a machine in Kent, which delivers it to a machine in London my girlfriend can login to. The communication pattern is represented by the thick lines in display 1.

From *gmdzi*, I can also login directly to a machine in Oxford, which mails directly to my girlfriend's machine. Thus another way of communicating with her is

---

Postal address: Department of Computer Science, University of Birmingham, Birmingham B15 2TT, England. *E-mail address*: D.V.Murphy@computer-science.birmingham.ac.uk.
Support from the Royal Society (via an ESEP fellowship at the GMD, Bonn) and the Australian Research Council is gratefully acknowledged.

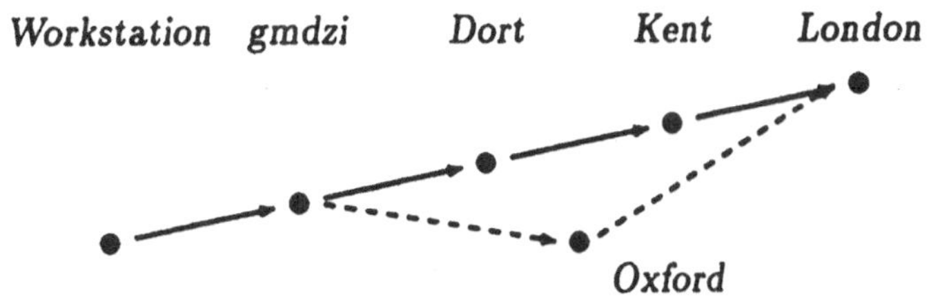

DISPLAY 1. Mailing From Sankt Augustin To London.

given by the dashed lines in display 1. The path via Oxford is quicker than the one via Dortmund and Kent during the daytime, but the Dortmund route is generally faster in the evening. Therefore, it makes sense to distinguish them.

This example motivates another concern: both *gmdzi* and *Dort* give me reasonable error messages should my mail fail, and I can do different things depending on where the failure happens (namely ringing up the systems manager in Sankt Augustin or mailing the postmaster in Dortmund). From my point of view, then, *gmdzi* and *Dort* are different places, as I do different things depending on whether *error* happens in one or the other.

I am not so fortunate with regard to things in England; failures in *Kent*, *London* or *Oxford* give little or no information to me, and I cannot tell where the error occurred. Thus it makes no sense for me to distinguish between the locations *Kent*, *London* and *Oxford*. My view of the network, then, is shown in display 2, where the open circles indicate observably distinct locations.

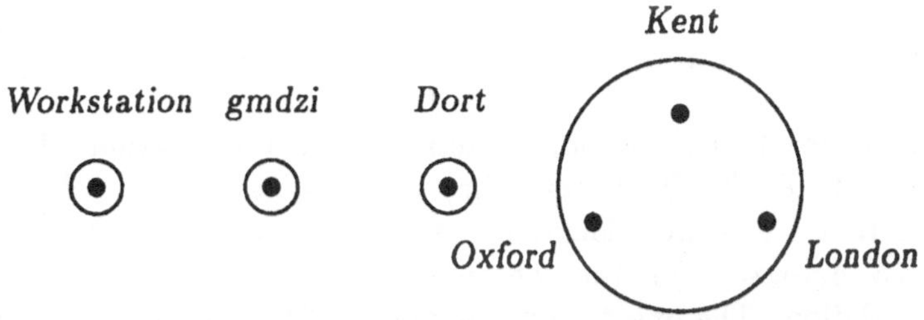

DISPLAY 2. My View of the Network.

In the following section a theory of concurrency that allows us to deal with the problems of location discussed above will be presented. We shall capture both the distinct identity of differently-distributed computations, and the similarity of happenings which cannot be observed to be in different places. Due to lack of space, all proofs are ommitted.

An Algebra of Located Processes

Suppose that a finite set A of actions is given. We begin by considering processes built from this set of actions with no distribution. These are the finite *regular* processes [8] which specify all finite nondeterministic automata. They are given by the following familiar syntax

$$LP ::= \text{NIL} \mid a.LP \mid LP + LP \mid \Phi(LP)$$

where $\Phi : A \rightarrow A$ is a relabeling. In what follows, the final *NIL* will often be omitted from processes; we write $a.b$ for $a.b.$*NIL*.

Processes built from this syntax are *local*, in that they describe the happenings in one place. To build distributed systems, we need a way of associating *locations* with processes. Consider as given a set X of locations—places where processes can be put. A distributed process is then either a local process at some location, or a (finite) parallel combination of such:

$$DP ::= LP_x \mid DP \,||\, DP$$

where $x \in X$. Typical members of *DP* will be written P, Q, and of *LP*, p and q.

Thus, the process which does a single action a at location x is just $(a.NIL)_x$, and this is different from the process that does an a at $y \neq x$, $(a.NIL)_y$. Similarly, the process which offers a choice between a and b at x in parallel with c at y is $(a+b)_x \,||\, c_y$. We have separated automata (*LP* processes) from networks (*DP* processes).

This syntax is fairly inexpressive — it does not, for instance, allow things in one place to affect those in another. Nevertheless, it is a good starting point for a theory of located concurrency which will be refined in latter sections. Moreover, it has the advantage that it does not allow choices to be distributed: it is rather hard to see how to implement the choice between eating lunch in Sankt Augustin and breakfast in London, since if one alternative is chosen, the other must be immediately withdrawn, and this requires instantaneous communication between separated locations [10].

Operational Semantics

The calculus *DP* of distributed processes will be given an operational semantics using a slightly enriched variant of the usual transition systems. First we give a standard semantics for *LP* as a rooted labeled transition system $(S, A, \Rightarrow, s_0)$, where S is a set of states, $s_0 \in S$ the root, and $\Rightarrow\, : S \times A \times S$ a transition relation over the actions A.

This is then extended to *DP* by tagging actions with the location where they happen. Thus we write $P \xrightarrow{a_x} Q$ to mean that the process P can evolve via the occurrence of an action $a \in A$ at a location $x \in X$ to become the process Q. A transition system $(S, A, \rightarrow, s_0, X)$ is then obtained, where $\rightarrow\, : S \times A \times X \times S$ is a *located* transition relation. The operational semantics of the calculus is defined as the smallest located transition system satisfying the rules given in display 3.

Standardly, we define bisimulation over the transition system given by the operational semantics:

Definition 1. A (strong) *located bisimulation* from a rooted located transition system $G = (S, A, \rightarrow, s_0, X)$ to another $H = (S', A', \rightarrow', s'_0, X)$ is a symmetric relation $R : S \times S'$ such that

(i) The roots are related; $s_0 \; R \; s'_0$.

(ii) If two states are related and a located transition is possible from one of them, then it is possible from the other: if we have $s \; R \; u$ and $s \xrightarrow{a_x} s'$ then $u \xrightarrow{a_x} u'$ and $s' \; R \; u'$.

If there exists a located bisimulation between the meanings of two processes $P, Q \in DP$, then we write $P \approx Q$.

We assume the obvious forgetful map from located transition systems to transition systems $Un(S, A, \rightarrow, s_0, X) = (S, A, \Rightarrow, s_0)$, writing $P \simeq Q$ if there is a strong bisimulation (in the usual sense of Park [12]) between $Un(P)$ and $Un(Q)$.

$$\text{PREF}\ \frac{}{a.p \stackrel{a}{\Longrightarrow} p}$$

$$\text{CL}\ \frac{p \stackrel{a}{\Longrightarrow} p'}{p+q \stackrel{a}{\Longrightarrow} p'} \qquad \text{CR}\ \frac{p \stackrel{a}{\Longrightarrow} p'}{q+p \stackrel{a}{\Longrightarrow} p'}$$

$$\text{REN}\ \frac{p \stackrel{a}{\Longrightarrow} p'}{\Phi(p) \stackrel{\Phi(a)}{\Longrightarrow} \Phi(p')}$$

$$\text{LOC}\ \frac{p \stackrel{a}{\Longrightarrow} p'}{p_x \stackrel{a_x}{\longrightarrow} p'_x}$$

$$\text{PL}\ \frac{P \stackrel{a_x}{\longrightarrow} P'}{P \,||\, Q \stackrel{a_x}{\longrightarrow} P' \,||\, Q} \qquad \text{PR}\ \frac{P \stackrel{a_x}{\longrightarrow} P'}{Q \,||\, P \stackrel{a_x}{\longrightarrow} Q \,||\, P'}$$

DISPLAY 3. The Operational Semantics of Distributed Processes.

Our operational semantics in terms of located transitions distinguishes processes whose behaviour with regard to where they perform their actions is different. For instance, it distinguishes not only $a_x \,||\, b_y$ from $(a.b + b.a)_x$ but also $a_x \,||\, b_y$ from $a_x \,||\, b_z$ thus capturing the difference between the places y and z.

This latter distinction may not be desirable; some observers may want to treat the locations y and z as essentially the same, perhaps for reasons outlined in the mail example above. To enable this ambiguity to be expressed, we give a *topology*, $\mathcal{O}X$ to the set X. The idea is that the basic opens $U \in \mathcal{O}X$ are *observably different* sets of locations, so that if no U contains y but not z, then y and z are not observably distinct. Thus my view of the network (shown in display 2) is captured by the topology on {*Workstation*, *gmdzi*, *Dort*, *Kent*, *London*, *Oxford*} specified by basic opens

{*Workstation*}, {*gmdzi*}, {*Dort*}, {*Kent*, *London*, *Oxford*}

Similarly, my girlfriend, who gets different error messages from *Kent*, *London* and *Oxford*, but for whom different sites across the channel are indistinguishable, works with an outlook given by the topology whose basic opens are

{*Workstation*, *gmdzi*, *Dort*}, {*Kent*}, {*London*}, {*Oxford*}

pictured in display 4.

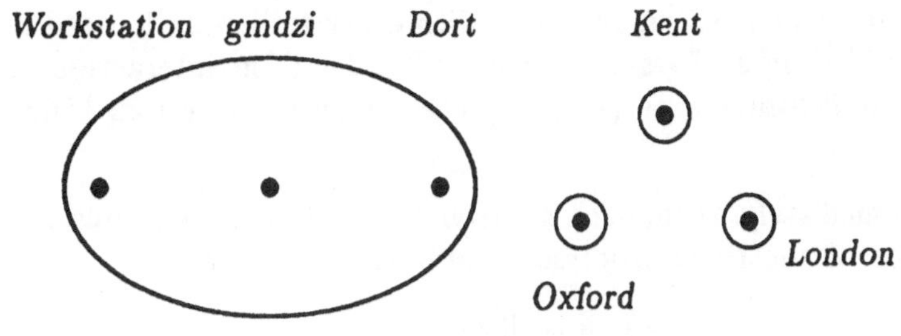

DISPLAY 4. My Girlfriend's View of the Network.

My girlfriend, then, is prepared to treat the three German locations *Workstation*, *gmdzi* and *Dort* as indistinguishable. Thus concurrency in those locations is of no interest to her; she can't distinguish an a in *Dort* in parallel with a b at *gmdzi* from $ab + ba$ at *gmdzi*, so she might as well interleave all the happenings in Germany.

Dealing with the Topology

In this subsection a denotational semantics for distributed processes over a set of locations X will be presented. This semantics will be parameterised by a topology $\mathcal{O}X$, representing which locations are to be viewed as distinguishable. We relate the two semantics under conditions on this topology.

We want to associate a labeled tree with a distributed process P. The first step is to characterise the trees associated with local processes. Let 0 be the tree with a single node, $\oplus$ the operation of joining two trees together at the root, and $a\triangleleft$ the operation of prefixing a single branch labeled a to a tree. Then the function $\mathcal{M} : LP \to \mathfrak{Tree}$, where $\mathfrak{Tree}$ is the domain of finite trees whose edges are labeled over some given set A, is defined by

$$\begin{aligned} \mathcal{M}(\textsc{Nil}) &= 0 & \qquad \mathcal{M}(a.p) &= a \triangleleft \mathcal{M}(p) \\ \mathcal{M}(p+q) &= \mathcal{M}(p) \oplus \mathcal{M}(q) & \qquad \mathcal{M}(\Phi(p)) &= \Phi(\mathcal{M}(p)) \end{aligned}$$

where $\Phi(T)$ for $T \in \mathfrak{Tree}$ relabels all edges a of T by $\Phi(a)$. It is well known that $\mathcal{M}(p)$ for $p \in LP$ gives the same tree as the operational semantics up to $\simeq$, so $\mathcal{M}$ is fully abstract over the class of processes LP.

Now, consider as given a topology $\mathcal{O}X$, and take some open set U in it. Let P be a distributed process over the locations X. We want to associate a labeled tree with the open U which represents P's behaviour at U. Clearly we can ignore all of the behaviour of P outside U, and interleave that at different $x \in U$. Thus $\mathcal{D} : DP \times \mathcal{O}X \to \mathfrak{Tree}$ is defined by

$$\begin{aligned} \mathcal{D}(p_x, U) &= \mathcal{M}(p) && \text{if } x \in U \\ \mathcal{D}(p_x, U) &= 0 && \text{otherwise} \\ \mathcal{D}(P \parallel Q, U) &= \mathcal{D}(P, U) \otimes \mathcal{D}(Q, U) && \end{aligned}$$

where $T \otimes T'$ interleaves the trees T and T' in the usual way [13].

Take $X = \{x, y, z\}$. Then, $\mathcal{D}(a_x, \{x\}) = a$, $\mathcal{D}(a_x, \{y\}) = 0$, $\mathcal{D}((a+b)_x \parallel c_z, \{x\}) = a \oplus b$, and $\mathcal{D}((a+b)_x \parallel c_z, \{x, z\}) = a \triangleleft c \,\oplus\, b \triangleleft c \,\oplus\, c \triangleleft (a \oplus b)$.

Definition 2. We say that two processes are location–equivalent up to $\mathcal{O}X$, and write $P \sim_{\mathcal{O}X} Q$, if for all $U \in \mathcal{O}X$, $\mathcal{D}(P, U) \simeq \mathcal{D}(Q, U)$.

Thus, with regard to a given notion of when two locations are the same, $\mathcal{O}X$, two processes are equivalent iff their contributions at each $U \in \mathcal{O}X$ are bisimilar.

Lemma 3. Let T_x be the transition system T with every transition located at x (so $s \xrightarrow{a_x} s'$ is an edge of T_x iff $s \xrightarrow{a} s'$ is an edge of T). Then the interaction between distribution and parallelism means $(p_x \parallel q_x)$ has a transition system bisimilar to $(\mathcal{M}(p) \otimes \mathcal{M}(q))_x$.

Corollary 4. Immediately, then, each nontrivial $P \in DP$ can be represented as a parallel composition of nontrivial processes in different locations

$$P_{x_1} \parallel \ldots \parallel P_{x_n}$$

where no $P_{x_i} \simeq 0$, and $i \neq j$ implies $x_i \neq x_j$.

Definition 5. We now define the locations of a process $P \in DP$ as those places it has a nontrivial contribution

$$\begin{aligned} \mathrm{loc}(p_x) &= \{x\} && \text{if } \mathcal{M}(p) \not\simeq 0 \\ &= \emptyset && \text{otherwise} \\ \mathrm{loc}(P \parallel Q) &= \mathrm{loc}(P) \cup \mathrm{loc}(Q) \end{aligned}$$

Proposition 6. If $P \approx Q$ then $\mathrm{loc}(P) = \mathrm{loc}(Q)$.

The previous machinery now allows us to show that certain observers correspond to the usual interleaving semantics for our process algebra, and others to the obvious completely distributed semantics:

Proposition 7. For any topology $\mathcal{O}X$, $P \approx Q \implies P \sim_{\mathcal{O}X} Q \implies P \simeq Q$.

Corollary 8. An easy corollary of proposition 7 is that if $\mathcal{O}X$ is indiscrete, then $\sim_{\mathcal{O}X}$ and $\simeq$ coincide and so we recover an interleaving semantics. Moreover, if the topology $\mathcal{O}X$ satisfies a certain *separation* property [3] then $\approx$ coincides with $\sim_{\mathcal{O}X}$:

Theorem 9. If $\mathcal{O}X$ is T_1, then $P \sim_{\mathcal{O}X} Q$ iff $P \approx Q$.

Finally, we treat the (rather unrealistic) case of a continuum of processes.

Proposition 10. Suppose that $F, G : X \to \mathfrak{Tree}$ are any two functions assigning trees to points with $F \not\simeq G$, and suppose that we can observe each function at an open via the evaluation map

$$F|_U = \bigotimes_{x \in U} F(x)$$

Then any topology both F and G are continuous in suffices to distinguish them.

Distributed Synchronisation

Thus far, we have distributed processes but no interaction between computations in different places. To add this feature, we extend the algebra LP of local processes with a new form of prefixing; $\widetilde{a^x}.P$ should be interpreted as 'await a communication identified by a from a process at x, and then behave like P'. Thus just as we interpret $P \xrightarrow{a} P' \xrightarrow{b} P''$ as 'if P does an a, it can then do a b', so we interpret $P \xrightarrow{\widetilde{a^x}} P' \xrightarrow{b} P''$ as 'if P is successfully communicated with by a process at x doing an a, then it can then do a b'. The idea, is that we will synchronise a process that wants an a from x with one that does an a at x, allowing the two processes to cooperate; see the examples overleaf.

The syntax of the calculus GP of global processes adds await prefixes to LP:

$$GP ::= \textsc{Nil} \mid a.GP \mid \widetilde{a^x}.GP \mid GP + GP \mid \Phi(GP)$$

where a ranges over A, x over X and $\tilde{a}$ over $\tilde{A}$. The elements of $\tilde{A} = \{\tilde{a} \mid a \in A\}$ can be thought of as actions which idle until triggered by a matching concurrent action [11] in a different place.

Semantics can be assigned to the calculus GP as a rooted labeled transition system $(S, E, \Rightarrow, s_0)$ much as before; $E = A \uplus (\tilde{A} \times X)$ is the set of all actions, including await-a-communication actions. The additional rule for await prefixes APref just fires them as usual, their special status as possible actions being representing by the distinguishing tilde.

We extend *DP* to a calculus *IP* of interacting distributed processes, with syntax

$$IP ::= GP_x \mid IP \,||\, IP \mid IP\backslash C$$

where $Act = (A \times X) \uplus (\tilde{A} \times X)$ and $C \subseteq Act$. Here $P\backslash C$ is the restriction of a set C of actions; no located action in C will be seen in the behaviour of $P\backslash C$.

$$\textsc{Pref}\ \frac{}{a.p \stackrel{a}{\Longrightarrow} p} \qquad \textsc{APref}\ \frac{}{\widetilde{a^x}.p \stackrel{\widetilde{a^x}}{\Longrightarrow} p}$$

$$\mathrm{CL}\ \frac{p \stackrel{f}{\Longrightarrow} p'}{p+q \stackrel{f}{\Longrightarrow} p'} \qquad \mathrm{CR}\ \frac{p \stackrel{f}{\Longrightarrow} p'}{q+p \stackrel{f}{\Longrightarrow} p'}$$

$$\textsc{Ren}\ \frac{p \stackrel{a}{\Longrightarrow} p'}{\Phi(p) \stackrel{\Phi(a)}{\Longrightarrow} \Phi(p')}$$

$$\textsc{Loc}\ \frac{p \stackrel{a}{\Longrightarrow} p'}{p_x \stackrel{a_x}{\longrightarrow} p'_x} \qquad \textsc{Pot}\ \frac{p \stackrel{\widetilde{a^x}}{\Longrightarrow} p'}{p_y \stackrel{\widetilde{a^x}}{\longrightarrow} p'_y}\ x \neq y$$

$$\mathrm{PL}\ \frac{P \stackrel{g}{\longrightarrow} P'}{P \,||\, Q \stackrel{g}{\longrightarrow} P' \,||\, Q} \qquad \mathrm{PR}\ \frac{P \stackrel{g}{\longrightarrow} P'}{Q \,||\, P \stackrel{g}{\longrightarrow} Q \,||\, P'}$$

$$\textsc{SyncL}\ \frac{P \stackrel{\widetilde{a^x}}{\longrightarrow} P' \quad Q \stackrel{a_x}{\longrightarrow} Q'}{P \,||\, Q \stackrel{\tau}{\longrightarrow} P' \,||\, Q'} \qquad \textsc{SyncR}\ \frac{P \stackrel{\widetilde{a^x}}{\longrightarrow} P' \quad Q \stackrel{a_x}{\longrightarrow} Q'}{Q \,||\, P \stackrel{\tau}{\longrightarrow} Q' \,||\, P'}$$

$$\textsc{Hide}\ \frac{P \stackrel{g}{\longrightarrow} P'}{P\backslash C \stackrel{g}{\longrightarrow} P'\backslash C}\ g \notin C$$

DISPLAY 5. The Operational Semantics of Interacting Processes.

The semantics of a process $P \in IP$ will be a tuple $(S, Act, \rightarrow, s_0, X)$, where $\rightarrow\ : S \times (Act \uplus \{\tau\}) \times X \times S$ is a located transition relation, the smallest one satisfying the rules of display 5. In that display, f ranges over $A \uplus (\tilde{A} \times X)$ and g over $Act \uplus \{\tau\}$.

The main novelty of the rules in display 5 is the two synchronisation rules SYNCL and SYNCR. These both define the result of a synchronisation as a special, silent action $\tau \notin A$. We need two rules as communication is blatantly asymmetric for the *IP* calculus; one process awaits what the other offers.*

We now present a few examples to develop the reader's intuition for the operational behaviour of *IP* processes. Note first that our rules do *not* allow synchronisations between processes in the same place; this is deliberate: very different resources

* This asymmetry in distinction to the pleasing symmetry of CCS, but seems unavoidable if a distributed setting is to be correctly modeled, as almost all common synchronisation protocols are inherently asymmetric. We could make things more CCS-like by making ~ a bijection, but then await and ordinary actions would have the same status, rather than awaits being passive and unobserved. The location of an await should then be observed.

are required for same–processor synchronisations and genuine communications, so the two situations should be distinguished. In fact, it is often more difficult to implement synchronisations between processes on the same processor than to deal with the distributed case; thus we ban the former by the sidecondition on rule POT.

EXAMPLE 1. Consider putting a process $\widetilde{a^x}.\textit{NIL}$ (which offers an a at x) in parallel with an uncooperative process, $b_x.\textit{NIL}$ say, allows us to see how $\widetilde{a^x}$ represents possible communications; the b_x is interleaved with the possibility of an a communication.

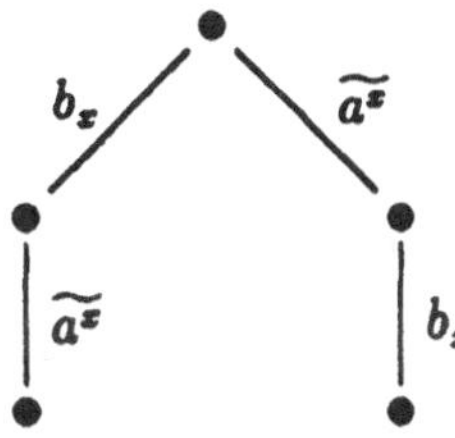

EXAMPLE 2. Now consider a process at y that awaits an a communication from a process at x, and then does a b. In parallel with a_x, we have

$$(\widetilde{a^x}.b)_y \parallel a_x \;=\; (a_x \lhd \widetilde{a^x} \lhd b_y) \oplus (\tau \lhd b_y) \oplus (\widetilde{a^x} \lhd (a_x \lhd b_y \oplus b_y \lhd a_x))$$

In other words, the cooperation either happens, or it doesn't.

EXAMPLE 3. We can force processes to cooperate using hiding as usual, as can be seen from the *IP* process

$$\big((\widetilde{a^x} + b)_y \parallel (a.b + c)_x\big) \setminus \{a_x, \widetilde{a^x}\}$$

which has the meaning shown opposite.

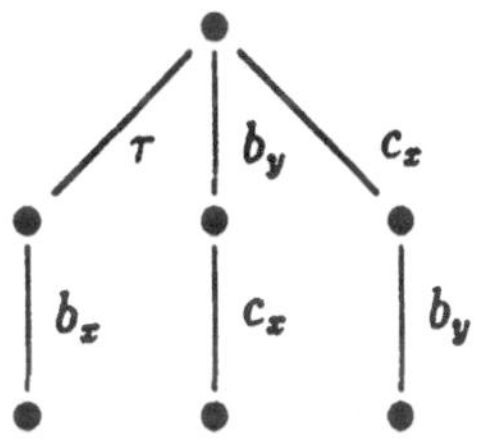

EXAMPLE 4. Our final example is a model of part of the mail system introduced earlier. My workstation sends a message to gmdzi, and then awaits a confirmation (in which case I get a *OK*) or an error message (in which case I get a *FAIL*):

$$\textit{WST} \stackrel{\text{def}}{=} \textit{SEND}.\left(\widetilde{\textit{CONF}}^{\textit{gmdzi}}.\textit{OK} + \widetilde{\textit{ERR}}^{\textit{gmdzi}}.\textit{FAIL}\right)$$

The *gmdzi* machine awaits my communication, then either passes it on to Dortmund (and listens for the result of so doing) or fails

$$\textit{GMD} \stackrel{\text{def}}{=} \widetilde{\textit{SEND}}^{\textit{Workstation}}.\left(\textit{SEND}.\left(\widetilde{\textit{CONF}}^{\textit{Dort}}.\textit{CONF} + \widetilde{\textit{ERR}}^{\textit{Dort}}.\textit{ERR}\right) + \widetilde{\textit{ERR}}\right)$$

It should now be clear how the first level of modelling of the network proceeds. To see the behaviour, consider a faulty machine in *Dort*: $\textit{UNIDO} \stackrel{\text{def}}{=} \widetilde{\textit{SEND}}^{\textit{gmdzi}}.\textit{ERR}$ which we place in parallel with the other processes, hiding the internal workings

$$\textit{NET} \stackrel{\text{def}}{=} (\textit{WST}_{\textit{Workstation}} \parallel \textit{GMD}_{\textit{gmdzi}} \parallel \textit{UNIDO}_{\textit{Dort}}) \setminus C$$

where $C = \{\textit{SEND}_{\textit{gmdzi}}, \textit{CONF}_{\textit{gmdzi}}, \textit{ERR}_{\textit{gmdzi}}, \textit{SEND}_{\textit{Dort}}, \textit{CONF}_{\textit{Dort}}, \textit{ERR}_{\textit{Dort}}, \widetilde{\textit{SEND}}^{\textit{Dort}}, \widetilde{\textit{CONF}}^{\textit{Dort}}, \widetilde{\textit{ERR}}^{\textit{Dort}}, \widetilde{\textit{SEND}}^{\textit{gmdzi}}, \widetilde{\textit{CONF}}^{\textit{gmdzi}}, \widetilde{\textit{ERR}}^{\textit{gmdzi}}\}$.

We then have that *NET* does a send at *Workstation*, followed by either three (a failure at *Dort*) or two (a failure at *gmdzi*) τs, followed by a *FAIL* at *Workstation*, much as expected.

Expanding Parallel Composition

In order to extend our previous results relating the denotational and operational semantics, we need to find a denotation for the $||$ of *IP*. This requires us to understand the structure of the transition systems generated by $P \,||\, Q$. Examining the four rules for parallel composition, PL, PR, SYNCL, and SYNCR, it is immediately clear that each corresponds to an operator. The first two rules correspond to the well–known left–merge and right–merge operators $\rfloor\!\rfloor$ and $\lfloor\!\lfloor$ [4]. The second two correspond to left and right versions of the synchronisation merge, $\rfloor$ and $\lfloor$, with the following defining rules

$$\frac{P \xrightarrow{\widetilde{a^x}} P' \qquad Q \xrightarrow{a_x} Q'}{P \rfloor Q \xrightarrow{\tau} P' \,||\, Q'} \qquad\qquad \frac{P \xrightarrow{\widetilde{a^x}} P' \qquad Q \xrightarrow{a_x} Q'}{Q \lfloor P \xrightarrow{\tau} Q' \,||\, P'}$$

THEOREM 11. Parallel composition decomposes thus:

$$P \,||\, Q = P \rfloor\!\rfloor Q + P \lfloor\!\lfloor Q + P \rfloor Q + P \lfloor Q$$

Recording Possible Behaviour

The expansion theorem gives us the key to extending the denotational semantics $\mathcal{D}$ for *DP* to one for *IP*; we have to record both the actual behaviour of a process (what actions it is capable of displaying by itself) and the possible behaviour in a bigger environment (what it might do if suitably triggered). For $p \in GP$ we define the meaning function using elements of *Act* straightforwardly:

$$\begin{aligned} \mathcal{M}(NIL) &= 0 & & \\ \mathcal{M}(a.p) &= a \lhd \mathcal{M}(p) & \mathcal{M}(\widetilde{a^x}.p) &= \widetilde{a^x} \lhd \mathcal{M}(p) \\ \mathcal{M}(p+q) &= \mathcal{M}(p) \oplus \mathcal{M}(q) & \mathcal{M}(\Phi(p)) &= \Phi(\mathcal{M}(p)) \end{aligned}$$

We then extend $\mathcal{D}$ to *IP*:

$$\begin{aligned} \mathcal{D}(p_x, U) &= \mathcal{M}(P) && \text{if } x \in U \\ &= \mathcal{M}(P) && \text{otherwise} \\ \mathcal{D}(P\backslash C, U) &= \pi_C(\mathcal{D}(P, U)) && \end{aligned}$$

where the operator π_C which prunes subtrees attached by edges with labels in C. This just leaves the case of $P \,||\, Q$ to define. Suppose the active and passive parts (i.e. the subtrees with initial edges labeled in A and in $\tilde{A} \times X$ respectively) are

$$\begin{aligned} \mathcal{D}_A(P,U) &= \textstyle\sum_i a_i.P_i & \mathcal{D}_A(Q,U) &= \textstyle\sum_j b_j.Q_j \\ \mathcal{D}_P(P,U) &= \textstyle\sum_i \widetilde{a^x}_i.P'_i & \mathcal{D}_P(Q,U) &= \textstyle\sum_j \widetilde{b^x}_j.Q'_j \end{aligned}$$

The expansion theorem gives us the clue as to the right way to combine these contributions to obtain $\mathcal{D}(P \,||\, Q,\, U)$; we have two terms representing the left and right merges, and two representing the two synchronisation merges, $\rfloor$ and $\lfloor$.

$$\begin{aligned} \mathcal{D}(P \,||\, Q, U) = & \sum_i a_i \lhd (\mathcal{D}(P_i \,||\, Q, U)) \oplus \sum_j b_j \lhd (\mathcal{D}(P \,||\, Q_j, U)) \\ & \oplus \tau \lhd \sum_{\widetilde{a^x}_i = b_j} \mathcal{D}(P'_i \,||\, Q_j, U) \oplus \tau \lhd \sum_{a_i = \widetilde{b^x}_j} \mathcal{D}(P_i \,||\, Q'_j, U) \end{aligned}$$

Thus the contribution at x with respect to an open U of $P \,||\, Q$ is the interleaving of the behaviour of the components together with synchronisations triggered by action at some other $y \in U$.

DEFINITION 12. We now extend our equivalences to $I\!P$: if there exists a located bisimulation between the meanings of two processes $P, Q \in I\!P$, then we write $P \approx Q$, while if there is a strong bisimulation between $Un(P)$ and $Un(Q)$ then we write $P \simeq Q$. Furthermore, we write $P \sim_{\mathcal{O}X} Q$ if for all $U \in X$, $\mathcal{D}(P, U) \simeq \mathcal{D}(Q, U)$. Results 7, 8 and 9 then extend naturally:

THEOREM 13. For any topology $\mathcal{O}X$, $P \approx Q \implies P \sim_{\mathcal{O}X} Q \implies P \simeq Q$. Moreover, $\sim_{\mathcal{O}X}$ coincides with $\approx$ if $\mathcal{O}X$ is a T_1 topology, and with $\simeq$ if it is indiscrete.

Concluding Remarks

We have presented a process algebra with an explicit notion of location. Two operational semantics,—an interleaving and a 'true concurrency' version,—have been presented, and a denotational semantics, parameterised by a notion of when two locations should be regarded as the same, has been shown to vary between these two operational extremes. The fully distributed setting corresponds to T_1 topologies, and the wholly interleaved one to indiscrete topologies.

Further Work

A notion of communication between distributed processes has been presented based on the idea of awaiting a communication. This notion allows us to discover an expansion theorem for parallel composition and hence to extend full abstraction to interacting processes. There is clearly, though, much to do before we have a comprehensive theory of located concurrency; in particular, it would be useful to investigate the obvious weak notions of equivalence that correspond to the strong bisimulations presented here.

It would also be interesting to investigate a theory of placing abstractly located processes such as $I\!P$-terms onto a network of processors with a fixed geometry. Such work should lead to a calculus of placing (with notions of 'most sequential placing', 'most distributed placing' and so on relative to a given topology) that would facilitate the layout of distributed algorithms on a wide variety of architectures.

It has been suggested that using a topology to specify an observer's concerns is over-complex, given that in most situations the set of locations is finite. There is some truth in this, but we justify ourselves on two grounds; firstly there is a long tradition of using topological ideas in theoretical computer science, and another connection is always welcome: secondly it will allow us, in a sequel, to discuss maps between located systems using continuous maps between spaces in a smooth fashion.

Related Work

The major process algebraic work on located concurrency theory is due to Boudol et al. [5]. Their calculus is similar to ours, although they do not enforce the absence of distributed choice. Our $\approx$ is the strong version of their 'location equivalence', and our notion of topology is motivated by similar concerns to their 'nice relations'. However, where they concentrate on axiomatisation and treat the subtleties of weak bisimulation, we deal with full abstraction and uncover a clean framework to describe observer's concerns. The most fundamental difference between our work and theirs, though, is that we assign locations statically to processes, while they do so dynamically. A variant of [5] with static locations has been presented by Aceto [1] and refined by Castellani [this volume] (our *DP* are Aceto's 'nets of automata'), but the differences are still considerable.

Various other calculi with more or less explicit notions of location have been presented in the process algebraic literature. The first was probably Castellani's distributed bisimulation [6] for CCS, which distinguishes processes on the basis of the local and global residues of an action. Aceto has recently shown this distributed bisimulation coincides with two other well-known equivalences for a simple calculus [2]. Also working with CCS, Kiehn has distinguished the local and global causes of an action [7], obtaining an equivalence coinciding with the location equivalence of Boudol et al. [5].

A rather different approach is taken by Montanari and Yankelevich [9], who lift the algebraic structure of terms to transitions, capturing spatial information via a kind of proven transition system. Various notions of observation of such systems then allow equivalences with differing spatial sensitivities to be formulated.

Acknowledgments
Conversations with Ilaria Castellani, Astrid Kiehn, Axel Poigné and especially Robin Knight have been most helpful while working on this material, as was a visit to HP Labs courtesy of Jeremy Gunawardena. Thanks, too, to the referees.

Bibliography

1. L. Aceto, *A static theory of localities*, Technical Report Number 1483, Inria, 1991.
2. L. Aceto, *Relating distributed, temporal and causal observations of simple processes*, manuscript, HP Science Center Pisa, submitted for publication, 1992.
3. A. Arkhangel'skii and V. Ponomarev, *Fundamentals of general topology*, D. Reidel, 1984.
4. J. Baeten and W. Weijland, *Process algebra*, Cambridge Tracts in Theoretical Computer Science, Volume 18, Cambridge University Press, 1990.
5. G. Boudol, I. Castellani, M. Hennessy, and A. Kiehn, *A theory of processes with locality*, Technical Report 13/91, Department of Computer Science, University of Sussex, 1991.
6. I. Castellani and M. Hennessy, *Distributed bisimulations*, Journal of the ACM, Volume 10 (1989), Pp. 887–911.
7. A. Kiehn, *Local and global causes*, Technical Report 342/23/91 A, Institut für Informatik, Technische Universität München, 1991.
8. R. Milner, *Communication and concurrency*, International series on computer science, Prentice Hall International, 1989.
9. U. Montanari and D. Yankelevich, *A Parametric Approach to Localities*, in the Proceedings of ICALP '92, (W. Kuich, Ed.), Springer-Verlag LNCS, Volume 623.
10. D. Murphy, *The physics of observation; a perspective for concurrency theorists*, Bulletin of the EATCS, Volume 44 (1991), Pp. 192–200.
11. ———, *Intervals and actions in a timed process algebra*, Arbeitspapiere der GMD 680, Gesellschaft für Mathematik und Dataverarbeitung, St. Augustin, 1992. Presented at MFPS '92 and submitted to Theoretical Computer Science.
12. D. Park, *Concurrency and automata on infinite sequences*, in Proceedings of Theoretical Computer Science 1981, Volume 104, Springer-Verlag LNCS, 1981.
13. G. Winskel, *Synchronization trees*, Theoretical Computer Science, Volume 34 (1985), Pp. 34–84.

The Boundary of Substitution Systems

Philippe Narbel
L.I.T.P, Institut Blaise Pascal, Paris 7
55-56, 1st fl., 2, Place Jussieu
75251 Paris
e-mail: narbel@litp.ibp.fr

Abstract. *The global limit set has been introduced in a preceding work as a generalization of the way of generating infinite words by substitution systems, i.e. by iterating a morphism on a finite alphabet. We prove here that the boundary set (the "adherence set") of a progressive substitution language is equal to its global limit set plus a simple set of words. This allows us to exhibit conditions to conclude that the full boundary is explicitly constructible, rationally codable and uncountable. The equivalence problem for boundaries is also shown decidable for iterated primitive morphisms.*

1 Introduction

Asymptotic sets of languages have become a familiar concept in formal language theory. They have appeared in at least two different main forms: first, as *limits* when rationality has been extended to infinite words; second, as *boundaries*[1] [BN80, Hea84b], used mostly to have a finer focus on the asymptotic structure of the words of a language. Originally, these asymptotic words have only one way to infinity but extensions to include both ways were done for limits in [NP82, Bea86] and for boundaries in [GN91, DL91]. One purpose of these asymptotic sets has been to reveal the characteristics of the devices which generate their corresponding language, as for instance automata and grammars. Another kind of process which leads to the use of asymptotic words consists of the infinite iteration of a morphism on a finite alphabet. This was initiated by Lindenmeyer for the so-called L-systems [RS80]. One of the most studied families has been the *D0L systems*, i.e. the deterministic context-free systems, which can be called also *substitutions systems*. Their one-way limits and boundaries were described respectively in [Sal81, Que87] and in [CIS82, Hea84a].

However, it happens that iteration of morphisms has almost never been used to generate two-way infinite asymptotic words but in a restricted way [Got63, CK71]. Also, the usual way for L-systems to generate asymptotic words seems to be too restrictive to be extended to handle n-dimensional substitution systems or geometrical substitution systems as *inflations* of tilings [GS87]. This called for some generalization which was initiated in [DB89] and formally attempted in [Nar93], where the *global limit set* was introduced. This set was shown to contain one-way and two-way infinite asymptotic words. Its definition was based on the topological view point as in [BN80, Hea84a]: a language is seen as a subset of a metric space whose distance involves comparison of factors, i.e. the

[1] In the literature, they have been called *adherence* sets, though the topological concept of *boundary* has been actually considered.

Cantor distance. Within this view, the boundary of a language is just the usual mathematical concept of boundary, and the global limit set one of its subset.

The aim of this paper is to describe the full boundary set of a language. The main result will be that, when the language is generated by iterations of a progressive morphism, the difference between its boundary and its global limit set is a simple set of words. This is a generalization of some of the results obtained in [CIS82, Hea84a] about the relationship between the limit and the boundary sets of D0L-systems. A consequence is that several characteristics of the boundary can be described through the results obtained for the limit set [Nar93]. We can conclude, for example, that whenever the morphism is primitive and aperiodic, the boundary can be uncountable and nevertheless rationally coded. This gives an instance of a link between a topological concept and rationality. Also, the boundary equivalence problem can be shown to be decidable in the primitive case using the decidability of the limit equivalence problem of $D0L$ systems [CIH84].

2 Definitions

Let Σ be a finite alphabet of symbols, and Σ^* be the set of all finite words plus the empty word 1, that is the free monoid generated by Σ using the concatenation product. The set Σ^+ is the semi-group $\Sigma^* \setminus \{1\}$. A right (respect. left, bi) infinite word is intuitively a word that has symbols to infinity to the right (respect. to the left, to both) direction. More formally, let $\mathbb{Z}$ denote the integers, $\mathbb{Z}^+$, $\mathbb{Z}^-$, the positive and negative integers with the zero. A *right infinite word* (respect. *left infinite word*) may be seen as a map $w : \mathbb{Z}^+ \to \Sigma$ (respect. $w : \mathbb{Z}^- \to \Sigma$). The bi-infinite words are more difficult to define since they do not have any default distinct point. So, consider an equivalence relation based on the *shift* function $\sigma(\widehat{w}(n))_{n\in\mathbb{Z}} = \widehat{w}(n+1)_{n\in\mathbb{Z}}$ where, if $\widehat{w}, \widehat{v} : \mathbb{Z} \to \Sigma$, then $\widehat{v} \sim \widehat{w}$ iff $\exists$ m such that $\widehat{v} \sim \sigma^m(\widehat{w})$. A *bi-infinite word* consists of a full $\sim_\sigma$-class in the set of all maps from $\mathbb{Z}$ to Σ. The set Σ^ω denotes the set of all the right-infinite words, the set ${}^\omega\Sigma$ of all the left-infinite, the set ${}^\omega\Sigma^\omega$ of all the bi-infinite ones and the infinitary set ${}^\infty\Sigma^\infty$ of all finite and right, left, bi-infinite ones. Any subset of ${}^\infty\Sigma^\infty$ is called a *language*.

However, the difficulty of handling bi-infinite words leads to the preferable use of *pointed words*, i.e. words with a fixed indexing. So, a *pointed bi-infinite word* is just a map $\widehat{w} : \mathbb{Z} \to \Sigma$. A *right* (respect. *left*) *infinite pointed words* are maps $\widehat{w} : \mathbb{Z}^+ \cup \{-m, .., -1\} \to \Sigma$ (respect. $\widehat{w} : \mathbb{Z}^- \cup \{1, .., m\} \to \Sigma$). In general, a hat will be added over the symbols representing pointed words or pointed languages. The image of $\widehat{w}(n)$ is abbreviated w_n and the symbol w_0 is called *the origin*. In the pointed infinitary set ${}^\infty\widehat{\Sigma}^\infty$, the *finite words* are maps on finite intervals of $\mathbb{Z}$ including the zero (the origin). We stress that in ${}^\infty\widehat{\Sigma}^\infty$ the finite pointed word $\widehat{w} = aa\mathbf{a}bbabbb$, for instance, is not equal to $\widehat{v} = aaabb\mathbf{a}bbb$, where the bold characters indicate the origins. The *length* $|\widehat{w}|$ of a word $\widehat{w}$ is equal to the cardinality of its source set. If $\widehat{w} : I \subseteq \mathbb{Z} \to \Sigma$ is a pointed word, and if $\widehat{v} : I' \subseteq I \to \Sigma$ where $I' = [k, l]$ is an interval in I containing the zero, then $\widehat{v}$ is a *pointed factor* of $\widehat{w}$ denoted by $\widehat{v} = [w_k, w_l] = w_k...w_l$ and $\widehat{v} \subseteq \widehat{w}$. Sometimes, the mixed notation $\widehat{v} \subset w$ is used, whenever w can be pointed so that the relation holds. The set of all pointed factors of a pointed word $\widehat{w}$ (respect. of a pointed language $\widehat{L}$) is denoted by $F(\widehat{w})$ (respect. by $F(\widehat{L})$).

Let us now define the metric structure of the infinitary pointed set ${}^{\infty}\widehat{\Sigma}^{\infty}$. In order to homogenize the infinitary set ${}^{\infty}\widehat{\Sigma}^{\infty}$, it can be assumed that all its words which are not yet bi-infinite have been padded to both infinities with some dummy symbol not already in Σ. Then, if $\widehat{u}, \widehat{v} \in {}^{\infty}\widehat{\Sigma}^{\infty}$ and an integer $p > 1$, it can be checked that,

$$d_p(\widehat{u},\widehat{v}) = \begin{cases} 0 & \text{iff } \widehat{u} = \widehat{v} \\ p^{-|\widehat{u},\widehat{v}|} & \text{otherwise} \end{cases} \quad \text{where} \quad |\widehat{u},\widehat{v}| = \begin{cases} 0 & \text{iff } u_0 \neq v_0 \\ 1 + max\{k \in \mathbb{N} \mid [u_{-k}, u_k] = [v_{-k}, v_k]\} \end{cases} \tag{1}$$

is a well-defined metric. The considered metric space is then the pair $({}^{\infty}\widehat{\Sigma}^{\infty}, d_p)$. Since the range of d_p is the discrete set $\{p^{-n}, n \in N\}$, this space can be seen *totally disconnected*. Using the fact that Σ is finite, it can be proved that it is *compact* and *perfect* [MH38]. This means that it is homeomorphic to the so-called middle-third Cantor set. Note that in this kind of space, the open sets are closed as well. Finally, it is *totally bounded*.

If $\widehat{L}$ is a pointed language included in $({}^{\infty}\widehat{\Sigma}^{\infty}, d_p)$, then $(\widehat{L})^-$ denotes its *closure* (or its *adherence*) and $\widehat{L}_c$ its *complement*. The *boundary* of $\widehat{L}$ is defined by

$$\partial\widehat{L} = (\widehat{L})^- \cap (\widehat{L}_c)^-.$$

A *Cauchy sequence* in $({}^{\infty}\widehat{\Sigma}^{\infty}, d_p)$ is a sequence (α_n) of words in ${}^{\infty}\widehat{\Sigma}^{\infty}$ (pointed words in sequences will be henceforth denoted by Greek letters) such that:

$$\forall\epsilon > 0 \,, \; \exists k \; \text{ such that } \; d_p(\alpha_n, \alpha_m) < \epsilon, \quad \forall n, m > k. \tag{2}$$

Let $\mathbf{CS}(\widehat{L})$ denote the set of all the Cauchy sequences formed by elements of $\widehat{L}$ and $\Im(\widehat{L})$ the set of all distinct limits of sequences in $\mathbf{CS}(\widehat{L})$, i.e.

$$\Im(\widehat{L}) = \{\widehat{w} \in (\widehat{L})^- \mid \exists(\alpha_n) \in \mathbf{CS}(\widehat{L}), \widehat{w} = lim_{n\to\infty}\alpha_n\}.$$

Here, we shall restrict our attention to languages $\widehat{L} \subset \widehat{\Sigma}^*$, with an infinite cardinality, i.e. a countable language of finite words. In this case, the following holds:

$$\partial\widehat{L} = \Im(\widehat{L}) \setminus \widehat{L},$$

which means also that $\partial\widehat{L}$ contains only infinite words, and that it is exactly the set of *completion points* of $\widehat{L}$. Note also that a completion of a totally bounded space is in fact a *compactification* [Eng89]. Finally, by compactness of $({}^{\infty}\widehat{\Sigma}^{\infty}, d_p)$, it can be figured out that $\partial\widehat{L} \neq \emptyset$.

Let us see what is the appearance of a Cauchy sequence (α_n) in $\mathbf{CS}(\widehat{L})$ whose limit is not already in $\widehat{L}$. Since it must converge to some infinite word, it can be assumed monotonic, i.e. $d_p(\alpha_{n+1}, \alpha_n) < d_p(\alpha_n, \alpha_{n-1})$ for all $n > 0$. Also, the very first element, α_0, can be assumed to be a single symbol in Σ (this allows us to uniquely define the origin for each limit word). Such a sequence can be then described by identifying and stacking longer and longer factors of the pointed words of (α_n), i.e. each α_n contains an $\alpha'_n \in F(\widehat{L})$ which gives a sequence (α'_n) with the property that $|\alpha'_{n+1}| > |\alpha'_n|$ for all n, and such that,

$$\begin{aligned} &\alpha_0 = \alpha'_0 \in \widehat{\Sigma}, \\ &\alpha_{n+1} = \delta_n \gamma_n \alpha'_n \beta_n \eta_n, \quad \text{where} \quad \alpha'_{n+1} = \gamma_n \alpha'_n \beta_n, \quad \text{and} \quad \beta_n, \gamma_n, \delta_n, \eta_n \in \Sigma^*. \end{aligned} \tag{3}$$

Such a sequence (α_n) is therefore associated to five sequences of factors indexed by their embeddings in (α_n). Note that the sequences (γ_n) and (β_n) are the ones which are identified, whereas (δ_n) and (η_n) do not contribute to reduce the distance. For instance, consider $\widehat{L}={}^*(a+b)\mathbf{a}(a+b)^*$, where the operator '+' is equivalent to an 'or', and the bold a is the fixed origin. Then,

$$\begin{array}{c} \vdots \\ bbbbbaaa\mathbf{a}bbaa \\ baa\mathbf{a}bbbb \\ a\mathbf{a}ba \\ \mathbf{a} \end{array}$$

is the beginning of a well-formed Cauchy sequence, where $\alpha'_0 = \mathbf{a}$, $\alpha'_1 = a\mathbf{a}b$, $\alpha'_2 = aa\mathbf{a}bb$,...

The boundary consists only of infinite words, but in view of the Cauchy sequences appearance, it can be seen that the limit words are of three different types:

left-infinite words in ${}^\omega\widehat{\Sigma}$ if $\exists\, n_0 < \infty$ such that $\beta_n \eta_n = 1$, $\forall n > n_0$,
right-infinite words in $\widehat{\Sigma}^\omega$ if $\exists\, n_0 < \infty$ such that $\delta_n \gamma_n = 1$, $\forall n > n_0$,
bi-infinite words in ${}^\omega\widehat{\Sigma}^\omega$ otherwise.

Hence, $\partial\widehat{L}$ decomposes itself in three sets respectively denoted by $L\partial\widehat{L}$, $R\partial\widehat{L}$ and $Bi\partial\widehat{L}$.

One special subset of $\partial\widehat{L}$ is the *global limit set* $Lim\widehat{L}$ of a language [Nar93]. Its inclusion into the boundary set can be shown by looking at its Cauchy sequences. These have the special property that their words must be entirely identified, i.e. there is no factor sequences which do not contribute to reduce the distance. This property can be explicited by a simplified sequence definition compared to (3): let $\alpha_n \in \widehat{L}$ for all n, with the property that $|\alpha_{n+1}| > |\alpha_n|$, then:

$$\begin{aligned} &\alpha_0 \in \widehat{\Sigma}, \\ &\alpha_{n+1} = \gamma_n \alpha_n \beta_n \text{ where } \beta_n, \gamma_n \in \Sigma^*, \end{aligned} \tag{4}$$

Again, the limit can be also decomposed in three subsets $LLim\widehat{L}$, $RLim\widehat{L}$ and $BiLim\widehat{L}$.

Whevener a non-pointed language $L \subset \Sigma^*$ is considered, its *pointed counterpart* $\widehat{L}$ consists of every different pointings of the words in L. For this kind of pointed languages, the following property holds:

$$\text{if } \widehat{w} \in \partial\widehat{L} \text{ then } \sigma^n(\widehat{w}) \in \partial\widehat{L}, \ \ \forall n \in \mathbb{Z}. \tag{5}$$

Also, the non-pointed boundary set ∂L can be defined as $(\partial\widehat{L}/\sim_\sigma)$. Note however that this is only a practical definition since the quotient space can be metrically and topologically non-separated [Con90].

In this paper, the languages on focus are the ones generated by a substitution system on a finite alphabet Σ. A map $\theta : \Sigma \to \Sigma^+$ is said to be a *morphism* when it is extended to ${}^\infty\Sigma^\infty$ by the following rule:

$$\theta(w) = \theta(\ldots s_i s_{i+1} s_{i+2} \ldots) = \ldots\theta(s_i)\theta(s_{i+1})\theta(s_{i+2})\ldots, \quad s_i \in \Sigma.$$

The nth iteration of the morphism is just the nth power of its composition. A *substitution system* is a pair (Σ, θ) whose language is given by

$$L_\theta = \{w \in \Sigma^+ \mid \quad \theta^n(s_j) = w, \;\; n \in \mathbb{Z}^+, \; s_j \in \Sigma\}. \tag{6}$$

The *growth* maps of L_θ are given by $g_j(n) = |\theta^n(s_j)|$ with $s_j \in \Sigma$. They can be explicitly obtained [RS80]. We also define the two related *extremal growth* maps $b_{min}(n) = \min_j\{g_j(n)\}$ and $b_{max}(n) = \max_j\{g_j(n)\}$. If $lim_{n\to\infty}|g_j(n)| = \infty$ for all j, then the morphism θ is said to be *progressive*. Note that being progressive ensures that $\widehat{L}_\theta$ is an infinite language included in $\widehat{\Sigma}^*$, and therefore that $\partial\widehat{L}_\theta \neq \emptyset$ holds. Note also that $\partial\widehat{L}_\theta \neq Lim\widehat{L}_\theta$ may happen. For instance, consider the following morphism: $\theta(a) = aa$, $\theta(b) = ab$, $\theta(c) = abab$. It can be checked that the word ${}^\omega \mathbf{aba}^\omega$ belongs to $Bi\partial\widehat{L}_\theta$ but not to $Bilim\widehat{L}_\theta$.

Recall that a D0L *system* [RS80] is a 3-tuple (Σ, θ, t) which generates a substitution language whose iterations start from a single word t in Σ^+, that is:

$$D0L_\theta(t) = \{w \in \Sigma^* \mid \quad \theta^n(t) = w, \;\; n \geq 0\}.$$

The justification of using the substitution language $\widehat{L}_\theta$ as defined in (6) is given by the fact, proved later on, that for any $t \in \Sigma^+$, the boundary set $\partial D0\widehat{L}_\theta(t)$ is a subset of $\partial\widehat{L}_\theta$ plus a set of words easily computed, whenever θ is progressive.

3 Relationships between the Boundary and the Limit Sets

This section presents a proof of the fact that the boundary set $\partial\widehat{L}_\theta$ is equal to an extension of $Lim\widehat{L}_\theta$, which is obtained by adding to it a set of bi-infinite words constructed by pasting some limit words of the non-pointed limit sets $LlimL_\theta$ and $RlimL_\theta$. This close relationship between the limit and the boundary sets will be proved by using their natural decomposition: first, the boundaries $L\partial\widehat{L}_\theta$ and $R\partial\widehat{L}_\theta$ will be shown equal to their counterpart limit sets $Llim\widehat{L}_\theta$ and $Rlim\widehat{L}_\theta$. This is simply derived from the results of [CIS82, Hea84a]. As a second step, the relationship between $Bi\partial\widehat{L}_\theta$ and $Bilim\widehat{L}_\theta$ will be investigated.

The complete recursive definition (4) of a Cauchy sequence is too general when one has to deal *apriori* with one-way infinite words only. If the sequence (α_n) is known to converge to a one-way infinite word, it can be easily redefined to be of the form (if the word is right-infinite):

$$\begin{aligned} &\alpha_0 \in \widehat{\Sigma}, \\ &\alpha_{n+1} = \alpha_n\beta_n \text{ where } \;\; \beta_n \in \Sigma^+. \end{aligned} \tag{7}$$

Thus, because of equation (5), the equivalence between $R\partial\widehat{L}_\theta$ and $Rlim\widehat{L}_\theta$ (respect. $L\partial\widehat{L}_\theta$ and $Llim\widehat{L}_\theta$) can be obtained by considering only words whose origins are at their finite extremities. For the one-way limit sets, this kind of words is canonical in their $\sim_\sigma$-classes and corresponds to the intuitive notion of an one-way infinite word. Now, a symbol a is said to be *right recurrent* (respect. *left recurrent*) if there exists some integer n such that $\theta^n(a) = av$ (respect. $\theta^n(a) = va$). The *order* of a recurrent symbol a is the least integer n giving the last equality. It is not hard to verify that,

Lemma 3.1 *Let θ be progressive. Then, the word $\widehat{w} = aw'$ belongs to $R\partial\widehat{L}_\theta$, with $a \in \Sigma$ (respect. $\widehat{u} = u'a \in L\partial\widehat{L}_\theta$), iff the symbol a is right-recurrent (respect. left-recurrent).*

The proof relies on the fact that Σ is finite; this means that there is a subsequence (ζ_n) of (α_n) such that $\zeta_n = \theta^{k_n}(s_0)$, for all n, for a fixed symbol s_0 in Σ, and for a growing sequence (k_n) of positive integers.

The last remarks lead to a direct translation into the original framework of the $D0L$-systems [CIS82, Hea84a]. Indeed, the definition (7) and the right/left recurrence notion are in accordance with the usual way to generate the unique one-way infinite words with $D0L$ systems. It can be proved then,

Proposition 3.1 *Let θ be progressive. Then,*

$$R\partial\widehat{L}_\theta = Rlim\widehat{L}_\theta \text{ and } L\partial\widehat{L}_\theta = Llim\widehat{L}_\theta.$$

Corollary 3.1 *Let θ be progressive. Then, $|R\partial L_\theta| = |\{a \text{ is right-recurrent}\}|$ and $|L\partial L_\theta| = |\{a \text{ is left-recurrent}\}|$.*

Note that the non-pointed limit set $LimL_\theta$ has been shown to be sometimes an uncountable set in [Nar93]. The corollary means then that the boundary set $Bi\partial\widehat{L}_\theta$ can be uncountable. This stresses the difference with the usual $D0L$ case.

Let us now investigate the relationship between $Bi\partial\widehat{L}_\theta$ and $Bilim\widehat{L}_\theta$. The method of proof is simply to take a word in the boundary with its corresponding converging sequence and to exhibit a sequence of growing subwords belonging to the language $\widehat{L}_\theta$. Since direct application of the morphism θ cannot handle the origin, the generation of these pointed subwords relies on the inverse correspondence θ^{-1}. It is not generally a map, but it happens to be sufficient to produce words in $\widehat{L}_\theta$: let $\widehat{w} \in \widehat{L}_\theta$ and the morphism image set be denoted by $\{\theta(\Sigma)\}$, there is then *at least one way* to write $\widehat{w} = ...x_i x_{i+1}...$, with the x_j's in $\{\theta(\Sigma)\}$ and where x_0 contains the origin:

$$\theta^{-1}(\widehat{w}) = ...s_i s_{i+1}... \in \widehat{L}_\theta \quad \text{with} \quad s_j \in \Sigma, \text{ and } \theta(s_j) = x_j. \tag{8}$$

It can be also iterated in $\widehat{L}_\theta$ by setting $\theta^{-m}(\widehat{w}) = ...s_i s_{i+1}...$, with $s_j \in \Sigma$, and $\widehat{w} = ...x_{i,m}x_{i+1,m}...$ with $x_{j,m} = \theta^m(s_j)$. If $\widehat{w} \in \widehat{\Sigma}$, we put $\theta^{-m}(\widehat{w}) = \widehat{w}$, $\forall m$.

The *right subfactor* after the origin of some word $\widehat{v}$ is denoted by v^ρ, and its *left subfactor* before the origin by v^λ, so that, $\widehat{v} = v^\lambda v_0 v^\rho$. The next lemma proves that any pointed word in $\widehat{L}_\theta$, which has both right and left factors sufficiently long, contains a pointed word in $\widehat{L}_\theta$ with minimum length:

Lemma 3.2 *Let $\widehat{v} \in F(\widehat{L}_\theta)$, and put*

$$m = \max_{i \in \mathbb{N}}\{|v^\rho| \geq (b_{max}(i) - 1) \quad and \quad |v^\lambda| \geq (b_{max}(i) - 1)\}.$$

Then, there exists $\widehat{u} \subseteq \widehat{v}$ such that $\widehat{u} \in \widehat{L}_\theta$ and $|\widehat{u}| \geq b_{min}(m)$.

Proof. If $\widehat{v} \in F(\widehat{L}_\theta)$, this means that there is at least a word $w \in L_\theta$ with $w \in \theta^n(s_j)$, $s_j \in \Sigma$, pointed such that $\widehat{v} \subseteq \widehat{w}$. According to the definition of the inversion, we get that $\forall p > 0$:

$$\theta^{-p}(\widehat{w}) = \theta^{-p}(...x_{-i,p}...x_{0,p}...x_{i,p}...) = ...\theta^{-p}(x_{-i,p})...\theta^{-p}(x_{0,p})...\theta^{-p}(x_{i,p})...$$

where the $x_{i,p}$'s belong to $\{\theta(\Sigma)\}$ and $x_{0,p}$ contains w_0. But, for all p,

$$0 < b_{min}(p) \leq |x_{i,p}| \leq b_{max}(p).$$

Since the factor $\widehat{v}$ is such that v^λ and v^ρ are longer than $b_{max}(m) - 1$, the factor $x_{0,p}$ must be completely included in the factor $\widehat{v}$ for all $p \leq m$. Hence, the searched word $\widehat{u}$ is $x_{0,m}$. $\diamond$

In the last proof, the word $\widehat{u} \in \widehat{L}_\theta$ is said to be *induced* by its *reference* word $\widehat{w}$. Note that the minimum length is attained whatever is the reference word. We call the longest subword in $\widehat{L}_\theta$ contained in $\widehat{w}$ and adjacent to $\widehat{u}$, the *residue word.*

Now, let the *pair set* $F^2(L_\theta)$ contains all factors of length two in words of L_θ. Then, the *pasted limit set* of L_θ is defined by pasting non-pointed one-way limit words:

$$PasL_\theta = \{w \in {}^\omega\Sigma^\omega | w = vu, \text{ where } v = v'x \in LlimL_\theta,\ u = yu' \in RlimL_\theta,\ xy \in F^2(L_\theta)\}. \tag{9}$$

The element $xy \in F^2(L_\theta)$ of the last definition is called a *pasting pair.* By corollary 3.1, the cardinality of $PasL_\theta$ is at most the finite number of pairs $s_i s_j$ such that s_i is left-recurrent and s_j is right-recurrent. Note that this set is very related to the pointed bi-infinite words obtained in the dynamical-oriented works [Got63, CK71]. The pointing of $PasL_\theta$ is obtained by setting the origin on a symbol which is at a finite shift power from the pasting pair. Since $F^2(L_\theta)$ is finite and computable, the set $Pas\widehat{L}_\theta$ is therefore easily obtained. Note that, if $\widehat{w}$ is in $Pas\widehat{L}_\theta$, then it can be written $\widehat{vu}$, where v and u are as in definition 9. Finally, if a factor $\widehat{v}$ of $\widehat{w}$ is such that $\widehat{w} = u\widehat{v}u'$ with $u, u' \in \Sigma^+$, then we denote this strict factor inclusion by $\widehat{v} \prec \widehat{w}$.

Theorem 3.1 *Let θ be progressive. Then, $\partial\widehat{L}_\theta = Lim\widehat{L}_\theta \cup Pas\widehat{L}_\theta$.*

Proof. Because of proposition 3.1, it is sufficient to consider only the bi-infinite asymptotic sets. Let us first show that $Bilim\widehat{L}_\theta \cup Pas\widehat{L}_\theta \subseteq Bi\partial\widehat{L}_\theta$. By definition, $Bilim\widehat{L}_\theta \subseteq Bi\partial\widehat{L}_\theta$. So, take a word $\widehat{w}$ in $Pas\widehat{L}_\theta$. This means that $\widehat{w} = \widehat{vu}$ where $v \in LlimL_\theta$ and $u \in RlimL_\theta$. Assume that the pasting pair of $\widehat{w}$ is given by $\gamma = ab \in F^2(L_\theta)$, so that $v = v'a$ and $u = bu'$. The claim is that $\widehat{w} = v'abu' \in \partial\widehat{L}_\theta$ (which is sufficient because of relation 5). From lemma 3.1, we know that the symbol a must be left-recurrent for some order p_1 and that b must be right-recurrent for some other order p_2. Define p as the least common multiple of p_1 and p_2, thus for all $n > 0$, there exists m, with $m > n$ such that,

$$\theta^{np}(ab) \prec \theta^{mp}(ab), \quad \forall m > n > 0. \tag{10}$$

Now, since $\gamma \in F^2(L_\theta)$, there is a word $\widehat{x} \in \widehat{L}_\theta$ such that $\gamma \subseteq \widehat{x}$. So, consider the following Cauchy sequence (α_n): let $\alpha_0 = a$ and let α_1 be $\widehat{x}$, pointed on the a of the γ it contains. Recursively, define $\alpha_{n+1} = \theta^{np}(x)$ to be pointed on the a in order that $\theta^{np}(\gamma)$ is identified with $\theta^{(n-1)p}(\gamma)$. This is possible because of equation 10. Since θ is progressive, (α_n) converges to $\widehat{w}$.

The other way, i.e. $Bi\partial\widehat{L}_\theta \subseteq Bilim\widehat{L}_\theta \cup Pas\widehat{L}_\theta$, is shown as follows: consider $\widehat{w}$ to be a word in $Bi\partial\widehat{L}_\theta$ and its corresponding sequence $(\alpha_n) \in \mathbf{CS}(\widehat{L}_\theta)$:

$$\alpha_{n+1} = \delta_n\gamma_n\alpha'_n\beta_n\eta_n, \quad \alpha'_{n+1} = \gamma_n\alpha'_n\beta_n.$$

Since $\widehat{w}$ is bi-infinite, there is a subsequence (ζ_n) of (α'_n) such that for all n, we have for the right and left subfactors of ζ_n:

$$|\zeta_n^\rho| \geq b_{max}(n) - 1 \quad \text{and} \quad |\zeta_n^\lambda| \geq b_{max}(n) - 1. \tag{11}$$

Hence, for each ζ_n, lemma 3.2 applies. This gives another sequence (μ_n) such that

$$\mu_n \subseteq \zeta_n \text{ and } \mu_n \in \widehat{L}_\theta, \quad \forall n.$$

Thus, the limit of (μ_n) belongs to $Lim\widehat{L}_\theta$. There are now two cases: First, if $|\mu_n^\rho|$ and $|\mu_n^\lambda|$ tend together to infinity when $(n \to \infty)$, then $\lim_{n\to\infty}\mu_n = \lim_{n\to\infty}\alpha_n$. Second, only one of $|\mu_n^\rho|$ or $|\mu_n^\lambda|$ tends to infinity. Notice that because of the multiple choice of the inducement of the word of the language by θ^{-1} in the lemma 3.2, the length $|\mu_n^\rho|$ or $|\mu_n^\lambda|$ can even diverge, because the property that $\mu_n \subseteq \mu_{n+1}, \forall n$ is not ensured. So, take a subsequence (μ'_n) of (μ_n) with $\mu'_n \subseteq \mu'_{n+1}$ forall n, so that one of the subfactors tends to infinity. Assume that it is the right subfactors. Then the sequence (μ'_n) is a well-defined sequence of $Rlim\widehat{L}_\theta$. Its left most symbol is therefore right-recurrent. Now, consider the sequence (ν_n) to be the residue words of (μ'_n), i.e. the words adjacent to the μ'_n. Since only the right side of $\widehat{w}$ is covered by (μ'_n), the length function $|(\nu_n)|$ tends to infinity. Hence, (ν_n) is a well-defined sequence of $Llim\widehat{L}_\theta$ which is just adjacent to (μ'_n). Again, its right most symbol must be left-recurrent. Therefore, the second case corresponds to a word belonging to $Pas\widehat{L}_\theta$ and the proof is completed. $\diamondsuit$

Furthermore, consider the non-pointed boundary language to be $(\partial\widehat{L}_\theta / \sim_\sigma)$:

Corollary 3.2 *Let θ be progressive. Then, $\partial L_\theta = LimL_\theta \cup PasL_\theta$.*

Hence, from the definition of the pasted set, only a finite set of words must be added to the limit set in order to give the full non-pointed boundary.

Before going further, let us present an example where the second case of the proof of theorem 3.1 occurs. Take the morphism $\theta(a) = aa$, $\theta(b) = ab$,

$\theta(c) = abab$ and consider the starting elements of a Cauchy sequence in $\mathbf{CS}(\widehat{L}_\theta)$ which converge to ${}^\omega abaaaaa^\omega$:

$$\vdots$$
$$aaaabaaaaaaa$$
$$aabaaaaa$$
$$aaaaa$$
$$\mathbf{a}$$

By applying the construction proposed in the proof of the theorem, the induced words (μ'_n) are invariably stopped by the symbol b and converge to $aaaaa^\omega$. Hence, the residue sequence (ν_n) gives ${}^\omega ab$. It is then easy to check that the symbol a is right-recurrent and the symbol b is left-recurrent. Since for any n and m the words of type ${}^m aba^n$ do not belong to L_θ, the word ${}^\omega abaaaaa^\omega$ strictly belongs to $Pas\widehat{L}_\theta$.

Let us also show the result about $D0L$-systems announced at the end of the introduction. Let the subalphabet Σ_t be the set of symbols which appear through the iteration of θ on the word $t \in \Sigma^+$, i.e. $\Sigma_t = \{s_i \in \Sigma \mid \exists\, n \geq 0,\ s_i \subset \theta^n(t)\}$. Then, let the morphism ν be the restriction of θ to Σ_t. This means that for $t \in \Sigma^+$, $D0\widehat{L}_\nu(t) = D0\widehat{L}_\theta(t)$ and that $\partial\widehat{L}_\nu \subset \partial\widehat{L}_\theta$. The set $PasD0\widehat{L}_\nu(t)$ contains the pasting pairs included in the language $D0\widehat{L}_\theta(t)$. So, the only difference with theorem 3.1 comes from the fact that, t being any word in Σ^+, it may imply that $Pas\widehat{L}_\nu$ is strictly included in $PasD0\widehat{L}_\nu(t)$. We can then conclude by the same arguments as in the proof of the theorem that,

Corollary 3.3 *Let θ be progressive. Then, $\partial D0\widehat{L}_\nu(t) = Lim\widehat{L}_\nu \cup PasD0\widehat{L}_\nu(t)$, for all $t \in \Sigma^+$.*

A coding of $Lim\widehat{L}_\theta$ by a rational language of one-way infinite words was described in [Nar93] whenever the morphism is *recognizable*. The recognizability property means that every word in $Lim\widehat{L}_\theta$ has a unique and locally determined factorization in $\{\theta(\Sigma)\}$, that is the inversion θ^{-1} is uniquely determined locally (see equation (8)). Moreover, it can be proved from [Mos90] that recognizability holds for a morphism whenever it is *primitive* and *aperiodic*, i.e. respectively, it exists an index $n < \infty$ such that all symbols of Σ are included in $\theta^n(s_i)$, for all $s_i \in \Sigma$; and, there is no word in $Lim\widehat{L}_\theta$ such that there exists an index p with $w_i = w_{i+p}$ for all i.

Therefore, in view of theorem 3.1, the remaining problem is to explicit what is in $Bilim\widehat{L}_\theta \cap Pas\widehat{L}_\theta$. The next proposition presents a sufficient condition to decide it (recall that $\prec$ denotes the strict factor inclusion):

Proposition 3.2 *Let θ be progressive. Let $\widehat{w}$ be a word in $Pas\widehat{L}_\theta$ such that its pasting pair is $\gamma \in F^2(L_\theta)$. Then $\widehat{w}$ belongs also to $Bilim\widehat{L}_\theta$ if and only if it exists $\widehat{u} \in \widehat{L}_\theta$ such that $\gamma \prec \widehat{u} \prec \theta^m(\gamma)$, with $m < \infty$.*

For instance, this allows us to conclude that the so-called Thue-Morse morphism $\theta(a) = ab$, $\theta(b) = ba$ is such that $Pas\widehat{L}_\theta \subset Bilim\widehat{L}_\theta$. In fact, from the last proposition, it can be then checked that,

Proposition 3.3 *Let θ be primitive. Then $Pas\widehat{L}_\theta \subset Bilim\widehat{L}_\theta$.*

Therefore, since proposition 3.3 can be interpreted as $\partial\widehat{L}_\theta = Lim\widehat{L}_\theta$ whenever the morphism is primitive, it can be concluded that,

Theorem 3.2 *Let θ be primitive and aperiodic. Then, the boundary set $\partial\widehat{L}_\theta$ can be coded by a rational language of one-way infinite words.*

Theorems 3.1 and 3.2 can be summed up by saying that compactification of the language $\widehat{L}_\theta$ included in $({}^\infty\widehat{\Sigma}^\infty, d_p)$ can be done by an automatic process whenever the morphism θ is primitive and aperiodic.

Finally, we present a decidability answer about the boundary equivalence problem in the class of the primitive morphisms, i.e. given two morphisms θ_1 and θ_2, one must decide whether the corresponding boundaries $\partial\widehat{L}_{\theta_1}$ and $\partial\widehat{L}_{\theta_2}$ are equal. Note that the usual one-way limits and boundaries of $D0L$ systems are finite sets, a fact which can be recovered by corollary 3.1. It is not the case for the full boundary which can be deduced to be uncountable whenever the morphism is recognizable [Nar93].

Theorem 3.3 *The boundary equivalence is decidable for primitive morphisms.*

Proof. By the result given in proposition 3.3, the limit sets can be directly used. Consider two morphisms θ_1 and θ_2 defined on the same alphabet Σ. From [CIH84, Hea84a] and lemma 3.1, the problem to know whether $Llim\widehat{L}_{\theta_1} = Llim\widehat{L}_{\theta_2}$ and $Rlim\widehat{L}_{\theta_1} = Rlim\widehat{L}_{\theta_2}$ is decidable. This allows us to make a few remarks. First, if a sequence (β_n) belongs to $\mathbf{CS}(L_{\theta_1})$ and has a limit in an one-way infinite limit set, say in $Rlim\widehat{L}_{\theta_1}$, then the result of [CIH84] implies that there is a sequence (β'_n) in $\mathbf{CS}(L_{\theta_2})$ which converges to the same word. Hence, for all indexes n there is an index k such that

$$\beta'_n \subseteq \beta_m, \quad \forall m > k. \tag{12}$$

Let us show then that $Bilim\widehat{L}_{\theta_1} = Bilim\widehat{L}_{\theta_2}$. Take a word $\widehat{w} \in Bilim\widehat{L}_{\theta_1}$ with its corresponding sequence (α_n). Since Σ is finite, we can assume that every $\alpha_n = \theta_1^{h_n}(s_0)$ with a fixed $s_0 \in \Sigma$ and powers $h_n \in \mathbb{Z}^+$. Consider then the sequence $(\beta_n) \in \mathbf{CS}(L_{\theta_1})$ defined such that $\beta_n = \theta_1^{np}(a)$ with a being right-recurrent of order p and which converges to a word in $Rlim\widehat{L}_{\theta_1}$. There is at least one such a since Σ is finite. Then, consider also its counterpart sequence $(\beta'_n) \in \mathbf{CS}(L_{\theta_2})$ which has the same limit. Because θ_1 is primitive and because equation (12) holds, there is an index l_1 and k_1 such that

$$s_0 \prec \beta'_{l_1} \subseteq \beta_{k_1}.$$

More generally, because of the construction of the sequences (β_n) and (β'_n), it can be concluded that for all n, there are indexes l_n and k_n such that

$$\theta_1^{h_n}(s_0) \prec \beta'_{l_n} \subseteq \beta_{k_n}.$$

Hence, define the sequence (ζ_n) in $\mathbf{CS}(L_{\theta_2})$ by identifying the β'_n such that all the iterates of $\theta_1^{h_n}(s_0)$ coincide in the same manner as in (α_n). But, this means that (ζ_n) converges to the same word as (α_n), though (ζ_n) is a sequence defined with words of the language L_{θ_2}. The conclusion is that if $Llim\widehat{L}_{\theta_1} = Llim\widehat{L}_{\theta_2}$, $Rlim\widehat{L}_{\theta_1} = Rlim\widehat{L}_{\theta_2}$ and if θ_1 and θ_2 are primitive on the same alphabet Σ, then $Bilim\widehat{L}_{\theta_1} = Bilim\widehat{L}_{\theta_2}$ holds. The result follows. ◇

References

[Bea86] Beauquier (D.). – *Automates de mots bi-infinis.* – Paris, 1986. Thèse d'Etat.

[BN80] Boasson (L.) and Nivat (N.). – Adherence of languages. *J. Comp. Syst. Sc.*, vol. 20, 1980, pp. 285–309.

[CIH84] Culik II (K.) and Harju (T.). – The ω–sequence equivalence problem for D0L systems is decidable. *Journal of the ACM*, vol. 31, number 2, april 1984, pp. 282–298.

[CIS82] Culik II (K.) and Salomaa (A.). – On infinite words obtained by iterating morphisms. *Theoretical Computer Science*, vol. 19, 1982, pp. 29–38.

[CK71] Coven (E.M.) and Keane (M.S.). – The structure of substitution minimal sets. *Transactions of the American Mathematical Society*, vol. 162, 1971, pp. 89–102.

[Con90] Connes (A.). – *Géométrie non Commutative.* – Interéditions, 1990.

[DB89] De Bruijn (N.G.). – Updown generation of Beatty sequences. *Indag. Math.*, vol. 51, 1989, pp. 385–407.

[DL91] Devolder (J.) and Litovsky (I.). – Finitely generated biω–languages. *Theoretical Computer Science*, vol. 85, 1991, pp. 33–52.

[Eng89] Engelking (R.). – *General Topology.* – Heldermann Verlag Berlin, 1989.

[GN91] Gire (F.) and Nivat (N.). – Langages algébriques de mots bi-infinis. *Theoretical Computer Science*, vol. 86, 1991, pp. 277–323.

[Got63] Gottschalk (W.H.). – Substitution minimal sets. *Transactions of the American Mathematical Society*, vol. 109, 1963, pp. 467–491.

[GS87] Grunbaum (B.) and Shephard (G.C.). – *Tilings and Patterns.* – Freeman and co., 1987.

[Hea84a] Head (T.). – Adherences of D0L languages. *Theoretical Computer Science*, vol. 31, 1984, pp. 139–149.

[Hea84b] Head (T.). – The adherences of languages as topological spaces. *In : Automata on infinite words.* pp. 147–163. – Springer Verlag. Lecture Notes in Comp. Sci. vol. 192.

[MH38] Morse (M.) and Hedlund (G.A.). – Symbolic dynamics. *American Journal of Mathematics*, vol. 60, 1938, pp. 815–866.

[Mos90] Mossé (B.). – *Puissances de mots et reconnaissabilité des points fixes d'une substitution.* – Technical report, PRC Math. Info., Université Aix-Marseille, 1990.

[Nar93] Narbel (P.). – The limit set of recognizable substitution systems. *In : STACS'93.* pp. 226–236. – Springer Verlag. Lecture Notes in Comp. Sci., 657.

[NP82] Nivat (N.) and Perrin (D.). – Ensembles reconnaissables de mots bi-infinis. *In : 14th ACM Symp. on Theory of Computing*, pp. 47–59.

[Que87] Queffelec (M.). – *Substitution Dynamical Systems - Spectral Analysis.* – Springer-Verlag, 1987, *Lecture Notes in Mathematics*, volume 1294.

[RS80] Rozenberg (G.) and Salomaa (A.). – *The mathematical theory of L systems.* – Academic press, 1980.

[Sal81] Salomaa (A.). – *Jewels of Formal Language Theory.* – Rockville, MD, Computer Science Press, 1981.

New Algorithms for Detecting Morphic Images of a Word

Jean NERAUD[1]

LIR, LITP, Université de ROUEN, Faculté des Sciences, Place E. Blondel, F-76128 MONT SAINT AIGNAN CEDEX, FRANCE

Abstract. We present efficient algorithms for two subcases of the general NP-complete problem [An 80] which consists in matching patterns with variables :
- matching an arbitrary one-variable pattern with constants
- matching a two-variable pattern

1 Introduction

Given a finite alphabet Σ, a recursive subset L of Σ^* (the free monoid generated by Σ) and a string $w \in \Sigma^*$, the pattern matching problem consists in deciding whether or not there exists a substring of w which belongs to L. In fact, this definition concerns many classical problems in the framework of combinatorics on words.

The most famous example is certainly the so-called "string matching" problem, which corresponds to L being a singleton. With this restriction, many famous fast algorithms have been implemented (cf e.g. [KMP 77], [BM 77], [GS 83]). Another classical question corresponds to the case where L is a finite language [AC 75], [FP 74], or more generally, the case where L is a regular set [T 68], [CN 89]. Approximate pattern matching is affected, and led to a great number of challenging combinatorical algorithms (e.g. [WS 78], [MM 89], [LV 88]).

Languages described by morphic images of words, a more general class of languages, are also concerned. Formally, given two finite disjoint sets, Σ, the alphabet of the "constants", and Δ, the alphabet of the "variables", we consider a word, say R, over $\Delta \cup \Sigma$. The language described by R, namely $L(R)$, is the set of all the words $\phi(R) \in \Sigma^*$, where ϕ stands for any morphism from $(\Delta \cup \Sigma)^*$ into Σ^* satisfying $\phi(a) = a$ for every "constant" $a \in \Sigma$. For instance, take $\Sigma = \{a, b, c, d\}$, $\Delta = \{X, Y, Z\}$ and consider the "pattern" $R = X^2aXYbZ \in (\Delta \cup \Sigma)^*$. By definition, the word $(ab)^2a^2bcbd$ belongs to $L(R)$. Indeed, such a string may be constructed by substituting ab to X, c to Y, and d to Z in the preceding word R.

From the point of view of the computational complexity, as shown in [An 80], given $R \in (\Delta \cup \Sigma)^*$, and given an arbitrary word $w \in \Sigma^*$, the problem of deciding whether $w \in \Sigma^* L(R) \Sigma^*$ or not is NP-complete. As shown by the author, it may be solved by applying a general $O(|w|^{2|\Delta|+1})$-time algorithm. Moreover, by examining each of the $O(|w|^2)$ factors of w, and by considering diophantine equations in the lengths of the candidates $\phi(X)$, with $X \in \Delta$, a $O(|w|^{|\Delta|+2})$-time improvement may be obtained.

With such conditions, it is of considerable interest to draw the limits of the class of "efficiently matchable" patterns, not only from a theoretical point of view, but also because matching morphic images has many important applications. Indeed, beside the classically mentioned examples of text editing, or Molecular Biology, matching patterns with variables is also an helpfull investigation tool in several domains of Computer Science, like Rewriting Theory, or in the framework of infinite words [B 92].

It is surprising that only the case of periodicities, i.e. the case where the pattern R is a one-variable word like X^5, has been studied in the litterature (cf e.g. [C 81], [AP 83], [ML 85] or [R 85]). The results led to construct $O(|w| ln|w|)$-time algorithms. In our paper, we broad this investigation by considering two new classes of patterns with variables :

- The first class corresponds to a one-variable pattern with constants. With such instances, each of the general algorithms runs in time $O(|w|^3)$. We establish that, given such a word R, and given arbitrary word w, deciding whether a factor of w is a morphic image of R may be done in time $O(|w|^2 ln|w|)$. It is of interest to notice that our method allows in fact to compute all the morphisms ϕ, and all the corresponding occurrences of the words $\phi(R)$.
- Our second result corresponds to the case of two-variable patterns. Given a two-element alphabet Δ, given R, an arbitrary word in Δ^*, like XYX^2YX, or $X^2Y^3X^5Y$,... and given an arbitrary word w, deciding whether or not the pattern R "appears" in w, may be done in time $O(|w|^5)$, or $O(|w|^4)$, by applying the general algorithms that we mentioned above. In our paper, we present a $O(|w|^2 ln^2|w|)$-time algorithm for solving the problem. Moreover, except in some simple special cases of pattern of type X^nY^p, or $X^n(YX)^p$, or X^nYX^p, where too many solutions may exist, our algorithm allows to compute all the morphisms which characterize the solutions.

Once more, the technics we introduce are a nice illustration of the prominent part of the combinatoric concept of "overlapping" when analizing such a type of problem [N 93].

2 Preliminaries

2.1 Definitions and notations

We adopt the standard notations of the free monoid theory : given a word w in Σ^* (the free monoid generated by Σ), we denote by $|w|$ its length, the empty word being the word of length 0. Given two words $u, w \in \Sigma^*$, we say that u is a *factor* (resp. *prefix*, *suffix*) of w iff we have $w \in \Sigma^* u \Sigma^*$ ($u\Sigma^*$, $\Sigma^* u$). If $w \in \Sigma^+ u \Sigma^+$ (resp. $u\Sigma^+$, $\Sigma^+ u$), we say that u is an *interior factor* (*proper*

prefix, *proper suffix*).
If u is a prefix (suffix) of w, we denote by $u^{-1}w$ (wu^{-1}) the unique word v such that $w = uv$ ($w = vu$). An *overlap* of w is a factor which is both proper prefix and suffix of w. Given a non-empty word $p \in \Sigma^*$, we say that it is a *period* of w iff w is a prefix of a word in p^+. In a classical way, the overlaps of w and its periods are in one-to-one correspondence (cf e.g. [KMP 77]).
The *primitive root* of a non-empty word w is the shortest word r such that $w \in r^+$ (if $w = r$, we say that w is a *primitive word*). As a direct consequence of the Defect theorem (cf e.g. [Lo 83] p. 6) :

Claim 1 *If x is a primitive word, then it cannot be an interior factor of x^2.* □

The following result is of folklore in the litterature :

Theorem 1 Fine and Wilf. *Let x, y be two words. Assume that two powers x^p, y^q of x and y have a common prefix of length at least equal to $|x| + |y| - g.c.d.(|x|, |y|)$. Then the words x and y are powers of a common word.* □

2.2 Matching morphic images of words : two problems

Consider the following general problem :

Instance : - Δ, the finite alphabet of "variables"
- Σ , a finite alphabet of "constants", with $\Delta \cap \Sigma = \emptyset$
- R, a word upon the alphabet $\Delta \cup \Sigma$ (the "pattern")
- $w \in \Sigma^*$, the "text"

Question : Decide whether or not there exists a non-erasing morphism

$$\phi : (\Delta \cup \Sigma)^* \to \Sigma^*$$

(i.e. with ε not in $\phi(\Sigma \cup \Delta)$), such that the two following conditions hold :
- for every letter $a \in \Sigma$, we have $\phi(a) = a$
- $\phi(R)$ is a factor of w.

As shown in [An 80], this problem is NP-complete. The preceding mapping ϕ will be called a *solution*. In our paper, we consider the two following restrictions :

- PAT1C, the restriction whose instance corresponds to Δ being a one-variable alphabet $\{X\}$.
- PAT2, with $\Delta = \{X, Y\}$, and $R \in \Delta^*$. With this last condition, every morphism solution ϕ may be identified to a substitution from Δ^* into Σ^*.

Moreover, we shall restrain our search to *unique decipherable morphisms*, i.e. morphisms such that $\phi(\Delta)$ is a unique decipherable set (otherwise, solving problem PAT2 simply comes down to detect periodicities)

2.3 Convention of implementation

From an algorithmic point of view, we represent words by linked lists of characters, sets of words being represented by lists of words. With this convention, union of sets will be interpreted as concatenation of lists. In a similar way, linear diophantine equations (inequations), will be represented by linked lists of integers.

3 Why a prominent part for overlaps

In this section, we present some technical results, which are the main feature of our algorithms. For short, given a solution ϕ, we shall see how the period of the word(s) in $\phi(\Delta)$ allows to determine the morphism ϕ itself.

3.1 One-variable prefix pattern with constants

This condition corresponds to the easiest case. Let $(u,v) \in \Sigma^* \times \Sigma^+$, and let $\phi : (\{X\} \cup \Sigma)^* \to \Sigma^*$ be a morphism. Consider the following condition :

Condition 1 *(i)* $\phi(a) = a$, *for every* $a \in \Sigma$
(ii) $\phi(X) \in (uv)^* u$

(we shall say that ϕ satisfies Condition 1 with respect to the pair of words (u,v))

Lemma 2. *Given the words* $w \in \Sigma^*$, $R \in (\{X\} \cup \Sigma)^*$, *and given a pair of words* $(u,v) \in \Sigma^* \times \Sigma^+$, *there exists a unique one-unknown linear diophantine constraint* (C), *such that the following property holds :*
For every morphism $\phi : (\{X\} \cup \Sigma)^* \to \Sigma^*$ *satisfying Condition 1 with respect to* (u,v), *if* $\phi(R)$ *is a prefix of* w, *then the length of the word* $\phi(X)$ *is solution of* (C).

Proof of Lemma 2 Let $(t_k)_{1 \le k \le n}$ be the unique sequence of words in Σ^* such that $R = t_0 X t_1 X t_2 ... X t_n$. For each integer $k \in [1,n]$, set $Sum(k) = |t_1| + ... + |t_k|$. Moreover, we set $w' = t_0^{-1} w$. Without loss of generality, assume that $(uv)^2$ is a prefix of $\phi(X)$ (otherwise, we directly get the word $\phi(X)$). Denote by i_{max} the greatest positive integer (if it exists) such that $\phi(X t_1 X ... X t_{i_{max}})$ is a prefix of a word in $(uv)^* u$, and set $z = |\phi(X)|$. According to Claim 1, for each integer $i \in [1, i_{max} - 1]$, the word t_i belongs to the set $(vu)^* v$.

- If $i_{max} = n$, then t_n is prefix of a word in $(vu)^* v$. According to Claim 1, the integer z is solution of the diophantine inequation :

$$nz + Sum(n-1) \le p|uv| + |u| \tag{1}$$

(where p stands for the greatest integer such that $(uv)^p u t_n$ is a prefix of w')

- Now, we assume that $i_{max} < n$ and we denote by p the greatest integer such that $(uv)^p$ is a prefix of w'. Once more, according to Claim 1 :
  - if $t_{i_{max}} \in (vu)^* v$, denote by q the greatest non-negative integer such that $(vu)^q$ is a prefix of $t_{i_{max}+1}$.
    If $(vu)^q v$ is not a prefix of $t_{i_{max}+1}$, then z is solution of the equation :

$$(i_{max} + 1)z + Sum(i_{max}) = (p-q)|uv| + |u| \tag{2}$$

    Otherwise, z is solution of the equation :

$$(i_{max} + 1)z + Sum(i_{max}) = (p-q)|uv| - |v| \tag{3}$$

- if $t_{i_{max}} \notin (vu)^*v$, we denote by q the greatest non-negative integer such that $(vu)^q$ is a prefix of $t_{i_{max}}$. As in the preceding case, if $(vu)^q v$ is a prefix of $t_{i_{max}}(uv)^2$, the integer z is solution of one of the two following equations :

$$i_{max}z + Sum(i_{max} - 1) = (p - q)|uv| + |u| \quad (4)$$

otherwise, z is solution of the equation :

$$i_{max}z + Sum(i_{max} - 1) = (p - q)|uv| - |v| \quad (5)$$

In each case, z is solution of a unique linear diophantine constraint. This completes the proof. □

Conversely, we notice that if (C) is an inequation, for every morphism ϕ satisfying Condition 1, if $\phi(X)$ is a solution of (I), then ϕ is solution of Problem PAT1C.

Example 1. Let $\Sigma = \{a, b\}$, $w = baa(aba)^{12}abb(aba)^2abba$. Consider the word $R = t_0Xt_1Xt_2Xt_3Xt_4$, with $t_0 = baa$, $t_1 = a$, $t_2 = aabaaba$, $t_3 = aabaabb$, $t_4 = \varepsilon$. Assume that we have $u = ab$, $v = a$. With the preceding notations, we obtain : $w' = t_0^{-1}w = (aba)^{12}abb(aba)^2abba$, $i_{max} = 2 < 4$, $t_{i_{max}} \in (vu)^*v$. Moreover, we have $q = 2$, and $(vu)^q v$ is not a prefix of $t_{i_{max}+1}$. Consequently, the length of z the corresponding candidate $\phi(X)$ satisfies the preceding equation (2) : $3|z| + 8 = (12 - 2)3 + 2$. We obtain $|\phi(X)| = 8$, which corresponds to the prefix $(aba)^2ab$ of w'. It is easy to verify that $\phi(R)$ is a prefix of w.

3.2 Two-variable prefix pattern : two technical lemma

Let $x, r, s \in \Sigma^*$ such that the following proprety holds :

Condition 2 *x and rs are primitive words, moreover $s \neq \varepsilon$.*

Given a morphism $\phi : \{X, Y\}^* \to \Sigma^*$, and given a tuple of non-negative integers (i, j, j', k), with $k \geq 1$, we consider the following condition :

Condition 3 *(i)* $\phi(X) = x^i$.
(ii) $\phi(Y) = x^j(rs)^k rx^{j'}$.
(iii) x is not a prefix, nor a suffix of the word $(rs)^k r$.

We say that the tuple (ϕ, i, j, j', k) satisfies Condition 3, with respect to the tuple of words (x, r, s).

Lemma 3. *Given the words $w \in \Sigma^*$, $R \in \{X, Y\}^*$ and given the tuple of words (x, r, s) of Condition 2, there exist :*
- a constant cardinality set of non-negative integers K,
- a finite family of intervals $(H_a)_{a \in A}$, with $H_a \subseteq [2, |w|]$, and $|A| \leq 2$,
- a corresponding finite family of linear diophantine systems $(S_a)_{a \in A}$,
such that, given a tuple (ϕ, i, j, j', k) satisfying Condition 3, if $\phi(R)$ is a prefix of w, then one at least of the two following conditions holds :
(i) We have $k \in K$
(ii) There exists an index $a \in A$, such that we have $k \in H_a$ and such that the tuple (i, j, j') is a solution of (S_a). □

The theorem of Fine and Wilf, and Claim 1 plays a prominent part in the proof, which consists in comparing the prefixes of the word w which belong to $(rs)^+r$, with the prefixes of R in a set of type $X^+(YX^{m_1})^+$, m_1 being a fixed positive integer. This comparison leads to an effective construction of the systems. Moreover, by construction :

Claim 2 *With the notation of Lemma 3 (ii), given $a \in A$, for every tuple (i, j, j', k) such that (i, j, j') is a solution of (S_a), and for every integer $k \in H_a$, the word $\phi(R)$ is a prefix of w.* □

Example 2. Let $\Sigma = \{a, b\}$, $R = XYX^2YX$ and $w = (ababa)^{20}b^2a^6bab^3$. Consider the tuple $(x, r, s) = (aba, baab, a)$.
We have $w = aba(rs)^{19}bab^2a^6bab^3$. With the notation of Condition 3 the word $x^{i+j}(rs)^k rx^{j'+2i+j}(rs)^k r$ is necessarily a prefix of $\phi(XYX^2Y)$. Moreover we have $x^{q_{cycle}} = srs$, with $q_{cycle} = 2$. In fact, the word R is of type $X^{m_0}(YX^{m_1})^{k_1}YX^{k_2}$, with $k_2 = 1$, $m_1 = 2$, and $k_1 = 1$, with $H = [2, 8]$, $(S)\ : i + j = 1,\ \ 2i + j + j' = q_{cycle}$.
It is easy to verify that each tuple $(i, j, j', k) = (1, 0, 0, k)$, with $0 \leq k \leq 8$ corresponds to a solution ϕ of our Problem PAT2.

In the case (i) of Lemma 3, the following result allows to decide whether Problem PAT2 has a solution or not :

Lemma 4. *With the condition (i) of Lemma 3, given the set K, and given an integer $k \in K$, there exists at most two linear diophantine systems (let (T) be their union)* [1] *such that the following property holds :*
For every morphism ϕ satisfying Condition 3, if $\phi(R)$ is a prefix of w, then the corresponding tuple (i, j, j') is a solution of (T).

Sketch of proof of Lemma 4. Let $k \in K$, and let $x = (rs)^k r$. There exists a unique sequence of non-negative integers $(k_0, ..., k_n)$ such that $R = X^{k_0}Y...YX^{k_n}$. In a similar way, given the solution ϕ, there exists a unique sequence of integers $(h_0, ..., h_n)$, such that $\phi(R) = x^{h_0}.y....yx^{h_n}$. Let p be the greatest integer such that x^p is a prefix of $(x^{h_0}...x^{h_{n-1}}y)^{-1}w$.
If $j + j' \geq 1$ the tuple (i, j, j') is solution of the linear system :

$$k_0 i + j = h_0,\ \ j' + k_a i + j = h_a (1 \leq a \leq n-1),\ \ j' + k_n i \leq p \qquad (6)$$

If $j = j' = 0$, a similar conclusion may be obtained by considering the sequences of positive integers (p_a) and (q_a), such that $R = Y^{p_1}.X^{q_1}....Y^{p_m}X^{q_n}$. □
Conversely, if (T) has a unique solution then, by computing the word $\phi(R)$, it will be easy to decide whether the corresponding morphism ϕ is solution of Problem (PAT2). Otherwise, as for Claim 2 :

Claim 3 *If (T) has more than one solutions (i, j, j') then all the corresponding morphisms ϕ are solutions of Problem (PAT2).* □

[1] In fact, each system is the conjonction of a system of at most $|R|$ equations, with at most one inequation.

4 Computing periods of the variables

4.1 One-variable pattern

Let $R = t_0Xt_1X...Xt_n$. We assume that $n > 1$ (otherwise, the problem may be trivially solved in time $O(|w|)$,by applying the classical KMP-algorithm [KMP 77]). Given an arbitrary word $t \in \Sigma^*$, denote by $LSQR(t)$ the set of the words $r \in \Sigma^*$ which satisfy the following condition :

Condition 4 *(i) r is a primitive word.*
(ii) r^2 is a prefix of t.

Clearly, a corresponding set $RSQR(t)$ may be defined by considering the suffixes of t. According to [CR 91] :

Claim 4 *We have $|LSQR(t)|, |RSQR(t)| \leq log_\Phi|t|$, where Φ stands for the golden ratio.* □

Let w_1 be a prefix of w, the input word, such that there exists a morphism ϕ with $\phi(R)$ a prefix of the word $w_2 = (w_1)^{-1}w$. Since the word $(t_1\phi(X))^2$ is a prefix of $w' = t_1(t_0^{-1}w_2)$, there exists a unique word $r \in LSQR(w')$ and a unique integer $k \geq 2$ such that $t_1\phi(X) = r^k$.
If $k = 1$ then we directly get $\phi(X) = t_1^{-1}r$. Consequently, we may assume that $k > 1$. With this condition , denote by v the shortest non-empty word such that $t_1 \in r^*v$, and set $u = rv^{-1}$. Clearly, the word $\phi(X)$ belongs to $(uv)^*u$, thus we obtain Condition 1.

From an algorithmic point of view, we shall decide whether there exists a solution for Problem (PAT1C) by applying an algorithm whose main scheme is the following :

Algorithm 1

```
answer ← FALSE
for each prefix w1 of w do {Step 1}
    w2 ← w1^{-1}.w; w2 ← t0^{-1}.w2;
    w' ← t1w2; L ← LSQR(w');
    for each word r in L do {Step 2}
        compute the words u, v as indicated above;
        compute constraint (C) (cf Lemma 2);
        if one solution then
            compute the candidate φ(X);
            if φ(R) is a prefix of w2 then answer ← TRUE endif
        else if more than one solution then answer ← TRUE endif
        endif
    endfor
endfor
endalgorithm
```

Complexity of Algorithm 1. In a classical way, the set $LSQR(w_2)$ will be computed by applying the KMP-algorithm [KMP 77]. Clearly, Loop {1} is applied $|w|$ times. According to Claim 4, Loop {2} is applied $O(ln|w|)$ times. Moreover, in Loop {2}, each instruction requires time $O(|w|)$ (If (C) corresponds to an inequation, we get $O(|w|)$ solutions). As a consequence :

Theorem 5. *Given a word $w \in \Sigma^*$, and a word $R \in (\{X\} \cup \Sigma)^*$, deciding whether Problem PAT1C has solution requires time $O(|w|^2 ln|w|)$.* □

It is easy to see that our algorithm allows to compute all the morphisms ϕ which are solution (and all the occurrences of the word $\phi(R)$ in our word w). Indeed, the processing applies an exhaustive search by examining all the prefixes of the word w.

4.2 Two-variable pattern

We shall explain how to obtain the condition of Section 3.2.

The case where the pattern has no interior factor of type YX^kY, with $k \geq 2$. It may be shown that the problem comes down to detect the morphic images of either a word of type $X^n(YX)^kY^p$ or a word of type $X^n(YX)^kX^p$. Moreover, with such conditions, by "intersecting" the periodicities problem PAT2 may be solved in time $O(|w|^2 ln^2|w|)$.

The general case. There exists a pair of words $(R_1, R_2) \in \Delta^*$ such that the following property holds :

Condition 5 *(i)* $R_1 \in \Delta^* Y$.
(ii) $R_2 \in X^k Y \Delta^*$, *with* $k \geq 2$.
(iii) $R = R_1 R_2$.
(Clearly, we may assume that R_1 is of minimal length)

With this notation, given a solution ϕ of Problem PAT2, there exists a unique pair of words (w_1, w_2) such that the following property holds :

Condition 6 *(i)* $\phi(R_1)$ *is a suffix of* w_1.
(ii) $\phi(R_2)$ *is a prefix of* w_2.
(iii) $w = w_1 w_2$.

By definition, there exists a word $x \in RSQR(w_1)$, such that $\phi(X) \in x^+$. Let y be the shortest word such that $\phi(Y) \in x^* y x^*$. By construction there exists a pair of non-negative integers (m, n), such that we have $w_2 w_1 = x^m y \Sigma^* y x^n$. More precisely, since x cannot be a prefix of y, and since it is a primitive word, according to the Theorem of Fine and Wilf, and according to Claim 1, we have in fact $(m, n) \in [p-2, p] \times [q-2, q]$, where p (resp. q) stands for the greatest integer such that x^p (x^q) is a prefix of w_2 (a suffix of w_1). Let F be the set of the corresponding words $x^{-m} w_2 w_1 x^{-n}$. By definition, the word y is an overlap of at least one element of F, say w'. According to [D 80], there exists a subset of $\Sigma^* \times \Sigma^+$, namely $OVL(w')$ such that each of the three following properties holds :

Condition 7 *(i) for each pair of words $(r, s) \in OVL(w')$, rs is a primitive word.*
(ii) for each overlap t of w', there exists a pair of words $(r, s) \in OVL(t)$ such that $t \in (rs)^+r$.
(iii) we have $OVL(w') \leq log_\Phi|w'|$.

As a consequence, given the prefix w_1 of w, the input word, we get $O(ln^2|w|)$ candidates (x, r, s). For each of them, we are in the conditions of Section 3.2 with the corresponding pair of words (R_2, w_2). Clearly, a similar condition holds with the reversed words of R_1, w_1. This leads to an algorithm for deciding whether or not Problem PAT2 has a solution :

Algorithm 2

```
answer ← FALSE ;
for each prefix w1 of the input word do
    E ← the set of the corresponding tuples (x, r, s) ;
    for each tuple (x, r, s) ∈ E do
        apply the construction of Section 3.2 with (R2, w2) ;
        {result : (U2), the union of a constant number of linear systems }
        do a similar operation with the reversed words of R1, w1 ; {result (U1) }
        (U) ← (U1) ∩ (U2) ;
        if one solution for (U) then
            if ( φ(R2) a prefix of w2 ) and (φ(R1) a suffix of w1 ) then
                answer ← TRUE
            else if more than one solution then answer ← TRUE endif
            endif
        endif
    endfor
endfor
endalgorithm
```

Theorem 6. *Given a word $w \in \Sigma^*$, and given a word $R \in \{X, Y\}^*$, deciding whether there exists a morphism ϕ such that $\phi(R)$ is a factor of w may be done in time $(O(|w|^2 ln^2|w|)$.* □

We finally notice that, except in the special case where $R = X^mY^nX^p$ (with $n \in \{0,1\}$), or $R = X^n(YX)^p$, (where $O(|w|^3$ solutions may exist), our methods allows to compute all the solutions ϕ of PAT2, with the corresponding occurrences of $\phi(R)$.

References

[An 80] Angluin D. Finding Patterns Common to a Set of Strings *Journ. of Computer and Syst. Sci.* 21, 46-62 (1980)

[AC 75] Aho A. and M. Coracick. Efficient String Matching : An Aid to Bibliographic Search, *Comm. ACM* (1975) Vol. 18, N.6, 333-340

[AP 83] Apostolico A. and F.P. Preparata. Optimal off-line detection of repetitions in a string, *Theoret. Comput. Sci.*, **22** (1983) 297-315

[B 92] Baker K. Open problems on avoidable and unavoidable patterns, manuscript (Université de Rouen, France)

[BM 77] Boyer R.S.and J.S. Moore, A fast string searching algorithm, *Comm. ACM* 20 (10) (1977) 62-72

[C 81] Crochemore M., An optimal algorithm for computing the repetitions in a word, *Information Proc. Letters*, 12 (1981), 244-250

[CN 89] Crochemore M. and J. Néraud. Unitary monoid with two generators : an algorithmic point of view, in : (*Proceedings of CAAP'90*), 1990

[CR 91] Crochemore M. and W. Rytter. Periodic prefixes of strings, in: (*Acts of Sequences'91*), 1991

[D 80] Duval J. P. Contribution à la combinatoire du monoide libre. Thèse de Doctorat d'Etat, Université de Rouen, 1980

[FP 74] Fisher M.J. and. M.S. Paterson, String Matching and other products, in : R.M. Karp ed., Complexity of Computation, *SIAM-AMS Proceedings*, Vol. 7 (Amer. Mathematical Soc. Providence, RI, 1974) 113-125

[GS 83] Galil Z., and J. Seiferas. Saving space in fast string-matching, *SIAM J. Comput.*, 1980, 417-438

[KMP 77] Knuth D, Morris J. and V. Pratt. Fast pattern matching in string, *SIAM J. Comput.* (1977) Vol. 6, N. 2, 323-350

[Lo 83] Lothaire M. "Combinatorics on words", Encyclopedia of Mathematics and appl., Addison Wesley Publish. Company (1983)

[LV 88] Landau G. and U. Vishkin. Fast string Matching with k Differences, *Journ. of Comput. and Sys. Sci* (1988) Vol 37, 63-78

[ML 85] Main G. and J. Lorentz. Linear time recognition of squarefree strings, in "Combinatoric Algorithms on Words", A. Apostolico and Z. Galil editor, NATO ASI, Springer Verlag, Berlin (1985)

[MM 89] Myers E. and W. Miller. Approximate matching of regular expressions, *Bulletin of Mathematical Biology* (1989) Vol. 51 (1), 5-37

[N 93] Néraud J. Deciding a finite set of words has rank at least two, to appear in *Theoretical Computer Science*, Vol. 109 (1993)

[R 85] Rabin O. Discovering repetitions in strings, in "Combinatoric Algorithms on Words", A. Apostolico and Z. Galil editor, NATO ASI, Springer Verlag, Berlin (1985)

[T 68] Thomson K. Regular Expression Search Algorithm, *Comm. of ACM* (1968) Vol 11, N. 6

[WS 78] Wagner R. and J. Seiferas. Correcting counter-automaton-recognizable languages, *SIAM J. Comput.* (1978) Vol. 7 (3)

Ignoring Nonessential Interleavings in Assertional Reasoning on Concurrent Programs*

Paweł Pączkowski

Instytut Matematyki, Uniwersytet Gdański,
ul. Wita Stwosza 57, 80-952 Gdańsk, POLAND
email: matpmp@halina.univ.gda.pl

Abstract. An approach allowing one to simplify assertional reasoning on concurrent programs is presented. In the adopted assertional framework, to verify such properties as partial correctness, mutual exclusion, or deadlock freedom, the inductive assertions method is applied to a labelled transition systems representing a program, where concurrency is modelled by action interleavings. In order to tackle the problem of state explosion a notion of reduction of the transition system representing a verified program is introduced, where some transitions and configurations that arise from nonessential interleavings of actions are ignored. To isolate nonessential interleavings, the trace equivalence, in the sense of Mazurkiewicz, is exploited. Decidability of verifying whether a given labelled transition system is a reduction is investigated.

1 Introduction

A well known, though undesirable effect of modelling concurrency by nondeterministic interleavings of actions is the state explosion in so constructed concurrent behaviours. Clearly, this affects reasoning on concurrent behaviours, where we have to characterize logically computations involving all interleavings of actions and, hence, all action interleavings have to be taken into account somehow.

There are, however, situations, where considering all action interleavings is not necessary. For example, a proof system for Hoare's CSP presented in [Apt et al. 80] contains the following rule

$$\frac{\{p\}\ S_1; S_2\ \{p_1\} \qquad \{p_1\}\ \alpha \parallel \bar{\alpha}\ \{p_2\} \qquad \{p_2\}\ T_1; T_2\ \{q\}}{\{p\}\ S_1; \alpha; T_1 \parallel S_2; \bar{\alpha}; T_2\ \{q\}}$$

where α, $\bar{\alpha}$ stand for a pair of matching communication commands and S_i, T_i, for $i = 1, 2$, are (possibly composite) commands not containing any communication commands. The interesting observation about this rule is that partial correctness properties of *sequential* compositions $S_1; S_2$ and $T_1; T_2$ allow one to derive a partial correctness property of the parallel composition of $S_1; \alpha; T_1$ and $S_2; \bar{\alpha}; T_2$. Thus, only one particular

* Part of the work presented here is included in author's PhD thesis. This research was supported in part by University of Gdańsk grant BW 5100-5-0091-2 and by CRIT IC 1010/II

order of executing atomic actions of concurrent statements S_1, S_2 and T_1, T_2 is analysed. Other orders are ignored. Such an approach is sound thanks to the convention that CSP processes work on disjoint sets of variables.

In this paper we generalize the idea exploited in the proof rule discussed above and present a technique for handling the state explosion in assertional reasoning on concurrent programs. Trace theory originated by Mazurkiewicz (see [Aalbersberg Rozenberg 88] or [Mazurkiewicz 88] for a survey) provides a convenient mathematical basis for this study.

The assertional framework we adopted for this research is derived from [Pączkowski 91] and can be viewed as a generalization of Floyd's method of program verification [Floyd 67]. In the full version of this paper [Pączkowski 93] and in [Pączkowski 91] it is argued that verifying such correctness properties of concurrent programs as partial correctness, mutual exclusion, or deadlock freedom, can be reduced to the following: characterize (by assertions) the states that can be reached at some crucial control points within the program. In an abstract setup that we consider in this paper, a verified program is represented by a finite labelled transition system (enriched with a modest extension) whose actions are provided with an input-output semantics. Configurations of this transition system represent control points within the program, the transitions correspond to atomic actions of the program. Concurrency is modelled by nondeterministic interleavings.

In order to tackle the problem of state explosion a notion of reduction of a transition system representing a verified program is introduced, where some transitions and configurations that arise from nonessential interleavings of actions are ignored. To isolate nonessential interleavings, an independence relation on actions is stipulated and the trace equivalence induced by the independence exploited.

A closely related approach to dealing with state explosion is presented in [Godefroid Wolper 91a, Godefroid Wolper 91b]. There, to reduce the cost of model checking and deadlock detection product automata representing uninterpreted concurrent behaviours are replaced with trace automata whose definition also appeals to Mazurkiewicz traces.

In this paper programs with interpreted actions are considered and an assertional framework is adopted. Moreover, we investigate decidability of verifying that a transition system is a 'trace equivalent' reduction of another one, which complements the approach of Godefroid and Wolper where an algorithm for constructing a trace automaton is given.

The paper is organized as follows. Section 2 contains preliminary definitions. In Sect. 3 an abstraction on typical assertional verification tasks is proposed and the inductive assertions method is adopted for the posed verification task. Section 4 contains a minimal background on trace theory which is needed in Sect. 5, where we present a method allowing one to avoid considering all action interleavings in assertional proofs. In Sect. 6 we discuss decidability issues. Section 7 contains examples. We end with some concluding remarks in Sect. 8.

2 Preliminaries

By a labelled transition system we will mean a structure $(Conf, Act, \longrightarrow)$ where $Conf$ is a set of configurations, Act is a set of actions and $\longrightarrow$ is a transition relation, $\longrightarrow \subset Conf \times Act \times Conf$.

An *extended labelled transition system* (abbreviated elts) is defined as a structure $(Conf, Act, \longrightarrow, \iota, E)$, where $(Conf, Act, \longrightarrow)$ is a labelled transition system, ι is a distinguished initial configuration, $i \in Conf$, and E is a set of distinguished configurations, $E \subset Conf$, whose role will be explained later. The set E will be called the *extension* of an elts. All elts's we are going to deal with in this paper will be finite, i.e. their sets of configurations and actions will be finite.

An elts $\mathcal{R} = (Conf_{\mathcal{R}}, Act_{\mathcal{R}}, \longrightarrow_{\mathcal{R}}, i_{\mathcal{R}}, E_{\mathcal{R}})$ is called a substructure (or a sub-elts) of an elts $\mathcal{T} = (Conf, Act, \longrightarrow, i, E)$ if $\mathcal{R}$ is included in $\mathcal{T}$ componentwise, i.e. $Conf_{\mathcal{R}} \subset Conf$, $Act_{\mathcal{R}} \subset Act$, $\longrightarrow_{\mathcal{R}} \subset \longrightarrow$, $i_{\mathcal{R}} = i$, $E_{\mathcal{R}} \subset E$.

We are going to consider elts's with interpreted actions. To this end, we assume a set of states called *States* and, for each action α of an elts, a relation on states denoted $[\![\alpha]\!]$, which describes the state change caused by an execution of α. For each transition $X \xrightarrow{\alpha} X'$ of an elts we define a corresponding set of *interpreted transitions* $\{\langle\sigma, X\rangle \xrightarrow{\alpha} \langle\sigma', X'\rangle \mid (\sigma, \sigma') \in [\![\alpha]\!]\}$. Interpreted transitions are members of the set $(States \times Conf) \times Act \times (States \times Conf)$.

A sequence of transitions of the form

$$\langle\sigma_0, X_0\rangle \xrightarrow{\alpha_1} \langle\sigma_1, X_1\rangle \xrightarrow{\alpha_2} \langle\sigma_2, X_2\rangle \xrightarrow{\alpha_3} \cdots \xrightarrow{\alpha_n} \langle\sigma_n, X_n\rangle \tag{1}$$

is called a *computation* of an elts $\mathcal{T}$ if $\langle\sigma_{i-1}, X_{i-1}\rangle \xrightarrow{\alpha_i} \langle\sigma_i, X_i\rangle$ are interpreted transitions induced by transitions $X_{i-1} \xrightarrow{\alpha_i} X_i$ of $\mathcal{T}$.

Definition 1. Let $\Sigma \subset States$. *A behaviour of an elts $\mathcal{T}$ from the set of starting states Σ*, denoted $Beh(\mathcal{T}, \Sigma)$, is the set of such computations (1) of $\mathcal{T}$ that $\sigma_0 \in \Sigma$. Pairs $\langle\sigma_i, X_i\rangle$ appearing in computations of a behaviour will be called *configurations of the behaviour.*

If X is a configuration of an elts $\mathcal{T}$ and Σ a set of states, let $|Beh(\mathcal{T}, \Sigma)|_X$ denote the set of states that can be reached at the configuration X by computations of $Beh(\mathcal{T}, \Sigma)$:

$$|Beh(\mathcal{T}, \Sigma)|_X = \{\sigma \in States \mid \langle\sigma, X\rangle \text{ is a configuration of } Beh(\mathcal{T}, \Sigma)\}$$

In the sequel we assume a first order assertion language $\mathcal{P}$ and some interpretation of $\mathcal{P}$. For an assertion p let $[\![p]\!]$ denote the set of states $\{\sigma \mid \sigma \models p\}$. We will write $\Sigma \models p$ when $\forall \sigma \in \Sigma \;\; \sigma \models p$, where $p \in \mathcal{P}$ and $\Sigma \subset States$.

3 An abstract schema of assertional reasoning

We consider an abstract assertional framework, where a concurrent program is represented by a finite elts whose configurations, transitions, and extension correspond, respectively, to control points within the program, noninterruptible actions of the program, and those control points that are crucial for establishing the desired property of the program. Concurrency is modelled by nondeterministic action interleavings.

Example 1. Figure 1 shows graphically an elts representing a program **while** $x \leq n$ **do begin** $x := x+1; s := s+x \;\|\; y := y+1$ **end**, where the assignments and evaluations of Boolean expressions, assumed to be noninterruptible actions of the program, are taken as the actions of the elts. (The distinction between solid and dotted arrows should be ignored at this point.) The initial configuration is denoted by i, $\{\varepsilon\}$ is taken as the extension. The unmarked arrows are assumed to be labelled as the arrows parallel to them. A structural method of transforming concurrent while-programs and CSP-like programs into labelled transition systems is described in [Pączkowski 91]. □

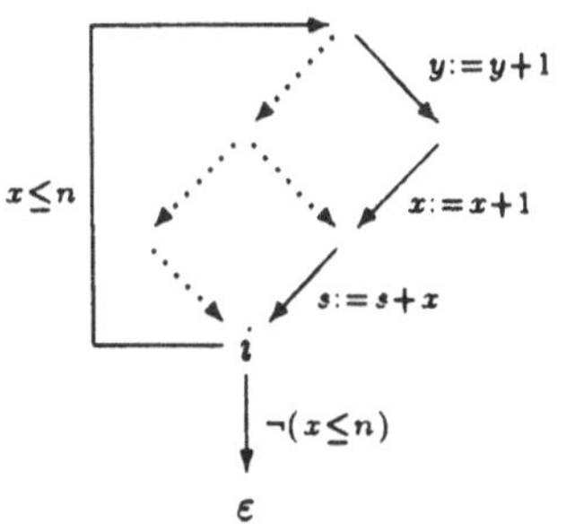

Fig. 1.

Providing the following characterization of selected reachable configurations of a behaviour of an elts will be adopted as a generic verification task.

Definition 2. Let $\mathcal{T} = (Conf, Act, \longrightarrow, \imath, E)$ be an elts and p an assertion defining a set of admissible starting states for computations of $\mathcal{T}$. We say that a family of assertions $\{p_X \mid X \in E\}$ *characterizes the behaviour* $Beh(\mathcal{T}, [\![p]\!])$ if for every $X \in E$ $|Beh(\mathcal{T}, [\![p]\!])|_X \models p_X$.

Partial correctness, mutual exclusion and deadlock freedom are examples of properties whose verification can be reduced to characterizing, in the sense of the definition above, a behaviour of an elts that represents the program under verification. When the extension of an elts representing a verified program consists just of the 'final' configuration, characterizing a behaviour by a family of assertions specializes clearly to a partial correctness property. For a mutual exclusion proof the extension E is assumed to consist of those configurations that violate the mutual exclusion property and *false* is taken as each p_X for $X \in E$ thus excluding reachability of those configurations. For a proof of deadlock freedom the extension E contains those configurations where a potential deadlock might occur and for each $X \in E$ p_X is an assertion specifying non-deadlocking states. A fuller motivation of the adopted generic verification task can be found in [Pączkowski 93].

We use an inductive assertions method for verifying that a family of assertions characterizes a behaviour of an elts. Let $\mathcal{T}$ be an elts with extension E and $\{p_X \mid X \in E\}$ a family of assertions. To prove that for each $X \in E$ $|Beh(\mathcal{T}, [\![p]\!])|_X \models p_X$ we associate assertions also with those configurations of $\mathcal{T}$ which do not belong to E in such a way that:

(P1) $p \supset p_\imath$, i.e. the assertion specifying the starting states for computations of $\mathcal{T}$ implies the assertion attached to the initial configuration of $\mathcal{T}$,

(P2) for each transition $X \xrightarrow{\alpha} X'$ of $\mathcal{T}$ the relation $[\![\alpha]\!]$ is partially correct with respect to assertions p_X and $p_{X'}$.

An elts together with a family of assertions attached to all its configurations will be called *an annotated elts*.

The following theorem ensures that the procedure described above is a sound method of verifying that a family of assertions characterizes a behaviour.

Theorem 3. *Let* $\mathcal{T} = (Conf, Act, \longrightarrow, \imath, E)$ *be an annotated elts, where* $\{p_X \mid X \in Conf\}$ *are the assertions attached to configurations of* $\mathcal{T}$*, and let* $p \in \mathcal{P}$*.*

If (P1) and (P2) above hold then $|Beh(\mathcal{T}, [\![p]\!])|_X \models p_X$ for every $X \in Conf$, thus also for every $X \in E$.

Proof. By the definition of $|Beh(\mathcal{T}, [\![p]\!])|_X$, for every $X \in Conf$ and every $\sigma \in |Beh(\mathcal{T}, [\![p]\!])|_X$ there is a computation

$$\langle \sigma_0, i \rangle \xrightarrow{\alpha_1} \langle \sigma_1, X_1 \rangle \xrightarrow{\alpha_2} \cdots \xrightarrow{\alpha_{n-1}} \langle \sigma_{n-1}, X_{n-1} \rangle \xrightarrow{\alpha_n} \langle \sigma, X \rangle \tag{2}$$

in $Beh(\mathcal{T}, [\![p]\!])$, where $\sigma_0 \models p$. Therefore, it is enough to show that for every computation (2) $\sigma \models p_X$. This is done by induction on the length of computation (2). If (2) has the length zero then $\sigma = \sigma_0$ and by (P1) $\sigma \models p_i$. If (2) has the length greater than zero then, by the induction hypothesis, we have $\sigma_{n-1} \models p_{X_{n-1}}$ which, by (P2), implies $\sigma \models X$. □

We assume that we have some means of verifying partial correctness of actions which is required in (P2) above. The actions correspond to basic primitives used in programming languages so it is fair to assume that we are able to reason about them. For example, if α is an assignment $x := t$ than $[\![\alpha]\!]$ is partially correct with respect to p_X and $p_{X'}$ if $p_X \supset p_{X'}[t/x]$.

We observe that, in effect, proving a property of a concurrent program involves attaching assertions to all configurations of an elts representing the program. The assertions attached to the configurations belonging to extension E are crucial for establishing the desired property of a program, the other assertions play an auxiliary role.

4 Trace equivalence

So far it has been assumed that elts's representing programs were so constructed that all interleavings of actions were taken into account. In order to capture the idea of nonessential interleavings we use trace equivalence.

Let us consider *Act*, the set of actions of an elts, as an alphabet over which words and languages can be formed. The primitive concept underlying trace equivalence is an independence relation on actions. Since we deal with interpreted actions, we impose an extra semantic condition on the independence relation which is normally not used in trace theory, where uninterpreted action systems are considered.

Definition 4. A symmetric relation I on actions, $I \subset Act \times Act$, satisfying

$$\text{if } \alpha I \beta \text{ then } [\![\alpha]\!][\![\beta]\!] = [\![\beta]\!][\![\alpha]\!] \tag{3}$$

will be called an *independence relation on actions*. When $\alpha I \beta$ we say that actions α and β are independent.

In trace theory, which is concerned with causal dependence of actions, it is customary to assume that the independence relation is irreflexive. We do not make this assumption; in fact, in our case I can be always assumed to be reflexive because (3) is trivially satisfied for $\beta \equiv \alpha$. However, the definition of trace equivalence is insensitive to this difference.

We note that although (3) appeals to the semantics of actions, syntactic considerations are often sufficient to establish that (3) holds. For example, if two assignments $x_1 := t_1$ and $x_2 := t_2$ obey the syntactic restriction that $x_1 \notin var(t_2)$, $x_2 \notin var(t_1)$ and $x_1 \not\equiv x_2$, where $var(t_i)$ denotes the set of variables appearing in the term t_i, then (3) obviously holds for the two assignments.

The definition of trace equivalence induced by an independence relation is routine:

Definition 5. Let $\sim'_I$ be a relation on Act^* defined

$$\sim'_I = \{(w\alpha\beta v, w\beta\alpha v) \mid \alpha I\beta, \quad w, v \in Act^*\}.$$

The reflexive transitive closure of $\sim'_I$, denoted $\sim_I$ is called the *trace equivalence* determined by I. Equivalence classes of Act^* with respect to $\sim_I$ are called traces and $[w]_I$ will denote the equivalence class of $w \in Act^*$.

The above definition implies that two words w, v over Act are trace equivalent if and only if there exists a sequence of words $w = w_1, w_2, \ldots, w_n = v$ such that w_{i+1} is obtained from w_i by a transposition of two adjacent independent actions.

The semantic condition (3) imposed on the independence relation extends to trace equivalent words as follows:

Proposition 6. *If* $\alpha_1 \ldots \alpha_n \sim_I \beta_1 \ldots \beta_n$ *then* $[\![\alpha_1]\!] \ldots [\![\alpha_n]\!] = [\![\beta_1]\!] \ldots [\![\beta_n]\!]$ □

5 Ignoring nonessential interleavings

Now we are ready to define a notion of *reduction*, which is a substructure of an elts where configurations and transitions arising from nonessential interleavings are ignored. Consider an elts $\mathcal{T} = (Conf, Act, \longrightarrow, i, E)$ and assume an independence relation I on Act, so that we can use the induced trace equivalence $\sim_I$.

Definition 7. A substructure $\mathcal{R}$ of $\mathcal{T}$ is called a *reduction* of $\mathcal{T}$ if

(R1) the extension of $\mathcal{R}$ coincides with the extension of $\mathcal{T}$,

(R2) for any sequence of transitions $i \xrightarrow{\alpha_1} X_1 \xrightarrow{\alpha_2} \cdots \xrightarrow{\alpha_n} X_n$ in $\mathcal{T}$, where $X_n \in E$, there is a sequence of transitions $i \xrightarrow{\alpha'_1}_{\mathcal{R}} X'_1 \xrightarrow{\alpha'_2}_{\mathcal{R}} \cdots \xrightarrow{\alpha_{n-1}'}_{\mathcal{R}} X'_{n-1} \xrightarrow{\alpha'_n}_{\mathcal{R}} X_n$ in $\mathcal{R}$ such that $\alpha_1 \ldots \alpha_n \sim_I \alpha'_1 \ldots \alpha'_n$.

Example 2. A reduction of the elts considered in Example 1 is shown in Fig. 1 using solid arrows. It is assumed that $x := x + 1$ and $s := s + x$ are independent of $y := y + 1$. □

The point in introducing the notion of reduction is that reductions adequately represent elts's for assertional proofs.

Theorem 8. *Let $\mathcal{R}$ be a reduction of $\mathcal{T}$. If a family of assertions $\{p_X \mid X \in E\}$ characterizes a behaviour $Beh(\mathcal{R}, [\![p]\!])$ then this family of assertions characterizes $Beh(\mathcal{T}, [\![p]\!])$.*

Proof. Follows from Lemma 9 below. □

If the independence relation is generous enough, then a reduction can be substantially smaller than the original elts. Since verifying that a family of assertions characterizes a behaviour of an elts involves attaching assertions to all configurations of the elts and analysing all its transition, the ability to use a reduction instead of the original elts can greatly simplify the verification task.

Lemma 9. *Let $\mathcal{R}$ be a reduction of $\mathcal{T}$ and let Σ be a set of states. For every X in E* $|Beh(\mathcal{R}, \Sigma)|_X = |Beh(\mathcal{T}, \Sigma)|_X$.

Proof. The inclusion $|Beh(\mathcal{R},\Sigma)|_X \subset |Beh(\mathcal{T},\Sigma)|_X$ is obvious. For the converse inclusion, take $\sigma \in |Beh(\mathcal{T},\Sigma)|_X$. By definition of behaviour there is a computation

$$\langle \sigma_0, i\rangle \xrightarrow{\alpha_1} \langle \sigma_1, X_1\rangle \xrightarrow{\alpha_2} \cdots \xrightarrow{\alpha_{n-1}} \langle \sigma_{n-1}, X_{n-1}\rangle \xrightarrow{\alpha_n} \langle \sigma, X\rangle$$

in $Beh(\mathcal{T},\Sigma)$, where $\sigma_0 \in \Sigma$. This means that

$$i \xrightarrow{\alpha_1} X_1 \xrightarrow{\alpha_2} \cdots \xrightarrow{\alpha_{n-1}} X_{n-1} \xrightarrow{\alpha_n} X$$

is a sequence of transitions in $\mathcal{T}$, $(\sigma_{j-1},\sigma_j) \in [\![\alpha_j]\!]$ for $j = 1,\ldots,n-1$ and $(\sigma_{n-1},\sigma) \in [\![\alpha_n]\!]$. By the definition of reduction there is a sequence of transitions

$$i \xrightarrow{\alpha_1'} X_1' \xrightarrow{\alpha_2'} \cdots \xrightarrow{\alpha_{n-1}'} X_{n-1}' \xrightarrow{\alpha_n'} X$$

in $\mathcal{R}$ such that $\alpha_1 \ldots \alpha_n \sim_I \alpha_1' \ldots \alpha_n'$. The latter equivalence gives $[\![\alpha_1]\!] \ldots [\![\alpha_n]\!] = [\![\alpha_1']\!] \ldots [\![\alpha_n']\!]$ so $Beh(\mathcal{R},\Sigma)$ contains a computation

$$\langle \sigma_0, i\rangle \xrightarrow{\alpha_1'} \langle \sigma_1', X_1'\rangle \xrightarrow{\alpha_2'} \cdots \xrightarrow{\alpha_{n-1}'} \langle \sigma_{n-1}', X_{n-1}'\rangle \xrightarrow{\alpha_n'} \langle \sigma, X\rangle$$

which proves that $\sigma \in Beh(\mathcal{R},\Sigma)$

We remark that a different notion of a reduced labelled transition system has been proposed in [Valmari Clegg 91]. Instead of trace equivalence the 'stubborn set' method has been used to define those reductions.

6 Decidability issues

Since elts's representing concurrent programs are large even for small programs, finding a reduction of an elts can be a cumbersome task. One solution, following an approach presented in [Godefroid 90], would be to provide a method for automated construction of a reduction, given an independence relation. Here we consider a different problem. One might wish to provide a particular substructure of an elts as a candidate for a reduction, for example, a substructure which is particularly easy to annotate with assertions. Then, an automated procedure for checking whether a given substructure of an elts is its reduction would be useful. It turns out, however, that a decision procedure for the property 'to be a reduction' does not exist for interesting independence relations.

Theorem 10. *Let us assume a fixed independence relation I. Transitivity of I is a necessary and sufficient condition for the existence of a procedure which given any elts and its substructure would decide whether the latter is a reduction of the former.*

Proof idea. It has been proved that transitivity of the assumed independence relation is a sufficient [Bertoni et al. 82] and necessary [Aalbersberg Hoogeboom 87] condition for decidability of the equality problem of regular trace languages. The following observation can be used to show that the problem of checking whether a substructure of an elts is its reduction is equivalent to the problem of equality of regular trace languages.

By definition, any regular trace language can be represented as $\{[w]_I \mid w \in L\}$, for some regular string language L. Let $\mathcal{R}$ be a substructure of $\mathcal{T}$. Consider $\mathcal{R}$ and $\mathcal{T}$ as finite automata, taking the (common) extension E as the set of final configurations. Condition (R2) of the definition of reduction is equivalent to requiring that trace languages determined by $\mathcal{R}$ and $\mathcal{T}$ are equal. □

In our case, where elts's represent concurrent programs and the independence relation on actions has to obey the semantic condition (3), transitivity does not hold as a rule. To see this, consider a program $S_1 \| S_2$ and assume that a_1, d_1 are atomic actions of S_1 and a_2 is an atomic action of S_2. Typically, while a_2 can be independent of both a_1 and d_1, the actions a_1, d_1 are not independent as they belong to the same process and operate on the same variables. But this violates the transitivity requirement.

Faced with the undecidability implied by Theorem 10 we propose tentatively a notion of strong reduction, where the condition (R2) of Definition 7 is replaced by a weaker requirement (SR3).

Definition 11. Let $\mathcal{R}$ be a substructure of $\mathcal{T} = (Conf, Act, \longrightarrow, i, E)$. $\mathcal{R}$ is called a *strong reduction* of $\mathcal{T}$ if

(SR1) the extension of $\mathcal{R}$ coincides with the extension E of $\mathcal{T}$,

(SR2) the set of configurations of $\mathcal{R}$ contains such a subset C that $E \subset C$ and every looping path in $\mathcal{T}$ has a configuration belonging to C

(SR3) for every path $X_0 \xrightarrow{\alpha_1} \cdots \xrightarrow{\alpha_n} X_n$ in $\mathcal{T}$, where $X_0, X_n \in C$ and $X_j \notin C$ for $0 < j < n$, there exists a path $X_0 \xrightarrow{\alpha'_1} X'_1 \xrightarrow{\alpha'_2} \cdots \xrightarrow{\alpha'_{n-1}} X'_{n-1} \xrightarrow{\alpha'_n} X_n$ in $\mathcal{R}$ such that $\alpha_1 \ldots \alpha_n \sim_I \alpha'_1 \ldots \alpha'_n$.

Example 3. The reduction presented in Example 2 is a strong reduction. $\{i, \varepsilon\}$ can be taken as the set C required to define a strong reduction. □

Proposition 12. *A strong reduction $\mathcal{R}$ of $\mathcal{T}$ is a reduction of $\mathcal{T}$. 'To be a strong reduction' is a decidable property.*

Proof idea. By the choice of C, any path of the form considered in (R2) can be presented as a concatenation of segments each of which has such a form as the paths considered in (SR3). As far as the second claim of the theorem is concerned we note that, by the choice of set C, verification of condition (SR3) involves considering only loop-free paths whose number has to be finite in a finite elts. □

An additional motivation for introducing strong reductions is that they adequately represent elts's for doing termination proofs by the application of the method presented in [Pączkowski 90]. However, the proposed notion of strong reduction is quite restrictive and defining a richer class of strong reductions for which the property 'to be a strong reduction' would still be decidable seems worthy of further research.

7 Examples

Example 4. Denote by S the program $Odd \| Even$, where

$$Odd \equiv \textbf{while } i \le n \textbf{ do begin } s := s + a_i \; ; \; i := i + 2 \textbf{ end}$$
$$Even \equiv \textbf{while } j \le n \textbf{ do begin } s := s + a_j \; ; \; j := j + 2 \textbf{ end}$$

S is partially correct with respect to the initial assertion $i = 1 \land j = 2 \land s = 0 \land n \ge 1$ and the final assertion $s = a_1 + a_2 + \cdots + a_n$. The summation is performed by two processes, *Odd* and *Even*, who are responsible for adding elements indexed with, respectively, odd and even numbers.

Figure 2 contains a pictorial presentation of an elts that represents the program S by taking all interleavings of actions into account and a reduction of this elts.

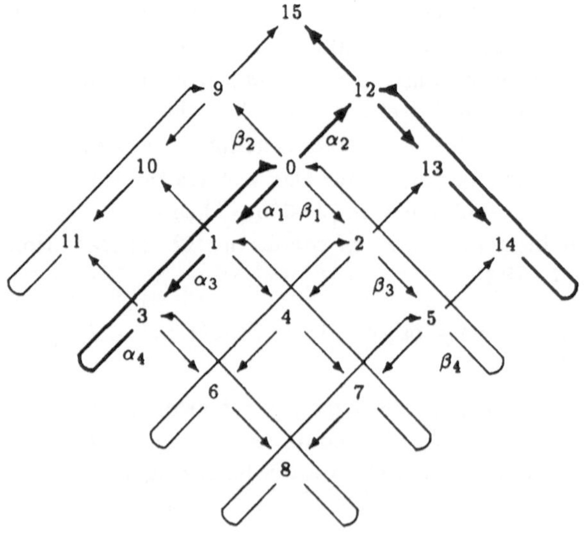

Fig. 2.

Actions α_1, α_2, α_3, α_4 represent the (noninterruptible) evaluations of tests and assignments $i \leq n$, $\neg(i \leq n)$, $s := s + a_i$, $i := i + 2$, respectively. Similarly, β_1, β_2, β_3, β_4 represent $j \leq n$, $\neg(j \leq n)$, $s := s + a_j$, $j := j + 2$, respectively. The independence relation that we take consists of pairs (α_i, β_j) and (β_j, α_i) for $i, j = 1, \ldots, 4$. Thick arrows represent transitions of the reduction, other arrows represent those transition that have been eliminated while constructing the reduction. The unmarked arrows are assumed to be labelled as the arrows drawn parallelly to them. The initial configuration is denoted by 0, the assumed extension that is suitable for a partial correctness proof contains the single 'final' configuration denoted by 15.

Thanks to the very rich independence relation a considerably smaller number of assertions and correctness checks along the transitions is required when the reduction rather than the unpruned representation of S is used for the partial correctness proof. □

Example 5. The presented method of eliminating nonessential interleavings was used to verify a program solving the problem of set partitioning [Dijkstra 82]. The program was written in a CSP-like language. The assumed independence relation was less regular in this case than in the example above.

The elts representing the program for set partitioning, where all possible interleavings were taken into account, contained 19 configurations and 28 transitions. A strong reduction of this elts which was constructed for a partial correctness proof contained 7 configurations and 14 transitions fewer. A strong reduction containing a different extension that was suitable for a deadlock freedom proof contained 6 configurations and 12 transitions fewer than the original elts. □

8 Concluding remarks

The presented technique is a modest step towards handling state explosion. Assertional proofs were simplified by letting one to avoid attaching assertions to those configurations of an elts representing a program that arise from nonessential interleavings and freeing one from checking partial correctness along transitions resulting from nonessential interleavings.

Analysis of the presented examples could be done by hand. For larger scale examples the use of some automated tools supporting the handling of transition systems involved seems necessary.

Acknowledgements

I would like to thank Werner Damm, Valery Nepomniaschy and Ernst-Rüdiger Olderog for helpful comments on this work. Comments of anonymous referees helped to improve the presentation.

References

[Aalbersberg Hoogeboom 87] I.J. Aalbersberg, H.J. Hoogeboom, *Decision problems for regular trace languages*, in: Proceedings ICALP 87, pp. 250-259, LNCS 267, 1987.

[Aalbersberg Rozenberg 88] I.J. Aalbersberg, G. Rozenberg, *Theory of traces*, Theoretical Computer Science 60, pp. 1-82 (1988).

[Apt et al. 80] K.R. Apt, N. Francez, W.P. de Roever, *A proof system for communicating sequential processes*, ACM TOPLAS 2(3), pp. 359-384 (1980).

[Bertoni et al. 82] A. Bertoni, G. Mauri, N. Sabadini, Equivalence and membership problems for regular trace languages, in: LNCS 140, pp. 61-71, 1982.

[Dijkstra 82] E.W. Dijkstra, *A correctness proof for communicating processes — A small exercise*, in: Selected writings on Computing: A Personal Perspective. Springer-Verlag, 1982.

[Floyd 67] R. W. Floyd, *Assigning meanings to programs*, in: Mathematical Aspects of Computer Science. (J.T. Schwartz, Ed.), pp. 19-32, Proceedings Symposium in Applied Mathematics, vol. 19, American Math. Soc., Providence, 1967.

[Godefroid 90] P. Godefroid, *Using partial orders to improve automatic verification methods*, in: Proc. Workshop on Computer Aided Verification, Rutgers, 1990.

[Godefroid Wolper 91a] P. Godefroid, P. Wolper, *A partial approach to model checking* in: Proceedings of the 6th LICS, pp. 406-415, 1991.

[Godefroid Wolper 91b] P. Godefroid, P. Wolper, *Using partial orders for the efficient verification of deadlock freedom and safety properties*, in: Proc. Workshop on Computer Aided Verification, 1991.

[Mazurkiewicz 88] A. Mazurkiewicz, *Trace semantics*, in: Advances in Petri Nets 1986, Part II, pp. 279-324, LNCS 255, 1987.

[Pączkowski 90] P. Pączkowski, *Proving termination of communicating programs*, in: Proceedings CONCUR'90, pp. 416-426, LNCS 458, 1990.

[Pączkowski 91] P. Pączkowski, Annotated Transition Systems for Verifying Concurrent Programs, PhD Thesis, CST-78-91 University of Edinburgh, 1991.

[Pączkowski 93] P. Pączkowski, Ignoring Nonessential Interleavings in Assertional Reasoning on Concurrent Programs, Preprint No. 90, Institute of Mathematics, University of Gdańsk, 1993.

[Valmari Clegg 91] A. Valmari, M. Clegg, *Reduced labelled transition systems save verification effort*, in: Proceedings of CONCUR'91, 1991.

CONSTANT TIME REDUCTIONS IN λ-CALCULUS

Michel Parigot & Paul Rozière

Equipe de logique — CNRS UA 753
45-55 5ème étage, Université Paris 7
2 place jussieu, 75251 PARIS Cedex 05, FRANCE
e-mail : parigot@logique.jussieu.fr
rozière@logique.jussieu.fr

Introduction

Lambda-calculus is a simple computational model which allows to represent both programs and data, such as natural numbers. In its typed versions, it corresponds exactly to intuitionistic proof systems, and expresses directly in this way the algorithmic content of proofs (the correspondence between intuitionistic proofs and typed terms is known as Curry-Howard isomorphism). In second order (or higher order) type systems, the data structures can be defined and the representation of the data in λ-calculus extracted from the definition.

It is well-known that all the recursive functions are representable as λ-terms. Even in typed λ-calculi, one has such completeness results, with respect to recursive functions whose termination is provable in an associated logical system (for strong enough type systems, one gets the functions whose termination is provable in usual mathematics). But all these completeness results are purely extensional: one gets a program which represents the function, but not necessary a program having the expected behavior, in term of time complexity for instance. Few theoretical works have been devoted to these intensional questions (see however [Co89], [Pa89]), though they are obviously important. In fact, even the simplest basic questions are very hard to answer.

From an intensional viewpoint, the choice of a representation of data is significant. Consider the case of the unary natural numbers. The usual representation is the *iterative* one (known as Church numerals), which corresponds to the standard second order definition of the set of natural numbers. Another natural representation is the *recursive* one (a variant of Scott numerals) which correspond to an alternative way of defining the set of natural numbers as a least fixed point. It is shown in [Pa89] that for the iterative representation of natural numbers, the addition is computable in a constant number of reduction steps, but the predecessor is not; for the recursive representation of natural numbers, the converse situation holds: the predecessor is computable in a constant number of reduction steps, but the addition is not. There is a third representation, the *mixed* one for which both predecessor and addition are computable in a constant number of reduction steps.

In this paper we analyze the *inherent complexity* of these three representations of natural numbers, our notion of complexity being the *number of reduction*

steps. More precisely, we characterize, for each of these representations, the class of functions from natural numbers to natural numbers which are computable in $< O(n)$ (it is shown to be, in each case, the same as the class of functions computable in $O(1)$).
For the iterative representation we get the functions which are ultimately of the form

$$a.m + b, \quad \text{with } a \in \mathbb{N} \text{ and } b \in \mathbb{N}$$

For the recursive representation we get the functions which are ultimately of the form

$$a.m + b, \quad \text{with } a = 0, 1 \text{ and } b \in \mathbb{Z}$$

For the mixed representation we get the functions which are ultimately of the form

$$a.m + b, \quad \text{with } a = \mathbb{N} \text{ and } b \in \mathbb{Z}$$

As a consequence of these characterizations, we get the results of [Pa89]. More precisely, we get the following complexity table for the basic operations on natural numbers:

	Iterative numerals	Recursive numerals	Mixed numerals
Successor	$O(1)$	$O(1)$	$O(1)$
Predecessor	$O(n)$	$O(1)$	$O(1)$
Addition	$O(1)$	$O(inf(n,p))$	$O(1)$

The method used to establish the results of this paper is rather general and can, in particular, be applied to prove similar results for data types other than the unary natural numbers. It makes use of labellings (and in particular internal labellings) in order to control the behavior of terms during reduction. It is a first step toward the solution of the following conjecture: in system F, there is no term which computes the comparison of two natural numbers n and p in $O(inf(n,p))$.

1 A simple labelled λ-calculus

In order to study the reduction process of a λ-term one needs to follow the behavior of particular subterms, in particular to know whether they "disappear" or not and, in the former case, to know precisely how they disappear. This is achieved by means of labellings. We need here only a very simple labelled λ-calculus (more general notions of labellings are presented in [Ba81])

The set Λ' of labelled λ-terms is defined by the following clauses:
(1) if x is a variable, then $x \in \Lambda'$;
(2) if x is a variable and $M \in \Lambda'$ then $\lambda x.M \in \Lambda'$;
(3) if $M, N \in \Lambda'$, then $(M)N \in \Lambda'$; [1]

[1] $(M)N$ stand for "M applied to N"

(4) if $M, N \in \Lambda'$ and i is a label, then $(M)_i N \in \Lambda'$.
The set Λ of λ-terms is just the subset of Λ' defined by clauses (i), (ii) and (iii). A term of the form $\lambda x.M$ is called an *abstraction*. A term of the form $(M)N$ (resp. $(M)_i N$) is called an *application* (resp. a *labelled application*). The *erasing* operation on terms, which replaces labelled applications $(M)_i N$ by the corresponding applications $(M)N$, transforms a labelled term into a term called its *trace*. A term of the form $(u)...(u)v$, with n occurrences of u is often denoted by$(u)^n v$.

The computation mechanism on λ-terms is the reduction of redexes. For labelled λ-terms, there are two kinds of redexes and therefore two basic rules of reduction:

$(\lambda x M)N \triangleright_1 M[N/x]$ $\qquad (\beta)$

$(\lambda x M)_i N \triangleright_1 M[N/x]$ $\qquad (\beta_i)$.

A *normal* term is term without subterms of the form $(\lambda x M)N$ or $(\lambda x M)_i N$. A *computation* of a term t is a reduction sequence of t which leads to a normal term, called the *result* of the computation. The erasing operation defines a one-one correspondence between reduction sequences on labelled terms and reduction sequences on their traces.

Let $t[(M)N]$ be a term with a distinguished subterm $(M)N$, which reduces to u. In order to describe precisely the behavior of $(M)N$ during the reduction of $t[(M)N]$, one considers $t' = t[(M)_i N]$, where i is a label not occurring in $t[(M)N]$. We have a corresponding reduction sequence between t' and u'. We call *heirs* of $(M)N$ in u, the traces of subterms of u' of the form $(M)_i N$. The subterm $(M)N$ is said to be *lost* during reduction, if it has no heir in u. It is said to be *used* if the rule β_i is used during the reduction of t'.

An important tool in the proofs which follow will be the use of *internal labels* (cf [Pa90]). One considers labels as constants in λ-terms. The definition of the set of terms is unchanged, constants being nothing else but variables which are not abstracted. To a labelled term t and a label i, one associates a term t^{-i} obtained by replacing inductively the subterms $(M)_i N$ of t by $((i)M)N$. Note that this operation has an effect on reduction: the reduction steps where the rule β_i is used, do not exist when the label i is replaced by the constant i; but if the rule β_i is not used, nothing is changed.

Lemma 1. *Let $t[(M)N]$ be a term with a distinguished subterm $(M)N$. Suppose that $t[(M)N]$ reduces to u. If $(M)N$ is not used during reduction, then $t[((c)M)N]$ reduces to a term u' which is obtained from u by replacing each heir $(M')N'$ of $(M)N$ by $((c)M')N'$. In particular, if $(M)N$ is lost and not used during reduction, then $t[((c)M)N]$ reduces to u.*

2 Recursive representation

We study in this section the recursive representation of natural numbers which is defined as follows: $\overline{0} := \lambda f \lambda x.x$ and $\overline{n+1} := \lambda f \lambda x.(f)\overline{n}$. For this representation there are terms which compute the successor and predecessor functions in a constant number of reduction steps: $\lambda n\lambda f\lambda x.(f)n$ for the successor function and $\lambda n.((n)\lambda x.x)\lambda f\lambda x.x$ for the predecessor function. There are also terms which compute the comparison, the difference and the addition of two numbers n and p in a number of reduction steps proportional to $inf(n,p)$. All these terms can be typed in a natural way in a type system with a least fixed point operator (cf [Pa92]).

We call *application of number p* in $\overline{n}$, the subterm $(f_p)\overline{n-p}$ when $\overline{n}$ is written

$$\lambda f_1\lambda x.(f_1)\ldots\lambda f_n\lambda x.(f_n)\lambda f\lambda x.x\ .$$

Proposition 2. *Let τ be a term which represents a function $\underline{\tau} : \mathbb{N} \to \mathbb{N}$.*
(1) If in the computation of $(\tau)\overline{n}$, the application of number p in $\overline{n}$ is lost and not used, then for all $m \geq p-1$, $\underline{\tau}(m) = \underline{\tau}(n)$.
(2) If in the computation of $(\tau)\overline{n}$, the application of number p in $\overline{n}$ is not lost and not used, then there exists an integer $b \in \mathbb{N}$ such that for all $m \geq p-1$, $\underline{\tau}(m) = m + b$.

Proof. (1) Because the application of number p in $\overline{n}$ is lost and not used, we have $(\tau)\lambda f_1\lambda x.(f_1)\ldots(f_{p-1})\lambda f_p\lambda x.((c)f_p)\overline{n-p} =_\beta (\tau)\overline{n}$. Replacing c by $\overline{m}$, for any natural number m, we get:

$$\forall m \in \mathbb{N},\ \ (\tau)\overline{p-1+m} =_\beta (\tau)\overline{n}\ .$$

(2) The term $(\tau)\lambda f_1\lambda x.(f_1)\ldots(f_{p-1})\lambda f_p\lambda x.((c)f_p)\overline{n-p}$ reduces to the term $\overline{\underline{\tau}(n)}$, where each heir $(M)N$ of the application of number p of $\overline{n}$ is replaced by $((c)M)N$. Suppose that the normal form of $(\tau)\overline{n}$ is $\lambda f_1\lambda x.(f_1)\ldots(f_q)\lambda f\lambda x.x$; necessarily the heirs of the application of number p are of the form $(f_j)N$, with $j \leq q$. Let k the smallest such j (in fact there is only one); replacing c by $\overline{m}$, for any natural number m, we get:

$$\forall m \in \mathbb{N},\ \ (\tau)\overline{p-1+m} =_\beta \overline{k+m}\ .$$

□

Theorem 3. *Let φ be a function from natural numbers to natural numbers. The following are equivalent.*
(1) There exist $n \in \mathbb{N}$ and $b \in \mathbb{Z}$ such that either (i) for all $m \geq n$, $\varphi(m) = b$, or (ii) for all $m \geq n$, $\varphi(m) = m + b$.
(2) There exist a term t which represents φ and a natural number n such that the computation of $(t)\overline{n}$ takes $< n$ reduction steps.
(3) There exist a term t which represents φ and a constant c such for all $n \in \mathbb{N}$, the computation of $(t)\overline{n}$ takes $< c$ reduction steps.

Proof. (1) $\Rightarrow$ (3) By composition b times of the successor (or the predecessor), one obtains a term τ_b which computes the function $n \to n+b$ in constant time. Using the term which computes the comparison in $O(inf(n,p))$, one obtains a term $test_q$ which computes in constant time the function φ_q defined by $\varphi_q(n,u,v) = u$ if $n = q$ and v otherwise. Now the required term t is in case (i)

$$\lambda x(((test_0)x)\overline{\varphi(0)})\ldots(((test_{n-1})x)\overline{\varphi(n-1)})\overline{\varphi(n)}\ ,$$

and in case (ii)

$$\lambda x(((test_0)x)\overline{\varphi(0)})\ldots(((test_{n-1})x)\overline{\varphi(n-1)})(\tau_b)x\ .$$

(3) $\Rightarrow$ (2) Obvious.
(2) $\Rightarrow$ (1) If the computation of $(\varphi)\overline{n}$ takes $< n$ steps of reduction, then one of the applications in $\overline{n}$ is not used during the reduction and proposition 2 applies. □

Corollary 4. *If a λ-term t represents the addition of two natural numbers, then the computation of $((t)\overline{n})\overline{p}$ takes at least $inf(n,p)$ reduction steps.*

Proof. Otherwise we get a term u which represents the function $2n$ and such that the computation of $(u)\overline{n}$ takes $< n$ reduction steps - which contradicts theorem 3. □

We have in fact the following generalization of theorem 3 using a straightforward generalization of proposition 2.

Theorem 5. *Let φ be a function $\mathbb{N}^k \to \mathbb{N}$. The following are equivalent.*
(1) There exist $n_1,\ldots,n_k \in \mathbb{N}$, $b \in \mathbb{N}$ and $i \in \{1\ldots k\}$ such that either (i) for all $m_1 \geq n_1,\ldots m_k \geq n_k$, $\varphi(m_1,\ldots,m_k) = b$ or (ii) for all $m_1 \geq n_1,\ldots m_k \geq n_k$, $\varphi(m_1,\ldots,m_k) = m_i + b$.
(2) There exist a term t which represents φ and a natural numbers $n_1,\ldots,n_k$ such that the computation of $(t)\overline{n_1}\ldots\overline{n_k}$ takes $< inf(n_1,\ldots,n_k)$ reduction steps.
(3) There exist a term t which represents φ and a constant c such for all $n_1,\ldots,n_k \in \mathbb{N}$, the computation of $(t)\overline{n_1}\ldots\overline{n_k}$ takes $< c$ reduction steps.

3 Mixed representation

We study in this section the mixed representation of natural numbers which is defined as follows: $\overline{n} = \lambda x\lambda f_1.(f_1)\lambda f_2.(f_2)\ldots\lambda f_n.(f_n)x$ (in particular, $\overline{0} := \lambda x.x$). For this representation there are terms which compute the successor, predecessor and addition functions in a constant number of reduction steps: $\lambda n\lambda x\lambda f.(f)(n)x$ for the successor function, $\lambda n\lambda x.((n)x)\lambda x.x$ for the predecessor function and $\lambda m\lambda n\lambda x.(n)(m)x$ for the addition. There are also terms which compute the comparison and the difference of two numbers n and p in a number of reduction steps proportional to $inf(n,p)$.
We call *application of number p* in $\overline{n}$, the subterm $(f_p)\lambda f_{p+1}.(f_{p+1})\ldots\lambda f_n.(f_n)x$ when $\overline{n}$ is written

$$\lambda x\lambda f_1.(f_1)\lambda f_2.(f_2)\ldots\lambda f_n.(f_n)x\ .$$

Note that $\lambda f_{p+1}.(f_{p+1})\ldots\lambda f_n.(f_n)x.$ is (the normal form of) $(\overline{n-p})x$.

Proposition 6. *Let τ be a term which represents the function $\underline{\tau}:\mathbb{N}\to\mathbb{N}$.*
(1) If in the computation of $(\tau)\overline{n}$, the application of number p in $\overline{n}$ is lost and not used, then for all $m \geq n-1$, $\underline{\tau}(m) = \underline{\tau}(n)$.

(2) If in the computation of $(\tau)\overline{n}$, the application of number p in $\overline{n}$ is not lost and not used, then there exist a natural number $k > 0$ and an integer $b \in \mathbb{Z}$ such that for all $m \geq n-1$, $\underline{\tau}(m) = k.m + b$ (k is the number of heirs of the application of number p in the result).

Proof. (1) Because the application of number p in $\overline{n}$ is lost and not used, we have $(\tau)\lambda x\lambda f_1.(f_1)\ldots\lambda f_p.((c)f_p)\lambda f_{p+1}.(f_{p+1})\ldots\lambda f_n.(f_n)x =_\beta (\tau)\overline{n}$. Replacing c by $\lambda f\lambda x.((\overline{m})x)f$, for any natural number m, we get:

$$\forall m \in \mathbb{N},\quad (\tau)\overline{n-1+m} =_\beta (\tau)\overline{n}\,.$$

(2) The term $(\tau)\lambda x\lambda f_1.(f_1)\ldots\lambda f_p.((c)f_p)\lambda f_{p+1}.(f_{p+1})\ldots\lambda f_n.(f_n)x$ reduces to the term $\overline{\underline{\tau}(n)}$, where each heir $(M)N$ of the application of number p of $\overline{n}$ is replaced by $((c)M)N$. Suppose that the normal form of $(\tau)\overline{n}$ is the term $\lambda x\lambda f_1.(f_1)\ldots\lambda f_q.(f_q)x$; necessarily the heirs of the application of number p are of the form $(f_j)N$, with $j \leq q$. Suppose there are k such heirs. Replacing c by $\overline{m}$, for any natural number m, we get:

$$\forall m \in \mathbb{N},\quad (\tau)\overline{n-1+m} =_\beta \overline{k(m-1)+\underline{\tau}(n)}\,.\qquad \square$$

Theorem 7. *Let φ be a function from natural numbers to natural numbers. The following are equivalent.*
(1) There exist $n \in \mathbb{N}$, $a \in \mathbb{N}$ and $b \in Z$ such that for all $m \geq n$, $\varphi(m) = a.m+b$.
(2) There exist a term t which represents φ and a natural number n such that the computation of $(t)\overline{n}$ takes $< n$ reduction steps.
(3) There exist a term t which represents φ and a constant c such for all $n \in \mathbb{N}$, the computation of $(t)\overline{n}$ takes $< c$ reduction steps.

Proof. (1) $\Rightarrow$ (3) Like for the proof of theorem 3 (using the fact that one has for this representation a term which computes the addition in a constant number of reduction steps).
(3) $\Rightarrow$ (2) Obvious.
(2) $\Rightarrow$ (1) If the computation of $(\varphi)\overline{n}$ takes $< n$ steps of reduction, then one of the applications in $\overline{n}$ is not used during the reduction and proposition 6 applies. $\square$

The generalization to k-ary functions concerning polynomials of degree 1 is left to the reader.

4 Iterative representation

We study in this section the iterative representation of natural numbers (also called Church numerals) which is defined as follows: $\overline{n} := \lambda f\lambda x.(f)^n x$, where $(f)^p x$ denotes $(f)\ldots(f)x$, where f occurs p times (in particular $\overline{0} := \lambda f\lambda x.x$). For this representation there are terms which compute the successor and addition functions in a constant number of reduction steps: $\lambda n\lambda f\lambda x.(f)((n)f)x$ for the successor function and $\lambda n\lambda p\lambda f\lambda x.((p)f)((n)f)x$ for the addition function. There

is also a term which computes the comparison of two numbers n and p in a number of reduction steps proportional to $inf(n,p)$, but it cannot be typed in system F_ω ([Kr87], [Ro89]).
We call *application of number* p in $\overline{n}$, the subterm $(f)^{n-p}x$ when $\overline{n}$ is written $\lambda f \lambda x.(f)^n x$.

Proposition 8. *Let τ be a term which represents a function $\underline{\tau} : \mathbb{N} \to \mathbb{N}$.*
(1) If in the computation of $(\tau)\overline{n}$, the application of number p in $\overline{n}$ is lost and not used, then for all $m \geq n-1$, $\underline{\tau}(m) = \underline{\tau}(n)$.
(2) If in the computation of $(\tau)\overline{n}$, the application of number p in $\overline{n}$ is not lost and not used, then there exist a natural number $k > 0$ and an integer $b \in \mathbb{Z}$ such that for all $m \geq n-1$, $\underline{\tau}(m) = k.m + b$ (k is the number of heirs of the application of number p in the result).

Proof. (1) Because the application of number p in $\overline{n}$ is lost and not used, we have $(\tau)\lambda f \lambda x.(f)^{p-1}((c)f)f^{n-p} =_\beta (\tau)\overline{n}$. Replacing c by $\overline{m}$, for any natural number m, we get:

$$\forall m \in \mathbb{N},\ (\tau)\overline{n-1+m} =_\beta (\tau)\overline{n}\,.$$

(2) The term $(\tau)\lambda f \lambda x.(f)^{p-1}((c)f)(f)^{n-p}x$ reduces to the term $\overline{\underline{\tau}(n)}$, where each heir $(M)N$ of the application of number p in $\overline{n}$ is replaced by $((c)M)N$. Because the normal form of $(\tau)\overline{n}$ is $\lambda f \lambda x.(f)^{\underline{\tau}(n)}x$, the heirs of the application of number p are of the form $(f)(f)^q x$. Suppose that there are k such heirs. Replacing c by $\overline{m}$, for any natural number m, we get:

$$\forall m \in \mathbb{N},\ (\tau)\overline{n-1+m} =_\beta \overline{k.(m-1)+\underline{\tau}(n)}\,. \qquad \square$$

The previous proposition doesn't give an optimal result: in case (2), the constant b is in fact necessarily ≥ 0. In order to prove this we will use the rigidity of the iterative structure of natural numbers, which is preserved during reduction.

We call *stable iteration of length* n a term of the form $(w_1)_1 \ldots (w_n)_n u$ with $w_1 =_\beta \ldots =_\beta w_n$ and w_i doesn't reduce to an abstraction.

Lemma 9. *Let $(w_1)\ldots(w_n)u$ and $(M)N$ be two subterms of a term t such that w_p is a subterm of N, for a certain $p \leq n$. Then either $(w_1)\ldots(w_n)u$ is a subterm of N or there exists $j < p$ such that $M = w_j$ and $N = (w_{j+1})\ldots(w_n)u$*

Proof. Obvious by induction on $\mathbb{N}$. $\square$

Lemma 10. *Let t a term which doesn't contain the labels $1,\ldots,n$ and s be the labelled term $\lambda f \lambda x.(f)_1 \ldots (f)_n x$. Suppose that $t[s/x]$ reduces to u, $(w)_i v$ is a subterm of u, for a certain $i \leq n$, and w doesn't reduce to an abstraction. Then w is an element of a stable iteration of length n in u.*

Proof. The proof is done by induction on the length of the reduction. If the length of the reduction is 0, then the result is given by the hypotheses of the lemma.

Suppose that the result is true for u and u reduces to u' in one step. We prove that the result is true for u'. We have $u = u_0[(\lambda x.M)_e N/y]$ (or $u = u_0[(\lambda x.M)N/y]$) and $u' = u_0[M[N/x]/y]$. Let $s' = (w'_p)_{e_p} v'$ be a subterm of u', for a certain $p \leq n$, such that w'_p doesn't reduce to an abstraction. We prove that w'_p is an element of a stable iteration of length n in u'.

(i) s' is a subterm of an occurrence of N in $M[N/x]$.

Then s' is also a subterm of N in u. By induction hypothesis, w'_p is an element of a labelled iteration of length n in u. Because $\lambda x.M$ cannot be an element of a stable iteration, the iteration itself is a subterm of N, by lemma 9. Therefore w'_p is an element of a stable iteration of length n in u'.

(ii) s' is a subterm of $M[N/x]$ and not a subterm of an occurrence of N in $M[N/x]$.

Then $s' = s[N/x]$ with s subterm of M distinct of x. Necessarily, s has the form $(w_p)_{e_p} v$ and w_p is not β-equivalent to an abstraction. By induction hypothesis, w_p is an element of a stable iteration $I = (w_1)_1 \ldots (w_n)_1 w$ of length n in u; necessarily, this iteration is a subterm of M. Let $I' = I[N/x] = (w'_1)_1 \ldots (w'_n)_1 w'$. All the w'_i are β-equivalent and, because w'_p 3 doesn't reduce to an abstraction, they also do not reduce to abstractions. Finally I' is a stable iteration of length n in u' and w'_p is an element of I'.

(iii) s' is not a subterm of $M[N/x]$.

Let $r = (\lambda x.M)_e N$ and $r' = M[N/x]$. We have $s' = (v_p[r'/y])_p v[r'/y]$ with v_p, v subterms of u_0. Consider now the subterm $(v_p[r/y])_p v[r/y]$ of u. By induction hypothesis, $v_p[r/y]$ is an element of a stable iteration of length n in u, $I = (v_1[r/y])_1 \ldots (v_n[r/y])_n t[r/y]$ with $v_1, \ldots, v_n, t$ subterms of u_0. Clearly $I' = (v_1[r'/y])_1 \ldots (v_n[r'/y])_n t[r'/y]$ is a stable iteration of length n in u' and $v_p[r'/y]$ is an element of I'. □

Proposition 11. *Let τ be a term which represents a function $\underline{\tau} : \mathbb{N} \to \mathbb{N}$. If in the computation of $(\tau)\overline{n}$, the application of number p in $\overline{n}$ is not used and has exactly k heirs in the result, then $\underline{\tau}(n) \geq k.n$.*

Proof. Let $u = \lambda f \lambda x (f)_1 \ldots (f)_n x$. The normal form w of $(\tau)\overline{n}$ contains k applications labelled p, which are all of the form $(f)_p v$ with v subterm of w. By lemma 10, it means that w contains k subterms of the form $(f)_1 \ldots (f)_n v$. All these subterms are necessarily distinct and therefore w contains at least $k.n$ occurrences of f. □

Theorem 12. *Let φ be a function from natural numbers to natural numbers. The following are equivalent.*

(1) There exist $n \in \mathbb{N}$, and $a, c \in \mathbb{N}$ such that for all $m \geq n$, $\varphi(m) = a.m + c$.

(2) There exist a term t which represents φ and a natural number n such that the computation of $(t)\overline{n}$ takes $< n$ reduction steps.

(3) There exist a term t which represents φ and a constant c such for all $n \in \mathbb{N}$, the computation of $(t)\overline{n}$ takes $< c$ reduction steps.

Proof. (1) ⇒ (3) The term $\tau_{a,c} = \lambda n \lambda f \lambda x (((\overline{a})(n)f)((\overline{c})f)x$ computes the function $n \to a.n + c$ in a constant number of steps (for $a, c \in \mathbb{N}$). One concludes as in the proof of theorem 3, using a composition with terms for the test functions.

(3) $\Rightarrow$ (2) Obvious.
(2) $\Rightarrow$ (1) If the computation of $(\varphi)\overline{n}$ takes $< n$ steps of reduction, then one of the applications in $\overline{n}$ is not used during the reduction and one can apply the propositions 8 and 11. □

Corollary 13. *If a λ-term t represents the predecessor function (or more generally, the subtraction of a given natural number), then the computation of $(t)\overline{n}$ takes at least n steps.*

Using the same technique, one can prove the following analogous of theorem 12 for k-ary functions.

Theorem 14. *Let φ be a function $\mathbb{N}^k \to \mathbb{N}$. The following are equivalent.*
(1) There exist $n_1, \ldots, n_k \in \mathbb{N}$, and $a_1, \ldots, a_k, b \in \mathbb{N}$ such that for all $m_1 \geq n_1, \ldots m_k \geq n_k$, $\varphi(m_1, \ldots, m_k) = a_1.m_1 + \ldots + a_k.m_k + b$.
(2) There exist a term t which represents φ and a k natural numbers $n_1, \ldots, n_k$ such that the computation of $(t)\overline{n_1} \ldots \overline{n_k}$ takes $< inf(n_1, \ldots, n_k)$ reduction steps.
(3) There exist a term t which represents φ and a constant c such for all $n_1, \ldots, n_k \in \mathbb{N}$, the computation of $(t)\overline{n_1} \ldots \overline{n_k}$ takes $< c$ reduction steps.

Proof (sketch). (1) $\Rightarrow$ (3) It is a consequence of theorem 12 ((1) $\Rightarrow$ (3)) and of the existence of a term for addition in a constant number of reduction steps.
(2) $\Rightarrow$ (1) In the term $(t)\overline{n_1} \ldots \overline{n_k}$ the application of number p in n_e is labelled by the pair of natural numbers (e, p).

It follows from (2) that there exits natural numbers $p_1 \leq n_1, \ldots, p_k \leq n_k$ such that in the computation of $(t)\overline{n_1} \ldots \overline{n_k}$ the applications (e, p_e), $1 \leq e \leq k$, are not used. From this we obtain a relaxed form of (1) (straightforward generalization of proposition 8) only for $b \in \mathbb{Z}$. We prove now that $b \in \mathbb{N}$.

We define stable iterations of length n as above but labelled with $(e, 1) \ldots (e, n)$ instead of $1, \ldots, n$. Lemma 10 can be stated as follows (changing only notations but not the result) : let τ a term which doesn't contain the labels $(e, 1), \ldots, (e, n)$ and s be the labelled term $\lambda f \lambda x.(f)_{(e,1)} \cdots (f)_{(e,n)}$. Suppose that $\tau[s/x]$ reduces to u, $(w)_{(e,i)}v$ is a subterm of u, for a certain $i \leq n$, and w doesn't reduce to an abstraction. Then w is an element of a stable iteration of length n in u labelled by $(e, 1), \ldots, (e, n)$. As in proposition 11 we deduce that $\varphi(n_1, \ldots, n_k) \geq a_1.n_1 + \ldots + a_k.n_k$ and then $b \geq 0$. □

References

[Ba81] H. Barendregt, The Lambda-Calculus, North-Holland, 1981.

[Co89] L. Colson, About primitive recursive algorithms, Proc. ICALP'89, Springer Lecture Notes in Computer Science, Vol. 372, pp 194-206.

[Gi72] J.Y. Girard, Interprétation fonctionnelle et élimination des coupures de l'arithmétique d'ordre supérieur, Thèse d'état, Université Paris 7,1972.

[Kr87] J.L. Krivine, Un algorithme non typable dans le système F, CRAS 304 Série I (1987), pp 123-126.

[Kr90] J.L. Krivine, Lambda-calcul types et modèles, Masson (1990).

[Pa89] M. Parigot, On the representation of data in lambda-calculus, Proc. CSL'89, Kaiserslautern, Springer Lecture Notes in Computer Science, Vol. 440, pp 309-321.

[Pa90] M. Parigot, Internal Labellings in Lambda-Calculus, Proc. MFCS 1990, Banská Bystrica, Springer Lecture Notes in Computer Science, Vol. 452, pp 439-445.

[Pa92] M. Parigot, Recursive programming with proofs, Proc. Coll. "Mathématique et Informatique", Marseille 1989, Theoretical Computer Science, Vol. 94, pp 335-356.

[Ro89] P. Rozière, Un résultat de non typabilité dans le système F_ω, CRAS 309 Série I (1989), pp 779-802.

Heterogeneous Unified Algebras

F. Parisi-Presicce, S. Veglioni

Dipartimento di Matematica Pura ed Applicata
Universitá degli Studi L'Aquila
67100 L'Aquila Italy

Abstract. The framework of Unified Algebras, recently developed for the axiomatic specification of ADT, is modified by introducing again the notion of sort as a classification mechanism for elements of a type. While retaining the idea of sorts as values, Heterogeneous Unified Algebras allow the distinction between certain sorts and the definition of subsorts by applying operations to them. A Specification Logic (which can be extended to an Institution using only injective signature morphisms) is defined, and initial algebra and free construction are shown to exist.

1 Introduction

In the algebraic approach to the specification of abstract data types, sorts play a major role, whether present in order to classify the values of the universe, or absent forcing to deal with partial operations and their difficulties. In Many Sorted Algebras (MSA), sorts are used not only as a first step toward abstraction, by classifying elements of the universe, but also to specify the functionalities of the operations and to limit the range of the variables in the axioms. The framework of MSA seems to be the most tractable and thus the one that received most attention, but also the one with major difficulties dealing with error values, inclusion of carriers corresponding to different sorts and overloading. MSA allow the carrier sets to overlap but not 'by design', i.e., not by specifying it. Order sorted algebras [6,11] allow to specify a partial order on the set of sorts to deal with subsort polymorphism, errors and exception-handling. The formalism for OSA signatures, algebras and axioms are very similar to those of MSA. An attempt at reconciling the two main approaches to OSA, namely Overloading OSA [5] and Universal OSA is due to Poigne [13].

A clear separation between elements and sorts has been challanged by Unified Algebras in [7,8]. In this framework, there is only one carrier whose values can represent either elements of data or classifications of elements into sorts. Operations can be applied to both elements and sorts, thus allowing to express directly some classifications of elements and to define explicitly sort constructors and equations on sorts. Since a unified specification is essentially an unsorted Horn clause specification, it supports initial algebra semantics. Some of the disadvantages of UA is that sort checking is in general undecidable and that, dealing with total operations, there are no type mismatch and static type checking can verify only the correct number of elements.

The view taken in this paper is that sort-independent semantics, for which the result of an operation depends only on the elements to which it is applied and not on their sorts, is important enough to question the adequacy of a 'sorted' world and to opt for a unique 'universe' in which operations with overlapping subdomain or codomain should be allowed to compose. On the other hand, some classifications are intended to be exclusive and ought to be represented by 'sorts'. The formalism of Heterogeneous Unified Algebras proposed here combines the 'universality' of UA with part of the classification typical of MSA.The main characteristics are :

- algebras are homogeneous and partially ordered, with operations of *sup* and *inf*, with a minimum value *nothing* and with particular values which act as sorts
- operations are total: when applied to arguments without a common sort, return the minimum value.
- unified treatment of elements and sorts, with operations extended also on sorts.

In the next section, we review the basic ideas on Unified Algebras. In section 3 we propose the model of Heterogeneous Unified Algebras with some of their properties, followed in section 4 by a treatment of parametrization which resembles MSA. In section 5, HUA are compared with a known institution to derive further properties. All proofs are omitted and can be found in [10].

2 Unified Algebras

We remind the readers [4] that an 'Institution' **I** consists of:

- a category $\mathbf{Sig}_I$ of signatures
- a functor $\mathbf{Mod}_I : \mathbf{Sig}_I \rightarrow \mathbf{Cat}^{op}$ which associates to each signature a category of models and to each signature morphism a functor between categories of models
- a functor $\mathbf{Sen}_I : \mathbf{Sig}_I \rightarrow \mathbf{Set}$ describing the set of sentences of each signature and how a signature morphism induces a translation of sentences
- a relation $\models_I$ of satisfaction between models and sentences

such that the following 'Satisfaction Condition' holds:
for any signature morphism $\phi : \Sigma \rightarrow \Sigma'$ in $\mathbf{Sig}_I$, for any sentence ax in $\mathbf{Sen}_I(\Sigma)$ and model M' in $\mathbf{Mod}_I(\Sigma)$, $M' \models_I \mathbf{Sen}_I(\phi)(ax) \iff \mathbf{Mod}_I(\phi)(M') \models_I ax$

The institution **UNI** for the specification of Unified Algebras, defined by Mosses in [7], is a specialization of **HH**, the institution for specifying abstract data types as homogeneous algebras using Horn formulas, in the sense that each signature is of the form $\Omega^0 = (OP, P^0)$, where OP contains the set OP^0 consisting of three operation symbols $\{$ *nothing*; $\_|\_$; $\_\&\_$ $\}$ and P^0 is the set of three predicate symbols $\{ \_=\_;\_\leq\_;\_:\_ \}$. The functor $\mathbf{Mod}_{UNI}(\Omega)$, which defines the unified algebras of a signature Ω, is also a specialization of the corresponding one of **HH**, being given by: $\mathbf{Mod}_{UNI}(\Omega) = \mathbf{Mod}_{HH}(\Omega, T^0)$, where T^0 is a fixed set of axioms dependent on Ω.

The main characteristics of unified algebras are:

- each unified algebra A is a homogeneous one with the structure of a distributive lattice on the carrier $|A|$, with a partial order $\_\leq\_$, the least value *nothing*, and the usual operations of $\_|\_$ (join) and $\_\&\_$ (meet). Moreover the carrier includes a distinguished subset of *elements* E_A defined with the help of a binary membership predicate $\_:\_$.

- both sorts and elements are *values* of the carrier $|A|$, so that operations can be applied to sorts and return sorts as results. Sort constraints are expressed as formulas: $x \leq y$ means that x is a subsort of y; while $x : y$ means that x is an element of sort y.

The specialization **UNI** maintains the main characteristics of **HH**, and in particular it is a liberal institution. The ability to distinguish sorts in a homogeneous structure like UA, imply the need to have a forgetful functor that 'forgets' in a stronger manner (the standard forgetful functor does not forget any value at all, only operations).

Now let $\phi : \Omega \rightarrow \Omega'$ be a unified signature morphism, and A' be in $\mathbf{Mod}_{UNI}(\Sigma')$. The application of the more forgetful functor $\_ \ddagger \phi$ (using mix-fix notation) to the algebra A' gives the algebra A defined by:

- $E_A = \{x \in E_{A'} \mid x \leq_{A'} \phi(t)_{A'}$ for some ground Ω-term $t\}$.
- $|A|$ is the least set that includes E_A such that when $\sigma \in \Omega_n$ and $x_1, ..., x_n \in |A|$ then also $\phi(\sigma)_{A'}(x_1, ..., x_n) \in |A|$.
- $\sigma_A = \phi(\sigma)_{A'}$ for each $\sigma \in \Omega$.

Furthermore, for each unified Ω'-homomorphism $h' : A' \rightarrow B'$, the morphism $h' \ddagger \phi$ is just h' regarded as an Ω-homomorphism from $A' \ddagger \phi$ to $B' \ddagger \phi$ ($A \ddagger \phi$ is a subalgebra of A).

Remarks:

- $A' \ddagger \phi$ is an algebra generated by its elements
- $A' \ddagger \phi \models ax \Leftarrow A' \models \phi(ax)$ with $\phi : Ax \rightarrow Ax'$ and $ax \in Ax$
- $A' \ddagger \phi \models axs \not\Rightarrow A' \models \phi(axs)$ with $\phi : \Sigma \rightarrow \Sigma'$.
- $\_ \ddagger \phi$ with respect to the identity does not leave unchanged the algebra.

For these reasons $\_ \ddagger \phi$ cannot be considered as the forgetful functor ($\mathbf{Mod}(\phi)$) of the institution, but we can recover its use with the introduction of *bounded-data-constraints*. A bounded-data-constraint consists of a theory and a signature morphism $< \phi : T'' \rightarrow T' , \theta : \Omega' \rightarrow \Omega >$, and it is verified by an Ω-algebra A if $A \ddagger \theta$ verifies T' and is naturally isomorphic to $F_\phi(A \ddagger \theta \ddagger \phi)$
where F_ϕ is the free functor of the institution (left-adjoint to $\mathbf{Mod}(\phi)$, indicated with $\_ \dagger \phi$).

The lack of 'institutionality' with the functor $\_ \ddagger$ leads to consequences such as:

- $A \ddagger \theta \models T$ does not imply $A \models \phi(T')$
- $F_\phi(A \ddagger \theta \ddagger \phi)$ is not free with respect to the more forgetful functor
- the functor $\_ \ddagger$ limits its effect to the elements, reducing the power of UA

3 Heterogeneous Unified Algebras

Heterogeneous Unified Algebras (HUA) are proposed as an answer to some problems just mentioned. While UA reflects some of the principles which originate the HUA, the total lack of 'sorts', i.e. the syntactical meaning of sort, creates other undesirable situations, in particular with regards to parametrized specifications. The first step towards HUA goes through the structure of the Power Algebras as complete boolean algebras. It is well known that completeness is a second order characteristic, which may prevent the existence of initial algebras. The next step, therefore, is to forego part of the expressivity of the algebra trying to recover initiality. The main characteristics of a heterogeneous unified algebra can be summarized as follows:

- the algebra is homogeneous, total and partially ordered, where the ordering reflects the inclusion of sorts
- the algebra has an heterogeneous nature, in view of particular values which act as sorts (we indicate such sorts as costant sorts or maximal sorts)
- the partially ordered carrier has a minimum value, denoted by *nothing*, which allows functions to be total, and a maximum value, which represents the entire universe.
- there are operations of *sup* and *inf* seen as the union and intersection of (sub) sorts
- the algebra treats elements and sorts in a unified way, allowing a universal structure.
- the sorts indicated in the signature do not share values but must classify the world into disjoint entities . If carriers corresponding to sorts not in the signature were allowed to overlap, the free constructions over such sorts would not exist
- inside the structure, it is possible to single out a core made up of 'sections' which have as maximal values the constant sorts. Each section is a boolean algebra (similar to unified algebras). Each core corresponds in a unique way to the algebra obtained by closing the domain with respect to the operation *sup*. The part not contained in the core contains no 'junk', nor 'confusion' with respect to sup.
- the operations are present in the signature with their functionality and when applied to values which are not related to the maximal sorts indicated in the functionality, return the same value they would return if applied to the minimum value
- operations not-strict are allowed.

This is realized by bringing back sorts into signatures; such sorts (constant sorts) play a different role than generated sorts and are characterized by absence of values in common ($inf(s, s') = minimum$). The algebra develops mainly below such sorts producing a structure in sections which is close to the structure of Many Sorted Algebras.

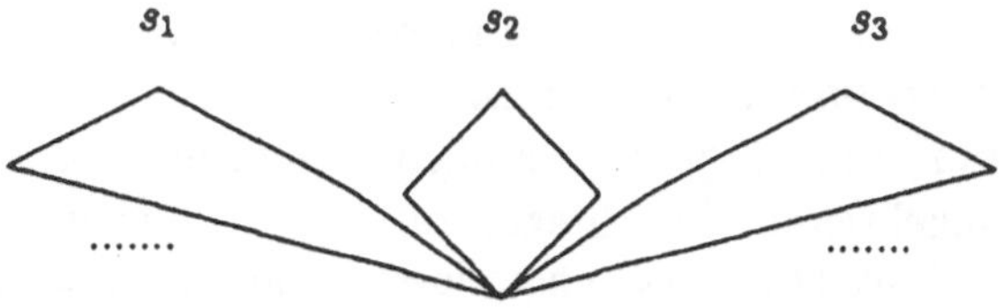

Finally, an HUA is obtained as the closure of such core with respect to *sup*, which represents non-deterministic choice of sorts of different nature. The whole structure is below a maximum value, which represents non deterministic choice among constant sorts.

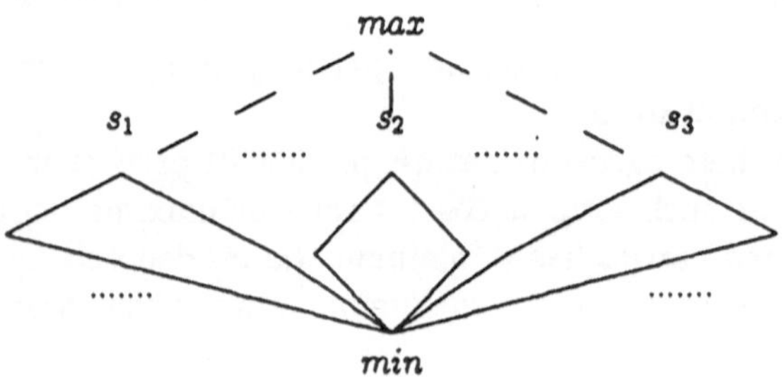

This work is based on the idea of constructing an HUA as a direct unique extension of a core. Now we define the logical structure of the Heterogeneuos Unified Algebras. The category of signatures $\mathbf{Sig}_{HUL}$ has the same objects and morphisms of $\mathbf{Sig}_{MSL}$ (many sorted-signatures and morphisms).

Definition 1. The category of heterogeneous unified signatures is $\mathbf{Sig}_{HUL}$, whose objects are pairs (S, OP), where S is a set of sorts and OP is a family of pairwise disjoint sets $OP = (OP_{w,s})_{w \in S^*, s \in S}$, and whose morphisms $\phi : (S, OP) \to (S', OP')$ are pairs of functions $(\phi_S : S \to S',\ \phi_{OP} : OP \to OP')$ with the usual property.

Every hu-signature and hu-morphism can be translated into a homogeneous signature and a homogeneous morphism respectively, by defining $\mathbf{Sig}_{H^0}$ as the category of homogeneous signatures and morphisms containing a fixed set $OP^0 = \{nothing, \_|\_\ , \_\&\_\}$ of operations, and a fixed set $P^0 = \{\_ = \_\ , \_ \leq \_\ , \_ : \_\}$ of predicates, and by letting the functor $SH : \mathbf{Sig}_{HUL} \to \mathbf{Sig}_H$ be defined by:

- $SH(S, OP) = (OP^0 \cup (OP_n)_{n \in N}, P^0)$ where OP_0 is the disjoint union of S and $\bigcup_{s \in S} OP_{\lambda,s}$ and $OP_n = \bigcup_{|w|=n} OP_{w,s}$ considering only the number of the arguments of the operations
- $SH(\phi : (S, OP) \to (S', OP')) : SH(S, OP) \to SH(S', OP')$ is the obvious extension of ϕ with $SH(\phi)(OP_0) = \phi_S(S) \cup (\bigcup \phi_{OP}(OP_{\lambda,s}))$, $SH(\phi)(OP_n) = \bigcup_{|w|=n} \phi_{OP}(OP_{w,s})$ and the identity on OP^0 e P^0.

If we let Axs_0 be the set of axioms in the top part of table 1 and denote by $\mathbf{Mod}_{H^0}$ the model functor of homogeneous signatures and predicates, then $\mathbf{Mod}_{H^0}(SH(S, OP), Axs_0)$ is a category of algebras which admits an initial model since the specification has only Horn axioms.

This initial algebra is a heterogeneous unified algebra in the sense described earlier. But not all the algebras in this category are heterogeneous unified ones since they may have values above the sorts. These 'unwanted' algebras can be eliminated by adding to Axs_0 some axioms Bx not in Horn form, to give algebras that are extensions of a core, without 'junk' or 'confusion'. We indicate now how these characteristics can be expressed in first order logic. Formally (with $S = \{s_1, ..., s_n\}$) the condition of no confusion can be expressed by the following formula not in Horn

$(S,OP) \longrightarrow ((OP_n)_{n \in N} \cup OP^0, P^0)$	
$x \leq y; y \leq x \Longrightarrow x = y$ $x \leq x$ $x \leq y; y \leq z \Longrightarrow x \leq z$	$x : x; x \leq y \Longrightarrow x : y$ $x : y \Longrightarrow x \leq y$ $x : y \Longrightarrow x : x$
$x \mid y = y \mid x$ $x \leq x \mid y$ $x \leq z; y \leq z \Longrightarrow x \mid y \leq z$	$x \& y = y \& x$ $x \& y \leq x$ $z \leq x; z \leq y \Longrightarrow z \leq x \& y$
$nothing \leq x, \quad x \leq all, \quad all = s_1 \mid s_2 \mid ... \mid s_n$ $(x \mid y) \& z = x \mid (y \& z) \quad (x \& y) \mid z = x \& (y \mid z)$ For all $\sigma \in OP_n, n \geq 1$, the n clauses: $x_i \leq x'_i \Longrightarrow \sigma(x_1, ..., x_i, ..., x_n) \leq \sigma(x_1, ..., x'_i, ..., x_n)$	
$\forall s_1, s_2 \in S, s_1 \neq s_2 \quad s_1 \& s_2 = nothing$ For all $\sigma \in OP_{w,s}$, the clauses: $\sigma(w) \leq s$ For all $\sigma \in OP_{s_1 ... s_n, s}$, the clauses: $\sigma(x_1, ..., x_n) = \sigma(x_1 \& s_1, ..., x_n \& s_n)$	

Table 1. Axs_0

form:

$(x_1 \leq s_1); ...; (x_i \leq s_i); (y \leq s_i); ...; (x_n \leq s_n); (x_i \neq nothing); (y \neq nothing) \Rightarrow (x_1 | ... | x_i | ... | x_n) \neq (x_1 | ... | y | ... | x_n)$

The condition of no junk can be expressed by the formula:

$$P(x) \Rightarrow \exists x_1, ... x_n : (x_1 \leq s_1); (x_n \leq s_n); (x_1 | ... | x_n = x)$$

where P is a predicate which is true if x is outside the core; we can think to P as: $P(x) \equiv \neg(x \leq s_1); \neg(x = s_1); ...; \neg(x \leq s_n); \neg(x = s_n)$.

The existential quantifier can be further eliminated by inserting n (hidden) operations as 'inverse to the sup': $\pi_1, ..., \pi_n$. The condition of no junk becomes then the equational formula $x = \pi_1(x) | ... | \pi_n(x)$.

Definition 2. The category HUA(Σ) of *Heterogeneuos Unified Algebras* over the signature Σ is HUA(Σ) = $\mathbf{Mod}_{H^0}(SH(\Sigma), Axs_1)$ where $Axs_1 = Axs_0 + Bx$.

Note that the hu-homorphisms map maximal sorts into maximal sorts, and the minimun into the minimum. This behavior resembles that of many-sorted homomorphisms.

Proposition 3. *The category* HUA(Σ) *has an initial algebra.*

For a given hu-signature $\Sigma = (S, OP)$, let $Sets_S$ be the category of families of sets indexed by S and index-preserving set functions. Given $\Sigma = (S, OP)$, let the forgetful functor D : HUA(Σ) $\rightarrow Sets_S$ be defined by associating to an algebra A the family $(A_s)_{s\in S}$ where $A_s = \{a \in A \mid a \leq s^A\}$ and to a homomorphism $h : A \to B$ the obvious family h_s of restrictions of h to A_s.

Given a signature $\Sigma = (S, OP)$ and a family $Y = (Y_s)_{s\in S}$ of pairwise disjoint sets of variables, consider the homogeneous algebra of $SH(S, OP)$-terms over the variables Y. By taking a quotient of this algebra with respect to the equivalence indexed by the sets of axioms Axs_0, we obtain a homogeneous algebra which is not a HUA. By taking the quotient with respect to Axs_1, instead, we obtain a set of minimal heterogeneous unified algebras which can be distinguished by the 'position' of the variables. From this set, we denote by $T_{OP}(Y)$ the one which satisfies: $y \in Y_s \Rightarrow y \leq s$, and call it the Σ-heterogeneous unified algebra of terms.

Definition 4. The Σ-heterogeneous unified algebra of terms $T_{OP}(Y)$, is obtained by choosing, from the homogeneous algebra of $SH(S, OP)$-terms over the family of variables Y quotiented with respect to Axs_1, the one which satisfies: $y \in Y_s \to y \leq s$.

Definition 5. Given $\Sigma = (S, OP)$, $Y = (Y_s)_{s\in S}$ and a Σ-heterogeneous unified algebra A, we define an assignment of variables any function $ass : Y \to A$ which satisfies: $y \in Y_S \to ass(y) \leq s^A$.

Proposition 6. *Given an assignment of variables $ass : Y \to A$, there exists a unique Σ-homomorphism extension $\overline{ass} : T_{OP}(Y) \to A$.*

Proposition 7. *The Σ-heterogeneous unified algebra of terms $T_{OP}(Y)$ is free on Y, in that the functor $Y \mapsto T_{OP}(Y)$ is left-adjoint of the forgetful functor* D.

Definition 8. The set HUS(Σ) of
Σ-heterogeneous unified sentences on Y is defined as the set of Horn formulas built on the terms of the free HUA $T_{OP}(Y)$ so that the following conditions are satisfied:
- an operation in OP is applied only to terms which respect its signature (e.g., $\sigma : s \to s'$ is applied only to terms below the constant sort s)
- the fixed operations and predicates can be applied only to pairs of terms below the same constant sort).

It is worth noticing that while this prevents us from expressing 'global' sentences, the expressive power of the structure is not really limited.

The definition of satisfaction relation $\models_{HUL} \subset |\mathbf{Mod}_{HUL}(\Sigma)| \times \mathbf{Sen}_{HUL}(\Sigma)$ is defined as usual.

Theorem 9. *The category* HUA($SPEC$) *of heterogeneous unified algebras of the specification $SPEC = (\Sigma, Ax)$ has an initial algebra.*

4 Parametrization with HUA

Now we define the notion of parametrization following the style in [1].

Definition 10. Given a morphism $\phi : \Sigma \to \Sigma'$, the induced morphism $(\tilde{\phi})$ to translate sentences is defined as $\mathbf{Sen}_{H^o}(SH(\phi))$.

The functor $\mathbf{Sen}_{HUL} : \mathbf{Sig}_{HUL} \to \mathbf{Set}$ is defined on objects as in Definition 8 and on morphisms as $\mathbf{Sen}_{H^o}(SH(\phi))$.

Definition 11. For each signature Σ the model functor $\mathbf{Mod}_{HUL} : \mathbf{Sig}_{HUL} \to \mathbf{Cat}^{op}$ gives the category HUA(Σ). Given a morphism $\phi : \Sigma \to \Sigma'$, the forgetful $\mathbf{Mod}_{HUL}(\phi) : \mathbf{Mod}_{HUL}(\Sigma') \to \mathbf{Mod}_{HUL}(\Sigma)$ is defined as follows:
- for $A' \in \mathbf{Mod}_{HUL}(\Sigma')$,
$|A| = |\mathbf{Mod}_{HUL}(\phi)(A')| = \{x \in |A'| : x \leq \phi(s)_{A'}\}$ and the operations OP^A and predicates P^{0A} are the restrictions to $|A| \subseteq |A'|$ of $\phi(OP)^{A'}$ and $\phi(P^0)^{A'}$
- for $B' \in \mathbf{Mod}_{HUL}(\Sigma')$ and $f' : A' \to B'$,
$\mathbf{Mod}_{HUL}(\phi)(f')$ is the restriction of f to $|A|$.

Note that if ϕ is not injective, then, to satisfy the axioms Axs_1, the parts in A below identified sorts must collapse to nothing.

We have now completed the definition of the quadruple **HUL** = (**Sig**, **Mod**, **Sen**, $\models$). Some important properties allow its use for parametrized specifications and parameter passing.

Definition 12. A specification morphism $\phi : (\Sigma, Axs) \to (\Sigma 1, Axs1)$ is a signature morphism $\phi : \Sigma \to \Sigma 1$ such that the induced functor $\mathbf{Sen}_{HUL}(\phi)$ satisfies $\mathbf{Sen}_{HUL}(\phi)(Axs) \subseteq Axs1$.

Remarks:

1. alternative definitions of specification morphisms [2] could be less restrictive, by allowing $\mathbf{Sen}_{HUL}(\phi)(Axs)$ to be derivable from $Axs1$, or more restrictive, by requiring an explicit correspondence between Axs and $Axs1$
2. the definition of specification morphism does not put constraints on the set Axs_1 of axioms defining the structure of HUA. In particular, for a non-injective signature morphism which maps, for example, s' and $s"$ to s, it is not required that the translation $s\&s = nothing$, of the Axs_1 axiom $s'\&s" = nothing$ be maintained.

Definition 13. The category of heterogeneous unified specifications $\mathbf{Spec}_{HUL}$ consists of the hu-specifications (Σ, Axs) as objects and of the specification morphisms of Definition 12 as morphisms.

Definition 14. A parametrized specification $PSPEC = (SPEC, SPEC1)$ consists of the formal parameter specification $SPEC = (S, OP, Axs)$ and the target specification $SPEC1 = SPEC + (S1, OP1, Axs1)$.

Proposition 15. *If $\phi : SPEC \to SPEC1$ is an injective specification morphism, then there exists a free functor $F_\phi : \mathrm{Mod}(SPEC) \to \mathrm{Mod}(SPEC1)$ as left-adjoint of the forgetful functor $\mathrm{Mod}(\phi)$.*

Definition 16. The semantics of a parametrized specification $PSPEC = (SPEC, SPEC1)$ is the free functor $F_\phi : \mathrm{Mod}(SPEC) \to \mathrm{Mod}(SPEC1)$, induced by the inclusion ϕ of $SPEC$ to $SPEC1$.

Having given the definition of parametrized specification and its semantics, we can now define parameter passing as in [1]. Given $PSPEC = (SPEC, SPEC1)$, and a specification morphism $h : SPEC \to SPEC'$, we define as parameter passing diagram the pushout diagram of h and ϕ. This definition is based on the following result:

Theorem 17. *The category of hu-specifications is closed under pushout.*

We can then define, as in [1], standard parameter passing by its *syntax* given by the parameter passing diagram; its *semantics* given by the semantics of $PSPEC$, $SPEC'$ and $SPEC1'$; and its *correctness* given by the conditions
$V_{p'}(T_{SPEC1'}) \simeq T_{SPEC'}$ (actual parameter protection)
$V_{h1}T_{SPEC1'} \simeq F \circ V_h(T_{SPEC'})$ (passing compatibility).

5 The specification Logic HUL

We are now going to show that **HUL** defines a specification logic by comparing it with the institution $\mathbf{HORN^0} = (\mathbf{Sig}_{HORN^0}, \mathbf{Mod}_{HORN^0}, \mathbf{Sen}_{HORN^0}, \models_{HORN^0}$), which differs from the institution $HORN$ in [4] because the signatures $\Sigma = (S, OP \cup OP^0, P^0)$ contain some fixed operation symbols $OP^0 = (OP^0_s)_{s\in S}$ with OP^0_s=$nothing_s$ $:\to s; s:\to s;$ $\_|_s\_, \_\&_s\_ : s, s \to s$ and some fixed binary predicate symbols $P^0 = (P^0_s)_{s\in S}$ with $P^0_s = \{\_ =_s \_; \_ :_s \_; \_ \leq_s \_\}$, and the signature morphisms maintain these symbols. The functor $\mathbf{Mod}_{HORN^0}$ associates to each signature Σ the category of algebras $\mathbf{Mod}_{HORN}(\Sigma, AX^0)$, and to each specification (Σ, Ax) the category $\mathbf{Mod}_{HORN}(\Sigma, Ax + AX^0)$. In terms of Specification Logics, the category $\mathbf{Spec}_{HORN^0}$ consists of specifications in $HORN$ which contain the fixed part AX^0. When the contest is clear, we will still write such signatures as $\Sigma = (S, OP)$ to simplify the notation.

Many of the properties of **HUL** depend on the strict relationship with $\mathbf{HORN^0}$.

Proposition 18. *1. The categories of signatures* $\mathbf{Sig}_{HUL}$ *and* $\mathbf{Sig}_{HORN^0}$ *are isomorphic.*
2. For $\Sigma = (S, OP)$ *an msh-signature, let* $AX^0 = \bigcup_{s\in S} AX^0_s$ *where* AX^0_s *is written in table 2.. Then* $\mathbf{Mod}_{HUL}(\Sigma) \simeq \mathbf{Mod}_{HORN^0}(\Sigma, AX^0)$.
3. The two functors $\mathbf{Sen}_{HUL}$ *and* $\mathbf{Sen}_{HORN^0}$ *are equivalent.*
4. The categories of specifications $\mathbf{Spec}_{HUL}$ *and* $\mathbf{Spec}_{HORN^0}$ *are equivalent.*
5. The relations $\models_{HUL}$ *and* $\models_{HORN^0}$ *are equivalent, in that they associate equivalent categories of models to each set of axioms.*

Theorem 19. **HUL** *is a Specification Logic.*

It remains to show the relationships existing between the forgetful functors.

Proposition 20. *If* $\phi : \Sigma \to \Sigma'$ *is an injective signature morphism, then* $\mathbf{Mod}_{HUL}(\phi)$ *is equivalent to the forgetful functor* $\mathbf{Mod}_{HORN^0}(\phi)$.

We can exploit liberality of $\mathbf{HORN^0}$ in the following.

Proposition 21. *By allowing in* $\mathbf{Sig}_{HUL}$ *only injective signature morphisms,* **HUL** *becomes a liberal institution. Furthermore it verifies both the amalgamation and the extension lemma.*

$\Sigma = (S, OP \bigcup OP^0, P^0)$	
$x \leq_s y; y \leq_s x \Longrightarrow x =_s y$ $x \leq_s x$ $x \leq_s y; y \leq_s z \Longrightarrow x \leq_s z$	$x :_s x; x \leq_s y \Longrightarrow x :_s y$ $x :_s y \Longrightarrow x \leq_s y$ $x :_s y \Longrightarrow x :_s x$
$x \vert_s y = y \vert_s x$ $x \leq_s x \vert_s y$ $x \leq_s z; y \leq_s z \Longrightarrow x \vert_s y \leq_s z$	$x \&_s y =_s y \&_s x$ $x \&_s y \leq_s x$ $z \leq_s x; z \leq_s y \Longrightarrow z \leq_s x \&_s y$
$x \leq_s s$ $nothing_s \leq_s x$ $(x \vert_s y) \&_s z =_s x \vert_s (y \&_s z)$ $(x \&_s y) \vert_s z =_s x \&_s (y \vert_s z)$ $\forall \sigma \in OP_{s_1 \ldots s_n, s}$, the n clauses: $x_i \leq_s x_i' \Longrightarrow \sigma(x_1, \ldots, x_i, \ldots, x_n) \leq_s \sigma(x_1, \ldots, x_i', \ldots, x_n)$	

Table 2. AX_s^0

The case of non injective morphisms distinguishes in a substantial manner this structure from that of many sorted algebras. In the case of identification of two sorts, the forgetful functor does not duplicate the elements. It is not possible to join the two parts of the algebra because in so doing the fixed axioms would be violated: hence the two 'domains' corresponding to the identified sorts would collapse to *nothing*. This behavior does not compromise the use of not injective morphisms for actualization of parameter, although passing compatibility may be violated.

6 Concluding Remarks

In the framework of Heterogeneous Unified Algebras the sorts in the signature become constant elements which can be referenced syntactically. It is a step closer to the Many Sorted Algebras than the Unified Algebras in which the lack of sorts creates problems with respect to the notions of institution, free functor and data constraints. The universe is in fact partitioned since parts of the structure relative to different sorts intersect only in the minimum element *nothing*. We think of HUA as a combination of the main positive characteristics of UA such as the constant *nothing*, the order on the sorts and the universality with the classification of MSA, removing from the latter the lack of a relationship among the sorts and from the former the lack of free constructions mentioned earlier. The condition imposed on HUA which prevents the intersection of parts that refer to different sorts has allowed some progress with respect to other structures which rely entirely on the principle of universality.

This work is intended as a first step toward the formulation of a structure to represent all those based on universality. Under investigation is the possibility of allowing intersections of parts of the algebras which refer to different sorts and free constructions with respect to instants of these named subsorts, and of exploiting the presence of complements in a universe which is in practice a coalesced sum of Boolean Algebras.

References

1. H.Ehrig, B.Mahr: Fundamentals of Algebraic Specification 1. Initial Semantics and Equations Springer, EATCS Monographs on Theoretical Computer Science **6** (1985).
2. H.Ehrig, F.Parisi-Presicce: Nonequivalence of categories for equational algebraic specifications. In Proc. 8th Workshop on Abstr. Data Types, 1991, Springer Lecture Notes in Computer Science 665 (1993) 222–235.
3. H.Ehrig, P.Pepper, F.Orejas: On Recent Trends in Algebraic Specification . In ICALP'89 Proc. Int. Coll. on Automata, Languages and Programming, Springer Lecture Notes in Computer Science **372** (1989) 263–288.
4. J.A.Goguen, R.M.Bustrall: Introducing Institutions. Proc. Logics of Programming Workshop, Springer Lecture Notes in Computer Science **164** (1984) 221–256.
5. J.A.Goguen, R.Diaconescu: A Short Oxford Survey of Order Sorted Algebras. Algebraic Specification Column, EATCS Bulletin **48** (1992) 120–133.
6. J.A.Goguen, J.Meseguer: Order Sorted Algebras I: equational deduction for multiple inheritance, overloading, exceptions and partial operations. Theoret. Comp. Sci. **105** (1992) 217–273.
7. P.D.Mosses: Unified Algebras and Institutions. In Proc. 4th IEEE Ann. Symp. on Logic in Computer Science, IEEE Press (1989) 304–312.
8. P.D.Mosses: Unified Algebras and Modules. In Proc. 16th Ann. ACM Symp. on Principles of Programming Languages, ACM (1989) 329–343.
9. P.D.Mosses: The use of sorts in algebraic specifications. In Proc. 8th Workshop on Abstr. Data Types, 1991, Springer Lecture Notes in Computer Science 665 (1993) 66–92.
10. F.Parisi-Presicce, S.Veglioni: Heterogeneous Unified Algebras. Tecnical Report **22**, Dip. di Matematica, Universit de L'Aquila (1992).
11. A.Poigné: Partial algebras, subsorting and dependent data types. In Proc. 5th Workshop on Abstract Data Types, 1987, Springer Lecture Notes in Computer Science **332** (1988) 208–234.
12. A.Poigné: Parametrization for order-sorted algebraic specifications. J. Comput. System Sci. vol. **40** no. 2 (1990) 229–268.
13. A.Poigné: Once more on order-sorted algebras. In Proc. Symposium on Math. Foundations of Computer Science, Springer Lecture Notes in Computer Science **520** (1991) 397–405.

A representation theorem for lambda abstraction algebras *

Don Pigozzi[1] and Antonino Salibra[2]

[1] Dep. Mathematics, Iowa State University, Ames, Iowa 50011, USA
[2] Dip. Informatica, University of Bari, Italy

Abstract. The concept of a *lambda abstraction algebra* (LAA) is designed to algebraize the untyped lambda calculus in the same way cylindric and polyadic algebras algebraize the first-order predicate logic. Like cylindric and polyadic algebras LAA's can be defined by true identities and thus form a variety in the sense of universal algebra. They provide a distinctly algebraic alternative to the highly combinatorial lambda calculus. A characteristic feature of LAA's is the algebraic reformulation of (β)-conversion as the definition of abstract substitution. The equational axioms of LAA's reflect (α)-conversion and Curry's recursive axiomatization of substitution in the lambda calculus. *Functional* LAA's arise from environment models or lambda models, the natural models of the lambda calculus. The main result of the paper is a stronger version of the functional representation theorem for locally finite LAA's, the algebraic analogue of the completeness theorem of lambda calculus.

1 Introduction

Although the axioms of lambda calculus are all in the form of equations, the lambda calculus is not a true equational theory since the variable-binding properties of lambda abstraction prevent variables in lambda calculus from operating as real algebraic variables.

The way in which lambda abstraction theory arises from the lambda calculus almost exactly parallels the way cylindric algebras are obtained from first-order logic. The axioms of first-order logic are like those of lambda calculus in that the formula-variables can not be substituted without restriction. In both cases the source of the problem is the way substitution for individuals is handled. By dealing with substitution at the level of the object language rather than the metalanguage, i.e., by abstracting it, a pure equational formalization of lambda calculus can be developed giving rise to the theory of lambda abstraction algebras. Like cylindric algebras, and in contrast to the lambda calculus, the axioms of lambda abstraction theory are pure identities (more accurately, they turn out to be equivalent to pure identities). Among the seven axioms, the first six constitute a recursive definition of the abstract substitution operator; they express

* The work of the first author was supported in part by National Science Foundation Grant #DMS 8805870. The work of the second author was supported in part by a NATO Senior Fellowship Grant of the Italian Research Council.

precisely the metamathematical content of β-conversion. The last axiom is an algebraic translation of α-conversion. The most significant feature of the axioms is that they are true identities in the sense that they continue to hold when arbitrary terms are substituted for the variables. Thus the theory of lambda abstraction algebras gives a pure equational theory of lambda calculus, and lambda abstraction algebras form a variety in the universal-algebraic sense.

There is a notion of a "natural" lambda abstraction algebra–the algebras that the axioms of lambda abstraction theory are intended to characterize. They correspond to functional polyadic algebras and, more loosely, to representable cylindric algebras; we call them ***functional lambda abstraction algebras***. Not surprisingly, they are closely related to the natural models of the lambda calculus, viz., the environment models [12] and functional β-models [7]. Functional lambda abstraction algebras are obtained by ***coordinatizing*** environment models by the variables in a natural way. The important point here is that, in contrast to environment models, the intentional form of the function corresponds to an actual element of the functional lambda abstraction algebra. In this sense functional lambda abstraction algebras are much richer than environment models, and this greater richness translates into a more algebraic theory.

The basic theory of lambda abstraction algebras is developed in [15]. The main result there may be viewed as the natural algebraic analogue of the completeness theorem for the lambda calculus. It is the functional representation theorem for locally finite lambda abstraction algebras: every locally finite lambda abstraction algebra $\mathcal{A}$ is isomorphic to a functional lambda abstraction algebra obtained by coordinatizing an environment model having the same carrier set as $\mathcal{A}$. This result corresponds to what is called the functional representation theorem for locally finite polyadic Boolean algebras (Halmos [9]). However the natural algebraic analogues of the completeness theorem for first-order logic are the stronger representation theorem for simple, locally finite polyadic Boolean algebras of infinite degree ([9], Thm. 17.3) and the representation theorem for locally finite cylindric algebras ([10], Thm. 3.2.8). The representation theorem for locally finite lambda abstraction algebras that corresponds to these results is the main result of the paper.

The rest of this paper is organized in the following manner. First, for the sake of self-containedness, in section 2 we recall basic definitions from [15] that are made use of in the rest of this paper. In section 3 we provide an algebraic version of the omega rule as defined in [3] and show that every locally finite LAA $\mathcal{A}$ satisfying the omega rule is isomorphic to a functional LAA obtained by coordinatizing the environment model associated with the zero-dimensional elements of $\mathcal{A}$. The strong representation theorem for locally finite LAA's is obtained in sections 4 and 5 by embedding every locally finite LAA into a locally finite LAA satisfying the omega rule.

Connections with Other Work. The theories of cylindric and polyadic algebras are two early contributions to the algebraization of quantifier logics and have greatly influenced our work. The main reference for cylindric algebras is

[10]; for polyadic algebras it is [9]. We also mention here Nemeti [13]. It contains an extensive survey of the various algebraic versions of quantifier logics.

Lambda abstraction algebras, as cylindric and polyadic algebras, can be also viewed as a contribution to the theory of abstract substitution. However, in lambda abstraction and cylindric algebras, abstract substitution is a defined operation, while in polyadic algebras it is one of the primitive notions. The importance of abstract substitution, and lambda abstraction, has been recognized for some time among computer scientists because it leads among other things to more natural term rewriting systems, which are useful in the analysis of processes of computations. See for example [1]. A pure theory of abstract substitution has been developed by Feldman and Pinter [8]. This work parallels ours in many respects and we acknowledge our indebtedness to it. Diskin in [4][6] has independently developed an algebraic framework for the untyped lambda calculus which parallels ours. The main result in [4] is a representation theorem equivalent to the one presented in [15].

There have been several attempts to reformulate the lambda calculus as a purely algebraic theory within the context of category theory: Obtulowicz and Wieger [14] via the *algebraic theories* of Lawvere; Adachi [2] via *monads*; Curien [5] via *categorical combinators*.

Finally, work connecting a theory of substitution in combination with abstract variable-binding operators has been recently done. See [16], [17], [18].

2 Basic notions and notations

In this first section we summarize, without proof, definitions and results from [15] that we will need in the subsequent part of the paper. Let I be a nonempty set. The similarity type of *lambda abstraction algebras of dimension* I is $(\cdot, (\lambda x : x \in I), (x : x \in I))$, where $\cdot$ is a binary operation symbol, λx is a unary operation symbol for every $x \in I$, and x is a constant symbol (i.e., nullary operation symbol) for every $x \in I$. The elements of I are to be thought of as the variables of lambda calculus although in their algebraic transformation they no longer play the role of variables in the usual sense. We will refer to them as *λ-variables*. The actual variables of the lambda abstraction theory will be referred to as *context variables* and denoted by the greek letters ξ, ν, and μ, possibly with subscripts. The terms of the language of lambda abstraction theory are called *λ-terms*. They are constructed in the usual way: every λ-variable x and context variable ξ is a λ-term; if t and s are λ-terms, then so are $t \cdot s$ and $\lambda x(t)$. An occurrence of a λ-variable x in a λ-term is *bound* if it falls within the scope of the operation symbol λx; otherwise it is *free*. The *free variables* of a λ-term are the λ-variables that have at least one free occurrence. A λ-term without any context variables is said to be *pure*. A λ-term without free variables is said to be *closed*. Λ_I and Λ_I^0 denote respectively the set of pure λ-terms and the set of closed pure λ-terms. Because of their similarity to the terms of the lambda calculus we use the standard notational conventions of the latter. The application operation symbol "·" is normally omitted, and the application of two terms is written as juxtaposition ts. When parentheses are omitted, association to the left is assumed. The left parenthesis delimiting the scope of a λ-abstraction is replaced with a period and the right parenthesis is omitted. For example, $\lambda x(ts)$ is written $\lambda x.ts$. Successive λ-abstractions $\lambda x \lambda y \lambda z \cdots$ are written $\lambda xyz \cdots$.

Definition 1. By a **lambda abstraction algebra of dimension** I we mean an algebraic structure of the form

$$\mathcal{A} := (A, \cdot^{\mathcal{A}}, (\lambda x^{\mathcal{A}} : x \in I), (x^{\mathcal{A}} : x \in I))$$

satisfying the following quasi identities for all $x, y, z \in I$ (subject to the indicated conditions) and all $\xi, \mu, \nu \in A$.

(β_1) $(\lambda x.x)\xi = \xi$;
(β_2) $(\lambda x.y)\xi = y, \quad x \neq y$;
(β_3) $(\lambda x.\xi)x = \xi$;
(β_4) $(\lambda xx.\xi)\mu = \lambda x.\xi$;
(β_5) $(\lambda x.\xi\mu)\nu = (\lambda x.\xi)\nu((\lambda x.\mu)\nu)$;
(β_6) $(\lambda y.\mu)z = \mu \rightarrow (\lambda xy.\xi)\mu = \lambda y.(\lambda x.\xi)\mu, \quad x \neq y, z \neq y$;
(α) $(\lambda y.\xi)z = \xi \rightarrow \lambda x.\xi = \lambda y.(\lambda x.\xi)y, \quad z \neq y$.

I is called the **dimension set** of $\mathcal{A}$. $\cdot^{\mathcal{A}}$ is called **application** and $\lambda x^{\mathcal{A}}$ is called **λ-abstraction** with respect to x.

The class of lambda abstraction algebras of dimension I is denoted by LAA_I. A LAA_I is *infinite dimensional* if I is infinite. (In the sequel of the paper every lambda abstraction algebra will be assumed infinite dimensional.)

We note here that in the presence of the other axioms, (β_6) and (α) are equivalent to identities. Thus LAA_I is a variety for every dimension set I.

Substitution and Dimension. When transformed into the equational language of lambda abstraction theory, (β)-conversion becomes the definition of abstract substitution. It takes the following form:

Definition 2. Let $\mathcal{A}$ be a LAA_I.

(i) $S^x_b(a) = (\lambda x.a)b$ for all $x \in I$ and $a, b \in A$.
(ii) $S^{\mathsf{x}}_{\mathsf{b}}(a) = S^{x_1}_{b_1}(\ldots(S^{x_n}_{b_n}(a))\ldots)$ for all $\mathsf{x} = x_1 \cdots x_n \in I^*$, $\mathsf{b} = b_1 \cdots b_n \in A^*$, and $a \in A$.

S is called the *(abstract) substitution operator.*

Definition 3. Let $\mathcal{A}$ be an LAA_I. Let $a \in A$ and $x \in I$. a is said to be **algebraically dependent on x** (over $\mathcal{A}$) if $(\lambda x.a)z \neq a$ for some $z \in I$; otherwise **a is algebraically independent of x** (over $\mathcal{A}$). The set of all $x \in I$ such that a is algebraically dependent on x over $\mathcal{A}$ is called the **dimension set** of a and is denoted by $\Delta^{\mathcal{A}}a$; thus

$$\Delta^{\mathcal{A}}a = \{\, x \in I : (\lambda x.a)z \neq a \text{ for some } z \in I \,\}.$$

a is **finite (infinite) dimensional** if Δa is finite (infinite).

In the following proposition we give some basic properties of dimension set.

Proposition 4. *Let $\mathcal{A} \in LAA_I$, $a, b \in A$, and $x \in I$.*

(i) $\Delta(ab) \subseteq \Delta a \cup \Delta b$.
(ii) $\Delta(\lambda x.a) = \Delta a \setminus \{x\}$.
(iii) $\Delta(S^x_b(a)) \subseteq (\Delta a \setminus \{x\}) \cup \Delta b)$.
(iv) $\Delta x \subseteq \{x\}$, *with equality holding if $\mathcal{A}$ is nontrivial.*

The metalogical aspect of the theory of LAA's leads in a most natural way to the concept of locally finite LAA.

Definition 5. A lambda abstraction algebra $\mathcal{A}$ is **locally finite** if it is of infinite dimension and every $a \in A$ is of finite dimension (i.e., $|\Delta a| < \omega$).

Examples of locally finite LAA's can be defined as follows. Recall that a set of equations between pure λ-terms closed under (α) and (β) conversion and under the rules of lambda calculus is called a lambda theory. It is very easy to check that any lambda theory Θ determines a congruence over the absolutely free algebra

$$\mathcal{A}_I := (\Lambda_I, \cdot^{\mathcal{A}_I}, (\lambda x^{\mathcal{A}_I} : x \in I), (x^{\mathcal{A}_I} : x \in I))$$

of pure λ-terms. Then $\mathcal{A}_I/\Theta$ is a locally finite LAA_I associated with the lambda theory Θ. The local finiteness is a direct consequence of the trivial fact that every λ-term is a finite string of symbols and hence contains only finitely many variables.

Functional Lambda Abstraction Algebras. Only recently has a general consensus developed as to what the models of the lambda calculus should be. (A brief but illuminating history of the process can be found in [12].) The notion of an *environment model* (the name is due to Meyer [12]) originated with Hindley and Longo [11]. Meyer describes them as "the natural, most general formulation of what might be meant by mathematical models of the untyped lambda calculus". We shall define environment models in terms of *functional* lambda abstraction algebras which are the natural models of the LAA axioms, i.e., the models the axioms were intended to characterize.

We begin by giving the formal definition of functional domain; environment models turn out to be special kinds of functional domains from whose coordinatization functional LAA's are constructed.

Definition 6. Let $\mathcal{V} = (V, \cdot^{\mathcal{V}}, \lambda^{\mathcal{V}})$ be a structure where V is a nonempty set, $\cdot^{\mathcal{V}}$ is a binary operation on V, and $\lambda^{\mathcal{V}} : V^V \xrightarrow{p} V$ is a partial function assigning elements of V to certain functions from V into itself. $\mathcal{V}$ is called a **functional domain** if for each f in the domain of $\lambda^{\mathcal{V}}$,

$$f(v) = (\lambda^{\mathcal{V}}(f)) \cdot^{\mathcal{V}} v, \qquad \text{for all } v \in V.$$

This definition of functional domain differs from the one in Meyer [12] but is easily seen to be equivalent.

If $p \in V^I$, $v \in V$, and $x \in I$, then $p(v/x) \in V^I$ is the mapping such that for all $y \in I$

$$p(v/x)_y := \begin{cases} v, & \text{if } y = x \\ p_y, & \text{otherwise} \end{cases}$$

Definition 7. Let $\mathcal{V} = (V, \cdot^{\mathcal{V}}, \lambda^{\mathcal{V}})$ be a functional domain and let I be a nonempty set. Let $V_I = \{ f : V^I \xrightarrow{p} V \}$, i.e., the set of all partial functions from V^I to V. By the **I-coordinatization** of $\mathcal{V}$ we mean the algebra

$$\mathcal{V}_I = (V_I, \cdot^{\mathcal{V}_I}, (\lambda x^{\mathcal{V}_I} : x \in I), (x^{\mathcal{V}_I} : x \in I)),$$

where for all $a, b : V^I \xrightarrow{p} V$, $x \in I$, and $p \in V^I$:

- $(a \cdot^{\mathcal{V}_I} b)(p) = a(p) \cdot^{\mathcal{V}} b(p)$, provided $a(p)$ and $b(p)$ are both defined; otherwise $(a \cdot^{\mathcal{V}_I} b)(p)$ is undefined.
- $(\lambda x^{\mathcal{V}_I}.a)(p) = \lambda^{\mathcal{V}}(< a(p(v/x)) : v \in V >)$, provided $< a(p(v/x)) : v \in V >$ is in the domain of $\lambda^{\mathcal{V}}$ (note this implies $a(p(v/x))$ is defined for all $v \in V$); otherwise $(\lambda x^{\mathcal{V}_I}.a)(p)$ is undefined.
- $x^{\mathcal{V}_I}(p) = p_x$.

A subalgebra $\mathcal{A}$ of total functions of $\mathcal{V}_I$, i.e., a subalgebra such that $(\lambda x^{\mathcal{V}_I}.a)(p)$ is defined for all $a \in A$ and $p \in V^I$, is called a *functional lambda abstraction algebra*. I is the *dimension set* of $\mathcal{A}$ and $\mathcal{V}$ is its *value domain*. The class of all functional lambda abstraction algebras of dimension I is denoted by $FLAA_I$. It is easy to prove that every $FLAA_I$ is a lambda abstraction algebra. In the sequel a subalgebra of $\mathcal{V}_I$ of total functions will be called a total subalgebra of $\mathcal{V}_I$.

Definition 8. An **environment model** is a functional domain $\mathcal{V}$ with the property that there exists at least one $FLAA_I$ with value domain $\mathcal{V}$, i.e., the coordinatization of $\mathcal{V}_I$ has at least one total subalgebra.

The functional β-models of the lambda calculus introduced by Krivine in [7] do seem to capture the essence of a functional locally finite LAA, but he apparently does not give it an algebraic structure. It is interesting to compare the conditions that Meyer and Krivine use to characterize respectively environment models and functional β-models among functional domains. In both cases it amounts to requiring that the coordinatization includes a total subalgebra. Meyer requires that the subalgebra be the image of the set of lambda terms under evaluation. Krivine like us specifices only the existence of a subalgebra, but since there is no explicit algebraic structure he can only describe subalgebras in terms of certain closure properties.

In our view the most natural models of the lambda calculus are functional LAA's which correspond via coordinatization exactly to environment models. This highlights the main difference between our approach and the traditional one to models of the lambda calculus: the latter focuses attention on functional domains while we focus on their coordinatization. It is possible to prove that our notion of environment model is equivalent to Meyer's [12] since both are equivalent to lambda models.

3 The Omega Rule

In this section we give a new proof of the representation theorem for locally finite LAA's that is simpler than the one in [15] but requires an additional assumption.

We introduce a notion which is directly defined in terms of the dimension set. An element a of a LAA $\mathcal{A}$ is called *zero-dimensional* if $\Delta a = \emptyset$. We denote the set of zero-dimensional elements by $Zd\mathcal{A}$. A metalogical interpretation of $Zd\mathcal{A}$ is clear: if $\mathcal{A}$ is a LAA associated with a lambda theory, then $Zd\mathcal{A}$ is the set of all those elements of $\mathcal{A}$ which are equivalence classes of closed pure λ-terms, i.e., terms without free variables.

The following is an algebraic version of the ω-rule as defined in Barendregt [3].

Definition 9. We say that a LAA_I $\mathcal{A}$ satisfies the *ω-rule* if, for all $a, b \in A$ and all $x \in I$, we have that

$$(\forall c \in Zd\mathcal{A} : S_c^x(a) = S_c^x(b)) \rightarrow a = b.$$

Let $\mathcal{A} = (A, \cdot^{\mathcal{A}}, \lambda x^{\mathcal{A}}, x^{\mathcal{A}})_{x\in I}$ be an arbitrary LAA_I satisfying the ω-rule. We define the functional domain $\mathcal{V} = (V, \cdot^{\mathcal{V}}, \lambda^{\mathcal{V}})$ *associated with the zero-dimensional elements of* $\mathcal{A}$ as follows: $V = Zd\mathcal{A}$ and $\cdot^{\mathcal{V}} = \cdot^{\mathcal{A}}$. The domain of $\lambda^{\mathcal{V}} : V^V \xrightarrow{p} V$ is

$$D_{\mathcal{A}} = \{ < S_v^x(a) : v \in V >: \lambda x^{\mathcal{A}}.a \in Zd\mathcal{A} \text{ and } x \in I\},$$

and for each function in this set we define

$$\lambda^{\mathcal{V}}(< S_v^x(a) : v \in V >) := \lambda x^{\mathcal{A}}.a.$$

It is possible to prove that $\lambda^{\mathcal{V}}$ is well-defined and that the structure $\mathcal{V}$ is a functional domain under the hypothesis that $\mathcal{A}$ satisfies the ω-rule.

The following is one of the main results of the paper.

Theorem 10. *Every locally finite LAA_I $\mathcal{A}$ satisfying the ω-rule is isomorphic to a functional lambda abstraction algebra. More precisely, $\mathcal{A}$ is isomorphic to a total subalgebra of the I-coordinatization of the functional domain associated with the zero-dimensional elements of $\mathcal{A}$.*

We recall here the completeness theorem for the lambda calculus as presented in [12]: every lambda theory Θ consists of precisely the equations valid in an environment model having Λ_I/Θ as carrier set. As a consequence of Thm. 10, we obtain the following version of the completeness theorem: Every lambda theory Θ closed under the ω-rule consists of precisely the equations valid in an environment model having Λ_I^0/Θ as carrier set.

As a corollary of Thm. 10 we get that every locally finite LAA is isomorphic to a functional LAA provided we can show that every locally finite LAA is embeddable in one satisfying the ω-rule. This is the subject of the next two sections.

4 Dilations

The process of going from an algebra to one of its dilations might be thought of as the adjunction of new λ-variables. It is not at all obvious that new variables can always be adjoined to an arbitrary lambda abstraction algebra; this section is devoted to the discussion of the most important special cases of the problem. The dilations of locally finite LAA's behave particularly well. In this regard their properties are similar to locally finite cylindric and polyadic algebras. Dilations do not exist in general, and if they do they may not be unique (up to isomorphism). But in the case of locally finite LAA_I's we have both existence and uniqueness. This result has the following consequence for the untyped lambda calculus: every lambda theory Θ over a set I of λ-variables admits a unique conservative extension to a lambda theory Θ' over a set J of λ-variables with $J \setminus I$ infinite. The results presented in this section will be the basis for the proof of the strong representation theorem for locally finite LAA's in the next section.

Let $\mathcal{A}$ be a LAA_J and $I \subseteq J$. By the *I-reduct* of $\mathcal{A}$ we mean the algebra

$$\mathbf{R}_I\mathcal{A} = (A, \cdot^{\mathcal{A}}, \lambda x^{\mathcal{A}}, x^{\mathcal{A}})_{x \in I}.$$

Clearly this is a LAA_I. Define $\mathrm{Nr}_I\mathcal{A} = \{a \in A : \Delta a \subseteq I\}$. This set forms a subuniverse of the I-reduct. By the *I-neat reduct* of $\mathcal{A}$ we mean the algebra $\mathbf{Nr}_I\mathcal{A}$ whose operations are corresponding operations of $\mathcal{A}$ restricted to $\mathrm{Nr}_I\mathcal{A}$. $\mathbf{Nr}_I\mathcal{A}$ is obviously a LAA_I.

Let $\mathcal{A}$ be a locally finite LAA_I and $I \subseteq J$. By a *J-dilation* of $\mathcal{A}$ we mean any locally finite LAA_J $\mathcal{B}$ such that $\mathcal{A} = \mathbf{Nr}_I\mathcal{B}$.

Let $\mathcal{A}$ be a locally finite LAA_I and let $J \supseteq I$, with $J \setminus I$ infinite. Extend the language of lambda abstraction algebras of dimension J by adjoining a new constant symbol $\overline{a}$ for each element $a \in A$. A pure λ-term in this extended language, denoted by $P_J(\mathcal{A})$, is called *a polynomial lambda term of dimension J over* $\mathcal{A}$. Denote by

$$\mathcal{P}_J(\mathcal{A}) := (P_J(\mathcal{A}), \cdot^{\mathcal{P}_J(\mathcal{A})}, (\lambda x^{\mathcal{P}_J(\mathcal{A})} : x \in J), (x^{\mathcal{P}_J(\mathcal{A})} : x \in J))$$

the term algebra over the set of polynomial lambda terms of dimension J over $\mathcal{A}$. The operations are defined in the expected way. For example, if $t, s \in P_J(\mathcal{A})$,

$$t \cdot^{\mathcal{P}_J(\mathcal{A})} s = \begin{cases} \overline{a \cdot^{\mathcal{A}} b}, & \text{if } t = \overline{a} \text{ and } s = \overline{b} \text{ with } a, b \in A \\ t \cdot s, & \text{otherwise.} \end{cases}$$

Clearly $\mathcal{A}$ is isomorphic to a subalgebra of the I-reduct of $\mathcal{P}_J(\mathcal{A})$. Moreover, $\mathcal{P}_J(\mathcal{A})$ has the following universal mapping property. Given any LAA_J $\mathcal{B}$ such that $\mathcal{A}$ is a subalgebra of $\mathbf{R}_I\mathcal{B}$, then there is a unique homomorphism $t \mapsto t^{\mathcal{B}}$ from $\mathcal{P}_J(\mathcal{A})$ into $\mathcal{B}$ such that $\overline{a}^{\mathcal{B}} = a$ for all $a \in A$.

For any set S, $S^\star$ is the set of all finite strings of elements of S without repetitions. A pair of sequences $\mathsf{x} = x_1 \cdots x_n$, $\mathsf{y} = y_1 \cdots y_n \in J^\star$ is called *a change of λ-variables from $J \setminus I$ to I*, denoted by $\mathsf{x} \mapsto \mathsf{y}$, if $x_i \in J \setminus I$ and $y_i \in I$ for all $i = 1, \cdots, n$. A change of λ-variables $\mathsf{x} \mapsto \mathsf{y}$ determines the following transformation on $P_J(\mathcal{A})$.

1. $(x_i)_{\mathsf{x}\mapsto\mathsf{y}} = y_i$;
2. $z_{\mathsf{x}\mapsto\mathsf{y}} = z$ if $z \neq x_i$ for all $i = 1, \cdots, n$;
3. $\overline{a}_{\mathsf{x}\mapsto\mathsf{y}} = \overline{a}$ for all $a \in A$;
4. $(tu)_{\mathsf{x}\mapsto\mathsf{y}} = (t_{\mathsf{x}\mapsto\mathsf{y}} u_{\mathsf{x}\mapsto\mathsf{y}})$;
5. $(\lambda z.t)_{\mathsf{x}\mapsto\mathsf{y}} = \lambda z_{\mathsf{x}\mapsto\mathsf{y}}.t_{\mathsf{x}\mapsto\mathsf{y}}$.

Define the set $O(t)$ of all the λ-variables *occurring* in $t \in P_J(\mathcal{A})$ as follows: (i) $O(x) = \{x\}$ for all $x \in J$; (ii) $O(\overline{a}) = \Delta^{\mathcal{A}}(a)$ for all $a \in A$; (iii) $O(tu) = O(t) \cup O(u)$; (iv) $O(\lambda x.t) = O(t) \cup \{x\}$. A change of λ-variables $\mathsf{x} \mapsto \mathsf{y}$ is *suitable for $t \in P_J(\mathcal{A})$* if the following two conditions hold:

(i) $\mathsf{x} = x_1 \cdots x_n$ contains all the λ-variables in $J \setminus I$ occurring in t;
(ii) $\mathsf{y} = y_1 \cdots y_n$ is a sequence of λ-variables in I such that $y_i \notin O(t)$ for all $i = 1, \cdots, n$.

Consider the following relation Θ over $P_J(\mathcal{A})$:

$$t \;\Theta\; u \text{ iff } t^{\mathcal{A}}_{\mathsf{x}\mapsto\mathsf{y}} = u^{\mathcal{A}}_{\mathsf{x}\mapsto\mathsf{y}} \text{ for all } \mathsf{x} \mapsto \mathsf{y} \text{ suitable for } t \text{ and } u.$$

Note $\overline{a}^{\mathcal{A}}_{\mathsf{x}\mapsto\mathsf{y}} = a$ for all $a \in A$. Thus $\overline{a} \;\Theta\; \overline{b}$ iff $a = b$.

Lemma 11. *The relation Θ is a congruence over $\mathcal{P}_J(\mathcal{A})$.*

Let $\mathcal{A}[J]$ be the quotient algebra $\mathcal{P}_J(\mathcal{A})/\Theta$, where $A[J] = P_J(\mathcal{A})/\Theta$, $\cdot^{\mathcal{A}[J]} = \cdot^{\mathcal{P}_J(\mathcal{A})/\Theta}$, and similarly for the other operators.

Two dilations $\mathcal{B}$ and $\mathcal{C}$ of a LAA_I $\mathcal{A}$ are *$\mathcal{A}$-isomorphic* if there exists an isomorphism $h : \mathcal{B} \to \mathcal{C}$ such that $h(a) = a$ for all $a \in A$.

Theorem 12. *Let $\mathcal{A}$ be a locally finite LAA_I. Then the algebra $\mathcal{A}[J]$ is a locally finite lambda abstraction algebra of dimension J which is the unique J-dilation of $\mathcal{A}$ up to isomorphism. $\mathcal{A}$ is isomorphic to the neat I-reduct of $\mathcal{A}[J]$.*

The algebra $\mathcal{A}[J]$ will be called *the J-dilation* of the locally finite LAA_I $\mathcal{A}$.

5 Representation of locally finite LAA's

Functional lambda abstraction algebras are the natural models of the LAA axioms, i.e., the models the axioms were intended to characterize. It is natural to conjecture that, if $\mathcal{A}$ is an arbitrary LAA_I, then it is possible to construct a functional domain $\mathcal{V}$ so that $\mathcal{A}$ is isomorphic to a functional LAA_I with value domain $\mathcal{V}$. The conjecture is not true in general. However, in [15] we proved that the conjecture is true at least for the lambda abstraction algebras of greatest interest in the lambda calculus, i.e., the locally finite lambda abstraction algebras. We provide a strong version of the representation theorem for locally finite lambda abstraction algebras based on the concept of dilation.

Proposition 13. *Let $\mathcal{A}$ be a locally finite LAA_I and let $J \setminus I$ be infinite. Then the I-reduct $\mathbf{R}_I\mathcal{A}[J]$ is a lambda abstraction algebra satisfying the ω-rule.*

The previous proposition has the following consequence for the untyped lambda calculus: every lambda theory Θ over a set I of λ-variables admits an extension to a lambda theory Θ' satisfying the ω-rule. Thus the extension of the lambda calculus by the ω-rule is consistent (see Barendregt [3] for another proof of this result).

The following is the promised strong representation theorem.

Theorem 14 (The Strong Representation Theorem for Locally Finite LAA's). *Every locally finite LAA_I $\mathcal{A}$ is isomorphic to a functional lambda abstraction algebra which is a total subalgebra of the I-coordinatization of the functional domain associated with the zero dimensional part of $\mathbf{R}_I\mathcal{A}[J]$ (with $J \setminus I$ infinite).*

Simple lambda abstraction algebras are the algebraic counterpart of Hilbert Post complete lambda theories (see Def. 4.1.22 in [3]). We recall that an algebra is simple if it admits only the two trivial congruences.

Theorem 15. *Let $\mathcal{A}$ be a locally finite LAA_I. Then the lattices of the congruences on $\mathcal{A}$ and on the applicative subreduct $\mathbf{Zd}\,\mathcal{A} = (Zd\mathcal{A}, \cdot^{\mathcal{A}})$ are isomorphic. Thus $\mathcal{A}$ is simple iff $\mathbf{Zd}\,\mathcal{A}$ is such.*

As a matter of terminology, by a congruence on a functional domain $\mathcal{V}$ we mean a congruence on the applicative reduct $(V, \cdot^{\mathcal{V}})$ of $\mathcal{V}$.

Corollary 16. *Every simple locally finite LAA_I $\mathcal{A}$ satisfying the ω-rule is isomorphic to a functional LAA_I whose value domain is simple.*

A *zero-dimensional subreduct* of a LAA_I $\mathcal{A}$ is an applicative algebra (i.e., an algebra with a binary operation) $\mathcal{C} = (C, \cdot^{C})$, such that $C \subseteq Zd\mathcal{A}$ and $\lambda xy.xy \in C$. This last condition has the consequence that $\lambda x.cx \in C$ for all $x \in I$ and $c \in C$. A *functional congruence* on a zero-dimensional subreduct $\mathcal{C}$ is a congruence Θ satisfying the following condition for all $a, b \in C$:

$$(\forall c \in C : ac \ \Theta \ bc) \rightarrow \lambda x.ax \ \Theta \ \lambda x.bx.$$

$\mathcal{C}$ is called *functionally simple* if it does not admit functional congruences.

Let $\mathcal{A}$ be a functional LAA_I with value domain $\mathcal{V}$. For every $v \in V$, we denote by $\overline{v}$ the constant map from V^I to V defined by $\overline{v}(p) = v$ for all $p \in V^I$. We say that $\mathcal{A}$ is a *rich* if the map $v \mapsto \overline{v}$ defines an embedding of $(V, \cdot^{\mathcal{V}})$ into the applicative reduct $(A, \cdot^{\mathcal{A}})$. Under this hypothesis it is possible to show that, for every functional congruence Θ on $\mathcal{V}$, $\mathcal{A}$ admits a homomorphic image $\mathcal{B}$ which is a rich functional LAA_I with value domain $\mathcal{V}/\Theta$.

Theorem 17 (The Representation Theorem for Simple Locally Finite LAA's). *Let $\mathcal{A}$ be a simple, locally finite LAA_I and let $\mathcal{B}$ be a rich functional LAA_I with value domain $\mathcal{V}$ such that $\mathcal{A}$ can be embedded into $\mathcal{B}$. Then, for every functional congruence Θ on $\mathcal{V}$, $\mathcal{A}$ is isomorphic to a functional LAA_I with value domain $\mathcal{V}/\Theta$. If the lattice of functional congruences on $\mathcal{V}$ admits a maximal functional congruence, then $\mathcal{A}$ is isomorphic to a functional LAA_I whose value domain is functionally simple.*

References

1. Abadi, M., Cardelli, L., Curien, P.L., Lévy, J.J.: Explicit substitutions. Proc. 17th Conference POPL, San Francisco (1990)
2. Adachi, T.: A categorical characterization of lambda calculus models. Research Report No. C-49, Dept. of Information Sciences, Tokio Institute of Technology (1983)
3. Barendregt, H.P.: The lambda calculus. Its syntax and semantics. Revised edition, North-Holland Publishing Co. (1985)
4. Beylin, I.D., Diskin, Z.B.: Lambda substitution algebras. This volume (1993)
5. Curien, P.L.: Categorical combinators, sequential algorithms and functional programming. Pitman Pub. (1986)
6. Diskin, Z.B.: Lambda term systems (submitted).
7. Krivine, J.L.: Lambda-Calcul, types et modèles, Masson, Paris (1990)
8. Feldman, N.: Axiomatization of polynomial substitution algebras. J. Symbolic Logic **47** (1982) 481–492
9. Halmos, P.: Homogeneous locally finite polyadic Boolean algebras of infinite degree. Fund. Math. **43** (1956) 255–325
10. Henkin, L., Monk, J.D., Tarski, A.: Cylindric algebras, Parts I and II. North-Holland Publishing Co. (1971, 1985)
11. Hindley, R., Longo, G.: Lambda-calculus models and extensionality. Zeit. f. Math. Logik u. Grund. der Math. **26** (1980) 289–310
12. Meyer, A.R.: What is a model of the lambda calculus ? Inform. Control **52** (1982) 87–122
13. Németi, I.: Algebraizations of quantifier logics. An introductory overview. Studia Logica **50** (1991) 485–569
14. Obtułowicz, A., Wiweger, A.: Categorical, functorial, and algebraic aspects of the type-free lambda calculus. Banach Center Publications **9**, Warsaw (1982)
15. Pigozzi, D., Salibra, A.: An introduction to lambda abstraction algebras (to appear).
16. Pigozzi, D., Salibra, A.: The abstract variable-binding calculus (Manuscript).
17. Pigozzi, D., Salibra, A.: Polyadic algebras over non-classical logics. In: Algebraic Logic, Banach Center Publications, Vol. 28, Polish Academy of Sciences (1993).
18. Salibra, A.: A general theory of algebras with quantifiers. In: Algebraic Logic, Proc. Conf. Budapest 1988 (Andréka, H., Monk, J.D., Németi, I., eds.), Colloq. Math. Soc. J. Bolyai **54** North-Holland Publishing Co. (1991) 573–620

On Saturated Calculi for a Linear Temporal Logic

Regimantas Pliuškevičius

Institute of Mathematics and Informatics,
Akademijos 4, Vilnius 2600, LITHUANIA

Abstract. A new type of finitary and infinitary sequential calculi (named the saturated ones) for a linear temporal logic are introduced. Non-logical axioms in saturated calculi are some sequents, indicating the saturation of the derivation process in these calculi. The finitary saturation suggests that "nothing new" can be obtained continuing the derivation process. An infinitary saturated calculus instead of an ω-type rule of inference has an infinite set of "saturated" sequents, but the form of these sequents is uniform. The saturation presents the unique deductive principle both for finitary and infinitary cases. The derivability in a finitary saturated calculus serves as a finitary completeness criterion for the first order linear temporal logic.

1 Introduction

In [7] it was proved that the first order temporal logic with unary predicate symbols and with equality and function symbols is incomplete, because Peano arithmetic can be embedded in this logic. Subtle incompleteness results for the first order linear temporal logic were obtained in [3, 4]. In [1] it was proved that the first order temporal logic becomes complete under nonstandard semantics and will allow new rules to define auxiliary predicates. In [2, 5, 6] it was proved that an infinitary calculus for the first order linear temporal logic with the ω-type rule of inference is complete under standard Kripke-style semantics.

It is well known that a propositional linear temporal logic is finitary complete and decidable. In a finitary complete discrete temporal sequential calculus the main rule of inference is the so-called invariant rule having the form of some analytic-cut-like rule in which the "cut formula" is called an invariant formula. The process of derivability in the finitary case differs, in principle, from the one in the infinitary complete case (with the ω-type rule of inference).

The purpose of this paper is to introduce a new type of finitary and infinitary sequential calculi (named the saturated ones) for a linear temporal logic with $\bigcirc$ (next) and $\Box$ (always). Non-logical axioms in saturated calculi are some sequents, indicating the saturation of the derivation process in these calculi. The saturation intuitively corresponds to a certain type of regularity in the derivations for the logic. The finitary saturation suggests that "nothing new" can be obtained continuing the derivation process. The derivability in a finitary saturated calculus serves as a finitary completeness criterion for the first order linear temporal logic. The "finitary saturation" replaces the invariant rule and is more "computer-aided" than latter. An infinitary saturated calculus in place of the ω-type rule of inference has an infinite set of "saturated" sequents, but the shape of these sequents is uniform. The attractive property of "saturation principle" (both finitary and infinitary) lies in presenting a unique deductive principle both for the finitary and infinitary cases. The non-logical axioms of a saturated calculus are constructed dependent on the specific peculiarities of the given sequent. Therefore for each given sequent (whose derivability requires the application of the induction-like postulate) a concrete saturated calculus is constructed. This property makes saturated calculi more effective than the traditional ones (based on the fixed induction-like postulate). For simplicity the saturated calculi are mainly described and founded for the so called class of Horn miniscoped formulas (in short: MH-formulas). MH-formulas have the form $((A \supset)^{\circ} \Box^{\circ} B)$, where A does not contain a positive occurrence of $\Box, (A \supset)^{\circ} \in \{\emptyset, A.\supset\}, \Box^{\circ} \in \{\emptyset, \Box\}, B$ does not contain a positive occurrence of $\Box$. Besides, in their temporal parts quantifiers enter only the formulas of the form $Q\overline{x}E(Q \in \{\forall, \exists\})$, where E is an elementary formula (i.e. $E = P(t_1, \ldots, t_n)$, P is an n-place ($n \geqslant 0$) predicate symbol). Despite of their simplicity, MH-formulas can express, for example, liveness properties of the form $\Box(A \supset \Diamond B)$. Some generalizations of saturated calculi to arbitrary formulas are described. The present paper extends the results of [5]. The proposed method of constructing saturated calculi can be applied to other non-classical logics containing the induction-like postulates.

2 Description of the Infinitary Sequential Calculus $G_{L\omega}$

The foundation of a saturated calculus (both finitary and infinitary) is carried out with the help of the infinitary calculus $G_{L\omega}$ containing the ω-type rule of inference. In $G_{L\omega}$ we consider arbitrary formulas (i.e. not only (MH-formulas) which are determined by means of logical symbols and a temporal operator $\Box$ as usual. Throughout the paper $\omega := \{0, 1, \ldots, n, \ldots\}$, the formula $\bigcirc \ldots \bigcirc A$ (k-times next A, $k \geqslant 1$) will be abbreviated as A^k (i.e. as a formula with the index k) more precisely: 1) if E is an elementary formula, $i, k \in \omega$, $k \neq 0$, then $(E^i)^k := E^{i+k}$ ($E^0 := E$); E^l ($l \geqslant 0$) will be called an atomic formula (if $l = 0$ then E^l becomes an elementary one); 2) $(A \odot B)^k := A^k \odot B^k if \odot \in \{\supset, \vee, \wedge\}$; 3) $(\sigma A)^k := \sigma A^k if \sigma \in \{\urcorner, \Box, \forall x, \exists x\}$.

A sequent is an expression of the form $\Gamma \rightarrow \Delta$, where Γ, Δ are arbitrary finite sets (not sequences or multisets) of formulas.

The calculus $G_{L\omega}$ is defined by the following postulates.

Axiom: $A \rightarrow A$.

Rules of inference:

1) temporal rules:

$$\frac{A, \Box A^1, \Gamma \rightarrow \Delta}{\Box A, \Gamma \rightarrow \Delta}(\Box \rightarrow) \qquad \frac{\{\Gamma \rightarrow \Delta, A^k\}_{k \in \omega}}{\Gamma \rightarrow \Delta, \Box A}(\rightarrow \Box_\omega),$$

where $k \in \omega$; here and below Γ^1 means $A_1^{k_1+1}, \ldots, A_n^{k_n+1}$, if $\Gamma = A_1^{k_1}, \ldots, A_n^{k_n}$, $n \geqslant 1$, $k_i \geqslant 0$, $1 \leqslant i \leqslant n$;

2) logical rules of inference consist of traditional invertible rules of inference;

3) structural rules: (W) (weakening); from the definition of a sequent it follows that $G_{L\omega}$ implicitly contains the structural rules "contraction" and "exchange".

Derivations in $G_{L\omega}$ are built up in a usual way (for the calculi with the ω-rule), i.e. in the form of an infinite tree (with finite branches); the height of a derivation D is an ordinal (defined in a traditional way). Let I be some calculus, then $I \vdash S$ means that the sequent S is derivable in S.

A derivation D in $G_{L\omega}$ will be called atomic if all axioms occurring in D have the form $E \rightarrow E$, where E is an atomic formula.

Lemma 2.1. An arbitrary derivation in $G_{L\omega}$ may be transformed into an atomic one.

Proof: using the rules of inference of $G_{L\omega}$.

Remark 2.1 All derivations in $G_{L\omega}$ will be regarded as atomic ones.

Lemma 2.2 (invertibility of the rules of inference of $G_{L\omega}$). If S_1 is a premise and S is the conclusion of any rule of inference of $G_{L\omega}$, different from (W)), then $G_{L\omega} \vdash S \Longrightarrow G_{L\omega} \vdash S_1$.

Proof: by induction on the height of the given atomic derivation.

Theorem 2.1 (a). The calculus $G_{L\omega}$ is sound and complete; (b) the cut rule is admissible in $G_{L\omega}$.

Proof: see [2, 5].

3 Description and Investigation of the Sequential Saturated Calculus Sat

Let $S = A_1, \ldots, A_n \rightarrow B_1, \ldots, B_m$, then $S^F = \bigwedge_{i=1}^{n} A_i \supset \bigvee_{i=1}^{m} B_i$. The sequent S will be called MH-sequent, if S^F is the MH-formula. First let us define the canonical form of MH-sequents (simply: sequents). A sequent S will be called primary, if $S = \Sigma_1, \Pi_1^1, \Box\Omega^1 \rightarrow \Sigma_2, \Pi_2^1, \Box\Delta^1$, where $\Sigma_i = \varnothing$ $(i = 1,2)$ or consists of logical formulas without indices; $\Pi_i^1 = \varnothing$ $(i = 1,2)$ or consists of logical formulas with indices; $\Box\Omega^1 = \varnothing$ or consists of arbitrary MH-formulas of the form $\Box A^1$; $\Box\Delta^1 = \varnothing$ or $\Box\Delta^1 = \Box B^1$ (as follows from the definition of MH-sequents, Ω^1, B^1 does not contain positive (with respect to S) occurrences of $\Box$). If $S = \Sigma_1, \Pi_1^1, \Box\Omega \rightarrow \Sigma_2, \Pi_2^1, \Box\Delta$ (i.e., it is not necessary that $\Box\Omega = \Box\Omega_1^1, \Box\Delta = \Box\Delta_1^1$) then such a sequent S will be called quasiprimary. It is clear that every primary sequent is the quasiprimary one. The sequent S will be called ordinary, if S contains both negative and positive occurrences of $\Box$. The sequent S will be called singular (simple) if S does not contain negative (positive, respectively) occurrences of $\Box$. An ordinary primary (quasiprimary) sequent will be called proper, if $S \neq \Box\Omega^1 \rightarrow \Box B^1$ $(S \neq \Box\Omega \rightarrow \Box B$, respectively).

The rules of inference of the calculus Sat consist of the rules of inference of $G_{L\omega}$, different from $(\rightarrow \Box_\omega)$, and the following three rules of inference:

$$\frac{\Pi_1 \rightarrow \Pi_2, A;\ \Pi_1 \rightarrow \Pi_2, \Box A^1}{\Pi_1 \rightarrow \Pi_2, \Box A}(\rightarrow \Box^1) \quad \frac{\Box\Gamma \rightarrow A}{\Box\Gamma \rightarrow \Box A}(\Box) \quad \frac{S_i^*}{S^*}(A)\ (i \in \{1,2\}),$$

where $\Pi_j \neq \varnothing$ $(j \in \{1,2\})$ and $\Pi_1 \neq \Box\Gamma$; $\Box\Gamma = \Box B_1, \ldots, \Box B_m$ $(m \geqslant 0)$; S^* is a primary sequent, i.e. $S^* = \Sigma_1, \Pi_1^1, \Box\Omega^1 \rightarrow \Sigma_2, \Pi_2^1, \Box\Delta^1$; $S_1^* = \Sigma_1 \rightarrow \Sigma_2$; $S_2^* = \Pi_1, \Box\Omega \rightarrow \Pi_2, \Box\Delta$.

Lemma 3.1 (invertibility of the rules of inference of Sat). All rules of inference of Sat, different from (W) and (A), are invertible in $G_{L\omega}$.

Proof. The invertibility of rules of inference, except $(\rightarrow \Box^1), (\Box), (A), (W)$. follows from invertibility of the rules of inference of $G_{L\omega}$. The invertibility of $(\rightarrow \Box^1)$ and $(\Box)$ follows from the fact that $G_{L\omega} \vdash \Box A \rightarrow A \wedge A^1$ and the admissibility of cut in $G_{L\omega}$.

Lemma 3.2 (disjunctional invertibility of (A)). Let S^* be the conclusion of (A), i.e. $S^* = \Sigma_1, \Pi_1^1, \Box\Omega^1 \rightarrow \Sigma_2, \Pi_2^1, \Box\Delta^1$, then $G_{L\omega} \vdash S^* \Longrightarrow \mathrm{Log} \vdash \Sigma_1 \rightarrow \Sigma_2$ or $G_{L\omega} \vdash \Pi_1, \Box\Omega \rightarrow \Pi_2, \Box\Delta$, where Log is the calculus obtained from $G_{L\omega}$ by dropping $(\rightarrow \Box_\omega), (\Box \rightarrow)$.

Proof: by induction on the height of the given atomic derivation.

A derivation in Sat is constructed bottom-up applying the rules of inference of Sat. As follows from Lemmas 3.1, 3.2 all sequents from the derivation in Sat are derivable in $G_{L\omega}$. Some sequents which indicate the saturation of a derivation process in Sat play the role of non-logical axioms in Sat. To define the class of these sequents let us define the tactic of constructing a derivation in Sat. First let us define the notion of reduction of a sequent. Let $\{i\}$ denote the set of rules of inference of Sat, let $S, S_1, \ldots, S_n$ denote sequents. The $\{i\}$-reduction (or briefly: reduction) of S to $S_1, \ldots, S_n$, denoted by $R(S)\{i\} \Longrightarrow \{S_1, \ldots S_n\}$ or briefly by $R(S)$, is defined to be a tree of sequents with the root S and leaves $S_1, \ldots, S_n$, and possibly some logical axioms such that every sequent in $R(S)$ different from S is an "upper sequent" of the rule of inference in $\{i\}$ whose "lower sequent" also belongs to $R(S)$.

Lemma 3.3. For each sequent S one can construct $R(S)\{i\} \Longrightarrow \{S_1, \ldots, S_n\}$, where $\forall i$ $(1 \leqslant i \leqslant)$ S_i is a primary (quasiprimary) sequent; $\{i\}$ is the set of rules of inference of Sat, different from (A), $(\Box)$ and the rules of inference for quantifiers (and $(\Box \rightarrow), (\rightarrow \Box^1)$, respectively); besides $G_{L\omega} \vdash S \Longrightarrow G_{L\omega} \vdash S_i$ $(i = 1, \ldots, n)$.

Proof: follows from Lemma 3.1.

The set of the primary (quasiprimary) sequents from Lemma 3.3 will be denoted by $P(S)$ ($QP(S)$, respectively). Let I_1 be the calculus obtained from $G_{L\omega}$ by dropping $(\rightarrow \Box_\omega)$; I_2 (I_3) be the calculus, obtained from I_1 by adding $(\Box)$ ($(\Box)$, (A) and dropping $(\Box \rightarrow)$, respectively).

Lemma 3.4. Let $G_{L\omega} \vdash S$, then (a) $I_1 \vdash S$, if S is simple; (b) $I_2 \vdash S$, if $S = \Box\Gamma \rightarrow \Box A$; (c) $I_3 \vdash S$, if S is singular.

Proof: applying Lemmas 3.1, 3.2.

Let S be a proper quasiprimary sequent, i.e. $S = \Sigma_1, \Pi_1^1 \rightarrow \Sigma_2, \Pi_2^1, \Box A$ (where $\Sigma_i \Pi_i^1, \Omega \neq \varnothing, i = 1,2$) let us define the notion of resolvent of the sequent S, denoted by $Re(S)$. Let us construct $P(S)$ and let $S_j^* \in P(S)$ $(j = 1, \ldots, n)$; let us apply (A) bottom-up to S_j^* and let S_{j2}^* be the "temporal" conclusion of this bottom-up application of (A) (i.e. $S_{j2}^* = \varnothing$, if $i = 1$ in (A))), then $Re(S) = \{S_{12}^*, \ldots, S_{n2}^*\} = \{S_1, \ldots, S_k\}$ $(k \leqslant n)$. We say that a sequent S absorbs the sequent S' (in symbols: $S' \prec S$ or $S \succ S'$), if S' can be obtained from S with the help of the structural rule of inference (W). Let us introduce the notion of the k-th resolvent of S, denoted by $Re^k(S)$ $(k \in \omega)$. $Re^k(S)$ is constructed as a tree of sequents, defined as follows $Re^0(S) = S$; $Re^1(S) = Re(S)$. Let $S_k \in Re^k(S)$, $S_l \in Re^l(S)$ $(k < l)$ and S_k, S_l belong to the same branch; we say that S_k is saturated, if $S_k \succcurlyeq S_l$ ($S' \preccurlyeq S''$ means $S' = S''$ or $S' \prec S''$); the sequent S_l is called the absorbed one. Let $S_k \in Re^k(S)$, we say that the sequent S_k is blocked if S_k is absorbed and $\forall j$ $(f-1 \leqslant j < k-1)$ $S_j \in Re^j(S)$ is either saturated, or absorbed, and S_f is saturated; f is some natural number and S_f, $S_{f+1}, \ldots, S_k$ belong to the same branch; otherwise $S_k \in Re^k(S)$ will be called non-blocked. Let $S_i \in Re^k(S)$ and S_i be non-blocked, then $Re^{k+1}(S) = \bigcup_i Re(S_i)$. A branch B in the derivation D will be called saturated if the leaf of B is blocked. A branch B in the derivation D in Sat will be called closed if the leaf of B is either a logical axiom or B is saturated. The derivation D in Sat of the sequent S will be called closed (in symbols: Sat $\vdash^D S$) if all the branches of D are closed. Let $S_i \in Re^k(S)$, then the sequent S_i is called the resolvent one. We say that the derivation D in Sat is constructed using the resolvent tactic, if D is constructed by generating $Re^k(S)$ $(k \in \omega)$. Let Sat $\vdash^D S$, then the set of saturated sequents from D will be denoted by Sat $\{S\}$. It is easy to see that the same sequent may have several sets Sat $\{S\}$.

Let us describe the tactic of constructing the derivation of an arbitrary MH-sequent S in Sat. Let us construct $QP(S) = \{S_1, \ldots, S_n\}$. If S_i is either simple or singular, or $S_i = \Box\Gamma \to \Box A$, then let us try to derive S_i in I_j $(j = 1, 2, 3)$ (see Lemma 3.4). If S_i is proper ordinary quasiprimary, then let us try to derive S_i by generating $Re^k(S_i)$.

Example 3.1. (a) Let $S = P(c), \Box\Omega_1, \Box\Omega_2, \Box\Omega_3 \to \Box Q(c)$, where $\Omega_1 = (\exists x P(x) \supset \forall x P^1(f(x)))$, $\Omega_2 = (\exists x P(x) \supset \forall x Q(x))$, $\Omega_3 = (\exists x Q(f(x)) \supset \forall x Q(x))$. Then $P(S) = P(c), \forall x P^1(f(x)), \Box\Omega_1^1, \Box\Omega_2^1, \Box\Omega_3^1, \Omega_2, \Omega_3 \to \Box Q^1(c)$. Let us generate $Re^k(S) : Re^o(S) = S$; $Re^1(S) = Re(S) = S^* = \forall x P(f(x)), \Box\Omega_1, \Box\Omega_2, \Box\Omega_3, \to \Box Q(c) = Re^2(S)$. Therefore Sat $\vdash S$ and Sat $\{S\} = S^*$. It is easy to verify that $G_{L\omega} \vdash Re^k(S), k \in \{0, 1\}$.
(b) Let $S = P, Q, \Box\Omega \to \Box Q$, where $\Box\Omega = \Box(P \supset P^2), \Box(P^2 \supset Q^2), \Box(Q^2 \supset Q^1)$. Then it is easy to verify that Sat $\vdash^D S$ and D contains two saturated branches and Sat $\{S\} = \{S_1, S_2\}$, where $S_1 = P, Q, \Box\Omega \to \Box Q$; $S_2 = P^1, Q^1, Q, \Box\Omega \to \Box Q$. In one branch $S_1 \succ P, Q, (P^1 \supset Q^1), (Q^1 \supset Q), \Box\Omega \to \Box Q = S_1'$ and $S_2 \succ P^1, Q^1, Q, (P \supset Q), \Box\Omega \to \Box Q = S_2'$. In another branch $S_1 \succ P, Q, P^1, (P^1 \supset Q^1), (Q^1 \supset Q), \Box\Omega \to \Box Q = S_3'$. If we do not notice that $S_i \succ S_i'$ $(i = 1, 2)$, then we get another Sat $\{S\} = \{S_1' S_2', S_3', S_4'\}$, where $S_4' = P, P^1, Q^1, Q, \Box\Omega \to \Box Q$.

Let us define the notion of a subformula and of a resolvent subformula of an arbitrary formula of the MH-sequent. The set of subformulas (resolvent subformulas) of formula A will be denoted by Sub (A) (RSub (A), respectively) To define the set RSub (A) an auxiliary set R^* Sub (A) is introduced. 1. Sub $(A) = R^*$ Sub $(A) = A$, if A is an elementary formula. 2. Sub $(A \odot B) = \{A \odot B\} \cup$ Sub $(A) \cup$ Sub (B); R^*Sub $(A \odot B) = \{A \odot B\} \cup R^*$ Sub $(A) \cup R^*$ Sub (B); where $\odot \in \{\supset, \wedge, \vee\}$; 3. Sub $(\neg A) = \{\neg A\} \cup$Sub (A); R^* Sub $(\neg A) = \{\neg A\} \cup R^*$ Sub (A). 4. Sub $(QxA(x)) = \{QxA(x)\} \cup$ Sub $\{A(t)\}$ $(Q \in \{\forall, \exists\}$, t is some term);

$R^* \operatorname{Sub}(QxA(x)) = \{QxA(x)\}$, if $A(x) = Q_1x_1, \ldots, Q_nx_nE(x, x_1, \ldots, x_n)$ ($Q_i \in \{\forall, \exists\}$, E is elementary), otherwise $R^* \operatorname{Sub}(QxA(x)) = \emptyset$. 5. $\operatorname{Sub}(A^1) = \{A^1\} \cup \{\operatorname{Sub}(A)\}$; $R^* \operatorname{Sub}(A^1) = \{A^1\} \cup R^* \operatorname{Sub}(A)$. 6. $\operatorname{Sub}(\Box A) = \{\Box A^k \mid k \in \omega\} \cup \operatorname{Sub}(A)$; $R^* \operatorname{Sub}(\Box A) = \{\Box A\} \cup R^* \operatorname{Sub}(A)$. Let A be a formula, then A^{-1} will mean the formula B^{k-1}, if $A = B^k$ ($k > 0$) and the empty word, otherwise. Let $R^* \operatorname{Sub}(A) = \{A_1, \ldots, A_n\}$, then $R\operatorname{Sub}(A) = \{A_1^{-1}, \ldots, A_n^{-1}\}$. Therefore, $\operatorname{Sub}(A)$ (where $\Box \in A$) is infinite, whereas $R\operatorname{Sub}(A)$ is finite, even in the non-propositional case.

Example 3.2. Let $A = \Box(\exists x P(x) \supset \forall x P^1(x))$, then $R\operatorname{Sub}(A) = \{A, \forall x P(x)\}$.

Let Sat$'$ be the calculus obtained from Sat by dropping ($\Box$).

Lemma 3.5. Let S be any proper ordinary quasiprimary sequent and let Sat$' \vdash^D S$, Then either $\forall i$ $(1 \leqslant i \leqslant n)$ $S_i \in \operatorname{Sat}\{S\} \Longrightarrow S_i = \Pi_i, \Box\Omega \to \Delta_i, \Box A$, where $\Pi_i, \Delta_i \in R\operatorname{Sub}(\Omega)$ or $S_i = \Box\Omega \to \Box A$; besides $G_{L\omega} \vdash S \Longrightarrow G_{L\omega} \vdash S_i$ ($i = 1, \ldots, n$) (a).

Proof. The form of S_i follows from the construction of D; the point (a) follows from Lemmas 3.1, 3.2.

Lemma 3.6. Let Sat$' \vdash^D S$ and $S' \in \operatorname{Sat}\{S\}$, let $Re(S') = \{S_1, \ldots, S_n\}$, then $\forall i$ $(1 \leqslant i \leqslant n)$ $S_i \preccurlyeq S^* \in \operatorname{Sat}\{S\}$.

Proof: follows from the definition of Sat$\{S\}$.

Example 3.3. Let S be the sequent from Example 3.1(b), and let Sat$\{S\} = \{S_1, S_2\}$, then $Re\{S_1\} = \{S_2\}$, $Re\{S_2\} = \{S_1', S_2'\}$, $S_1' \prec S_1$, $S_2' \prec S_2$. Let Sat$\{S\} = \{S_1', S_2', S_3', S_4',\}$, then $Re\{S_1'\} = S_2'$; $Re\{S_2'\} = \{S_1', S_4'\}$; $Re\{S_3'\} = S_4'$; $Re\{S_4'\} = S_3'$.

4 Description of an Invariant Calculus IN and the Equivalence of $G_{L\omega}$, Sat and IN for MH-sequents

To clarify the role of saturated sequents let us introduce an "invariant" calculus IN. The postulates of the calculus IN are obtained from the calculus $G_{L\omega}$ replacing ($\to \Box_\omega$) by ($\to \Box^1$), ($\Box$) and the following one:

$$\frac{\Gamma, \Box\Omega \to \Delta, R; \quad R \to R^1; \quad R \to A}{\Gamma, \Box\Omega \to \Delta, \Box A} (\to \Box),$$

where (1) $\Gamma, \Box\Omega \to \Delta, \Box A \in \operatorname{Sat}\{S\} = \{\Pi_1, \Box\Omega \to \Delta_1, \Box A, \ldots, \Pi_n, \Box\Omega \to \Delta_n, \Box A\}$ and S is a proper ordinary quasiprimary sequent such that Sat$' \vdash S$; (2) $R = \bigvee_{i=1}^{n} (\Pi_i^\wedge \wedge \neg\Delta_i^\vee) \wedge \Box\Omega$ (where $\Gamma^\wedge(\Gamma^\vee)$ means the conjunction (disjunction) of formulas from Γ); the formula R is called an invariant formula.

Remark 4.1. It is clear that the rule of inference ($\to \Box$) destroys the subformula property (it becomes an analytic cut-like rule of inference after adding the conditions (1), (2), see above), which is restored by means of saturation principle. The rule of inference ($\to \Box$) corresponds to the induction-like axiom $A \wedge \Box(A \supset A^1) \supset \Box A$.

Lemma 4.1. In the calculi $G_{L\omega}$, IN the following rule of inference: $\frac{\Gamma \to \Delta}{\Gamma^1 \to \Delta^1}$ (+1) is admissible.

Proof: by induction on the height of the derivation of sequent $\Gamma \to \Delta$.

Lemma 4.2. In the calculus $G_{L\omega}$ the rule of inference $(\to \Box)$ is admissible.

Proof. Using the premises of $(\to \Box)$, admisibility of cut and $(+1)$ in $G_{L\omega}$ and by induction on k we get $G_{L\omega} \vdash \Gamma \to \Delta, A^k$ $(k \in \omega)$. Hence, by $(\to \Box_\omega)$ we get $G_{L\omega} \vdash \Gamma \to \Delta, \Box A$.

Lemma 4.3. The rule of inference $(\to \Box^1)$ is admissible in $G_{L\omega}$.

Proof. Using the fact that $G_{L\omega} \vdash A \wedge \Box A^1 \to \Box A$ and the admissibility of cut in $G_{L\omega}$.

Lemma 4.4. The rule of inference $(\Box)$ is admissible in $G_{L\omega}$.

Proof. Using Lemma 4.1 and $(\Box \to), (W)$ we get $G_{L\omega} \vdash \Box\Gamma \to A \Longrightarrow G_{L\omega} \vdash \Box\Gamma \to A^k$ $(k \in \omega)$. Therefore by $(\to \Box_\omega)$ we get $G_{L\omega} \vdash \Box\Gamma \to \Box A$.

Theorem 4.1. $IN \vdash S \Longrightarrow G_{L\omega} \vdash S$.

Proof: follows from Lemmas 4.2, 4.3, 4.4.

Now, we prove that Sat $\vdash S \Longrightarrow G_{L\omega} \vdash S$. Let I be the calculus obtained from $G_{L\omega}$ replacing $(\to \Box_\omega)$ by $(\to \Box^1)$.

Lemma 4.5. Let Sat $\vdash S$ and $S_i = \Pi_i, \Box\Omega \to \Delta_i, \Box A \in \text{Sat}\,\{S\}$ $(i = 1, \ldots, n)$, then $\forall i$ $(1 \leqslant i \leqslant n)$ $I \vdash \Pi_i, \Box\Omega \to \Delta_i, B \Longrightarrow I \vdash \Pi_i, \Box\Omega \to \Delta_i, B^1$.

Proof: using Lemma 3.5 (see Lemma 2.7 [5])

Lemma 4.6. Let Sat $\vdash^D S$ and $S_p \in D$ $(p = 1, 2, \ldots)$ then $G_{L\omega} \vdash S_p$ for all p.

Proof: using Lemmas 3.6, 4.5 (see Lemma 3.4 [5]).

Theorem 4.3. Sat $\vdash S \Longrightarrow G_{L\omega} \vdash S$.

Proof: follows from Lemma 4.6.

Now, we prove that Sat $\vdash S \Longleftrightarrow IN \vdash S$.

Lemma 4.7. Let Sat$'$ $\vdash S$, then $I \vdash R \to R^1$, where R is the invariant formula, as indicated above.

Proof: Taking R instead of B in Lemma 4.5 and using logical rules of inference we get $I \vdash \Pi_i$, $\Box\Omega \to \Delta_i$, R (for each i), therefore, by Lemma 4.5 we get $I \vdash S_i^* = \Pi_i$, $\Box\Omega \to \Delta_i$, R^1. Applying logical rules of inference to S_i^* we get $I \vdash R \to R^1$.

Lemma 4.8. Let Sat$'$ $\vdash^D S$ then $\forall i$ $(1 \leqslant i \leqslant n)$ $IN \vdash S_i = \Pi_i, \Box\Omega \to \Delta_i, \Box A \in$ Sat $\{S\}$.

Proof. It is obvious that Prop $\vdash \Pi_i, \Box\Omega \to \Delta_i, R$ (1). From Lemma 4.7 we get $I \vdash R \to R^1$ (2). From the construction of D it follows that $I_1 \vdash \Pi_i, \Box\Omega \to \Delta_i, A$ $(3_i')$ $(i = 1, \ldots, n)$ (where I_1 is the calculus obtained from $G_{L\omega}$ by dropping $(\to \Box_\omega)$). Applying $(\neg \to), (\wedge \to), (\vee \to)$ to $(3_i')$ we get $I_1 \vdash R \to A$ (3). Applying $(\to \Box)$ to (1), (2), (3) we get $IN \vdash S_i$.

Lemma 4.9. Let Sat $\vdash^D S$ and $IN \vdash^{D_1} S^*$, where S^* is any leaf of D, then $IN \vdash^{D_2} S$, besides if D_1 does not contain the applications of $(\to \Box)$, then D_2 does not contain the applications of $(\to \Box)$, either.

Proof: follows from the fact that any application of (A) can be replaced by the application of $(+1)$ (which is admissible in IN) and (W).

Lemma 4.10. $IN \vdash S \Longrightarrow$ Sat $\vdash S$.

Proof: follows from the definition of Sat.

Theorem 4.4. Sat $\vdash S \Longleftrightarrow IN \vdash S$.

Proof. The part $\Longrightarrow$ follows from Lemmas 3.6, 4.8, 4.9 the part $\Longleftarrow$ follows from Lemma 4.10.

Lemma 4.11 Let $S = \Pi, \Box\Omega \rightarrow \Delta, \Box A$ be a proper ordinary quasiprimary sequent, then $G_{L\omega} \vdash S \Longrightarrow \mathrm{Sat}' \vdash S$.

Proof. Starting from S let us construct a tree D of sequents, applying the resolvent tactic. Since $G_{L\omega} \vdash S$, (as follows from Lemmas 3.1, 3.2), all the sequents from D are derivable in $G_{L\omega}$. Let us consider the bottom-up applications of $(\rightarrow \Box^1)$ in D. As $G_{L\omega} \vdash S$, then (as follows from Lemma 3.1 and since S is the MH-sequent), we conclude that the left premise of $(\rightarrow \Box^1)$ is derivable in I_1 (where I_1 is obtained from $G_{L\omega}$ by dropping $(\rightarrow \Box_\omega)$). Next, let us consider the bottom-up applications of (A) in D. Since $G_{L\omega} \vdash S$, (as follows from Lemma 3.2) the "logical premise' of (A) is derivable in Log (where Log is obtained from $G_{L\omega}$ by dropping $(\rightarrow \Box_\omega), (\Box \rightarrow)$). Relative to the "temporal" applications of (A) we conclude that the "temporal" premise of (A) is of the form $\Gamma, \Box\Omega \rightarrow \Theta, \Box A$. Therefore, applying the resolvent tactic to S (and using the fact that $G_{L\omega} \vdash S$) we can reduce S either to logical axioms or to the sequents $S_i = \Pi_i, \Box\Omega \rightarrow \Delta_i, \Box A$, (in separate case $\Pi_i, \Delta_i = \varnothing$) where $\Pi_i, \Delta_i \in R\mathrm{Sub}(\Omega)$. Since the set $R\mathrm{Sub}(\Omega)$ is finite, starting from some k-th resolvent $Re^k(S)$ we shall generate the resolvent sequents which were obtained previously in a branch of the tree D and each resolvent sequent is either saturated or absorbed in each branch of the tree D. Therefore each "resolvent branch" of D is saturated and (each "resolvent branch") can be cut off getting the blocked leaf. The leaves of other branches of D are logical axioms. Thus, we get $\mathrm{Sat}' \vdash^D S$. Let us note that we described the worst way of obtaining the derivation in Sat$'$ of the proper quasiprimary sequent S. Instead of getting the equality $S^{**} = S^*$ (which yields that S^* is saturated and S^{**} is absorbed) we can get (in some cases earlier) the inequality $S^{**} \prec S^*$. Therefore it is not necessary to generate resolvent sequents, containing all possible resolvent subformulas of the formulas from Ω.

Theorem 4.5. Let S be an MH-sequent, then $G_{L\omega} \vdash S \Longleftrightarrow \mathrm{Sat} \vdash S$.

Proof. The part $\Longrightarrow$ follows from Lemmas 3.3, 4.11, the part $\Longleftarrow$ follows from Theorem 4.2.

Theorem 4.6. Let S be an MH-sequent, then $G_{L\omega} \vdash S \Longleftrightarrow IN \vdash S$.

Proof: follows from Theorems 4.4, 4.5.

5 Some Generalizations of Saturated Sequential Calculi to Arbitrary Sequents

5.1 The Case of Arbitrary M-sequents

Let us consider the construction of a saturated calculus for arbitrary miniscoped sequents (abbreviated: M-sequents). Let Sat^M (IN^M) be the calculus obtained from Sat (IN, respectively) replacing the rule of inference $(\Box)$ by the following ones:

$$\frac{S_1; \ldots; S_n}{\Box\Gamma \rightarrow \Box A_1, \ldots, \Box A_n}\ (\Box_n),$$

where $S_1 = \Box\Gamma \to A_1, \Box A_2, \ldots, \Box A_n; \ldots; S_n = \Box\Gamma \to \Box A_1, \ldots, \Box A_{n-1}, A_n$.
Theorem 5.1. Let S be an M-sequent, then $G_{L\omega} \vdash S \iff \mathrm{Sat}^M \vdash S \iff IN^M \vdash S$.
Proof: analogously as in Theorems 4.4, 4.5, 4.6.

5.2 The Case of □-∀-sequents

For simplicity let us consider non-miniscoped sequents without ∃, without positive occurrences of ∀, containing only monadic predicates, only one function symbol and only one constant (as follows from [4] this class of temporal sequents is incomplete. Moreover the sequents satisfy the following (so called □∀-restriction: all negative occurrences of ∀ enter only the scope of □. The rules of inference of the saturated calculus Sat^*_λ ($\lambda \in \{\varnothing, \infty\}$) are obtained from those of Sat^M, replacing (A) by a new one, in which $S^* = \Sigma_1, \forall\overline{x}\Delta, \Pi^1_1, \Box\Omega^1 \to \Sigma_2, \Pi_2, \Box\nabla^1$, where $\forall\overline{x}\Delta = \varnothing$ or consists of formulas of the type $\forall x A$, where $\Box \notin A$ and A contains indices; $S^*_1 = \Sigma_1 \to \Sigma_2$; $S^*_2 = \Pi_1, \Box\Omega \to \Pi_2, \Box\nabla$. The calculus Sat^*_∞ for □-∀-sequents is infinitary, (in general) therefore besides the finitary saturated sequents, we must formulate so called ω-saturated sequents. Let $S' = \Gamma, \Pi(b), \Box\Omega \to \Delta, \Theta(b) \in Re^k(S)$, we say that the sequent S' is l-saturated if there exists $S^* \in Re^p(S)$ $(p > k)$ such that S^*, S' belongs to the same branch and $S^* = \Gamma, \Sigma_1, \Pi(f^{c\cdot l}(b)), \ldots, f^{c\cdot l-(l+1)}(b)) \to \Delta, \Sigma_2, \Theta(f^{c\cdot l}(b), \ldots, f^{c\cdot l-(l+1)}(b))$, S^* is called l-absorbed. (where c is a natural number; $f^m(b) = f(b)$, if $m = 1$ and $f^m(b) = f(f^{m-1}(b))$, if $m > 1$). If the sequent S' is l-saturated l-absorbed for all $c \in \omega$, then S' will be called ω-saturated, (ω-absorbed, respectively). Let $S_k \in Re^k(S)$, we say that the sequent S_k is ω-blocked if S_k is ω-absorbed and $\forall j$ $(f - 1 \leqslant j < k - 1)$ $S_j \in Re^j(S)$ is either ω-saturated, or ω-absorbed and S_f is ω-saturated . A branch B in the derivation D will be called ω-saturated if the leaf of B is ω-blocked. A branch in the derivation D in Sat^*_λ will be called λ-closed if the leaf of B is either closed (see section 3), or ω-saturated. A derivation D in Sat^*_λ will be called closed if all branches are λ-closed. The notation $\mathrm{Sat}^* \vdash^D S$ ($\mathrm{Sat}^*_\infty \vdash^D S$) will mean that the closed derivation of S does not contain (contain, respectively) a ω-saturated branch.
Theorem 5.2. Let S be a □-∀-sequent, then $G_{L\omega} \vdash S \iff \mathrm{Sat}^*_\lambda \vdash S$ ($\lambda \in \{\varnothing, \infty\}$).
Proof: using the proofs of Theorems 4.4, 4.5.

Let IN^* be the calculus obtained from IN^M replacing Sat^M by Sat^* in the restriction (1) of formulating $(\to \Box)$.
Theorem 5.3 (completeness criterion for □-∀-sequents). Let S be a □-∀-sequent, then $\mathrm{Sat}^* \vdash S \Longrightarrow (\forall M \models S \Longleftarrow IN^* \vdash S)$.
Proof: using the proofs of Theorems 1.1, 4.4, 4.5, 4.6.
Example 5.2. Let $S = P(b), \Box\Omega \to \Box A$, where $\Box\Omega = \Box\forall x(P(x) \supset (P^1(x) \vee P^1(f^2(x)))), \Box\forall x(P(x) \supset Q(x)), \Box\forall x(Q(f^2(x)) \supset Q(x))$; $A = Q(b) \vee P(b)$. Then it is easy to verify that $\mathrm{Sat}^*_\infty \vdash^D S$ and D contains one ω-closed branch involving one type of ω-closed sequents of the form $S_{c+1} = P(f^{(c+1)\cdot 2}(b)), \Box\Omega \to \Box A$ $(c \in \omega)$ and an infinite set of finitary saturated branches involving one type of finitary

saturated sequents of the form $S_c = P(f^{c \cdot 2}(b)), \square\Omega \to \square A$ $(c \in \omega)$.

5.3 Some Ways of Removing the □∀-restriction

In some cases the □∀-restriction may be eliminated using a logical equivalence and Barcan formula $\square \forall x A(x) \equiv \forall x \square A(x)$. In the opposite case it is necessary to add to the calculus Sat^*_λ the rule of inference (which is admissible, but not invertible in $G_{L\omega}$) of the following type:

$$\frac{A(t), \Gamma \to \Delta; \quad \square B(t), \forall x(A(x) \vee \square B(x))^1, \Gamma \to \Delta}{\forall x(A(x) \vee \square B(x)), \Gamma \to \Delta} (\forall^1 \to)$$

References

1. H. Andreka, I. Nemeti, J.Sain: On the strength of temporal proofs. In: A. Kreczmar, G. Mirkowska (eds.): Mathematical foundations of computer science 1989. Lecture Notes in Computer Science 379. Berlin: Spriger 1989, pp. 135–144.
2. H. Kawai: Sequential calculus for a first order infinitary temporal logic. Zeitshr. für Math. Logik und Grundlagen der Math. 33, 423–432 (1987).
3. F. Kroger: On the interpretability of arithmetic in temporal logic. Theoretical Computer Science 73, 47–60 (1990).
4. S. Merz: Decidability and incompleteness results for first-order temporal logic of linear time. Journal of Applied Non-classical Logics 2, 139–156 (1992).
5. R. Pliuškevičius: Completeness criterion for the Horn-like first order linear temporal logic. Proceedings of conference on applied logic. Logic at Work, Amsterdam 1992.
6. A. Szalas: A complete axiomatic characterization of first-order temporal logic of linear time, Theoretical Computer Science 54, 199–214 (1987).
7. A. Szalas: Concerning the semantic consequence relation in first-order temporal logic. Theoretical Computer Science 47, 329–334 (1986).

The Snack Powerdomain for Database Semantics

Hermann Puhlmann

Technische Hochschule Darmstadt, Fachbereich Mathematik
Schlossgartenstrasse 7, D–64289 Darmstadt, Germany
e-mail: puhlmann@mathematik.th-darmstadt.de

Abstract. Recently the use of domain theory for a semantics of databases has been proposed. To model set-valued structures in this framework, a powerdomain construction will be needed. As an appropriate construction a modification of the recently introduced snack powerdomain is investigated and shown to be a free algebra. Moreover, the construction preserves bounded completeness. A slight modification of the snack powerdomain yields the scone powerdomain which, additionally, is distributive. Both constructions promise to bear fruit in the domain theoretic approach to database semantics.

1 Introduction

Think of a database relation which lists the books of a library. Typically, not all books of the library will be listed in the relation, e. g. the books which are older than the computer system might not be included. On the other hand, we may take it for granted that every book which is listed in the database relation is actually owned by the library. Hence this database contains less elements than one would like.

Now imagine that you want to borrow one of the library's books. In order to do so you will have to produce your library card. Your library card gives evidence that you are a registered user of the library, and without being registered you can't possibly borrow a book. There may, however, be people who are still registered with the library whilst not being users of the library any more, so this database contains more than is desired.

These two examples illustrate two opposite ways in which database relations may reflect the real world. The situation can be expressed nicely if we think of database tuples as elements of a partially ordered set. Motivations of this are the treatment of null values (see e. g. [11]) and orderings on complex objects (see e. g. [1]). While the ordered sets occuring in these approaches are of a very special kind, Buneman, Jung and Ohori in [3] allowed for much more general ordered sets, namely they proposed that a database tuple is an element of a Scott domain.

We follow this idea and in this work search for structures in which database relations together with their intended meaning can be represented. We will then see how we can deal with the initial examples.

2 Mathematical Concepts

We assume the reader's familiarity with the usual concepts of domain theory as they can be found in [5] or [6]. Note that in this paper "domain" means algebraic cpo and Scott domains enjoy the additional property of bounded completeness.

The powerdomain orders $\leq^\sharp$ and $\leq^\flat$ will play an important role in this paper. They are preorders on the powerset of an ordered set $(P, \leq)$ and defined by

$$A \leq^\sharp B :\Leftrightarrow \forall b \in B\ \exists a \in A : a \leq b \qquad (\Leftrightarrow \uparrow A \supseteq \uparrow B)\ ,$$
$$A \leq^\flat B :\Leftrightarrow \forall a \in A\ \exists b \in B : a \leq b \qquad (\Leftrightarrow \downarrow A \subseteq \downarrow B)\ .$$

On upper sets (for $\leq^\sharp$) and lower sets (for $\leq^\flat$) as well as on antichains (in either case) these are order relations. It is often convenient to think in terms of upper and lower sets and we will comment on this at the appropriate place.

3 Sandwiches and Mixes

Using the powerdomain orderings $\leq^\flat$ and $\leq^\sharp$ we can express the two ways of interpreting relations which we encountered in our initial examples. As it was proposed in [3], we treat relations as antichains in an ordered space $\mathcal{D}$ of partial descriptions. The total descriptions of real world objects are maximal elements of $\mathcal{D}$ and the sets we want to describe by a relation are subsets of the set of maximal elements of $\mathcal{D}$. Then for the relation R listing the library's books and the set M of all books which the library really owns we have the relation $R \leq^\flat M$. That is, for every tuple r in R there must be a book owned by the library which is described by r. If we want to improve the approximation of M we can do so by making the information about books which we already covered by tuples in R more accurate, e. g. by giving not only author and title of a book but also the publisher or adding tuples for books which are not yet in the database. The resulting relation R' will be related to R by $R \leq^\flat R'$.

The relation Q about users of the library approximates the set N of all library users in a different way. There we have $Q \leq^\sharp N$ which means that for every library user n there must be a tuple in Q which describes n. What will be a better approximation of N than Q?—The descriptions of the library users will have to be improved, and outdated entries, i. e. those which do not correspond to users of the library, should be removed. It becomes clearer if we say what it means for Q to be not as good as a relation Q': For every tuple in Q' we can find a less accurate tuple in Q. So Q and Q' are related by $Q \leq^\sharp Q'$.

So we have two interpretations of relations, each coming along with a powerdomain ordering. Will we therefore have to calculate in different ordered spaces according to the intended meaning of a relation? And what will we do when a relation is meant in both interpretations simultaneously?—A solution was suggested by Buneman, Davidson and Watters in [2] and Gunter in [4]: Rather than giving an antichain together with the intended meaning, i. e. ordering, they take pairs of antichains, one antichain for each ordering. A set S of real world objects is described by such a pair (U, L) of antichains if $U \leq^\sharp S$ and $L \leq^\flat S$. In order to have a description of S in the $\leq^\sharp$-sense only, one chooses L to be the empty set. Since $\emptyset \leq^\flat S$ for all $S \subseteq \mathcal{D}$, the condition imposed by the second component of the pair becomes vacuously satisfied. Similarly we can express that only a match in the $\leq^\flat$-sense is requested by setting the first component to $\{\bot\}$ where $\bot$ is the least element of $\mathcal{D}$.

We can think of pairs (U, L) in the following way: The U-part restricts the described real-world set S to be a subset of $\uparrow U \cap \mathcal{D}_{\max}$. The L-part gives information about elements of S we know of because every element of L must be an approximation of an element in S. So the slogan is:

"Everything must be in U (precisely: in $\uparrow U$) and L gives examples."

In [2] and [4] there are additional conditions on the pairs (U, L). In [2] the condition is that $\uparrow l \cap \uparrow U \neq \emptyset$ for all $l \in L$. In other words, every element of L must approximate some real world object which is also approximated by U. The pairs (U, L) which satisfy this condition are called *sandwiches.* In [4] the condition is that $L \subseteq \uparrow U$, which means the examples can only live inside the allowed set $\uparrow U$. Pairs which satisfy this condition are called *mixes.* Sandwiches as well as mixes are ordered by the product ordering $\leq^\sharp \times \leq^\flat$, i. e. the U-parts are ordered by $\leq^\sharp$ and the L-parts by $\leq^\flat$. These are the orderings we found to be appropriate before.

Given a domain $\mathcal{D}$, the sandwich and the mixed powerdomain are respectively defined as ideal completions of the ordered sets of sandwiches and mixes on the set $K(\mathcal{D})$ of compact elements of $\mathcal{D}$. So these powerdomains are defined in the spirit of the classical powerdomains, viz. the upper and the lower powerdomain, which are ideal completions of $K(\mathcal{D})$ ordered by $\leq^\sharp$ and $\leq^\flat$, respectively (see [9]).

In order to get increasingly good approximations to a *set* of maximal elements of $\mathcal{D}$ one might be tempted to take an increasing sequence of elements in a powerdomain which has a maximal element as limit. The upper and lower powerdomain, however, have no interesting maximal elements. In the upper powerdomain only the empty set is maximal, in the lower powerdomain only the entire set $\mathcal{D}$. Both, the mixed and the sandwich powerdomain overcome this deficiency. We can think of their maximal elements as being of the form (M, M) where M is a subset of $\mathcal{D}_{\max}$ (More precisely, these elements are the ideals of mixes or sandwiches (U, L) where $U, L \subseteq K(\mathcal{D})$ and $U \leq^\sharp M$, $L \leq^\flat M$.)

In this regard, sandwiches and mixes are what is needed for modelling databases. The sandwich powerdomain, however, has no nice algebraic characterisation. The drawback of mixes is that they do not preserve bounded completeness as the example in Fig. 1 illustrates. Remembering our slogan that everything must be within U and L gives examples, we can say which sets the two mixes in this example approximate. In this special case the U-parts do not impose any restrictions to the approximated sets since $\mathcal{D}_{\max} \subseteq \uparrow\{x, y\}$ and $\mathcal{D}_{\max} \subseteq \uparrow\{w, z\}$. The set $\{x, y\}$ seen as the L-part says that the approximated set must at least contain (a or c) and (b or d). The L-part $\{w, z\}$ requires (a or b) and (c or d) to be in the set. Given the information that both mixes approximate the same set of maximal elements, one wants to combine this information. It turns out

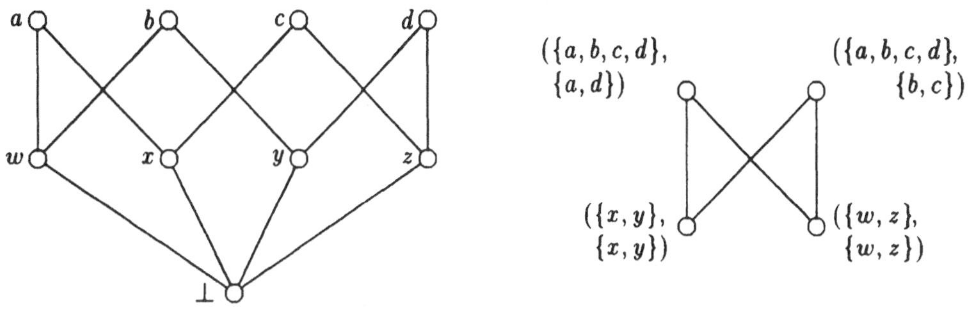

The domain on the left is bounded complete but the two mixes $(\{x, y\}, \{x, y\})$ and $(\{w, z\}, \{w, z\})$ have two minimal upper bounds.

Fig. 1. Mixes do not preserve bounded completeness

that each of the minimal upper bounds indicated in Fig. 1 is too restrictive. The information we really have is that the approximated set must contain (a and d) or (b and c). We might as well leave it as a conjunction of disjunctions: (a or c) and (b or d) and (a or b) and (c or d).

In the following section we define *snacks* which allow exactly for this: The L-part consists of clusters each of which gives information about the approximated set in a disjunctive manner.

4 Snacks on Ordered Sets

Let $(P, \leq)$ be an ordered set. By $\mathcal{U}(P)$ we denote the set of all finitely generated upper subsets of P ordered by the Smyth or upper order ($\leq^\sharp$) which is, for upper sets, just the superset relation. To exclude the empty set we write $\mathcal{U}_{\neq\emptyset}(P)$. The set of finite antichains of $\mathcal{U}_{\neq\emptyset}(P)$, ordered by the Hoare or lower order ($\leq^\flat$), is denoted by $\mathcal{L}(\mathcal{U}_{\neq\emptyset}(P))$.

The set of generators of a set $U \in \mathcal{U}(P)$ is denoted by $\mathcal{MIN}(U)$, the set of $\leq^\sharp$-maximal upper sets in a set L of upper sets by $\mathcal{MAX}^\sharp(L)$.

One might ask why we use upper sets in the case of the Smyth ordering but antichains in the case of the Hoare ordering. The answer is easy. Both ways ensure that we have order relations rather than preorders which we would have if we took arbitrary subsets. We use the Smyth ordering on a first level where we take subsets of a poset P. From our point of view, the most intuitive way of thinking about the Smyth order is to regard it as the superset relation on the upper sets which are generated by the sets we want to compare. Of course, for the Hoare ordering the same argument holds stated for the lower sets. However, we take the Hoare ordering on a second level where we want to order sets of upper sets. Now, what is a lower set of upper sets? Here, we think, it is easier to take the generators rather than the lower set and we therefore use antichains.

Definition 1. A *snack* on an ordered set P is a pair $(U, L) \in \mathcal{U}(P) \times \mathcal{L}(\mathcal{U}_{\neq\emptyset}(P))$ such that $L_i \subseteq U$ for all $L_i \in L$. The set of all snacks on P is denoted by $\mathcal{S}^0(P)$.

The set $\mathcal{S}^0(P)$ is itself an ordered set. The order is inherited from $\mathcal{U}(P) \times \mathcal{L}(\mathcal{U}_{\neq\emptyset}(P))$, i. e. $(U, L) \leq (U', L')$ iff $U \supseteq U'$ and for all $L_i \in L$ there is a $L'_j \in L'$ such that $L_i \supseteq L'_j$.

Snacks were originally proposed by Buneman and studied by Ngair in [7]. Ngair, however, allowed the second component of a snack to be the set containing the empty set. This led to what he called "anomalous snacks". These complicate the algebraic (and the semantic) treatment. Ngair also addressed semantic issues of database operations while we want to investigate the algebraic structure of the set of snacks on a poset.

4.1 Operations on Snacks

Our aim is to use snacks for defining a powerdomain construction. Powerdomains, generally, are defined as ideal completions of constructions over the set of compact elements of the underlying domain which are characterised algebraically as free algebras over the domain.

For this characterisation, appropriate operations are needed. These often appear like rabbits pulled out of the hat. We are afraid that we cannot entirely avoid this phenomenon. But we can try to motivate the choice of operations by looking how we can operate on snacks.

As in the case of the other powerdomains we have a formal union for snacks. It is basically the componentwise union. Since this need not yield the required antichain of upper sets in the second component we have to make it one by taking the $\leq^\sharp$-maximal (i. e. the $\subseteq$-minimal) upper sets only. Formally we define this operation as

$$\uplus : \mathcal{S}^0(P) \times \mathcal{S}^0(P) \to \mathcal{S}^0(P) \ ,$$
$$(U, L) \uplus (U', L') := (U \cup U', \mathcal{MAX}^\sharp(L \cup L')) \ .$$

This is a monotone semilattice-operation with unit element $(\emptyset, \emptyset)$.

A semilattice operation on a set induces an order relation. Note that the order derived from the formal union has nothing to do with the order we defined on snacks. In so far, the situation is as for mixes or sandwiches. But unlike these constructions, snacks have binary meets. Therefore we get another semilattice operation $\wedge$, and the order induced by $\wedge$ is in fact the given order on snacks. The operation $\wedge$ is given by

$$\wedge : \mathcal{S}^0(P) \times \mathcal{S}^0(P) \to \mathcal{S}^0(P) \ ,$$
$$(U, L) \wedge (U', L') := (U \cup U', \mathcal{MAX}^\sharp(\{L_i \cup L'_j | L_i \in L, L'_j \in L'\})) \ .$$

It is easy to see that $\wedge$ is well-defined and yields the infimum w. r. t. the given order.

The two semilattice operations $\wedge$ and $\uplus$ are connected via the distributivities $(r \uplus s) \wedge t = (r \wedge t) \uplus (s \wedge t)$ and $(r \wedge s) \uplus t = (r \uplus t) \wedge (s \uplus t)$.

We promised that snacks will have bounded binary suprema if the underlying poset has. This is indeed the case. But since suprema are not needed for the algebraic characterisation we will postpone this issue to Sect. 6.

4.2 Building Snacks out of Singletons

The set of all snacks on a poset P can be seen as generated by the elements of P in the sense that in $\mathcal{S}^0(P)$ there are representatives for the elements of P such that every snack can be built out of those by means of $\uplus$, $\wedge$ and $(\emptyset, \emptyset)$. The representatives are called *singletons* and produced by the snack-singleton function

$$\{\!|\cdot|\!\} : P \to \mathcal{S}^0(P) ,$$
$$\{\!|x|\!\} := (\uparrow x, \{\uparrow x\}) .$$

Note that the singleton function is monotone.

A way of constructing snacks out of singletons is shown in the following lemma.

Lemma 2. *Let* $s = (U, \{L_1, \ldots, L_k\})$ *be a snack on* P. *Furthermore, let the generators of* U *be* $u_1, \ldots, u_n$ *and those of* L_i $(i = 1, \ldots, k)$ *be* $l_{i1}, \ldots, l_{im_i}$. *Then* s *can be built out of singletons as*

$$s = (\{\!|u_1|\!\} \wedge \ldots \wedge \{\!|u_n|\!\} \wedge (\emptyset, \emptyset)) \uplus$$
$$(\{\!|l_{11}|\!\} \wedge \ldots \wedge \{\!|l_{1m_1}|\!\}) \uplus \ldots \uplus (\{\!|l_{k1}|\!\} \wedge \ldots \wedge \{\!|l_{km_k}|\!\}) . \quad \square$$

5 The Snack Algebra

From the previous section we have a collection of operations on snacks and equations for the operations. We now drop the specific structure of snacks on a poset and only keep the algebraic properties:

Definition 3. A *snack algebra* is a set S together with two semilattice operations $\uplus, \wedge : S \times S \to S$ and a constant $e \in S$ such that for all $r, s, t \in S$:

$$e \uplus r = r \qquad (S1)$$
$$(r \uplus s) \wedge t = (r \wedge t) \uplus (s \wedge t) \qquad (S2)$$
$$((r \wedge s) \uplus t) \wedge (r \uplus t) = (r \wedge s) \uplus t \qquad (S3)$$

Each semilattice operation induces an order on S. The order which is induced by $\uplus$ is not relevant for our purposes. The order induced by $\wedge$ as a meet-semilattice operation will be denoted by $\leq$, and we will consider S as a poset $(S, \leq)$. In this setting it is clear that $\wedge$ is monotone and $(S3)$ says that also $\uplus$ is monotone w.r.t. this order.

Note that instead of (S3) we could equivalently have included the distributivity $(r \wedge s) \uplus t = (r \uplus t) \wedge (s \uplus t)$ in the definition.

Structures with two semilattice operations are sometimes called bisemilattices or quasi-lattices, and various distributivities on them have been studied [10, 8]. In this line of thought a snack-algebra is a distributive bisemilattice with a unit element e for one of the semilattice operations. For our purposes, however, the idea of a snack-algebra being a poset is more important. So we include the axiom of monotonicity (S3) into the definition.

Definition 4. The category with snack algebras as objects and snack algebra homomorphisms (i. e. functions preserving $\uplus$, $\wedge$, and e) as morphisms is denoted by **SnAlg**.

Looking at snacks on a poset P and the operations $\uplus$ and $\wedge$ on $\mathcal{S}^0(P)$ we observe that $\mathcal{S}^A(P) := (\mathcal{S}^0(P), \uplus, \wedge, (\emptyset, \emptyset))$, i. e. the set of snacks over P together with the operations and the constant $(\emptyset, \emptyset)$, is a snack algebra. We already proved that both operations are semilattice operations, $\wedge$ is the infimum w. r. t. the given order on snacks, i. e. the snack-ordering is induced by $\wedge$, the snack $(\emptyset, \emptyset)$ is the unit for $\uplus$, and the distributivity law (S2) holds. As we pointed out before, monotonicity for $\uplus$ is seen easily.

Before establishing our main result in Theorem 9 we will now present some auxiliary lemmata.

Lemma 5. *For any r, s in a snack-algebra we have $(r \uplus s) \wedge e = (r \wedge s) \wedge e$.* □

Lemma 6. *Let P be a poset, $U \in \mathcal{U}(P)$ and $\{u_1, \dots, u_n\}$ the set of generators for U. For every finite subset $\{u_1, \dots, u_n, v_1, \dots, v_m\}$ of U and every monotone mapping f from P into a snack algebra Q we have*

$$f(u_1) \wedge \dots \wedge f(u_n) = f(u_1) \wedge \dots \wedge f(u_n) \wedge f(v_1) \wedge \dots \wedge f(v_m) \ . \quad \square$$

Lemma 7. *Let $A = \{a_1, \dots, a_n\}$ and $B = \{b_1, \dots, b_m\}$ be subsets of an ordered set P such that $A \leq^\sharp B$. Furthermore, let Q be an ordered set with binary meets, and $f : P \to Q$ monotone. Then $f(a_1) \wedge \dots \wedge f(a_n) \leq f(b_1) \wedge \dots \wedge f(b_m)$.* □

Lemma 8. *Let P be a poset, $U = \{u_1, \dots, u_n\}$, $L' = \{l'_1, \dots, l'_k\}$ and $L = \{l_1, \dots, l_m\}$ be subsets of P with $U \leq^\sharp L' \leq^\sharp L$ and f a monotone function from P to a snack algebra Q. Then*

$$\begin{aligned} &(e \wedge f(u_1) \wedge \dots \wedge f(u_n)) \uplus (f(l'_1) \wedge \dots \wedge f(l'_k)) \uplus (f(l_1) \wedge \dots \wedge f(l_m)) \\ = &(e \wedge f(u_1) \wedge \dots \wedge f(u_n)) \uplus (f(l_1) \wedge \dots \wedge f(l_m)) \ . \end{aligned}$$

Proof. One proves the equality by showing "$\leq$" and "$\geq$". The calculations make use of Lemma 7 and the monotonicity and idempotence of $\uplus$. □

We are now ready to prove our main result. It says that the snacks on a poset P form a free algebra generated by P. We formulate this in the language of category theory, and V denotes the forgetful functor from the category of snack algebras and snack algebra homomorphisms to the category of posets and monotone functions.

Theorem 9. *For any poset P, the snack algebra $\mathcal{S}^A(P)$ is freely generated by P, that is, for any snack algebra Q and any monotone $f : P \to VQ$, there is a unique morphism $f^+ : \mathcal{S}^A(P) \to Q$ such that Vf^+ makes the following diagram commutative:*

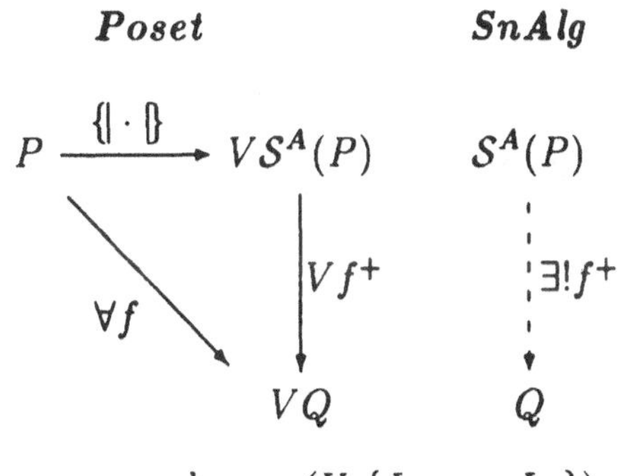

Proof. By Lemma 2, every snack $s = (U, \{L_1, \ldots, L_k\})$ can be uniquely represented by the generators $u_1, \ldots, u_n$ of U and the generators $l_{i1}, \ldots, l_{im_i}$ of L_i $(i = 1, \ldots, k)$. If the desired morphism f^+ exists then

$$f^+(s) = (e \wedge f(u_1) \wedge \ldots \wedge f(u_n)) \uplus$$
$$(f(l_{11}) \wedge \ldots \wedge f(l_{1m_1})) \uplus \ldots \uplus (f(l_{k1}) \wedge \ldots \wedge f(l_{km_k})) \ .$$

It must be shown that f^+ as defined above is a snack-algebra morphism. Obviously, $(\emptyset, \emptyset)$ is mapped to e. That f^+ also preserves $\uplus$ and $\wedge$ is seen by some easy computations using the above lemmata. □

The result of Theorem 9 can easily be carried from posets to domains.

Definition 10. Let D be a domain and $K(D)$ the set of compact elements of D. Then the ideal completion

$$\mathcal{S}^P D := \mathsf{idl}(\mathcal{S}^0(K(D)))$$

is the *snack powerdomain* over D.

Definition 11. A continuous snack algebra is a snack algebra $(D, \uplus, \wedge, e)$ where D is a domain and $\uplus$ and $\wedge$ are continuous. A continuous snack algebra homomorphism is a continuous homomorphism between snack algebras. The category of continuous snack algebras and continuous snack algebra homomorphisms is denoted by **CSnAlg**.

Again, we have a singleton function $\{\!|\cdot|\!\}_{\mathsf{idl}} : D \to \mathcal{S}^P D$. If $x \in K(D)$ then $\{\!|x|\!\}_{\mathsf{idl}}$ is the principal ideal in $\mathcal{S}^0(K(D))$ which is generated by the snack $(\uparrow x, \{\uparrow x\})$. If x is not compact then $\{\!|x|\!\}_{\mathsf{idl}}$ is the union of the singletons of those compact elements which approximate x.

This is the general recipe for extending a monotone mapping which is given on the compact elements to a continuous mapping between algebraic domains. With this in mind the following is a straightforward corollary to Theorem 9. We denote the forgetful functor from the category **CSnAlg** to the category **Dom** of domains and continuous functions by V.

Corollary 12. *For any domain D, the continuous snack algebra $\mathcal{S}^P(D)$ is freely generated by D, that is, for any continuous snack-algebra E and any continuous $f : D \to VE$ there is a unique morphism $f^+ : \mathcal{S}^P D \to E$ such that Vf^+ makes the following diagram commutative:*

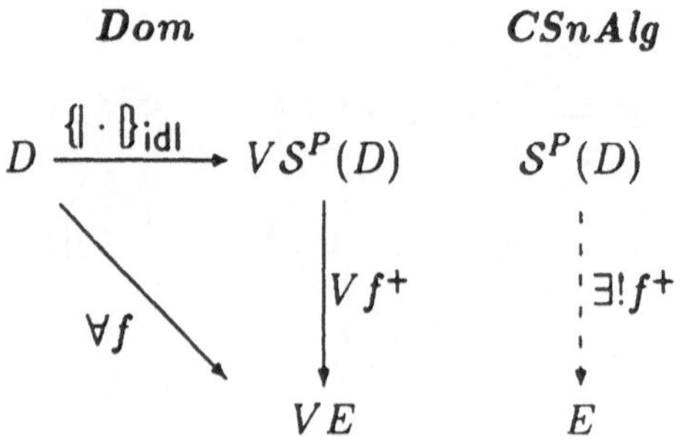

□

6 Bounded Suprema and Scones

In order to use a powerdomain construction for database semantics in the spirit of [3], the construction must carry Scott domains to Scott domains. In particular, bounded completeness must be preserved. It was first observed by Buneman and Gunter that the snack powerdomain does this, and the following construction can also be found in [7]:

Lemma 13. *If the poset P has bounded binary suprema then so has $\mathcal{S}^0(P)$.*
For two bounded snacks $s = (U, L)$ and $s' = (U', L')$ on P, the supremum is calculated as $s \vee s' = (U \cap U', \mathcal{MAX}^{\natural}(\{L_i \cap U \cap U' \mid L_i \in L \cup L'\}))$.

Proof. Crucial for the proof is the fact that the occuring intersections of finitely generated upper sets are again finitely generated. This is the case since the generators of the intersections can be calculated as binary suprema of the generators of the sets which are intersected. The boundedness of s and s' comes in when we have to show that the occuring intersections are not empty. □

Corollary 14. *For a Scott domain D the snack powerdomain $\mathcal{S}^P(D)$ is again a Scott domain.* □

There is still something unfortunate about the snack powerdomain: It does not preserve distributivity. If we relax the condition $L_i \subseteq U$ in the definition of a snack to $L_i \cap U \neq \emptyset$ we get a structure which we call *scone*. The *scone powerdomain* is defined in analogy to the snack powerdomain. For scones we have very similar operations as for snacks, and distributivity comes in for free:

Theorem 15. *The scone powerdomain of a Scott domain is a distributive Scott domain.* □

Many techniques in the domain theoretic database framework of [3] depend on the distributivity of the domains used. The database operations therefore might be calculated using scones. Intersecting the L-parts of a scone with its

U-part yields a snack which describes the same sets and is easier to read. So scones might be used for doing calculations whereas snacks are for displaying information to the user. Of course, a detailed investigation of the applicability of these structures is still to be done.

Acknowledgements

It is my desire and pleasure to thank Achim Jung who set me on the snack's trail and never failed to guide my work. Klaus Keimel made numerous helpful suggestions to improve the presentation of the material. Andrea Schalk and Philipp Sünderhauf carefully listened to me in our tea-time discussions and repeatedly helped me out of cul-de-sacs in my proofs. Through his interest in database semantics, Peter Buneman encouraged my work, and when he and Leonid Libkin visited Darmstadt in October 1992, I had most interesting and fruitful discussions with them. Finally, I thank the anonymous referees for their comments on the paper.

References

1. F. Bancilhon and S. Khoshafian. A calculus for complex objects. In *Proceedings of the 5th ACM SIGACT/SIGMOD Symposium on Principles of Database Systems*, pages 53–59, Cambridge, Massachusetts, March 1986.
2. P. Buneman, S. Davidson, and A. Watters. A semantics for complex objects and approximate queries. Technical Report MS-CIS-87-99, University of Pennsylvania, November 1988.
3. P. Buneman, A. Jung, and A. Ohori. Using powerdomains to generalize relational databases. *Theoretical Computer Science*, 91:23–55, 1991.
4. Carl A. Gunter. The mixed powerdomain. *Theoretical Computer Science*, 103(2):311–334, September 1992.
5. Carl A. Gunter. *Semantics of Programming Languages. Structures and Techniques.* MIT Press, 1992.
6. A. Jung. *Cartesian Closed Categories of Domains*, volume 66 of *CWI Tracts*. Centrum voor Wiskunde en Informatica, Amsterdam, 1989.
7. Teow-Hin Ngair. *Convex Spaces as an Order-theoretic Basis for Problem Solving.* PhD thesis, Department of Computer and Information Science, University of Pennsylvania, July 1992.
8. J. Płonka. On distributive quasi-lattices. *Fundamenta Mathematicae*, 60:191–200, 1967.
9. G. D. Plotkin. Post-graduate lecture notes in advanced domain theory (incorporating the "Pisa Notes"). Dept. of Computer Science, Univ. of Edinburgh, 1981.
10. A. Romanowska. On bisemilattices with one distributive law. *Algebra Universalis*, 10:36–47, 1980.
11. Yannis Vassiliou. Null values in data base management: A denotational semantics approach. In *Proceedings of the ACM SIGMOD Int. Conf. Management of Data, Boston*, 1979.

Verifying Properties of Module Construction in Type Theory*

Bernhard Reus Thomas Streicher

Institut für Informatik, Universität München,
Leopoldstraße 11B, D-8000 München 40 (Germany)
[reus|streiche]@informatik.uni-muenchen.de

Abstract. This paper presents a comparison between algebraic specifications-in-the-large and a type theoretical formulation of modular specifications, called *deliverables*. It is shown that the laws of module algebra can be translated to laws about deliverables which can be proved correct in type theory. The adequacy of the *Extended Calculus of Constructions* as a possible implementation of type theory is discussed and it is explained how the reformulation of the laws is influenced by this choice.

1 Introduction

One of the most popular approaches to formal specification of software systems, i.e. *programs* or *program modules*, is the theory of algebraic specifications. This approach is suitable for *programming-in-the-small*, i.e. the specification of *one single* (program) module. For building bigger structured software systems – the discipline of so-called *programming-in-the-large* – it is necessary to deal with modules and their specifications as objects, because one wants to define functions and predicates on modules and their specifications. Admittedly in specification languages such as *CLEAR* [Burstall, Goguen 80] or *ASL* [Wirsing 86] one has certain *built-in linguistic constructs* for forming compound specifications but they *cannot be considered as functions operating on specifications* as there is not something like a "sort or type of specifications". ([Leszczylowski, Wirsing 91] is a proposal in this direction.) Furthermore one has no linguistic means to refer to modules as objects but only can formulate predicates on modules, i.e. specifications.

It seems to be fairly difficult to extend the formalism of algebraic specifications in a simple way to the needs of programming-in-the-large. One reason is that types of modules cannot be expressed without employing the concepts of *dependent types* and *sums of families of types* whereas the type systems used in algebraic specifications do not go beyond simply typed λ-calculus. The other reason is that in the theory of algebraic specifications a specification does not simply denote a class of modules but also contains some syntactic information (e.g. about the underlying signature or the special form of the axioms in order to guarantee properties such as confluence or termination). Therefore a drastic

* This work was partially sponsored by the BMFT-project KORSO

shift of concepts has to take place in order to adapt specifications to the needs of programming-in-the-large. A very appealing suggestion into this direction has been made and forcefully argued for by R. Burstall and J. McKinna when they introduced their concept of *deliverables* [Burstall, McKinna 92] [McKinna 92]. We will show that certain relations between specifications as formulated by the laws of so called *module algebra* [Bergstra et al. 89] can be reformulated in the framework of deliverables. Thereby we heavily build on previous work by Zhaohui Luo, see [Luo 91], where he has expressed the fundamental operations on specifications as operations on deliverables. The version of type theory we have chosen is the *Extended Calculus of Constructions* (XCC) [Luo 89] as it provides an infinite type hierarchy with sums and it is implemented within the LEGO system [Luo et al. 92]. Having a good implementation available is of great importance as machine assistance is indispensable for constructing and checking completely formalized proofs.
The paper is organized as follows: Section 2 introduces the concept of a *deliverable* as a generalization of a specification-in-the-small, whereas Section 3 discusses them in-the-large and provides also their algebraic counterparts. The main contribution of this paper is contained in Section 4 where the module algebra axioms and their type theoretic analagons are presented. A proof sketch for one of the laws is included in order to illustrate the typical problems arising from the choice of XCC. Finally Section 5 contains a short summary.

2 Specifications as deliverables

In this section we recap the basic intuition behind the *generalization from algebraic specifications to deliverables* [Luo 91]. Essentially there is nothing new in our presentation maybe with the exception that we put more emphasis on explaining the differences between algebraic specifications and their generalization to deliverables.
In the theory of algebraic specifications a specification is given by two components [Wirsing 89]. A signature indicating the *type of modules under consideration* and a collection of axioms expressing which *laws the intended subcollection of this type of modules should satisfy.*
Now the *deliverable approach* starts from the idea of *model class semantics.* The basic assumption is that the collection of specified objects is more fundamental than the form of its syntactic presentation. Therefore a type of modules is not represented indirectly by a syntactic object like a signature but more generally by an *arbitrary type.* Typically the modules of a signature with one sort s and one unary function symbol $f : S \rightarrow S$ are represented by the type

$$(\sum S : Type)\ S \rightarrow S\ .$$

For the notation see e.g. [Nordström et al. 90]. Thus under presence of dependent types and sums of families of types there is no longer need for signatures as objects. Due to α-conversion for bound variables problems with renamings of signatures disappear.

Now a deliverable in the sense of Burstall and McKinna is a pair $\langle S, P\rangle$ where S is a type and $P : S \to Prop$ is a predicate on S. Accordingly, the deliverable approach amounts to manipulating and stating relations between the model classes corresponding to algebraic specifications in a constructive way.
An important aspect of deliverables is that they organize into a category *DELIV* whose underlying type of objects is

$$(\sum S : Type)\ S \to Prop\ .$$

The type $S \to Prop$ will be abbreviated by $Pred(S)$ throughout the paper. Note that since S itself can be a type we must avoid circular types. But absence of circularity is checked automatically by the implementation of XCC in the LEGO system.
A morphism from deliverable $D1 = \langle S_1, P_1\rangle$ to deliverable $D2 = \langle S_2, P_2\rangle$ is a function $f : S_1 \to S_2$ together with the proof that the proposition

$$(\Pi x : S_1)\ P_1(x) \to P_2(f(x)) \quad (*)$$

holds. The most important aspect of deliverable morphisms is that they provide an adequate concept of refinement step as a morphism f from D_1 to D_2 can be interpreted as a *verified constructive implementation of* D_2 *in terms of* D_1. This notion of implementation is most adequate as it makes a clear distinction between the *computational part* of an implementation - the function $f : S_1 \to S_2$ and the proof of the *verification condition* (*) expressing that f transforms modules of type S_1 satisfying requirement P_1 into modules of type S_2 satisfying requirement P_2.
This suggests a methodology of program development by stepwise implementation – employing n-ary deliverable morphisms instead of only unary ones – where in each implementation step one already gets some part of the final functional program. Thus the deliverable approach provides an accurate mathematical setting for the *stepwise refinement method* for program development.
The module algebra axioms state equalities between specifications which can be used to flatten structural specifications, i.e. expand compound specifications to explicit form. Structured specifications are used in particular for building reusable specifications. As well known from the theory of algebraic specifications the operations *derive*, *translate*, *sum*, and *union* are not defined on the collection of *all* algebraic specifications but are inherently partial. This *partiality is made explicit by an appropriate typing* of these operators. Furthermore in the formulation of the laws of module algebra there is a certain amount of *informality* in expressing *side conditions of syntactic nature*, i.e. assumptions about signatures and signature morphisms. In the deliverable formulation these informalities are avoided by expressing these side conditions by typing assumptions.

3 Specification-in-the-large

We have already seen how specifications can be expressed as deliverables. But in this framework not only specifications-in-the-small can be represented adequately but also specifications-in-the-large. Most of the relevant specification

operations have been already described in [Luo 91]. After a short summary of these operations we will take a closer look on the module algebra axioms of [Bergstra et al. 89] and check if they correspondingly hold for deliverables.
First, let us recall the basic notions of algebraic specifications of [Wirsing 89]: We assume the class *Sign* of all signatures to be given. The collection *Spec* of all specifications consists of all pairs $\langle \Sigma, C \rangle$ of a signature Σ together with a class C of Σ-algebras, which is, in general, closed under isomorphism (i.e. for all $A, B \in Alg(\Sigma)$, $A \in C$ and $A \cong B$ implies $B \in C$).

$$Spec =_{def} \{\langle \Sigma, C \rangle \mid \Sigma \in Sign \text{ and } C \subseteq Alg(\Sigma)\ \}.$$

On *Spec* we define two projection functions: the left projection

$$sig : Spec \to Sign,\ \ sig(\langle \Sigma, C \rangle) =_{def} \Sigma \text{ for all } \langle \Sigma, C \rangle \in Spec,$$

assigns to each element of *Spec* its signature, the right projection whereas

$$Mod : Spec \to \{C \subseteq Alg(\Sigma) \mid \Sigma \in Sign\},$$
$$Mod(\langle \Sigma, C \rangle) =_{def} C \text{ for all } \langle \Sigma, C \rangle \in Spec,$$

assigns to each element of *Spec* its class of models.
For any signature Σ and any set E of Σ-formulas, the basic specification $< \Sigma, E >$ denotes the pair $\langle \Sigma, Alg(\Sigma, E) \rangle$ in *Spec*, i.e.

$$sig(< \Sigma, E >) =_{def} \Sigma,$$
$$Mod(< \Sigma, E >) =_{def} \{A \in Alg(\Sigma) \mid A \models E\}.$$

Thus, the models of $< \Sigma, E >$ are all Σ-algebras which satisfy E.

3.1 Specification Operations

union The definition of $\cup$ for algebraic specifications requires that the signatures of both arguments are equal i.e. $sig(sp_1) = sig(sp_2)$:

$$sig(sp_1 \cup sp_2) =_{def} \Sigma,$$
$$Mod(sp_1 \cup sp_2) =_{def} Mod(sp_1) \cap Mod(sp_2).$$

The corresponding type theoretical definition is :

$$union : \Pi\ S{:}Type.\ Pred(S) \to Pred(S) \to Pred(S)$$
$$union\ (S)(SP_1)(SP_2) =_{def} \lambda\ x{:}S.\ SP_1(x) \wedge SP_2(x)$$

where, as mentioned in Section 2, *Pred* abbreviates $\lambda S{:}Type.\ S \to Prop$. The intersection of model classes corresponds to the conjunction of formulas that the module x must fulfil. The condition that both specifications must have the same signature is expressed by the type of *union*.

derive The ASL-operation '**derive from**' is the basis for renaming specifications and exporting subsignatures. For any signature Σ', any signature morphism $\sigma : \Sigma \to \Sigma'$ and any class C' of Σ'-algebras, it is defined as follows:

$$sig(\textit{derive from } \langle \Sigma', C' \rangle \textit{by } \sigma) =_{def} \Sigma \ ,$$
$$Mod(\textit{derive from } \langle \Sigma', C' \rangle \textit{by } \sigma) =_{def} \{A|_\sigma \in Alg(\Sigma) \mid A \in C'\} \ ,$$

where $A|_\sigma$ is the σ-reduct of A.
The type theoretic definition is:

$$derive : \Pi \ S', S{:}Type. \ Pred(S') \to (S' \to S) \to Pred(S)$$
$$derive \ (S') \ (S) \ (SP') \ (\varphi) =_{def} \lambda x{:}S. \exists s{:}S'. (SP' \ s) \wedge \varphi(s) = x$$

If we read $\{A|_\sigma \in Alg(\Sigma) \mid A \in Mod(SP')\}$ as $\{A \in Alg(\Sigma) \mid \exists B \in Alg(\Sigma'). B \in Mod(SP') \wedge \sigma(B) = A\}$ then "$A \in Alg(\Sigma)$" corresponds to the typing "$x{:}S$", "$\exists B \in Alg(\Sigma')$" corresponds to "$\exists s{:}S'$" and "$B \in Mod(SP') \wedge \sigma(B) = A$" corresponds to "$(SP' \ s) \wedge \varphi(s) = x$".
Note the difference between the signature morphism σ and the induced structure map φ which goes in the opposite direction.

export The *export* operation can be defined in terms of *derive*:

$$\textit{export } \Sigma_0 \textit{ from } \langle \Sigma, C \rangle =_{def} \textit{derive from } \langle \Sigma, C \rangle \textit{by } i,$$

where i is the canonical injection $i : \Sigma_0 \to \Sigma$, $i(x) =_{def} x$.
In the deliverables approach we write for the first projection:

$$derive \ (\textstyle\sum x{:}A.B(x)) \ (A) \ (P) \ (\pi_1)$$

instead of the *export*.

translate The algebraic definition is:

$$sig(translate\langle \Sigma, C \rangle \textit{ with } \sigma) =_{def} \Sigma',$$
$$Mod(translate\langle \Sigma, C \rangle \textit{ with } \sigma) =_{def} \{A \in Alg(\Sigma') | \ A|_\sigma \in C\},$$

where $A|_\sigma$ is the Σ-reduct of A w.r.t. $\sigma : \Sigma \to \Sigma'$.
The type theoretic definition is:

$$translate : \Pi \ S', S{:}Type. \ Pred(S) \to (S' \to S) \to Pred(S')$$
$$translate(S')(S)(SP)(\varphi) =_{def} \lambda x{:}S'. SP(\varphi(s))$$

The condition "$A \in Alg(\Sigma')$" corresponds to the typing "$x{:}S'$" and "$A|_\sigma \in C$" corresponds to "$SP(\varphi(s))$". In algebraic specifications the function φ in *translate* and *derive* can only be generated by a signature morphism which is characterized by its ability to forget, identify, and rename. Since renaming is irrelevant due to α-conversion for simulating this approach one should use for φ only functions that project and identify (via λ-abstraction and products with identical components). In the type theoretic approach *translate* and *derive* are therefore stronger than their algebraic counterparts because φ is not necessarily induced by a signature morphism.

sum: putting two specifications together In algebraic specifications the +-operator is needed in order to build the amalgamation of two specifications, i.e. for gluing them together at a common interface. More formally, given specifications $sp_1, sp_2 \in Spec$ with signatures $sig(sp_1) = \Sigma_1$ and $sig(sp_2) = \Sigma_2$,

$$sp_1 + sp_2 =_{def} (\textit{translate } sp_1 \textit{ with } in_1) \cup (\textit{translate } sp_2 \textit{ with } in_2).$$

where $in_k : \Sigma_k \to \Sigma_1 \cup \Sigma_2$ embed the signatures into their union by $in_k(x) = x$ for $x \in \Sigma_k$, $k = 1, 2$.
The $\cup$ on signatures has no direct analogue on the module level, since in the deliverables approach there are no more signatures indicating the *shared parts*. Therefore the common interface must be given explicitly. This shared part must of course fulfil the requirements of both specifications. *sum* is the type theoretic counterpart of +. Note that S denotes the *shared part*.

$$\begin{aligned} &sum : \Pi S{:}Type.\Pi R,T{:}S \to Type.\ Pred(\Sigma s{:}S.R(s)) \to Pred(\Sigma s{:}S.T(s)) \to \\ &\qquad Pred(\Sigma s{:}S.\,(R(s) \times T(s))) \\ &sum\ S\ R\ T\ SP_1\ SP_2 =_{def} \lambda x{:}(\Sigma s{:}S.\,R(s) \times T(s)). \\ &\qquad SP_1\langle \pi_1(x), \pi_1(\pi_2(x))\rangle \wedge SP_2\langle \pi_1(x), \pi_2(\pi_2(x))\rangle \end{aligned}$$

Now "$\sum s{:}S.\,R(s) \times T(s)$" corresponds to "$\Sigma_1 \cup \Sigma_2$" "$S$" corresponds to "$\Sigma_1 \cap \Sigma_2$", "$R$" to "$\Sigma_1$", "$T$" to "$\Sigma_2$" and "$\lambda x{:}\sum s{:}S.\,R(s) \times T(s).\ \langle \pi_1(x), \pi_k(\pi_2(x))\rangle$" is the projection function that corresponds to "in_k". *sum* combines two principles, namely adding axioms (union) and enlarging the signature or structure type and can therefore serve to derive all other notions of combining specifications like *enrich* or *parallel* [Luo 91].

3.2 Equality on deliverables on the same underlying type

Before equations on specifications as deliverables can be proved one has to fix an equality on them, i.e. on *Pred*(S). Certainly we want to consider two deliverables as equal if the predicates are equivalent (and not if they are equal intensionally), i.e.:

$$\begin{aligned} &equal : \Pi\ S{:}Type.\ Pred(S) \to Pred(S) \to Prop \\ &equal\ (S)(D_1)(D_2) = \forall\ s{:}S.\ D_1(s) \Longleftrightarrow D_2(s) \end{aligned}$$

For the sake of readability we write $SP_1 =_S SP_2$ instead of *equal* $(S)(SP_1)(SP_2)$ and $=$ without decoration denotes Leibniz equality, i.e.
$(a = b : S) =_{def} \forall P{:}Pred(S).\ P(a) \Rightarrow P(b)$.

4 Module algebra axioms in type theory

4.1 The module algebra

Let us recall the Module Algebra specification of [Bergstra et al. 89] in the form of [Wirsing 89] omitting the type theoretically unnecessary operation *reach* (*reach* can be rephrased as a second order formulae expressing the induction principle). Therefore, the equations concerning *reach* (T2) and (S2) are left out and (E1) is omitted, because it corresponds to the definition of deliverables.

spec *BSA* $\equiv$
extend *SIGNATURE, FORMULA* **by**
sort *spec*
constructors $< .,. >$: *signature, set formula* $\rightarrow$ *spec*,
$. \cup .$: *spec, spec* $\rightarrow$ *spec*,
export . from . : *signature, spec* $\rightarrow$ *spec*,
translate . with . : *spec, signature morphism* $\rightarrow$ *spec*
functions *sig* : *spec* $\rightarrow$ *signature*
axioms $\forall$ sp, sp_1, sp_2 : *spec*, $\Sigma, \Sigma_1, \Sigma', \Sigma''$: *signature*,
E_1, E_2 : set formula, $\sigma : \Sigma \rightarrow \Sigma'$, $\sigma' : \Sigma' \rightarrow \Sigma''$: signature morphism.

(U1) $sp \cup sp = sp$,
(U2) $sp_1 \cup sp_2 = sp_2 \cup sp_1$,
(U3) $sp \cup (sp_1 \cup sp_2) = (sp \cup sp_1) \cup sp_2$,
(U4) $< \Sigma, E_1 > \cup < \Sigma, E_2 > = < \Sigma, E_1 \cup E_2 >$,

(E2) $\Sigma = sig(sp_1) \cap sig(sp_2) \Rightarrow$
$(export\ \Sigma\ from\ sp_1) \cup (export\ \Sigma\ from\ sp_2) = export\ \Sigma\ from\ (sp_1 + sp_2)$,
(E3) $\Sigma \subseteq \Sigma_1 \subseteq sig(sp) \Rightarrow$
$export\ \Sigma\ from(export\ \Sigma_1\ from\ sp) = export\ \Sigma\ from\ sp$,
(E4) $export\ sig(sp)\ from\ sp = sp$,

(T1) $translate < \Sigma, E > with\ \sigma = < \Sigma', \sigma^*(E) >$,
(T3) $translate\ (sp_1 \cup sp_2)\ with\ \sigma = (translate\ sp_1\ with\ \sigma) \cup (translate\ sp_2\ with\ \sigma)$,
(T4) $translate\ (export\ \Sigma\ from\ sp)\ with\ \sigma = export\ \Sigma'\ from\ (translate\ sp\ with\ \hat{\sigma})$,
(T5) $translate\ (translate\ sp\ with\ \sigma)\ with\ \sigma' = translate\ sp\ with\ \sigma' \circ \sigma$,

(αh) $\Sigma_1 \subseteq \Sigma = sig(sp) \wedge \sigma|_{\Sigma_1} = id \wedge injective(\sigma) \Rightarrow$
$export\ \Sigma_1\ from\ sp = export\ \Sigma_1\ from\ (translate\ sp\ with\ \sigma)$,
endspec

Here $\hat{\sigma} : sig(sp) \rightarrow \Sigma' \cup (sig(sp) \backslash \Sigma)$ denotes the extension of σ to $sig(sp)$ defined by $\hat{\sigma}(x) = \sigma(x)$, if $x \in \Sigma$, and x, otherwise. σ^* is the obvious extension of σ to formulas over $dom(\sigma)$.

4.2 The analogue of the laws in type theory

We will now present the corresponding formulation in a type theoretic framework. Therefore, we assume the following context:

$A,D,E : Type,\ \ B,C : A \rightarrow Type,$
$S_B =_{def} \sum x{:}A.B(x),\ \ S_C =_{def} \sum x{:}A.C(x),\ \ \ S_{AC} =_{def} \sum x{:}A.\ B(x) \times C(x),$
$F : S_B \rightarrow Type,\ \ S_{BF} =_{def} \sum y{:}S_B.F(y).$

Let $\hat{P}$ be an abbreviation for $\lambda s{:}S_B.\ P\langle \pi_1(s), \pi_2(s)\rangle$ (analogously for Q) and use the following definitions of *comp* and *join*, where *comp* denotes function composition and *join* takes two functions and yields a function on the dependent sum of their domains, since the domain of the second function may depend on the domain of the first one.

$comp : \Pi\ A,B,C{:}Type.\ (A \rightarrow B) \rightarrow (B \rightarrow C) \rightarrow (A \rightarrow C)$

$comp\ A\ B\ C\ f\ g\ s =_{def} g(f(s))$
$comp_{\pi_1} =_{def} comp(S_{BF})(S_B)(A)(\pi_1)(\pi_1)$

$join : \Pi X,Y{:}Type.\ \Pi B{:}X \rightarrow Type.\ \Pi C{:}Y \rightarrow Type.\ \Pi f{:}(X \rightarrow Y).(\Pi x{:}X.B(x) \rightarrow C(f(x)) \rightarrow (\sum x{:}X.B(x)) \rightarrow (\sum y{:}Y.C(y))$
$join\ X\ Y\ B\ C\ f\ g =_{def} \lambda z{:}(\sum x{:}X.B(x)).\ \langle f(\pi_1(z)), g(\pi_1(z))(\pi_2(z)) \rangle$

Let *Co* denote this context of definitions. For the logical connectives we use the usual second-order definitions, see e.g. [Luo 91].

(U1) $\forall SP{:}Pred(A).\ (union\ A\ SP\ SP) =_A SP$

(U2) $\forall SP_1,SP_2 :Pred(A).\ (union\ A\ SP_1\ SP_2) =_A (union\ A\ SP_2\ SP_1)$

(U3) $\forall\ SP_1,SP_2,SP_3 :Pred(A).$
$(union\ A\ SP_1\ (union\ A\ SP_2\ SP_3)) =_A (union\ A\ (union\ A\ SP_1\ SP_2)\ SP_3)$

(E2) $\forall P{:}Pred(S_B).\forall R{:}Pred(S_C).$
$(union\ A\ (derive\ S_B\ A\ \hat{P}\ \pi_1)\ (derive\ S_C\ A\ \hat{R}\ \pi_1)) =_A$
$(derive\ S_{AC}\ A\ (sum\ A\ B\ C\ \hat{P}\ \hat{R})\ \pi_1)$

(E3) $\forall P{:}Pred(S_{BF}).$
$(derive\ S_B\ A\ (derive\ S_{BF}\ S_B\ P\ \pi_1)\ \pi_1) =_A (derive\ S_{BF}\ A\ P\ comp_{\pi_1})$

(E4) $\forall P{:}Pred(A).\ (derive\ A\ A\ P\ (\lambda s{:}A.s)) =_A P$

(T3) $\forall P,R{:}Pred(A).\ \forall f{:}D \rightarrow A.$
$(translate\ D\ A\ (union\ A\ P\ R)\ f) =_D (union\ D\ (translate\ D\ A\ P\ f)\ (translate\ D\ A\ R\ f))$

(T4) $\forall P{:}Pred(S_B).\ \forall f{:}D \rightarrow A.$
$(translate\ D\ A\ (derive\ S_B\ A\ P\ \pi_1)\ f\ ) =_D (derive\ (\sum x{:}D.B(f(x)))\ D$
$(translate\ \sum x{:}D.B(f(x))\ S_B\ \hat{P}\ (join\ f\ \lambda x{:}D.\lambda y{:}B(f(x)).y))\ \pi_1)$

(T5) $\forall P{:}Pred(A).\ \forall f{:}D \rightarrow A.\ \forall g{:}E \rightarrow D.$
$(translate\ E\ D\ (translate\ D\ A\ P\ f)\ g) =_E (translate\ E\ A\ P\ f(comp\ E\ D\ A\ g\ f))$

(ah) $\forall P{:}Pred(S_B).\ (\forall x{:}A.\ \forall y{:}B(x).\ P\langle x,y\rangle \Rightarrow \exists z{:}F\langle x,y\rangle.true\ ) \Rightarrow$
$(derive\ S_B\ A\ P\ \pi_1) =_A (derive\ S_{BF}\ A\ (translate\ S_{BF}\ S_B\ \hat{P}\ \pi_1)\ comp_{\pi_1})$

Remark: The axioms (U4) and (T1) reflect the definition of the operators *union* and *translate* as deliverables and are therefore omitted in the type theoretic formulation. Note that instead of quantifying over signature morphisms due to the generality of deliverables we have to quantify over functions between the underlying types (in the opposite direction).
The choice of the projection functions and the bracketing of the products is, of course, arbitrary. In this presentation we always chose the first projection, π_1, and the structure types are bracketed correspondingly, e.g. $\sum x{:}A.B(x)$. But the projected type may be only a part of A which is independent from the other components of the (sum) type. The projection would then yield some arbitrary

components of the dependent sum. Still it can be written as composition of first and second projection functions. As a consequence, the stated equations are valid up to canonical isomorphism between product types.
(E2)–(E4): The *export* operation used in *BSA* must be expressed via *derive* using a projection function (cf. Section 3.1) Therefore, conditions on the signatures (formulated via *sig* in *BSA*) have to be expressed by the appropriate typing condition on the level of deliverables. In (E2) the information "$\Sigma = sig(sp_1) \cap sig(sp_2)$" is expressed by the fact that sp_1 has structure type $\sum x{:}A.B(x)$ and sp_2 has type $\sum x{:}A.C(x)$ (with B, C arbitrary families of structure types). Since Σ corresponds to A, which is the common part, the antecedent is modelled adequately.
Note that using $\hat{P}$ and $\hat{R}$ in formulating the liftings of (E2),(T4) and (αh) has become necessary in order to prove them in XCC where one cannot prove

$$P(s) \Leftrightarrow P(\langle \pi_1(s), \pi_2(s) \rangle)$$

uniformly in P and s. In other words the Leibniz equality of s and $\langle \pi_1(s), \pi_2(s) \rangle$ is not provable in XCC. If we would use type theory with an eliminator for $\sum$-types à la Martin-Löf, cf. [Nordström et al. 90], then the invariance of predicates under surjective pairing would become provable and we could avoid the tricky use of $\hat{P}$. In the next Section we will give a proof sketch for (E2) and explain this in more detail.
(T3) and (T5) are obvious. (T4) can avoid the $\hat{\sigma}$ construction of *BSA* using a concrete function on structures which is built via *join*.
When lifting (αh) the first part of the premiss expressing the subsignature assumption is again formulated by assuming that the type under consideration is a $\sum$-type of the appropriate form. The assumption of injectivity of σ is mirrored by the use of the projection π_1.

4.3 Provability in XCC

Theorem: The properties (U1)–(αh) from Section 4.2. are provable in XCC relative to the context *Co*.
Proof: All the propositions obtained by lifting the module algebra laws to type theory (instead of being assumed as axioms like in the module algebra) have been proved in the LEGO system, using the *Extended Calculus of Constructions* XCC [Luo 89]. The complete proof as LEGO-code is available from the authors. □
Once the module algebra laws are *adequately* rephrased in type theory the proofs are not very difficult. There are, however, some subtle difficulties when proving them in XCC. The next Section discusses these problems in detail.

Proof of (E2) In the following let $\pi_{x.y}(u)$ be an abbreviation for $\pi_y(\pi_x(u))$. To give an impression of how proofs look like we present a proof sketch for (E2). It is appropriate to give the motivation for using "split" versions of predicates and explain some difficulties arising from intensional type theory. If we expand the definitions of the operations then (E2) becomes

$\forall P{:}S_B.\ \forall Q{:}S_C.\ \forall s{:}A.$
$[\exists\ t{:}S_B.\ P\ \langle\pi_1(t),\ \pi_2(t)\rangle \wedge (s{=}\ \pi_1(t))] \wedge$
$[\exists\ u{:}S_C.\ R\ \langle\pi_1(u),\ \pi_2(u)\rangle \wedge (s{=}\ \pi_1(u))]$
$\Leftrightarrow$
$[\exists\ w{:}S_{AC}.\ (\ P\ \langle\pi_1(w),\ \pi_{1.2}(w)\rangle \wedge R\ \langle\pi_1(w),\ \pi_{2.2}(w)\rangle\ ) \wedge (s = \pi_1(w))].$

The "$\Leftarrow$"-direction is easy to verify. The "realizers" for the existential quantifiers of the left hand side are $t =_{def} \langle\pi_1(w), \pi_{1.2}(w)\rangle$ and $u =_{def} \langle\pi_1(w), \pi_{2.2}(w)\rangle$. The "$\Rightarrow$"-direction is a little more complicated to prove. We suspect the realizer of the existential quantifier to be $w =_{def} \langle\pi_1(t),\ \langle\pi_2(t),\ \pi_2(u)\rangle\rangle$. But although $\pi_1(t) = s = \pi_1(u)$ holds, the type checker is not able to accept this tuple as well-typed. This is because in an intensional type theory like XCC the rule:

$$\frac{A{:}Type;\ B,C{:}A \to Type;\ x{:}S_B;\ y{:}S_C\ \vdash \pi_1(x) = \pi_1(y) \in A}{A{:}Type;\ B,C{:}A \to Type;\ x{:}S_B;\ y{:}S_C\ \vdash \langle x, \pi_2(y)\rangle \in S_{AC}}$$

is not derivable, i.e. from the fact that one can prove $\pi_1(x) = \pi_1(y)$ with Leibniz equality, it does not automatically follow that the types $C(\pi_1(x))$ and $C(\pi_1(y))$ are intensionally equal. But since we are only proving existence and need not construct the object explicitly, we can fortunately employ the following **Lemma:**

$\forall P{:}Pred(S_B).\ \forall x{:}A.$
$(\exists y{:}(B\ x).\ P\langle x,\ y\rangle) \Leftrightarrow (\exists\ s{:}S_B.\ P\ \langle\pi_1(s), \pi_2(s)\rangle)\ \wedge\ (x = \pi_1(s))$

the proof of which has been provided by Z.Luo [Luo 92]. Now after discharging P, R and s we can apply the "$\Leftarrow$" direction of this lemma twice (instantiating P with P and x with s) and yield as new goal

$\exists y_1{:}B(s).\ P\ \langle s,\ y_1\rangle \wedge \exists y_2{:}C(s).\ R\ \langle s,\ y_2\rangle \Rightarrow [\exists\ w{:}S_{AC}.\ldots]$

This Lemma is used to "decompose" the objects t into two separated objects s and y_1 with dependent types. The same holds for u with s and y_2. Sinced both objects y_1 and y_2 depend on the same s one can now build the realizer of the existential quantifier on the right hand side of the implication, namely w $=_{def} \langle s, \langle y_1, y_2\rangle\rangle$ and the rest of the proof is straightforward.
Note that the splitting of P and Q is necessary because after substituting the realizer for w we get the proposition $P\ \langle\pi_1(t), \pi_2(t)\rangle$ which cannot be deduced from $P(t)$.

5 Conclusion

We have expressed the laws of module algebra formally in type theory. As in XCC one cannot derive surjective pairing for dependent sum types, the module algebra laws had to be presented in a "split" version, i.e. instead of $P(s)$ one writes $P(\langle\pi_1(s), \pi_2(s)\rangle)$. But this problem could be avoided by chosing Martin Löf's notation of dependent sum types or alternatively by introducing surjective pairing into XCC as an axiom.

As type theory can express in a quite flexible way functions and predicates on deliverables it does not seem necessary to restrict to the specific laws of module algebra. Nevertheless these can be considered as a starting point for a library of theorems useful for the composition of large systems and might be reused for implementation proofs.

Acknowledgement: Thanks to Martin Wirsing for moral support, for reading several drafts and for his comments on how to improve readability. Thanks also to Zhaohui Luo for answering detailed questions about XCC and for the enlightenment he gave us with the proof of what we have baptized *Luo's Lemma*.

References

[Bergstra et al. 89] J. A. Bergstra, J. Heering, P. Klint: *Algebraic Specification.* ACM Press, New York, 1989.

[Burstall, Goguen 80] R.M. Burstall, J.A. Goguen: *The semantics of Clear, a specification language.* In: Proc. Advanced Course on Abstract Software Specification. LNCS 86, Springer, Berlin, 1980.

[Burstall, McKinna 92] R.M. Burstall, J. McKinna: *Deliverables: a categorical approach to program development in type theory.* ECS-LFCS-92-242, Dept. of Computer Science, Edinburgh University, 1992.

[McKinna 92] J. McKinna: *Deliverables: a categorical approach to program development in type theory.* Ph.D. Thesis, University of Edinburgh, 1992. Also appeared as ECS-LFCS-92-247, Dept. of Computer Science, Edinburgh University, 1992.

[Leszczylowski, Wirsing 91] J. Leszczylowski, M. Wirsing: *Polymorphism, Parameterisation and Typing: An Algebraic Specification Perspective.* In: Proc. of the 8th STACS, LNCS 480, Springer, Berlin, p. 1 - 15.

[Luo 89] Z. Luo: *ECC, an Extended Calculus of Constructions.* In: Proc. of the Fourth Ann. Symp. on Logic in Computer Science (LICS), IEEE, 1989, p. 385 - 395.

[Luo 91] Z. Luo: *Program specification and data refinement in type theory.* In: Proc. of the Fourth Intern. Conf. on the Theory and Practice of Software Development (TAPSOFT), LNCS 493, Springer, Berlin, 1991, p. 143 - 168.

[Luo 92] Z. Luo: private communication.

[Luo et al. 92] Z. Luo, R. Pollack: *LEGO Proof Development System: Users' Manual.* ECS-LFCS-92-211, Dept. of Computer Science, Edinburgh University, 1992.

[Nordström et al. 90] B. Norström, K. Petersson, J.M. Smith: *Programming in Martin Löf's Type Theory.* Oxford University Press, Oxford, 1990.

[Streicher 88] T. Streicher: *Semantics of Type Theory - Correctness, Completeness and Independence Results.* Birkhäuser, Boston, 1991.

[Streicher, Wirsing 90] T. Streicher, M. Wirsing: *Dependent types considered necessary for algebraic specification languages.* In: Recent Trends in Data Type Specification, Proc. 7th International Workshop on Specification of Abstract Data Types. LNCS 534, Springer, Berlin, 1990, p. 323 - 340.

[Wirsing 86] M. Wirsing: *Structured algebraic specifications: a kernel language.* Theoretical Computer Science 42, 1986, p. 123 - 249.

[Wirsing 89] M. Wirsing: *Algebraic Specification* in J.V.Leeuwen (ed.): Handbook of Theoretical Computer Science, Volume B, Elsevier, Amsterdam, 1990, p. 675 - 788.

On Time-Space Trade-Offs in Dynamic Graph Pebbling

Peter Ružička, Juraj Waczulík
Institute of Computer Science, Comenius University
842 15 Bratislava, Slovakia

Abstract

Pebble game on dynamic graphs is studied as an abstract model for the incremental computations. We investigate how the time T and/or the space S is changing according to the number m of insert-edge/delete-edge operations on directed acyclic graphs of size n. Time-space trade-off of the form $T = \Theta\left(\frac{n^2}{S+m}\right)$ is given for standard pebble game on permutation graphs. In the case of the minimal space pebbling an extreme (superpolynomial) explosion of the time is related to the unit change of the graph size. If the space is from a small interval then a superpolynomial time increase is also achieved relative to the small (slowly increasing) change of the graph size.

1 Introduction

Pebbling directed acyclic graphs (dags) introduces a relevant paradigm in programming. Dags introduce an abstract computational structure of a given problem and pebbling simulates the computation of this problem on the computational graph. Pebbling mediates us to model a variety of computations and it enables us to investigate the relationship between space and time complexity of these computations by means of a simple combinatorial one-person game played with pebbles on dags according to given rules. There are well known exploitations of pebbling in gaining time-space trade-offs for concrete computational problems in various areas of computer science, including compilers, operating and database systems, parallel and distributed systems, and programming methodology. In the theory of pebbling well known results have been obtained concerning space complexity and time-space trade-offs for pebbling special classes of dags. A survey of principal classical results and basic applications of pebbling is given by Pippenger (1980). Recent pebbling results are presented by Wilber (1988) and Venkateswaran, Tompa (1989).

Recently much effort has been devoted to the investigation of the incremental and dynamic computations. There has been a growing interest in the incremental/dynamic problems on graphs. In these problems one would like to answer queries on graphs that are undergoing a sequence of updates (insertions of edges in case of incremental updates and insertions&deletions of edges in case of fully dynamic ones). Mostly, the goal of the incremental/dynamic graph algorithms is to update efficiently the solution of the problem after incremental/dynamic changes, rather than having to recompute it each time. Seldom, the objective is to express the computational complexity of problems on graphs relative to the size of incremental/dynamic changes.

We propose to study pebble game on incrementally changing dags, acting as an abstract model of incremental computations. With respect to pebble game incremental ("insert-edge") and decremental ("delete-edge") changes on dags are symmetrical. We are interested in characterizing the relationship between time-space complexity of pebbling and the size of incremental changes on dags. We investigate the asymptotical behaviour of the time T for various values of the space S, size n and distance m parameters. The basic question is to determine the classes of dags $\mathcal{G}$ and $\mathcal{D}$ such that there is (as great as possible) asymptotical difference in time–space trade–offs relative to (as small as possible) distance m of the classes.

We present the following trade–off results for the standard pebble game. There are classes $\mathcal{B}, \mathcal{G}, \mathcal{H}$ of dags of the size n such that

- for space $S(n)$ in the interval $\langle 2, n\rangle$ it holds $T(n, S(n)) = \Theta\left(\frac{n^2}{S(n)+m}\right)$ on a class of dags of the distance m from $\mathcal{B}$, where $2 \leq S(n) + m \leq n$;
- for the minimal space $S_{min}(n)$ in the range $\omega(1)$ to $O\left(n^{\frac{1}{3}}\right)$ needed to pebble dags from $\mathcal{G}' \subset \mathcal{G}$ it holds $T(n, S_{min}(n)) = 2^{\Theta\left(S_{min}(n) \cdot log \frac{n}{S_{min}(n)}\right)}$ on $\mathcal{G}'$, but there exists a class of dags $\mathcal{D}$ of the distance 1 from $\mathcal{G}'$ such that $T(n, S_{min}(n)) = O(n^2)$ on $\mathcal{D}$;
- for the minimal space $S_{min}(n)$ in the range $\alpha(n)$ (the inverse of the Ackermann function) to $\left(\frac{\sqrt{n}}{\alpha(n)}\right)^{1-\epsilon}$ needed to pebble dags from $\mathcal{H}' \subset \mathcal{H}$ it holds $T(n, S(n)) = 2^{\Omega(S(n) \cdot log(c(n) \cdot S(n)))}$ on $\mathcal{H}'$ for $S(n)$ in the interval $\langle S_{min}(n), S_{min}(n) + c^{1-\epsilon}(n)\rangle$, but there exists a class of dags $\mathcal{D}$ of the distance $c(n)$ from $\mathcal{H}'$, where $c(n)$ in the range $\alpha(n)$ to $\left(\frac{\sqrt{n}}{\alpha(n)}\right)^{1-\epsilon}$, $0 < \epsilon < 1$ and satisfies the condition $c^2(n) \cdot S^2(n) + c(n) \cdot S^3(n) = \Theta(n)$ such that $T(n, S_{min}(n)) = 2^{O(c(n) \cdot log S_{min}(n))}$ on $\mathcal{D}$.

2 Preliminaries

Throughout this paper we consider only directed acyclic graphs (**dags**) with in–degree bounded by 2. An input (an output) vertex in a dag is a vertex with in–degree (out–degree) equal to 0. A dag having exactly one output vertex is called rooted, and its output vertex is the root of the dag. The **size** of a dag G (denoted as $size(G)$) is equal to the number of its edges. By $\mathcal{G}_n$ is denoted a subclass of dags of size n from the class $\mathcal{G}$.

Standard pebble game is played on dags using the following rules:

S-1: A pebble can be removed from an arbitrary vertex of the graph.

S-2: If all direct predecessors of a vertex v are covered by pebbles, then a pebble can be laid also on the vertex v.

S-3: If all direct predecessors of a vertex v are covered by pebbles, then a pebble can be moved from some predecessor of the vertex v onto the vertex v.

A **configuration** C is a set of vertices, comprising just those vertices of a dag that have pebbles on them.

A **transition** is an ordered pair of configurations, the second one of which follows from the first one according to one of the rules **S-i**.

A **computation** $\mathcal{C}$ is a sequence of configurations, where each successive pair forms a transition.

A **complete computation** is one that begins and ends with the empty configuration and every vertex has a pebble on it at some configuration in the computation.

We shall be particularly interested in the **time** $T_{\mathcal{C}}$ and the **space** $S_{\mathcal{C}}$ of the complete computation $\mathcal{C}$. $T_{\mathcal{C}}$ is the number of transitions of the computation $\mathcal{C}$, and $S_{\mathcal{C}}$ is the maximum number of pebbles on the graph over all configurations in the computation $\mathcal{C}$.

The **minimal space** $S_{\min}(G)$ of the standard pebble game on the dag G is the minimum of S_C over all complete computations C on the dag G.
The **time** $T(G,S)$ of the standard pebble game **with** S **pebbles** on the dag G is the minimum of T_C over all complete computations C with S pebbles on the dag G. We use notation $T(G)$ if value S is evident.

Assume $\mathcal{G}$ is a class of dags. The **minimal space function** $S_{\min}(n)$ of a class $\mathcal{G}$ is the maximum of $S_{\min}(G)$ over all dags G in the subclass $\mathcal{G}_n$. The **time function** $T(n,S)$ of a class $\mathcal{G}$ is the maximum of $T(G,S)$ over all dags G in the subclass $\mathcal{G}_n$.

A dag G is called **open** in a configuration C if and only if there is a path in G from an input to an output such that it does not contain a vertex in C. Otherwise G is called **closed** in C.

A dag D is of the **distance** m from a dag G (denoted as $d(G,D) = m$) if and only if D is obtained from G by deleting m edges.
The class of dags $\mathcal{D}$ is of the distance m from the class of dags $\mathcal{G}$ (denoted as $d(\mathcal{G},\mathcal{D}) = m$) if and only if for each $G \in \mathcal{G}$ there is $D \in \mathcal{D}$ and vice versa such that $d(G,D) = m$.

3 General Case

In this section we investigate the asymptotical behaviour of the time T for various values of the space S, size n and distance m parameters. In case of an arbitrary space (it means at least the minimal pebbling space and not more than the number of vertices) a general trade-off result is given for the standard pebble game on permutation graphs. We start with the proposition that there are no time-space differences for standard pebble game on classes of dags of constant distance.

Proposition 3.1 *Let G and D be dags, $S_{\min}$ be the minimal space for D and $d(G,D) = m$. Then for every space S, $S_{\min} \leq S \leq n$ it holds*

$$T(G, S+m) \leq (m+1) \cdot T(D,S)$$

Proof: We show that for every complete computation C on D with S_C pebbles and time T_C we are able to construct the complete computation C' for G with time $T_{C'} \leq (m+1) \cdot T_C$ and space $S_{C'} \leq S_C + m$.

G is obtained from D by inserting m edges. Let these edges be outgoing from vertices $w_1, \ldots, w_k$, $k \leq m$. Let G_{w_i} denote the rooted subgraph of G with the root w_i. Without loss of generality we can assume the vertices are ordered such that for every $1 \leq i \leq k$ the subgraph G_{w_i} doesn't contain vertices w_j for $j > i$. Such ordering exists because G is an acyclic graph.

The computation C' is the following. At first the permanent pebbles are successively laid on the vertices $w_1, \ldots, w_k$ using at most m additional permanent pebbles. Afterwards, direct (step-by-step) simulation of the complete computation C is used for pebbling G. The preprocessing needs at most $S_C + m$ pebbles and time at most $m \cdot T_C$. Hence, the complete computation C' on G needs time less than $(m+1) \cdot T_C$ and space at most $S_C + m$. □

Consequence 3.2 *Let $\mathcal{G}$ and $\mathcal{D}$ be arbitrary classes of dags satisfying $d(\mathcal{G},\mathcal{D}) = O(1)$. Then the time-space trade-offs for standard pebble games on $\mathcal{G}$ and $\mathcal{D}$ are asymptotically equal.*

Let π_n be a permutation on n elements. The **permutation graph** Π_n (corresponding to the permutation π_n) is the graph $\Pi_n = (V, E)$ such that

$$\begin{aligned} V &= \{u_1, \ldots, u_n, v_1, \ldots, v_n\} \\ E &= \{(u_i, u_{i+1}), (v_i, v_{i+1}) \mid 1 \le i \le n-1\} \cup \{(u_i, v_{\pi_n(i)}) \mid 1 \le i \le n\} \end{aligned}$$

The size of Π_n is $3 \cdot n - 2$ and $S_{\min} = 2$ for $n \ge 2$. Permutation graph pebbling has been studied by Lengauer, Tarjan (1982).

Theorem 3.3 *Let $\mathcal{P} = \{\Pi_n \mid n > 1\}$. For $\mathcal{P}$ and an arbitrary m, $0 \le m \le n$, there is a class $\mathcal{D}$, $d(\mathcal{P}, \mathcal{D}) = m$, such that for the space $S(n)$ in the interval $\langle 2, n\rangle$ an upper bound on the time-space trade-off is given by*

$$T(n, S(n)) = O\left(\frac{n^2}{S(n) + m}\right).$$

Proof: Let Π_n be a permutation graph for some permutation π_n. For an arbitrary m, $0 \le m < n$, construct the dag D_n of the distance m from Π_n such that edges $(u_{j \cdot k-1}, u_{j \cdot k})$ for $1 \le j \le m$, $k = \lfloor \frac{n}{m} \rfloor$ are deleted from Π_n (for technical reasons (u_0, u_1) equals to (u_{n-1}, u_n)).

We derive the complete computation on D_n with $S(n)$, $2 \le S(n) + m \le n$, pebbles using $O(\frac{n^2}{S(n)+m})$ pebble placements.

Reserve one pebble for u-path (a path of u-vertices) and one pebble for v-path (a path of v-vertices) in D_n. Moreover, situate the rest $S(n) - 2$ pebbles as permanent ones equidistantly on segments of u-vertices $u_{j \cdot k}, \ldots, u_{j \cdot (k+1)-1}$. This can be done in the following manner. Let $p = \lceil \log(\frac{S(n)-2}{m+1} + 1) + 1 \rceil$. Initialy, denote "boundary" the set of vertices u_1, u_n, $u_{j \cdot k}$ for $j = 1, 2, \ldots, m$. Repeat p times the following step: for all neighbour "boundary" vertices u_i, u_j place a permanent pebble on the "free" vertex $u_{\lceil (j-i)/2 \rceil}$ and add this vertex to the "boundary". So there are $S(n) + m + 1$ segments, divided by "boundary" vertices, each of the length $O(\frac{n}{S(n)+m})$. The complete computation now performs in the following way. The idea is to pebble v-path in the increasing order of indices. If a pebble lays on the vertex v_i, then move a pebble from v_i to v_{i+1} immediately after pebbling the path from the nearest left-side "boundary" vertex to u_j satisfying $\pi_n(j) = i + 1$. The complete computation on D_n with $S(n)$ pebbles needs $O(n)$ pebble placements on v-vertices and $O(\frac{n^2}{(S(n)+m)^2})$ pebble placements on each segment of u-vertices. There are $S(n) + m + 1$ segments and thus altogether $O(\frac{n^2}{S(n)+m})$ pebble placements on u-vertices are needed. □

Suppose $0 \le n = 2^h$ and the set to be permuted be $N = \{i \mid 0 \le i < n\}$. Let $f : N \longrightarrow \{0,1\}^h$ be the bijective mapping where $f(i)$ is the binary string of the length h representing the number i. Let $f(i) = b_0 \ldots b_{h-1}$. The bit reversal of i (denoted by rev(i)) is defined to be the number j such that $f(j) = b_{h-1} \ldots b_0$. The **bit reversal permutation graph** B_n is the graph $B_n = (V, E)$ such that

$$\begin{aligned} V &= \{u_1, \ldots, u_n, v_1, \ldots, v_n\} \\ E &= \{(u_i, u_{i+1}), (v_i, v_{i+1}) \mid 1 \le i \le n-1\} \cup \{(u_i, v_{rev(i)}) \mid 1 \le i \le n\} \end{aligned}$$

Theorem 3.4 *Let $\mathcal{B} = \{B_n \mid n > 1\}$. For $\mathcal{B}$ and an arbitrary m, $0 \le m \le n-2$, there is a class of dags $\mathcal{D}$, $d(\mathcal{B}, \mathcal{D}) = m$, such that a lower bound on the time-space trade-off is given by*

$$T(n, S(n)) = \Omega\left(\frac{n^2}{S(n) + m}\right).$$

Proof: In fact, we prove precisely $T\cdot(S(n)+m) > \frac{n^2}{16}$. The proof is trivial for $S(n)+m > \frac{n}{4}$. Assume $S(n)+m \leq \frac{n}{4}$. Take $k = \lceil\log(S(n)+m)\rceil + 1$ and $h = \lceil\log n\rceil$. Let D_n is obtained from B_n by deleting m edges from the u-path. Assume the v-path is divided into 2^{h-k} segments, each of the length 2^k . Let t_p be the time at which a pebble is laid on the last unpebbled vertex of the p-segment for the first time. The immediate predecessors of all vertices in the p-segment divide the u-path into $2^k - 1$ intervals of the length 2^{h-k}. At the time $t_p - 1$ at most $S(n)-1$ pebbles are on u-vertices. Thus, at least $2^k-1-(S(n)+m-1) \geq S(n)+m$ intervals are pebble-free and without deleted edge at t_p-1. All of them have to be pebbled before t_p. This takes at least $(S(n)+m)\cdot 2^{h-k} > \frac{n}{4}$ pebble placements. Therefore $t_p - t_{p-1} > \frac{n}{4}$ for $1 < p \leq 2^{h-k}$ and thus each complete computation on D_n needs time at least

$$2^{h-k}\cdot\frac{n}{4} > \frac{n^2}{16\cdot(S(n)+m)}$$

what proves the theorem. □

We conclude the section with two comments. The lower bound argument presented in the proof of Theorem 3.4 holds also for a broader class of dags of distance m from B_n such that m edges of the form either (u_i, u_{i+1}) or $(u_i, v_{rev(i)})$ are deleted from B_n. The idea is to "simulate the missing edge" by placing a permanent pebble on the vertex u_i at the beginning of the complete computation (counting zero time for such a placement). Note that deleted edges of the form (v_i, v_{i+1}) can influence the asymptotical time complexity.

As a consequence of Theorem 3.3 and 3.4 it follows time hierarchy for space restricted dynamic pebbling formulated in the following way. Let $S(n)$ be a space function, $m_i(n)$ $(i = 1, 2, \ldots)$ be distance functions in the range from $S(n)$ to n satisfying $\lim_{n\to\infty}\frac{m_i(n)}{m_{i+1}(n)} = 0$ and $T_i(n) = \Theta(\frac{n^2}{m_i(n)})$ be time functions. Then there is a dag D_n^i, $d(B_n, D_n^i) = m_i(n)$ such that D_n^i is pebbleable simultaneously in time $T_i(n)$ and space $S(n)$, but is not pebbleable in $T_{i+1}(n)$ with space $S(n)$.

4 Special Cases

In the previous section we investigated the behaviour of time with respect to space, size and distance. If there are no restrictions put on space then we are able to exhibit only quadratic differences in time on special classes of dags. Now, the basic question is to determine the classes of dags $\mathcal{G}$ and $\mathcal{D}$ such that there are (as great as possible) differences in time relative to (as small as possible) distances of $\mathcal{G}$ and $\mathcal{D}$ for space restricted standard pebble game.

4.1 Pebbling in minimal space

We show there are superpolynomial time differences in the space optimal pebbling on a class of dags of distance 1. We start with presenting an upper bound on time complexity for pebbling a graph of size n pebbleable in space $S(n)$.

Proposition 4.1 *Let G be a rooted dag of size n and $S_{min}(n)$ be the minimal number of pebbles needed to pebble G. Then for the pebbling time $T(G, S(n))$ it holds*

$$T(G,S(n)) = 2^{O(S_{min}(n)\cdot\log\frac{n}{S_{min}(n)})}.$$

Proof: Each complete computation on rooted dags can be transformed onto the complete computation without repeated configurations.

Consider a complete computation with $S_{min}(n)$ pebbles in the form starting with $S_{min}(n)$ applications of the rule S-2, following with a sequence of applications either of the rule S-3 or of the rule S-1 followed by the rule S-2 and ending with $S_{min}(n)$ applications of the rule S-1. The worst-case time complexity of these three phases of computation is estimated as follows

$$T(G,S(n)) \leq S_{min}(n) + 2 \cdot \binom{n}{S_{min}(n)} + S_{min}(n)$$

Applying Stirling formula $n! \geq \sqrt{2\pi n}(\frac{n}{e})^n$ one can obtain the time estimation

$$T(G,S(n)) \leq \frac{\sqrt{2}}{\sqrt{\pi S_{min}(n)}} 2^{S_{min}(n)\log\frac{e \cdot n}{S_{min}(n)}} + 2S_{min}(n)$$

and hence the asymptotical time-space trade-off formula is in the form

$$T(G,S(n)) = 2^{O\left(S_{min}(n)\cdot\log\frac{n}{S_{min}(n)}\right)}$$

□

In a special case when a graph can be pebbled with a constant number of pebbles the following upper bound on time complexity can be derived.

Consequence 4.2 *Let G be a rooted dag of size n and S_{min} be a constant. Then for the pebbling time $T(G,S)$ it holds*

$$T(G,S) = O(n^{S_{min}}).$$

Now, we present a lower bound on the time complexity of the minimal space pebbling up to space $O(n^{\frac{1}{3}})$ which is asymptotically tight to the upper bound given in Proposition 4.1. Basic to all graphs discussed later in this paper is the pyramidal graph $P_k = (V,E)$, where

$$\begin{aligned} V &= \{u_{i,j} \mid 1 \leq i \leq j \leq k\} \\ E &= \{(u_{i,j}, u_{s,t}) \mid s = i+1, t \in \{j, j+1\}, 1 \leq i \leq j, s \leq t \leq k\} \end{aligned}$$

Lemma 4.3 (Cook,1974) *The pyramidal graph P_k has the size equal to $\Theta(k^2)$ and the minimal number of pebbles $S_{min}(P_k) = k$.*

Consider three graph components P_i, R_j, H_l, where P_i is a pyramidal graph with i inputs, 1 output and $\frac{1}{2}i(i+1)$ vertices, R_j is a chain graph with 1 input, 1 output and j vertices and H_l is a combination graph with 1 input, 2 outputs and $l\cdot(l+2)$ vertices. A pyramid P_i is a fragment of rectangular grid. A chain R_j is a sequence of j vertices, where there is an edge oriented from the k-th vertex to the $(k+1)$-st one. A combination graph H_l consists of two pyramids P_l and a chain R_{2l}, where the i-th vertex of R_{2l} for $1 \leq i \leq l$ is connected with oriented edge to the i-th input vertex of the first pyramid and the j-th vertex of R_{2l} for $l+1 \leq j \leq 2\cdot l$ is connected with oriented edge to the $(j-l)$-th input vertex of the second pyramid.

A class of dags $\mathcal{G} = \{G_{k,m} \mid k \geq 1, m \geq 2\}$ is defined as follows. $G_{k,m}$ is constructed from k levels $U_1^m, \ldots, U_k^m$, where the i-th level U_i^m for $1 \leq i \leq k$ consists of the serial connection of three components P_i, R_m, H_i (i.e. the output of P_i is connected with oriented edge to the input of R_m and the output of R_m is connected to the input of H_i). The graph $G_{k,m}$ is completed using the following connections of levels U_i^m and U_{i+1}^m for $1 \leq i < k$. The first (second) output of U_i^m is connected with oriented edges to all odd (even) vertices in the component R_m of the level U_{i+1}^m (see the Fig.1).

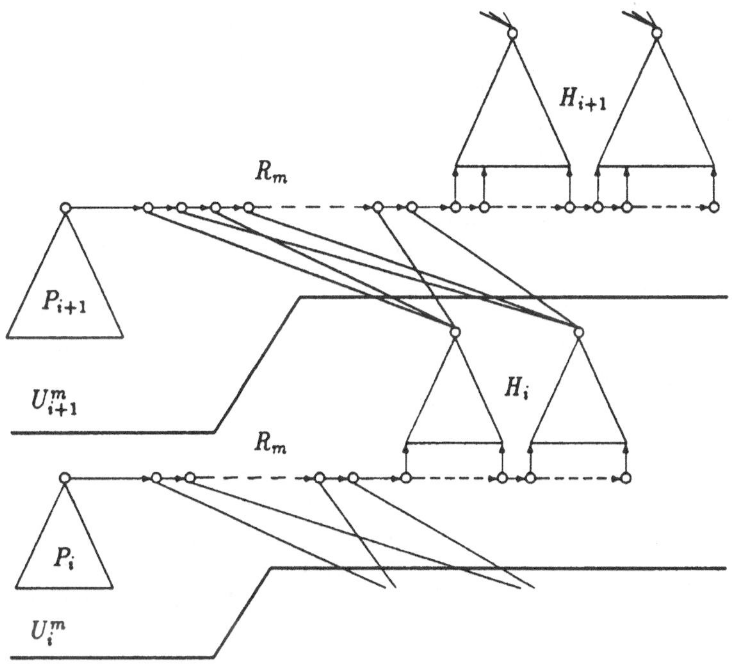

Fig.1: The i-th and $(i+1)$-st level of the graph $G_{k,m}$.

Proposition 4.4 *The class of graphs $\mathcal{G}$ satisfies two conditions:*

$$size(G_{k,m}) = \Theta(k^3 + k \cdot m) \text{ and } S_{min}(G_{k,m}) = k$$

Proof: The size of the graph $G_{k,m}$ fulfils the following recurrent relation

$$\begin{aligned} size(G_{k,m}) &= a_1 \cdot m + size(G_{k-1,m}) \\ size(G_{1,m}) &= a_2 \cdot m \end{aligned}$$

for k in the range $\omega(1)$ to $O(\sqrt{m})$ and some constants $a_1, a_2 > 1$ and

$$\begin{aligned} size(G_{k,m}) &= a_3 \cdot k^2 + size(G_{k-1,m}) \\ size(G_{1,m}) &= a_2 \cdot m \end{aligned}$$

for k in the range $\Omega(\sqrt{m})$ to $m^{O(1)}$ and a constant $a_3 > 1$.
The solution is in the form $size(G_{k,m}) = \Theta(k^3 + k \cdot m)$.

The second condition follows directly from the Lemma 4.3. □

Consider the k-th level of the graph $G_{k,2m}$. Denote by u the output of the first component P_k of the k-th level U_k^{2m} and by $x_0, x_1, \ldots, x_{2m-1}$ the sequence of vertices in the second component R_{2m} of the k-th level U_k^{2m} and by α_0, α_1 the output vertices of the $(k-1)$-st level U_{k-1}^{2m} in the graph $G_{k,2m}$. The following proposition holds:

Lemma 4.5 *In any computation on the graph $G_{k,2m}$ using k pebbles for pebbling the vertex x_i, $0 \leq i < 2m-1$, there have to be an open path from an input vertex to x_{i+1} leading through $\alpha_{(i+1) \bmod 2}$ at some configuration in which x_i is pebbled.*

Proof: Case $i = 0$: Consider an arbitrary computation C_k on $G_{k,2m}$ with k pebbles starting with the empty configuration and ending with a configuration in which a pebble is placed on x_0 for the first time. From Lemma 4.3 it follows that k pebbles are needed to close all path from inputs to the vertex u in the pyramid P_k. Furthermore, if there is at least one path open in the pyramid P_k, then k pebbles are needed to close P_k. Two important facts follow.

Firstly, a computation on the subgraph $G_{k-1,2m}$ of the graph $G_{k,2m}$ can be started after closing the pyramid P_k with the root u because at the moment of closing the pyramid the subgraph $G_{k-1,2m}$ is empty.

Secondly, the pyramid P_k with the root u have to be maintained as closed during the rest of the computation C_k because otherwise the computation C_k would be repeated from some configuration before closing P_k. At least 1 pebble has to be permanently placed on P_k in order to maintain the pyramid P_k closed. Thus all path to x_0 through u are closed. Hence, there have to be a configuration (say C) in the computation C_k such that 1 pebble is placed on the pyramid P_k with the root u and $k-1$ pebbles are placed on the pyramid P_{k-1} with the root α_0. That means that in the configuration C there is no pebble on a path leading to the vertex α_1 in $G_{k-1,2m}$ and thus there is an open path to x_{i+1} through α_1.

Case $0 < i < 2m-1$: It is sufficient to repeat arguments used in case $i = 0$ with the addition that permanent pebble keeping closeness of P_k lay on x_{i-1} and thus it maintains all path to x_i through x_{i-1} closed. □

Theorem 4.6 *Let $S(n)$ be a space function in the range $\omega(1)$ to $O(n^{\frac{1}{3}})$.*

1. *There is a subclass $\mathcal{G}' = \{G \mid size(G) = n, n \geq n_0\}$ of $\mathcal{G}$ with the minimal space function $S(n)$ such that it holds*
$$T(n, S(n)) = 2^{\Omega(S(n)\cdot\log\frac{n}{S(n)})}.$$
2. *There is a class of dags $\mathcal{D}$, $d(\mathcal{G}', \mathcal{D}) = 1$ with the minimal space function $S(n)$ such that it holds*
$$T(n, S(n)) = O\left(n^2\right).$$

Proof:

Part 1: Consider the graph $G_{k,2m}$ with $k = S(n)$ and $m = c_1 \cdot \frac{n}{S(n)}$ for appropriately chosen constant $c_1 > 0$. We show that the following recurrent relation holds:

$$\begin{aligned} T(G_{k,2m}) &\geq b_1 \cdot m \cdot T(G_{k-1,2m}) \\ T(G_{1,2m}) &\geq b_2 \cdot m \end{aligned}$$

It follows directly that the relation holds for $i = 1$. Suppose now that it holds also for $i = 2, 3, \ldots, k-1$. Following Lemma 4.5 the time needed for pebbling $G_{k,2m}$ with $k = S(n)$ pebbles is lower estimated by

$$T(G_{k,2m}) \geq d_2 \cdot k^2 + \sum_{i=0}^{2m-1} [\frac{d_1}{2} T(G_{k-1,2m})]$$

because for pebbling the pyramid P_k it is needed $\frac{1}{2}k(k+1)$ computational steps and for pebbling $2m$ vertices $x_0, x_1, \ldots, x_{2m-1}$ it is needed $\frac{1}{2}d_1 \cdot T(G_{k-1,2m})$ computational steps. Choose $k = S(n) = c_2 \cdot n^{\frac{1}{3}}$ for appropriate constant $c_2 > 0$. So we have $m = c_3 \cdot n^{\frac{2}{3}}$ and thus for appropriately chosen constant $b_1 > 0$ it is obtained

$$T(G_{k,2m}) \geq b_1 \cdot m \cdot T(G_{k-1,2m}).$$

Solving the recurrent relation one can obtain $T(G_{k,2m}) \geq d \cdot m^k$ for some constant $d > 0$ and thus for $G \equiv G_{S(n),n/S(n)}$ it holds

$$T(G) = 2^{\Omega(S(n) \log \frac{n}{S(n)})}.$$

Part 2: In order to obtain the class $\mathcal{D}$ with the required property it is sufficient to remove the edge outgoing from the output vertex of the component R_m at the level U_{k-2}^m in the graph $G_{k,m}$. The analysis is straightforward. □

If the minimum space necessary to pebble a dag is a constant, then minimum space and polynomial time are achieved simultaneously and the polynomial time is asymptotically tightly expressed in the following form.

Consequence 4.7 *Let k be a constant. Then $T(G_{k,m}, k) = \Theta(m^k)$.*

4.2 Nonminimal space pebbling

In this section we show there are also significant time differences in the nonoptimal space restricted pebbling on dags of close distance.

We present a class of dags $\mathcal{H}$ and prove a superpolynomial lower bound on its pebbling time with restricted number of pebbles. $\mathcal{H} = \{H_{k,c} \mid k \geq 1, c \geq 1\}$ is defined inductively for fixed c (the number of output vertices). The basis graph $H_{1,c}$ is an input vertex with c outgoing edges leading to c various output vertices. The graph $H_{i+1,c}$ is composed of c pyramidal graphs P_{i+1} and the subgraph $H_{i,c}$. These $c+1$ components are connected to c chains in the following manner. Consider each chain is divided into $i+1$ sections of vertices. Each section consists of $2 \cdot c$ vertices. The output vertex of the j-th pyramid P_{i+1} is connected to the j-th vertex of each section of all chains. The j-th output of the subgraph $H_{i,c}$ is connected to the $(c+j)$-th vertex of each section of all chains. Thus each output vertex of either the pyramid P_{i+1} or the subgraph $H_{i,c}$ has totally $c \cdot (i+1)$ outgoing edges (see the Fig.2).

Proposition 4.8 *The class of graphs $\mathcal{H}$ satisfies two conditions:*

$$size(H_{k,c}) = \Theta(c^2 \cdot k^2 + c \cdot k^3) \text{ and } S_{\min}(H_{k,c}) = k.$$

Let $\alpha(n)$ be the inverse of Ackermann function, ε be in $(0,1)$, $\beta(n) = \left(\frac{\sqrt{n}}{\alpha(n)}\right)^{1-\varepsilon}$ and n_0 be sufficiently large integer.

Theorem 4.9 *Let $S(n)$ be a space function in the range $\alpha(n)$ to $\beta(n)$ and $c(n)$ be in the range from 2 to $\beta(n)$ such that the condition $c^2(n) \cdot S^2(n) + c(n) \cdot S^3(n) = \Theta(n)$ is satisfied.*

1. *There is a subclass of dags $\mathcal{H}' = \{H \mid size(H) = n, n \geq n_0\}$ of $\mathcal{H}$ with the minimal space function $S(n)$ such that it holds*

$$T(n, S(n) + c^{1-\varepsilon}(n)) = 2^{\Omega(S(n) \cdot \log(c(n) \cdot S(n)))}.$$

2. *There is a class of dags $\mathcal{D}$, $d(\mathcal{H}', \mathcal{D}) = c(n)$, with the minimal space function $S(n)$ such that it holds*

$$T(n, S(n)) = 2^{O(c(n) \cdot \log S(n))}.$$

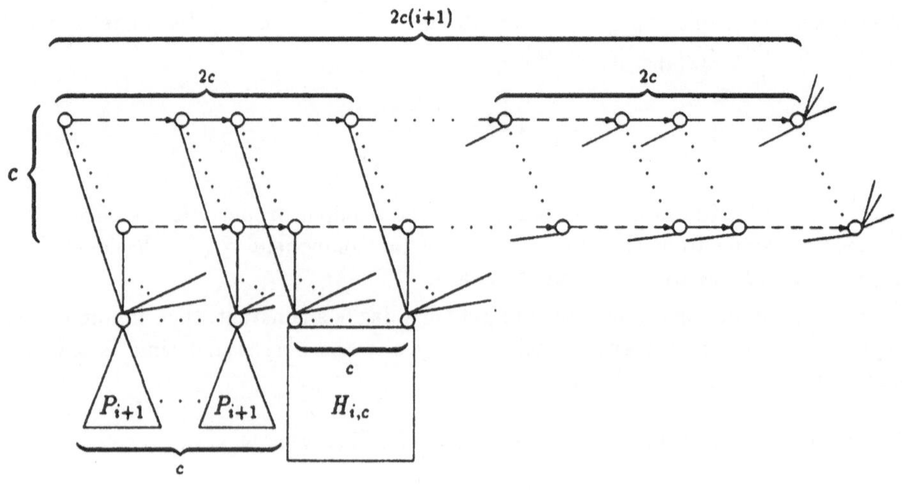

Fig.2: The $(i+1)$-st level of the graph $H_{k,c}$.

Proof:

Part 1: First, consider $H_{k,c}$ for fixed c. From (Carlson,Savage,1982) for the time needed to pebble $H_{k,c}$ using $k+e$ pebbles it follows

$$T(H_{k,c}, k+e) \geq \left(\frac{c-e}{e+1}\right)^k \cdot k!$$

where $0 \leq e \leq c-3$ and $c^2 \cdot k^2 + c \cdot k^3 = \Theta(n)$. Thus, if we choose $e = c^{1-\varepsilon} - 1$ $(0 < \varepsilon < 1)$, then pebbling $H_{k,c}$ requires time at least $(c^\varepsilon - 2)^k \cdot k!$. Hence, when $c(n)$ is not fixed and space is chosen from the interval ranging between $\alpha(n)$ and $\beta(n)$, the time necessary to pebble the graph is at least

$$2^{\Omega(S(n) \cdot \log(c(n) \cdot S(n)))}.$$

Part 2: Assume $H_{k,c}$ for fixed c.

1. Consider a subgraph consisting of $k - 2 \cdot c$ "low levels" of $H_{k,c}$: Linear time according to the subgraph size is sufficient to pebble $k - 2 \cdot c$ "low levels" of $H_{k,c}$, since $2 \cdot c$ pebbles can be laid on c pyramid outputs and c subgraphs outputs.

2. Consider a subgraph consisting of c "middle levels" of $H_{k,c}$: Quadratic time with respect to the subgraph size is sufficient to pebble c "middle levels" of $H_{k,c}$, since c pebbles are laid on c subgraph outputs and c pyramids are repebbled k times for each of the c levels.

3. Consider c "upper levels" of $H_{k,c}$: Delete $4 \cdot c$ edges from $H_{k,c}$ leading to c pyramid outputs and c subgraph outputs of the level $k-c$ in $H_{k,c}$. Consider now the component $D_{k,c}$ consisting of c "upper levels" of $H_{k,c}$. Time $T(D_{k,c}, k)$ sufficient to pebble this component using k pebbles is

$$\begin{aligned} T(D_{k,c}, k) &\leq k \cdot c \cdot T(D_{k-1,c}, k) + k^3 \cdot c \\ T(D_{k-c,c}, k) &\leq k \cdot c + k^3 \cdot c \end{aligned}$$

and thus for D consisting of the components $D_{k,c}$ and $H_{k-c,c}$ it follows

$$T(D,k) \leq \frac{c^2 \cdot k^2 \cdot k!}{(k-c)!}.$$

Hence, for nonfixed $c(n)$ the upper bound on $T(D,S(n))$ is given as

$$T(D,S(n)) = 2^{O(c(n)\cdot \log S(n))}.$$

□

5 Conclusions

We have exhibited classes of graphs on which we proved interesting properties concerning asymptotical relationship among parameters T, S, n and m. Similar results can be obtained for other variants of the pebble game, i.e. for the nondeterministic black-and-white pebble game.

An interesting problem (with respect to Theorem 4.6) might be to find a class of dags of size n which can be pebbled in space $\omega(n^{\frac{1}{3}})$, requires exponential time for the minimal space pebbling and is not tolerant according to the "edge deleting". A possible attempt to solve this problem is to construct a class of rooted dags of size n requiring polynomial time for pebbling in the minimal space $\omega(\sqrt{n})$.

References

1. Carlson, D.A. and Savage, J. E.: Extreme Time-Space Tradeoffs for Graphs with Small Space Requirements. IPL 14, 1982, pp. 223-227

2. Cook, S.A.: An Observation on Time-Storage Trade Off. JCSS 9, 1974, pp. 308-316

3. Lengauer, T. and Tarjan, R.E.: Asymptotically Tight Bounds on Time-Space Trade-Offs in a Pebble Game. JACM 29, 1982, pp. 1087-1130

4. Pippenger, N.: Pebbling. 5th IBM MFCS, Tokyo, 1980, p. 19

5. Venkateswaran, H. and Tompa, M.: A New Pebble Game that Characterizes Parallel Complexity Classes. SIAM J. Computing 18, 1989, pp. 533-549

6. Wilber, R.: White Pebbles Help. JCSS 36, 1988, pp. 108-124

Deterministic Behavioural Models for Concurrency

*Vladimiro Sassone** *Mogens Nielsen***
*Glynn Winskel***

*Dipartimento di Informatica, Università di Pisa, Italy
**Computer Science Department, Aarhus University, Denmark

EXTENDED ABSTRACT

This paper offers three candidates for a deterministic, noninterleaving, behaviour model which generalizes Hoare traces to the noninterleaving situation. The three models are all proved equivalent in the rather strong sense of being equivalent as categories. The models are: deterministic labelled event structures, generalized trace languages in which the independence relation is context-dependent, and deterministic languages of pomsets.

Introduction

Models for concurrency can be classified according to whether they can represent the structure of *systems* or just their *behaviours* (*Behaviour or System model*); whether they can faithfully take into account the difference between *concurrency* and *nondeterminism* (*Interleaving or Noninterleaving model*); and, finally, whether they can represent the *branching structure* of processes, i.e., the points in which choices are taken, or not (*Linear or Branching Time model*).

In [9], the authors studied a range of models based on such a classification. The classification has the shape of a *cube* whose vertices are *categories* of models—corresponding to the eight classes of models obtained by varying the parameters above—and whose edges establish formal relationships between such categories. More precisely, the edges of the cube are special forms of *adjunctions*, namely *reflections* or *coreflections*, which express translations between models.

Generally speaking, the model chosen to represent a class is a canonical and universally accepted representative of that class. For the behavioural models they are Hoare languages [3] for interleaving, linear-time models, synchronization trees [12] for interleaving, branching-time models, and labelled event structures [13] for noninterleaving, branching-time models. However, for the class of noninterleaving, linear-time models, there does not, at present, seem to be an obvious choice of a corresponding canonical model.

The choice taken in [9] is deterministic labelled event structures, i.e. labelled event structures where the enabling relation between configurations and events is deterministic in the sense that whenever two events with the same label are enabled at a common configuration they are the same. The following is an example of such an event structure, together with its domain of configurations.

*This author has been supported by a grant of the Danish Research Academy.

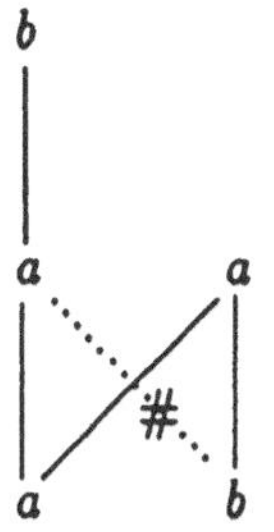

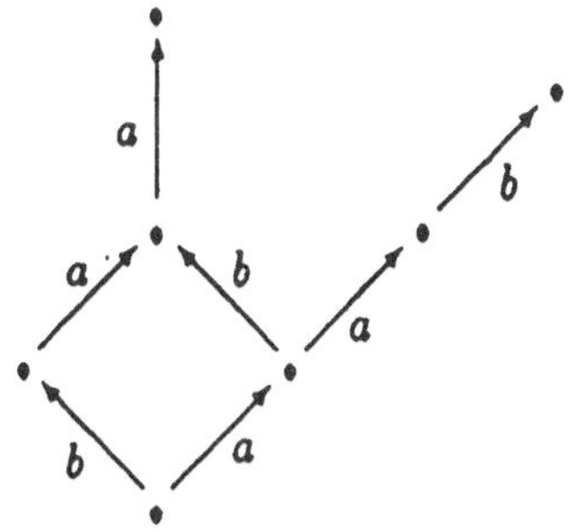

The choice of deterministic labelled event structures is based, by analogy, on the observation that Hoare trace languages may be viewed as deterministic synchronization trees, and that labelled event structures are a canonical generalization of synchronization trees within noninterleaving models. In this paper we investigate the relationship between this model and two of the most-studied, noninterleaving generalizations of Hoare languages in the literature: the pomsets of Pratt [8], and the traces of Mazurkiewicz [6].

Pomsets, an acronym for *p*artial *o*rdered *m*ulti*sets*, are labelled partial ordered sets. A noninterleaving representation of a system can be readily obtained by means of pomsets simply by considering the (multiset of) labels occurring in the run ordered by the *causal dependency* relation inherited from the events. The system itself is then represented by a set of pomsets. For instance, the labelled event structure given in the example discussed above can be represented by the following set of pomsets.

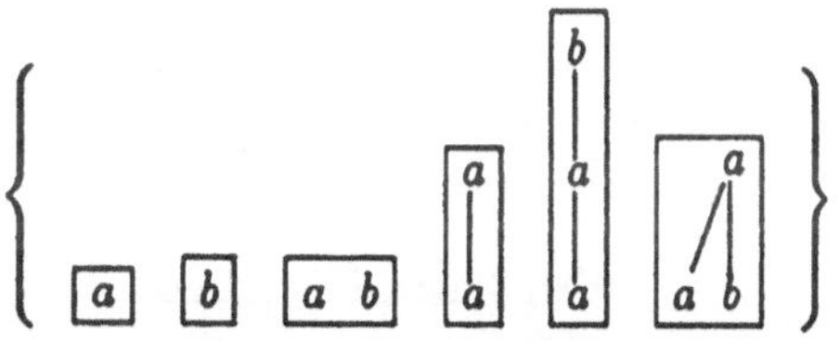

A simple but conceptually relevant observation about pomsets is that strings can be thought of as a particular kind of pomsets, namely those pomsets which are *finite* and *linearly* ordered. In other words, a pomset $\boxed{a_1 < a_2 < \cdots < a_n}$ represents the string $a_1a_2 \cdots a_n$. On the other side of such correspondence, we can think of (finite) pomsets as a generalization of the notion of word (string) obtained by relaxing the constraint which imposes that the symbols in a word be linearly ordered. This is why in the literature pomsets have also appeared under the name *partial words* [2]. The analogy between pomsets and strings can be pursued to the point of defining languages of partial words, called *partial languages*, as prefix-closed—for a suitable extension of this concept to pomsets—sets of pomsets on a given alphabet of labels.

Since our purpose is to study linear-time models which are deterministic, we shall consider only pomsets without *autoconcurrency*, i.e., pomsets such that all the elements carrying the same label are linearly ordered. Following [11], we shall refer to this kind of pomsets as *semiwords* and to the corresponding languages as *semilanguages*. We shall identify a category <u>**dSL**</u> of deterministic semilanguages equivalent to the category of deterministic labelled event structures. Although pomsets have been studied extensively (see e.g. [8, 1, 2]), there are few previous results about formal relationships of pomsets with other models for concurrency.

Mazurkiewicz trace languages [6] are defined on an alphabet L together with a symmetric irreflexive binary relation I on L, called the *independence relation.* The relation I induces an equivalence on the strings of L^* which is generated by the simple rule

$$\alpha ab\beta \simeq \alpha ba\beta \quad \text{if} \quad a\, I\, b,$$

where $\alpha, \beta \in L^*$ and $a, b \in L$. A trace language is simply a subset M of L^* which is prefix-closed and $\simeq$-closed, i.e., $\alpha \in M$ and $\alpha \simeq \beta$ implies $\beta \in M$. It represents a system by representing all its possible behaviours as the sequences of (occurrences of) events it can perform. Since the independence relation can be taken to indicate the events which are *concurrent* to each other, the relation $\simeq$ does nothing but relate runs of the systems which differ only in the order in which independent events occur.

However, Mazurkiewicz trace languages are too abstract to describe faithfully labelled event structures. Consider for instance the labelled event structure shown earlier. Clearly, any trace language with alphabet $\{a, b\}$ able to describe such a labelled event structure must be such that $ab \simeq ba$. However, it cannot be such that $aba \simeq aab$. Thus, we are forced to move from the well-known model of trace languages. We shall introduce here a new notion of *generalized Mazurkiewicz trace language*, in which the independence relation is *context-dependent.* For instance, the event structure shown in the above picture will be represented by a trace language in which a is independent from b *at* ϵ, i.e., after the empty string, in symbols $a\, I_\epsilon\, b$, but a is *not* independent from b *at* a, i.e., after the string a has appeared, in symbols $a\, \not{I}_a\, b$. In particular, we shall present a category <u>**GTL**</u> of generalized trace languages which is equivalent to the category <u>**dLES**</u> of deterministic labelled event structures. We remark that a similar idea of generalizing Mazurkiewicz trace languages has been considered also in [4].

Summing up, we present the chain of equivalences

$$\underline{\mathbf{dSL}} \simeq \underline{\mathbf{dLES}} \simeq \underline{\mathbf{GTL}}$$

which, besides identifying models which can replace <u>**dLES**</u> in our classification, also introduce interesting *deterministic behavioural models* for concurrency and formalizes their mutual relationships. This being an extended abstract, all the proofs are omitted; however, all the translations between models are presented in full. The reader interested in the complete treatment is referred to [10].

1 Preliminaries

One of the most studied noninterleaving models for concurrency is that of *event structures* [7, 12]. Their first class objects are events, assumed to be the *atomic computational steps,* which are related to each other by cause/effect and conflict relationships.

DEFINITION 1.1 *(Labelled Event Structures)*
A labelled event structure is a structure $ES = (E, \#, \leq, \ell, L)$ consisting of a set of events E partially ordered by $\leq$; a symmetric, irreflexive relation $\# \subseteq E \times E$, the conflict relation, such that

$$\{ e' \in E \mid e' \leq e \} \text{ is finite for each } e \in E,$$

$$e \# e' \leq e'' \text{ implies } e \# e'' \text{ for each } e, e', e'' \in E;$$

a set of labels L and a labelling function $\ell: E \rightarrow L$. For an event $e \in E$, define $\lfloor e \rfloor = \{ e' \in E \mid e' \leq e \}$. Moreover, we write $\mathrel{W}$ for the relation $\# \cup \{ (e,e) \mid e \in E_{ES} \}$.
A labelled event structure morphism from ES_0 to ES_1 is a pair of partial functions (η, λ), where $\eta: E_{ES_0} \rightarrow E_{ES_1}$ and $\lambda: L_{ES_0} \rightarrow L_{ES_1}$ are such that

$$(i)\ \lfloor \eta(e) \rfloor \subseteq \eta(\lfloor e \rfloor), \quad (ii)\ \eta(e) \mathrel{W} \eta(e') \Rightarrow e \mathrel{W} e', \quad (iii)\ \lambda \circ \ell_{ES_0} = \ell_{ES_1} \circ \eta.$$

*Taking composition to be the componentwise composition of partial functions, defines the category **LES** of labelled event structures.*

The computational intuition behind event structures is very simple: an event e can occur when all its *causes*, i.e., $\lfloor e \rfloor \setminus \{e\}$, have occurred and no event which it is in conflict with has already occurred. This is formalized by the following notion of *configuration*.

DEFINITION 1.2 *(Configurations)*
Given a labelled event structure ES, define the configurations of ES to be those subsets $c \subseteq E_{ES}$ which are

Conflict Free: $\forall e_1, e_2 \in c$, *not* $e_1 \# e_2$

Left Closed: $\forall e \in c\ \forall e' \leq e,\ e' \in c$

Let $\mathcal{L}(ES)$ denote the set of configurations of ES. We say that event e is enabled at a configuration c, in symbols $c \vdash e$, if (i) $e \notin c$; (ii) $\lfloor e \rfloor \setminus \{e\} \subseteq c$; and (iii) $e' \in E_{ES}$ and $e' \# e$ implies $e' \notin c$.

DEFINITION 1.3 *(Deterministic Event Structures)*
A labelled event structure ES is deterministic if and only if for any $c \in \mathcal{L}(ES)$, and for any pair of events $e, e' \in E_{ES}$, whenever $c \vdash e$, $c \vdash e'$ and $\ell(e) = \ell(e')$, then $e = e'$.
*This defines the category **dLES** as a full subcategory of **LES**.*

Configurations of event structures may be viewed as *labelled partial orders* on L, i.e., as triples $(E, \leq, \ell)$, where E is a set, $\leq \subseteq E^2$ a partial order relation; and $\ell: E \rightarrow L$ is a *labelling* function. We say that a labelled partial order $(E, \leq, \ell)$ is *finite* if E is so.

DEFINITION 1.4 *(Partial Words)*
A partial word on L is an isomorphism class of finite labelled partial orders. Given a finite labelled partial order p we shall denote with $[\![p]\!]$ the partial word which contains p. We shall also say that p represents the partial word $[\![p]\!]$.

A semiword is a partial word which does not exhibit autoconcurrency, i.e., such that all its subsets consisting of elements carrying the same label are linearly ordered. This is a strong simplification. Indeed, given a labelled partial order p representing a semiword on L and any label $a \in L$, such hypothesis allows us to talk *unequivocally* of the first element labelled a, of the second element labelled a, ..., the n-th element labelled a. In other words, we can represent p unequivocally as a (strict) partial order whose elements are pairs in $L \times \omega$, (a, i) representing the i-th element carrying label a. Thus, we are led to the following definition, where for n a natural number, $[n]$ denote the initial segment of length n of $\omega \setminus \{0\}$, i.e., $[n] = \{1, \ldots, n\}$.

Definition 1.5 *(SemiWords)*
A (canonical representative of a) semiword on an alphabet L is a pair $x = (A_x, <_x)$ where

- $A_x = \bigcup_{a \in L} \left(\{a\} \times [n_a^x]\right)$, *for some $n_a^x \in \omega$, and A_x is finite;*
- $<_x$ *is a transitive, irreflexive, binary relation on A_x such that*

  $$(a,i) <_x (a,j) \quad \textit{if and only if} \quad i < j,$$

  where $<$ is the usual (strict) ordering on natural numbers.

The semiword represented by x is $\left[\!\left[(A_x, \leq, \ell)\right]\!\right]$, where $(a,i) \leq (b,j)$ if and only if $(a,i) <_x (b,j)$ or $(a,i) = (b,j)$, and $\ell\big((a,i)\big) = a$. However, exploiting in full the existence of such an easy representation, from now on, we shall make no distinction between x and the semiword which it represents. In particular, as already stressed in Definition 1.5, with abuse of language, we shall refer to x as a semiword. The set of semiwords on L will be indicated by $SW(L)$. The usual set of words (strings) on L is (isomorphic to) the subset of $SW(L)$ consisting of semiwords with *total* ordering.

A standard ordering used on words is the prefix order $\sqsubseteq$, which relates α and β if and only if α is an initial segment of β. Such idea is easily extended to semiwords in order to define a prefix order $\sqsubseteq\, \subseteq SW(L) \times SW(L)$. Consider x and y in $SW(L)$. Following the intuition, for x to be a prefix of y, it is necessary that the elements of A_x are contained also in A_y with the same ordering. Moreover, since new elements can be added in A_y only "on the top" of A_x, no element in $A_y \setminus A_x$ may be less than an element of A_x. This is formalized by saying

$$x \sqsubseteq y \quad \text{if and only if} \quad A_x \subseteq A_y \ \text{ and } \ <_x \,=\, <_y \cap A_x^2$$
$$\text{and} \ <_y \cap ((A_y \setminus A_x) \times A_x) = \emptyset.$$

It is quickly realized that $\sqsubseteq$ is a partial order on $SW(L)$ and that it coincides with the usual prefix ordering on words.

Example 1.6 *(Prefix Ordering)*
As a few examples of the prefix ordering of semiwords, it is

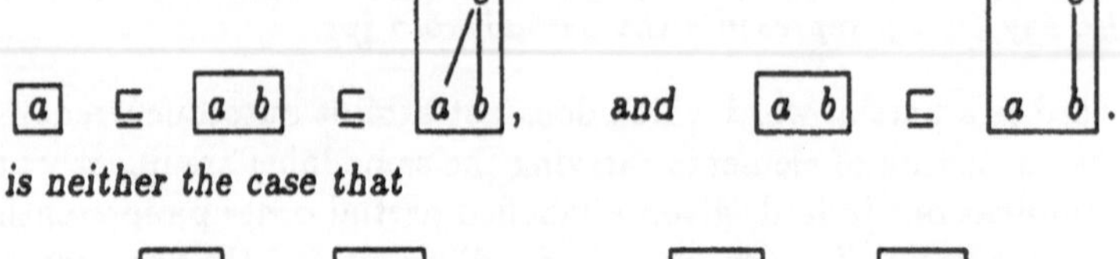

However, it is neither the case that

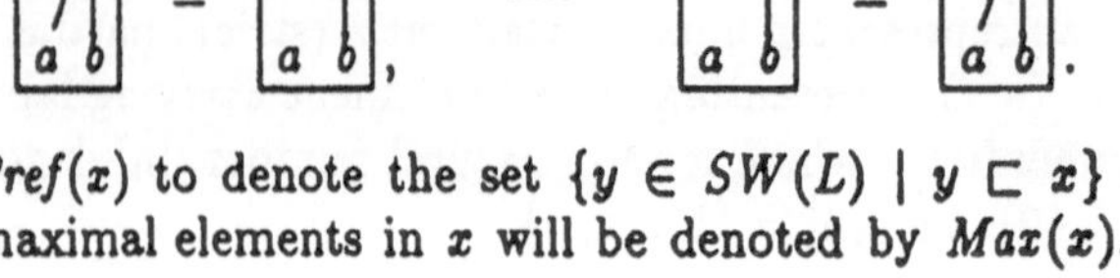

We shall use $Pref(x)$ to denote the set $\{y \in SW(L) \mid y \sqsubset x\}$ of *proper prefixes* of x. The set of maximal elements in x will be denoted by $Max(x)$. Semiwords with a maximum element play a key role in our development. For reasons that will be clear later, we shall refer to them as to *events*.

Another important ordering is usually defined on semiwords: the *"smoother than"* order, which takes into account that a semiword can be extended just by relaxing its ordering. More precisely, x is smoother than y, in symbols $x \preccurlyeq y$, if x imposes more order contraints on the elements of y. Formally,

$$x \preccurlyeq y \qquad \text{if and only if} \qquad A_x = A_y \quad \text{and} \quad <_x \supseteq <_y.$$

It is easy to see that $\preccurlyeq \subseteq SW(L) \times SW(L)$ is a partial order. In the following, we shall use $Smooth(x)$ to denote the set of *smoothings* of x, i.e., the set $\{y \in SW(L) \mid y \preccurlyeq x\}$.

EXAMPLE 1.7 *(Smoother than Ordering)*
The following few easy situations exemplify the smoother than ordering of semiwords.

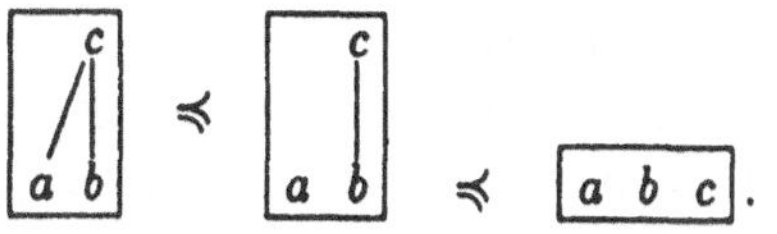

On the other hand, neither

2 Semilanguages and Event Structures

Semilanguages are a straightforward generalization of Hoare languages to prefix-closed subsets of $SW(L)$.

DEFINITION 2.1 *(SemiLanguages)*
A semilanguage is a pair (SW, L)*, where* L *is an alphabet and* SW *is a set of semiwords on* L *which is*

Prefix closed: $y \in SW$ *and* $x \sqsubseteq y$ *implies* $x \in SW$;
Coherent: $Pref(x) \subseteq SW$ *and* $|Max(x)| > 2$ *implies* $x \in SW$.

Semilanguage (SW, L) *is deterministic if*

$$x, y \in SW \text{ and } Smooth(x) \cap Smooth(y) \neq \emptyset \quad \text{implies} \quad x = y.$$

In order to fully understand this definition, we need to appeal to the intended meaning of semilanguages. A semiword in a semilanguage describes a (partial) run of a system in terms of the observable properties (labels) of the events which have occurred, together with the causal relationships which rule their interactions. Thus, the *prefix closedness* clause captures exactly the intuitive fact that any *initial segment* of a (partial) computation is itself a (partial) computation of the system.

In this view, the *coherence* axiom can be interpreted as follows. Suppose that there is a semiword x whose proper prefixes are in the language. i.e., they are runs of the system, and suppose that $|Max(x)| > 2$. This means that, given any pair of maximal elements in x, there is a computation of the system in which the corresponding events have both occurred. Then, in this case, the coherence axiom asks for x to be a possible computation of the system, as well. In other words, we can look at coherence as to the axiom which forces a set of events to be conflict free if it is *pairwise* conflict free, as in [7] for *prime event structures* and in [6] for *proper trace languages.*

To conclude our discussion about Definition 2.1, let us analyze the notion of *determinism.* Remembering our interpretation of semiwords as runs of a system, it is easy to realize how the existence of distinct x and y such that $Smooth(x) \cap Smooth(y) \neq \emptyset$ would imply nondeterminism. In fact, if there were two different runs with a common linearization, then there would be two different computations exhibiting the same observable behaviour, i.e., in other words, two *non equivalent* sequences of events with the same strings of labels.

Also the notion of morphisms of semilanguage can be derived smoothly as extension of the existing one for Hoare languages.

Any $\lambda: L_0 \rightharpoonup L_1$ determines a partial function $\hat{\lambda}: SW(L_0) \rightharpoonup SW(L_1)$ which maps x to its *relabelling* through λ, if this represents a semiword, and is undefined otherwise. Consider now semilanguages (SW_0, L_0) and (SW_1, L_1), and suppose for $x \in SW_0$ that $\hat{\lambda}$ is defined on x. Although one could be tempted to ask that $\hat{\lambda}(x)$ be a semiword in SW_1, this would by far too strong a requirement. In fact, since in $\hat{\lambda}(x)$ the order $<_x$ is strictly preserved, morphisms would always strictly preserve causal dependency, and this would be out of tune with the existing notion of morphisms for event structures, in which sequential tasks can be simulated by *"more concurrent"* ones. Fortunately enough, we have an easy way to ask for the existence of a more concurrent version of $\hat{\lambda}(x)$ in SW_1. It consists of asking that $\hat{\lambda}(x)$ be a smoothing of some semiword in SW_1.

DEFINITION 2.2 *(Semilanguage Morphisms)*
Given the semilanguages (SW_0, L_0) and (SW_1, L_1), a partial function $\lambda: L_0 \rightharpoonup L_1$ is a morphism $\lambda: (SW_0, L_0) \rightarrow (SW_1, L_1)$ if[1]

$$\forall x \in SW_0 \qquad \hat{\lambda}\downarrow x \quad \text{and} \quad \hat{\lambda}(x) \in Smooth(SW_1).$$

It is worth observing that, if (SW_1, L_1) is *deterministic*, there can be *at most one* semiword in SW_1, say x_λ, such that $\hat{\lambda}(x) \in Smooth(x_\lambda)$. In this case, we can think of $\lambda: (SW_0, L_0) \rightarrow (SW_1, L_1)$ as mapping x to x_λ.

EXAMPLE 2.3
Given $L_0 = \{a, b\}$ and $L_1 = \{c, d\}$, consider the deterministic semilanguages below.

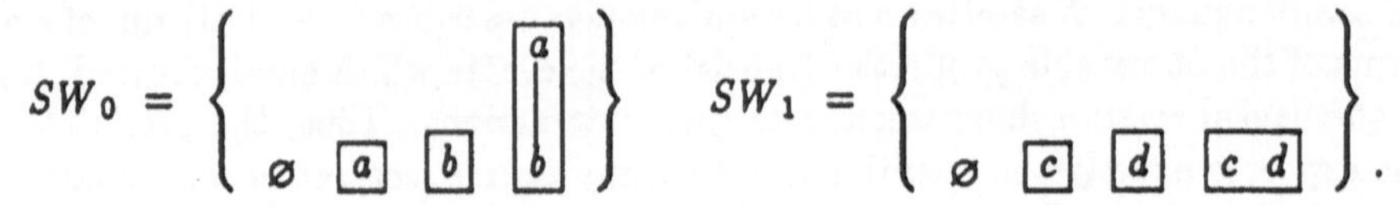

[1]Here, and in the following, we use $f\downarrow x$ to mean that a partial function f is defined on argument x.

Then, the function λ which maps a to c and b to d is a morphism from (SW_0, L_0) to (SW_1, L_1). For instance,

$$\hat{\lambda}\left(\boxed{\begin{array}{c}a\\|\\b\end{array}}\right) = \boxed{\begin{array}{c}c\\|\\d\end{array}} \preccurlyeq \boxed{c\ \ d}.$$

Observe that the function $\lambda': L_0 \to L_1$ which sends both a and b to c is not a morphism since $\hat{\lambda}$ applied to $\boxed{b < a}$ gives $\boxed{c < c}$ which is not the smoothing of any semiword in SW_1, while $\lambda'': L_1 \to L_0$ which sends both c and d to a is not a morphism from (SW_1, L_1) to (SW_0, L_0) since $\hat{\lambda}$ is undefined on $\boxed{c\ \ d}$.

It can be shown that semilanguages and their morphisms, with composition that of partial functions, form a category whose full subcategory consisting of deterministic semilanguages will be denoted by **dSL**. In the following, we shall define *translation* functors between **dLES** and **dSL**.

Given a deterministic semilanguage (SW, L) define $dsl.dles\big((SW, L)\big)$ to be the structure $(E, \leq, \#, \ell, L)$, where

- $E = \left\{e \;\middle|\; e \in SW,\ e \text{ is an } \textit{event}\text{, i.e., } e \text{ has a maximum element}\right\}$;
- $\leq\; = \;\sqsubseteq \cap E^2$;
- $\# = \left\{(e, e') \in E^2 \;\middle|\; e \text{ and } e' \text{ are incompatible wrt } \sqsubseteq\right\}$;
- $\ell(e)$ is the label of the maximum element of e.

THEOREM 2.4
$dsl.dles\big((SW, L)\big)$ is a deterministic labelled event structure.

Consider now a deterministic labelled event structure $DES = (E, \leq, \#, \ell, L)$. Define $dles.dsl(DES)$ to be the structure (SW, L), where

$$SW = \left\{ \left[\!\left[(c, \leq \cap\, c^2, \ell_{|c})\right]\!\right] \;\middle|\; c \text{ is a } \textit{finite} \text{ configuration of } DES \right\}.$$

THEOREM 2.5
$dles.dsl(DES)$ is a deterministic semilanguage.

It can be shown that $dsl.dles$ and $dles.dsl$ extend to functors which when composed with each other yield functors naturally isomorphic to identity functors. In other words, they form an *adjoint equivalence* [5, chap. III, pg. 91], i.e., an adjunction which is both a *reflection* and a *coreflection*. It is worthwhile noticing that this implies that the mappings $dsl.dles$ and $dles.dsl$ constitute a bijection between deterministic semilanguages and isomorphism classes of deterministic labelled event structures—isomorphism being identity up to the names of events.

THEOREM 2.6
*The categories **dSL** and **dLES** are equivalent.*

In fact, dropping the axiom of coherence in Definition 2.1 we get semilanguages equivalent to labelled *stable event structures* [12].

3 Trace Languages and Event Structures

Generalized trace languages extend trace languages by considering an independence relation which may vary while the computation is progressing. Of course, we need a few axioms to guarantee the consistency of such an extension.

DEFINITION 3.1 *(Generalized Trace Languages)*
A generalized trace language is a triple (M, I, L), where L is an alphabet, $M \subseteq L^*$ is a prefix-closed and $\simeq$-closed set of strings, $I: M \to 2^{L \times L}$ is a function which associates to each $s \in M$ a symmetric and irreflexive relation $I_s \subseteq L \times L$, such that

I is consistent: $s \simeq s'$ *implies* $I_s = I_{s'}$;
M is I-closed: $a\, I_s\, b$ *implies* $sab \in M$;
I is coherent: (i) $a\, I_s\, b$ *and* $a\, I_{sb}\, c$ *and* $c\, I_{sa}\, b$ *implies* $a\, I_s\, c$,
(ii) $a\, I_s\, c$ *and* $c\, I_s\, b$ *implies* ($a\, I_s\, b$ *if and only if* $a\, I_{sc}\, b$);

where $\simeq$ is the least equivalence relation on L^* such that $sabu \simeq sbau$ if $a\, I_s\, b$.

As in the case of trace languages, we have an equivalence relation $\simeq$ which equates those strings representing the same computation. Thus, I must be consistent in the sense that it must associate the same independence relation to $\simeq$-equivalent strings. In order to understand the last two axioms, the following picture shows in terms of computations ordered by prefix the situations which those axioms forbid. There, the dots represent computations, the labelled edges represent the prefix ordering, and the dotted lines represent the computations forced in M by the axioms.

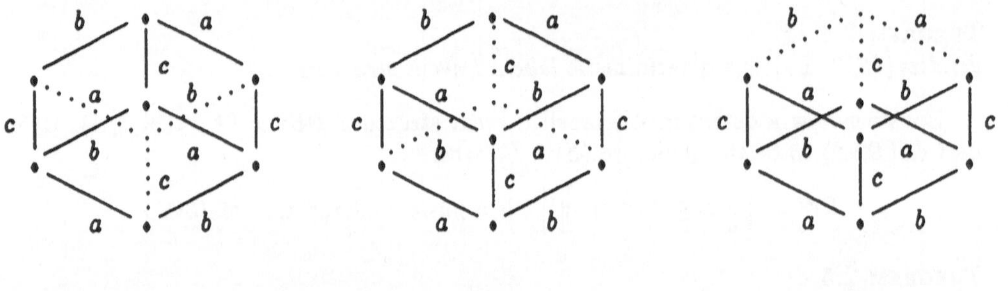

It is easy to see that axiom (i) rules out the situation described by just the solid lines in (A)—impossible for stable event structures, while axiom (ii) eliminates cases (B)—which is beyond the descriptive power of ***general event structures*** [12] and (C)—impossible for event structures with ***binary*** conflict. They narrow down to those orderings of computations arising from prime event structures.

DEFINITION 3.2 *(Generalized Trace Language Morphisms)*
Given the generalized trace languages (M, I, L) and (M', I', L'), a partial function $\lambda: L \to L'$ is a morphism $\lambda: (M, I, L) \to (M', I', L')$ if

λ preserves words: $s \in M$ *implies* $\lambda^*(s) \in M'$;
λ respects independence: $a\, I_s\, b$ *and* $\lambda\downarrow a$, $\lambda\downarrow b$ *implies* $\lambda(a)\, I'_{\lambda^*(s)}\, \lambda(b)$;

where λ^* *is inductively defined by* $\lambda^*(\epsilon) = \epsilon$ *and* $\lambda^*(sa) = \begin{cases} \lambda^*(s)\lambda(a) & \text{if } \lambda\downarrow a \\ \lambda^*(s) & \text{otherwise.} \end{cases}$

Generalized trace languages and their morphisms, under the usual composition of partial functions, form the category <u>**GTL**</u>.

A derived notion of event in generalized trace languages can be captured by the relation $\sim$ defined as the least equivalence such that

$$a\, I_s\, b \;\text{ implies }\; sa \sim sba \qquad\text{and}\qquad s \simeq s' \;\text{ implies }\; sa \sim s'a.$$

The events occurring in $s \in M$, denoted by $Ev(s)$, are the $\sim$-classes a representative of which occurs as a non empty prefix of s, i.e., $\Big\{[u]_\sim \;\Big|\; u \text{ is a non empty prefix of } s\Big\}$. It can be shown that $s \simeq s'$ if and only if $Ev(s) = Ev(s')$. Extending the notation, we shall write $Ev(M)$ to denote the events of (M, I, L), i.e., the $\sim$-equivalence classes of non empty strings in M.

Now, given a generalized trace language (M, I, L) define $gtl.dles\Big((M, I, L)\Big)$ to be the structure $(Ev(M), \leq, \#, \ell, L)$, where

- $[s]_\sim \leq [s']_\sim$ if and only if $\forall u \in M,\ [s']_\sim \in Ev(u)$ implies $[s]_\sim \in Ev(u)$;
- $[s]_\sim \# [s']_\sim$ if and only if $\forall u \in M,\ [s]_\sim \in Ev(u)$ implies $[s']_\sim \notin Ev(u)$;
- $\ell\Big([s]_\sim\Big) = a$ if and only if $s = s'a$.

THEOREM 3.3
$gtl.dles\Big((M, I, L)\Big)$ *is a deterministic labelled event structure.*

On the other hand, in order to define a generalized trace language from a deterministic labelled event structure $DES = (E, \leq, \#, \ell, L)$, consider

$$M = \Big\{\ell^*(e_1 \cdots e_n) \;\Big|\; \{e_1, \ldots, e_n\} \subseteq E \;\text{ and }\; \{e_1, \ldots, e_{i-1}\} \vdash e_i,\ i = 1, \ldots, n\Big\}.$$

Since DES is deterministic, any $s \in M$ identifies unequivocally a string of events $Sec(s) = e_1 \cdots e_n \in E^*$ such that $\{e_1, \ldots, e_{i-1}\} \vdash e_i$, $i = 1, \ldots, n$, and $\ell^*(e_1 \cdots e_n) = s$. Now, for any $s \in M$, take $I_s = \Big\{(a, b) \;\Big|\; sab \in M,\ Sec(sab) = xe_0e_1 \text{ and } e_0\ co\ e_1\Big\}$. Then, define (M, I, L) to be $dles.gtl(DES)$.

THEOREM 3.4
$dles.gtl(DES)$ *is a generalized trace language.*

As in the case treated in the previous section, *dles.gtl* and *gtl.dles* extend to functors between <u>**GTL**</u> and <u>**dLES**</u> which form an adjoint equivalence. Such an equivalence restricts to an isomorphism of generalized trace languages and isomorphism classes of deterministic labelled event structures.

THEOREM 3.5
Categories <u>**GTL**</u> *and* <u>**dLES**</u> *are equivalent.*

The result extends to labelled *stable event structures* by dropping the *'only if'* implication in part *(ii)* of the coherence axiom of Definition 3.1. Of course, it follows from Theorem 2.6 and Theorem 3.5 that <u>**dSL**</u> and <u>**GTL**</u> are equivalent. In the full paper [10], we also define direct translations between such categories.

References

[1] J. GISCHER. The Equational Theory of Pomsets. *Theoretical Computer Science*, n. 61, pp. 199–224, 1988.

[2] J. GRABOWSKI. On Partial Languages. *Fundamenta Informaticae*, n. 4, pp. 428–498, 1981.

[3] C.A.R. HOARE. *Communicating Sequential Processes*. Englewood Cliffs, 1985.

[4] P.W. HOOGERS, H.C.M KLEIJN, AND P.S. THIAGARAJAN. A Trace Semantics for Petri Nets. In Proceedings of *ICALP '92*, LNCS, n. 623, pp. 595–604, Springer Verlag, 1992.

[5] S. MACLANE. *Categories for the Working Mathematician*. GTM, Springer-Verlag, 1971.

[6] A. MAZURKIEWICZ. Basic Notions of Trace Theory. In *lecture notes for the REX summer-school in temporal logic*, LNCS, n. 354, pp. 285–363, Springer-Verlag, 1988.

[7] M. NIELSEN, G. PLOTKIN, AND G. WINSKEL. *Petri nets, Event Structures and Domains, part 1*. Theoretical Computer Science, n. 13, pp. 85–108, 1981.

[8] V. PRATT. Modeling Concurrency with Partial Orders. *International Journal of Parallel Processing*, n. 15, pp. 33–71, 1986.

[9] V. SASSONE, M. NIELSEN, AND G. WINSKEL. *A Classification of Models for Concurrency.* To appear as Technical Report Daimi, Computer Science Department, Aarhus University, 1993. Extended abstract to appear in Proceedings of *CONCUR '93*.

[10] V. SASSONE, M. NIELSEN, AND G. WINSKEL. *Deterministic Behavioural Models for Concurrency.* To appear as Technical Report Daimi, Computer Science Department, Aarhus University, 1993.

[11] P. STARKE. Traces and Semiwords. LNCS, n. 208, pp. 332–349, Springer-Verlag, 1985.

[12] G. WINSKEL. Event Structures. In *Advances in Petri nets*, LNCS, n. 255, pp. 325–392, Springer-Verlag, 1987.

[13] G. WINSKEL, AND M. NIELSEN. Models for Concurrency. To appear in the *Handbook of Logic in Computer Science*. A draft appears as DAIMI PB 429, 1992.

Real-Time Refinement: Semantics and Application

David Scholefield, Hussein Zedan, He Jifeng†

Formal Systems Research Group
Department of Computer Science
University of York, Heslington, York (UK)
†Programming Research Group
Oxford University, Keble Road, Oxford (UK)

Abstract. A formal framework for a calculus of real-time systems is presented. Specifications and program statements are combined into a single language called TAM (the Temporal Agent Model), that allows the user to express both functional and timing properties. A specification oriented semantics for TAM is given, along with the definition of a refinement relation and a calculus which is sound with respect to that relation. A simple real-time program is also developed using the calculus.

1 Introduction

In most formal development methods there are at least two languages involved, one for the specification task, and one for the design task (often the translation to implementation is ignored, or considered to be trivial). However, an inherent problem with such a 'multi language' approach is the lack of method by which suitable designs are arrived at. A combination of experience and guess-work must be used in order to formulate a design, and then verification – a time consuming task – is undertaken. If the verification fails then the design task is undertaken again. This cycle is undergone repeatedly until verification is achieved.

To overcome this problem we have developed the 'Temporal Agent Model' (TAM) which is a theory centered around a wide-spectrum language in which both specifications and executable programs can be intermixed. A real-time functional specification in TAM is transformed step-by-step into a mixed program containing both specification fragments and executable code. Such transformations continue until a completely executable program is produced which is guaranteed correct with respect to the original specification. The program may then be analysed by run-time schedulability and allocation tools in the usual manner, and executed.

The paper introduces extensions to first-order predicate logic to cover time, a wide spectrum language with a specificational semantics, a refinement calculus, and an example of program development.

2 The TAM Language

We define a real-time system as a collection of concurrently executing computation agents which communicate asynchronously via time-stamped shared data areas called shunts. The time-stamps refer to the time of the last write to that shunt. Shunts may have only one writer but any number of readers, they may also be hidden from agents. Time is global i.e. a single clock is accessible to every agent and shunt. The time domain is discrete, linear, and modelled naturally by the positive integers. There is a unique 'first time' instant from which we assume all systems will measure their execution release times for all agents.

The syntax of TAM can be defined recursively in terms of agents by:

$$\mathcal{A} ::= x := e \mid x \leftarrow s \mid x \rightarrow s \mid w : \Phi \mid \amalg \mid (x : T)\mathcal{A} \mid \mathcal{A}|\mathcal{A}$$
$$\mid [S]\mathcal{A} \mid \mathcal{A};\mathcal{A} \mid \bigsqcup_{i \in I} g_i \Rightarrow \mathcal{A}_i \mid \mathcal{A} \triangleright_t^s \mathcal{A} \mid \mu_n^S \mathcal{A}$$

Where x is a computation variable, s is a shunt, T is a type name, t is a time (constant), I is some finite indexing set, g_i are boolean expressions on variables and shunts, e is an expression on variables and shunts, and S is a set of times.

$x := e$ performs an assignment of the value of the expression e (defined over the variables of the agent) to the variable x. The values of the variables and shunts found in e are taken at the release time of the assignment.

$x \leftarrow s$ performs an input from the shunt s into the variable x. The type of x must therefore be a positive (time-stamp) and value pair. The input is asynchronous and therefore requires no prior write to the shunt.

$x \rightarrow s$ performs an asynchronous output to the shunt s from the variable (or constant value) x. The time of the output is used as the time-stamp and is expected to be provided by the run-time environment.

$w : \Phi$ is an agent which specifies required behaviour on variables and shunts. This statement form is discussed in detail below.

$\amalg$ is an agent which may terminate at any arbitrary time but does not change any variable or shunt.

$(x : T)\mathcal{A}$ introduces the unique variable or shunt x and associates it with the agent $\mathcal{A}$. Initially x has an undefined value (any value from the appropriate type set T). The variable or shunt is destroyed (unavailable) after the termination of $\mathcal{A}$.

$\mathcal{A}|\mathcal{A}$ executes the two agents concurrently. The two agents are released at the same time, and the concurrent composition terminates some arbitrary time after both agents terminate. We use the shorthand $\prod_{i \in I} \mathcal{A}_i$ to denote indexed concurrent composition (I is finite).

$[S]\mathcal{A}$ is an agent which has a duration (a duration of an agent is the interval over which the agent may be active, i.e. the difference between the release time and the termination time) equal to one of the values in the set S. This agent form therefore describes a *deadline* which we expect to be respected by the run-time execution environment.

$\mathcal{A};\mathcal{A}$ defines the sequential composition of the two agents. Immediately upon termination of the first agent the second agent will be released.

$\bigsqcup^t_{i\in I} g_i \Rightarrow \mathcal{A}_i$ evaluates all of the guards g_i (which are boolean expressions on local state space), and executes one of the agents corresponding to a true guard. If no guards evaluate to true, then the choice agent terminates correctly. The constant t defines an execution window within which the guards are evaluated; it is assumed that the evaluation will take t time units. The indexing set I is always finite.

$\mathcal{A} \triangleright^s_t \mathcal{A}$ is a timeout operator. The agent waits for a write to occur on shunt s for t time units. As soon as the shunt is written to the right-hand agent is released. If after t time units no write has occurred then the left-hand agent is released.

$\mu^S_n \mathcal{A}$ executes the agent $\mathcal{A}$ in sequence n times. Each execution of $\mathcal{A}$ is also given the deadline set S.

We impose the syntactic restriction that no agent may share its variables with a concurrently executing agent, and only one concurrently executing agent may write to any given shunt. An agent is assumed to have an implicit alphabet of local variables, input shunts, and output shunts.

A system is described by a single agent which is assumed, without loss of generality, to be released at time 0. For all but the most simple of systems the agent will have internal concurrency, and all sub-agents will be released at time 0. Later release times are then specified by the use of the delay agent.

The syntax of the specification agent in TAM is defined as:

$$w : \Phi$$

where w is a set of variables and shunts which may be changed during the behaviour defined by the specification, and Φ is a timed logic formula which describes that behaviour. The timed logic formula is a first-order logic formula with simple extensions to cover time. We add terms of the form $x@t$ to denote the value of the variable (or shunt) x at time t. We will also use the notation $s@t.ts$ to refer to the time stamp found in shunt s at time t, and similarly, $s@t.v$ for the value. Quantification is allowed for any timed variable or shunt over any finite execution interval (this allows the syntactic replacement of a timed variable or shunt with a unique free variable for each time unit in the interval).

In a specification $w : \Phi$, the formula Φ may contain two unique free variables $t, t' \in Time$ which represent the time at which the behaviour described by Φ starts (t), and the time at which the behaviour terminates (t').

3 Semantics

We start by defining some useful predicates on shunts and variables. The predicate '*stable*' asserts that the shunt s will not be changed during the given interval:

Definition Stable (shunts) $stable(s, n, m) =_{def} \bigwedge_{\sigma\in[n+1,m]} s@\sigma = s@(\sigma - 1)$

Similarly for variables:

Definition Stable (variables) $stable(x, n, m) =_{def} x@m = x@n$

In addition, the definitions for *stable* are extended to sets of variables or shunts.

The predicate *write* asserts that a given value is written to a shunt within an interval, and that the shunt remains stable at all other times within the interval:

Definition Write $write(x, s, n, m) =_{def}$

$$\bigvee_{\sigma \in [n+1,m]} stable(s, n, \sigma - 1) \wedge s@\sigma = (\sigma, x@n) \wedge stable(s, \sigma, m)$$

We also define an operator for dividing formulae into time consecutive subformulae:

Definition Chop. Given two timed logic formulae $\mathcal{A}$ and $\mathcal{B}$, then,

$$\mathcal{A} \frown \mathcal{B} =_{def} \exists m \in [t, t'](\mathcal{A}[m/t'] \wedge \mathcal{B}[m/t])$$

The semantics of an agent are now given by a timed logic formula. The specification statement is defined in this manner also, giving a natural interpretation for a refinement relation. Note that we use the notation $\tilde{w}$ to denote those variables and shunts which are owned by the specification agent, but which do not appear in the frame (i.e. if W is the alphabet of the specification statement $w : \Phi$, then $\tilde{w} =_{def} W/w$). Where variables appear dashed (x') we assume $x@t$'.

Definition Semantics. See figure 1.

4 Refinement

A refinement relation is defined as a *strengthening* of a timed formula. This models the intuitive definition of refinement as a lessening of nondeterminism.

Definition Refinement.

$$\mathcal{A} \sqsubseteq \mathcal{B} \quad \text{iff} \quad [\![\mathcal{B}]\!] \Rightarrow [\![\mathcal{A}]\!]$$

Composite agents allow us to describe complex behaviour by combining simpler agents. However, in system development we would not want to complicate the refinement method in proportion to the complexity of the desired system, instead we would wish to have simple refinement obligations which could be trivially composed in order to discharge complex refinement proof obligations.

In [6] the importance of compositionality of proof systems for concurrent real-time formalisms is discussed. We assert that the refinement calculus for TAM is compositional i.e. systems can be sub-divided into subsystems which may then be refined in isolation, and recomposed to give a system which is a refinement of the original specification. This form of compositionality is clearly dependent upon the fact that

$$[\![w:\Phi]\!] =_{def} stable(\tilde{w},t,t') \wedge \Phi \qquad [\![\amalg]\!] =_{def} \emptyset : t \leq t'$$

$$[\![x := e]\!] =_{def} [\![\{x:T\} : (t < t' \wedge x' = e@t)]\!]$$

$$[\![x \rightarrow s]\!] =_{def} [\![\ \{s : [Time \times T]\} : write(x,s,t,t')]\!]$$

$$[\![x \leftarrow s]\!] =_{def} [\![\{x : [Time \times T]\} : t < t' \wedge \exists m \in (t,t'](x' = s@m)]\!]$$

$$[\![(x:T)\mathcal{A}]\!] =_{def} \exists x \in T([\![\mathcal{A}]\!]) \qquad [\![\mathcal{A}|\mathcal{B}]\!] =_{def} ([\![\mathcal{A}]\!]\frown[\![\amalg]\!]) \wedge ([\![\mathcal{B}]\!]\frown[\![\amalg]\!])$$

$$[\![[S]\mathcal{A}]\!] =_{def} [\![\mathcal{A}]\!] \wedge t' - t \in S \qquad [\![\mathcal{A};\mathcal{B}]\!] =_{def} [\![\mathcal{A}]\!]\frown[\![\mathcal{B}]\!]$$

$$[\![\bigsqcup_{i \in I}^{n} g_i \Rightarrow \mathcal{A}_i]\!] =_{def} [\![(\emptyset : t' = t+n)]\!]\frown(((\bigwedge_{i \in I} \neg g_i@t) \wedge [\![\amalg]\!]) \vee \bigvee_{i \in I}(g_i@t \wedge [\![\mathcal{A}_i]\!])))$$

$$[\![\mathcal{A} \triangleright_n^s \mathcal{B}]\!] =_{def} \text{Nowrite}\frown[\![\mathcal{A}]\!] \ \vee \ \text{Input}\frown[\![\mathcal{B}]\!]$$

$$[\![\mu_{n+1}^S\mathcal{A}]\!] =_{def} ([\![[S]\mathcal{A}]\!])\frown[\![\mu_n^S\mathcal{A}]\!] \qquad [\![\mu_0^S\mathcal{A}]\!] =_{def} [\![\amalg]\!]$$

$$\text{Nowrite} =_{def} [\![\emptyset : t' = t + d \ \wedge \ s@t.ts < t \ \wedge \ stable(s,t,t')]\!]$$

$$\text{Input} =_{def} [\![\emptyset : t' \in [t, t+d] \ \wedge \ s@t'.ts = t' \ \wedge \ \forall m \in [t,t')(s@m = s@t)]\!]$$

Fig. 1. TAM Semantics

refinement can not introduce non-local variables and therefore break the interference constraint on concurrent systems, and similarly can not introduce unrestricted shunts. The property of compositionality also holds for the other agent constructors (deadline, sequence, variable declaration, shunt restriction, timeout, guards and recursion), and in the refinement calculus compositionality equates to the property of *monotonicity*.

We therefore assert the following theorem:

Theorem (Refinement Monotonicity) [9]

If $\mathcal{A} \sqsubseteq \mathcal{B}$ then for any context $\mathcal{C}(\_)$ we have $\mathcal{C}(\mathcal{A}) \sqsubseteq \mathcal{C}(\mathcal{B})$

There are at least sixty useful refinement laws which are sound with respect to the above definition of refinement (for a full list see [9] [10]), however, here we only list the three which are used in the following example.

The first allows for a strengthening of the specification predicate:

(strengthen) $w : \Phi \sqsubseteq w : \Psi$ if $\Psi \Rightarrow \Phi$

The second allows for new variable or shunt introduction:

(introduce) $w : \Phi \sqsubseteq (x : T)(w \cup \{x : T\} : \Phi)$

The third rule allows for variable or shunt elimination:

(eliminate) $w \cup \{s : T\} : \Phi \wedge stable(s, t, t') \sqsubseteq w - \{s : T\} : \Phi$

(if s does not appear free in Φ).

5 Example

The TAM calculus has been used on a number of case studies [9] [10], and here we present only a short example. We refine a specification for a simple comparator which reads data from two input shunts, and if the data values are equal, outputs the value 'true' on an output shunt, otherwise it outputs the value 'false'. The comparator must take no longer than 10 time units to compare and output.

In addition we are provided with an axiom which asserts that the value of the input shunts will remain stable during the comparator period:

$A.1 \quad =_{def} \; stable(\{in_1, in_2\}, t, t')$

Given three shunts:

$$\begin{array}{lll} \mathrm{in}_1, \mathrm{in}_2 & : [Time \times T] & \text{(input shunts)} \\ \mathrm{out} & : [Time \times bool] & \text{(output shunt)} \end{array}$$

then the comparator can be specified by:

$$\begin{array}{lll} \mathrm{Spec} & =_{def} & \{out : [Time \times bool]\} : t' \leq t + 10 \wedge \\ & & (\bigwedge_{\sigma \in [t, t']} in_1 @ \sigma.v = in_2 @ \sigma.v) \Rightarrow write(out, \text{'}true\text{'}, t, t') \wedge \\ & & (\neg \bigwedge_{\sigma \in [t, t']} in_1 @ \sigma.v \neq in_2 @ \sigma.v) \Rightarrow write(out, \text{'}false\text{'}, t, t') \end{array}$$

Rule *strengthen* allows us to strengthen the specification, and we do so by constraining the termination time further, and appealing to the axiom A.1 to allow us to choose an arbitrary time at which to test the input shunts.

$$\begin{array}{ll} \mathrm{Spec} \sqsubseteq & \{out : [Time \times bool]\} : t' = t + 10 \wedge \\ & (in_1 @ (t+1).v = in_2 @ (t+1).v) \Rightarrow write(out, \text{'}true\text{'}, t, t') \wedge \\ & (in_1 @ (t+1).v \neq in_2 @ (t+1).v) \Rightarrow write(out, \text{'}false\text{'}, t, t') \end{array}$$

The rule *strengthen* also allows us to introduce existentially quantified variables which will be used to hold the values read from the shunts.

$$\begin{array}{ll} \mathrm{Spec} \sqsubseteq & \{out : [Time \times bool]\} : t' = t + 10 \wedge \\ & \exists r_1, r_2 : [Time \times T](\\ & r_1 @ (t+1) = in_1 @ (t+1) \wedge \end{array}$$

$$r_2@(t+1) = in_2@(t+1) \wedge$$
$$(r_1@(t+1).v = r_2@(t+1).v) \Rightarrow write(out, \text{'true'}, t, t') \wedge$$
$$(r_1@(t+1).v \neq r_2@(t+1).v) \Rightarrow write(out, \text{'false'}, t, t'))$$

We can move the existential quantification out of the specification and declare them as local variables by *introduction*. In addition we can introduce an intermediate time value m (we are clearly moving towards sequential composition).

$$\begin{aligned}\text{Spec} \sqsubseteq\ & (r_1, r_2 : [Time \times T])\{r_1, r_2 : [Time \times T],\ out : [Time \times bool]\} : \\ & \exists m \in [t, t'](t' = m + 9\ \wedge\ m = t + 1 \wedge \\ & r_1@m = in_1@m\ \wedge\ r_2@m = in_2@m \wedge \\ & (r_1@m.v = r_2@m.v) \Rightarrow write(out, \text{'true'}, t, t') \wedge \\ & (r_1@m.v \neq r_2@m.v) \Rightarrow write(out, \text{'false'}, t, t'))\end{aligned}$$

The next refinement allows us to reduce the period during which the write may take place. This is sound by *strengthen*.

$$\begin{aligned}\text{Spec} \sqsubseteq\ & (r_1, r_2 : [Time \times T])\{r_1, r_2 : [Time \times T],\ out : [Time \times bool]\} : \\ & \exists m \in [t, t'](t' = m + 9\ \wedge\ m = t + 1\ \wedge\ r_1@m = in_1@m \wedge \\ & r_2@m = in_2@m\ \wedge\ stable(out, t, m) \wedge \\ & (r_1@m.v = r_2@m.v) \Rightarrow write(out, \text{'true'}, m, t') \wedge \\ & (r_1@m.v \neq r_2@m.v) \Rightarrow write(out, \text{'false'}, m, t'))\end{aligned}$$

We can now refine to a sequential agent (by def ;).

$$\text{Spec} \sqsubseteq (r_1, r_2 : [Time \times T])\mathcal{A}; \mathcal{B}$$

where

$$\mathcal{A} =_{def} \{r_1, r_2 : [Time \times T],\ out : [Time \times bool]\} :$$

$$stable(out, t, t')\ \wedge\ t' = t + 1\ \wedge\ r_1@t' = in_1@t'\ \wedge\ r_2@t' = in_2@t'$$

$$\begin{aligned}\mathcal{B} =_{def}\ & \{r_1, r_2 : [Time \times T],\ out : [Time \times bool]\} : \\ & (r_1@t.v = r_2@t.v) \Rightarrow write(out, \text{'true'}, t, t') \wedge \\ & (r_1@t.v \neq r_2@t.v) \Rightarrow write(out, \text{'false'}, t, t') \wedge \\ & t' = t + 9\end{aligned}$$

We refine these two agents independently. Firstly, the shunt *out* can be removed from $\mathcal{A}$ by *eliminate*.

$$\mathcal{A} \sqsubseteq \{r_1, r_2 : [Time \times T]\} : t' = t + 1\ \wedge\ r_1@t' = in_1@t'\ \wedge\ r_2@t' = in_2@t'$$

The agent $\mathcal{A}$ is eventually refined down to a concurrent agent which reads the two shunts (the full proof of this refinement can be found in [10]).

$$\mathcal{A} \sqsubseteq [1](r_1 \leftarrow in_1 \mid r_2 \leftarrow in_2)$$

The agent $\mathcal{B}$ is now refined by removing the variables from the frame (appealing to *strengthen* by adding a conjunct asserting their stability).

$$\begin{aligned}\mathcal{B} \sqsubseteq \{out : [Time \times bool]\} : & (r_1@t.v = r_2@t.v) \Rightarrow write(out, \text{`}true\text{'}, t, t') \wedge \\ & (r_1@t.v \neq r_2@t.v) \Rightarrow write(out, \text{`}false\text{'}, t, t') \wedge \\ & t' = t + 9\end{aligned}$$

We can reduce the writing interval still further (which will allow us to refine to a guarded agent) by appealing to *strengthen*.

$$\begin{aligned}\mathcal{B} \sqsubseteq \{out : [Time \times bool]\} : & (r_1@t.v = r_2@t.v) \Rightarrow write(out, \text{`}true\text{'}, t+2, t') \wedge \\ & (r_1@t.v \neq r_2@t.v) \Rightarrow write(out, \text{`}false\text{'}, t+2, t') \wedge \\ & t' = t + 9 \ \wedge \ stable(out, t, t+2)\end{aligned}$$

The constraint on the termination time can be replaced by a deadline (by def $[\{S\}]\mathcal{A}$).

$$\begin{aligned}\mathcal{B} \sqsubseteq [\{9\}]\{out : [Time \times bool]\} : & (r_1@t.v = r_2@t.v) \Rightarrow write(out, \text{`}true\text{'}, t+2, t') \wedge \\ & (r_1@t.v \neq r_2@t.v) \Rightarrow write(out, \text{`}false\text{'}, t+2, t') \wedge \\ & stable(out, t, t+2)\end{aligned}$$

We may now refine to a guard (by def guards).

$$\begin{aligned}\mathcal{B} \sqsubseteq [\{9\}]\ & (r_1.v = r_2.v) \Rightarrow \{out : [Time \times bool]\} : write(out, \text{`}true\text{'}, t, t') \\ & \sqcup_2 \\ & (r_1.v \neq r_2.v) \Rightarrow \{out : [Time \times bool]\} : write(out, \text{`}false\text{'}, t, t')\end{aligned}$$

Both guarded agents are then refined by writes.

$$\begin{aligned}\mathcal{B} \sqsubseteq [\{9\}]\ & (r_1.v = r_2.v) \Rightarrow \text{`}true\text{'} \rightarrow out \\ & \sqcup_2 \\ & (r_1.v \neq r_2.v) \Rightarrow \text{`}false\text{'} \rightarrow out\end{aligned}$$

Placing the two agents back in sequence gives us the final program:

$$\begin{aligned}Spec \sqsubseteq (r_1, r_2 : [Time \times T])\ & [\{1\}](r_1 \leftarrow in_1 \mid r_2 \leftarrow in_2)\ ; \\ & [\{9\}](\ r_1 = r_2 \Rightarrow \text{`}true\text{'} \rightarrow out \\ & \qquad \sqcup_2 \\ & \qquad r_1 \neq r_2 \Rightarrow \text{`}false\text{'} \rightarrow out\ )\end{aligned}$$

6 Conclusions

TAM is unique in providing a wide-spectrum development language for real-time systems in which abstract specifications can be refined down to concrete executable programs. Wide-spectrum languages for non real-time systems have been studied extensively, for example in the SETL language [11], and the CIP project [4], wide-spectrum languages based upon predicate logic are given transformation rules which allow refinement in a manner similar to TAM.

The utility of a wide-spectrum language can be clearly seen in the refinement method used by Morgan in his calculus [8] [7]. In this language, the concrete syntax is provided by an extended version of Dijkstra's Guarded Command Language [5]. The abstract specification syntax is provided by a statement form:

$w : [pre, post]$

where 'w' (called the 'frame') defines the scope of the specification, i.e. those state variables which may be changed by the behaviour defined by the specification , and 'pre' and '$post$' are first-order predicate logic formulae which describe the relationship between the program state before the 'execution' of the specification statement, and after the termination of the specification statement respectively. The specification statement can therefore be viewed as a description of the minimum requirements on the behaviour of any concrete statement which may replace it during refinement.

Similarly, in Back and Wright's wide-spectrum language [2], the concrete code is a version of Dijkstra's Guarded Command Language and a statement, called an *assert statement*, is denoted $\{b\}$, where b is a formula on the local state. The assert statement will terminate correctly if the local state satisfies the formulae when 'executed', and will abort otherwise.

Original work by Back [1], and later [2], can be seen as the first investigation of adding specification statements to programming languages to aid in the process of verification.

The common factor of both Morgan and Back and Wright's languages is that they are transformational; they describe computations which have all input data available at the start of execution, and provide the result at the time of termination. This restriction provides the basis for the 'shape' of Morgan's specification statement – it describes a relationship between initial and final states. In real-time systems we are interested in *reaction*, i.e. input and output during the execution of an agent. In addition, we are interested in the time at which the inputs and outputs occur; our specification statement for real-time systems reflects these requirements.

In future we aim to produce a toolkit which automates the refinement process by providing theorem guiders for the discharging of refinement proof obligations. In addition, language extensions for supporting programming in the large are being investigated.

7 References

[1] R. J. R. Back, **"Correctness Preserving Program Refinements: Proof Theory and Applications"**, Tract 131, Mathematisch Centrum, Amsterdam. 1980.

[2] R. J. R. Back, **"A Calculus Of Refinements for Program Derivations"**, Acta-Informatica, 25, p593-624. 1988. .

[3] R. J. R. Back, J Wright, **"Refinement Concepts Formalised In Higher Order Logic"**, BCS-FACS, Vol 2, No.3. 1990.

[4] The CIP Language Group, **"The Munich Project CIP. Vol1"**, LNCS-183. 1985.

[5] E. Dijkstra, **"A Discipline Of Programming"**, Prentice-Hall. 1976.

[6] J. Hooman, **"Specification and Compositional Verification of Real-Time Systems"**, Ph.D. Thesis, Technical University of Eindhoven. 1991.

[7] C. Morgan, **"Programming From Specifications"**. Prentice-Hall International, C.A.R. Hoare Series,. 1990.

[8] C. Morgan, K. Robinson, P. Gardiner, **"On The Refinement Calculus"**, Oxford University Technical Report PRG-70, Oct 1988.

[9] D. J. Scholefield, **"A Refinement Calculus for Real-Time Systems"**, Department of Computer Science D.Phil Thesis. University of York. July 1992.

[10] D. J. Scholefield, H.S.M. Zedan, J. He, **"A Specification Oriented Semantics for Real-Time Refinement**, Theoretical Computer Science, (to appear). 1993.

[11] E. Schonberg, D. Shields, **"From Prototype To Efficient Implementation: a Case Study Using SETL and C"**. Courant institute of mathematical sciences, Dept of computer science, New York University. 1985.

Deciding Testing Equivalence for Real-Time Processes with Dense Time

Bernhard Steffen[1] and Carsten Weise[2]

[1] Lehrstuhl für Programmiersysteme, University of Passau, Germany
Email: steffen@fmi.uni-passau.de
[2] Lehrstuhl für Informatik I, Aachen University of Technology, Germany
Email: carsten@i1.informatik.rwth-aachen.de

Abstract. We present a decision algorithm for testing equivalence of real-time systems with a dense time domain. Real-time systems are modelled by timed graphs, while the decision algorithm uses "mutually refined" timer region graphs. The mutual refinement is important for the synchronization of the timers of different real-time systems. Key to our decision algorithm is the fact that - despite the dense time domain - testing can be reduced to Π-bisimulation in very much the same way as in the untimed case.

1 Introduction

Recently, many real-time process calculi have been proposed (see [BB90, Che92, MT90, NSY91, RR88, Yi91]) adding a time-domain to existing calculi to model real-time systems. Well-studied are extensions with discrete time-domains, which at first sight seem to be sufficient, because typical (technical) applications have discrete time steps. However, even if one is only interested in discrete-time systems, dense-time models are superior because they allow to abstract from the specific granularity of the time domain. This is particularly important in the light of refinement: in a discrete-time model the refinement of a system may require a complete redesign whenever the granularity of the time domain changes. In contrast, dense-time models support such refinements. Unfortunately, there is a price to pay for this generality: whereas most of the decidabilty results for finite state systems straightforwardly carry over to discrete time models, decidability of equivalence relations such as bisimulation and testing equivalence is not at all obvious for dense-time calculi, which, owing to their time domain, have infinitely many states.

Timer region graphs [ACD90] build a powerful framework for decision algorithms for real-time processes modelled as *timed graphs* [ACD90, NSY91], a generalization of labelled transition systems, which admits dense time domains. The real-time calculus $\mathbf{ATP}_D$ [NSY91] e.g. can be translated into timed graphs.

In this paper we present a decision procedure for *testing equivalence* of timed graphs. Testing equivalence is a natural equivalence for real-time systems as it can be seen as a "black-box"-theory. The decision procedure uses an (untimed!) Π-bisimulation between *deterministic* graphs, built from *mutually refined timer region graphs*. Timer region graphs are a finite abstraction of *timer evaluation graphs*, which constitute an operational semantics of timed graphs and thus are the model of reference for the correctness of the algorithm.

After a short preliminary section, we recall the definition of timed graphs and the derived notion of timer evaluation graphs in Sect. 3. Timer region graphs are presented in Sect. 4. Afterwards extended and refined graphs are defined in Sect. 5 and testing equivalence for real-time systems in Sect. 6. The following Sect. 7, where the decision procedure is developed, is the heart of the paper. Finally Sect. 8 discusses related work and direction to future work respectively.

2 Preliminaries

We briefly recall some basic notions.

Definition 1. A *(rooted) labelled transition system* $(N, n_0, L, \longrightarrow)$ consists of a (possibly infinite) set of nodes N, a start node $n_0 \in N$, a (possibly infinite) set of labels L, also called the alphabet, and a transition relation $\longrightarrow \subseteq N \times L \times N$.
Two paths in a labelled transition system (LTS) are called *equivalent* iff they represent the same sequence of labels. An LTS is *finite* if N and $\longrightarrow$ are finite. In the following we often write $n \xrightarrow{\ell} n'$ instead of $(n, \ell, n') \in \longrightarrow$.

Proposition 2. *Given an LTS $(N, n_0, L, \longrightarrow)$ and an equivalence relation $\Pi \subseteq N \times N$, there exists a relation $\sim_\Pi \subseteq N \times N$ called Π-*bisimulation *which is characterized by $\sim_\Pi \subseteq \Pi$ and $n \sim_\Pi m$ iff for all labels $\ell \in L$:*

$$(1)\ n \xrightarrow{\ell} n' \quad \text{then } \exists m' \in N : m \xrightarrow{\ell} m' \wedge n' \sim_\Pi m'$$

$$(2)\ m \xrightarrow{\ell} m' \quad \text{then } \exists n' \in N : n \xrightarrow{\ell} n' \quad \wedge n' \sim_\Pi m' \ .$$

An $N \times N$-bisimulation is a (strong) bisimulation $\sim$ in the sense of [Mil89]. The notion of *weak transition relation* reflects the intuition that τ is internal to a process:
Definition 3. Given an LTS $(N, n_0, L, \longrightarrow)$, and a special symbol $\tau \in L$, the *weak* transition relation $\Rightarrow$ is the least relation defined by: $\overset{\epsilon}{\Rightarrow} = \xrightarrow{\tau^*}$, $\overset{a}{\Rightarrow} = \xrightarrow{\tau^*}\xrightarrow{a}\xrightarrow{\tau^*}$.

3 Timed and Timer Evaluation Graphs

We represent real-time processes as *timed graphs* [ACD90][NSY91], a generalization of LTS's. Timed graphs build a very general framework for the specification and analysis of real-time processes, in which many real-time operators – e.g. the full calculus $\mathbf{ATP}_D$[NSY91] – can be modelled. They are parameterized with a finite set of timers $\mathcal{T}$, which can be seen as variables over a (possibly dense) time domain. We will use $\mathbb{R}^0_+$, the set of non-negative real numbers, as our time domain. Then $\mathbb{R}_+ := \mathbb{R}^0_+ \setminus \{0\}$ and d typically denotes elements from the time domain.

Transitions of timed graphs are labelled with an action a, an enabling condition b and a set $R \in 2^{\mathcal{T}}$ of timers reset after performing the transition. Enabling conditions are expressions giving upper and lower bounds for timer values. As usual, the constants occurring in these expressions are restricted to come from $\mathbb{N}$, the set of natural numbers. This fact is central for our decidability results[3]. We further assume a finite set $\mathcal{A}$ of visible actions and an internal action τ, and define $\mathcal{A}_\tau := \mathcal{A} \cup \{\tau\}$.

[3] Note that a process where all constants are rational can be turned into one with integer constants by multiplication of all constants with an appropriate value.

Definition 4.

1. The set $Cond(\mathcal{T})$ of conditions over a timer set $\mathcal{T} = \{T_1, \ldots, T_k\}$ is recursively defined by:
   (a) $T_i < c, T_i \leq c, T_i > c, T_i \geq c \in Cond(\mathcal{T})$ for $c \in \mathbb{N}, T_i \in \mathcal{T}$
   (b) $e \wedge f, e \vee f \in Cond(\mathcal{T})$ for $e, f \in Cond(\mathcal{T})$
   with the usual interpretation of operators.
2. A *Timed Graph* (or process) P is a finite LTS $(M, m_0, A_\tau \times Cond(\mathcal{T}) \times 2^{\mathcal{T}}, \rightarrow)$, where $\mathcal{T}$ is a finite timer set, with an associated boolean function $div : M \rightarrow \{\texttt{tt}, \texttt{ff}\}$ representing divergence potential: if $div(m)$ is true, the process can idle in m for any amount of time.– A node m without outgoing arcs always satisfies $div(m)$.

If not stated otherwise, we will assume in the following that a timed graph P is represented as $(M_P, m_0^P, A_\tau \times Cond(\mathcal{T}_P) \times 2^{\mathcal{T}_P}, \rightarrow_P)$.

Intuitively, one can think of timers as synchronously running clocks, initially reset and increasing – with intermediate resets – as time progresses. The values of the timers steer the execution of timed graphs by means of enabling conditions.

In order to arrive at the usual (unconditional) notion of transition systems, we make this dependency explicit by introducing timer evaluation graphs:

Definition 5.

1. A *timer evaluation* is a function $\nu : \mathcal{T} \rightarrow \mathbb{R}_+^0$, which associates each timer variable with its current value. The set of all timer evaluations for a given set $\mathcal{T}$ is denoted by $\mathcal{V}(\mathcal{T})$. The *inital evaluation* ν_0 has all timers reset.
2. For some evaluation ν and some $d \in \mathbb{R}_+^0$, we denote by $\nu + d$ the evaluation where d is added to each timer value, i.e. $\forall T_i \in \mathcal{T} : (\nu + d)(T_i) := \nu(T_i) + d$
3. $[R \mapsto 0]\nu$, where $R \subseteq \mathcal{T}$, denotes the function that resets each timer $T_i \in R$ in the argument timer evaluation, i.e. $[R \mapsto 0]\nu(T_i) := \begin{cases} \nu(T_i) & T_i \notin R \\ 0 & T_i \in R \end{cases}$
4. For a given timed graph P with alphabet A_τ its *timer evaluation graph* $TE(P, \mathcal{T}_P)$ is the LTS $(M_P \times \mathcal{V}(\mathcal{T}_P), <m_0^P, \nu_0>, A_\tau \cup \mathbb{R}_+, \rightarrow)$, where the transition relation is given by
   (a) $<m, \nu> \xrightarrow{a} <m', [R \mapsto 0]\nu'>, a \in A_\tau$, if the transition a from m to m' is enabled under the timer evaluation ν (i.e. its condition is true for ν), and the timers in R are reset along this transition,
   (b) $<m, \nu> \xrightarrow{d} <m, \nu + d>, d \in \mathbb{R}_+$, if either there is some action enabled from m at or after $\nu + d$ or $div(m)$ is true.

The weak transition relation for timer evaluation graphs requires the following additional rules: $n \overset{0}{\Rightarrow} n$ and $\forall d, d' \in \mathbb{R}_+ : n \overset{d}{\Rightarrow}\overset{d'}{\Rightarrow} n'$ then $n \overset{d+d'}{\Longrightarrow} n'$.

4 Timer Region Graphs

Timer evaluation graphs have infinitely many states: every state of the timed graph corresponds to as many states as there are timer evaluations. However,a finite partitioning of the set of timer evaluations leads to timer region graphs ([ACD90]), which are concrete enough to decide bisimilarity and testing. This partitioning is motivated by three observations, based on the fact that there are only finitely many conditions in a timed graph, considering integer bounds only:

1. Conditions cannot distinguish between timer values with identical integral parts.
2. For a given timed graph, there exists a maximal value c to which timers are compared. Thus the conditions cannot distinguish between timer values in (c, ∞).
3. The order between the fractional parts of different timers is important, because the largest fractional part becomes integral next, which may cause the enabling of a transition. Further it is important if a timer has an integral value.

Assuming the maximal constant c to be given from now on, this leads to the following definition (see [ACD90]):

Definition 6. The equivalence relation $\simeq_c$ on $\mathcal{V}(\mathcal{T})$ is defined by:

$$\nu \simeq_c \nu' :\Leftrightarrow \begin{cases} (a)\ \forall T_i \in \mathcal{T}: & \lfloor \nu(T_i) \rfloor = \lfloor \nu'(T_i) \rfloor \quad \vee \quad \nu(T_i) > c \wedge \nu'(T_i) > c \\ (b)\ \forall T_i, T_j \in \mathcal{T}: & fract(\nu(T_i)) \leq fract(\nu(T_j)) \Leftrightarrow \\ & fract(\nu'(T_i)) \leq fract(\nu'(T_j)) \\ (c)\ \forall T_i \in \mathcal{T}: & fract(\nu(T_i)) = 0 \Leftrightarrow fract(\nu'(T_i)) = 0 \end{cases}$$

where $fract(z)$ and $\lfloor z \rfloor$ denote the fractional and integral part of z. A *timer region* is a $\simeq_c$-equivalence class written $[\nu]_c$. Moreover, $\Gamma_c(\mathcal{T})$ denotes the set of all regions and γ_0 the *initial region* $[\nu_0]_c$, which is the singleton set $\{\nu_0\}$.

The index c will be dropped if clear from the context. To clearly identify the timer set we will sometimes denote the inital region by $\gamma_0(\mathcal{T})$. Obviously, the set of regions is finite. Moreover, as motivated by the three observations, conditions cannot distinguish between the evaluations within a region.

Lemma 7. *For some condition $b \in Cond(\mathcal{T})$ with all constants smaller than or equal to c holds:* $\forall \gamma \in \Gamma_c(\mathcal{T}): \ (\exists \nu \in \gamma : b[\nu] = \mathtt{tt}) \ \Leftrightarrow \ (\forall \nu \in \gamma : b[\nu] = \mathtt{tt})$.

For regions, a successor-function can be defined by (see [ACD90]):

Definition 8. $succ : \Gamma_c(\mathcal{T}) \rightarrow \Gamma_c(\mathcal{T})$ is defined by $\gamma' = succ(\gamma)$ iff

$$\gamma \neq \gamma' \text{ and } \forall \nu \in \gamma : \exists d \in \mathbb{R}_+ : \nu + d \in \gamma' \wedge \forall d' < d : \nu + d' \in \gamma \cup \gamma' \ .$$

A timer region graph is defined as the abstraction of the timer evaluation graph, where two states $<m, \nu_1>$ and $<m, \nu_2>$ are identified iff $\nu_1 \simeq \nu_2$. By this, time steps of different duration are represented by the same transition, which we will label with the special symbol χ. The set $\mathcal{A} \cup \{\chi\}$ is written as $\mathcal{A}_\chi$. The well-definedness of the timer region graph defined below is a consequence of Lemma 7:

Definition 9. Given a timer evaluation graph $TE(P, \mathcal{T}_P)$ with transition relation $\rightarrow_P$, its corresponding *timer region graph* $TR(P, \mathcal{T}_P)$ is the LTS $(M_P \times \Gamma(\mathcal{T}_P)$, $<m_0^P, \gamma_0>, A_\tau \cup \{\chi\}, \rightarrow)$, where the transition relation is given by:

$$\forall a \in A_\tau : <m, \gamma> \xrightarrow{a} <m', [R \mapsto 0]\gamma> \Leftrightarrow \exists \nu \in \gamma : <m, \nu> \xrightarrow{a}_P <m', [R \mapsto 0]\nu>$$

$$<m, \gamma> \xrightarrow{\chi} <m, \gamma' = succ(\gamma)> \Leftrightarrow \exists \nu \in \gamma, d \in \mathbb{R}_+ : \nu + d \in \gamma' \wedge <m, \nu> \xrightarrow{d}_P <m, \nu + d>$$

As $\Gamma(\mathcal{T})$ depends only on the set of timer bounds, the nodes and transitions of the timer region graph can directly be constructed from the underlying timed graph.

5 Extended Graphs

Central to our decision procedure is the synchronization of the timers of P and Q, gained by extending the processes' timer sets $\mathcal{T}_P$ and $\mathcal{T}_Q$ to $\mathcal{T}_P \cup \mathcal{T}_Q$. This generalizes the idea in [ACD90], where only one timer was added for the decision procedure.

Definition 10. Let $TE(P, \mathcal{T}_P)$ be a timer evaluation graph, $\mathcal{T}_e$ a set of timers disjoint from $\mathcal{T}_P$, and $\mathcal{T} := \mathcal{T}_P \cup \mathcal{T}_e$. The *extended timer evaluation graph* $TE(P, \mathcal{T})$ is the LTS $(M_P \times \mathcal{V}(\mathcal{T}), <m_0^P, \gamma_0>, A_\tau \cup \mathbb{R}_+, \rightarrow)$ where $\rightarrow$ is given by:

$$\forall R' \subseteq \mathcal{T}, \mu \in \mathcal{V}(\mathcal{T}) \text{ where } \mu|_{\mathcal{T}_P} = \nu :$$

$$\text{if } <m,\nu> \xrightarrow{a}_P <m', [R \mapsto 0]\nu>, a \in A_\tau \text{ then } <m,\mu> \xrightarrow{a} <m', [R \cup R' \mapsto 0]\mu>$$

$$\text{if } <m,\nu> \xrightarrow{d}_P <m, \nu + d>, d \in \mathbb{R}_+ \text{ then } <m,\mu> \xrightarrow{d} <m, [R' \mapsto 0](\mu + d)>$$

The $\mathcal{T}_e$-extension $TR(P, \mathcal{T})$ of a given region graph $TR(P, \mathcal{T}_P)$ is the region graph rising from the extension $TE(P, \mathcal{T})$ of the underlying timer evaluation graph $TE(P, \mathcal{T}_P)$.

Note that the new timers increase synchronously with the original timers, but may be reset at any time. Extended timer evaluation graphs have the following simple property:

Lemma 11. *For nodes of the $\mathcal{T}_e$-extension $TE(P, \mathcal{T})$ of the timer evaluation graph $TE(P, \mathcal{T}_P)$ holds: $\forall m \in M_P, \nu, \nu' \in \mathcal{V}(\mathcal{T})$: $\nu|_{\mathcal{T}_P} = \nu'|_{\mathcal{T}_P}$ implies $<m,\nu> \sim <m,\nu'>$.*

Thus the newly introduced timers do not influence the behaviour of the transition system. The following lemma shows that extended graphs are useful for the comparison of processes. For *mutually extended timer evaluation graphs* of two processes P and Q, i.e. the graph for P is extended by the timer set of Q and vice versa, comparison of states with identical evaluations only is necessary:

Lemma 12. *Let $TE(P, \mathcal{T}_P)$ and $TE(Q, \mathcal{T}_Q)$ be timer evaluation graphs with disjoint timer sets, and $TE(P, \mathcal{T})$ and $TE(Q, \mathcal{T})$ their mutually extended timer evaluation graphs, both defined over $\mathcal{T} := \mathcal{T}_P \cup \mathcal{T}_Q$. Moreover, let $<m,\nu>$ and $<m',\nu'>$ be states of $TE(P, \mathcal{T}_P)$ and $TE(Q, \mathcal{T}_Q)$ resp. and $\mu \in \mathcal{V}(\mathcal{T})$ be the unique evaluation satisfying $\mu|_{\mathcal{T}_P} = \nu \wedge \mu|_{\mathcal{T}_Q} = \nu'$. Then $<m,\nu> \sim <m',\nu'>$ if and only if $<m,\mu> \sim <m',\mu>$, where $<m,\mu>$ and $<m',\mu>$ are states of $T(P, \mathcal{T})$ and $T(Q, \mathcal{T})$, respectively.*

To see that extending region graphs is a kind of refinement, we extend the usual restriction for timer evaluations to timer regions:

Definition 13. Let $\gamma \in \Gamma(\mathcal{T} \cup \mathcal{T}_e)$ be some timer region. Then the restriction of γ to $\mathcal{T}$ is defined by: $\gamma|_{\mathcal{T}} := \{\nu|_{\mathcal{T}} \in \mathcal{V}(\mathcal{T}) \mid \nu \in \gamma\}$.

The set of regions $\gamma \in \Gamma(\mathcal{T} \cup \mathcal{T}_e)$ which all have the same restricted region $\gamma|_{\mathcal{T}}$ – i.e. which only differ in the values of the timers from $\mathcal{T}_e$ – can be seen as a refinement of $\gamma|_{\mathcal{T}}$. In the same way, a node $<m,\gamma'>$ from a a timer region graph $TR(P, \mathcal{T}_P)$ is refined by the set $\{<m,\gamma> \mid \gamma \in \Gamma(\mathcal{T}), \gamma|_{\mathcal{T}_P} = \gamma'\}$ contained in the $\mathcal{T}_e$-extension $TR(P, \mathcal{T} = \mathcal{T}_P \cup \mathcal{T}_e)$ of $TR(P, \mathcal{T}_P)$. Therefore, extended region graphs are also called *refined region graphs*, and mutually extended region graphs – where one graph is extended by the timer set of the other – are called *mutually refined region graphs*.

Note that in refined region graphs, nodes generally have several outgoing χ-arcs. By the extension the number of consecutive χ-transitions in general increases, as the new timers may introduce additional crossings of integer bounds not present before.

6 Testing for Real-Time Systems

Testing equivalence is a well-known notion of equivalence for process algebras [DH84]. Processes are distinguished by special processes, called experimenters (or testers), capable of an action ω (*success*). The testee may only engage in observable actions the experimenter is willing to accept, while both perform invisible actions independently. A test is succesful if the tester arrives at an accepting ω-transition.

Definition 14. Let a process P and an experimenter E be given with start states $p_0 = <m_0^P, \nu_0^P>$, $e_0 = <m_0^E, \nu_0^E>$ and arbitrary states p and e of their timer evaluation graphs $TE(P, \mathcal{T}_P)$ and $TE(E, \mathcal{T}_E)$.

1. An expression $p||e$ is called a *configuration*. A configuration is *successful* if $e \xrightarrow{\omega}$.
2. The *configuration transition relation* $\longrightarrow$ is defined by:

$$\forall \ell \in A \cup \mathbb{R}_+ : \begin{array}{c} p \xrightarrow{\ell}_P p', e \xrightarrow{\ell}_E e' \text{ then } p||e \xrightarrow{\ell} p'||e' \text{ and} \\ p \xrightarrow{\tau}_P p' \text{ then } p||e \xrightarrow{\tau} p'||e \text{ and } e \xrightarrow{\tau}_E e' \text{ then } p||e \xrightarrow{\tau} p||e' \;. \end{array}$$

3. A maximal (possibly infinite) sequence $p_1||e_1 \xrightarrow{\ell_1} p_2||e_2 \xrightarrow{\ell_2} \ldots$ is called a *computation* of $p_1||e_1$, if there is a constant $g \in \mathbb{R}_+$ so that for every ℓ_i from $\mathbb{R}_+$ we have $\ell_i \geq g$. The computation is called *successful*, if for some i, $p_i||e_i$ is successful.
4. $p\,\text{MAY}\,e$ if $p||e$ has a successful computation. $P\,\text{MAY}\,E$ if $p_0\,\text{MAY}\,e_0$.
5. $p\,\text{MUST}\,e$ if every computation of $p||e$ is successful. $P\,\text{MUST}\,E$ if $p_0\,\text{MUST}\,e_0$.

Note that this is the definition of testing given in [CZ91], generalized to an uncountable set of labels $A_\tau \cup \mathbb{R}_+$, and with an additional lower bound on time-steps in a computation. This last requirement supresses *Zeno-computations* (see [NSY91]) which have infinitly many time-steps but never reach a given bound. The testing-preorders and -equivalences are defined by:

Definition 15.

$P \begin{Bmatrix} \leq_{may} \\ \leq_{must} \end{Bmatrix} Q$ iff for every experimenter E, $P \begin{Bmatrix} \text{MAY} \\ \text{MUST} \end{Bmatrix} E$ implies $Q \begin{Bmatrix} \text{MAY} \\ \text{MUST} \end{Bmatrix} E$.

$P \leq_{test} Q$ iff $P \leq_{may} Q$ and $P \leq_{must} Q$.

$P \cong_i Q$ iff $P \leq_i Q$ and $Q \leq_i P$, where $i \in \{may, must, test\}$.

In the untimed case, testing equivalence can be characterized by means of the language and so-called acceptance sets of the processes. A similar characterization can be given here. This requires the introduction of words and languages for timed graphs. The usual approach to describe languages of real-time processes with dense time domains are so-called *dense words*. These are simply mappings from the time domain into the words over the alphabet, giving for each point in time the word accepted by the process. We will use a slightly different notion based on relative time, which we call a *timed word*[4]. An example is $a(3).b(2).c(0).f(1)$, which means that an a was observed at time 3, a b and then a c at time 5, and an f at time 6.

[4] This notion is similar to timed traces [RR88]

Definition 16. Given an alphabet $\mathcal{A}$, the set of *timed actions* of $\mathcal{A}$ over $\mathbb{R}_+^0$ is defined by: $\mathcal{A}(\mathbb{R}_+^0) := \{a(d) | a \in \mathcal{A}, d \in \mathbb{R}_+^0\}$. The set of *timed words* is $\mathcal{A}(\mathbb{R}_+^0)^*$, which is the set of finite words over $\mathcal{A}(\mathbb{R}_+^0)$. The function $\Delta : \mathcal{A}(\mathbb{R}_+^0)^* \to \mathbb{R}_+^0$, measuring the duration of a timed word, is defined by: $\Delta(\varepsilon) := 0$, $\Delta(a(d).w) := d + \Delta(w)$. The domain for real-time languages is

$$Dom(\mathcal{A}, \mathbb{R}_+^0) := \{<w,d> | w \in \mathcal{A}(\mathbb{R}_+^0)^*, d \in \mathbb{R}_+^0, d \geq \Delta(w)\} \ .$$

The weak transition relation $\Rightarrow$ is extended to words from $\mathcal{A}(\mathbb{R}_+^0)^*$ and $Dom(\mathcal{A}, \mathbb{R}_+^0)$:

$$\begin{array}{llll} d, d' \in \mathbb{R}_+, a \in \mathcal{A}, w \in \mathcal{A}(\mathbb{R}_+^0)^* : & & n \stackrel{a}{\Rightarrow} n' & \text{then } n \stackrel{a(0)}{\Longrightarrow} n' \\ n \stackrel{d}{\Rightarrow}\stackrel{a}{\Rightarrow} n' & \text{then } n \stackrel{a(d)}{\Longrightarrow} n' & n \stackrel{a(d)}{\Longrightarrow}\stackrel{w}{\Rightarrow} n' & \text{then } n \stackrel{a(d).w}{\Longrightarrow} n' \\ n \stackrel{w}{\Longrightarrow} n' & \text{then } n \stackrel{<w,\Delta(w)>}{\Longrightarrow} n' & n \stackrel{<w,d>}{\Longrightarrow}\stackrel{d'}{\Rightarrow} n' & \text{then } n \stackrel{<w,d+d'>}{\Longrightarrow} n' \end{array}$$

We define the following terms over timed words:

Definition 17. Let $TE(P,T)$ be some (extended) timer evaluation graph. Then

1) $\mathcal{L}(P) := \{<w,d> \in Dom(\mathcal{A}, \mathbb{R}_+^0) \mid n_0 \stackrel{<w,d>}{\Longrightarrow}\}$ 2) $S_\chi(n) := \{d \in \mathbb{R}_+ | n \stackrel{d}{\Rightarrow}\}$

3) $S_\chi(P, <w,d>) := \bigcap_{n \in X} S_\chi(n)$ where $X := \{n | n_0 \stackrel{<w,d>}{\Longrightarrow} n\}$

4) $der(n,a) := \{n' \mid n \stackrel{a}{\rightarrow} n'\}$ 5) $S(n) := \{a \mid a \in \mathcal{A}_\tau, n \stackrel{a}{\rightarrow}\}$

6) $Acc(P, <w,d>) := \{S(n') \mid n_0 \stackrel{<w,d>}{\Longrightarrow} n', n' \not\stackrel{\tau}{\rightarrow}\}$

7) $M \subset\subset N \Leftrightarrow \forall S \in M \; \exists S' \in N : S' \subseteq S$

Then $\mathcal{L}(P)$ is the *language* of P, $der(n,a)$ is the set of *a-derivatives* of n, $S(.)$ the set of *enabled actions* and $Acc(.,.)$ is an *acceptance set*.

All these notions are well-known from the theory of untimed testing, execpt for S_χ. The set $S_\chi(P, <w,d>)$ describes all time moves possible from states reachable via $<w,d>$, and thus can be seen as a "time acceptance set". Such a set is either empty or an interval of the form $[0,d)$ or $[0,d]$. - Enabled actions of evaluation and region graphs are related in the following way:

Lemma 18. *Enabled actions of an (extended) evaluation graph $TE(P,T)$ and its region graph are related by:* $\forall <m,\nu> \in M_P \times \mathcal{V}(T) : S(<m,\nu>) = S(<m,[\nu]>) \setminus \{\chi\}$.

7 Deciding Timed Testing

The point of our decision procedure is the reduction of testing to *untimed Π-bisimulation* for finite state transition systems, which is known to be decidable [KS83]. This approach is similar to the one presented by [CH89]. However the presence of time requires a far more sophisticated reduction. We start by giving a characterization for the "timed" may- and must-preorders, where we assume that processes cannot engage in arbitrary long sequences of τ-moves[5]. Note that this is almost the characterization of the untimed case, with an additional clause for $S_\chi(.)$:

[5] The case of infinite τ-moves can be treated along the lines of [CH89].

Theorem 19 A Characterization of Testing.

$$P \leq_{may} Q \iff \mathcal{L}(P) \subseteq \mathcal{L}(Q)$$

$$P \leq_{must} Q \iff \forall <w,d> \in Dom(\mathcal{A}, D) : Acc(Q, <w,d>) \subset\subset Acc(P, <w,d>) \\ \wedge S_\chi(Q, <w,d>) \supseteq S_\chi(P, <w,d>)$$

While the basic idea of the proof is the same as in the untimed case, there is a subtle problem arising from the real numbers. For a given timed word, e.g. $u = <a(e).b(\pi - e), 4>$, it is not generally possible to construct a test accpeting this and only this word. This can be solved by considering the mutually refined timer region graphs of P and Q, as shown in the extended version of the paper ([SW93]).

The decision procedure uses *deterministic timer region graphs*, constructed in the standard way by first building the τ-closure and then applying a powerset-construction to the set of states. We recall the standard definition of deterministic transition systems:

Definition 20. Let an LTS $TS = (N, n_0, L \cup \{\tau\}, \rightarrow_1)$ and its weak transition relation $\Rightarrow$ be given. The *τ-closure* of a subset $X \subseteq N$ of nodes is defined by $X^\tau := \{n' \in N | n \in X, n \stackrel{\varepsilon}{\Rightarrow} n'\}$. The *deterministic transition system* of TS is the LTS $(\{X \subseteq N | X = X^\tau\}, \{n_0\}^\tau, L, \rightarrow)$, with the transition relation given by

$$\forall \ell \in L : X \stackrel{\ell}{\rightarrow} X' \Leftrightarrow X' = (\bigcup_{n \in X} der(n, \ell))^\tau \ .$$

Definition 21. The *deterministic timer region graph* $D(P,T)$ is the deterministic transition system built from $TR(P,T)$.

Deterministic timer region graphs (deterministic graphs for short) are LTS's over A_χ. Thus (untimed) bisimulation is defined on them. For the correctness of our decision algorithm, which is based on deterministic graphs, we need the following notion:

Definition 22. Let $TE(P, T_P)$ be a timer evaluation graph, $D(P,T)$ be the deterministic graph of its refined region graph $TR(P,T)$, p be some (finite) path in $D(P,T)$ from the start state X_0 to some state X and $<m,\nu>$ be a state of $TE(P, T_P)$ and $<w,d> \in Dom(\mathcal{A}, \mathbb{R}_+^0)$. Then $<m,\nu>$ is *reachable via* $<w,d>$ *consistent with* p iff there is $\mu \in \mathcal{V}(T) : \mu|_{T_P} = \nu$ and $<m,[\mu]> \in X$ and either

1. p consists of the start state only, i.e. $p = X_0$ and $<w,d> = <\varepsilon, 0>$, or
2. the state X is reached via $\ell \in A_\chi$, i.e. there is a path p' with $p = p' \stackrel{\ell}{\rightarrow} X$ and a state $<m',\nu'>$ reachable in $TE(P)$ via $<w',d'>$ consistent with p', and either
   (a) $\ell = a \in \mathcal{A}$, $<m',\nu'> \stackrel{a}{\Rightarrow} <m,\nu>$, $<w,d> = <w'.a(d' - \Delta(w')), d'>$, or
   (b) $\ell = \chi$, $\exists d'' \in \mathbb{R}_+ : <m',\nu'> \stackrel{d''}{\Longrightarrow} <m,\nu>$, $<w,d> = <w', d' + d''>$.

Intuitively this definition means that we can find a path in $TE(P, T_P)$ with a labelling "adding up" to $<w,d>$ leading from the start state to $<m,\nu>$ which "touches" each node of the path p. – Now we are ready to relate nodes of two deterministic graphs constructed from mutually refined timer region graphs:

Lemma 23. *Let $D(P,\mathcal{T})$ and $D(Q,\mathcal{T})$ be the determinstic graphs of two mutually refined timer region graphs with common timer set $\mathcal{T} := \mathcal{T}_P \cup \mathcal{T}_Q$ and X_P, X_Q be states in them reachable by equivalent paths p and q resp. Then we have*

(1) $\forall <m,\gamma>\in X_P, <m',\gamma'>\in X_Q : \exists \gamma'' :$
$\gamma''|_{\mathcal{T}_P} = \gamma|_{\mathcal{T}_P} \ \wedge\ \gamma''|_{\mathcal{T}_Q} = \gamma'|_{\mathcal{T}_Q} \ \wedge\ <m,\gamma''>\in X_P, <m',\gamma''>\in X_Q$

(2) $\forall <m,\nu>\in M_P \times \mathcal{V}(\mathcal{T}_P), <w,d>\in Dom(\mathcal{A}, \mathbb{R}^0_+)$,
where $<m,\nu>$ *is reachable via* $<w,d>$ *consistent with* p :
(*i*) $\exists \mu \in \mathcal{V}(\mathcal{T}) : \mu|_{\mathcal{T}_P} = \nu \wedge <m,[\mu]>\in X_P$
(*ii*) $Acc(Q,<w,d>) = \{\, S(<m',[\mu']>) \setminus \{\chi\} \mid \mu'|_{\mathcal{T}_P} = \nu, <m',[\mu']>\in X_Q, <m',\mu'>\overset{\tau}{\not\to}\}$

(3) *as (2) with the roles of P and Q exchanged*

Here (1) states that whenever we have two states of the timer region graphs in the two nodes of the deterministic graph (reachable by equivalent paths), we can find states in the same nodes which are comparable regarding their timing behaviour (i.e. have the same region) and include the same sets of states from the timer evaluation graphs (upto bisimilarity) according to Lemma 11.

Then (2)(*i*) says that every node of the deterministic graph represents all states reachable via $<w,d>$ consistent with a path leading to it. From (2)(*ii*) follows that $Acc(Q,<w,d>)$ can be computed by looking at the node X_Q only. This is central for the decision of MUST-equivalence. Then MAY-equivalence is decidable because (untimed) bisimulation on deterministic graphs is decidable:

Theorem 24. *Let $D(P,\mathcal{T})$ and $D(Q,\mathcal{T})$ be mutually refined deterministic region graphs of two processes P and Q. Then $D(P,\mathcal{T}) \sim D(Q,\mathcal{T}) \iff P \cong_{may} Q$.*

The proof of this theorem uses the characterization of $\cong_{may}$ given in Theorem 19:

$$D(P,\mathcal{T}_Q) \sim D(Q,\mathcal{T}_P) \iff \mathcal{L}(P) = \mathcal{L}(Q)$$

The if-case follows directly from the construction of deterministic graphs and part (1) of Lemma 23, while the converse case uses (2)(*i*) and (3)(*i*) from Lemma 23.

As in the untimed case [CH89], we define a relation Π on states of deterministic graphs, which resembles the definition of $\subset\!\subset$ and reduces $\cong_{must}$ to Π-bisimulation on deterministic graphs:

Definition 25. Let $D(P,\mathcal{T}_Q)$ and $D(Q,\mathcal{T}_P)$ be deterministic graphs of two mutually refined timer region graphs, and N_P and N_Q their respective sets of nodes. Then the relation $\Pi \subseteq N_P \times N_Q$ is defined by $\quad (X_P, X_Q) \in \Pi \quad \iff$

(1) $\forall <m,\gamma>\in X_P, <m,\gamma>\overset{\tau}{\not\to}: \exists <m',\gamma'>\in X_Q$ where $<m',\gamma'>\overset{\tau}{\not\to}$ and $\gamma|_{\mathcal{T}_P} = \gamma'|_{\mathcal{T}_P}$ so that $S(<m',\gamma'>) \subseteq S(<m,\gamma>)$ and

(2) $\forall <m',\gamma'>\in X_Q, <m',\gamma'>\overset{\tau}{\not\to}: \exists <m,\gamma>\in X_P$ where $<m,\gamma>\overset{\tau}{\not\to}$ and $\gamma'|_{\mathcal{T}_Q} = \gamma|_{\mathcal{T}_Q}$ so that $S(<m,\gamma>) \subseteq S(<m',\gamma'>)$

Now we can relate MUST-equivalence to Π-bisimulation on deterministic graphs:

Theorem 26. *Given two processes P and Q with there mutual refined deterministic timer region graphs $D(P, T_Q)$ and $D(Q, T_P)$, and the relation Π as above, we have*

$$D(P, T_Q) \sim_\Pi D(Q, T_P) \quad \Longleftrightarrow \quad P \cong_{must} Q \ .$$

The proof is based on the characterization of $\cong_{must}$ given in Theorem 19:

$$D(P, T_Q) \sim_\Pi D(Q, T_P) \quad \Longleftrightarrow \quad \forall <w, d> \in Dom(\mathcal{A}, D):$$
$$Acc(P, <w, d>) \subset\subset Acc(Q, <w, d>) \text{ and } S_\chi(P, <w, d>) \supseteq S_\chi(Q, <w, d>) \text{ and}$$
$$Acc(Q, <w, d>) \subset\subset Acc(P, <w, d>) \text{ and } S_\chi(Q, <w, d>) \supseteq S_\chi(P, <w, d>)$$

Using part (2) and (3) of Lemma 23, this can be seen from the definition of Π. If we forget about the χ's in the sets $S(<m, \gamma>)$ of the definition of Π, it should be clear that Π then models the relation $\subset\subset$ for acceptance sets. Further $S_\chi(.)$ is directly related to the number of consecutive χ-actions in the timer region graph, and therefore the proposition on $S_\chi(.)$ is guaranteed by the presence or absence of χ's in the $S(.)$-sets of the timer region graph.

Theorem 26 reduces the verification of $P \cong_{must} Q$ to Π-bisimulation-checking for the deterministic mutually refined timer region graphs of P and Q. Note that this requires the decision of *untimed* bisimulation only.

The decidabilty of $\cong_{may}$ and $\cong_{must}$ leads to the main result of the paper:

Theorem 27. *$P \cong_{test} Q$ is decidable for arbitrary timed graphs P and Q.*

8 Related and Future Work

Recently, more and more real-time process calculi have been proposed, most of them being extensions of untimed calculi with some time domain, see e.g. [BB90], [Yi91], [RR88]. When using a discrete time domain, most decidability results carry over from the untimed to the timed case. This is not easily seen with dense time. Some decision algorithms have been proposed for bisimulation [Fok91][HLY91]. However, these algorithms cannot handle parallelism or recursion. The only exception is the algorithm presented in [Cer92], which, similarly to our approach, uses timer regions. In fact, our approach using mutually refined timer region graphs can also be used to decide bisimulation. An application of this is given in [MW93]. In contrast to [Cer92], the resulting algorithm does not require the construction of a product graph (cf. [SW93]). – Testing equivalence has been defined for real-time processes in [CZ91] for a discrete time calculus. We do not know of any paper dealing with testing equivalence for dense time calculi.

We have presented an algorithm that decides testing equivalence for timed graphs, which is based on the comparison of different timed graphs via mutually refined timer region graphs. Our presentation focused on processes without infinite τ-sequences. Application to processes with infinite τ-sequences is straightforward along the lines of [CH89]. As may be expected, our algorithm is doubly exponential in the number of timers. The first exponential blow-up is caused by the construction of the timer region graphs, which is also necessary for bisimulation on timed graphs, and the second blow-up, which also arises in the untimed case, is caused by the powerset construction yielding the deterministic graph.

It can be shown that testing is a congruence for the real-time calculus $\mathbf{ATP}_D$, with the usual problem with non-deterministic choice [DH84]. Thus our algorithm may be used as a part of a compositional analysis tool for real-time systems. We plan to implement our algorithm in the Concurrency Workbench [CPS93].

We would like to thank Rance Cleaveland for pointing out the importance of testing equivalence and several unnamed referees for comments on this paper.

References

[ACD90] R. Alur, C. Courcoubetis, D. Dill. *Model-Checking for Real-Time Systems.* 5th LICS'90, Philadelphia, PA., Jun. '90, pp. 414-425

[BB90] J.C.M. Baeten, J.A. Bergstra. *Real-Time Process Algebra.* Report P8916b University of Amsterdam, Mar. '90.

[Cer92] K. Čerāns. *Decidability of Bisimulation Equivalences for Parallel Timer Processes.* in: Proceedings of CAV '92

[Che92] L. Chen. *An Interleaving Model for Real-Time Systems.* Report ECS-LFCS-91-184, Univ. of Edinburgh, Nov. '91

[CPS93] R. Cleaveland, J. Parrow, B. Steffen. *The Concurrency Workbench.* ACM Trans. on Prog. Lang. and Systems, Vol. 15, No. 1, Jan. '93, pp. 36-72.

[CH89] R. Cleaveland, M. Hennessy. *Testing Equivalences as a Bisimulation Equivalence.* in: LNCS 407, Springer, New York '89, pp. 11-23

[CZ91] R. Cleaveland, A. Zwarico. *A Theory of Testing for Real-Time.* 6th LICS '91, Amsterdam, The Netherlands, Jul. '91, pp. 110-119

[DH84] R. DeNicola, M. Hennessy. *Testing Equivalences for Processes.* Theoretical Computer Science No. 24, '84, pp. 83-133

[Fok91] W.J. Fokkink. *Normal forms in real-time process algebra.* Report CS-R9194, CWI, Amsterdam, '91.

[HLY91] U. Holmer, K.G. Larsen, W. Yi. *Deciding Properties of Regular Real-Timed Processes.* Report R 91-20 University of Aalborg, Jun. '91.

[KS83] P.C. Kanellakis, S.A. Smolka. *CCS Expressions, Finite State Processes,and Three Problems of Equivalence.* 2nd PODC, Montreal, Aug. '83, pp. 228-240

[MW93] T. Margaria, C. Weise. *Continuous Real Time Models in Practice.* to appear in: Proc. 5th Euromicro Worksh. on Real-Time Systems, Oulu(Finland), Jun. '93

[Mil89] R. Milner. *Communication and Concurrency.* Prentice-Hall, New York '89

[MT90] F. Moller, C. Tofts. *A Temporal Calculus of Communicating Systems.* in: LNCS 458, Springer, New York '90, pp. 401-415

[NSY91] X. Nicollin, J. Sifakis, S. Yovine. *From ATP to Timed Graphs and Hybrid Systems.* REX Workshop "Real-Time: Theory in Practice", Jun. '91

[RR88] G.M. Reed, A.W. Roscoe. *A Timed Model for Communicating Sequential Processes.* Journal of Theoretical Computer Science 58, '88, pp. 249-261

[SW93] B. Steffen, C. Weise. *Deciding Equivalences for Dense Real-Time Processes.* Technical Report. Aachener Informatik Berichte, Aachen University of Technology, to appear.

[Yi91] W. Yi. *CCS + Time = an Interleaving Model for Real-Time Systems.* in: LNCS 510, Springer, New York '91, pp. 217-228

A Calculus for Higher Order Procedures with Global Variables

Werner Stephan and Andreas Wolpers

German Research Center for Artificial Intelligence (DFKI), Stuhlsatzenhausweg 3, W-6600 Saarbrücken 11, Germany

Abstract. An arithmetically complete axiom system for full Algol-like higher order procedures with mode depth one is presented. To show soundness, a translation of the calculus into a variant of Dynamic Logic is defined. The completeness proof is outlined.

1 Introduction

The treatment of higher-order procedures (procedures can be passed as parameters) in Hoare-like axiomatic systems has been a major research topic since the problem of sound and relatively complete calculi for 'simple' procedures was solved in a satisfactory way, [Coo78], [Old81], [Apt81]. All known proof systems for higher-order procedures extend the basic formalism introduced by Hoare, [Hoa69], in a significant way. Olderog, [Old84], and Damm and Josko, [DJ83], extend the language for pre- and postconditions while the form of the partial correctness assertions is maintained. In contrast to that German, Clarke and Halpern, [GCH89], extend the expressiveness of the underlying logic adhering to a first-order language for pre- and postconditions.

The formalisms mentioned above are restricted to a sub-language (often called L4) where no global variables are allowed. It is known [Cla84] that without this restriction there can be no sound and relatively complete, [Coo78], calculus like the one presented in [GCH89] for L4. Although the existence of a uniform enumeration procedure (for partial correctness assertions) for arithmetical universes is obvious (from which the existence of a Hoare-logic follows, [Lip77]), we are still left with the problem of exhibiting a Hoare-like axiomatic system for this kind of language. In this paper we present a formalism for an unrestricted language with mode depth one. It is presented as a derived calculus in the context of a more general framework. We think that the generalisation to arbitrary finite mode depths is obvious.

The difficulties in treating global variables come from the fact that an unbounded number of variables that are not "visible" on the surface can be accessed by procedures that are passed as parameters. We discuss this problem and the main ideas of our solution by a small example.

As an example we present the implementation of a stack machine that reads an input string of commands, like `push(a)`, `push(b)`, `push(c)`, `assign(1,e)`, `top`, `pop`, `top`, and produces an output string representing a trace of the selected top elements. The execution of the command assign(1, e) changes the second element in the stack to e. The output for the above example is c, e.

Conceptually in our implementation a stack consists of two procedures named *eval* and *assign*. The procedure *eval* evaluates the remaining input string starting with the current state of the stack machine. It uses a global procedure *scan* that analyses and shortens the input string given by the global variable *in*, and gives back (in call by reference parameters) two strings and a natural number. We use the notation $p \Leftarrow \lambda(x : y : q).\alpha$ for the declaration of a procedure p with call-by-value, call-by-reference, and procedural parameters x, y, and q, respectively. $\lambda(x : y : q).\alpha$ is called an ***abstraction***. Sans serif typeface is used for strings. By ε we denote the empty string and $\cdot$ is the concatenation operator.

```
proc {global procedure declarations}
  scan ⇐ ...
  assign0 ⇐ λ(i, y ::).in := ε
  eval0 ⇐ λ(::).in := ε
  push ⇐ λ(: t : e, a). {body of push}
     proc   {local procedure declarations}
        assign ⇐ λ(i, x ::). if i ≡ 0 then t := x else a(i − 1, x ::) fi;
        eval ⇐ λ(::). var opc, arg, ind init ε, ε, 0
                 begin scan(: opc, arg, ind :);
                        case opc of
                          top: out := out · , · t; eval(::)
                          pop: e(::)
                          assign: assign(ind, arg ::); eval(::)
                          push: var top init arg
                                          begin push(: top : eval, assign) end
                          ε: skip
                        esac end
     begin eval(::) end
begin {main program}
  var top init ε begin push(: top : eval0, assign0) end
end
```

A procedure call $assign(i, x ::)$ sets the i-th element of the stack to x. The procedure *push* creates a new stack given a new top element (as a call-by-reference parameter) and two procedural parameters that represent the old stack.

For sake of readability we have used a **case**-Statement for some nested conditionals. In order to specify the behaviour of our stack machine, we use the abstract data type *Stack* with the operations *empty*, *push*, *pop*, and *top*. In addition we may define an operation *assign* of type $Stack \times String \times Nat \rightarrow Stack$ and an operation *depth* of type $Stack \rightarrow Nat$. A *trace* is the sequence of values of the top elements of the stack at those times where a top operation is encountered in the input string. The first command of a string is given by $opc(s)$, $arg(s)$, and $ind(s)$, where $ind : String \rightarrow Nat$. If an error occurs then $opc(s) = \varepsilon$. The operation $cut : String \rightarrow String$ cuts off the first command. The specification of the main program is

$$in = in0 \wedge out = \varepsilon\{\text{main program}\}out = trace(empty, in0) \ .$$

The problem in proving this assertion is straightforward. The only "visible" variables on the top level are *in* and *out*. On the level of the *push* procedure we have in addition the formal (call-by-reference) parameter t. Following [GCH89] the behaviour of *push* is described relative to the behaviour of the procedures that are passed as arguments. The basic idea of our approach is to use so-called *read procedures* that are added to the program. In our example we have to add

$$\begin{array}{l} read0 \Leftarrow \lambda(:st:).st := empty \quad \text{and} \\ read1 \Leftarrow \lambda(:st:).\mathbf{var}\ s\ \mathbf{init}\ empty\ \mathbf{begin}\ r(:s:);\ st := push(t,s)\ \mathbf{end} \end{array}$$

to the outer and inner declaration, respectively. Also, we have to add a formal procedure parameter r to *push* and pass *read0* in the call of *push* in the main program and *read1* in the recursive call. Using abstractions of read procedures in extended pre- and postconditions we are now able to specify the behaviour for example of an anonymous application of the abstraction γ_{eval} of the *eval* procedure in an environment e by

$$in \equiv in0 \wedge out \equiv out0 \wedge (e|\gamma_{read1}) \equiv st0\{e|\gamma_{eval}[::]\}out \equiv out0 \cdot trace(in0, st0) \ .$$

The assertion $(e|\gamma_{read1}) \equiv st0$ intuitively means that the read abstraction γ_{read1} when executed in e terminates and yields the value $st0$. Clearly, the modified program is equivalent to the old one since the read procedures are never called.

After having discussed some characteristic rules of our calculus, we provide a translation of formulas of our system into a variant of Dynamic Logic that achieves uninterpreted reasoning by using some fixed auxiliary data structures and additional constructs, [HRS89], [Ste89]. Based on this translation one can prove that all rules are derived rules of the general axiomatic system, thereby establishing the soundness of our calculus. In this paper we will only discuss the translation into Dynamic Logic (DL).

In the last section we will outline the completeness proof for arithmetical universes.

2 The Proof System

We start with the syntax of our programming language.

2.1 The Syntax of Programs

The basic vocabulary is given by a signature and two sets, X and P, the sets of *program variables* and *procedure identifiers*, respectively. We use x, y, z as typical elements of X and p, q, r as typical elements of P. *Commands*, (procedure) *declarations*, and *abstractions* ($\gamma \in Abs$) are given by the following rules.

$$\begin{array}{rcl} \alpha & ::= & \mathbf{skip} \mid \mathbf{abort} \mid x := \sigma \mid \alpha_0;\alpha_1 \mid \mathbf{if}\ \epsilon\ \mathbf{then}\ \alpha_0\ \mathbf{else}\ \alpha_1\ \mathbf{fi} \mid \\ & & \mathbf{var}\ x\ \mathbf{init}\ \sigma\ \mathbf{begin}\ \alpha\ \mathbf{end} \mid \mathbf{proc}\ \delta\ \mathbf{begin}\ \alpha\ \mathbf{end} \mid p(\bar{\sigma} : \bar{z} : \bar{r}) \\ \delta & ::= & p_1 \Leftarrow \gamma_1; \ldots; p_n \Leftarrow \gamma_n \qquad \gamma ::= \lambda(\bar{x} : \bar{y} : \bar{q}).\alpha \end{array}$$

The syntax of boolean expressions ϵ and algebraic expressions σ is as usual. Procedures take three (possibly empty) lists of parameters, a list of call-by-value parameters, a list of call-by-reference parameters, and a list of procedure parameters. Of course there are additional type (sort) constraints that in particular guarantee correct parameter transmissions.

2.2 Environments and Metavariables

Our proof system is derived from a variant of Dynamic Logic (DL) that incorporates certain meta-level concepts into the object language. In particular this concerns the use of *names* for program variables (metavariables) as first class objects, and the auxiliary data structure of *environments*. Here we will discuss only those aspects of this formalism that are relevant for the derived calculus to be presented here. We will also use some special notation to make things more readable.

Metavariables, we will use $\mathsf{x}, \mathsf{y}, \mathsf{z}$ as typical elements, are variables that (in a given state) denote names of program variables. They are necessary to treat local variables properly in the context of statements that can access an unbounded number of variables. To simplify our syntax we allow metavariables to occur in algebraic expressions instead of ordinary program variables. In these places metavariables can *always* be thought of as being automatically dereferenced. We use ι for a syntactic entity consisting of a metavariable or a program variable.

Apart from program variables, metavariables and variables for environments (e) there are *logical* variables (u, v, w). All symbols except program variables are *rigid* in the sense that they cannot be changed by programs. To evaluate the various kinds of expressions we use states (denoted by s) that map variables to their meaning.

Environments are *partial mappings* that map variable names to variable names and procedure names to *closures* consisting of an (inner) environment and an abstraction. Environments have a finite domain and are finitely nested. In arithmetical universes environments can thus be encoded into domain values. The set of environments Env is given by $Env = \bigcup\{Env_i \mid i \in \omega\}$, where the sets Env_i of environments of depth at most i and C_i of closures of depth at most i are defined inductively as follows. Let $Id = X \cup P$.

$$\begin{aligned} Env_0 &= \{env : Id \rightarrow X \mid Dom(env) \subset X \text{ finite } \} \\ Env_{i+1} &= \{env : Id \rightarrow X \cup C_i \mid env(X) \subset X,\ env(P) \subset C_i, \text{ and } Dom(env) \text{ finite}\} \\ C_i &= \bigcup\{(Env_j \times Abs) \mid j \leq i\} \ . \end{aligned}$$

$\emptyset$ denotes the *empty environment*, that is undefined for all arguments. Let $env =_{id} env'$ denote the fact that the environments env and env' agree on all identifiers except perhaps id. $mod(env, id, d)$ is the environment env' satisfying $env' =_{id} env$ and $env'(id) = d$. For $\delta = p_1 \Leftarrow \gamma_1; \ldots; p_n \Leftarrow \gamma_n$ the set of *approximating environments* w.r.t. δ and a given environment env, $Ap_\delta(env) = \{ap_\delta(n, env) \mid n \geq 0\}$, is inductively defined as follows

$$\begin{aligned} ap_\delta(0, env) &= \emptyset \ , \\ ap_\delta(n+1, env) &= mod(mod(env, p_1, (ap_\delta(n, env), \gamma_1)), \ldots, p_n, (ap_\delta(n, env)|\gamma_n)\ldots). \end{aligned}$$

Let Ran denote the range of a function. Then $Ran^*(env) \subseteq X \cup (Env \times Abs)$ is defined as $Ran^*(env) = Ran(env) \cup \bigcup\{Ran^*(env') | env(p) = (env'|\gamma) \text{ for some } p \text{ and } \gamma\}$.

Expressions for environments (as used in the context of our derived calculus) are built up by the following laws

$$\rho \ ::= \ \perp \mid e \mid \rho[x/\iota] \mid \rho_0[p/(\rho_1 \mid \gamma)] \ .$$

Given a state s expressions are evaluated by $[\![\cdot]\!]_s$. For name expressions we have $[\![\mathsf{x}]\!]_s = s(\mathsf{x})$, $[\![x]\!]_s = x$, and for environment expressions

$$[\![\bot]\!]_s = \emptyset \qquad [\![\rho[x/\iota]]\!]_s = mod([\![\rho]\!]_s, x, [\![\iota]\!]_s)$$
$$[\![e]\!]_s = s(e) \qquad [\![\rho_0[p/(\rho_1 \mid \gamma)]]\!]_s(p) = mod([\![\rho_0]\!]_s, p, ([\![\rho_1]\!]_s \mid \gamma)) \ .$$

In some cases we have to constrain metavariables and environment variables. In place of a metavariable x we may use $(\mathsf{x} \downarrow new(\rho_1, \ldots, \rho_n))$. A state s satisfies this constraint iff $s(\mathsf{x}) \notin \bigcup_{i=1}^{n} Ran^*([\![\rho_i]\!]_s)$. In place of an environment variable e_0 we may use $(e_0 \downarrow e_1, \ldots, e_n : Ap_\delta(\rho))$, where the sequence $e_1, \ldots, e_n$ may be empty. A state s satisfies this constraint iff $s(e_i) \in Ap_\delta([\![\rho]\!]_0)$ for $0 \leq i \leq n$ and $depth(s(e_j)) < depth(s(e_{j+1}))$ for $0 \leq j < n$. The variables $e_1, \ldots, e_n$ are used to control the induction process. In the following we will restrict ourselves to states satisfying the constraints contained in a given formula φ, that is we set $\mathcal{I} \approx_s \varphi$ iff s does not satisfy all constraints in φ or $\mathcal{I} \models_s \varphi$. In our variant of DL the constraints mentioned above can be expressed as formulas.

2.3 Extended Partial Correctness Assertions

Pre- and postconditions of *extended partial correctness assertions* are first-order formulas where apart from program variables, logical variables, metavariables, and ordinary constants, atomic terms are made up from closures $(\rho|\gamma)$, where γ is an abstraction of a read procedure. Abstractions of read procedures have to be of the form $\lambda(: z :).\alpha$, where in α the program variable z occurs only on the left hand side of assignments. Intuitively a closure in pre- and postconditions stands for the value returned by the corresponding call. A formal interpretation will be given in section 3 by a translation into a DL formula. As can be seen from this translation, these closures have to be regarded as flexible symbols. Quantification in pre- and postconditions is allowed only on logical variables.

Extended partial correctness assertions (epca's) are of the form

$$\mu ::= \theta\{\rho|\alpha\}\xi \mid \theta\{\rho|\gamma[\bar{v} : \bar{\iota} : \bar{\chi}]\}\xi \ ,$$

where the elements of $\bar{\chi}$ are syntactic closures of the form $(\rho|\gamma)$. Assertions $\theta\{\rho|\bullet\}\xi$ have to be read: 'If θ holds and the execution of $\bullet$ in the environment ρ terminates, then ξ holds afterwards.' $\rho|\gamma[\bar{v} : \bar{\iota} : \bar{\chi}]$ denotes the ***anonymous application*** of the abstraction γ to the argument lists $\bar{v}$, $\bar{\iota}$, and $\bar{\chi}$ in ρ.

Formulas of the top level axiomatic system are of the form $\Gamma \Rightarrow \Delta \rightarrow \mu$. Δ is a finite set of epca's of the form $\theta\{e|\gamma[\bar{v} : \bar{\iota} :]\}\xi$. The basic technique to prove assertions about higher-order procedures, also used in [GCH89], is to prove (by induction) an implication where the premise is made up of assumptions about the procedure parameters. In our case this is done by using assertions of the form $\theta \quad \{e|\gamma[\bar{v} : \bar{\iota} :]\} \quad \xi$. Note that for a fixed program there is only a finite number of abstractions. However, for the full language this is not sufficient. Since assignments can cause side effects on the "hidden" variables, we may have to add assertions about the global behaviour of assignments as additional assumptions to Δ. In simple cases we can do without this additional apparatus. Assumptions about assignments are of the form $\theta\ \{\bot|\gamma_{assign}[v:\iota:]\}\ \xi$, where $\gamma_{assign} = \lambda(x:y:).y := x$. This is necessary to formulate the assumptions about assignments in a sufficiently general way. Note that these abstractions differ only in the sorts of the bound variables.

Γ contains inductive assumptions and general statements that have been proved by induction. The elements of Γ are of the form $\Delta \rightarrow \mu$. The scope of program variables and ordinary logical variables is the epca they occur in. They are thought to be universally quantified. The scope of metavariables, environment variables, and logical variables marked by '$\tilde{}$' is $\Delta \rightarrow \mu$. Again, these variables are thought to be universally quantified in Γ. See 3 for a more formal treatment of hidden quantifiers.

2.4 The Top Level Calculus

Our calculus is layered in the sense that there is a top level calculus for formulas $\Gamma \Rightarrow \Delta \rightarrow \mu$ and two sub-calculi, one for proving implications between pre- and postconditions and one for proving assertions about environment expressions. For example $\rho[x] \hookrightarrow \iota$ expresses the fact that the identifier x is mapped to ι by ρ . We present only some rules of the complete calculus. In all rules, variables occurring only in the premises must be *fresh* variables.

$$\vdash \rho[\bar{y}] \hookrightarrow \bar{\iota}: \quad \frac{\Gamma \Rightarrow \Delta \rightarrow \theta[\bar{\rho}] \wedge \sigma[\bar{\iota}] \equiv \mathsf{x}\ \ \{\rho[x/(\mathsf{x}\downarrow new(\rho,\bar{\rho}))]|\alpha\}\ \ \xi[\bar{\rho}]}{\Gamma \Rightarrow \Delta \rightarrow \theta[\bar{\rho}]\ \ \{\rho|\mathbf{var}\ x\ \mathbf{init}\ \sigma[\bar{y}]\ \mathbf{begin}\ \alpha\ \mathbf{end}\}\ \ \xi[\bar{\rho}]} \quad \textbf{var declaration}$$

$$\frac{\Gamma \Rightarrow \Delta \rightarrow \theta\ \ \{(e\downarrow : Ap_\delta(\rho))|\alpha\}\ \ \xi}{\Gamma \Rightarrow \Delta \rightarrow \theta\ \ \{\rho|\mathbf{proc}\ \delta\ \mathbf{begin}\ \alpha\ \mathbf{end}\}\ \ \xi} \quad \textbf{proc declaration}$$

$$\theta \vdash \theta', \xi' \vdash \xi: \quad \frac{\Gamma \Rightarrow \Delta \rightarrow \theta'\ \ \{\rho|\alpha\}\ \ \xi'}{\Gamma \Rightarrow \Delta \rightarrow \theta\ \ \{\rho|\alpha\}\ \ \xi} \quad \text{Consequence}$$

$$\frac{\Gamma \cup \Gamma' \Rightarrow \Delta \rightarrow \theta\ \ \{\rho_0|\alpha_0\}\ \ \xi \qquad \ldots\ \Gamma \cup \Gamma'' \Rightarrow \eta_i(e,\bar{e})\ \ldots}{\Gamma \Rightarrow \Delta \rightarrow \theta\ \ \{\rho_0|\alpha_0\}\ \ \xi} \quad \text{Recursion}$$

$\vdash \rho_0[p] \hookrightarrow (\rho_1|\gamma) \wedge \rho_0[\bar{r}] \hookrightarrow \bar{\chi} \wedge \rho_0[\bar{z}] \hookrightarrow \bar{\iota}_0 \wedge \rho_0[\bar{x}_1] \hookrightarrow \bar{\iota}_1, \ldots, \rho_0[\bar{x}_n] \hookrightarrow \bar{\iota}_n$:

$$\frac{\ldots\ \Gamma \Rightarrow \Delta \rightarrow \mu_i\ \ldots \qquad \Gamma \Rightarrow \Delta \cup \Delta' \rightarrow \theta \wedge v_1 \equiv \sigma_1[\bar{\iota}_1] \wedge \ldots \wedge v_n \equiv \sigma_n[\bar{\iota}_n]\ \ \{\rho_1|\gamma[v_1,\ldots,v_n:\bar{\iota}_0:\bar{\chi}]\}\ \ \xi}{\Gamma \Rightarrow \Delta \rightarrow \theta\ \ \{\rho_0|p(\sigma_1[\bar{x}_1],\ldots\sigma_n[\bar{x}_n]:\bar{z}:\bar{r})\}\ \ \bar{\xi}} \quad \text{Calls}$$

where $\Delta' = \{\mu_1,\ldots,\mu_n\}$

We present a simplified version of the recursion rule where only a single recursive procedure is declared. Assume we have reached a situation $\Gamma \Rightarrow \Delta \rightarrow \theta\ \ \{\rho_0|\alpha_0\}\ \ \xi$, where for a declaration $\delta = p \Leftarrow \lambda(x:y:q).\alpha$ we have $\rho_0 = (e\downarrow\bar{e} : Ap_\delta(\rho_1))\ldots$. Then, for $\eta_i(*,\bullet) = \Delta \cup \Delta_i \rightarrow \theta_i\ \{(*\downarrow\bullet : Ap_\delta(\rho_1)) \mid \gamma[\bar{v}_i:\bar{\iota}_i:\bar{\chi}_i]\}\ \xi_i$ and a fresh environment variable $\tilde{e}$, we can start to prove additional general statements $\eta_i(\tilde{e},\bar{e})$. The elements of the list $\bar{\chi}_i$ are of the form $(e' \mid \gamma')$. The set Δ_i contains assumptions about $(e' \mid \gamma')$. $\Gamma' = \bigcup\{\eta_i(\tilde{e},\bar{e}) \mid 1 \leq i \leq n\}$ is the set of additional assumptions. The statements $\eta_i(\tilde{e}, \bar{e})$ have to be proved by induction (on the depth of approximations). The set of additional inductive assumptions is $\Gamma'' = \bigcup\{\eta_i(\tilde{e}, e \cdot \bar{e}) \mid 1 \leq i \leq n\}$.

Recall that metavariables, environment variables, and marked logical variables, are implicitly quantified in Γ.

The rule for procedure calls includes the separation of assumptions concerning abstractions that are passed as parameters (or mimic assignments). We thus obtain implications that can be matched with a previously proven general statement about the abstraction γ or the inductive hypothesis.

The sub-calculus for manipulating environment expressions proves, for example, statements like $(e \downarrow \bar{e} : Ap_{p \Leftarrow \gamma}(\rho))[p] \hookrightarrow (e' \downarrow e \cdot \bar{e} : Ap_{p \Leftarrow \gamma}(\rho))$, where e' is a fresh environment variable.

3 Translation into Dynamic Logic

The variant of DL we have used achieves uninterpreted reasoning in a finitary calculus by using auxiliary data structures like environments, and variables for names (of program variables) together with additional constructs like $(e|\alpha)$ and $(e|\gamma[\ldots])$. In this way meta-level proofs in infinitary systems like Salwicki's Algorithmic Logic, [Sal70], and uninterpreted DL, see for example [Gol82], can be carried out in a formal way. The main application of this kind of reasoning is the derivation of (sound) special purpose proof rules, which was a main motivation for DL, as introduced by Pratt, [Pra76]. The rules presented above are an example of such a derived calculus. The logic is implemented in the KIV system, [HRS89], that follows the paradigm of Tactical Theorem Proving to realize in practice sound extensions of the basic formalism.

The most interesting part of the translation concerns the read procedures in the epca's. Formulas $\theta[(\rho_1|\gamma_1), \ldots, (\rho_m|\gamma_m)]\{\pi\}\xi[(\rho'_1|\gamma'_1), \ldots, (\rho'_n|\gamma'_n)]$ are translated into

$$\exists u_1, \ldots, u_n.(\bigwedge_{i=1}^{m} \forall z_i.\langle \rho_i|\gamma_i[: z_i :]\rangle z_i \equiv u_i) \wedge \theta[u_1, \ldots, u_n]) \rightarrow$$
$$[\pi] \; \exists u_1, \ldots, u_n.(\bigwedge_{i=1}^{n} (\forall z_i.\langle \rho'_i|\gamma'_i[: z_i :]\rangle z_i \equiv u_i) \wedge \xi[u_1, \ldots, u_n])$$

where $u_1, \ldots, u_n$ is a pairwise distinct vector of fresh logical variables.

The sub-calculus used in the rule of consequence is a sequent calculus (like it underlies the KIV system), with sequents that in this special case are of the form $\varphi_1, \ldots, \varphi_m \vdash \psi_1, \ldots, \psi_n$ where the φ_i and ψ_i are either DL formulas $\langle\pi\rangle\varphi$ or first-order formulas. Note that in general program formulas occur on both sides of '$\vdash$', that is we prove implications between diamond formulas (total correctness assertions). A typical situation is that on the left hand side we have an assertion about a closure $(e|\gamma)$ that is bound to some procedure name in the environment of a closure occurring on the right hand side. As long as we do not use local recursive procedures in the read abstractions (which was necessary neither for the examples we did nor for the completeness proof), we can do with a rather simple set of rules. We were able to prove that this system is powerful enough to solve the deduction problems occurring if one follows the fixed 'strategy' for introducing read abstractions that is used in the completeness proof. Up to now we have no completeness result for the sub-calculus in its own.

Sequents without DL formulas are called ***verification conditions***. The sub-calculus mentioned above ends up with such sequents. In verification conditions metavariables can be replaced by fresh logical variables, since different metavariables are always assumed to stand for different programming variables.

The translation of $\Gamma \Rightarrow \Delta \to \mu$, necessary to obtain the soundness of the top level system, is mainly a matter of quantification. We only present the structure of this quantification and discuss it on semantic grounds. Let $s =_M s'$ iff s and s' differ at most on M. $\Gamma \Rightarrow \Delta \to \mu$ becomes $\dot\forall_{\mathsf{x}}.\forall_e.(\bigwedge \Gamma^* \to (\bigwedge \Delta^* \to \forall_x.\forall_u.tr(\mu)))$, where $tr(\mu)$ is the translation presented above and, for example, $\mathcal{I} \models_s \forall_x.\psi$ iff $\mathcal{I} \models_{s'} \psi$ for all s' such that $s =_X s'$. In the case of metavariables we need an additional constraint. $\mathcal{I} \models_s \dot\forall_{\mathsf{x}}.\psi$ iff $\mathcal{I} \models_{s'} \psi$ for all s' such that $s' =_{\mathsf{X}} s$, $s'(\mathsf{x}) = s'(\mathsf{y})$ implies $\mathsf{x} = \mathsf{y}$, and $s'(\mathsf{x}) \neq x$ for all x occurring in ψ. The sets Δ^* and Γ^* are defined as $\Delta^* = \{\forall_x.\forall_u.tr(\mu) \mid \mu \in \Delta\}$ and $\Gamma^* = \{\dot\forall_{\mathsf{x}}.\forall_e.\forall_{\bar u}.(\bigwedge \Delta^* \to \forall_x.\forall_u.tr(\mu)) \mid \Delta \to \mu \in \Gamma\}$.

4 Completeness

Our completeness proof uses and extends the basic ideas from [GCH89].

4.1 Semantics

The semantics of units is defined by a set of equations (standard model conditions). It is much easier to formulate these equations for so-called semantic (or partially evaluated) units where all expressions containing only rigid variables have already been evaluated. To this end, we define $[\rho|\gamma]_s = ([\![\rho]\!]_s|\gamma)$, and

$$s\,[\rho \mid \alpha]\,t \quad \text{iff} \quad s\,[[\rho]_s \mid \alpha]\,t \text{ , and}$$
$$s\,[\rho \mid \gamma[\bar v : \bar\imath : \bar\chi]]\,t \quad \text{iff} \quad s\,[[\rho]_s \mid \gamma[[\bar v]_s : [\bar\imath]_s : [\bar\chi]_s]]\,t\ .$$

In the following we will use a for elements of our domain A. Examples of equations for (semantic) units are

$[\![env \mid \alpha_0; \alpha_1]\!] = [\![env \mid \alpha_0]\!] \circ [\![env \mid \alpha_1]\!]$,

$[\![env \mid \lambda(\bar x : \bar y : \bar q).\alpha\ [\bar a : \bar z : \bar{cl}]]\!] = (\bar x'/\bar a) \circ [\![mod^*(mod^*(mod^*(env, \bar x, \bar x'), \bar y, \bar z), \bar q, \bar{cl}) \mid \alpha]\!]$,

where $(\bar x'/\bar a)$ is the assignment relation , and $\bar x'$ are new variables .

$[\![env \mid \textbf{proc}\ \delta\ \textbf{begin}\ \alpha\ \textbf{end}]\!] = \bigcup\{[\![env' \mid \alpha]\!] \mid env' \in Ap_\delta(env)\}$.

Applying the equations from left to right for each unit we get an infinitely branching *semantic* tree. One can prove that semantic trees are of finite depth. A semantic tree for a command α with free variables $x_1, \ldots, x_n$ is a tree with $([\![\bot[x_1/x_1] \ldots [x_n/x_n]]\!]|\alpha)$ at the root. For a given α there are (infinitely) many semantic trees that differ in the choice of the new variables. One can prove (by induction on the structure of semantic trees) that the semantics is nevertheless uniquely defined.

4.2 Schemes

The first observation is that all units in semantic trees for a command α that are applications of abstractions can be described by a *finite* number of so-called schemes. In the case of applications, there is one kind of scheme for 'normal' (or 'recursive') calls and one for calls of procedures that have been passed as parameters

$$\begin{aligned}
\pi^0[\bar{e}, \bar{x}] &: \ \rho[\bar{e}, \bar{x}] \| \alpha \\
\pi^1[\bar{e}, \bar{e}', \bar{\mathsf{x}}, \bar{\mathsf{x}}', \bar{v}] &: \ \rho[\bar{e}, \bar{\mathsf{x}}] \mid \gamma[\bar{v} : \bar{\mathsf{x}}' : \bar{\chi}[\bar{e}']] \\
\pi^2[e, \bar{\mathsf{x}}', \bar{v}] &: \ e \mid \gamma[\bar{v} : \bar{\mathsf{x}}' :] .
\end{aligned}$$

Theorem 1. *For all commands α there is a finite number of of schemes such that in all semantic trees for α all units are instances of some scheme.*

Of course, not every instance of a scheme (belonging to α) is a unit in a tree for α.

4.3 Read Procedures and Encodings of Environments

The read procedures used in the completeness proof are defined according to a certain strategy. They collect the values of the "hidden" variables into structured list in accordance with the structure of the environment in which the procedures are executed. Based on these read procedures, and an encoding $enc(\cdot)$ of tuples of environments and information about the "visible" program variables, we can define a simulating procedure (of mode depth zero) for procedures passed as arguments. This simulating procedure serves as a basis for showing the existence of the pre- and postconditions required for the completeness proof.

4.4 The Completeness Proof

The completeness proof relies on the fact that the behaviour of all instances of a scheme can be described in an exhaustive way by pre- and postconditions. For applications, the preconditions will be of the forms $\theta^1 = \bar{u} \equiv \bar{\chi} \wedge \bar{w} \equiv \mathsf{x} \wedge \bar{w}' \equiv \bar{\mathsf{x}}'$ and $\theta^2 = \bar{u} \equiv \bar{\chi} \wedge \bar{w} \equiv \mathsf{x}'$. For a given scheme the closures $\chi = (\rho|\gamma_{read})$ occurring in θ^1 and θ^2 can be chosen in such a way that for all instances of that scheme in the semantic tree they collect the values of all "hidden" variables. Restricting ourselves to application schemes, we get the following results.

Theorem 2. *For application schemes $\pi^1[\bar{e}, \bar{e}', \bar{\mathsf{x}}, \bar{\mathsf{x}}', \bar{v}]$, $\pi^2[e, \bar{\mathsf{x}}', \bar{v}]$ and pre- and postconditions $\theta^1(\bar{e}, \bar{e}', \bar{\mathsf{x}}, \bar{\mathsf{x}}')$, $\theta^2(e, \bar{\mathsf{x}}')$ as indicated above there are postconditions $\xi^1(\tilde{u}, \bar{e}, \bar{e}', \bar{\mathsf{x}}, \bar{\mathsf{x}}', \bar{v})$, $\xi^2(\tilde{u}, e, \bar{\mathsf{x}}', \bar{v})$ such that for each instance $\pi^1[\overline{env}, \overline{env}', \bar{x}, \bar{x}', \bar{a}]$ and $\pi^2[env, \bar{x}', \bar{a}]$ in the semantic trees for a given command we have*

1. *for all t*

$$\models_{s'} tr(\theta^1) \text{ and } s \, [\![\pi^1[\overline{env}, \overline{env}', \bar{x}, \bar{x}', \bar{a}]]\!] \, t \text{ for some } s$$
$$\textit{iff} \models_{t'} tr(\xi^1),$$

where $s(\bar{e}/\overline{env}) \circ (\bar{e}'/\overline{env}') \circ (\bar{\mathsf{x}}/\bar{x}) \circ (\bar{\mathsf{x}}'/\bar{x}')s'$
and $t(\bar{e}/\overline{env}) \circ (\bar{e}'/\overline{env}') \circ (\bar{\mathsf{x}}/\bar{x}) \circ (\bar{\mathsf{x}}'/\bar{x}') \circ (\bar{v}/\bar{a}) \circ (\tilde{u}/enc(\overline{env}, \overline{env}', \bar{x}, \bar{x}'))t'$

2. for all t

$$\models_{s'} tr(\theta^2) \text{ and } s\,[\![\pi^2[env, \bar{x}', \bar{a}]]\!]\,t \text{ for some } s \text{ iff } \models_{t'} tr(\xi^2),$$
$$\text{where } s(e/env) \circ (\bar{\mathrm{x}}'/\bar{x}')s' \text{ and } t(e/env) \circ (\bar{\mathrm{x}}'/\bar{x}') \circ (\bar{v}/\bar{a}) \circ (\tilde{u}/enc(env, \bar{x}'))t'$$

In order to construct a formal derivation using the pre- and postconditions from above we must be able to formulate appropriate implications.

Theorem 3. *The pre- and postconditions from theorem 2 can be used to formulate* valid *implications* $\bigwedge \Delta \rightarrow \theta^1\{\rho \mid \gamma[\bar{v} : \bar{\mathrm{x}} : \bar{\chi}]\}\xi^1$, *where the formulas in* Δ *are of the form* $\theta^2\{e \mid \gamma'[\bar{v}' : \bar{\mathrm{x}}' :]\}\xi^2$.

References

[Apt81] K. R. Apt. Ten years of Hoare's Logic, a Survey—Part 1. *ACM TOPLAS*, 3:431–483, 1981.

[Cla84] E.M. Clarke. The Characterization Problem for Hoare Logics. In C. A. R. Hoare and J. C. Shepherdson, editors, *Mathematical Logic and Programming Languages*, International Series in Computer Science. Prentice Hall, 1984.

[Coo78] S. A. Cook. Soundness and Completeness of an Axiom System for Program Verification. *SIAM Journal of Computing*, 7(1):70–90, 1978.

[DJ83] Werner Damm and Bernhard Josko. A Sound and Relatively* Complete Axiomatization of Clarke's Language L_4. *Acta Informatica*, 20:59–101, 1983.

[GCH89] Steven M. German, Edmund M. Clarke, and Joseph Y. Halpern. Reasoning about Procedures as Parameters in the Language L4. *Information and Computation*, pages 265–359, 1989.

[Gol82] Robert Goldblatt. *Axiomatising the Logic of Computer Programming*. Number 130 in Lecture Notes in Computer Science. Springer-Verlag, 1982.

[Hoa69] Charles Anthony Richard Hoare. An Axiomatic Basis for Computer Programming. *Communications of hte ACM*, 12:576–580, 583, 1969.

[HRS89] Maritta Heisel, Wolfgang Reif, and Werner Stephan. A Dynamic Logic for Program Verification. In *Proceedings of Logic at Botik*, number 363 in Lecture Notes in Computer Science, pages 134–145. Springer Verlag, 1989.

[Lip77] R. J. Lipton. A Necessary and Sufficient Condition for the Existance of Hoare Logics. In *Proceedings, 18th IEEE Symposium on Foundations of Computer Science*, pages 1–6, 1977.

[Old81] Ernst Rüdiger Olderog. Sound and Complete Hoare-like Calculi based on Copy Rules. *Acta Informatica*, 16:161–197, 1981.

[Old84] Ernst-Rüdiger Olderog. Correctness of Programs with Pascal-like Procedures without Global Variables. *Theoretical Computer Science*, 30:49–90, 1984.

[Pra76] V. R. Pratt. Semantical Considerations on Floyd-Hoare Logic. In *Proc. 17th IEEE Symp. on Foundations of Computer Science*, pages 109–121, October 1976.

[Sal70] A. Salwicki. Formalised Algorithmic Languages. *Bull. Acad. Pol. Sci., Ser. Sci. Math. Astron. Phy.*, pages 227–232, 1970.

[Ste89] Werner Stephan. Axiomatisierung rekursiver Prozeduren in der Dynamischen Logik. Habilitationsschrift, Universität Karlsruhe, 1989.

Variable Substitution with Iconic Combinators

David Stevens

School of Computer Science, University of Birmingham, Edgbaston, Birmingham, B15 2TT, United Kingdom

Abstract. An iconic notation for representing combinators is introduced, and the extensions of combinatory logic needed to work with it are developed. It is shown how entire sets of combinators can be defined by just two rules for constructing and interpreting iconic names. The two rules – a general abstraction rule and a general reduction rule – are the only ones required to perform variable substitution. It is suggested that techniques using the results reported here may have advantages over other combinator-based methods for implementing a functional language.

1 Introduction

In this paper, I present the theoretical background to a whole family of related algorithms for variable substitution. Although a new notation for combinators is used, nevertheless some of the established concepts and methods of combinatory logic underpin the approach. I therefore begin by outlining the traditional way of handling substitution in combinatory logic. This leads on to a statement of the issue addressed by the new approach.

1.1 Variable Substitution

Single variable substitution can be expressed as

$$[p/x]Q = ([x]Q)p \ .$$

The left side reads 'the substitution of p for all occurrences of variable x in expression Q'. The formula states that this is equal to a new expression derived from the original one, applied as a function to argument p. The new expression $[x]Q$ is 'the x-abstract of Q'. The formula also shows that substitution viewed as a process splits naturally into two distinct stages:

Stage 1 (Abstraction). *Obtain $[x]Q$. This is done by repeatedly rewriting Q until all occurrences of x have been removed. Rewriting at this stage is specified by a list – usually ordered – of variable abstraction rules; each rule eliminates one or more occurrence of x, and almost all of them introduce a combinator into the abstract.*

Stage 2 (Reduction). *Apply $[x]Q$ to p, then simplify to normal form. The simplification is done by repeatedly rewriting $([x]Q)p$ until no more simplification is possible. Rewriting at this stage is specified by a set of combinator reduction rules; the set contains a rule for each combinator mentioned in the abstraction rules of Stage 1.*

The x-abstract serves as an intermediary between the stages. It is an expression in which x does not occur. Instead it contains a collection of combinators that together provide a recipe for rebuilding Q with p in place of x.

About the simplest way to specify variable substitution is to take Curry's list of abstraction rules (abf) (Curry and Feys 1958, p. 190), together with reduction rules for the combinator set $\{\mathbf{S}, \mathbf{K}, \mathbf{I}\}$. Space does not allow its performance to be illustrated, but the resulting algorithm is commonly observed to result in many abstraction steps at Stage 1, large abstracts, and many reduction steps at Stage 2; all in all, it leads to a very longwinded substitution process.

The usual way to decrease the length of the process is to extend the specification with abstraction and reduction rules for a few – carefully selected – new combinators. Curry, for example, suggests adding rules for **B** and **C** (Curry and Feys 1958, p. 190), and in another context, for **W** (Curry 1964, p. 135). The idea is that these combinators make the process somewhat less tedious over a wide range of expressions. The technique has its drawbacks however. Each addition to the specification slightly increases its complexity; that must be weighed against the benefits offered. Another difficulty is the lack of a definite criterion for deciding when to call off the search for useful combinators.

The need to balance complexity of specification against length of substitution process poses a fundamental dilemma for the traditional approach. A solution is suggested by returning to the $(abf) + \{\mathbf{S}, \mathbf{K}, \mathbf{I}\}$ specification. This has two desirable qualities: all its combinators are logically necessary, and there is little or no redundancy among its rules. An extended specification possesses neither quality. Particular combinators are included, while others are left out, for empirical – rather than logical – reasons. Repetitious rules are inevitable because the combinators are treated as unrelated to each other, whereas a well known result implies that the new ones could all have been expressed in terms of the original ones **S**, **K** and **I**.

1.2 Variable Substitution with Iconic Combinators

The theory of iconic combinators is an approach to variable substitution that makes available all possible combinators suited to doing the job – rather than an empirically justified subset. Furthermore, the theory formalizes certain kinds of connection between combinators – rather than treating them as independent entities. The theory aims to achieve a specification with no empirically justified parts, and no redundancy; a brief substitution process; and compact abstracts.

I first show how to construct the *iconic representation of a combinator*; the central concepts of a *reference expression* and a *structural description* are also explained. I follow this by stating formal rules for the two stages of substitution: a *general abstraction rule* for eliminating variables, and a *general reduction rule* for simplifying combinatory expressions. My intention is to focus on the logical form of the notation and substitution scheme, so detailed consideration of practical applications would be out of place. However, I do give a brief outline – in the conclusion – of benefits that the results put forward might be expected to bring to a functional language implementation.

2 Iconic Representations

2.1 Reference Expressions

A set of iconic combinators is defined in relation to a reference expression of predetermined shape and size. Any shape and size would do, but to simplify things attention is limited to 'square' expressions of 1 level, 2 levels, 3 levels, and so on. We will thus be dealing with an infinite sequence of combinator sets.

The 1 level reference expression is $t_1\,t_2$, where the t_j are terms in the expression. A typical combinator defined on this expression has abstract representation $\Phi_1\Phi_2$. Conventions for the concrete representation will be given shortly. The 2 level expression is $t_1\,t_2\,(t_3\,t_4)$, and a typical representation $\Phi_1\Phi_2\Phi_3\Phi_4$; the 3 level expression is $t_1\,t_2\,(t_3\,t_4)\,(t_5\,t_6\,(t_7\,t_8))$, and a typical representation $\Phi_1\Phi_2\Phi_3\Phi_4\Phi_5\Phi_6\Phi_7\Phi_8$; and so on.

2.2 Structural Descriptions

A structural description of a target expression is a term-by-term encoding made relative to an abstraction variable and a reference expression. Terms in the target are categorized on the independent dimensions of structure (they are either atoms or combinations), and content (they either do, or do not, contain the abstraction variable). This leads to four non-overlapping two-dimensional categories. It is irrelevant however, whether a term not containing the variable is an atom or a combination, so two of the categories can be merged. The further category of non-existent terms is also needed, because there might be no term in the target to correspond with a term appearing in the reference expression.

Table 1 lays out the final result. The first column lists the categories. For each category, the column headed 'term specifiers' shows symbols(s) used to denote general term(s) in the target, and the column headed 'iconic letter' shows a letter drawn from the alphabet $\{\mathbf{o},\mathbf{k},\mathbf{i},\mathbf{s}\}$. The structural description of the target is built up from the iconic letters.

Table 1. Term specifiers and iconic letters

Category	Term specifiers	Iconic letter
Non-existent term	ϵ	**o**
Any term *not* containing abstraction variable	$E, E_1, E_2, \ldots$	**k**
Instance of abstraction variable	x	**i**
Combination containing abstraction variable	$X, X_1, X_2, \ldots$	**s**

To illustrate how the notation is used, consider a target expression that has been categorized as

$$E_1\,x\,(E_2\,x)\ .$$

Relative to the 2 level reference expression $t_1 t_2 (t_3 t_4)$, the structure of the target is described by the sequence of iconic letters $\langle \mathbf{k}, \mathbf{i}, \mathbf{k}, \mathbf{i} \rangle$, since t_1 and t_3 do not contain the abstraction variable, while t_2 and t_4 are instances of the variable. Relative to the 1 level reference expression $t_1\ t_2$, the description is $\langle \mathtt{s}, \mathtt{s} \rangle$, since both t_1 and t_2 are combinations containing the abstraction variable.

A structural description of any target expression, relative to any variable and any reference expression, can be encoded by the same method. Some restrictions on parsing the target are stated in Sect. 3.1; they include a convention for handling non-existent terms. From now on, I will refer to the structural description of an expression – encoded in the manner illustrated – as the *iconic representation of a combinator*; for notational convenience the letters in the representation will be concatenated into a string.

3 The Elimination of Variables

Table 2 shows the variable abstraction rules for 1, 2 and 3 level reference expressions. As before, x is the abstraction variable, the t_j are terms in the reference expression, and the Φ_j are letters in an iconic representation. There is just one general abstraction rule for each reference expression.

Table 2. General 1, 2 and 3 level abstraction rules

Level	Variable abstraction rule
1	$[x]\, t_1\, t_2 \Rightarrow \Phi_1 \Phi_2\, a_1\, a_2$
2	$[x]\, t_1\, t_2\, (t_3\, t_4) \Rightarrow \Phi_1 \Phi_2 \Phi_3 \Phi_4\, a_1\, a_2\, a_3\, a_4$
3	$[x]\, t_1\, t_2\, (t_3\, t_4)\, (t_5\, t_6\, (t_7\, t_8)) \Rightarrow \Phi_1 \Phi_2 \ldots \Phi_8\, a_1\, a_2\, \ldots\, a_8$

Iconic encoding clauses
If $t_j = \epsilon$ then $\Phi_j = \mathtt{o}$ and $a_j = \epsilon$, if $t_j = E$ then $\Phi_j = \mathbf{k}$ and $a_j = t_j$, if $t_j = x$ then $\Phi_j = \mathtt{i}$ and $a_j = \epsilon$, if $t_j = X$ then $\Phi_j = \mathtt{s}$ and $a_j = [x]\, t_j$.

All the rules use the same iconic encoding clauses. A separate clause applies to each category of a term t_j in the reference expression. Each clause defines how to generate one iconic letter Φ_j. The clauses also state what to do with the term t_j itself: it is either discarded (if categorized as ϵ or x), copied into the abstract (if categorized as E), or subjected to a recursive application of the rule (if categorized as X).

The general abstraction rules extend the earlier method for encoding a structural description of an expression. Each – perhaps recursive – application of a rule eliminates all occurrences of a single variable x, and so allows a complete description to be made of a parameterized version of the expression. In other words, applying one of the general abstraction rule to an expression Q yields the function $[x]Q$.

As an example, consider eliminating z from $a\ z\ (b\ z\ c)$. Using the 1 level abstraction rule, derivation of the z-abstract proceeds as follows:

$$\begin{aligned}[z]\, a\, z\, (b\, z\, c) &\Rightarrow \mathbf{ss}\, ([z]\, a\, z)([z]\, b\, z\, c) \ , \\ &\Rightarrow \mathbf{ss}\, (\mathbf{ki}\, a)\, ([z]\, b\, z\, c) \ , \\ &\Rightarrow \mathbf{ss}\, (\mathbf{ki}\, a)\, (\mathbf{sk}\, ([z]\, b\, z)\, c) \ , \\ &\Rightarrow \mathbf{ss}\, (\mathbf{ki}\, a)\, (\mathbf{sk}\, (\mathbf{ki}\, b)\, c) \ .\end{aligned}$$

Using the 2 level rule gives a different z-abstract:

$$\begin{aligned}[z]\, a\, z\, (b\, z\, c) &\Rightarrow \mathbf{kisk}\, a\, ([z]\, b\, z)\, c \ , \\ &\Rightarrow \mathbf{kisk}\, a\, (\mathbf{koio}\, b)\, c \ .\end{aligned}$$

Using the 3 level rule gives a third z-abstract in a single step:

$$[z]\, a\, z\, (b\, z\, c) \Rightarrow \mathbf{koiokiko}\, a\, b\, c \ .$$

3.1 Computational Model

A general abstraction rule provides a formal specification for variable elimination. To complete the definition, a computational model explaining how to interpret the specification must also be given. It is evident from the earlier descriptions and examples that the model is a term rewriting system which applies a general rewrite rule: any permutation of term categories in a target expression can be matched by an appropriate instantiation of the reference expression on the left side of the rule. Three principles embodying restrictions on parsing the target are now stated; these are needed to avoid ambiguities.

Principle 1. *When inserting non-existent terms into the target to make it correspond one-to-one with the reference expression, the term u must be expanded to $u\,\epsilon$, rather than to $\epsilon\, u$.*

For instance, using the 2 level abstraction rule:

$$[z]\, a\, (b\, z) = [z]\, a\, \epsilon\, (b\, z) \Rightarrow \mathbf{koki}\, a\, b \ .$$

Principle 2. *A term that does not contain the abstraction variable must be left unanalyzed.*

In $[z]\ a\ b\ (c\ d)$, the entire expression should be treated as an instance of term category E, rather than being analyzed as $E_1\ E_2$ for the 1 level abstract, or as $E_1\ E_2\ (E_3\ E_4)$ for the 2 level abstract. So, for instance, the 1 level abstract is

$$[z]\, a\, b\, (c\, d) \Rightarrow \mathbf{ko}\, (a\, b\, (c\, d)) \ .$$

Principle 3. *A combination that contains the abstraction variable must be analyzed as deeply as possible, limited only by the number of levels in the reference expression.*

Following this principle minimizes occurrences of iconic letter **s**. The principle is also a prescription to minimize the number of internal abstractions, because one is needed whenever **s** is encoded into the iconic representation. For instance, the 2 level abstract $[z]\, a\, (b\, z)$ was given above - correctly - as **koki** $a\, b$, rather than as **koso** $a\, ([z]\, b\, z)$ leading on to **koso** a (**koio** b).

The restrictions embodied in the three principles inhibit certain categorizations of a target expression. The corresponding instantiations of the reference expression will therefore never be used, even though they are implicit in the general abstraction rule.

4 Combinator Reduction

Table 3 shows the combinator reduction rules for 1, 2 and 3 level reference expressions. The Φ_j are letters in an iconic representation, the a_j are the leading arguments of the combinator, and x is its final argument. The t_j denote terms in the expression resulting from carrying out the reduction step; this expression has the same shape and size as the reference expression. As for variable abstraction, there is just one general reduction rule for each reference expression.

Table 3. General 1, 2 and 3 level reduction rules

Level	Combinator reduction rule
1	$\Phi_1\Phi_2\, a_1\, a_2\, x \rightarrow t_1\, t_2$
2	$\Phi_1\Phi_2\Phi_3\Phi_4\, a_1\, a_2\, a_3\, a_4\, x \rightarrow t_1\, t_2\, (t_3\, t_4)$
3	$\Phi_1\Phi_2 \ldots \Phi_8\, a_1\, a_2 \ldots a_8\, x \rightarrow t_1\, t_2\, (t_3\, t_4)\, (t_5\, t_6\, (t_7\, t_8))$

Iconic decoding clauses
If $\Phi_j = \mathbf{o}$ then $a_j = \epsilon$ and $t_j = \epsilon$,
if $\Phi_j = \mathbf{k}$ then $t_j = a_j$,
if $\Phi_j = \mathbf{i}$ then $a_j = \epsilon$ and $t_j = x$,
if $\Phi_j = \mathbf{s}$ then $t_j = a_j\, x$.

All the rules use the same iconic decoding clauses. A separate clause applies to each letter that can occur in the iconic representation. Each clause defines how to generate a single term t_j in the resultant expression. The clauses also state the assumption made about argument a_j: it is assumed either to be absent (if $\Phi_j = \mathbf{o}$ or $\Phi_j = \mathbf{i}$), or to be present (if $\Phi_j = \mathbf{k}$ or $\Phi_j = \mathbf{s}$). The arity of a combinator - the number of arguments needed for a reduction - can therefore be calculated from its iconic representation, by the formula

$$arity\,(\Phi_1\Phi_2..\Phi_n) = \sum_{j=1}^{n}(\text{if } \Phi_j = \mathbf{k} \text{ or } \Phi_j = \mathbf{s} \text{ then } 1 \text{ otherwise } 0) + 1 \ .$$

Curry states a fundamental requirement for the correctness of a substitution scheme:

$$[x/x]Q = ([x]Q)x \rightarrow Q \ , \qquad (1)$$

for all variables x, and all expressions Q (Curry and Feys 1958, p. 188); he proves the correctness of his $(abf) + \{\mathbf{S}, \mathbf{K}, \mathbf{I}\}$ specification by induction on the size of Q. The same proof method can be used to demonstrate that the reduction rules of Table 3 are exact inverses of the abstraction rules of Table 2, and that the iconic method therefore satisfies (1).

The example started earlier can now be continued. For the expression $az(bzc)$, we derived the 1 level z-abstract $\mathbf{ss}\,(\mathbf{ki}\,a)\,(\mathbf{sk}\,(\mathbf{ki}\,b)\,c)$. Applying the abstract to argument p, and simplifying using the general 1 level reduction rule gives:

$$\begin{aligned} \mathbf{ss}\,(\mathbf{ki}\,a)\,(\mathbf{sk}\,(\mathbf{ki}\,b)\,c)\,p &\rightarrow \mathbf{ki}\,a\,p\,(\mathbf{sk}\,(\mathbf{ki}\,b)\,c\,p), && \text{reducing } \mathbf{ss} \ , \\ &\rightarrow a\,p\,(\mathbf{sk}\,(\mathbf{ki}\,b)\,c\,p), && \text{reducing } \mathbf{ki} \ , \\ &\rightarrow a\,p\,(\mathbf{ki}\,b\,p\,c), && \text{reducing } \mathbf{sk} \ , \\ &\rightarrow a\,p\,(b\,p\,c), && \text{reducing } \mathbf{ki} \ . \end{aligned}$$

Using the 2 level z-abstract and reduction rule gives:

$$\begin{aligned} \mathbf{kisk}\,a\,(\mathbf{koio}\,b)\,c\,p &\rightarrow a\,p\,(\mathbf{koio}\,b\,p\,c), && \text{reducing } \mathbf{kisk} \ , \\ &\rightarrow a\,p\,(b\,p\,c), && \text{reducing } \mathbf{koio} \ . \end{aligned}$$

In the 3 level case only one reduction step is needed:

$$\mathbf{koiokiko}\,a\,b\,c\,p \rightarrow a\,p\,(b\,p\,c) \ .$$

The 1, 2 and 3 level z-abstracts are different combinatory expressions. When applied to an argument however, they all simplify to the same result: the substitution of the argument for all occurrences of z in the original expression. In other words, the abstracts are equivalent when treated as functions; an idea often expressed by saying they are equal in extension. It would be a disaster for the substitution scheme if this were not so.

4.1 Specific Reduction Rules

A single general reduction rule is sufficient to work with all the combinators in an iconic set. It is however possible to explicitly instantiate the general rule for a particular iconic representation. The result is a rule applying to that representation only; in other words, a specific reduction rule. Table 4 contains the complete set of specific rules covered by the general 1 level rule in Table 3. The results show that some of the combinators generated by the new scheme are the familiar ones of combinatory logic.

Table 4. Specific 1 level reduction rules

Iconic representation	Specific reduction rule	Traditional name
oo	**oo** $y \rightarrow \epsilon$	
ok	**ok** $f\ y \rightarrow f$	**K**
oi	**oi** $y \rightarrow y$	**I**
os	**os** $f\ y \rightarrow f\ y$	$\mathbf{Z}_1$
ko	**ko** $f\ y \rightarrow f$	**K**
kk	**kk** $f\ g\ y \rightarrow f\ g$	
ki	**ki** $f\ y \rightarrow f\ y$	$\mathbf{Z}_1$
ks	**ks** $f\ g\ y \rightarrow f\ (g\ y)$	**B**
io	**io** $y \rightarrow y$	**I**
ik	**ik** $f\ y \rightarrow y\ f$	**T**
ii	**ii** $y \rightarrow y\ y$	**M**
is	**is** $f\ y \rightarrow y\ (f\ y)$	**O**
so	**so** $f\ y \rightarrow f\ y$	$\mathbf{Z}_1$
sk	**sk** $f\ g\ y \rightarrow f\ y\ g$	**C**
si	**si** $f\ y \rightarrow f\ y\ y$	**W**
ss	**ss** $f\ g\ y \rightarrow f\ y\ (g\ y)$	**S**

The traditional names **K**, **I**, **B**, **C**, **W** and **S** are widely used to refer to combinators with the properties shown here. The names **T**, **M** and **O** might not be so well known. They are used, for example, by Smullyan (1985) in his ornithology of combinators, to name respectively the *Thrush*, *Mockingbird* and *Owl*. $\mathbf{Z}_1$ is used as a name for the lambda term $\lambda fy.fy$ in Church's system for representing the natural numbers (Church 1941, p. 28).

The three principles of the computational model for iconic abstraction prevent certain representations from being generated. The corresponding instantiations of the general reduction rule are consequently never used. For the 1 level reference expression, the representations prevented from occurring are **ok**, **oi**, **os**, **kk** and **so**. The appearance in Table 4 of more than one iconic representation for the same combinator is thus of no practical significance.

5 Conclusion

An infinite sequence of substitution algorithms can be defined on the square reference expression. Two rules are needed to specify the L level algorithm:

For stage 1. *The general L level abstraction rule is required to specify the elimination of variables.*

For stage 2. *The general L level reduction rule is required to specify the simplification of combinatory expressions.*

The sets of iconic encoding and decoding clauses are also needed; these remain the same at all levels.

The aims of the theory were stated to be a specification with no empirically justified parts, and no redundancy; a brief substitution process; and compact abstracts. It is clear that the first aim has been achieved, but what of the second and third? For any expression Q, and any variable x, an algorithm exists that achieves both, no matter how small the required length of process or size of abstract. For instance, at the 'level of full penetration' (Stevens 1992a), substitution is carried out in just two steps: one abstraction step, and one reduction step. At that level, the abstract contains a single 'fully penetrating' combinator.

The set of combinators made available by the L level algorithm is complete, in the sense that the associated set of iconic representations contains a full structural description of every parameterized expression with complexity up to that of the L level reference expression. The search for useful new combinators can thus be called off, or at least replaced by the much simpler task of choosing a suitable reference expression.

5.1 Extensions

There are three fairly obvious ways to extend the substitution scheme. One is to change the shape of the reference expression. Keeping shape constant, and varying size – 1 level, 2 levels, and so on – defines a new infinite sequence of combinator sets. Stevens (1992a) examines algorithms based on 'non-square' reference expressions. The change affects the efficiency with which structural descriptions – the iconic representations of combinators – are encoded.

Another possibility is to enlarge the alphabet of iconic letters. For instance, **b** and **c** could be added, to encode by a single letter the term combinations categorized as $E\ X$ and $X\ E$ respectively. This idea has not been followed up, because iconic encoding and decoding would become much more complicated, to little or no overall advantage.

The third possibility is to add one or more new abstraction rules to the specification for variable elimination. For example, Stevens (1992b) adds the 'eta abstraction' rule $[x]\ E\ x \Rightarrow E$. The new rule suppresses generation of the iconic representation **ki** – and its aliases at other levels – thus decreasing the length of abstracts. The specific reduction rule for **ki** is $\mathbf{ki}\ f\ y \rightarrow f\ y$, so the outcome is clearly a desirable one.

5.2 A Practical Application

Turner (1979b) shows how results from combinatory logic can be applied to implementing a functional language. His first substitution algorithm uses the combinator set $\{\mathbf{S}, \mathbf{K}, \mathbf{I}\}$. With a view to improving efficiency, he also reports a second and third algorithm. The second is taken directly from Curry's work, and uses the set $\{\mathbf{S}, \mathbf{K}, \mathbf{I}, \mathbf{B}, \mathbf{C}\}$. For the third, Turner (1979a) adds his own 'long-reach' combinators $\mathbf{S}'$, $\mathbf{B}'$ and $\mathbf{C}'$, to give the set $\{\mathbf{S}, \mathbf{K}, \mathbf{I}, \mathbf{B}, \mathbf{C}, \mathbf{S}', \mathbf{B}', \mathbf{C}'\}$. Following Turner's lead, others have put forward different combinator sets. For example, Joy et al. (1985) suggest $\{\mathbf{S},\mathbf{I},\mathbf{B},\mathbf{C},\mathbf{J},\mathbf{S}',\mathbf{B}',\mathbf{C}',\mathbf{J}'\}$, while Stevens (1991) reports an algorithm using $\{\mathbf{S},\mathbf{K},\mathbf{I},\mathbf{B},\mathbf{C},\mathbf{W},\mathbf{S}',\mathbf{B}',\mathbf{C}',\mathbf{W}'\}$.

All the algorithms just mentioned provide a set of combinators that is defined in advance, and for practical reasons is quite small. An iconic substitution algorithm also provides a pre-defined set, but there is no reason in this case why the set must be small. It ought to be possible to use a bigger than normal set, without filling up memory with a large collection of specific rules, or slowing down code generation (the abstraction stage) by repeated searching through a long list of abstraction rules. The bigger set of combinators should lead to more compact object code (the abstracts), and perhaps to faster or less memory-intensive program evaluation (the reduction stage). Stevens (1992b) puts these ideas to the test, examining the use of various iconic algorithms in a functional language implementation, and discussing the conclusions that can be drawn.

Acknowledgements

I would like to thank Antoni Diller for reading earlier versions of this paper, for his helpful comments, and for many interesting discussions. I am also grateful to the Science and Engineering Research Council for financial support during the period of my research.

References

Church, A.: The Calculi of Lambda Conversion. Princeton University Press (1941)

Curry, H.B.: The Elimination of Variables by Regular Combinators. In Bunge, (ed): The Critical Approach to Science and Philosophy. Collier-Macmillan (1964) 127–143

Curry, H.B., Feys, R.: Combinatory Logic vol I. Amsterdam, North-Holland (1958)

Joy, M.S., Rayward-Smith, V.J., Burton, F.W.: Efficient Combinator Code. Computer Languages **10** (1985) 211–224

Smullyan, R.: To Mock a Mockingbird. Alfred A. Knopf (1985)

Stevens, D.: Bracket Abstraction Algorithm D. CSIRP-91-4, School of Computer Science, University of Birmingham (1991)

Stevens, D.: Variable Substitution with Iconic Combinators. CSIRP-92-6, School of Computer Science, University of Birmingham (1992a)

Stevens, D.: Compiling a Functional Language with Iconic Combinators. CSIRP-92-7, School of Computer Science, University of Birmingham (1992b)

Turner, D.A.: Another Algorithm for Bracket Abstraction. The Journal of Symbolic Logic **44** (1979a) 267–270

Turner, D.A.: A New Implementation Technique for Applicative Languages. Software – Practice and Experience **9** (1979b) 31–49

Feature Constraints with First-Class Features

Ralf Treinen*

German Research Center for Artificial Intelligence (DFKI), Stuhlsatzenhausweg 3, 66123 Saarbrücken, Germany, email: treinen@dfki.uni-sb.de

Abstract. Feature Constraint Systems have been proposed as a logical data structure for constraint (logic) programming. They provide a record-like view to trees by identifying subtrees by keyword rather than by position. Their atomic constraints are finer grained than in the constructor-based approach. The recently proposed *CFT* [15] in fact generalizes the rational tree system of Prolog II.
We propose a new feature constraint system *EF* which extends *CFT* by considering features as first class values. As a consequence, *EF* contains constraints like $x[v]w$ where v is a variable ranging over features, while *CFT* restricts v to be a fixed feature symbol.
We show that the satisfiability of conjunctions of atomic *EF*-constraints is NP-complete. Satisfiability of quantifier-free *EF*-constraints is shown to be decidable, while the $\exists^*\forall^*\exists^*$ fragment of the first order theory is undecidable. .

1 Introduction

Feature constraints provide records as logical data structure for constraint (logic) programming. Their origins are the feature descriptions from computational linguistics (see [13] for references) and Aït-Kacis's ψ-terms [1] which have been employed in the logic programming language Login [2]. Smolka [13] gives a unified logical view of most earlier feature formalisms and presents an expressive feature logic.

The predicate logic view to feature constraints, which has been pioneered by [13], laid the ground for the development of the constraint systems *FT* [3, 5] and *CFT* [15]. The latter constraint system subsumes Colmerauer's classical rational tree constraint system [6], but provides for finer grained and more expressive constraints. An efficient implementation of tests for satisfiability and entailment in *CFT* has been given in [16]. In fact, satisfiability of *CFT*-constraints can be tested in at most quadratic time, and for a mildly restricted case in quasi-linear time. *CFT* is the theoretical base for the constraint system of Oz [14].

CFT's standard model consists of so-called *feature trees*, that is possibly infinite trees where the nodes are labeled with *label symbols* and the edges are labeled with *feature symbols*. The labels of the edges departing from a node,

* Supported by the Bundesminister für Forschung und Technologie (contract ITW 9105), the Esprit Basic Research Project ACCLAIM (contract EP 7195) and the Esprit Working Group CCL (contract EP 6028).

called the *features of* that node, are pairwise distinct. The atomic constraints of *CFT* are equations, *label constraints* Ax ("x has label A"), *feature constraints* $x[f]y$ ("y is a child of x via feature f") and *arity constraints* $x\{f_1, \ldots, f_n\}$ ("x has exactly the features $f_1, \ldots, f_n$"). A rational tree constraint $x \doteq K(y_1, \ldots, y_n)$ can now be expressed in *CFT* as

$$Kx \wedge x\{1, \ldots, n\} \wedge x[1]y_1 \wedge \ldots x[n]y_n .$$

Note that in *CFT* we can express the fact that x has the child y at feature f by $x[f]y$, which is inconvenient to express in the rational tree constraint system if the signature is finite and impossible if the signature is infinite. *CFT*'s atomic constraints are finer grained and hence lead to an elegant and powerful tree constraint system for logic programming. A complete axiomatization of *FT* (that is *CFT* without arity constraints) has been given in [5], while the question of complete axiomatizability of *CFT* is still open.

In this paper we are concerned with an extension of *CFT* which is desirable for logic programming and which also leads to a more basic view of feature constraints. Instead of considering a family of binary feature constraints $x[f]y$, indexed by feature symbols f, we consider features to be first-class values and introduce *one* ternary variable feature constraint $x[v]y$, where v ranges over a distinguished sort of feature symbols. The interesting point is that we now get quantification over features for free from predicate logic, which leads to a dramatic gain in expressiveness. In contrast to [15], we can now for instance express the fact that y is a direct subtree of x by $\exists v\, x[v]y$, and an arity constraint $x\{f_1, \ldots, f_n\}$ can now be seen as a mere abbreviation for

$$\forall v(\exists y\, x[v]y \leftrightarrow \bigvee_{i=1}^{n} v \doteq f_i) .$$

It turns out that in certain (intended) models the subtree relation between feature trees is expressible (see Section 7).

Feature descriptions with first class features have already been considered by Johnson [10]. In contrast to our work, Johnson was not concerned with quantifiers or with arity constraints.

After fixing the constraint system *EF* in Section 2, we will address the problem of satisfiability of positive constraints, that is of conjunctions of atomic constraints, in Section 4. Although redundant for the first order theory, we keep the arity constraint since adding arity constraints to the fragment of *quantifier-free* formulae still leads to a gain in expressiveness. For the same reason, we add a constraint $x[t]\uparrow$ standing for $\forall y(\neg x[t]y)$. We present a nondeterministic algorithm with polynomial complexity. Note that, as an easy corollary of [4], satisfiability of constraints *without* arity constraints is decidable using the algorithm of [15]. In Section 5 we show the problem to be NP-hard, which results in positive constraint satisfaction to be NP-complete.

Section 6 extends the solvability result to conjunctions of positive and negative atomic constraints, which yields decidability of the $\exists^*$ fragment of the first order theory. We finally show in Section 7 that the canonical models of *EF* have undecidable first order theories. The proofs missing in this paper can be found in [18].

2 F- and EF-Constraints

In this section we define the constraint system F and its extension EF. We assume a fixed set FEA of *feature symbols*, ranged over by f, g, and a fixed set LAB of *label symbols*[2], ranged over by A, B. The language of the constraint system F has two sorts, *feat* and *tree*, and an infinite supply of variables of each sort. We use letters x, y, z to denote *tree* variables and letters v, w for *feat* variables. The constant symbols of sort *feat* are the elements of FEA, there are no further constant or function symbols. The predicate symbols are, besides equality $\doteq$:

- A unary predicate symbol L of type *tree* for each $L \in LAB$. We use prefix notation Lx for the so-called *label constraints*.
- A ternary predicate symbol $\cdot[\cdot]\cdot$ of sort *tree* $\times$ *feat* $\times$ *tree*. We use mixfix notation $x[v]y$ for the so-called *feature constraints*.

A *constraint* μ is a possibly empty conjunction of literals, where we identify as usual a finite multiset with the conjunction of its members. A constraint ϕ is a *clause* if ϕ contains no equation. A constraint γ (resp. clause ψ) which contains no negated atom is called *positive*. We write $\tilde{\forall}\mu$ ($\tilde{\exists}\mu$) for the universal (existential) closure of μ.

Now we describe three F-structures which are candidates for being "natural" models. All three structures are in fact models of the axiom system presented in Section 3.

The structure $\mathfrak{I}$ consists of all finite and infinite feature trees (see Section 1). $\mathfrak{F}$ consists of all finitely branching (but probably infinite) feature trees. The structure $\mathfrak{R}$ consists of all finitely branching (probably infinite) feature trees which have only finitely many subtrees (these trees are called *rational*). In the three models, a constraint Ax holds if and only if the root of x is labeled with label symbol A, and $x[v]y$ holds if y is a child of x via feature v. Hence, we have the substructure relationship $\mathfrak{R} \subset \mathfrak{F} \subset \mathfrak{I}$.

A first indication of the great expressivity of F is the fact that these models are not elementarily equivalent, in contrast to the situation of constructor trees where the model of all infinite trees and the model of rational trees cannot be distinguished by a single first order logic sentence [12]. This can be seen as follows: We take $y \prec x$ as an abbreviation for $\exists v\; x[v]y$ (read: y is a child of x). Now the formula

$$fs(x) \quad := \quad \exists y\Big(x \prec y \wedge \forall z_1, z_2\,(z_2 \prec z_1 \wedge z_1 \prec y \rightarrow z_2 \prec y)\Big)$$

[2] Labels have been called *sorts* in earlier publications on feature constraint systems (e.g. [15]). We changed this name here in order to avoid confusion with the sorts of predicate logic.

expresses in some sense that x can be "flattened". The reader will easily verify that $\forall x\, fs(x)$ is valid in $\mathfrak{R}$, but neither in $\mathfrak{F}$ nor in $\mathfrak{I}$. In fact, $fs(x)$ holds in $\mathfrak{F}$ iff x is rational, and holds in $\mathfrak{I}$ iff x has at most *cardinality*(*FEA*) many subtrees.

The following formula expresses that there is a feature tree x with an infinite sequence of children which have themselves a strictly increasing number of children. This formula holds in $\mathfrak{I}$ but not in $\mathfrak{F}$.

$$\exists x(\exists y\, y \prec x \wedge \forall y_1(y_1 \prec x \rightarrow \exists y_2(y_2 \prec x \wedge \forall z(z \prec y_1 \rightarrow z \prec y_2) \wedge \\ \exists z(\neg z \prec y_1 \wedge z \prec y_2))))$$

Note that the formula $\exists x \forall y\, y \prec x$ does not hold in $\mathfrak{I}$, since the cardinality of $\mathfrak{I}$ is strictly greater than the cardinality of *FEA*.

We now extend the constraint system F to the system *EF* by adding the following predicates symbols:

- A unary predicate symbol F for each finite subset F of *FEA*. We use postfix notation xF for the so-called *arity constraints*.
- A binary predicate symbol $\cdot[\cdot]\uparrow$ of sort *tree* $\times$ *feat*.

The additional predicate symbols of *EF* have explicit definitions in F:

$$x\{f_1, \ldots, f_n\} \leftrightarrow \forall v(\exists y\, x[v]y \leftrightarrow \bigvee_{i=1}^{n} v \doteq f_i) \qquad x \neq y$$
$$x[v]\uparrow \leftrightarrow \neg\exists y\, x[v]y \qquad x \neq y$$

Hence, we will consider $\mathfrak{I}$, $\mathfrak{F}$ and $\mathfrak{R}$ to be *EF*-structures as well as F-structures.

3 An Axiomatization

In this section we give a system *EF* of axioms which describe the "intended" structures of *EF*. We begin with five straightforward axiom schemes.

$(D)\ \neg f \doteq g$	$f, g \in FEA;\ f \neq g$
$(L)\ \forall x\, \neg(Ax \wedge Bx)$	$A \neq B$
$(F)\ \forall x, y, z, v(x[v]y \wedge x[v]z \rightarrow y \doteq z)$	
$(A)\ \forall x(x\{f_1, \ldots, f_n\} \leftrightarrow \forall v(\exists y\, x[v]y \leftrightarrow \bigvee_{i=1}^{n} v \doteq f_i))$	$x \neq y$
$(U)\ \forall x, v(x[v]\uparrow \leftrightarrow \neg\exists y\, x[v]y)$	$x \neq y$

One possible model of these axioms interprets all relation symbols as the empty relation. We wish to exclude those models by requiring that certain clauses like

$$Ax \wedge x\{f, g\} \wedge x[f]x \wedge x[g]y \wedge y[v]\uparrow \wedge y[w]z \qquad (1)$$

have a solution. In order to state the axiom scheme for the satisfiability of clauses we need some more definitions. A *solved positive clause* is a positive clause ϕ which satisfies:

1. if $Ax \wedge Bx \subseteq \phi$ then $A = B$;
2. if $x[t]y \wedge x[t]z \subseteq \phi$ then $y = z$;
3. if $x[t]y \wedge x[s]\uparrow \subseteq \phi$ then $t \neq s$;
4. if $xF \wedge xG \subseteq \phi$ then $F = G$;
5. if $xF \wedge x[t]y \subseteq \phi$ then $t \in F$;
6. if $xF \wedge x[t]\uparrow \subseteq \phi$ then t is a variable.

For example (1) is a solved positive clause. A variable x is *constrained* in a clause ϕ if ϕ contains a constraint of the form Ax, xF, $x[t]y$ or $x[t]\uparrow$. We say that ϕ *constrains* x *at* t if ϕ contains $x[t]y$, $x[t]\uparrow$ or xF with $t \in F$. We use $C\phi$ to denote the set of constrained variables of ϕ. For a clause ϕ we define

$$\Delta\phi := \{\neg(v \doteq t) \mid \phi \text{ constrains some } x \text{ at } v \text{ and at } t,\ v \neq t,\ v \text{ is a variable}\}$$

Now the axiom scheme stating satisfiability of solved positive clauses reads

(Con) $\tilde{\forall}(\Delta\phi \rightarrow \exists C\phi\, \phi)$	if ϕ is a solved positive clause

Taking the solved positive clause (1), we obtain the axiom

$$\forall z, v, w\Big(\neg v \doteq w \rightarrow \exists x, y(Ax \wedge x\{f,g\} \wedge x[f]x \wedge x[g]y \wedge y[v]\uparrow \wedge y[w]z)\Big)$$

Note that $\neg v \doteq w$ is satisfiable in every model of the axioms, hence (1) is satisfiable in every model of the axioms.

Taking the clause (1) we know that in the three structures of Section 2 the solution to x is unique if y and z are fixed. This is what the last axiom scheme expresses. We write $\hat{\exists}x\Psi$ (read: "there is at most one x such that Ψ") as an abbreviation for

$$\forall y_1, y_2\Big(\Psi[x \mapsto y_1] \wedge \Psi[x \mapsto y_2] \rightarrow y_1 \doteq y_2\Big)$$

and accordingly for sets of variables. This quantifier has the important property that for all formulas Φ, Ψ

$$\tilde{\forall}\exists X(\Phi \wedge \Psi) \wedge \tilde{\forall}\hat{\exists}X\Psi \models \tilde{\forall}(\Psi \rightarrow \Phi)\,.$$

A variable x is *determined* in a clause ϕ if ϕ contains a label constraint Ax, an arity constraint xF and for each $f \in F$ a feature constraint of the form $x[f]y$. We use $\mathcal{D}\phi$ to denote the set of all determined variables in ϕ. If for instance ϕ is the clause from (1) then $\mathcal{D}\phi = \{x\}$. The axiom scheme on the uniqueness of solutions reads

(Det) $\tilde{\forall}\hat{\exists}\mathcal{D}\phi\, \phi$	if ϕ is a solved positive clause

Taking for ϕ the clause from (1) we get the following instance of (Det):

$$\forall y, z, v, w\, \hat{\exists}x(Ax \wedge x\{f,g\} \wedge x[f]x \wedge x[g]y \wedge y[v]\uparrow \wedge y[w]z)$$

Lemma 1. *The structures* $\mathfrak{I}$, $\mathfrak{F}$ *and* $\mathfrak{R}$ *are models of EF.*

4 Satisfiability of Positive *EF*-Constraints is in NP

In this section, we present a nondeterministic algorithm which decides satisfiability of positive constraints in the models of *EF*. The algorithm consists of a rewrite relation $\Rightarrow_P$ such that in all models of *EF* the constraint γ is equivalent to the disjunction of its $\Rightarrow_P$-normal forms. Every $\Rightarrow_P$-irreducible form is either $\perp$ or of the form $\delta \wedge \phi$, where δ is an idempotent substitution, ϕ is solved positive constraint and $\delta\phi = \phi$. In the following, χ ranges over variables of sort *feat* or *tree*, and t ranges over arbitrary terms (that is, variables or feature symbols). The first set of rules ensures that the equational part of an irreducible constraint (if different from $\perp$) is an idempotent substitution which is applied to the remainder.

(P1) $$\frac{t \doteq t \wedge \phi}{\phi}$$

(P2) $$\frac{\chi \doteq t \wedge \phi}{\chi \doteq t \wedge \phi[\chi \leftarrow t]} \quad \chi \in \mathcal{V}\phi, \chi \neq t$$

(P3) $$\frac{f \doteq v \wedge \phi}{v \doteq f \wedge \phi}$$

(P4) $$\frac{f \doteq g \wedge \phi}{\perp} \quad f \neq g$$

The rules of the second set coincide with the conditions of the definition of solved positive clauses. *(P5)* guarantees condition 1, *(P6)* condition 2, *(P7)* condition 3, *(P8)* condition 4, *(P9)* and *(P10)* condition 5 and *(P11)* and *(P12)* condition 6.

(P5) $$\frac{Ax \wedge Bx \wedge \phi}{\perp} \quad A \neq B$$

(P6) $$\frac{x[t]y \wedge x[t]z \wedge \phi}{y \doteq z \wedge x[t]z \wedge \phi}$$

(P7) $$\frac{x[t]y \wedge x[t]\uparrow \wedge \phi}{\perp}$$

(P8) $$\frac{xF \wedge xG \wedge \phi}{\perp} \quad F \neq G$$

(P9) $$\frac{xF \wedge x[f]y \wedge \phi}{\perp} \quad f \notin F$$

(P11) $$\frac{xF \wedge x[f]\uparrow \wedge \phi}{\perp} \quad f \in F$$

(P12) $$\frac{xF \wedge x[f]\uparrow \wedge \phi}{xF \wedge \phi} \quad f \notin F$$

(P10) $$\frac{xF \wedge x[v]y \wedge \phi}{v \doteq f \wedge xF \wedge x[f]y \wedge \phi[v \leftarrow f]} \quad f \in F$$

Note that only the rule (P10) is indeterministic by allowing for an arbitrary choice of v among the members of F. The rewriting relation $\Rightarrow_P$ defined by the above rewrite system is terminating, as the reader easily verifies. A theory T is *satisfaction complete* [9] if for every positive constraint γ either $T \models \tilde{\exists}\gamma$ or $T \models \neg\tilde{\exists}\gamma$ holds. Hence we obtain

Theorem 2. *EF is satisfaction complete. A positive constraint γ is satisfiable in EF iff there is an $\Rightarrow_P$-irreducible form of γ different from $\perp$.*

Note that the length of a rewriting sequence starting from γ is polynomial in the size of γ.

Corollary 3. *Satisfiability of positive EF-constraints is decidable in NP time.*

5 Satisfiability of Positive *EF*-Constraints is NP-hard

We employ a reduction of the Minimum Cover Problem [8] to the satisfiability problem of positive *EF*-constraints. Since the Minimum Cover Problem is known to be NP-complete and since our reduction is polynomial, this will prove satisfiability to be NP-hard.

The Minimum Cover Problem reads as follows:

Given a collection $S_1, \ldots, S_n$ of finite sets and a natural number $k \leq n$. Is there a subset $I \subseteq \{1, \ldots, n\}$ with *cardinality*$(I) \leq k$ such that

$$\bigcup_{j \in I} S_j = \bigcup_{i=1}^{n} S_i \quad ?$$

Let an instance $(S_1, \ldots, S_n; k)$ of the Minimum Cover Problem be given. We define $U := \bigcup_{i=1}^{n} S_i$ and for any $u \in U$: $\delta_u := \{j \mid i \in S_j\}$. Without loss of generality, we assume that $1, \ldots, n \in FEA$. We construct a constraint $\Psi := \Psi_1 \wedge \Psi_2 \wedge \Psi_3$ that is satisfiable if and only if the instance of the minimum cover problem has a solution. We use variables x_u $(u \in U)$ for the elements of U and variables $z_1, \ldots, z_n$ to denote the sets $S_1, \ldots, S_n$. The first formula Ψ_1 requires that z_j is a direct subtree of x_u if and only if $u \in S_j$:

$$\bigwedge_{u \in U} x_u \delta_u \quad \wedge \quad \bigwedge_{u \in U} \bigwedge_{j \in \delta_u} x_u[j] z_j \ .$$

The choice of an appropriate set I is now expressed as an assignment of labels to the variables z_i. The idea is to assign the label IN to the variable z_i if $i \in I$, and OUT otherwise. The formula Ψ_2 expresses the fact that at least $n - k$ of the z_i have the label IN. It is defined as

$$\exists x \left(x\{1, \ldots, n\} \quad \wedge \quad \bigwedge_{i=1}^{n} x[i] z_i \quad \wedge \quad \bigwedge_{i=1}^{n-k} \exists v, y (x[v] y \wedge \text{OUT}\, y \wedge y\{i\}) \right) .$$

The arity constraints for y forces for each i a different choice of y. The formula Ψ_3 expresses the fact that each x_i has an immediate subtree with label IN, which according to the definition of Ψ_1 must be one of the z_i.

$$\bigwedge_{i \in U} \exists v, z (x_i[v] z \wedge \text{IN}\, z) \ .$$

The length of the formula Ψ is in fact linear in the size of the representation of the minimum cover problem. Hence, together with Corollary 3 we obtain

Theorem 4. *Satisfiability of positive EF-constraints is NP-complete.*

6 Satisfiability of Constraints

In this section, we extend the results of Section 4 to conjunction of positive and negative atomic constraints. This extension is complicated by the fact that the independence of constraints[6] does not hold in our case, in contrast to the constraint system *CFT* of [15]. A counter example to the Independence Property[3] is

$$x\{f,g\} \wedge x[f]y \wedge x[g]z \wedge Ay \wedge Bz \wedge x[v]x' \models_{EF} Ax' \vee Bx'$$

but the left hand side does not imply any of the two disjuncts alone.

We define a rewrite system $\Rightarrow_N$ by the rules of Section 4 plus the following ones:

(N1) $\dfrac{Ax \wedge \neg Ax \wedge \phi}{\bot}$ (N2) $\dfrac{Ax \wedge \neg Bx \wedge \phi}{Ax \wedge \phi}$ $A \neq B$

(N3) $\dfrac{xF \wedge \neg xF \wedge \phi}{\bot}$ (N4) $\dfrac{xF \wedge \neg xG \wedge \phi}{xF \wedge \phi}$ $F \neq G$

(N5) $\dfrac{\neg x[t]y \wedge \phi}{x[t]\uparrow \wedge \phi}$ (N6) $\dfrac{\neg x[t]y \wedge \phi}{\exists z(x[t]z \wedge \neg z \doteq y) \wedge \phi}$ z new

(N7) $\dfrac{\neg x[t]\uparrow \wedge \phi}{\exists z\, x[t]z \wedge \phi}$ z new

---

A clause ψ is a *solved clause*, if its positive part is a solved positive clause, if ψ does not contain constraints of the form $\neg x[t]\uparrow$ or $\neg x[t]y$, and if ψ contains a negative label (resp. arity) constraint for x, then it does not contain a positive label (resp. arity) constraint for x. Note that every $\Rightarrow_N$-normal form of a constraint is either $\bot$ or of the form $\delta \wedge \phi \wedge \tau$, where δ is an idempotent substitution with $\delta(\phi \wedge \tau) = \phi \wedge \tau$ (hence, δ is not relevant for satisfiability), ϕ is a solved clause and τ is a conjunction of negated equations.

Lemma 5. *Every solved clause is satisfiable in every model of EF.*

We still need a criterion whether some solved clause together with some inequations is satisfiable. We say that a set η of equations is *complete wrt.* ψ, if $\eta \models x \doteq y$ and $x[t]x' \wedge y[t]y' \subseteq \psi$ imply that $\eta \models x' \doteq y'$. Given ψ and η, it is easy to compute a set η' of equations such that $\psi \wedge \eta \models\!\!\dashv_{EF} \psi \wedge \eta'$. ($\eta'$ can be seen as the equational part of the congruence closure of $\eta \wedge \psi$, see [15].)

Lemma 6. *Let ψ be a solved clause, and η be complete wrt. ψ. If $\psi \wedge \eta$ is satisfiable in EF and if $\mathcal{V}\eta \subseteq \mathcal{D}\psi$, then $\psi \models_{EF} \eta$.*

A clause ψ is called *saturated* if $xF \in \psi$ and $f \in F$ imply that $x[f]y \in \psi$ for some y. Every solved clause can be transformed into an equivalent saturated clause, with existentially quantifiers for the new variables.

---

[3] A constraint system is *independent* [6] if: $\phi \wedge \neg\phi_1 \wedge \ldots \wedge \neg\phi_2$ is satisfiable iff $\phi \wedge \neg\phi_i$ is satisfiable for every i This equivalent to: $\phi \models \phi_1 \vee \ldots \phi_n$ iff $\phi \models \phi_i$ for some i.

Lemma 7. *Let ψ be a solved and saturated clause and let $\eta_1, \ldots, \eta_n$ be conjunctions of equations such that for every i: $\mathcal{V}\eta_i \not\subseteq \mathcal{D}\psi$. Then.*

$$\models_{EF} \tilde{\exists}\left(\psi \wedge \neg\eta_1 \wedge \ldots \wedge \neg\eta_n\right).$$

Theorem 8. *EF is complete for Σ_1, that is for every quantifier-free formula w, either $\models_{EF} \tilde{\exists} w$ or $\models_{EF} \neg\tilde{\exists} w$.*

It is decidable whether for an quantifier-free w: $\models_{EF} \tilde{\exists} w$.

Proof. We transform a given quantifier-free formula w into disjunctive normal form and test every disjunction (i.e., constraint) μ for satisfiability as follows: We compute all $\Rightarrow_N$-normal forms of μ. (Note that $\Rightarrow_N$ is terminating.) μ is satisfiable iff one of its normal forms ν is. If $\nu = \bot$, then ν is of course not satisfiable in any model of *EF*. Otherwise we extend ν to an (modulo new variables) equivalent saturated clause ν'. If there is an inequation $\neg x \doteq y$ in ν' such that all variables of the completion of $x \doteq y$ wrt. ν' are determined in ν', then ν' is by Lemma 5 and Lemma 6 not satisfiable in any model of *EF*. Otherwise, by Lemma 7, ν' is satisfiable in every model of *EF*.

7 Undecidability of the First Order Theory

In this section we will just give the key argument why the first order theories of the $\mathfrak{I}$, $\mathfrak{F}$ and $\mathfrak{R}$ are undecidable. For complete proofs we refer to [18]. Venkataraman [19] has shown that the first order theory of constructor trees with the subterm relation is undecidable (see also [17]). Since feature constraints are in fact even more expressive than constructor tree constraints, it suffices to show that we can express the subterm relation between feature trees as a first order logic formula. To be more specific, we don't have to code the subterm relation in its full generality. It is sufficient that for each structure under consideration there is a set *Rep* of feature trees that contains at least the rational feature trees such that $s(x, y)$ holds iff $x \in Rep$ und y is a subtree of x:

$$\begin{aligned} s(x,y) := \exists z \quad & (x \prec z \wedge \forall x_1, x_2\, (x_2 \prec x_1 \wedge x_1 \prec z \rightarrow x_2 \prec z)) \wedge \\ & \forall z\, ((x \prec z \wedge \forall x_1, x_2\, (x_2 \prec x_1 \wedge x_1 \prec z \rightarrow x_2 \prec z)) \\ & \quad \rightarrow y \prec z) \end{aligned}$$

With a direct coding of the Post Correspondence Problem into the three theories along the technique given in [17] we can show (see [18]):

Theorem 9. *The $\exists^*\forall^*\exists^*$ fragment of the first order theories of the structures $\mathfrak{R}$, $\mathfrak{I}$ and $\mathfrak{F}$ are undecidable.*

I am grateful to Gert Smolka for discussions on an earlier version of this paper and to an anonymous referee for useful comments.

References

1. H. Aït-Kaci. An algebraic semantics approach to the effective resolution of type equations. *Theoretical Comput. Sci.*, 45:293–351, 1986.
2. H. Aït-Kaci and R. Nasr. LOGIN: A logic programming language with built-in inheritance. *Journal of Logic Programming*, 3:185–215, 1986.
3. H. Aït-Kaci, A. Podelski, and G. Smolka. A feature-based constraint system for logic programming with entailment. In *Int. Conf. on 5th Generation Computer Systems*, pages 1012–1021, 1992.
4. R. Backofen. On the decidability of functional uncertainty. In *Rewriting Techniques and Applications*, LNCS, 1993. Springer-Verlag.
5. R. Backofen and G. Smolka. A complete and recursive feature theory. In *Proc. of the 31 th ACL*, Columbus, Ohio, 1993. Complete version as DFKI Research Report RR-92-30.
6. A. Colmerauer. Equations and inequations on finite and infinite trees. In *2nd Int. Conf. on 5th Generation Computer Systems*, pages 85–99, 1984.
7. H. Comon. Unification et disunification. Théorie et applications, 1988. Doctoral Thesis, Institut National Polytechnique de Grenoble.
8. M. R. Garey and D. S. Johnson. *Computers and Intractability. A Guide to the Theory of NP-Completeness.* W. H. Freeman and Company, New York, 1979.
9. J. Jaffar and J.-L. Lassez. Constraint logic programming. In *14th POPL*, pages 111–119, Munich, Germany, Jan. 1987. ACM.
10. M. Johnson. *Attribute-Value Logic and the Theory of Grammar.* CSLI Lecture Notes 16. Center for the Study of Language and Information, Stanford University, CA, 1988.
11. E. Kounalies, D. Lugiez, and L. Potier. A solution of the complement problem in associative-commutative theories. In *MFCS 1991*, LNAI, vol. 520, pages 287–297, Springer-Verlag.
12. M. J. Maher. Complete axiomatizations of the algebras of finite, rational and infinite trees. In *Third LICS*, pages 348–357. 1988.
13. G. Smolka. Feature constraint logics for unification grammars. *Journal of Logic Programming*, 12:51–87, 1992.
14. G. Smolka, M. Henz, and J. Würtz. Object-oriented concurrent constraint programming in Oz. Research Report RR-93-16, Deutsches Forschungszentrum für Künstliche Intelligenz, Stuhlsatzenhausweg 3, D-W-6600 Saarbrücken, Germany, Apr. 1993.
15. G. Smolka and R. Treinen. Records for logic programming. In K. Apt, editor, *Proceedings of the Joint International Conference and Symposium on Logic Programming*, pages 240–254, 1992.
16. G. Smolka and R. Treinen. Records for logic programming. Research Report RR-92-23, Deutsches Forschungszentrum für Künstliche Intelligenz, Stuhlsatzenhausweg 3, D-W-6600 Saarbrücken, Germany, Aug. 1992.
17. R. Treinen. A new method for undecidability proofs of first order theories. *Journal of Symbolic Computation*, 14(5):437–457, Nov. 1992.
18. R. Treinen. On feature constraints with variable feature access. Research Report RR-93-21, Deutsches Forschungszentrum für Künstliche Intelligenz, Stuhlsatzenhausweg 3, D-W-6600 Saarbrücken, Germany, 1993.
19. K. N. Venkataraman. Decidability of the purely existential fragment of the theory of term algebra. *J. ACM*, 34(2):492–510, Apr. 1987.

Between Min Cut and Graph Bisection *

Dorothea Wagner[1], Frank Wagner[2]

[1] Fachbereich Mathematik, Technische Universität Berlin, Straße des 17. Juni 136, W-1000 Berlin 12, Germany, wagner@math.tu-berlin.de
[2] Institut für Informatik, Fachbereich Mathematik, Freie Universität Berlin, Arnimallee 2-6, W-1000 Berlin 33, Germany, wagner@math.fu-berlin.de

Abstract. We investigate a class of graph partitioning problems whose two extreme representatives are the well-known Min Cut and Graph Bisection problems. The former is known to be efficiently solvable by flow techniques, the latter to be $\mathcal{NP}$-complete. The results presented in this paper are

- a monotony result of the type "The more balanced the partition we look for has to be, the harder the problem".
- a complexity result clarifying the status of a large part of intermediate problems in the class.

Thus we show the existence and partly localize an "efficiency border" between the two extremes.

1 Introduction

Partitioning the vertex set of a graph such that the two parts are connected by few cut edges is of great interest since it is (or would be) the foundation for divide-and-conquer algorithms for a lot of problems. For the special case of planar graphs [Lipton and Tarjan]'s Planar Separator Theorem (which deals with the closely related problem of partitioning a graph by taking away few vertices) is a key tool for a large class of algorithms on planar graphs. For general graphs two main results are known:

Using flow methods [Ford and Fulkerson] showed how to partition a graph efficiently such that two given vertices, a source and a sink, are in different parts and the number of cut edges is minimum . For divide-and-conquer approaches this Min Cut algorithm is not very helpful since it yields only a method to cut a graph into pieces of possibly very different sizes resulting in a linear recursion depth.

A bisection of the graph into two equal-sized parts with provably few cut edges would be the most desirable thing, but the corresponding decision problem MINIMUM BISECTION was shown to be $\mathcal{NP}$-hard by [Garey, Johnson and Stockmeyer]. In order to have a logarithmical recursion depth it would suffice to guarantee that both pieces contain at least a constant fraction of the vertices. The complexity status of this weaker version of the bisection problem was open so far.

In this paper we discuss the family of all such problems defined formally as:

* This work was done while the authors were with Lehrstuhl für angewandte Mathematik insbesondere Informatik, Rheinisch-Westfälische Technische Hochschule Aachen.

Definition 1. Let f be a function from the positive integers to the positive reals. An *f-balanced-bipartition (f-BB)* of a graph is a pair V_1, V_2 of subsets of its vertex set such that

- $V_1 \cup V_2 = V$, $V_1 \cap V_2 = \emptyset$ and
- $|V_i| \geq f(|V|)$.

The *(cut-)size* $|V_1, V_2|$ of such an f-BB is the number of edges between V_1 and V_2. An f-BB is *minimum* if its cutsize is minimum. f is called the *balancing function*.

Remarks: Throughout the paper we assume $f(n) \leq \frac{n}{2}$, since else an f-BB cannot exist.
We assume that the balancing function is efficiently computable. This is no real restriction for who wants to know a balanced bipartition, if she or he can't even compute the balancing function?

The decision problem MIN f-BB is:

Instance: A graph G and a bound K.
Question: Does G have an f-BB of size at most K?

In this context MINIMUM BISECTION equals MIN $\frac{n}{2}$-BB and MIN CUT (without specified source and sink) is exactly MIN 1-BB.

In Section 2 we will present a monotony result of the form "The more balanced the partition we look for has to be, the harder the problem" .

This motivates us to present in Sections 3 and 4 the results of our search for the "efficiency border":

Even MIN αn^{ϵ}-BB is $\mathcal{NP}$-hard for arbitrarily small positive constants α and ϵ; this includes, for $\epsilon = 1$, the problem of cutting off at least a constant fraction of the vertices. MIN C-BB is efficiently solvable for every constant C and the complexity status of MIN $\log n$-BB is left as an open problem.

2 A Monotony Result

A precise formulation of the dependency between the size of the balancing function and the complexity of the corresponding problem is the following theorem.

Theorem 2 Monotony. *Let g be a balancing function with the property*

$$0 \leq g(n)\text{-}g(n\text{-}1) \leq 1.$$

Then $f(n) \leq g(n)$ for all n implies
MIN f-BB is polynomially reducible to MIN g-BB.

Remark: The technical condition $0 \leq g(n) - g(n-1) \leq 1$ guarantees that g is monotonically increasing. i.e. the larger the graph the larger the pieces we look for, and that the speed of increase is bounded. There are a lot of functions which fulfill this property e. g. constants, $\log n$, n^{ϵ} (for $\epsilon \in [0,1]$). Of course, we have to consider the admissible balancing functions, which correspond to these functions, i.e. if F is of one of these types we consider $f := \min\{\frac{n}{2}, F\}$.

Proof. Let G, K be an instance of f-BB. By adding an appropriate number δ of isolated vertices we reduce it to an instance H, K of g-BB.

$$V_H := V_G \cup V_\delta \text{ with } |V_\delta| = \delta, \quad E_H := E_G$$

δ is chosen such that (if $|V_G| =: n$ and $|V_H| =: m$)

$$m - \lceil g(m) \rceil = n - \lceil f(n) \rceil.$$

Such a δ always exists, since for $r(\ell) := \ell - \lceil g(\ell) \rceil$ we have

(i) $r(n) = n - \lceil g(n) \rceil \leq n - \lceil f(n) \rceil$
(ii) $r(2n) = 2n - \lceil g(2n) \rceil \geq n \geq n - \lceil f(n) \rceil$
(iii) $r(\ell) - r(\ell - 1) = 1 - (\lceil g(\ell) \rceil - \lceil g(\ell - 1) \rceil) \in \{0, 1\}$

So, there is an $m \in [n, \ldots, 2n]$ with $m - \lceil g(m) \rceil = r(m) = n - \lceil f(n) \rceil$. We always choose the smallest such m.

Here we need that the balancing functions are efficiently computable. Because of this the reduction is polynomial.

Now, let G have an f-BB V_G^1, V_G^2 of size at most K. Let w.l.o.g. $|V_G^1| \geq |V_G^2|$:

Case 1: $|V_G^2| \geq g(m)$.
Then $V_H^1 := V_G^1 \cup V_\delta$, $V_H^2 := V_G^2$ defines a g-BB of the same size.
Case 2: $|V_G^2| < g(m)$.
Then $V_H^1 := V_G^1 \cup V_\delta^1$, $V_H^2 := V_G^2 \cup V_\delta^2$ is a g-BB of the kind we look for if $|V_\delta^2| = \delta_2 := \lceil g(m) \rceil - |V_G^2|$, since:

a) $\delta_2 \leq \lceil g(m) \rceil - \lceil f(n) \rceil = m - n = \delta$.
b) By construction $|V_H^2| \geq g(m)$.
c) It remains to show that $|V_H^1| \geq g(m)$:
Assume otherwise, then $m - \lceil g(m) \rceil = |V_H^1| < g(m) \leq \frac{m}{2}$, so $\lceil g(m) \rceil > \frac{m}{2}$.
But this is only possible if m is odd and $\lceil g(m) \rceil = \frac{m+1}{2}$.
Now $\frac{m-1}{2} \geq \lceil g(m-1) \rceil \geq \lceil g(m) \rceil - 1 = \frac{m-1}{2}$, so $\lceil g(m-1) \rceil = \frac{m-1}{2}$.
Then $m - 1 - \lceil g(m-1) \rceil = m - \lceil g(m) \rceil$, which because of the minimality of m and δ means that $m = n$, so $\lceil f(n) \rceil = \frac{n+1}{2} > \frac{n}{2}$. But this contradicts the existence of an f-BB.

Let on the other hand (H, K) be a YES-instance of MIN g-BB. Then $V_G^i := V_H^i \cap V_G$ is an f-BB of G of the same size, since
$|V_G^i| \leq |V_H^i| \leq m - \lceil g(m) \rceil = n - \lceil f(n) \rceil \quad (i = 1, 2)$
$\Longrightarrow |V_G^i| \geq \lceil f(n) \rceil \quad (i = 1, 2)$.

Notice that this theorem is useful in two directions:

- If we show the $\mathcal{NP}$-completeness of MIN f-BB for a concrete function f the same holds for every larger function.
- If we can decide MIN g-BB efficiently, we can do so for every smaller balancing function. This part is even constructive in the sense that we can efficiently construct a g-BB of minimum size, if we can do so for the balancing function f.

3 Approaching the efficiency border from below

Finding a minimum 1-BB can be done by solving the classical Max Flow problem [Ford and Fulkerson] for all pairs of source/sink. This gives an $\mathcal{O}(n^5)$ polynomial algorithm.

A much faster solution was given by [Hao and Orlin] who essentially reduce it to one application of the preflow push algorithm by [Goldberg and Tarjan] yielding a $\mathcal{O}(nm \log n^2/m)$ algorithm.

A minimum C-BB (C constant) can be constructed by taking C of the vertices as one sink and C as one source and solving the Max Flow problem on the (slightly) smaller graphs for all possible choices of the $2C$ vertices. Since there are $\mathcal{O}(n^{2C})$ ways of choosing the source/sink-vertices this gives a $\mathcal{O}(n^{2C+3})$ polynomial algorithm. This method stops to work for MIN $\log n$-BB as it would lead to an $\mathcal{O}(n^{2\log n+3})$ algorithm which is superpolynomial.

4 ...from above

The smallest balancing function f for which we are able to show the $\mathcal{NP}$-completeness of MIN f-BB is an arbitrary small but positive power of n. An interesting point is that the reduction is from MINIMUM BISECTION, i.e. it runs in a sense in the opposite direction of Theorem 2.

Theorem 3 Upper bound. *MIN αn^ϵ-BB is $\mathcal{NP}$-complete for $\alpha, \epsilon > 0$.*

Proof. We assume that $\epsilon \leq 1$ and $\alpha \leq \frac{1}{4}$. Theorem 2 shows that the theorem also holds for larger α as long as αn^ϵ fulfills the technical condition.

Given an instance G, K of MINIMUM BISECTION we reduce it to an instance H, L of MIN f-BB such that

G, K is a YES-instance of MINIMUM BISECTION

$\Leftrightarrow H, L$ is a YES-instance of MIN f-BB.

First we handle two extreme cases: If n is odd (i.e. no bisection can exist) let $H = K_1$ and $L = 0$; if n is even and $K \geq \frac{n^2}{4}$ (i.e. such a bisection always exist since the maximum cut size is at most $\frac{n^2}{4}$) let $H = K_2$ and $L = 1$.

So let n be even and $K < \frac{n^2}{4}$: H is now constructed from G by substituting each vertex of G by a K_{n^2}, a "small" clique on n^2 vertices. Each edge of G is simulated by an edge between the corresponding cliques. To ensure the close relation between bisections of G and f-BB's of H we add as an enforcer an additional "large" clique on $\left\lfloor \left(\frac{n^3}{2\alpha}\right)^{1/\epsilon} \right\rfloor - n^3$ vertices. The small cliques are called $H_1, \ldots, H_n$ — corresponding to $v_1, \ldots, v_n$, the vertices of G — the large one H_0. H_0 is connected to each of the small cliques by n additional edges which are chosen avoiding parallel edges, such that H is also a simple graph. The bound L is chosen as $K + \frac{n^2}{2}$. Is H_0 large enough to avoid parallel edges?

$$|V_{H_0}| = \left\lfloor \left(\frac{n^3}{2\alpha}\right)^{1/\epsilon} \right\rfloor - n^3 \geq \left\lfloor (2n^3)^{1/\epsilon} \right\rfloor - n^3$$
$$\geq 2n^3 - n^3 = n^3$$

This is of course large enough. The construction of H guarantees that a bisection of G induces an $\alpha|V_H|^\epsilon$-BB of H where the smaller part contains exactly $\lceil\alpha|V_H|^\epsilon\rceil$ vertices:

A bisection of G where one half consists of the vertices $v_{j_1},\ldots,v_{j_{n/2}}$ induces a partition of V_H into, say V_H^1 and V_H^2 with

$$V_H^1=\bigcup_{j=1}^{\frac{n}{2}} V_{H_{j_l}} \text{ and } V_H^2=V_H\backslash V_H^1.$$

Since

$$\alpha|V_H|^\epsilon=\alpha\left\lfloor\left(\frac{n^3}{2\alpha}\right)^{1/\epsilon}\right\rfloor^\epsilon$$

we have

$$\frac{n^3}{2}\geq\alpha|V_H|^\epsilon\geq\alpha\left(\left(\frac{n^3}{2\alpha}\right)^{1/\epsilon}-1\right)^\epsilon$$

$$\geq\alpha\left(\frac{n^3}{2\alpha}-1\right),$$

since $(x-1)^\epsilon\geq x^\epsilon-1$, if $\epsilon\leq1$ and $x\geq1$,

$$=\frac{n^3}{2}-\alpha.$$

Thus $\lceil\alpha|V_H|^\epsilon\rceil=\frac{n^3}{2}$, since $\alpha<1$.

So a bisection of G of size at most K induces an $\alpha|V_H|^\epsilon$-BB of H of size $K+\frac{n}{2}\cdot n$ (the simulated edges of G plus the edges between the $V_{H_{j_l}}$ and V_{H_0}) which is at most L.

On the other hand assume that H, L is a YES-instance for MIN $\alpha|V_H|^\epsilon$-BB. Let V_H^1, V_H^2 be a $\alpha|V_H|^\epsilon$-BB of size at most L. Observe that none of the cliques (small or large) of H can have vertices in both V_H^1 and V_H^2, because otherwise the cut-size would be at least n^2-1 (every clique has at least n^2 vertices) which is larger than

$$L=K+\frac{n^2}{2}<\frac{n^2}{4}+\frac{n^2}{2}=\frac{3}{4}n^2.$$

Thus V_H^1, V_H^2 induces canonically a bipartition V_G^1, V_G^2 of G. We will modify V_H^1, V_H^2 without increasing its size such that the induced bipartition is a bisection:

If $|V_G^1|=|V_G^2|$ we are done, so let w.l.o.g. $|V_G^2|>|V_G^1|$. It follows that $V_{H_0}\subseteq V_H^1$, since else $|V_H^1|<\frac{n}{2}\cdot n^2=\lceil\alpha|V_H|^\epsilon\rceil$ in contradiction to our assumption.

Define M as $\{i\in\{1,\ldots,n\}\mid V_{H_i}\subseteq V_H^2\}$, let M_2 be an arbitrary subset of M of order $n/2$ and let M_1 be $M\setminus M_2$. Now we move the small cliques with indices from M_1 from V_H^2 to V_H^1 thus producing a new bipartition W_H^1, W_H^2 which obviously induces a bisection of G. What about the cut size?

Let $C := \bigcup_{i \in M_1} V_{H_i}$ be the set of moved vertices, then

$$|W_H^1, W_H^2| - |V_H^1, V_H^2| = |C, W_H^2| - |V_H^1, C| =: \alpha.$$

Now

$$|C, W_H^2| \leq \frac{|C|}{n^2} \cdot \frac{|W_H^2|}{n^2} = |M_1| \cdot \frac{n}{2},$$

since the only edges between C and W_H^2 (both consisting only of small cliques) are the simulated edges of G.

In addition we have $|V_H^1, C| \geq |V_{H_0}, C| = |M_1| \cdot \frac{n}{2}$ and thus $\alpha \leq 0$, which means that the cutsize was not increased by our modification.

For the size of the induced bisection W_G^1, W_G^2 of G we have

$$\begin{aligned} |W_G^1, W_G^2| &= |W_H^1 \setminus V_{H_0}, W_H^2| \\ &= |W_H^1, W_H^2| - |V_{H_0}, W_H^2| \\ &\leq L - \frac{n^2}{2} = K. \end{aligned}$$

Thus we have shown the claimed existence of a bisection of G of the right size. What about the efficiency of the reduction. The only critical point is the size of the large clique, which is $\mathcal{O}(n^{3/\epsilon})$ which is a polynomial in the input size as long as ϵ is constant.

If we try to show the $\mathcal{NP}$-completeness for smaller balancing functions along the same lines as in the proof of Theorem 3, we fail since an analogous reduction would no longer be polynomial: adding a large clique for the case of MIN $\log n$-BB would yield a graph H of size $e^{n^3/2}$.

5 Conclusion

Using similar methods as [Bui, Chaudhuri, Leighton and Sipser] we can extend the hardness results to three-regular graphs.

The concluding figure is a graphical summary of the results in this paper.

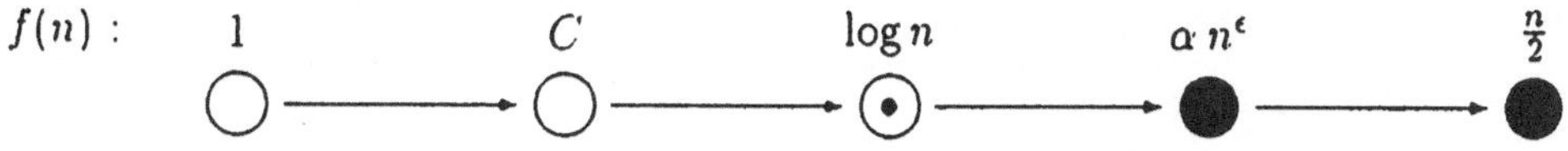

where ○ stands for $\in \mathcal{P}$,

● for $\mathcal{NP}$ – *complete*

and ⊙ for open.

References

[Bui, Chaudhuri, Leighton and Sipser] T. N. Bui, S. Chaudhuri, F. T. Leighton, M. Sipser: *Graph Bisection Algorithms With Good Average Case Behaviour*, Combinatorica **7** (1987) 171–191

[Ford and Fulkerson] L. R. Ford, D. R. Fulkerson: *Maximal Flow Through a Network*, Canadian J. Math. **8** (1956) 399–404

[Garey, Johnson and Stockmeyer] M. J. Garey, D. S. Johnson, L. Stockmeyer: *Some Simplified $\mathcal{NP}$-complete Graph Problems*, Theoretical Computer Science **1** (1976) 237–267

[Goldberg and Tarjan] A. V. Goldberg, R. E. Tarjan: *A New Approach to the Maximum Flow Problem*, Journal of the ACM **35** (1988) 921–940

[Hao and Orlin] J. Hao, J. B. Orlin: *A Faster Algorithm for Finding the Minimum Cut in a Graph*, Proceedings of the third ACM-SIAM Symposium on Discrete Algorithms (SODA'91) (1991) 165–174

[Lipton and Tarjan] R. J. Lipton, R. E. Tarjan: *A Planar Separator Theorem*, SIAM J. on Applied Math. **36** (1979) 177–189

Paths and Cycles in Finite Periodic Graphs

Egon Wanke

Institut für methodische Grundlagen (I1.Leng), Gesellschaft für Mathematik und Datenverarbeitung, W-5205 Sankt Augustin 1, Germany

Abstract. We consider *finite periodic graphs* G^m defined by nonnegative integer vectors m and directed graphs G whose edges are labeled with *integer vector-weights*. G^m has a vertex (u, x) for each vertex u of G and each nonnegative integer vector x less than or equal to m. G^m has an edge from (u, x) to $(v, x + z)$ if and only if G has an edge from u to v with vector weight z.
We analyze the complexity and present algorithms for finding paths and cycles in finite periodic graphs. The present paper shows that path and cycle problems on finite periodic graphs are PSPACE-complete under various restrictions, but solvable in polynomial time if the vector weights of the edges are bounded.

1 Introduction and summary

A d-dimensional *periodic graph* G^∞ is an infinite graph finitely described by a so-called *static graph* G. A static graph is a directed graph with d-dimensional integer vector-weights associated with the edges. The periodic graph G^∞ is obtained by repeating the static graph G in a d-dimensional orthogonal grid. That is, a copy of the vertex set of G is placed at each point in the integral lattice $\mathbb{Z}^d$. Then for each edge of G from u to v with vector-weight y, the copy of u at each lattice point z is connected with the copy of v at lattice point $z + y$. The connection-pattern between the vertices is the same everywhere in the grid. For an integer vector m, the *finite periodic graph* G^m is the subgraph of the infinite periodic graph G^∞ induced by the vertices associated with nonnegative lattice points less than or equal to m.

The static/periodic graph model has been considered by several authors with respect to various applications. For example, Karp, Miller, and Winograd [8] analyze systems of *uniform recurrence equations* modeled by infinite periodic graphs. Rao [12] considers *regular iterative algorithms* and maps certain finite parts of periodic graphs onto finite processor arrays for possible systolic computations. Iwano and Steiglitz [7], Kosaraju and Sullivan [10], and Cohen and Megiddo [2] show that cycles in infinite periodic graphs can be found in polynomial time. Orlin provides in [11] some polynomial time algorithms for graph problems on one-dimensional infinite periodic graphs. Kodialam and Orlin develop in [9] an efficient algorithm to determine the strongly connected components of an infinite periodic graph. Cohen and Megiddo [3] develop a linear time algorithm for testing planarity of infinite periodic graphs. Backes, Schwiegelshohn, and Thiele [1] analyze the structure of longest paths in infinite periodic graphs represented by strongly connected static graphs. Höfting and Wanke [5, 6] present a polynomial time solution for the minimum cost path problem in infinite periodic graphs with bounded dimensions.

A close look to the motivation of the research cited below manifests that solving problems on infinite periodic graphs is just a simple way to escape the problem for finite periodic graphs. For example, when modeling dependencies of variables in recurrence equations, the equations usually are defined for a certain (finite) domain. Finding a cycle in an infinite periodic graph does not imply that some recurrence equations are not well-defined for their (finite) domain. Also, in VLSI-design, regular grid structured circuits are always finite. The existence of infinitely many distinct paths with decreasing costs between two specified vertices in an infinite periodic graph does not imply the existence of a negative temporal correlation between the corresponding two points in some finite circuit. In infinite graphs a negative circulation is not always a negative cycle; see [5, 6].

We analyze the complexity of path and cycle problems in finite periodic graphs. Each finite periodic graph G^m is given in a succinct description. G^m has $n \cdot \prod_{i=1}^{d} m_i + 1$ vertices when G has n vertices and $m = (m_1, \ldots, m_d)$. Path problems in finite periodic graphs G^m defined by recurrence equations are strongly related to scheduling problems for computations. Whereby, the existence of a cycle in G^m proofs that the equation is not well-defined for the domain specified by m.

Problem	Restriction				Complexity
	$\|V\|$	vector weights	m	d	
Path and Cycle	–	–	–	–	in PSPACE
	1	$\{-1,0,1\}^d$	$\{0,1\}^d$	–	PSPACE-hard
	–	–	–	2	PSPACE-hard
	1	–	–	4	PSPACE-hard
	1	–	–	1	NP-hard
Cycle	–	$\{-1,0,1\}^d$	–	$O(1)$	in P
Path	–	$\{-1,0,1\}$	–	1	in P

The full paper shows that paths and cycle problems in finite periodic graphs are PSPACE-complete even if (1) the static graphs have one vertex and the vector-weights of the edges are $0,1$-vectors, (2) the vector-weights of the edges are 2-dimensional, or (3) the static graphs have one vertex and the vector-weights are 4-dimensional. The path and cycle problem are NP-hard even if the static graphs have one vertex and the vector-weights of the edges are 1-dimensional. The cycle problem is solvable in polynomial time if the dimension of the vector-weights are bounded by a constant and the integers in the vector-weights are represented unary. The path problem is solvable in polynomial time if the edges of the static graphs are labeled by unary represented integers.

The table above summarizes our results. Due to space restrictions, we concentrate our attention to the PSPACE completeness of the unrestricted path and cycle problem.

2 The problem

Static graphs are systems $G = (V, E)$, where V is a finite set of *vertices*, $E \subseteq V \times \mathbb{Z}^d \times V$ is a finite set of labeled *edges*, where (u, x, v) is an edge from u to v associated with a d-dimensional *transit vector* x of integers. We allow *multiple edges*, i.e. several edges with the same source and target vertex.

The d-dimensional 0-vector and 1-vector is denoted by $\mathbf{0}^d$ and $\mathbf{1}^d$, respectively. We simply write $\mathbf{0}$ and $\mathbf{1}$ if the dimension is clear from the context. For a d-dimensional nonnegative integer vector $m \in \mathbb{Z}^d$, let $[m] = \{m' \in \mathbb{Z}^d | \mathbf{0} \leq m' \leq m\}$ be the set of all nonnegative d-dimensional integer vectors less than or equal to m.

The *finite periodic graph* G^m associated with a d-dimensional static graph $G = (V, E)$ and a d-dimensional nonnegative integer vector m is defined by the vertex set $V \times [m]$ and the set of all edges from (u, z) to $(v, z + x)$, where (u, x, v) is from E.

A *path* p of *length* n from a vertex u to a vertex v is an alternating sequence $(u_1, e_1, u_2, \ldots, u_n, e_n, u_{n+1})$ of vertices and edges such that $u_1 = u$, $u_{n+1} = v$, and e_i is an edge from u_i to u_{i+1} for $i = 1, \ldots, n$. A path of length ≥ 1 from a vertex u to itself is called a *cycle*. A path or cycle is called *simple* if all vertices in the path or cycle are pairwise distinct. We define $tran(e) = x$ if e is an edge with transit vector x, and $tran(p) = \sum_{i=1}^{n} tran(e_i)$, if p is a path in some static graph.

We consider the following problem definitions:

Name: BOUNDED-PATH

Instance: A d-dimensional static graph G, two vertices u, v of G, a nonnegative *source vector* $s \in \mathbb{Z}^d$, a nonnegative *target vector* $t \in \mathbb{Z}^d$, and a nonnegative *bound vector* $m \in \mathbb{Z}^d$.

Question: Is there a path from (u, s) to (v, t) in G^m?

Name: BOUNDED-CYCLE

Instance: A d-dim. static graph G and a nonnegative *bound vector* $m \in \mathbb{Z}^d$.

Question: Is there a cycle in G^m?

Note that each path from (u, s) to (v, t) in some finite periodic graph G^m corresponds to some path $p = (u_1, e_1, u_2, \ldots, u_n, e_n, u_{n+1})$ from $u_1 = u$ to $u_{n+1} = v$ in the static graph G, and vice versa, such that $s + tran(p) = t$ and $\mathbf{0} \leq s + \sum_{i=1}^{k} tran(e_i) \leq m$ for $k = 1, \ldots, n$. Each cycle in G^m corresponds to some cycle $p = (u_1, e_1, u_2, \ldots, u_n, e_n, u_{n+1})$ in G, and vice versa, such that $tran(p) = \mathbf{0}$ and $\mathbf{0} \leq \sum_{i=1}^{k} tran(e_i) \leq m$ for $k = 1, \ldots, n$.

3 The solution

The following simple nondeterministic algorithm proofs that BOUNDED-PATH and BOUNDED-CYCLE belong to the complexity class PSPACE: "Guess edge by edge a path p from (u, s) to (v, t) in G^m." This can be done in polynomial space, because for each guessed sub-path p' only its last vertex (u', m') need to be stored. Since PSPACE is closed under nondeterminism BOUNDED-PATH and BOUNDED-CYCLE belong to PSPACE.

The proof of the PSPACE hardness of BOUNDED-PATH is done by a reduction from QUANTIFIED BOOLEAN FORMULAS (QBF). Quantified Boolean formulas are build from *Boolean variables* $X = \{x_1, x_2, \dots, x_n\}$, the operators "$\vee$", "$\wedge$", "$\neg$", "parentheses", and the quantifications "$\exists x$" and "$\forall x$" for variables x over X. Let E be a Boolean expression over X build with the operators "$\vee$", "$\wedge$", "$\neg$", and "parentheses". The problem whether a quantified Boolean formula $Q_1 x_1 : Q_2 x_2 : \cdots Q_n x_n : E$ is true is PSPACE-complete; see for example [4].

Without loss of generality, we assume that

1. n is even,
2. $Q_1, Q_3, \dots, Q_{n-1}$ are existential quantifications,
3. $Q_2, Q_4, \dots, Q_n$ are universal quantifications, and
4. E is in 3-*conjunctive normal form*, i.e,
   (a) E is of the form $E_1 \wedge E_2 \wedge \cdots \wedge E_m$ (E_i is called a *clause* over X) and
   (b) each E_i is of the form $y_{i1} \vee y_{i2} \vee y_{i3}$. ($y_{ij}$ is called a *literal* over X), and
   (c) each y_{ij} is either x or $\neg x$, where x is a variable over X.

QBF with the restrictions above is called QUANTIFIED 3-SATISFIABILITY (Q3SAT). Since the quantifications for the variables are fixed, the instance for Q3SAT merely consists of the pair (X, E).

Theorem 1. *BOUNDED PATH is PSPACE-hard, even if* $m = 1$.

Proof. Let (X, E) be an instance of Q3SAT, where $X = \{x_1, x_2, \dots, x_n\}$ and $E = \{E_1 \wedge E_2 \wedge \cdots \wedge E_m\}$.

Let G' denote the static graph which we will design from the instance (X, E). The dimension $d = n$ of G' is the number of variables in X. Each vector $z \in \{0, 1\}^n$ will be considered as an assignment $A(z)$ of the variables in X. The assignments $x_i = 0$ and $x_i = 1$ are represented by 0 and 1, respectively, at coordinate i in z. For example, $A((0, 1, 0, 1))$ yields the assignment $x_1 = 0, x_2 = 1, x_3 = 0, x_4 = 1$.

For any j, $1 \leq j \leq n$, let $\alpha(n, j)$ and $\alpha(n, -j)$ denote the vector $(t_1, t_2, \dots, t_n)$, where $t_j = 1$ or $t_j = -1$, respectively, and $t_i = 0$ for $i \neq j$. For example, $\alpha(4, 3) = (0, 0, 1, 0)$ and $\alpha(5, -2) = (0, -1, 0, 0, 0)$.

We define G' hierarchically. That is, we abbreviate certain subgraphs of G' by labeled edges. The subgraphs may be build from further subgraphs which are again declared by labeled edges. Figure 1 shows an example for $n = 4$. An edge from a vertex u to a vertex v labeled with

"if $x_i = 1$" defines a vertex w, an edge from u to w with transit vector $\alpha(n, -i)$, and an edge from w to v with transit vector $\alpha(n, i)$.

"if $x_i = 0$" defines a vertex w, an edge from u to w with transit vector $\alpha(n, i)$, and an edge from w to v with transit vector $\alpha(n, -i)$.

“set $x_i = 1$” defines a vertex w, an edge from u to w with transit vector $\mathbf{0}$, an edge from u to w with transit vector $\alpha(n,i)$, and an edge from w to v labeled with “if $x_i = 1$”.

“set $x_i = 0$” defines a vertex w, an edge from u to w with transit vector $\mathbf{0}$, an edge from u to w with transit vector $\alpha(n,-i)$, and an edge from w to v labeled with “if $x_i - 0$”.

“(A) then (B)” where A and B are arbitrary edge labels, defines a vertex w and an edge from u to w labeled with A and an edge from w to v labeled with B.

“(A) or (B)” where A and B are arbitrary edge labels, defines an edge from u to v labeled with A and an edge from u to v labeled with B.

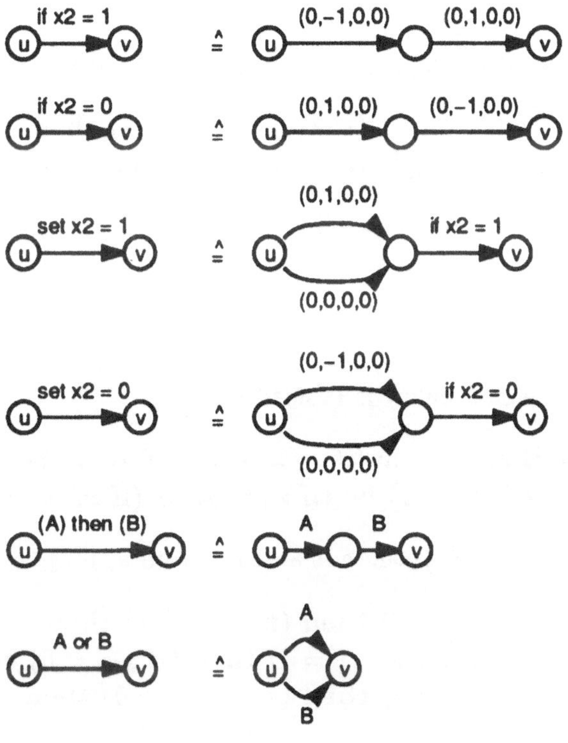

Fig. 1. Some labeled edges and the subgraphs which they define for $n = 4$.

The following facts are immediate from the subgraph definitions. Let $G(A)$ be the static graph defined by an edge from u to v labeled with A. Let $s = (s_1, \ldots, s_n)$ and $t = (t_1, \ldots, t_n)$, then there is a path from (u,s) to (v,t) in

$G(\textbf{if } x_i = 1)^{\mathbf{1}}$ if and only if $s_i = 1$ and $t = s$,
$G(\textbf{if } x_i = 0)^{\mathbf{1}}$ if and only if $s_i = 0$ and $t = s$,
$G(\textbf{set } x_i = 1)^{\mathbf{1}}$ if and only if $t_i = 1$ and $t_j = s_j$ for all $j \neq i$,
$G(\textbf{set } x_i = 0)^{\mathbf{1}}$ if and only if $t_i = 0$ and $t_j = s_j$ for all $j \neq i$,

$G((A)$ **then** $(B))^{\mathbf{1}}$ if and only if there is a path from (u,s) to (w,r) in $G(A)^{\mathbf{1}}$ and a path from (w,r) to (v,t) in $G(B)^{\mathbf{1}}$ for some $r \in \{0,1\}^d$,
$G((A)$ **or** $(B))^{\mathbf{1}}$ if and only if there is a path from (u,s) to (v,t) in $G(A)^{\mathbf{1}}$ or a path from (u,s) to (v,t) in $G(B)^{\mathbf{1}}$.

The static graph G' is defined by $n/2+3$ vertices $u_1, u_2, \ldots, u_{n/2+3}$ and the subgraphs defined by the following $n+2$ labeled edges (for an example of G' see figure 2.):

1. For each $i \in \{1, \ldots, n/2\}$, there is an edge from u_i to u_{i+1} labeled with

$$(\textbf{set } x_{2 \cdot i-1} = 1) \textbf{ or } (\textbf{set } x_{2 \cdot i-1} = 0).$$

These edges control the assignment with respect to the existential quantifications.

2. There is an edge from $u_{n/2+1}$ to $u_{n/2+2}$ labeled with

$$((\textbf{if } Y_{11}) \textbf{ or } (\textbf{if } Y_{12}) \textbf{ or } (\textbf{if } Y_{13})) \textbf{ then } \cdots$$
$$\cdots \textbf{ then } ((\textbf{if } Y_{m1}) \textbf{ or } (\textbf{if } Y_{m2}) \textbf{ or } (\textbf{if } Y_{m3})),$$

where Y_{ij} is "$x_l = 1$" or "$x_l = 0$" if $y_{ij} = x_l$ or $y_{ij} = \neg x_l$, respectively. This edge verifies the chosen assignment of the variables. For example, if

$$E = (x_1 \vee \neg x_2 \vee x_3) \wedge (x_2 \vee \neg x_3 \vee x_4)$$

then the edge defined below is labeled with

$$((\textbf{if } x_1 = 1) \textbf{ or } (\textbf{if } x_2 = 0) \textbf{ or } (\textbf{if } x_3 = 1)) \textbf{ then}$$
$$((\textbf{if } x_2 = 1) \textbf{ or } (\textbf{if } x_3 = 0) \textbf{ or } (\textbf{if } x_4 = 1)).$$

3. For each $i \in \{1, \ldots, n/2\}$, there is an edge from $u_{n/2+2}$ to u_{i+1} labeled with

$$(\textbf{if } x_n = 1) \textbf{ then } (\textbf{if } x_{n-2} = 1) \textbf{ then } \cdots$$
$$\cdots \textbf{ then } (\textbf{if } x_{2 \cdot i+2} = 1) \textbf{ then } (\textbf{if } x_{2 \cdot i} = 0) \textbf{ then}$$
$$(\textbf{set } x_n = 0) \textbf{ then } (\textbf{set } x_{n-2} = 0) \textbf{ then } \cdots$$
$$\cdots \textbf{ then } (\textbf{set } x_{2 \cdot i+2} = 0) \textbf{ then } (\textbf{set } x_{2 \cdot i} = 1).$$

These edges control the assignment of the variables with respect to the universal quantifications.

4. There is an edge from $u_{n/2+2}$ to $u_{n/2+3}$ labeled with

$$(\textbf{if } x_n = 1) \textbf{ then } (\textbf{if } x_{n-2} = 1) \textbf{ then } \cdots$$
$$\cdots \textbf{ then } (\textbf{if } x_4 = 1) \textbf{ then } (\textbf{if } x_2\textbf{=1}) \textbf{ then}$$
$$(\textbf{set } x_n = 0) \textbf{ then } (\textbf{set } x_{n-1} = 0) \textbf{ then } \cdots$$
$$\cdots \textbf{ then } (\textbf{set } x_2 = 0) \textbf{ then } (\textbf{set } x_1 = 0).$$

This edge verifies whether the choices of the universal quantifications are complete.

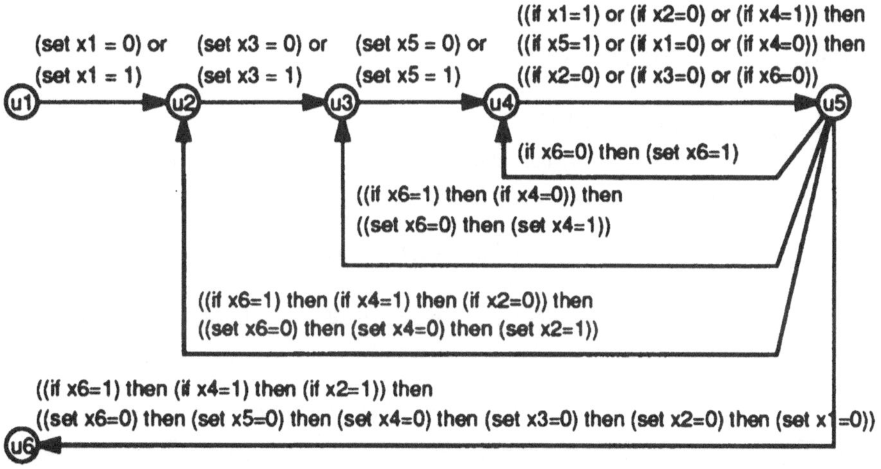

Fig. 2. The static graph G' for $X = \{x_1, x_2, x_3, x_4, x_5, x_6\}$ and $E = \{\{x_1, \neg x_2, x_4\}, \{x_5, \neg x_1, \neg x_4\}, \{\neg x_2, \neg x_3, \neg x_6)\}\}$.

Let p be any path from $(u_1, \mathbf{0})$ to $(u_{n/2+3}, \mathbf{0})$ in $G'^{\mathbf{1}}$. The coordinates $2, 4, \cdots, n$ in the transit vectors associated with the vertices along the path p are counted up in a binary fashion (forced by the edges of type 3), whereby only the coordinates associated with succeeding existential quantifications can arbitrarily be changed (edges of type 1). The last edge (of type 4) additionally forces all coordinates to zero. Since each defined assignment is verified to satisfy the expression E (forced by the edge of type 2), the Boolean formula

$$\exists x_1 : \forall x_2 : \ldots \forall x_{n-1} : \exists x_n : E$$

has to be true. Conversely, if the Boolean formula below is true there exists such a path p in $G'^{\mathbf{1}}$. □

Theorem 1 shows also that BOUNDED-PATH is PSPACE complete if the source and target vector are zero. Obviously, BOUNDED-PATH is PSPACE-complete for any fixed source and target vector. Thus it remains PSPACE-complete if only the static graph G is given as instance,

Theorem 2. *BOUNDED-CYCLE is PSPACE-hard, even if the bound vector* $m = \mathbf{1}$.

Proof. Use the same construction as in the proof of theorem 1 and add to G' a new edge e' from v to u with transit vector $\mathbf{0}$. Let G'' be the static graph G' with the additional edge e'. If there is a path from $(u, \mathbf{0})$ to $(v, \mathbf{0})$ in $G'^{\mathbf{1}}$ then there is a cycle in $G''^{\mathbf{1}}$. Conversely, if there is a cycle in $G''^{\mathbf{1}}$, then the cycle has to pass an edge defined by e', because the coordinates defined by the universal quantifications are counted up in a binary fashion along the cycle in $G'^{\mathbf{1}}$. If the cycle passes an edge defined by e', then it contains a path from $(u, \mathbf{0})$ to $(v, \mathbf{0})$, which is also contained in $G'^{\mathbf{1}}$. □

BOUNDED-PATH and BOUNDED-CYCLE are also PSPACE-hard for static graphs with exactly one vertex, even if $m = \mathbf{1}^d$. The proof is shown in the full paper. Under such restrictions the structure of the static graphs are unimportant for the complexity of the problems. In fact, the BOUNDED-PATH and BOUNDED-CYCLE problem for static graphs with exactly one vertex seems to be a pure combinatorial problem which can also be formulated as follows:

Name: BOUNDED INTEGER VECTOR SUM

Instance: A nonnegative integer d, some integer vectors $x_1, \dots, x_n \in \mathbb{Z}^d$ and three nonnegative integer vectors $s, t, m \in \mathbb{Z}^d$.

Question: Is there a nonempty sequence of integers $j_1, \dots, j_k \in \{1, \dots, n\}$ such that $s + \sum_{i=1}^{k} x_{j_i} = t$ and for each $k' = 1, \dots, k : \mathbf{0} \le m_1 + \sum_{i=1}^{k'} x_{j_i} \le m$?

Corollary 3. *BOUNDED INTEGER VECTOR SUM is PSPACE-complete for any fixed* $s, t \in \{0, 1\}^d$ *and* $m = \mathbf{1}^d$.

We have also shown that BOUNDED-PATH is log-space hard for PSPACE, even if the static graphs are 2-dimensional, and that BOUNDED-PATH and BOUNDED-CYCLE for d-dimensional static graphs are log-space transformable to BOUNDED-PATH for $d + 2$-dimensional static graphs with exactly one vertex. That is, we have also proved the PSPACE-completeness of the following restricted version of BOUNDED INTEGER VECTOR SUM.

Corollary 4. *BOUNDED INTEGER VECTOR SUM is log-space complete for PSPACE even if* $d = 4$.

It is left open whether BOUNDED-PATH or BOUNDED-CYCLE are PSPACE-hard for 3-dimensional integer vectors and static graphs with one vertex. It is also left open whether BOUNDED-PATH or BOUNDED-CYCLE are PSPACE-complete for 1-dimensional static graphs. On the other side, BOUNDED-PATH in 1-dimensional static graphs with one vertex is NP-hard. This shows a trivial transformation from SUBSET SUM, which is NP-complete. An instance for SUBSET SUM is a set of nonnegative integers $x_1, \dots, x_n, m$. It has a solution if and only if there is a sequence of indices $i_1, \dots, i_k \in \{1, \dots, n\}$ such that $\sum_{j=1}^{k} x_{i_j} = m$. Let G be a 1-dimensional static graph G with one vertex w and edges with transit integers x_i for $i = 1, \dots, n$. Then there is a path from $(w, 0)$ to (w, m) in G^m if and only if $x_1, \dots, x_n, m$ has a solution for SUBSET SUM.

If G contains an additional edge with transit integer $-m$, then there is a cycle in G^m if and only if $x_1, \dots, x_n, m$ has a solution for SUBSET SUM. That is, BOUNDED-CYCLE is also NP-hard for 1-dimensional static graphs with one vertex.

Corollary 5. *BOUNDED-PATH and BOUNDED-CYCLE are NP-hard, even if the static graphs are 1-dimensional and have one vertex.*

It is also left open, whether or not BOUNDED-PATH and BOUNDED-CYCLE are members of NP, if the static graphs are 1-dimensional and have one vertex. However, the next paragraphs shows that these problems are weak-NP-hard, because they are solvable in polynomial time if the integers $x_1, \ldots, x_n$ are given unary.

From [5, 6], we know that infinite periodic graphs G^∞ defined by static graphs $G = (V, E)$ have cycles if and only if they have cycles with at most

$$|V|^2 + |V| \cdot (2 \cdot (|V| \cdot \|tran\| + |V|^2 \cdot \|tran\|) + 1)^d \tag{1}$$

edges, where $\| \ \|$ is the maximum norm. For any fixed dimension d and for unary represented transit vectors, the cycle length above is polynomially in the size of the static graph.

Let $b(G)$ be the cycle length specified above in equation 1 for a static graph G. If any coordinate i in the bound vector $m = (m_1 \ldots, m_d)$ is less than $b(G)$, then G can be *folded down* in polynomial time with respect to coordinate i. For instance, let $m' = (m_1, \ldots, m_{i-1}, m_{i+1}, \ldots, m_d)$. Then we can label the edges in $G^{(<0^{i-1}>, m_i, <0^{d-i}>)}$ such that the result is a $d-1$-dimensional static graph H with $H^{m'} = G^m$. The transit vectors of the edges in H are those from the edges in G without coordinate i. That is, the cycle problem for infinite periodic graphs is polynomially transformable to the cycle problem in infinite periodic graphs by folding down all coordinates i where m_i is less than $b(G)$. Since the cycle problem in infinite periodic graphs takes polynomial time, we obtain the following corollary.

Corollary 6. *BOUNDED-CYCLE is in P, if the dimension of the static graphs is fixed and the transit vectors of the edges are represented unary.*

A solution seems to be more difficult for BOUNDED-PATH. We do not know whether or not BOUNDED-PATH can be solved in polynomial time, if the dimension of the static graphs are fixed and the transit vectors are represented unary. However, for the 1-dimensional case, we know a polynomial time solution.

Theorem 7. *BOUNDED-PATH is in P, if the static graphs are 1-dimensional and the integer vectors are represented unary.*

Proof. Let $G = (V, E)$ be a 1-dimensional static graph with unary represented transit integers. First transform G such that all transit integers are either -1, 0, or 1. This can be done in polynomial time by substituting each edge with nonzero transit integer x with a path of length abs(t), where all edges in the path have transit vector 1 or -1, respectively.

Let $[G^n]$ for some nonnegative integer n be the graph with vertex set $V \times \{0, n\}$ and an edge from (u, x) to (v, y) if and only if there is a path from (u, x) to (v, y) in G^n. $[G^n]$ for some integer n can easily be constructed with $[G^{\lceil n/2 \rceil}]$ and $[G^{\lfloor n/2 \rfloor}]$ in time $O(|V|^3)$. That is, $[G^n]$ can be constructed in time $O(\lceil \log(n) \rceil \cdot |V|^3)$.

To solve the BOUNDED-PATH problem for two vertices u, v, and three non-negative integers s, t, m, we combine $[G^s]$, $[G^{t-s}]$, and $[G^{m-t}]$ to a graph H with vertex set $V \times \{0, s, t, m\}$ such that there is a path from (u, s) to (v, t) in H if and only if there is a path from (u, s) to (v, t) in G^m. H can be constructed in polynomial time. There is an edge from (u, x) to (u, y) in H if and only if

1. $0 \leq x, y \leq s$ and $[G^s]$ has an edge from (u, x) to (v, y), or
2. $s \leq x, y \leq t$ and $[G^{t-s}]$ has an edge from $(u, x-s)$ to $(v, y-s)$, or
3. $t \leq x, y \leq m$ and $[G^{m-t}]$ has an edge from $(u, x-t)$ to $(v, y-t)$. □

References

1. W. Backes, U. Schwiegelshohn, and L. Thiele. Analysis of free schedule in periodic graphs. *Proceedings of the fourth annual ACM Symposium on Parallel Algorithms and Architectures*, pages 333–343, 1992.
2. E. Cohen and N. Megiddo. Strongly polynomial-time and NC algorithms for detecting cycles in dynamic graphs. In *Annual ACM Symposium on Theory of Computing*, pages 523–534, 1989.
3. E. Cohen and N. Megiddo. Recognizing properties of periodic graphs. The Victor Klee Festschrift, Honorary Volume of Applied Geometry and Discrete Mathematics, 1990.
4. M.R. Garey and D.S. Johnson. *Computers and Intractability, A Guide to the Theory of NP-Completeness.* W.H. Freeman and Company, San Francisco, 1979.
5. F. Höfting and E. Wanke. Minimum cost path in periodic graphs. Technical Report 99, Universität-Gesamthochschule-Paderborn, Paderborn, FRG, April 1992.
6. F. Höfting and E. Wanke. Polynomial algorithms for minimum cost paths in periodic graphs. *Proceedings of the fourth annual ACM-SIAM Symposium on Discrete Algorithms*, pages 493–499, 1993.
7. K. Iwano and K. Steiglitz. Testing for cycles in infinite graphs with periodic structure. In *Proceedings of Annual ACM Symposium on Theory of Computing '87*, pages 46–55, 1987.
8. R.M. Karp, R.E. Miller, and A. Winograd. The organization of computations for uniform recurrence equations. *Journal of the ACM*, 14(3):563–590, July 1967.
9. M. Kodialam and J.B. Orlin. Recognizing strong connectivity in (dynamic) periodic graphs and its relation to integer programming. *Proceedings of the second annual ACM-SIAM Symposium on Discrete Algorithms*, pages 131–135, 1991.
10. S.R. Kosaraju and G.F. Sullivan. Detecting cycles in dynamic graphs in polynomial time. In *Proceedings of Annual ACM Symposium on Theory of Computing '88*, pages 398–406, 1988.
11. J.B. Orlin. Some problems on dynamic/periodic graphs. In W.R. Pulleyblank, editor, *Progress in Combinatorial Optimization*, pages 273–293. Academic Press, Orlando, Florida, 1984.
12. S.K. Rao. *Regular iterative algorithms and their implementations on processor arrays.* PhD thesis, Department of Electrical Engineering, Stanford University, 1985.

Learning Decision Lists from Noisy Examples *

Jilei Yin
Computer Center
Fudan University, Shanghai 200433 China

Zhu Hong†
Computer Science Dept.
Fudan University, Shanghai 200433 China

Abstract In this paper we solve an open problem raised by Rivest [3], show that the concept class k-Decision lists can be learned, in the sense of Valiant, from random examples with classification noise.

1 Introduction

Rivest introduced a new representation for Boolean functions [3] decision lists, and proved that the function class k-DL (the set of decision lists with conjunctive clauses of size k at each decision) can be learned from examples under Valiant's learning model. In this model, one is given examples under drawn from some probability distribution over some domain space, say $\{0,1\}^n$, and classified according to some target function $f : x \in \{0,1\}^n \rightarrow \{1,0\}$ as positive or negtive. The learning problem is, only given access to classified examples, and the knowledge that $f \in \Gamma$ for some class Γ of functions, to produce some hypothesis function g which agrees with f on the classification of almost all examples drawn from the probability distribution.

k-DL properly includes k-CNF and k-DNF [4] so dicision list is a powerful method for representing Boolean functions. As an open problem, Rivest asked "Can the functions in k-DL be learned efficiently when the supplied examples are 'noisy'?" (i.e. the classifications of the given examples may be erroneous with some probability η). This paper shows that the answer to this problem is 'yes'.

---

*This work was supported by National Sciences Foundation of China under Grant 6907330

†On visit Institute for Informatics, Slovak Academy of Sciences, partially supported by Grant of Slovak Academy of Sciences No.88, by EC Cooperative Action IC 1000 Algorithms for Future Technologies

2 Valiant's Learning Model and Decision Lists

We define a concept as a set of objects, where objects are described in terms of a set of *attribute-value* pairs.

We assume that there are n Boolean attributes to be considered, and denote the set of such attributes as $V_n = \{v_1, v_2, \ldots, v_n\}$. In a particular learning situation, each attribute will have a preassigned meaning. We use an assignment – a mapping form V_n to $\{0,1\}^n$, to describe an object, hence an objcet is indentical to an element in $X_n = \{0,1\}^n$. Obviously Boolean functions are very suitable to describe such concepts. A Boolean function f is the description of a concept c if $x \in c \Leftrightarrow f(x) = 1, x \in X_n$. Hence concepts are equivalent to Boolean functions.

Each Boolean formula defines a corresponding Boolean function from X_n to $\{0,1\}$ in a natural way . In the following, we do not distinguish between formulae and the Boolean functions they represent (so that we mihgt consider the size of a concept , e.g. the length of its formula).

$L_n = \{v_1, \neg v_1, v_2, \neg v_2, \ldots, v_n, \neg v_n\}$

$C(n,k)$ ={all terms (conjunctions) with at most k literals drawn from L_n }, k is considered a constant here.

A decision list on n variables is a list of pairs $L = (f_1, a_1), (f_2, a_2), \ldots, (f_t, a_t)$, where for each $i, 1 \leq i \leq t, a_i \in \{0,1\}$, and $f_i \in C(n,k)$, except that the last function f_t is always TRUE. The Boolean function represented by L is defined by letting $L(x) = a_j$, where $x \in X_n$, j is the least index such that $f_j(x) = 1$. This class of functions is called $k - DL(n)$. $k - DL = \bigcup_{n \geq 1} k - DL(n)$.

The following is a quick and formal sketch of Valiant's learning model.

An example of a concept f (f is also considered a Boolean function) is the description of an object(an assignment), together with its classification: $(x, f(x))$. A Boolean function f is consistent with (x,a) if $f(x) = a$.

Generally, we have the knowledge that the function we want to learn is in Γ for some class Γ of functions. Γ is called hypothesis space (or concept class). The goal of learning algorithms is to learn which concept in Γ is actually being used to classify the examples it has seen.

We partition Γ according to the number n of attributes upon which the concept depends:

$$\Gamma = \bigcup_{n \geq 1} \Gamma(n)$$

$\Gamma(n)$ is the subset of Γ that depends on V_n.

For any integer $n \geq 1$, we assume there exists a fixed but arbitrary probability distribution D_n defined on the set X_n, unknown to the learner. A learning algorithm can obtain information about a concept f (assume that $f \in \Gamma(n)$) by asking an oracle $EX(f)$. When $EX(f)$ is called, it returns an example (x,a) of f, where x is drawn from X_n according to D_n. Each call of $EX()$ is independent.

We define the discrepancy $diff(f,g)$ between two function f and g in Γ(n) as the probability (according to D_n) that f and g differ:

$$diff(f,g) = \sum_{\{x|f(x)\neq g(x)\}} D_n(x)$$

DEFINITION: A concept class Γ *is learnable (from examples)* iff there exists a learning algorithm A (possibly randomized) such that for any n ($n \geq 1$ is an integer), any $1 > \varepsilon, \delta > 0$, any probability distribution D_n on X_n, and any concept $f \in \Gamma(n)$, A accesses f only via $EX(f)$, and satisfies:
1. Poly-time:A's running time is polynomially bounded in $1/\epsilon, 1/\delta, n$, and the size of f.
2. Robustness: A outputs a concept g in $\Gamma(n)$such that , with probability at least $1-\delta$, $diff(f,g) < \epsilon$.

Valiant's learning model captures many learning situations encounted in practice. This model overcomes two of the weaknesses of the preceeding learning models [1]: the lack of robustness, and the lack of a complexity measure. Notise that this model is distribution-free, which means the results will hold for any probability distribution governing the generation of examples. So it can be expected that performance bounds derived under this model will hold in a wide variety of real-world learning-situation.

3 Learning Decision Lists from Noisy Examples

It's unresonable to require all the supplied examples are correct in practice. The algorithm to learn $k-DL$, given in [3], is sensitive to even very small amounts of noise in examples. In this section, we first give a formal definition of noise and then show that for $k-DL$ a substantial rate of errors can be accommodated.

We have a new oracle $EX_\eta()$ instead of $EX()$. At each call of $EX_\eta()$,$EX()$ is called first, then with probability η, the sign of the example presented by $EX()$ is changed. Each call of $EX_\eta()$ is independent. This kind of noise is also called Classification Noise. Without loss of generality, we can assume $\eta < 1/2$ [2].

From the algorithm given in [3], we know that the learnability of $1-DL$ is essential to that of $k-DL$. To simplify the proof, we just show that $1-DL$ can be learned from noisy examples.

Let S be a set of examples w.r.t a function $f \epsilon 1-DL(n)$, S_0^i denotes the set of examples $(x, f(x))$ in S such that $x=(x_1,...,x_n)$ and $x_i=0$, and S_1^i denotes the set of $(x, f(x))$ such that $x_i=1$. v_i is *informative* on S if both S_0^i and S_1^i are nonempty.

A decision list can be partitioned into a head and a tail, a head is a pair and tail is either empty or again a decision list.

Algorithm *Noise*(S)

INPUT: S is a set of examples obtained from $EX_\eta(f)$, its cardinality will be specified in the theorem, f is some unknown target function in $1-DL(n)$.

OUTPUT: a function g in $1 - DL$ that satisfies $Pr\{diff(f,g) > \epsilon\} < \delta$

```
1.      IV={informative variables on S}
        if IV = then begin
           POS = {(x,a) ∈ S | a = 1};
           NEG = {(x,a) ∈ S | a = 0};
           if | POS | / | S |≥ 1/2 then Return((true, 1))
                                  else Return((true, 0))
        end
2.      for each variable v_i in IV do begin
           if |S_0^i|/M < ε'/2 then (1) T' = Noise(S_1^i)
/* M is an integer whose value will be specified in the theorem */
                                    (2) Return(¬v_i, *), T')
           if |S_1^i|/M < ε'/2 then (1) T' = Noise(S_0^i)
/* '*' stands either 0 or 1 which means this value is not important */
                                    (2) Return((v_i, *), T')
        end
3.      for each variable v_i in IV do begin
           error(x_i, 0, 0) =| (x, 1) ∈ S_0^i | / | S_0^i |
/* empirical-inconsistent rate of (¬v_i, 0) */
           error(x_i, 0, 1) =| (x, 0) ∈ S_0^i | / | S_0^i |
           error(x_i, 1, 0) =| (x, 1) ∈ S_1^i | / | S_1^i |
           error(x_i, 1, 1) =| (x, 0) ∈ S_0^i | / | S_1^i |
        end
        Select (x_j, a, b) whose error is the smallest
        T' = Noise(S_{1-a}^j)
        if a = 0 then Return((¬v_j, b), T')
                 else Return((v_j, b), T')
```

Obviously functions produced by Noise are in $1 - DL$, and the running time of $Noise(S)$ is $O(n \mid S \mid)$.

THEOREM. Let $\eta < 1/2$ be an upper bound of the rate of classification noise. For any $0 < \epsilon, \delta < 1$, integer $n \geq 1$, any target function f in $1 - DL(n)$ and any distribution D_n on X_n, when

$$\mid S \mid \geq max\{\frac{32(n+1) * ln(16(n+1)^2/\delta)}{\epsilon^3 * (1-2\eta)^2}, \frac{8(n+1)^2 * ln(2(n+1)/\delta)}{\epsilon^2}\}$$

and $M = \mid S \mid$, the function g produced by Noise(S) satisfies

$$Pr\{diff(f,g) > \epsilon\} < \delta$$

Examples in S are drawn from $EX_\eta(f)$ independently.

In the proof to follow, we will use a special case of Höeffding's inequality. Consider a Bernoulli random variable with probability p of having the value 1 ("success"), and $1-p$ of having value 0 ("fail"). In m independent trails, if p' denotes the empirical rate of success, we have

$$Pr\{| p-p' |\geq s\} \leq 2e^{-2s^2m} \quad for 0 \leq p,s \leq 1$$

let $\delta = 2e^{-2s^2m}$, we have

$$Pr\{| p-p' |\geq \sqrt{\frac{ln(2/\delta)}{2m}}\} \leq \delta$$

Proof: For every f in $1-DL(n)$, f has an equivalent form which has at most $n+1$ pairs. $x \in X_n$ *arrives at* the pair (f_i, a_i) in decision list L if i is the least index in L such that $f_i(x)=1$.

Set $\delta' = \delta/(8(n+1)^2), \gamma = \delta/(n+1), N = 4n, \epsilon' = \epsilon/2(n+1)$. In the following, we will prove that any pair (f_i, a_i) in L produced by Noise(S) satisfies either

$Pr\{Pr\{(f_i, a_i)$ outputs a wrong answer according to target function f | a random object x arrives at $(f_i, a_i)\} > \epsilon/2\} < \gamma$ (1)

or

$Pr\{Pr\{$ a random object $x \in X_n$ arrives at $(f_i, a_i)\} > \epsilon'\} < \gamma$ (2)

We can view a random object arriving at a certain pair as a Bernoulli trail. The total number of trials $| S |$ is larger than

$$\frac{8(n+1)^2 * ln(2(n+1)/\delta)}{\epsilon^2} = \frac{2ln(2/\gamma)}{(\epsilon')^2}$$

If a pair (f_i, a_i) is output at the second step of the algorithm, due to Höeffding's inequality, (f_i, a_i) satisfies either

$Pr\{|$ the probability of a random object arrives at $(\neg v_j, a_i) - | S_0^j | /M |> \epsilon'/2\} < \gamma \ (f_i = \neg v_j)$ (3)

$Pr\{|$ the probability of a random object arrives at $(v_j, a_i) - | S_1^j | /M |> \epsilon'/2\} < \gamma \ (f_i = v_i)$ (4)

Without loss of generality, we assume that (f_i, a_i) satisfies (3). Notise that $| S_0^j | /M \leq \epsilon'/2$, hence (f_i, a_i) satisfies (2).

Consider two pairs:$(f_1, a_1), (f_2, a_2)$ (we assume that the partial decision lists before these two pairs are the same): one correct and one with error $\epsilon/2$, i.e. $Pr\{f(x) \neq a_1 \mid$ a random object x arrives at $(f_1, a_1)\} = 0$ and $Pr\{f(x) \neq a_2 \mid$ a random object x arrives at $(f_2, a_2)\} = \epsilon/2$ (such (f_1, a_1) always exists).

When (x, a) is drawn from $EX_\eta(f)$, however,

$$Pr\{a \neq a_1 \mid x \text{ arrives at } (f_1, a_1,)\} = \eta$$

and

$$Pr\{a \neq a_2 \mid x \text{ arrives at } (f_2, a_2)\} = (1-\epsilon/2)\eta + \epsilon(1-\eta)/2 \ = \eta + (1-2\eta)\epsilon/2$$

(f_2, a_2) is called an $\epsilon/2$-bad pair.

For each pair, do at least $\epsilon' * M/2 \geq \frac{8ln(2/\delta')}{\epsilon^2(1-2\eta)^2}$ independent Bernoulli trials respectively. Let p_i' be the empirical failure rate (i.e. the inconsistent rate) of pair (f_i, a_i).
$Pr\{p_1' \geq p_2'\} \leq Pr\{p_1' - \eta \geq (1-2\eta)\epsilon/4\} + Pr\{(\eta + (1-2\eta)\epsilon/2 - p_2' \geq (1-2\eta)\epsilon/4\} \leq Pr\{| p_1' - \eta | \geq (1-2\eta)\epsilon/4\} + Pr\{| (\eta + (1-2\eta)\epsilon/2) - p_2' | \geq (1-2\eta)\epsilon/4\} \leq 2\delta'$

If a pair is output at the third step of the algorithm, all the competitors have at least $\epsilon' * M/2$ Bernoulli trials(each example in S_0^i is a trial for $(\neg v_i, 0) and (\neg v_i, 1)$). There at most N candidates, so the probability that the algorithm outputs an $\epsilon/2$-bad pair at the third step is less than $2N\delta' < \gamma$. Hence pairs produced at the third step satisfy (1).

From the previous discussion, it's obvious that the pair output at the first step of the algorithm satisfies either (1) or (2).

With probability at least $1-(n+1)\gamma = 1-\delta$, all pairs satisfy either

$Pr\{(f_i, a_i)$ outputs a wrong answer according to target function f | a random object x arrives at $(f_i, a_i)\} \leq \epsilon/2$ (5)

or

$Pr\{$ a random object $x \in X_n$ arrives at $(f_i, a_i)\} \leq \epsilon'$ (6)

We denote the set of pairs satisfy (6) as α, and the set β consists of the remain pairs, so that the probability that a decision list L produced by the algorithm outputs a wrong answer is

$\sum_{(f_i,a_i)\in L} Pr\{$ a random object x arrives at pair $(f_i, a_i)\}$ *
$Pr\{(f_i, a_i)$ outputs a wrong answer | a random object x arrives at $(f_i, a_i)\}$
$\leq \sum_{(f_i,v_i)\in\alpha} \epsilon' + \epsilon/2 \sum_{(f_i,v_i)\in\beta} Pr\{$ a random object arrives at $(f_i, a_i)\}$
$\leq (n+1)\epsilon' + \epsilon/2 = \epsilon.$

Hence we have: The concept class $k-DL$ can be efficiently learned from noisy examples. The sample complexity and the running time is polynomial in $1/\epsilon, 1/\delta, 1/\eta$, and n^k. It's not necessary to know the exact noise rate ,we just need its upper bound. The only request on the bound of the noise rate η is $\eta < 1/2$, which is better than Rivest expected.

References

1. Angluin,D. Smith,C. (1983) Indutive inference: Theory and Methods ACM Comp. Surveys,15(3),237-270.
2. Laird,P.D. (1988) Learning from Good and Bad Data. Kluwer Academic Publishers.
3. Rivest ,R.L. (1987) Learning Decision Lists. Machine Learning,2:229-246.
4. Valiat,L.G. (1984) A theory of the learnable . Comm. ACM, 27,No.11, 1134-1142.

Analytic Tableaux for Finite and Infinite Post Logics

Nicolas Zabel

LIFIA–IMAG
46, Av. Félix Viallet
F 38 031 Grenoble Cedex - FRANCE
e-mail: zabel@lifia.imag.fr(uucp)

Abstract. We present a tableau-based calculus for the finite and infinite valued Post logics, well suited for automated deduction. We use a possible world semantics and a prefixed tableau calculus based on it. The formula prefixes contain arithmetical expressions and variables. The world-information is handled by solving constraints which express ordering problems. Some hints of future work are given.

1 Introduction

This article investigates the automation of finite and infinite Post logics for their own. They are used in Computer Sciences. For example, by using Post logic as the logic *underlying* Salwicki's algorithmic logic, properties of programmes can naturally be expressed even if the programming language includes such features as **CASE** and **LABEL/GOTO** and allows recursivity.

The structure of the paper is the following. First, in section 2, we recall the logical matrix defining these logics and give some background informations. The semantical issues of this section will be followed by the definition of a Kripke style possible world semantics, using as a starting point, the imbedding of Dummett's LC [2] in the Post logic. The tableau calculus we study in sections 3 to 4 can be seen as an enhancement of the Gentzen calculus. Section 5 gives some hints of future work.

The proofs are omitted and we have to refer a forthcoming long version of the present paper.

2 Definition of Post Logics

We shall recall only the features needed in this work.

The number of truth values distinguishes one Post logic from another; the semantics of the connectives are very close. Therefore as long as no confusion may arise, we shall speak about the Post logics, having in mind any Post logic.

The set of logical symbols of Post logics contains the binary connectives $\dot{\wedge}$, $\dot{\vee}$, $\dot{\Rightarrow}$, the unary ones $\dot{\neg}$, $(\mathsf{D}_{\dot{1}+i})_{i\in\omega}$, the constants $(\mathbf{e}_i)_{i\leq\omega}$, and the quantifiers $\forall$, $\exists$. Albeit the logical symbols are countably many, the set of well formed formulas, called $\mathcal{W}ff$, remains primitive recursive.

The Post logic containing all symbols listed above is called logic of order **sup** ω. The Post logic of order $m < \omega$ does not contain the symbols D_{1+n} and e_n for $n \geq m$.

The Post logic of order m (i. e. with $m+1$ truth values) is obtained from LC by adding new connectives. LC is an extension of intuitionistic logic obtained by adding the axiom $p \dot{\Rightarrow} q \dot{\vee} q \dot{\Rightarrow} p$. The Post logic of order 0 is the classical logic.

We adopt the following notations: Σ and Ω denote the signature of the function symbols and of the predicate symbols. We assume that Σ contains at least a constant symbol. $\mathcal{V}_f$, $\mathcal{V}_b$ are the infinite set of free and bound variables; an element of $\mathcal{V}_f$ will always been written with a dot (e. g. $\dot{x}$).

Figure 1 contains the semantics of the connectives and the quantifiers.

- the carrier of the algebra is $\mathcal{P}_m = \{n : n < m\} \cup \{\omega\}$; the truth values are ordered according to the natural order of the ordinal numbers.
- the sole designated element is ω.
-

symbol op	$\dot{\wedge}$	$\dot{\vee}$	e_i	$\forall$	$\exists$
semantics $\Re$op	min	max	i	min	sup

symbol op	$\dot{\Rightarrow}$	$\dot{\neg}$	D_j
semantics $\Re$op	$\langle i,j\rangle \mapsto \omega$ if $i \leq j$ $\langle i,j\rangle \mapsto j$ otherwise	$n \mapsto \Re_{\dot{\Rightarrow}}(n,0)$	$n \mapsto \omega$ if $n \geq j$ $n \mapsto 0$ otherwise

For $0 < j < 1+m$ (Recall $\omega = 1+\omega$), and $i \in \mathcal{P}_m$.

Fig. 1. Matrix Semantics of the Post logic of order m (P_m)

Definition 2.1. A *P_m-interpretation $\widehat{\Im}$ with domain (of discourse) $\mathcal{I}$* is a mapping with the following profile:

$$\widehat{\Im} : \begin{array}{rcll} \Sigma_n & \longrightarrow & \mathcal{I}^{\mathcal{I}^n} & \text{for all } n \in \mathbb{N} \\ \Omega_n & \longrightarrow & \mathcal{P}_m{}^{\mathcal{I}^n} & \text{for all } n \in \mathbb{N} \\ \mathcal{V}_f & \longrightarrow & \mathcal{I} & \end{array}$$

$\widehat{\Im}$ P_m-satisfies F iff $\widehat{\Im}(F) = \omega$.

A variant $\widehat{\Im}'$ of an P_m-interpretation $\widehat{\Im}$ of domain $\mathcal{I}$ is an P_m-interpretation of domain $\mathcal{I}$ such that for all $P \in \Omega$, $\widehat{\Im}'(P) = \widehat{\Im}(P)$.

If $\widehat{\Im}$ is an P_m-interpretation of domain $\mathcal{I}$, $\dot{x}$ a free variable in $\mathcal{V}_f$ and a an element of $\mathcal{I}$, $\widehat{\Im}^{\dot{x}}_a$ is the P_m-interpretation such that $\widehat{\Im}^{\dot{x}}_a(\dot{x}) = a$ and for all $X \in \Sigma \cup \Omega \cup \mathcal{V}_f \setminus \{\dot{x}\}$, $\widehat{\Im}^{\dot{x}}_a(X) = \widehat{\Im}(X)$. $\Diamond$

The Kripke frames (see [7, 1]) for Post logics are similar to those for LC (see e. g. [10]). They supply another semantics of the Post logics.

Definition 2.2. A *Kripke frame for the Post logic of order m* is a couple $\langle (\Im_i)_{i<m}, \Im_\omega \rangle$ verifying the following conditions:

- $m \leq \omega$, $\mathcal{P}_m$ denotes (as before) the set $\{n \in \mathbb{N} : n < m\} \cup \{\omega\}$
- $(\Im_i)_{i<m}$ a sequence of interpretations with same domain of individuals $\mathcal{I}$ such that:
  - (*Constant domain hypothesis*) for all $i, j \in \mathcal{P}_m$, $X \in \Sigma \cup \mathcal{V}_f$, $\Im_i(X) = \Im_j(X)$
  - (*Bivalence*) for all $i \in \mathcal{P}_m$, $n \in \mathbb{N}$, $P \in \Omega_n$, $\Im_i(P)$ is a mapping in $\mathcal{I}^n \rightarrow \{0, 1\}$
  - (*Monotonicity*) for all $i < j \in \mathcal{P}_m$, $n \in \mathbb{N}$, $P \in \Omega_n$,

$$\{a \in \mathcal{I}^n : \Im_{i,P}(a) = 0\} \subseteq \{a \in \mathcal{I}^n : \Im_{j,P}(a) = 0\}$$

  - (*Inductive definition of an evaluation*) for all $i \in \mathcal{P}_m$
    (*i_1*) $\Im_i(F_1 \dot\wedge F_2) = 1$ iff $\Im_i(F_1) = \Im_i(F_2) = 1$,
    (*i_2*) $\Im_i(F_1 \dot\vee F_2) = 1$ iff $\Im_i(F_1) = 1$ or $\Im_i(F_2) = 1$,
    (*i_3*) $\Im_i(F_1 \dot\Rightarrow F_2) = 1$ iff for all $j \leq i$, $\Im_j(F_2) = 1$, whenever $\Im_j(F_1) = 1$,
    (*i_4*) $\Im_i(\forall x \cdot F(x)) = 1$ iff for all $a \in \mathcal{I}$, $\Im_{i\,a}^{\,x}(F(\dot x)) = 1$,
    (*i_5*) $\Im_i(\exists x \cdot F(x)) = 1$ iff there is $a \in \mathcal{I}$ such that $\Im_{i\,a}^{\,x}(F(\dot x)) = 1$,
    (*i_6*) for all $j < m$, $\Im_i(\mathsf{D}_{j+1}\, F) = 1$ iff $\Im_j(F) = 1$,
    (*i_7*) $\Im_i(\mathsf{e}_{j+1}) = 1$ iff $j \geq i$ and $\Im_i(\mathsf{e}_0) = 0$

For all $i \in \mathcal{P}_m$ and $F \in \mathcal{W}ff$, we shall note $i \Vdash\!\!- F$ iff $\Im_i(F) = 1$, and $i -\!\!\dashv\!| F$ iff $\Im_i(F) = 0$.

A formula F is satisfied by this Kripke frame iff $\Im_\omega(F) = 1$. ◊

Let F be a formula of $\mathcal{W}ff$. There is a P_m Kripke frame $\langle(\Im_i)_{i<m}, \Im_\omega\rangle$ satisfying F iff there is a P_m-interpretation $\widehat{\Im}$ such that $\widehat{\Im}(F) = \omega$.

3 A Constraint Based Calculus

The calculus developed in this section, decomposes the reasoning into two distinct parts: the reasoning in the Post logic and the reasoning about the possible world.

In the sequel, $\mathcal{A}_f$ denotes a set of countably many free variables, called *arithmetical variables*. The signature for the order theory is $\langle 0, \omega, s, \prec\rangle$, where 0 and ω are constants (they denote respectively the ordinals 0 and ω, s is a unary symbol (for the successor function), and $\prec$ is the predicate for the order relation. For $i, j < m$, $i \dot- j =_{df} \begin{cases} 0 & \text{if } i \leq j < m \\ i - j & \text{otherwise} \end{cases}$ The signs are taken in $\mathcal{C}_{\mathcal{A}_f} = \{i + k \Vdash\!\!-, i + k -\!\!\dashv\!|, i \Vdash\!\!-, i -\!\!\dashv\!|, \omega \Vdash\!\!-, \omega -\!\!\dashv\!| \; : \; k \in \mathcal{A}_f, i < m\}$. Notice that the equality and the *less than or equal* relation between two arithmetical expressions are seen as an abbreviation: $a \preceq b$ iff $a \prec b+1$; $a = b$ iff $a \preceq b \wedge b \preceq a$. The set $\mathcal{E}$ of all elementary constraints is then defined by $\{i \prec j, i+k \prec j+k', i \prec j + k', i + k \prec j, i + k \prec j, i + k \prec m : i, j < m, k, k' \in \mathcal{A}_f\}$. $\mathcal{E}$ does not contain expressions where ω is on the lefthand side of $\prec$ nor terms like $\omega + k$. Due to this choice, we are forced to distinguish sometimes the decomposition rules for $\omega \Vdash\!\!- \ldots$ and $\omega -\!\!\dashv\!| \ldots$

$$cpa_1 : \frac{i''\Vdash F\dot{\wedge} G}{\begin{array}{c} i''\Vdash F \\ i''\Vdash G \end{array}} \qquad cpa_2 : \frac{i''\dashv\!\| F\dot{\wedge} G}{i''\dashv\!\| F \;\mid\; i''\dashv\!\| G}$$

$$cpa_3 : \frac{i''\Vdash F\dot{\vee} G}{i''\Vdash F \;\mid\; i''\Vdash G} \qquad cpa_4 : \frac{i''\dashv\!\| F\dot{\vee} G}{\begin{array}{c} i''\dashv\!\| F \\ i''\dashv\!\| G \end{array}}$$

$$cpa_5 : \frac{i''\Vdash F\dot{\Rightarrow} G}{0\dashv\!\| F \;\Bigg|\; i''\Vdash G \;\Bigg|\; \begin{array}{c} 0\preceq k, k\prec i'' \\ k\Vdash G \\ k+1\dashv\!\| F \end{array}} \qquad cpa_6 : \frac{i''\dashv\!\| F\dot{\Rightarrow} G}{\begin{array}{c} 0\preceq k, k\prec 1+i'' \\ k\Vdash F \\ k\dashv\!\| G \end{array}}$$

$$cpa_7 : \frac{i''\Vdash \dot{\neg} F}{0\dashv\!\| F} \qquad cpa_9 : \frac{i''\Vdash \mathsf{D}_{j+1}\, F}{j\Vdash F} \qquad cpa_{11} : \frac{i'\Vdash \mathsf{e}_j}{i'\preceq j}$$

$$cpa_8 : \frac{i''\dashv\!\| \dot{\neg} F}{0\Vdash F} \qquad cpa_{10} : \frac{i''\dashv\!\| \mathsf{D}_{j+1}\, F}{j\dashv\!\| F} \qquad cpa_{12} : \frac{i'\dashv\!\| \mathsf{e}_j}{j\preceq i'}$$

Fig. 2. Connective Rules of CPA_m (First Part)

In the sequel, we shall use the conventions summed up in the following table:

The symbol	denotes elements from
k	$\mathcal{A}_f$
i	$\mathcal{P}_m \setminus \{\omega\}$
i''	$\mathcal{Q} = \mathcal{P}_m \cup \{i+k : 0 \le i < m, k \in \mathcal{A}_f\}$
i'	$\mathcal{Q} \setminus \{\omega\}$

If τ is a mapping from $\mathcal{A}_f$ either to $\mathcal{A}_f$ or to **sup** ω, τ is extended to signed formulas and to sets of signed formulas as usually: $\tau(i+k\Vdash F)$ is by definition $i+\tau(k)\Vdash F$ and if $\mathcal{F}$ is a set of signed formulas, $\tau(\mathcal{F})$ is defined as $\{\tau(sF) : sF \in \mathcal{F}\}$.

3.1 The Semantic Tableau Calculus CPA_m

Definition 3.3. The *Post semantic tableau calculus* (CPA_m) whose language is defined as follows: the signs of the formulas are the elements of the set $C_{\mathcal{A}_f}$; the formal language is the set of all signed formulas and the symbol $\times$ (which points out an inconsistency); the inference rules are given in Figures 2 to 5.

A *parameter* is a constant in Σ_0 which does not occur in S. We assume that there are infinitely many parameters. A parameter, an arithmetical variable are defined to be *new* if and only if they do not occur in the formulas of the branch to which the unique conclusion of the inference introducing this object, is added.

for any term t

$$cpa_{13} : \frac{i'' \Vdash \forall x \cdot F(x)}{i'' \Vdash F(t)}$$

$$cpa_{15} : \frac{i'' \dashv\!\vert\, \exists x \cdot F(x)}{i'' \dashv\!\vert\, F(t)}$$

For any new parameter a

$$cpa_{14} : \frac{i'' \dashv\!\vert\, \forall x \cdot F(x)}{i'' \dashv\!\vert\, F(a)}$$

$$cpa_{16} : \frac{i'' \Vdash \exists x \cdot F(x)}{i'' \Vdash F(a)}$$

Fig. 3. Quantifier Rules of CPA_m (Second Part)

$$cpa_{17} : \frac{\omega \Vdash \mathbf{e}_j}{\times} \qquad cpa_{18} : \frac{\omega \dashv\!\vert\, \mathbf{e}_\omega}{\times}$$

$$cpa_{19} : \frac{i'' \Vdash F \qquad j'' \dashv\!\vert\, F}{i'' \preceq j''} \qquad cpa_{20} : \frac{\omega \Vdash F \qquad j' \dashv\!\vert\, F}{\times}$$

Fig. 4. "Closure" Rules of CPA_m (Third Part)

An *ordering constraint* O corresponds to each branch, this constraint determined by the sequence of inferences which have allowed to move from the root to the leaf of the branch. $\Diamond$

The rule cpa_{13} needs some comments. By definition of **sup** ω, $\Im(\forall x \cdot F(x)) =$ **sup** ω, if there is a sequence $a_0, \ldots, a_n, \ldots$ with $\Im(F(a_n)) = n$ for all $n \in \mathbb{N}$ (even if for no individual a, $\Im(F(a)) =$ **sup** ω). Nevertheless, if a is a new parameter, $F(a)$ can be added to the set of formulas without loss of consistency.

3.2 Soundness of CPA_m

In the sequel, instead of proving the unsatisfiability in P_ω, the unsatisfiability in P_n, with a large enough n is proven; this claim will be established below. As the value n does not occur explicitly (but in the rather rhetorical caution *the used indices are less than* n: this constraint will be satisfied when n is chosen large enough).

The sequence of the Post logics of increasing order verifies a property that should be brought together with the *finite frame property* [6, Chapter 8]. In our context, this property, called *finite model property* in [5], essentially states that for verifying the P_ω-unsatisfiability of a formula (or of a finite set of formulas), it is sufficient to be sure of the unsatisfiability in some Post logic of finite order.

Proposition 3.4. *Let $\mathcal{F}$ be a* finite *set of formulas (signed or not). $\mathcal{F}$ is P_ω-unsatisfiable iff there is $n < \omega$ such that $\mathcal{F}$ is P_n-unsatisfiable.*

$$cpa_{21}: \frac{i'' \prec 0}{\times} \qquad cpa_{22}: \frac{i+k \prec k}{\times}$$

$$\text{if } 1+i+j \geq m$$

$$cpa_{23}: \frac{i+k \prec j+k'}{i \dot{-} j+k \prec j \dot{-} i+k'} \qquad cpa_{24}: \frac{k+i \prec k' \quad j+k' \prec k''}{\times}$$

$$cpa_{25}: \frac{i+k \prec k' \quad k' \prec 1+j+k''}{i \dot{-} j+k \prec j \dot{-} i+k''} \qquad cpa_{26}: \frac{k \prec 1+i+k' \quad j+k' \prec k''}{j \dot{-} i+k \prec i \dot{-} j+k''}$$

$$\text{if } 1+i+j < m \qquad \text{if } i+j < m$$

$$cpa_{27}: \frac{k+i \prec k' \quad j+k' \prec k''}{1+i+j+k \prec k''} \qquad cpa_{28}: \frac{k \prec 1+i+k' \quad k' \prec j+k''}{k \prec i+j+k''}$$

Fig. 5. Constraints Rules of CPA_m (Fourth and Last Part)

Theorem Soundness . *Let τ be an arbitrary instantiation of the arithmetical variables. For any instance of an inference rule in Figures 2 to 5, if the premise(s) is/are satisfied by an interpretation $\Im$, then there is an instantiation τ' which differs from τ only on the new arithmetical variables, such that the instance of the conclusions got by using τ' of at least one alternative are all satisfied by a variant of $\Im$, that differs from $\Im$ by the values assigned to new parameters.*

3.3 Completeness of CPA_m

For proving the completeness of CPA_m, we adapt the standard technique of analytic consistency property devised for classical logic [9]. Such a property is here a collection of sets of signed formulas and elementary contraints. If we consider these sets as "knowledge states", an analytic consistency property is a deductively closed collection of knowledge states, where "deductive" refers to the inference rules of CPA_m.

Definition 3.5. We shall call *constraint based analytic consistency property* any family $\mathcal{P}ca$ of ordered pairs of sets of signed formulas and elementary contraints from $\mathcal{E}$ such that the following statements, where $T \in \mathcal{P}ca$, S is the set of all signed formulas of T and O the set of all elementary constraints of T, hold

- S and O satisfy: if $i+k \Vdash\!\!- e_j \in S$, then $T \cup \{i+k \prec j\} \in \mathcal{P}ca$ (similar for $i+k \!-\!\!\dashv\!\!| e_j \in S$), if $k+i \Vdash\!\!- Ps, k'+j \!-\!\!\dashv\!\!| Ps \in S$ then $\{k'+j \prec k+i\} \cup T \in \mathcal{P}ca$
- S fulfils the following conditions which mimic the inference rules of *Figures 2* and *3*. For example: *ν-formula cases*:
  if $i+k \Vdash\!\!- F \Rightarrow G \in S$, then either $\{0 \!-\!\!\dashv\!\!| F\} \cup T \in \mathcal{P}ca$ or $\{i+k \Vdash\!\!- G\} \cup T \in \mathcal{P}ca$
  or for any new k', $\{k' \Vdash\!\!- F, k'+1 \!-\!\!\dashv\!\!| G\} \cup \{0 \preceq k', k' \prec i+k\} \cup T \in \mathcal{P}ca$

- O fulfils the following conditions:
  ($PcaO_1$) Base case:
  $i' \prec 0, i+k \prec k, m+j' \prec i' \notin O$
  ($PcaO_2$) Simplification by the constants:
  if $i+k \prec j+k' \in O$, then $\{i \dot{-} j + k \prec j \dot{-} i + k'\} \cup T \in \mathcal{P}ca$

  ($PcaO_3$) Transitivity of $\prec$ compare with *Figure 5.*

If a constraint based analytic consistency property contains only one element H, H is called a *Hintikka set.* ◊

We verify that the extension lemmas (renaming, finite character [3]) remain valid in this new setting.

Lemma Hintikka's lemma for CPA$_m$ *Let H be a Hintikka set of P_m, $m \leq \omega$. The set O of all elementary constraints in H is satisfiable in $\langle \mathcal{P}_m, < \rangle$. Further, let τ be a solution of O. $\tau(H \setminus O)$ is satisfiable.*

Theorem 3.6. *Any element of a constraint based analytic consistency property is satisfiable.*

Proof. Any analytic consistency property can be extended to an analytic consistency property of finite character. By application of Tuckey's lemma, any element S of the analytic consistency property may be extended to a maximal element in that property. The previous lemma states the satisfiability of the maximal elements of any analytic consistency property, and this implies the satisfiability of S. *q.e.d.*

Proposition 3.7. *The family of all the sets consistent for the finite tableaux in CPA$_m$, $\mathcal{P}$, is a constraint based analytic consistency property.*

Proof. By a case analysis. The verification of the points *(PcaO)* is trivial.

Among the other items, we choose to examplify the verification with the ν-formula case: the same arguments occur in the case of the β-formulas and of the π-formulas.

Let $i' \dashv\!\!| F \Rightarrow G \in S$ with $i' \in Q$, $P \in \mathcal{P}$. Assume that $P \cup \{0 \dashv\!\!| F\} \notin \mathcal{P}$, $P \cup \{i' |\!\!\vdash G\} \notin \mathcal{P}$, $P \cup \{k |\!\!\vdash G, k+1 \dashv\!\!| F, 0 \prec k, k \prec i'\} \notin \mathcal{P}$.

Hence for all these cases, there is a finite closed tableau, however then there is also a closed tableau for P which cannot be in $\mathcal{P}$. *q.e.d.*

3.4 A Fast Test for the Satisfiability of O

In the proof of the model existence theorem, the application of the inference rules stated in the clauses *(PcaO)* leads in practice to fastly growing search spaces.

Nevertheless it is worth to keep these rules in the tableau calculus although it would be enough to know that the constraints are satisfiable. When we know exactly which constraints are necessary for closing a branch, it is possible to apply strategies shortening the tableau proof. The set O is translated into a graph.

4 Skolemisation

In order to use easily the unification, it must be possible to skolemise a formula. As far as the skolemisation concerns the individuals, it is easier to use P_m-interpretations.

[8] has shown that the Post logic of order ω admitted a prenex normal form and that the formulas of type $\mathsf{D}_i\, F$ could be skolemised. We shall handle skolemisation in a different way and propose in addition to the usual way of skolemising, a lazy skolemisation. The case of finite Post logics is treated easily. By using *Proposition 3.4*, we conclude about the Post logic of order **sup** ω.

In the Post logics, all the connectives are monotonic. The argument of $\Re_{\dot{\neg}}$ and the first one of $\Re_{\Rightarrow}$ are antitonic, i. e. when $i, j, k \in \mathcal{P}_m$ and $i \leq j$, $\Re_{\dot{\neg}}(i) \geq \Re_{\dot{\neg}}(j)$ and $\Re_{\Rightarrow}(i,k) \geq \Re_{\Rightarrow}(j,k)$. All the other arguments of all remainding connectives and quantifiers are isotonic. Therefore any position is isotonic (also called a positive position) or antitonic (negative position). Furthermore because of monotonicity of all the elements of the logical signature, a lazy skolemisation, albeit possible, is less interesting; it is possible to skolemise the formula, without having to put it into a normal form, *before* searching (for example) a refutation by tableaux.

4.1 Skolemisation in the Finite Post Logics

Proposition 4.8. *Let $\widehat{\Im}$ be a P_m-interpretation with domain $\mathcal{I}$, F be a closed formula, p denote a position of F, $\boldsymbol{x} : \langle x_1, \ldots, x_n \rangle$ be the sequence of all the variables bound by a quantifier whose scope embodies p, f a n-ary function symbol not occurring in Σ.*

When p is an isotonic position of F and when $F|_p$ is $\exists x \cdot G(x)$,
$\widehat{\Im}(F) = \omega$, iff there is an expansion $\widehat{\Im}'$ of $\widehat{\Im}$, verifying $\widehat{\Im}'(F[G(f\boldsymbol{x})]_p) = \omega$.

When p is an antitonic position and when $F|_p$ is $\forall x \cdot G(x)$,
$\widehat{\Im}(F) = \omega$, iff there is an expansion $\widehat{\Im}'$ de $\widehat{\Im}$, verifying $\widehat{\Im}'(F[G(f\boldsymbol{x})]_p) = \omega$.

Proof. Cf. [11]. *q.e.d.*

We can verify that by skolemising a formula, in a tableau refutation no parameter is introduced, for the corresponding decomposition rules do not apply. The symbol f is called Skolem function symbol. By "top–down" skolemisation, Skolem functions of smaller arity are introduced than by "bottom–up" skolemisation. We call F', the result of "top-down" skolemisation of F, the skolemised from of F.

4.2 Skolemisation in the Post Logic of Order sup ω

Proposition 4.9. *Let F be a formula of $\mathcal{W}ff$. F is P_ω-unsatisfiable, iff the skolemised F' is F too.*

Proof. F is P_ω-unsatisfiable, iff there is $n < \omega$ such that F is P_n-unsatisfiable (*Proposition 3.4*) iff F' is also P_n-unsatisfiable (*Proposition 4.8*), iff F' is P_ω-unsatisfiable (*Proposition 3.4*). *q.e.d.*

Hence for proving the soundness of the skolemisation, we have used its soundness for the finite valued logics; the fact that P_ω is the "limit" of $(\mathsf{P}_i)_{i<\omega}$ extends the claim to the infinite case. This technique is very general and could also be applied to other sequences of logics.

5 Conclusion and Future Work

We have presented a sound and complete calculus for Post logics and stressed out that by handling explicitly and symbolically the truth values, the calculus CPA_m obtained is more efficient and produces more readable refutations. We have presented also an alternative way of treating existential quantifiers at isotonic positions, when unification is used, without any use of normal forms. CPA_m seems sufficiently interesting for devising strategies (like "factorisation" and restrictions to the applyability of *(cpa$_{19}$)* and *(cpa$_{20}$)* with unification). Another topic of future work will be on the study of the inclusion of graded equality in this framework.

Acknowledgement

The author wishes to thank Ricardo Caferra for reading and discussing the drafts of this article.

References

1. B. Dahn. KRIPKE-style semantics for some many-valued propositional calculi. *Bulletin de l'Académie Polonaise des Sciences. série des sciences math., astr. et phys.*, XXII(2):99–102, 1974.
2. Michael Dummett. A propositional calculus with denumerable matrix. *Journal of Symbolic Logic*, 24(2):97–106, June 1959.
3. Melvin Fitting. *First Order Logic and Automated theorem Proving*. Springer-Verlag, 1990.
4. Reiner Hähnle. *Automated Theorem Proving in Multiple Valued Logics*. Oxford University Press. to appear.
5. Ronald Harrop. Some structure results for propositional calculi. *Journal of Symbolic Logic*, 30(3):271–292, September 1965.
6. G. E. Hughes and M. J. Cresswell. *A Companion to Modal Logic*. Methuen, 1984.
7. L. Maksimova and D. Vakarelov. Semantics for ω^+-valued predicate calculi. *Bulletin de l'Académie Polonaise des Sciences. série des sciences math., astr. et phys.*, XXII(8):765–771, 1974.
8. Ewa Orlowska. The Gentzen style axiomatisation of ω^+-valued logic. *Studia Logica*, 35(4):433–445, 1976.

9. Raymond M. Smullyan. *First Order Logic*. Springer-Verlag, 1968.
10. Dirk van Dalen. Intuitionistic logic. In Dov Gabbay and Franz Günthner, editors, *Handbook of Philosophical Logic, vol. III: Alternatives of classical logic*, volume 166 of *Synthese Library*, chapter III.4, pages 225–240. Reidel, 1986.
11. Nicolas Zabel. *Nouvelles techniques de déduction automatique en logiques polyvalentes du premier ordre, finies et infinies*. Thèse, Institut National Polytechnique de Grenoble, 1992. forthcoming.

Subject index

Object orientedness

Parallel computation and algorithms

Process equivalence

Real time

Semantics

Author index

Springer-Verlag and the Environment

Lecture Notes in Computer Science

For information about Vols. 1–639
please contact your bookseller or Springer-Verlag

Vol. 640: C. Sledge (Ed.), Software Engineering Education. Proceedings, 1992. X, 451 pages. 1992.

Vol. 641: U. Kastens, P. Pfahler (Eds.), Compiler Construction. Proceedings, 1992. VIII, 320 pages. 1992.

Vol. 642: K. P. Jantke (Ed.), Analogical and Inductive Inference. Proceedings, 1992. VIII, 319 pages. 1992. (Subseries LNAI).

Vol. 643: A. Habel, Hyperedge Replacement: Grammars and Languages. X, 214 pages. 1992.

Vol. 644: A. Apostolico, M. Crochemore, Z. Galil, U. Manber (Eds.), Combinatorial Pattern Matching. Proceedings, 1992. X, 287 pages. 1992.

Vol. 645: G. Pernul, A M. Tjoa (Eds.), Entity-Relationship Approach – ER '92. Proceedings, 1992. XI, 439 pages, 1992.

Vol. 646: J. Biskup, R. Hull (Eds.), Database Theory – ICDT '92. Proceedings, 1992. IX, 449 pages. 1992.

Vol. 647: A. Segall, S. Zaks (Eds.), Distributed Algorithms. X, 380 pages. 1992.

Vol. 648: Y. Deswarte, G. Eizenberg, J.-J. Quisquater (Eds.), Computer Security – ESORICS 92. Proceedings. XI, 451 pages. 1992.

Vol. 649: A. Pettorossi (Ed.), Meta-Programming in Logic. Proceedings, 1992. XII, 535 pages. 1992.

Vol. 650: T. Ibaraki, Y. Inagaki, K. Iwama, T. Nishizeki, M. Yamashita (Eds.), Algorithms and Computation. Proceedings, 1992. XI, 510 pages. 1992.

Vol. 651: R. Koymans, Specifying Message Passing and Time-Critical Systems with Temporal Logic. IX, 164 pages. 1992.

Vol. 652: R. Shyamasundar (Ed.), Foundations of Software Technology and Theoretical Computer Science. Proceedings, 1992. XIII, 405 pages. 1992.

Vol. 653: A. Bensoussan, J.-P. Verjus (Eds.), Future Tendencies in Computer Science, Control and Applied Mathematics. Proceedings, 1992. XV, 371 pages. 1992.

Vol. 654: A. Nakamura, M. Nivat, A. Saoudi, P. S. P. Wang, K. Inoue (Eds.), Parallel Image Analysis. Proceedings, 1992. VIII, 312 pages. 1992.

Vol. 655: M. Bidoit, C. Choppy (Eds.), Recent Trends in Data Type Specification. X, 344 pages. 1993.

Vol. 656: M. Rusinowitch, J. L. Rémy (Eds.), Conditional Term Rewriting Systems. Proceedings, 1992. XI, 501 pages. 1993.

Vol. 657: E. W. Mayr (Ed.), Graph-Theoretic Concepts in Computer Science. Proceedings, 1992. VIII, 350 pages. 1993.

Vol. 658: R. A. Rueppel (Ed.), Advances in Cryptology – EUROCRYPT '92. Proceedings, 1992. X, 493 pages. 1993.

Vol. 659: G. Brewka, K. P. Jantke, P. H. Schmitt (Eds.), Nonmonotonic and Inductive Logic. Proceedings, 1991. VIII, 332 pages. 1993. (Subseries LNAI).

Vol. 660: E. Lamma, P. Mello (Eds.), Extensions of Logic Programming. Proceedings, 1992. VIII, 417 pages. 1993. (Subseries LNAI).

Vol. 661: S. J. Hanson, W. Remmele, R. L. Rivest (Eds.), Machine Learning: From Theory to Applications. VIII, 271 pages. 1993.

Vol. 662: M. Nitzberg, D. Mumford, T. Shiota, Filtering, Segmentation and Depth. VIII, 143 pages. 1993.

Vol. 663: G. v. Bochmann, D. K. Probst (Eds.), Computer Aided Verification. Proceedings, 1992. IX, 422 pages. 1993.

Vol. 664: M. Bezem, J. F. Groote (Eds.), Typed Lambda Calculi and Applications. Proceedings, 1993. VIII, 433 pages. 1993.

Vol. 665: P. Enjalbert, A. Finkel, K. W. Wagner (Eds.), STACS 93. Proceedings, 1993. XIV, 724 pages. 1993.

Vol. 666: J. W. de Bakker, W.-P. de Roever, G. Rozenberg (Eds.), Semantics: Foundations and Applications. Proceedings, 1992. VIII, 659 pages. 1993.

Vol. 667: P. B. Brazdil (Ed.), Machine Learning: ECML – 93. Proceedings, 1993. XII, 471 pages. 1993. (Subseries LNAI).

Vol. 668: M.-C. Gaudel, J.-P. Jouannaud (Eds.), TAPSOFT '93: Theory and Practice of Software Development. Proceedings, 1993. XII, 762 pages. 1993.

Vol. 669: R. S. Bird, C. C. Morgan, J. C. P. Woodcock (Eds.), Mathematics of Program Construction. Proceedings, 1992. VIII, 378 pages. 1993.

Vol. 670: J. C. P. Woodcock, P. G. Larsen (Eds.), FME '93: Industrial-Strength Formal Methods. Proceedings, 1993. XI, 689 pages. 1993.

Vol. 671: H. J. Ohlbach (Ed.), GWAI-92: Advances in Artificial Intelligence. Proceedings, 1992. XI, 397 pages. 1993. (Subseries LNAI).

Vol. 672: A. Barak, S. Guday, R. G. Wheeler, The MOSIX Distributed Operating System. X, 221 pages. 1993.

Vol. 673: G. Cohen, T. Mora, O. Moreno (Eds.), Applied Algebra, Algebraic Algorithms and Error-Correcting Codes. Proceedings, 1993. X, 355 pages 1993.

Vol. 674: G. Rozenberg (Ed.), Advances in Petri Nets 1993. VII, 457 pages. 1993.

Vol. 675: A. Mulkers, Live Data Structures in Logic Programs. VIII, 220 pages. 1993.

Vol. 676: Th. H. Reiss, Recognizing Planar Objects Using Invariant Image Features. X, 180 pages. 1993.

Vol. 677: H. Abdulrab, J.-P. Pécuchet (Eds.), Word Equations and Related Topics. Proceedings, 1991. VII, 214 pages. 1993.

Vol. 678: F. Meyer auf der Heide, B. Monien, A. L. Rosenberg (Eds.), Parallel Architectures and Their Efficient Use. Proceedings, 1992. XII, 227 pages. 1993.

Vol. 679: C. Fermüller, A. Leitsch, T. Tammet, N. Zamov, Resolution Methods for the Decision Problem. VIII, 205 pages. 1993. (Subseries LNAI).

Vol. 680: B. Hoffmann, B. Krieg-Brückner (Eds.), Program Development by Specification and Transformation. XV, 623 pages. 1993.

Vol. 681: H. Wansing, The Logic of Information Structures. IX, 163 pages. 1993. (Subseries LNAI).

Vol. 682: B. Bouchon-Meunier, L. Valverde, R. R. Yager (Eds.), IPMU '92 – Advanced Methods in Artificial Intelligence. Proceedings, 1992. IX, 367 pages. 1993.

Vol. 683: G.J. Milne, L. Pierre (Eds.), Correct Hardware Design and Verification Methods. Proceedings, 1993. VIII, 270 Pages. 1993.

Vol. 684: A. Apostolico, M. Crochemore, Z. Galil, U. Manber (Eds.), Combinatorial Pattern Matching. Proceedings, 1993. VIII, 265 pages. 1993.

Vol. 685: C. Rolland, F. Bodart, C. Cauvet (Eds.), Advanced Information Systems Engineering. Proceedings, 1993. XI, 650 pages. 1993.

Vol. 686: J. Mira, J. Cabestany, A. Prieto (Eds.), New Trends in Neural Computation. Proceedings, 1993. XVII, 746 pages. 1993.

Vol. 687: H. H. Barrett, A. F. Gmitro (Eds.), Information Processing in Medical Imaging. Proceedings, 1993. XVI, 567 pages. 1993.

Vol. 688: M. Gauthier (Ed.), Ada-Europe '93. Proceedings, 1993. VIII, 353 pages. 1993.

Vol. 689: J. Komorowski, Z. W. Ras (Eds.), Methodologies for Intelligent Systems. Proceedings, 1993. XI, 653 pages. 1993. (Subseries LNAI).

Vol. 690: C. Kirchner (Ed.), Rewriting Techniques and Applications. Proceedings, 1993. XI, 488 pages. 1993.

Vol. 691: M. Ajmone Marsan (Ed.), Application and Theory of Petri Nets 1993. Proceedings, 1993. IX, 591 pages. 1993.

Vol. 692: D. Abel, B.C. Ooi (Eds.), Advances in Spatial Databases. Proceedings, 1993. XIII, 529 pages. 1993.

Vol. 693: P. E. Lauer (Ed.), Functional Programming, Concurrency, Simulation and Automated Reasoning. Proceedings, 1991/1992. XI, 398 pages. 1993.

Vol. 694: A. Bode, M. Reeve, G. Wolf (Eds.), PARLE '93. Parallel Architectures and Languages Europe. Proceedings, 1993. XVII, 770 pages. 1993.

Vol. 695: E. P. Klement, W. Slany (Eds.), Fuzzy Logic in Artificial Intelligence. Proceedings, 1993. VIII, 192 pages. 1993. (Subseries LNAI).

Vol. 696: M. Worboys, A. F. Grundy (Eds.), Advances in Databases. Proceedings, 1993. X, 276 pages. 1993.

Vol. 697: C. Courcoubetis (Ed.), Computer Aided Verification. Proceedings, 1993. IX, 504 pages. 1993.

Vol. 698: A. Voronkov (Ed.), Logic Programming and Automated Reasoning. Proceedings, 1993. XIII, 386 pages. 1993. (Subseries LNAI).

Vol. 699: G. W. Mineau, B. Moulin, J. F. Sowa (Eds.), Conceptual Graphs for Knowledge Representation. Proceedings, 1993. IX, 451 pages. 1993. (Subseries LNAI).

Vol. 700: A. Lingas, R. Karlsson, S. Carlsson (Eds.), Automata, Languages and Programming. Proceedings, 1993. XII, 697 pages. 1993.

Vol. 701: P. Atzeni (Ed.), LOGIDATA+: Deductive Databases with Complex Objects. VIII, 273 pages. 1993.

Vol. 702: E. Börger, G. Jäger, H. Kleine Büning, S. Martini, M. M. Richter (Eds.), Computer Science Logic. Proceedings, 1992. VIII, 439 pages. 1993.

Vol. 703: M. de Berg, Ray Shooting, Depth Orders and Hidden Surface Removal. X, 201 pages. 1993.

Vol. 704: F. N. Paulisch, The Design of an Extendible Graph Editor. XV, 184 pages. 1993.

Vol. 705: H. Grünbacher, R. W. Hartenstein (Eds.), Field-Programmable Gate Arrays. Proceedings, 1992. VIII, 218 pages. 1993.

Vol. 706: H. D. Rombach, V. R. Basili, R. W. Selby (Eds.), Experimental Software Engineering Issues. Proceedings, 1992. XVIII, 261 pages. 1993.

Vol. 707: O. M. Nierstrasz (Ed.), ECOOP '93 – Object-Oriented Programming. Proceedings, 1993. XI, 531 pages. 1993.

Vol. 708: C. Laugier (Ed.), Geometric Reasoning for Perception and Action. Proceedings, 1991. VIII, 281 pages. 1993.

Vol. 709: F. Dehne, J.-R. Sack, N. Santoro, S. Whitesides (Eds.), Algorithms and Data Structures. Proceedings, 1993. XII, 634 pages. 1993.

Vol. 710: Z. Ésik (Ed.), Fundamentals of Computation Theory. Proceedings, 1993. IX, 471 pages. 1993.

Vol. 711: A. M. Borzyszkowski, S. Sokołowski (Eds.), Mathematical Foundations of Computer Science 1993. Proceedings, 1993. XIII, 782 pages. 1993.

Vol. 712: P. V. Rangan (Ed.), Network and Operating System Support for Digital Audio and Video. Proceedings, 1992. X, 416 pages. 1993.

Vol. 713: G. Gottlob, A. Leitsch, D. Mundici (Eds.), Computational Logic and Proof Theory. Proceedings, 1993. XI, 348 pages. 1993.

Vol. 714: M. Bruynooghe, J. Penjam (Eds.), Programming Language Implementation and Logic Programming. Proceedings, 1993. XI, 421 pages. 1993.

Vol. 715: E. Best (Ed.), CONCUR'93. Proceedings, 1993. IX, 541 pages. 1993.

Vol. 716: A. U. Frank, I. Campari (Eds.), Spatial Information Theory. Proceedings, 1993. XI, 478 pages. 1993.

Vol. 717: I. Sommerville, M. Paul (Eds.), Software Engineering – ESEC '93. Proceedings, 1993. XII, 516 pages. 1993.